PHARMACEUTICAL STATISTICS

PHARMACEUTICAL STATISTICS

T.E. Gopala Krishna Murty
M.Pharm, Ph.D
Professor and Principal
Department of Pharmaceutics
Bapatla College of Pharmacy,
Bapatla, Guntur Dist. (A.P.)

P. Srinivasa Babu
M.Pharm, Ph.D
Professor and Principal
Department of Pharmaceutics
Vignan Pharmacy College
Vadlamudi, Guntur Dist. (A.P.)

P. Seshagiri Rao
M.Sc., M.Phil
Associate Professor
Department of Mathematics
Vignan Pharmacy College
Vadlamudi, Guntur Dist. (A.P.)

PharmaMed Press
An imprint of Pharma Book Syndicate
A Unit of BSP Books Pvt. Ltd.
4-4-309/316, Giriraj Lane,
Sultan Bazar, Hyderabad - 500 095.

Published by

PharmaMed Press

An imprint of Pharma Book Syndicate

A unit of BSP Books Pvt. Ltd.

4-4-309/316, Giriraj Lane, Sultan Bazar, Hyderabad - 500 095.
Phone: 040-23445605, 23445688; Fax: 91+40-23445611
e-mail: info@pharmamedpress.com

ISBN: 978-93-85433-79-5 (HB)

PREFACE

Statistics is an area of applied mathematics concerned with the data collection, analysis, interpretation and presentation.

Statistics is used in almost every field of research in social sciences, climate research, business and pharmaceutical research...

Many universities have introduced statistics as one of the subjects in the curriculum. Many books are available in the market but they have focused mainly on Mathematics and moreover the language used is very difficult to understand the subject.

Hence the need to compile a book very useful to the pharmacists has arisen and this book covers all the subjects related to statistical analysis of pharmaceutical sciences.

This book is written in a colloquial style and is an ongoing effort to make the basic knowledge easily understandable to the pharmaceutical community.

The main strength of the book is its straightforward and clear explanations of all covered topics and the very well-done diagrams which help readers better visualize key concepts. The text contains a large number of instructive and clearly explained numerical examples making the book very easy to self-study.

We have taken a lot of efforts for the ideal printing of the book but some types may have crept in.

Any suggestions and criticism are always welcome and the modifications will be made in the next edition.

We trust that the readers will welcome the edition and we are sure that the book will certainly meet all your requirements.

CONTENTS

CHAPTER 3

DATA PRESENTATION

CHAPTER 4

PROBABILITY AND DISTRIBUTIONS

CHAPTER 5

CORRELATION AND REGRESSION

CHAPTER 8

STATISTICAL INFERENCE: SIGNIFICANCE OF TESTING

CHAPTER 9

ANALYSIS OF VARIANCE (ANOVA)

CHAPTER 10

NON PARAMETRIC METHODS

CHAPTER 11

STATISTICAL QUALITY CONTROL

CHAPTER 12

FACTORIAL DESIGNS

CHAPTER 13

OPTIMIZATION TECHNIQUES AND SCREENING DESIGNS

CHAPTER 1

BASIC CONCEPTS

Statistics is the science which deals with the methods of collecting, classifying, presenting, comparing and interpreting numerical data collected on any sphere of inquiry. Knowledge on statistical concepts is desired for the development, production, evaluation and marketing of various pharmaceutical dosage forms. Statistics depends on the measurement of continuous and discrete variables. Discontinuous or discrete variable is lacking continuity and it may take several values. For example, the number of students in several classes in a school may vary from 50 to 100, taking several discrete integer values. A continuous variable takes on values which can be measured to any depth. Statistical analysis requires the generation of data and subsequent statistical treatment. During the collection of data, there is a possible presence of either determinate errors or indeterminate errors. Determinant errors are constant errors; unsuspected and corrected once they are identified. They are present in each case. For example, presence of impurities in active pharmaceutical active ingredient is inevitable and depends upon the raw materials used and the conditions employed for the synthesis. The affect of these impurities on the quality, performance and safety of the finished dosage form may vary. Such errors must be eliminated before studying the significance of indeterminate errors. Indeterminate errors occur by accident or chance and vary from one experiment to another experiment. They cannot be corrected due to the fluctuations that occur in all measurements.

Determinate and indeterminate errors can be explained with regard to the determination of the bulk density of the sample. If the errors in density measurement are due to the transfer of a sample in to the container or in adequate working of the bulk density apparatus, then those errors are designated as determinate errors. However the errors are due to the fluctuations in temperature during the repeated experimentation and hence they are termed as indeterminate errors.

The outcome of statistical decision depends upon the selected sample. Statistics involves the examination of the relatively small sample and making decision about the population. The sample should be carefully chosen, subsequently modified (if necessary) so that the sample has the same characteristics as the bulk material. During the dissolution of tablet dosage form, if the samples are collected from different positions of the dissolution rate testing apparatus at different time intervals then the statistician cannot find the source of variation in amount of drug dissolved because it may be due to the sample position or due to the time. In this experiment, the data is graphically represented by plotting a graph in between time (X-axis) and amount of drug dissolved (Y-axis). The validity or reproducibility of the parameter represented on Y-axis depends upon the sample collected at regular intervals of dissolution study.

1.1 Types of Statistics

The statistical procedures are divided into descriptive and inferential. Descriptive statistics describes the observed/collected data. It represents the procedure used to organize, categorize and summarize the data collected from the experiment. Such statistics accurately, efficiently and effectively represent the observed outcome in a clear and understandable manner. Examples include average of outcome, pie charts, bar graphs or other visual representations.

Inferential statistics make predictions about a large population based on a sample from those populations. It uses the statistical tests such as ANOVA, t-test, chi-square test.

Inferential statistics involves the following steps:

1. Identification/establishment of a research problem
2. Establishment of a hypothesis
3. Selection of suitable statistical test
4. Selection of data
5. Collection of required data
6. Performing the statistical tests
7. Making the decision based on the test results.

1.2 Parameters and Statistics

Statistical data usually involves a relatively small portion of data. This data is numerically manipulated to take decisions about the population. Parameters are characteristics of population and statistics are characteristics of samples. Parameters are usually represented by Greek symbols (μ, σ, ψ) and statistics are denoted by letters ($\overline{X}$, S^2, and r).

***Example*:**

A pharmaceutical company produces 20,000 Paracetamol tablets per batch. To meet pharmacopoeial standards, these tablets are subjected to various quality control tests such as weight variation, drug content (assay), hardness, disintegration time, friability and dissolution tests. To determine population parameter, all the formulated tablets should be weighed and from this either average weight or range of weights are calculated. However it is very tedious. If all the tablets are subjected to disintegration tests to determine the average disintegration time of 20,000 tablets (whole batch), it destroys all tablets (destructive testing) which is not desirable. Thus calculation of population parameters may be either impractical or impossible. So performance of a statistical analysis is preferred to make a statement about population.

Statistical analysis of the average weight can be explained as follows:

20 tablets should be collected by following appropriate sampling procedure. The average weight of 20 tablets is calculated. With some statistical manipulation of this data, we can get the actual average weight of entire batch of 20,000 tablets at the selected confidence interval (usually 95%). If the sample is carefully and accurately collected then the observed/claim (descriptive statistic) is 100 % accurate. Similarly disintegration and dissolution tests can be performed on 6 randomly collected tablets.

1.3 Sampling and Independent Observations

Random sampling from a population is essential for any inferential test. In random sampling, each individual member/observation has an equal chance of being selected for the sample. For example, in a batch of 20,000 tablets each tablet is having theoretically an equal chance for selection. The second requirement of inferential testing is that the observations are independent of each other. The sample should not affect the outcome of any other sample. For example, the disintegration time of the 6th tablet is not influenced by the disintegration time of other 5 tablets which are subjected to disintegration test. An independent result must represent an outcome not dependent on the result of any other observation.

1.4 Variables and Variation

A variable is any measurable property that can vary from one observation to another. These variables are identified and controlled in a well designed process/experiment. The variables shall be controlled to formulate a product having consistent quality. During the developmental stage, all probable variables which may directly/indirectly influence the quality, safety and effectiveness of a drug/drug delivery system should be identified. In process optimization studies, the process variables such as temperature, flow rate, speed of coating pan etc, and formulation variables like concentration of excipients should be optimized. The assay of a drug is influenced by the variables such as the method employed (gravimetric/volumetric/spectrophotometric), instrument used, analyst, sample, unidentified, uncontrollable errors (noise). In validation of a process, the process is splitted into different steps and the variables affecting each step are clearly defined.

The observation could have an infinite number of variables. For example, the hardness (observation) of a tablet is influenced by the variables like binder; moisture content of granules, granule size and size distribution, compression force etc. The variables should be controlled to achieve reproducible results.

The number of possible variables is limited only on imagination. Variables may be discrete or continuous. The variable should be thoroughly determined before selecting the appropriate statistical test.

1.4.1 Discrete Variables

A discrete variable is characterized by jumps and gaps between one value and the next. These variables are placed in specific, finite number of categories or classifications. Variables that can only take on a finite number of values are called "discrete variables. All qualitative variables are discrete. Some quantitative variables are discrete, such as performance rated as 1, 2, 3, 4 and 5 temperature rounded to the nearest degree. A type of variable, also called a categorical or nominal variable, which has a finite number of possible values that do not have an inherent order. For example, hair color would be a discrete variable, because it can only have a limited number of values, such as red, brown, and black, that does not occur in any particular order.

The solid dosage forms can be categorized as powders, tablets and capsules. The tablets can be classified based on passage or failure of a specific assay or dissolution requirement. These variables are also referred as qualitative, category or nominal. They can be differentiated as above and below the midpoint of distribution. Levels of discrete variable are exhaustive when the variable accounts for all possible outcomes. For example, capsule size 0-5 are exhaustive for humans; whereas capsule no 2 is not exhaustive for the human use because there are capsules other than the size 2. The levels of discrete variable may be set as mutually exclusive when the categories are not having the same number in common with each other. For example, the weight variation tolerances for uncoated tablets are categorized as average weight < 130 mg, 130-324 mg and more than 324 mg. Such average weights are not mutually exclusive because the average weights 130 and 324 mg are included in two of the discrete groups. To represent a mutually exclusive and exhaustive set of categories, the average weight group should be as follows: less than 130 mg, 131-324 mg and greater than 325mg.

1.4.2 Continuous Variables

Continuous variables also referred as quantitative variables. Discrete variables are usually based on counting, whereas continuous variables are involving measurements. Examples include viscosity, surface tension, blood glucose levels, systolic blood pressure. Theoretically a difference can be found in continuous variable value depending upon the sensitivity of the instrument used. For example, the plasma concentration of a sample may be analyzed by UV and HPLC techniques. The concentration of a drug detected from UV method may be 24 µg/ml whereas the concentration observed from HPLC method may be 24.6 µg/ml. Similarly in analytical balance of 10 mg sensitivity, the weight of the sample is 70 mg and the same sample weight may be 72, 72.4, 72.46 mg respectively from the balances having a sensitivity of 1mg, 0.1 mg and 0.01 mg respectively. Therefore, any measurement result for a continuous variable actually represents an interval from half a unit below or above the value.

Occasionally continuous variables cannot be easily measured but can be ranked in order of magnitude. The experimental results may be expressed as above or below a certain value (like midpoint). For example the temperature is usually represented as high, medium or low. Similarly, the pharmaceutical dosage forms are evaluated for a particular quality control test and the results are mentioned as passes/fails. In analgesic studies, pain severity is assigned with the scores like no pain = 0, slight pain = 1, moderate pain = 2 and severe pain = 3. Similarly the scores can be allotted to the taste of the drug.

1.4.3 Measurement of Variables

The variables are measured in terms of scales. In case of **nominal scale**, observations are classified qualitatively based on a characteristic. They differ in nature and cannot be arranged in a meaningful order. E.g.: medicines in a shop are arranged into tablets, capsules, liquids.

In **ordinal scale,** the items in a population are divided into two or more categories and the categories are arranged on a scale by some attribute, that is a ordinal scale. The difference between the two observations is not considered but their relative position is considered based on ranking with numbers. E.g. medicines in a shop are arranged in order of their acceptance by patients: very well accepted, accepted, tolerated, rejected. Such scale is not precise but extremely important in non parametric tests. Nominal and ordinal scales are also referred as non-metric scales.

The third type of scale is **interval scale** where the difference between each level of scale is equal. For example, the particle size is determined by optical microscopy and the observed data is represented as the number of particles belongs to a particular interval.

The fourth type of scale is **ratio scale** where the numerical value of one observation is divided with the other observation value. E.g.: the AUC value of formulation A is 100 μg ml/hr and the second formulation AUC is 50 μg ml/hr. Then the data can be expressed as the AUC value observed from the first formulation is twice the AUC value of second formulation. The interval and ratio scales are referred as metric scales.

1.4.4 Independent and Dependent Variables

Independent variables are qualitative (either continuous or discrete). The independent variables should be identified before conducting an experiment (predictor variable) and controlled during experiment. The variables which are measured against the independent variable are called dependent variables. During research work, the dependent variables are measured and compared based on different levels or categories of independent variables. For example the hardness of a tablet is the independent variable, if all the variables influencing the hardness are properly controlled. The hardness of such tablets may be tested with three different hardness testers (Monsanto, Pfizer and Strong Cobb). The variations in hardness observed from these testers are dependent variable as it is influenced by the device. The results are different even though the hardness of the tablet is same. Such extraneous factors which may influence the dependent variable are known as confounding or nuisance variable.

1.5 Applications of Statistics in Pharmacy

- Analytical statistics such as correlation, regression is used to establish the functional relation between variables like time and amount of drug dissolved.
- Descriptive statistical parameters such as mean and standard deviation are used to express the data generated from the replicated experiments to confirm the reproducibility.
- Inferential statistical tests such as t test, ANOVA are used in bioavailability and bioequivalence studies.
- Applied statistics is used in quality control tests.
- Statistical treatment of data is required for validation of a process.
- Statistical quality control charts are used to determine the quality of the product.
- Statistical knowledge is essential for validation of analytical technique.
- A useful tool in the design of experiments like bioavailability studies.
- Useful in interpretation of experimental results.
- Statistical usage is recommended by the drug regulatory agencies.
- Collection of true representative sample from the bulk needs the concept of statistics.

CHAPTER 2

MEASURES OF CENTRAL TENDENCY AND MEASURES OF DISPERSION

Central tendency is a descriptive statistics involving the continuous variables. The center of dispersion and the dispersion of observations within the distribution are important in inferential statistics such as ANOVA and t-test. The center of dispersion is generally represented with mode, median and mean. The dispersion of the observed values from the central value is denoted with the range, variance and standard deviation.

2.1 Measures of Central Tendency

Measures of central tendency can be used when the collected data is depicted as ordinal, interval or ratio scales. The word average or the term, measures of central tendency has been defined by various authors as given below.

Simpson and Kofka observe that "A measure of central tendency is a typical value around which other figures congregate". Ya Lun Chou states that "an average is a typical value in the sense that it is sometimes employed to represent all the individual values in a series of a variable".

2.1.1 The Objectives of Average

- To arrive at a single value that is representative of the characteristics of the entire mass of data.
- To facilitate comparison.
- To trace precise relationship.
- To know about the universe from the sample
- To help in decision making

2.1.2 Characteristics of a Good Average

- It should be rigidly defined.
- Its definition should be in the form of a mathematical formula.
- It should be easy to calculate and simple to follow.
- It should be based on all observations in the series.
- It should not be affected much by a few extreme values.
- It should be capable of being used in further statistical comparison or processing.
- It should possess sampling stability.

2.1.3 Types of Averages

Measures of central tendency or averages are usually of either mathematical or positional. Arithmetic mean ($\bar{x}$), geometric mean (G.M) and harmonic mean (H.M) fall under mathematical mean while mean (Md) and mode (Mo) are representing positional average.

2.1.4 Arithmetic Average (Mean)

The mean for a large number of population or universe is known as the universe or population mean (μ).The mean value is calculated by summing the observed data and dividing with the number of observations. Compared with the other measures of central tendency, mean allows the consideration of data spreading and use the extreme values for calculation.

Illustration: An anti diabetic drug was administered to a group of persons consisting of diabetic individuals and persons with diabetic but not yet diagnosed. Then the average blood glucose level computed from that data indicates population average.

S No of the Patient	Blood Glucose Level (mg/dl)
1	84
2	86
3	86
4	88
5	85
6	84

***Solution*:** The average blood glucose level

$$= \frac{84+86+86+88+85+84}{6}$$

$$= \frac{513}{6} = 85.5 \text{ mg/dl}$$

Most of the cases population mean is not advisable to measure due to the problems in identifying all the population suffering with disease, more time requirement, highly expensive, unethical and impracticable. So a sample of n objects is taken from a population to estimate the properties of population. The average value of such observed data represents the sample mean ($\bar{x}$) and expressed as $\bar{x} = \frac{(\Sigma x_i)}{n}$. The value of $\bar{x}$ approaches to μ, as the number of measurements is increased.

In a pharmaceutical industry, thousands of tablets are produced per batch and the average weight of the tablet is determined with 20 tablets. The mean weight (average weight) observed from the individual weights of 20 tablets indicates sample mean.

Illustration: Determine the average weight of 20 tablets randomly collected from a batch of 20,000 tablets.

S No	Weight (mg)
1	99
2	98
3	97
4	96
5	98
6	100
7	101
8	97
9	103
10	96
11	98
12	100
13	102
14	99
15	103
16	102
17	99
18	100
19	98
20	99

Solution: Average weight $= \frac{(\Sigma x_i)}{n} = \frac{1985}{20} = 99.25$ mg

Mean Value from the Frequency Data

Frequency distribution data is one of the descriptive statistical analyses used to simplify and summarize the collected data in the form of table. In frequency distribution data, the collected observations (variables) are classified into discrete variable. For example, the particle size of 100 particles are measured and listed in the following Table 2.1.

Table 2.1 The particle size data of suspended particles observed before subjecting to freeze thaw cycle

2	4	6	5	7	9	23	10	8	7
4	5	7	8	3	6	9	12	17	9
6	8	21	12	14	3	6	8	10	22
5	19	7	13	8	10	22	14	8	21
7	10	12	8	6	15	24	7	10	13
6	9	13	15	18	12	10	14	16	14
10	8	6	7	9	21	12	15	14	17
12	11	18	4	7	8	13	16	16	13
16	12	10	18	9	10	10	22	24	15
14	10	11	6	7	9	24	15	17	16

Guidelines for selection of intervals in a frequency table.

1. Intervals should be selected in relation to the nature of data. The intervals such as 0-2 and 26-28 are meaningless from the above data, as the observations are not fitted in this range.
2. Half of the selected intervals that contain least frequency should contain at least 10% of the data.
3. Eight to twenty intervals are preferable.

Sometimes the pharmaceutical scientist has to calculate the mean value from the frequency data. While analyzing the particle size of the dispersed particle in the suspension or the globule size in an emulsion by employing microscopic technique, the following type of data is generated (i.e., projection of 100 particles under microscope). Such data is very confusing and meaningful interpretation is difficult. Interpretation of changes in particle size due to ageing or exposure to various stress conditions is highly difficult. Such changes can be easily solved by following frequency mean calculations. The above table can be presented as follows to get some meaningful interpretation.

Table 2.2 Frequency distribution data using 6 class intervals

Class Interval	Frequency	% Frequency
2-5	9	9
6-9	33	33
10-13	25	25
14-17	19	19
18-21	7	7
22-25	7	7

Table 2.3 The mean particle size of the dispersed particles

Class Interval (x)	Frequency (f)	Midpoint of Class Interval (m)	fm
2-5	9	3.5	31.5
6-9	33	7.5	247.5
10-13	25	11.5	287.5
14-17	19	15.5	294.5
18-21	7	19.5	136.5
22-25	7	23.5	164.5
Total	100	81	1162

$$\bar{x} = \frac{\Sigma fm}{\Sigma f} = \frac{1162}{100} = 11.62\ \mu$$

Table 2.4 Frequency distribution data using 8 class intervals

Class interval	Frequency	% Frequency
2-4	6	6
5-7	23	23
8-10	24	24
11-13	14	14
14-16	16	16
17-19	7	7
20-22	6	6
23-25	4	4

Table 2.5 The mean particle size of the dispersed particles

Class Interval (x)	Frequency (f)	Midpoint of Class Interval (m)	fm
2-4	6	3	18
5-7	23	6	138
8-10	24	9	216
11-13	14	12	168
14-16	16	15	240
17-19	7	18	126
20-22	6	21	126
23-25	4	24	96
Total	100		1128

$$\bar{x} = \frac{\Sigma fm}{\Sigma f} = \frac{1128}{100} = 11.28\ \mu$$

Table 2.6 The particle size data of suspended particles observed after subjecting to freeze thaw cycle

18	15	12	22	18	21	13	20	7	11
10	24	23	9	24	7	15	12	18	21
22	6	3	23	12	16	9	19	8	17
9	20	21	10	18	22	11	13	10	12
24	11	7	11	5	14	20	5	16	14
22	14	19	6	8	19	13	16	17	8
8	7	12	5	12	16	24	9	23	12
23	22	9	17	21	6	18	10	14	9
6	16	23	7	9	13	6	24	6	21
24	22	10	21	23	12	22	17	11	6

Table 2.7 Frequency distribution data using 6 class intervals

Class Interval	Frequency	% Frequency
2-5	3	3
6-9	24	24
10-13	22	22
14-17	15	15
18-21	17	17
22-25	19	19

Table 2.8 The mean particle size of the dispersed particles

Class Interval (x)	Frequency (f)	Midpoint of Class Interval (m)	fm
2-5	3	3.5	10.5
6-9	24	7.5	180
10-13	22	11.5	253
14-17	15	15.5	232.5
18-21	17	19.5	331.5
22-25	19	23.5	446.5
Total	100		1454

$$\bar{x} = \frac{\Sigma fm}{\Sigma f} = \frac{1454}{100} = 14.54\ \mu$$

From the above data, the average particle size of the dispersion observed before and after subjecting to freeze thaw cycle was found to be 11.62 μ and 14.54 μ respectively. The changes in particle size are characterized only with the treatment of the observed data as above.

Merits of Arithmetic Mean

- It is rigidly defined so that different interpretations by different persons are not possible.
- It is easy to understand and easy to calculate.
- It is considered as more representative of distribution, as it takes all values into consideration.
- It is possible to calculate even in absence of some data.
- It gives weight to all items in direct proportion to their size.
- It is less affected by fluctuations of sampling.
- It provides a good basis for comparison.

Demerits of Arithmetic Mean

- It cannot be determined by inspection nor by graphically.
- Not suitable to study the qualitative phenomena due to non numerical measurements.
- It is very much affected by extreme values.

2.1.5 Mode

The mode is simply the value which is most popular or observed with the greatest frequency. According to Croxton and Cowden "The mode of a distribution is the value at the point around which the items tend to be most heavily concentrated". In the words of

A.M. Tuttle, "Mode is the value which has the greater frequency density in its immediate neighborhood". It is illustrated with the following example.

The size of the observed particles is 2, 4, 5, 8, 6, 5, 4, 2, 5, 6, 8, 5, 6, 4, 5, 8, 5, 6, 4 and 5.

The above data can be categorized as follows:

Particle Size (μ)	Frequency (f)
2	2
4	4
5	7
6	4
8	3

In the above example higher frequency values are observed for the particle size 5. So the mode is 5.

The mode value can be observed from the following histogram.

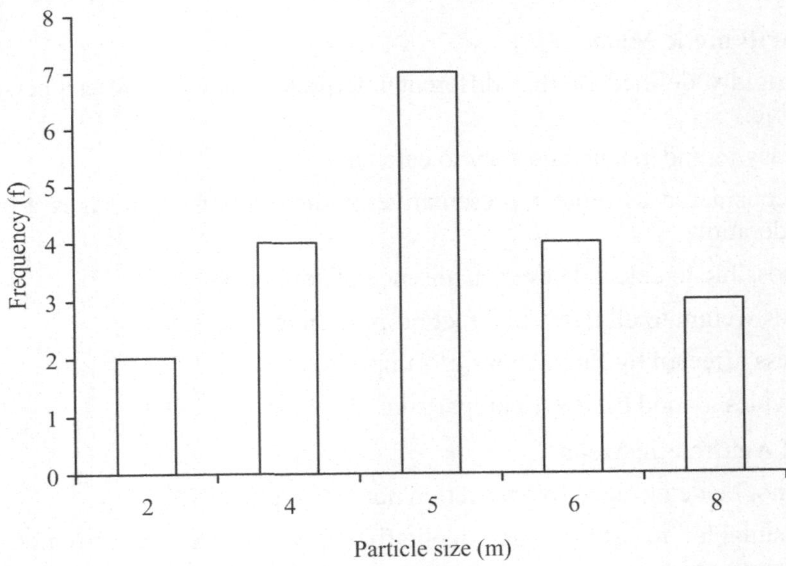

Mode values are useful only when the data is divided into categories. A distribution may be multi mode i.e., having different values that have same greatest relative frequency. An example of bimodal distribution observed from the plasma concentrations associated with oral and parental injection is shown in the following histogram.

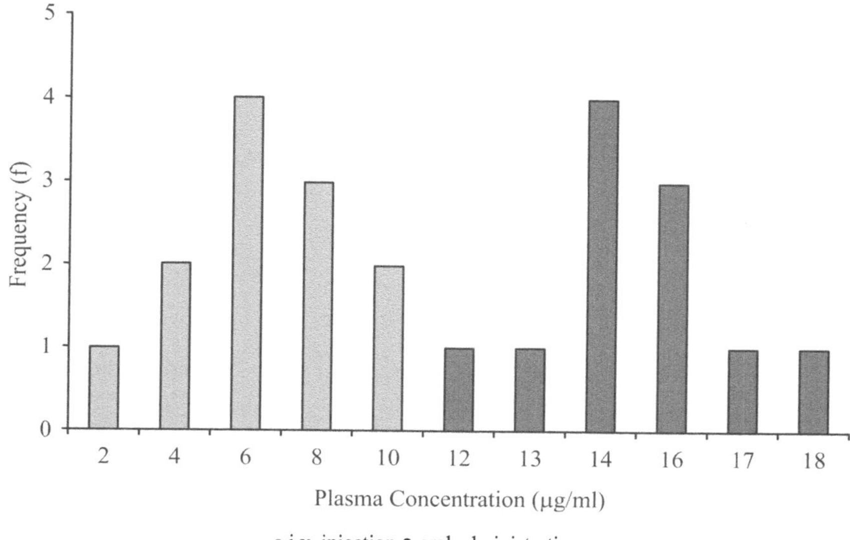

The first histogram (left side histogram) represents the i.v. injection and the other histogram represents oral route where low plasma concentrations are seen at the same time (1 hour of post administration).

The plasma concentration values observed from i.v. injection is 14, 13, 16, 12, 17, 16, 14, 12, 18, 14, 16 and 14.The mode is 16.

The plasma concentration observed after oral administration is 4, 6, 2, 8, 10, 6, 4, 6, 8, 10, 6 and 8. The mode value is 14.

Mode of Grouped Data

In case of grouped data, the computation of mode can be studied under the unusual classification of discrete and continuous series. In a uni-modal distribution where the highest concentration is in a single discrete value there should not be any difficulty in locating modal value or a class –interval containing this value by inspection.

Illustration: Mode in the following series

Particle Size (μ)	Frequency (f)
15	40
20	50
30	60
40	80
50	70
60	65
70	35

In the above example two highest frequency values are 30 and 40. The first impression will be that the mode lies in value 40 because the highest frequency concentration of 80 is in it, but when we add the neighboring frequencies as described below:

60	80
80	70
70	65
-----	-----
210	215
-----	-----

We find that the mode lies in the class value of 50 with 70 frequencies because of the influence of the frequency in the neighboring groups. If the totals are equal or the difference is not significant, resort to grouping is required.

Steps in grouping table:

- Prepare a grouping table with 6 columns
- Write down the frequency against the respective term in column 1.
- Group the frequency in twos, starting from the top and enter in column 2. Found the totals and identify the highest value.
- Group the frequency in twos, leaving the first frequency and enter in column 3. Found the totals and identify the highest value.
- Group the frequency in threes, starting from the top and enter in column 4. Found the totals and identify the highest value.
- Group the frequency in threes, leaving the first frequency and enter in column 5. Found the totals and identify the highest value.
- Group the frequency in threes, leaving the first and second frequency and enter in column 6. Found the totals and identify the highest value.
- Construct the analysis table to identify the item which repeats largest number of times.

Illustration: Determine the Mode from the following series

Size	20	40	60	80	100	120	140
Frequency	8	15	23	32	16	4	2

Grouping Table

Size	**Frequency**	**Grouping**				
		2	**3**	**4**	**5**	**6**
20	8	23		46		

Table *Contd...*

40	15		38		70	
60	23	**55**				**71**
80	32		**48**	**52**		
100	16	20			22	
120	4		6			
140	2					

Analysis Table

Column	Size of items containing maximum frequency					
1	80					
2	80	60				
3	80		100			
4	80		100	120		
5	80	60			40	
6	80	60	100			
No. of times the item occurs	6	3	3	1	1	

Since item 80 occurs 6 times, it is mode value.

Interpolation Formulae

The exact value of mode in the case of continuous frequency distribution can be obtained by following the interpolation formulae.

$$M_o = l_1 + \frac{\Delta_1}{\Delta_1 + \Delta_2} \times h \quad \text{or} \quad l_2 - \frac{\Delta_2}{\Delta_1 + \Delta_2} \times h$$

where l_1 and l_2 are the real lower and upper limits of the model class

h = magnitude of the modal class = $l_2 - l_1$

Δ_1 = $f_1 - f_0$

Δ_2 = $f_1 - f_2$

f_1 = frequency of the modal class

f_2 = frequency of the class preceding the modal class

f_0 = frequency of the class succeeding the modal class

Illustration: The particle size range and its distribution are furnished below. Find the mode.

Size range	0-10	10-20	20-30	30-40	40-50
Frequency	12	26	30	18	14

Size Range	Frequency
0-10	12
10-20	26 f_2
20-30	30 f_1
30-40	18 f_0
40-50	14

Highest frequency is 30

∴ Corresponding class = Mode class = 20 – 30

∴ $l_1 = 20$

$f_1 = 30, f_2 = 26, f_0 = 18$

$\Delta_1 = f_1 - f_0 = 30 - 18 = 12$

$\Delta_2 = f_1 - f_2 = 30 - 26 = 4$

N = magnitude of modal class = $l_2 - l_1 = 30 - 20 = 10$

$$\therefore \quad \text{Mode} = l_1 + \frac{\Delta_1}{\Delta_1 + \Delta_2} \times h$$

$$= 20 + \frac{12}{12+4} \times 10$$

$$= 20 + \frac{120}{16}$$

$$= 20 + 7.5 = 27.5$$

∴ Mode = 27.5 μ

Location of Mode Graphically

From the frequency distribution, mode can also be located graphically by using the following steps.

- Draw the histogram of the given distribution.
- Join the top right corner of the highest rectangle by a straight line to the top right corner of the proceeding rectangle.
- Join the top left corner of the highest rectangle by a straight line to the top left corner of the rectangle on the right.
- Draw a perpendicular on the horizontal axis from the point of intersection of the two diagonal lines.
- Indicate the mode from the foot of the perpendicular.

Illustration: The particle size range and its distribution are furnished below. Find the mode graphically.

Size range	0-10	10-20	20-30	30-40	40-50
Frequency	16	24	32	17	11

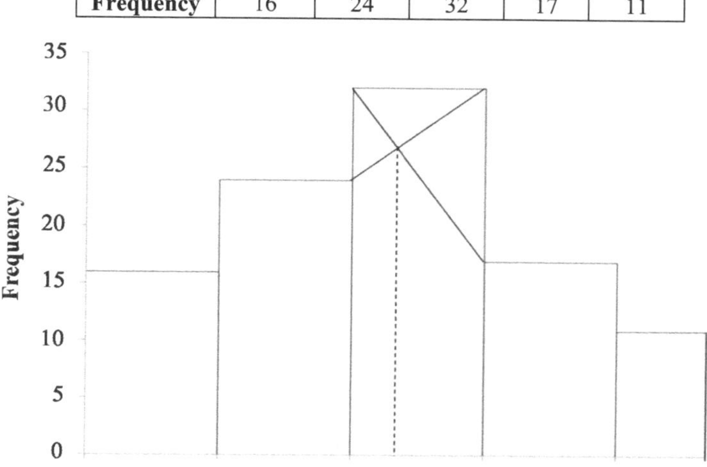

Advantages of Mode

- It is readily understood and easily calculated.
- It is not affected by extreme observations.
- It can be conveniently located even if the class interval of unequal distribution is provided.

Disadvantages of Mode

- It is not based on all observations
- It is not capable of further mathematical treatment
- In many cases, it is impossible to get definite value of mode
- It is affected by sampling fluctuations

2.1.6 Median

Median is the value of middle item of a series arranged in an ascending or a descending order of magnitudes. Secrist states that "Median of a series is the value of the item actual or estimate when a series is arranged in order of magnitude which divides the data in to two parts". L.R.Connor defines median as 'that value of a variable which divides the group in to two equal parts, one part comprising all values greater and the other all values less than the median". The median is the center point of the distribution. It divides the data in to two equal parts.

It is calculated with the following steps.

1. Collect the data.
2. Arrange the data in a rank order.
3. Observe the central value in case of odd number of observations and the mean of the two central values in case of even observations and express it as the median value.

Illustration: Calculate the median from the following data 5, 7, 6, 8, 7, 5, 7, 6 and 7.

Above data is rearranged as 5, 5, 6, 6, 7, 7, 7, 7 and 8.

The central value (median) of the above data is 7.

In the above illustration, the number of items was odd and there was no difficulty in locating the middle items and its corresponding value.

In case the number of items in a series is even, the two middle values can be taken as the median but conventionally an arithmetic average of the two middle values is taken as the median. This is illustrated with the following example.

Calculate the median from the following observations 2, 4, 5, 6, 2, 4, 5, 7, 5 and 2.

The above data is orderly arranged as 2, 2, 2, 4, 4, 5, 5, 5, 6 and 7.

In the above example, the central values are 4 and 5 and the median is the average of these two values i.e., 4.5

Example: The disintegration time observed from two batches of tablets formulated with two disintegrating agents is given below. Observe the better disintegrating agent base on median value.

Batch 1	6	8	4	9	10	6
Batch 2	12	10	8	13	9	10

The above data is orderly arranged as

Batch 1	4	6	6	8	9	10
Batch 2	8	9	10	10	12	13

The median of batch 1 is $6 + 8/2 = 7$

The median of batch 2 is $10 + 10/2 = 10$

The disintegrating agent present in Batch 1 tablet is better as it offered low disintegration time.

Median value is a better estimate than mode but is not affected by extreme values. Two sets of data may have same median but differ in their extreme values. However these extreme values are also important for inferential statistics to derive the meaningful results from the observed data.

For example the median values from the following two sets are equal.

The disintegration time observed from six tablets of first batch is 12, 11, 12, 11, 13, 12 respectively and the observed values from the second batch are 5, 8, 14, 10, 14 and 15.

In both cases the median value is 12 but the observed extreme values from the first set are 11 and 13 which are very close to the central value. However the extreme values of second batch are 5 and 15. The first batch is more preferable than the second batch as the extreme values are distributed very narrowly towards the central value and indicating good consistency.

Median value is defined by its position in orderly arranged values of a set. Sometimes it is desirable to describe the particular observation with respect to its position in a set of data. Most effectively it is represented in terms of percentile score.

$$\text{Percentile} = \frac{\text{number of values less than the given value}}{\text{total number of values}}$$

For example; 12,000 students appeared for GPAT examination. 11,500 students got less marks than Krishna. Then the percentile score of Krishna is 11500/12000 = 0.9583; usually it is multiplied with 100 then the value becomes 95.83.

Computation of median in a continuous series

The exact value of median in the case of continuous frequency distribution can be obtained by the following interpolation formulae.

$$\mathrm{M_d} = l_1 + \frac{\frac{1}{2}N - C}{f} \times h \qquad \text{or}$$

$$\mathrm{M_d} = l_2 - \frac{\frac{1}{2}N - C}{f} \times h$$

where, $\mathrm{M_d}$ = The value of the median

l_1 and l_2 = The lower and the upper limit of the class in which the median lies

f = The frequency of the median class

C = The cumulative frequency of the class preceding the median class or the sum of frequencies of all classes lower then the median class

h = The class interval of the median class, i.e., $l_2 - l_1$

Illustration: The particle size range and its distribution are furnished below. Find the median.

Size Range	0-10	10-20	20-30	30-40	40-50
Frequency	12	20	38	16	14

Solution: Given distribution is

Size Range	Frequency	Cumulative Frequency
0-10	12	12
10-20	20	32
20-30	38	70
30-40	16	86
40-50	14	100 = N

$\therefore$ N = 100, N/2 = 50

The cumulative frequency just more than 50 is 70. So corresponding class is 20-30 it is median class.

$\therefore$ $l_1 = 20$, its frequency is 38

$\therefore$ $f = 38$, and c = proceeding cumulative frequency = 32;

h = the median class interval = 30 – 20 = 10

$$\therefore \quad \text{Median } M_d = l_1 + \frac{\frac{1}{2}N - C}{f} \times h$$

$$= 20 + \frac{50 - 32}{38} \times 10$$

$$= 20 + \frac{18}{38} \times 10$$

$$= 20 + \frac{180}{38} = 20 + 4.73 = 24.73\ \mu$$

Advantages of Median

- It is easy to calculate and is readily understood
- It is not at all affected by the extremes
- The median value can be located graphically.
- It is the only average that can be used in qualitative analysis

Disadvantages of Median

- It cannot be exactly determined in case of even number of observations.
- It is not suitable satisfactorily for algebraic treatment.
- It is not based on all items of the series
- It is much affected by fluctuations of sampling

2.1.7 Comparison of Mode, Median and Mean Values

In a symmetrical distribution mean, median and mode are identical and have the same value. However, in actual life most distributions are not symmetrical. In moderately asymmetrical distributions, the values of mean, median and mode have the following empirical relationship given by Karl Pearson:

- Mean – Mode = 3(Mean – Median)
- Mean = Mode + $\left(\frac{3}{2}\right)$ (Median – Mode) = $\left(\frac{1}{2}\right)$ (3Median – Mode)
- Mode = 3 Median – 2 Mean
- Median = Mode + $\left(\frac{3}{2}\right)$ (Mean – Mode) = $\left(\frac{1}{2}\right)$ (2 Mean + Mode)

Mode, median and mean values can be compared from the following data.

Sample 1	Sample 2	Sample 3
20	30	20
30	50	30
50	20	20
30	30	50
20	20	30

In the above example mean = median = 30 and mode = 50

Example: Calculate the mean, mode and median from the following data

Sample 1	Sample 2	Sample 3	Sample 4
10	10	10	5
10	15	10	10
20	15	20	20
30	30	25	25
30	30	35	30

The above example can be summarized as follows

Sample	Mean	Median	Mode
1	20	20	30
2	20	15	30
3	20	20	35
4	18	20	30

Sample 1 & 2 have same mean and mode but have different median

Sample 1 & 3 have same mean, median but have different mode

Sample 1 & 4 have same median & mode and have different mean

Illustration: Calculate the mean, median and mode from the following data.

Size Range	0-4	4-8	8-12	12-16	16-20	20-24	24-28
Frequency	6	12	21	32	19	7	3

Solution: Given distribution is

Size Range	Frequency (f_i)	Mid values (x_i)	f_i x_i	Cumulative Frequency
0-4	6	2	12	6
4-8	12	6	72	18
8-12	21	10	210	39
12-16	32	14	448	71
16-20	19	18	342	90
20-24	7	22	154	97
24-28	3	26	78	100=N
	100		1316	

$$\text{Mean} = \bar{x} + \frac{\Sigma f_i x_i}{\Sigma f_i} = \frac{1316}{100} = 13.16$$

Mean $$\bar{x} + \frac{\Sigma f_i x_i}{\Sigma f_i} = \frac{1316}{100} = 13.16$$

and here N = 100, $\therefore \ \frac{N}{2} = 50$

The cumulative frequency just more than 50 is 71.

$\therefore$ Median class = 12 – 16 $\therefore l_1 = 12$

and h = 16 – 12 = 4

Median frequency = f = 32

c = preceding cumulative frequency = 39

$$\therefore \qquad \text{Median} = l_1 + \left(\frac{\frac{N}{2} - c}{f} \right) h$$

$$= 12 + \left(\frac{50 - 39}{32} \right) 4 = 12 + \frac{(-11)}{8}$$

$$= 12 - 1.37$$

$$= 10.63$$

So we have mean = 13.16 μ

Median = 10.63 μ

∴ Mode = 3(median) – 2 (mean)

= 3(10.63) – 2 (10.63)

= 10.63

2.1.8 Geometric Mean

The geometric mean of a series containing N observations is the Nth root of the product of values. If there are two values, the square root of the product of two values is the geometric mean. It is most appropriate average to be used when it is desired to give more weightage to smaller items and small weightage to larger items. Geometric mean is used when the data is positively skewed i.e., most of the values are located to the left side of the distribution compared to the right side and accordingly the calculated mean values are slightly pulled to the right. It is showed with the following example.

Illustration: The plasma concentrations observed from 12 patients is given below. Calculate the mean value.

Subject	Plasma concentration (μg/ml)
1	1.1
2	1.5
3	2.1
4	1.5
5	3.7
6	1.6
7	1.7
8	1.4
9	2.6
10	1.7
11	1.6
12	7.1

The above data after arranging in ascending order,

Subject	Plasma concentration (µg/ml)
1	1.1
8	1.4
2	1.5
4	1.5
6	1.6
11	1.6
7	1.7
10	1.7
3	2.1
9	2.6
5	3.7
12	7.1

The graphical representation of data in normal distribution (i.e., plasma concentration Vs frequency) is as follows.

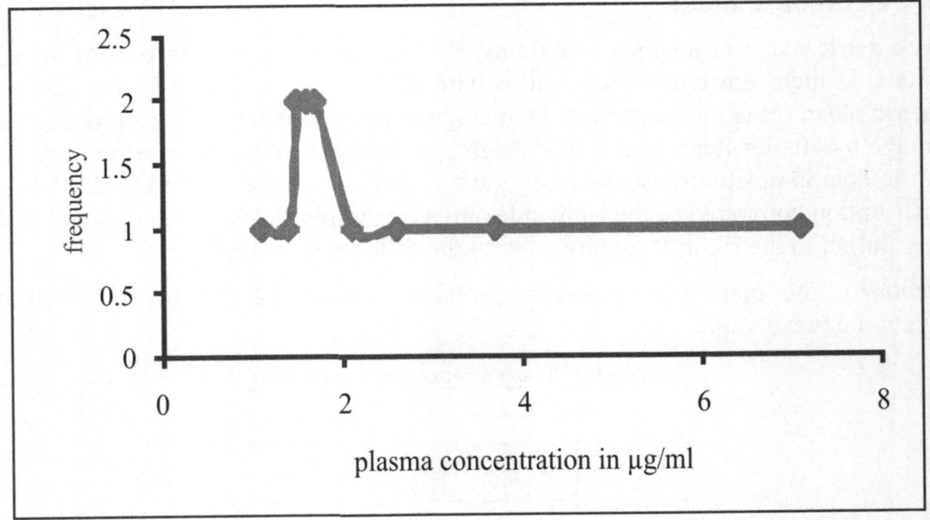

The above graph is showing positively skewed. So the plasma concentrations are transformed to logarithmic value and the following log normal distribution is constructed (log plasma concentration Vs frequency). This graph is not positively skewed.

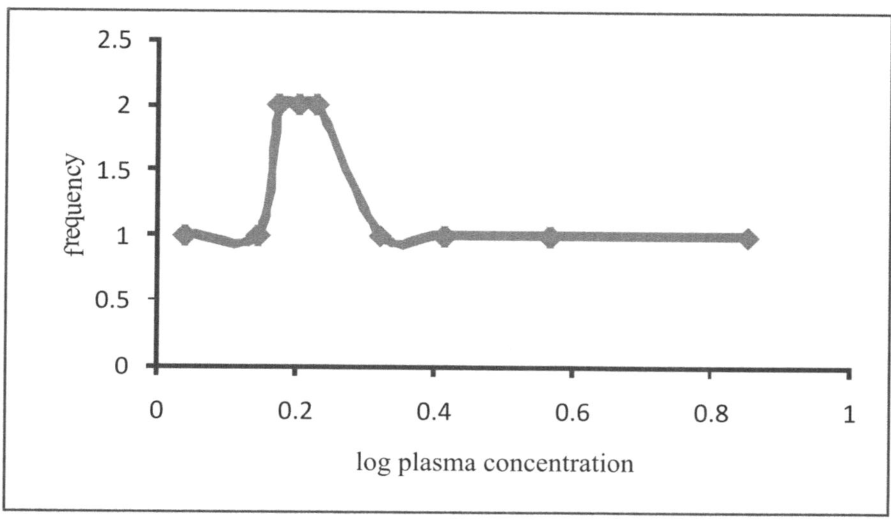

Calculation of Geometric Mean

1. Arrange the data in ascending order
2. Transform the data into log values
3. Calculate the logarithmic mean value(i.e. summing the log values and dividing with the number of observations)
4. Take the anti log to the log mean value and express it as geometric mean value.

The log transformation of the given data results the following table.

Subject	Plasma concentration (μg/ml)	Log transformation
1	1.1	0.041
8	1.4	0.146
2	1.5	0.176
4	1.5	0.176
6	1.6	0.204
11	1.6	0.204
7	1.7	0.23
10	1.7	0.23
3	2.1	0.3222
9	2.6	0.414
5	3.7	0.568
12	7.1	0.851
Total	27.6	3.5622
Mean	2.3	0.2968

Geometric mean X_G = anti log (0.2968)

= 1.98 μ

Advantages of Geometric Mean

- It is rigidly defined and its value is precise
- It is based on all observations of a series
- It is not much affected by the presence of extremely small or large observations
- It is suitable for further mathematical treatment
- It is not much affected by the fluctuations of sampling

Disadvantages of Geometric Mean

- It is neither easy to calculate nor simple to understand
- It may be a value which does not exist in the series
- If any value in a series is zero, the geometric mean would also be zero

2.1.9 Harmonic Mean (H.M)

Harmonic mean of a series is the reciprocal of the arithmetic average of the reciprocal of the values of its various items.

The arithmetic mean of the reciprocals of *N* values $X_1, X_2, X_3, , ,X_n$ is

$$\frac{(1/X_1)+(1/X_2)+......+(1/X_n)}{n}$$

The harmonic mean is the reciprocal of the arithmetic mean:

$$\text{H.M.} = \frac{n}{(1/X_1)+(1/X_2)+......+(1/X_n)} = \text{Reciprocal}\left[\frac{\Sigma(1/X)}{n}\right]$$

Illustration: Find the harmonic mean from the following data.

9.46, 87.58, 152.46, 0.28, 0.06

Solution: Given details 9.46, 87.58, 152.46, 0.28, 0.06

x_i	$1/x_i$
9.46	0.1057
87.58	0.0114
152.46	0.0065
0.28	3.5714
0.06	16.666

$$\Sigma\frac{1}{x_i} = 20.361$$

$$\therefore \qquad H.M = \frac{n}{\Sigma \frac{1}{x}} = \frac{5}{20.361} = 0.2455$$

Advantages of Harmonic Mean

- It is rigidly defined and is based on all observations
- It is capable of further algebraic treatment
- It is not much affected by fluctuations of sampling

Disadvantages of Harmonic Mean

- It is not readily understood and is difficult to compute
- It is only a summary figure

2.2 Measures of Dispersion

Dispersion is an important measure sought for describing the characteristics of variability in data. It finds out how individual values fall apart, on an average from the representative values. Brooks and Dick define "dispersion is the degree of the scatter or variation of the variable about a central value". According to A.L.Bowley "dispersion is the measure of the variations of the item." The different measures of central tendency are mean, mode and median. Measure of dispersion allows the spreading of the data around the average value. The central value may be same in two or more distribution but may differ with respect to dispersion. For example consider the two series.

1.	5	10	15	20	75
2.	25	25	30	20	

The two series have same mean 25,but the variation from mean to first series are 20, 15, 0, 5, – 50; and the second series are 0, 0, –5, 5 respectively. These two variations are different. The measure of dispersion helps us in studying the important characteristics of the distribution. It measures the extent to which there are differences between individual observations and central value. The dispersion may be defined as the measure of variations of variable.

2.2.1 Objectives of Measuring Dispersion

Measures of dispersion are calculated to serve the following purposes.

- To judge the reliability of measures of central tendency
- To compare two or more series with regard to their variability
- To control the variability
- To facilitate the use of other statistical measure

2.2.2 Methods of Measuring Dispersion

1. Range
2. Mean deviation
3. Variance and standard deviation

1. Range: The range is the simplest method of studying the dispersion. It is defined as the difference between the highest and the lowest values of observations of the distribution.

E.g.: Find the range of distribution for the data 5, 10, 18, 19, 23, 27.

$$\text{Range } R = X_n - X_1 = 27 - 5 = 22$$

where, X_n = last item and X_1 = first item of a series arranged in the ascending order of magnitude.

Coefficient of Range

$$\text{Coefficient of range } = \frac{X_n - X_1}{X_n + X_1} = \frac{27-5}{27+5} = \frac{22}{32} = \frac{11}{16}$$

Illustration: Process control record of automatic compression of a tablet from four tablet compression machines is given below. Determine the range and average range of 4 machines (Tablet label weight 300 mg).

Time the tablet is sampled	Weight of the tablet(mg)					
	Machine A	Machine B	Machine C	Machine D	$\bar{x}$	R
8.30 am	296	294	298	301	297.25	7
10.30 am	292	290	302	292	294.0	12
12.45 pm	298	288	303	298	296.75	15
2.20 pm	302	304	294	302	300.5	10
4.30 pm	305	306	292	306	302.25	14
6.00 pm	295	301	296	305	299.25	10

Solution: Grand mean $= \dfrac{297.25 + 294 + 296.75 + 300.5 + 302.25 + 299.25}{6} = 298.37$

$$\text{Average range } R = \frac{7+12+15+10+14+10}{6} = 11.33$$

Merits and Demerits of Range

- It is easily calculated and readily understood
- It is not based on all the observations
- It is very much affected by the fluctuations of sampling
- It is not suitable for mathematical treatment
- It cannot be computed from frequency distributions

2. Mean Deviation or Average Deviation: Clearly range is a position average which is not based on all observations. So it does not show any scatter of observations from an average. Mean deviation is a measure of dispersion based on all items. The mean deviation is the arithmetic mean of the deviations of the individual values from the average of the given data. Mean deviation can be computed from any average, *i.e.* mean, median and mode. Mean is generally preferred because it is mathematically a sound measure of central value.

Mean deviation about mean $= \frac{\Sigma|x - \bar{x}|}{n}$ (For series of individual values)

Where $\bar{x}$ = mean, x – variable, n = number of observations

Mean deviation about mean $= \frac{\Sigma\, f|x - \bar{x}|}{\Sigma\, f}$ (For discrete data)

Mean deviation about mean $= \frac{\Sigma\, f|m - \bar{x}|}{\Sigma\, f}$ (For continuous data)

When m is mid values of class inverse

Mean deviation about median $= \frac{\Sigma\, |x - M|}{n}$ (For series of individual values)

When M = median of series

Mean deviation about median $= \frac{\Sigma\, f\,|x - M|}{\Sigma\, f}$ (For discrete data)

Mean deviation about median $= \frac{\Sigma\, f\,|m - M|}{\Sigma\, f}$ (For continuous data)

m is mid value of class interval, M = median.

Coefficient of Mean Deviation: Mean deviation calculated from any measure of central tendency is an absolute measure. When, it is divided by the average used to calculate it, we can get coefficient of mean deviation. It will give a relative measure of dispersion suitable to compare two or more series.

$$\text{Coefficient of M.D.} = \frac{\text{M.D.}}{\text{Mean or Median or Mode}}$$

Illustration: Calculate the mean deviation of mean, mode and median from the following data

The C_{max} values observed from the 6 subjects is 12.4, 14.6, 13.8, 10.4, 15.3 and 9.8. Calculate the mean, median, mode and the coefficient of mean deviation.

Solution: Given data is 12.4, 14.6, 13.8, 10.4, 15.3, 9.8

$$\text{A.M.} = \frac{\Sigma x_i}{n} = \frac{12.4+14.6+13.8+10.4+15.3+9.8}{6} = \frac{76.3}{6} = 12.7166$$

Arrange the data in increasing order. We get 9.8, 10.4, 12.4, 13.8, 14.6, 15.3

Here number of observations is 6 which is even

$$\therefore \quad \text{Median} = \frac{1}{2}\left[\frac{6}{2}\text{th term} + \frac{6+2}{2}\text{th term}\right]$$

$$= \frac{1}{2}[\text{3rd term} + \text{4th term}]]$$

$$= \frac{1}{2}\ [12.4+13.8] = 13.1$$

And mode = 3 (median) – 2(mean)

$= 3(13.1) - 2\ (12.7166)$

$= 39.3\text{-}25.4332$

$= 13.87$

$$\text{The coefficient of mean deviation} = \frac{\text{mean deviation}}{\text{mean}}$$

To find mean deviation about mean

x_i	$x_i - \bar{x}$	$\lvert x_i - \bar{x}\rvert$
12.4	– 0.3166	0.3166
14.6	1.8834	1.8834
13.8	1.0834	1.0834
10.4	2.3166	2.3166
15.	2.5834	2.5834
9.8	– 2.9166	2.9166
		11.1

$$\therefore \quad \text{Mean deviation about mean} = \frac{\sum\lvert x_i - \bar{x}\rvert}{n} = \frac{11.1}{6} = 1.85$$

$$\text{The coefficient of mean deviation} = \frac{\text{mean deviation}}{\text{mean}} = \frac{1.85}{12.7166} = 0.14547$$

Illustration: The absorbance values of a same solution observed from two spectrophotometers is given below. Suggest a suitable spectrophotometer based on the mean deviation.

Absorbance observed from spectrophotometer 1 is 0.104, 0.108. 0.112, 0.117, 0.106 and 0.114

Absorbance observed from spectrophotometer 2 is 0.110, 0.108, 0.120, 0.095, 0.104 and 0.116

Solution: To find the mean deviation of absorbance data observed from spectrophotometer 1.

x_i	$x_i - \bar{x}$	$\lvert x_i - \bar{x} \rvert$
0.104	– 0.006	0.006
0.108	– 0.002	0.002
0.112	0.002	0.002
0.117	0.007	0.007
0.106	– 0.004	0.004
0.114	0.004	0.004
		0.025

$$\therefore \bar{x} = \frac{\Sigma x_i}{n} = \frac{0.104 + 0.108 + 0.112 + 0.117 + 0.106 + 0.114}{6} = \frac{0.661}{6} = 0.1101 \cong 0.11$$

$$\text{Now the mean deviation about mean} = \frac{\Sigma \lvert x_i - \bar{x} \rvert}{n} = \frac{0.025}{6} = 0.004$$

To find the mean deviation about mean for the absorbance data observed from spectrophotometer 2 is 0.110, 0.108, 0.120, 0.095, 0.104, 0.116

x_i	$x_i - \bar{x}$	$\lvert x_i - \bar{x} \rvert$
0.110	0.0012	0.0012
0.108	– 0.0008	0.0008
0.120	– 0.0112	0.0112
0.095	– 0.0138	0.0138
0.104	– 0.0048	0.0048
0.116	0.0072	0.0072
0.653		0.0380

$$\text{A.M} = \frac{\Sigma \lvert x_i - \bar{x} \rvert}{n} = \frac{0.025}{6} = 0.1088$$

$$\therefore \quad \Sigma \lvert x_i - \bar{x} \rvert = 0.0380$$

Now the mean deviation about mean

$$= \frac{\Sigma|x_i - \bar{x}|}{n} = \frac{0.0380}{6} = 0.0063$$

The mean deviation values of spectrophotometer 1 is less than the mean deviation of spectrophotometer 2. Hence spectrophotometer 1 is suitable

Illustration: Find out the coefficient of mean deviation by using mean from the following data.

Globule Size	0.5	5-10	10-15	15-20	20-25	25-30	30-35
Frequency	4	16	25	30	14	8	3

Solution: Given data is arranged as follows.

Globule size	**Frequency** f_i	**Mid values** (x_i)	$x_i - 17.5$	$\frac{x_i - 17.5}{5} = y_i$	$y_i x_i f_i$
0-5	4	2.5	– 15	– 3	– 12
5-10	16	7.5	– 10	– 2	– 32
10-15	25	12.5	– 5	– 1	– 25
15-20	30	17.5	0	0	0
20-25	14	22.5	5	1	14
25-30	8	27.5	10	2	16
30-35	3	32.5	15	3	9
	100				– 30

$$\therefore \quad \text{Mean} = A + \frac{\Sigma f_i y_i}{\Sigma f_i} \times n$$

$$= 17.5 + \frac{-30}{100} \times 5 = 17.5 - \frac{150}{100} = 17.5 - 1.5 = 16$$

x_i	$x_i - \bar{x}$	$\lvert x_i - \bar{x}\rvert$	f_i	$\lvert x_i - \bar{x}\rvert x f_i$
2.5	– 13.5	13.5	4	94
7.5	– 8.5	8.5	16	136
12.5	– 3.5	3.5	25	87.5
17.5	1.5	1.5	30	45
22.5	6.5	6.5	14	91
27.5	11.5	11.5	8	92
32.5	16.5	16.5	3	49.5
				595

$$\therefore \quad \text{Mean deviation about mean} = \frac{\Sigma f_i |x_i - \bar{x}|}{\Sigma f_i} = \frac{595}{100} = 5.95$$

Advantages of Mean Deviation

- Mean deviation is easy to calculate among the measures of dispersion
- It is more comprehensive measure of variation calculated by using all observations
- It is less affected by extreme values compared to standard deviation

Disadvantages of Mean Deviation

- It ignores the positive and negative signs of deviation
- Is not suitable for further algebraic treatment
- It gives accurate results only when deviations are taken from median.

3. Variance and Standard Deviation

Standard Deviation: Standard deviation is the most commonly used absolute measure of dispersion. It was first introduce by Karl Pearson in 1893. It may be defined as the positive square root of the arithmetic mean of the squares of the deviations of given observations from their arithmetic mean

Standard deviation is defined as the square root of the variance.

The following formulae is used to find standard deviation of series of individual observation

$$s = \sqrt{\frac{\Sigma\,(x-\bar{x})^2}{(n-1)\text{ or } n}} = \sqrt{\frac{\Sigma\,(d\bar{x})^2}{(n-1)\text{ or } n}}$$

where $d\bar{x} = x - \bar{x}$ and $(d\bar{x})^2 = (x-\bar{x})^2$

(or)

$$s = \sqrt{\frac{\Sigma x^2 - \frac{(\Sigma x)^2}{n}}{n \text{ or } (n-1)}} \qquad \begin{bmatrix} n-1 \text{ for } & n \leq 30 \\ n \text{ for } & n > 30 \end{bmatrix}$$

Variance: variance is also called mean square deviation. The term was first introduced by R.A. Fisher. This is an important measure in the quantitative analysis of the data in the biological, agricultural and medical sciences.

To calculate variance

1. Find deviation of variables from the mean
2. Square the deviation
3. Add the above
4. Divide the above sum by (n – 1). When n is number of observation.

Sample variance $= s^2 = \frac{\Sigma(x-\bar{x})^2}{n-1}$ or $\frac{\Sigma(d\bar{x})^2}{n-1}$

Population variance $\sigma^2 = \frac{\Sigma(x-\mu)^2}{N}$

where $\bar{x}$ is arithmetic mean (or) The variance of series individual values is

$$s^2 = \frac{\Sigma x^2 - \frac{(\Sigma x)^2}{n}}{(n-1) \text{ or } n}$$

(**Note:** We use n–1 for number of observation less than 30 n for number of observations more than 30)

In the case of frequency distribution (discrete series)

Variance $= s^2 = \frac{\Sigma f x^2 - \frac{(\Sigma fx)^2}{\Sigma f}}{\Sigma f}$

For the continuous data the variance is obtained by $s^2 = \frac{\Sigma fm^2 - \frac{(\Sigma fm)^2}{\Sigma f}}{\Sigma f}$

[where m is mid values of class intervals]

Illustration: Determination of the standard deviation of the dispersed particles of a suspension

Class interval (x)	Frequency (f)	Midpoint of class interval (m)	Fm	m²	fm²
2-5	9	3.5	31.5	12.25	110.25
6-9	33	7.5	247.5	56.25	1856.25
10-13	25	11.5	287.5	132.25	3306.25
14-17	19	15.5	294.5	240.25	4564.75
18-21	7	19.5	136.5	380.25	2661.75
22-25	7	23.5	164.5	552.25	3865.75
Total	100	81	1162	1373.5	16365

$$s = \sqrt{\frac{\Sigma fm^2 - \frac{(\Sigma fm)^2}{\Sigma f}}{\Sigma f}} = \sqrt{\frac{16365 - \frac{(1162)^2}{100}}{100}}$$

$$= \sqrt{\frac{2862.56}{100}} = \sqrt{28.62} = 5.35$$

Advantages of Standard Deviation

- The standard deviation is calculated by using all observations.
- It is suitable for further mathematical treatment
- It is least affected by the sampling fluctuations
- It is useful to determine the reliability of the means of two or more different series.

Disadvantages of Standard Deviation

- It is neither easy to understand nor simple to calculate
- It is affected by the extreme values

Symbols used for sample and population measures of central tendency:

	Sample Statistic	Population Parameter
Mean	$\overline{X}$	μ
Variance	s^2	σ^2
Standard deviation	s	σ
Number of observations	n	N

Relationship between S.D. and Range

Range R is less efficient measure of variability compared to standard deviation but there is a relationship between these two values. Standard deviation is obtained from the range by divisor that depends on sample size.

Divisors for Estimating Standard Deviation from Range

Sample size (N)	Divisor (D)
2	1.1
3	1.7
7	2.7
10	3.1
30	4.0
100	5.0

Example: The hardness values 3 tablets are 5.4, 5.5, and 5.6 kg/m^2. Here the mean hardness value is 5.5 and the standard deviation is 0.1. The range is 0.2. The divisor for 3 is 1.7. Theoretically calculated s.d from the range = 0.2/1.7 = 0.11, which is equal to the practically observed standard deviation.

Solved Examples

1. Calculate the standard deviation of following data 2, 6, 9, 11, 15, 23, 30.

***Solution*:** Here the number of observations = n = 7

$$\Sigma x = 2 + 6 + 9 + 11 + 15 + 23 + 30 = 96$$

$$\Sigma x^2 = 2^2 + 6^2 + 9^2 + 11^2 + 15^2 + 23^2 + 30^2 = 1896$$

$$\therefore \quad \text{Standard deviation s} = \sqrt{\frac{\Sigma x^2 - \frac{(\Sigma x)^2}{n}}{n-1}}$$

$$= \sqrt{\frac{1896 - \frac{(96)^2}{7}}{6}}$$

$$= \sqrt{\frac{1896 - \frac{9216}{7}}{6}}$$

$$= \sqrt{\frac{1896 - 1316.57}{6}}$$

$$= \sqrt{\frac{579.45}{6}} = \sqrt{96.57} = 9.83$$

2. Calculate the standard deviation of 8, 10, 10, 10, 12, 13, 15, 15, 17 and 20

***Solution*:** Here the number of observations = 10

Variable x	x^2
8	64
10	100
10	100
10	100
12	144
13	169
15	225
15	225
17	289
20	400
$\Sigma x = 130$	$\Sigma x^2 = 1816$

$$\text{Standard deviation s} = \sqrt{\frac{\Sigma x^2 - \frac{(\Sigma x)^2}{n}}{n-1}}$$

$$= \sqrt{\frac{1816 - \frac{(130)^2}{10}}{10-1}}$$

$$= \sqrt{\frac{1816-1690}{9}}$$

$$= \sqrt{\frac{126}{9}}$$

$$= \sqrt{14}$$

$$= 3.74$$

Coefficient of Variation/Relative Standard Deviation (RSD): We know that standard deviation is an absolute measure of dispersion. It is expressed in terms of units in which original data are collected. It is clear that the standard deviation of the height of plants cannot compare with the standard deviation of weight of the plants because they are expressed in different units. That is heights in cm and weights in grams. So the standard deviation must be converted in to a relative measure for the purpose of comparison. "The relative measure of dispersion is known as coefficient of variation". Coefficient of variation is denoted by C.V and is defined as

$$\text{C.V.} = \frac{S}{\bar{x}} \times 100$$

where S is standard deviation and $\bar{x}$ is mean.

Coefficient of variation is the best measure of comparing the variability or consistency of two or more samples.

The smaller the coefficient of variation, the results are more consistent and greater coefficient of variation shows more variable results or less homogeneous results.

If we observe the standard deviation from the following data, it cannot be compared as the weight expressed in mg and the hardness is expressed in kg/m^2 (both are not having same units).

Example:

S. No.	Weight (mg)	Hardness (kg/m^2)
1	98	4.2
2	96	4.6
3	94	4.8
4	95	4.4
5	98	4.5
6	97	4.7

Similarly the standard deviation observed from the tablets containing different doses of same active ingredient cannot be compared and sometimes that comparison may give false impression

***Example*:** The potency observed from the tablets containing 100 mg and 50 mg strength is given below.

Data 1	Data 2
100	50
99	49
98	49

For data 1, standard deviation = 1 and mean = 99

$$\text{Coefficient of variation} = \frac{\text{Standard deviation}}{\text{Mean}} \times 100$$

$$= \frac{1}{99} \times 100 = 1.01$$

For data 2, standard deviation = 0.5773 and Mean = 49.33

$$\text{Coefficient of variation} = \frac{\text{Standard deviation}}{\text{Mean}} \times 100$$

$$= \frac{0.5773}{49.33} \times 100 = 1.17$$

The difference between standard deviations is more (1 – 0.5773 = 0.4227) compared to difference between coefficient of variations (1.17038 – 1.01 = 0.16038). So coefficient of variation is more preferable to compare.

***Example*:** From the following data compute the values of coefficient of variation. Compare the two series for the data consistency.

Series A: $\bar{x} = 30.5$ $S = 15$

Series B: $\bar{x} = 25$ $S = 7.5$

Solution:

For series A

$$C.V = \frac{S}{\overline{X}} \times 100$$

$$= \frac{15}{30.5} \times 100$$

C.V. = 49.18%

For series B

$$C.V = \frac{S}{\overline{X}} \times 100$$

$$= \frac{15}{30.5} \times 100$$

C.V = 30%

The above data shown that the contents of series A are more variable and less consistent as compared to the data of series B which is more consistent.

Example: The following table provides the scores of two bats men A and B in series of innings

A	12	115	6	73	7	19	119	36	84	29
B	47	12	16	42	4	51	37	48	13	0

Who is the better scorer and who is more consistent?

Solution: For Batsman A

Score: 12 115 6 73 7 19 119 36 84 29

$\Sigma x = 500$

Mean of score of Batsman A

$$\overline{x} = \frac{\Sigma x}{n} = \frac{500}{10} = 50$$

$$\Sigma x^2 = (12)^2 + (115)^2 + (6)^2 + (73)^2 + (7)^2 + (19)^2 + (119)^2 + (36)^2 + (84)^2 + (29)^2$$

$$= 42498$$

Standard deviation of

$$\text{Batsman A} = s = \sqrt{\frac{\Sigma x^2 - \frac{(\Sigma x)^2}{n}}{n-1}} = \sqrt{\frac{42498 - 25000}{9}}$$

$$= \sqrt{\frac{17498}{9}} = \sqrt{1944.22} = 44.09$$

$$C.V = \frac{s}{\overline{x}} \times 100 = \frac{44.09 \times 100}{50} = 88.18$$

For Batsman B

Score: 47 12 16 42 4 51 37 48 13 0

Mean of score of Batsman A

$$\bar{x} = \frac{\Sigma x}{n} = \frac{270}{10} = 27$$

Again,

$$\Sigma x^2 = (47)^2 + (12)^2 + (16)^2 + (42)^2 + (4)^2 + (51)^2 + (37)^2 + (48)^2 + (13)^2 + (0)^2$$
$$= 10832$$

Standard deviation of score of Batsman B =

$$S = \sqrt{\frac{\Sigma x^2 - \frac{(\Sigma x)^2}{n}}{n-1}}$$

$$= \sqrt{\frac{10832 - \frac{(270)^2}{10}}{9}}$$

$$= \sqrt{\frac{10832 - 7290}{9}}$$

$$= \sqrt{\frac{3542}{9}} = \sqrt{393.55} = 19.83$$

$$C.V = \frac{S}{\bar{x}} \times 100$$

$$= \frac{19.83 \times 100}{27} = 73.44$$

$\therefore$ C.V of player A = 88.18

C.V. of player B = 73.44

C.V. of A is more than the coefficient of variation of player B.

$\therefore$ Batsman A is more consistent and player A is better scorer.

2.2.3 Precision and Accuracy

Accuracy is regarding the agreement of the obtained result with the actual, true result. We call it as nearness to true value. Precision is the extent of reproducibility of the result.

Example: Determine the system precision from the following absorbance values obtained after analyzing the same solution repeatedly six times.

0.611, 0.613, 0.620, 0.618, 0.625, 0.614

Solution:

$$\text{mean } \bar{x} = \frac{3.701}{6} = 0.61$$

$$SD = \sqrt{\frac{\Sigma x^2 - \frac{(\Sigma x)^2}{n}}{n-1}}$$

$$= \sqrt{2.283035 - \frac{(3.701)^2}{6}}$$

$$= \sqrt{\frac{2.283035 - 2.282900167}{5}}$$

$$= \sqrt{2.69666 \times 10^{-5}} = 0.005193$$

$$= 5.193 \times 10^{-3}$$

$$\text{Relative standard deviation} = \frac{\text{Standard deviation}}{\text{mean}} \times 100$$

$$= \frac{0.005193}{0.6168} \times 100 = 0.84$$

It is less than 2% and hence the data is having good precision.

The accuracy expresses the closeness of agreement between the value which is accepted either as a conventional true value or an accepted reference value and the value found. This is something termed trueness. Accuracy studies for drug substance and drug product are recommended to be performed at 80,100 and 120% levels. At each recommended level studied, replicate samples are evaluated. The RSD of the replicates will provide the analysis variation or how precise the test method is the mean of the replicates, expressed as % label claim, indicates how accurate the test method is:

Example: Determine the accuracy of the analytical method from the following amounts observed by analyzing six samples containing 50% level.

50.4, 48.7, 47.9, 49.6, 48.7, 47.9

$$\bar{x} = \frac{\Sigma x_i}{n} = \frac{293.2}{6} = 48.87;$$

$$\text{S.D.} = \frac{\sqrt{(2.3409)+(0.0289)+(0.9409)+(0.5329)+(0.0289)+0.9409}}{5} = 0.981$$

$$\text{Coefficient of variation} = \frac{\text{SD}}{\text{mean}} \times 100 = 2.0073$$

Mean is nearer to 50 and coefficient of variation is 2. The data exhibits both precision and accuracy.

2.2.4 Standard Error of Statistics

The average amount of the variability of observations of a sampling distribution is called standard error. The standard deviation used to measure the variability of the values of statistics from sample to sample is called the standard error of that statistics. Thus the standard deviation and standard error have same meaning and same connection but are used in different cases and different circumstances. Both of them are used to measure the variability of observations. Standard error is used to measure the variability of values of statistics computed from samples of same size drawn from population. Where as standard deviation is used to measure the variation of observations of the population itself.

In statistics there are two types of standard errors

1. Standard error of sample mean
2. Standard error of proportion

1. **Standard error of sample mean:** It is the standard deviation of the sampling distribution of sample means. Standard error of sample means is denoted by σ_x and is defined as $\sigma_x = \frac{\sigma}{\sqrt{n}}$ where σ is standard deviation of population and n is the sample size.
2. **Standard error of proportion:** Standard error of proportion is computed from the proportions of all possible samples of the same size drawn from a population and denoted by S_p and is defines as

$$S_P = \sqrt{\frac{P(1-P)}{n}}$$

where P = Population proportion and n = sample size.

Standard error of sample proportion

Case (i)

If the population size is infinitely large or the sample is drawn with replacement, then the Standard error of sample proportion is calculated with the following formulae.

$$S.E\ (p) = \sqrt{\frac{Pq}{n}}$$

If the value of population proportion (P) is known, where q = 1 – P, n = the sample size

Case (ii)

If the population size is finite and the samples are drawn without replacement, then the Standard error of sample proportion is converted with the formulae

$$S.E\ (p) = \sqrt{\frac{Pq}{n}} \times \sqrt{\frac{N-n}{N-1}}$$

If P is known and where q = 1 – P, N = size population and n = size of sample.

Uses of standard error of statistic

1. Standard error is used to measure the variability of the values of a statistics from its mean.
2. Standard error is used to test the hypothesis and to draw a statistical conclusion from it.
3. Standard error is used to set up the confidence limits with in which the population parameter may lie.

Estimation: Estimation of popular parameters like mean, standard deviation, variance, correlation coefficient etc., from corresponding sample statistic is one of the important problems of statistical inference. Estimates are found to differ from the true value of the population parameter but the tolerable divergence between the estimates and the true value of the population parameter may be specified before handling the data.

There are two types of estimations

1. Point estimation
2. Interval estimation

Point Estimation: The estimate of a population parameter given by a single number is called point estimation of the parameter. In point estimation a single statistic is used to provide an estimate of population parameter.

In a point estimation we find a statistic which may be used for or replace an un known parameter of the population for all practical purposes.

A best estimator must possess the following properties.

(i) Unbiased ness
(ii) Consistency
(iii) Efficiency
(iv) Sufficiency

Interval estimation: In some of the cases the point estimation is not desirable and we are interested in finding such limits with in which with and known probability or to a known degree of reliance, the value of the population parameter is expected to lie. This type of process of estimation is called the interval estimation.

The interval estimation of the population parameter is the estimation of population parameter by an interval around the point. In the interval estimation the estimate for the parameter lies between two limits. The two limits with in which the estimate for the parameter lies are known as confidence limits or fiducial limits and the interval bounded by these two limits or confidence interval. The probability that we associate with an interval is called confidence level.

Algorithm to compute confidence limits for a population parameter

(i) Compute or select an appropriate sample statistic t.

(ii) Obtain SE (t), the standard error of sampling distribution for the sample statistic t.

(iii) Select the confidence level and corresponding to that level of confidence and the critical value.

Interval estimation for large samples: For large samples, the underlying distribution of standardized variant corresponding to the sampling distribution of sample statistic 't' will be asymptotically normal i.e.,

$$Z = \frac{t - E(t)}{S.E.(t)} \sim N(0,1)$$

Confidence interval estimate of the mean: Suppose the mean population is μ and mean samples $\overline{x}$ and assume that sample mean is normal or large. The interval estimate of population mean μ by the sample mean $\overline{x}$ of sampling distribution of means is given by following algorithm.

Algorithm

(i) Calculate $\overline{x}$ or consider $\overline{x}$

(ii) Select confidence level and corresponding to that specific level of confidence, note the use of confidence coefficient z.

(iii) Calculate the standard error (S.E ($\overline{x}$)) with the help of given result.

Case (i)

The standard deviation of the normal population is known

$$\text{S.E}\ (\overline{x}) = \frac{\sigma}{\sqrt{n}}$$

when n = sample size.

Case (ii)

The standard deviation of population is not known

$$\text{S.E}\ (\bar{x}) = \frac{s}{\sqrt{n-1}}$$

where n = sample size and s = standard deviation of sampling distribution of sample.

Determination of confidence interval

Case (i)

Standard deviation σ is known and population is normal or any population with large n.

$$\text{Confidence interval} = \left[\bar{x} - \text{SE}\ (\bar{x}) \times Z_\alpha,\ \bar{x} + \text{SE}\ (\bar{x}) \times Z_\alpha\right]$$

$$= \left[\bar{x} - \frac{\sigma}{\sqrt{n}} \times Z_\alpha,\ \bar{x} + \frac{\sigma}{\sqrt{n}} \times Z_\alpha\right] \left[\because\ \text{SE}\ (\bar{x}) = \frac{\sigma}{\sqrt{n}}\right]$$

Case (ii)

Standard deviation is not known and population with large 'n'. Confidence interval is

$$= \left[\bar{x} - \text{SE}\ (\bar{x}) \times Z_\alpha,\ \bar{x} - \text{SE}(\bar{x}) \times Z_\alpha\right]$$

$$= \left[\bar{x} - \frac{s}{\sqrt{n-1}} \times Z_\alpha,\ \bar{x} + \frac{s}{\sqrt{n-1}} \times Z_\alpha\right] \left(\because\ \text{S.E}\ (\bar{x}) = \frac{s}{\sqrt{n-1}}\right)$$

i.e.,

$$= \bar{x} - \frac{s}{\sqrt{n-1}} \times Z_\alpha \le \mu \le \bar{x} + \frac{s}{\sqrt{n-1}} \times Z_\alpha$$

Select the value of Z_α from the following table

Confidence level (1 – α)%	Value of confidence coefficient Z_α
90%	1.64
95%	1.96
98%	2.33
99%	2.58
other	3.00

Confidence interval estimate of the proportion

Population Proportion: The population proportion is the ratio of the number of elements possessing characteristic to the total number of elements in the population and it is denoted by P.

$$\therefore \qquad P = \frac{\text{Number of elements possessing the characteristic}}{\text{Total number of elements in population}}$$

Sample Proportion: Sample proportion is the ratio of number of elements possessing a characteristic to the total number of elements 'n' in sample and it is denoted by p.

$$\therefore \quad p = \frac{\text{Number of elements possessing a characteristic}}{\text{Total number of elements in sample}}$$

Case (i)

If the sampling is from infinite population

Confidence interval = [p – Z × SE (p), p + Z × SE (p)]

where Z is confidence coefficient and S.E (p) is the standard error of $p=\sqrt{pq/n}$

Case (ii)

If the sampling is from finite population of size N

Confidence interval = [p – Z × SE (p), p + Z × SE (p)]

$$SE\ (p) = \sqrt{(N-n)pq / N(n-1)}$$

Case (iii)

If the sampling is from two different proportions

$$SE\ (p_1p_2) = \sqrt{p_1q_1 / n_1 + p_2q_2 / n_2}$$

Solved Problems

1. A Random sample of size 100 has mean 20, the population variance being 25. Find the interval estimate of population mean with a confidence level of (i) 95% (ii) 99%.

***Solution*:**

(i) 95% confidence level ⇒ $Z_\alpha = 1.96$

Confidence Interval

$$= \left[\bar{x} - S.E(\bar{x}) \times Z_\alpha,\ \bar{x} + S.E(\bar{x}) \times Z_\alpha\right]$$

$$= \left[\bar{x} - \frac{\sigma}{\sqrt{n}} \times Z_\alpha,\ \bar{x} + \frac{\sigma}{\sqrt{n}} \times Z_\alpha\right]$$

$$= \left[20 - \frac{5}{\sqrt{100}} \times 1.96,\ 20 + \frac{5}{\sqrt{100}} \times 1.96\right]$$

$$= \left[20 - 0.5 \times 1.96,\ 20 + 0.5 \times 1.96\right]$$

$$= \left[20 - 0.98,\ 20 + 0.98\right] = [19.02, 20.98]$$

(ii) For 99% confidence level $\Rightarrow Z_\alpha = 2.58$

Confidence interval

$$= \left[\bar{x} - S.E(\bar{x}) \times Z_\alpha,\ \ \bar{x} + S.E(\bar{x}) \times Z_\alpha\right]$$

$$= [20 - 0.5 \times 2.58,\ \ 20 + 0.5 \times 2.58]$$

$$= [20 - 1.29,\ 20 + 1.29] = [18.71, 21.29]$$

2. A sample of size 9 from a normal population gave $\bar{x} = 16$ and $S_x^2 = 10$. Find a 99% interval for population mean.

***Solution*:** Given $\bar{x} = 16$, $n = 9$, $S_n^2 = 10$ $S_x = \sqrt{10} = s$

$$S.E.\ (x) = \frac{s}{\sqrt{n-1}} = \frac{\sqrt{10}}{\sqrt{9-1}} = \frac{\sqrt{10}}{\sqrt{8}} = 1.118$$

Also $t_{0.01}$ for 8 degrees of freedom = 3.36. (n is small i.e., less than 30)

Confidence interval $= \left[\bar{x} - S.E \times t_{0.01},\ \bar{x} + S.E \times t_{0.01}\right]$

$$= [16 - 1.118 \times 3.36,\ 16 + 1.118 \times 3.36]$$

$$= [12.24, 19.76]$$

Exercises

1. Find the standard deviation of the following data 37, 43, 48, 29, 57, 63.

Ans. 11.5

2. Find the standard deviation of the following data

Number of Branches	8	11	14	17	20	23
Number of Plants	2	4	6	10	6	3

Ans. 4.03

3. From the following figures find the standard deviation

Marks	0-10	10-20	20-30	30-40	40-50	50-60
Frequency	2	4	6	10	6	3

Ans. 15.71

4. Find the coefficient of variation of the following data 63, 63, 66, 67, 68, 69, 70, 62, 71, 71.

Ans. 5.07%

5. Prices of particular commodity in five years in two cities are given below.

Price in City A	Price in City B
10	20
20	22
18	19
12	23
15	16

From the above data, find the city which had more stable prices

Ans. City A, C.V = 24.6; for City B, C.V = 12.25

∴ City B had more stable price than city A.

6. The absorbance values observed from a spectrophotometer are 0.104, 0. 106, 0.108, 0.105, 0.106 and 0.106. Calculate the mean, standard deviation and R.S.D.

Ans. Mean = 0.106, s.d. = 0.0013, R.S.D = 1.25.

7. The peak areas observed from a chromatograph are 14400, 14980, 15046, 13975, 14295 and 15200. Calculate the mean, standard deviation and R.S.D.

Ans. Mean = 14649.33, s.d. = 492.41, R.S.D = 0.034

8. The plasma concentrations of a drug observed from 6 subjects are 2.4, 2.8, 3.1, 2.7, 2.2 and 2.9 μg/ml respectively. Calculate the mean, standard deviation and R.S.D.

Ans. Mean = 2.68, s.d = 0.33, R.S.D = 12.35

9. Find the globule size of an emulsion from the following data.

Ans. 71 μ

Range	No. of globules
100-200	2
200-300	5
300-500	6
500-1000	3
1000-2000	2
2000-3000	2

CHAPTER 3

DATA PRESENTATION

In any statistical investigation, the first task of the statistician after data collection is the arrangement of the data in suitable table or graphical representation. The pharmaceutical scientists perform several experiments during preformulation, formulation and evaluation of pharmaceutical dosage forms. An abundant data is generated from these experiments. Such data should be stored for regulatory requirement, literature and preserved for future use. Data can be communicated verbally or described in a written format. Such representations are less informative and interpretation is more difficult.

Most commonly data is summarized in the form of either tables or graphs. Such descriptive statistics is more beneficial to describe and explaining the research data. For interpretation of the results of inferential statistics also such graphs and tables are useful.

3.1 Tabulation of Data

Tabulation is one of the most important devices of presenting the data in a condensed and readily comprehensive form. The statistical data is arranged in vertical columns and horizontal rows.

3.1.1 Components of a Table

An ideal table should contain the following components.

- *Table number:* A table should be numbered for ease of identification and reference.
- *Title:* It should be placed at the top of the table.
- *Captions:* It refers to the headings of vertical column.
- *Stubs:* It refers to the headings of horizontal rows.
- *Body:* It contains the numerical information.
- *Prefatory notes and foot note:* Prefatory note is placed below the title as a part of the title. A foot note is keyed with the title, column or row heading and represented with the symbol. It gives explanation to the symbols and placed at the foot of the table.
- *Source note:* It refers to the source of information.

3.1.2 Types of Tables

The tables are categorised based on the elements covered (simple and complex tables), objective (reference or summary) and nature of data (original or derived). In simple table, the data is classified with respect to single characteristic and the data is grouped in to different classes with respect to characteristics in case of complex or manifold table. A general purpose table contains raw data covering variety of data in detailed way. For example, the table may contain the information regarding the entire quality control data of a batch of tablets. A specific purpose table presents data relating to the specific problem. It is exemplified with the table containing the information related to dissolution data only. Original table contains the information collected from the original sources as such without any treatment. The parameters such as mean and standard deviation are calculated from the origin data and the resulting data is used for preparation of derived table.

Illustration: The tabulation of the research results in original table is the simplest and least informative. For example, as a part of quality control test, 20 tablets will be selected from the entire batch and then they will be weighed and documented in the form of Table 3.1.

Table 3.1 Results of the 20 Tablets Weight

S. No.	Weight of Tablet	S. No.	Weight of tablet
1	250	11	242
2	246	12	248
3	252	13	250
4	257	14	255
5	250	15	249
6	248	16	250
7	255	17	252
8	248	18	255
9	250	19	250
10	256	20	246

The data represented in Table 3.1 is simple but useless as it is not allowing interpretation. This data can be analysed by ranking the data as shown in Table 3.2. The tablets can be ranked based on their weight from the smallest to the largest.

Table 3.2 Rank Ordering of the Tablets Based on Weight

Rank	Weight (mg)	Rank	Weight (mg)
1	242	11	250
2	246	12	250
3	246	13	250
4	248	14	252
5	248	15	252
6	248	16	255

Table 3.2 *Contd...*

Rank	Weight (mg)	Rank	Weight (mg)
7	249	17	255
8	250	18	255
9	250	19	256
10	250	20	257

From the above table, the central value of the distribution (around 250 mg) and the spread of the outcome (smallest weight 242 mg and the largest weight 257 mg) are easily observed.

The above descriptive statistics is better organized and summarized with the reduction of data. The continuous variable (weight of the tablet) can be grouped into categories as weight less than 250 mg, weight equal to 250 mg and weight greater than 250 mg to form discrete variable.

Weight of Tablet	Number of Occurrences/Frequency	Percentage
< 250 mg	7	35
= 250	6	30
> 250 mg	7	35

One of the common problems associated with the conversion of continuous variable in to a discrete variable is the loss of information. In the basic data, 9 different classes 242, 246, 248, 249, 250, 252, 255, 256 and 257 are there where as in the descriptive data; it is reduced to 3 classes.

Similarly the assay results, disintegration time and amount of drug dissolved, particle size of a powder, globule size of emulsion can be measured and tabulated.

3.2 Stem and Leaf Plot

It was developed by John Tukey in 1960s for visual representation of the continuous data. It is also called as stem plot. It is one of the compact way to summarize and tabulating the large amount of data. It can be illustrated with the following example.

Table 3.3 Globule Size Data of Castor Oil Emulsion

44	65	18	40	26	32	60	44	36	19
26	38	30	32	28	41	54	42	38	24
30	43	49	58	42	37	46	47	34	29
58	30	52	31	47	36	32	39	40	32
32	54	65	43	51	29	29	28	42	36
27	19	42	55	34	30	32	26	44	38
19	33	35	36	20	44	46	38	46	40
60	48	32	44	32	19	45	19	48	42
25	39	28	29	47	50	42	24	39	54
33	50	32	31	45	38	40	29	43	59

Guidelines to construct the stem and leaf plot

1. Identification of the first integer digits of the data
2. Arranging the first digit in ascending order vertically to form the stem
3. Add the second digit to one of the integer of the stem to form leaves
4. In case of three digit number, the stem may be either single or double digit and accordingly the leaf is double or single digit.

In the above example the globule size is in between 18-65 μ. So the stem is ranging from 1 to 6. The first value in this table is 44 so we place 4 next to 4 in the stem column. The next value is 26 so we place 6 next to 2 in the stem column and so on to construct the following stem and leaf diagram.

Stem	Leaves
1	9 9 8 9 9 9
2	6 7 5 8 9 6 8 0 9 9 8 6 4 9 4 9
3	0 2 3 8 0 3 9 0 5 2 2 2 1 6 1 4 2 2 7 6 0 8 2 2 9 8 6 8 4 9 2 6 8
4	4 3 8 9 2 0 3 4 2 7 7 5 1 4 6 6 5 2 0 4 2 7 0 2 4 6 8 2 0 2
5	8 4 0 2 8 5 1 0 4 4 9
6	0 5 5 0

The above data is easily visualised and without any manipulation, the frequency of globules can be recorded. The first digit of the stem is 1 and the second digit or the digit of leaf may be in between 1 to 9. So the class interval is 10-19. The frequency is nothing but the number of leaves. In the above example, the number of leaves for first column is 6. So the frequency of globules having a size range of 10 to 19 is 6. From the above observations, the following frequency table can be constructed.

Size Range	Frequency
10-19	6
20-29	16
30-39	33
40-49	30
50-59	11
60-69	4

3.3 Back to Back Stem and Leaf Plot

A back to back stem and leaf plot is used to compare the two sets of data visually. For example, particle size data observed from two suspensions containing same drug and formulated with two suspending agents is given below.

Table 3.4 Particle Size Data of Paracetamol Suspension Formulated with 2% Tragacanth (Formulation A)

19	24	28	35	42	20	18	26	32	40
23	29	34	40	39	24	26	30	35	44
28	35	41	45	54	30	28	34	38	36
32	40	44	50	60	32	30	37	42	38
34	45	48	54	67	36	38	40	46	51
42	49	52	57	70	18	40	45	50	57
54	57	60	59	58	24	42	48	54	60
61	54	71	62	42	19	44	50	58	32
42	59	49	70	36	30	46	52	60	38
46	60	56	58	29	32	50	59	24	40

Table 3.5 Particle Size Data of Paracetamol Formulated with 2% Sodium CMC (Formulation B)

26	30	29	24	30	18	32	17	24	26
24	28	27	26	28	24	26	27	30	32
30	36	30	28	24	22	23	30	28	34
32	34	17	30	26	27	25	32	26	37
34	18	22	32	27	25	28	34	40	24
35	26	24	34	19	20	30	36	38	20
40	38	27	36	20	24	32	30	36	18
24	34	30	28	22	26	34	28	29	25
19	36	32	19	24	30	36	26	32	27
20	40	34	24	32	33	38	24	35	36

The above data can be represented as Back to Back Stem and Leaf Plot as follows.

Formulation A		**Formulation B**
9 8 9 8	1	98799878
3 8 4 9 8 9 0 4 4 6 8 6 4	2	64408697247468848467024427
		504663587864486964057
245459602602080472586828	3	024506484600240246020320246
		802460086252476
22605914890522024605826040	4	000
4749260479848002904817	5	
10020700	6	
100	7	

The above Back to Back Stem and Leaf Plot the data can be further simplified as follows.

Size Range	Frequency	
	Formulation A	Formulation B
10-19	4	8
20-29	13	47
30-39	24	42
40-49	26	3
50-59	22	-
60-69	8	-
70-79	3	-

Thus by visualizing the above table, the particle size observed from the formulation B is less compared to formulation A. As more frequency is observed for smaller particle size in case of formulation B.

3.4 Graphical Representation of Data

In statistics, graphs are used either to transmit the collected information or to establish a correlation in between the variables.

One of the simplest data presentations is with bar graphs or block diagram. They are drawn with spaces between the bars symbolizing the discontinuity among the level of variables. It is shown with the following example. Twenty tablets of metformin were analyzed for the drug content and observed % drug content of the labelled claim is shown in the Table 3.6.

Table 3.6 Percent Assay Results of Metformin

S No	% Drug Content	S No	% Drug Content
1	96	11	96
2	94	12	102
3	89	13	104
4	101	14	106
5	104	15	112
6	88	16	98
7	86	17	87
8	99	18	113
9	100	19	89
10	97	20	97

The above information can be categorized as three mutually exclusive and exhaustive levels as less than 90 mg and 90-110 mg and more than 110 mg. So the above data is reduced as follows:

Category	Frequency	Percentage
< 90	5	25
90 – 110	13	65
> 110	2	10

The above data is presented in the form of following graph (Fig. 3.1) by taking % potency on the horizontal plane (x-axis or abscissa) and the frequency on vertically (y-axis or ordinate).

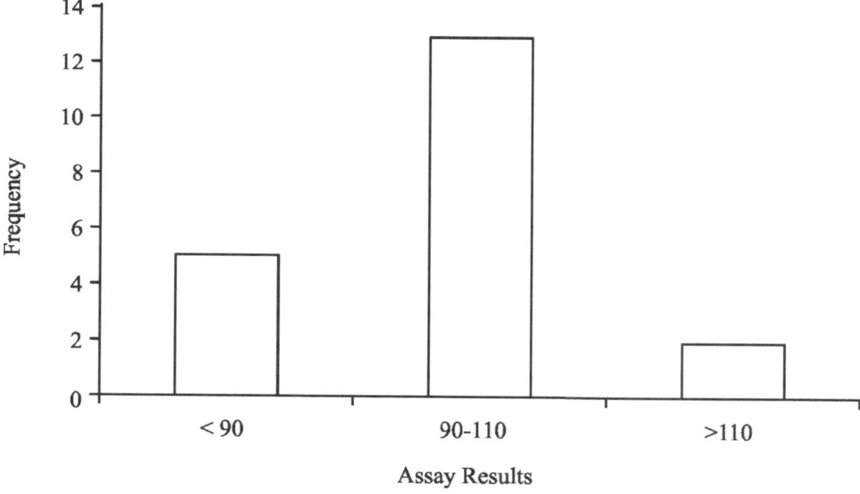

Fig. 3.1 Bar Graph showing the Frequency of Assay Results.

Pictograms are similar to bar charts and meant to provide the same type of information but the bars are replaced with proportional number of icons (Fig. 3.2).

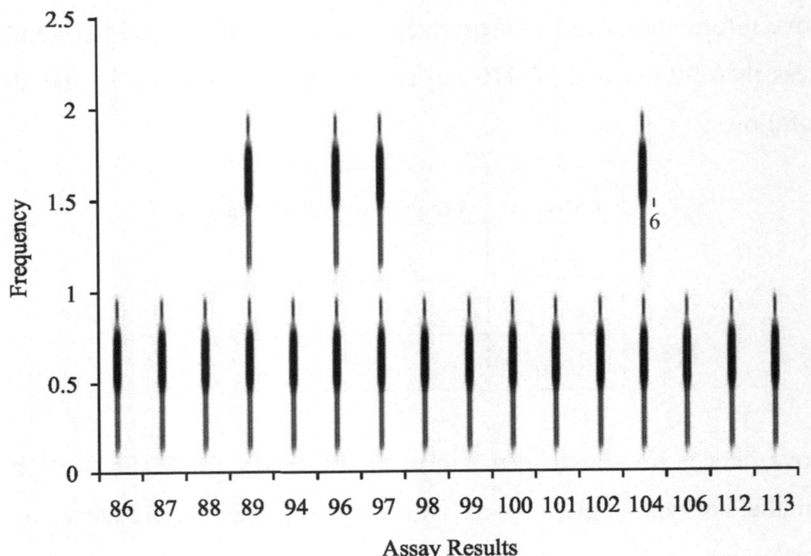

Fig. 3.2 Pictogram showing the Frequency of Assay Results.

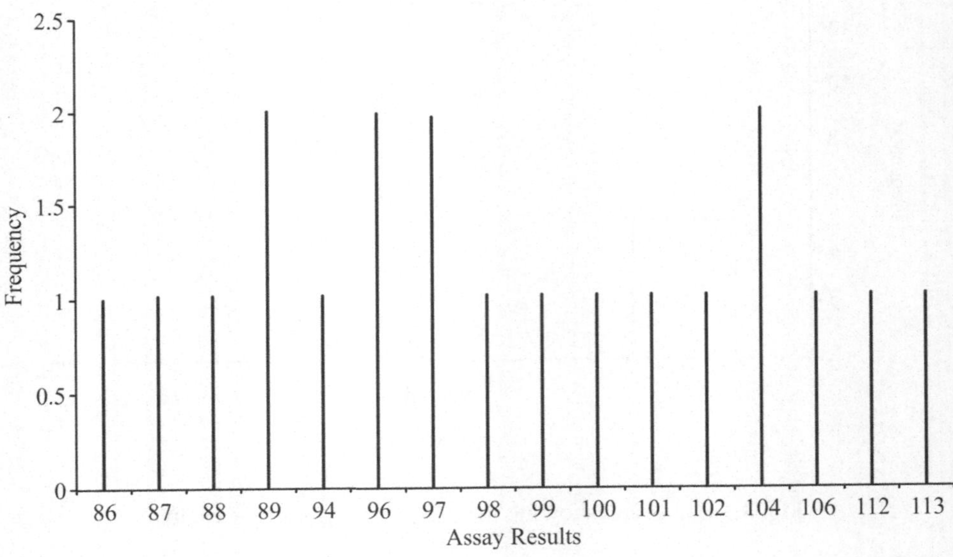

Fig. 3.3 Line Chart showing the Frequency of Assay Results.

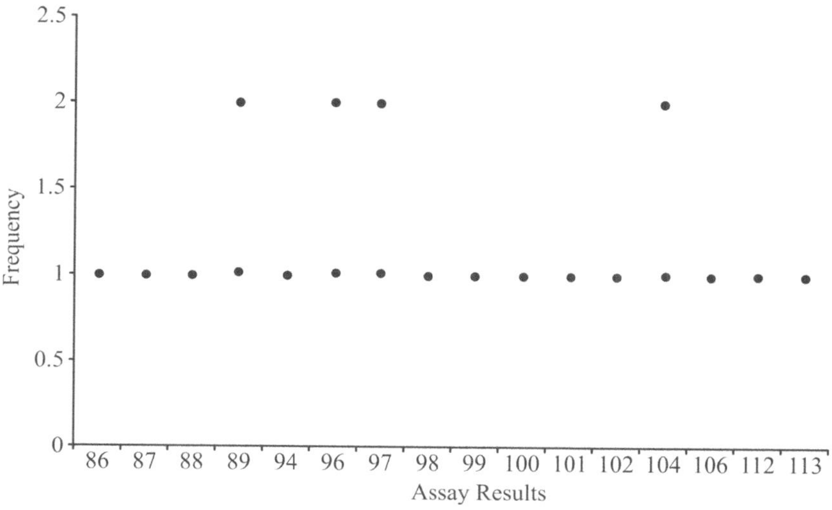

Fig. 3.4 Point Plot showing the Frequency of Assay Results.

The data given in the Table 3.6 can also be represented in the form of pictograms, line chart, point plot and pie chart. In line chart (Fig. 3.3), thin lines are used to represent the frequency associated with each level of the discrete variable. Point plots also referred to as dot diagram are constructed by representing the frequency with number of points or dots (Fig. 3.4). Pie chart allows the observation and comparison of a discrete variable in relationship to that variable as a whole. Pie chart is more convenient and effective when the data can be divided into parts. The pie chart corresponding to the data of Table 3.6 is shown in Fig. 3.5. The circle (pie) represents 100% results and each segment (slice) has an area proportional to the area of the circle representing the contribution of particular segment.

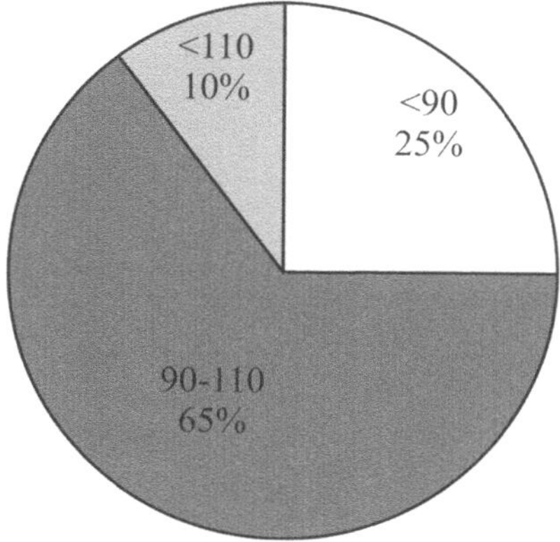

Fig. 3.5 Pie chart showing Assay Results.

3.4.1 Histograms

Histogram is one of the most popular ways of representing and summarizing the data. They are used to display the distribution of larger sample sizes of a continuous variable. The frequency or proportion of observation in each class interval is plotted as a bar or rectangle. Thus each bar represents a single value or range of values within the width of the bar. For proper interpretation of the data, the class intervals should be of equal width. The choice of intervals for histogram depends on the nature of the data, the distribution of the data and the purpose of presentation. The number of class intervals proposed is differing from one author to another. Sturges Rule provides more useful information to determine the number of intervals based on the sample size (N):

$$\text{Intervals (K)} = 1 + 3.32 \log_{10}(N)$$

$$\text{Width (w)} = (\text{Range})/K$$

Number of intervals and width for the data given in Table 3.5

N = 100 range = 71 – 18 = 53

$$\text{Intervals} = 1 + 3.32 \log_{10}(100)$$

$$= 1 + 3.32(2) = 1 + 6.62 = 7.62 = 8 \text{ intervals}$$

$$\text{Width} = 53/8 = 6.62\ (7)$$

The data presented in Table 3.7 is divided in to 8 intervals according to Sturges Rule and showed in the following Table.

Table 3.7 Class Intervals Created using Sturges Rule

Interval	Frequency
17-23	6
24-30	15
31-37	15
38-44	21
45-51	15
52-58	14
59-65	10
66-72	4

The number of intervals for the data given in table 3.8 is 8, the range = 40 – 16 = 24 and the width of the interval is 24/8 = 3. The frequency table constructed according to Sturges Rule is shown in Table 3.8.

Table 3.8 The Class Intervals Created using Sturges Rule

Interval	Frequency
17-19	8
20-22	7
23-25	16
26-28	22
29-31	13
32-34	18
35-37	10
38-40	6

The histograms related to the data given in Table 3.7 and 3.8 are showed as histograms in Fig. 3.6 and 3.7 respectively.

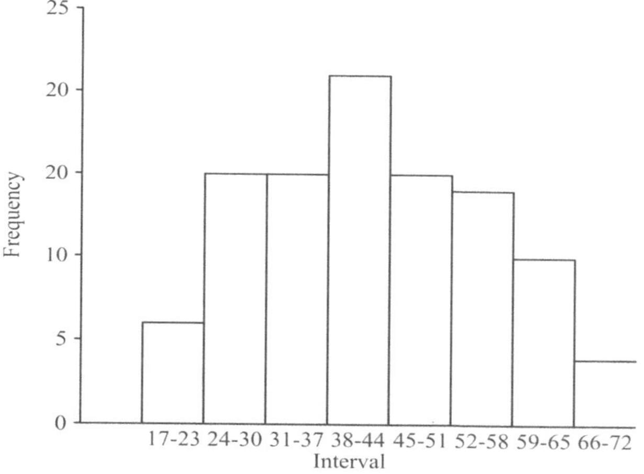

Fig 3.6 Histograms for data presented in Table 3.7.

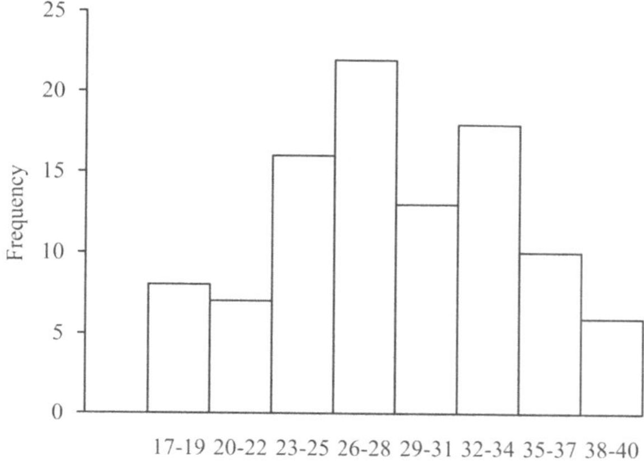

Fig. 3.7 Histograms for data presented in Table 3.8.

Histograms can also be used to observe the differences between formulations. The particle size frequency data observed from two formulations is presented and showed in Table 3.9 and Fig. 3.8.

Table 3.9 The Class Intervals Created using Sturges Rule

Interval	Frequency	
	Sample 1	**Sample 2**
17-19	8	4
20-22	7	10
23-25	16	20
26-28	22	25
29-31	13	10
32-34	18	15
35-37	10	10
38-40	6	6

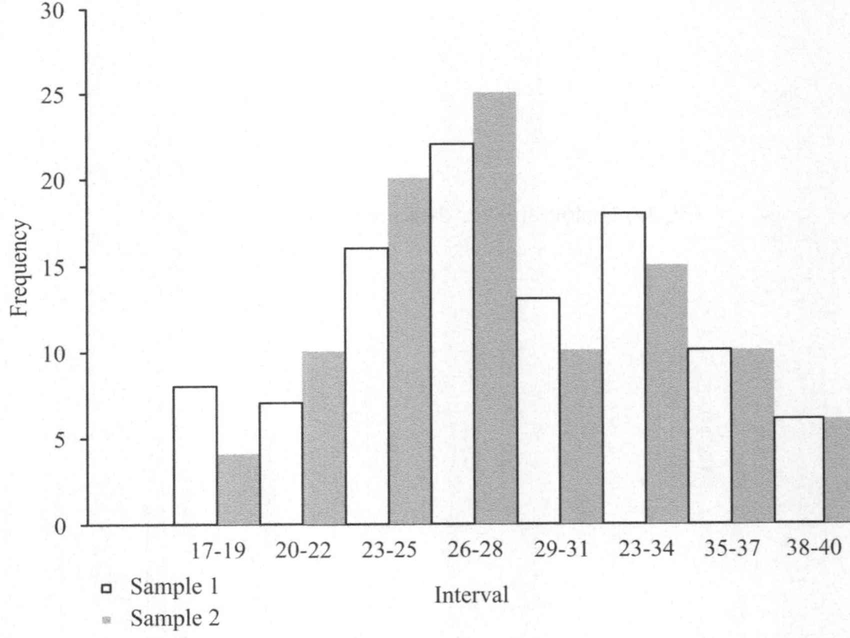

Fig. 3.8 Histograms for data presented in Table 3.9.

Illustration

Table 3.10 In vitro performance parameters of Beclomethasone dipropionate DPI formulated with overages

n = 3	mcg of Beclomethasaone dipropionate per dose				% based on total dose			
Overages	0%	10%	15%	20%	0	10	15	20
Drug deposited in capsules	7.6 ± 1.1	9.4 ± 0.9	10.1 ± 1.2	13.6 ± 0.9	3.84	4.38	4.44	5.68
drug collected in device	17.0 ± 1.2	18.6 ± 0.7	20.6 ± 1.5	23.4 ± 1.3	8.60	8.68	9.06	9.77
Emitted dose	173.1 ± 6.2	186.4 ± 5.9	196.7 ± 5.4	202.6 ± 6.4	87.56	86.94	86.50	84.56
Total dose	197.7 ± 6.9	214.4 ± 6.5	227.4 ± 6.0	239.6 ± 7.0	100%			

The above data is represented in the following figure.

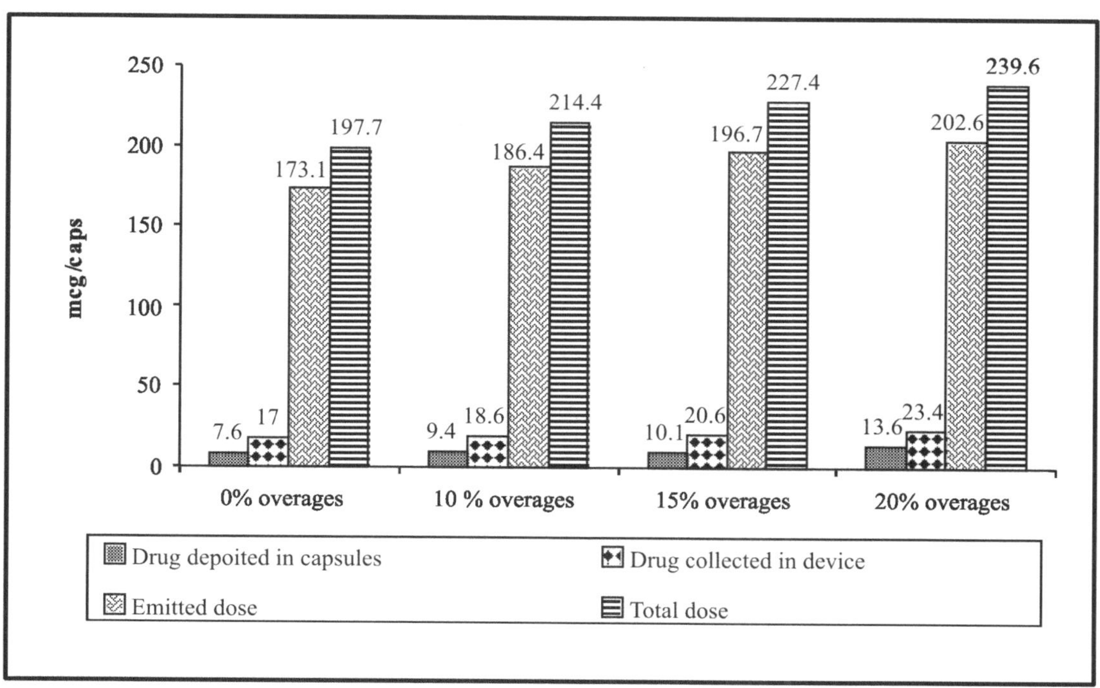

Fig. 3.9 Histograms correlating emitted dose and overages of Beclomethasone dipropionate DPI 200 mcg

3.4.2 Construction and Labeling of Graphs

Effective graphical representation needs knowledge of statistical principles, computer and drafting competency. The general graph should be executed to demonstrate the experimental data clearly without any ambiguity. The following recommendations can be followed to construct the graph.

1. ***Title of the graph:*** Proper title should be given to the graph. The reader must know the purpose of the graph from the title. For example: the dissolution profiles of Diclofenac Sodium sustained release formulations formulated with different polymers. The title may be placed below or above the graph.
2. ***Labeling the axes:*** The parameters with their units should be clearly indicated on x and y axis. The variable parameter of the experiment should be on x-axis and the observed result with such changes should be indicated on y-axis. For example, in beer's plot the absorbance values are recorded (y-axis) for variable concentrations (x-axis). In this example the unit for concentration is μg/ml and the absorbance is dimensionless number. It is shown in Table 3.11 and Fig. 3.10.

Table 3.11 Calibration Curve Data for Piroxicam

Concentration (μg/ml)	Absorbance
0	0
10	0.104
20	0.206
30	0.309
40	0.417
50	0.523

The above data is represented in the following figure

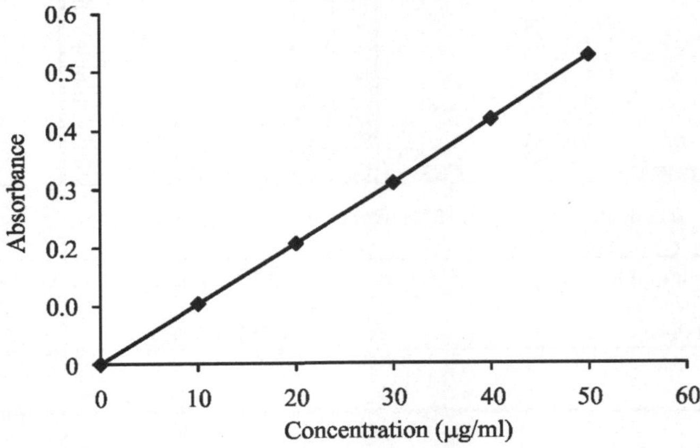

Fig. 3.10 Calibration Curve Data for Piroxicam.

3. ***Assigning numerical values to the selected scale:*** To achieve a fair picture, proper scaling and appropriate proportion of the scale is required. This is especially required when the graph is intended to differentiate/compare more than one formulation. The data presented in Table 3.12 is showed in Fig. 3.11. In this example the graphs were plotted by taking different proportion of scales and accordingly the elegancy of the graphs are affected. More clarity is observed in Fig. 3.11(a).

Table 3.12 Dissolution Data of Atorvastatin

Time (min)	% Drug Dissolved
0	0
5	36.24
10	58.86
15	74.92
20	86.35
25	94.58
30	98.64

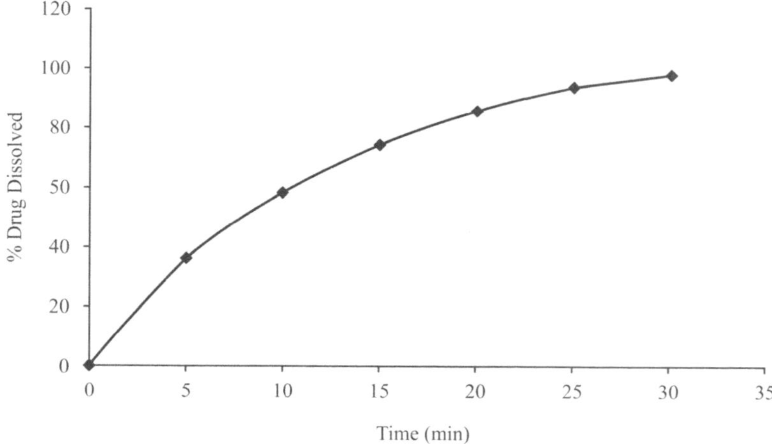

(a) Scale x-axis: 5 units; y-axis: 20 units

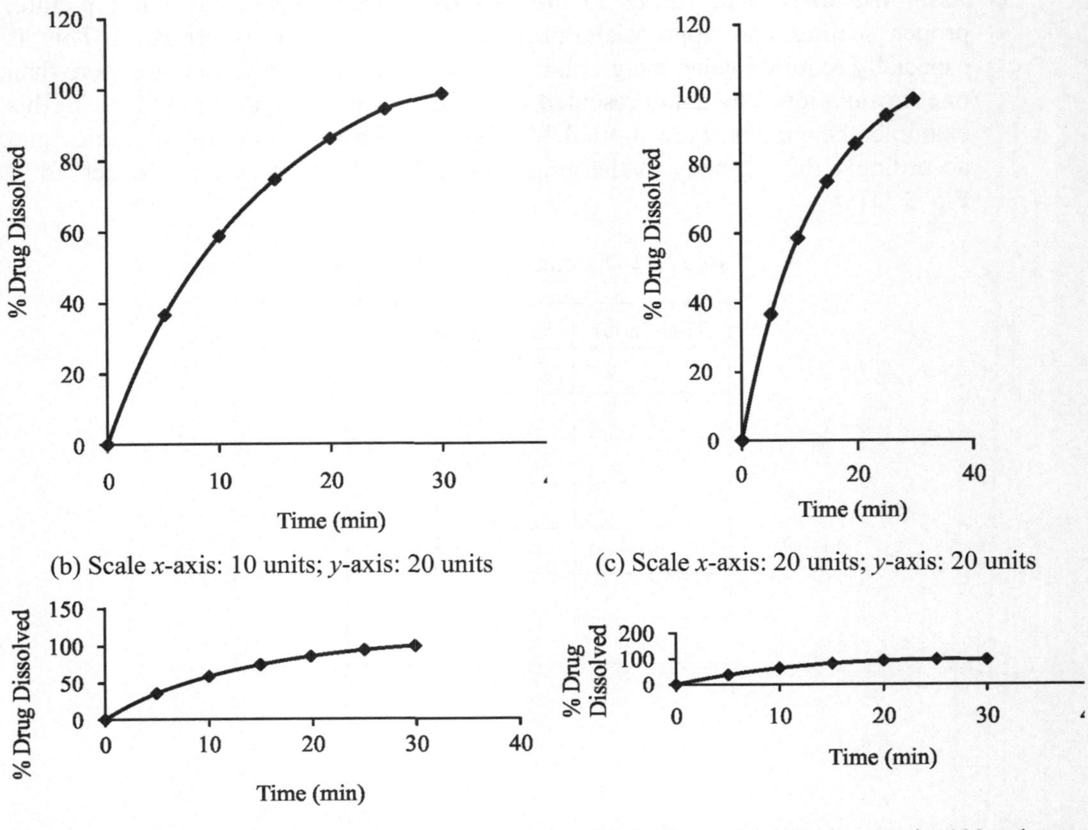

(b) Scale *x*-axis: 10 units; *y*-axis: 20 units

(c) Scale *x*-axis: 20 units; *y*-axis: 20 units

(d) Scale *x*-axis: 10 units; *y*-axis: 50 units

(e) Scale *x*-axis: 10 units; *y*-axis: 100 units

Fig. 3.11 Dissolution Profiles of Atorvastatin.

4. ***Indication of standard deviation:*** To confirm the reproducibility of the result, the experiment will be repeatedly performed and the result will be expressed in the form of mean and standard deviation. Then such standard deviation/standard error/percentage should be displayed in the graph and the number of replications of that experiment should be mentioned in the graph.

   E.g. The plasma concentration profiles of Nifedipine observed from the rabbit.

Table 3.13 Plasma Concentration Data of Nifedipine Observed from Rabbit

Time (hr)	Plasma Concentration (µg/ml) $(\bar{x} \pm s.d.)$
0.5	32.42 ± 0.8
1.0	38.87 ± 1.2
2.0	46.26 ± 1.6
4.0	46.67 ± 1.8

Table 3.13 *Contd...*

Time (hr)	Plasma Concentration (μg/ml) $(\bar{x} \pm s.d.)$
6.0	42.81 ± 1.5
10.0	36.66 ± 1.2
12.0	23.45 ± 0.9
16.0	13.33 ± 0.6
20.0	1.08 ± 0.02

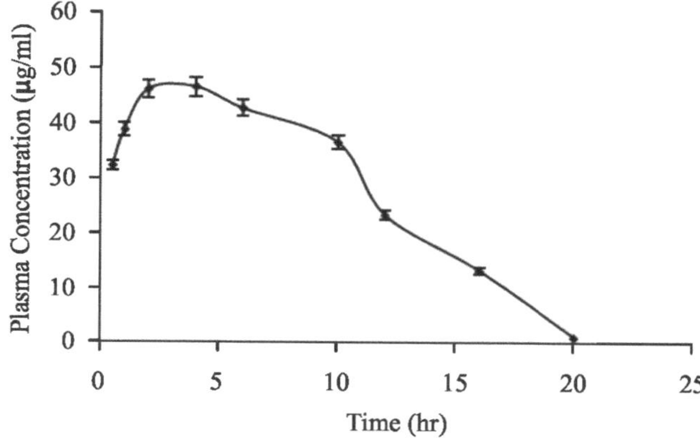

Fig. 3.12 Plasma Concentration Profiles of Nifedipine.

The experiments were conducted in triplicate.

5. ***Number of curves on same graph:*** Sometimes the data may be generated from different formulations containing same drug and same formulation stored at different conditions. In such situations, the visual representation of data collected from all formulations or at different conditions on same graph will give good clarity. These curves are useful whenever the difference in the experimental result is noticed. If the data observed from any two formulations are super-impossible, then such graphs are useless and more confusing.

   E.g.: The dissolution data observed from four formulations are showed in Table 3.14 and 3.15 and the corresponding figures are shown in Fig. 3.13 and Fig. 3.14.

Table 3.14 Drug Release Data of Metformin Hydrochloride Sustained Release Formulations Prepared with Different Polymers

Time (hr)	% Drug Released			
	Formulation A	**Formulation B**	**Formulation C**	**Formulation D**
0	0	0	0	0
1	9.24	14.14	4.86	19.26
2	19.56	29.26	9.91	36.74
3	28.78	43.82	14.27	55.92
5	48.67	74.52	26.12	94.19

Table 3.14 *Contd...*

Time (hr)	% Drug Released			
	Formulation A	Formulation B	Formulation C	Formulation D
6	57.79	89.24	31.24	99.82
7	69.36	99.16	36.58	-
8	76.42	-	41.82	-
9	85.71	-	46.79	-
10	98.42	-	52.48	-

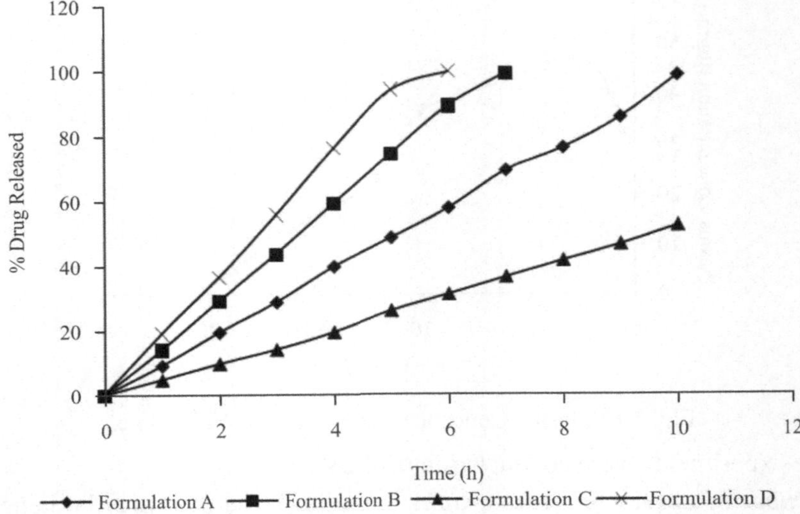

Fig. 3.13 Drug Release Profiles of Metformin Hydrochloride Sustained Release Formulations Prepared with Different Polymers.

Table 3.15 Drug Release Data of Metformin Hydrochloride Sustained Release Tablets Stored at Different Temperatures

Time (hr)	% Drug Released			
	Room Temperature	40 °C	50 °C	60 °C
0	0	0	0	0
1	9.5	10.02	10.46	11.16
2	19.26	19.82	20.12	21.24
3	30.18	30.26	31.09	32.46
4	39.82	40.72	41.26	42.57
5	51.16	52.26	53.39	54.19
6	60.74	61.56	62.16	63.89
7	72.19	73.24	74.36	75.92
8	83.24	84.39	85.42	86.56
9	91.56	92.46	93.19	94.82
10	99.3	99.32	98.92	99.4

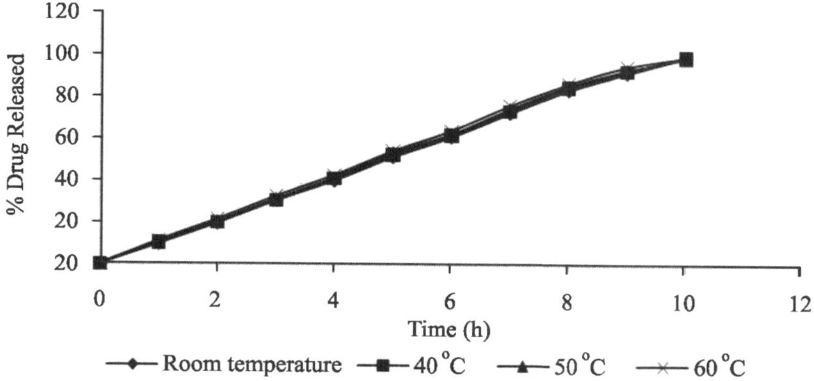

Fig. 3.14 Drug Release Profiles of Metformin Hydrochloride Sustained Release Tablets Stored at Different Temperatures.

In the above example more clarity is observed and the differences are witnessed in Fig. 3.13 where as the Fig. 3.14 is not differentiating the data and hence not suitable.

6. ***Differentiation of curves with different symbols:*** Sometimes the same graph may contain more than one curve. In such cases each curve should be denoted with different symbol or with different colours and the legend of this figure should indicate the symbol along with the information related to that curve (Fig. 3.13).

7. ***Manner of connectivity of the points:*** Sometimes the points are connected as a smooth curve or connected with a straight line. To establish the functional relationship between two parameters, the points are connected as a smooth curve. For illustration, there exists a relationship between temperature and solubility/miscibility. Such graphs are denoted with smooth curve.

Table 3.16 Miscibility Data of Phenol in Water at Different Temperatures

Concentration of Phenol (% v/v)	**Miscibility Temperature (°C)**
25	64
27	65
30	66
35	68
40	68
45	66
50	60

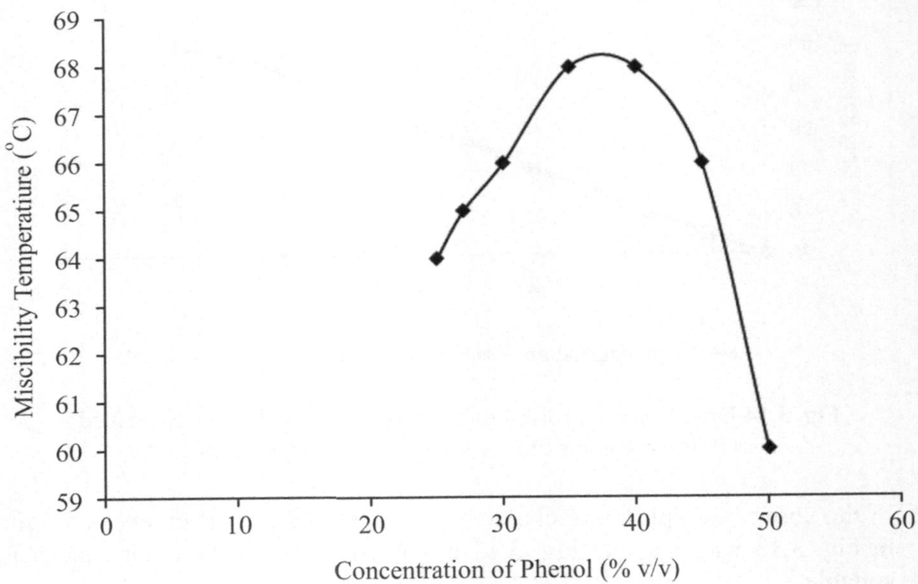

Fig. 3.15 Miscibility Profile of Phenol in Water at Different Temperatures.

Similarly there exists a relationship between time and plasma concentration and hence it is also showed graphically as a smooth curve (Fig. 3.16). Sometimes the graph is used to calculate the area under the curve. In such cases, to obtain the trapezoids the points are connected with straight lines (Fig. 3.16).

Table 3.17 Plasma Concentration Data of an antibiotic

Time (hr)	Plasma Concentration (μg/ml)
0.5	3.7
1	5.8
2	7.3
3	6.8
4	6.2
6	4.8
8	3.5
12	1.9
18	0.7
24	0.3

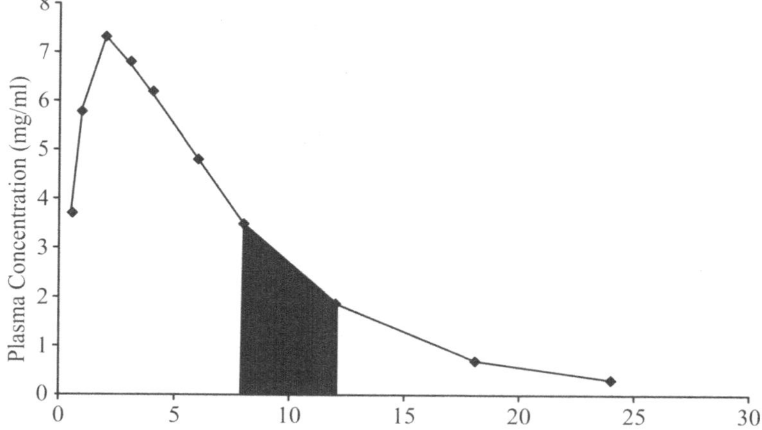

Fig. 3.16 Plasma Concentration Profiles of an antibiotic.

8. *Visual display of two or more continuous variables*

Scatter Plot

These plots are used to show the relationship between two continuous variables. They cover the ranges of the two variables. They are used to represent the paired observations. They are also known as correlation plots and useful for correlation and regression. They are also used in inferential tests. In these linear graphs, all the points are not connected, only those points lied on the linear line are connected. It is exemplified in the following figure. The paired observations (5, 41.24) are not fitted in the linear line and hence that pair was omitted in the graph.

E.g.: Dissolution data of Theophylline sustained release formulations is presented in Table 3.18.

Table 3.18 Dissolution Data of Theophylline Sustained Release Formulations

Time (hr)	Amount of Drug Dissolved (mg)
0	0
1	9.86
2	18.92
3	29.72
4	37.56
5	41.24
6	58.79
7	69.65
8	81.26
9	90.74
10	98.49

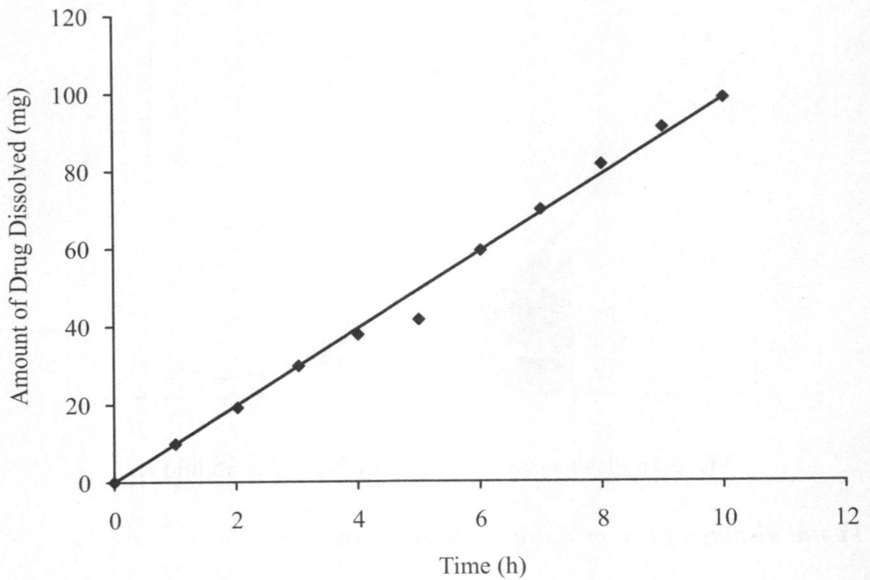

Fig. 3.17 Dissolution Profiles of Theophylline Sustained Release Formulations

Sequence Plot

As a measure of in process quality control, the quality of the pharmaceutical dosage form is observed continuously during the production. Such recorded observations of continuous variable can be represented on vertical axis and the time of observations can be arranged chronologically on x-axis to form sequence plot. It is showed in Table 3.19 and Fig. 3.18.

Table 3.19 The Average Weight of the Tablets Collected from a Rotary Tablet Machine at Various Time Intervals

Time	Average Weight of Tablet (mg)
08:30 AM	98.42
09:45 AM	99.16
10:20 AM	100.09
11:35 AM	97.58
12:50 PM	99.39
01:40 PM	98.58
02:20 PM	101.24
03:15 PM	99.86
04:00 PM	98.74

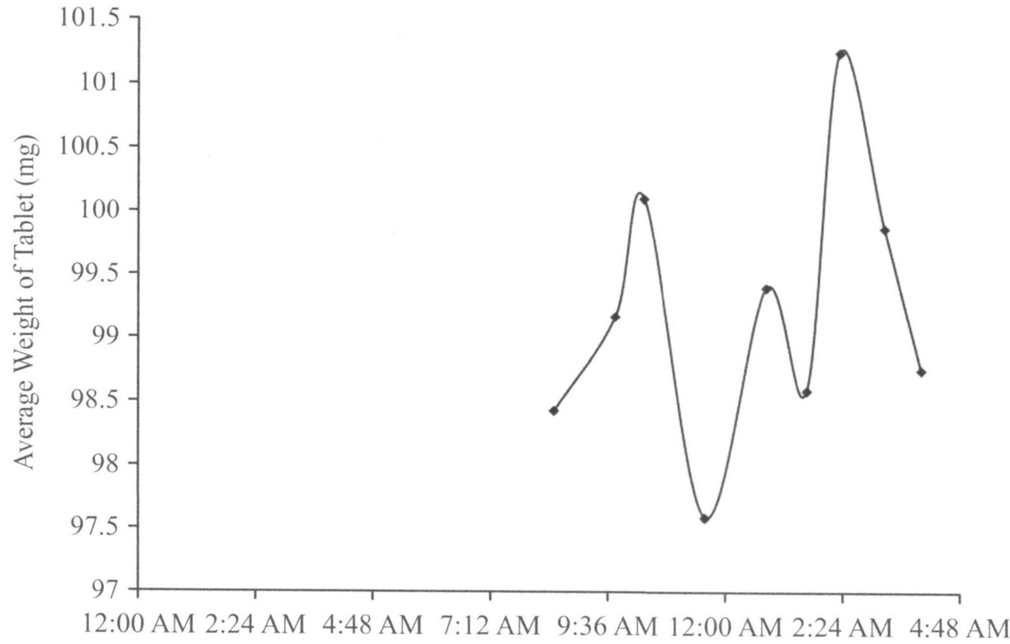

Fig. 3.18 The average weight of the tablets collected from a rotary tablet machine at various time intervals.

Semi-Log Plots

Sometimes the data should be represented on semi-log plots. In these plots the scale of *y*-axis is represented on log values. It is showed in the following Table 3.20 and Fig. 3.19.

Table 3.20 Dissolution Data of Glipizide Tablets

Time (h)	% Drug Undissolved
0	100
5	40
10	20
15	8
20	3
25	1

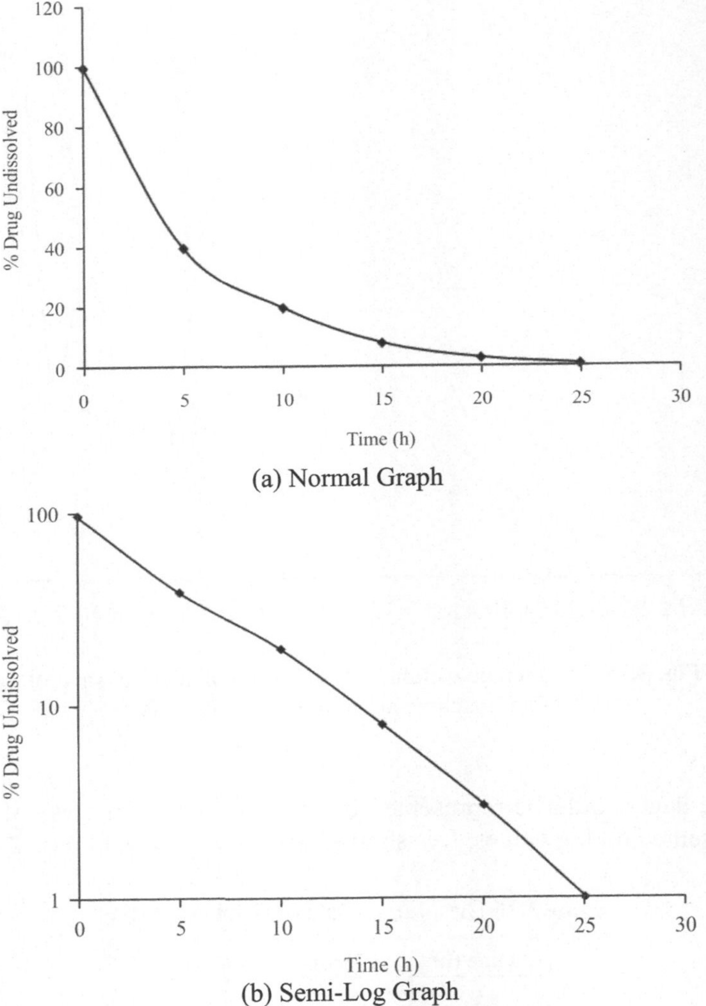

Fig. 3.19 Dissolution Profiles of Glipizide Tablets.

Probit plots

The particle size distribution of a sample is expressed as normal, log normal and as probit plots. In normal distribution plots, a graph is plotted between the particle size and frequency. Log normal distribution graph is plotted between log particle size and frequency. Log probability plots are plotted between log particle size and cumulative percent frequency on probability scale. The author can transform percent to probits by observing table 3.22.It is exemplified from the table 3.21. Alternatively the data can be represented as histogram and cumulative frequency plots. The corresponding graphs are given in Fig. 3.20-3.24.

Table 3.21 Particle Size and it's a Frequency Data

Size Range	Frequency	% Over Size*	% Under Size*
10-20	20	80	20
20-30	30	50	50
30-40	40	10	90
40-50	6	4	96
50-60	3	1	99
60-70	1	0	100

* % over size indicates the cumulative % of particles having a size more than the stated size
* % under size indicates the cumulative % of particles having a size less than the stated size

The above data is represented in the form of following graphs.

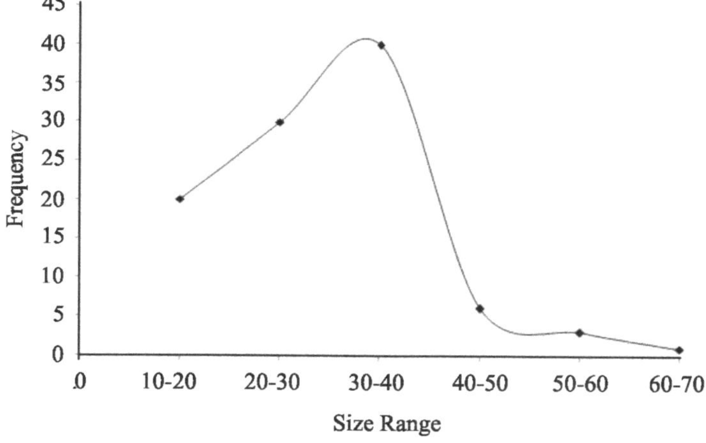

Fig 3.20 Normal distribution graph.

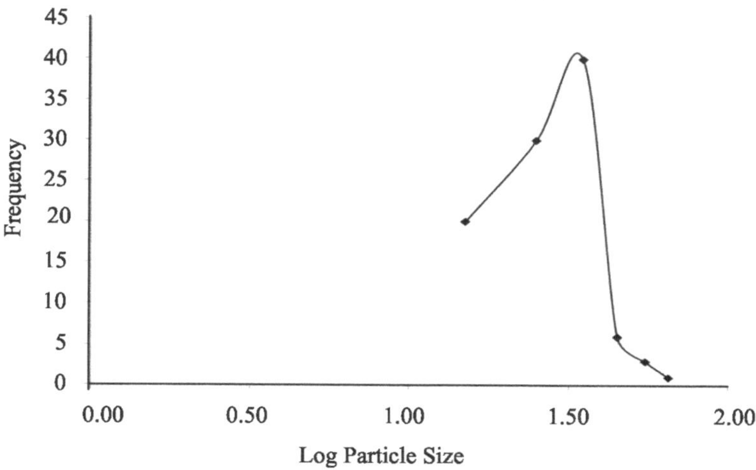

Fig 3.21 Log normal distribution graph.

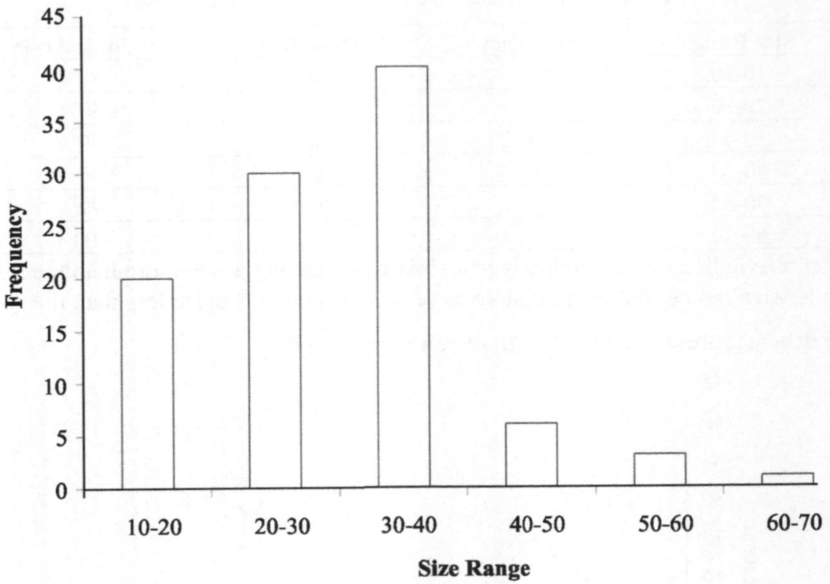

Fig 3.22 Histogram.

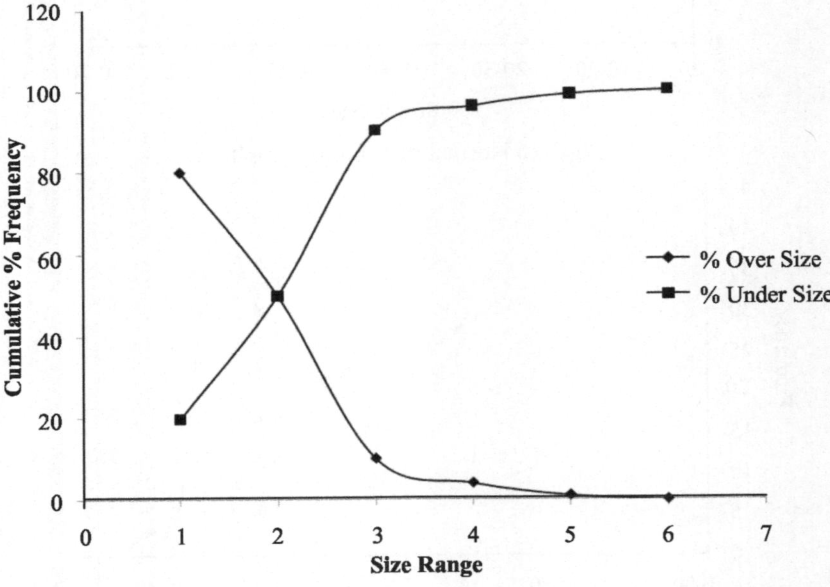

Fig 3.23 Cumulative % frequency size distribution graph.

Table 3.22 Transformation of Percentages to Probits

%	0	1	2	3	4	5	6	7	8	9
0	-	2.67	2.95	3.12	3.25	3.36	3.45	3.52	3.59	3.66
10	3.72	3.77	3.82	3.87	3.92	3.96	4.01	4.05	4.08	4.12
20	4.16	4.19	4.23	4.26	4.29	4.33	4.36	4.39	4.42	4.45
30	4.48	4.50	4.53	4.56	4.59	4.61	4.64	4.67	4.69	4.72
40	4.75	4.77	4.80	4.82	4.85	4.87	4.90	4.92	4.95	4.97
50	5.00	5.03	5.05	508	5.10	5.13	5.15	5.18	5.20	5.23
60	5.25	5.28	5.31	5.33	5.36	5.39	5.41	5.44	5.47	5.50
70	5.52	5.55	5.58	5.61	5.64	5.07	5.71	5.74	5.77	5.81
80	5.84	5.88	5.92	5.95	5.99	6.04	6.08	6.13	6.18	6.23
90	6.28	6.34	0.41	0.48	6.55	6.64	6.75	6.88	7.05	7.33
–	0.0	0.1	0.2	0.3	0.4	0.5	0.6	0.7	0.8	0.9
99	7.33	7.37	7.41	7.40	7.51	7.58	7.65	7.75	7.88	8.09

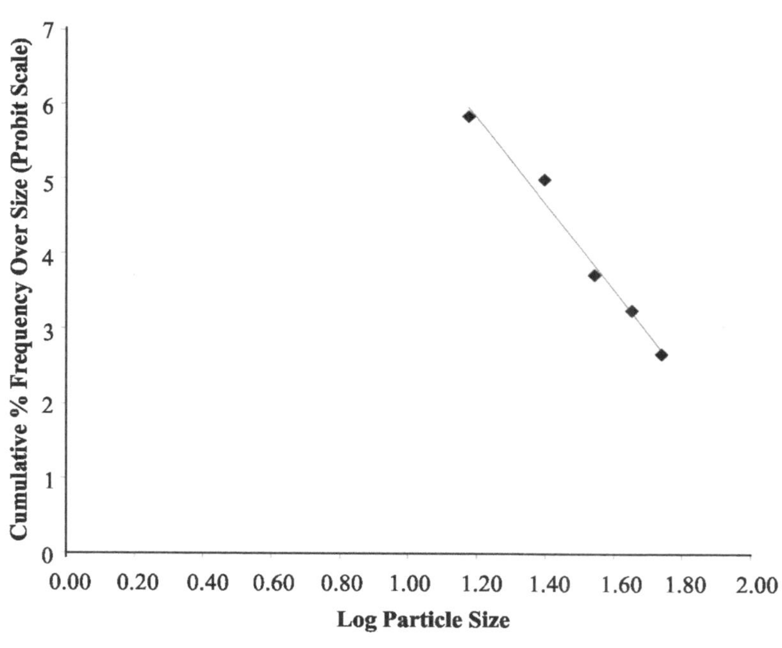

Fig. 3.24 Probability distribution graph.

CHAPTER 4

PROBABILITY AND DISTRIBUTIONS

The theory of probability enables a person to make educated guesses on matters where the full facts are not known or there is a uncertainty about the outcome.

4.1 Some Basic Definitions

Random Experiment

A Random Experiment is an experiment that can be repeated any number of times under same (identical) conditions in which

(i) All possible outcomes of the experiment are known before conducting the experiment (in advance).

(ii) The actual outcome in a particular case is not known in advance.

***Example* 1:** Suppose consider a simple experiment that is tossing a coin (unbiased). We have only two possible outcomes: Head (H) and Tail (T). In a particular trail, one does not know in advance the outcome. This experiment can be done any number of times under identical conditions.

***Example* 2:** Selecting a card from a pack of 52 playing cards. Here we have 52 possible outcomes. In a particular trial, one does not know in advance the outcome. This experiment can be done any number of times under identical conditions.

Simple Event

In a Random Experiment, any possible outcome is called simple event.

***In Example* 1,** there are two simple events namely H, T.

***In Example* 2,** there are 52 simple out came. Therefore there are 52 simple events exists.

Sample Space

The set of all possible simple events of a random experiment is called the sample space associated with the random experiment. It is denoted by 'S'. In the random experiment of tossing a coin, the sample space S = {H, T}. In the random experiment of throwing a dice, the sample space S = {1, 2, 3, 4, 5, 6}.

Note: If S is the sample space of a random experiment if

(i) Every element of S is an outcome of the experiment.

(ii) Every performance of the experiment results in an out came that corresponds to exactly one element of sample space S.

Event

A subset E of sample space S of a random experiment is called an event. That is, a set of elementary events (simple events) of a random experiment is called an event.

Example: In a rolling dice, S = {1, 2, 3, 4, 5, 6}, E_1 = {1, 2, 3}, E_2 = {3, 5} are events of S.

Note:

1. An event E is said to be happen (or occur), if an out come of the experiment belongs to E otherwise we say that E has not happened (or not occurred).
2. The complement of an event E is denoted by $\overline{E}$ or E^C is the event given by $\overline{E} = S - E$ which is called complementary event of E.
3. The Set S and empty set Ø are trivial subset of set S are events called certain (definite) event and impossible event respectively.

   For Example, in the experiment of 'tossing a coin' the sample space S = {H, T} where H stands for Head and T stands for Tail. The event of occurrence of either H or T is a certain (definite) event and the occurrence of both H and T is impossible event.

Mutually Exclusive Events

Two or more events are said to be mutually exclusive events, if the occurrence of one of the events prevents the occurrence of remaining events. So, the events E_1, E_2, E_3...E_n are said to be mutually exclusive if $E_i \cap E_j = \emptyset$ for $i \neq j$ & $1 \leq i, j \leq n$.

Example: In the experiment of tossing a coin, the events "happening of head (H)" and "happening of tail (T)" are mutually exclusive events. Since the event happening of head (H) prevents the occurrence of happening of tail (T). So the two events are mutually exclusive events.

Mutually Exhaustive Events

Two or more events are said to be mutually exhaustive events if the performance of the experiment always results in the occurrence of at least one of them. The events E_1, E_2, E_3...E_n are said to be mutually exhaustive events if $E_1 \cup E_2 \cup E_3 \ldots \cup E_n = S$.

Example: In the experiment of throwing a die, the events $E_1 = \{1, 3, 6\}$, $E_2 = \{2, 4, 5\}$ are exhaustive events since $E_1 \cup E_2 = S = \{1, 2, 3, 4, 5, 6\}$.

Equally Likely Events

Two or more events are said to be equally likely events if the chance of occurrence of each is equal i.e., there is no reason to expect one of them to happen in preference to other.

Example: In the experiment of tossing a coin, the chance of happening of 'head' is equal to the chance of happening of 'tail'. So the two events are equally likely events.

Classical or Mathematical Definition of Probability

If a random experiment results in n equally likely, exclusive, exhaustive events and m of them are favorable to happening of an event E, then the probability of occurrence of E or simply the probability of E denoted by P (E) is defined by

$$P(E) = \frac{m}{n}$$

Note:

1. For any event $0 \le P(E) \le 1$.
2. $P(E) + P(E^C) = 1$

   $$\text{Since } P(E^C) = \frac{n-m}{n} = 1 - \frac{m}{n}$$

   $$\therefore P(E) + P(E^C) = \frac{m}{n} + 1 - \frac{m}{n} = 1$$

3. If the outcomes of a random experiment are not equal, then the probability of an event in such an experiment is not defined.

   For Example the probability of a student pass an examination is not 1/2 since the out came of pass and failure in examination are not equally likely.
4. If the random experiment contains infinitely many outcomes than this definition cannot be applied to find the probability of an event in such an experiment.

Solved Problems

1. A die is thrown. What is the probability of getting an odd number?

Solution:

The total number of elementary events of sample space S = 6

$\because \quad S = \{1, 2, 3, 4, 5, 6\}$

$\therefore \quad n(S) = 6$

Let E be the event that occurrence of odd number when die is thrown.

$\therefore \quad E = \{1, 3, 5\}, n(E) = 3$

$$\therefore \quad P(E) = \frac{n(E)}{n(S)} = \frac{3}{6} = \frac{1}{2}$$

2. Three dice are thrown. What is the probability of getting same number on the three faces?

***Solution*:**

Let S be the sample space of experiment of throwing three dice.

$\therefore \quad n(S) = 6 \times 6 \times 6 = 6^3$

Let E be the event that the three faces show same number.

$\therefore \quad E = \{(1, 1, 1), (2, 2, 2), (3, 3, 3), (4, 4, 4), (5, 5, 5), (6, 6, 6)\}$

$\therefore \quad n(E) = 6$

$$\therefore \quad P(E) = \frac{n(E)}{n(S)} = \frac{6}{6^3} = \frac{1}{6^2} = \frac{1}{36}$$

3. A bag contains 3 Red, 4 Black, 5 White balls. What is the probability that two balls are drawn simultaneously from the bag are red and black ball?

***Solution*:**

Total number of balls = 3 + 4 + 5 =12

The number of ways of selecting two balls from these 12 is ${}^{12}C_2$.

$\therefore \quad n(S) = {}^{12}C_2$

Out of 3 red balls, one red ball can be selected in ${}^{3}C_1$ ways and out of 4 Black balls, one Black ball can be selected in ${}^{4}C_1$ ways. Let E be the event that getting red and black ball in a draw.

$\therefore \quad n(E) = 3 \times 4 = 12$

$$\therefore \quad P(E) = \frac{n(E)}{n(S)} = \frac{12}{12_{C_2}} = \frac{12}{\left(\frac{12 \times 11}{2}\right)} = \frac{2}{11}$$

4. If two cards are drawn from a well shuffled pack of card, what is the probability that one is king and other is queen?

Solution:

The number of ways of selecting two cards out of from 52 cards = ${}^{52}C_2$

$\therefore$ n (S) = ${}^{52}C_2$

Let E be the event that occurrence one king and one queen from the selection.

$\therefore \quad n(E) = {}^{4}C_1 \times {}^{4}C_1 = 16$ (In 52 cards there are 4 kings and 4 queens)

$$\therefore \quad P(E) = \frac{n(E)}{n(S)} = \frac{16}{{}^{52}C_2} = \frac{16}{\left(\frac{52 \times 51}{2}\right)} = \frac{16 \times 2}{52 \times 51} = \frac{8}{13 \times 51} = \frac{8}{663}$$

5. In a class, there are 60 boys and 40 girls. Half of the boys and half of the girls know cricket. Find the probability of the event that the person selected from the class is either a girl or a boy knowing cricket.

Solution:

Let E_1 be the event that the selected person is a girl with cricket knowledge.

$\therefore \quad n(E_1) = 20.$

Let E_2 be the event that the selected person is a boy with cricket knowledge.

$\therefore \quad n(E_2) = 30.$

Let E be the event that selected person is a girl or a boy knowing cricket knowledge.

$\therefore \quad n(E) = n(E_1) + n(E_2) = 20 + 30 = 50$

Also n(S) = total number of students = 60 + 40 = 100

$$\therefore \quad P(E) = \frac{n(E)}{n(S)} = \frac{50}{100} = \frac{1}{2}$$

6. If two numbers are randomly selected from 10 consecutive natural numbers, find the probability that the sum of two numbers is

(i) An even number and

(ii) An odd number.

Solution:

We know that sum of any two even numbers or sum of any two odd numbers is an even number and sum of one even and one odd number is an odd number. Here the number of consecutive numbers is 10 in which 5 are even and 5 are odd. The number of ways of selecting 2 numbers from 10 is ${}^{10}C_2$. $\therefore$ n(S) = ${}^{10}C_2$.

(i) Let E_1 be the event that sum of selecting two numbers from the 10 numbers is even. The number of ways of selecting two even numbers from 5 even numbers is ${}^5C_2$. The number of ways of selecting two odd numbers from 5 odd numbers is ${}^5C_2$.

$$\therefore \quad n(E_1) = {}^5C_2 + {}^5C_2 = 10 + 10 = 20$$

$$\therefore \quad P(E_1) = \frac{n(E_1)}{n(S)} = \frac{20}{{}^{10}C_2} = \frac{20}{\left(\frac{10\times 9}{2}\right)} = \frac{4}{9}$$

(ii) Let E_2 be the event that sum of selected two numbers is odd.

$$\therefore \quad n(E_2) = {}^5C_1 \times {}^5C_1 = 25$$

$$P(E_2) = \frac{n(E_2)}{n(S)} = \frac{25}{{}^{10}C_2} = \frac{25}{\left(\frac{10\times 9}{2}\right)} = \frac{50}{10\times 9} = \frac{5}{9}$$

or
$$P(E_2) = 1 - P(E_1) = 1 - \frac{4}{9} = \frac{5}{9}$$

7. A box contains 20 mangoes in which 5 are damaged. If 5 mangoes are selected at random from the box, find the probability of the event that

(i) None of them is damaged.

(ii) Only one of them is damaged.

(iii) At least one of them is damaged.

***Solution*:**

The number of ways of selecting 5 mangoes out of 20 is ${}^{20}C_5$.

(i) Let E_1 be the event that selects 5 mangoes in which none of them is damaged.

$$\therefore \quad n(E_1) = {}^{(20-5)}C_5 = {}^{15}C_5$$

$$\therefore \quad P(E_1) = \frac{n(E_1)}{n(S)} = \frac{{}^{15}C_5}{{}^{20}C_5}$$

(ii) Let E_2 be the event that selects 5 mangoes in which only one of them is damaged.

$\therefore$ $n(E_2) = {}^{15}C_4 \times {}^{5}C_1$ (i.e., 4 are good selected from 15 good mangoes and 1 is selected from 5 damaged mangoes)

$$\therefore \quad P(E_2) = \frac{{}^{15}C_4 \times {}^{5}C_1}{{}^{20}C_5}$$

(iii) Let E_3 be event that at least one of them is damaged.

$$\therefore \quad P(E_3) = 1 - P(\text{all the selected 5 mangoes are good}) = 1 - \frac{{}^{15}C_5}{{}^{20}C_5}$$

8. A Pair of dice is thrown. What is the probability that sum of on the faces are 11?

***Solution*:**

A pair of dice is thrown.

$\therefore$ $n(S) = 6 \times 6 = 36$

Let E be the event that sum on the two faces is 11.

$\therefore$ $E = \{(5, 6), (6, 5)\} \Rightarrow n(E) = 2$

$$\therefore \quad P(E) = \frac{n(E)}{n(S)} = \frac{2}{36} = \frac{1}{18}$$

4.2 Relative Frequency (or Statistical or Empirical) Definition of Probability

Suppose a random experiment is conducted n times. Out of which an event E occurs m(n) times then the ratio $r_n = \frac{m(n)}{n}$ is called n^{th} relative frequency of event. Therefore, the sequence $r_1, r_2, r_3, \ldots r_n \ldots$ if r_n tends to a definite limit say l as $n \to \infty$. i.e., $\lim_{n\to\infty} r_n = l$ is defined to be probability of event E and we write $P(E) = \lim_{n\to\infty} r_n = l$.

Solved Examples

1. A fair coin is tossed 400 times. Find the probability of getting head on odd number of times?

***Solution*:**

The total number of points on sample space = 2^{400}.

Let E be the event of getting a head an odd number of times.

$$\therefore \quad n(E) = {}^{400}C_1 + {}^{400}C_3 + {}^{400}C_5 + \dots\dots\dots + {}^{400}C_{399} = \frac{2^{400}}{2} = 2^{399}$$

$$\text{Required Probability} = P(E) = \frac{2^{399}}{2^{400}} = \frac{1}{2}$$

2. Out of 1, 00,000 new born babies 76,282 survived till the age 20. Find the probability that a new born baby survives till 20.

***Solution*:** Here m = 76,282; n = 100000.

$$\text{Required Probability} = \frac{76282}{100000} = 0.76282$$

Axiomatic Approach of Probability

Let S be the sample space of random experiment, then the set of all subsets of S is called the power set of S and is denoted by P(s). Observe that the set of all possible events of this experiment is power set P(S).

Let S be the sample space of a random experiment and function P: P(s) → R satisfies the following axioms is called a probability function.

(i) $P(E) \geq 0$ for all $E \in P(s)$ (axiom of non negative)

(ii) $P(s) = 1$ (axiom of certainty)

(iii) If $E_1, E_2 \in P(s)$ and $E_1 \cap E_2 = \emptyset$ then $P(E_1 \cup E_2) = P(E_1) + P(E_2)$ (axiom of additivity)

For each $E \in P(s)$, then the real number P(E) is called probability of event E. If E = {a} then we write P (a) instead of P ({a}).

Note:

1. $P(\emptyset) = 0$ for any sample space S, $S \cup \emptyset = S$, $S \cap \emptyset = \emptyset$,

   $P(S) = P(S \cup \emptyset) = P(S) + P(\emptyset) = P(S) + 0 = 1 + 0 = 1.$

2. If S is countably infinite and $\{E_n\}$ is sequence of mutually exclusive events then $P\left(\bigcup_{n=1}^{\infty} E_n\right) = \sum_{n=1}^{\infty} P(E_n)$

   i.e., $P(E_1 \cup E_2 \cup \dots\dots) = P(E_1) + P(E_2) + \dots.$

3. If E_1, E_2 are any two events of a sample space such that $E_1 \subseteq E_2$ then

   $P(E_2 - E_1) = P(E_2) - P(E_1).$

Addition Theorem on Probability

If E_1, E_2 are any two events of a random experiment and P is probability function then $P(E_1 \cup E_2) = P(E_1) + P(E_2) - P(E_1 \cap E_2)$.

Here $(E_1 \cup E_2)$ means occurrence event either E_1 or E_2 and $E_1 \cap E_2$ means occurrence of event both E_1 and E_2.

Solved Problems

1. A and B are events with $P(A) = 0.4$, $P(B) = 0.5$ and $P(A \cap B) = 0.3$. Find the probability that

(i) A does not occur.

(ii) Either A or B.

(iii) Neither A nor B.

(iv) A occurs but B does not occur.

Solution:

(i) $P(\bar{A}) = 1 - P(A) = 1 - 0.4 = 0.6$

(ii) $P(A \cup B) = P(A) + P(B) - P(A \cap B) = 0.4 + 0.5 - 0.3 = 0.6$

(iii) $P(\bar{A} \cap \bar{B}) = P(\overline{A \cup B}) = 1 - P(A \cup B) = 1 - 0.6 = 0.4$

(iv) $P(A \cap \bar{B}) = P(A) - P(A \cap B) = 0.4 - 0.3 = 0.1$

2. A card is drawn from well shuffled pack of 52 playing cards. What is the probability that the drawn card is a spade or a king?

Solution:

Let E_1 be the event that selected card is spade

$$\therefore \quad P(E_1) = \frac{^{13}C_1}{^{52}C_1} = \frac{13}{52}$$

Let E_2 be the event that selected card is king

$$\therefore \quad P(E_2) = \frac{^{4}C_1}{^{52}C_2} = \frac{4}{52}$$

$$P(E_1 \cap E_2) = P\text{(selected card is spade King)} = \frac{1}{52}$$

$$P(E_1 \cup E_2) = P(E_1) + P(E_2) - P(E_1 \cap E_2) = \frac{13}{52} + \frac{4}{52} - \frac{1}{52} = \frac{16}{52} = \frac{4}{13}$$

3. The probability for a contractor to get a road contract is $\frac{2}{3}$ and to get a building contract is $\frac{5}{9}$. The probability to get at least one contract is $\frac{4}{5}$. Find the probability that he gets both contracts?

Solution:

Let E_1 be the event of getting road contract and E_2 be the event of getting building contract.

$$\therefore \quad P(E_1) = \frac{2}{3},\ P(E_2) = \frac{5}{9},\ P(E_1 \cup E_2) = \frac{4}{5}$$

$$P(E_1 \cup E_2) = P(E_1) + P(E_2) - P(E_1 \cap E_2)$$

$$P(E_1 \cap E_2) = P(E_1) + P(E_2) - P(E_1 \cup E_2) = \frac{2}{3} + \frac{5}{9} - \frac{4}{5} = \frac{19}{45}$$

4. A committee of 30 members, each member is proficient either in pharmaceutics or in pharma chemistry or in both. 21 are proficient in pharmaceutics and 17 are proficient in pharma chemistry. Find the probability that a person selected from the committee is proficient in both.

Solution:

Let M be the event that selected person is proficient in pharmaceutics and P be the event that selected person is proficient in pharma chemistry.

$$\therefore \quad P(M) = \frac{21}{30},\ P(P) = \frac{17}{30}$$

$P(M \cup P) = 1$ ($\because$ The person is either proficient in pharmaceutics or pharma chemistry or in both).

$$P(M \cap P) = P(M) + P(P) - P(M \cup P) = \frac{21}{30} + \frac{17}{30} - 1 = \left(\frac{38}{30}\right) - 1 = \frac{8}{30} = \frac{4}{15}$$

5. If one ticket is randomly selected from tickets numbered 1 to 20 then find the probability that the number on the ticket is

(i) A multiple of 5 or 7

(ii) A multiple of 3 or 5.

Solution:

(i) Let A be the event that selected number from 1 to 20 is multiple of 5.

Let B be the event that selected number is multiple of 7.

A = {5, 10, 15, 20}, B = {7, 14}, A∩B = Ø,

$$P(A\cup B) = P(A) + P(B) - P(A\cap B) = \frac{4}{20} + \frac{2}{20} - P(\emptyset) = \frac{6}{20} - 0 = \frac{6}{20} = \frac{3}{10}$$

(ii) Let A be the event that selected number is divisible by 3.

Let B be the event that selected number is divisible by 5.

A= {3, 6, 9, 12, 15, 18}, B ={5, 10, 15, 20}, A∩B = {15},

$$P(A\cup B) = P(A) + P(B) - P(A\cap B) = \frac{6}{20} + \frac{4}{20} - \frac{1}{20} = \frac{9}{20}$$

6. A, B, C are three newspapers from a city. 20% of population read A, 16% read B, 14% read C, 8% read both A and B, 5% read both A and C, 4% read both B and C and 2% read all the three. Find the percentage of the population who read at least one newspaper.

Solution:

Let A, B, C be the events that selecting A, B, C newspapers respectively.

$$\therefore \quad P(A) = \frac{20}{100},\ P(B) = \frac{16}{100},\ P(C) = \frac{14}{100},\ P(A\cap B) = \frac{8}{100},\ P(A\cap C) = \frac{5}{100},$$

$$P(B\cap C) = \frac{4}{100},\ P(A\cap B\cap C) = \frac{2}{100}.$$

$P(A\cup B\cup C) =$

$P(A) + P(B) + P(C) - P(A\cap B) - P(A\cap C) - P(B\cap C) + P(A\cap B\cap C)$

$$= \frac{20}{100} + \frac{16}{100} + \frac{14}{100} - \frac{8}{100} - \frac{5}{100} - \frac{4}{100} + \frac{2}{100} = \frac{50}{100} - \frac{17}{100} + \frac{2}{100} = \frac{35}{100}$$

The percentage of the population who read at least one newspaper is

$$\frac{35}{100} \times 100 = 35$$

Conditional Event

Let A and B are two events of a random experiment. If the event B occurs after the occurrence of event A then the event "happening of B after happening of A" is called a conditional event and is denoted by B/A. Similarly A/B stands for the event happening of A after happening of B.

Conditional Probability

Suppose A and B are two events of a sample space S and $P(A) \neq 0$ then the probability of B after happening of A is called Conditional Probability of B given A and is denoted by P(B/A) and is defined as $P\left(\frac{B}{A}\right) = \frac{P(B \cap A)}{P(A)}$, $P(A) \neq 0$.

Similarly $$P\left(\frac{A}{B}\right) = \frac{P(A \cap B)}{P(B)}, \quad P(B) \neq 0.$$

Solved Examples

1. A pair of dice is rolled. What is the probability that the sum of two dice is 7 and neither die shows 2.

Solution:

Given a pair of dice are rolled.

$\therefore \quad n(S) = 6 \times 6 = 36.$

Let A be the event that neither die show 2.

$$\therefore \quad P(A) = \frac{5}{6} \times \frac{5}{6} = \frac{25}{36}$$

Let B be the event that sum on the two faces is 7.

$B = \{(1, 6), \{2, 5),(3, 4),(4, 3),(5, 2),(6, 1)\} \Rightarrow n(B) = 6.$

Now $n(A \cap B) = 4$, $P(A \cap B) = \frac{4}{36}$

$$\therefore \quad P\left(\frac{B}{A}\right) = \frac{P(A \cap B)}{P(A)} = \frac{\frac{4}{36}}{\frac{25}{36}} = \frac{4}{25}$$

2. A pair of dice is rolled. What is the probability that the neither dice shows 2 and their sum is 7.

Solution:

$$P\left(\frac{A}{B}\right) = \frac{P(A \cap B)}{P(B)} = \frac{\frac{4}{36}}{\frac{6}{36}} = \frac{4}{6} = \frac{2}{3}$$ (Using above solution)

3. Suppose that an unbiased pair of dice is rolled. Let A denote event that same number show on the either die. Let B denote the event that sum is greater than 7. Find (i) $P\left(\frac{A}{B}\right)$ and (ii) $P\left(\frac{B}{A}\right)$.

Solution:

Given A is the event that same number show on the either die. n (A) = 6, $P(A) = \frac{6}{36}$

B be the event that sum greater than 7 i.e., sum is 8 or 9 or 10 or 11 or 12.

For above number of possibilities n(B) = 5 + 4 + 3 + 2 + 1 = 15.

$A \cap B = \{(4, 4), (5, 5), (6, 6)\} \Rightarrow n(A \cap B) = 3$

(i) $$P\left(\frac{A}{B}\right) = \frac{P(A \cap B)}{P(B)} = \frac{\frac{3}{36}}{\frac{15}{36}} = \frac{3}{15} = \frac{1}{5}$$

(ii) $$P\left(\frac{B}{A}\right) = \frac{P(A \cap B)}{P(A)} = \frac{\frac{3}{36}}{\frac{6}{36}} = \frac{3}{6} = \frac{1}{2}$$

Independent Events

Two events A and B are said to be independent events if the happening of each one does not depend upon the other. i.e., $P\left(\frac{A}{B}\right) = P(A)$ and similarly $P\left(\frac{B}{A}\right) = P(B)$.

$$\therefore \frac{P(A \cap B)}{P(B)} = P(A) \text{ and } \frac{P(B \cap A)}{P(A)} = P(B) \Rightarrow P(A).P(B) = P(A \cap B).$$

$\therefore$ Two events A and B are said to be independent of $P(A \cap B) = P(A).P(B)$.

Note:

1. If $E_1, E_2, E_3 \ldots E_n$ are independent then

   $P(E_1 \cap E_2 \cap E_3 \ldots \cap E_n) = P(E_1).P(E_2).P(E_3) \ldots P(E_n)$.

2. If A and B are not independent then we say they are dependent and

   $P(A \cap B) \neq P(A).P(B)$.

Solved Problems

1. A card is selected at random from a pack of 52 cards. Let A be the event that the card is a face card and let B be the event that the card is a heart. Show that A and B are independent.

Solution:

Given A be event that the selected card is a face card.

$\therefore \quad n(A) = 4 \times 3 = 12$ (Kings-4, Queens-4, Jacks-4)

$P(A) = \frac{12}{36} = \frac{1}{3}$

Let B be the event that selected card is heart.

$n(B) = 13.$

$P(B) = \frac{13}{52} = \frac{1}{4}.$

$A \cap B = \{\text{King heart, Queen heart, Jack heart}\}$, $n(A \cap B)$

$= 3,\ P(A \cap B) = \frac{3}{36} = \frac{1}{12}.$

Now $P(A).P(B) = \frac{1}{3} \cdot \frac{1}{4} = \frac{1}{12}.$

$\therefore$ $P(A \cap B) = P(A).P(B)$

$\therefore$ A, B are independent.

2. Suppose A and B are two independent events with

$P(A) = \frac{1}{2}$ and $P(B) = \frac{1}{3}$, then find $P(A - B)$, $P(B - A)$, $P((A - B) \cup (B - A))$.

Solution:

Given A and B are independent events with $P(A) = \frac{1}{2}$ and $P(B) = \frac{1}{3}$

$P(A - B) = P(A) - P(A \cap B)$

$= P(A) - P(A).P(B)$ (Since A and B are independent Events)

$= \frac{1}{2} - \frac{1}{2} \cdot \frac{1}{3} = \frac{1}{2} - \frac{1}{6} = \frac{4}{12} = \frac{1}{3}$

$P(B - A) = P(B) - P(B \cap A)$

$= P(B) - P(A).P(B)$ (Since A and B are independent Events)

$$= \frac{1}{3} - \frac{1}{2} \cdot \frac{1}{3} = \frac{1}{3} - \frac{1}{6} = \frac{1}{6}$$

$P((A - B) \cup (B - A)) = P(A - B) + P(A - B)$

(since A—B and B—A are mutually inclusive events).

$= P(A) - P(A).P(B) + P(B) - P(A).P(B)$

$= P(A) + P(B) - 2.P(A).P(B)$

$$= \frac{1}{2} + \frac{1}{3} - 2\left(\frac{1}{2}\right)\left(\frac{1}{3}\right)$$

$$= \frac{1}{2} + \frac{1}{3} - \frac{1}{3} = \frac{1}{2}$$

3. The probability that A hit the target is 1/2 and the probability that B hit the target is 1/3. If each of them fired once what's the probability that the target will be hit at least once.

Solution:

$$P(A) = \frac{1}{2}, \ P(B) = \frac{1}{3}, \ P(A \cup B) = ?$$

$P(A \cap B) = P(A).P(B)$ (Since A and B are independent Events).

$\therefore \quad P(A \cup B) = P(A) + P(B) - P(A \cap B)$ ($\because$ by Addition theorem)

$= P(A) + P(B) - P(A).P(B)$

$$= \frac{1}{2} + \frac{1}{3} - \left(\frac{1}{2} \cdot \frac{1}{3}\right)$$

$$= \frac{1}{2} + \frac{1}{3} - \frac{1}{6} = \frac{4}{6} = \frac{2}{3}$$

4. A Problem is given to three students A, B, C. The chances of solving the problem from these 3 students was found to be $\frac{1}{2}$, $\frac{1}{3}$ and $\frac{1}{4}$ respectively. Then find the probability that the problem will be solved.

Solution:

Given $P(A) = \frac{1}{2}$, $P(B) = \frac{1}{3}$ and $P(C) = \frac{1}{4}$

$$P(A \cup B \cup C) = 1 - P\left(\overline{A \cup B \cup C}\right)$$

$$= 1 - P\left(\overline{A} \cap \overline{B} \cap \overline{C}\right) \qquad (\because \text{By demorgans law})$$

$$= 1 - P(\overline{A}).P(\overline{B}).P(\overline{C})$$

($\because$ Since A , B, C are independent, so $\overline{A}$, $\overline{B}$, $\overline{C}$ are also independent)

$$= 1 - \left(1 - \frac{1}{2}\right)\left(1 - \frac{1}{3}\right)\left(1 - \frac{1}{4}\right)$$

$$= 1 - \frac{1}{2} \times \frac{2}{3} \times \frac{3}{4} = 1 - \frac{1}{4} = \frac{3}{4}$$

5. Suppose there are 10 boys and 5 girls in a class. If we choose three children one after another in succession at random, then what is the probability that all the three are boys?

Solution:

Let A_i be the event that choosing a boy in its trial = 1, 2, 3 now we have to find $P(A_1 \cap A_2 \cap A_3)$.

$$P(A_1) = \frac{10}{15}, P\left(\frac{A_2}{A_1}\right) = \frac{9}{14}, P\left(\frac{A_3}{A_1 \cap A_2}\right) = \frac{8}{13}$$

Hence $\quad P(A_1 \cap A_2 \cap A_3) = P(A_1).P\left(\frac{A_2}{A_1}\right).P\left(\frac{A_3}{A_1 \cap A_2}\right)$

($\because$ By multiplication theorem, that is if A and B are any two events $P(B) \neq 0$, $P(A) \neq 0$ then

$$P(A \cap B) = P(A).P\left(\frac{B}{A}\right) = P(B).P\left(\frac{A}{B}\right))$$

$$= \frac{10}{15} \times \frac{9}{14} \times \frac{8}{13} = \frac{24}{91}$$

6. An urn contains 10 red balls, 10 white balls. Suppose two balls are drawn one after another without replacement. Find the probability that second ball is white. Given that first ball is red.

Solution:

After selecting red ball from the Urn, it contains 9 red, 10 white balls.

Then the probability of getting white $= \dfrac{10}{(9+10)} = \dfrac{10}{19}$

7. Bag B_1 contains 4 white and 2 black balls. Bag B_2 contains 5 white and 3 black balls. A bag is drawn at random and a ball is chosen at random from it. What is the probability that it is white?

Solution:

Let E_1, E_2 denote two events of choosing bags B_1 and B_2 respectively.

$$\therefore \quad P(E_1) = P(E_2) = \frac{1}{2}$$

Let W be the event that the ball chosen from selected bag is white.

$$P\left(\frac{W}{E_1}\right) = \frac{4}{6}, \quad P\left(\frac{W}{E_2}\right) = \frac{5}{8}$$

Now $W = (W \cap E_1) \cup (W \cap E_2)$.

$$P(W) = P((W \cap E_1) \cup (W \cap E_2))$$

$$= P(W \cap E_1) + P(W \cap E_2)$$

$$= P(E_1).P\left(\frac{W}{E_1}\right) + P(E_2).P\left(\frac{W}{E_2}\right)$$

$$= \frac{1}{2} \times \frac{4}{6} + \frac{1}{2} \times \frac{5}{8} = \frac{1}{3} + \frac{5}{16} = \frac{16+15}{48} = \frac{31}{48}$$

8. A and B are independent events such that the probability of both the events to occur is 1/6 and the probability of both the events do not occur is 1/3. Find P(A).

Solution:

$$P(A \cap B) = \frac{1}{6}, \; P\left(\overline{A} \cap \overline{B}\right) = \frac{1}{3}, \; P(A).P(B) = \frac{1}{6}, \; P(\overline{A}).P(\overline{B}) = \frac{1}{3}$$

Let $P(A) = x, \; P(B) = y$

$$\therefore \quad P(\overline{A}) = 1 - x, \; P(\overline{B}) = 1 - y$$

$$xy = \frac{1}{6}$$

$$(1 - x)(1 - y) = \frac{1}{3}$$

$$1 - x - y + xy = \frac{1}{3}$$

$$1 - x - y + \frac{1}{6} = \frac{1}{3}$$

$$x + y = \frac{7}{6} - \frac{1}{3} = \frac{7-2}{6} = \frac{5}{6}$$

$$x + y = \frac{5}{6},\ xy = \frac{1}{6}$$

$$(x - y)^2 = (x + y)^2 - 4xy = \frac{25}{36} - \frac{4}{6} = \frac{25-24}{36} = \frac{1}{36}$$

$$x - y = \pm\frac{1}{6},\ x + y = \frac{5}{6}$$

if,

$$x - y = \frac{1}{6}$$

$$x + y = \frac{5}{6}$$

$$2x = \frac{6}{6} = 1$$

$$x = \frac{1}{2}$$

if,

$$x - y = -\frac{1}{6}$$

$$x + y = \frac{5}{6}$$

$$2x = \frac{4}{6} = \frac{2}{3}$$

$$x = \frac{1}{3}$$

$\therefore$ $$P(A) = \frac{1}{2} \text{ or } \frac{1}{3}$$

4.3 Baye's Theorem

Suppose E_1, E_2, $E_3 \ldots E_n$ are n mutually exclusive and exhaustive events of a random experiment with $P(E_i) \neq 0$ for i=1, 2, 3,…n then for any event A of random experiment with $P(A) \neq 0$,

$$P\left(\frac{E_k}{A}\right) = \frac{P(E_k).P\left(\frac{A}{E_k}\right)}{\sum_{i=1}^{n} P(E_i).P\left(\frac{A}{E_i}\right)}, \qquad k = 1, 2, 3 \ldots n$$

Solved Problems

1. A shopkeeper buys a particular type of electrical bulbs from three manufacturers M_1, M_2 and M_3. He buys 25% of his requirement from M_1, 45% from M_2 and 30% from M_3. Based on the past experience he found that 2% of M_3 bulbs are defective, where as only 1% of M_1 and type M_2 are defective. If a bulb is chosen by him at random is defective, find the probability that it was of type M_3.

Solution:

Let E be the event that the bulb chosen is defective .We have to find out $P\left(\frac{M_3}{E}\right)$

And given $P(M_1) = 0.25$, $P(M_2) = 0.45$, $P(M_3) = 0.3$

$$P\left(\frac{E}{M_1}\right) = 0.01,\ P\left(\frac{E}{M_2}\right) = 0.01,\ P\left(\frac{E}{M_3}\right) = 0.02.$$

Now from the Baye's Theorem,

$$P\left(\frac{M_3}{E}\right) = \frac{P(M_3)\cdot P\left(\frac{E}{M_3}\right)}{P(M_1)\cdot P\left(\frac{E}{M_1}\right) + P(M_2)\cdot P\left(\frac{E}{M_2}\right) + P(M_3)\cdot P\left(\frac{E}{M_3}\right)}$$

$$= \frac{0.02 \times 0.3}{(0.25 \times 0.01) + (0.45 \times 0.01) + (0.30 \times 0.20)} = 0.46$$

2. Suppose that an urn B_1 contains 2 black, 3 white and another urn B_2 has 3 black and 4 white balls. One urn is selected at random and a ball is drawn. If the ball is found to be white, then what is the probability that the urn chosen was B_2.

Solution:

Let W be the event that the ball chosen from selected urn is white and let B_1, B_2 are events of selected urns B_1 and B_2.

$\therefore \quad P(B_1) = P(B_2) = \frac{1}{2}, P\left(\frac{W}{B_1}\right) = \frac{3}{5}, P\left(\frac{W}{B_2}\right) = \frac{4}{7}$

By Baye's Theorem,

$$P\left(\frac{B_2}{W}\right) = \frac{P(B_2)\cdot P\left(\frac{W}{B_2}\right)}{P(B_1)\cdot P\left(\frac{W}{B_1}\right) + P(B_2)\cdot P\left(\frac{W}{B_2}\right)}$$

$$= \frac{\frac{1}{2}\times\frac{4}{7}}{\left(\frac{1}{2}\times\frac{3}{5}\right)+\left(\frac{1}{2}\times\frac{4}{7}\right)} = \frac{\frac{4}{7}}{\frac{3}{5}+\frac{4}{7}}$$

$$= \frac{\frac{4}{7}}{\frac{21+20}{35}} = \frac{20}{41}$$

3. Three urns have the following composition of balls.

Urn 1: 1 white and 2 black

Urn2 : 2 white and 3 black

Urn 3: 3 white and 4 black

One of the urns is selected at random and a ball is drawn. It turns out to be white. Find the probability that it came from urn 3.

Solution:

Let B_1, B_2, B_3 are the events of selecting urns 1, 2, 3 respectively.

Let W be the event of choosing white ball from the selected urn.

$\therefore \quad P(B_1) = P(B_2) = P(B_3) = \frac{1}{2}$

$P\left(\frac{W}{B_1}\right) = \frac{1}{3}, P\left(\frac{W}{B_2}\right) = \frac{2}{5}, P\left(\frac{W}{B_3}\right) = \frac{3}{7}$

$$\therefore \quad P\left(\frac{B_3}{W}\right) = \frac{P(B_3)\cdot P\left(\frac{W}{B_3}\right)}{P(B_1)\cdot P\left(\frac{W}{B_1}\right) + P(B_2)\cdot P\left(\frac{W}{B_2}\right) + P(B_3)\cdot P\left(\frac{W}{B_3}\right)}$$

$$= \frac{\frac{1}{2} \times \frac{3}{7}}{\left(\frac{1}{2} \times \frac{1}{3}\right) + \left(\frac{1}{2} \times \frac{2}{5}\right) + \left(\frac{1}{2} \times \frac{3}{7}\right)} = \frac{\frac{3}{7}}{\frac{1}{3} + \frac{2}{5} + \frac{3}{7}}$$

$$= \frac{\frac{3}{7}}{\frac{35 + 42 + 45}{15 \times 7}} = \frac{3 \times 15}{122} = \frac{45}{122}$$

4. There are 3 black and 4 white balls in one bag, 4 black and 3 white balls in second bag. A die is rolled and first bag is selected if the dice shows up 1 or 3, and the second bag for the rest. One bag is selected in this manner and a ball is drawn from it and the ball is found to be white. What is the probability that it is coming from first bag.

Solution:

Let B_1, B_2 be events that selecting first and second bags respectively.

$$P(B_1) = \frac{2}{6} = \frac{1}{3}, \; P(B_2) = \frac{4}{6} = \frac{2}{3}$$

Let W be the event of choosing ball from the selected bag is white.

$$P\left(\frac{W}{B_1}\right) = \frac{4}{7}, \; P\left(\frac{W}{B_2}\right) = \frac{3}{7}$$

Now
$$P\left(\frac{B_1}{W}\right) = \frac{P(B_1) \cdot P\left(\frac{W}{B_1}\right)}{P(B_1) \cdot P\left(\frac{W}{B_1}\right) + P(B_2) \cdot P\left(\frac{W}{B_2}\right)}$$

$$= \frac{\frac{1}{3} \times \frac{4}{7}}{\left(\frac{1}{3} \times \frac{4}{7}\right) + \left(\frac{2}{3} \times \frac{3}{7}\right)} = \frac{1 \times 4}{1 \times 4 + 2 \times 3} = \frac{4}{10} = \frac{2}{5}$$

4.4 Probability Distributions

Random Variable

Suppose S is a sample space associated with a random experiment, then the function $X: S \rightarrow R$ is called a random variable and R is set of real numbers.

Example: Let S be the sample space of experiment of rolling a fair dice.

Then $X: S \rightarrow R$ is given by

$$\left.\begin{aligned} X(n) &= 0 \text{ if n is odd} \\ &= 1 \text{ if n is even} \end{aligned}\right\} \text{ is a random variable.}$$

Here $S = \{1, 2, 3, 4, 5, 6\}$

$\therefore$ $X(1) = 0, X(2) = 1, X(3) = 0, X(4) = 1, X(5) = 0, X(6) = 1$

$\therefore$ $X^{-1}(0) = \{1, 3, 5\}, X^{-1}(1) = \{2, 4, 6\}$

Note: $X^{-1}(P(R)) = P(S)$ where P(S) stands for power set of S.

Probability Function Induced by Random Variable X

Suppose S is a sample space of random experiment, $P: P(S) \to R$ is probability function and $X: S \to R$ is random variable then the function P^1 defined on P(R) by $P^1(Y) = P(X^{-1}(Y))$ for each $y \in P(R)$ is a probability function on P(R) have P^1 is called probability function induced by X.

Example: Let S be sample space of the random experiment of rolling a fair dice. Define

$$X: S \to R \text{ by } \quad X(n) = 0 \text{ if n is odd}$$
$$= 1 \text{ if n is even.}$$

Then we find P^1.

Solution: Here $S = \{1, 2, 3, 4, 5, 6\}$

$$X(1) = X(3) = X(5) = 0 \Rightarrow X^{-1}(0) = \{1, 3, 5\}$$
$$X(2) = X(4) = X(6) = 1 \Rightarrow X^{-1}(1) = \{2, 4, 6\}$$
$$P^1(0) = P(X^{-1}(0)) = P(\{1, 3, 5\}) = \frac{1}{6} + \frac{1}{6} + \frac{1}{6} = \frac{3}{6} = \frac{1}{2}$$
$$P^1(1) = P(X^{-1}(1)) = P(\{2, 4, 6\}) = \frac{1}{6} + \frac{1}{6} + \frac{1}{6} = \frac{3}{6} = \frac{1}{2}$$
$$= \begin{cases} \frac{1}{2} \text{ of 0 or } 1 \in y \text{ but not both} \\ 1 \text{ of } \{0,1\} \in y \end{cases}$$

Probability Distribution Function

Suppose S is the sample space in a random experiment. Let $X: S \to R$ is a random variable then to each $x \in R$ the event $\{a \in S: X(a) \le x\}$ is denoted by $X \le x$.

i.e.,
$$(X \le x) = \{a \in S: X(a) \le x\}$$
$$= \{a \in S: X(a) \in (-\infty, x]\}$$
$$= X^{-1}((-\infty, x]) \text{ for each } x \in R$$

Suppose x is a random variable connected with random experiment and P is a probability function associated with it then $F: R \to R$ defined by $F(x) = P(X \le x)$ for each $x \in R$ is called probability distribution function of X.

Note:

(i) $0 \leq F(x) \leq 1$ for all $x \in R$

(ii) $x_1 \leq x_2 \Rightarrow F(x_1) \leq F(x_2)$

(iii) $\lim_{n \to \infty} F(x) = 1$ and $\lim_{n \to \infty} F(x) = 0$

4.4.1 Discrete Random Variable – Continuous Random Variable

Suppose X: $S \rightarrow R$ is a random variable. If the range of X is either finite or countable infinite then X is called a discrete random variable.

A random variable which can take all real values in an interval (a, b) is called as continuous variable.

4.4.2 Binomial Distribution

Suppose if we conduct an experiment with two possible outcomes then it is called Bernoulli trial. In general a possible outcome of a Berrnoulli trial is called either a success (S) or Failure (F).

If p is the probability of success and q is the probability of failure then it follows that $p + q = 1$. Many of the problems can be solved by determining the probability of success in n independent. Bernoullis trials with probability of success p and probability of failure $q = 1 - p$, considered as a function of r we call this function the binomial distribution.

A discrete random variable X is said to follow a binomial distribution (or simply a binomial variable with parameters n and p) where $0 < p < 1$ if

$$P(X = r) = {}^nC_r \times p^r \times q^{n-r}, \; n \in \{0, 1, 2, 3, \ldots.. n\}$$

If X is a binomial variable with parameters n, p then it is also written as X~B(n, p).

Binomial Distribution follows under the experimental conditions:

1. The number n of trials is finite.
2. Each trial results in two mutually exclusive outcomes termed as success and failure.
3. The n trials are independent of each other.
4. The probability of p of success is constant for each trial.
5. If the number of trials in any attempt is n and if there are N such sets(attempts) then frequency function of the binomial distribution is given by

$$N\,P(r) = N \times {}^nC_r \times p^r \times q^{n-r} \;\; r \in \{0, 1, 2, \ldots.n\}.$$

   i.e., the number of sets in which get exactly r comes

$$= N \times {}^nC_r \times p^r \times q^{n-r} \;\; x \in \{0, 1, 2 \ldots.n\}.$$

Definition: If X~B (n, p), then the mean μ and variance σ^2 of X are equal to np and npq respectively.

Solved Problems

1. 10 coins are tossed simultaneously. Find the probability of getting 6 heads.

Solution:

In the experiment of tossing a coin the probability of getting head $=\frac{1}{2}$,

Probability of getting tail $=\frac{1}{2}$,

$\therefore$ Then the probability of getting r heads in a random throw of 8 coins.

$$= P(X=r) = {}^8C_r \cdot \left(\frac{1}{2}\right)^r \cdot \left(\frac{1}{2}\right)^{8-r}$$

$$\therefore \quad P(X=6) = {}^8C_6 \cdot \left(\frac{1}{2}\right)^6 \cdot \left(\frac{1}{2}\right)^{8-6}$$

$$= {}^8C_2 \cdot \left(\frac{1}{2}\right)^6 \cdot \left(\frac{1}{2}\right)^2$$

$$= \frac{8\times 7}{2} \times \left(\frac{1}{2}\right)^8 = \frac{28}{2^8} = \frac{7}{2^6} = \frac{7}{64}$$

2. The mean and variance of a binomial distribution are 4 and 3 respectively. Find the distribution and find $p(X \geq 1)$.

Solution:

Given mean = np = 4 and variance = npq = 3 $\Rightarrow \frac{npq}{np} = \frac{3}{4} \Rightarrow q = \frac{3}{4}$

$\therefore \quad p = 1 - q = 1 - \frac{3}{4} = \frac{1}{4} \qquad \therefore np = 4.$

$\Rightarrow \quad n.\frac{1}{4} = 4 \Rightarrow n = 16$

$\therefore \quad X \sim B(n, p) = B(16, \frac{1}{2})$

$$\text{Now } p(X \geq 1) = 1 - p(X < 1)$$

$$= 1 - p(X = 0)$$

$$= 1 - {}^{16}C_0 \cdot \left(\frac{1}{4}\right)^0 \cdot \left(\frac{3}{4}\right)^{16} = 1 - \left(\frac{3}{4}\right)^{16}$$

3. The difference between the mean and variance of a binomial variable is 5/9 then, find the probability for the event of 2 successes when the experiment is conducted 5 times.

Solution:

Given $np - npq = \frac{5}{9}$ and n=5

Then $np(1-q) = \frac{5}{9} \Rightarrow 5\,p(p) = \frac{5}{9}$ $(\because n = 5,\ 1-q = p)$

$$\Rightarrow \quad p^2 = \frac{1}{9} \Rightarrow p = \frac{1}{3}$$

$(\because p > 0)$

$$\therefore \quad q = 1 - p = 1 - \frac{1}{3} = \frac{2}{3}$$

Now $p\,(X = r) = {}^{n}C_r \cdot (p)^r \cdot (q)^{n-r}$

$$p\,(X = 2) = {}^{5}C_2 \cdot \left(\frac{1}{3}\right)^2 \cdot \left(\frac{2}{3}\right)^3$$

$$= 10 \times \frac{2^3}{3^5} = \frac{80}{243}$$

4. If given that 10% of electric bulbs are manufactured by a company are defective. In the sample of 20 bulbs, find the probability that more than 1 is defective.

Solution:

Let P be the probability of selected bulb is defective.

$$P = \frac{10}{100} = \frac{1}{10} \text{ and } q = 1 - \frac{1}{10} = \frac{9}{10}$$

Now $P(X > 1) = 1 - P(X \leq 1)$

$= 1 - \{P(X=0) + P(X=1)\}$

$$= 1 - \left\{ {}^{20}C_0 \left(\frac{1}{10}\right)^0 \left(\frac{9}{10}\right)^{10} + {}^{20}C_1 \left(\frac{1}{10}\right)^1 \left(\frac{9}{10}\right)^9 \right\}$$

$$= 1 - \left\{ \frac{9^{10}}{10^{10}} + \frac{20 \times 9^9}{10^{10}} \right\} = 1 - \frac{9^9(9+20)}{10^{10}} = 1 - \frac{(29) \cdot 9^9}{10^{10}}$$

5. On an average, rain fall is 12 days in every 30 days. Find the probability that rain will fall on just three days of a given week.

Solution:

From given data, the probability of rainy day = $\frac{12}{30}=\frac{4}{10}=\frac{2}{5}=p$

$$q = 1-p = 1-\frac{2}{5}=\frac{3}{5}$$

For n = 7, P(X = 3) = ?

$$P(X=3) = {}^7C_3 \cdot p^3 \cdot q^{7-3} = {}^7C_3 \cdot \left(\frac{2}{5}\right)^3 \cdot \left(\frac{3}{5}\right)^4$$

6. One in 9 ships is likely to be wrecked when they are set to sail, When 6 ships are sail find the probability for (i) At least one will arrive safely (ii) Exactly 3 will arrive safely.

Solution:

Probability of ship is wrecked is $\frac{1}{9}$

Probability that ship arrive safely is $1-\frac{1}{9}=\frac{8}{9}$

$\therefore \quad q=\frac{1}{9}, p=\frac{8}{9}, n=9$

(i) $P(X \geq 1) = 1 - P(X < 0)$

$$= 1 - {}^9C_0 \cdot (p)^0 \cdot (q)^{9-0}$$

$$= 1 - {}^9C_0 \cdot \left(\frac{8}{9}\right)^0 \cdot \left(\frac{1}{9}\right)^9 = 1-\left(\frac{1}{9}\right)^9$$

(ii) $P(X=3) = {}^9C_3 \cdot \left(\frac{8}{9}\right)^3 \cdot \left(\frac{1}{9}\right)^6$

7. For a binomial distribution mean is 6 and standard deviation is $\sqrt{2}$. Find the first two terms of distribution.

Solution:

Mean = np = 6, variance = npq = (standard deviation)^2 = $(\sqrt{2})^2 = 2$.

$$q = \frac{npq}{np}=\frac{2}{6}=\frac{1}{3} \text{ and } p = 1-q=\frac{2}{3} \text{ and } np = 6 \Rightarrow n\cdot\frac{2}{3}=6 \Rightarrow n=9$$

$$\therefore \quad P(X=0) = {}^9C_0 \cdot \left(\frac{2}{3}\right)^0 \cdot \left(\frac{1}{3}\right)^9 ; P(X=1) = {}^9C_1 \cdot \left(\frac{2}{3}\right)^1 \cdot \left(\frac{1}{3}\right)^8$$

$$\therefore \quad \text{The first two terms are } {}^9C_0 \cdot \left(\frac{1}{3}\right)^9, \ {}^9C_1 \cdot \frac{2}{3^9}$$

i.e., $\frac{1}{3^9}, \frac{18}{3^9}$

4.4.3 Poisson Distribution

In Binomial Distribution the number of trials 'n' is known and finite. But sometimes it is not possible. If we know the number of times the event occurred but not knowing how many times it did not occur, then Binomial Distribution is not possible. In such cases we compute the binomial probabilities approximately by using Poisson distribution.

The following are some of the experiments which may be analyzed using Poisson distribution.

(i) The number of telephone calls received at a telephone exchange in a given time interval.

(ii) The number of printing errors in a page of a book.

(iii) The number of defective dosage forms in a batch.

(iv) The number of road/rail accidents reported in a city in a given period of time (say a day).

Discrete Random variables X (taking values 0, 1, 2, 3….n) following Poisson distribution of its probability function is of the form

$$P(X=r) = \frac{e^{-\lambda} \cdot \lambda^r}{r!} \qquad \lambda(>0) \text{ is a parameter.}$$

Poisson distribution can be used under following experimental conditions:

(i) Each trial results in two mutually exclusive outcomes termed as success and failure.

(ii) The number of such trials n is sufficiently large.

(iii) The trials are independent of each other.

(iv) The probability of success is very small.

(v) If X is a Poisson variable with parameter λ, then the mean and variance of X are equal to λ.

Solved Problems

1. A Poisson variable satisfies P(X = 0) = P(X = 1). Find P(X = 3).

Solution:

Given $P(X = 0) = P(X = 1)$

$$\Rightarrow \quad \frac{e^{-\lambda}\lambda^0}{01} = \frac{e^{-\lambda}\lambda^1}{1!}$$

$$\Rightarrow \quad e^{-\lambda} = e^{-\lambda}\lambda$$

$$\Rightarrow \quad \lambda = 1$$

$$\therefore \quad P(X = 3) = \frac{e^{-\lambda}\lambda^3}{3!} = \frac{e^{-1}\cdot(1)^3}{3!} = \frac{1}{e\cdot(3!)} = \frac{1}{6e}$$

2. In a city 10 accidents take place in a span of 50 days assuming that number of accidents follows the Poisson distribution. Find the probability that there will be 3 or more accidents in a day.

Solution:

The average number of accidents per day $\lambda = \frac{10}{50} = 0.2$

$$P(X \geq 3) = 1 - P(X < 3)$$

$$= 1 - P(P(X = 0) + P(X = 1) + P(X = 2))$$

$$= 1 - \left(\frac{e^{-\lambda}\lambda^0}{0!} + \frac{e^{-\lambda}\lambda^1}{1!} + \frac{e^{-\lambda}\lambda^2}{2!}\right)$$

$$= 1 - \left(e^{-0.2} + \frac{e^{-0.2}(0.5)}{1!} + \frac{e^{-0.2}(0.5)^2}{2!}\right)$$

$$= 1 - e^{-0.2}\left(1 + \frac{1}{2} + \frac{1}{8}\right) = 1 - e^{-0.2}\left(\frac{8+4+1}{8}\right)$$

$$= 1 - \frac{13}{8}e^{-0.2}$$

3. The number of persons joining a cinema ticket counter in a minute has Poisson distribution with parameter 6. Find the probability that (i) No one joins the queue in a particular minute. (ii) Two or more persons join the queue in a minute.

Solution:

Given $\lambda = 6$.

(a) $P(X = 0) = \frac{e^{-\lambda}\lambda^0}{0!} = \frac{e^{-6}(6)^0}{0!} = e^{-6}$

(b) $P(X > 2) = 1 - P(X < 2)$

$$= 1 - \{P(X = 0) + P(X = 1)\}$$

$$= 1 - \left\{\frac{e^{-\lambda}\lambda^0}{0!} + \frac{e^{-\lambda}\lambda^1}{1!}\right\} = 1 - e^{-\lambda}\left(1 + \frac{\lambda}{1}\right)$$

$$= 1 - e^{-6}(1+6) = 1 - 7e^{-6}$$

4. In a book of 400 pages there are 200 typographical errors. Assuming that the number of errors per page follows the Poisson law, find the probability that a random of sample of 3 pages will contain the typographical error.

Solution:

The average number of errors per page in book is $\lambda = \frac{200}{400} = 0.5$

The probability that there are r errors per page is $P(X = r) = \frac{e^{-\lambda}\lambda^r}{r!}$

$\therefore$ $P(X = 0) = \frac{e^{-\lambda}\lambda^0}{0!} = e^{-0.5}$

$\therefore$ The required probability that a random sample of 3 pages will contain errors is

$$[P(X = 0)]^3 = [e^{-0.5}]^3 = e^{-0.15}$$

5. In a large consignment of apples 3% are rotten. What is the probability that certain of 48 apples will contain less than two rotten apples?

Solution:

$$P = \frac{3}{100}, \; n = 48, \; \therefore \; \lambda = np = \frac{3}{100} \times 48 = \frac{36}{25} = 1.44$$

Required probability

$$P(X < 2) = P(X = 0) + P(X = 1)$$

$$= \frac{e^{-\lambda}\lambda^0}{0!} + \frac{e^{-\lambda}\lambda^1}{1!} = e^{-1.44}(1 + 1.44) = 2.44e^{-1.44}$$

6. On an average 1 in 100 tablets manufactured by a firm are defective. If tablets are supplied in packets of 5 each, find the probability that a packet has at least one defective tablet. If 1000 packets are tested, in how many packets would you expect defective tablets?

Solution:

Given an average 1 in 100 tablets manufactured by a firm are defective and tablets are supplied in a packet of 5 each.

$$\therefore \quad \lambda = \frac{1}{100} \times 5 = \frac{5}{100} = 0.05$$

$$\therefore \quad P(X \geq 1) = 1 - P(X < 0) = 1 - \frac{e^{-\lambda}\lambda^0}{0!} = 1 - e^{-0.05}$$

The number of packets can we expect defective tablets = $(1000).(1 - e^{-0.05})$

7. In a certain manufacturing process 5% of the items produced turn out to be defective. Find the probability that in a sample of 40 items at most 2 will be defective (Given $e^{-2} = 0.132$).

Sol:

$$\lambda = np = 40 \times \frac{5}{100} = \frac{200}{100} = 2$$

$$P(X \leq 2) = \{P(X = 0) + P(X = 1) + P(X = 2)\}$$

$$= e^{-\lambda}\left\{\frac{\lambda^0}{0!} + \frac{\lambda^1}{1!} + \frac{\lambda^2}{2!}\right\} = e^{-0.2}\{1 + 2 + 2\} = 0.66$$

4.4.4 Normal Distribution

Normal Distribution is a distribution of continuous data. It is a distribution of normal phenomenon such as

(i) Weight (of people working at some work, of products produced by same machine)

(ii) Height (of animals, plants and buildings)

(iii) Length (various maker of car)

(iv) Time (taken to complete a standard work).

A random variable X is said to follow a normal distribution with parameters μ (called mean) and σ^2 (called variance). If its probability density function f(x) is given by

$$f(x) = \frac{1}{\sigma\sqrt{2\pi}} e^{-\frac{(x-\mu)^2}{2\sigma^2}} \qquad \begin{cases} -\infty < \mu < \infty \\ \sigma > 0 \\ -\infty < x < \infty \end{cases}$$

where $e = 2.7183$, $\sqrt{2}\pi = 2.5060$

The graph of f(x) is a bell shaped curve symmetrical about mean μ this curve is known as Normal probability curve.

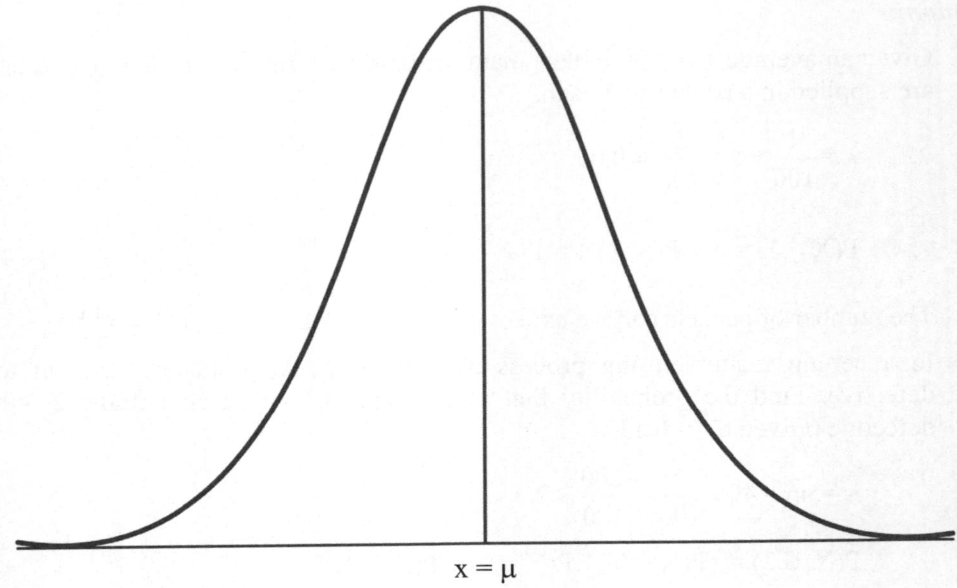

Standard Normal Distribution

If $X \sim N(\mu, \sigma^2)$ then $z = \dfrac{X-\mu}{\sigma}$ is a standard normal variant with mean ($\mu = 0$) and standard deviation ($\sigma = 1$) and we write $z \sim N(0, 1)$.

Probability function of the standard normal variant z is given by

$$f(z) = \frac{1}{\sqrt{2\pi}} e^{-\frac{z^2}{2}} \qquad -\infty < z < \infty$$

The curve $y = f(z)$ is known as standard normal curve. The shape of the curve is same as that of the normal curve.

The standard normal curve helps us to find the area within two assigned limits under the curve.

The area between the standard normal curve drawn at two assigned limits a and b will give the proportion of cases for which the values of z lie between a and b. Hence the area between the two assigned limits a and b under the standard normal curve will represent the probability that z will be between a and b. It is denoted by $P(a \leq Z \leq b)$.

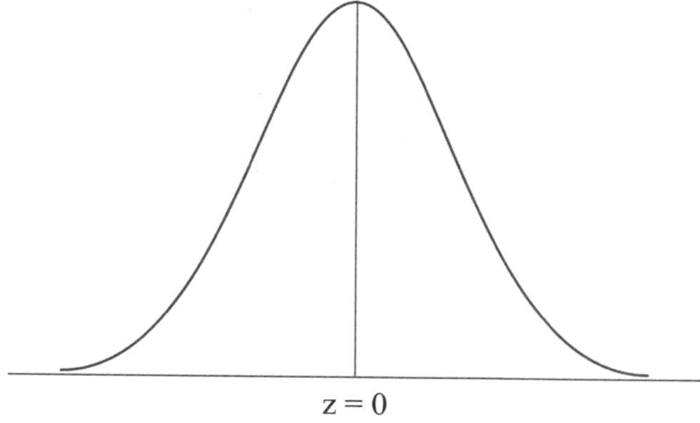

Properties of Normal Curve

The normal probability curve with mean μ and standard deviation σ has the following properties:

1. The equation of curve is $y = f(x) = \frac{1}{\sigma\sqrt{2\pi}} e^{-\frac{\left(\frac{x-\mu}{\sigma}\right)^2}{2}}$ $\quad -\infty \leq x \leq \infty$
2. It is bell shaped curve. The top of the bell is directly above the mean.
3. The curve is symmetric above the line $x = \mu$ (mean) and $-\infty \leq x \leq \infty$.
4. Mean, mode, median are coincide at $x = \mu$ as the distribution is symmetrical.
5. *x*-axis is asymptote (target at ∞) to the curve.
6. The point of inflection of the curve are at $x = \mu + \sigma$ and $x = \mu - \sigma$ and the curve changes from concave to convex at $x = \mu + \sigma$ and $x = \mu - \sigma$.
7. The total area under the normal curve is equal to unity and the percentage distribution of area under the normal curve is given below.
   (i) Above 68% of area fall between $\mu - 2\sigma$ and $\mu + \sigma$.
   (ii) Above 95.5% of area fall between $\mu - 2\sigma$ and $\mu + \sigma\mu$.
   (iii) Above 99.7% of area fall between $\mu - 3\sigma$ and $\mu + 3\sigma$.
8. The normal distribution curve is uni model (has a single peak).
9. Since mean = median =μ, the coordinate at $x = \mu$ (or $z = 0$) divides the whole area into two equal parts the area to the right of the ordinate as well as to the left of the ordinate at $x = \mu$ (or $z = 0$) is 0.5.
10. No portion of the curve lies below the *x*–axis as f(x) being probability can never be negative.

Fitting a Normal Distribution by Method of Area

The total area under the normal curve is 1 and it represents the total number of frequencies. The mean of the distribution is located at the center of the distribution and divide the whole normal curve into two equal parts.

Working Rule:

(a) Write the class intervals (column 1).

(b) Write the lower limits x of class interval (column 2).

(c) Find the deviation of the lower limits from the mean of the series $d = x - \mu$ (column 3).

(d) Find $\frac{x-\mu}{\sigma} = z$ (column 4).

(e) Write the area corresponding to $z = \frac{x-\mu}{\sigma}$ (by normal table), (column 5).

(f) Find the area under normal curve between the successive values of z. This can be obtained by subtracting the successive area obtained in column 5 when the corresponding zs have the same sign and adding them when the z s have the opposite sign (column 6).

(g) Find the expected frequency by multiplying the relative frequency obtained in column 6 by the number of observations (column 7).

Solved Problems

1. In a sample of 1000 cases, the mean of certain test is 14 and standard deviation is 2.5. Assuming the distribution is normal. Find

(a) How many students score between 12 and 15?

(b) How many score above 18?

(c) How many score below 8?

Solution:

Given mean = 14 and standard deviation = 2.5

$$\text{(i) } P(12 < x < 15) = P\left(\frac{12-\mu}{\sigma} < \frac{x-\mu}{\sigma} < \frac{15-\mu}{\sigma}\right)$$

$$= P\left(\frac{12-14}{2.5} < z < \frac{15-14}{2.5}\right)$$

$$= P\left(\frac{-2}{2.5} < z < \frac{1}{2.5}\right) = P(-0.8 < z < 0.4)$$

$= (\text{Area between } z = 0 \text{ to } z = -0.8) + (\text{Area between } z = 0 \text{ to } z = 0.4)$

$= (\text{Area between } z = 0 \text{ to } z = 0.8) + (\text{Area between } z = 0 \text{ to } z = 0.4)$

$= 0.2881 + 0.1554 = 0.4435$

$\therefore$ The required number of students $= 1000 \times 0.4435 = 443.5 \simeq 444$ (Approx)

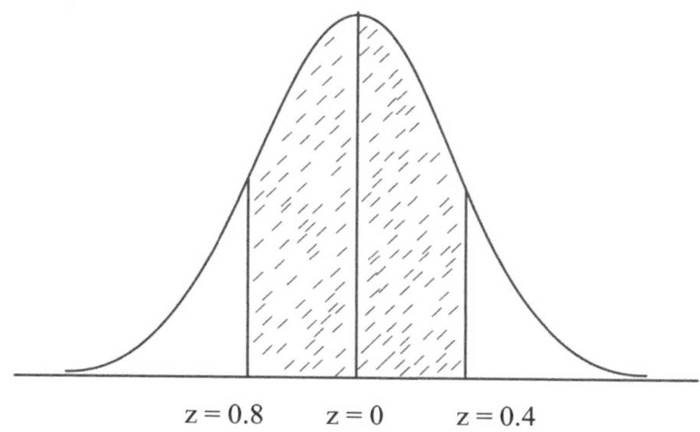

(ii) $P(x > 18) = P\left(\frac{x-\mu}{\sigma} > \frac{18-\mu}{2.5}\right)$

$= P\left(z > \frac{18-14}{2.5}\right) = P\left(z > \frac{4}{2.5}\right) = P(z > 1.6)$

$= 0.5 - (\text{Area between } z = 0 \text{ to } z = 1.6)$

$= 0.5 - 0.4452 = 0.0548$

$\therefore$ The required number of students $= 1000 \times 0.0548 = 54.8 \simeq 55$ (Approx)

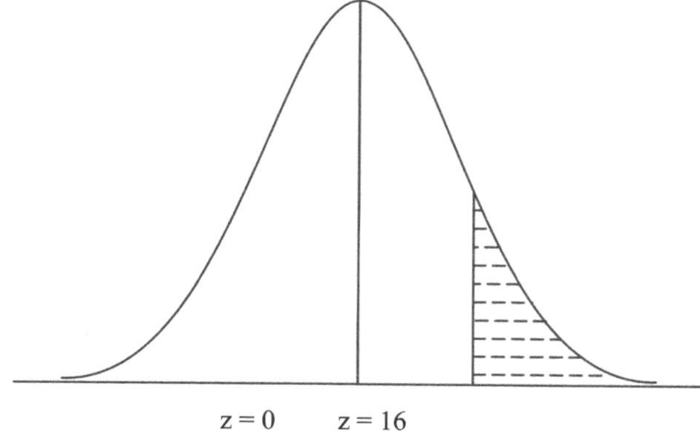

$$\text{(iii) } P(x < 8) = P\left(\frac{x-\mu}{\sigma} < \frac{8-\mu}{\sigma}\right)$$

$$= P\left(z < \frac{8-14}{2.5}\right)$$

$$= P(z < -2.4)$$

$$= 0.5 - (\text{Area between } z = 0 \text{ to } z = -2.4)$$

$$= 0.5 - 0.4918 = 0.0082$$

∴ The required number of students = $1000 \times 0.0082 = 8.2 \simeq 8$ (Approx)

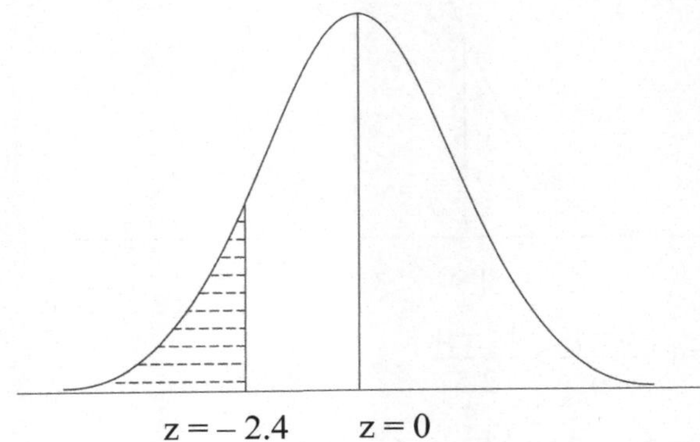

2. Assuming the mean height of student is μ and following exactly normal distribution with 68.22 inches mean and variance 10.8 inches squares. How many students in a college of 1000 students would you expect to be over 6 feet tall?

Solution:

1 feet = 12 inches. ∴ 6 feet = 72 inches.

So first to find

$$P(x > 72) = P\left(\frac{x-\mu}{\sigma} > \frac{72-\mu}{\sigma}\right)$$

$$= P\left(z > \frac{72-68.22}{10.8}\right)$$

$$= P(z > 0.35)$$

$$= 0.5 - (\text{Area between } z = 0 \text{ to } z = 0.35)$$

$$= 0.5 - 0.3749 = 0.1251$$

∴ The expected number of students having height more than 6 feet

$$= 1000 \times 0.1251 = 125.1 \simeq 125$$

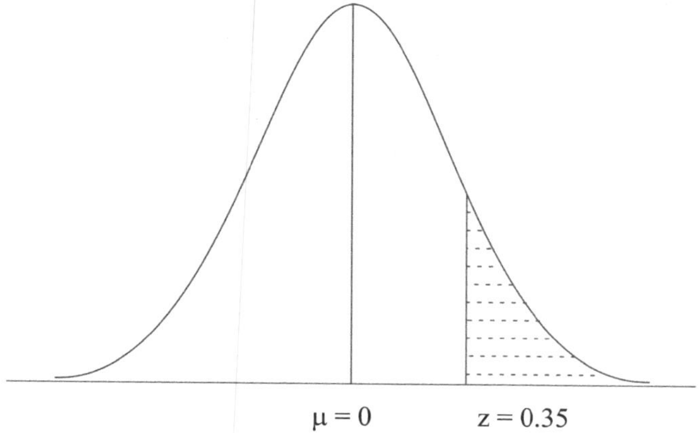

3. In a distribution of exactly normal 7% of the items are under 35 and 89% are under 63. What are the mean and standard deviation of the distribution?

Solution:

$P(x < 35) = 0.07$

Let μ be mean and σ be standard deviation

$$\Rightarrow P\left(\frac{x-\mu}{\sigma} < \frac{35-\mu}{\sigma}\right) = 0.07$$

$$\Rightarrow P\left(z_1 < \frac{35-\mu}{\sigma}\right) = 0.07$$

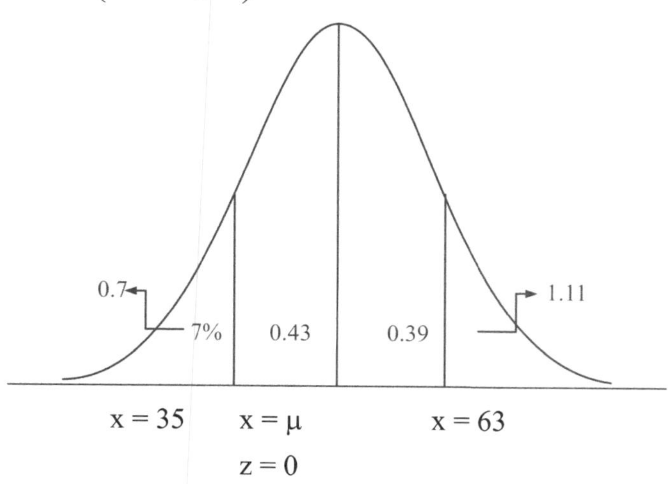

$P(x < 63) = 0.89$

$$\Rightarrow P\left(\frac{x-\mu}{\sigma} < \frac{63-\mu}{\sigma}\right) = 0.89$$

$$\Rightarrow P\left(z_2 < \frac{63-\mu}{\sigma}\right) = 0.89$$

$$\Rightarrow P\left(0 < z_2 < \frac{63-\mu}{\sigma}\right) = 0.39$$

$$\text{and } P\left(0 < z < \frac{35-\mu}{\sigma}\right) = 0.43$$

$$\therefore \quad \frac{35-\mu}{\sigma} = -1.48 \quad \text{and} \quad \frac{63-\mu}{\sigma} = 1.23$$

Solving the two equations we get $\mu = 50.3$ and $\sigma = 10.33$.

4. A normal distribution with $\mu = 50$ and $\sigma = 10$ is given. Find the value of x that has

(a) 13% of the area to the left.

(b) 14% of the area to the right.

Solution:

Given $\mu = 50$ and $\sigma = 10$

(i) $P(x < x_1) = 0.13.$

$$= P\left(\frac{x-\mu}{\sigma} < \frac{x_1-\mu}{\sigma}\right) = 0.13$$

$$= P\left(z < \frac{x_1-50}{10}\right) = 0.13$$

$$= P\left(0 < z < \frac{x_1-50}{10}\right) = 0.37$$

$$= \frac{x_1-50}{10} = -1.103 = \frac{x_1-50}{10} = -1.103$$

$x_1 = 50 - 11.03 = 38.97$

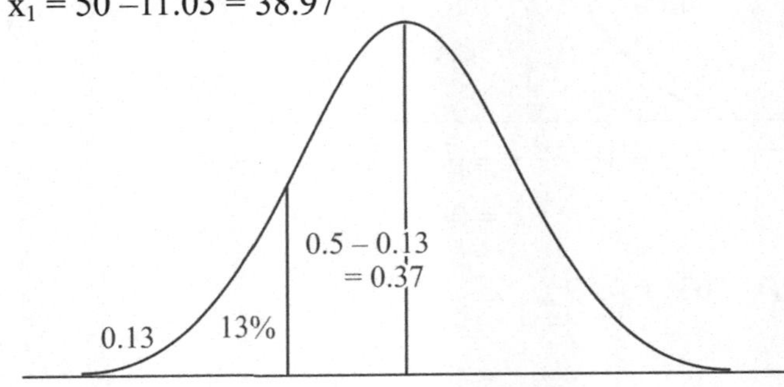

(ii) $P(x > x_2) = 0.14.$

$$= P\left(\frac{x-\mu}{\sigma} > \frac{x_2 - 50}{10}\right) = 0.14$$

$$= P\left(0 \le z < \frac{x_2 - 50}{10}\right) = 0.36$$

$$= \frac{x_2 - 50}{10} = 1.09$$

$X_2 = 50 + 10.9 = 60.9$

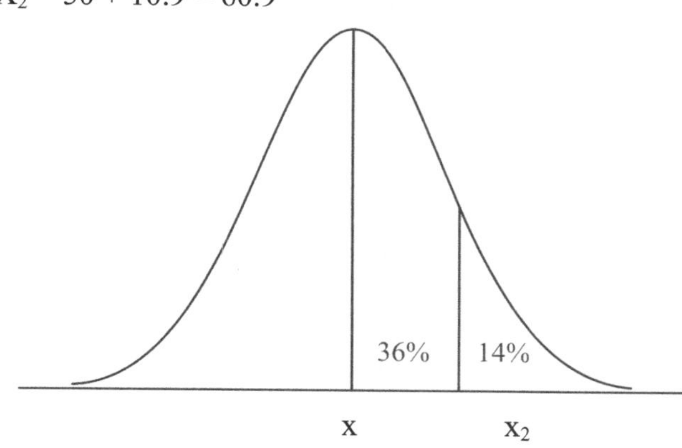

5. In a large number of children, 55% are under 60 units in height, 40% are between 60 and 65 units. Assuming a normal distribution, find the mean height and standard deviation.

Solution:

Given $P(x < 60) = 0.55$

$$P\left(\frac{x-\mu}{\sigma} < \frac{60-\mu}{\sigma}\right) = 0.55$$

$$P\left(z < \frac{60-\mu}{\sigma}\right) = 0.55$$

$$\Rightarrow 0.5 + P\left(0 < z < \frac{60-\mu}{\sigma}\right) = 0.55$$

$$\Rightarrow \frac{60-\mu}{\sigma} = 0.103$$

and $P(60 < x < 65) = 0.4$

$$\Rightarrow P\left(\frac{60-\mu}{\sigma} < \frac{x-\mu}{\sigma} < \frac{65-\mu}{\sigma}\right) = 0.4$$

$$\Rightarrow P\left(\frac{60-\mu}{\sigma} < z < \frac{65-\mu}{\sigma}\right) = 0.4$$

$$\Rightarrow P\left(0 < z < \frac{65-\mu}{\sigma}\right) = 0.45$$

$$(0.4 + 0.05 = 0.45)$$

$$\therefore \quad \frac{65-\mu}{\sigma} = 1.605$$

$$60 - \mu = 0.103\sigma$$

$$\Rightarrow \quad 65 - \mu = 1.605\sigma$$

$$5 \quad = 1.562\sigma$$

$$\sigma = \frac{5}{1.506} = 3.3288$$

$$\therefore \quad \mu = 65 - (1.605).(3.3288) = 59.6714$$

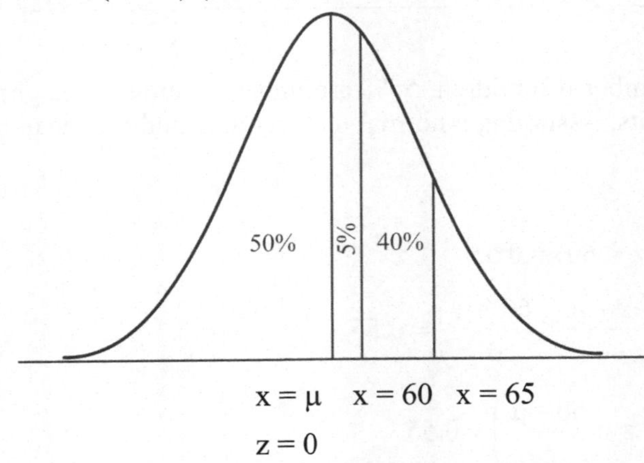

6. In a normal distribution, 31% of the items are under 45 and 8% items are over 64. Find the mean and standard deviation of the distribution.

Solution:

$$\text{If } x = 45, \ \frac{x-\mu}{\sigma} = \frac{45-\mu}{\sigma}$$

$$P(x \le 45) = P\left(z \le \frac{45-\mu}{\sigma}\right) = 0.31$$

$$= 0.5 - P\left(z = 0 \text{ to } z = \frac{45-\mu}{\sigma}\right) = 0.31$$

$$\Rightarrow \quad P\left(0 \le z \le \frac{45-\mu}{\sigma}\right) = 0.5 - 0.31 = 0.19$$

$$\Rightarrow \quad \frac{45-\mu}{\sigma} = -\ 0.5 \qquad \text{(From Table)}$$

$$= 45 - \mu = -\ 0.5\sigma \qquad \text{.....(1)}$$

Again $P(x > 64) = 0.08$

$$P\left(\frac{x-\mu}{\sigma} > \frac{64-\mu}{\sigma}\right) = 0.08$$

$$\Rightarrow \quad 0.5 - P\left(0 < z < \frac{64-\mu}{\sigma}\right) = 0.08$$

$$\Rightarrow \quad P\left(0 < z < \frac{64-\mu}{\sigma}\right) = 0.5 - 0.08 = 0.42$$

$$\Rightarrow \quad \frac{64-\mu}{\sigma} = 1.401$$

$$= 64 - \mu = 1.401\sigma \qquad \text{.....(2)}$$

$\therefore$ Solving $45 - \mu = -\ 0.5\sigma$ and $64 - \mu = 1.401\sigma$

We get $\mu = 50$ and $\sigma = 10$

$\therefore$ Mean $\mu = 50$ and S D $\sigma = 10$

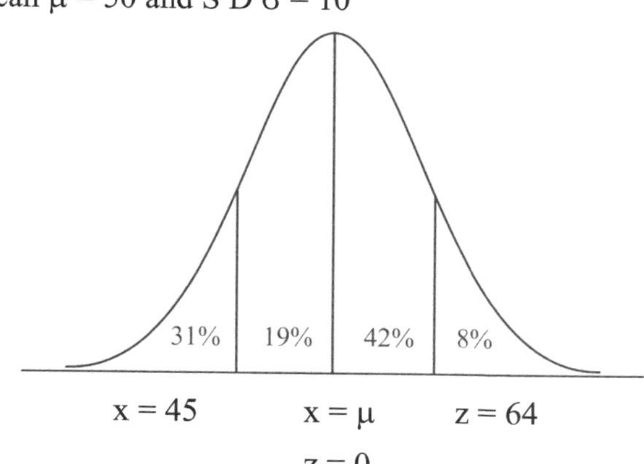

CHAPTER 5

CORRELATION AND REGRESSION

Introduction

In many biological experiments, there exists a certain relationship between two continuous variables. The changes in one variable result a change in the other variable. Correlation measures the closeness of the relation between the two variables. For example the dose of drug and response; Time and amount of drug dissolved.

5.1 Correlation

Correlation is used in statistical analysis to determine the degree or extent to which two variables fluctuate with reference to each other. The word relationship indicates that there is some connection to measure the closeness of the relationship between the variables.

According to Croxton and Cowden, "If the relationship is of quantitative nature, the appropriate statistical tool to discover and measure that relation and express it in the form of a brief formula is correlation".

According to Simpson and Kafka, "Correlation analysis deals with the association between two or more variables".

According to A.M. Tuttle, "Correlation is an analysis of the co-variation between two variables".

According to L.R. Connor, "If two or more quantities vary in sympathy so that movements in the one tend to be accompanied by corresponding movements in the other, then they are said to be correlated".

Although Karl Parson was the first to establish the mathematical formula but Sir Francis Galton was the first to develop the technique to obtain it graphically.

So, the correlation refers to the study of relationship between the two variables.

Correlation analysis helps to determining the degree of relationship between two or more variables but it does not indicate a cause and effect of relationship between two variables. It explains only co variation.

5.1.1 Types of Correlation

On the basic nature of the relationship between the variables i.e., the direction in which changes take place in the variables or the ratio by which they change correlation may be

(i) Positive or negative correlation

(ii) Linear or non linear correlation

(iii) Simple or multiple correlations

(iv) Partial or total

(i) ***Positive or negative correlation*****:** If both the variables tend to move together in the same direction so that an increase in the value of one variable is associated with an increase in the value of the other variable, and decrease in the value of one variable is associated with a decrease in the value of the other variable. This type of correlation is called positive correlation or direct correlation.

Example 1: The relationship between concentration and absorbance exhibits positive correlation. It can be observed from the following data.

Concentration(μg/ml)	Absorbance
2	0.204
4	0.408
6	0.612
8	0.816
10	1.02

In negative correlation, the increase in one variable results a decrease in another variable. It is showed in example 2. The potency of a drug present in a drug delivery system decreases with time.

Example 2: Time and the amounts of drug remained in the dosage form exhibits negative correlation and it is showed in the following example:

Time (months)	Potency (mg)
0	100
3	98
6	96
8	94
10	92
12	90

(ii) ***Linear or non linear correlation*:** If the variation in the values of two variables has a constant ratio, the correlation between them is linear.

Example 3:

Variable (x)	10	20	30	40	50
Variable (y)	20	40	60	80	100

From the above example it is clear that the ratio of two variables i.e., the ratio of change between them is same.

When all the plotted points (x_i, y_i) (x_i is value of one variable and y is corresponding value of another variable) lie exactly on a straight line then the correlation said to be perfect linear correlation.

In a non linear correlation, the amount of change in one variable does not bear a constant ratio to amount of change in the other relative variable. For example if we double the time interval of dissolution, it may not necessarily double the amounts of drug dissolved. The graph of non-linear relationship will form a curve. It is also called curve linear relationship.

(iii) ***Simple and Multiple Correlations*:**

Simple correlation: If we study only two variables relationship, then it is described as simple correlation.

E.g.: Amounts of drug dissolved and hardness

Multiple correlations: If we study three or more variables relationship, then such correlation is described as multiple correlations. Multiple correlations consist of the measurement of relationship between a dependent variable and two or more independent variables.

E.g.: The study of relationship between amount of drug release with that of hardness and granule size.

(iv) ***Partial and total Correlations*:** These two correlations are the examples of multiple correlations.

Partial Correlation: The study of two variables excluding some other variables called partial correlation.

E.g.: In the study of relationship between potency of drug and humidity and temperature, excluding the effect of humidity on potency is a partial correlation.

The *total correlation* is based on all the relevant variables, which, however, is normally not feasible.

E.g.: The study of relationship between potency of drug and temperature as well as humidity.

5.1.2 Methods of Studying Correlation

(i) Scatter or dot diagram method

(ii) Graphical method

(iii) Karl pearsons coefficient of correlation method

(i) ***Scatter or dot diagram method*:** A scatter diagram is a tool for analyzing relationships between two variables. One variable is plotted on the horizontal axis and the other is plotted on the vertical axis. The pattern of their intersecting points can graphically show relationship patterns. Most often a scatter diagram is used to prove or disprove cause-and-effect relationships. It is a very simple method to identify whether there is a relationship between two variables x and y or not. Let the variables x and y be plotted along the x and y-axis in a coordinate plane by taking suitable scale corresponding to every ordered pair (x_i, y_i) there corresponds a point or dot in the coordinate plane.

The diagram of dots or points so obtained is called a scatter diagram or dot diagram. With the help of this diagram, we get an idea about the relationship between the two variables x and y.

Scatter diagrams will generally show one of six possible correlations between the variables:

Strong Positive Correlation: The value of Y clearly increases as the value of X increases.

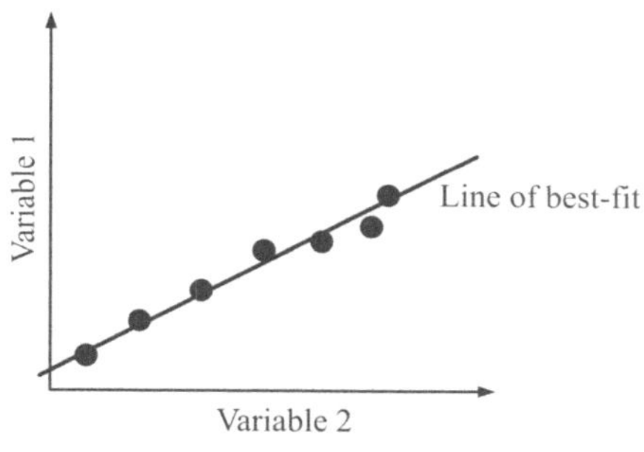

Positive Correlation

Strong Negative Correlation: The value of Y clearly decreases as the value of X increases.

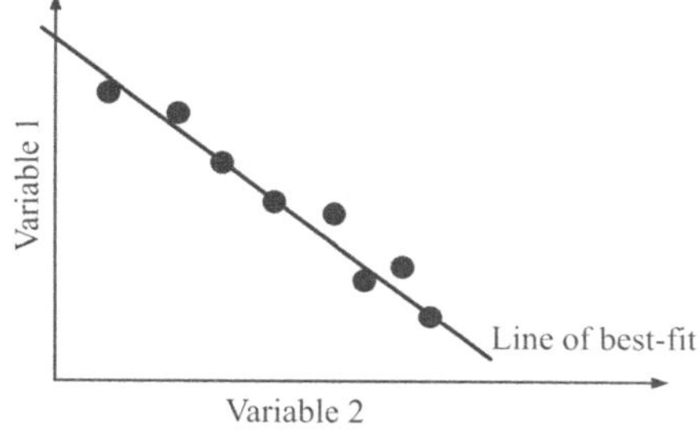

Negative Correlation

Weak Positive Correlation: The value of Y increases slightly as the value of X increases.

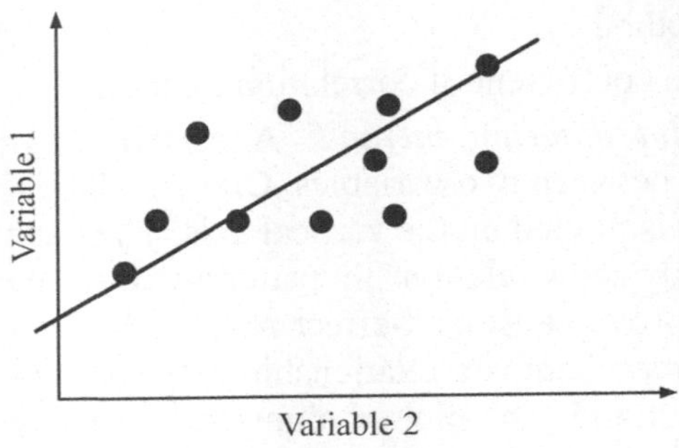

Weak Positive Correlation

Weak Negative Correlation: The value of Y decreases slightly as the value of X increases.

Complex Correlation: The value of Y seems to be related to the value of X, but the relationship is not easily determined.

No Correlation: There is no demonstrated connection between the two variables.

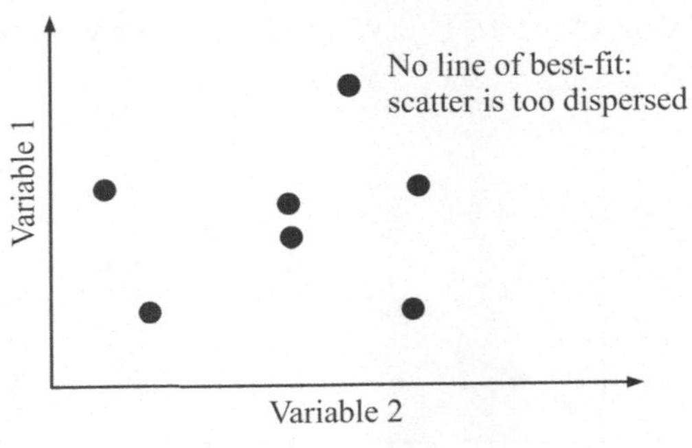

No Correlation

(ii) ***Graphical Method*:** It is a simple method of finding the correlation between two variables. The individual values of two variables x and y are plotted on a graph paper to obtain the relation between the variable x and other variable y. If both curves move in the same direction, i.e., parallel to each other either upward or downward; the correlation is said to be positive. If they are moving in the opposite direction then the correlation is said to be negative. This type of graphical method is used in environmental biology and general studies.

(iii) ***Correlation coefficient*:** The above two methods (i) Scatar diagram (ii) graphical method are only indicate a tendency that two variables x and y have direct relationship even though, these methods do not provide any numerical measure of correlation. Unless the relationship is established with a quantitative expression the relationship above will be no use. So the degree of relationship can be established by calculating a coefficient called correlation coefficient which provides a quantitative measure of the degree of closeness between the two variables. This coefficient of correlation is developed by Karl Pearson.

The karl Pearson's coefficient is denoted by r and is defined as

$$r = \frac{\Sigma(x-\bar{x})(y-\bar{y})}{\sqrt{\Sigma(x-\bar{x})^2}\sqrt{\Sigma(y-\bar{y})^2}}$$

where $\bar{x}$ and $\bar{y}$ are means of two variables x and y and $(x-\bar{x})$, $(y-\bar{y})$ are the deviations of values of variables from respected means. The above formula can also be written as

$$r = \frac{\Sigma dx\, dy}{\sqrt{\Sigma dx^2}\sqrt{\Sigma dy^2}}$$

$$r = \frac{\Sigma xy - \frac{1}{n}(\Sigma x\, \Sigma y)}{\sqrt{\left[\Sigma x^2 - \frac{(\Sigma x)^2}{n}\right]}\sqrt{\Sigma y^2 - \frac{(\Sigma y)^2}{n}}}$$

It has also been indicated as product moment formulae of coefficient of correlation.

Rough table

x	**y**	**xy**	**x^2**	**y^2**
Σx	Σy	Σxy	Σx^2	Σy^2

So calculate Σx, Σy, Σxy, Σx^2, Σy^2 and substitute in above formula we obtain Karl Pearson`s coefficient of correlations, here 'n' is number of observations in each variable.

If r = 1 we conclude that the degree of correlation between two variables is perfect positive correlation and if r = – 1, we conclude that relation is perfect negative correlation. Note the value of r is always lies between –1 to 1. If r = 0 thus we conclude that there is no correlation between the two variables.

Example: Calculate the Pearson`s correlation coefficient from the following data

Subject	weight (kg)	Height (m)
1	90	1.92
2	80	1.76
3	70	1.54
4	60	1.37
5	50	1.18

Weight (kg) (x)	Height (m) (y)	xy	x^2	y^2
90	1.92	172.8	8100	3.6864
80	1.76	140.8	6400	3.0976
70	1.54	107.8	4900	2.3716
60	1.37	82.2	3600	1.8769
50	1.18	59	2500	1.3924
Σx=350	Σy=7.77	Σxy=562.6	25500	12.4249

∴ Coefficient of correlation

$$\therefore \text{ Calculated } r = \frac{\Sigma xy - \frac{1}{n}\Sigma x \Sigma y}{\sqrt{\Sigma x^2 - \frac{(\Sigma x)^2}{n}}\sqrt{\Sigma y^2 - \frac{(\Sigma y)^2}{n}}}$$

$$= \frac{562.6 - \frac{1}{5}.(350)(7.77)}{\sqrt{25500 - \frac{(350)^2}{5}}\sqrt{12.4249 - \frac{(7.77)^2}{5}}}$$

$$= \frac{562.6 - 543.9}{\sqrt{1000}\ \sqrt{0.35032}}$$

$$= \frac{18.7}{31.6227 \times 0.598} = \frac{18.7}{18.71431}$$

$$= 0.\ 99$$

$$= 0.1$$

5.1.3 Statistical Significance of a Correlation Coefficient

A positive or negative correlation, the correlation coefficient value should be more than 0.99 in process validation of pharmaceutical industry. The rough guide for correlation is as follows.

<0.2	slight or almost negligible relationship
0.2-0.4	low correlation
0.4-0.7	moderate correlation
0.7-0.9	high correlation
>0.9	very high correlation

Significance test of correlation coefficient

In the significance test, we want to decide more specifically whether a correlation coefficient (r) is large enough to suggest the existence of correlation between paired values of population or whether the obtained r can be attributed to change.

In order to test the significance of deviation of the estimate of correlation coefficient (r) from zero we may apply t-test with n – 2 degree of freedom. The formula of t is defined as

$$t = \frac{r\sqrt{n-2}}{\sqrt{1-r^2}}$$

The significance of an observed correlation can be, tested with reference to the table't' value.

If the calculated t value exceeds the table t value for $p = 0.05$ or 0.01 we have reason to believe that sample indicate a real correlation between two variables.

Important steps for significant test

(i) Fix p either 0.05 or 0.01

(ii) Find degree of freedom n – 2

(iii) Find tabled value of t probably level 0.05 or 0.01 with degree of freedom n – 2.

(iv) Compare above with calculated value of t

(v) Calculated t > tables t ⇒ real correlation

(vi) Calculated t < tabled t ⇒ no correlation

The rate of drug release from a sustained release formulation may follow either first order or zero order release. Similarly the rate of drug decomposition may follow either first order or zero order decomposition. In order to identify the rate of release, the correlation coefficient should be calculated and compared with each other and the significance of deviation is detected by applying't' test.

***Example*:**

The drug release data from a tablet containing 200 mg of active ingredient is furnished in the following table. Compare the correlation coefficients of first order and zero order release.

Time (hrs)	Amount of drug release (in mg)
0	0
1	19.86
2	39.04
3	59.74
4	78.82
5	99.36
6	121.18
7	139.86
8	158.72
9	182.37
10	198.56

Solution:

Time (x)	Amount of drug released (y)	% drug release	% drug unreleased	Log % drug in unreleased (z)
0	0	0	100	2
1	19.86	9.93	90.07	1.9546
2	39.04	19.52	80.48	1.9057
3.	59.74	29.87	70.13	1.8459
4.	78.82	39.41	60.59	1.7824
5.	99.36	49.68	50.32	1.7017
6.	121.18	60.59	39.41	1.5956
7.	139.86	69.93	30.07	1.4781
8.	158.72	79.36	20.64	1.3147
9.	182.37	91.185	8.815	0.9452
10.	198.56	99.28	0.72	– 0.1427

To find the zero order correlation coefficient

x	y
0	0
1	19.86
2	39.04
3	59.74
4	78.82
5	99.36
6	121.18
7	139.86
8	158.72
9	182.37
10	198.56

$$r = \frac{\Sigma xy - \frac{1}{n}\Sigma x \Sigma y}{\sqrt{\Sigma x^2 - \frac{(\Sigma x)^2}{n}}\sqrt{\Sigma y^2 - \frac{(\Sigma y)^2}{n}}}$$

$\Sigma xy = 7692.03$, $\Sigma x = 55$, $\Sigma y = 1097.51$,

$\Sigma y^2 = 153694.7517$

$\Sigma x^2 = 385$

$$r = \frac{7692.03 - \frac{1}{11}(55)(1097.51)}{\sqrt{385 - \frac{(55)^2}{11}}\sqrt{153694.7517 - \frac{(1097.51)^2}{11}}}$$

$$= \frac{7692.03 - 5487.55}{\sqrt{110}\sqrt{44192.18806}}$$

$$= \frac{2204.48}{10.49 \times 210.2193}$$

$$= \frac{2204.48}{2205.2} = 0.9996$$

$\therefore$ r is very vey nearer to 1

$\therefore$ The drug release follows zero order

$$t = \frac{r\sqrt{n-2}}{\sqrt{1-r^2}}$$

$$= \frac{0.9996\sqrt{9}}{\sqrt{1-(0.9996)^2}}$$

$$= \frac{0.9996 \times 3}{0.0228} = \frac{2.9988}{0.0228} = 131.52$$

The table t-value for the degree of freedom 9 (11 – 2) at 0.05 level of significance is 2.26. The calculated t- value is greater than table value. So there is real correlation.

To find the coefficient of correlation of first order

Time (hrs)	Log % drug unreleased
0	0
1	1.9546
2	1.9057
3	1.8459
4	1.7824
5	1.7017
6	1.5956
7	1.4781
8	1.3147
9	0.9452
10	-0.427

$\Sigma x = 55, \quad \Sigma x^2 = 385,$

$\Sigma y = 14.3812$

$\Sigma xy = 64.4595$

$\Sigma y^2 = 24.3051545$

Coefficient of Correlation

$$r = \frac{\Sigma xy - \frac{1}{n}\Sigma x \Sigma y}{\sqrt{\Sigma x - \frac{(\Sigma x)^2}{n}}\sqrt{\Sigma y^2 - \frac{(\Sigma y)^2}{n}}}$$

$$= \frac{64.4595 - \frac{1}{11}(55)(14.3812)}{\sqrt{385 - \frac{(55)^2}{11}}\sqrt{(24.3056) - \frac{(14.3812)^2}{11}}}$$

$$= \frac{-7.4455}{\sqrt{110}\sqrt{5.0388}}$$

$$= \frac{-7.4465}{10.4880 \times 2.3460}$$

$$= \frac{-7.4465}{24.6050} = -0.3026$$

and $$t = \left|\frac{r\sqrt{n-2}}{\sqrt{1-r^2}}\right| = \left|\frac{-0.3026\sqrt{10-2}}{\sqrt{1-(0.3026)^2}}\right|$$

$$= \frac{0.3026\sqrt{8}}{\sqrt{1-(0.3026)^2}}$$

$$= \frac{0.3026 \times 2.828427}{\sqrt{1-0.9156676}}$$

$$= \frac{0.3026 \times 2.828427}{\sqrt{0-90843324}}$$

$$= \frac{0.85588}{0.9531176}$$

$$= 0.89$$

The table t-value for the degree of freedom 9 (11 – 2) at 0.05 level of significance is 2.26. The calculated t- value is less than table value. So there is no correlation.

In some cases the data may not show any correlation, however if it is splitted into two sets then it may show correlation. Initially drug release may follow first order followed by zero order. In case of non-linear pharmacokinetics, the absorption or distribution or metabolism or excretion initially follows first order and then it may exhibit zero order.

E.g. The drug release data observed from the tablet containing 100 mg of drug

Time (hrs)	**Amount of drug release (in mg)**
0	0
1	20
2	36
3	48
4	56
5	66
6	76
7	85
8	95

Calculate

(i) r for zero order

(ii) r for first order

(iii) r for first order up to 4 hours, zero order from 4-8 hours.

Solution:

Time hrs	Amount of drug release (mg)	% drug release	% drug unreleased	Log % drug unreleased
0	0	0	100	2
1	20	20	80	1.9030
2	36	36	64	1.8061
3	48	48	52	1.71600
4	56	56	44	1.6434
5	66	66	34	1.5314
6	76	76	24	1.3802
7	85	85	15	1.17609
8	95	95	5	0.69897

(i) To find r for zero order

Time x	Amount drug release y
0	0
1	20
2	36
3	48
4	56
5	66
6	76
7	85
8	95

$\Sigma x = 28$

$\Sigma x^2 = 204$

$\Sigma y = 482$

$\Sigma y^2 = 33518$

$\Sigma xy = 2601$

$$r = \frac{\Sigma xy - \frac{1}{n}\Sigma x \Sigma y}{\sqrt{\Sigma x^2 - \frac{(\Sigma x)^2}{n}}\sqrt{\Sigma y^2 - \frac{(\Sigma y)^2}{n}}}$$

$$= \frac{2601 - \frac{1}{9}.28 \times 482}{\sqrt{204 - \frac{(28)^2}{9}}\sqrt{33518 - \frac{(482)^2}{9}}}$$

$$= \frac{1101.44}{10.81 \times 87.77} = 0.160$$

t- Value for the above r value is $t=\dfrac{r\sqrt{n-2}}{\sqrt{1-r^2}}$ =0.43

The table t-value for the degree of freedom 7 (9 – 2) at 0.05 level of significance is 2.36. The calculated t- value is less than table value. So there is no correlation.

(ii) To find r for first order

Time hrs (x)	Log% drug unreleased (y)
0	2
1	1.9030
2	1.8061
3	1.71600
4	1.6434
5	1.5314
6	1.3802
7	1.17609
8	0.69897

$\Sigma x = 28$

$\Sigma x^2 = 204$

$\Sigma y = 13.85516$

$\Sigma y^2 = 22.65071052$

$\Sigma xy = 46.99639$

$$r=\frac{\Sigma xy-\frac{1}{n}\Sigma x\Sigma y}{\sqrt{\Sigma x^2-\frac{(\Sigma x)^2}{n}}\sqrt{\Sigma y^2-\frac{(\Sigma y)^2}{n}}}$$

$$=\frac{46.99639-\frac{1}{9}.(28)(13.85516)}{\sqrt{204-\frac{(28)^2}{9}}\sqrt{(22.6507)-\frac{(13.8516)^2}{9}}}$$

$$=\frac{3.891447778}{(10.81)(1.149436643)}$$

$$=\frac{3.891447878}{12.42541011}$$

$$=0.31$$

t- Value for the above r value is $t = \dfrac{r\sqrt{n-2}}{\sqrt{1-r^2}}$ =0.86

The table t-value for the degree of freedom7 (9 – 2) at 0.05 level of significance is 2.36. The calculated t- value is less than table value. So there is no correlation.

(iii)

(a) To find r for first order up to 4 hours

Time (x)	Log % drug unreleased (y)
0	2
1	1.9030
2	1.8061
3	1.71600
4	1.6434

$\Sigma x = 10$

$\Sigma x^2 = 30$

$\Sigma y = 9.0685$

$\Sigma y^2 = 16.5\ 288\ 25\ 77$

$\Sigma xy = 17.2368$

$$r = \frac{\Sigma xy - \frac{1}{n}\Sigma x . \Sigma y}{\sqrt{\Sigma x^2 - \frac{(\Sigma x)^2}{n}}\sqrt{\Sigma y^2 - \frac{(\Sigma y)^2}{n}}}$$

$$= \frac{17.2368 - \frac{1}{5}\times 10 \times (9.0685)}{\sqrt{30 - \frac{(10)^2}{5}}\sqrt{(16.5288) - \frac{(9.0685)^2}{5}}}$$

$$= \frac{-0.9002}{3.1622 \times 0.285064} = \frac{-0.9002}{0.901429}$$

$= -\ 0.9986$

$$\text{Now } t = \frac{\left|r\sqrt{n-2}\right|}{\sqrt{1-r^2}} = \frac{\left|-0.9986\sqrt{5-2}\right|}{\sqrt{1-(-0.9986)^2}}$$

$$= \frac{0.9986 \times 1.7320}{\sqrt{1-0.99720}} = \frac{1-7295752}{0.0529} = 32.695$$

The table t-value for the degree of freedom 3 (5 – 2) at 0.05 level of significance is 3.18. The calculated t-value is more than table value. So there is real correlation.

(b) To find zero order from 4 to 8 hours

$$\Sigma x = 30$$
$$\Sigma x^2 = 190$$
$$\Sigma y = 378$$
$$\Sigma y^2 = 29518$$
$$\Sigma xy = 2365$$

Time (x)	Log % drug release (y)
4	56
5	66
6	76
7	85
8	95

$$r = \frac{\Sigma xy - \frac{1}{n}\Sigma x \Sigma y}{\sqrt{\Sigma x^2 - \frac{(\Sigma x)^2}{n}}\sqrt{\Sigma y^2 - \frac{(\Sigma y)^2}{n}}}$$

$$= \frac{2365 - \frac{1}{5}.(3.0)(378)}{\sqrt{190 - \frac{(30)^2}{5}}\sqrt{29518 - \frac{(378)^2}{5}}}$$

$$= \frac{97}{3.1628 \times 30.68} = 0.999642$$

and $$t = \frac{r\sqrt{1-2}}{\sqrt{1-r^2}} = \frac{0.999642.\sqrt{3}}{\sqrt{1-(0.999642)^2}}$$

$$= \frac{1.73143}{\sqrt{0.00071}} = \frac{1.73143}{0.02675}$$

$$= 64.72$$

The table t-value for the degree of freedom 3 (5 – 2) at 0.05 level of significance is 3.18. The calculated t- value is more than table value. So there is real correlation.

5.1.4 Pair Wise Correlations Involving more than Two Variables

When there are more than two continuous variables affecting each data point, it is possible to calculate pair wise correlation. If we are evaluating three continuous variables on the same subject, we can calculate the correlation coefficients r_{xy}, r_{xz}, r_{yz}. It is evaluated by creating a table referred as inter correlation matrix (correlation matrix). The correlation matrix arranges the correlation coefficients in a square with equal number of rows and columns.

Example of an inter correlation matrix:

Variables	X	Y	Z
X	r_{xx}	r_{xy}	r_{xz}
Y	r_{xy}	r_{yy}	r_{yz}
Z	r_{xz}	r_{yz}	r_{zz}

Abbreviated inter correlation matrix:

Variables	Y	Z
X	r_{xy}	r_{xz}
Y	1	r_{yz}

To illustrate the use of an inter correlation matrix, consider the data presented in the following table.

Leveled variables from the four conditions in a dissolution study:

Compressional force	Hardness (kg/cm^2)	Disintegration time (min)	Dissolution rate (min^{-1})
4	2	5	0.502
6	3	8	0.368
8	4	10	0.274
10	5	15	0.169
12	6	20	0.126

Layout of correlation matrix:

Variables	Compression force	Hardness	Disintegration time	Dissolution rate
Compression force(a)	1.00	r_{ab}	r_{ac}	r_{ad}
Hardness(b)	r_{ab}	1.00	r_{bc}	r_{bd}
Disintegration time(c)	r_{ac}	r_{bc}	1.00	r_{cd}
Dissolution rate(d)	r_{ad}	r_{bd}	r_{cd}	1.00

To calculate r_{ab}

Compression force (a) = x	Hardness (b) = y	xy	X^2	Y^2
4	2	8	16	4
6	3	18	36	9
8	4	32	64	16
10	5	50	100	25
12	6	72	144	36
40	**20**	**180**	**340**	**90**

$$\mathbf{r_{ab}} = \frac{\Sigma \mathbf{xy} - \frac{1}{\mathbf{n}}(\Sigma \mathbf{x} \Sigma \mathbf{y})}{\sqrt{\Sigma \mathbf{x}^2 - \frac{(\Sigma \mathbf{x})^2}{\mathbf{n}}}\sqrt{\Sigma \mathbf{y}^2 - \frac{(\Sigma \mathbf{y})^2}{\mathbf{n}}}} = \frac{180 - \frac{1}{5}(40)\ (20)}{\sqrt{340 - \frac{(40)^2}{5}}\sqrt{90 - \frac{(20)^2}{5}}}$$

$$= \frac{180 - 160}{\sqrt{340 - 320}\sqrt{10}} = \frac{20}{\sqrt{20}\,.\sqrt{10}} = \frac{20}{\sqrt{200}} = 1.4142$$

It is more than 1 takes it as 1

To calculate r_{ac}

Compression force (a) x	Disintegration time (c) y	xy	x^2	y^2
4	5	20	16	25
6	8	48	36	64
8	10	80	64	100
10	15	150	100	225
12	20	240	144	400
40	**58**	**538**	**360**	**894**

$$\therefore r_{ac} = \frac{\Sigma xy - \frac{1}{n}\Sigma x \Sigma y}{\sqrt{\Sigma x^2 - \frac{(\Sigma x)^2}{n}}\sqrt{\Sigma y^2 - \frac{(\Sigma y)^2}{n}}}$$

$$= \frac{538 - \frac{1}{5}\ .\ (40)\ (58)}{\sqrt{360 - \frac{(40)^2}{5}}\sqrt{814 - \frac{(58)^2}{5}}}$$

$$= \frac{74}{\sqrt{40} \times \sqrt{141.2}} = \frac{74}{75.153} = 0.9846$$

To Calculate r_{ad}

Compression force (a) = x	Dissolution rate (y)	xy	x^2	y^2
4	0.502	2.008	16	0.252004
6	0.368	2.208	36	0.135424
8	0.274	2.192	64	0.075076
10	0.169	1.69	100	0.028561
12	0.126	1.512	144	0.015876
38	1.439	9.61	360	0.506941

$$\therefore r_{ad} = \frac{\Sigma xy - \frac{1}{n}\Sigma x \Sigma y}{\sqrt{\Sigma x^2 - \frac{(\Sigma x)^2}{n}}\sqrt{\Sigma y^2 - \frac{(\Sigma y)^2}{n}}}$$

$$= \frac{9.61 - \frac{1}{5}(38)\,(1.439)}{\sqrt{360 - \frac{(38)^2}{5}}\sqrt{(0.506941) - \frac{(1.439)^2}{5}}} = -0.9872$$

Similarly we can find r_{bc}, r_{bd}, r_{cd}

$r_{bc} = 0.9846$, $r_{bd} = -0.9872$

$r_{cd} = -0.9512$

Correlation matrix: from the above data correlation coefficients are calculated and presented in the following table.

Variables	Compression force	Hardness	Disintegration time	Dissolution rate
Compression force	1.00	1.00	0.9846	– 0.9872
Hardness	1.00	1.00	0.9846	– 0.9872
Disintegration time	0.9846	0.9846	1.00	– 0.9512
Dissolution rate	– 0.9872	– 0.9872	– 0.9512	1.00

From the above results, it is concluded as follows:

1. The correlation existing between compressional force and hardness is strongly positive.
2. The correlation existing between compressional force and disintegration time is strongly positive.

3. The correlation existing between compressional force and dissolution rate is strongly negative.
4. The correlation existing between hardness and disintegration time is strongly positive.
5. The correlation existing between hardness and dissolution rate is strongly negative.
6. The correlation existing between disintegration time and dissolution rate is strongly negative.

5.1.5 Partial Correlations

Partial correlation shows that correlation between two continuous variables, while removing the effect of other continuous variables. It extracts the common effect of one variable from the relationship between two other variables of interest. In case of three continuous variables x, y, z, the correlation between variables x, y by eliminating the effect of variable z ($r_{xy,z}$) is calculated with the following formula

$$r_{yx,z} = \frac{r_{yx} - (r_{xz})(r_{yz})}{\sqrt{(1-r_{xz}^2)(1-r_{yz}^2)}}$$

In the above example, if we are interested on correlation of disintegration time (c) and dissolution rate (d) by removing the effect of hardness (b) on compression force (d), then the partial correlation is as follows:

$$r_{cd,a} = \frac{r_{cd} - (r_{ca})(r_{da})}{\sqrt{1-(r_{ca})^2}\sqrt{1-(r_{da})^2}}$$

$$= \frac{-0.9512 - (0.9846)(-0.9872)}{\sqrt{(1-(0.9846)^2(1-(-0.9872)^2)}}$$

$$= \frac{-0.9512 + 0.97199712}{\sqrt{(0.03056284)(0.02543616)}}$$

$$= \frac{0.02079712}{0.027880522} = 0.745937253$$

Problem: The range of compression force and disintegration time of the tablets is given in the following table. Find the coefficient of correlation between compression force and the number of tablets passing the disintegration time.

Compression force	Disintegration time	
	No. tested	No. passed
0-2	6	5
2-4	6	4
4-6	6	3
6-8	6	1

Solution:

Compression fource	Mid value (x)	Deviation from assumed av 5	Disinte-gration time No. passed (y)	Deviation from assumed are 3 (dy)	$(dx)^2$	$(dy)^2$	dx dy
0-2	1	-4	5	2	16	4	8
2-4	3	-2	4	1	4	1	-2
4-6	5	0	3	0	0	0	0
6-8	7	2	1	-2	4	4	-4
N = 4		$\Sigma dx = -4$		$\Sigma dy = 1$	$\Sigma dx^2 = 24$	$\Sigma dy^2 = 9$	$\Sigma dxdy = 2$

$$\text{Now } r = \frac{\Sigma dxdy - \dfrac{\Sigma dx \times \Sigma dy}{N}}{\sqrt{\left\{\Sigma dx^2 - \dfrac{(\Sigma dx)^2}{N}\right\}\left\{\Sigma dy^2 - \dfrac{(\Sigma dy)^2}{N}\right\}}}$$

$$\frac{2 - \dfrac{(-4)(1)}{4}}{\sqrt{\left\{24 - \dfrac{(-4)^2}{4}\right\}\left\{9 - \dfrac{(1)^2}{4}\right\}}} = \frac{3}{\sqrt{(20)\left(\dfrac{35}{4}\right)}} = \frac{3}{\sqrt{5 \times 35}}$$

$$= \frac{3}{13.22} = 0.2269$$

5.1.6 Standard Error of the Coefficient of Correlation

The more commonly used measure of reliability of the value of Pearsonian coefficient of correlation is the standard error of the coefficient of correlation. An amount added to and deducted from the coefficient of correlation produces a range within which coefficients of correlation of other groups collected from same series at random will fall, with a probability of 67.45%. It is calculated with the following formulae.

$$\text{S.E.}(r) = \frac{1 - r^2}{\sqrt{N}}$$

r = observed correlation coefficient.

N = number of pairs of observation.

Coefficient of determination

The coefficient of determination (r^2) is the square of the coefficient of correlation. It is more useful and readily comprehensible measure for indicating the percentage variation in the dependent variable which is accounted for by the independent variable. It gives the ratio of the explained variance to the total variance and it is calculated with the following formulae.

$$r^2 = \text{Explained Variance/Total Variance}$$

The coefficient of determination is a much more useful measure of the linear co variation of two variables. For example, if the value of r is 0.8, we cannot conclude that the 80% of the variation in the dependent variable is due to the variation in the independent variable. The coefficient of determination in this case is 0.64 which implies that only 64% of the variation in the dependent variable has been explained by the independent variables.

5.1.7 Correlation in Bivariate Frequency Table

When the number of observations in a bivariate distribution is fairly large, the data can be classified according to two measurements in a two-way frequency table to make the presentation concise.

The formulae for calculating the correlation coefficient is given below.

$$r = \frac{\Sigma fdxdy - \dfrac{\Sigma fdx \times \Sigma fdy}{N}}{\sqrt{\left\{\Sigma fdx^2 - \dfrac{(\Sigma fdx)^2}{N}\right\}\left\{\Sigma fdy^2 - \dfrac{(\Sigma fdy)^2}{N}\right\}}}, \quad \text{where } N = \Sigma f$$

Illustration: The data presented in the following table shows a negative correlation between granule size and dissolution rate. Calculate r value.

Dissolution rate	**Granule size**			
	3-5	5-7	7-9	9-11
2-4			2	4
4-6		1	2	3
6-8		2	1	1
8-10	2	4		

Granule size (x-series)

Dissolution rate y				3-5	5-7	7-9	9-11	Total			
Class				3-5	5-7	7-9	9-11				
Midpoint				4	6	8	10				
				– 4	– 2	0	+ 2	Total	fdy	fdy^2	fdx dy
		dy	dx	– 2	– 1	0	+ 1	f			
2-4	3	– 4	– 2	-	-	2(0)	4(– 8)	6	– 12	24	– 8
4-6	5	– 2	– 1	-	1(1)	2(0)	3(– 3)	6	– 6	6	– 2
6-8	7	0	0	-	2(0)	1(0)	1(0)	4	0	0	0
8-10	9	2	1	2 (– 4)	4(– 4)	-	-	4	4	4	– 8
Total (f)				2	7	5	8	22	$\Sigma fdy = – 4$	$\Sigma fdy^2 = 34$	$\Sigma fdxdy = – 18$
fdx				– 4	– 7	0	8	$\Sigma fdx = – 3$			
fdx^2				8	7	0	8	$\Sigma fdx^2 = 23$			
fdxdy				– 4	– 3	0	– 11	$\Sigma fdxdy = – 18$			

$$\therefore \quad r = \frac{\Sigma fdxdy - \dfrac{\Sigma fdx \times \Sigma fdy}{N}}{\sqrt{\left(\Sigma fdx^2 - \dfrac{(\Sigma fdx)^2}{N}\right)\left(\Sigma fdy^2 - \dfrac{(\Sigma fdy)^2}{N}\right)}}$$

$$= \frac{(-18) - \dfrac{(-3)(-14)}{22}}{\sqrt{\left(23 - \dfrac{(-3)^2}{22}\right)\left(34 - \dfrac{(-14)^2}{22}\right)}} = \frac{-18 - 1.9090}{\sqrt{(22.59)(25.090)}}$$

$$= \frac{(-19.9090)}{23.8072} = -0.836$$

5.2 Regression Analysis

The statistical method which helps us to estimate the unknown value of one variable from the known value of the related variable is called Regression. The variables whose values are known is referred as the independent variable and the variable whose values are influenced by the independent variables are known as the dependent variables.

The literal meaning of the word regression is return or going back.

According to M.M. Blair, Regression is the measure of average relationship between two or more variables in terms of the original units of the data.

According to Ya-Lu-Chow, Regression analysis attempts to establish the nature of the relationship between the variables. It allows the study of the functional relationship between the variables and thereby provides a mechanism for prediction or forecasting.

So we conclude that regression analysis is a statistical device and with the help of regression analysis we can estimate or predict the unknown values of one variable from known values of another variable.

5.2.1 Differences between Correlation and Regression

Correlation	Regression
Correlation is the relationship between two or more variables which vary in sympathy with the other in the same or the opposite direction	Regression means going back and it is a mathematical measure showing the average relation between two variables
Correlation find out the degree of relationship between two variables and not the cause and effect of variables	Regression indicates the cause and effect relationship between the variables and establishes a functional relationship.
The range of coefficient of correlation ship lies between (–1) and 1	Regression coefficient is an absolute figure
It has limited applications	Regression has wider applications

5.2.2 Regression Lines

The average relationship between two variables is described by the regression lines. When the exact value of one variable is given, the most probable value of other variable is shown in the regression line. If there are two relative variables then there will be two regression lines one of each variable with reference to other. Suppose there are two variables x and y, there will be one regression line for y with reference to x. This line is known as regression line of y on x. Another line is for x with reference to y. It is called regression line of x on y.

(i) ***Line of regression of x on y*****:**

Line of regression of x on y gives the best estimate for the values of x for particular value of y and its equation is defined as

$$x - \bar{x} = r.\frac{\sigma x}{\sigma y}(y - \bar{y})$$

Here, $\bar{x}$, $\bar{y}$ are means of x series and y-series respectively, Σx, Σy are standard deviations of x and y series respectively and r is correlation coefficient between x and y.

The line of regression of x on y also written as x = a + by

where $\frac{-a}{b}$ is the intercept of line and $\frac{1}{b}$ is the slope of line x on y.

The regression coefficient as x on y is

$$b = r.\,\frac{\sigma x}{\sigma y}$$

$$\therefore \qquad b = \frac{\Sigma xy - \frac{1}{n}\Sigma x.\Sigma y}{\sqrt{\Sigma x^2 - \frac{(\Sigma x)^2}{n}}\sqrt{\Sigma y^2 - \frac{(\Sigma y)^2}{n}}} \times \frac{\sqrt{\Sigma x^2 - \frac{(\Sigma x)^2}{n}}}{\sqrt{\frac{\Sigma y^2 - \frac{(\Sigma y)^2}{n}}{n}}}$$

$$\therefore \qquad b = \frac{\Sigma xy - \frac{1}{n}\Sigma x\,\Sigma y}{\left(\Sigma y^2 - \frac{(\Sigma y)^2}{n}\right)}$$

We obtain b by above formula

Again we have x = a + by

$\therefore \quad \Sigma x = \Sigma a + \Sigma by$

$\therefore \quad \Sigma x = na + b\Sigma y$

Dividing on both sides by 'n' we get

$$\frac{(\Sigma x)}{n} = \frac{na}{n} + b.\frac{\Sigma y}{n}$$

$\Rightarrow \quad \bar{x} = a + b\bar{y}$

$\therefore \quad a = \bar{x} - b\bar{y}$

$\therefore$ We obtain constant a by the formula

$$a = \bar{x} - b\bar{y}$$

where $\bar{x}$ is mean of variable x, $\bar{y}$ is mean of variable y

Algorithm to find Regression equation of x on y

Step 1. Let required regression equation of x on y is x = a + by where b is regression coefficient.

2. Obtain regression coefficient of x on y i.e., b by

$$\frac{\Sigma xy - \frac{1}{n}\Sigma x \, \Sigma y}{\left(\Sigma y^2 - \frac{(\Sigma y)^2}{n}\right)}$$

3. Obtain 'a' from $\bar{x} = a + b\bar{y}$, i.e., $a = \bar{x} - b\bar{y}$

4. Substitute a and b in x = a + by

(i) **Line of regression of y on x:** Line of regression of y on x gives the best estimate for the values of y for particular value of x and its equation is defined as $y - \bar{y} = r.\frac{\sigma y}{\sigma x}(x - \bar{x})$

where $\bar{x}$, $\bar{y}$ are the means of variables x and y respectively, Σx, Σy are standard deviation of x and y, r is the coefficient of correlation between x and y.

The line of regression of y as x is also written as y = a + bx where a is intercept of line and b is slope of line, here b is called regression coefficient of y on x and it is obtained by

$$b = r\frac{\sigma y}{\sigma x}$$

$$= \frac{\Sigma xy - \frac{1}{n}\Sigma x.\Sigma y}{\sqrt{\Sigma x^2 - \frac{(\Sigma x)^2}{n}}\sqrt{\Sigma y^2 - \frac{(\Sigma y)^2}{n}}} \times \frac{\sqrt{\frac{\Sigma y^2 - \frac{(\Sigma y)^2}{n}}{n}}}{\sqrt{\frac{\Sigma x^2 - \frac{(\Sigma x)^2}{n}}{n}}}$$

$$\therefore \qquad b = \frac{\Sigma xy - \frac{1}{n}\Sigma x \Sigma y}{\Sigma x^2 - \frac{(\Sigma x)^2}{n}}$$

Again we have,

$$y = a + bx$$

$$\Rightarrow \qquad \Sigma y = \Sigma a + \Sigma bx$$

$$\Rightarrow \qquad \Sigma y = na + b\Sigma x$$

where n is number of observations

$$\Rightarrow \qquad \frac{\Sigma y}{n} = a + b\frac{\Sigma x}{n} \Rightarrow \overline{y} = a + b\overline{x}$$

$$\therefore \qquad a = \overline{y} - b\overline{x}$$

So substitute a, b in y = a + bx we obtain regression of y on x

Algorithm to find regression equation of y on x

Step 1: Let the required equation y = a + bx

Step 2: b is obtained by $\dfrac{\Sigma xy - \frac{1}{n}\Sigma x \Sigma y}{\left(\Sigma x^2 - \frac{(\Sigma x)^2}{n}\right)}$

Step 3: obtained a by $\overline{y} - b\overline{x}$

Step 4: substitute a, b in y = a + bx

Note:

1. If b_{xy}, b_{yx} are coefficients of regression of x on y and y on x respectively then the coefficient of correlation between x and y is $r = \sqrt{b_{xy}.b_{yx}}$
2. The two regression lines x on y and y on x always intersect at the means $(\overline{x}, \overline{y})$
3. If r = 0 then the regression coefficient is zero.
4. The regression lines become identical if r = ± 1. It follows from the regression equations that $x = \overline{x}$ and $y = \overline{y}$, and these lines are perpendicular to each other.
5. If one of the regression coefficient is greater than unity thus the other is less than unity since, suppose by x > 1, also we know that $r^2 \leq 1$ and $r^2 = b_{yx}\, b_{xy}$.

$$\Rightarrow \qquad b_{yx} \,.\, b_{xy} \leq 1$$

$$b_{xy} \leq \frac{1}{b_{yx}} < 1 \text{ (by x > 1)}$$

6. Both the regression coefficients will have same sign that is either both are positive or both are negative.
7. The sign of correlation is same as that of regression coefficients.

5.2.3 Method of Least Squares

In this method we obtain the equation of curve which fits best a given set of observations. The least square method is only based on the assumption that the sum of squares of differences between the estimated values and the actual observed values is very low (minimum).

Fitting a Straight line

The best fitting straight line can be obtained by means of an equation which shows the relation between two variables.

For obtaining best values of the constants in the estimating equation, a set of equations depending on the number of constants in the estimating equation are required to be solved. The system of equations required to be solved for obtaining the values of constants in the estimations equation is known as normal equations.

If the estimate equation of straight line is $y = a + bx$ then its corresponding normal equations are

$$\Sigma y = na + b\Sigma x$$

$$\Sigma xy = a\Sigma x + b\Sigma x^2$$

where

x = independent variable

y = Dependent variable

a and b are constants, Σ = sign of summation

n = number of pairs of observations.

Fitting a Parabola

Take the estimate equation of parabola for given values is $y = ax^2 + bx + c$ where a, b, c are constants these are obtained by solving the corresponding normal equation.

$$\Sigma y = a\Sigma x^2 + b\Sigma x + cn$$

$$\Sigma xy = a\Sigma x^3 + b\Sigma x^2 + c\Sigma x$$

$$\Sigma x^2 y = a\Sigma x^4 + b\Sigma x^3 + c\Sigma x^2$$

Solved Examples

1. The regression equations of two variables x and y are x = 0.7 y + 5.2, y = 0.3 x + 2.8. Find the mean of variables and coefficient of correlation between them.

Solution: Given regression equations of two variables x and y are

$$x = 0.7y + 5.2 \quad .....(1)$$

$$y = 0.3x + 2.8 \quad .....(2)$$

The point of interaction of (1) and (2) given the means of variables x and y

(1) ⇒ $x - 0.7\,y - 5.2 = 0$

(2) ⇒ $0.3\,x - y + 2.8 = 0$

$$\Rightarrow \quad \begin{matrix} & x & & y & & 1 & \\ -0.7 & & -5.2 & & 1 & & -0.7 \\ -1 & & 2.8 & & 0.3 & & -1 \end{matrix}$$

$$\Rightarrow \quad \frac{x}{(-0.7)(2.8)-(1)(5.2)} = \frac{y}{(-5.2)(0.3)-2.8} = \frac{1}{-1+(0.3)(0.7)}$$

$$\frac{x}{-7.16} = \frac{y}{-4.36} = \frac{1}{-0.79}$$

$$\Rightarrow \quad x = \frac{-7.16}{-0.79} = 9.06,\; y = \frac{-4.36}{-0.77} = 5.52$$

∴ Mean of variable x = 9.06

Mean of variable y = 5.52

and we have x = 0.7y + 5.2

∴ Regression coefficient of x on y = b_{xy} = 0.7

$y = 0.3x + 2.8$

∴ Regression coefficient of x on y = b_{xy} = 0.7, y = 0.3x + 2.8

∴ Regression coefficient of y on x = b_{yx} = 0.3

∴ Coefficient of correlation between the variables

$$x \text{ and } y \text{ is } r = \sqrt{b_{xy}.b_{yx}} = \sqrt{0.7 \times 0.3}$$

$$= \sqrt{0.21} = 0.4582$$

$$\approx 0.46$$

2. Find the regression equations from the following data.

x	6	2	10	4	8
y	9	11	5	8	7

Solution:

x	y	xy	x^2	y^2
6	9	54	36	81
2	11	22	4	121
10	5	50	100	25
4	8	32	16	64
8	7	56	64	49
$\Sigma x = 30$	40	214	220	340

n = 5

To find the regression essential equation of x on y

Let the required estimation be x = a + by

Regression coefficient of x on y = b

$$= \frac{\Sigma xy - \frac{1}{n}\Sigma x\, \Sigma y}{\left(\Sigma y^2 - \frac{(\Sigma y)^2}{n}\right)}$$

$$b = \frac{214 - \frac{1}{5}.30 \times 40}{340 - \frac{(40)^2}{5}} = \frac{214 - 240}{340 - 320} = \frac{-26}{20} = -1.3$$

$\therefore$ $b = -1.3$

and $\bar{x} = \frac{\Sigma x}{n} = \frac{30}{5} = 6,\ \bar{y} = \frac{\Sigma y}{n} = \frac{40}{5} = 8$

and $\bar{x} = a + b\bar{y} = 1,\ a = \bar{x} - b\bar{y}$

$= 6 - (-1.3)\ 8$

$= 6 + (1.3)\ (8) = 16.4$

$\therefore$ Regression equation is x = a + by i.e., x = 16.4 + (–1.3)y

To find the regression equation of y on x

Let the required equation be y = a + bx

where b = regression coefficient of y on x

$$= \frac{\Sigma xy - \frac{1}{n}\Sigma x.\Sigma y}{\Sigma x^2 - \frac{(\Sigma x)^2}{n}}$$

$$= \frac{(-26)}{220 - \frac{(30)^2}{5}} = \frac{-26}{220-180} = \frac{-26}{40} = -0.65$$

and $\bar{y} = a + b\bar{x}$

$\Rightarrow$ $a = \bar{y} - b\bar{x} = 8 - (-0.65)(6)$

$= 8 + 0.65 \times 6 = 11.9$

$\therefore$ Regression equation is

$y = a + bx$ i.e., $y = 11.9 + (-0.65)\ x$

i.e., $y = 11.9 - 0.65\ x$

3. Find the regression lines from the following data.

Age of husband	18	19	20	21	22	23	24	25	26	27
Age of wife	17	17	18	18	19	19	19	20	21	22

Solution:

Age of husband (x)	**Age of wife (y)**	**xy**	**x^2**	**y^2**
18	17	306	324	289
19	17	323	361	289
20	18	360	400	324
21	18	378	441	324
22	19	418	484	361
23	19	437	529	361
24	19	456	576	361
25	20	500	625	400
26	21	546	676	441
27	22	594	729	484
$\Sigma x = 225$	$\Sigma y = 190$	$\Sigma xy = 4318$	$\Sigma x^2 = 5145$	$\Sigma y^2 = 3634$ and $n = 10$

To find regression equation of x on y

Let the required be

$$x = a + by$$

where
$$b = \frac{\Sigma xy - \frac{1}{n}\Sigma x.\Sigma y}{\Sigma y^2 - \frac{(\Sigma y)^2}{n}}$$

$$= \frac{4318 - \frac{1}{10}(225)(190)}{3634 - \frac{(190)^2}{10}}$$

$$= \frac{43}{24} = 1.79$$

and also $\bar{x} = a + b\bar{y}$

$\therefore$ $a = \bar{x} - b\bar{y}$

$$= \frac{225}{10} - 1.79 \times \frac{190}{10}$$

$$a = 22.5 - 1.79 \times 19 = -11.51$$

$\therefore$ Regression equation of x on y is $x = a + by$

i.e., $x = -11.51 + 1.79\,y$

To find the regression equation of y on x

Let the required equation be $y = a + bx$

where
$$b = \frac{\Sigma xy - \frac{1}{n}\Sigma x\,\Sigma y}{\Sigma x^2 - \frac{(\Sigma x)^2}{n}}$$

$$= \frac{43}{5145 - \frac{(225)^2}{10}}$$

$$= \frac{43}{82.5} = 0.521212$$

$$\approx 0.52$$

and $\bar{y} = a + b\bar{x} \quad \Rightarrow a = \bar{y} - b\bar{x}$

$$= \frac{190}{10} - 0.52 \times \frac{225}{10}$$

$$= 19 - 0.52 \times 22.5$$

$$a = 7.3$$

∴ The regression equation of y on x is

$$y = 7.3 + 0.52\ x$$

4. In a partially destroyed record the following data are available.

Variance of

$$x = 25$$

Regression equation of

$$x \text{ on } y: 5x - y = 22$$

Regression equation of y on

$$x: 64\ x - 45\ y = 24$$

Find

(a) Mean value of x and y
(b) Coefficient of correlation between x and y
(c) Standard deviation of y

Solution:

(a) The means of variables x and y i.e., $\bar{x}$, $\bar{y}$ are points of intersection of Regression lines

∴ $\bar{x}$, $\bar{y}$ lies on both the lines

∴ $5\bar{x} - \bar{y} - 22 = 0$

$64\bar{x} - 45\bar{y} - 24 = 0$

$\bar{x}$		$\bar{y}$		**1**
– 1	– 22	5	– 1	
– 45	– 24	64	– 45	

$$\frac{\bar{x}}{24 - (45)(22)} = \frac{\bar{y}}{(-22)(64) + (24)(5)} = \frac{1}{(5)(-45) + 64}$$

$$\bar{x} = \frac{24 - (45)(22)}{5(-45) + 64},\quad \bar{y} = \frac{(-22)(64) + (24)(5)}{5(-45) + 64}$$

⇒ $\bar{x} = 6 \qquad \bar{y} = 8$

(b) The regression equation of x on y is

$$5x - y = 22$$

i.e., $5x = y + 22$

$\Rightarrow \quad x = \frac{22}{5} + \frac{y}{5}$

$\Rightarrow \quad b_{xy} = 1/5$

The regression equation of y on x is

$$64x - 45y = 24$$

$\Rightarrow \quad 45y = 64x - 24$

$\Rightarrow \quad y = \frac{-24}{45} + \frac{64}{45}x$

$\Rightarrow \quad b_{yx} = \frac{64}{45}$

$\therefore$ Coefficient of correlation between x and y is

$$r = \sqrt{b_{xy} \times b_{yx}} = \sqrt{\frac{1}{5} \times \frac{64}{45}} = \sqrt{\frac{64}{225}}$$

$$= 8/15$$

$\therefore \quad r = 8/15$

(c) Variance of x is $x^2 = 25$

$\therefore \quad \sigma_x = 5$

We know that $b_{yx} = \frac{\sigma_y}{\sigma_x}.r$

$$\therefore \quad \sigma_y = \left(\frac{b_{yx}.\sigma_x}{r}\right) = \frac{\frac{64}{45} \times 5}{8/15}$$

$$= \frac{8}{3} \times 5 = \frac{40}{3}$$

$$= 13.33$$

Hence standard deviation of y = 13.33

5. It is given that the mean value of R variable x is 20 and the mean value of y is 45 and the regression coefficient of y on x is 4, and that of x on y is $\frac{1}{9}$. Then find.

(i) The coefficient of correlation

(ii) The standard deviation of x if the standard deviation of y is 12

(iii) Also write down the equation of regression lines

Solution: (i) Given that $\overline{x} = 20$, $\overline{y} = 45$, $b_{yx} = 4$, $b_{xy} = \frac{1}{9}$

∴ Coefficient of correlation

$$r = \sqrt{b_{yx}.b_{xy}}$$

$$\therefore \quad r = \sqrt{4.\tfrac{1}{9}} = \tfrac{2}{3}$$

$$\therefore \quad r = \tfrac{2}{3}$$

(ii) and

$$\sigma_x = 12,$$

We have $b_{yx} = \frac{\sigma y}{\sigma x}.r$

$$\therefore \quad \sigma_x = \frac{\sigma y}{b_{yx}} \times r = \frac{12}{4} \times \frac{2}{3} = 2$$

$$\sigma_x = 2$$

(iii) The regression equation of y on x is

$$y - = \frac{\sigma_y}{\sigma_x}.r(x - \overline{x})$$

$$\therefore \quad y - 45 = \frac{12}{2} \times \frac{2}{3}\ (x - 20)$$

$$\therefore \quad y - 45 = 4(x - 20)$$

$$\therefore \quad 4x - y - 35 = 0$$

i.e., $\quad y = 4x - 35$

The regression equation of x or y

$$x - \overline{x} = \frac{\sigma x}{\sigma y}(r)(y - \overline{y})$$

$$\therefore \quad x - 20 = \frac{2}{12}\left(\frac{2}{3}\right)(y-45)$$

$$\therefore \quad x - 20 = \frac{1}{9}(y-45)$$

$$\Rightarrow \quad x = 20 + y/9 - 5$$

The required equation of x on y

$$x = 15 + y/9$$

6. From the following data, find two regression lines

	x	y
Arithmetic mean	36	85
Standard deviation	11	8

Correlation coefficient between x and y is 0.66

Solution: Given, $\bar{x} = 36,\ \bar{y} = 85$

$\sigma_x = 11,\ \sigma_y = 8,\ r = 0.66$

$\therefore$ Regression coefficient of x on y = $b_{xy} = \frac{\sigma x}{\sigma y} \times r$

$$= \frac{11}{8} \times 0.66$$

Regression coefficient of y on x = by x $= \frac{\sigma y}{\sigma x} \times r$

$$= \frac{8}{11} \times 0.66$$

$$= 8 \times 0.6 = 4.8$$

$\therefore$ The regression equation of x on y is

$$x - \bar{x} = \frac{\sigma y}{\sigma x} r(y - \bar{y}) \Rightarrow x - 36 = 11/8 \times 0.66(y-85)$$

The regression equation of y on x is

$$y - \bar{y} = \frac{\sigma y}{\sigma x} r(x - \bar{x})$$

$$\Rightarrow y - 85 = 4.8\ (x - 36)$$

7. Find the concentration of the drug present in a solution whose absorbance value is 0.36. The standard calibration curve for the given drug is given below.

Concentration (μ g/ml)	Absorbance
2	0.104
4	0.208
6	0.310
8	0.415
10	0.519

Solution: Suppose corresponding equation is y = a + bx

Where
$$b = \frac{\Sigma xy - \frac{1}{n}\Sigma x \Sigma y}{\Sigma x^2 - \frac{(\Sigma x)^2}{n}}$$

$$\Sigma x = 30 \; \Sigma x^2 = 220 \qquad \Sigma y = 1.556$$

$$\Sigma xy = 11.41$$

$$\Rightarrow \quad b = \frac{11.41 - \frac{1}{5}(30)1.556}{\left((220) - \frac{(30)^2}{5}\right)}$$

$$= \frac{2.074}{40} = 0.05185$$

Again $y = a + bx \Rightarrow \Sigma y = \Sigma a + \Sigma bx$

$\Rightarrow \quad \Sigma y = na + b\Sigma x$

$\Rightarrow \quad \bar{y} = a + b.\bar{x}$

Now, $\bar{x} = \frac{30}{6} = 5$

$$\bar{y} = \frac{1.556}{5} = 0.3112$$

$\Rightarrow \quad 0.3112 = a + (0.05185)\,(5)$

$\Rightarrow \quad a = 0.3112 - (0.05185)\,(5)$

$= 0.05195$

∴ The regression equation of y on x is

$$y = a + bx$$

i.e. $y = (0.05195) + (0.05185)\ x$

When $y = 0.36$ then

$$0.36 = 0.05195 + (0.05185)\ x$$

$$\Rightarrow \quad x = \frac{0.36 - 0.05195}{0.05185} = 0.30808$$

$$x = 5.941176\ \mu g/ml$$

Or

Let the required equation is

$$x = a + by$$

where

$$b = \frac{\Sigma xy - \frac{1}{n}\Sigma x\,\Sigma y}{\Sigma y^2 - \frac{(\Sigma y)^2}{n}}$$

$$\Sigma y^2 = 0.591766$$

$$\Rightarrow \quad b = \frac{11.41 - \frac{1}{5}(30)(1.556)}{(0.591766) - \frac{(1.556)^2}{5}}$$

$$= \frac{2.074}{0.1075388}$$

$$= 19.78$$

Also $\bar{x} = a + b\bar{y}$

$$\Rightarrow \quad a = \bar{x} - b\bar{y}$$

$$5 - (19.2860)\ (0.3112)$$

$$= -\ 1.0018032$$

Now $x = (-\ 1.0018\ 032) + (19.2860)\ y$

When $y = 0.36$

$$x = (-\ 1.00\ 18032) + (19.2860)\ (0.36)$$

$$= 5.94115668$$

8. The dissolution samples were collected at regular one hour intervals and the samples were analyzed for the drug content. However at 8th hour the sample was unable to be collected. Determine the amount of drug dissolved at 8th hour.

Time (x) (hr)	**Amount of drug dissolved (mg) (y)**
1	5
1	10
3	15
4	20
5	25
6	30
7	35
8	-
9	45

Solution:

$$\Sigma x = 37$$

$$\Sigma y = 185$$

$$\Sigma x^2 = 221$$

$$\Sigma xy = 1105$$

$$\bar{y} = \Sigma y / 8 = \frac{185}{8} = 23.125$$

$$\bar{x} = \Sigma x / 8 = 37/8 = 4.625$$

Let corresponding regression essential is

$$y = a + bx$$

When $$b = \frac{\Sigma xy - \frac{1}{n}\Sigma x \Sigma y}{\Sigma x^2 - \frac{(\Sigma x)^2}{n}}$$

$$\Rightarrow \quad b = \frac{1105 - \frac{1}{8}.(37)(185)}{221 - \frac{(37)^2}{8}}$$

$$= \frac{249.375}{49.875} = 5$$

Again $\bar{y} = a + b\bar{x}$

$$23.125 = a + 5(4.625)$$

$$a = 23.125 - 5(4.625)$$

$$= 0$$

Now $y = a + bx$

i.e., $y = 0 + 5x$

$\therefore$ $y = 5x$

At $x = 8, y = 5 \times 8 = 40$

9. Fit a straight line of y on x from given data by using method of least square.

x	0	1	2	3	4	5	6
y	2	1	3	2	4	3	5

x	y	xy	x^2
0	2	0	0
1	1	1	1
2	3	6	4
3	2	6	9
4	4	16	16
5	3	15	25
6	5	30	36
$\Sigma x = 21$	$\Sigma y = 20$	$\Sigma xy = 74$	91

$n = 7$

Solution:

Let the required equation of line is $y = a + bx$

Now the corresponding normal equations are

$$\Sigma y = na + b\Sigma x, \quad \Sigma xy = a\Sigma x + b\Sigma x^2$$

$\Rightarrow$ $20 = 7a + b.21$ $\Rightarrow$ $7a + 21b = 20$ .....(1)

and $74 = 21a + 91b$ $\Rightarrow$ $21a + 91b = 94$ .....(2)

Solving (1) and (2)

Now (1) × 3 $\Rightarrow$ $21\ a + 63\ b = 60$

(2) $\Rightarrow$ $21\ a + 91\ b = 74$

$\Rightarrow$ $-28\ b = -14$

$\Rightarrow$ $b = \dfrac{14}{28} = 0.5$

Substitute b = 0.5 in (1)

$$7a + 21 \times 0.5 = 20$$

$$7a = 20 - 10.5 = 9.5$$

$$a = \frac{9.5}{7} = 1.357$$

$$\therefore \quad y = 1.357\,x + 0.5$$

5.2.4 Regression Equation in Group Frequency Distribution

Like correlation analysis, the regression analysis of a grouped frequency distribution requires a two-way correlation table in which values of one of variable is represented horizontally in the first row and the values of other variable is represented vertically in the first column. The formulae for calculating the regression coefficients are:

$$b_{xy} = \left\{ \frac{\Sigma fdx\ dy - \dfrac{(\Sigma fdx)\times(\Sigma fdy)}{N}}{\Sigma fdy^2 - \dfrac{(\Sigma fdy)^2}{N}} \right\} \times \frac{c_x}{c_y}$$

$$b_{yx} = \left\{ \frac{\Sigma fdx\ dy - \dfrac{(\Sigma fdx)\times(\Sigma fdy)}{N}}{\Sigma fdx^2 - \dfrac{(\Sigma fdx)^2}{N}} \right\} \times \frac{c_y}{c_x}$$

where c_x= width of the class-interval of the *X*-variable and

c_y= width of the class-interval of *Y*-variable

Example:

Age of the patient	AUC value			
	100-150	150-200	200-250	250-300
20-30		2	2	6
30-40		3	4	3
40-50		5	3	2
50-60	2	6	2	

AUC (x-series)

Age of patient ((y))

	x			100-150	150-200	200-250	250-300	Total			
	Midpoint			125	175	225	275				
				-100	– 50	0	+ 50	Total	fdy	fdy^2	fdx dy
	Mid value		dx / dy	– 2	– 1	0	+ 1	F			
20-30	25	– 20	– 2	-	2(4)	2(0)	6 (– 12)	10	– 20	40	– 8
30-40	35	– 10	– 1	-	3(3)	4(0)	3 (– 3)	10	– 10	10	0
40-50	45	0	0	-	5(0)	3(0)	2(0)	10	0	0	0
50-60	55	10	+ 1	2 (– 4)	6(– 6)	2(0)	-	10	10	10	– 10
Total (f)				2	16	11	11	40	$\Sigma fdy = -20$	$\Sigma fdy^2 = 60$	$\Sigma fdxdy = -18$
fdx				– 4	– 16	0	11	$\Sigma fdx = -9$			
fdx^2				8	16	0	11	$\Sigma fdx^2 = 35$			
fdxdy				– 4	1	0	– 15	$\Sigma fdxdy = -18$			

$\therefore$ $\Sigma f = 40$ $\Sigma fdx = -9$, $\Sigma fdx^2 = 35$, $\Sigma fdxdy = -18$

$\Sigma fdy = -20$, $\Sigma fdy^2 = 60$, $cx = 50$, $cy = 10$

Now

$$b_{xy} = \left\{ \frac{\Sigma fdxdy - \dfrac{(\Sigma fdx)(\Sigma fdy)}{N}}{\Sigma fdy^2 - \dfrac{(\Sigma fdy)^2}{N}} \right\} \times \frac{c_x}{c_y}$$

$$= \left\{ \frac{-18 - \dfrac{1}{40}(-9)(-20)}{60 - \dfrac{(-20)^2}{40}} \right\} \left(\frac{50}{10} \right)$$

$$= \frac{(-18 - 4.5)}{(60 - 10)}(5) = \frac{-22.5}{50} \times 5 = \frac{-22.5}{10}$$

$$= -2.25$$

$$b_{yx} = \left\{ \frac{\Sigma fdxdy - \dfrac{(\Sigma fdx) \times (\Sigma fdy)}{N}}{\Sigma fdy^2 - \dfrac{(\Sigma fdy)^2}{N}} \right\} \times \frac{c_y}{c_x}$$

$$= \frac{(-18 - \dfrac{(-9)(-20)}{40}}{\left(35 - \dfrac{(-9)^2}{40}\right)} \times \frac{10}{50} = \frac{(-18 - 4.5)}{\left(35 - \dfrac{81}{40}\right)} \times \frac{1}{5}$$

$$= \frac{-22.5}{\left(35 - \dfrac{81}{40}\right)} \times \frac{1}{5}$$

$$= \frac{-4.5}{35 - 2.025} \times \frac{-4.5}{32 - 975} = -\ 0.136$$

$\therefore$ The regression Coefficients are -2.25, -0.136

Solved Examples

1. Determine the correlation coefficient between x and y of the following data

x	– 3	– 2	– 1	1	2	3
y	9	4	1	1	4	9

Solution:

The correlation coefficient between x and y is

$$r = \frac{\Sigma xy - \frac{1}{n}\Sigma x \Sigma y}{\sqrt{\Sigma x^2 - \frac{(\Sigma x)^2}{n}}\sqrt{\Sigma y^2 - \frac{(\Sigma y)^2}{n}}}$$

x	y	xy	x^2	y^2
– 3	9	– 27	9	81
– 2	4	8	4	16
– 1	1	-1	1	1
1	1	1	1	1
2	4	8	4	16
3	9	27	9	81
$\Sigma x=0$	$\Sigma y=28$	$\Sigma xy=0$	$\Sigma x^2=28$	$\Sigma y^2=196$

$$\therefore \quad r = \frac{0 - \frac{1}{6}(0)(28)}{\sqrt{28-0}\sqrt{196 - \frac{(28)^2}{6}}} = 0$$

$\therefore \quad r = 0$

$\therefore$ The correlation coefficient is zero

2. The following is the data on head and body weights of 10 insects

Head weight (mg)	18	20	25	28	19	27	30	34	36	33
Body weight (mg)	52	54	60	62	59	63	65	70	78	77

Compute the correlation coefficient between head & body weights of insects and test its significance

Solution:

Head weight (x)	Body weight (y)	x^2	y^2	xy
18	52	324	2704	936
20	54	400	2916	1080
25	60	625	3600	1500
28	62	784	3844	1736
19	59	361	3481	1121
27	63	729	3969	1701
30	65	900	4225	1950
34	70	1156	4900	2380
36	78	1296	6084	2808
33	77	1089	5929	2541
Σx=270	Σy=640	$\Sigma x^2=7664$	$\Sigma y^2=41652$	Σxy=17753

$$\text{Coefficient of correlation} = \frac{\Sigma xy - \frac{1}{n}\Sigma x\Sigma y}{\sqrt{\Sigma x^2 - \frac{(\Sigma x)^2}{n}}\sqrt{\Sigma y^2 - \frac{(\Sigma y)^2}{n}}}$$

$$= \frac{17753 - \frac{1}{10}(270)(640)}{\sqrt{\left(7664 - \frac{(270)^2}{10}\right)}\sqrt{\left(41652 - \frac{(640)^2}{10}\right)}}$$

$$= \frac{17753 - 270 \times 640}{\sqrt{(374)(692)}} = \frac{473}{\sqrt{258808}} = \frac{473}{508.7317} = 0.9297$$

3. From the following data obtain two regression equations

X	6	2	10	4	8
Y	9	11	5	8	7

Solution:

To find the regression equation of x on y

Where $$b = \frac{\Sigma xy - \frac{1}{n}\Sigma x\Sigma y}{\Sigma y^2 - \frac{(\Sigma y)^2}{n}}$$

and $\bar{x} = a + b\bar{y}$

$\Rightarrow$ $a = \bar{x} - b\bar{y}$

X	y	x^2	y^2	xy
6	9	36	81	54
2	11	4	121	22
10	5	100	25	50
4	8	16	64	32
8	7	64	49	56
$\Sigma x = 30$	$\Sigma y = 40$	$\Sigma x^2 = 220$	$\Sigma y^2 = 340$	$\Sigma xy = 214$

$$\therefore \quad b = \frac{\Sigma xy - \frac{1}{n}\Sigma x \Sigma y}{\Sigma y^2 - \frac{(\Sigma y)^2}{n}}$$

$$b = \frac{214 - \frac{1}{5}.(30)(40)}{340 - \frac{1}{5}(40)^2}$$

$$= \frac{214 - 240}{340 - \frac{1600}{5}} = \frac{-26}{340 - 320} = \frac{-26}{20}$$

$$= \frac{-2.6}{2} = -1.3$$

Here n = number of observations = 5

$$\bar{x} = \frac{\Sigma x}{n} = \frac{30}{5} = 6$$

$$\bar{y} = \frac{\Sigma y}{n} = \frac{40}{5} = 8$$

$\therefore$ a = $\bar{x} - b\bar{y}$

= 6 – (– 1.3) (8)

= 6 + 10.4

= 16.4

∴ Required equation is

x = a + by

i.e x = 16.4 + (– 1.3) y i.e x = 16.4 – 1.3y

To find the regression equation of y on x suppose the required equation is

Y = a + b x

where $$b = \frac{\Sigma xy - \frac{1}{n}\Sigma x \Sigma y}{\Sigma x^2 - \frac{(\Sigma x)^2}{n}}$$

$$= \frac{214 - \frac{1}{5} \times 30 \times 40}{220 - \frac{1}{5} \times (30)^2} = \frac{214 - 240}{220 - 180} = \frac{-26}{40}$$

$$= \frac{-13}{20} = \frac{-1.3}{2} = -0.65$$

and $\bar{y} = a + b\bar{x}$

⇒ a = $\bar{y} - b\bar{x}$ = 8 – (– 0.65) (6)

= 8+3.90 = 11.90

y = 11.90 + (– 0.65) x

4. The following data was collected from patients with respect to their age (x) and number of days of convalescence (y) in fever hospital

Age	64	56	50	48	46	35	38	31
Days	18	12	10	9	6	5	8	4

Fit the regression equation of number of days of convalescence and age of patient and test the regression coefficient at probability level 0.05 and estimate the days of convalescence when the age is 60.

Solution:

Given data is

Age (x)	**Days (y)**	**xy**	**x^2**	**y^2**
64	18	1152	4096	324
56	12	672	3136	144
50	10	500	2500	100
48	9	432	2304	81
46	6	276	2116	36
35	5	175	1225	25
38	8	304	1444	64
31	4	124	961	16
$\Sigma x = 368$	$\Sigma y = 72$	$\Sigma xy = 3635$	$\Sigma x^2 = 17782$	$\Sigma y^2 = 790$

$$y = a + bx$$

where $$b = \frac{\Sigma xy - \frac{1}{n}\Sigma x \Sigma y}{\Sigma x^2 - \frac{(\Sigma x)^2}{n}}$$

$$= \frac{3635 - \frac{1}{8}(368)(72)}{17782 - \frac{(368)^2}{8}}$$

$$= \frac{3635 - 3312}{17782 - 16928} = \frac{323}{854}$$

$$= 0.37822014$$

$$= 0.378$$

and $$\bar{y} = a + b\bar{x}$$

$\Rightarrow$ $$a = \bar{y} - b\bar{x}$$

$$\bar{y} = \frac{\Sigma y}{n} = \frac{640}{8} = 80$$

$$\bar{x} = \frac{\Sigma x}{n} = \frac{270}{8} = 33.75$$

$\therefore$ $$a = 80 - (0.378)(33.75)$$

$$= 80 - 12.7575 = 67.2425$$

∴ Regression equation is

$$y = a+bx$$

i.e $y = (67.2425) + (0.378)x$

Exercise I

1. Calculate the correlation coefficient from the following data.

x : 12	9	8	10	11	13	7
y : 14	8	6	9	11	12	3

Ans. r = 0.95

2. Ten students got the following percentage of marks in mathematics and statistics.

Marks in Maths	39	65	62	90	82	75	25	98	36	78
Marks in Statistics	47	53	58	86	62	68	60	91	51	84

Calculate the coefficient of correlation

Ans. 0.78

3. Calculate the coefficient of correlation between x and y variable from the following data.

x	50	55	65	64	60	63	62	61	65	70
y	110	112	117	120	110	112	113	112	121	125

Ans. r = 0.782, t = 3.55

The calculated t value is higher than the tabled value for both 5% (2.31) and 1% (3.36) level of probability. So, the coefficient of correlation is highly significant between the variables x and y.

4. Find the coefficient of correlation in the following data.

x	8	10	15	17	20	22	24	25
y	25	30	32	35	37	40	42	45

Ans. 0.983

Exercise II

1. Find out a regression line y on x from the data given below.

x	1	2	3	4	5	6	7	8	9
y	1	3	2	5	5	4	7	6	9

Ans. y = 0.67 x + 0.67

2. Find the regression equation of following data.

x	1	2	3	4	5	6	7
y	2	4	7	6	5	6	3

Ans. x = 0.69 y + 0.54; y = 0.39 x + 3.46

3. Two random variables have the regression lines with equations 3x + 2y = 26, and 6x + y = 31. Find the mean values and correlation coefficient between x and y.

Ans. $\overline{x} = 4, \ \overline{y} = 13, \ r = 0.43$

4. In a laboratory experiment on correlation research study the equation to the two regression lines were found to be 2x – y + 1 = 0 and 3x – 2y + 7 = 0. Find the mean of x and y also workout the values of regression coefficient of correlation between the two variables x and y.

Ans. $\overline{x} = 5, \ \overline{y} = 11, \ r = 0.866$

5. Fit a straight line to the following data by the method of least squares.

x	0	1	2	3	4
y	1	1.8	3.3	4.5	6.3

Ans. y = 11.96 – 4.29 x

6. Calculate the correlation coefficient and represent the regression equation from the following data. Determine the concentration corresponding to the absorbance 0.245.

Concentration (µg/ml)	Absorbance
2	0.103
4	0.209
6	0.315
8	0.424
10	0.535

Ans. r = 0.999, regression equation y = 0.054 × –0.007 where y = absorbance, x = concentration, the concentration corresponding to the absorbance 0.245 = 4.55 µg/ml

7. Determine the correlation coefficients for zero & first order equations from the following data. Report the amount of drug dissolved from the dosage form at 7th hour based on the best fitted regression equation. The strength of the dosage form is 100 mg.

Time (hr)	Amount of drug dissolved (mg)
1	9.2
2	17.16
4	35.72
6	56.38
8	72.92
10	99.84

Hint: Zero order, amount of drug dissolved vs., first order time log % drug un dissolved vs. time.

Ans. r = 0.997 for zero order, regression equation y=9.89x – 2.57 where y = amount of drug dissolved, x = time. r = – 0.801 for first order. Amount of drug dissolved at 7th hour according to the best fitted line (zero order) = 66.66mg

8. Report the correlation coefficients of Higuchi & Peppas plots from the following data.

Time (hr)	Amount of drug dissolved (mg)
1	7.4
2	15.8
4	27.5
6	40.92
8	52.49
10	65.34

Hint: Higuchi, amount of drug dissolved vs. square root of time. Peppas, log percent drug dissolved vs. log time.

Ans. r = 0.996 for Higuchi, r = 0.999 for Peppas

CHAPTER 6

TRANSFORMATIONS AND OUTLIERS

Most of the cases, the statistical data are assumed to be normally distributed with equal variance. The statistical inference with non equal variance (heteroscedascity) is more problematic in t tests, ANOVA and regression. The variance heterogeneity and/or lack of normality can be corrected by a data transformation.

Observations of unusual results or getting the results that are violating the common assumptions are called outliers. During the statistical treatment of data, the data generated due to the errors or mistakes should be omitted to get better statistical interpretation. These unusual values may or may not explainable. The prudent experimenter will make every effort to find the cause for errors and make attempts to modify the data or analysis. If the cause is not detected, then the scientific judgment with regard to the discrepancy of these results is required. The observation that is extreme and not representing to the bulk of data is referred as outlier and it is detected with an appropriate statistical test.

6.1 Transformations

Each data point is transformed to other value by applying log, square root, reciprocal etc. such changes are made prior to statistical analysis. Such transformations are made in following cases.

1. The data is not normally distributed.
2. The variance is not constant within treatment groups.
3. The linear relationship exists only with the data transformation.

6.1.1 The Logarithmic Transformation

The most commonly used transformation in pharmaceutical sciences is the log transformation. Either logs to the base 10 ($\log_{10}$) or the base e ($\log_e$/ln) can be used. These transformations are used when the standard deviation is proportional to mean value.

Applications

1. Particle size analysis of powder

The particle size distribution of the pharmaceutical powder samples is not following the normal distribution.

The following features are observed in normal distribution.

(a) The distribution is symmetric around the mean (equally distributed around the mean)

(b) The mean and mode are equal

(c) 68% of the population lie $\pm$ 1σ from the mean

(d) 95% of the population lie $\pm$ 3σ from the mean

(e) 99.7% of the population lie $\pm$ 3σ from the mean

Example 1: The particle size distribution data observed from a pharmaceutical powder sample is given in table 6.1. The data is treated statistically with normal distribution and log normal distribution (log transformation of the data) to assess the distribution pattern of the sample

Table 6.1 Particle size distribution data of the powder sample

Size range	Mean size	Log mean size	Number of particles	Percent frequency
2-4	3	0.4771	3	1
4-6	5	0.6989	48	16
6-8	7	0.8450	96	32
8-10	9	0.9542	72	24
10-12	11	1.0413	45	15
12-14	13	1.1139	21	7
14-16	15	1.1760	9	3
16-18	17	1.2304	4	1.3
18-20	19	1.2787	2	0.6
			300	

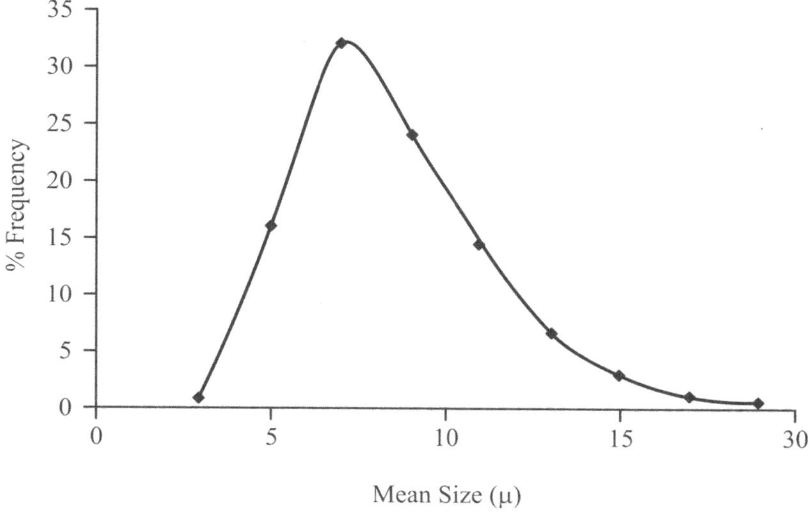

Fig. 6.1 Particle Size Data before Transformation.

The above data is represented as normal distribution (mean size vs % frequency, Fig 6.1) and log normal distribution (% frequency vs log mean size, Fig 6.2). A typical bell shaped curve is observed in log normal distribution. The above example clearly indicating that the log transformation of mean particle size is well essential to describe the data more appropriately.

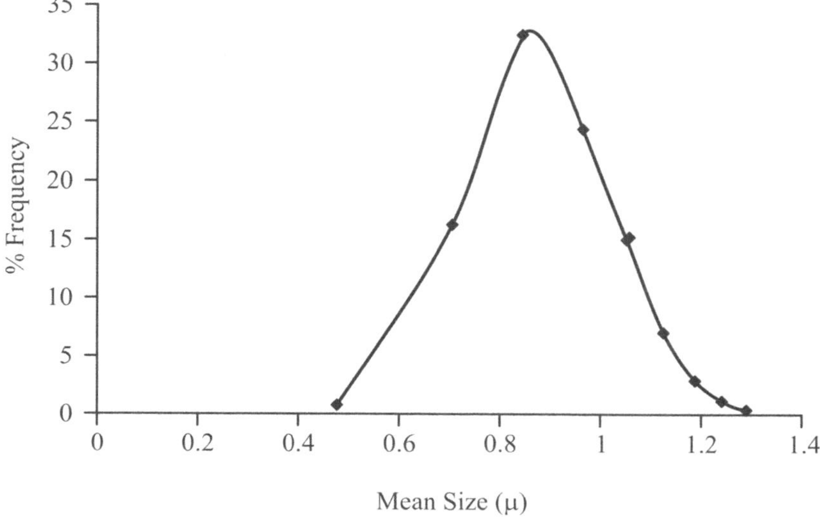

Fig. 6.2 Particle Size Data after Transformation

2. Interpretation of rheological data

The logarithmic plots of shear stress and shear rate relationships are used to illustrate the power law and calculation of its parameters.

Power law equation, $S = A\,D^n$

Logarithm of both sides of equation results

$\text{Log } S = \log A + n \log D$

where S = Shear stress
D = Shear rate
A= Appropriate proportionality constant and n = power index.

For pseudo plastic materials, n is less than 1 and greater than 1for dilatants materials. A plot of log S against log D results in a straight line of slope η and intercept log A. The logarithmic plots of stress and shear rate is convenient to provide an overall summary of data observed at various shear rates. In conventional graphs, low shear rate is excessively compacted in presence of high shear rates. These log-log plots are the most useful plots to observe both low and high shear rate within one graph.

Example 2: The shear rate and shear stress observed from a fluid is given below. Report its flow behavior.

Shear Rate (D) (sec^{-1})	**Shear Stress (S) ($dyne/cm^2$)**
13.20	13.20
26.40	26.38
39.60	39.62
52.80	52.79
66.00	65.98

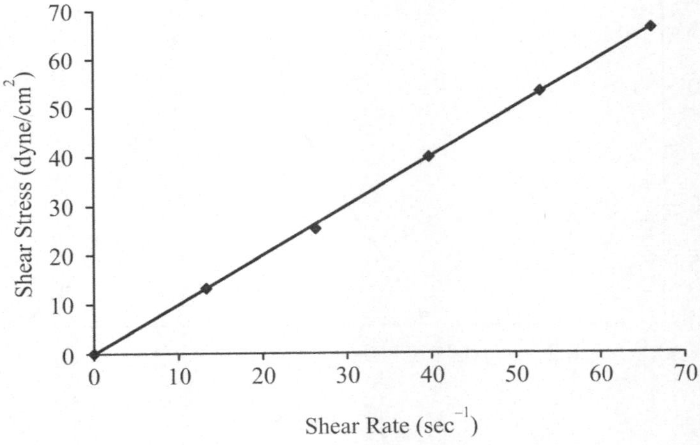

Fig. 6.3 Rheogram of a Newtonian fluid

The plot of shear rate against shear stress is linear (Fig 6.3) and passing through the origin (intercept 0). Hence it is Newtonian flow.

Example 3: The shear rate and shear stress observed from a fluid is given below. Report its flow behavior.

Shear Rate (D) (sec^{-1})	Shear Stress (S) ($dyne/cm^2$)
13.20	8.00
26.40	9.30
39.60	10.45
52.80	11.60

The plot of shear rate against shear stress is linear (Fig 6.4) after reaching to its yield value (y-intercept). Hence it is plastic flow.

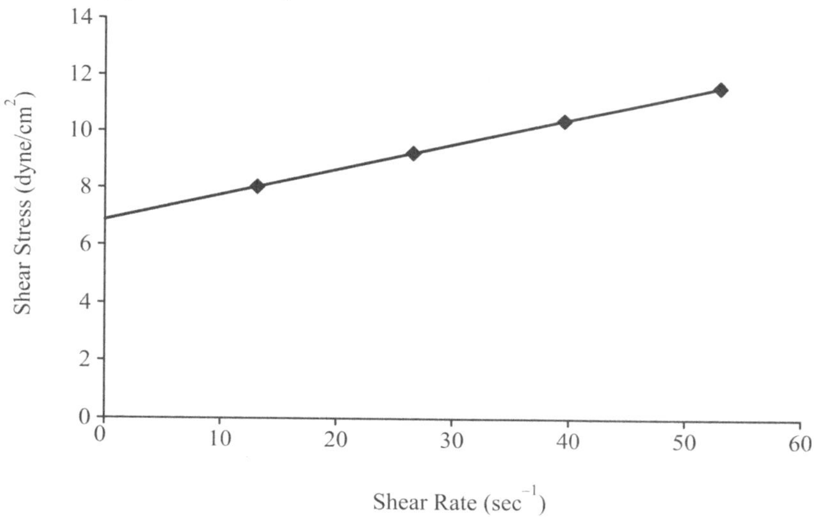

Fig. 6.4 Rheogram indicating Plastic Flow

Example 4: The observed shear stress from a fluid exposed to different shear rate is depicted in the following table. Interpret the data.

Shear rate (D) (sec^{-1})	Shear stress (S) ($dyne/cm^2$)
13.20	13.78
26.40	22.29
39.60	25.66
52.80	27.52
66.00	31.79

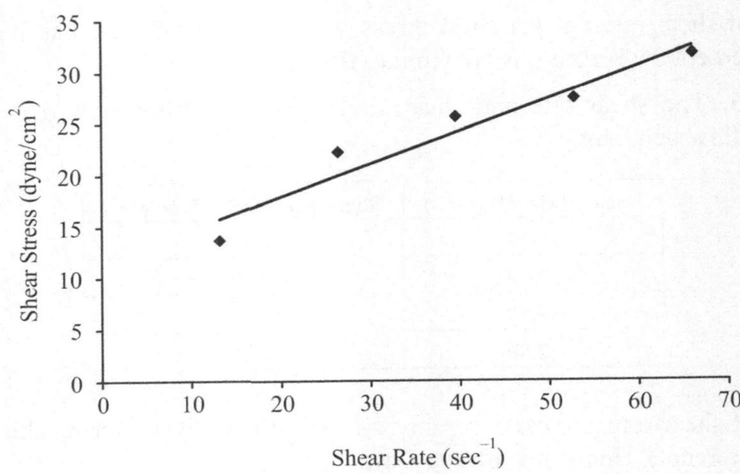

Fig 6.5 Rheogram before Log Transformation

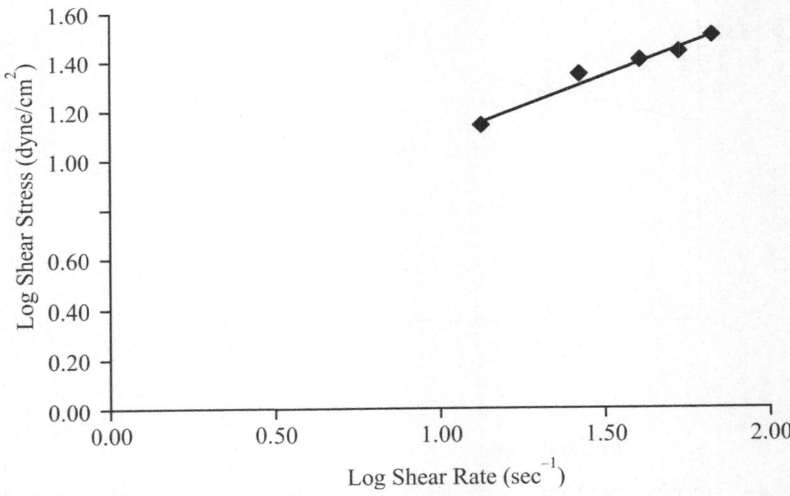

Fig 6.6 Rheogram after Log Transformation

In Fig. 6.5 and 6.6, more linearity is observed in the Fig. 6.6 and thus the data indicates that the log transformation is essential. The values of shear rate and shear stress are best fitted in the power law and the power index value is 0.5229 i.e., less than 1. Hence it is pseudo plastic flow.

Example 5: The shear rate and shear stress observed from a fluid is given below. Report its flow behavior.

Shear rate (D) (sec^{-1})	Shear stress (S) ($dyne/cm^2$)
13.20	0.781
26.40	1.266
39.60	1.362
52.80	1.502
66.00	1.754

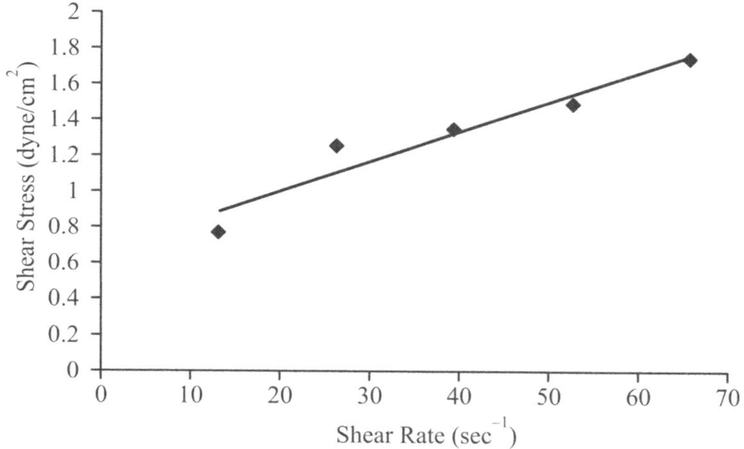

Fig 6.7 Rheogram before Log Transformation

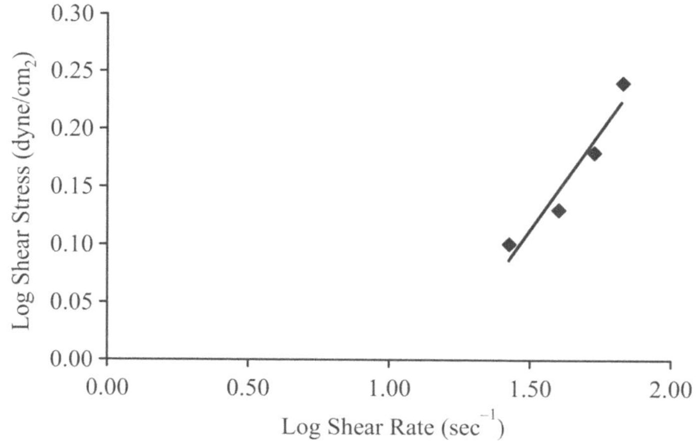

Fig 6.8 Rheogram after Log Transformation

In Fig. 6.7 and 6.8, more linearity is observed in the Fig. 6.8 and thus the data indicates that the log transformation is essential The values of shear rate and shear stress are best fitted in the power law and the power index value is1.278 i.e., more than 1. Hence it is dilatant flow.

3. Determination of molecular weight of a polymer

Mark-Houwink equation $[\eta] = K\ M^a$ is often used to determine the molecular weight of a polymer.

$$\log [\eta] = \log K + a \log M$$

where η = Intrinsic viscosity/limiting viscosity number

M = Molecular weight of the polymer

a and K are constants for a given polymer solvent combination at a given temperature.

The relationship existing between the molecular weight of a polymer and intrinsic viscosity is shown in the Fig 6.9.

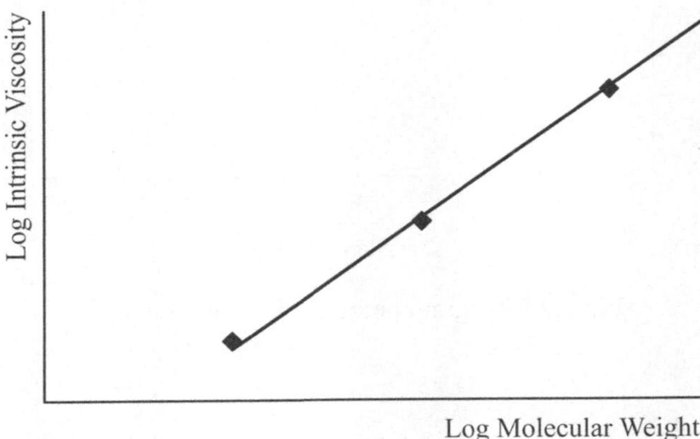

Fig 6.9 Mark-Houwink Plot

4. Determination and decomposition rate constant

The drugs and the corresponding dosage forms undergo decomposition either by zero order or with first order. The plots before transmitting in to logarithmic form are shown in Fig 6.10. More often the active pharmaceutical ingredient of a pharmaceutical dosage form undergo first order decomposition i.e., concentration dependent. To get linearity and to determine the decomposition rate constant, a graph is plotted in between time and log % drug undecomposed (Fig 6.11). The first order process is usually expressed in logarithmic form

$$\ln c = \ln c_0 - kt \quad \text{or} \quad \log c = \log c_0 - \frac{kt}{2.303}$$

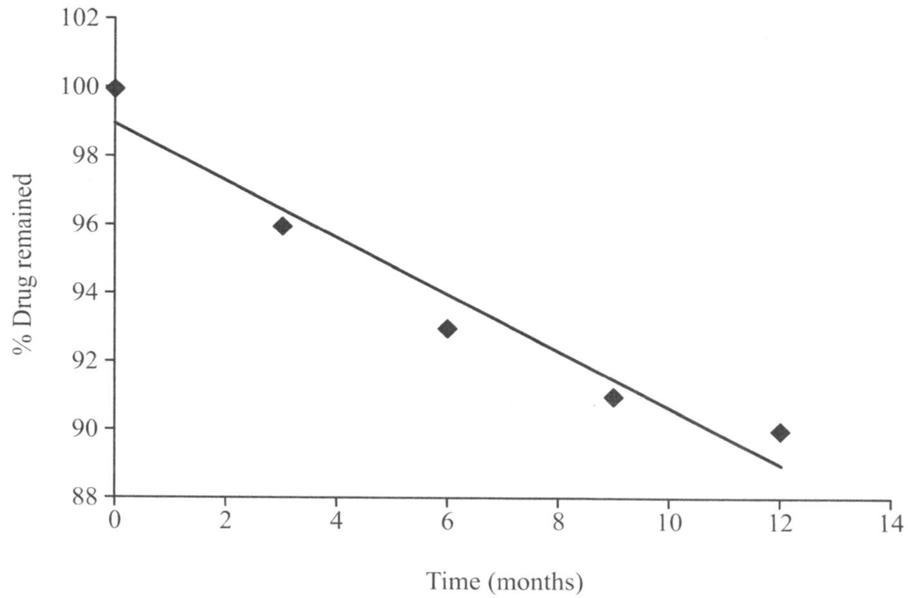

Fig: 6.10 Decomposition Profile before Log Transformation

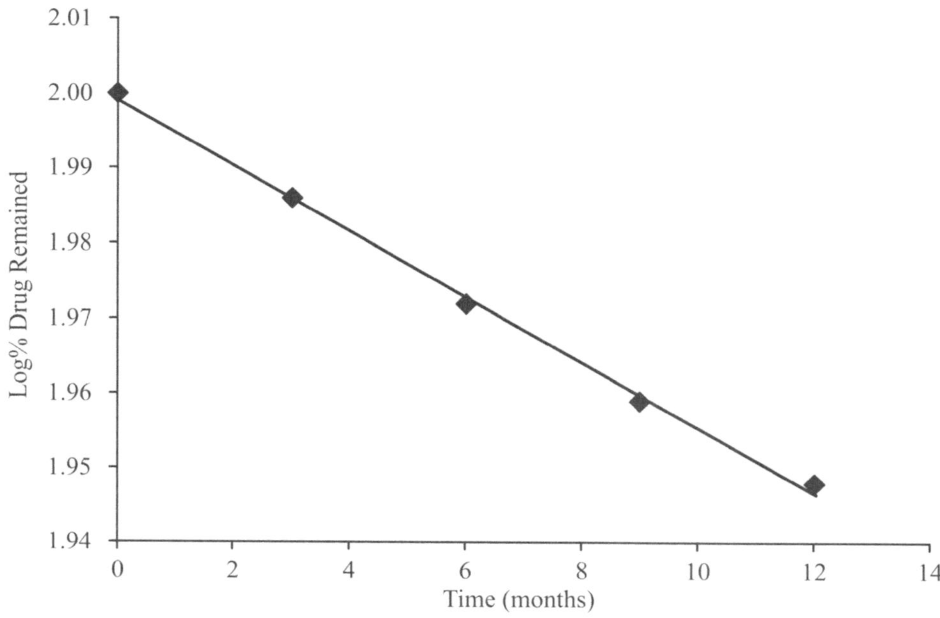

Fig 6.11 Decomposition Profile after Log Transformation

5. Determination of dissolution rate constant

The drug release from the conventional dosage forms such as tablets, capsules usually follows first order kinetics and the dissolution rate is computed from the plot drawn in between time and log % drug undissolved (Fig 6.12).

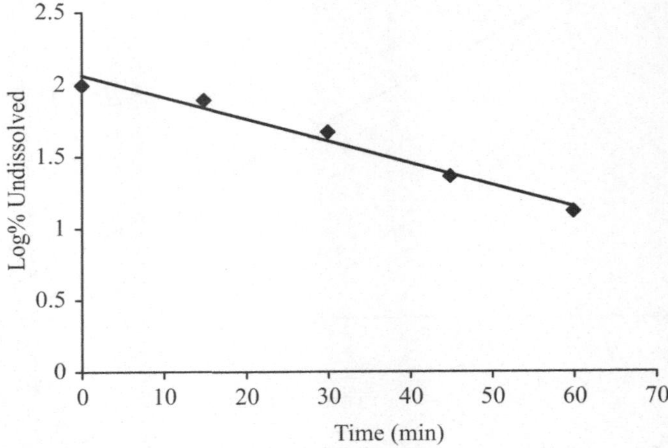

Fig 6.12 First Order Plot for Dissolution

6. Identification of mechanism of drug release

To identify the mechanism of drug release from a controlled drug delivery system, koresmeyer peppas plots are constructed. These plots are constructed among the log percent drug release and log time (Fig 6.13). The slope of the regression line is employed to identify the mechanism of drug release.

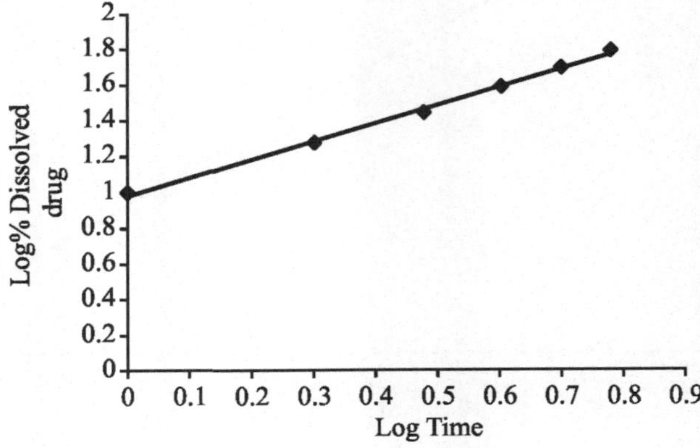

Fig 6.13 Peppas Plots

7. Analysis of residuals

Prior to performing transformations or statistical tests, the observed data/graph is examined by exploratory data analysis (EDA). A significant aspect of EDA is examination of residuals. Residuals are the deviations of observed data from the list to the statistical model. It is usually employed to determine absorption rate constant from bioavailability studies.

Method of Residuals:

The method of residuals (also known as feathering or peeling) is a useful procedure for fitting a curve to the experimental data of a drug when the drug does not clearly follow a one-compartment model.

Example 6: 100 mg of a drug was administered by rapid IV injection to a 70 kg, healthy adult male. Blood samples were taken periodically after the administration of drug, and the plasma fraction of each sample was assayed for drug. The following data were obtained.

Time (hr)	Plasma conc. (μg/ml)
0.25	44.00
0.5	33.00
1.0	19.00
1.5	14.00
2.0	10.00
4.0	6.50
8.0	2.80
12.0	1.20
16.0	0.52

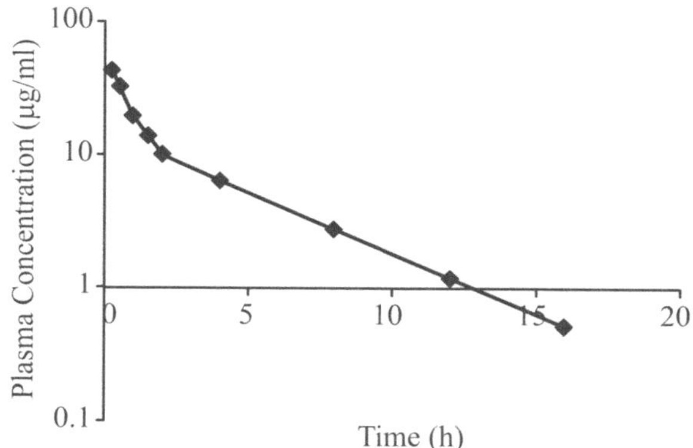

Fig 6.14 Semi log Plots of Plasma Concentration Profiles.

When these data are plotted on semi logarithmic graph paper, a curved line is observed (Fig. 6.14). The curved-line relationship between the logarithm of the plasma concentration and time indicates that the drug is distributed in more than one compartment.

8. To equalize the variance

To show the linearity over a range of drug concentrations, the experimental data is usually collected at different levels. More often the standard deviation of the data increased with the mean value of the estimated drug concentration. Such alterations are not useful for statistical inference and to calculate the correlation coefficient in between known drug concentration (x-axis) and mean assay (y-axis). In such cases, to minimize the variance and to get good linearity the data on both sides is transformed logarithmically and plotted the graph in between log mean drug assay and log known drug concentration (Fig 6.15). Low standard deviations are observed from these graphs compared to the graphs drawn before log transformation (Fig 6.16).

Example 7: The results of the drug content at three different levels.

25 mg		50 mg		75 mg	
Assay	**Log Assay**	**Assay**	**Log Assay**	**Assay**	**Log Assay**
24	1.3802	54	1.7323	80	1.9030
26	1.4149	56	1.7481	74	1.8692
28	1.4471	58	1.7634	82	1.9138
30	1.4771	46	1.6627	72	1.8573
25	1.3979	55	1.7403	84	1.9242
27	1.4313	47	1.6720	68	1.8325

Average:	26.66	1.4247	52.66	1.7198	76.66	1.8833
S.d.:	2.160	0.0348	4.966	0.042	6.282	0.0359

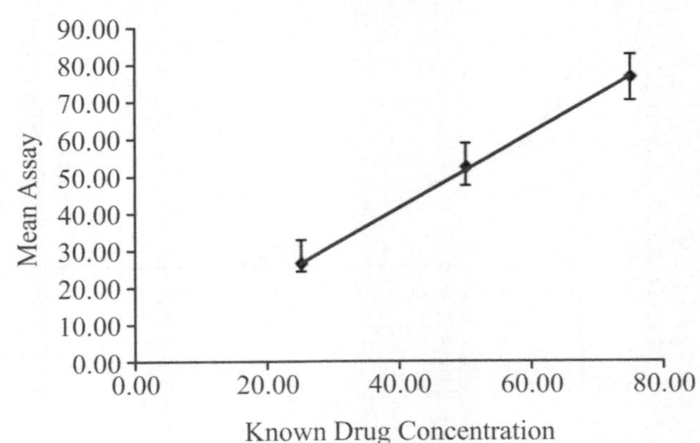

Fig 6.15 Means of Assay Values

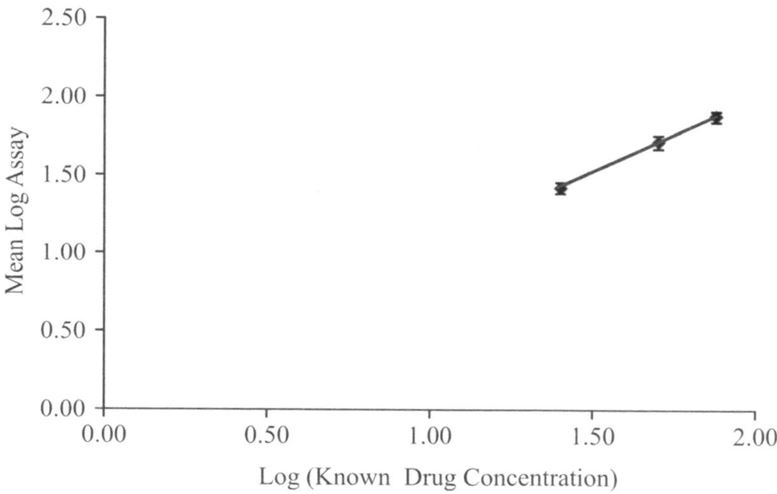

Fig 6.16 Log Transformed Means of Assay Values

Problems during the usage of Variance

Interpretation of data reported in logarithmic form is problematic and more often the formal report of the experimental result need back transformation (anti log).

Example 8: The rate constant for the decomposition of drug at 50 °C is 1.26 hr $^{-1}$ and 60 °C is 3.84 hr $^{-1}$. What is the activation energy and frequency factor?

E_a may be obtained from the following equation.

$$\log\frac{k_2}{k_1} = \left(\frac{E_a}{2.303R}\right)\times\frac{(T_2 - T_1)}{T_2T_1}$$

$$\log\frac{3.84}{1.26} = \left(\frac{E_a}{2.303\times 1.97}\right)\times\left[\frac{(333-323)}{333\times 323}\right]$$

E_a = 23613.61848 k cal/mole

At 50°C, the frequency factor A is calculated from the following equation.

$$\log\ K = \log\ A - \left[\left(\frac{E_a}{2.303}\right)\times\left(\frac{1}{RT}\right)\right]$$

$$\log\ 1.26 = \log\ A - \left[\left(\frac{23613.61848}{2.303\times 1.987}\right)\times\frac{1}{323}\right]$$

A = 1.40928 × 10^{16} sec $^{-1}$

The second problem is log transformation of data changes the variance. In case of small variance of original data, the variance of ln transformed values is having a variance equal to the coefficient of variation of original data.

Example 9: Calculate the mean and standard deviation before and after getting log transformation

X	ln X
80	4.3820
84	4.4308
76	4.3307
72	4.2766
75	4.3174
82	4.4067

Mean	:	78.16	Mean	:	4.357
Standard deviation	:	4.578	Standard deviation	:	0.0587
C.V(s.d /mean)	:	0.0585	C.V(s.d /mean)	:	0.0134

In the above example, the coefficient of variation of the original data is equal to the standard deviation of the transformed data.

6.1.2 Square Root Transformation

The square root transformation is useful in cases where the variance (s.d^2) is proportional to mean. To get equal variance, the parameters like blood cell content are subjected to square root transformation to get equal variance.

Example 10: The RBC count observed from 6 subjects during 2 weeks is given below.

Subject	RBC count			
	1st week		2nd week	
	Normal	Square root transformation	Normal	Square root transformation
1	4.5	2.12	4.8	2.19
2	5.4	2.32	5.2	2.28
3	5.6	2.36	5.9	2.42
4	5.8	2.40	6.4	2.52
5	6.2	2.48	5.9	2.42
6	6.4	2.52	6.6	2.56

Mean :	5.65	2.366	5.8	2.4
Standard deviation :	0.674	0.141	0.689	0.14

In the above example, the variance is equal after square root transforming of the data. Square root transformations are useful for construction of Higuchi plots. These plots are constructed in between amount of drug release and square root of time.

Example 11: The drug release data observed from a controlled release formulation is shown in the following table. The data is represented as such (Fig 6.17) and after transforming in to square root (Fig 6.18). More linearity is observed after transforming in to square root.

Time (h)	$\sqrt{\text{Time}}$	Amount of drug release (mg)
5	2.236	40
10	3.162	56.2
15	3.872	67.8
20	4.472	79.3
25	5.000	88.6

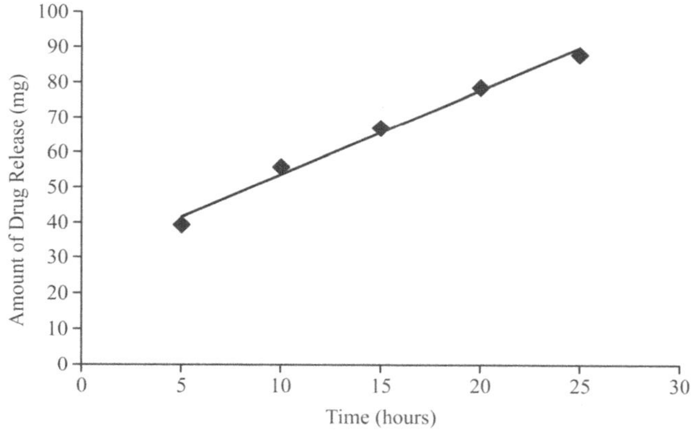

Fig. 6.17 Dissolution Profiles before Square Root Transformation

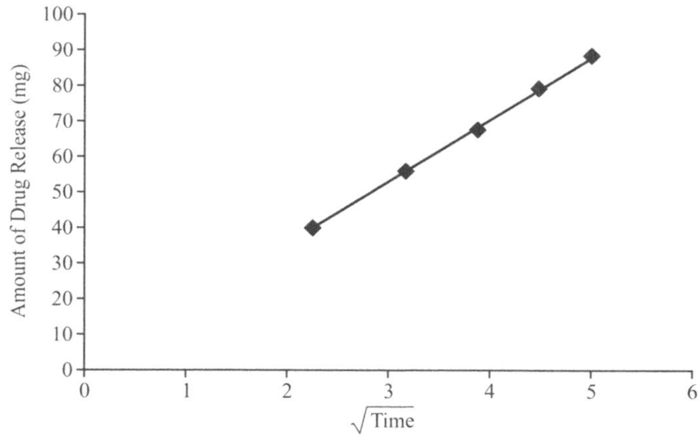

Fig 6.18 Dissolution Profiles after Square Root Transformation (Higuchi Plots)

Casson and the Fitch plots are also required square root transformation. These plots are used to visualize the yield value. In Fitch plot (Fig. 6.19) the stress is plotted against the square root of shear rate and in Casson plot (Fig. 6.20), the square root of stress is plotted against square root of shear rate.

Example 12: Rheological data observed from a fluid

Shear rate (cm^{-1})	**$\sqrt{\text{Shear rate}}$ $\left(\sqrt{cm^{-1}}\right)$**	**Stress (dyne/cm^2)**	**$\sqrt{\text{Stress}}$ $\left(\sqrt{dyne/cm^2}\right)$**
1	1	200	14.142
4	2	400	20
9	3	600	24.49
16	4	800	28.28
25	5	1000	31.62

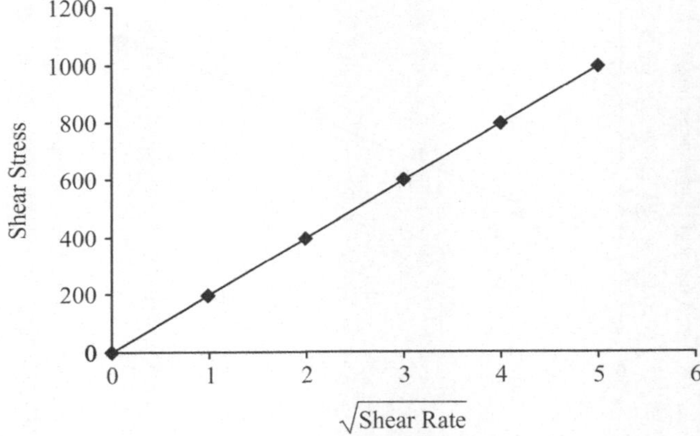

Fig. 6.19 Fitch Plot

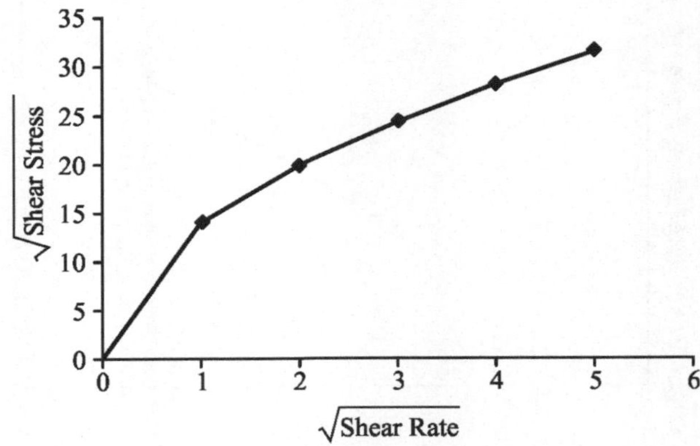

Fig. 6.20 Casson Plot

6.1.3 Reciprocal Transformations

Transformations are used when the standard deviation is proportional to the square of mean. Such transformations are more often used in chemical kinetics. The log decomposition rate constant is proportional to absolute temperature according to Arrhenius equation. To determine the decomposition rate constant at room temperature and to calculate shelf life of a dosage form these plots are useful (Fig 6.21).

Example 13: The decomposition rate constant of a drug observed at different temperatures is given below. Calculate K_{25}.

Temperature (°C)	K
40	0.00011
50	0.00026
60	0.00078
70	0.0016

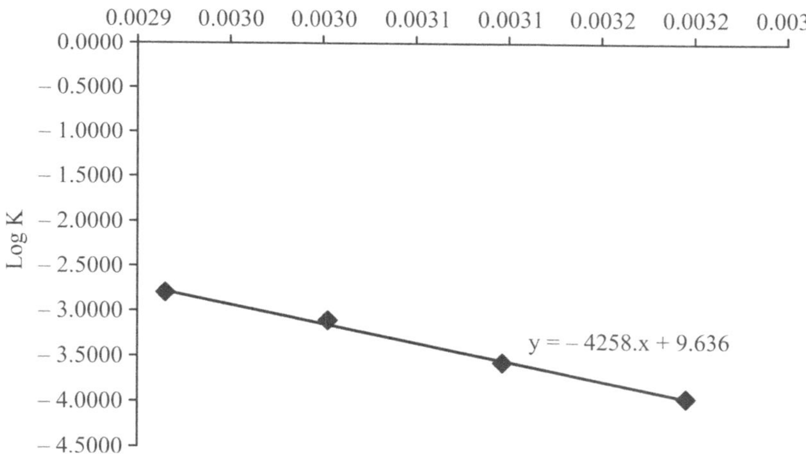

Fig 6.21 Arrhenius Plot

The k value at 25 °C is calculated from the above regression equation

$$\text{Log } k = (-4258 \times 0.003356) + 9.636 = -4.65259$$

$$k = 2.2254 \times 10^{-5}$$

6.1.4 The Arcsin Transformations for Proportions

The Arcsin is the inverse sine function ($\sin^{-1}$). When the data contains proportions that are not having homogeneity of variance, this transformation is used. If the proportions under consideration are varying from one observation to another, the variance of proportions pq/N is also varied. To apply Arcsin transformation, each proportion should be based on same number of observations, N. If the proportion is very small (<0.03) or very greater (>0.97) these transformations are not preferred. Arcsin transformation = Arcsin. $\sqrt{p} = \sin^{-1}\sqrt{p}$ and back transformation as sin (arcsin)^2.

Example 14: The incidence of the adverse affect observed from the 25 subjects with the administration of a new drug during a clinical study is given below.

Time	Numbers of subjects showing adverse reaction	Proportion of subjects showing adverse reaction	Arcsin P
1	2	0.08	4.588
2	6	0.24	13.88
3	4	0.16	9.206
4	8	0.32	18.66
5	6	0.24	13.88
6	3	0.12	6.892

6.2 Outliers

Outliers refer to relatively small or large values that are different from the main data. Considerable judgment should be exercised before discarding the suspected values. If the cause of an outlier is known, it is advised to include the reason(s) for the omission of the aberrant value in experimental results. In most cases, extreme values are observed without obvious causes. Then it is difficult to identify the cause, as it may occur either due to error of observation or it may occur routinely in experiments. The presence of an outlier or omission of an outlier from the data will change the validity of interpretation and conclusion of experimental results. If the reason for the outlier is not known, then it is handled with suitable statistical criteria. Statistical tests for the presence of outliers are based on the assumption that the data have a normal distribution. If the suspected outlier is not having any possible occurrence in an experiment, then it can be omitted without any statistical test. For example, the first order decomposition rate constants observed from six formulations are 0.104, 0.106, 0.108, 0.107, 0.105, and 1.03. In this example, the 6th value is considered as outlier and it can be discarded without confusion. As there is no scope for getting a first order rate constant values exceeding 1.

Statistical approaches to make a decision about outlier.

The following approaches are useful to make a better decision about outlier.

1. Dixon's test

Dixon criteria for testing an extreme mean can be used to decide whether the suspected extreme value belongs with the main data or not.

Steps involved:

(a) Order the experimental data numerically from low to high

(b) Calculate the ratio of the difference of extreme value from one of the nearest neighboring values to the range of observations.

Example: If the observations are denoted as X_1, X_2, X_3....X_N, for N observations and X_1 is the extreme value, then the r is calculated with the following formulae.

$$r = \frac{X_2 - X_1}{X_N - X_1};$$

(c) The formula for r value depends on sample size which is shown in Table 6.1.

Table 6.2 Dixon's Criteria for Rejecting Outliers

K		Significance	level
		5%	1%
3	$r_{10} = (X_2–X_1)/(X_k–X_1)$ if smallest value is suspected;	0.941	0.988
4		0.765	0.889
5	$=(X_k–X_{k-1})/(X_k–X_1)$ if largest value is suspected	0.642	0.780
6		0.560	0.698
7		0.507	0.637
8	$r_{11} = (X_2–X_1)/(X_{k-1}–X_1)$ if smallest value is suspected;	0.554	0.683
9		0.512	0.635
10	$= (X_k–X_{k-1})/(X_k–X_2)$ if largest value is suspected	0.477	0.597
11	$r_{21} = (X_3–X_1)/(X_{k-1}–X_1)$ if smallest value is suspected;	0.576	0.679
12		0.546	0.642
13	$= (X_k–X_{k-2})/(X_k–X_2)$ if largest value is suspected	0.521	0.615
14	$r_{22} = (X_3–X_1)/(X_{k-2}–X_1)$ if smallest value is suspected;	0.546	0.641
15		0.525	0.616
16	$= (X_k–X_{k-2})/(X_k–X_3)$ if largest value is suspected	0.507	0.595
17		0.490	0.577
18		0.475	0.561
19		0.462	0.547
20		0.450	0.535
21		0.440	0.524
22		0.430	0.514
23		0.421	0.505
24		0.413	0.497
25		0.406	0.489

Table 6.3 Critical Values of T for a Two Sided Test at the 5% Level of Significance (Test of Outliers)

Sample size	T
3	1.155
4	1.481
5	1.715
6	1.887
7	2.020
8	2.126
9	2.215
10	2.290
11	2.355
12	2.412
13	2.462
14	2.507
15	2.549
16	2.585
17	2.620
18	2.651
19	2.681
20	2.709
25	2.822
30	2.908
35	2.979
40	3.036
50	3.128
100	3.383

(d) Compare the calculated r value with the tabulated value (Table 6.2). If the ratio is greater than or equal to tabulated value, the observation is considered to be an outlier.

Example: The hardness of 6 tablets is 4.2, 4.5, 4.7, 4.3, 4.6 and 8.9 kg/cm^2.

$$x_n = 8.9$$

$$\bar{x} = \frac{4.2+4.5+4.7+4.3+4.6+8.9}{6}$$

$$= \frac{31.2}{6} = 5.2$$

$$s = \sqrt{\frac{\Sigma(x-\bar{x})^2}{N-1}}$$

$$s = \sqrt{\frac{(4.2-5.2)^2+(4.5-5.2)^2+(4.7-5.2)^2+(4.3-5.2)^2+(4.6-5.2)^2+(8.9-5.2)^2}{5}}$$

$$= \sqrt{\frac{(-1)^2+(-0.7)^2+(-0.5)^2+(-0.9)^2+(-0.6)^2+(3.7)^2}{5}}$$

$$= \sqrt{\frac{16.6}{5}} = \sqrt{3.32} = 1.82208$$

$$\therefore T = \frac{x_n-\bar{x}}{s} = \frac{8.9-5.2}{1.82208} = \frac{3.7}{1.82208} = 2.03064$$

∴ Calculated T=2.03

Tabled T = 1.887

Calculated T > Tabled T

∴ The suspected value 8.9 is outlier

2. The T procedure(Grubb's test)

The ratio in T procedure is calculated as follows

$$T = \frac{X_n - \overline{X}}{S}$$

where, X_n is the smallest or largest value which is suspected as an outlier; $\overline{X}$ is mean and S is standard deviation.

The calculated T_n value should be compared with table value (from Table 6.3) and to consider the suspected value as outlier, the table value should be less than the calculated value.

WINSORIZING

In this method, the extreme values (both high & low values) are changed to the values of their closed neighbors. The mean of such values is calculated and compared with the mean of the untransformed data. This procedure protects the outlier values and very little information is lost.

Eg: Blood glucose values observed from 12 subjects

Subject	1	2	3	4	5	6	7	8	9	10	11	12
Blood Glucose Level	120	116	118	121	124	114	117	118	140	116	118	120

Calculate the mean before omitting the outlier and after omitting the outlier 140

Winorizing

Mean before omitting the outlier 140

$$= \frac{120+116+118+121+124+114+117+118+140+116+118+120}{12}$$

$$= \frac{1442}{12} = 120.16$$

Mean after omitting the outlier 140

$$= \frac{120+116+118+121+124+114+117+118+116+118+120}{11}$$

$$= \frac{1302}{11} = 118.36$$

Approaches to be followed when the Outliers are rejected:

The following approaches are used when the outliers are rejected by following the statistical analysis.

1. Repeat the experiment to obtain the replaced value.
2. Estimate the missing value by statistical methods.
3. Analyze the data without the discarded value.

CHAPTER 7

SAMPLE CHOOSING

Samples are the units of population representing the best estimation of true parameters of population. Validation of an excellent experiment or investigation is essentially depends on the sample choosing. Statistical inference of experimental results depends on sampling procedure. The samples should be collected systematically with careful thoughts. The sampling procedure may be more or less difficult depending on the experimental situation. Some of the sample collection cases in pharmaceutical or medicinal sciences include the selection of patients for clinical trial, analysis of a sample from a bulk container, collection of a dosage form for quality control, selection of a blend to test for homogeneity etc. Entire population (100% sampling) is never done, as it is costly, time consuming and may yield erroneous observations. If entire pharmaceutical dosage forms are tested, then the manufacturer cannot release the dosage forms into the market. So it is advised to collect appropriate sample and subject to experimental evaluation to get a meaningful interpretation about entire batch.

The samples should be randomly selected *i.e.,* the probability of selection of an individual unit (sample) is having equal (same) chance.

Factors to be considered for selection of a sample

1. *Nature of population*: In some cases the population is easily identified as in case of dosage forms, where as in other cases like diabetic patients it is more difficult to identify and pick up the sample.
2. *Cost of sampling*: The quantity of the sample collected for an experiment may be restricted by its cost.
3. *Time of sampling*: The time required for collection of sample is decided by the objective of an experiment. If the objective is analysis of finished dosage form, then the samples are collected once the dosage forms are completely manufactured. In some cases, the objective may be monitoring the quality of dosage forms produced continuously from a machine, in such cases samples should be collected from the population at regular intervals (time consuming).

4. *Ease of sampling*: The sampling procedure should allow the collection of sample more easily as some times it is not practically possible.
5. *Desired accuracy and precision*: The accuracy and precision of the experimental result depends on the sampling procedure and the sample size.

7.1 Sampling Procedures

1. **Sampling by questionnaire:** The response of a drug can be collected from the patients get administered with that drug by asking them to fill the questionnaires. The results so obtained can be tabulated and interpretive. In this sampling, the results may probably bias because some patients may not return/respond to the questionnaire. The characteristics of responding and non responding patients are different. Some cases the patient may busy to answer and some cases the patients may feel little or no benefit with the drug. The other reasons for bias include the way in which the question is asked, the order in which questions are asked and the psychological interaction between the interviewer and respondent.
2. **Random sampling:** The results of biological experiments are more often depends on animal, laboratories and technicians and all these variables contribute an error in experimental result. The concurrent conditions such as temperature and humidity – sometimes contribute experimental variability. The importance of random sampling can be illustrated with the following example.

   A clinical pharmacist wish to examine and compare the activity of two drugs on 40 rats, so that the first 20 animals caught are put on drug 1 and the last 20 are put on drug 2. The first 20 are less lively compared to last 20 and may differ in size, weight and activity. So the results are biased and the decision on the drug was made based on the method of choosing the animals, but the intention of the experiment is to select the drug based on its activity.

   In order to eliminate the subjective influences on experimental results, a table of random numbers or computer generated random numbers can be used.

Guidelines to use Random Tables

1. Assign numbers to the patients.
2. Allot the patients to the first group by reading across the random table.
3. Initial numbers of random table are allotted to first group and the remaining patients are allotted to the second group.
4. A zero will respond to the patient 10, 20…etc.
5. If the number is repeated, then skip the number and proceed to the next one.
6. Ignore the number which exceeds the largest number.

Table 7.1 A short table of random numbers

39	61	09	51	68	81	26	30	51	20	61	41	25
89	35	48	61	72	10	84	34	10	44	72	94	77
37	98	37	56	40	30	70	31	75	03	68	32	15
20	55	68	05	53	73	60	28	96	48	91	81	18

Example 1: Allotment of 30 animals into two groups (15 each)

The animal numbers allotted for first group based on random table are 39, 9, 26, 30, 20, 25, 35, 10, 34, 37, 40, 31, 3, 32, and 15.

We were not included the numbers 61, 51, 68, 81, 52 etc as the animals with that numbers are not available. The remaining numbers are allotted to second group.

Example 2: Allotment of 5 animals to each group from a total of 10 animals

First group 3, 9, 6, 1 and 10

Second group the remaining numbers (2, 4, 5, 7 and 8)

3. **Selective sampling:** It can be considered as non-probability sampling method. It is convenient, easy and more reliable than random sampling. Selective sampling allows the sample collection to represent all the units in the population.

   (a) **Systematic sampling:** It is a process in which every n^{th} object is selected. For example, in a pharmaceutical industry, thousands of tablets are produced during a day and it is not possible to take every dosage form. The time for sample collection (i.e. every one hour or third hour etc.) is randomly selected and the samples are collected at the chosen time interval and analyzed. In this method, every sample does not have an equal chance of being selected.

   (b) **Stratified sampling:** In this sampling technique, the population is divided into groups (strata) having similar characteristics and then the objects can be randomly selected from each group. For example, in a clinical study the research scientist may wish to include 25% of alcoholics in control and experimental groups. Then the volunteers are categorized as alcoholics and non-alcoholics. Then 10 out of 40 volunteers (total volunteers including alcoholics and non-alcoholics) are randomly selected for experimental group and an additional 10 alcoholics are randomly selected for control group. Similarly, two groups of 30 non alcoholics are selected randomly to complete the study design.

   (c) **Cluster sampling:** It is also known as multistage sampling. This sampling technique is employed when there are several primary units (individual unit dosage forms) which are clustered (secondary units) in a bottle. It is not possible to analyze the quality of a bulk active pharmaceutical ingredient (API) from each container (the number of such containers are more). The best procedure is to select the suitable number of containers randomly (secondary units) and then extraction of sample randomly (top, middle and bottom) for analysis.

The estimate of the variance of the mean in cluster sampling is calculated with the following equation

$$(1 - f_1)\, S_1^2/n \;+ [S_2^2/(nm)]\,(f_1\,(1 - f_2))$$

f_1 and f_2 are the sampling fractions of the primary and secondary units

S_1^2 and S_2^2 are the variance among the primary and secondary units

n = number of primary units

m = number of samples collected from primary unit

Eg: Determine the variance in drug content observed from 10 tablets picked from 25 bottles (containing 250 tablets) collected from 1000 bottles.

f_1 = 25 bottles/ 1000 bottles = 0.025

f_2 = 10 tablets/ 250 tablets = 0.04

n= 25, m=10

If the variances in this example, S_1^2 and S_2^2 are 3 and 9 respectively, the estimated variance of the mean of 250 tablets sampled from 25 bottles (10 tablets per bottle) is

$$\frac{(1-0.025)3}{25} + \left[\frac{9}{25 \times 10}\right] (0.025)\,(1 - 0.04)$$

$$\frac{2.925}{25} + 0.036 \times 0.025 \times 0.96$$

$$0.117 + 0.000864 = 0.1178$$

7.2 Precision, Accuracy and Bias

It is desired to collect the sample which is representing the true population. It should be evidenced with good precision, accuracy and free from bias.

Precision measures the variability of a group of measurements and reflected by a small relative standard deviation (RSD). The experimental results observed from a group of samples collected from a population should be compacted (less scatter). Such closeness of the observed data indicates good precision. For e.g.: The potency of the 10 tablets was tested from a container containing 250 tablets (labeled strength 100 mg). The observed potencies are 92, 93, 91, 90, 94, 92, 94, 93, 90 and 92 mg respectively.

The mean and standard deviations are 92.1 and 1.449 respectively. The observed low RSD (1.573) indicates good precision but the values are not near to the actual value. It is observed in Fig 7.1 and 7.3 as the samples are very close to each other.

Accuracy is concerned with correctness of the results and refers how closely the sample data represents the true value of the population. In another case, the observed potencies are 93, 107, 94, 105, 96, 104, 97, 103, 98 and 102 mg and the actual value is 100. Here, the mean and standard deviations are 99.9 and 4.9091 respectively indicating good accuracy as the average value of the observed values is near to the actual value but the high RSD (4.914) indicates that the values are not precise.

The sample should be precise and accurate. For example, the observed potencies are 99,100,101,98,99,100,100,98,101 and 100 mg and the actual value is 100. The mean and standard deviations are 99.6 and 1.074 respectively. The RSD is 1.078. These values indicate good precision and accuracy as the values are nearer to the actual value and the observed values are very close to each other (low RSD).

In Fig. 7.1, all the observed values are close each other but differ with true value. So Fig. 7.1, representing precise but not the accurate data. In Fig. 7.2, the observed data is neither close with each other nor near to the true value. So this fig represents the data which is neither precise nor accurate. In Fig. 7.3, the observed data is close to each other and near to true value and thus indicating both precision and accuracy.

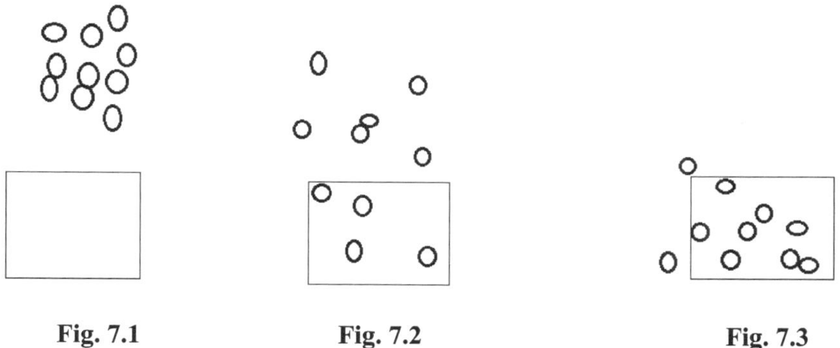

Fig. 7.1 **Fig. 7.2** **Fig. 7.3**

The lack of accuracy is Bias. Bias can be thought of as systematic error (constant error) in the measurement. For example if the sample collected from the top layer of a bulk and analyzed for its particle size or density, then the particle size and density are not the true representations of the population. The smaller particles and low density material is usually find in upper layer. Such bias is due to incorrect sampling, in appropriate experimental design, inadequate blending or blunders in recording the data. The reasons for systematic error should be identified and controlled to have good accuracy in experimental results.

7.3 Reliability and Validity of Sampling Data

Reliability indicates the consistency in experimental results and deals with the amount of errors associated with the observed values. To generate the reliable data, all sources of error and their magnitude should be known to the experimenter. The errors may be either constant errors or random (chance errors). To confirm the reproducibility, the samples are tested twice to establish the strong relationship between successive measurements and it is referred as test-retest reliability. In another method of testing, the subjects are grouped into odd items and even items and then the results are interpreted to determine the relationship. Such test is referred as split-half reliability.

Validity indicates that the observed data represents the true measurement. To be a valid sample, the sample must be a reliable sample. So the degree of validity is limited by its degree of reliability. Randomness of experiment sampling is essential to get valid experimental results.

7.4 Sampling and Non-sampling Errors

The errors involved in the collection, processing and analysis of data may be broadly classified under sampling and non sampling errors.

Sampling errors: It is due to the fact that only a part of the population has been used to estimate the population parameters and draw inferences about them. These errors are due to the following reasons.

- Faulty selection of the sample.
- Inappropriate substitution.
- Faulty demarcation of sampling units.
- Variability of the population

Non-sampling errors: The non-sampling errors primarily arise at the stages of observation, approximation and processing of the data. These errors are due to the following reasons.

- Faulty planning.
- Errors in response.
- Non-response bias.
- Errors in design of experiment.
- Errors in compilation.
- Publication errors.

7.5 Objectives of Sampling

The objectives of the sampling are as follows:

- To obtain information about the population.
- To set up the limit of accuracy of the population parameter.
- To test the significance of the population characteristics.

7.6 Sampling Distribution of a Statistic

Sampling distribution of a statistic refers to the distribution of the various values, which can be assumed by that statistic computed from the various samples of the same size randomly drawn from the population. The statistical measures of statistic like mean, standard deviation may be computed for each sample and such values may be arranged as a frequency distribution known as the sampling distribution. The population may contain either finite or infinite number of sampling units. Sampling may be done with or without sample replacement. If the population size is N and the sample size is n, then the total number of possible samples in case of replacement of samples is equal to Nn. If the sampling is done without replacement, then the possible number of samples will be NC_n.

CHAPTER 8

STATISTICAL INFERENCE: SIGNIFICANCE OF TESTING

8.1 Introduction

The parameters observed from the sample are usually employed to estimate the true population parameters. The sample mean and variance are the predictors of true mean and more often called as point estimates. Scientific experiments are conducted either to estimate some parameter or property of a system or to compare the experiments. The reliability of the sample statistic as an estimate of the true parameter depends on the confidence interval. A confidence interval is an interval width in which the true value lies. The width of the interval depends on properties of population, the estimated sample parameters and the degree of the desired probability. Most of the statistical problems are concerned with normal distribution and deals with the sampling of the means and the statistician is interested in confidence intervals of means.

Confidence intervals are calculated with the following formulae.

$$\text{P\% confidence interval for Z distribution} = Z = \frac{\bar{x} - \mu}{\frac{\sigma}{\sqrt{n}}}$$

where $\bar{x}$ = sample mean

n = sample size

σ = population standard deviation

z_p = Normal deviate corresponding to (p+1)/2 percentile of cumulative standard normal distribution

μ = Population mean

P% confidence interval for t distribution when σ is unknown and estimated from the sample.

$$\bar{x} \pm \frac{ts}{\sqrt{n}}$$

The value of t depends on degree of freedom (n –1).

Example:

(i) Calculate the 90% confidence interval for an N =10, mean value 20, and s.d is 0.3. The t value from the table for a degree of freedom 9 is 1.83.

Solution:

$$\text{P\% confidence interval} = \bar{x} \pm \frac{ts}{\sqrt{n}} = 20 \pm \frac{1.83 \times 0.3}{\sqrt{10}}$$

$$= 20 \pm \frac{0.549}{3.1622} = 20 \pm 0.17$$

(ii) Calculate the confidence intervals at 90%, 95% and 99% from the following data;

The assay results of 10 tablets were 98.4, 99.6, 99.9, 101.2, 102.6, 97.6, 98.9, 99.7, 100.5 and 101.2 respectively.

Solution: The assay results of 10 tablets were 98.4, 99.6, 99.9, 101.2, 102.6, 97.6, 98.9, 99.7, 100.5 and 102.2 respectively

$$\therefore \quad \bar{x} = \frac{98.4 + 99.6 + \ldots + 102.2}{10} = \frac{999.6}{10} = 99.96$$

$$\text{Standard deviation} = \sqrt{\frac{\Sigma\left(x - \bar{x}\right)^2}{n-1}}$$

$$= \sqrt{\frac{(98.4 - 99.96)^2 + (99.6 - 99.96)^2 + \ldots\ldots + (102.2 - 99.96)^2}{10-1}}$$

$$= \sqrt{\frac{2.4336 + 0.1296 + 0.0036 + 1.5376 + 6.9696 + 5.5696 + 1.1236 + 0.0676 + 0.2916 + 5.0176}{9}}$$

$$= \sqrt{\frac{23.144}{9}} = \sqrt{2.5715} = 1.60$$

For n = 10, degrees of freedom = n – 1= 9

The values of 't', for different confidence intervals are given in the following table.

%	t	**Confidence interval** $\bar{x} \pm \frac{ts}{\sqrt{n}}$
90	1.83	$99.96 \pm \frac{1.83 \times 1.60}{\sqrt{10}} = 99.96 \pm 0.9259$
95	2.26	$99.96 \pm \frac{2.26 \times 1.60}{\sqrt{10}} = 99.96 \pm 1.143479$
99	3.25	$99.96 \pm \frac{2.25 \times 1.60}{\sqrt{10}} = 99.96 \pm 1.64438$

So, the confidence intervals are

(99.03, 100.89)

(99.82, 101.10)

(98.32, 101.60)

(iii) Calculate the confidence intervals at 90%, 95% and 99% from the following data; 102.4, 103.7, 100.6, 98.5, 99.4, 97.9, 97.5, 98.2, 98.6 and 99.0

Solution: Given data is 102.4, 103.7, 100.6, 98.5, 99.4, 97.9, 97.5, 98.2, 98.6, 99.0, n = 10

$$\text{Mean} = \bar{x} = \frac{995.8}{10} - 99.58$$

$$\text{Standard deviation} = \sqrt{\frac{(102.4 - 99.58)^2 + \ldots\ldots\ldots + (99.0 - 99.58)^2}{9}}$$

$$= \sqrt{\frac{37.516}{9}} = \sqrt{4.16844} = 2.04$$

For n = 10, degrees of freedom = 10 – 1 = 9

The values of 't', for different confidence intervals are given in the following table.

%	t	**Confidence interval** $\bar{x} \pm \frac{ts}{\sqrt{n}}$
90	1.83	$99.58 \pm \frac{1.83 \times 2.041}{\sqrt{10}} = 99.58 \pm 1.1811$
95	2.26	$99.58 \pm \frac{2.26 \times 2.041}{\sqrt{10}} = 99.58 \pm 1.458$
99	3.25	$99.58 \pm \frac{3.25 \times 2.041}{\sqrt{10}} = 99.58 \pm 2.0976$

8.2 Hypothesis Testing

Hypothesis testing is the process of inferring from a sample (a small representative proportion of total population) whether or not to accept a certain statement about a population or populations. It is also referred to as significance testing. The statistical test results have two possible outcomes, either we cannot reject null hypothesis (H_0) or we reject null hypothesis in favor of alternative hypothesis (H_1). Null hypothesis is stated as no real difference in the outcomes. The type of null hypothesis will depend up on the type of variables and the researcher is interested in measuring the outcomes. The alternative hypothesis states that the differences in outcomes. These two hypotheses are mutually exclusive and exhaustive. There are two possible errors associated with hypothesis testing. Type 1 error (level of significance) is the probability of rejecting a true null hypothesis. It uses the symbol $\propto$ or p. The probability of all outcomes less type 1 error $(1 - \propto)$ is known as confidence level. The probability of accepting a false null hypothesis is type 2 error (β). In hypothesis testing, the statistician should try to minimize $\propto$ and maximize $1-\beta$. Probability of rejecting a false null hypothesis is called power $(1-\beta)$.

The possible outcome of hypothesis testing includes:

1. $1-\propto$: Do not reject (H_0) hypothesis when in fact $\mu_1 = \mu_2$ is true
2. $\propto$: Reject (H_0) hypothesis when in fact $\mu_1 = \mu_2$ is true
3. $1-\beta$: Reject (H_0) hypothesis when in fact $\mu_1 = \mu_2$ is false.
4. β: Do not reject null hypothesis when in fact $\mu_1 = \mu_2$ is false.

The acceptance region is that area in a statistical distribution where the outcomes will not lead to a rejection of the hypothesis under test. In contrast, the rejection region, or critical region, represents outcomes in a statistical distribution, which lead to the rejection of the hypothesis under test and acceptance of the alternative hypothesis. In other words, outcomes in the acceptance region could occur as a result of random or chance error.

The differences observed in between two values may be statistically significant or statistically not insignificant. The data obtained from clinical trials, bioavailability studies, stability studies and during product development is treated statistically to identify whether the differences are statistically significant or not. Common probability distributions such as z, t, χ^2 and F are used for testing the significance. These tests compare the sample and population means or two sample means or sample and population proportion or two sample proportions or the variances between two samples. This test of significance helps us to determine whether the observed differences are due to chance or not.

Steps involved in Statistical Inference:

1. ***Statement of null hypothesis:*** In null hypothesis, there is no difference between statistic and population parameter or the standard and observed values. Null hypothesis is traditionally written as $H_0 : \mu = \mu_0$ (hypothetical mean).

***Examples of null hypothesis*:**

(i) The drug content of a batch of tablet is 75 mg.
Null hypothesis H_0: $\mu = 75$ mg

(ii) The disintegration time of uncoated tablet is less than or equal to 15 minutes.
Null hypothesis H_0: $\mu \leq 15$ minutes

(iii) Oral hypoglycemic drug reduces the blood glucose level to below 80 mg.
Null hypothesis H_0: $\mu < 80$ mg

2. ***Statement of alternative hypothesis:*** This is the alternate and opposite of null hypothesis. It may be one side alternative or two side alternative. The previous example 1, the drug content from a batch of tablet is 75 mg. In one side alternative hypothesis, the alternative statement is the average drug content should not be less than 75 mg and it can be expressed as

$$H_A: \mu < 75 \text{ mg}$$

In two side alternative hypothesis, the alternative statement is the average drug content should not be either less than or greater than 75 mg.

$$H_A: \mu \neq 75 \text{ mg}$$

In the field of pharmacy, both alternative hypotheses are used. In case of disintegration time of uncoated tablets I.P., the alternative hypothesis is one sided as the tablets having the disintegration time more than 15 minutes are rejected.

In case of content uniformity test, the acceptance criterion is $\pm$ 15% of average drug content. It indicates the rejection of dosage containing the drug more than or less than 15% of the average drug content. So, in this example, two sided alternative hypothesis is required.

3. ***Choosing the level of significance:*** Level of significance ($\propto$ or type 1 error) is the probability of rejection of the null hypothesis.

Example: $P < 0.01$ or highly significant at 1% level refers to $\propto$ error. If the hypothesis is tested at 1% level means the chances of rejection of the null hypothesis which is true is 1%. Two most commonly used level of significances are 0.05 (5%) and 0.01 (1%) and the confidence intervals are 95% and 99% respectively. The confidence of getting correct decision in case of $P = 0.01$ is 99% and in case of $P = 0.05$, the confidence of 95%. The $\propto$ error takes place due to the rejection of the null hypothesis which is actually true. The error may also happen due to the acceptance of null hypothesis which is not actually correct. It is known as type 2 error or β.

4. ***Choosing appropriate sample:*** Sufficient number of independent samples should be collected from the bulk to represent the similar mean and standard derivation values with that of bulk samples. The meaningful valid statistical interpretation is possible only if the sample is a true representative of population.

5. ***Calculation of statistic values:*** The statistic values such as t static and Z static values should be computed by using the appropriate mathematical formulae depending up on the number of samples.
6. ***Decision:*** The decision to accept or to reject the null hypothesis will be made after comparing the calculated value and table value. If the table value is more than the calculated value, then the null hypothesis is true and the null hypothes statement is "there is no difference in tested samples". If the calculated value is more than table value then the differences are statistically significant.

8.3 Parametric Tests

The parametric procedures include such tests as the t-tests, analysis of variance (ANOVA or F-test), correlation and linear regression. In addition to the requirements that the samples must be randomly selected from their population and independently measured, two additional "parameters" must be met. First, it must be assumed that the sample is drawn from a population whose distribution approximates that of a normal distribution. Second, when two or more distributions are being compared, there must be homogeneity of variance or homoscedasticity (sample variances must be approximately equal).

8.4 Z test

The Z test is applied when sample size is more than 30. Z test can be applied to compare the mean of single population or to compare the means from two independent populations. It may be one tailed or two tailed test. If we consider only one side of a decision then we call it one tailed test, however if we are considering both sides of test, then it is called as two tailed test.

8.4.1 Test of Significance of a Mean and Large Sample

The following steps are included in this test.

- *Null Hypothesis*: There is no significant differences between sample and population mean.

$$H_o : \bar{x} = \mu$$

- *Alternative hypothesis*: Two samples are not having same mean.

$$H_1 : \bar{x} \neq \mu$$

- *Computation of test statistic*: The test statistic is

$$Z = \frac{\bar{x} - \mu}{\frac{\sigma}{\sqrt{n}}}$$

- *Level of significance*: Generally fixed at 5%. Alternatively 1% and 2% may also be used.

- *Critical value*: The critical values of the statistic Z is obtained by referring the table areas under the Normal curve-Z value. These values are given in the following table.

Critical value	Level of significance				
	1%	2%	4%	5%	10%
Two- tailed test	± 2.576	± 22.326	± 22.054	± 21.960	± 21.645
Right- tailed test	+ 2.326	+ 2.054	+ 1.751	+1.645	+ 1.282
Left- tailed test	– 2.326	– 2.054	– 1.751	– 1.645	– 1.282

- *Decision:* A decision should be arrived as to whether the null hypothesis is to be accepted or rejected.

Example

The manufacturer's claim for a product is the strength of tablet is 40 mg with the standard deviation of 1.5 mg. 40 tablets were collected randomly from a batch and analyzed for the drug content. The average drug content was found to be 38 mg per tablet. Analyze statistically and report whether the tablets contains labeled amount of drug?

Formulae $$Z = \frac{\overline{x} - \mu}{\frac{\sigma}{\sqrt{n}}}$$

In the above problem, one tailed test is used when the objective is the rejection of tablets having potency less than 40 mg ($\mu < 40$ mg) and if we want to reject the tablets having potency less than 40 mg as well as greater than 40 mg then two tailed test is used.

Solution:

Null hypothesis: $\mu = 40$; alternative hypothesis $\mu \neq 40$ i.e., $\mu > 40$ or $\mu < 40$ and given standard deviation = 1.5 mg = σ

The number of tablets were collected from a batch = 40 = n.

The average drug content is 38 mg/tablet = $\overline{x}$; $\mu = 40$

Now the test statistic

$$Z = \frac{\overline{x} - \mu}{\frac{\sigma}{\sqrt{n}}}$$

$$= \frac{38 - 40}{\frac{(1.5)}{\sqrt{40}}} = \frac{-2\sqrt{40}}{1.5}$$

$$= \frac{-2 \times 6.32455532}{1.5}$$

$$= \frac{-12.64911064}{1.5}$$

$$= -8.4327$$

$$|Z|_{cal} = 8.4327$$

The tabled value of Z with level of probability 0.05 (in two tailed test) is 1.96.

The calculated Z value is greater than the tabled value

The null hypothesis is rejected i.e., $\mu = 40$

Note: In the above problem one tailed test is used when the object is the rejection of tablets having potency less than 40 mg ($\mu < 40$ mg). Here we take null hypothesis as $\mu = 40$ mg and tabled Z value in one tail (left) with level of problem 0.05 is –1.645. So, null hypothesis rejected.

8.4.2 Test of Significance of Difference between Two Means of Large Samples

Sometimes the samples are obtained from two different sources. Then to know the differences in their mean due to chance and whether the samples are belong to same population, this test is used. To perform this test, the following data should be collected.

	Population1	**Population2**
Population mean	μ_1	μ_2
Population s.d	σ_1	σ_2
Sample mean	$\overline{X}_1$	$\overline{X}_2$
Sample s.d	s_1	s_2
Sample size	n_1	n_2

It involves the following steps.

- *Null hypothesis*: The two samples have been drawn from same parent population or the two samples are drawn from different populations but have same mean and equal standard deviation.

  $$H_0 : \mu_1 = \mu_2$$

- *Alternative hypothesis*: Two samples are not having same mean.

  $$H_1 : \mu_1 \neq \mu_2$$

- *Standard error:* The standard error of the sampling distribution of two samples of size n_1 and n_2 is calculated with the following equation.

  $$S.E\left(\overline{X}_1 - \overline{X}_2\right) = \sqrt{\left(\frac{\sigma_1^2}{n_1} + \frac{\sigma_2^2}{n_2}\right)}$$

- *Computation of test statistic:* The test statistic is calculated with the following equation.

$$Z = \frac{\overline{X}_1 - \overline{X}_2}{\sqrt{\left(\frac{\sigma_1^2}{n_1} + \frac{\sigma_2^2}{n_2}\right)}} \quad \text{Known variance}$$

$$Z = \frac{\overline{X}_1 - \overline{X}_2}{\sqrt{\left(\frac{s_1^2}{n_1} + \frac{s_2^2}{n_2}\right)}} \quad \text{Unknown variance}$$

The other steps such as level of significance, critical value and decision to be taken in testing the significance of difference between two means will be same as test of significance of a mean and large sample.

Example: The mean blood glucose levels observed from a group of 150 persons from town A was 90 mg/dl with an s.d of 4 and the mean and s.d. of blood glucose levels noticed from 200 persons of town B is 120 and 5 respectively. Test whether there is any significant difference between the mean blood glucose levels of the persons representing two towns A and B.

Solution:

Given mean blood glucose level observed from group of 150 persons from town A was 90 m/dl with standard deviation 4.

$\therefore$ $n_1 = 150$, $\bar{x}_1 = 90$, $s_1 = 4$ and

the mean and standard deviation of blood glucose levels notices from 200 persons of town B is 120 and 5, respectively.

$\therefore$ $n_2 = 200$, $\bar{x}_2 = 120$, $s_2 = 5$

Null Hypothesis: Suppose there is no significance difference between the blood glucose levels of persons representing two towns A and B. i.e., $\mu_1 = \mu_2$

Alternative hypothesis $\mu_1 \neq \mu_2$

Test statistic
$$Z = \frac{\overline{x_1} - \overline{x_2}}{\sqrt{\frac{s_1^2}{n_1} + \frac{s_2^2}{n_2}}}$$

$$= \frac{90 - 120}{\sqrt{\frac{(4)^2}{150} + \frac{(5)^2}{200}}}$$

$$= \frac{-30}{\sqrt{\frac{16}{50}+\frac{25}{200}}} = \frac{-30}{\sqrt{0.2316}} = -62.32$$

$$\therefore |Z| = 62.32$$

∴ Calculated Z value = 62.32

Tabled Z value at the level of significance 0.05 is 1.96

Null hypothesis is rejected

∴ There is a significant difference between glucose levels of persons representing two towns A and B.

8.4.3 Test of Significance of Difference between Two Standard Deviations

Sometimes we have to determine whether the two standard deviations of two samples come from same population or not. Then the following procedure is followed.

- *Null hypothesis*: The two samples standard deviations do not differ significantly.

  $$H_0 : \sigma_1 = \sigma_2$$

- *Alternative hypothesis*: Two samples are not having same standard deviation.

  $$H_1 : \sigma_1 \neq \sigma_2$$

- *Computation of test statistic:* The test statistic is calculated with the following equation.

$$Z = \frac{s_1 - s_2}{\sqrt{\frac{\sigma_1^2}{2n_1}+\frac{\sigma_2^2}{2n_2}}} = \frac{s_1 - s_2}{\sqrt{\frac{s_1^2}{2n_1}+\frac{s_2^2}{2n_2}}}$$

The other steps such as level of significance, critical value and decision to be taken in testing the significance of difference between two means will be same as test of significance of a mean and large sample.

Example: A reputed Pharma company is producing tablets from two different machines. A test is carried out to ascertain the variability of the machine and the results are as follows.

	Machine1	**Machine2**
No. of tablets in sample	100	120
Average weight	320	346
Standard deviation	2.5	3.1

Solution: Given data is

	Machine1	**Machine2**
No. of tablets in sample	100	120
Average weight	320	346
S.D	2.5	3.1

So here $s_1 = 2.5$ $s_2 = 3.1$

$n_1 = 100,$ $n_2 = 120$

Null Hypothesis: The two samples standard deviations do not differ significant by

$$H_0 : \sigma_1 = \sigma_2$$

Alternative hypothesis: Two samples are not having same standard deviation

$$H_1 : \sigma_1 \neq \sigma_2$$

$$\therefore \text{test statistic} = Z = \frac{s_1 - s_2}{\sqrt{\frac{\sigma_1^2}{2n_1} + \frac{\sigma_2^2}{2n_2}}} = \frac{s_1 - s_2}{\sqrt{\frac{s_1^2}{2n_1} + \frac{s_2^2}{2n_2}}} = \frac{2.5 - 3.1}{\sqrt{\frac{(2.5)^2}{200} + \frac{(3.1)^2}{240}}}$$

$$\frac{-0.6}{\sqrt{\frac{(2.5)^2}{200} + \frac{(3.1)^2}{240}}}$$

$$= \frac{-0.6}{\sqrt{0.07129}}$$

$$= \frac{-0.6}{0.2670} = -2.2471$$

$$|Z| = 2.2471$$

$\therefore$ Calculated z value = 2.2471

Tabled z value at level of significance

0.05 is 1.96

$\therefore$ Calculated z value > tabled z value

$\therefore$ Null hypothesis is rejected

The two sample standard deviations are significant

Binomial Distribution

Normal distribution is an example of continuous probability distribution. Though the experimental data is not exactly fitted in to the normal distribution, but in certain pharmaceutical experiments the data may be discontinuous and data may not be properly fitted in to normal distribution then alternative distributions such as binomial distribution is required.

Examples

1. In order to meet pharmacopoeia standards, the manufacture may prepare formulations with different compositions and preference will be given only for one formulation.
2. Checking the performance of a drug after treatment (whether the drug is active or not).

If the outcome of experimental result is in the form of accept/reject, passed/failed, survival/death, improvement/worst then the binomial distribution is required. The probability of the one of the possible outcome (p) and the number of experiments or trials conducted (n) are required for binomial distribution. If one of the possible outcome is p, then the possibility of other outcome is 1–p and denoted with q. If the values of $np = nq$ and equal to or greater than 5 then the distribution is close to normal distribution. The normal and binomial distributions are equal if the p value and n values are higher.

8.4.4 Test of Significance of a Sample Proportion

In this method, the standard error of sample proportion is used in place of standard error of sample mean and the other steps are similar to testing of significance of sample mean. It is described as follows.

- *Null hypothesis*: The sample is drawn from population with proportion P.

  $$H_0 : p = P_0$$

- *Alternative hypothesis*: The sample is not drawn from population with proportion P_0.

  $$H_1 : p \neq P_0$$

- *Standard error of proportion*: It is calculated with the following formulae.

  $$S.E.(p) = \sqrt{\frac{P(1-P)}{n}}$$

- *Computation of test statistic:* The test statistic is calculated with the following equation.

  $$Z = \frac{p - P_0}{\sqrt{\frac{P_0 q_0}{n}}}$$

The other steps such as level of significance, critical value and decision to be taken in testing the significance of difference between two means will be same as test of significance of difference between two sample means.

Example

10% of population from a place was reported to be diabetic. A drug was administered to 75 persons and their blood glucose levels were observed and then 5 persons out of 75 were found to be diabetic. Whether the drug is anti diabetic or not.

Solution:

Null Hypothesis: There is no effect of the drug on the persons i.e., NP = 5 here 10% population from a place were reported to be diabetic

∴ Hypothetical proportion $P_0 = \frac{10}{100} = 0.1 \Rightarrow q_0 = 1 - 0.1 = 0.9$

Given 5 persons out of 75 were found to be diabetic.

∴ observed proportion $= p = \frac{5}{75} = 0.06$

∴ The value of Z for normal distribution

$$= Z = \frac{p - P}{\sqrt{\frac{P_0 q_0}{n}}}$$

$$= \frac{0.06666 - 0.1}{\sqrt{\frac{(0.1)\ (0.9)}{75}}} = \frac{-\ 0.03333334}{\sqrt{0.0012}}$$

$$= -\ 0.96$$

$$|Z| = 0.96$$

Z_{tab} at the level of Probability = 0.05 = 1.96

∴ $t_{cal} < t_{tab}$

Null hypothesis is accepted.

8.4.5 Comparison of a Sample Proportion to a known Proportion

The sample proportion can be compared with the known proportion by treating the data with the following formulae.

$$Z = \frac{|p - P_0|}{\sqrt{\frac{P_0 q_0}{n}}}$$

p = Observed proportion

P_0 = Hypothetical Proportion

n = Number of observations

$q_0 = 1 - P_0$

Because of discrete nature of binomial data, a correction factor is recommended to improve the normal approximation. The correction often called Yates continuity correction is made with the following formulae.

$$Z = \frac{|p - P_0| - \frac{1}{2n}}{\sqrt{\frac{P_0 q_0}{n}}}$$

Example:

The content uniformity test data obtained from a pharma company indicated that 2% of products are outside the limits. Thirty out of 600 tablets from a fresh batch were found to be out side the limits. What is its statistical significance?

Solution:

Null hypothesis $H_0 = \mu$ i.e., there is no significance difference between, the observed content uniformity and the claimed content uniformity

Observed proportion (sample proportion)

$$= \frac{30}{600} = \frac{3}{60} = \frac{1}{20} = 0.05$$

$$\therefore \qquad p = 0.05$$

$$\text{Hypothetical proportion} = \frac{2}{100} = 0.02 = P_0$$

$$\therefore \qquad q_0 = 1 - P_0 = 1 - 0.02 = 0.08$$

$$\therefore \quad \text{Test statistic} = Z = \frac{|p - P_0| - \frac{1}{2N}}{\sqrt{P_0 q_0 / N}}$$

Here $\qquad N = 600$

$$\therefore \qquad Z = \frac{|0.05 - 0.02| - \frac{1}{2(600)}}{\sqrt{\frac{(0.02)(0.08)}{600}}} = \frac{0.03 - \frac{1}{1200}}{\sqrt{\frac{0.0016}{600}}}$$

$$= \frac{0.02916666}{\sqrt{0.000002626}}$$

$$= \frac{0.02916666}{0.001632993}$$

$$= 17.86$$

Calculated $Z = 17.86$

and tabled z value at the level of probability 0.05 is 1.96

∴ z calculated > z tabled

∴ Null hypothesis is rejected

∴ There is a significance difference between the hypothetical proportion and observed proportion

8.4.6 Comparison of Two Proportions

Sometimes the statistical differences between two sample proportions should be determined and then the following steps are used.

- *Null hypothesis*: There is no significant difference between the two population proportions.

  $$H_0 : P_1 = P_2$$

- *Alternative hypothesis*: The samples are not drawn from same population.

  $$H_1 : P_1 \neq P_2$$

- *Computation of test statistic:* If the data is obtained from the two groups and it is proposed to compare with either standard or hypothetical value, then the following formulae is used.

$$Z = \frac{|P_1 - P_2| - \frac{1}{n}}{\sqrt{P_0 q_0 \left(\frac{1}{n_1} + \frac{1}{n_2}\right)}}$$

The other steps such as level of significance, critical value and decision to be taken in testing the significance of difference between two means will be same as test of significance of difference between two sample means.

Example:

The number of tablets accepted and rejected after modification in the composition is given below. The data should be treated statistically to determine whether the modification is significant or not. The following table shows the number of tablets accepted and rejected after modification in composition.

	Accepted	Rejected	Total
Unmodified	40	60	100
Modified	70	30	100
Total	110	90	200

Solution:

Null hypothesis H_0: There is no significant difference between acceptance and rejection after modification in the compositions of tablets.

Total number of tablets = 200 = N

From the given data $P_0 = \frac{110}{200} = 0.55$

$q_0 = 1 - 0.55 = 0.45$

P_1 = proportion of accepted tablets before modification = $\frac{40}{200} = 0.3636$

P_2 = proportion of accepted tablets after modification = $\frac{70}{110} = 0.6363$

$n_1 = 100, \qquad n_2 = 100$

∴ Test statistic $Z = \frac{|P_1 - P_2| - \frac{1}{n}}{\sqrt{P_0 q_0\left(\frac{1}{n_1} + \frac{1}{n_2}\right)}}$

$$= \frac{|0.3636 - 0.6363| - \frac{1}{(200)}}{\sqrt{(0.55)(0.45)\left(\frac{1}{100} + \frac{1}{100}\right)}}$$

$$= \frac{0.2763 - \frac{1}{200}}{\sqrt{(0.55)(0.45)(0.01 + 0.01)}}$$

$$= \frac{0.2713}{0.070356236}$$

$= 3.85$

$|Z|_{cal} = 3.85$

Z_{tab} = 1.96 at level of probability 0.05.

∴ $Z_{cal} > Z_{tab}$

∴ Null hypothesis is rejected.

∴ The modification has a significant affect.

8.5 't' test

If we take a very large number of samples from a population and calculate the mean for each sample and then plot the frequency distribution of these means, the resulting sampling distribution is student's t-distribution. The 't' distribution is having similar shape to the normal distribution; however it may be more spreader with high tail area. The t distribution is close to the normal distribution when the sample size is increased as shown in the following figure.

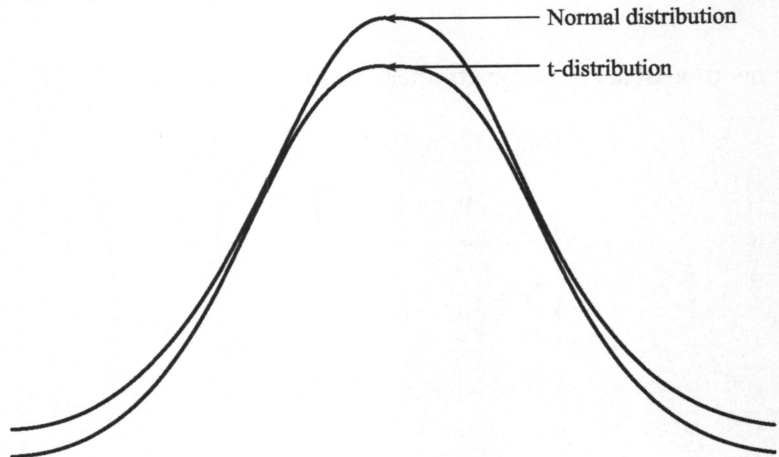

The number of t distributions is defined by the degree of freedom. The degree of freedom is number of samples minus one (n – 1). 't' distribution attains more closeness towards the normal distribution, if the sample size approaches to 30.

For interpretation of the't' test, t-tables are used. The properties of the normal curves are constant for all normal curves. (Z- Test), however the t-distribution varies according to the number of items present in the sample. These tables will give the value up to the sample size 30 for one tailed and two tailed tests with different levels of significance. The one tailed and the two tailed t distribution is presented in the following figures.

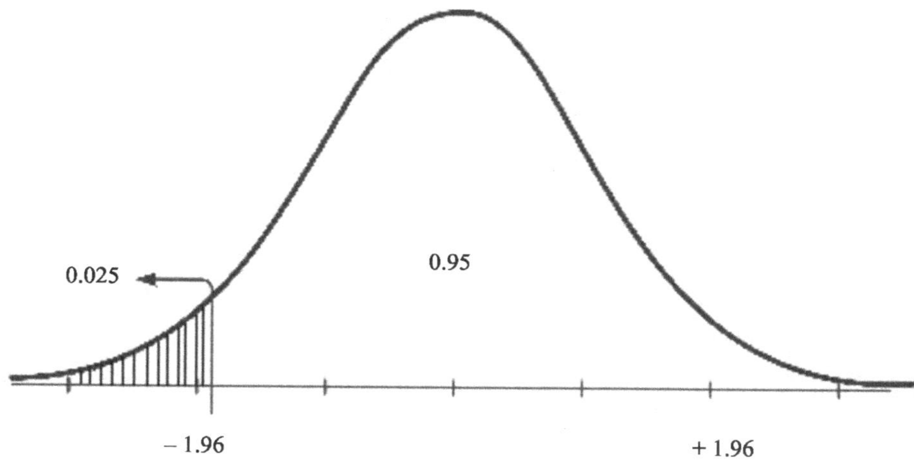

Representing the one tailed t-test

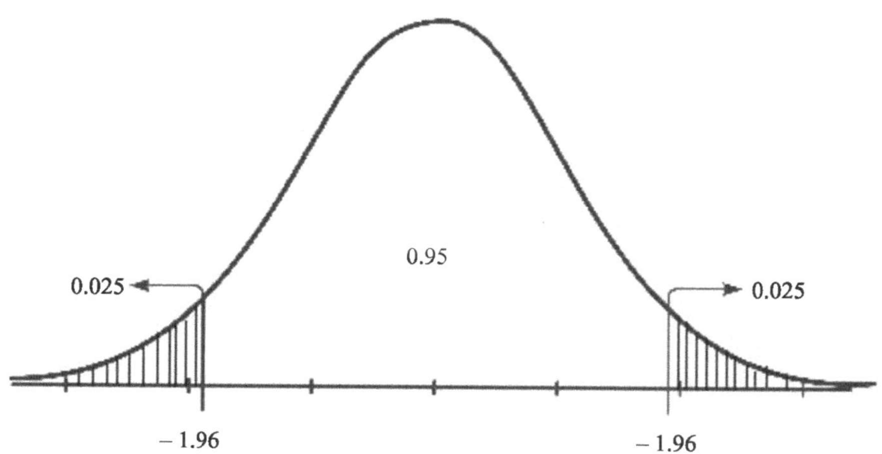

Representing the two tailed t-test

The following assumptions are made in the student's t-test.

- The parent proportion from which sample drawn is normal.
- The sample is drawn by random sampling method.
- The population standard deviation is not known.

The critical values of the t-table can be found from the Table 8.1.

Table 8.1 Critical Values of the *t* Distribution

ν	α(2): 0.50 α(l): 0.25	0.20 0.10	0.10 0.05	0.05 0.025	0.02 0.01	0.01 0.005	0.005 0.0025	0.002 0.001	0.001 0.0005
1	1.000	3.078	6.314	12.706	31.821	63.657	127.321	318.309	636.619
2	0.816	1.886	2.920	4.303	6.965	9.925	14.089	22.327	31.599
3	0.765	1.638	2.353	3.182	4.541	5.841	7.453	10.215	12.924
4	0.741	1.533	2.132	2.776	3.747	4.604	5.598	7.173	8.610
5	0.727	1.476	2.015	2.571	3.365	4.032	4.773	5.893	6.869
6	0.718	1.440	1.943	2.447	3.143	3.707	4.317	5.208	5.959
7	0.711	1.415	1.895	2.365	2.998	3.499	4.029	4.785	5.408
8	0.706	1.397	1.860	2.306	2.896	3.355	3.833	4.501	5.041
9	0.703	1.383	1.833	2.262	2.821	3.230	3.690	4.297	4.781
10	0.700	1.372	1.812	2.228	2.764	3.169	3.581	4.144	4.587
11	0.697	1.363	1.796	2.201	2.718	3.106	3.497	4.025	4.437
12	0.695	1.356	1.782	2.179	2.681	3.055	3.428	3.930	4.318
13	0.694	1.350	1.771	2.:so	2.650	3.012	3.372	3.«2	4.221
14	0.692	1.345	1.761	2.145	2.624	2.977	3.326	3.787	4.140
15	0.691	1.341	1.753	2.131	2.602	2.947	3.286	3.733	4.073
16	0.690	1.337	1.746	2.120	2.583	2.921	3.252	3.686	4.015
17	0.689	1.333	1.740	2.110	2.567	2.898	3.222	3.646	3.965
18	0.688	1.330	1.734	2.101	2.552	2.878	3.197	3.610	3.922
19	0.688	1.328	1.729	2.093	2.539	2.861	3.174	3.579	3.883
20	0.687	1.325	1.725	2.086	2.528	2.845	3.153	-3.552	3.850
21	0.686	1.323	1.721	2.080	2.518	2.831	3.135	3.527	3.819
22	0.686	1.321	1.717	2.074	2.508	2.819	3.119	3.505	3.792
23	0.685	1.319	1.714	2.069	2.500	2.807	3.104	3.485	3.768
24	0.685	1.318	1.711	2.064	2.492	2.797	3.091	3.467	3.745
25	0.684	1.316	1.708	2.060	2.485	2.787	3.078	3.450	3.725
26	0.684	1.315	1.706	2.056	2.479	2.779	3.067	3.435	3.707
27	0.684	1.314	1.703	2.052	2.473	2.771	3.057	3.421	3.690
28	0.683	1.313	1.701	2.048	2.467	2.763	3.047	3.408	3.674
29	0.683	1.311	1.699	2.045	2.462	2.756	3.038	3.396	3.659
30	0.683	1.310	1.697	2.042	2.457	2.750	3.030	3.385	3.646
31	0.682	1.309	1.696	2.040	2.453	2.744	3.022	3.375	3.633
32	0.682	1.309	1.694	2.037	2.449	2.738	3.015	3.365	3.622
33	0.682	1.308	1.692	2.035	2.445	2.733	3.008	3.356	3.611
34	0.682	1.307	1.691	2.032	2.441	2.728	3.002	3.348	3.601
35	0.682	1.306	1.690	2.030	2.438	2.724	2.996	3.340	3.591
36	0.681	1.306	1.688	2.028	2.434	2.719	2.990	3.333	3.582
37	0.681	1.305	1.687	2.026	2.431	2.715	2.985	3.326	3.574

Table 8.1 *Contd...*

ν	α(2): 0.50 α(l): 0.25	0.20 0.10	0.10 0.05	0.05 0.025	0.02 0.01	0.01 0.005	0.005 0.0025	0.002 0.001	0.001 0.0005
38	0.681	1.304	1.686	2.024	2.429	2.712	2.980	3.319	3.566
39	0.681	1.304	1.685	2.023	2.426	2.708	2.976	3.313	3.558
40	0.681	1.303	1.684	2.021	2.423	2.704	2.971	3.307	3.551
41	0.681	1.303	1.683	2.020	2.421	2.701	2.967	3.301	3.544
42	0.680	1.302	1.682	2.018	2.418	2.698	2.963	3.296	3.538
43	0.680	1.302	1.681	2.017	2.416	2.695	2.959	3.291	3.532
44	0.680	1.301	1.680	2.015	2.414	2.692	2.956	3.286	3.526
45	0.680	1.301	1.679	2.014	2.412	2.690	2.952	3.281	3.520
46	0.680	1.300	1.679	2.013	2.410	2.687	2.949	3.277	3.515
47	0.680	1.300	1.678	2.012	2.408	2.685	2.946	3.273	3.510
48	0.680	1.299	1.677	2.011	2.407	2.682	2.943	3.269	3.505
49	0.680	1.299	1.677	2.010	2.405	2.680	2.940	3.265	3.500
50	0.679	1.299	1.676	2.009	2.403	2.678	2.937	3.261	3.496
52	0.679	1.298	1.675	2.007	2.400	2.674	2.932	3.255	3.488
54	0.679	1.297	1.674	2.005	2.397	2.670	2.927	3.248	3.480
56	0.679	1.297	1.673	2.003	2.395	2.667	2.923	3.242	3.473
58	0.679	1.296	1.672	2.002	2.392	2.663	2.918	3.237	3.466
60	0.679	1.296	1.671	2.000	2.390	2.660	2.915	3.232	3.460
61									
62	0.678	1.295	1.670	1.999	2.388	2.657	2.911	3.227	3.454
64	0.678	1.295	1.669	1.998	2.386	2.655	2.908	3.223	3.449
66	0.678	1.295	1.668	1.997	2.384	2.652	2.904	3.218	3.444
68	0.678	1.294	1.668	1.995	2.382	2.650	2.902	3.214	3.439
70	0.678	1.294	1.667	1.994	2.381	2.648	2.899	3.211	3.435
72	0.678	1.293	1.666	1.993	2.379	2.646	2.896	3.207	3.431
74	0.678	1.293	1.666	1.993	2.378	2.644	2.894	3.204	3.427
76	0.678	1.293	1.665	1.992	2.376	2.642	2.891	3.201	3.423
78	0.678	1.292	1.665	1.991	2.375	2.640	2.889	3.198	3.420
80	0.678	1.292	1.664	1.990	2.374	2.639	2.887	3.195	3.416
82	0.677	1.292	1.664	1.989	2.373	2.637	2.885	3.193	3.413
84	0.677	1.292	1.663	1.989	2.372	2.636	2.883	3.190	3.410
86	0.677	1.291	1.663	1.988	2.370	2.634	2.881	3.188	3.407
88	0.677	1.291	1,662	1.987	2.369	2.633	2.880	3.185	3.405
90	0.677	1.291	1,662	1.987	2.368	2.632	2.878	3.183	3.402
92	0.677	1.291	1.662	1.986	2.368	2.630	2.876	3.181	3.399
94	0.677	1.291	1.661	1.986	2.367	2.629	2.875	3.179	3.397
96	0.677	1.290	1.661	1.985	2.366	2.628	2.873	3.177	3.395
98	0.677	1.290	1.661	1.984	2.365	2.627	2.872	3.175	3.393
100	0.677	1.290	1.660	1.984	2.364	2.626	2.871	3.174	3.390
105	0.677	1.290	1.659	1.983	2.362	2.623	2.868	3.170	3.386

Table 8.1 *Contd...*

ν	α(2): 0.50 α(1): 0.25	0.20 0.10	0.10 0.05	0.05 0.025	0.02 0.01	0.01 0.005	0.005 0.0025	0.002 0.001	0.001 0.0005
110	0.677	1.289	1.659	1.982	2.361	2.621	2.865	3.166	3.381
115	0.677	1.289	1.658	1.981	2.359	2.619	2.862	3.163	3.377
120	0.677	1.289	1.658	1.980	2.358	2.617	2.860	3.160	3.373
125	0.676	1.288	1.657	1.979	2.357	2.616	2.858	3.157	3.370
130	0.676	1.288	1.657	1.978	2.355	2.614	2.856	3.154	3.367
135	0.676	1.288	1.656	1.978	2.354	2.613	2.854	3.152	3.364
140	0.676	1.288	1.656	1.977	2.353	2.611	2.852	3.149	3.361
145	0.676	1.287	1.655	1.976	2.352	2.610	2.851	3.147	3.359
150	0.676	1.287	1.655	1.976	2.351	2.609	2.849	3.145	3.357
160	0.676	1.287	1.654	1.975	2.350	2.607	2.846	3.142	3.352
170	0.676	1.287	1.654	1.974	2.348	2.605	2.844	3.139	3.349
180	0.676	1.286	1.653	1.973	2.347	2.603	2.842	3.136	3.345
190	0.676	1.286	1.653	1.973	2.346	2.602	2.840	3.134	3.342
200	0.676	1.286	1.653	1.972	2.345	2.601	2.839	3.131	3.340
250	0.675	1.285	1.651	1.969	2.341	2.596	2.832	3.123	3.330
300	0.675	1.284	1.650	1.968	2.339	2.592	2.828	3.118	3.323
350	0,675	1.284	1.649	1.967	2.337	2.590	2.825	3.114	3.319
400	0.675	1.284	1.649	1.966	2.336	2.588	2.823	3.111	3.315
450	0.675	1.283	1.648	1.965	2.335	2.587	2.821	3.108	3.312
500	0.675	1.283	1.648	1.965	2.334	2.586	2.820	3.107	3.310
600	0.675	1.283	1.647	1.964	2.333	2.584	2.817	3.104	3.307
700	0.675	1.283	1.647	1.963	2.332	2.583	2.816	3.102	3.304
800	0.675	1.283	1.647	1.963	2.331	2.582	2.815	3.100	3.303
900	0.675	1.282	1.647	1.963	2.330	2.581	2.814	3.099	3.301
1000	0.675	1.282	1.646	1.962	2.330	2.581	2.813	3.098	3.300
∞	0.6745	1.2816	1.6449	1.9600	2.3263	2.5758	2.8070	3.0902	3.2905

Applications of t-test

8.5.1 Comparison of Sample Mean with Population Mean

It involves the following steps.

- *Null hypothesis*. The given random sample is drawn from the normal population with mean μ_0.

  $$H_0 : \mu = \mu_0$$

- *Alternative hypothesis* There is a significant difference between sample and population means.

  $$H_1 : \mu \neq \mu_0$$

- *Computation of statistic.* To observe the differences in sample and population mean, the following formulae is used.

$$t = \frac{\overline{x} - \mu_0}{\frac{s}{\sqrt{n-1}}}$$

- *Critical value.* The critical value of t for n–1 degrees of freedom at the desired level of significance should be determined from the t-tables.
- *Decision.* If the calculated t-value is less than the tabulated t-value, then the null hypothesis is accepted and if it is greater than the table value then the null hypothesis is rejected and alternative hypothesis is accepted.

Example:

Six tablets are randomly collected and subjected to disintegration test. The observed disintegration time for these tablets was 10, 12, 8, 10, 9 and 11 minutes respectively. The populations mean disintegration time was 15 minutes. To observe the disintegration time of the mean sample is statically differed significantly from the population disintegration time, it is treated as follows.

It is one tail t test, as we are considering about the disintegration time equal to or less than15 minutes but not greater/lesser than 15 minutes.

Solution:

Null hypothesis $H_0 : \mu = 15$ min, alternative $H_1 : \mu > 15$

The population disintegration time is 15 minutes. $\therefore\ \mu_0 = 15$

The disintegration time for six tablets is 10, 12, 8, 10, 9 and 11 minutes.

$$\therefore \qquad n = 6$$

$$\therefore \qquad \overline{x} = \frac{10+12+8+10+9+11}{6} = 10;$$

Standard deviation

$$S = \sqrt{\frac{(15-10)^2 + (12-10)^2 + (8-10)^2 + (9-10)^2 + (11-10)^2}{6-1}}$$

$$\left(\because\ S = \sqrt{\frac{\Sigma(x-\overline{x})^2}{n-1}}\right)$$

$$= \sqrt{\frac{25+4+4+1+1}{5}} = \sqrt{\frac{35}{5}} = \sqrt{7} = 2.64$$

The test statistic $t = \dfrac{\bar{x} - \mu_0}{\dfrac{s}{\sqrt{n-1}}}$

$$= \frac{10-15}{\dfrac{2.6457}{\sqrt{6-1}}} = \frac{-5}{\left(\dfrac{2.6457}{\sqrt{5}}\right)}$$

$$= \frac{-5 \times \sqrt{5}}{2.6457} = \frac{-11.18033989}{2.6457} = -4.2258 \Rightarrow |t| = 4.22$$

Tabled value of at the level of probability 0.05 with degrees of freedom n – 1 = 6 – 1 = 5 (right tailed) is equal to tabled value of t at the level of probability 0.05 × 2 = 0.01 with degree of freedom n – 1 = 5 in two tailed test = 2.015.

∴ Calculated t value is greater than the tabled value

∴ Null hypothesis is rejected

∴ $\mu = 15$ is rejected

∴ $\mu > 15$

8.5.2 Comparison of Two Sample Means

This test is used to test the significance of differences between two means of small samples (< 30). It should be used under following assumptions.

- The samples are normally distributed.
- The samples are random and independent of each other.
- Population variances are equal and unknown.

The following steps are involved.

- ***Null hypothesis***: The given random samples are drawn from the normal population and their means are same.

  $$H_0 : \bar{x}_1 = \bar{x}_2$$

- ***Alternative hypothesis:*** There is a significant difference between two sample means.

  $$H_1 : \bar{x}_1 \neq \bar{x}_2$$

- ***Computation of statistic:*** The variance is computed from the sample data and assumed that the variance is same for both groups. To compare the means of such samples the following formulae is used.

$$t = \frac{\bar{x}_1 - \bar{x}_2}{s\sqrt{\frac{1}{n_1} + \frac{1}{n_2}}}$$

$$S = \sqrt{\frac{(n_1 - 1)\, s_1^2 + (n_2 - 1)\, s_2^2}{n_1 + n_2 - 2}}$$

- *Critical value.* The critical value of t for $n_1 + n_2 - 2$ degrees of freedom at the desired level of significance should be determined from the t-tables.
- *Decision.* If the calculated t-value is less than the tabulated t-value, then the null hypothesis is accepted and if it is greater than the table value then the null hypothesis is rejected and alternative hypothesis is accepted.

Example: To observe the precision of analytical method, the experiments may be performed by two analysts with two spectrophotometers at different laboratories.

The observed absorbance values of a drug solution from two spectrophotometers are furnished below. Whether the differences in absorbance values noticed from these instruments is significant or not.

S.No.	Spectrophotometer 1	Spectrophotometer 2
1	0.234	0.240
2	0.236	0.243
3	0.240	0.245
4	0.242	0.246
5	0.237	0.242
6	0.235	0.244

Solution:

Null hypothesis H_0: There is no significant difference between the absorbance values of drug solution from the two spectrophotometers.

Mean of observations of spectrophotometer 1 is

$$\bar{x}_1 = \frac{1.424}{6} = 0.23733$$

Mean of observations of spectrophotometer 2 is

$$\bar{x}_2 = \frac{1.46}{6} = 0.243333$$

Pooled standard deviation

$$S = \sqrt{\frac{(n_1 - 1)\, s_1^2 + (n_2 - 1)\, s_2^2}{n_1 + n_2 - 2}}$$

$$s_1^2 = \frac{\Sigma x_1^2 - \frac{(\Sigma x_1)^2}{n_1}}{n_1 - 1}$$

$$= \frac{0.33801 - \frac{(1.424)^2}{6}}{5}$$

$$= \frac{0.33801 - 0.337962666}{5}$$

$$= \frac{0.000047333}{5} = 0.000009466$$

$$s_2^2 = \frac{\Sigma x_2^2 - \frac{(\Sigma x_2)^2}{n_2}}{n_2 - 1}$$

$$= \frac{0.35529 - \frac{(1.46)^2}{6}}{5}$$

$$= \frac{0.35529 - 0.355266666}{5}$$

$$= \frac{0.0000046668}{5}$$

∴ $$S = \sqrt{\frac{(6-1).\ (0.000009466) + (6-1)\ (0.0000046668)}{6+6-2}}$$

$$= \sqrt{\frac{0.000009466 + 0.00002333}{2}}$$

$$= \sqrt{0.0000070664} = 0.0026583$$

∴ Test statistic $$t = \frac{\bar{x}_1 - \bar{x}_2}{s\sqrt{\frac{1}{n_1} + \frac{1}{n_2}}}$$

$$= \frac{0.23733 - 0.243333}{0.0026583\sqrt{\frac{1}{6} + \frac{1}{6}}}$$

$$= \frac{-\ 0.006003}{0.001534770} = -\ 3.9093$$

$$|t|_{cal} = 3.90$$

t_{tab} at the level of probability 0.05 with degree of freedom $n_1 + n_2 - 2 = 10$ is 2.228

∴ $t_{cal} > t_{tab}$

∴ Null hypothesis is rejected

∴ There is significance difference between the absorbance values of drug solution observed from the two spectrophotometers.

Example:

An herbal extract was administrated to one group of animals and the other group of animals received marketed formulation. The percent glucose level reduction values observed at 4th hour of post administration are given in the following table. The difference in blood glucose levels is significant or not.

% blood glucose reduction		
S.No.	**Herbal formulation**	**Marketed formulation**
1	30	40
2	26	42
3	28	38
4	29	39
5	28	41
6	29	42

Solution:

Null hypothesis: H_0: There is no significance difference between the blood glucose levels in two groups.

From given data

$$\overline{x}_1 = \frac{30+26+28+29+28+29}{6} = \frac{170}{6} = 28.3333$$

$$\overline{x}_2 = \frac{40+42+38+39+41+42}{6} = \frac{242}{6} = 40.3333$$

$$s_1^2 = \frac{\Sigma x_1^2 - \frac{(\Sigma x_1)^2}{n_1}}{n_1 - 1}$$

$$= \frac{4826 - \frac{(28900)}{6}}{6-1} = \frac{4826 - 4816.666}{5} = 1.86$$

$$s_2^2 = \frac{\Sigma x_2^2 - \frac{(\Sigma x_2)^2}{n_2}}{n_2 - 1}$$

$$= \frac{9774 - \frac{58564}{6}}{5} = \frac{9774 - 9760.6666}{5}$$

$$= \frac{13.33333}{5} = 2.66$$

$$S = \sqrt{\frac{(n_1 - 1)\ S_1^2 + (n_2 - 1)\ S_2^2}{n_1 + n_2 - 2}}$$

$$= \sqrt{\frac{5.(1.86666) + 5.(2.6666)}{6 + 6 - 2}}$$

$$= \sqrt{\frac{5\ (4.53326)}{10}} = \sqrt{\frac{4.53326}{2}}$$

$$= \sqrt{2.26663} = 1.50$$

Test statistic $$t = \frac{\overline{x_1} - \overline{x_2}}{s\sqrt{\frac{1}{n_1} + \frac{1}{n_2}}}$$

$$= \frac{28.3333 - 40.3333}{1.505533\sqrt{\frac{1}{6} + \frac{1}{6}}}$$

$$= \frac{-12}{1.505533 \times 0.57735269}$$

$$= \left|\frac{-12}{0.869}\right| = 13.80$$

$\therefore$ $t_{cal} = 13.80$

t_{tab} at the level of probability 0.05 with degree of freedom $n_1 + n_2 - 2 = 12 - 2 = 10$ is 2.228

$\therefore$ $t_{cal} > t_{tab}$

$\therefore$ Null hypothesis is rejected

$\therefore$ there is a significant difference between the glucose levels in two groups.

8.5.3 Comparison of Two Samples Means having Different Variance

If the variations existing between the two groups are not same and the ratio of the two variations is exceeding 4, then the problem can be solved with Behrens- Fisher procedure. It involves the following steps.

- *Null hypothesis.* The given random samples are drawn from the normal population and their means are same.

  $$H_0 : \bar{x}_1 = \bar{x}_2$$

- *Alternative hypothesis* There is a significant difference between two sample means.

  $$H_1 : \bar{x}_1 \neq \bar{x}_2$$

- *Computation of statistic.* The variance is computed from the sample data and assumed that the variance is same for both groups. To compare the means of such samples the following formulae is used.

  $$t = \frac{|\bar{x}_1 - \bar{x}_2|}{\sqrt{\frac{s_1^2}{n_1} + \frac{s_2^2}{n_2}}}$$

- *Critical value.* The critical value of t for $n_1 + n_2 - 2$ degrees of freedom at the desired level of significance should be determined from the t-tables.
- *Decision.* If the calculated t-value is less than the tabulated t-value, then the null hypothesis is accepted and if it is greater than the table value then the null hypothesis is rejected and alternative hypothesis is accepted.

Example: The blood sugar values observed from the two treatments A and B are given below. Observe whether the differences in these two treatments are significant or not.

Treatment A	Treatment B
140	90
160	92
130	96
152	88
170	90
120	86
138	93
142	95
156	90
164	84

In the above problem, the variance ratio is more than 4 and it is treated statistically with modified procedure as follows.

for treatment A

$$\bar{x}_1 = \frac{140+160+130+152+170+120+138+142+156+164}{10}$$

$$= 147.2$$

$$s_1^2 = \frac{(140-147.2)^2 + (160-147.2)^2 + \ldots\ldots + (147.2-164)^2}{10-1}$$

$$= \frac{51.84+163.84+295.84+\ldots\ldots+77.44+282.24}{9}$$

$$= 251.73$$

For treatment B

$$\bar{x}_2 = \frac{90+92+96+88+90+86+93+95+90+84}{10} = 90.4$$

$$s_2^2 = \frac{(90.4-90)^2 + (92-90.4)^2 + \ldots\ldots + (84-90.4)^2}{9}$$

$$= \frac{0.16+2.56+31.36+5.76+0.16+19.36+6.76+21.16+0.16+40.96}{9}$$

$$= \frac{128.4}{9} = 14.26$$

In this problem, the ratio of s_1^2 to s_2^2 is exceeding 4, so the following formula is used.

$$t = \frac{|\bar{x}_1 - \bar{x}_2|}{\sqrt{\frac{s_1^2}{n_1} + \frac{s_2^2}{n_2}}} = \frac{|147.2-90.4|}{\sqrt{\frac{251.73}{10} + \frac{14.26}{10}}} = \frac{56.8}{\sqrt{26.6}}$$

$$= \frac{56.8}{5.157} = 10.66$$

$t_{cal} = 11.013$

t_{tab} at level of probability 0.05 with d.f 18 = 2.10

$\therefore$ $t_{cal} > t_{tab}$

$\therefore$ The treatment is significant

8.5.4 Paired t-Test

If the two sets of measurements are collected from the same set of formulations, animals (before or after treatment), twins or from two analytical methods, then that data is called paired observations, use of t-test for such paired observations is called as paired t-test. It involves following steps.

- *Null hypothesis.* There is no significant difference in the variable before and after the test or alternatively the mean of the samples observed before and after the treatment are same.

  $$H_0 : \bar{x}_1 = \bar{x}_2$$

- *Alternative hypothesis* There is a significant difference between two sample means.

  $$H_1 : \bar{x}_1 \neq \bar{x}_2$$

- *Computation of statistic.* The variance is computed from the sample data and assumed that the variance is same for both groups. To compare the means of such samples the following formulae is used.

  $$\text{Test statistic } t = \frac{d}{\frac{s}{\sqrt{n}}}$$

- *Critical value.* The critical value of t for n–1 degrees of freedom at the desired level of significance should be determined from the t-tables.
- *Decision.* If the calculated t-value is less than the tabulated t-value, then the null hypothesis is accepted and if it is greater than the table value then the null hypothesis is rejected and alternative hypothesis is accepted.

Problem

The systolic blood pressure levels of a hypertensive patient observed before and after drug treatment are furnished below. To observe the statistical differences in blood pressure with drug treatment, the data is treated as follows.

Patient No	Before treatment	After treatment
1	140	120
2	142	123
3	145	122
4	148	118
5	146	120
6	144	122

Solution:

Patient No	Before treatment A	After treatment B	d = A – B
1	140	120	20
2	142	123	19
3	145	122	23
4	148	118	30
5	146	120	26
6	144	122	22

$$\bar{d} = \frac{\Sigma d}{n} = \frac{140}{6} = 23.3333$$

S = standard deviation of 'd'

$$= \sqrt{\frac{\Sigma d^2 - \frac{(\Sigma d)^2}{n}}{n-1}}$$

$$= \sqrt{\frac{3350 - \frac{(140)^2}{6}}{5}} = \sqrt{\frac{3350 - 3266.666667}{5}}$$

$$= \sqrt{\frac{83.3333}{5}} = \sqrt{16.6666} = 4.08$$

Null hypothesis H_0: There is no significance difference between the blood pressure of the patient before and after the drug treatment.

Test statistic $$t = \frac{\bar{d}}{s\sqrt{\frac{1}{N}}} = \frac{\bar{d}}{\frac{s}{\sqrt{N}}} = \frac{23.3333}{\frac{(4.082482905)}{\sqrt{6}}}$$

$$= \frac{23.3333 \times \sqrt{6}}{4.082482905} = \frac{57.1546702}{4.082482905} = 13.99$$

∴ $t_{calculated}$ value = 13.99

The tabled value of t at the level of probability 0.05 with degrees of freedom = n – 1 = 6 – 1 = 5 is 2.571.

∴ $t_{cal} > t_{tab}$

∴ The null hypothesis is rejected

∴ There is a significant difference between the blood pressure of the patient before and after the drug treatment.

Problem:

The disintegration time observed from tablets before and after incorporation of disintegrating agents (without changing other factors) is showed in the following table. Report the statistical differences in disintegration time observed due to the presence of disintegrant.

Tablet No	Without disintegrant	With disintegrant
1	20	15
2	22	14
3	23	13
4	24	12
5	25	13
6	23	13

Solution:

Null hypothesis H_0: There is no significance difference between the disintegration times (min) of tablets formulated with and without disintegrant.

Tablet No	Without disintegrant (A)	With disintegrant (B)	d = A – B
1	20	15	5
2	22	14	8
3	23	13	10
4	24	12	12
5	25	13	12
6	23	13	10

$$\bar{d} = \frac{\Sigma d}{n} = \frac{57}{6} = 9.5$$

s = Standard deviation of

$$d = \sqrt{\frac{\Sigma d^2 - \frac{(\Sigma d)^2}{n}}{n-1}}$$

$$= \sqrt{\frac{577 - \frac{3249}{6}}{5}}$$

$$= \sqrt{\frac{577 - 541.5}{5}}$$

$$= \sqrt{\frac{35.5}{5}} = \sqrt{7.1} = 2.664582$$

Test statistic $t = \dfrac{\bar{d}}{\dfrac{s}{\sqrt{n}}}$

$$= \frac{(9.5)\sqrt{6}}{2.664582} = \frac{23.27015}{2.664582} = 8.73$$

and the tabled value of t at the level of probability 0.05 with degrees of freedom n – 1 = 5 is 2.571

∴ $t_{cal} > t_{tab}$

∴ The null hypothesis is rejected

∴ There is significant difference between the disintegration of drug in two cases.

Problem:

The percent drug content observed from a formulation prepared with and without antioxidant after six months of storage is given below. Whether the antioxidant significantly affected the stability or not?

S.No	With antioxidant	Without antioxidant
1	98	84
2	97	82
3	96	85
4	98	86
5	99	85
6	98	87

Solution:

Null hypothesis: There is no significant difference between the drug content values observed from a formulations prepared with and without antioxidant after six months of storage.

S.No	With antioxidant A	Without antioxidant B	d = A – B
1	98	84	14
2	97	82	15
3	96	85	11
4	98	86	12
5	99	85	14
6	98	87	11

$$\bar{d} = \frac{\Sigma d}{n} = \frac{77}{6} = 12.833$$

Standard deviation of

$$d = \sqrt{\frac{\Sigma d^2 - \frac{(\Sigma d)^2}{n}}{n-1}}$$

$$= \sqrt{\frac{1003 - \frac{5929}{6}}{5}}$$

$$= \sqrt{\frac{1003 - 988.1666667}{5}}$$

$$= \sqrt{\frac{14.83333}{5}} = \sqrt{2.966} = 1.72$$

Test statistic $$t = \frac{d}{\frac{s}{\sqrt{n}}}$$

$$= \frac{12.833}{\frac{1.72}{\sqrt{6}}}$$

$$= \frac{12.833 \times \sqrt{6}}{1.72}$$

$$= \frac{31.43430}{1.72}$$

$$= 18.27$$

∴ $t_{cal} = 18.27$; t_{tab} at the level of 0.05 problem with degrees of freedom $n - 1 = 5$ is 2.571.

∴ $t_{cal} > t_{tab}$

∴ Null hypothesis is rejected.

∴ There is significant difference between the drug contents in the both cases.

8.6 Comparison of Variances (F-test)

To compare the variances of two samples and to verify whether they are equal or not, we perform this test. In this test, the ratio of two variances is computed and denoted with F. The F-test is based on F-distribution, which may be defined as a distribution of two independent Chi-square variants each divided by the corresponding degrees of freedom. In other words:

$$F = \frac{\chi_1^2/v_1}{\chi_2^2/v_2} \qquad (v_1 > v_2)$$

The range of F values is from zero to infinity. The shape of the F-distribution curve depends upon the degrees of freedom. This test is based on the following assumptions.

- The populations for each sample must be normally distributed with same mean and variance.
- All samples must be independent and randomly distributed.
- The largest to smallest variance ratio should be equal to or greater than 1.
- The F value cannot be negative.

If the variance ratio is significant, then t-test is not applicable. In one sided variance test, the hypothesis is "the variance of first sample $\left(s_1^2\right)$ is equal to or less than the variance of second sample $\left(s_2^2\right)$". To accept this null hypothesis, S_1^2 should be smaller than s_2^2, if $\left(s_1^2\right)$ is larger than $\left(s_2^2\right)$ test of significance is required and the calculated value is compared with the table value of F distribution with the degree of freedom $n_1 - 1/n_2 - 2$. n_1 and n_2 are the sample sizes of variations observed. If the ratio $\left(s_1^2/s_2^2\right)$ exceeds the table value at the specified level of significance, then the variations are considered as significant.

If the sample from which larger variances are resulted is not known, then the two sided test is used. If a drug is estimated by two methods and the variance observed from which method is large is not known then two sided test is used. This test involves the following steps.

- *Null hypothesis.* There is no significant difference in the variances.

  $H_0 : \sigma_1^2 = \sigma_2^2$

- *Alternative hypothesis.* There is a significant difference between two sample variances.

  $H_1 : \sigma_1^2 \neq \sigma_2^2$

- *Computation of statistic.* It is calculated with the formulae $F = s_1^2/s_2^2$
- *Critical value.* The critical value of F for $n_1 - 1/n_2 - 1$ degrees of freedom at the desired level of significance should be determined from the F tables (table 8.2).
- *Decision.* If the calculated F-value is less than the tabulated F-value, then the null hypothesis is accepted and if it is greater than the table value then the null hypothesis is rejected and alternative hypothesis is accepted.

Table 8.2 (a) Critical Values of the *F* Distribution Numerator DF = 1

Denom. DF	α(2): 0.50 α(l): 0.25	0.20 0.10	0.10 0.05	0.05 0.025	0.02 0.01	0.01 0.005	0.005 0.0025	0.002 0.001	0.001 0.0005
1	5.83	39.9	161.	648.	4050.	16200.	64800.	405000.	1620000.
2	2.57	8.53	18.5	38.5	98.5	199.	399.	999.	2000.
3	2.02	5.54	10.1	17.4	34.1	55.6	89.6	167.	267.
4	1.81	4.54	7.71	12.2	21.2	31.3	45.7	74.1	106.
5	1.69	4.06	6.61	10.0	16.3	22.8	31.4	47.2	63.6
6	1.62	3.78	5.99	8.81	13.7	18.6	24.8	35.5	46.1
7	1.57	3.59	5.59	8.07	12.2	16.2	21.1	29.2	37.0
8	1.54	3.46	5.32	7.57	11.3	14.7	18.8	25.4	31.6
9	1.51	3.36	5.12	7.21	10.6	13.6	17.2	22.9	28.0
10	1.49	3.29	4.96	6.94	10.0	12.8	16.0	21.0	25.5
11	1.47	3.23	4.84	6.72	9.65	12.2	15.2	19.7	23.7
12	1.46	3.18	4.75	6.55	9.33	11.8	14.5	18.6	22.2
13	1.45	3.14	4.67	6.41	9.07	11.4	13.9	17.8	21.1
14	1.44	3.10	4.60	6.30	8.86	11.1	13.5	17.1	20.2
15	1.43	3.07	4.54	6.20	8.68	10.8	13.1	16.6	19.5
16	1.42	3.05	4.49	6.12	8.53	10.6	12.8	16.1	18.9
17	1.42	3.03	4.45	6.04	8.40	10.4	12.6	15.7	18.4
18	1.41	3.01	4.41	5.98	8.29	10.2	12.3	15.4	17.9
19	1.41	2.99	4.38	5.92	8.18	10.1	12.1	15.1	17.5
20	1.40	2.97	4.35	5.87	8.10	9.94	11.9	14.8	17.2
21	1.40	2.96	4.32	5.83	8.02	9.83	11.8	14.6	16.9
22	1.40	2.95	4.30	5.79	7.95	9.73	11.6	14.4	16.6
23	1.39	2.94	4.28	5.75	7.88	9.63	11.5	14.2	16.4
24	1.39	2.93	4.26	5.72	7.82	9.55	11.4	14.0	16.2
25	1.39	2.92	4.24	5.69	1.11	9.48	11.3	13.9	16.0
26	1.38	2.91	4.23	5.66	7.72	9.41	11.2	13.7	15.8
27	1.38	2.90	4.21	5.63	7.68	9.34	11.1	13.6	15.6
28	1.38	2.89	4.20	5.61	7.64	9.28	11.0	13.5	15.5
29	1.38	2.89	4.18	5.59	7.60	9.23	11.0	13.4	15.3
30	1.38	2.88	4.17	5.57	7.56	9.18	10.9	13.3	15.2
35	1.37	2.85	4.12	5.48	7.42	8.98	10.6	12.9	14.7
40	1.36	2.84	4.08	5.42	7.31	8.83	10.4	12.6	14.4
45	1.36	2.82	4.06	5.38	7.23	8.71	10.3	12.4	14.1
50	1.35	2.81	4.03	5.34	7.17	8.63	10.1	12.2	13.9
60	1.35	2.79	4.00	5.29	7.08	8.49	9.96	12.0	13.5
70	1.35	2.78	3.98	5.25	7.01	8.40	9.84	11.8	13.3
80	1.34	2.77	3.96	5.22	6.96	8.33	9.75	11.7	13.2
90	1.34	2.76	3.95	5.20	6.93	8.28	9.68	11.6	13.0
100	1.34	2.76	3.94	5.18	6.90	8.24	9.62	11.5	12.9
120	1.34	2.75	3.92	5.15	6.85	8.18	9.54	11.4	12.8
140	1.33	2.74	3.91	5.13	6.82	8.14	9.48	11.3	12.7
160	1.33	2.74	3.90	5.12	6.80	8.10	9.44	11.2	12.6
180	1.33	2.73	3.89	5.11	6.78	8.08	9.40	11.2	12.6
200	1.33	2.73	3.89	5.10	6.76	8.06	9.38	11.2	12.5
300	1.33	2.72	3.87	5.07	6.72	8.00	9.30	11.0	12.4
500	1.33	2.72	3.86	5.05	6.69	7.95	9.23	11.0	12.3
∞	1.32	2.71	3.84	5.02	6.64	7.88	9.14	10.8	12.1

Table 8.2 (b) Critical Values of the *F* Distribution Numerator DF = 2

Denom. DF	α(2): 0.50 α(l): 0.25	0.20 0.10	0.10 0.05	0.05 0.025	0.02 0.01	0.01 0.005	0.005 0.0025	0.002 0.001	0.001 0.0005
1	7.50	49.5	200.	800.	5000.	20000.	80000.	500000.	2000000.
2	3.00	9.00	19.0	39.0	99.0	199.	399.	999.	2000.
3	2,28	5.46	9.55	16.0	30.8	49.8	79.9	149.	237.
4	2.00	4.32	6.94	10.6	18.0	26.3	38.0	61.2	87.4
5	1.85	3.78	5.79	8.43	13.3	18.3	25.0	37.1	49.8
6	1.76	3.46	5.14	7.26	10.9	14.5	19.1	27.0	34.8
7	1.70	3.26	4.74	6.54	9.55	12.4	15.9	21.7	27.2
8	1.66	3.11	4.46	6.06	8.65	11.0	13.9	18.5	22.7
9	1.62	3.01	4.26	5.71	8.02	10.1	12.5	16.4	19.9
10	1.60	2.92	4.10	5.46	7.56	9.43	11.6	14.9	17.9
11	1.58	2.86	3.98	5.26	7.21	8.91	10.8	13.8	16.4
12	1.56	2.81	3.89	5.10	6.93	8.51	10.3	13.0	15.3
13	1.55	2.76	3.81	4.97	6.70	8.19	9.84	12.3	14.4
14	1.53	2.73	3.74	4.86	6.51	7.92	9.47	11.8	13.7
15	1.52	2.70	3.68	4.77	6.36	7.70	9.17	11.3	13.2
16	1.51	2.67	3.63	4.69	6.23	7.51	8.92	11.0	12.7
17	1.51	2.64	3.59	4.62	6.11	7.35	8.70	10.7	12.3
18	1.50	2.62	3.55	4.56	6.01	7.21	8.51	10.4	11.9
19	1.49	2.61	3.52	4.51	5.93	7.09	8.35	10.2	11.6
20	1.49	2.59	3.49	4.46	5.85	6.99	8.21	9.95	11.4
21	1.48	2.57	3.47	4.42	5.78	6.89	8.08	9.77	11.2
22	1.48	2.56	3.44	4.38	5.72	6.81	7.96	9.61	11.0
23	1.47	2.55	3.42	4.35	5.66	6.73	7.86	9.47	10.8
24	1.47	2.54	3.40	4.32	5.61	6.66	7.77	9.34	10.6
25	1.47	2.53	3.39	4.29	5.57	6.60	7.69	9.22	10.5
26	1.46	2.52	3.37	4.27	5.53	6.54	7.61	9.12	10.3
27	1.46	2.51	3.35	4.24	5.49	6.49	7.54	9.02	10.2
28	1.46	2.50	3.34	4.22	5.45	6.44	7.48	8.93	10.1
29	1.45	2.50	3.33	4.20	5.42	6.40	7.42	8.85	9.99
30	1.45	2.49	3.32	4.18	5.39	6.35	7.36	8.77	9.90
35	1.44	2.46	3.27	4.11	5.27	6.19	7.14	8.47	9.52
40	1.44	2.44	3.23	4.05	5.18	5.07	6.99	8.25	9.25
45	1.43	2.42	3.20	4.01	5.1 1	5.97	6.86	8.09	9.04
50	1.43	2.41	3.18	3.97	5.06	5.90	6.77	7.96	8.88
60	1.42	2.39	3.15	3.93	4.98	5.79	6.63	7.77	8.65
70	1.41	2.38	3.13	3.89	4.92	5.72	6.53	7.64	8.49
80	1.41	2.37	3.11	3.86	4.88	5.67	6.46	7.54	8.37
90	1.41	2.36	3.10	3.84	4.85	5.62	6.41	7.47	8.28
100	1.41	2.36	3.09	3.83	4.82	5.59	6.37	7.41	8.21
120	1.40	2.35	3.07	3.80	4.79	5.54	6.30	7.32	8.10
140	1.40	2.34	3.06	3.79	4.76	5.50	6.26	7.26	8.03
160	1.40	2.34	3.05	3.78	4.74	5.48	6.22	7.21	7.97
180	1.40	2.33	3.05	3.77	4.73	5.46	6.20	7.18	7.93
200	1.40	2.33	3.04	3.76	4.71	5.44	6.17	7.15	7.90
300	1.39	2.32	3.03	3.73	4.68	5.39	6.11	7.07	7.80
500	1.39	2.31	3.01	3.72	4.65	5.35	6.06	7.00	7.72
∞	1.39	2.30	3.00	3.69	4.61	5.30	5.99	6.91	7.60

Table 8.2 (c) Critical Values of the *F* Distribution Numerator DF = 3

Denom. DF	**α(2): 0.50**	**0.20**	**0.10**	**0.05**	**0.02**	**0.01**	**0.005**	**0.002**	**0.001**
	α(1): 0.25	**0.10**	**0.05**	**0.025**	**0.01**	**0.005**	**0.0025**	**0.001**	**0.0005**
1	8.20	53.6	216.	864.	5400.	21600.	86500.	540000.	2160000.
2	3.15	9.16	19.2	39.2	99.2	199.	399.	999.	2000.
3	2.36	5.39	9.28	15.4	29.5	47.5	76.1	141.	225.
4	2.05	4.19	6.59	9.98	16.7	24.3	35.0	56.2	80.1
5	1.88	3.62	5.41	7.76	12.1	16.5	22.4	33.2	44.4
6	1.78	3.29	4.76	6.60	9.78	12.9	16.9	23.7	30.5
7	1.72	3.07	4.35	5.89	8.45	10.9	13.8	18.8	23.5
8	1.67	2.92	4.07	5.42	7.59	9.60	12.0	15.8	19.4
9	1.63	2.81	3.86	5.08	6.99	8.72	10.7	13.9	16.8
10	1.60	2.73	3.71	4.83	6.55	8.08	9.83	12.6	15.0
11	1.58	2.66	3.59	4.63	6.22	7.60	9.17	11.6	13.7
12	1.56	2.61	3.49	4.47	5.95	7.23	8.65	10.8	12.7
13	1.55	2.56	3.41	4.35	5.74	6.93	8.24	10.2	11.9
14	1.53	2.52	3.34	4.24	5.56	6.68	7.91	9.73	11.3
15	1.52	2.49	5.29	4.15	5.42	6.48	7.63	9.34	10.8
16	1.51	2.46	3.24	4.08	5.29	6.30	7.40	9.01	10.3
17	1.50	2.44	3.20	4.01	5.19	6.16	7.21	8.73	9.99
18	1.49	2.42	3.16	3.95	5.09	6.03	7.04	8.49	9.69
19	1.49	2.40	3.13	3.90	5.01	5.92	6.89	8.28	9.42
20	1.48	2.38	3.10	3.86	4.94	5.82	6.76	8.10	9.20
21	1.48	2.36	3.07	3.82	4.87	5.73	6.64	7.94	8.99
22	1.47	2.35	3.05	3.78	4.82	5.65	6.54	7.80	8.82
23	1.47	2.34	3.03	3.75	4.76	5.58	6.45	7.67	8.66
24	1.46	2.33	3.01	3.72	4.72	5.52	6.36	7.55	8.51
25	1.46	2.32	2.99	3.69	4.68	5.46	6.29	7.45	8.39
26	1.45	2.31	2.98	3.67	4.64	5.41	6.22	7.36	8.27
27	1.45	2.30	2.96	3.65	4.60	5.36	6.16	7.27	8.16
28	1.45	2.29	2.95	3.63	4.57	5.32	6.10	7.19	8.07
29	1.45	2.28	2.93	3.61	4.54	5.28	6.05	7.12	7.98
30	1.44	2.28	2.92	3.59	4.51	5.24	6.00	7.05	7.89
35	1.43	2.25	2.87	3.52	4.40	5.09	5.80	6.79	7.56
40	1.42	2.23	2.84	3.46	4.31	4.98	5.66	6.59	7.33
45	1.42	2.21	2.81	3.42	4.25	4.89	5.55	6.45	7.15
50	1.41	2.20	2.79	3.39	4.20	4.83	5.47	6.34	7.01
60	1.41	2.18	2.76	3.34	4.13	4.73	5.34	6.17	6.81
70	1.40	2.16	2.74	3.31	4.07	4.66	5.26	6.06	6.67
80	1.40	2.15	2.72	3.28	4.04	4.61	5.19	5.97	6.57
90	1.39	2.15	2.71	3.26	4.01	4.57	5.14	5.91	6.49
100	1.39	2.14	2.70	3.25	3.98	4.54	5.11	5.86	6.43
120	1.39	2.13	2.68	3.23	3.95	4.50	5.05	5.78	6.34
140	1.38	2.12	2.67	3.21	3.92	4.47	5.01	5.73	6.28
160	1.38	2.12	2.66	3.20	3.91	4.44	4.98	5.69	6.23
180	1.38	2.11	2.65	3.19	3.89	4.42	4.95	5.66	6.19
200	1.38	2.11	2.65	3.18	3.88	4.41	4.94	5.63	6.16
300	1.38	2.10	2.63	3.16	3.85	4.36	4.88	5.56	6.08
500	1.37	2.09	2.62	3.14	3.82	4.33	4.84	5.51	6.01
∞	1.37	2.08	2.61	3.12	3.78	4.28	4.77	5.42	5.91

Table 8.2 (d) Critical Values of the *F* Distribution Numerator DF = 4

Denom. DF	α(2): 0.50 α(1): 0.25	0.20 0.10	0.10 0.05	0.05 0.025	0.02 0.01	0.01 0.005	0.005 0.0025	0.002 0.001	0.001 0.0005
1	8.58	55.8	225.	900.	5620.	22500.	90000.	562000.	2250000.
2	3.23	9.24	19.2	39.2	99.2	199.	399.	999.	2000.
3	2.39	5.34	9.12	15.1	28.7	46.2	73.9	137.	218.
4	2.06	4.11	6.39	9.60	16.0	23.2	33.3	53.4	76.1
5	1.89	3.52	5.19	7.39	11.4	15.6	21.0	31.1	41.5
6	1.79	3.18	4.53	6.23	9.15	12.0	15.7	21.9	28.1
7	1.72	2.96	4.12	5.52	7.85	10.1	12.7	17.2	21.4
8	1.66	2.81	3.84	5.05	7.01	8.81	10.9	14.4	17.6
9	1.63	2.69	3.63	4.72	6.42	7.96	9.74	12.6	15.1
10	1.59	2.61	3.48	4.47	5.99	7.34	8.89	11.3	13.4
11	1.57	2.54	3.36	4.28	5.67	6.88	8.25	10.3	12.2
12	1.55	2.48	3.26	4.12	5.41	6.52	7.76	9.63	11.2
13	1.53	2.43	3.18	4.00	5.21	6.23	7.37	9.07	10.5
14	1.52	2.39	3.11	3.89	5.04	6.00	7.06	8.62	9.95
15	1.51	2.36	3.06	3.80	4.89	5.80	6.80	8.25	9.48
16	1.50	2.33	3.01	3.73	4.77	5.64	6.58	7.94	9.08
17	1.49	2.31	2.96	3.66	4.67	5.50	6.39	7.68	8.75
18	1.48	2.29	2.93	3.61	4.58	5.37	6.23	7.46	8.47
19	1.47	2.27	2.90	3.56	4.50	5.27	6.09	7.27	8.23
20	1.47	2.25	2.87	3.51	4.43	5.17	5.97	7.10	8.02
21	1.46	2.23	2.84	3.48	4.37	5.09	5.86	6.95	7.83
22	1.45	2.22	2.82	3.44	4.31	5.02	5.76	6.81	7.67
23	1.45	2.21	2.80	3.41	4.26	4.95	5.67	6.70	7.52
24	1.44	2.19	2.78	3.38	4.22	4.89	5.60	6.59	7.39
25	1.44	2.18	2.76	3.35	4.18	4.84	5.53	6.49	7.27
26	1.44	2.17	2.74	3.33	4.14	4.79	5.46	6.41	7.16
27	1.43	2.17	2.73	3.31	4.11	4.74	5.40	6.33	7.06
28	1.43	2.16	2.71	3.29	4.07	4.70	5.35	6.25	6.97
29	1.43	2.15	2.70	3.27	4.04	4.66	5.30	6.19	6.89
30	1.42	2.14	2.69	3.25	4.02	4.62	5.25	6.12	6.82
35	1.41	2.11	2.64	3.18	3.91	4.48	5.07	5.88	6.51
40	1.40	2.09	2.61	3.13	3.83	4.37	4.93	5.70	6.30
45	1.40	2.07	2.58	3.09	3.77	4.29	4.83	5.56	6.13
50	1.39	2.06	2.56	3.05	3.72	4.23	4.75	5.46	6.01
60	1.38	2.04	2.53	3.01	3.65	4.14	4.64	5.31	5.82
70	1.38	2.03	2.50	2.97	3.60	4.08	4.56	5.20	5.70
80	1.38	2.02	2.49	2.95	3.56	4.03	4.50	5.12	5.60
90	1.37	2.01	2.47	2.93	3.53	3.99	4.45	5.06	5.53
100	1.37	2.00	2.46	2.92	3.51	3.96	4.42	5.02	5.48
120	1.37	1.99	2.45	2.89	3.48	3.92	4.36	4.95	5.39
140	1.36	1.99	2.44	2.88	3.46	3.89	4.32	4.90	5.33
160	1.36	1.98	2.43	2.87	3.44	3.87	4.30	4.86	5.29
180	1.36	1.98	2.42	2.86	3.43	3.85	4.27	4.83	5.26
200	1.36	1.97	2.42	2.85	3.41	3.84	4.26	4.81	5.23
300	1.35	1.96	2.40	2.83	3.38	3.80	4.21	4.75	5.15
500	1.35	1.96	2.39	2.81	3.36	3.76	4.17	4.69	5.09
∞	1.35	1.94	2.37	2.79	3.32	3.72	4.11	4.62	5.00

Table 8.2 (e) Critical Values of the *F* Distribution Numerator DF = 5

Denom. DF	α(2): 0.50 α(1): 0.25	0.20 0.10	0.10 0.05	0.05 0.025	0.02 0.01	0.01 0.005	0.005 0.0025	0.002 0.001	0.001 0.0005
1	8.82	57.2	230.	922.	5760.	23100.	92200.	576000.	2310000.
2	3.28	9.29	19.3	39.3	99.3	199.	399.	999.	2000.
3	2.41	5.31	9.01	14.9	28.2	45.4	72.6	135.	214.
4	2.07	4.05	6.26	9.36	15.5	22.5	32.3	51.7	73.6
5	1.89	3.45	5.05	7.15	11.0	14.9	20.2	29.8	39.7
6	1.79	3.11	4.39	5.99	8.75	11.5	14.9	20.8	26.6
7	1.71	2.88	3.97	5.29	7.46	9.52	12.0	16.2	20.2
8	1.66	2.73	3.69	4.82	6.63	8.30	10.3	13.5	16.4
9	1.62	2.61	3.48	4.48	6.06	7.47	9.12	11.7	14.1
10	1.59	2.52	3.33	4.24	5.64	6.87	8.29	10.5	12.4
11	1.56	2.45	3.20	4.04	5.32	6.42	7.67	9.58	11.2
12	1.54	2.39	3.11	3.89	5.06	6.07	7.20	8.89	10.4
13	1.52	2.35	3.03	3.77	4.86	5.79	6.82	8.35	9.66
14	1.51	2.31	2.96	3.66	4.69	5.56	6.51	7.92	9.11
15	1.49	2.27	2.90	3.58	4.56	5.37	6.26	7.57	8.66
16	1.48	2.24	2.85	3.50	4.44	5.21	6.05	7.27	8.29
17	1.47	2.22	2.81	3.44	4.34	5.07	5.87	7.02	7.98
18	1.46	2.20	2.77	3.38	4.25	4.96	5.72	6.81	7.71
19	1.46	2.18	2.74	3.33	4.17	4.85	5.58	6.62	7.48
20	1.45	2.16	2.71	3.29	4.10	4.76	5.46	6.46	7.27
21	1.44	2.14	2.68	3.25	4.04	4.68	5.36	6.32	7.10
22	1.44	2.13	2.66	3.22	3.99	4.61	5.26	6.19	6.94
23	1.43	2.11	2.64	3.18	3.94	4.54	5.18	6.08	6.80
24	1.43	2.10	2.62	3.15	3.90	4.49	5.11	5.98	6.68
25	1.42	2.09	2.60	3.13	3.85	4.43	5.04	5.89	6.56
26	1.42	2.08	2.59	3.10	3.82	4.38	4.98	5.80	6.46
27	1.42	2.07	2.57	3.08	3.78	4.34	4.92	5.73	6.37
28	1.41	2.06	2.56	3.06	3.75	4.30	4.87	5.66	6.28
29	1.41	2.06	2.55	3.04	3.73	4.26	4.82	5.59	6.21
30	1.41	2.05	2.53	3.03	3.70	4.23	4.78	5.53	6.13
35	1.40	2.02	2.49	2.96	3.59	4.09	4.60	5.30	5.85
40	1.39	2.00	2.45	2.90	3.51	3.99	4.47	5.13	5.64
45	1.38	1.98	2.42	2.86	3.45	3.91	4.37	5.00	5.49
50	1.37	1.97	2.40	2.83	3.41	3.85	4.30	4.90	5.37
60	1.37	1.95	2.37	2.79	3.34	3.76	4.19	4.76	5.20
70	1.36	1.93	2.35	2.75	3.29	3.70	4.11	4.66	5.08
80	1.36	1.92	2.33	2.73	3.26	3.65	4.05	4.58	4.99
90	1.35	1.91	2.32	2.71	3.23	3.62	4.01	4.53	4.92
100	1.35	1.91	2.31	2.70	3.21	3.59	3.97	4.48	4.87
120	1.35	1.90	2.29	2.67	3.17	3.55	3.92	4.42	4.79
140	1.34	1.89	2.28	2.66	3.15	3.52	3.89	4.37	4.74
160	1.34	1.88	2.27	2.65	3.13	3.50	3.86	4.33	4.69
180	1.34	1.88	2.26	2.64	3.12	3.48	3.84	4.31	4.66
200	1.34	1.88	2.26	2.63	3.11	3.47	3.82	4.29	4.64
300	1.33	1.87	2.24	2.61	3.08	3.43	3.77	4.22	4,56
500	1.33	1.86	2.23	2.59	3.05	3.40	3.73	4.18	4.51
∞	1.33	1.85	2.21	2.57	3.02	3.35	3.68	4.10	4.42

Table 8.2 (f) Critical Values of the *F* Distribution Numerator DF = 6

Denom. DF	α(2): 0.50 α(1): 0.25	0.20 0.10	0.10 0.05	0.05 0.025	0.02 0.01	0.01 0.005	0.005 0.0025	0.002 0.001	0.001 0.0005
1	8.98	58.2	234.	937.	5860.	23400.	93700.	586000.	2340000.
2	3.31	9.33	19.3	39.3	99.3	199.	399.	999.	2000.
3	2.42	5.28	8.94	14.7	27.9	44.8	71.7	133.	211.
4	2.08	4.01	6.16	9.20	15.2	22.0	31.5	50.5	71.9
5	1.89	3.40	4.95	6.98	10.7	14.5	19.6	28.8	38.5
6	1.78	3.05	4.28	5.82	8.47	11.1	14.4	20.0	25.6
7	1.71	2.83	3.87	5.12	7.19	9.16	11.5	15.5	19.3
8	1.65	2.67	3.58	4.65	6.37	7.95	9.83	12.9	15.7
9	1.61	2.55	3.37	4.32	5.80	7.13	8.68	11.1	13.3
10	1.58	2.46	3.22	4.07	5.39	6.54	7.87	9.93	11.7
11	1.55	2.39	3.09	3.88	5.07	6.10	7.27	9.05	10.6
12	1.53	2.33	3.00	3.73	4.82	5.76	6.80	8.38	9.74
13	1.51	2.28	2.92	3.60	4.62	5.48	6.44	7.86	9.07
14	1.50	2.24	2.85	3.50	4.46	5.26	6.14	7.44	8.53
15	1.48	2.21	2.79	3.41	4.32	5.07	5.89	7.09	8.10
16	1.47	2.18	2.74	3.34	4.20	4.91	5.68	6.80	7.74
17	1.46	2.15	2.70	3.28	4.10	4.78	5.51	6.56	7.43
18	1.45	2.13	2.66	3.22	4.01	4.66	5.36	6.35	7.18
19	1.44	2.11	2.63	3.17	3.94	4.56	5.23	6.18	6.95
20	1.44	2.09	2.60	3.13	3.87	4.47	5.11	6.02	6.76
21	1.43	2.08	2.57	3.09	3.81	4.39	5.01	5.88	6.59
22	1.42	2.06	2.55	3.05	3.76	4.32	4.92	5.76	6.44
23	1.42	2.05	2.53	3.02	3.71	4.26	4.84	5.65	6.30
24	1.41	2.04	2.51	2.99	3.67	4.20	4.76	5.55	6.18
25	1.41	2.02	2.49	2.97	3.63	4.15	4.70	5.46	6.07
26	1.41	2.01	2.47	2.94	3.59	4.10	4.64	5.38	5.98
27	1.40	2.00	2.46	2.92	3.56	4.06	4.58	5.31	5.89
28	1.40	2.00	2.45	2.90	3.53	4.02	4.53	5.24	5.80
29	1.40	1.99	2.43	2.88	3.50	3.98	4.48	5.18	5.73
30	1.39	1.98	2.42	2.87	3.47	3.95	4.44	5.12	5.66
35	1.38	1.95	2.37	2.80	3.37	3.81	4.27	4.89	5.39
40	1.37	1.93	2.34	2.74	3.29	3.71	4.14	4.73	5.19
45	1.36	1.91	2.31	2.70	3.23	3.64	4.05	4.61	5.04
50	1.36	1.90	2.29	2.67	3.19	3.58	3.98	4.51	4.93
60	1.35	1.87	2.25	2.63	3.12	3.49	3.87	4.37	4.76
70	1.34	1.86	2.23	2.59	3.07	3.43	3.79	4.28	4.64
80	1.34	1.85	2.21	2.57	3.04	3.39	3.74	4.20	4.56
90	1.33	1.84	2.20	2.55	3.01	3.35	3.70	4.15	4.50
100	1.33	1.83	2.19	2.54	2.99	3.33	3.66	4.11	4.45
120	1.33	1.82	2.18	2.52	2.96	3.28	3.61	4.04	4.37
140	1.32	1.82	2.16	2.50	2.93	3.26	3.58	4.00	4.32
160	1.32	1.81	2.16	2.49	2.92	3.24	3.55	3.97	4.28
180	1.32	1.81	2.15	2.48	2.90	3.22	3.53	3.94	4.25
200	1.32	1.80	2.14	2.47	2.89	3.21	3.52	3.92	4.22
300	1.32	1.79	2.13	2.45	2.86	3.17	3.47	3.86	4.15
500	1.31	1.79	2.12	2.43	2.84	3.14	3.43	3.81	4.10
∞	1.31	1.77	2.10	2.41	2.80	3.09	3.37	3.74	4.02

Table 8.2 (g) Critical Values of the *F* Distribution Numerator DF = 7

Denom. DF	α(2): 0.50 α(1): 0.25	0.20 0.10	0.10 0.05	0.05 0.025	0.02 0.01	0.01 0.005	0.005 0.0025	0.002 0.001	0.001 0.0005
1	9.10	58.9	237.	948.	5930.	23700.	94900.	593000.	2370000.
2	3.34	9.35	19.4	39.4	99.4	199.	399.	999.	2000.
3	2.43	5.27	8.89	14.6	27.7	44.4	71.0	132.	209.
4	2.08	3.98	6.09	9.07	15.0	21.6	31.0	49.7	70.7
5	1.89	3.37	4.88	6.85	10.5	14.2	19.1	28.2	37.6
6	1.78	3.01	4.21	5.70	8.26	10.8	14.0	19.5	24.9
7	1.70	2.78	3.79	4.99	6.99	8.89	11.2	15.0	18.7
8	1.64	2.62	3.50	4.53	6.18	7.69	9.49	12.4	15.1
9	1.60	2.51	3.29	4.20	5.61	6.88	8.36	10.7	12.8
10	1.57	2.41	3.14	3.95	5.20	6.30	7.56	9.52	11.2
11	1.54	2.34	3.01	3.76	4.89	5.86	6.97	8.66	10.1
12	1.52	2.28	2.91	3.61	4.64	5.52	6.51	8.00	9.28
13	1.50	2.23	2.83	3.48	4.44	5.25	6.15	7.49	8.63
14	1.49	2.19	2.76	3.38	4.28	5.03	5.86	7.08	8.11
15	1.47	2.16	2.71	3.29	4.14	4.85	5.62	6.74	7.68
16	1.46	2.13	2.66	3.22	4.03	4.69	5.41	6.46	7.33
17	1.45	2.10	2.61	3.16	3.93	4.56	5.24	6.22	7.04
18	1.44	2.08	2.58	3.10	3.84	4.44	5.09	6.02	6.78
19	1.43	2.06	2.54	3.05	3.77	4.34	4.96	5.85	6.57
20	1.43	2.04	2.51	3.01	3.70	4.26	4.85	5.69	6.38
21	1.42	2.02	2.49	2.97	3.64	4.18	4.75	5.56	6.21
22	1.41	2.01	2.46	2.93	3.59	4.11	4.66	5.44	6.07
23	1.41	1.99	2.44	2.90	3.54	4.05	4.58	5.33	5.94
24	1.40	1.98	242	2.87	3.50	3.99	4.51	5.23	5.82
25	1.40	1.97	2.40	2.85	3.46	3.94	4.44	5.15	5.71
26	1.39	1.96	2.39	2.82	3.42	3.89	4.38	5.07	5.62
27	1.39	1.95	2.37	2.80	3.39	3.85	4.33	5.00	5.53
28	1.39	1.94	2.36	2.78	3.36	3.81	4.28	4.93	5.45
29	1.38	1.93	2.35	2.76	3.33	3.77	4.24	4.87	5.38
30	1.38	1.93	2.33	2.75	3.30	3.74	4.19	4.82	5.31
35	1.37	1.90	2.29	2.68	3.20	3.61	4.02	4.59	5.04
40	1.36	1.87	2.25	2.62	3.12	3.51	3.90	4.44	4.85
45	1.35	1.85	2.22	2.58	3.07	3.43	3.81	4.32	4.71
50	1.34	1.84	2.20	2.55	3.02	3.38	3.74	4.22	4.60
60	1.33	1.82	2.17	2.51	2.95	3.29	3.63	4.09	4.44
70	1.33	1.80	2.14	2.47	2.91	3.23	3.56	3.99	4.32
80	1.32	1.79	2.13	2.45	2.87	3.19	3.50	3.92	4.24
90	1.32	1.78	2.11	2.43	2.84	3.15	3.46	3.87	4.18
100	1.32	1.78	2.10	2.42	2.82	3.13	3.43	3.83	4.13
120	1.31	1.77	2.09	2.39	2.79	3.09	3.38	3.77	4.06
140	1.31	1.76	2.08	2.38	2.77	3.06	3.35	3.72	4.01
160	1.31	1.75	2.07	2.37	2.75	3.04	3.32	3.69	3.97
180	1.31	1.75	2.06	2.36	2.74	3.02	3.30	3.67	3.94
200	1.30	1.75	2.06	2.35	2.73	3.01	3.29	3.65	3.92
300	1.30	1.74	2.04	2.33	2.70	2.97	3.24	3.59	3.85
500	1.30	1.73	2.03	2.31	2.68	2.94	3.20	3.54	3.80
∞	1.29	1.72	2.01	2.29	2.64	2.90	3.15	3.47	3.72

Table 8.2 (h) Critical Values of the *F* Distribution Numerator DF = 8

Denom. DF	α(2): 0.50 α(1): 0.25	0.20 0.10	0.10 0.05	0.05 0.025	0.02 0.01	0.01 0.005	0.005 0.0025	0.002 0.001	0.001 0.0005
1	9.19	59.4	239.	957.	5980.	23900.	95700.	598000.	2390000.
2	3.35	9.37	19.4	39.4	99.4	199.	399.	999.	2000.
3	2.44	5.25	8.85	14.5	27.5	44.1	70.5	131.	208.
4	2.08	3.95	6.04	8.98	14.8	21.4	30.6	49.0	69.7
5	1.89	3.34	4.82	6.76	10.3	14.0	18.8	27.6	36.9
6	1.78	2.98	4.15	5.60	8.10	10.6	13.7	19.0	24.3
7	1.70	2.75	3.73	4.90	6.84	8.68	10.9	14.6	18.2
8	1.64	2.59	3.44	4.43	6.03	7.50	9.24	12.0	14.6
9	1.60	2.47	3.23	4.10	5.47	6.69	8.12	10.4	12,4
10	1.56	2.38	3.07	3.85	5.06	6.12	7.33	9.20	10.9
11	1.53	2.30	2.95	3.66	4.74	5.68	6.74	8.35	9.76
12	1.51	2.24	2.85	3.51	4.50	5.35	6.29	7.71	8.94
13	1.49	2.20	2.77	3.39	4.30	5.08	5.93	7.21	8.29
14	1.48	2.15	2.70	3.29	4.14	4.86	5.64	6.80	7.78
15	1.46	2.12	2.64	3.20	4.00	4.67	5.40	6.47	7.37
16	1.45	2.09	2.59	3.12	3.89	4.52	5.20	6.19	7.02
17	1.44	2.06	2.55	3.06	3.79	4.39	5.03	5.96	6.73
18	1.43	2.04	2.51	3.01	3.71	4.28	4.89	5.76	6.48
19	1.42	2.02	2.48	2.96	3.63	4.18	4.76	5.59	6.27
20	1.42	2.00	2.45	2.91	3.56	4.09	4.65	5.44	6.09
21	1.41	1.98	2.42	2.87	3.51	4.01	4.55	5.31	5.92
22	1.40	1.97	2.40	2.84	3.45	3.94	4.46	5.19	5.78
23	1.40	1.95	2.37	2.81	3.41	3.88	4.38	5.09	5.65
24	1.39	1.94	2.36	2.78	3.36	3.83	4.31	4.99	5.54
25	1.39	1.93	2.34	2.75	3.32	3.78	4.25	4.91	5.43
26	1.38	1.92	2.32	2.73	3.29	3.73	4.19	4.83	5.34
27	1.38	1.91	2.31	2.71	3.26	3.69	4.14	4.76	5.25
28	1.38	1.90	2.29	2.69	3.23	3.65	4.09	4.69	5.18
29	1.37	1.89	2.28	2.67	3.20	3.61	4.04	4.64	5.11
30	1.37	1.88	2.27	2.65	3.17	3.58	4.00	4.58	5.04
35	1.36	1.85	2.22	2.58	3.07	3.45	3.83	4.36	4.78
40	1.35	1.83	2.18	2.53	2.99	3.35	3.71	4.21	4.59
45	1.34	1.81	2.15	2.49	2.94	3.28	3.62	4.09	4.45
50	1.33	1.80	2.13	2.46	2.89	3.22	3.55	4.00	4.34
60	1.32	1.77	2.10	2.41	2.82	3.13	3.45	3.86	4.19
70	1.32	1.76	2.07	2.38	2.78	3.08	3.37	3.77	4.08
80	1.31	1.75	2.06	2.35	2.74	3.03	3.32	3.70	4.00
90	1.31	1.74	2.04	2.34	2.72	3.00	3.28	3.65	3.94
100	1.30	1.73	2.03	2.32	2.69	2.97	3.25	3.61	3.89
120	1.30	1.72	2.02	2.30	2.66	2.93	3.20	3.55	3.82
140	1.30	1.71	2.01	2.28	2.64	2.91	3.17	3.51	3.77
160	1.29	1.71	2.00	2.27	2.62	2.88	3.14	3.48	3.73
180	1.29	1.70	1.99	2.26	2.61	2.87	3.12	3.45	3.70
200	1.29	1.70	1.98	2.26	2.60	2.86	3.11	3.43	3.68
300	1.29	1.69	1.97	2.23	2.57	2.82	3.06	3.38	3.61
500	1.28	1.68	1.96	2.22	2.55	2.79	3.03	3.33	3.56
∞	1.28	1.67	1.94	2.19	2.51	2.74	2.97	3.27	3.48

Table 8.2 (i) Critical Values of the *F* Distribution Numerator DF = 9

Denom. DF	α(2): 0.50 α(1): 0.25	0.20 0.10	0.10 0.05	0.05 0.025	0.02 0.01	0.01 0.005	0.005 0.0025	0.002 0.001	0.001 0.0005
1	9.26	59.9	241.	963.	6020.	24100.	96400.	602000.	2410000.
2	3.37	9.38	19.4	39.4	99.4	199.	399.	999.	2000.
3	2.44	5.24	8.81	14.5	27.3	43.9	70.1	130.	207.
4	2.08	3.94	6.00	8.90	14.7	21.1	30.3	48.5	69.0
5	1.89	3.32	4.77	6.68	10.2	13.8	18.5	27.2	36.3
6	1.77	2.96	4.10	5.52	7.98	10.4	13.4	18.7	23.9
7	1.69	2.72	3.68	4.82	6.72	8.51	10.7	14.3	17.8
8	1.63	2.56	3.39	4.36	5.91	7.34	9.03	11.8	14.3
9	1.59	2.44	3.18	4.03	5.35	6.54	7.92	10.1	12.1
10	1.56	2.35	3.02	3.78	4.94	5.97	7.14	8.96	10.6
11	1.53	2.27	2.90	3.59	4.63	5.54	6.56	8.12	9.48
12	1.51	2.21	2.80	3.44	4.39	5.20	6.11	7.48	8.66
13	1.49	2.16	2.71	3.31	4.19	4.94	5.76	6.98	8.03
14	1.47	2.12	2.65	3.21	4.03	4.72	5.47	6.58	7.52
15	1.46	2.09	2.59	3.12	3.89	4.54	5.23	6.26	7.11
16	1.44	2.06	2.54	3.05	3.78	4.38	5.04	5.98	6.77
17	1.43	2.03	2.49	2.98	3.68	4.25	4.87	5.75	6.49
18	1.42	2.00	2.46	2.93	3.60	4.14	4.72	5.56	6.24
19	1.41	1.98	2.42	2.88	3.52	4.04	4.60	5.39	6.03
20	1.41	1.96	2.39	2.84	3.46	3.96	4.49	5.24	5.85
21	1.40	1.95	2.37	2.80	3.40	3.88	4.39	5.11	5.69
22	1.39	1.93	2.34	2.76	3.35	3.81	4.30	4.99	5.55
23	1.39	1.92	2.32	2.73	3.30	3.75	4.22	4.89	5.43
24	1.38	1.91	2.30	2.70	3.26	3.69	4.15	4.80	5.31
25	1.38	1.89	2.28	2.68	3.22	3.64	4.09	4.71	5.21
26	1.37	1.88	2.27	2.65	3.18	3.60	4.03	4.64	5.12
27	1.37	1.87	2.25	2.63	3.15	3.56	3.98	4.57	5.04
28	1.37	1.87	2.24	2.61	3.12	2.52	3.93	4.50	4.96
29	1.36	1.86	2.22	2.59	3.09	3.48	3.89	4.45	4.89
30	1.36	1.85	2.21	2.57	3.07	3.45	3.85	4.39	4.82
35	1.35	1.82	2.16	2.50	2.96	3.32	3.68	4.18	4.57
40	1.34	1.79	2.12	2.45	2.89	3.22	3.56	4.02	4.38
45	1.33	1.77	2.10	2.41	2.83	3.15	3.47	3.91	4.25
50	1.32	1.76	2.07	2.38	2.78	3.09	3.40	3.82	4.14
60	1.31	1.74	2.04	2.33	2.72	3.01	3.30	3.69	3.98
70	1.31	1.72	2.02	2.30	2.67	2.95	3.23	3.60	3.88
80	1.30	1.71	2.00	2.28	2.64	2.91	3.17	3.53	3.80
90	1.30	1.70	1.99	2.26	2.61	2.87	3.13	3.48	3.74
100	1.29	1.69	1.97	2.24	2.59	2.85	3.10	3.44	3.69
120	1.29	1.68	1.96	2.22	2.56	2.81	3.06	3.38	3.62
140	1.29	1.68	1.95	2.21	2.54	2.78	3.02	3.34	3.57
160	1.28	1.67	1.94	2.19	2.52	2.76	3.00	3.31	3.54
180	1.28	1.67	1.93	2.19	2.51	2.74	2.98	3.28	3.51
200	1.28	1.66	1.93	2.18	2.50	2.73	2.96	3.26	3.49
300	1.27	1.65	1.91	2.16	2.47	2.69	2.92	3.21	3.42
500	1.27	1.64	1.90	2.14	2.44	2.66	2.88	3.16	3.37
∞	1.27	1.63	1.88	2.11	2.41	2.62	2.83	3.10	3.30

Table 8.2 (j) Critical Values of the *F* Distribution Numerator DF = 10

Denom. DF	α(2): 0.50 α(1): 0.25	0.20 0.10	0.10 0.05	0.05 0.025	0.02 0.01	0.01 0.005	0.005 0.0025	0.002 0.001	0.001 0.0005
1	9.32	60.2	242.	969.	6060.	24200.	96900.	606000.	2420000.
2	3.38	9.39	19.4	39.4	99.4	199.	399.	999.	2000.
3	2.44	5.23	8.79	14.4	27.2	43.7	69.8	129.	206.
4	2.08	3.92	5.96	8.84	14.5	21.0	30.0	48.1	68.3
5	1.89	3.30	4.74	6.62	10.1	13.6	18.3	26.9	35.9
6	1.77	2.94	4.06	5.46	7.87	10.3	13.2	18.4	23.5
7	1.69	2.70	3.64	4.76	6.62	8.38	10.5	14.1	17.5
8	1.63	2.54	3.35	4.30	5.81	7.21	8.87	11.5	14.0
9	1.59	2.42	3.14	3.96	5.26	6.42	7.77	9.89	11.8
10	1.55	2.32	2.98	3.72	4.85	5.85	6.99	8.75	10.3
11	1.52	2.25	2.85	3.53	4.54	5.42	6.41	7.92	9.24
12	1.50	2.19	2.75	3.37	4.30	5.09	5.97	7.29	8.43
13	1.48	2.14	2.67	3.25	4.10	4.82	5.62	6.80	7.81
14	1.46	2.10	2.60	3.15	3.94	4.60	5.33	6.40	7.31
15	1.45	2.06	2.54	3.06	3.80	4.42	5.10	6.08	6.91
16	1.44	2.03	2.49	2.99	3.69	4.27	4.90	5.81	6.57
17	1.43	2.00	2.45	2.92	3.59	4.14	4.73	5.58	6.29
18	1.42	1.98	2.41	2.87	3.51	4.03	4.59	5.39	6.05
19	1.41	1.96	2.38	2.82	3.43	3.93	4.46	5.22	5.84
20	1.40	1.94	2.35	2.77	3.37	3.85	4.35	5.08	5.66
21	1.39	1.92	2.32	2.73	3.31	3.77	4.26	4.95	5.50
22	1.39	1.90	2.30	2.70	3.26	3.70	4.17	4.83	5.36
23	1.38	1.89	2.27	2.67	3.21	3.64	4.09	4.73	5.24
24	1.38	1.88	2.25	2,64	3.17	3.59	4.03	4.64	5.13
25	1.37	1.87	2.24	2.61	3.13	3.54	3.96	4.56	5.03
26	1.37	1.86	2.22	2.59	3.09	3.49	3.91	4.48	4.94
27	1.36	1.85	2.20	2.57	3.06	3.45	3.85	4.41	4.86
28	1.36	1.84	2.19	2.55	3.03	3.41	3.81	4.35	4.78
29	1.35	1.83	2.18	2.53	3.00	3.38	3.76	4.29	4.71
30	1.35	1.82	2.16	2.51	2.98	3.34	3.72	4.24	4.65
35	1.34	1.79	2.11	2.44	2.88	3.21	3.56	4.03	4.39
40	1.33	1.76	2.08	2.39	2.80	3.12	3.44	3.87	4.21
45	1.32	1.74	2.05	2.35	2.74	3.04	3.35	3.76	4.08
50	1.31	1.73	2.03	2.32	2.70	2.99	3.28	3.67	3.97
60	1.30	1.71	1.99	2.27	2.63	2.90	3.18	3.54	3.82
70	1.30	1.69	1.97	2.24	2.59	2.85	3.11	3.45	3.71
80	1.29	1.68	1.95	2.21	2.55	2.80	3.05	3.39	3.64
90	1.29	1.67	1.94	2.19	2.52	2.77	3.01	3.34	3.58
100	1.28	1.66	1.93	2.18	2.50	2.74	2.98	3.30	3.53
120	1.28	1.65	1.91	2.16	2.47	2.71	2.94	3.24	3,46
140	1.28	1.64	1.90	2.14	2.45	2.68	2.90	3.20	3.42
160	1.27	1.64	1.89	2.13	2.43	2.66	2.88	3.17	3.38
180	1.27	1.63	1.88	2.12	2.42	2.64	2.86	3.14	3.35
200	1.27	1.63	1.88	2.11	2.41	2.63	2.84	3.12	3.33
300	1.26	1.62	1.86	2.09	2.38	2.59	2.80	3.07	3.27
500	1.26	1.61	1.85	2.07	2.36	2.56	2.76	3.02	3.22
∞	1.25	1.60	1.83	2.05	2.32	2.52	2.71	2.96	3.14

Table 8.2 (k) Critical Values of the *F* Distribution Numerator DF = 11

Denom. DF	α(2): 0.50 α(1): 0.25	0.20 0.10	0.10 0.05	0.05 0.025	0.02 0.01	0.01 0.005	0.005 0.0025	0.002 0.001	0.001 0.0005
1	9.37	60.5	243.	973.	6080.	24300.	97300.	608000.	2430000.
2	3.39	9.40	19.4	39.4	99.4	199.	399.	999.	2000.
3	2.45	5.22	8.76	14.4	27.1	43.5	69.5	129.	205.
4	2.08	3.91	5.94	8.79	14.5	20.8	29.8	47.7	67.8
5	1.89	3.28	4.70	6.57	9.96	13.5	18.1	26.6	35.5
6	1.77	2.92	4.03	5.41	7.79	10.1	13.1	18.2	23.2
7	1.69	2.68	3.60	4.71	6.54	8.27	10.4	13.9	17.2
8	1.63	2.52	3.31	4.24	5.73	7.10	8.73	11.4	13.8
9	1.58	2.40	3.10	3.91	5.18	6.31	7.63	9.72	11.6
10	1.55	2.30	2.94	3.66	4.77	5.75	6.86	8.59	10.1
11	1.52	2.23	2.82	3.47	4.46	5.32	6.29	7.76	9.05
12	1.49	2.17	2.72	3.32	4.22	4.99	5.85	7.14	8.25
13	1.47	2.12	2.63	3.20	4.02	4.72	5.50	6.65	7.63
14	1.46	2.07	2.57	3.09	3.86	4.51	5.21	6.26	7.13
15	1.44	2.04	2.51	3.01	3.73	4.33	4.98	5.94	6.73
16	1.43	2.01	2.46	2.93	3.62	4.18	4.79	5.67	6.40
17	1.42	1.98	2.41	2.87	3.52	4.05	4.62	5.44	6.12
18	1.41	1.95	2.37	2.81	3.43	3.94	4.48	5.25	5.89
19	1.40	1.93	2.34	2.76	3.36	3.84	4.35	5.08	5.68
20	1.39	1.91	2.31	2.72	3.29	3.76	4.24	4.94	5.50
21	1.39	1.90	2.28	2.68	3.24	3.68	4.15	4.81	5.35
22	1.38	1.88	2.26	2.65	3.18	3.61	4.06	4.70	5.21
23	1.37	1.87	2.24	2.62	3.14	3.55	3.99	4.60	5.09
24	1.37	1.85	2.22	2.59	3.09	3.50	3.92	4.51	4.98
25	1.36	1.84	2.20	2.56	3.06	3.45	3.85	4.42	4.88
26	1.36	1.83	2.18	2.54	3.02	3.40	3.80	4.35	4.79
27	1.35	1.82	2.17	2.51	2.99	3.36	3.75	4.28	4.71
28	1.35	1.81	2.15	2.49	2.96	3.32	3.70	4.22	4.63
29	1.35	1.80	2.14	2.48	2.93	3.29	3.66	4.16	4.56
30	1.34	1.79	2.13	2.46	2.91	3.25	3.61	4.11	4.50
35	1.33	1.76	2.07	2.39	2.80	3.12	3.45	3.90	4.25
40	1.32	1.74	2.04	2.33	2.73	3.03	3.33	3.75	4.07
45	1.31	1.72	2.01	2.29	2.67	2.96	3.25	3.64	3.94
50	1.30	1.70	1.99	2.26	2.63	2.90	3.18	3.55	3.83
60	1.29	1.68	1.95	2.22	2.56	2.82	3.08	3.42	3.68
70	1.29	1.66	1.93	2.18	2.51	2.76	3.00	3.33	3.58
80	1.28	1.65	1.91	2.16	2.48	2.72	2.95	3.27	3.50
90	1.28	1.64	1.90	2.14	2.45	2.68	2.91	3.22	3.44
100	1.27	1.64	1.89	2.12	2.43	2.66	2.88	3.18	3.40
120	1.27	1.63	1.87	2.10	2.40	2.62	2.83	3.12	3.33
140	1.27	1.62	1.86	2.09	2.38	2.59	2.80	3.08	3.28
160	1.26	1.61	1.85	2.07	2.36	2.57	2.78	3.05	3.25
180	1.26	1.61	1.84	2.07	2.35	2.56	2.76	3.02	3.22
200	1.26	1.60	1.84	2.06	2.34	2.54	2.74	3.00	3.20
300	1.26	1.59	1.82	2.04	2.31	2.51	2.70	2.95	3.14
500	1.25	1.58	1.81	2.02	2.28	2.48	2.66	2.91	3.09
∞	1.25	1.57	1.79	1.99	2.25	2.43	2.61	2.84	3.01

Table 8.2 (l) Critical Values of the *F* Distribution Numerator DF = 12

Denom. DF	α(2): 0.50 α(1): 0.25	0.20 0.10	0.10 0.05	0.05 0.025	0.02 0.01	0.01 0.005	0.005 0.0025	0.002 0.001	0.001 0.0005
1	9.41	60.7	244.	977.	6110.	24400.	97700.	611000.	2440000.
2	3.39	9.41	19.4	39.4	99.4	199.	399.	999.	2000.
3	2.45	5.22	8.74	14.3	27.1	43.4	69.3	128.	204.
4	2.08	3.90	5.91	8.75	14.4	20.7	29.7	47.4	67.4
5	1.89	3.27	4.68	6.52	9.89	13.4	18.0	26.4	35.2
6	1.77	2.90	4.00	5.37	7.72	10.0	12.9	18.0	23.0
7	1.68	2.67	3.57	4.67	6.47	8.18	10.3	13.7	17.0
8	1.62	2.50	3.28	4.20	5.67	7.01	8.61	11.2	13.6
9	1.58	2.38	3.07	3.87	5.11	6.23	7.52	9.57	11.4
10	1.54	2.28	2.91	3.62	4.71	5.66	6.75	8.45	9.94
11	1.51	2.21	2.79	3.43	4.40	5.24	6.18	7.63	8.88
12	1.49	2.15	2.69	3.28	4.16	4.91	5.74	7.00	8.09
13	1.47	2.10	2.60	3.15	3.96	4.64	5.40	6.52	7.48
14	1.45	2.05	2.53	3.05	3.80	4.43	5.12	6.13	6.99
15	1.44	2.02	2.48	2.96	3.67	4.25	4.88	5.81	6.59
16	1.43	1.99	2.42	2.89	3.55	4.10	4.69	5.55	6.26
17	1.41	1.96	2.38	2.82	3.46	3.97	4.52	5.32	5.98
18	1.40	1.93	2.34	2.77	3.37	3.86	4.38	5.13	5.75
19	1.40	1.91	2.31	2.72	3.30	3.76	4.26	4.97	5.55
20	1.39	1.89	2.28	2.68	3.23	3.68	4.15	4.82	5.37
21	1.38	1.87	2.25	2.64	3.17	3.60	4.06	4.70	5.21
22	1.37	1.86	2.23	2.60	3.12	3.54	3.97	4.58	5.08
23	1.37	1.84	2.20	2.57	3.07	3.47	3.89	4.48	4.96
24	1.36	1.83	2.18	2.54	3.03	3.42	3.83	4.39	4.85
25	1.36	1.82	2.16	2.51	2.99	3.37	3.76	4.31	4.75
26	1.35	1.81	2.15	2.49	2.96	3.33	3.71	4.24	4.66
27	1.35	1.80	2.13	2.47	2.93	3.28	3.66	4.17	4.58
28	1.34	1.79	2.12	2.45	2.90	3.25	3.61	4.11	4.51
29	1.34	1.78	2.10	2.43	2.87	3.21	3.56	4.05	4.44
30	1.34	1.77	2.09	2.41	2.84	3.18	3.52	4.00	4.38
35	1.32	1.74	2.04	2.34	2.74	3.05	3.36	3.79	4.13-
40	1.31	1.71	2.00	2.29	2.66	2.95	3.25	3.64	3.95
45	1.30	1.70	1.97	2.25	2.61	2.88	3.16	3.53	3.82
50	1.30	1.68	1.95	2.22	2.56	2.82	3.09	3.44	3.71
60	1.29	1.66	1.92	2.17	2.50	2.74	2.99	3.32	3.57
70	1.28	1.64	1.89	2.14	2.45	2.68	2.92	3.23	3.46
80	1.27	1.63	1.88	2.11	2.42	2.64	2.87	3.16	3.39
90	1.27	1.62	1.86	2.09	2.39	2.61	2.83	3.11	3.33
100	1.27	1.61	1.85	2.08	2.37	2.58	2.80	3.07	3.28
120	1.26	1.60	1.83	2.05	2.34	2.54	2.75	3.02	3.22
140	1.26	1.59	1.82	2.04	2.31	2.52	2.72	2.98	3.17
160	1.26	1.59	1.81	2.03	2.30	2.50	2.69.	2.95	3.14
180	1.25	1.58	1.81	2.02	2.28	2.48	2.67	2.92	3.11
200	1.25	1.58	1.80	2.01	2.27	2.47	2.66	2.90	3.09
300	1.25	1.57	1.78	1.99	2.24	2.43	2.6!	2.85	3.02
500	1.24	1.56	1.77	1.97	2.22	2.40	2.58	2.81	2.97
∞	1.24	1.55	1.75	1.94	2.18	2.36	2.53	2.74	2.90

Table 8.2 (m) Critical Values of the *F* Distribution Numerator DF =13

Denom. DF	α(2): 0.50 α(l): 0.25	0.20 0.10	0.10 0.05	0.05 0.025	0.02 0.01	0.01 0.005	0.005 0.0025	0.002 0.001	0.001 0.0005
1	9.44	60.9	245.	980.	6130.	24500.	98000.	613000.	2450000.
2	3.40	9.41	19.4	39.4	99.4	199.	399.	999.	2000.
3	2.45	5.21	8.73	14.3	27.0	43.3	69.1	128.	204.
4	2.08	3.89	5.89	8.71	14.3	20.6	29.5	47.2	67.1
5	1.89	3.26	4.66	6.49	9.82	13.3	17.9	26.2	34.9
6	1.77	2.89	3.98	5.33	7.66	9.95	12.8	17.8	22.7
7	1.68	2.65	3.55	4.63	6.41	8.10	10.1	13.6	16.8
8	1.62	2.49	3.26	4.16	5.61	6.94	8.51	11.1	13.4
9	1.58	2.36	3.05	3.83	5.05	6.15	7.43	9.44	11.3
10	1.54	2.27	2.89	3.58	4.65	5.59	6.66	8.32	9.80
11	1.51	2.19	2.76	3.39	4.34	5.16	6.09	7.51	8.74
12	1.49	2.13	2.66	3.24	4.10	4.84	5.66	6.89	7.96
13	1.47	2.08	2.58	3.12	3.91	4.57	5.31	6.41	7.35
14	1.45	2.04	2.51	3.01	3.75	4.36	5.03	6.02	6.86
15	1.43	2.00	2.45	2.92	3.61	4.18	4.80	5.71	6.47
16	1.42	1.97	2.40	2.85	3.50	4.03	4.61	5.44	6.14
17	1.41	1.94	2.35	2.79	3.40	3.90	4.44	5.22	5.86
18	1.40	1.92	2.31	2.73	3.32	3.79	4.30	5.03	5.63
19	1.39	1.89	2.28	2.68	3.24	3.70	4.18	4.87	5.43
20	1.38	1.87	2.25	2.64	3.18	3.61	4.07	4.72	5.25
21	1.37	1.86	2.22	2.60	3.12	3.54	3.98	4.60	5.10
22	1.37	1.84	2.20	2.56	3.07	3.47	3.89	4.49	4.97
23	1.36	1.83	2.18	2.53	3.02	3.41	3.82	4.39	4.84
24	1.36	1.81	2.15	2.50	2.98	3.35	3.75	4.30	4.74
25	1.35	1.80	2.14	2.48	2.94	3.30	3.69	4.22	4.64
26	1.35	1.79	2.12	2.45	2.90	3.26	3.63	4.14	4.55
27	1.34	1.78	2.10	2.43	2.87	3.22	3.58	4.08	4.47
28	1.34	1.77	2.09	2.41	2.84	3.18	3.53	4.01	4.40
29	1.33	1.76	2.08	2.39	2.81	3.15	3.49	3.96	4.33
30	1.33	1.75	2.06	2.37	2.79	3.11	3.45	3.91	4.27
35	1.32	1.72	2.01	2.30	2.69	2.98	3.29	3.70	4.02
40	1.31	1.70	1.97	2.25	2.61	2.89	3.17	3.55	3.85
45	1.30	1.68	1.94	2.21	2.55	2.82	3.08	3.44	3.71
50	1.29	1.66	1.92	2.18	2.51	2.76	3.01	3.35	3.61
60	1.28	1.64	1.89	2.13	2.44	2.68	2.91	3.23	3.46
70	1.27	1.62	1.86	2.10	2.40	2.62	2.84	3.14	3.36
80	1.27	1.61	1.84	2.07	2.36	2.58	2.79	3.07	3.29
90	1.26	1.60	1.83	2.05	2.33	2.54	2.75	3.02	3.23
100	1.26	1.59	1.82	2.04	2.31	2.52	2.72	2.99	3.19
120	1.26	1.58	1.80	2.01	2.28	2.48	2.67	2.93	3.12
140	1.25	1.57	1.79	2.00	2.26	2.45	2.64	2.89	3.07
160	1.25	1.57	1.78	1.99	2.24	2.43	2.62	2.86	3.04
180	1.25	1.56	1.77	1.98	2.23	2.42	2.60	2.83	3.01
200	1.24	1.56	1.77	1.97	2.22	2.40	2.58	2.82	2.99
300	1.24	1.55	1.75	1.95	2.19	2.37	2.54	2.76	2.93
500	1.24	1.54	1.74	1.93	2.17	2.34	2.50	2.72	2.88
∞	1.23	1.52	1.72	1.90	2.13	2.29	2.45	2.66	2.81

Table 8.2 (n) Critical Values of the *F* Distribution Numerator DF = 14

Denom. DF	α(2): 0.50 α(1): 0.25	0.20 0.10	0.10 0.05	0.05 0.025	0.02 0.01	0.01 0.005	0.005 0.0025	0.002 0.001	0.00 0.0005
1	9.47	61.1	245.	983.	6140.	24600.	98300.	614000.	2460000.
2	3.41	9.42	19.4	39.4	99.4	199.	399.	999.	2000.
3	2.45	5.20	8.71	14.3	26.9	43.2	69.0	128.	203.
4	2.08	3.88	5.87	8.68	14.2	20.5	29.4	46.9	66.8
5	1.89	3.25	4.64	6.46	9.77	13.2	17.8	26.1	34.7
6	1.76	2.88	3.96	5.30	7.60	9.88	12.7	17.7	22.6
7	1.68	2.64	3.53	4.60	6.36	8.03	10.1	13.4	16.6
8	1.62	2.48	3.24	4.13	5.56	6.87	8.43	10.9	13.3
9	1.57	2.35	3.03	3.80	5.01	6.09	7.35	9.33	11.1
10	1.54	2.26	2.86	3.55	4.60	5.53	6.58	8.22	9.67
11	1.51	2.18	2.74	3.36	4.29	5.10	6.02	7.41	8.62
12	1.48	2.12	2.64	3.21	4.05	4.77	5.58	6.79	7.84
13	1.46	2.07	2.55	3.08	3.86	4.51	5.24	6.31	7.23
14	1.44	2.02	2.48	2.98	3.70	4.30	4.96	5.93	6.75
15	1.43	1.99	2.42	2.89	3.56	4.12	4.73	5.62	6.36
16	1.42	1.95	2.37	2.82	3.45	3.97	4.54	5.35	6.03
17	1.41	1.93	2.33	2.75	3.35	3.84	4.37	5.13	5.76
18	1.40	1.90	2.29	2.70	3.27	3.73	4.23	4.94	5.53
19	1.39	1.88	2.26	2.65	3.19	3.64	4.11	4.78	5.33
20	1.38	1.86	2.22	2.60	3.13	3.55	4.00	4.64	5.15
21	1.37	1.84	2.20	2.56	3.07	3.48	3.91	4.51	5.00
22	1.36	1.83	2.17	2.53	3.02	3.41	3.82	4.40	4.87
23	1.36	1.81	2.15	2.50	2.97	3.35	3.75	4.30	4.75
24	1.35	1.80	2.13	2.47	2.93	3.30	3.68	4.21	4.64
25	1.35	1.79	2.11	2.44	2.89	3.25	3.62	4.13	4.54
26	1.34	1.77	2.09	2.42	2.86	3.20	3.56	4.06	4.46
27	1.34	1.76	2.08	2.39	2.82	3.16	3.51	3.99	4.38
28	1.33	1.75	2.06	2.37	2.79	3.12	3.46	3.93	4,30
29	1.33	1.75	2.05	2.36	2.77	3.09	3.42	3.88	4.24
30	1.33	1.74	2.04	2.34	2.74	3.06	3.38	3.82	4.18
35	1.31	1.70	1.99	2.27	2.64	2.93	3.22	3.62	3.93
40	1.30	1.68	1.95	2.21	2.56	2.83	3.10	3.47	3.76
45	1.29	1.66	1.92	2.17	2.51	2.76	3.02	3.36	3.63
50	1.28	1.64	1.89	2.14	2.46	2.70	2.95	3.27	3.52
60	1.27	1.62	1.86	2.09	2.39	2.62	2.85	3.15	3.38
70	1.27	1.60	1.84	2.06	2.35	2.56	2.78	3.06	3.28
80	1.26	1.59	1.82	2.03	2.31	2.52	2.73	3.00	3.20
90	1.26	1.58	1.80	2.02	2.29	2.49	2.69	2.95	3.14
100	1.25	1.57	1.79	2.00	2.27	2.46	2.65	2.91	3.10
120	1.25	1.56	1.78	1.98	2.23	2.42	2.61	2.85	3.03
140	1.24	1.55	1.76	1.96	2.2!	2.40	2.58	2.81	2.99
160	1.24	1.55	1.75	1.95	2.20	2.38	2.55	2.78	2.95
180	1.24	1.54	1.75	1.94	2.18	2.36	2.53	2.76	2.93
200	1.24	1.54	1.74	1.93	2.17	2.35	2.52	2.74	2.91
300	1.23	1.53	1.72	1.91	2.14	2.31	2.47	2.69	2.84
500	1.23	1.52	1.71	1.89	2.12	2.28	2.44	2.64	2.79
∞	1.22	1.50	1.69	1.87	2.08	2.24	2.39	2.58	2.72

Table 8.2 (o) Critical Values of the *F* Distribution Numerator DF = 15

Denom. DF	$\alpha(2)$: 0.50	0.20	0.10	0.05	0.02	0.01	0.005	0.002	0.001
	$\alpha(1)$: 0.25	0.10	0.05	0.025	0.01	0.005	0.0025	0.001	0.0005
1	9.49	61.2	246.	985.	6160.	24600.	98500.	616000.	2460000.
2	3.41	9.42	19.4	39.4	99.4	199.	399.	999.	2000.
3	2.46	5.20	8.70	14.3	26.9	43.1	68.8	127.	203.
4	2.08	3.87	5.86	8.66	14.2	20.4	29.3	46.8	66.5
5	1.89	3.24	4.62	6.43	9.72	13.1	17.7	25.9	34.5
6	1.76	2.87	3.94	5.27	7.56	9.81	12.7	17.6	22.4
7	1.68	2.63	3.51	4.57	6.31	7.97	9.98	13.3	16.5
8	1.62	2.46	3.22	4.10	5.52	6.81	8.35	10.8	13.1
9	1.57	2.34	3.01	3.77	4.96	6.03	7.28	9.24	11.0
10	1.53	2.24	2.85	3.52	4.56	5.47	6.51	8.13	9.56
11	1.50	2.17	2.72	3.33	4.25	5.05	5.95	7.32	8.52
12	1.48	2.10	2.62	3.18	4.01	4.72	5.52	6.71	7.74
13	1.46	2.05	2.53	3.05	3.82	4.46	5.17	6.23	7.13
14	1.44	2.01	2.46	2.95	3.66	4.25	4.89	5.85	6.65
15	1.43	1.97	2.40	2.86	3.52	4.07	4.67	5.54	6.26
16	1.41	1.94	2.35	2.79	3.41	3.92	4.47	5.27	5.94
17	1.40	1.91	2.31	2.72	3.31	3.79	4.31	5.05	5.67
18	1.39	1.89	2.27	2.67	3.23	3.68	4.17	4.87	5.44
19	1.38	1.86	2.23	2.62	3.15	3.59	4.05	4.70	5.24
20	1.37	1.84	2.20	2.57	3.09	3.50	3.94	4.56	5.07
21	1.37	1.83	2.18	2.53	3.03	3.43	3.85	4.44	4.92
22	1.36	1.81	2.15	2.50	2.98	3.36	3.76	4.33	4.78
23	1.35	1.80	2.13	2.47	2.93	3.30	3.69	4.23	4.66
24	1.35	1.78	2.11	2.44	2.89	3.25	3.62	4.14	4.56
25	, 1.34	1.77	2.09	2.41	2.85	3.20	3.56	4.06	4.46
26	1.34	1.76	2.07	2.39	2.81	3.15	3.50	3.99	4.37
27	1.33	1.75	2.06	2.36	2.78	3.11	3.45	3.92	4.29
28	1.33	1.74	2.04	2.34	2.75	3.07	3.40	3.86	4.22
29	1.32	1.73	2.03	2.32	2.73	3.04	3.36	3.80	4.15
30	1.32	1.72	2.01	2.31	2.70	3.01	3.32	3.75	4.09
35	1.31	1.69	1.96	2.23	2.60	2.88	3.16	3.55	3.85
40	1.30	1.66	1.92	2.18	2.52	2.78	3.04	3.40	3.68
45	1.29	1.64	1.89	2.14	2.46	2.71	2.96	3.29	3.55
50	1.28	1.63	1.87	2.11	2.42	2.65	2.89	3.20	3.45
60	1.27	1.60	1.84	2.06	2.35	2.57	2.79	3.08	3.30
70	1.26	1.59	1.81	2.03	2.31	2.51	2.72	2.99	3.20
80	1.26	1.57	1.79	2.00	2.27	2.47	2.67	2.93	3.12
90	1.25	1.56	1.78	1.98	2.24	2.44	2.63	2.88	3.07
100	1.25	1.56	1.77	1.97	2.22	2.41	2.60	2.84	3.02
120	1.24	1.55	1.75	1.94	2.19	2.37	2.55	2.78	2.96
140	1.24	1.54	1.74	1.93	2.17	2.35	2.52	2.74	2.91
160	1.24	1.53	1.73	1.92	2.15	2.33	2.49	2.7!	2.88
180	1.23	1.53	1.72	1.91	2.14	2.31	2.48	2.69	2.85
200	1.23	1.52	1.72	1.90	2.13	2.30	2.46	2.67	2.83
300	1.23	1.51	1.70	1.88	2.10	2.26	2.42	2.62	2.77
500	1.22	1.50	1.69	1.86	2.07	2.23	2.38	2.58	2.72
∞	1.22	1.49	1.67	1.83	2.04	2.19	2.33	2.51	2.65

Table 8.2 (p) Critical Values of the *F* Distribution Numerator DF = 16

Denom. DF	α(2): 0.50 α(1): 0.25	0.20 0.10	0.10 0.05	0.05 0.025	0.02 0.01	0.01 0.005	0.005 0.0025	0.002 0.001	0.001 0.0005
1	9.52	61.3	246.	987.	6170.	24700.	98700.	617000.	2470000.
2	3.41	9.43	19.4	39.4	99.4	199.	399.	999	2000.
3	2.46	5.20	8.69	14.2	26.8	43.0	68.7	127.	202.
4	2.08	3.86	5.84	8.63	14.2	20.4	29.2	46.6	66.2
5	1.88	3.23	4.60	6.40	9.68	13.1	17.6	25.8	34.3
6	1.76	2.86	3.92	5.24	7.52	9.76	12.6	17.4	22.3
7	1.68	2.62	3.49	4.54	6.28	7.91	9.91	13.2	16.4
8	1.62	2.45	3.20	4.08	5.48	6.76	8.29	10.8	13.0
9	1.57	2.33	2.99	3.74	4.92	5.98	7.21	9.15	10.9
10	1.53	2.23	2.83	3.50	4.52	5.42	6.45	8.05	9.46
11	1.50	2.16	2.70	3.30	4.21	5.00	5.89	7.24	8.43
12	1.48	2.09	2.60	3.15	3.97	4.67	5.46	6.63	7.65
13	1.46	2.04	2.51	3.03	3.78	4.41	5.11	6.16	7.05
14	1.44	2.00	2.44	2.92	3.62	4.20	4.84	5.78	6.57
15	1.42	1.96	2.38	2.84	3.49	4.02	4.61	5.46	6.18
16	1.41	1.93	2.33	2.76	3.37	3.87	4.42	5.20	5.86
17	1.40	1.90	2.29	2.70	3.27	3.75	4.25	4.99	5.59
18	1.39	1.87	2.25	2.64	3.19	3.64	4.11	4.80	5.36
19	1.38	1.85	2.21	2.59	3.12	3.54	3.99	4.64	5.16
20	1.37	1.83	2.18	2.55	3.05	3.46	3.89	4.49	4.99
21	1.36	1.81	2.16	2.51	2.99	3.38	3.79	4.37	4.84
22	1.36	1.80	2.13	2.47	2.94	3.31	3.71	4.26	4.71
23	1.35	1.78	2.11	2.44	2.89	3.25	3.63	4.16	4.59
24	1.34	1.77	2.09	2.41	2.85	3.20	3.56	4.07	4.48
25	1.34	1.76	2.07	2.38	2.81	3.15	3.50	3.99	4.39
26	1.33	1.75	2.05	2.36	2.78	3.11	3.45	3.92	4.30
27	1.33	1.74	2.04	2.34	2.75	3.07	3.40	3.86	4.22
28	1.32	1.73	2.02	2.32	2.72	3.03	3.35	3.80	4.15
29	1.32	1.72	2.01	2.30	2.69	2.99	3.31	3.74	4.08
30	1.32	1.71	1.99	2.28	2.66	2.96	3.27	3.69	4.02
35	1.30	1.67	1.94	2.21	2.56	2.83	3.11	3.48	3.78
40	1.29	1.65	1.90	2.15	2.48	2.74	2.99	3.34	3.61
45	1.28	1.63	1.87	2.11	2.43	2.66	2.90	3.23	3.48
50	1.27	1.61	1.85	2.08	2.38	2.61	2.84	3.14	3.38
60	1.26	1.59	1.82	2.03	2.31	2.53	2.74	3.02	3.23
70	1.26	1.57	1.79	2.00	2.27	2.47	2.67	2.93	3.13
80	1.25	1.56	1.77	1.97	2.23	2.43	2.62	2.87	3.06
90	1.25	1.55	1.76	1.95	2.21	2.39	2.58	2.82	3.00
100	1.24	1.54	1.75	1.94	2.19	2.37	2.55	2.78	2.96
120	1.24	1.53	1.73	1.92	2.15	2.33	2.50	2.72	2.89
140	1.23	1.52	1.72	1.90	2.13	2.30	2.47	2.68	2.84
160	1.23	1.52	1.71	1.89	2.11	2.28	2.44	2.65	2.81
180	1.23	1.51	1.70	1.88	2.10	2.26	2.42	2.63	2.78
200	1.23	1.51	1.69	1.87	2.09	2.25	2.41	2.61	2.76
300	1.22	1.49	1.68	1.85	2.06	2.21	2.36	2.56	2.70
500	1.22	1.49	1.66	1.83	2.04	2.19	2.33	2.52	2.65
∞	1.21	1.47	1.64	1.80	2.00	2.14	2.28	2.45	2.58

Table 8.2 (q) Critical Values of the *F* Distribution Numerator DF =17

Denom. DF	α(2): 0.50 α(1): 0.25	0.20 0.10	0.10 0.05	0.05 0.025	0.02 0.01	0.01 0.005	0.005 0.0025	0.002 0.001	0.001 0.0005
1	9.53	61.5	247.	989.	6180.	24700.	98900.	618000.	2470000.
2	3.42	9.43	19.4	39.4	99.4	199.	399.	999.	2000.
3	2.46	5.19	8.68	14.2	26.8	42.9	68.6	127.	202.
4	2.08	3.86	5.83	8.61	14.1	20.3	29.1	46.5	66.0
5	1.88	3.22	4.59	6.38	9.64	13.0	17.5	25.7	34.2
6	1.76	2.85	3.91	5.22	7.48	9.71	12.5	17.4	22.1
7	1.67	2.61	3.48	4.52	6.24	7.87	9.85	13.1	16.3
8	1.61	2.45	3.19	4.05	5.44	6.72	8.23	10.7	12.9
9	1.57	2.32	2.97	3.72	4.89	5.94	7.16	9.08	10.8
10	1.53	2.22	2.81	3.47	4.49	5.38	6.40	7.98	9.38
11	1.50	2.15	2.69	3.28	4.18	4.96	5.84	7.17	8.34
12	1.47	2.08	2.58	3.13	3.94	4.63	5.40	6.57	7.57
13	1.45	2.03	2.50	3.00	3.75	4.37	5.06	6.09	6.97
14	1.44	1.99	2.43	2.90	3.59	4.16	4.79	5.71	6.49
15	1.42	1.95	2.37	2.81	3.45	3.98	4.56	5.40	6.11
16	1.41	1.92	2.32	274	3.34	3.83	4.37	5.14	5.79
17	1.39	1.89	2.27	2.67	3.24	3.71	4.21	4.92	5.52
18	1.38	1.86	2.23	2.62	3.16	3.60	4.07	4.74	5.29
19	1.37	1.84	2.20	2.57	3.08	3.50	3.94	4.58	5.09
20	1.37	1.82	2.17	2.52	3.02	3.42	3.84	4.44	4.92
21	1.36	1.80	2.14	2.48	2.96	3.34	3.74	4.31	4.77
22	1.35	1.79	2.11	2.45	2.91	3.27	3.66	4.20	4.64
23	1.35	1.77	2.09	2.42	2.86	3.21	3.58	4.10	4.52
24	1.34	1.76	2.07	2.39	2.82	3.16	3.52	4.02	4.41
25	1.33	1.75	2.05	2.36	2.78	3.11	3.46	3.94	4.32
26	1.33	1.73	2.03	2.34	2.75	3.07	3.40	3.86	4.23
27	1.33	1.72	2.02	2.31	2.71	3.03	3.35	3.80	4.15
28	1.32	1.71	2.00	2.29	2.68	2.99	3.30	3.74	4.08
29	1.32	1.71	1.99	2.27	2.66	2.95	3.26	3.68	4.02
30	1.31	1.70	1.98	2.26	2.63	2.92	3.22	3.63	3.96
35	1.30	1.66	1.92	2.18	2.53	2.79	3.06	3.43	3.72
40	1.29	1.64	1.89	2.13	2.45	2.70	2.95	3.28	3.54
45	1.28	1.62	1.86	2.09	2.39	2.62	2.86	3.17	3.41
50	1.27	1.60	1.83	2.06	2.35	2.57	2.79	3.09	3.31
60	1.26	1.58	1.80	2.01	2.28	2.49	2.69	2.96	3.17
70	1.25	1.56	1.77	1.97	2.23	2.43	2.62	2.88	3.07
80	1.25	1.55	1.75	1.95	2.20	2.39	2.57	2.81	3.00
90	1.24	1.54	1.74	1.93	2.17	2.35	2.53	2.76	2.94
100	1.24	1.53	1.73	1.91	2.15	2.33	2.50	2.73	2.89
120	1.23	1.52	1.71	1.89	2.12	2.29	2.45	2.67	2.83
140	1.23	1.51	1.70	1.87	2.10	2.26	2.42	2.63	2.78
160	1.23	1.50	1.69	1.86	2.08	2.24	2.40	2.60	2.75
180	1.22	1.50	1.68	1.85	2.07	2.22	2.38	2.58	2.72
200	1.22	1.49	1.67	1.84	2.06	2.21	2.36	2.56	2.70
300	1.22	1.48	1.66	1.82	2.03	2.17	2.32	2.50	2.64
500	1.21	1.47	1.64	1.80	2.00	2.14	2.28	2.46	2.59
∞	1.21	1.46	1.62	1.78	1.97	2.10	2.23	2.40	2.52

Table 8.2 (r) Critical Values of the *F* Distribution Numerator DF = 18

Denom. DF	α(2): 0.50 α(1): 0.25	0.20 0.10	0.10 0.05	0.05 0.025	0.02 0.01	0.01 0.005	0.005 0.0025	0.002 0.001	0.001 0.0005
1	9.55	61.6	247.	990.	6190.	24800.	99100.	619000.	2480000.
2	3.42	9.44	19.4	39.4	99.4	199.	399.	999.	2000.
3	2.46	5.19	8.67	14.2	26.8	42.9	68.5	127.	202.
4	2.08	3.85	5.82	8.59	14.1	20.3	29.0	46.3	65.8
5	1.88	3.22	4.58	6.36	9.61	13.0	17.4	25.6	34.0
6	1.76	2.85	3.90	5.20	7.45	9.66	12.4	17.3	22.0
7	1.67	2.61	3.47	4.50	6.21	7.83	9.79	13.1	16.2
8	1.61	2.44	3.17	4.03	5.41	6.68	8.18	10.6	12.8
9	1.56	2.31	2.96	3.70	4.86	5.90	7.11	9.01	10.7
10	1.53	2.22	2.80	3.45	4.46	5.34	6.35	7.91	9.30
11	1.50	2.14	2.67	3.26	4.15	4.92	5.79	7.11	8.27
12	1.47	2.08	2.57	3.11	3.91	4.59	5.36	6.51	7.50
13	1.45	2.02	2.48	2.98	3.72	4.33	5.02	6.03	6.90
14	1.43	1.98	2.41	2.88	3.56	4.12	4.74	5.66	6.43
15	1.42	1.94	2.35	2.79	3.42	3.95	4.51	5.35	6.04
16	1.40	1.91	2.30	2.72	3.31	3.80	4.32	5.09	5.72
17	1.39	1.88	2.26	2.65	3.21	3.67	4.16	4.87	5.45
18	1.38	1.85	2.22	2.60	3.13	3.56	4.02	4.68	5.23
19	1.37	1.83	2.18	2.55	3.05	3.46	3.90	4.52	5.03
20	1.36	1.81	2.15	2.50	2.99	3.38	3.79	4.38	4.86
21	1.36	1.79	2.12	2.46	2.93	3.31	3.70	4.26	4.71
22	1.35	1.78	2.10	2.43	2.88	3.24	3.62	4.15	4.58
23	1.34	1.76	2.08	2.39	2.83	3.18	3.54	4.05	4.46
24	1.34	1.75	2.05	2.36	2.79	3.12	3.47	3.96	4.35
25	1.33	1.74	2.04	2.34	2.75	3.08	3.41	3.88	4.26
26	1.33	1.72	2.02	2.31	2.72	3.03	3.36	3.81	4.17
27	1.32	1.71	2.00	2.29	2.68	2.99	3.31	3.75	4.10
28	1.32	1.70	1.99	2.27	2.65	2.95	3.26	3.69	4.02
29	1.31	1.69	1.97	2.25	2.63	2.92	3.22	3.63	3.96
30	1.31	1.69	1.96	2.23	2.60	2.89	3.18	3.58	3.90
35	1.29	1.65	1.91	2.16	2.50	2.76	3.02	3.38	3.66
40	1.28	1.62	1.87	2.11	2.42	2.66	2.90	3.23	3.49
45	1.27	1.60	1.84	2.07	2.36	2.59	2.82	3.12	3.36
50	1.27	1.59	1.81	2.03	2.32	2.53	2.75	3.04	3.26
60	1.26	1.56	1.78	1.98	2.25	2.45	2.65	2.91	3.11
70	1.25	1.55	1.75	1.95	2.20	2.39	2.58	2.83	3.01
80	1.24	1.53	1.73	1.92	2.17	2.35	2.53	2.76	2.94
90	1.24	1.52	1.72	1.91	2.14	2.32	2.49	2.71	2.88
100	1.23	1.52	1.71	1.89	2.12	2.29	2.46	2.68	2.84
120	1.23	1.50	1.69	1.87	2.09	2.25	2.41	2.62	2.78
140	1.22	1.50	1.68	1.85	2.07	2.22	2.38	2.58	2.73
160	1.22	1.49	1.67	1.84	2.05	2.20	2.35	2.55	2.70
180	1.22	1.48	1.66	1.83	2.04	2.19	2.34	2.53	2.67
200	1.22	1.48	1.66	1.82	2.03	2.18	2.32	2.51	2.65
300	1.21	1.47	1.64	1.80	1.99	2.14	2.28	2.46	2.59
500	1.21	1.46	1.62	1.78	1.97	2.11	2.24	2.41	2.54
∞	1.20	1.44	1.60	1.75	1.93	2.06	2.19	2.35	2.47

Table 8.2 (s) Critical Values of the F Distribution Numerator DF = 19

Denom. DF	α(2): 0.50 α(1): 0.25	0.20 0.10	0.10 0.05	0.05 0.025	0.02 0.01	0.01 0.005	0.005 0.0025	0.002 0.001	0.001 0.0005
1	9.57	61.7	248.	992.	6200.	24800.	99200.	620000.	2480000.
2	3.42	9.44	19.4	39.4	99.4	199.	399.	999.	2000.
3	2.46	5.19	8.67	14.2	26.7	42.8	68.4	127.	201.
4	2.08	3.85	5.81	8.58	14.0	20.2	28.9	46.2	65.7
5	1.88	3.21	4.57	6.34	9.58	12.9	17.4	25.5	33.9
6	1.76	2.84	3.88	5.18	7.42	9.62	12.4	17.2	21.9
7	1.67	2.60	3.46	4.48	6.18	7.79	9.74	13.0	16.1
8	1.61	2.43	3.16	4.02	5.38	6.64	8.13	10.5	12.8
9	1.56	2.30	2.95	3.68	4.83	5.86	7.06	8.95	10.7
10	1.53	2.21	2.79	3.44	4.43	5.31	6.31	7.86	9.23
11	1.49	2.13	2.66	3.24	4.12	4.89	5.75	7.06	8.20
12	1.47	2.07	2.56	3.09	3.88	4.56	5.32	6.45	7.43
13	1.45	2.01	2.47	2.96	3.69	4.30	4.98	5.98	6.84
14	1.43	1.97	2.40	2.86	3.53	4.09	4.70	5.60	6.37
15	1.41	1.93	2.34	2.77	3.40	3.91	4.47	5.29	5.98
16	1.40	1.90	2.29	2.70	3.28	3.76	4.28	5.04	5.66
17	1.39	1.87	2.24	2.63	3.19	3.64	4.12	4.82	5.40
18	1.38	1.84	2.20	2.58	3.10	3.53	3.98	4.63	5.17
19	1.37	1.82	2.17	2.53	3.03	3.43	3.86	4.47	4.97
20	1.36	1.80	2.14	2.48	2.96	3.35	3.76	4.33	4.80
21	1.35	1.78	2.11	2.44	2.90	3.27	3.66	4.21	4.65
22	1.35	1.77	2.08	2.41	2.85	3.21	3.58	4.10	4.52
23	1.34	1.75	2.06	2.37	2.80	3.15	3.50	4.00	4.41
24	1.33	1.74	2.04	2.35	2.76	3.09	3.44	3.92	4.30
25	1.33	1.73	2.02	2.32	2.72	3.04	3.38	3.84	4.21
26	1.32	1.71	2.00	2.29	2.69	3.00	3.32	3.77	4.12
27	1.32	1.70	1.99	2.27	2.66	2.96	3.27	3.70	4.04
28	1.31	1.69	1.97	2.25	2.63	2.92	3.22	3.64	3.97
29	1.31	1.68	1.96	2.23	2.60	2.88	3.18	3.59	3.9!
30	1.31	1.68	1.95	2.21	2.57	2.85	3.14	3.53	3.85
35	1.29	1.64	1.89	2.14	2.47	2.72	2.98	3.33	3.61
40	1.28	1.61	1.85	2.09	2.39	2.63	2.87	3.19	3.44
45	1.27	1.59	1.82	2.04	2.34	2.56	2.78	3.08	3.31
50	1.26	1.58	1.80	2.01	2.29	2.50	2.71	2.99	3.21
60	1.25	1.55	1.76	1.96	2.22	2.42	2.61	2.87	3.06
70	1.24	1.54	1.74	1.93	2.18	2.36	2.54	2.78	2.96
80	1.24	1.52	1.72	1.90	2.14	2.32	2.49	2.72	2.89
90	1.23	1.51	1.70	1.88	2.11	2.28	2.45	2.67	2.83
100	1.23	1.50	1.69	1.87	2.09	2.26	2.42	2.63	2.79
120	1.22	1.49	1.67	1.84	2.06	2.22	2.37	2.58	2.73
140	1.22	1.48	1.66	1.83	2.04	2.19	2.34	2.54	2.68
160	1.22	1.48	1.65	1.82	2.02	2.17	2.32	2.51	2.65
180	1.21	1.47	1.64	1.81	2.01	2.15	2.30	2.48	2.62
200	1.21	1.47	1.64	1.80	2.00	2.14	2.28	2.46	2.60
300	1.21	1.46	1.62	1.77	1.97	2.10	2.24	2.41	2.54
500	1.20	1.45	1.61	1.76	1.94	2.07	2.20	2.37	2.49
∞	1.20	1.43	1.59	1.73	1.90	2.03	2.15	2.31	2.42

Table 8.2 (t) Critical Values of the *F* Distribution Numerator DF = 20

Denom. DF	α(2): 0.50 α(l): 0.25	0.20 0.10	0.10 0.05	0.05 0.025	0.02 0.01	0.01 0.005	0.005 0.0025	0.002 0.001	0.001 0.0005
1	9.58	61.7	248.	993.	6210.	24800.	99300.	621000.	2480000.
2	3.43	9.44	19.4	39.4	99.4	199.	399.	999.	2000.
3	2.46	5.18	8.66	14.2	26.7	42.8	68.3	126.	201.
4	2.08	3.84	5.80	8.56	14.0	20.2	28.9	46.1	65.5
5	1.88	3.21	4.56	6.33	9.55	12.9	17.3	25.4	33.8
6	1.76	2.84	3.87	5.17	7.40	9.59	12.3	17.1	21.8
7	1.67	2.59	3.44	4.47	6.16	7.75	9.70	12.9	16.0
8	1.61	2.42	3.15	4.00	5.36	6.61	8.09	10.5	12.7
9	1.56	2.30	2.94	3.67	4.81	5.83	7.02	8.90	10.6
10	1.52	2.20	2.77	3.42	4.41	5.27	6.27	7.80	9.17
11	1.49	2.12	2.65	3.23	4.10	4.86	5.71	7.01	8.14
12	1.47	2.06	2.54	3.07	3.86	4.53	5.28	6.40	7.37
13	1.45	2.01	2.46	2.95	3.66	4.27	4.94	5.93	6.78
14	1.43	1.96	2.39	2.84	3.51	4.06	4.66	5.56	6.31
15	1.41	1.92	2.33	2.76	3.37	3.88	4.44	5.25	5.93
16	1.40	1.89	2.28	2.68	3.26	3.73	4.25	4.99	5.61
17	1.39	1.86	2.23	2.62	3.16	3.61	4.09	4.78	5.34
18	1.38	1.84	2.19	2.56	3.08	3.50	3.95	4.59	5.12
19	1.37	1.81	2.16	2.51	3.00	3.40	3.83	4.43	4.92
20	1.36	1.79	2.12	2.46	2.94	3.32	3.72	4.29	4.75
21	1.35	1.78	2.10	2.42	2.88	3.24	3.63	4.17	4.60
22	1.34	1.76	2.07	2.39	2.83	3.18	3.54	4.06	4.47
23	1.34	1.74	2.05	2.36	2.78	3.12	3.47	3.96	4.36
24	1.33	1.73	2.03	2.33	2.74	3.06	3.40	3.87	4.25
25	1.33	1.72	2.01	2.30	2.70	3.01	3.34	3.79	4.16
26	1.32	1.71	1.99	2.28	2.66	2.97	3.28	3.72	4.07
27	1.32	1.70	1.97	2.25	2.63	2.93	3.23	3.66	3.99
28	1.31	1.69	1.96	2.23	2.60	2.89	3.19	3.60	3.92
29	1.31	1.68	1.94	2.21	2.57	2.86	3.14	3.54	3.86
30	1.30	1.67	1.93	2.20	2.55	2.82	3.11	3.49	3.80
35	1.29	1.63	1.88	2.12	2.44	2.69	2.95	3.29	3.56
40	1.28	1.61	1.84	2.07	2.37	2.60	2.83	3.14	3.39
45	1.27	1.58	1.81	2.03	2.31	2.53	2.74	3.04	3.26
50	1.26	1.57	1.78	1.99	2.27	2.47	2.68	2.95	3.16
60	1.25	1.54	1.75	1.94	2.20	2.39	2.58	2.83	3.02
70	1.24	1.53	1.72	1.91	2.15	2.33	2.51	2.74	2.92
80	1.23	1.51	1.70	1.88	2.12	2.29	2.46	2.68	2.85
90	1.23	1.50	1.69	1.86	2.09	2.25	2.42	2.63	2.79
100	1.23	1.49	1.68	1.85	2.07	2.23	2.38	2.59	2.75
120	1.22	1.48	1.66	1.82	2.03	2.19	2.34	2.53	2.68
140	1.22	1.47	1.65	1.81	2.01	2.16	2.31	2.49	2.64
160	1.21	1.47	1.64	1.80	1.99	2.14	2.28	2.47	2.60
180	1.21	1.46	1.63	1.79	1.98	2.12	2.26	2.44	2.58
200	1.21	1.46	1.62	1.78	1.97	2.11	2.25	2.42	2.56
300	1.20	1.45	1.61	1.75	1.94	2.07	2.20	2.37	2.49
500	1.20	1.44	1.59	1.74	1.92	2.04	2.17	2.33	2.45
∞	1.19	1.42	1.57	1.71	1.88	2.00	2.12	2.27	2.37

Table 8.2 (u) Critical Values of the F Distribution Numerator DF = 22

Denom. DF	α(2): 0.50 α(1): 0.25	0.20 0.10	0.10 0.05	0.05 0.025	0.02 0.01	0.01 0.005	0.005 0.0025	0.002 0.001	0.001 0.0005
1	9.61	61.9	249.	995.	6220.	24900.	99600.	622000.	2490000.
2	3.43	9.45	19.5	39.5	99.5	199.	399.	999.	2000.
3	2.46	5.18	8.65	14.1	26.6	42.7	68.2	126.	201.
4	2.08	3.84	5.79	8.53	14.0	20.1	28.7	45.9	65.3
5	1.88	3.20	4.54	6.30	9.51	12.8	17.2	25.3	33.6
6	1.76	2.83	3.86	5.14	7.35	9.53	12.3	17.0	21.7
7	1.67	2.58	3.43	4.44	6.11	7.69	9.62	12.8	15.9
8	1.61	2.41	3.13	3.97	5.32	6.55	8.02	10.4	12.6
9	1.56	2,29	2.92	3.64	4.77	5.78	6.95	8.80	10.5
10	1.52	2.19	2.75	3.39	4.36	5.22	6.20	7.71	9.06
11	1.49	2.11	2.63	3.20	4.06	4.80	5.64	6.92	8.04
12	1.46	2.05	2.52	3.04	3.82	4.48	5.21	6.32	7.27
13	1.44	1.99	2.44	2.92	3.62	4.22	4.87	5.85	6.68
14	1.42	1.95	2.37	2.81	3.46	4.01	4.60	5.48	6.21
15	1.41	1.91	2.31	2.73	3.33	3.83	4.37	5.17	5.83
16	1.39	1.88	2.25	2.65	3.22	3.68	4.18	4.91	5.52
17	1.38	1.85	2.21	2.59	3.12	3.56	4.02	4.70	5.25
18	1.37	1.82	2.17	2.53	3.03	3.45	3.88	4.51	5.03
19	1.36	1.80	2.13	2.48	2.96	3.35	3.76	4.35	4.83
20	1.35	1.78	2.10	2.43	2.90	3.27	3.66	4.21	4.67
21	1.35	1.76	2.07	2.39	2.84	3.19	3.56	4.09	4.52
22	1.34	1.74	2.05	2.36	2.78	3.12	3.48	3.98	4.39
23	1.33	1.73	2.02	2.33	2.74	3.06	3.41	3.89	4.27
24	1.33	1.71	2.00	2.30	2.70	3.01	3.34	3.80	4.17
25	1.32	1.70	1.98	2.27	2.66	2.96	3.28	3.72	4.07
26	1.32	1.69	1.97	2.24	2.62	2.92	3.22	3.65	3.99
27	1.31	1.68	1.95	2.22	2.59	2.88	3.17	3.58	3.91
28	1.31	1.67	1.93	2.20	2.56	2.84	3.13	3.52	3.84
29	1.30	1.66	1.92	2.18	2.53	2.80	3.08	3.47	3.77
30	1.30	1.65	1.91	2.16	2.51	2.77	3.04	3.42	3.71
35	1.28	1.62	1.85	2.09	2.40	2.64	2.89	3.22	3.48
40	1.27	1.59	1.81	2.03	2.33	2.55	2.77	3.07	3.31
45	1.26	1.57	1.78	1.99	2.27	2.47	2.68	2.96	3.18
50	1.25	1.55	1.76	1.96	2.22	2.42	2.62	2.88	3.08
60	1.24	1.53	1.72	1.91	2.15	2.33	2.52	2.75	2.94
70	1.23	1.51	1.70	1.88	2.11	2.28	2.45	2.67	2.84
80	1.23	1.49	1.68	1.85	2.07	2.23	2.39	2.61	2.77
90	1.22	1.48	1.66	1.83	2.04	2.20	2.35	2.56	2.71
100	1.22	1.48	1.65	1.81	2.02	2.17	2.32	2.52	2.67
120	1.21	1.46	1.63	1.79	1.99	2.13	2.28	2.46	2.60
140	1.21	1.45	1.62	1.77	1.97	2.11	2.24	2.42	2.56
160	1.21	1.45	1.61	1.76	1.95	2.09	2.22	2.39	2.52
180	1.20	1.44	1.60	1.75	1.94	2.07	2.20	2.37	2.50
200	1.20	1.44	1.60	1.74	1.93	2.06	2.19	2.35	2.48
300	1.20	1.43	1.58	1.72	1.89	2.02	2.14	2.30	2.41
500	1.19	1.42	1.56	1.70	1.87	1.99	2.11	2.26	2.37
∞	1.18	1.40	1.54	1.67	1.83	1.95	2.05	2.19	2.30

Table 8.2 (v) Critical Values of the *F* Distribution Numerator DF = 24

Denom. DF	α(2): 0.50 α(1): 0.25	0.20 0.10	0.10 0.05	0.05 0.025	0.02 0.01	0.01 0.005	0.005 0.0025	0.002 0.001	0.001 0.0005
1	9.63	62.0	249.	997.	6230.	24900.	99800.	623000.	2490000.
2	3.43	9.45	19.5	39.5	99.5	199.	399.	999.	2000.
3	2.46	5.18	8.64	14.1	26.6	42.6	68.1	126.	200.
4	2.08	3.83	5.77	8.51	13.9	20.0	28.7	45.8	65.0
5	1.88	3.19	4.53	6.28	9.47	12.8	17.1	25.1	33.4
6	1.75	2.82	3.84	5.12	7.31	9.47	12.2	16.9	21.5
7	1.67	2.58	3.41	4.41	6.07	7.64	9.56	12.7	15.8
8	1.60	2.40	3.12	3.95	5.28	6.50	7.95	10.3	12.5
9	1.56	2.28	2.90	3.61	4.73	5.73	6.89	8.72	10.4
10	1.52	2.18	2.74	3.37	4.33	5.17	6.14	7.64	8.96
11	1.49	2.10	2.61	3.17	4.02	4.76	5.58	6.85	7.95
12	1.46	2.04	2.51	3.02	3.78	4.43	5.16	6.25	7.19
13	1.44	1.98	2.42	2.89	3.59	4.17	4.82	5.78	6.60
14	1.42	1.94	2.35	2.79	3.43	3.96	4.55	5.41	6.13
15	1.41	1.90	2.29	2.70	3.29	3.79	4.32	5.10	5.75
16	1.39	1.87	2.24	2.63	3.18	3.64	4.13	4.85	5.44
17	1.38	1.84	2.19	2.56	3.08	3.51	3.97	4.63	5.18
18	1.37	1.81	2.15	2.50	3.00	3.40	3.83	4.45	4.95
19	1.36	1.79	2.11	2.45	2.92	3.31	3.71	4.29	4.76
20	1.35	1.77	2.08	2.41	2.86	3.22	3.61	4.15	4.59
21	1.34	1.75	2.05	2.37	2.80	3.15	3.51	4.03	4.44
22	1.33	1.73	2.03	2.33	2.75	3.08	3.43	3.92	4.31
23	1.33	1.72	2.01	2.30	2.70	3.02	3.35	3.82	4.20
24	1.32	1.70	1.98	2.27	2.66	2.97	3.29	3.74	4.09
25	1.32	1.69	1.96	2.24	2.62	2.92	3.23	3.66	4.00
26	1.31	1.68	1.95	2.22	2.58	2.87	3.17	3.59	3.92
27	1.31	1.67	1.93	2.19	2.55	2.83	3.12	3.52	3.84
28	1.30	1.66	1.91	2.17	2.52	2.79	3.07	3.46	3.77
29	1.30	1.65	1.90	2.15	2.49	2.76	3.03	3.41	3.70
30	1.29	1.64	1.89	2.14	2.47	2.73	2.99	3.36	3.64
35	1.28	1.60	1.83	2.06	2.36	2.60	2.83	3.16	3.41
40	1.26	1.57	1.79	2.01	2.29	2.50	2.72	3.01	3.24
45	1.26	1.55	1.76	1.96	2.23	2.43	2.63	2.90	3.11
50	1.25	1.54	1.74	1.93	2.18	2.37	2.56	2.82	3.01
60	1.24	1.51	1.70	1.88	2.12	2.29	2.46	2.69	2.87
70	1.23	1.49	1.67	1.85	2.07	2.23	2.39	2.61	2.77
80	1.22	1.48	1.65	1.82	2.03	2.19	2.34	2.54	2.70
90	1.22	1.47	1.64	1.80	2.00	2.15	2.30	2.50	2.64
100	1.21	1.46	1.63	1.78	1.98	2.13	2.27	2.46	2.60
120	1.21	1.45	1.61	1.76	1.95	2.09	2.23	2.40	2.53
140	1.20	1.44	1.60	1.74	1.93	2.06	2.19	2.36	2.49
160	1.20	1.43	1.59	1.73	1.91	2.04	2.17	2.33	2.45
180	1.20	1.43	1.58	1.72	1.90	2.02	2.15	2.31	2.43
200	1.19	1.42	1.57	1.71	1.89	2.01	2.13	2.29	2.41
300	1.19	1.41	1.55	1.69	1.85	1.97	2.09	2.24	2.35
500	1.18	1.40	1.54	1.67	1.83	1.94	2.05	2.20	2.30
∞	1.18	1.38	1.52	1.64	1.79	1.90	2.00	2.13	2.23

Table 8.2 (w) Critical Values of the F Distribution Numerator DF = 26

Denom. DF	α(2): 0.50 α(1): 0.25	0.20 0.10	0.10 0.05	0.05 0.025	0.02 0.01	0.01 0.005	0.005 0.0025	0.002 0.001	0.001 0.0005
1	9.64	62.1	249.	999.	6240.	25000.	99900.	624000.	2500000.
2	3.44	9.45	19.5	39.5	99.5	199.	399.	999.	2000.
3	2.46	5.17	8.63	14.1	26.6	42.6	68.0	126.	200.
4	2.08	3.83	5.76	8.49	13.9	20.0	28.6	45.6	64.9
5	1.88	3.18	4.52	6.26	9.43	12.7	17.1	25.0	33.3
6	1.75	2.81	3.83	5.10	7.28	9.43	12.1	16.8	21.4
7	1.67	2.57	3.40	4.39	6.04	7.60	9.50	12.7	15.7
8	1.60	2.40	3.10	3.93	5.25	6.46	7.90	10.2	12.4
9	1.55	2.27	2.89	3.59	4.70	5.69	6.84	8.66	10.3
10	1.52	2.17	2.72	3.34	4.30	5.13	6.09	7.57	8.89
11	1.48	2.09	2.59	3.15	3.99	4.72	5.54	6.78	7.87
12	1.46	2.03	2.49	3.00	3.75	4.39	5.11	6.19	7.12
13	1.44	1.97	2.41	2.87	3.56	4.13	4.77	5.72	6.53
14	1.42	1.93	2.33	2.77	3.40	3.92	4.50	5.35	6.07
15	1.40	1.89	2.27	2.68	3.26	3.75	4.27	5.04	5.69
16	1.39	1.86	2.22	2.60	3.15	3.60	4.09	4.79	5.37
17	1.38	1.83	2.17	2.54	3.05	3.47	3.92	4.57	5.11
18	1.36	1.80	2.13	2.48	2.97	3.36	3.79	4.39	4.89
19	1.35	1.78	2.10	2.43	2.89	3.27	3.67	4.23	4.70
20	1.35	1.76	2.07	2.39	2.83	3.18	3.56	4.09	4.53
21	1.34	1.74	2.04	2.34	2.77	3.11	3.47	3.97	4.38
22	1.33	1.72	2.01	2.31	2.72	3.04	3.38	3.86	4.25
23	1.32	1.70	1.99	2.28	2.67	2.98	3.31	3.77	4.14
24	1.32	1.69	1.97	2.25	2.63	2.93	3.24	3.68	4.03
25	1.31	1.68	1.95	2.22	2.59	2.88	3.18	3.60	3.94
26	1.31	1.67	1.93	2.19	2.55	2.84	3.13	3.53	3.85
27	1.30	1.65	1.91	2.17	2.52	2.79	3.08	3.47	3.78
28	1.30	1.64	1.90	2.15	2.49	2.76	3.03	3.41	3.71
29	1.29	1.63	1.88	2.13	2.46	2.72	2.99	3.35	3.64
30	1.29	1.63	1.87	2.11	2.44	2.69	2.95	3.30	3.58
35	1.27	1.59	1.82	2.04	2.33	2.56	2.79	3.10	3.35
40	1.26	1.56	1.77	1.98	2.26	2.46	2.67	2.96	3.18
45	1.25	1.54	1.74	1.94	2.20	2.39	2.59	2.85	3.05
50	1.24	1.52	1.72	1.91	2.15	2.33	2.52	2.76	2.95
60	1.23	1.50	1.68	1.86	2.08	2.25	2.42	2.64	2.81
70	1.22	1.48	1.65	1.82	2.03	2.19	2.35	2.56	2.71
80	1.22	1.47	1.63	1.79	2.00	2.15	2.30	2.49	2.64
90	1.21	1.45	1.62	1.77	1.97	2.12	2.26	2.44	2.58
100	1.21	1.45	1.61	1.76	1.95	2.09	2.23	2.41	2.54
120	1.20	1.43	1.59	1.73	1.92	2.05	2.i8	2.35	2.48
140	1.20	1.42	1.57	1.72	1.89	2.02	2.15	2.31	2.43
160	1.19	1.42	1.57	1.70	1.88	2.00	2.12	2.28	2.40
180	1.19	1.41	1.56	1.69	1.86	1.98	2.10	2.26	2.37
200	1.19	1.41	1.55	1.68	1.85	1.97	2.09	2.24	2.35
300	1.18	1.39	1.53	1.66	1.82	1.93	2.04	2.18	2.29
500	1.18	1.38	1.52	1.64	1.79	1.90	2.01	2.14	2.24
∞	1.17	1.37	1.50	1.61	1.76	1.86	1.95	2.08	2.17

Table 8.2 (x) Critical Values of the *F* Distribution Numerator DF = 28

Denom. DF	α(2): 0.50 α(1): 0.25	0.20 0.10	0.10 0.05	0.05 0.025	0.02 0.01	0.01 0.005	0.005 0.0025	0.002 0.001	0.001 0.0005
1	9.66	62.2	250.	1000.	6250.	25000.	100000.	625000.	2500000.
2	3.44	9.46	19.5	39.5	99.5	199.	399.	999.	2000.
3	2.46	5.17	8.62	14.1	26.5	42.5	67.9	126.	200.
4	2.08	3.82	5.75	8.48	13.9	19.9	28.5	45.5	64.7
5	1.88	3.18	4.50	6.24	9.40	12.7	17.0	24.9	33.2
6	1.75	2.81	3.82	5.08	7.25	9.39	12.1	16.7	21.3
7	1.66	2.56	3.39	4.38	6.02	7.57	9.45	12.6	15.6
8	1.60	2.39	3.09	3.91	5.22	6.43	7.86	10.2	12.3
9	1.55	2.26	2.87	3.58	4.67	5.65	6.80	8.60	10.2
10	1.51	2.16	2.71	3.33	4.27	5.10	6.05	7.52	8.82
11	1.48	2.08	2.58	3.13	3.96	4.68	5.49	6.73	7.81
12	1.46	2.02	2.48	2.98	3.72	4.36	5.07	6.14	7.05
13	1.43	1.96	2.39	2.85	3.53	4.10	4.73	5.67	6.47
14	1.42	1.92	2.32	2.75	3.37	3.89	4.46	5.30	6.01
15	1.40	1.88	2.26	2.66	3.24	3.72	4.23	4.99	5.63
16	1.39	1.85	2.21	2.58	3.12	3.57	4.05	4.74	5.32
17	1.37	1.82	2.16	2.52	3.03	3.44	3.89	4.53	5.05
18	1.36	1.79	2.12	2.46	2.94	3.33	3.75	4.34	4.83
19	1.35	1.77	2.08	2.41	2.87	3.24	3.63	4.18	4.64
20	1.34	1.75	2.05	2.37	2.80	3.15	3.52	4.05	4.47
21	1.33	1.73	2.02	2.33	2.74	3.08	3.43	3.93	4.33
22	1.33	1.71	2.00	2.29	2.69	3.01	3.35	3.82	4.20
23	1.32	1.69	1.97	2.26	2.64	2.95	3.27	3.72	4.08
24	1.31	1.68	1.95	2.23	2.60	2.90	3.20	3.63	3.98
25	1.31	1.67	1.93	2.20	2.56	2.85	3.14	3.56	3.89
26	1.30	1.66	1.91	2.17	2.53	2.80	3.09	3.49	3.80
27	1.30	1.64	1.90	2.15	2.49	2.76	3.04	3.42	3.72
28	1.29	1.63	1.88	2.13	2.46	2.72	2.99	3.36	3.65
29	1.29	1.62	1.87	2.11	2.44	2.69	2.95	3.31	3.59
30	1.29	1.62	1.85	2.09	2.41	2.66	2.91	3.26	3.53
35	1.27	1.58	1.80	2.02	2.30	2.53	2.75	3.06	3.30
40	1.26	1.55	1.76	1.96	2.23	2.43	2.64	2.91	3.13
45	1.25	1.53	1.73	1.92	2.17	2.36	2.55	2.80	3.00
50	1.24	1.51	1.70	1.89	2.12	2.30	2.48	2.72	2.90
60	1.23	1.49	1.66	1.83	2.05	2.22	2.38	2.60	2.76
70	1.22	1.47	1.64	1.80	2.01	2.16	2.31	2.51	2.66
80	1.21	1.45	1.62	1.77	1.97	2.11	2.26	2.45	2.59
90	1.21	1.44	1.60	1.75	1.94	2.08	2.22	2.40	2.53
100	1.20	1.43	1.59	1.74	1.92	2.05	2.19	2.36	2.49
120	1.20	1.42	1.57	1.71	1.89	2.01	2.14	2.30	2.42
140	1.19	1.41	1.56	1.69	1.86	1.99	2.11	2.26	2.38
160	1.19	1.40	1.55	1.68	1.85	1.97	2.08	2.23	2.35
180	1.19	1.40	1.54	1.67	1.83	1.95	2.06	2.21	2.32
200	1.18	1.39	1.53	1.66	1.82	1.94	2.05	2.19	2.30
300	1.18	1.38	1.51	1.64	1.79	1.90	2.00	2.14	2.24
500	1.17	1.37	1.50	1.62	1.76	1.87	1.97	2.10	2.19
∞	1.17	1.35	1.48	1.59	1.72	1.82	1.91	2.03	2.12

Table 8.2 (y) Critical Values of the F Distribution Numerator DF = 30

Denom. DF	α(2): 0.50 α(1): 0.25	0.20 0.10	0.10 0.05	0.05 0.025	0.02 0.01	0.01 0.005	0.005 0.0025	0.002 0.001	0.001 0.0005
1	9.67	62.3	250.	1000.	6260.	25000.	100000.	626000.	2500000.
2	3.44	9.46	19.5	39.5	99.5	199.	399.	999.	2000.
3	2.47	5.17	8.62	14.1	26.5	42.5	67.8	125.	200.
4	2.08	3.82	5.75	8.46	13.8	19.9	28.5	45.4	64.6
5	1.88	3.17	4.50	6.23	9.38	12.7	17.0	24.9	33.1
6	1.75	2.80	3.81	5.07	7.23	9.36	12.0	16.7	21.2
7	1.66	2.56	3.38	4.36	5.99	7.53	9.41	12.5	15.5
8	1.60	2.38	3.08	3.89	5.20	6.40	7.82	10.1	12.2
9	1.55	2.25	2.86	3.56	4.65	5.62	6.76	8.55	10.2
10	1.51	2.16	2.70	3.31	4.25	5.07	6.01	7.47	8.76
11	1.48	2.08	2.57	3.12	3.94	4.65	5.46	6.68	7.75
12	1.45	2.01	2.47	2.96	3.70	4.33	5.03	6.09	7.00
13	1.43	1.96	2.38	2.84	3.51	4.07	4.70	5.63	6.42
14	1.41	1.91	2.31	2.73	3.35	3.86	4.42	5.25	5.95
15	1.40	1.87	2.25	2.64	3.21	3.69	4.20	4.95	5.58
16	1.38	1.84	2.19	2.57	3.10	3.54	4.01	4.70	5.27
17	1.37	1.81	2.15	2.50	3.00	3.41	3.85	4.48	5.01
18	1.36	1.78	2.11	2.44	2.92	3.30	3.71	4.30	4.78
19	1.35	1.76	2.07	2.39	2.84	3.21	3.59	4.14	4.59
20	1.34	1.74	2.04	2.35	2.78	3.12	3.49	4.00	4.42
21	1.33	1.72	2.01	2.31	2.72	3.05	3.40	3.88	4.28
22	1.32	1.70	1.98	2.27	2.67	2.98	3.31	3.78	4.15
23	1.32	1.69	1.96	2.24	2.62	2.92	3.24	3.68	4.03
24	1.31	1.67	1.94	2.21	2.58	2.87	3.17	3.59	3.93
25	1.31	1.66	1.92	2.18	2.54	2.82	3.11	3.52	3.84
26	1.30	1.65	1.90	2.16	2.50	2.77	3.06	3.44	3.75
27	1.30	1.64	1.88	2.13	2.47	2.73	3.00	3.38	3.68
28	1.29	1.63	1.87	2.11	2.44	2.69	2.96	3.32	3.61
29	1.29	1.62	1.85	2.09	2.41	2.66	2.92	3.27	3.54
30	1.28	1.61	1.84	2.07	2.39	2.63	2.88	3.22	3.49
35	1.27	1.57	1.79	2.00	2.28	2.50	2.72	3.02	3.25
40	1.25	1.54	1.74	1.94	2.20	2.40	2.60	2.87	3.08
45	1.24	1.52	1.71	1.90	2.14	2.33	2.51	2.76	2.96
50	1.23	1.50	1.69	1.87	2.10	2.27	2.45	2.68	2.86
60	1.22	1.48	1.65	1.82	2.03	2.19	2.35	2.55	2.71
70	1.21	1.46	1.62	1.78	1.98	2.13	2.28	2.47	2.62
80	1.21	1.44	1.60	1.75	1.94	2.08	2.22	2.41	2.54
90	1.20	1.43	1.59	1.73	1.92	2.05	2.18	2.36	2.49
100	1.20	1.42	1.57	1.71	1.89	2.02	2.15	2.32	2.44
120	1.19	1.41	1.55	1.69	1.86	1.98	2.11	2.26	2.38
140	1.19	1.40	1.54	1.67	1.84	1.96	2.07	2.22	2.33
160	1.18	1.39	1.53	1.66	1.82	1.93	2.05	2.19	2.30
180	1.18	1.39	1.52	1.65	1.81	1.92	2.03	2.17	2.27
200	1.18	1.38	1.52	1.64	1.79	1.91	2.01	2.15	2.25
300	1.17	1.37	1.50	1.62	1.76	1.87	1.97	2.10	2.19
500	1.17	1.36	1.48	1.60	1.74	1.84	1.93	2.05	2.14
∞	1.16	1 .34	1 .46	1.57	1.70	1.79	1.88	1.99	2.07

Table 8.2 (z) Critical Values of the *F* Distribution Numerator DF = 40

Denom. DF	**α(2): 0.50** **α(1): 0.25**	**0.20** **0.10**	**0.10** **0.05**	**0.05** **0.025**	**0.02** **0.01**	**0.01** **0.005**	**0.005** **0.0025**	**0.002** **0.001**	**0.001** **0.0005**
1	9.71	62.5	251.	1010.	6290.	25100.	101000.	629000.	2510000.
2	3.45	9.47	19.5	39.5	99.5	199.	399.	999.	2000.
3	2.47	5.16	8.59	14.0	26.4	42.3	67.5	125.	199.
4	2.08	3.80	5.72	8.41	13.7	19.8	28.2	45.1	64.1
5	1.88	3.16	4.46	6.18	9.29	12.5	16.8	24.6	32.7
6	1.75	2.78	3.77	5.01	7.14	9.24	11.9	16.4	21.0
7	1.66	2.54	3.34	4.31	5.91	7.42	9.26	12.3	15.2
8	1.59	2.36	3.04	3.84	5.12	6.29	7.68	9.92	12.0
9	1.54	2.23	2.83	3.51	4.57	5.52	6.62	8.37	9.94
10	1.51	2.13	2.66	3.26	4.17	4.97	5.88	7.30	8.55
11	1.47	2.05	2.53	3.06	3.86	4.55	5.33	6.52	7.55
12	1.45	1.99	2.43	2.91	3.62	4.23	4.91	5.93	6.81
13	1.42	1.93	2.34	2.78	3.43	3.97	4.57	5.47	6.23
14	1.41	1.89	2.27	2.67	3.27	3.76	4.30	5.10	5.77
15	1.39	1.85	2.20	2.59	3.13	3.58	4.08	4.80	5.40
16	1.37	1.81	2.15	2.51	3.02	3.44	3.89	4.54	5.09
17	1.36	1.78	2.10	2.44	2.92	3.31	3.73	4.33	4.83
18	1.35	1.75	2.06	2.38	2.84	3.20	3.59	4.15	4.61
19	1.34	1.73	2.03	2.33	2.76	3.11	3.47	3.99	4.42
20	1.33	1.71	1.99	2.29	2.69	3.02	3.37	3.86	4.25
21	1.32	1.69	1.96	2.25	2.64	2.95	3.27	3.74	4.11
22	1.31	1.67	1.94	2.21	2.58	2.88	3.19	3.63	3.98
23	1.31	1.66	1.91	2.18	2.54	2.82	3.12	3.53	3.87
24	1.30	1.64	1.89	2.15	2.49	2.77	3.05	3.45	3.76
25	1.29	1.63	1.87	2.12	2.45	2.72	2.99	3.37	3.67
26	1.29	1.61	1.85	2.09	2.42	2.67	2.93	3.30	3.59
27	1.28	1.60	1.84	2.07	2.38	2.63	2.88	3.23	3.51
28	1.28	1.59	1.82	2.05	2.35	2.59	2.84	3.18	3.44
29	1.27	1.58	1.81	2.03	2.33	2.56	2.79	3.12	3.38
30	1.27	1.57	1.79	2.01	2.30	2.52	2.76	3.07	3.32
35	1.25	1.53	1.74	1.93	2.19	2.39	2.60	2.87	3.09
40	1.24	1.51	1.69	1.88	2.11	2.30	2.48	2.73	2.92
45	1.23	1.48	1.66	1.83	2.05	2.22	2.39	2.62	2.79
50	1.22	1.46	1.63	1.80	2.01	2.16	2.32	2.53	2.69
60	1.21	1.44	1.59	1.74	1.94	2.08	2.22	2.41	2.55
70	1.20	1.42	1.57	1.71	1.89	2.02	2.15	2.32	2.45
80	1.19	1.40	1.54	1.68	1.85	1.97	2.10	2.26	2.38
90	1.19	1.39	1.53	1.66	1.82	1.94	2.06	2.21	2.32
100	1.18	1.38	1.52	1.64	1.80	1.91	2.02	2.17	2.28
120	1.18	1.37	1.50	1.61	1.76	1.87	1.98	2.11	2.21
140	1.17	1.36	1.48	1.60	1.74	1.84	1.94	2.07	2.17
160	1.17	1.35	1.47	1.58	1.72	1.82	1.92	2.04	2.13
ISO	1.16	1.34	1.46	1.57	1.71	1.80	1.90	2.02	2.11
200	1.16	1.34	1.46	1.56	1.69	1.79	1.88	2.00	2.09
300	1.15	1.32	1.43	1.54	1.66	1.75	1.84	1.94	2.02
500	1.15	1.31	1.42	1.52	1.63	1.72	1.80	1.90	1.98
∞	1.14	1.30	1.39	1.48	1.59	1.67	1.74	1.84	1.90

Table 8.2 (aa) Critical Values of the *F* Distribution Numerator DF = 50

Denom. DF	α(2): 0.50 α(1): 0.25	0.20 0.10	0.10 0.05	0.05 0.025	0.02 0.01	0.01 0.005	0.005 0.0025	0.002 0.001	0.001 0.0005
1	9.74	62.7	252.	1010.	6300.	25200.	101000.	630000.	2520000.
2	3.46	9.47	19.5	39.5	99.5	199.	399.	999.	2000.
3	2.47	5.15	8.58	14.0	26.4	42.2	67.4	125.	198.
4	2.08	3.80	5.70	8.38	13.7	19.7	28.1	44.9	63.8
5	1.88	3.15	4.44	6.14	9.24	12.5	16.7	24.4	32.5
6	1.75	2.77	3.75	4.98	7.09	9.17	11.8	16.3	20.8
7	1.66	2.52	3.32	4.28	5.86	7.35	9.17	12.2	15.1
8	1.59	2.35	3.02	3.81	5.07	6.22	7.59	9.80	11.8
9	1.54	2.22	2.80	3.47	4.52	5.45	6.54	8.26	9.81
10	1.50	2.12	2.64	3.22	4.12	4.90	5.80	7.19	8.43
11	1.47	2.04	2.51	3.03	3.81	4.49	5.25	6.42	7.43
12	1.44	1.97	2.40	2.87	3.57	4.17	4.83	5.83	6.69
13	1.42	1.92	2.31	2.74	3.38	3.91	4.50	5.37	6.11
14	1.40	1.87	2.24	2.64	3.22	3.70	4.23	5.00	5.66
15	1.38	1.83	2.18	2.55	3.08	3.52	4.00	4.70	5.29
16	1.37	1.79	2.12	2.47	2.97	3.37	3.81	4.45	4.98
17	1.36	1.76	2.08	2.41	2.87	3.25	3.65	4.24	4.72
18	1.34	1.74	2.04	2.35	2.78	3.14	3.52	4.06	4.50
19	1.33	1.71	2.00	2.30	2.71	3.04	3.40	3.90	4.31
20	1.32	1.69	1.97	2.25	2.64	2.96	3.29	3.77	4.15
21	1.32	1.67	1.94	2.21	2.58	2.88	3.20	3.64	4.00
22	1.31	1.65	1.91	2.17	2.53	2.82	3.12	3.54	3.88
23	1.30	1.64	1.88	2.14	2.48	2.76	3.04	3.44	3.76
24	1.29	1.62	1.86	2.11	2.44	2.70	2.98	3.36	3.66
25	1.29	1.61	1.84	2.08	2.40	2.65	2.91	3.28	3.57
26	1.28	1.59	1.82	2.05	2.36	2.61	2.86	3.21	3.49
27	1.28	1.58	1.81	2.03	2.33	2.57	2.81	3.14	3.41
28	1.27	1.57	1.79	2.01	2.30	2.53	2.76	3.09	3.34
29	1.27	1.56	1.77	1.99	2.27	2.49	2.72	3.03	3.28
30	1.26	1.55	1.76	1.97	2.25	2.46	2.68	2.98	3.22
35	1.24	1.51	1.70	1.89	2.14	2.33	2.52	2.78	2.98
40	1.23	1.48	1.66	1.83	2.06	2.23	2.40	2.64	2.82
45	1.22	1.46	1.63	1.79	2.00	2.16	2.31	2.53	2.69
50	1.21	1.44	1.60	1.75	1.95	2.10	2.24	2.44	2.59
60	1.20	1.41	1.56	1.70	1.88	2.01	2.14	2.32	2.45
70	1.19	1.39	1.53	1.66	1.83	1.95	2.07	2.23	2.35
80	1.18	1.38	1.51	1.63	1.79	1.90	2.02	2.16	2.28
90	1.18	1.36	1.49	1.61	1.76	1.87	1.98	2.11	2.22
100	1.17	1.35	1.48	1.59	1.74	1.84	1.94	2.08	2.18
120	1.16	1.34	1.46	1.56	1.70	1.80	1.89	2.02	2.11
140	1.16	1.33	1.44	1.55	1.67	1.77	1.86	1.98	2.06
160	1.15	1.32	1.43	1.53	1.66	1.75	1.83	1.95	2.03
180	1.15	1.32	1.42	1.52	1.64	1.73	1.81	1.92	2.00
200	1.15	1.31	1.41	1.51	1.63	1.71	1.80	1.90	1.98
300	1.14	1.29	1.39	1.48	1.59	1.67	1.75	1.85	1.92
500	1.14	1.28	1.38	1.46	1.57	1.64	1.71	1.80	1.87
∞	1.13	1.26	1.35	1.43	1.52	1.59	1.65	1.73	1.79

Table 8.2 (ab) Critical Values of the *F* Distribution Numerator DF = 60

Denom. DP	α(2): 0.50 α(1): 0.25	0.20 0.10	0.10 0.05	0.05 0.025	0.02 0.01	0.01 0.005	0.005 0.0025	0.002 0.001	0.001 0.0005
1	9.76	62.8	252.	1010.	6310.	25300.	101000.	631000.	2530000.
2	3.46	9.47	19.5	39.5	99.5	199.	399.	999.	2000.
3	2.47	5.15	8.57	14.0	26.3	42.1	67.3	124.	198.
4	2.08	3.79	5.69	8.36	13.7	19.6	28.0	44.7	63.6
5	1.87	3.14	4.43	6.12	9.20	12.4	16.6	24.3	32.4
6	1.74	2.76	3.74	4.96	7.06	9.12	11.7	16.2	20.7
7	1.65	2.51	3.30	4.25	5.82	7.31	9.12	12.1	15.0
8	1.59	2.34	3.01	3.78	5.03	6.18	7.54	9.73	11.8
9	1.54	2.21	2.79	3.45	4.48	5.41	6.49	8.19	9.72
10	1.50	2.11	2.62	3.20	4.08	4.86	5.75	7.12	8.34
11	1.47	2.03	2.49	3.00	3.78	4.45	5.20	6.35	7.35
12	1.44	1.96	2.38	2.85	3.54	4.12	4.78	5.76	6.61
13	1.42	1.90	2.30	2.72	3.34	3.87	4.44	5.30	6.04
14	1.40	1.86	2.22	2.61	3.18	3.66	4.17	4.94	5.58
15	1.38	1.82	2.16	2.52	3.05	3.48	3.95	4.64	5.21
16	1.36	1.78	2.11	2.45	2.93	3.33	3.76	4.39	4.91
17	1.35	1.75	2.06	2.38	2.83	3.21	3.60	4.18	4.65
18	1.34	1.72	2.02	2.32	2.75	3.10	3.47	4.00	4.43
19	1.33	1.70	1.98	2.27	2.67	3.00	3.35	3.84	4.24
20	1.32	1.68	1.95	2.22	2.61	2.92	3.24	3.70	4.08
21	1.31	1.66	1.92	2.18	2.55	2.84	3.15	3.58	3.93
22	1.30	1.64	1.89	2.14	2.50	2.77	3.07	3.48	3.81
23	1.30	1.62	1.86	2.11	2.45	2.71	2.99	3.38	3.69
24	1.29	1.61	1.84	2.08	2.40	2.66	2.92	3.29	3.59
25	1.28	1.59	1.82	2.05	2.36	2.61	2.86	3.22	3.50
26	1.28	1.58	1.80	2.03	2.33	2.56	2.81	3.15	3.42
27	1.27	1.57	1.79	2.00	2.29	2.52	2.76	3.08	3.34
28	1.27	1.56	1.77	1.98	2.26	2.48	2.71	3.02	3.27
29	1.26	1.55	1.75	1.96	2.23	2.45	2.67	2.97	3.21
30	1.26	1.54	1.74	1.94	2.21	2.42	2.63	2.92	3.15
35	1.24	1.50	1.68	1.86	2.10	2.28	2.47	2.72	2.91
40	1.22	1.47	1.64	1.80	2.02	2.18	2.35	2.57	2.75
45	1.21	1.44	1.60	1.76	1.96	2.11	2.26	2.46	2.62
50	1.20	1.42	1.58	1.72	1.91	2.05	2.19	2.38	2.52
60	1.19	1.40	1.53	1.67	1.84	1.96	2.09	2.25	2.38
70	1.18	1.37	1.50	1.63	1.78	1.90	2.01	2.16	2.28
80	1.17	1.36	1.48	1.60	1.75	1.85	1.96	2.10	2.20
90	1.17	1.35	1.46	1.58	1.72	1.82	1.92	2.05	2.15
100	1.16	1.34	1.45	1.56	1.69	1.79	1.89	2.01	2.10
120	1.16	1.32	1.43	1.53	1.66	1.75	1.84	1.95	2.04
140	1.15	1.31	1.41	1.51	1.63	1.72	1.80	1.91	1.99
160	1.15	1.30	1.40	1.50	1.61	1.69	1.77	1.88	1.95
180	1.14	1.29	1.39	1.48	1.60	1.68	1.75	1.85	1.93
200	1.14	1.29	1.39	1.47	1.58	1.66	1.74	1.83	1.90
300	1.13	1.27	1.36	1.45	1.55	1.62	1.69	1.78	1.84
500	1.13	1.26	1.35	1.42	1.52	1.58	1.65	1.73	1.79
∞	1.12	1.24	1.32	1.39	1.47	1.53	1.59	1.66	1.71

Table 8.2 (ac) Critical Values of the *F* Distribution Numerator DF = 70

Denom. DF	α(2): 0.50 α(1): 0.25	0.20 0.10	0.10 0.05	0.05 0.025	0.02 0.01	0.01 0.005	0.005 0.0025	0.002 0.001	0.001 0.0005
1	9.77	62.9	252.	1010.	6320.	25300.	101000.	632000.	2530000.
2	3.46	9.48	19.5	39.5	99.5	199.	399.	999.	2000.
3	2.47	5.15	8.57	14.0	26.3	42.1	67.2	124.	198.
4	2.08	3.79	5.68	8.35	13.6	19.6	28.0	44.6	63.4
5	1.87	3.14	4.42	6.11	9.18	12.4	16.6	24.3	32.3
6	1.74	2.76	3.73	4.94	7.03	9.09	11.7	16.1	20.6
7	1.65	2.51	3.29	4.24	5.80	7.28	9.07	12.1	14.9
8	1.59	2.33	2.99	3.77	5.01	6.15	7.49	9.67	11.7
9	1.54	2.20	2.78	3.43	4.46	5.38	6.45	8.13	9.65
10	1.50	2.10	2.61	3.18	4.06	4.83	5.71	7.07	8.28
11	1.46	2.02	2.48	2.99	3.75	4.41	5.16	6.30	7.29
12	1.44	1.95	2.37	2.83	3.51	4.09	4.74	5.71	6.55
13	1.41	1.90	2.28	2.70	3.32	3.84	4.41	5.26	5.98
14	1.39	1.85	2.21	2.60	3.16	3.62	4.14	4.89	5.53
15	1.38	1.81	2.15	2.51	3.02	3.45	3.91	4.59	5.16
16	1.36	1.77	2.09	2.43	2.91	3.30	3.73	4.34	4.85
17	1.35	1.74	2.05	2.36	2.81	3.18	3.57	4.13	4.60
18	1.34	1.71	2.00	2.30	2.72	3.07	3.43	3.95	4.38
19	1.33	1.69	1.97	2.25	2.65	2.97	3.31	3.79	4.19
20	1.32	1.67	1.93	2.20	2.58	2.88	3.20	3.66	4.03
21	1.31	1.65	1.90	2.16	2.52	2.81	3.11	3.54	3.88
22	1.30	1.63	1.88	2.13	2.47	2.74	3.03	3.43	3.76
23	1.29	1.61	1.85	2.09	2.42	2.68	2.95	3.34	3.64
24	1.28	1.60	1.83	2.06	2.38	2.63	2.89	3.25	3.54
25	1.28	1.58	1.81	2.03	2.34	2.58	2.83	3.17	3.45
26	1.27	1.57	1.79	2.01	2.30	2.53	2.77	3.10	3.37
27	1.27	1.56	1.77	1.98	2.27	2.49	2.72	3.04	3.29
28	1.26	1.55	1.75	1.96	2.24	2.45	2.67	2.98	3.22
29	1.26	1.54	1.74	1.94	2.21	2.42	2.63	2.92	3.16
30	1.25	1.53	1.72	1.92	2.18	2.38	2.59	2.87	3.10
35	1.23	1.49	1.66	1.84	2.07	2.25	2.43	2.67	2.86
40	1.22	1.46	1.62	1.78	1.99	2.15	2.31	2.53	2.70
45	1.21	1.43	1.59	1.74	1.93	2.08	2.22	2.42	2.57
50	1.20	1.41	1.56	1.70	1.88	2.02	2.15	2.33	2.47
60	1.19	1.38	1.52	1.64	1.81	1.93	2.05	2.21	2.33
70	1.18	1.36	1.49	1.60	1.75	1.86	1.97	2.12	2.22
80	1.17	1.34	1.46	1.57	1.71	1.82	1.92	2.05	2.15
90	1.16	1.33	1.44	1.55	1.68	1.78	1.88	2.00	2.09
100	1.16	1.32	1.43	1.53	1.66	1.75	1.84	1.96	2.05
120	1.15	1.31	1.41	1.50	1.62	1.71	1.79	1.90	1.98
140	1.14	1.29	1.39	1.48	1.60	1.68	1.76	1.86	1.93
160	1.14	1.29	1.38	1.47	1.58	1.65	1.73	1.83	1.90
180	1.14	1.28	1.37	1.46	1.56	1.64	1.71	1.80	1.87
200	1.13	1.27	1.36	1.45	1.55	1.62	1.69	1.78	1.85
300	1.13	1.26	1.34	1.42	1.51	1.58	1.64	1.72	1.78
500	1.12	1.24	1.32	1.39	1.48	1.54	1.60	1.68	1.73
∞	1.11	1.22	1.29	1.36	1.43	1.49	1.54	1.60	1.65

Table 8.2 (ad) Critical Values of the *F* Distribution Numerator DF = 80

Denom. DF	α(2): 0.50 α(1): 0.25	0.20 0.10	0.10 0.05	0.05 0.025	0.02 0.01	0.01 0.005	0.005 0.0025	0.002 0.001	0.001 0.0005
1	9.78	62.9.	253.	1010.	6330.	25300.	101000.	633000.	2530000.
2	3.46	9.48	19.5	39.5	99.5	199.	399.	999.	2000.
3	2.47	5.15	8.56	14.0	26.3	42.1	67.1	124.	198.
4	2.08	3.78	5.67	8.33	13.6	19.5	27.9	44.6	63.3
5	1.87	3.13	4.41	6.10	9.16	12.3	16.5	24.2	32.2
6	1.74	2.75	3.72	4.93	7.01	9.06	11.6	16.1	20.5
7	1.65	2.50	3.29	4.23	5.78	7.25	9.04	12.0	14.8
8	1.59	2.33	2.99	3.76	4.99	6.12	7.46	9.63	11.6
9	1.54	2.20	2.77	3.42	4.44	5.36	6.42	8.09	9.61
10	1.50	2.09	2.60	3.17	4.04	4.80	5.68	7.03	8.23
11	1.46	2.01	2.47	2.97	3.73	4.39	5.13	6.26	7.25
12	1.44	1.95	2.36	2.82	3.49	4.07	4.71	5.68	6.51
13	1.41	1.89	2.27	2.69	3.30	3.81	4.38	5.22	5.94
14	1.39	1.84	2.20	2.58	3.14	3.60	4.11	4.86	5.49
15	1.37	1.80	2.14	2.49	3.00	3.43	3.89	4.56	5.12
16	1.36	1.77	2.08	2.42	2.89	3.28	3.70	4.31	4.81
17	1.35	1.74	2.03	2.35	2.79	3.15	3.54	4.10	4.56
18	1.33	1.71	1.99	2.29	2.70	3.04	3.40	3.92	4.34
19	1.32	1.68	1.96	2.24	2.63	2.95	3.28.	3.76	4.15
20	1.31	1.66	1.92	2.19	2.56	2.86	3.18	3.62	3.99
21	1.30	1.64	1.89	2.15	2.50	2.79	3.08	3.50	3.84
22	1.30	1.62	1.86	2.11	2.45	2.72	3.00	3.40	3.72
23	1.29	1.61	1.84	2.08	2.40	2.66	2.93	3.30	3.60
24	1.28	1.59	1.82	2.05	2.36	2.60	2.86	3.22	3.50
25	1.28	1.58	1.80	2.02	2.32	2.55	2.80	3.14	3.41
26	1.27	1.56	1.78	1.99	2.28	2.51	2.74	3.07	3.33
27	1.26	1.55	1.76	1.97	2.25	2.47	2.69	3.00	3.25
28	1.26	1.54	1.74	1.94	2.22	2.43	2.64	2.94	3.18
29	1.25	1.53	1.73	1.92	2.19	2.39	2.60	2.89	3.12
30	1.25	1.52	1.71	1.90	2.16	2.36	2.56	2.84	3.06
35	1.23	1.48	1.65	1.82	2.05	2.22	2.40	2.64	2.83
40	1.22	1.45	1.61	1.76	1.97	2.12	2.28	2.49	2.66
45	1.21	1.42	1.57	1.72	1.91	2.05	2.19	2.38	2.53
50	1.20	1.40	1.54	1.68	1.86	1.99	2.12	2.30	2.43
60	1.18	1.37	1.50	1.63	1.78	1.90	2.02	2.17	2.29
70	1.17	1.35	1.47	1.59	1.73	1.84	1.94	2.08	2.18
80	1.16	1.33	1.45	1.55	1.69	1.79	1.89	2.01	2.11
90	1.16	1.32	1.43	1.53	1.66	1.75	1.84	1.96	2.05
100	1.15	1.31	1.41	1.51	1.63	1.72	1.81	1.92	2.01
120	1.14	1.29	1.39	1.48	1.60	1.68	1.76	1.86	1.94
140	1.14	1.28	1.38	1.46	1.57	1.65	1.72	1.82	1.89
160	1.13	1.27	1.36	1.45	1.55	1.62	1.69	1.79	1.85
180	1.13	1.27	1.35	1.43	1.53	1.61	1.67	1.76	1.83
200	1.13	1.26	1.35	1.42	1.52	1.59	1.66	1.74	1.80
300	1.12	1.24	1.32	1.39	1.48	1.55	1.61	1.68	1.74
500	1.11	1.23	1.30	1.37	1.45	1.51	1.56	1.63	1.68
∞	1.10	1.21	1.27	1.33	1.40	1.45	1.50	1.56	1.60

Table 8.2 (ae) Critical Values of the F Distribution Numerator DF = 90

Denom. DF	α(2): 0.50 α(1): 0.25	0.20 0.10	0.10 0.05	0.05 0.025	0.02 0.01	0.01 0.005	0.005 0.0025	0.002 0.001	0.001 0.0005
1	9.79	63.0	253.	1010.	6330.	25300.	101000.	633000.	2530000.
2	3.46	9.48	19.5	39.5	99.5	199.	399.	999.	2000.
3	2.47	5.15	3.56	14.0	26.3	42.0	67.1	124.	197.
4	2.08	3.78	5.67	8.33	13.6	19.5	27.9	44.5	63.2
5	1.87	3.13	4.41	6.09	9.14	12.3	16.5	24.2	32.1
6	1.74	2.75	3.72	4.92	7.00	9.00	11.6	16.1	20.4
7	1.65	2.50	4.92	4.22	5.77	7.23	9.01	12.0	14.8
8	1.59	2.32	2.98	3.75	4.97	6.10	7.44	9.60	11.6
9	1.53	2.19	2.76	3.41	4.43	5.34	6.40	8.06	9.57
10	1.49	2.09	2.59	3.16	4.03	4.79	5.66	7.00	8.20
11	1.46	2.01	2.46	2.96	3.72	4.37	5.11	6.23	7.21
12	1.43	1.94	2.36	2.81	3.48	4.05	4.69	5.65	6.48
13	1.41	1.89	2.27	2.68	3.28	3.79	4.36	5.19	5.91
14	1.39	1.84	2.19	2.57	3.12	3.58	4.09	4.83	5.45
15	1.37	1.80	2.13	2.48	2.99	3.41	3.86	4.53	5.09
16	1.36	1.76	2.07	2.40	2.87	3.26	3.68	4.28	4.78
17	1.34	1.73	2.03	2.34	2.78	3.13	3.52	4.07	4.53
18	1.33	1.70	1.98	2.28	2.69	3.02	3.38	3.89	4.31
19	1.32	1.68	1.95	2.23	2.61	2.93	3.26	3.73	4.12
20	1.31	1.65	1.91	2.18	2.55	2.84	3.16	3.60	3.96
21	1.30	1.63	1.88	2.14	2.49	2.77	3.06	3.48	3.81
22	1.29	1.62	1.86	2.10	2.43	2.70	2.98	3.37	3.69
23	1.29	1.60	1.83	2.07	2.39	2.64	2.90	3.28	3.57
24	1.28	1.58	1.81	2.03	2.34	2.58	2.84	3.19	3.47
25	1.27	1.57	1.79	2.01	2.30	2.53	2.78	3.11	3.38
26	1.27	1.56	1.77	1.98	2.26	2.49	2.72	3.04	3.30
27	1.26	1.54	1.75	1.95	2.23	2.45	2.67	2.98	3.22
28	1.26	1.53	1.73	1.93	2.20	2.41	2.62	2.92	3.15
29	1.25	1.52	1.72	1.91	2.17	2.37	2.58	2.86	3.09
30	1.25	1.51	1.70	1.89	2.14	2.34	2.54	2.81	3.03
35	1.23	1.47	1.64	1.81	2.03	2.20	2.38	2.61	2.80
40	1.21	1.44	1.60	1.75	1.95	2.10	2.26	2.47	2.63
45	1.20	1.41	1.56	1.70	1.89	2.03	2.17	2.36	2.50
50	1.19	1.39	1.53	1.67	1.84	1.97	2.10	2.27	2.40
60	1.18	1.36	1.49	1.61	1.76	1.88	1.99	2.14	2.25
70	1.17	1.34	1.46	1.57	1.71	1.81	1.92	2.05	2.15
80	1.16	1.33	1.44	1.54	1.67	1.77	1.86	1.98	2.08
90	1.15	1.31	1.42	1.52	1.64	1.73	1.82	1.93	2.02
100	1.15	1.30	1.40	1.50	1.61	1.70	1.78	1.89	1.97
120	1.14	1.28	1.38	1.47	1.58	1.66	1.73	1.83	1.90
140	1.13	1.27	1.36	1.45	1.55	1.62	1.70	1.79	1.86
160	1.13	1.26	1.35	1.43	1.53	1.60	1.67	1.75	1.82
180	1.13	1.26	1.34	1.42	1.51	1.58	1.65	1.73	1.79
200	1.12	1.25	1.33	1.41	1.50	1.56	1.63	1.71	1.77
300	1.11	1.23	1.31	1.38	1.46	1.52	1.58	1.65	1.70
500	1.11	1.22	1.29	1.35	1.43	1.48	1.53	1.60	1.65
∞	1.10	1.20	1.26	1.31	1.38	1.43	1.47	1.52	1.56

Table 8.2 (af) Critical Values of the *F* Distribution Numerator DF = 100

Denom. DF	α(2): 0.50 α(1): 0.25	0.20 0.10	0.10 0.05	0.05 0.025	0.02 0.01	0.01 0.005	0.005 0.0025	0.002 0.001	0.001 0.0005
1	9.80	63.0	253.	1010.	6330.	25300.	101000.	633000.	2530000.
2	3.47	9.48	19.5	39.5	99.5	199.	399.	999.	2000.
3	2.47	5.14	8.55	14.0	26.2	42.0	67.1	124.	197.
4	2.08	3.78	5.66	8.32	13.6	19.5	27.9	44.5	63.2
5	1.87	3.13	4.41	6.08	9.13	12.3	16.5	24.1	32.1
6	1.74	2.75	3.71	4.92	6.99	9.03	11.6	16.0	20.4
7	1.65	2.50	3.27	4.21	5.75	7.22	8.99	12.0	14.8
8	1.58	2.32	2.97	3.74	4.96	6.09	7.42	9.57	11.6
9	1.53	2.19	2.76	3.40	4.41	5.32	6.38	8.04	9.54
10	1.49	2.09	2.59	3.15	4.01	4.77	5.64	6.98	8.17
11	1.46	2.01	2.46	2.96	3.71	4.36	5.09	6.21	7.19
12	1.43	1.94	2.35	2.80	3.47	4.04	4.67	5.63	6.45
13	1.41	1.88	2.26	2.67	3.27	3.78	4.34	5.17	5.88
14	1.39	1.83	2.19	2.56	3.11	3.57	4.07	4.81	5.43
15	1.37	1.79	2.12	2.47	2.98	3.39	3.85	4.51	5.06
16	1.36	1.76	2.07	2.40	2.86	3.25	3.66	4.26	4.76
17	1.34	1.73	2.02	2.33	2.76	3.12	3.50	4.05	4.50
18	1.33	1.70	1.98	2.27	2.68	3.01	3.36	3.87	4.28
19	1.32	1.67	1.94	2.22	2.60	2.91	3.24	3.71	4.10
20	1.31	1.65	1.91	2.17	2.54	2.83	3.14	3.58	3.93
21	1.30	1.63	1.88	2.13	2.48	2.75	3.04	3.46	3.79
22	1.29	1.61	1.85	2.09	2.42	2.69	2.96	3.35	3.66
23	1.29	1.59	1.82	2.06	2.37	2.62	2.89	3.25	3.55
24	1.28	1.58	1.80	2.02	2.33	2.57	2.82	3.17	3.45
25	1.27	1.56	1.78	2.00	2.29	2.52	2.76	3.09	3.36
26	1.27	1.55	1.76	1.97	2.25	2.47	2.70	3.02	3.27
27	1.26	1.54	1.74	1.94	2.22	2.43	2.65	2.96	3.20
28	1.25	1.53	1.73	1.92	2.19	2.39	2.60	2.90	3.13
29	1.25	1.52	1.71	1.90	2.16	2.36	2.56	2.84	3.06
30	1.25	1.51	1.70	1.88	2.13	2.32	2.52	2.79	3.01
35	1.23	1.47	1.63	1.80	2.02	2.19	2.36	2.59	2.77
40	1.21	1.43	1.59	1.74	1.94	2.09	2.24	2.44	2.60
45	1.20	1.41	1.55	1.69	1.88	2.01	2.15	2.33	2.47
50	1.19	1.39	1.52	1.66	1.82	1.95	2.08	2.25	2.37
60	1.18	1.36	1.48	1.60	1.75	1.86	1.97	2.12	2.23
70	1.16	1.34	1.45	1.56	1.70	1.80	1.90	2.03	2.13
80	1.16	1.32	1.43	1.53	1.65	1.75	1.84	1.96	2.05
90	1.15	1.30	1.41	1.50	1.62	1.71	1.80	1.91	1.99
100	1.14	1.29	1.39	1.48	1.60	1.68	1.76	1.87	1.95
120	1.14	1.28	1.37	1.45	1.56	1.64	1.71	1.81	1.88
140	1.13	1.26	1.35	1.43	1.53	1.60	1.67	1.76	1.83
160	1.13	1.26	1.34	1.42	1.51	1.58	1.64	1.73	1.79
180	1.12	1.25	1.33	1.40	1.49	1.56	1.62	1.70	1.76
200	1.12	1.24	1.32	1.39	1.48	1.54	1.60	1.68	1.74
300	1.11	1.22	1.30	1.36	1.44	1.50	1.55	1.62	1.67
500	1.10	1.21	1.28	1.34	1.41	1.46	1.51	1.57	1.62
∞	1.09	1.18	1.24	1.30	1.36	1.40	1.44	1.49	1.53

Table 8.2 (ag) Critical Values of the *F* Distribution Numerator DF = 120

Denom. DF	α(2): 0.50 α(1): 0.25	0.20 0.10	0.10 0.05	0.05 0.025	0.02 0.01	0.01 0.005	0.005 0.0025	0.002 0.001	0.001 0.0005
1	9.80	63.1	253.	1010.	6340.	25400.	101000.	634000.	2540000.
2	3.47	9.48	19.5	39.5	99.5	199.	399.	999.	2000.
3	2.47	5.14	8.55	13.9	26.2	42.0	67.0	124.	197.
4	2.08	3.78	5.66	831	13.6	19.5	27.8	44.4	63.1
5	1.87	3.12	4.40	6.07	9.11	12.3	16.4	24.1	32.0
6	1.74	2.74	3.70	4.90	6.97	9.00	11.6	16.0	20.3
7	1.65	2.49	3.27	4.20	5.74	7.19	8.96	11.9	14.7
8	1.58	2.32	2.97	3.73	4.95	6.06	7.39	9.53	11.5
9	1.53	2.18	2.75	3.39	4.40	5.30	6.35	8.00	9.49
10	1.49	2.08	2.58	3.14	4.00	4.75	5.61	6.94	8.12
11	1.46	2.00	2.45	2.94	3.69	4.34	5.07	6.18	7.14
12	1.43	1.93	2.34	2.79	3.45	4.01	4.65	5.59	6.41
13	1.41	1.88	2.25	2.66	3.25	3.76	4.31	5.14	5.84
14	1.39	1.83	2.18'	2.55	3.09	3.55	4.04	4.77	5.39
15	1.37	1.79	2.11	2.46	2.96	3.37	3.82	4.47	5.02
16	1.35	1.75	2.06	2.38	2.84	3.22	3.63	4.23	4.72
17	1.34	1.72	2.01	2.32	2.75	3.10	3.47	4.02	4.46
18	1.33	1.69	1.97	2.26	2.66	2.99	3.34	3.84	4.25
19	1.32	1.67	1.93	2.20	2.58	2.89	3.22	3.68	4.06
20	1.31	1.64	1.90	2.16	2.52	2.81	3.11	3.54	3.90
21	1.30	1.62	1.87	2.11	2.46	2.73	3.02	3.42	3.75
22	1.29	1.60	1.84	2.08	2.40	2.66	2.93	3.32	3.62
23	1.28	1.59	1.81	2.04	2.35	2.60	2.86	3.22	3.51
24	1.28	1.57	1.79	2.01	2.31	2.55	2.79	3.14	3.41
25	1.27	1.56	1.77	1.98	2.27	2.50	2.73	3.06	3.32
26	1.26	1.54	1.75	1.95	2.23	2.45	2.68	2.99	3.24
27	1.26	1.53	1.73	1.93	2.20	2.41	2.62	2.92	3.16
28	1.25	1.52	1.71	1.91	2.17	2.37	2.58	2.86	3.09
29	1.25	1.51	1.70	1.89	2.14	2.33	2.53	2.81	3.03
30	1.24	1.50	1.68	1.87	2.11	2.30	2.49	2.76	2.97
35	1.22	1.46	1.62	1.79	2.00	2.16	2.33	2.56	2.73
40	1.21	1.42	1.58	1.72	1.92	2.06	2.21	2.41	2.56
45	1.20	1.40	1.54	1.68	1.85	1.99	2.12	2.30	2.44
50	1.19	1.38	1.51	1.64	1.80	1.93	2.05	2.21	2.34
60	1.17	1.35	1.47	1.58	1.73	1.83	1.94	2.08	2.19
70	1.16	1.32	1.44	1.54	1.67	1.77	1.87	1.99	2.09
80	1.15	1.31	1.41	1.51	1.63	1.72	1.81	1.92	2.01
90	1.15	1.29	1.39	1.48	1.60	1.68	1.76	1.87	1.95
100	1.14	1.28	1.38	1.46	1.57	1.65	1.73	1.83	1.90
120	1.13	1.26	1.35	1.43	1.53	1.61	1.68	1.77	1.83
140	1.13	1.25	1.33	1.41	1.50	1.57	1.64	1.72	1.78
160	1.12	1.24	1.32	1.39	1.48	1.55	1.61	1.69	1.75
180	1.12	1.23	1.31	1.38	1.47	1.53	1.59	1.66	1.72
200	1.11	1.23	1.30	1.37	1.45	1.51	1.57	1.64	1.69
300	1.10	1.21	1.28	1.34	1.41	1.46	1.51	1.58	1.62
500	1.10	1.19	1.26	1.31	1.38	1.42	1.47	1.53	1.57
∞	1.08	1.17	1.22	1.27	1.32	1.36	1.40	1.45	1.48

Table 8.2 (ah) Critical Values of the *F* Distribution Numerator DF = 140

Denom. DF	α(2): 0.50 α(1): 0.25	0.20 0.10	0.10 0.05	0.05 0.025	0.02 0.01	0.01 0.005	0.005 0.0025	0.002 0.001	0.001 0.0005
1	9.81	63.1	253.	1010.	6340.	25400.	101000.	634000.	2540000.
2	3.47	9.48	19.5	39.5	99.5	199.	399.	999.	2000.
3	2.47	5.14	8.55	13.9	26.2	42.0	67.0	124.	197.
4	2.08	3.77	5.65	8.30	13.5	19.4	27.8	44.4	63.0
5	1.87	3.12	4.39	6.06	9.10	12.3	16.4	24.0	31.9
6	1.74	2.74	3.70	4.90	6.96	8.98	11.5	15.9	20.3
7	1.65	2.49	3.26	4.19	5.72	7.18	8.94	11.9	14.7
8	1.58	2.31	2.96	3.72	4.93	6.05	7.37	9.50	11.5
9	1.53	2.18	2.74	3.38	4.39	5.28	6.33	7.97	9.46
10	1.49	2.08	2.57	3.13	3.98	4.73	5.59	6.92	8.09
11	1.46	2.00	2.44	2.94	3.68	4.32	5.05	6.15	7.11
12	1.43	1.93	2.33	2.78	3.44	4.00	4.63	5.57	6.38
13	1.41	1.87	2.25	2.65	3.24	3.74	4.29	5.11	5.81
14	1.39	1.82	2.17	2.54	3.08	3.53	4.02	4.75	5.36
15	1.37	1.78	2.11	2.45	2.95	3.36	3.80	4.45	5.00
16	1.35	1.75	2.05	2.37	2.83	3.21	3.61	4.20	4.69
17	1.34	1.71	2.00	2.31	2.73	3.08	3.45	3.99	4.44
18	1.33	1.69	1.96	2.25	2.65	2.97	3.32	3.81	4.22
19	1.32	1.66	1.92	2.19	2.57	2.87	3.20	3.66	4.03
20	1.31	1.64	1.89	2.15	2.50	2.79	3.09	3.52	3.87
21	1.30	1.62	1.86	2.10	2.44	2.71	3.00	3.40	3.73
22	1.29	1.60	1.83	2.07	2.39	2.65	2.92	3.29	3.60
23	1.28	1.58	1.81	2.03	2.34	2.59	2.84	3.20	3.49
24	1.27	1.57	1.78	2.00	2.30	2.53	2.77	3.11	3.38
25	1.27	1.55	1.76	1.97	2.26	2.48	2.71	3.03	3.29
26	1.26	1.54	1.74	1.94	2.22	2.43	2.66	2.96	3.21
27	1.26	1.53	1.72	1.92	2.18	2.39	2.60	2.90	3.13
28	1.25	1.51	1.71	1.90	2.15	2.35	2.56	2.84	3.06
29	1.24	1.50	1.69	1.88	2.12	2.32	2.51	2.79	3.00
30	1.24	1.49	1.68	1.86	2.10	2.28	2.47	2.74	2.94
35	1.22	1.45	1.61	1.77	1.98	2.15	2.31	2.53	2.71
40	1.21	1.42	1.57	1.71	1.90	2.05	2.19	2.39	2.54
45	1.19	1.39	1.53	1.66	1.84	1.97	2.10	2.27	2.41
50	1.18	1.37	1.50	1.63	1.79	1.91	2.03	2.19	2.31
60	1.17	1.34	1.46	1.57	1.71	1.81	1.92	2.06	2.16
70	1.16	1.32	1.42	1.53	1.65	1.75	1.84	1.96	2.06
80	1.15	1.30	1.40	1.49	1.61	1.70	1.79	1.90	1.98
90	1.14	1.28	1.38	1.47	1.58	1.66	1.74	1.84	1.92
100	1.14	1.27	1.36	1.45	1.55	1.63	1.70	1.80	1.87
120	1.13	1.26	1.34	1.42	1.51	1.58	1.65	1.74	1.80
140	1.12	1.24	1.32	1.39	1.48	1.55	1.61	1.69	1.75
160	1.12	1.23	1.31	1.38	1.46	1.52	1.58	1.66	1.71
180	1.11	1.22	1.30	1.36	1.45	1.50	1.56	1.63	1.68
200	1.11	1.22	1.29	1.35	1.43	1.49	1.54	1.61	1.66
300	1.10	1.20	1.26	1.32	1.39	1.44	1.49	1.55	1.59
500	1.09	1.18	1.24	1.29	1.35	1.40	1.44	1.49	1.53
∞	1.08	1.16	1.20	1.25	1.30	1.33	1.37	1.41	1.44

Table 8.2 (ai) Critical Values of the *F* Distribution Numerator DF = 200

Denom. DF	α(2): 0.50 α(1): 0.25	0.20 0.10	0.10 0.05	0.05 0.025	0.02 0.01	0.01 0.005	0.005 0.0025	0.002 0.001	0.001 0.0005
1	9.82	63.2	254.	1020.	6350.	25400.	102000.	635000.	2540000.
2	3.47	9.49	19.5	39.5	99.5	199.	399.	999.	2000.
3	2.47	5.14	8.54	13.9	26.2	41.9	66.9	124.	197.
4	2.08	3.77	5.65	8.29	13.5	19.4	27.7	44.3	62.9
5	1.87	3.12	4.39	6.05	9.08	12.2	16.4	24.0	31.8
6	1.74	2.73	3.69	4.88	6.93	8.95	11.5	15.9	20.2
7	1.65	2.48	3.25	4.18	5.70	7.15	8.90	11.8	14.6
8	1.58	2.31	2.95	3.70	4.91	6.02	7.33	9.45	11.4
9	1.53	2.17	2.73	3.37	4.36	5.26	6.29	7.93	9.40
10	1.49	2.07	2.56	3.12	396	4.71	5.56	6.87	8.04
11	1.46	1.99	2.43	2.92	3.66	4.29	5.01	6.10	7.06
12	1.43	1.92	2,32	2.76	3.41	3.97	4.59	5.52	6.33
13	1.40	1.86	2.23	2.63	3.22	3.71	4.26	5.07	5.76
14	1.38	1.82	2.16	2.53	3.06	3.50	3.99	4.71	5.31
15	1.37	1.77	2.10	2.44	2.92	3.33	3.77	4.41	4.95
16	1.35	1.74	2.04	2.36	2.81	3.18	3.58	4.16	4.64
17	1.34	1.71	1.99	2.29	2.71	3.05	3.42	3.95	4.39
18	1.32	1.68	1.95	2.23	2.62	2.94	3.28	3.77	4.17
19	1.31	1.65	1.91	2.18	2.55	2.85	3.16	3.61	3.98
20	1.30	1.63	1.88	2.13	2.48	2.76	3.06	3.48	3.82
21	1.29	1.61	1.84	2.09	2.42	2.68	2.96	3.36	3.68
22	1.28	1.59	1.82	2.05	2.36	2.62	2.88	3.25	3.55
23	1.28	1.57	1.79	2.01	2.32	2.56	2.81	3.16	3.44
24	1.27	1.56	1.77	1.98	2.27	2.50	2.74	3.07	3.34
25	1.26	1.54	1.75	1.95	2.23	2.45	2.68	2.99	3.24
26	1.26	1.53	1.73	1.92	2.19	2.40	2.62	2.92	3.16
27	1.25	1.52	1.71	1.90	2.16	2.36	2.57	2.86	3.09
28	1.25	1.50	1.69	1.88	2.13	2.32	2.52	2.80	3.02
29	1.24	1.49	1.67	1.86	2.10	2.29	2.48	2.74	2.95
30	1.24	1.48	1.66	1.84	2.07	2.25	2.44	2.69	2.89
35	1.22	1.44	1.60	1.75	1.96	2.11	2.27	2.49	2.66
40	1.20	1.41	1.55	1.69	1.87	2.01	2.15	2.34	2.49
45	1.19	1.38	1.51	1.64	1.81	1.93	2.06	2.23	2.36
50	1.18	1.36	1.48	1.60	1.76	1.87	1.99	2.14	2.26
60	1.16	1.33	1.44	1.54	1.68	1.78	1.88	2.01	2.11
70	1.15	1.30	1.40	1.50	1.62	1.71	1.80	1.92	2.00
80	1.14	1.28	1.38	1.47	1.58	1.66	1.74	1.85	1.93
90	1.13	1.27	1.36	1.44	1.55	1.62	1.70	1.79	1.86
100	1.13	1.26	1.34	1.42	1.52	1.59	1.66	1.75	1.82
120	1.12	1.24	1.32	1.39	1.48	1.54	1.60	1.68	1.74
140	1.11	1.22	1.30	1.36	1.45	1.51	1.56	1.64	1.69
160	1.11	1.21	1.28	1.35	1.42	1.48	1.53	1.60	1.65
180	1.10	1.21	1.27	1.33	1.41	1.46	1.51	1.57	1.62
200	1.10	1.20	1.26	1.32	1.39	1.44	1.49	1.55	1.60
300	1.09	1.18	1.23	1.28	1.35	1.39	1.43	1.48	1.52
500	1.08	1.16	1.21	1.25	1.31	1.35	1.38	1.43	1.46
∞	1.07	1.13	1.17	1.21	1.25	1.28	1.30	1.34	1.36

Table 8.2 (aj) Critical Values of the F Distribution Numerator DF = ∞

Denom. DF	α(2): 0.50 α(1): 0.25	0.20 0.10	0.10 0.05	0.05 0.025	0.02 0.01	0.01 0.005	0.005 0.0025	0.002 0.001	0.001 0.0005
1	9.85	63.3	254.	1020.	6370.	25500.	102000.	637000.	2550000.
2	3.48	9.49	19.5	39.5	99.5	199.	399.	999.	2000.
3	2.47	5.13	8.53	13.9	26.1	41.8	66.8	123.	196.
4	2.08	3.76	5.63	8.26	13.5	19.3	27.6	44.0	62.6
5	1.87	3.11	4.37	6.02	9.02	12.1	16.3	23.8	31.6
6	1.74	2.72	3.67	4.85	6.88	8.88	11.4	15.7	20.0
7	1.65	2.47	3.23	4.14	5.65	7.08	8.81	11.7	14.4
8	1.58	2.29	2.93	3.67	4.86	5.95	7.25	9.33	11.3
9	1.53	2.16	2.71	3.33	4.31	5.19	6.21	7.81	9.26
10	1.48	2.06	2.54	3.08	3.91	4.64	5.47	6.76	7.91
11	1.45	1.97	2.40	2.88	3.60	4.23	4.93	6.00	6.93
12	1.42	1.90	2.30	2.72	3.36	3.90	4.51	5.42	6.20
13	1.40	1.85	2.21	2.60	3.17	3.65	4.18	4.97	5.64
14	1.38	1.80	2.13	2.49	3.00	3.44	3.91	4.60	5.19
15	1.36	1.76	2.07	2.40	2.87	3.26	3.69	4.31	4.83
16	1.34	1.72	2.01	2.32	2.75	3.11	3.50	4.06	4.52
17	1.33	1.69	1.96	2.25	2.65	2.98	3.34	3.85	4.27
18	1.32	1.66	1.92	2.19	2.57	2.87	3.20	3.67	4.05
19	1.30	1.63	1.88	2.13	2.49	2.78	3,08	3.51	3.87
20	1.29	1.61	1.84	2.09	2.42	2.69	2.97	3.38	3.71
21	1.28	1.59	1.81	2.04	2.36	2.61	2.88	3.26	3.56
22	1.28	1.57	1.78	2.00	2.31	2.55	2.80	3.15	3.43
23	1.27	1.55	1.76	1.97	2.26	2.48	2.72	3.05	3.32
24	1.26	1.53	1.73	1.94	2.21	2.43	2.65	2.97	3.22
25	1.25	1.52	1.71	1.91	2.17	2.38	2.59	2.89	3.13
26	1.25	1.50	1.69	1.88	2.13	2.33	2.54	2.82	3.05
27	1.24	1.49	1.67	1.85	2.10	2.29	2.48	2.75	2.97
28	1.24	1.48	1.65	1.83	2.06	2.25	2.44	2.69	2.90
29	1.23	1.47	1.64	1.81	2.03	2.21	2.39	2.64	2.84
30	1.23	1.46	1.62	1.79	2.01	2.18	2.35	2.59	2.78
35	1.20	1.41	1.56	1.70	1.89	2.04	2.18	2.38	2.54
40	1.19	1.38	1.51	1.64	1.80	1.93	2.06	2.23	2.37
45	1.18	1.35	1.47	1.59	1.74	1.85	1.97	2.12	2.23
50	1.16	1.33	1.44	1.55	1.68	1.79	1.89	2.03	2.13
60	1.15	1.29	1.39	1.48	1.60	1.69	1.78	1.89	1.98
70	1.13	1.27	1.35	1.44	1.54	1.62	1.69	1.79	1.87
80	1.12	1.24	1.32	1.40	1.49	1.56	1.63	1.72	179
90	1.12	1.23	1.30	1.37	1.46	1.52	1.58	1.66	1.72
100	1.11	1.21	1.28	1.35	1.43	1.49	1.54	1.62	1.67
120	1.10	1.19	1.25	1.31	1.38	1.43	1.48	1.54	1.59
140	1.09	1.18	1.23	1.28	1.35	1.39	1.43	1.49	1.53
160	1.08	1.16	1.21	1.26	1.32	1.36	1.40	1.45	1.49
180	1.08	1.15	1.20	1.24	1.30	1.33	1.37	1.42	1.45
200	1.07	1.14	1.19	1.23	1.28	1.31	1.35	1.39	1.42
300	1.06	1.11	1.15	1.18	1.22	1.25	1.27	1.30	1.33
500	1.05	1.09	1.11	1.14	1.16	1.18	1.20	1.23	1.24
∞	1.00	1.00	1.00	1.00	1.00	1.00	1.00	1.00	1.00

Example:

The disintegration times of tablets observed from the two batches is given below. Whether the variation is significant or not?

Batch A	Batch B
1	30
2	26
3	28
4	29
5	28
6	29

Solution:

$$A = \frac{21}{6} = 3.5 = \overline{x}_1$$

Standard deviation of batch 1

$$S_1 = \sqrt{\frac{\Sigma x^2 - \frac{(\Sigma x)^2}{n}}{n-1}}$$

$$= \sqrt{\frac{91 - \frac{(21)^2}{6}}{5}}$$

$$= \sqrt{\frac{91 - 73.5}{5}}$$

$$= \sqrt{\frac{17.5}{5}}$$

$$= \sqrt{3.5}$$

$$= 1.87$$

Mean of tablets batch $B = \frac{170}{6} = 28.33$

Standard deviation of batch 2

$$S_2 = \sqrt{\frac{4826 - \frac{(170)^2}{6}}{5}}$$

$$= \sqrt{\frac{4826 - \frac{28900}{6}}{5}}$$

$$= \sqrt{\frac{4826 - 4816.67}{5}}$$

$$= \sqrt{\frac{9.33}{5}} = \sqrt{1.866} = 1.37$$

F statistic $= \frac{3.497}{1.877} = 1.86$

$F_{cal} = 1.86$

F_{tab} value with degree of freedom $n_1 - 1 = 5$ and $n_2 - 1 = 5$ at the level 1 of probability 0.05 is 5.05

∴ $F_{cal} < F_{tab}$

∴ Null hypothesis is accepted.

8.7 Chi-square Test of Significance

The earlier tests are based on the assumption that the samples are drawn from normal population. But, in many situations it is not possible to make any assumption about the parent distribution from which samples have been drawn. In such cases, distribution free Chi-square test is used. The chi square tests are involved only with discrete variables. In the goodness-of-fit test there is one discrete variable. For the test of independence, two discrete variables are compared: one is usually independent (i.e., experimental vs. control group) and the other variable is dependent upon the first (i.e., cured vs. died). Chi square tests evaluate the importance of the difference between what is expected and what is observed; Chi Square *(χ^2)* can best be thought of as a discrepancy statistic. It analyzes the difference between observed values and those values that one would normally expect to occur. It is calculated by determining the difference between the frequencies actually observed in a sample data set and the expected frequencies based on probability. Some textbooks classify it as a nonparametric procedure because it is not concerned with distributions about a central point and does not require assumptions of homogeneity or normality.

Significance of difference between two distributions is tested with chi-square (χ^2) distribution. The chi-square statistic is calculated with the following formulae

$$\chi^2 = \Sigma \frac{(f_o - f_E)^2}{f_E}$$

f_o = observed count, f_E = expected count

The χ^2 test can be applied to study the influence of either 2 or 3 parameters.

Like the z-, t- and F-tests, critical values for the chi square distribution are presented in tabular form. Like the t- and F-distributions, there is not one single distribution, but a set of distributions. The characteristics of each distribution are dependent on the number of associated degrees of freedom. The first column on the left of table indicates the number of degrees of freedom and the remaining columns are the critical chi square values at different acceptable levels of Type I error ($\propto$). The decision rule is written similar to previous tests. For example, assume we are dealing with four degrees of freedom and wish to have a 95% level of confidence. The decision rule would be: With $\propto = 0.05$ reject H_0 if $\chi^2 > \chi^2_4\ (0.05) = 9.448$. If the calculated statistic derived from the formula is larger than the critical (9.448), the null hypothesis (the observed and expected values are the same) is rejected.

This distribution is having the following properties.

- Mean of χ^2 distribution = Degrees of freedom = υ
- Standard deviation of χ^2 distribution is the square root of 2 υ
- Median of χ^2 distribution divides the curve into two equal parts.
- Mode of χ^2 distribution is equal to $\upsilon - 2$.
- χ^2 curve is always positively skewed.
- Chi-square values increase with degrees of freedom.
- The range of Chi- square values are 0-∞.
- When $\upsilon > 30$, then $\sqrt{2}\,\chi^2 - \sqrt{2}\,\upsilon - 1$ follows normal distribution.

Chi-square test is used in following conditions.

- The frequencies must be absolute.
- The total number of observations must be > 50.
- The samples should be independent to each other.
- It is dependent on degrees of freedom and on sample data.
- The expected frequency of any cell or item must not be less than 5.
- The data should be expressed in original units.

Different types of Chi-square distribution

Various types of Chi-square distribution are as follows.

- Test of goodness of fit
- Test of independence.
- Test of population variance

8.7.1 Test of Goodness of Fit

The χ^2 test of goodness of fit is used to test if the deviations between observed and theoretical values can be attributed to chance or are due to some inadequacy of the theory to fit the observed data. It involves the following steps.

- *Null hypothesis.* There is no significant differences between the observed and expected value.
- *Computation of the expected value.*
- *Computation of χ^2 value.* It is calculated with the following equation.

$$\chi^2 = \sum_{i=1}^{n}\left[\frac{(O_i - E_i)^2}{E_i}\right]$$

$$= \frac{(O_1 - E_1)^2}{E_1} + \frac{(O_2 - E_2)^2}{E_2} + \ldots\ldots + \frac{(O_n - E_n)^2}{E_n}$$

- *Calculate degree of freedom* (n – 1).
- *Decision.* If the calculated χ^2 value is less than the tabulated χ^2 value, then the null hypothesis is accepted and if it is greater than the table value then the null hypothesis is rejected and alternative hypothesis is accepted. The table values are given in Table 8.3.

Table 8.3 Critical Values of the Chi-Square Distribution

ν	α: 0.999	0.995	0.99	0.975	0.95	0.90	0.75	0.50	0.25	0.10	0.05	0.025	0.01	0.005	0.001
1	0.000	0.000	0.000	0.001	0.004	0.016	0.102	0.455	1.323	2.706	3.841	5.024	6.635	7.879	10.828
2	0.002	0.010	0.020	0.051	0.103	0.211	0.575	1.386	2.773	4.605	5.991	7.378	9.210	10.597	13.816
3	0.024	0.072	0.115	0.216	0.352	0.584	1.213	2.366	4.108	6.251	7.815	9.348	11.345	12.838	16.266
4	0.091	0.207	0.297	0.484	0.711	1.064	1.923	3.357	5.385	7.779	9.488	11.143	13.277	14.860	18.467
5	0.210	0.412	0.554	0.831	1.145	1.610	2.675	4.351	6.626	9.236	11.070	12.833	15.086	16.750	20.515
6	0.381	0.676	0.872	1.237	1.635	2.204	3.455	5.348	7.841	10.645	12.592	14.449	16.812	18.548	22.458
7	0.599	0.989	1.239	1.690	2.167	2.833	4.255	6.346	9.037	12.017	14.067	16.013	18.475	20.278	24.322
8	0.857	1.344	1.646	2.180	2.733	3.490	5.071	7.344	10.219	13.362	15.507	17.535	20.090	21.955	26.124
9	1.152	1.735	2.088	2.700	3.325	4.168	5.899	8.343	11.389	14.684	16.919	19.023	21.666	23.589	27.877
10	1.479	2.156	2.558	3.247	3.940	4.865	6.737	9.342	12.549	15.987	18.307	20.483	23.209	25.188	29.588
11	1.834	2.603	3.053	3.816	4.575	5.578	7.584	10.341	13.701	17.275	19.675	21.920	24.725	26.757	31.264
12	2.214	3.074	3.571	4.404	5.226	6.304	8.438	11.340	14.845	18.549	21.026	23.337	26.217	28.300	32.909
13	2.617	3.565	4.107	5.009	5.892	7.042	9.299	12.340	15.984	19.812	22.362	24.736	27.688	29.819	34.528
14	3.041	4.075	4.660	5.629	6.571	7.790	10.165	13.339	17.117	21.064	23.685	26.119	29.141	31.319	36.123
15	3.483	4.601	5,229	6.262	7.261	8.547	11.037	14.339	18.245	22.307	24.996	27.488	30.578	32.801	37.697
16	3.942	5.142	5.812	6.908	7.962	9.312	11.912	15.338	19.369	23.542	26.296	28.845	32.000	34.267	39.252
17	4.416	5.697	6.408	7.564	8.672	10.085	12.792	16.338	20.489	24.769	27.587	30.191	33.409	35.718	40.790
18	4.905	6.265	7.015	8.231	9.390	10.865	13.675	17.338	21.605	25.989	28.869	31.526	34.805	37.156	42.312
19	5.407	6.844	7.633	8.907	10.117	11.651	14.562	18.338	22.718	27.204	30.144	32.852	36.191	38.582	43.820
20	5.921	7.434	8.260	9.591	10.851	12.443	15.452	19.337	23.828	28.412	31.410	34.170	37.566	39.997	45.315
21	6.447	8.034	8.897	10.283	11.591	13.240	16.344	20.337	24.935	29.615	32.671	35.479	38.932	41.401	46.797
22	6.983	8.643	9.542	10.982	12.338	14.041	17.240	21.337	26.039	30.813	33.924	36.781	40.289	42.796	48.268
23	7.529	9.260	10.196	11.689	13.091	14.848	18.137	22.337	27.141	32.007	35.172	38.076	41.638	44.181	49.728

Table 8.3 *Contd....*

ν	α: 0.999	0.995	0.99	0.975	0.95	0.90	0.75	0.50	0.25	0.10	0.05	0.025	0.01	0.005	0.001
24	8.085	9.886	10.856	12.401	13.848	15.659	19.037	23.337	28.241	33.196	36.415	39.364	42.980	45.559	51.179
25	8.649	10.520	11.524	13.120	14.611	16.473	19.939	24.337	29.339	34.382	37.652	40.646	44.314	46.928	52.620
26	9.222	11.160	12.198	13.844	15.379	17.292	20.843	25.336	30.435	35.563	38.885	41.923	45.642	48.290	54.052
27	9.803	11.808	12.879	14.573	16.151	18.114	21.749	26.336	31.528	36.741	40.113	43.195	46.963	49.645	55.476
28	10.391	12.461	13.565	15.308	16.928	18.939	22.657	27.336	32.620	37.916	41.337	44.461	48.278	50.993	56.892
29	10.986	13.121	14.256	16.047	17.708	19.768	23.567	28.336	33.711	39.087	42.557	45.722	49.588	52.336	58.301
30	11.588	13.787	14.953	16.791	18.493	20.599	24.478	29.336	34.800	40.256	43.773	46.979	50.892	53.672	59.703
31	12.196	14.458	15.655	17.539	19.281	21.434	25.390	30.336	35.887	41.422	44.985	48.232	52.191	55.003	61.098
32	12.811	15.134	16.362	18.291	20.072	22.271	26.304	31.336	36.973	42.585	46.194	49.480	53.486	56.328	62.487
33	13.431	15.815	17.074	19.047	20.867	23.110	27.219	32.336	38.058	43.745	47.400	50.725	54.776	57.648	63.870
34	14.057	16.501	17.789	19.806	21.664	23.952	28.136	33.336	39.141	44.903	48.602	51.966	56.061	58.964	65.247
35	14.688	17.192	18.509	20.569	22.465	24.797	29.054	34.336	40.223	46.059	49.802	53.203	57.342	60.275	66.619
36	15.324	17.887	19.233	21.336	23.269	25.643	29.973	35.336	41.304	47.212	50.998	54.437	58.619	61.581	67.985
37	15.965	18.586	19.960	22.106	24.075	26.492	30.893	36.336	42.383	48.363	52.192	55.668	59.893	62.883	69.346
38	16.611	19.289	20.691	22.878	24884	27.343	31.815	37.335	43.462	49.513	53.384	56.896	61.162	64.181	70.703
39	17.262	19.996	21.426	23.654	25.695	28.196	32.737	38.335	44.539	50.660	54.572	58.120	62.428	65.476	72.055
40	17.916	20.707	22.164	24.433	26.509	29.051	33.660	39.335	45.616	51.805	55.758	59.342	63.691	66.766	73.402
41	18.576	21.421	22.906	25.215	27.326	29.907	34.585	40.335	46.692	52.949	56.942	60.561	64.950	68.053	74.745
42	19.239	22.138	23.650	25.999	28.144	30.765	35.510	41.335	47.766	54.090	*58.124*	61. 777	66.206	69.336	76.084
43	19.906	22.859	24.398	26.785	28.965	31.625	36.436	42.335	48.840	55.230	59.304	62.990	67.459	70.616	77.419
44	20.576	23.584	25.148	27.575	29.787	32.487	37,363	43.335	49.913	56.369	60.481	64.201	68.710	71.893	78.750
45	21.251	24.311	25.901	28366	30.612	33.350	38.291	44.335	50.985	57.505	61.656	65.410	69.957	73.166	80.077
46	21.929	25.041	26.657	29.160	31.439	34.215	39.220	45.335	52.056	58.641	62.830	66.617	71.201	74.437	81.400
47	22.610	25.775	27.416	29.956	32.268	35.081	40.149	46.335	53.127	59.774	64.001	67.821	72.443	75.704	82.720

Table 8.3 *Contd....*

ν	α: 0.999	0.995	0.99	0.975	0.95	0.90	0.75	0.50	0.25	0.10	0.05	0.025	0.01	0.005	0.001
48	23.295	26.511	28.177	30.755	33.098	35.949	41.079	47.335	54.196	60.907	65.171	69.023	73.683	76.969	84.037
49	23.983	27.249	28.941	31.555	33.930	36.818	42.010	48.335	55.265	62.038	66.339	70.222	74.919	78.231	85.351
50	24.674	27.991	29.707	32.357	34.764	37.689	42.942	49.335	56.334	63.167	67.505	71.420	"76.154	79.490	86.661
51	25.368	28.735	30.475	33.162	35.600	38.560	43.874	50.335	57.401	64.295	68.669	72.616	77.386	80.747	87.968
52	26.065	29.481	31.246	33.968	36.437	39.433	44.808	51.335	58.468	65.422	69.832	73.810	78.616	82.001	89.272
53	26.765	30.230	32.019	34.776	37.276	40.308	45.741	52.335	59.534	66.548	70.993	75.002	79.843	83.253	90.573
54	27.468	30.981	32.793	35.586	38.116	41.183	46.676	53.335	60.600	67.673	72.153	76.192	81.069	84.502	91.872
55	28.173	31.735	33.570	36.398	38.958	42.060	47.610	54.335	61.665	68.796	73.311	77.380	82.292	85.749	93.168
56	28.881	32.491	34.350	37.212	39.801	42.937	48.546	55.335	62.729	69.919	74.468	78.567	83.513	86.994	94.461
57	29.592	33.248	35.131	38.027	40.646	43.816	49.482	56.335	63.793	71.040	75.624	79.752	84.733	88.236	95.751
58	30.305	34.008	35.913	38.844	41.492	44.696	50.419	57.335	64.857	72.160	76.778	80.936	85.950	89.477	97.039
59	31.021	34.770	36.698	39.662	42.339	45.577	51.356	58.335	65.919	73.279	77.931	82.117	87.166	90.715	98.324
60	31.738	35.535	37.485	40.482	43.188	46.459	52.294	59.335	66.981	74,397	79.082	83.298	88.379	91.952	99.607
61	32.459	36.301	38.273	41.303	44.038	47.342	53.232	60.335	68.043	75.514	80.232	84.476	89.591	93.186	100.888
62	33.181	37.068	39.063	42.126	44.889	48.226	54.171	61.335	69.104	76.630	81.381	85.654	90.802	94.419	102.166
63	33.906	37.838	39.855	42.950	45.741	49.111	55.110	62.335	70.165	77.745	82.529	86.830	92.010	95.649	103.442
64	34.633	38.610	40.649	43.776	46.595	49.996	56.050	63.335	71.225	78.860	83.675	88.004	93.217	96.878	104.716
65	35.362	39.383	41.444	44.603	47.450	50.883	56.990	64.335	72.285	79.973	84.821	89.177	94.422	98.105	105.988
66	36.093	40.158	42.240	45.431	48.305	51.770	57.931	65.335	73.344	81.085	85.965	90.349	95.626	99.330	107.258
67	36.826	40.935	43.038	46.261	49.162	52.659	58.872	66.335	74.403	82.197	87.108	91.519	96.828	100.554	108.526
68	37.561	41.713	43.838	47.092	50.020	53.548	59.814	67.335	75.461	83.308	88.250	92.689	98.028	101.776	109.791
69	38.298	42.494	44.639	47.924	50.879	54.438	60.756	68.334	76.519	84.418	89.391	93.856	99.228	102.996	111.055
70	39.036	43.275	45.442	48.758	51.739	55.329	61.698	69.334	77.577	85.527	90.531	95.023	100.425	104.215	112.317
71	39.777	44.058	46.246	49.592	52.600	56.221	62.641	70.334	78.634	86.635	91.670	96.189	101.621	105.432	113.577
72	40.520	44.843	47.051	50.428	53.462	57.113	63.585	71.334	79.690	87.743	92.808	97.353	102.816	106.648	114.835

Table 8.3 *Contd....*

ν	α: 0.999	0.995	0.99	0.975	0.95	0.90	0.75	0.50	0.25	0.10	0.05	0.025	0.01	0.005	0.001
73	41.264	45.629	47.858	51.265	54.325	58.006	64.528	72.334	80.747	88.850	93.945	98.516	104.010	107.862	116.092
74	42.010	46.417	48.666	52.103	55.189	58.900	65.472	73.334	81.803	89.956	95.081	99.678	105.202	109.074	117.346
75	42.757	47.206	49.475	52.942	56.054	59.795	66.417	74.334	82.858	91.061	96.217	100.839	106.393	110.286	118.599
76	43.507	47.997	50.286	53.782	56.920	60.690	67.362	75.334	83.913	92.166	97.351	101.999	107.583	111.495	119.850
77	44.258	48.788	51.097	54.623	57.786	61.586	68.307	76.334	84.968	93.270	98.484	103.158	108.771	112.704	121.100
78	45.010	49.582	51.910	55.466	58.654	62.483	69.252	77.334	86.022	94.374	99.617	104.316	109.958	113.911	122.348
79	45.764	50.376	52.725	56.309	59.522	63.380	70.198	78.334	87.077	95.476	100.749	105.473	111.144	115.117	123.594
80	46.520	51.172	53.540	57.153	60.391	64.278	71.145	79.334	88.130	96.578	101.879	106.629	112.329	116.321	124.839
81	47.277	51.969	54.357	57.998	61.261	65.176	72.091	80.334	89.184	97.680	103.010	107.783	113.512	117.524	126.083
82	48.036	52.767	55.174	58.845	62.132	66.076	73.038	81.334	90.237	98.780	104.139	108.937	1 14.695	118,726	127.324
83	48.796	53.567	55.993	59.692	63.004	66.976	73.985	82.334	91.289	99.880	105.267	110.090	115.876	119.927	128.565
84	49.557	54.368	56.813	60.540	63.876	67.876	74.933	83.334	92.342	100.980	106.395	111.242	117.057	121.126	129.804
85	50.320	55.170	57.634	61.389	64.749	68.777	75.881	84.334	93.394	102.079	107.522	112.393	118.236	122.325	131.041
86	51.085	55.973	58.456	62.239	65.623	69.679	76.829	85.334	94.446	103.177	108.648	113.544	119.414	123.522	132.277
87	51.850	56.777	59.279	63.089	66.498	70.581	77.777	86.334	95.497	104.275	109.773	114.693	120.591	124.718	133.512
88	52.617	57.582	60.103	63.941	67.373	71.484	78.726	87.334	96.548	105.372	110.898	115.841	121.767	125.913	134.745
89	53.386	58.389	60.928	64.793	68.249	72.387	79.675	88.334	97.599	106.469	112.022	116.989	122.942	127.106	135.978
90	54.155	59.196	61.754	65.647	69.126	73.291	80.625	89.334	98.650	107.565	113.145	118.136	124.116	128.299	137.208
91	54.926	60.005	62.581	66.501	70.003	74.196	81.574	90.334	99.700	108.661	1 14.268	119.282	125.289	129.491	138.438
92	55698	60.815	63.409	67.356	70.882	75.100	82.524	91.334	100.750	109.756	115.390	120.427	126.462	130.681	139.666
93	56.472	61.625	64.238	68.211	71.760	76.006	83.474	92.334	101.800	110.850	116.511	121.571	127.633	131.871	140.893
94	57.246	62.437	65.068	69.068	72.640	76.912	84.425	93.334	102.850	111.944	117.632	122.715	128.803	133.059	142.119
95	58.022	63.250	65.898	69.925	73.520	77.818	85.376	94.334	103.899	113.038	118.752	123.858	129.973	134.247	143.344
96	58.799	64.063	66.730	70.783	74.401	78.725	86.327	95.334	104.948	114.131	119.871	125.000	131.141	135.433	144.567
97	59.577	64.878	67.562	71.642	75.282	79.633	87.278	96.334	105.997	115.223	120.990	126.141	132.309	136.619	145.789

Table 8.3 *Contd....*

ν	α: 0.999	0.995	0.99	0.975	0.95	0.90	0.75	0.50	0.25	0.10	0.05	0.025	0.01	0.005	0.001
98	60.356	65.694	68.396	72.501	76.164	80.541	88.229	97.334	107.045	116.315	122.108	127.282	133.476	137.803	147.010
99	61.137	66.510	69.230	73.361	77.046	81.449	89.181	98.334	108.093	117.407	123.225	128.422	134.642	138.987	148.230
100	61.918	67.328	70.065	74.222	77.929	82.358	90.133	99.334	109.141	118.498	124.342	129.561	135.807	140.169	149.449
101	62.701	68.146	70.901	75.083	78.813	83.267	91.085	100.334	110.189	119.589	125.458	130.700	136.971	141.351	150.667
102	63.484	68.965	71.737	75.946	79.697	84.177	92.038	101.334	111.236	120.679	126.574	131.838	138.134	142.532	151.884
103	64.269	69.785	72.575	76.809	80.582	85.088	92.991	102.334	112.284	121.769	127.689	132.975	139.297	143.712	153.099
104	65.055	70.606	73.413	77.672	81.468	85.998	93.944	103.334	113.331	122.858	128.804	134.111	140.459	144.891	154.314
105	65.841	71.428	74.252	78.536	82.354	86.909	94.897	104.334	114.378	123.947	129.918	135.247	141.620	146.070	155.528
106	66.629	72.251	75.092	79.401	83.240	87.821	95.850	105.334	115.424	125.035	131.031	136.382	142.780	147.247	156.740
107	67.418	73.075	75.932	80.267	84.127	88.733	96.804	106.334	116.471	126.123	132.144	137.517	143.940	148.424	157.952
108	68.207	73.899	76.774	81.133	85.015	89.645	97.758	107.334	117.517	127.211	133.257	138.651	145.099	149.599	159.162
109	68.998	74724	77.616	82.000	85.903	90.558	98.712	108.334	118.563	128.298	134.369	139.784	146.257	150.774	160.372
110	69.790	75.550	78.458	82.867	86.792	91.471	99.666	109.334	119.608	129.385	135.480	140.917	147.414	151.948	161.581
111	70.582	76.377	79.302	83.735	87.681	92.385	100.620	110.334	120.654	130.472	136.591	142.049	148.571	153.122	162.788
112	71.376	77.204	80.146	84.604	88.570	93.299	101.575	111.334	121.699	131.558	137.701	143.180	149.727	154.294	163.995
113	72.170	78.033	80.991	85.473	89.461	94.213	102.530	112.334	122.744	132.643	138.811	144.311	150.882	155.466	165.201
114	72.965	78862	81.836	86.342	90.35 1	95.128	103.485	113.334	123.789	133.729	139.921	145.441	152.037	156.637	166.406
115	73.761	79.692	82.682	87.213	91.242	96.043	104.440	114.334	124.834	134.8L3	141.030	146.571	153.191	157.808	167.610
116	74.558	80.522	83.529	88.084	92.134	96.958	105.396	115.334	125.878	135.898	142.138	147.700	154.344	158.977	168.813
117	75.356	81.353	84.377	88.955	93.026	97.874	106.352	116.334	126.923	136.982	143.246	148.829	155.496	160.146	170.016
118	76.155	82.185	85.225	89.827	93.918	98.790	107.307	117.334	127.967	138.066	144.354	149.957	156.648	161.314	171.217
119	76.955	83.018	86.074	90.700	94.811	99.707	108.263	118.334	129.011	139.149	145.461	151.084	157.800	162.481	172.418
120	77.755	83.852	86.923	91.573	95.705	100.624	109.220	119.334	130055	140.233	146.567	152.211	158.950	163.648	173.617
121	78.557	84.686	87.773	92.446	96.598	101.541	110.176	120.334	131.098	141.315	147.674	153.338	160.100	164.814	174.816
122	79.359	85.521	88.624	93.320	97.493	102.458	111.133	121.334	132.142	142.398	148.779	154.464	161.250	165.980	176.014

Table 8.3 *Contd....*

123	80.162	86.356	89.475	94.195	98.387	103.376	112.089	122.334	133.185	143.480	149.885	155.589	162.398	167.144	177.212
124	80.965	87.192	90.327	95.070	99.283	104.295	113.046	123.334	134.228	144.562	150.989	156.714	163.546	168.308	178.408
125	81.770	88.029	91.180	95.946	100.178	105.213	114.004	124.334	135.271	145.643	152.094	157.839	164.694	169.471	179.604
126	82.575	88.866	92.033	96.822	101.074	106.132	114.961	125.334	136.313	146.724	153.198	158.962	165.841	170.634	180.799
127	83.381	89.704	92.887	97.698	101.971	107.051	115.918	126.334	137.356	147.805	154.302	160.086	166.987	171.796	181.993
128	84.188	90.543	93.741	98.576	102.867	107.971	116.876	127.334	138.398	148.885	155.405	161.209	168.133	172.957	183.186
129	84.996	91.383	94.596	99.453	103.765	108.891	117.834	128.334	139.440	149.965	156.508	162.331	169.278	174.118	184.379
130	85.804	92.223	95.451	100.331	104.662	109.811	118.792	129.334	140.482	151.045	157.610	163.453	170.423	175.278	185.571
131	86.613	93.063	96.307	101.210	105.560	110.732	119.750	130.334	141.524	152.125	158.712	164.575	171.567	176.438	186.762
132	87.423	93.904	97.163	102.089	106.459	111.652	120.708	131.334	142.566	153.204	159.814	165.696	172.711	177.597	187.953
133	88.233	94.746	98.021	102.968	107.357	112.573	121.667	132.334	143.608	154.283	160.915	166.816	173.854	178.755	189.142
134	89.044	95.588	98.878	103.848	108.257	113.495	122.625	133.334	144.649	155.361	162.016	167.936	174.996	179.913	190.331
135	89.856	96.431	99.736	104.729	109.156	114.417	123.584	134.334	145.690	156.440	163.116	169.056	176.138	181.070	191.520
136	90.669	97.275	100.595	105.609	110.056	115.338	124.543	135.334	146.731	157.518	164.216	170.175	177.280	182.226	192.707
137	91.482	98.119	101.454	106.491	110.956	116.261	125.502	136.334	147.772	158.595	165.316	171.294	178.421	183.382	193.894
138	92.296	98.964	102.314	107.372	111.857	117.183	126.461	137.334	148.813	159.673	166.415	172.412	179.561	184.538	195.080
139	93.111	99.809	103.174	108.254	112.758	118.106	127.421	138.334	149.854	160.750	167.514	173.530	180.701	185.693	196.266
140	93.926	100.655	104.034	109.137	113.659	119.029	128.380	139.334	150.894	161.827	168.613	174.648	181.840	186.847	197.451

Example: The performance of an anti inflammatory formulation is evaluated by observing the edema. It is graded as strong edema (4), moderate edema (3), slight edema (2), very slight edema (1) and no edema (0). The performance of the formulation shows the proportions 0:1:2:3:4. As a check on the performance of the formulation, 400 persons were examined and the data is given in the following table.

Grade	**4**	**3**	**2**	**1**	**0**
frequency	10	30	50	70	240

Solution:

Null hypothesis: There is no significance difference between expected frequencies and observed frequencies

Grade	4	3	2	1	0
Observed frequency	$\frac{0}{10}\times 400 = 0$	$\frac{1}{10}\times 400 = 40$	$\frac{2}{10}\times 400 = 80$	$\frac{3}{10}\times 400 = 120$	$\frac{4}{10}\times 400 = 160$
Expected (E_i) frequency	10	30	50	70	240
$(O_i - E_i)$	– 10	10	30	50	– 80
$(O_i - E_i)^2$	100	100	900	2500	6400
$\frac{(O_i - E_i)^2}{E_i}$	$\frac{100}{10} = 10$	$\frac{100}{30} = 3.33$	$\frac{900}{50} = 18$	$\frac{2500}{70} = 35.71$	$\frac{6400}{240} = 26.66$

$$\therefore\ \chi^2 = \sum_{i=1}^{5} \frac{(O_i - E_i)^2}{E_i} = 10 + 3.33 + 18 + 35.71 + 26.66 = 93.7$$

$\therefore$ Calculated $\chi^2 = 93.7$

Tabled χ^2 value at level of probability 0.05 with degrees of freedom n – 1 = 5 – 1 = 4 is 0.711

$\therefore$ Calculated $\chi^2 >$ tabled χ^2

$\therefore$ Null hypothesis is rejected

There is a significance difference between observed frequencies and expected frequencies

8.7.2 Test of Independence

When the categories of events analysed fall into two rows and two columns a two-way classification table is used. For more than two rows and columns, contingency table is used.

χ^2 test for two parameters

Two groups of persons received anti hypertensive formulation and placebo respectively. The blood pressures from these two groups were measured and the number of hypertensive patients was showed in the following table. Whether the anti hypertensive formulation is differed from placebo significantly or not.

	Number with hypertension	Number without hypertension	Total
Drug	20	80	100
Placebo	40	60	100
Total	60	140	200

Solution:

Null hypothesis: There is no significant difference between the anti hypersensitive (drug) and placebo.

	Number with hypertension	Number without hypertension	Total
Drug	20	80	100
Placebo	40	60	100
Total	60	140	200

To calculate the expected frequencies

	Number with hypertension	Number without hypertension
Drug	$\frac{100 \times 60}{200} = 30$	$\frac{100 \times 140}{200} = 70$
Placebo	$\frac{100 \times 60}{200} = 30$	$\frac{100 \times 140}{200} = 70$

Calculation of χ^2 value from the observed and expected frequencies

		Observed frequency f_o	Expected frequency f_e	$f_o - f_e$	$(f_o - f_e)^2$	$\frac{(f_o - f_e)^2}{f_e}$
Drug	Hypertension	20	30	–10	100	$\frac{100}{30}$
	Without hypertension	80	70	–10	100	$\frac{100}{70}$
Placebo	Hypertension	40	30	–10	100	$\frac{100}{30}$
	Without hypertension	60	70	–10	100	$\frac{100}{70}$

$$\chi^2 = \Sigma \frac{(f_o - f_e)^2}{f_e}$$

$$= \frac{100}{30} + \frac{100}{70} + \frac{100}{30} + \frac{100}{70} = 9.523809524$$

Degree of freedom (D.f) = (c – 1) (r – 1) = (2 – 1) (2 – 1) = 1 × 1 = 1

The tabled value of X^2 at level of probability 0.05 with degrees of freedom 1 is 3.84

$\therefore$ $\chi^2_{cal} > \chi^2_{tab}$

$\therefore$ Null hypothesis is rejected

$\therefore$ There is mean difference between drug and placebo.

χ^2 test for three parameters

Three groups of persons were administered with placebo, herbal drug and synthetic compound. The blood pressures were observed and the patients are characterized as per the following table. Observe the differences are significant or not.

	Number of patient with hypertension	**Number of patient without hypertension**	**Total**
Placebo	50	50	100
Herbal formulation	40	60	100
Synthetic compound	20	80	100
Total	110	190	300

Solution: Null hypothesis: There is no significant difference in observed blood pressure of the three groups of persons.

	Number of patients with hypertension	**Number of patients without hypertension**
Placebo	$\frac{100 \times 110}{300} = \frac{110}{3} = 37$	$\frac{100 \times 190}{300} = \frac{190}{3} = 63$
Herbal formulation	$\frac{100 \times 110}{300} = \frac{110}{3} = 37$	$\frac{100 \times 190}{300} = \frac{190}{3} = 63$
Synthetic compound	$\frac{100 \times 110}{300} = \frac{110}{3} = 37$	$\frac{100 \times 190}{300} = \frac{190}{3} = 63$

Calculation of x^2 from observed and calculated frequencies:

f_o = observed frequency f_e = expected frequency

		f_o	f_e	$f_o - f_e$	$(f_o - f_e)^2$	$\frac{(f_o - f_e)^2}{f_e}$
Placebo	No. of Patients with hypertension	50	37	13	169	4.5675
	No. of Patients without hypertension	50	63	–13	169	2.6825
Herbal formulation	No. of Patients with hypertension	40	37	3	9	0.24324
	No. of Patients without hypertension	60	63	–3	9	0.14285
Synthetic compound	No. of Patients with hypertension	20	37	–17	289	7.81081
	No. of Patients without hypertension	80	63	17	289	4.58730

$$\chi^2 = \Sigma \frac{(f_o - f_e)^2}{f_e} = 20.0342$$

Degree of freedom $= (c - 1)(r - 1) = (2 - 1)(3 - 1) = 2$

∴ Tabled values of x^2 at level of probability 0.05 with degrees of freedom 2 is 5.991

∴ $\chi^2_{cal} > \chi^2_{tab}$

∴ Null hypothesis is rejected.

Continuity correction in χ^2 test

The continuity correction is applied using the following rule. If the fractional part of the difference O – E is greater than 0 but less than or equal to 0.5, delete the fractional part. If the fractional part is greater than 0.5 or 0, then reduce the fractional part to 0.5.

For the above problem the correction values are 4.5, 2.5, 0, 0, 7.5, and 4.5

$$\chi^2 = 4.5+2.5+0+0+7.5+4.5 = 19$$

Degree of freedom $= (c–1)(r–1) = (2–1)(3–1) = 2$

The tabled value of χ^2 at the level of probability 0.05 with degrees of freedom 2 is 5.991.

∴ $\chi^2_{cal} > \chi^2_{tab}$

∴ Null hypothesis is rejected

8.7.3 Test for the Population Variance

This test is used to test that the populations from which the sample is drawn is from normal distribution. It involves the following steps.

- *Null hypothesis*. There is no significant difference between the observed and expected variance.

$$H_0 : \sigma_1^2 = \sigma_2^2$$

- *Computation of χ^2 value.* It is calculated with the following equation.

$$\chi^2 = ns^2/n\sigma^2$$

- *Calculate degree of freedom* (n – 1).
- *Decision.* If the calculated χ^2 value is less than the tabulated χ^2 value, then the null hypothesis is accepted and if it is greater than the table value then the null hypothesis is rejected and alternative hypothesis is accepted.

Example: The weight of ten tablets is 120, 122, 123, 121, 122, 123, 124, 120, 123 and 125 mg respectively. Can we say that the above samples are drawn from the population having variance of 1.5 mg?

Solution:

Null hypothesis H_0 : There is no significance difference between observed & expected variance

$$\bar{x} = \frac{120+122+123+\ldots+125}{10}$$

$$= \frac{1223}{10} = 122.3$$

Weight (x)	$\mathbf{x - \bar{x}}$	$(\mathbf{x - \bar{x}})^2$
120	–2.3	5.29
122	-0.3	0.09
123	0.7	0.49
121	–1.3	1.69
122	–0.3	0.09
123	0.7	0.49
124	1.7	2.89
120	–2.3	5.29
123	0.7	0.49
125	2.7	7.29
		$\Sigma(x-\bar{x})^2 = 24.1$

$$\text{Standard deviation of sample} = s = \sqrt{\frac{\Sigma(x-\bar{x})^2}{n-1}} = \sqrt{\frac{24.1}{9}}$$

$$= \sqrt{2.6777} = 1.6363$$

Now $$\chi^2 = \frac{ns^2}{n\sigma^2} = \frac{10\times 2.6777}{10\times 1.5} = \frac{2.6777}{1.5} = 1.78513$$

Degree of freedom = n – 1 = 10 – 1 = 9

∴ Tabled χ^2 at the level of probability0.05 with degrees of freedom 9 = 3.33

∴ Calculated χ^2 (=1.78513) < tabled χ^2 (3.33)

∴ Null hypothesis is accepted

∴ Samples are drawn from the population having variance 1.5 mg

Exercise :

1. In a survey conducted by a national institute on the occurrence of a common side effect, headache, with two brands x and y containing same drug, the results with brand 'x' are yes in 206 individuals and no with 18062 individuals. The results with brand 'y' are yes in 159 individuals and no in 18201 individuals. Apply an appropriate statistical test to assess the differences in occurrence of headache.

Answer :

Chi-square			
Chi-square, df	2.008,1		
P value	0.1565		
P value summary	ns		
One- or two-sided	Two-sided		
Statistically significant? (alpha<0.05)	No		
Data analyzed	**Headache**	**No headache**	**Total**
brand X	206	18062	18268
Brand Y	159	16201	16360
Total	365	34263	34628

2. The following are the fasting and post prandial blood glucose levels of 6 patients. Use appropriate test to determine if there is any significant difference between them.

Fasting	post prandial
90	140
83	132
98	158
82	121
99	144
100	156

Answer:

Paired t test	
P value	< 0.0001
P value summary	***
Are means signif. different? (P < 0.05)	Yes
One- or two-tailed P value?	Two-tailed
	t = 16.23 df = 5
Number of pairs	6
How big is the difference?	
Mean of differences	– 49.83
95% confidence interval	– 57.73 to – 41.94
R squared	0.9814
How effective was the pairing?	
Correlation coefficient (r)	0.9098
P Value (one tailed)	0.0059
P value summary	**
Was the pairing significantly effective?	Yes

3. The given data are the intelligent quotients (IQ) of two groups of children fed with two different health drinks for 6 months. Determine if the difference in IQ is due to the difference in health drinks.

Mindstar	Intellisure
41	22
53	34
61	29
46	19
68	39
49	26

Answer:

Unpaired t test	
P value	0.0006
P value summary	***
Are means signif. different? (P < 0.05)	Yes
One- or two-tailed P value?	Two-tailed
t = 4.880 df = 10	
How big is the difference?	
Mean ± SEM of column A	53.00 ± 4.074 N = 6
Mean ± SEM of column B	28.17 ± 3.049 N = 6
Difference between means	24.83 ± 5.089
95% confidence interval	13.50 to 36.17
R squared	0.7043
F test to compare variances	
F, DF numerator, Df denominator	1.786, 5, 5
P value	0.5399
P value summary	ns
Are variances significantly different?	No

CHAPTER 9

ANALYSIS OF VARIANCE (ANOVA)

Analysis of variance is the statistical procedure employed to detect the differences (if any) existing in between two or more samples. The t-test was appropriate for the one or two sample cases (one or two levels of the discrete independent variable), the F-test or one-way analysis of variance can be expanded to k levels of the independent variable. The calculation involves an *analysis of variance* of the individual sample means around a central grand mean. Like the t-test, the dependent variable represents data that is in continuous distribution. The analysis of variance is also referred to as the F-test, The ANOVA represents a variety of techniques used to identify and measure sources of variation within a collection of observations, hence the name analysis of variance. The F distribution is presented in the following figure.

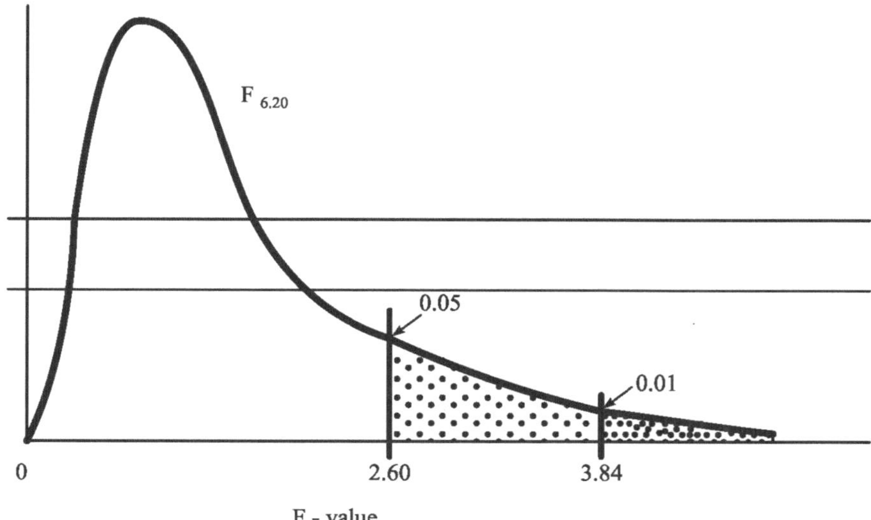

The calculated F statistic must be positive. Therefore, unlike the t-test, we are only interested in positive values to the extreme of our critical value. Similar to the t-distribution, the F-distribution is a series of curves, whose shape differs based on the degrees of freedom. The decision to accept or reject the null hypothesis, based on the shape of the F-distribution, is dependent on both the total sample size and the number of levels associated with the discrete independent variable. As the number of degrees of freedom gets larger, the F distribution will approach the shape of a normal distribution.

It is connected to experimental design. The experimental data generated from different formulations, different subjects and analytical methods is usually treated with ANOVA to get better interpretation. The concept of ANOVA may be simple or complex depending upon the design of experiment. For example, in the experiment " studies on the influence of composition on dissolution rate of tablet dosage form", the experimental results are influenced by the parameters such as composition of the tablet, efficiency/reproducibility of tablet machine, performance of the dissolution rate testing apparatus, the analytical method etc. If the objective of the experiment is restricted to study the influence of composition of tablet on dissolution rate, then the ideally designed experiment should eliminate the influence of the parameters that are expected to influence the dissolution rate other than the factor which is under examination. Thus the meaningful interpretation of experimental results is possible from ANOVA only with the well designed experiment.

In ANOVA, the total variance of the data is separated in to factors that are expected to influence the experimental outcome. In bioavailability experiments, the plasma concentrations of drug may be varied due to the differences in subjects, order of administration, due to chance and due to differences in formulations. The final analysis of variation should separate these factors from the experimental errors (errors by chance). The hypothesis testing depends on this separation.

It is mainly used to test the hypothesis of equality between variances and equality amongst the means. Generally the variance between samples is greater than the variance within samples so that the F value is always greater than 1 otherwise the variances should be interchanged.

9.1. One-way Analysis of Variance

The same assumptions for the t-test hold true for this procedure namely normality and homogeneity of variance. Under one-way, the influence of any one factor is considered. One-way ANOVA is used when we wish to test the equality of treatment means in experiments where two or more treatments are randomly assigned to different independent experimental units. The total variability in a set of data in one-way ANOVA is separated in to

- The deviation of group mean from the overall mean.
- The random error in the experiment which is assumed to be normal with mean zero and variance σ^2.

To obtain an F value we need two estimates of the variance. It is necessary to examine the variability (analysis of variance) of observations within groups as well as between groups.

For example, if we want to study the influence of two binders on dissolution rate, the results can be treated with t-test. In t-test, the differences in the mean dissolution rate values observed between the binders are studied. The dissolution studies are repeatedly performed six times to validate the dissolution method by confirming low variations in dissolution rate. It indicates the possibility of variations in dissolution rate caused by the differences between the binder, within the binder and experimental errors. Such different sources of variation cannot be detected and meaningful interpretation of the results is not possible with simple t-test. Similarly, t-test cannot be applicable to study the influence of three binders or more on dissolution rate. In such cases, the results are analyzed with ANOVA. This test is based on the following assumptions.

- The errors are normal with constant variance.
- The errors are independent.

Steps involved in one-way ANOVA

1. Statement of null hypothesis

   E.g: There is no difference in treatments

   $H_0 : \mu_1 = \mu_2 = \mu_3 = \mu_{k=\mu}$

2. Selection of level of confidence (α)

   The confidence levels should be decided to observe the table value and to compare the table and observed values. More often it is decided as 0.05.

3. Calculate the total sum of squares from the observed data. It is calculated using all of the data, ignoring the treatment grouping. This is calculated with the formulae.

   $SST = \Sigma x^2 - (\Sigma x)^2 / N$

   where $(\Sigma x)^2 / N$ is known as correction term

   N = total number of observations

4. Calculation of sum of squares between groups (treatments, columns). It is calculated with the following formulae.

   $$SSC = \frac{(\Sigma x_{c_1})^2}{n_1} + \frac{(\Sigma x_{c_2})^2}{n_2} + \frac{(\Sigma x_{c_3})^2}{n_3} + \frac{(\Sigma x_{c_4})^2}{n_4} - \frac{(\Sigma xi)^2}{N}$$

   SSC = sum of squares between the groups

   Σx_1 = sum of the first group, n_1 = number of observations in first group

   Σx_i = sum of the ith (last group) x_i = number of observations in last group

5. Calculate the sum of squares within groups (rows)
   It is computed with the following formulae
   SSE = SST – SSC
6. Construction of ANOVA table

   Construct the ANOVA table with the following format

Source of variation	Degree of freedom	Sum of squares	Mean squares	F
Between group	C – 1	SSC	MSC = SSC/C – 1	F = MSC/MSE
Error	N – C	SSE	MSE = SSE/N – C	
Total	N – 1	SST		

   C = number of groups; N = no. of observations
   MSC = Mean sum of square between groups
   MSE = Mean square within groups
7. Refer the F table with the degree of freedom n_1/n_2; n_1 = d.f of MSC; n_2 = d.f of MSE
8. **Interpretation**

   If the calculated value is equal to or greater than the table value at the desired level of significance, then null hypothesis is rejected and conclusion will be the differences between treatment means is significant.

Example

A drug was estimated by 4 analysts repeatedly for 3 times and the observed drug content is given below. Whether the differences from analyst-analyst is significant or not at 0.05 level of significance?

Assay results from different analysts

Trial	% drug content			
	Analyst 1	Analyst 2	Analyst 3	Analyst 4
1	98	97	98	99
2	99	96	97	96
3	100	98	99	97

Solution:

Null Hypothesis: There is no significant difference between the percent drug content observed by the four analysts

$$\text{Now, } \Sigma x^2 = (98)^2 + (99)^2 + (100)^2 + (97)^2 + (96)^2 + (98)^2 + (98)^2 + (97)^2 + (99)^2$$
$$+ (99)^2 + (96)^2 + (97)^2 = 9607 + 9801 + 10,000 + 9409 + 9216$$
$$+ 9604 + 9604 + 9409 + 9801 + 9801 + 9216 + 9409 = 114874$$

$$\Sigma x = 99 + 98 + 100 + 97 + 96 + 98 + 98 + 97 + 99 + 99 + 96 + 97 = 1174$$

$$N = 12$$

$$\therefore \quad \frac{(\Sigma x)^2}{N} = \frac{(1174)^2}{12} = 11\,4856.333\,333$$

$$\therefore \quad SST = \Sigma x^2 - \frac{(\Sigma x)^2}{N} = 17.666\,667$$

$$SSC = \left(\frac{(\Sigma xc_1)^2}{n_1} + \frac{(\Sigma xc_2)^2}{n_2} + \frac{(\Sigma xc_3)^2}{n_3} + \frac{(\Sigma xc_4)^2}{n_4}\right) - \frac{(\Sigma x)^2}{N}$$

$$= \frac{(297)^2}{3} + \frac{(291)^2}{3} + \frac{(294)^2}{3} + \frac{(292)^2}{3} - (114856.333333)$$

$$= \frac{88209 + 84681 + 86436 + 85264}{3} - 114856.33$$

$$= 114\,863.333 - 114856.333 = 7$$

ANOVA Table

Source of Variation	D.f	Sum of squares	Mean squares	F-Ratio
Between groups	C – 1 = 4 – 1 = 3	SSC = 7	MSC = SSC/C – 1 = 2.33	MSC/MSE = 2.33/1.33 = 1.75
Error	N – C = 12 – 4 = 8	SSE = 10.67	MSE = SSE/N – C = 1.33	
Total	11	SST = 17.67		

9.1.1 ANOVA vs t-test

The correct analysis of the data containing more than 2 groups is the application of ANOVA rather than t-test. To analyze the data given in the above problem (mean differences), six t-tests are required to notice the differences between 1&2, 1&3, 1&4, 2&3, 2&4 and 3&4 analysts. It is time consuming, complicated and the interpretation of results is difficult. If the differences between 1&2 and 3&4 analysts is significantly differed and the differences detected among other cases is insignificant then the question will arise that out of 1&2 and 3&4 analysts which difference is more significant. However if the overall variation is not insignificant from ANOVA, then there is no meaning to conduct individual t-tests. If significant differences are observed from ANOVA, then to detect the differences observed in between which two treatments are significant t-test is required.

9.1.2. Multiple Comparisons in ANOVA

If significant differences are observed from ANOVA then to determine which of the treatment differ, a separate test is required to compare each pair of treatments. More significant results are obtained with multiple comparison results compared to ANOVA. If the level of significance is 0.05, then it indicates that the 5 times out of 100, a difference will be found to be significant when the treatments are identical. In the same experiment if more than one pair of treatments are tested for significance in the same experiment then the chances of getting significant differences are more than 5% though the treatments are identical.

Multiple comparisons are performed with the following approaches.

1. **Least significant difference method (LSD):** The simplest method of multiple comparison approach for same size in all groups is constructed with the following formula.

$$LSD = t\sqrt{2s^2/N}$$

N = number of samples in a group

$$t = \frac{\overline{X}_1 - \overline{X}_2}{\sqrt{s^2\left(\frac{1}{N_1} + \frac{1}{N_2}\right)}}$$

S^2 = with in mean square variance

t = table value at α level of significance with the degree of freedom

If the sample sizes from the treatments (groups) are not equal then the following formula is used.

$$LSD = t\sqrt{s^2\left(\frac{1}{N_1} + \frac{1}{N_2}\right)}$$

N_1 = Sample size in first group

N_2 = Sample size in second group

Example: The hardness of tablets observed from tablets prepared with four different binders is given below. Compare the variances in hardness attributed by binders and report wether it is significant or not.

	Hardness			
Trial	**Binder A**	**Binder B**	**Binder C**	**Binder D**
1	4	5.2	5.1	4.9
2	3.8	4.9	5.0	5.1
3	3.6	4.8	4.9	5.2

Solution:

Null hypothesis: There is no significance difference between the hardness attributed by binders.

$$\text{Correction term} = \frac{(\Sigma x_i)^2}{N} = \frac{(56.5)^2}{12} = \frac{3192.25}{12} = 266.020$$

$$\text{Total sum of squares} = \text{SST} = \Sigma x^2 - \frac{(\Sigma x)^2}{N} = 269.57 - 266.02 = 3.55$$

Between sum of squares

$$\text{SSC} = \frac{(\Sigma x_{c_1})^2}{n_1} + \frac{(\Sigma x_{c_2})^2}{n_2} + \frac{(\Sigma x_{c_3})^2}{n_3} + \frac{(\Sigma x_{c_4})^2}{n_4} - \frac{(\Sigma xi)^2}{N}$$

$$= \frac{(11.4)^2}{3} + \frac{(14.9)^2}{3} + \frac{(15)^2}{3} + \frac{(15.2)^2}{3} - 266.020$$

$$= (129.96 + 222.01 + 225 + 231.04/3) - 266.02$$

$$= (808.01/3) - 266.02 = 269.34 - 266.02 = 3.32$$

Within sum of squares = Total sum of squares – between sum of squares

$$= 3.55 - 3.32 = 0.23$$

ANOVA Table

Source of Variation	**d.f**	**Sum of squares**	**Mean squares**	**F**
Between groups	C-1 = 4–1 = 3	SSC= 3.32	3.32/3 = 1.11	1.11/0.029 = 37.91
Within groups	N-C = 12 – 4 = 8	SSE = 0.23	0.23/8 = 0.029	
Total	N-1 = 11	3.55		

The calculated F value = 37.91

The tabled F value at the level of Probability 0.05 with degrees of freedom

$n_1 = 3$, and $n_2 = 8$ is 4.07

∴ Calculated F value > tabled F value

∴ Null hypothesis is rejected

There is a significance difference between hardness of tablets attributed by binders

Now $$\text{LSD} = t\sqrt{s^2\left(\frac{1}{\frac{1}{N_1}+\frac{1}{N_2}}\right)} = t\sqrt{s^2/\frac{2}{N}}$$

where s^2 is within mean square variance and t is tabled value of F at the α level with approximate degree of freedom (d.f = the number of degrees of freedom related to the WMS of the ANOVA)

$\therefore$ $s^2 = 0.02916666$; N = 3; t = tabled value of t at the level of probability 0.05 with degrees of freedom 8 = 2.31

$\therefore$ $$\text{L.S.D} = 2.31\sqrt{\frac{2\times 0.02916666}{3}} = 0.32$$

The Average hardness of tablets from the binders

Binder 1 $= \bar{A} = 3.8$, for binder 2 $= \bar{B} = 4.96$

For binder 3 $= \bar{C} = 5$ for binder 4 $= \bar{D} = 5.06$

Now $|\bar{A}-\bar{B}| = 1.16$, $|\bar{A}-\bar{C}| = 1.2$, $|\bar{A}-\bar{D}| = 1.26$

$|B-\bar{C}| = 0.04$ $|\bar{B}-\bar{D}| = 0.1$ $|\bar{C}-\bar{D}| = 0.06$

$\therefore$ $|\bar{A}-\bar{B}|$, $|\bar{A}-\bar{C}|$, $|\bar{A}-\bar{D}|$ are more than 0.322(L.S.D. value)

The average hardness of tablets containing the binders B, C, and D is significantly greater than the hardness of tablet containing binder A.

2. **Turkey's multiple range tests:** In this test, the error rate α is fixed at 5% level. Larger differences between treatments are required compared to LSD to get significance in Turkey's test. The LSD has greater power compared to Turkey's test. The differences are calculated with the following formulae.

$$|\text{Difference}| = Q\sqrt{\frac{s^2}{N}}$$

where Q values can be found from studentized range and the values are given in table 9.1. It depends on number of treatments, d.f and α level. Any difference in treatment means exceeding $Q\sqrt{\frac{s^2}{N}}$ value is significant.

If the number of samples is different from group-group, then the following formulae is used.

Table 9.1 Turkey's **Multiple Range test** Values of 'Q'

d.f	Number of treatments																		
	2	**3**	**4**	**5**	**6**	**7**	**8**	**9**	**10**	**11**	**12**	**13**	**14**	**15**	**16**	**17**	**18**	**19**	**20**
1	18.0	27.0	32.8	37.2	40.5	43.1	45.4	47.3	49.1	50.6	51.9	53.2	54.3	55.4	56.3	57.2	58.0	58.8	59.6
2	6.09	8.33	9.80	10.89	11.73	12.43	13.03	13.54	13.99	14.39	14.75	15.08	15.38	15.65	15.91	16.14	16.36	16.57	16.77
3	4.50	5.91	6.83	7.51	8.04	8.47	8.85	9.18	9.46	9.72	9.95	10.16	10.35	10.52	10.69	10.84	10.98	11.12	12.24
4	3.93	5.04	5.76	6.29	6.71	7.06	7.35	7.60	7.83	8.03	8.21	8.37	8.52	8.67	8.80	8.92	9.03	9.14	9.24
5	3.64	4.60	5.22	5.67	6.03	6.33	6.58	6.80	6.99	7.17	7.32	7.47	7.60	7.72	7.83	7.93	8.03	8.12	8.21
6	3.46	4.34	4.90	5.31	5.63	5.89	6.12	6.32	6.49	6.65	6.79	6.92	7.04	7.14	7.24	7.34	7.43	7.51	7.59
7	3.34	4.16	4.68	5.06	5.35	5.59	5.80	5.99	6.15	6.29	6.42	6.54	6.65	6.75	6.84	6.93	7.01	7.08	7.16
8	3.26	4.04	4.53	4.89	5.17	5.40	5.60	5.77	5.92	6.05	6.18	6.29	6.39	6.48	6.57	6.65	6.73	6.80	6.87
9	3.20	3.95	4.42	4.76	5.02	5.24	5.43	5.60	5.74	5.87	5.98	6.09	6.19	6.28	6.36	6.44	6.51	6.58	6.65
10	3.15	3.88	4.33	4.66	4.91	5.12	5.30	5.46	5.60	5.72	5.83	5.93	6.03	6.12	6.20	6.27	6.34	6.41	6.47
11	3.11	3.82	4.26	5.58	4.82	5.03	5.20	5.35	5.49	5.61	5.71	5.81	5.90	5.98	6.06	6.14	6.20	6.27	6.33

Table 9.1 *Contd...*

d.f	Number of treatments																		
	2	**3**	**4**	**5**	**6**	**7**	**8**	**9**	**10**	**11**	**12**	**13**	**14**	**15**	**16**	**17**	**18**	**19**	**20**
13	3.06	3.73	4.15	4.46	4.69	4.88	5.05	5.19	5.32	5.43	5.53	5.63	5.71	5.79	5.86	5.93	6.00	6.06	6.11
14	3.03	3.70	4.11	4.41	4.64	4.83	4.99	5.13	5.25	5.36	5.46	5.56	5.64	5.72	5.79	5.86	5.92	5.98	6.03
15	3.01	3.67	4.08	4.37	4.59	4.78	4.94	5.08	5.20	5.31	5.40	5.49	5.57	5.65	5.72	5.79	5.85	5.91	5.96
16	3.00	3.65	4.05	4.34	4.56	4.74	4.90	5.03	5.15	5.26	5.35	5.44	5.52	5.59	5.66	5.73	5.79	5.84	5.90
17	2.98	3.62	4.02	4.31	4.52	4.70	4.86	4.99	5.11	5.21	5.31	5.39	5.47	5.55	5.61	5.68	5.74	5.79	5.84
18	2.97	3.61	4.00	4.28	4.49	4.67	4.83	4.96	5.07	5.17	5.27	5.35	5.43	5.50	5.57	5.63	5.69	5.74	5.79
19	2.96	3.59	3.98	4.26	4.47	4.64	4.79	4.92	5.04	5.14	5.23	5.32	5.39	5.46	5.53	5.59	5.65	5.70	5.75
20	2.95	3.58	3.96	4.24	4.45	4.62	4.77	4.90	5.01	5.11	5.20	5.28	5.36	5.43	5.50	5.56	5.61	5.66	5.71
30	2.89	3.48	3.84	4.11	4.30	4.46	4.60	4.72	4.83	4.92	5.00	5.08	5.15	5.21	5.27	5.33	5.38	3.43	5.48
40	2.86	3.44	3.79	4.04	4.23	4.39	4.52	4.63	4.74	4.82	4.90	4.98	5.05	5.11	5.17	5.22	5.27	5.32	5.36
120	2.80	3.6	3.69	3.92	4.10	4.24	4.36	4.47	4.56	4.64	4.71	4.78	4.84	4.90	4.95	5.00	5.04	5.09	5.13
∞	2.77	3.32	3.63	3.86	4.03	4.17	4.29	4.39	4.47	4.55	4.62	4.68	4.74	4.80	4.84	4.89	4.93	4.97	5.01

$$|\text{Difference}| = Q\sqrt{\frac{s^2}{\frac{2N_1N_2}{N_1+N_2}}}$$

N_1 and N_2 are the sample sizes of groups 1 & 2 respectively.

For above problem, the value of Q in the Studentized range for four treatments and degrees of freedom 8 is 4.53.

$$Q\sqrt{\frac{S^2}{N}} \text{ Value for the above problem}$$

$$= 4.53 \times \sqrt{\frac{0.02916666}{3}}$$

$$= 4.53 \times 0.0986 = 0.447$$

In this test also, the mean differences in hardness values observed from the tablets A&B, A&C and A&D is exceeding the calculated value. So these mean differences in hardness values are statistically significant.

3. **Scheffe method:** To compare the large number of treatments within a shorter interval, Scheffe method is used. In this method also the α rate is set as 5%. To compare the two means, the following formula is used.

$$= \sqrt{S^2(k-1)F\left(\frac{1}{N_1}+\frac{1}{N_2}\right)}$$

S^2 = Appropriate variance estimated from one way ANOVA (MSW)

k = no. of treatments

F = ANOVA (F) table value with numerator G – 1 and the denominator N–G at the degree of freedom 0.05, N_1 and N_2 are the sample sizes of two groups.

For the above problem $S^2 = 0.0291666$

$K-1 = 4-1 = 3$, $N_1 = N_2 = 3$

$$\therefore \quad \sqrt{S^2(k-1)F\left(\frac{1}{\frac{1}{N_1}+\frac{1}{N_2}}\right)}$$

$$= \sqrt{(0.0291666)(3)\left(\frac{1}{3}+\frac{1}{3}\right)}$$

$$= \sqrt{(0.0291666)(3)\left(\frac{2}{3}\right)}$$

$$= \sqrt{0.0291666 \times 2}$$

$$= \sqrt{0.0583332 \times 2} = 0.24$$

Mean of first sample = 3.8 = $\overline{A}$

Mean of second sample = 4.97= $\overline{B}$

Mean of third sample = 5 = $\overline{C}$

Mean of fourth sample = 5.07 = $\overline{D}$

$$\left|\overline{A}-\overline{B}\right| = 1.16,\ \left|\overline{A}-\overline{C}\right| = 1.2,\ \left|\overline{A}-\overline{D}\right| = 1.26$$

$$\left|\overline{B}-\overline{C}\right| = 0.04,\ \left|\overline{B}-\overline{D}\right| = 0.1,\ \left|\overline{C}-\overline{D}\right| = 0.06$$

$\left|\overline{A}-\overline{B}\right|,\ \left|\overline{A}-\overline{C}\right|,\ \left|\overline{A}-\overline{D}\right|$ Values are more than 0.24

The average hardness observed from tablets containing binders B, C, and D is significantly greater than the hardness of tablet formulated with binder A.

4. **Newman-Keuls Test:** This test utilizes the Turkey's factor Q in a sequential fashion. The means to be compared are arranged in order of their magnitude.

   In the above example, the means are arranged according to its magnitude as follows:

   Calculate the differences needed to get significance of 2, 3 … n means where n is the total number of treatment means.

   For the above data, the mean hardness values observed from four tablets containing binder A, binder B, binder C and binder D are 3.8, 4.9, 5, and 5.06 respectively. To apply the test, compute the difference needed for significance for the comparison of 2, 3, .n means (where n is the total number of treatment means). The given experiment consists of 4 treatments.

∴ We will obtain differences needed for significance for 2, 3 and 4 means.

Initially, consider the first two means using Turkey's test $Q\sqrt{\frac{S^2}{N}}$

With 2 treatments and 8 d.f, the table value of $Q = 3.26$

$$\therefore \quad Q\sqrt{\frac{S^2}{N}} = 3.26\sqrt{\frac{0.0291666}{3}} = 3.26 \times 0.098601217 = 0.32$$

(N = number of tablet used in each treatment)

For three treatments and 8 d.f, the table value of $Q = 4.04$

$$Q\sqrt{\frac{S^2}{N}} = 4.04\sqrt{\frac{0.0291666}{3}} = 4.04 \times 0.0986$$

$$= 0.398$$

For four treatments and 8 d.f, the table value of $Q = 4.53$

$$Q\sqrt{\frac{S^2}{N}} = 4.53 \times 0.98601217 = 0.45$$

Note that the last value 0.45 is same value as that obtained for the turkey test.

Thus the differences needed for 2, 3 and 4 means to be considered significantly are 0.32, 0.40, and 0.45. It can be represented as follows.

Number of treatments	2	3	4
Critical difference	0.32	0.40	0.45

The four order means are

A	**B**	**C**	**D**
3.8	4.96	5	5.06

The above notation is standard. Any two means connected by the same under scored line are not significantly different. Two means not connected by the underscored line are significantly different. In the above example, theoretically we can connect the means A, B and C or the means B, C and D with underscore. Examination of the under scored line in this example shows that only two means not connected are 3.8 and 5.06, corresponding to the binder A and D.

The determination of significant and non significant differences follows. The difference between treatments A and D covering 4 means is equal to 1.26 which exceeds 0.45, resulting in a significant difference. Since the difference of means of B and D is 0.1 which is less than the critical value of 0.40 for three means. This is described by the underscore. We need not compare B and C or C and D since these will not be considered different based on the underscore.

The difference between treatments A and C is 1.2, which is more than the critical value of 0.40 for 3 means. We need to compare A & B and B & C since these will be considered different based on the differences. The mean difference between A & B is 1.16, which is more than the critical value of 0.32 for 2 means and hence it can be concluded that the mean differences between A&B is significant. The mean difference between B & C is 0.04, which is less than the critical value of 0.32 for 2 means and hence it can be concluded that the mean differences between B&C is not significant. So we cannot connect the means A, B and C with underscore. Finally the treatments B, C, D are not significant and A is significantly differing from B, C and D.

5. **Dunnett's test:** In certain pharmaceutical calculations, the observed data should be compared against the standard data. Such examples are furnished below.
    1. Comparison of dissolution data of different formulations against the approved formulation.
    2. Comparison of stability data collected after exposure of drug to different conditions against initial potency.
    3. Comparison of bioavailability of tested formulations with the bioavailability of standard approved formulation.
    4. Clinical response observed from different formulations against a control (not treated with drug).

    In such cases, the required comparison is between A versus B, A versus C and A versus D. Where A is the standard and B, C, D are the test products. Dunnett reported multiple comparison procedure for treatments versus control. It is calculated with the following formulae.

$$D' = t'\sqrt{S^2\left(\frac{1}{N_1}+\frac{1}{N_2}\right)}$$

t' = Two sided t studentized range and these values are given in table 9.2

N_1 and N_2 = numbers of observations in control and formulating

Example

In the above problem 'Binder A' is taken as standard ; then the comparison should be made between Binder A vs. Binder B; Binder A vs. Binder C, Binder A vs. Binder D. Dunnett devised multiple comparison procedure for treatments versus control is defined as

$$D^1 = t^1 \sqrt{S^2\left(\frac{1}{N_1}+\frac{1}{N_2}\right)}$$

In the present example, the number of treatments to be compared to the control, is equal to 3 and degrees of freedom = 8 for a two sided test at the 0.05 level the value of t' is 2.88 (including control, the number of groups are 4).

Table 9.2 Dunnett Table

Number of Groups Including Control Group

d.f	alpha	2	3	4	5	6	7	8	9	10
5	0.05	2.57	3.03	3.29	3.48	3.62	3.73	3.82	3.9	3.97
	0.01	4.03	4.63	4.98	5.22	5.41	5.56	5.69	5.8	5.89
6	0.05	2.45	2.86	3.1	3.26	3.39	3.49	3.57	3.64	3.71
	0.01	3.71	4.21	4.51	4.71	4.87	5	5.1	5.2	5.28
7	0.05	2.36	2.75	2.97	3.12	3.24	3.33	3.41	3.47	3.53
	0.01	3.5	3.95	4.21	4.39	4.53	4.64	4.74	4.82	4.89
8	0.05	2.31	2.67	2.88	3.02	3.13	3.22	3.29	3.35	3.41
	0.01	3.36	3.77	4	4.17	4.29	4.4	4.48	4.56	4.62
9	0.05	2.26	2.61	2.81	2.95	3.05	3.14	3.2	3.26	3.32
	0.01	3.25	3.63	3.85	4.01	4.12	4.22	4.3	4.37	4.43
10	0.05	2.23	2.57	2.76	2.89	2.99	3.07	3.14	3.19	3.24
	0.01	3.17	3.53	3.74	3.88	3.99	4.08	4.16	4.22	4.28
11	0.05	2.2	2.53	2.72	2.84	2.94	3.02	3.08	3.14	3.19
	0.01	3.11	3.45	3.65	3.79	3.89	3.98	4.05	4.11	4.16
12	0.05	2.18	2.5	2.68	2.81	2.9	2.98	3.04	3.09	3.14
	0.01	3.05	3.39	3.58	3.71	3.81	3.89	3.96	4.02	4.07
13	0.05	2.16	2.48	2.65	2.78	2.87	2.94	3	3.06	3.1
	0.01	3.01	3.33	3.52	3.65	3.74	3.82	3.89	3.94	3.99
14	0.05	2.14	2.46	2.63	2.75	2.84	2.91	2.97	3.02	3.07
	0.01	2.98	3.29	3.47	3.59	3.69	3.76	3.83	3.88	3.93
15	0.05	2.13	2.44	2.61	2.73	2.82	2.89	2.95	3	3.04
	0.01	2.95	3.25	3.43	3.55	3.64	3.71	3.78	3.83	3.88
16	0.05	2.12	2.42	2.59	2.71	2.8	2.87	2.92	2.97	3.02
	0.01	2.92	3.22	3.39	3.51	3.6	3.67	3.73	3.78	3.83
17	0.05	2.11	2.41	2.58	2.69	2.78	2.85	2.9	2.95	3
	0.01	2.9	3.19	3.36	3.47	3.56	3.63	3.69	3.74	3.79
18	0.05	2.1	2.4	2.56	2.68	2.76	2.83	2.89	2.94	2.98
	0.01	2.88	3.17	3.33	3.44	3.53	3.6	3.66	3.71	3.75
19	0.05	2.09	2.39	2.55	2.66	2.75	2.81	2.87	2.92	2.96
	0.01	2.86	3.15	3.31	3.42	3.5	3.57	3.63	3.68	3.72
20	0.05	2.09	2.38	2.54	2.65	2.73	2.8	2.86	2.9	2.95
	0.01	2.85	3.13	3.29	3.4	3.48	3.55	3.6	3.65	3.69
24	0.05	2.06	2.35	2.51	2.61	2.7	2.76	2.81	2.86	2.9
	0.01	2.8	3.07	3.22	3.32	3.4	3.47	3.52	3.57	3.61
30	0.05	2.04	2.32	2.47	2.58	2.66	2.72	2.77	2.82	2.86
	0.01	2.75	3.01	3.15	3.25	3.33	3.39	3.44	3.49	3.52
40	0.05	2.02	2.29	2.44	2.54	2.62	2.68	2.73	2.77	2.81
	0.01	2.7	2.95	3.09	3.19	3.26	3.32	3.37	3.41	3.44
60	0.05	2	2.27	2.41	2.51	2.58	2.64	2.69	2.73	2.77
	0.01	2.66	2.9	3.03	3.12	3.19	3.25	3.29	3.33	3.37

The critical difference

$$= 2.88\sqrt{(0.0292)\left(\frac{1}{3}+\frac{1}{3}\right)}$$

$$= 2.88\sqrt{0.0292x\frac{2}{3}} = 2.88\sqrt{0.0584/3} = 2.88\sqrt{0.195} = 2.88x0.139 = 0.402$$

We have $\left|\overline{A}-\overline{B}\right| = 1.16$, $\left|\overline{A}-\overline{C}\right| = 1.2$, $\left|\overline{A}-\overline{D}\right| = 1.26$

$\therefore$ $\left|\overline{A}-\overline{B}\right|$, $\left|\overline{A}-\overline{C}\right|$, $\left|\overline{A}-\overline{D}\right|$ are more than 0.94

$\therefore$ The treatments B, C, D are, differ significantly from standard method.

9.1.3 Multiple Correlated Outcomes

Some of the Pharmaceutical experiments are influenced by several factors. Among these factors, two or more factors may be correlated. For example, the dissolution rate of a drug is influenced by the following factors and there is a possible correlation among these factors.

1. The concentration of binder.
2. Concentration of disintegrant
3. Hardness of the tablet
4. Compression force employed during compression of the tablet and
5. Concentration of diluents.

In this example, hardness and compression force are correlated. In this experiment, if the dissolution rate is studied with each independent variable separately (at 0.01 level of significance) , then the overall level of significance for 5 variables will be the sum of each individual level of significance. The other problem is if all the variables are correlated, then the selected α level (0.05) is applicable to all the variables or not. The overall level of significance is adjusted in case of correlation with the formulae.

$$\alpha = 1-(1-\gamma)^m$$

α = overall level of significance

$m = k^{1-\overline{R}}$

m = 1 in case of perfect correlation among the variables then $\alpha = \gamma$

k = No. of outcome variables

$\overline{R}$ = average correlation = $\Sigma\dfrac{R_i}{k-1}$

$\gamma = 1-(1-\alpha)^{1/m}$

Example

The dissolution rate constant obtained from a formulation is correlated with the above five variables. The correlation matrix is given below. Solve the α level for testing each outcome r.

Correlation matrix for five variables affecting dissolution rate

	1	**2**	**3**	**4**	**5**
1	1.00	0.68	0.86	0.72	0.84
2	0.68	1.00	0.84	0.68	0.72
3	0.86	0.84	1.00	0.78	0.76
4	0.72	0.68	0.78	1.00	0.82
5	0.84	0.72	0.76	0.82	1.00

Correlation matrix for five variables effecting dissolution rate

$$r_1 = (0.68 + 0.86 + 0.72 + 0.84) / 4 = \frac{3.1}{4} = 0.775$$

$$r_2 = (0.68 + 0.84 + 0.68 + 0.72) / 4 = \frac{2.92}{4} = 0.73$$

$$r_3 = (0.86 + 0.84 + 0.78 + 0.76) / 4 = \frac{3.24}{4} = 0.81$$

$$r_4 = (0.72 + 0.68 + 0.78 + 0.82) / 4 = \frac{3}{4} = 0.75$$

$$r_5 = (0.84 + 0.72 + 0.76 + 0.82) / 4 = \frac{3.14}{4} = 0.785$$

The average correlation is

$$(0.775 + (0.73 + 0.84 + 0.75 + 0.785)/5 = \frac{3.85}{5} = 0.77$$

$$\therefore\ m = K^{1-0.77} = 5^{1-0.77} = 5^{0.23} = 1.447981.793$$

$\therefore$ The level for each variable is adjusted to

$$\gamma = 1 - (1-\alpha)^{1/m} = 1 - (1 - 0.05)^{\frac{1}{1.447981793}}$$

$$= 1 - 0 - 95^{\frac{1}{1.44798}}$$

$$= 1 - 0.96$$

$$= 0.035$$

Testing of the individual outcome variables should be performed at the 0.035 level

9.1.4 One-way ANOVA with Unequal Sample Sizes

There is a possible death of animals during bioavailability/bioequivalence or clinical studies. The number of trials conducted for different variables may not be the same due to mistakes/inadequate procedures or due to unavoidable circumstances. Then the ANOVA should be conducted with the existing data itself.

Example

The C_{max} values noticed from 4 formulations observed from 6 volunteers is given below. Whether the differences are significant or not?

C_{max} (μg/ml) values observed from 4 formulations ($F_1 - F_4$).

Subject No	F_1	F_2	F_3	F_4
1	2.4	2.8	3.1	2.7
2	2.6	2.9	3.2	2.9
3	2.5	3.0	2.8	3.0
4	2.6	2.7	3.0	2.8
5	2.7	-	-	2.9
6	2.5	-	-	-
Total	**15.3**	**11.4**	**12.1**	**14.3**

Null Hypothesis: There is no significant difference between C_{max} values noticed from 4 formulations observed from 6 volunteers.

$$\text{Total sum of squares} = \Sigma x^2 - \frac{(\Sigma x_i)^2}{N}$$

$$= ((2.4)^2 + (2.6)^2 + \ldots. + (2.9)^2) - \frac{(53.1)^2}{19}$$

$$= 149.25 - 148.4005265$$

$$= 0.85$$

BSS = between sum of squares

$$= \frac{(\Sigma x_{c_1}^2)}{n_{c_1}} + \frac{(\Sigma x_{c_2}^2)}{n_{c_2}} + \frac{(\Sigma x_{c_3})^2}{n_{c_3}} + \frac{(\Sigma x_{c_4})^2}{n_{c_4}} - \frac{(\Sigma x_i)^2}{N}$$

$$= \frac{(15.3)^2}{6} + \frac{(11.4)^2}{4} + \frac{(12.1)^2}{4} + \frac{(14.3)^2}{5} - 148.4005263$$

$$= (39.015 + 32.49 + 36.6025 + 40.898) - 148.40$$

$$= 0.605$$

Within sum of squares = SSE = SST – SSC

= 0.849 – 0.605

= 0.244

ANOVA Table

Source of Variation	d.f.	Sum of Squares	Mean Squares	F
Between groups	C – 1 = 4 – 1 = 3	SSC = 0.605	SSC/C–1 = 0.202	0.202/0.016 = 12.37
Within groups (Error)	N – C = 19 – 4 = 15	SSE = 0.244	SSE/N –C= 0.016	
Total	18	0.85		

The calculated F value = 12.37

Tabled F value at the level of probability 0.05 with degrees of freedom for $n_1 = 3$, and $n_2 = 15$ is 3.29

∴ F calculated value > F tabled value

∴ Null hypothesis is rejected

Hence there is a significant difference between the items

9.2 Two-way ANOVA

In one-way ANOVA, the influence of one factor on different sample groups is studied. Under two-way ANOVA, the influence of two factors is studied. Besides these two factors, there may be sampling variation which is characterized as residual variation. The sum of squares of variations in column (SSC), plus the sum of squares of variations in rows (SSR) plus the sum of squares as the residual due to errors (SSE) make up the total sum of squares of variations (SST).

$$SST = SSC + SSR + SSE$$

Example

1. If a drug is estimated from same batch of formulation by different analytical methods, the source of variation is the analytical method employed for the estimation of drug. However if different analytical methods are used for the estimation of drug belong to different batches of formulation, and then the source of variation are analytical method and batch. In first case one way ANOVA is adequate however two-way ANOVA is required for the second case.

2. If the raw material is supplied by different suppliers and estimated by different methods then also two-way ANOVA is used, as it is influenced by the supplier and the analytical method.
3. If the shelf life is determined for different formulations (differences in composition, batches, physical properties of drug etc.,) packed in different packaging materials, then two-way ANOVA is preferred as the shelf life is varied due to the sources of variation is formulation and packaging materials.
4. The dissolution rate of different formulations may be tested with different apparatus, and then two-way ANOVA is used as the dissolution rate is varied from formulation and apparatus.
5. The absorbance values may be observed by different analysts from different spectrophotometers. Then also two-way ANOVA is preferable.

Problem

The T_{50} values observed from different formulations with 3 different dissolution rate testing apparatus. Treat the data statistically.

Dissolution rate testing apparatus	**T_{50} values of formulations (min)**		
	1	**2**	**3**
1	15	12	20
2	14	13	19
3	13	12	18

Solution:

***Null Hypothesis*:** There is no significance difference between T_{50} values observed from formulations containing same drug and same strength and subjected to dissolution rate testing with different apparatus.

Dissolution rate testing apparatus	**T_{50} values of formulations** 1	2	3	**Total**
1	15	12	20	47
2	14	13	19	46
3	13	12	18	43
Total	42	37	57	136 = N

$$\text{The total sum of squares} = SST = \Sigma x^2 - C.T = \Sigma x^2 - \frac{(\Sigma x_i)^2}{N}$$

$$= 2132 - \frac{(136)^2}{9} = 2132 - 2055.1111 = 76.88$$

Column sum of squares = (SSC) = $\dfrac{\left(\Sigma x_{c_i}\right)^2}{n_{c_1}} - CT$

$$= \frac{(42)^2}{3} + \frac{(37)^2}{3} + \frac{(57)^2}{3} - 2055.1111 = \frac{6382}{3} - 2055.1111$$

$$= 72.22222$$

Row sum of squares (SSR) = $-\dfrac{6174}{3} - 2055.11 = 2.89$

Residual = SSE= SST – SSC– SSR

$$= 76.88 - 72.22 - 2.89$$

$$= 1.77$$

ANOVA Table

Source of Variation	d.f	S.S	M.S	F
Between columns	c – 1 = 2	SSC = 72.22	SSC/C –1 = 36.11	$F_{2,4}$ = 36.11/0.44 = 81.66
Between rows	r –1 = 2	SSR = 2.89	SSR /r – 1 = 1.44	$F_{2,4}$ = 1.44/0.44 = 3.27
Residual	(c–1) (r–1) = 4	SSE = 1.77	0.44	
Total	8	76.88	379977	

Calculated F value for T_{50} values = 81.66 and the tabled F value at the level of probability 0.05 with d.f at $n_1 = 2$, $n_2 = 4$ is 6.94

∴ $F_{cal} > F_{tab}$

There is significant difference between T_{50} values of formulation

Calculated F value for Dissolution rate of testing apparatus is 3.27 and tabled F value at the level of probability 0.05 with degrees of freedom 2, 4 is 6.94

∴ $F_{calculated}$ value < F_{table} value

∴ there is no significant difference between dissolution rates observed from different dissolution rate testing apparatus.

9.2.1 Fixed and Random Effects in the Two-way ANOVA

The proper test of significance in the two-way design depends on the model and the presence of interaction among the selected factors. A fixed category includes all the treatments of interest. Out of two factors, one factor may be fixed and the other may be selected randomly. For example, the pharmacokinetic parameters of a drug may vary from product to product (A-D) and differ from subject-subject. In some subjects, product A may show better results and some other patients may be more benefited with other products. In such cases, the objective may be fixed to study the differences in pharmacokinetic parameters due to the variations in formulations and then the subjects may be selected randomly.

If the tablets are formulated with disintegrants and super disintegrating agents, then the dissolution rate is expected to be more from the formulations containing superdisintegrating agent as they are more efficient. Here the variation from the formulations is fixed. So for ANOVA, the treatments are fixed as long as the objective is to observe the differences between the selected formulations. If the objective is to draw a conclusion about the differences in disintegrants (variations in populations) then the formulations should be selected randomly i.e., the formulations should be formulated with variety of disintegrants (not fixed disintegrants). If the data is collected from fixed no of items (not selected randomly) of a formulation (belong to some treatment/column), then the variation observed within a treatment is restricted to these samples tested and not applicable to obtain a valid conclusion about that batch. So the samples (rows) i.e., subjects in clinical study, tablets for dissolution studies should be selected randomly and treatments (columns) are fixed. The objective is to study the variations in disintegrants, then the formulations are fixed and the tablets from the batches are selected randomly (One-way Anova).

9.2.2 Two-way ANOVA with Replication

In case of objective to study the variations in formulations and batches the question will arise whether the collected samples of a formulation are true representatives of the selected batch or not? Then in order to confirm it, replication studies are preferable. In this study, replication studies (dissolution study) should be conducted with another tablet collected from the same batch. It is best to have equal number of replications for each cell of the two-way design. The experimenter should be sure that the experiment is properly replicated.

Example

The C_{max} values observed from a replicated crossover study of the test product were compared with the standard product. Whether the differences are significant or not?

	Test Product		**Standard product**	
	Trial (1)	**Trial (2)**	**Trial (1)**	**Trial (2)**
1	2.4	2.3	3.0	2.8
2	2.5	2.4	2.9	2.7
3	2.3	2.6	3.1	3.0
4	2.4	2.2	2.8	2.7
5	2.5	2.5	3.2	3.0
6	2.5	2.4	3.1	3.1

Solution: In the above problem, it is considered that the products (test and standard) are fixed and the subjects are random. The analysis of the data results in one new term in the ANOVA, that due to the within-cell SS. The within-cell SS represents the variability due to replications.

Subject	C_{max} values (μg /ml)				
	Test product		Standard product		Row total
	Trial (1)	Trial (2)	Trial (1)	Trial (2)	
1	2.4	2.3	3.0	2.8	10.5
2	2.5	2.4	2.9	2.7	10.5
3	2.3	2.6	3.1	3.0	11.0
4	2.4	2.2	2.8	2.7	10.1
5	2.5	2.5	3.2	3.0	11.2
6	2.5	2.4	3.1	3.1	11.1
Column total	14.6	14.4	18.1	17.3	64.4

Correction term= $(64.4)^2/24=4147.36/24=172.81$

Total sum of squares = SST $= \Sigma x^2 - C.T$
$= (2.4)^2 + (2.3)^2 + (2.5)^2 + (2.4)^2 + (3.1)^2 + (3.1)^2 - 172.81$
$= 174.96 - 172.81 = 2.15$

Column sum of squares = SSC $= \Sigma\, C_J^2/ Rr$ = Product sum of squares

(C_j is the total sum of observations in column j, R is number of rows and r is the number of replicates)

$= (29^2 + 35.4^2/12) - 172.81 = 174.51 - 172.81 = 1.70$

Row sum of squares = SSR = subject SS

$= (\sum R_i /C_r) - C.T = \{((7.7)^2+(7.8)^2+....+(8)^2) / 4\} - 172.81$

$= (692.16/4) - 172.81 = 173.04 - 172.81 = 0.23$

Where R_i is the sum of observations in row i, C is number of columns and r is number of replications.

Within-cell sum of squares $= \Sigma(x_1-x_2)^2$

x_1 and x_2 are the observed values from trial 1 and trial 2 of each cell respectively.

$[(2.4-2.3)^2 /2] + [(2.5-2.4)^2 /2] + [(2.3-2.6)^2 /2] +...+ [(3.1-3.1)^2 /2] = 0.127$

Residual C×R sum of squares = SST – SSC – SSR-SSW

$= 2.15 - 1.70 - 0.23 - 0.127 = 0.093$

Analysis Table

Source of variation	d.f	S.S	M.S	F
Between columns	C–1 = 1	1.7	1.7	$F_{1,5}$ = 91.40
Between rows	R – 1 = 5	0.23	0.046	$F_{5,12}$ = 4.34
Residual	(C–1) (R – 1) = 5	0.093	0.0186	$F_{5,12}$ = 1.75
Within cells	R*C*(N – 1) = 12	0.127	0.0106	

N=Number of replications

Test of Significance: To test for differences among products (H_0: $\mu_A = \mu_B$)

The F distribution with 1 and 5 d.f. has table value at level of probability 0.05 = 6.608

∴ F calculated value > F table value

∴ Null hypothesis is rejected

∴ Products are significantly different at 5% level

Test of Significance: To test for differences among subjects

The F distribution with 5 and 12 d.f. has table value at level of probability 0.05 = 3.106

∴ F calculated value > F table value

Subjects are significantly different at 5% level

The F test (C*R mean square/within mean square) is a test of interaction. In the absence of interaction (subject*drug product), the C*R mean square would equal the within mean square on the average. A value of the ratio sufficiently larger than one indicates interaction. In this example, the F ratio is 1.75, 0.0186/0.0106; this is not significant at the 5% level ($F_{5,\ 12}$ =3.106). The absence of a subject* drug product interaction in this experiment suggests that subjects are similar in their ability to distinguish the two products.

Missing Data

Even incase of a well designed experiment, due to certain factors, the data may be missed. The missed value may not cause much difference in one-way ANOVA and the interpretation is correct though it is collected with the missed value. In case of two-way ANOVA if the calculation is completed without the complete data then the interpretation is not correct. The missing value can be calculated with the following formulae.

$$r_{ij} = \frac{rr_i + cr_i - y}{(r-1)(c-1)}$$

r = number of rows,

c = number of columns

r_{ij} = observation in the ith row and j^{th} column

r_i = total i^{th} row, r_J is the total J^{th} column, y is grand total of all observations.

Example

The C_{max} values observed with the administration of different doses of drug to the patients having different age is given below. Find out the differences C_{max} due to age and dose is significant or not.

Patient	C_{max} values (μg/ml)		
	Low dose	Medium dose	High dose
Children	30	40	50
Adult	20	–	35
Old	20	30	40

Solution:

Patient	Low dose	Medium dose	High dose	Total
Children	30	40	50	120
Adult	20	-	35	55
Old	20	30	40	90
Total	70	70	125	265

The missing value in the second column and 2nd row is obtained from below formulae and the obtained value is substituted in the above table. Then the problem is solved as follows.

$$r_{ij} = \frac{(ry_i + cy_j - y)}{(r-1)(c-1)} = \frac{3 \times 55 + 3 \times 70 - 265}{2 \times 2} = 27.5$$

Patient	Low dose	Medium dose	High dose	Total
Children	30	40	50	120
Adult	20	27.5	35	82.5
Old	20	30	40	90
Total	70	97.5	125	292.5

Total sum of squares = SST = $\Sigma x^2 - CT$

$$= (30)^2 + (20)^2 + (20)^2 + \ldots\ldots\ldots + (40)^2 - \left(\frac{(292.5)^2}{9}\right)$$

$$= 10281.25 - 9506.25$$

$$= 775$$

$$\text{Column sum of squares SSC} = \frac{\left(\Sigma^x C_i\right)^2}{{}^n C_1} - CT$$

$$= \frac{(70)^2 + (97.5)^2 + (125)^2}{3} - C.F$$

$$= 10010.41667 - 9506.25$$

$$= 504.167$$

$$\text{Row sum of squares} = SSR = \frac{\sum R_i^2}{C} - C.F$$

$$= \frac{(120)^2}{3} + \frac{(82.5)^2}{3} + \frac{(90)^2}{3} - 9506.25$$

$$= 9768.75\text{-}9506.25 = 262.5$$

Residual sum of squares = SST – SSC – SSR = 8.33

Source of Variance	**D.F**	**S.S**	**M.S = $\frac{S.S}{D.f}$**	**F**
Between Columns	C – 1 = 2	504.167	252.083	$\frac{252.0833334}{2.083333} = 121.0002$
Between rows	R – 1 = 2	262.5	131.25	$\frac{131.25}{2} = 65.625$
Residual	(C – 1) (R – 1) = 4	8.33	2.083	
Total	8	775		

Test Significance:

Null hypothesis: There is no significance difference between C_{max} due to age.

Calculated F value = 121.0002

Tabled F value with level of significance 0.05and with degrees of freedom 2, 4 is 6.944

Calculated F value > table F value

∴ The null hypothesis is rejected

∴ There is a significance difference in C_{max} values due to age.

Null hypothesis: There is no significance difference between C_{max} due to dose.

Calculated F value = 65.62

Tabled F value with level of significance 0.05and with degrees of freedom 2, 4 is 6.944

Calculated F value > table F value

∴ The null hypothesis is rejected

∴ There is a significant difference in C_{max} values due to dose.

9.3 Three-way ANOVA

Three-way ANOVA is used to observe the statistical differences observed from three different sources. This test is more often employed in bioavailability studies, as the sources of variations are from formulation, subject and order of administration of pharmaceutical dosage forms.

***Example*:** The area under blood level vs. time curves observed from the tablet and solution formulation containing a new drug is given below. Find the analysis of variance accounts for order of administration, between subjects and between formulations.

Subject	Tablet	Solution	Order of administration
1	72	86	S-T
2	68	81	T-S
3	64	83	S-T
4	70	84	S-T
5	67	82	T-S
6	65	80	T-S
7	68	83	T-S
8	66	81	S-T
9	67	80	T-S
10	68	82	S-T

S-T solution first and tablet second

T-S tablet first and solution second

Solution:

Subject	Tablet	Solution	Order of administration	
1	72	86	S-T	$R_1 = 158$
2	68	81	T-S	$R_2 = 149$
3	64	83	S-T	$R_3 = 147$
4	70	84	S-T	$R_4 = 154$
5	67	82	T-S	$R_5 = 149$
6	65	80	T-S	$R_6 = 145$
7	68	83	T-S	$R_7 = 151$
8	66	81	S-T	$R_8 = 147$
9	67	80	T-S	$R_9 = 147$
10	68	82	S-T	$R_{10} = 150$
	$C_1 = 675$	$C_2 = 822$	$O_1 751$ $O_2 = 746$	

Variance accounts for order of administration with 1 d.f as well as subject and formulations variance, the error sum of squares is the total sum of squares minus the sum of squares due to order, subjects and formulations. The following formulae is used in three-way ANOVA.

Between subjects SSR $= \dfrac{\Sigma R_i^2}{2} - C.F$

Between formulation SSC $= \dfrac{\Sigma C_i^2}{10} - C.F$

Between order of treatments SSO $= \dfrac{\Sigma O_i^2}{10} - C.F$

where o_1 is the sum of results of first visit and o_2 is the sum of result for second visit.

Total sum of squares $= \Sigma x_i^2 - CT$

Now C.T = Correction term $= \dfrac{(\Sigma x_i)^2}{N}$

$$= \frac{(675+822)^2}{20} = \frac{(1497)^2}{20} = 112050.45$$

$$\Sigma x_i^2 = 113211$$

∴ Total sum of squares = 113211 – 112050.45

= 1160.55

Between subjects SSR = $((158)^2 + (149^2) + \ldots .. + (150)^2)/2$ – C.T

= (224235/2) – 112050.45

= 112117.5 – 112050.45

= 67.05

Between formulations SSC = $[(675)^2 + (822)^2]/10$=113130.9-112050.45=1080.45

Between order of administration SSO = $[(751)^2 + (746)^2]/10$ – C.T

= 112051.7 – 112050.45 = 1.25

ANOVA Table

Source of variance	D.F	S.S	M.S	F
Subjects	9	67.05	67.05/9=7.45	$F_{9,8}$=7.45/1475 = 0.005
Formulation	1	1080.45	1080.45/1=1080.45	$F_{1,8} = \dfrac{1080.45}{1475} = 0.732$
Order	1	1.25	1.25/1=1.25	
Error	8	11.8	11.8/8=1475	$F_{1,8} = \dfrac{1.25}{1475} = 0.000847$
Total	19	1160.55		

The observed AVC values for the formulation and order of administration are not significant ($F_{1, 8} = 5.32$ at the 5% level) suggesting that the values of total absorption from both formulations are similar. The test for the subjects is not significant (the table value $F_{9, 8} = 3.38$ and the calculated value is 0.005). A significant order effect would occur as the average results of first visit differed from the average results of second visit. Such a difference could be caused by systematic differences in the experimental procedure. A significant order effect does not invalidate the experiment.

9.4 Analysis of Covariance (ANCOVA)

Analysis of covariance is the application of ANOVA with regression to improve precision in comparison of treatments. The accuracy of ANCOVA is higher than ANOVA.

Example

To compare the stability of two formulations, the formulations are stored at 40 °C, 75% RH for 6 months and potency observed after 6 months is given below. Compare the two formulations.

Formulation 1		Formulation 2	
Initial potency	**Potency after storage**	**Initial potency**	**Potency after storage**
98	94	95	94
96	93	94	93
97	94	96	95
96	92	97	96
98	96	96	95
97	94	95	95

In the above case, the differences in stability of 2 formulations may be significant; however the final potency is also influenced by the covariable initial potency. The above validity is more significantly detected only if the initial variations are also included in the calculation.

ANOVA comparing formulations based on potency

Observed after storage:

Formulation 1	Formulation 2
94	94
93	93
94	95
92	96
96	95
94	95
Total 563	568

$$\Sigma x = 563 + 568 = 1131$$

$$\frac{(\Sigma x)^2}{N} = \frac{(1131)^2}{12} = 106596.75$$

$$\Sigma x^2 = (94)^2 + (93)^2 + (94)^2 + ... + 95)^2 = 106613$$

Total sum of squares = $SST = \Sigma x^2 - \frac{(\Sigma x)^2}{N}$

$$= 106613 - 106596.75 = 16.25$$

Between (Column) Sum of Squares = SSC

$$= \frac{(563)^2 + (568)^2}{6} - \frac{(\Sigma x)^2}{N}$$

$$= 106598.8333 - 106596.75 = 2.083$$

With in sum of squares = Total sum of squares – Between sum of squares

$$= 16.25 - 2.083 = 14.167$$

ANOVA Table

Source of Variance	d.f	S.S	M.S	F
Between formulations	1	2.083333	$\frac{2.083333}{1} = 2.08$	$\frac{2.083333}{1.4166667} = 1.47$
With in formulations	10	14.166667	$\frac{14.166667}{10} = 1.42$	
Total	11	16.25		

ANOVA Comparing formulations based on initial potency

Formulation 1	Formulation 2
98	95
96	94
97	96
96	97
98	96
97	95
Total 582	573

$$\Sigma x = 582 + 573 = 1155$$

$$\frac{(\Sigma x)^2}{N} = \frac{(1155)^2}{12} = 111168.75$$

$$\Sigma x^2 = (98)^2 + (96)^2 + ... + (95)^2 = 111185$$

$$\Sigma x^2 - \frac{(\Sigma x)^2}{N} = 111185 - 111168.75 = 16.25$$

Total sum of squares =SST= $\Sigma x^2 - \frac{(\Sigma x)^2}{N}$ =16.25

Between (Column) Sum of Squares = SSC = $\frac{(582)^2 + (573)^2}{6} - 111168.75 = 6.75$

With in sum of squares = Total sum of squares – Between sum of squares

= 16.25 – 6.75 = 9.5

ANOVA Table

Source of Variance	d.f	S.S	M.S	F
Between formulations	1	6.75	$\frac{6.75}{1} = 6.75$	$\frac{6.75}{0.95} = 7.1052$
With in formulations	10	9.5	$\frac{9.5}{10} = 0.95$	
Total	11	16.25		

With in formulation 1, potency after storage sum of squares

$$= (94^2 + 93^2 + ... + 94^2) - \frac{(563)^2}{6}$$

$$= 52837 - 52828.16667 = 8.83333$$

With in formulation 1, initial potency sum of squares

$$= \left((98)^2 + (96)^2 + (97)^2 + ... + (97)^2\right) - \frac{(582)^2}{6}$$

$$= 56458 - 56454 = 4$$

With in formulation 2, potency after storage sum of squares

$$= (94)^2 + (93)^2 + ... + (95)^2 - \frac{(568)^2}{6}$$

$$= 53776 - 53770.66667$$

$$= 5.3333333$$

With in formulation 2 Potency before storage

$$= (95)^2 + (94)^2 + ... + (95)^2 - \frac{(573)^2}{6}$$

$$= (54727) - 54721.5 = 5.5$$

Formulation 1:

Initial potency(x)	Potency after Storage(y)	$x - \bar{x}$	$y - \bar{y}$	$(x - \bar{x})(y - \bar{y})$
98	94	1	0.17	0.17
96	93	– 1	– 0.83	0.83
97	94	0	0.17	0
96	92	– 1	– 1.83	1.83
98	96	1	2.17	2.17
97	94	0	0.17	0
Total 582	563			5

$\bar{x} = 97$ $\bar{y} = 93.83$

Formulation 2:

Initial potency (x)	Potency after Storage (y)	$x - \bar{x}$	$y - \bar{y}$	$(x - \bar{x})(y - \bar{y})$
95	94	– 0.5	– 0.666	0.33
94	93	– 1.5	– 1.666	2.499
96	95	0.5	0.334	0.167
97	96	1.5	1.334	2.001
96	95	0.5	0.334	0.167
95	95	– 0.5	0.334	– 0.167
Total 573	**568**			**5**

Between cross product of sum of squares

Initial Potency (x)	Potency after Storage (y)	Initial Potency (x^1)	Potency After Storage (y^1)	$x - \bar{x}^1$	$y - \bar{y}^1$	$(x - \bar{x}^1)(y - \bar{y}^1)$
98	94	95	94	2.5	– 0.666	– 1.665
96	93	94	93	0.5	– 1.666	– 0.833
97	94	96	95	1.5	– 0.666	– 0.999
96	92	97	96	0.5	– 2.666	– 0.333
98	96	96	95	2.5	1.334	3.335
97	94	95	95	1.5	– 0.666	– 0.999
$\bar{X} = 97$	$\bar{Y} = 93.83$	$\bar{X}^1 = 95.5$	$\bar{Y}^1 = 94.666$			Total 2.494

Detailed Computation for ANOVA For Above data

Source	(1) d.f	(2) Y = Potency after storage	(3) X=Potency before storage	(4) After × before	(5) slope	(6) regss	(7) D.f	(8) Res S.S
Within Formulation 1	5	8.83	4	5	1.25	13.80	4	-
Within Formulation 2	5	5.33	5.5	5	0.91	4.41	4	4.97
Separate regressions	-	-	-	-	-	-	8	0.93
With in methods	10	14.17	9.5	10	1.05	10.57	9	
Between methods	1	2.08	6.75	2.50	-		-	3.64
Total	11	16.25	16.25	12.50	3.21			-

Column (1) & (2) are simple ANOVA for potency after storage

Column (1) & (3) are simple ANOVA for potency before storage

Column (4) is the cross product S.S $\Sigma(x-\bar{x})(y-\bar{y})$

Column (5) is computed as Column (4)/Column (3) [Potency after storage is Y variable, potency before storage is X variable]

Column (6) is the Column (3) $\times$Column (5) Squared

Column (7) is the d.f for residual (Column (8))

Column (8) is the Column (2) – Column (6)

Exercise:

1. In a state level cricket tournament the runs scored by the top 6 batsmen of four teams with similar performance are tabulated. Perform adequate test to determine if there is any significant difference observed between the groups when compared with previous state champions (challengers). Pick out the best team based upon the scores of top 6 batsmen and rank them accordingly.

Answer:

One-way analysis of variance	
P value	0.0021
P value summary	**
Are means signif. different? ($P < 0.05$)	Yes
Number of groups	5
F	5.727
R squared	0.4782

Table *Contd...*

ANOVA Table	SS	df	MS			
Treatment (between columns)	42000	4	10500			
Residual (within columns)	45840	25	1834			
Total	87840	29				
Dunnett's Multiple Comparison Test	Mean Diff.	q	Significant? P < 0.05?	Summary	95% CI of diff	Rank
challegers vs dare devils	106.3	4.301	Yes	***	41.85 to 170.8	4
challegers vs warriors	83.67	3.384	Yes	**	19.19 to 148.1	3
challegers vs riders	79.17	3.202	Yes	*	14.69 to 143.6	2
challegers vs chargers	41.33	1.672	No	ns	– 23.15 to 105.8	1

2. The following are the changes in serum HDL levels on four intervals in 30 day treatment with an allopathic and ayurvedic drug. Apply appropriate statistical test to report the effective treatment.

	Allopathic			**Ayurvedic**		
Day 0	45.7	56.8	37.4	54.3	43.3	48.3
Day 10	65.6	63.3	43.2	79	85.6	58.2
Day 20	74.5	71.5	59.5	112.5	124.5	89.1
Day 30	76.4	98.3	79.4	122.1	138.6	124.2

Answer:

Two way ANOVA					
Two-way RM ANOVA	**Matching by cols**				
Source of Variation	**Df**	**Sum-of-squares**	**Mean square**	**F**	
Interaction	3	1755	584.9	8.014	0.0034
Time	3	11950	3984	54.59	< 0.0001
Treatment	1	3955	3955	12.13	0.0253
Subjects (matching)	4	1305	326.1	4.468	0.0193
Residual	12	875.8	72.99		

Chapter 10

Non-Parametric Methods

10.1 Introduction

Most of the statistical tests are based on the assumption that the sample is following a known distribution (*e.g.*, normal or binomial). The results obtained from this assumption may be deviated from the actual value, if the sample is not fitted in theses distributions. In such cases distribution from non-parametric methods are used. The Z, t and F tests made inferences about the population mean(s), non-parametric tests make inferences about the population median(s). In non-parametric tests the distribution is continuous, and the data is symmetrical and independent. In these types of tests, we rank each observation from our data set. Although, we are not specifying which distribution our data is from, it must be from a continuous distribution. Each distribution must be described by a density curve that allows observations to take any value in some interval.

Thus when there is a doubt about the distribution of the parent population, a non parametric method should be used. In many situations particularly in social and behavioral sciences observations are difficult or impossible to take on numerical scales. Non parametric tests are well suited under such situations.

These methods involve the analysis of ranks and in doing such we decrease the accuracy of our information (changing from the raw data to a relative ranking). Most non-parametric tests require that the data be ranked on an ordinal scale. Ranking involves assigning the value 1 to the smallest observation, 2 to the second smallest, and continuing this process until n is assigned to the largest observation. We may obscure true differences and make it difficult to identify differences which are significant. In other words, non-parametric tests require differences to be larger if they are to be found significant.

Non-parametric tests assume that the variance in the group is equal and applicable for nominal variables. Non-parametric statistical tests can be useful when dealing with extremely small sample sizes or when the requirements of normality and homo-scedasticity cannot be met or assumed. These tests are simple to calculate, but are traditionally less powerful and the researcher needs to evaluate the risk of a Type II error.

Often referred to as distribution-free statistics, non parametric statistical tests do not make any assumptions about the population distribution. Nonparametric tests are relatively simple to calculate. Their speed and convenience offers a distinct advantage over conventional tests.

Ranking the data

Data	rank
22	1
28	5
26	3
25	2
27	4

In the case of ties, the average of the rank values is assigned to each tied observation.

Data	rank
25	2.5
28	7.5
30	10
24	1
26	5
25	2.5
26	5
28	7.5
29	9
26	5

In this example there were two 25s (ranks 2, and 3) with an average rank of 2.5, three26s (ranks4, 5 and 6) with an average rank of 5 and two 28s (ranks 7, and 8) with an average rank of 7.5

When comparing sets of data from different groups or different treatment levels (levels of the independent variable), ranking involves all of the observations regardless of the discrete level in which the observation occurs:

Group A (n=6)		Group B (n=7)		Group C (n=6) Total (N=20)	
Data	**rank**	**Data**	**rank**	**Data**	**rank**
16	5	20	14	19	11.5
18	9	19	11.5	24	19
17	6.5	15	3.5	23	18
12	1	17	6.5	20	14
14	2	18	9	18	9
15	3.5	22	17	20	14
		21	16	25	20

Accuracy of the ranking process may be checked in two ways. First, the last rank assigned should be equal to the total N. The second way to check the accuracy of the ranking procedure is the fact that the sum of all the summed ranks should equal to $N(N+1)/2$, where N equals the total number of observations:

General Procedure

1. **Observing the nominal variables:** The first step in non parametric tests is the observation of certain variable in the test performed for example age, weight, height, male/female colour, taste, oedama.
2. **Generation of ordinal data:** In ordinal data, the data is arranged in an arbitrary scale like arranging the data based on greater than the predetermined, /fixed or specified value and accordingly they can be scored.

   E.g.: In evaluation of rubber closures, the extracts are injected to the abdominal region of the rabbit and observed for oedema at the site of injection. In this test, the nominal data is observation of animals for oedama and ordinal data is based on categorization of oedema depending upon sevearity of oedema like no oedema, very slight oedema, moderate oedama and severe oedema and accordingly scoring is allotted.
3. **Ranking based on the observed values:** Based on the numerical values of the data, the data is arranged as per ranks. The animal which is having higher oedama (based on diameter) will receive higher rank and the animal not having oedema is receiving the lowest rank. These ranks are used to analyze the observed data from non parametric tests.

Advantages

1. Less complication in calculation
2. Rapid and easy
3. Easy to understand
4. More efficient than parametric tests, if the data is not distributed normally.

Disadvantages

1. Less power compared to parametric tests. Power indicates the efficacy test to recognize the differences existing if any.
2. Complete information provided by the sample is not utilized.
3. Poor interpretation.
4. Large sample requirement to obtain comparable results with parametric tests, if both parametric and non parametric tests are applicable.

10.2 Sign Test

Sign test is the simplest of the non-parametric tests. It is a test of equality of medians of two comparative groups. It can be applied to ranked or high level data such as continuous interval and ratio type data. This test is used for the paired data, pairs are matched and the differences of the measurements for each pair are tabulated. The differences are categorized with regard to the sign of the difference. The number of times one treatment has a higher value than the other treatment is counted. Ties are not counted. Depending up on the sample size, either normal or binomial probabilities are used. If the test shows significance, it indicates that the two comparative populations have different medians at the α level of significance. If positive and negative signs are observed to occur with equal frequency, then the treatments have a similar median. If the positive or negative signs are predominant, then one treatment has a higher median than the other.

10.2.1 Sign Test following Normal Probability

***E.g.*:** Twenty tablets from a batch were collected randomly and use the sign test to observe the statistical significances between the observed and the mean weight.

Mean weight = 100 mg

Weight of 20 tablets is 102, 103, 102, 103, 99, 101, 96, 100, 98, 101, 102, 100, 98, 99, 103, 102, 104, 103, 100, and 98

Calculation

1. Selection of Sample Size:

   Sample size = Total number of samples – The number of samples having weight equal to mean weight.

   In this example, the sample size is 17 as the total number is 20 and the tablets with the mean weight 100 mg are 3.

2. Null hypothesis $\mu = 100$
3. Alternative hypothesis $\mu < 100$
4. Level of significance 5%
5. Calculation: Replacing the each value greater than 100 with plus sign and each value less than 100 with a minus sign, the sample values yields

   ++++ –+– – ++– –++++ –.

The criteria is based on the probability of setting either + or –ve signs.

The number of positive or negative signs required for a sample size of 6 to 20 for significance at the 5% level of significance for sign test is given in Table 10.1. The table value for the sample size 17 is13. In the above problem, the number of positive signs is 11 and the negative signs are 6. The observed number of signs is less than table value. So the differences are not significant. The null hypothesis is accepted i.e. the mean weight of the tablets is equal to100 mg.

Table 10.1 Number of Positive or Negative Signs Needed for Significance for the Sign Test

Sample size	Number of positive or negative signs for significance	
	5% Level	1% Level
6	6	--
7	7	--
8	8	8
9	8	9
10	9	10
11	10	11
12	10	11
13	11	12
14	12	13
15	12	13
16	13	14
17	13	15
18	14	15
19	15	16
20	15	17

If the frequency of getting +ve and –ve signs is equal then it can be concluded samples are having equal median. If they are not equal, then the statistical test is based on the binomial distribution and the probability of outcome can be calculated from the table of binomial distribution for small sample size and the sample size greater than 20, the normal approximation test is used with the following formulae.

$$Z = \frac{|p - 0.5| - 1/(2N)}{0.5/\sqrt{N}}$$

P = observed proportion

N = Sample size excluding the ties.

$\therefore$ here $P = \frac{11}{17}$, N = 17

$= 0.64$

$$\therefore \quad Z = \frac{|0.647058823 - 0.5| - \frac{1}{2\times 17}}{\frac{0.5}{\sqrt{17}}} = \frac{0.14708823 - 0.029411764}{0.121267812} = 0.97$$

Z tabled value at the level of probability 0.05 = 1.96

∴ Calculated Z value is less than tabled Z value

∴ Null hypothesis is accepted

The calculation can be simplified as follows.

Z= [|number of +'s- number of –'s| –1] /√ number of +'s – number of –'s

Z= [|11– 6| –1] /√ 11– 6 = 4/√5 = 4/2.24 = 1.79

Z tabled value at the level of probability 0.05 = 1.96

∴ Calculated Z value is less than tabled Z value

∴ Null hypothesis is accepted

10.2.2 Paired Sample Sign Test

This test is applied if the paired data is used and the procedure is described below.

1. Replace the pair of sample values with +ve sign if the initial/first value is greater than final/second value and with –ve sign if the first value is smaller than second value. Ignore if the difference is zero.
2. Proportion of positive responses should be calculated and compared with the null hypothesis of equality of treatments ($H_o : P_o = 0.5$) using one sample binomial test.

***E.g.*:** The drug solution was divided in to two equal parts and one part is estimated by HPLC and the other part with UV method for 6 times and the observed drug content is furnished below. Whether the efficacy of method A and B are equal or not?

S.No	Drug Content (mg/tablet)		
	HPLC method	**UV method**	**Difference**
1.	100	98	+2
2.	102	99	+3
3.	103	97	+6
4.	102	100	+2
5.	101	99	+2
6.	102	99	+3

Null hypothesis: $\overline{x}_1 = \overline{x}_2$

Alternative hypothesis: $\overline{x}_1 \neq \overline{x}_2$

From the above table, 6 positive differences were observed and the sample size is 6. The observed proportion P is 1 and the hypothetical proportion P_o is 0.5.

$$Z = \frac{[1-0.5] - \frac{1}{12}}{\frac{0.5}{\sqrt{6}}} = \frac{0.5 - 0.083}{0.204} = 2.04$$

Since Z is greater 1.96, the difference is significant at 5% level. The drug content estimated from the HPLC method is more than the UV method, suggesting the HPLC method is more suitable than UV method.

10.2.3 Sign Test for Large Samples

Sign test for large sample is applied when np and n (1–p) are greater than 5 and Z value is calculated by the following formulae

$$Z = \frac{P - nP_0}{\sqrt{nP_0(1-P_0)}}$$

$P_0 = 0.5$; P = observed probability

Example

The weight of tablets collected from a compression station observed before and after some adjustment of a rotary tablet machine is given below. Whether the adjustment affects the weight of a tablet or not?

S.No	Weight before adjustment	Weight after adjustment
1	100	102
2	98	100
3	102	99
4	103	105
5	102	102
6	103	101
7	100	102
8	99	101
9	100	102
10	102	100
11	100	99
12	98	100
13	99	98
14	100	103
15	97	99
16	96	98
17	98	100
18	98	98
19	102	104
20	102	102

S.No	Weight before adjustment(a)	Weight after adjustment (b)	Difference d (a – b)
1	100	102	–2
2	98	100	–2
3	102	99	+3
4	103	105	–2
5	102	102	0
6	103	101	2
7	100	102	–2
8	99	101	–2
9	100	102	–2
10	102	100	2
11	100	99	1
12	98	100	–2
13	99	98	1
14	100	103	-3
15	97	99	–2
16	96	98	–2
17	98	100	–2
18	98	98	0
19	102	104	–2
20	102	102	0

$P_0 = 0.5$

$n = 20 - 3 = 17$

No. of –ve signs = 12

+ ve signs = 5

$$\therefore \text{ observed probability} = \frac{12}{17} = 0.705882352$$

$$\therefore \qquad Z = \frac{|P - nP_0|}{\sqrt{nP_0(1-P_0)}} = \frac{|0.705882352 - 17 \times 0.5|}{\sqrt{17 \times 0.5 \times 0.5}}$$

$$= \frac{7.794117648}{\sqrt{4.25}} = \frac{7.794117648}{2.06155293}$$

$$= 3.780$$

∴ Calculated Z value = 3.780

Tabled Z value at the level of probability 0.05 is 1.96

∴ Calculated Z value is greater than the tabled z value

∴ Null hypothesis is rejected

The efficacy of the tablet station is not same with the adjustment and they are significantly differed.

10.3 Wilcoxon Signed Rank Test

This test is used to improve the sensitivity of sign test in comparison of means. In this test, the magnitude of difference between the paired results is taken into consideration in addition to the sign. This test is used to compare the two treatments in a paired design. The following steps are involved in this test.

1. Differentiate the treatment pairs.
2. Discard the pair having zero difference.
3. Rank the untied pairs in order of magnitude disregarding sign.
4. Allot the average rank to the differences of equal magnitude.
5. Reassign the ranks based on the sign of original difference.
6. Sum the ranks with like signs.
7. Compare the smaller rank sum value with the table value.
8. The smaller rank sum value should be equal to or less than table value (table 10.2) for the means to be statistically significant.
9. For sample size exceeding 20, a normal approximation is available to compare population means with the following formulae.

$$Z = \left| \frac{T - N(N+1)/4}{\sqrt{N(N+1)(2N+1)/24}} \right|$$

Table 10.2 Values Leading to Significance for the Wilcoxon Signed Rank Test (Two-Sided Test)

Sample size, *N*	**5% Level**	**1% Level**
6	0	--
7	2	--
8	3	0
9	5	1
10	8	3
11	10	5
12	13	7
13	17	10
14	21	13
15	25	16
16	30	19
17	35	23
18	40	28
19	46	32
20	52	37

Example

The disintegration time of six randomly selected tablets is 14, 13, 12, 15, 14, 15 minutes respectively. Use the signed rank test at 0.05 level of significance to find out the differences in observed and claimed disintegration time (15 minutes.) is statistically significant or not.

Null hypothesis $\mu = 15$

Alternative hypothesis $\mu \neq 15$

Level of significance $\alpha = 0.05$

Calculation

Subtract 15 minutes from each value we get $\overset{1}{-1} - \overset{3}{2}, -\overset{4}{3}, 0, -\overset{2}{1}, 0$ and the ranks of the magnitude of difference are 1.5, 3, 4, and 1.5

In this example, the tablet No 1 and 5 are having equal disintegration time and received 1.5 ranks (average of 1st and second rank). The disintegration time of 4th and 6th tablet is equal to the theoretical mean and it is discarded from the study, as the difference is zero. All tablets received negative ranks, as the difference is negative in all cases and the sum of the response is =1.5 + 3 + 4 + 1.5=10

Here

$T = 10$

$N = 4$

The table value for the sample size 4 and the level of significance 5% from binomial distribution is more than the calculated value and hence null hypothesis is accepted. Alternatively it is calculated assuming normal distribution as follows.

$$Z = \left| \frac{T - N(N+1)/4}{\sqrt{N(N+1)(2N+1)/24}} \right| = \frac{10 - 4(5)/4}{\sqrt{4(4+1)(2 \times 4 + 1)/24}}$$

$$= \frac{5}{\sqrt{4 \times 5 \times 9/24}} = \frac{5}{\sqrt{7.5}} = \frac{5}{2.7386} = 1.8257$$

The calculated z value = 1.8257, tabled z value = 1.96 at level of 0.05

$\therefore$ Null hypothesis is accepted.

10.4 Wilcoxon Matched-pairs Test

The Wilcoxon matched-pairs test is similar to the matched-pair t-test. To accomplish this test, a traditional pre-post test (before-after) table is constructed and the differences are calculated similar to the matched-pair t-test. The absolute differences (regardless of sign, positive or negative) are then ranked from smallest to largest. A T-value is calculated for the sum of the ranks associated with the *least frequent* sign (positive or negative). If all the signs are positive or negative then the T-value would be zero and no ranks would be associated with the least frequent sign. A unique aspect of this test is that a certain amount of data may be ignored. If a difference is zero, the difference is neither positive nor negative; therefore, a sign cannot be assigned. Thus, data associated with no differences are eliminated and the number of pairs (n) is reduced appropriately. The hypotheses for the Wilcoxon matched-pairs test are not concerned with mean differences, as seen with the t-test

Example The onset of action observed for two formulations from the same 8 subjects is given below. Determine the statistical differences in onset of action observed between test and reference products.

Subject	**Onset of action(min)**	
	Test product	**Reference product**
1	5.4	6.3
2	5.7	6.1
3	5.2	6.7
4	5.6	6.2
5	5.2	6.7
6	5.8	6.4
7	5.6	6.1
8	5.2	6.7

Solution:

Subject	**Onset of action(min)**		
	Test product A	**Reference product B**	**Difference B – A**
1	5.4	6.3	0.9
2	5.7	6.1	0.4
3	5.2	5.4	0.2
4	5.6	5.2	– 0.4
5	5.2	6.7	1.5
6	5.8	5.4	– 0.4
7	5.6	6.1	0.5
8	5.2	6.7	1.5

Ranking the differences by ignoring the sign.

Subject	Value	Rank	Assigned rank	Assigned rank with sign
3	0.2	1	1	1
2	0.4	2	3	3
4	-0.4	3	3	-3
6	-0.4	4	3	-3
7	0.5	5	5	5
1	0.9	6	6	6
5	1.5	7	7.5	7.5
8	1.5	8	7.5	7.5

The sum of ranks with positive signs = 30

The sum of ranks with negative sign = 6

From the table 10.2, the value of the smaller of the two rank sums required for significance at the 5% level for sample size 8 is 3. The sample size is equal to the number of pairs-number of ties. In our example, the smaller rank sum is 6(more than the table value), therefore the difference is insignificant. For larger samples (size>20), a normal approximation is used to compare the population means. This formulae works well for smaller sample sizes. In our example, N = 8 and T = 30 (either the large or smaller rank sum can be used. The Z value is calculated with the following formulae.

$$Z = \frac{|T - N(N+1)/4|}{\sqrt{[N(N+1/2)(N+1)]/12}}$$

$$Z = \frac{30 - 8(8+1)/4}{\sqrt{8(8+1/2)(8+1)/12}} = 1.68$$

The calculated Z value = 1.68, tabled Z value = 1.96 at level of 0.05

∴ Null hypothesis is accepted

10.5 Wilcoxon Rank Sum Test (Test for Differences between Two Independent Groups)

If two treatments are to be compared where the observations have been obtained from two independent groups, then this test is used. It is alternative to the two independent sample t-tests. This test is also known as Mann-Whitney test or U test. This test is applicable if the data is ordinal (the observations can be ordered). It is based on the assumption that the distributions are symmetrical and the mean and medians are distributed equally. In this test, the magnitude of the difference is also taken in to consideration in addition to the sign.

Procedure

1. Rank the pooled observations from both groups regardless of group designation.
2. Give equal rank to the identical observations.

3. Return the observations to their respective treatment groups
4. Replace the observations with the corresponding ranks.
5. Sum the ranks in each group.
6. If the sample size is equal in two groups, then the sum of the rank in either group can be used. If the sample size is less in one groups, then the sum of the rank in that group should be used
7. For moderate sized samples, the statistical test for equality of the distribution means may be estimated using the normal distribution. For sample size less than 10, follow the table 10.3 for significance. If the sample size is equal to greater than 10 significance is calculated based on normal distribution by using the following formulae.

$$Z = \frac{|T - N_1(N_1 + N_2 + 1)/2|}{\sqrt{N_1 N_2 (N_1 + N_2 + 1)/12}}$$

where T = sum of ranks for the smaller sample size.

N_1= smaller sample size.

N_2= larger sample size.

Table 10.3 Critical Values for Wilcoxon Rank Sum Test ($\alpha = 0.05$)

Size of larger sample	Size of smaller sample (*M*)						
	M = 3	4	5	6	7	8	9
M	5, 16	11, 25	18, 37	26, 52	37, 68	49, 87	63, 108
M + 1	6, 18	12, 28	19, 41	28, 56	39, 73	51, 93	66, 114
M + 2	6, 21	12, 32	20, 45	29, 61	41, 78	54, 98	68, 121
M + 3	7, 23	13, 35	21, 49	31, 65	43, 83	56, 104	71, 127
M + 4	7, 26	14, 38	22, 53	32, 70	45, 88	58, 110	74, 133
M + 5	8, 28	15, 41	24, 56	34, 74	46, 94	61, 115	77, 139
M + 6	8, 31	16, 44	25, 60	36, 78	48, 99	63, 121	79, 146
M + 7	9, 33	17, 47	26, 64	37, 83	50, 104	65, 127	82, 152
M + 8	10, 35	17, 51	27, 68	39, 87	52, 109	68, 132	85, 158
M + 9	10, 38	18, 54	29, 71	41, 91	54, 114	70, 138	88, 164
M + 10	11, 40	19, 57	30, 75	42, 96	56, 119	72, 144	90, 171
M + 15	13, 53	24, 72	36, 94	50, 118	66, 144	84, 172	104, 202
M + 20	16, 65	28, 88	42, 113	58, 140	76, 169	96, 200	118, 223
M + 25	18, 78	32, 104	48, 132	66, 162	86, 194	108, 228	132, 264

Table 10.4 Critical Difference for Significance (α = 0.05) Comparing All Possible Paris of Treatments for Nonparametric One-way ANOVA

N (for each treatment)	Number of treatments				
	3	4	5	6	7
3	15	23	30	37	45
4	24	35	46	57	69
5	33	48	63	79	96
6	43	63	83	104	125
7	54	79	105	131	158
8	66	96	128	160	192
9	79	115	152	190	229
10	92	134	178	223	268
11	106	155	205	257	309
12	121	176	233	292	352
13	136	199	263	329	397
14	152	222	294	368	444
15	169	246	326	408	492
16	186	271	359	449	542
17	203	296	393	492	593
18	221	323	428	536	646
19	240	350	464	581	700
20	259	378	501	627	756
21	278	406	538	674	814
22	298	435	577	723	872
23	319	465	617	773	932
24	340	496	657	824	994
25	361	527	699	875	1056

Table 10.5 Critical Differences for Significance (α = 0.05) Comparing All Possible Pairs of Treatments for Nonparametric Two-way ANOVA

N (for each treatment)	Number of Treatments				
	3	4	5	6	7
3	6	8	10	13	15
4	7	10	12	15	18
5	8	11	14	17	20
6	9	12	15	18	22
7	9	13	16	20	24
8	10	14	17	21	25
9	10	14	18	23	27
10	11	15	19	24	28
11	11	16	20	25	30
12	12	16	21	26	31
13	12	17	22	27	32
14	13	18	23	28	34
15	13	18	24	29	35
16	13	19	24	30	36
17	14	19	25	31	37

Table 10.5 *Contd...*

N (for each treatment)	Number of Treatments				
	3	**4**	**5**	**6**	**7**
18	14	20	26	32	38
19	14	20	27	33	39
20	15	21	27	34	40
21	15	21	28	35	41
22	16	22	29	35	42
23	16	22	29	36	43
24	16	23	30	37	44
25	17	23	31	38	45

Table 10.6 Critical Values for Number of Runs at the 5% Level of Significance

Sample size, N	Two-sided test Lower number	Upper number	One-sided test Lower number
10	2	9	3
12	3	10	3
14	3	12	4
16	4	13	5
18	5	14	6
20	6	15	6
22	7	16	7
24	7	18	8
26	8	19	9
28	9	20	10
30	10	21	11
32	11	22	11
34	11	24	12
36	12	25	13
38	13	26	14
40	14	27	15

Table 10.7 Probability of Getting at Least One Run of Given Size for N Samples

N	5% Level	1% Level
10	5	--
20	7	8
30	8	9
40	9	10
50	10	11

Example:

The C_{max} values of two formulations are given below, whether difference in their C_{max} values is significant at 0.05 level or not.

Subject	Test product		Reference product	
	C_{max} values (ng/ml)	Rank	C_{max} values (ng/ml)	Rank
1	2.4	5	2.3	2
2	2.5	7	2.3	2
3	2.4	5	2.7	12
4	2.6	9	2.8	14.5
5	2.7	12	2.6	9
6	2.8	14.5	2.4	5
7	2.6	9	2.3	2
8	2.7	12	2.9	16
Sum of ranks		64.5		62.5

Solution:

Null hypothesis: There is no significance difference between the C_{max} values of test product and reference product.

Here

N_1 & $N_2 = 8$

T = sum of ranks for the smaller sample size = 62.5

$$Z = \frac{|T - N_1(N_1 + N_2 + 1)/2|}{\sqrt{N_1 N_2 (N_1 + N_2 + 1)/12}}$$

$$Z = \frac{\left|62.5 - \left(8 \times \frac{17}{2}\right)\right|}{\sqrt{8 \times 8 \times \frac{17}{12}}} = \frac{|62.5 - 68|}{9.52} = 0.58$$

Calculated Z value = 0.58

∴ Tabled Z value at level of probability 0.05 = 1.96

∴ $T_{cal} < T_{tab}$ ∴ Null hypothesis is accepted.

10.6 Kruskal-Wallis Test (One-way ANOVA)

The H test or Kruskal-Wallis test is used to test the null hypothesis i.e. independent samples came from identical populations or not. This test is also known as H test. The H value is calculated with the following formulae.

$$H = \chi^2_{k-1} = \frac{12}{N(N+1)}\left(\Sigma \frac{R_t^2}{n_i}\right) - 3(N+1)$$

N = Total number of observations in all groups

R_i = Sum of the ranks in i^{th} group

n_i = Number of observations in i^{th} group

k = Number of groups

If each sample has at least five observations, the sampling distribution is very close to chi-square distribution with k–1 degrees of freedom.

In one-way ANOVA,we are testing to see if our population means are equal. In the Kruskal-Wallis test, we test to see if our population medians are equal. The idea of this test is to rank all the responses from all the groups together and then apply one-way ANOVA to the ranks rather than to the original observations. This test is applied when the treatments are more than two. It is assumed that the variance in two groups is equal. Equal number of sample sizes is preferable but not compulsory. It is an extension of the rank sum test to more than two treatments. If the average of at least two comparative treatments is different, then the significant differences are observed with this test. The analysis consists of combining all the data like Wilcoxcon rank sum test corresponding to each observation are retained and summed for each group. It is approximately distributed as Chi-square with the degree of freedom k–1. k is number of treatments. The Chi-square is applicable if the number of observations in each group is greater than 5.

The critical values for this test can be find from the table 10.4. In page 340, above the example incorporate the sentence the critical values for this test can be find from the table 10.5. In page 346, above the example incorporate the sentence the critical values for this test can be find from the tables 10.6 and 10.7.

Example

Tablets of drug x were prepared by three different techniques such as wet granulation, drug granulation and direct compression techniques. The disintegration time observed from the tablets prepared by these methods is given below.

Direct compression: 8, 7, 6, 7, 9, 10

Wet granulation: 11, 12, 13, 12, 13, 10

Dry granulation: 15, 14, 16, 15, 17, 18

Null hypothesis: Population means are identical

Alternative hypothesis: Population means are not equal

Level of significance $\alpha = 0.05$

Criterion

Reject the null hypothesis if H > 5.991, which is the value of $\psi^2\,0.05$ for 2 degrees of freedom.

The ranks allotted to the tablets prepared with direct compression technique based on observed disintegration time $R_1 = 4, 2.5, 1, 2.5,5, 6.5$.

Sum = 4 + 2.5 + 1 + 2.5 + 5 + 6.5 = 21.5

The ranks allotted to the tablets prepared with wet granulation technique based on observed disintegration time $R_2 = 8, 9.5, 11.5, 9.5, 11.5, 6.5$.

Sum = 8 + 9.5 + 11.5 + 9.5 + 11.5 + 6.5 = 56.5

The ranks allotted to the tablets prepared with dry granulation technique based on observed disintegration time $R_3 = 14.5, 13, 16, 14.5, 17, 18$

Sum = 93

$$H = \frac{12}{18\times19}\left[\frac{(21.5)^2}{6}+\frac{(56.5)^2}{6}+\frac{(93)^2}{6}\right]-3\times19$$

$$= \frac{12}{18\times19}\left[\frac{462.25+3192.25+8649}{6}\right]-3\times19$$

$$= \frac{12}{18\times19}\left(\frac{12303.5}{6}\right)-3\times19$$

$$= 71.9502924-57$$

$$= 14.95$$

Decision

$H = 14.95 > 5.991$

Reject the null hypothesis.

Example

The blood glucose levels observed after three hours of post administration from three groups, control and two different formulations of antidiabetic drug is given below. Determine the significance of difference between formulations with Kruskal-Wallis test at 0.05 level of significance.

Subject No	Blood Glucose level		
	Control	**Formulation 1**	**Formulation 2**
1	130	90	86
2	136	92	88
3	132	93	87
4	134	91	88
5	133	90	90
6	135	89	89

Solution:

Subject No	Control	Rank	Formulation 1	Rank	Formulation 2	Rank
1	130	13	90	8	86	1
2	136	18	92	11	88	3.5
3	132	14	93	12	87	2
4	134	16	91	10	88	3.5
5	133	15	90	8	90	8
6	135	17	89	5.5	89	5.5
		$93 = R_1$		$54.5 = R_2$		$23.5 = R_3$

$$\chi^2_{k-1} = \frac{12}{N(N+1)}\left(\Sigma\frac{R_t^2}{n_i}\right) - 3(N+1)$$

$$= \frac{12}{18.719}\left[\frac{R_1^2}{6} + \frac{R_2^2}{6} + \frac{R_3^2}{6}\right] - 3(19)$$

$$= \frac{12}{18.19}\left[\frac{(93)^2 + (54.5)^2 + (23.5)^2}{6}\right] - 3(19)$$

$$= \frac{1}{9(19)}\left[(93)^2 + (54.5)^2 + (23.5)^2\right] - (3.19)$$

$$= \frac{12171.5}{171} - (3.19)$$

$$= 71.17836257 - 3(19)$$

$$= 14.17$$

The value of Chi Square with 2 d.f must be equal to or greater than 5.99 to be significant at the 5% level.

If statistically significant differences are observed, then we have to find out which treatments are different. To perform pair wise comparisons, the number of observations per treatment must be same. The difference between the sums of ranks observed from two different groups should be calculated. In this example, the difference between the control group and one of the formulations or alternatively the difference in sum of ranks observed from the two formulations should be calculated. If the observed differences in each case is higher than the Chi-square distribution, then the significant differences are existing among control and other two formulations. In some cases, the differences observed from control and formulation 1 might be significant where as in other cases the differences observed from control and formulation 2 is significant or other possibility is the differences in two formulations is significant.

Correction: To increase the observed value of Chi-square, the following correction is suggested. This correction increases the degree of significance, if the null hypothesis is rejected and it may result the statistically significant differences when the original Chi-square missed the significance.

$$\text{correction} = \frac{x^2}{1 - \Sigma\left(t_i^3 - t_i\right)\left(N^3 - N\right)}$$

t_i = number of tied observed in group i

N = total number of observations

The correction for Chi-square to the above problem is

The above answer after correction results

$$= \frac{14.178362}{1-(\frac{6+6+24}{5814})} = \frac{14.178362}{1-0.00619195} = \frac{14.178362}{0.993808049}$$

$$= 14.26$$

10.7 Friedman Test (Two-way Analysis of Variance)

This test is applied when the data is in the form of a two-way ANOVA design and data should be suitable for ranking and data should be from more than two groups. In this test, treatments are ranked within each block (columns). If the sample size is sufficiently large, then Chi-square distribution can be used to approximate the test of significance. The Chi-square test is

$$\chi^2_{c-1} = \frac{12}{rc(c+1)}\left(\Sigma R_i^2\right) - 3r(c+1)$$

d.f = c – 1

r = number of rows

c = number of columns

R_i = sum of the ranks in i^{th} group

The ranks should be allotted within each block (subject, formulation, method etc). If the ties are observed in the assigned ranks, then the average rank should be assigned to the tied observations.

Example

The disintegration time observed from six formulations each containing different disintegrants is given below. This difference in disintegration time due to disintegrants is significant or not?

	Disintegration time (minutes)		
S.No	**Disintegrant A**	**Disintegrant B**	**Disintegrant C**
1	8	12	15
2	7	11	16
3	6	10	17
4	7	11	15
5	5	10	18
6	6	9	17

Solution:

In the above problem, the lowest disintegration time observed from formulation 1 is with disintegrant A is 8 minuts so it is assigned 1st rank. The next highest value observed is from the formulation containing disintegrant B is 12 so it is assigned rank 2 and formulation containing disintegrant C is 15 and assigned rank 3. The same procedure is followed for other groups. The assigned ranks are shown in the following table.

	Disintegration time (minutes)					
Tablet formulation	**Disintegrant A**	**Rank**	**Disintegrant B**	**Rank**	**Disintegrant C**	**Rank**
1	8	1	12	2	15	3
2	7	1	11	2	16	3
3	6	1	10	2	17	3
4	7	1	11	2	15	3
5	5	1	10	2	18	3
6	6	1	9	2	17	3
R_i		6		12		18

$$= [12/6\times3(3+1)]\ (6^2+12^2+18^2) - 3\times6(3+1) = 12$$

χ^2 tabled value with level of Probability 0.05 and having d.f = c – 1 = 2 is 5.99

There is significance difference between disintegrans

$$\chi^2_{\text{cal}} > \chi^2_{\text{table}}$$

Modified Friedman Test

To get approximate F distribution with c – 1, (c – 1) (r – 1) degree of freedom and to get better comparison compared to Chi-square distribution for the Friedman test, this modification is applied. In this case, the statistic parameter T_2 is calculated with the following formulae.

$$T_2 = \frac{(r-1)\left[B_2 - rc(c+1)^2/4\right]}{A_2 - B_2}$$

$A_2 = \sum x_i^2$ where x_i are the individual ranks. If there are no ties, then A_2 is calculated with the following equation.

$$A_2 = cr\ (c + 1)\ (2c + 1)/6;\ B_2 = \frac{1}{r}\Sigma\left(c_i^2\right)$$

C = number of columns and r= number of rows.

C_i = sum of observations in column i

The above problem is on subjecting to modified Friedman test results

$$A_2 = cr(c + 1)\ (2c + 1)/6 = 3 \times 6\ (3 + 1)\ (2 \times 3 + 1)/6$$

$$= 3 \times 4 \times (7) = 84$$

$$B_2 = \frac{1}{r}\Sigma(c_i)^2 = \frac{1}{6}\left[(6)^2 + (12)^2 + (18)^2\right] = \frac{1}{6}[504] = 84$$

$$\therefore \quad T_2 = \frac{(6-1)\left[84 - 3.6\frac{(3+1)^2}{4}\right]}{84-84} = \infty$$

10.8 Multiple Comparisons for the Modified Friedman Test

In previous case, as the differences in disintegrants was found to be significant then the differences between which two super disintegrants is significant is detected by the following formulae.

$$[C_j - c_i] > t\sqrt{[2r(A_2 - B_2)]/[(r-1)(c-1)]}$$

t = tabled t value with (r – 1) (c – 1) d.f at the specified α level (usually 0.05)

For the above problem, the table t-value for the degree of freedom (6-1) (3-1) =10 at 5% level of significance for two tailed test is 2.23.

$$[C_J - C_I] > 2.23\sqrt{2 \times 5(84 - 84)/(6-1)(3-1)} = 0$$

Any difference in rank sum greater than or equal to 0 is significant at the 0.05 level. For the above problem, the rank sums observed from disintegrants A&B, A&C and B&C are significantly differ and indicating that the differences in disintegration time observed from the tablets containing three disintegrants is significant.

Dunns multiple comparison test

Kruskal-wallis test and friedman test are nonparametric tests that compares three or more unpaired groups (non gaussian distribution). Analysis of such data requires a post test for effective comparison between the groups. Dunns test is one such post test that makes pair wise comparisons among the groups. It compares the difference in the sum of ranks between two columns with the expected average difference and calculates a‘t’ value for each pair of columns. These‘t’ values suggests, if difference is statistical significant or not?

In the Dunns test, the difference in the mean ranks between the groups is compared pair wise and a standard deviation for each of these comparisons is calculated. By dividing the absolute difference in ranks between each pair by its standard deviation, a‘t’ statistic value is generated for each pair, which are compared with the standard normal *z* distribution. If the resulting statistic at α/2 (0.05) is significantly large, the null hypothesis of no difference between the ranks is rejected. Multiple comparisons are made only if a significant overall test has been observed at the usual α (0.05). This is to reduce the error that occurs in nonparametric tests.

$$T_{ij} = \frac{|D_{ij}|}{\sigma_{ij}} = \frac{|\bar{R}_i - \bar{R}_j|}{\sigma_{ij}} = \frac{|\bar{R}_i - \bar{R}_j|}{\sqrt{\frac{N(N+1)}{12}\left[\frac{1}{n_i} + \frac{1}{n_j}\right]}}$$

D_{ij} = Mean rank difference

σ_{ij} = Standard deviation of differences

$\left|\overline{R}_i - \overline{R}_j\right|$ = the absolute value of difference in mean ranks between group and group j

N = total sample size

n_i = sample size for group i

n_j = sample size for group j

10.9 Quade Test

If more number of ties is observed in Friedman, then to improve the sensitivity of the test Quade test is used. The range of values (largest- smallest rank for each row) is calculated for each block (row) and the blocks are ranked depending up on range of values within a block. These ranks are denoted as Q_1..... Q_r, where r is the number of rows (blocks). Calculate the 'T' statistic with the following formulae.

$$T = \frac{(r-1)B}{A-B}$$

$$A = \Sigma s_{ij}^2 ; B = \Sigma s_i^2 / r$$

$s_{ij} = Q_i\,[\,R\,(x_j) - (k+1)/2]$

$s_i = \Sigma\, s_{ij}$ for each treatment

k = number of columns

x_j = rank of each observation

The example shown in Friedman test is solved as follows.

Range of row1=15 – 8 = 7

Range of row2=16 – 7 = 9

Range of row3=17 – 6 = 11

Range of row4=15 – 7 = 8

Range of row5=18 – 5 = 13

Range of row6=17 – 6 = 11

$Q_1 = 1$

$Q_2 = 3$

$Q_3 = 4.5$

$Q_4 = 2$

$Q_5 = 6$

$Q_6 = 4.5$

The calculation of S_{11} is as follows.

$S_{11} = Q_1 [R(x_{11}) - (k+1)/2]$

$Q_1 = 1$

$R(x_{11})$ = the rank assigned to the observation representing the first row of the column 1 = 1

k = number of columns = 3

$$1[1 - (3 + 1)/2] = - 1$$

Similarly S_{12}, S_{13}, S_{21}, S_{22}, S_{23}, S_{31}… S_{63}, are calculated and given in the following table.

Tablet formulation	Disintegrant A	Disintegrant B	Disintegrant C	Range	Rank
1	– 1	0	1	7	1
2	– 3	0	3	9	3
3	– 4.5	0	4.5	11	4.5
4	– 2	0	2	8	2
5	– 6	0	6	13	6
6	– 4.5	0	4.5	11	4.5
R_i	– 21	0	21		

$A = \Sigma(S_i)^2 = 0$

$B = \Sigma S_i^2/r = (-21)^2 + (0)^2 + (21)^{2/}6 = 147$

The test statistic is $T = \dfrac{(r-1)B}{A-B} = 5 \times 147/0 - 147 = -5$

The table F value with (c – 1) 2 and (r – 1) (c – 1) 10 degree of freedom at 0.05 level is 4.1. The calculated value is greater than table value. Therefore at least two disintegrants are significantly different. Multiple comparisons can be made if the F test shows significance. The difference between the sums of any two treatments is significant, if the difference value exceeds

$$t \times \left[\frac{2r(A-B)}{(r-1)(c-1)}\right]^{1/2}$$

The table t-value for the degree of freedom (r – 1) (c – 1) 10 at the 0.05 alpha levels is 2.23. Substitution of this data in the above formulae results

$$2.23[12(0 - 147)/5 \times 2]^{½} = 29.62$$

We conclude that the disintegrant C gives higher results compared to A & B disintegrants.

10.10 Rank Correlation

There are many occasions where it is not possible to measure the variable under consideration quantitatively or where the statistical series is composed of items, the exact magnitude of which cannot be ascertained. For example, it may be possible to judge the taste of the formulation but it may be difficult to give them a numerical grade in terms of taste. In such cases, it is not possible to measure the relationship between the attributes by coefficient of correlation. If the two series are ranked according to the two attributes, the correlation coefficient between their ranks is called rank correlation. When continuous data are measured on an ordinal scale or if assumptions of population normality cannot apply to the data, the Spearman rank-order correlation, or Spearman rho, offers an alternate procedure to the correlation coefficient. Similar to other nonparametric tests, a procedure is used to rank the order of the observations for both variables and then the difference between the two ranks makes up part of the test statistic.

In this test, Ranks are allotted separately to x' and y' by following the same order either low to high or high to low. If any same values are observed, then they are assigned with the mean of the rank and jointly given for both observations. The rank correlation is calculated with the following formulae.

$$\text{Spearman Rank Order Correlation} = \rho = 1 - \frac{6\left(\Sigma d^2\right)}{n^3 - n}$$

p = Spearman's rank correlation coefficient

n = number of pairs

d = difference between the pair of ranks

Unlike the correlation coefficient, which is concerned with the means for both the X and Y variables, here the investigator is interested in the correlation between the rankings.

Example

The time and the % drug dissolved data is given below. Calculate the rank correlation.

Time (hrs)	% drug release	Ranks	Ranks of time	d	d^2
1	15	7	1	6	36
2	29	6	2	4	16
3	44	5	3	2	4
4	61	4	4	0	0
5	74	3	5	–2	4
6	88	2	6	–4	16
7	99	1	7	–6	36
					$\Sigma d^2 = 112$

$$r_s = 1 - \frac{6.(112)}{7(7^2 - 1)} = 1 - \frac{6 \times 112}{7 \times 48} = 1 - 2 = -1$$

The rank correlation = –1

10.11 Test of Randomness: Runs

A run is a succession of identical sign (weight, hardness, absorbance value etc.,) followed and proceeded by different sign which are having higher/lower than median value.

Example

Consider the following table which is representing the weight of 20 tablets.

S.No	Weight of the tablet (mg)	S.No	Weight of the tablet (mg)
1	104	11	104
2	103	12	105
3	98	13	106
4	97	14	97
5	96	15	96
6	105	16	98
7	106	17	99
8	104	18	105
9	99	19	107
10	98	20	106

The median value from the above table is 103.5 ($\because \frac{103+104}{2}$) mg. Assign the +ve sign to those tablets having weight higher than median value and –ve sign to those tablets having weight less than median value. The above tablets can be arranged as follows after assigning the sign.

$$\underbrace{+}\underbrace{----}\underbrace{+++}\underbrace{--}\underbrace{+++}\underbrace{----}\underbrace{+++}$$

The numbers of runs required for sample No. 20, as per the table value is 6. The observed runs are 7 (greater than table value) which indicate that the runs are randomly distributed.

The above underlined portion constitutes the run. The total number of runs appearing in an arrangement gives an indication of randomness. If the runs are less than there is a possibility of grouping/clustering. If the number of runs is more, then there is a possibility of cyclic pattern or repetition. The sequence is random or not can be observed from statistical tables. If the number of runs is less than or equal to the lower number of greater than or equal to upper value, the sequence is considered as non random at that level of significance. The sample size (N) is the number of values above and below the median.

For sample sizes of 40 or more a normal distribution is applicable and Z value is calculated with the formulae

$$Z = \frac{|r-(N/2+1)|}{\sqrt{N(N-2)/4(N-1)}}$$

r = number of runs

N = sample size

Here the number of runs = 7, Sample size (N) = 20

$$Z = \frac{r-(N/2+1)}{\sqrt{N(N-2)/4(N-1)}} = \frac{\left|7-(\frac{20}{2}+1)\right|}{\sqrt{(20)(20-2)/4(20-1)}}$$

$$= \frac{4}{\sqrt{(20\times18)/4\times19}}$$

$$= \frac{4}{\sqrt{5\times18/19}}$$

$$= \frac{4}{\sqrt{4.73}}$$

$$= \frac{4}{2.176} = 1.838$$

The observed Z value is less than table value and indicates random distribution

Comparison of parametric and non parametric tests is given in the following table.

Setting	Normal test	Rank test
One Sample	One-sample *t* test	Wilcoxon signed rank test
Matched pairs	Apply one-sample test to differences within pairs	Wilcoxon signed rank test
Two independent samples	Two-sample *t* test	Wilcoxon rank sum test
Several independent samples	One-way ANOVA *F* test	Kruskal-Wallis test

Exercise

1. Treatment of five patients with an anti hypertensive drug has revealed has shown the following changes in the systolic BP. Report the effective antihypertensive drug.

Patient ID	Before	After
AA	154	110
AB	168	125
GV	146	142
AD	135	120
AE	145	112

Answer:

Wilcoxon signed rank test	
P value	0.0313
P value summary	*
Are medians signif. different? (P < 0.05)	Yes
One- or two-tailed P value?	One-tailed
Sum of positive ranks –15	negative ranks – 0
Sum of signed ranks (W)	15

2. The given data is the time taken by three laboratory animals to complete a task i.e., to reach a hidden platform. Perform adequate statistical test to determine the best species to complete the task in short period.

Mice	Rat	Hamster
63	37	23
73	27	46
76	24	56
60	40	51

Answer:

Kruskal-Wallis test			
P value	0.0183		
P value summary	*		
Do the medians vary signif. (P < 0.05)	Yes		
Number of groups	3		
Kruskal-Wallis statistic	8		
Dunn's Multiple Comparison Test	**Difference in rank sum**	**Significant? P < 0.05?**	**Summary**
mice vs rat	7	Yes	*
mice vs hamster	5	No	ns
rat vs hamster	–2	No	ns

3. The following are the marks obtained by five students in mathematics in different mid term exams of a year. Use appropriate test to see if any change in marks occurred and report their performance.

	Term 1	**Term 2**	**Term 3**
GS	54	43	44
JM	23	34	22
HM	45	65	38
JW	54	77	55
PS	45	46	62

Answer:

Friedman test			
P value	0.6914		
P value summary	ns		
Are means significantly different? (P < 0.05)	No		
Number of groups	3		
Friedman statistic	1.2		
Dunn's Multiple Comparison Test	**Difference in rank sum**	**Significant ? P < 0.05?**	**Summary**
Term 1 vs. Term 2	– 3	No	ns
Term 1 vs. Term 3	0	No	ns
Term 2 vs. Term 3	3	No	ns

4. The following is the amount of money spent by men and women on purchase of five different chocolates. Determine the order of choice.

Men	**Women**
32	94
28	85
16	79
31	62
26	72

Answer:

Mann Whitney test	
P value	0.0079
P value summary	**
Are medians significantly different ? (P < 0.05)	Yes
One- or two-tailed P value?	Two-tailed
Sum of ranks in column A - 15	Sum of ranks in column B - 40
Mann-Whitney U	0.000

CHAPTER 11

STATISTICAL QUALITY CONTROL

11.1 Introduction

Pharmaceutical industry, being an integral part of health care system, plays a key role in the manufacture of various pharmaceuticals. The pharmaceutical products must follow certain standards that enable the product to conform to the desired quality, safety, efficacy and purity as per regulatory norms. The quality control department always strives for the production of quality items that conform to the specific standards. In order to develope such products; it is necessary to test each and every product. However, inpractical sense testing each and every product is almost a cumbersome process. Therefore, applying statistical techniques to the data obtained from sampling enables to estimate the quality of the entire batch. Thus, application of statistical techniques in quality control is refered to as statistical quality control.

The terms quality control and statistical quality control differ from one another. Quality control does not define the sample size and sampling techniques; where as the statistical quality control defines both in order to avoid the bias occuring with the quality control.

11.2 Statistical Quality Control

Statistical quality control refers to the statistical techniques employed for the maintenance of uniform quality in a continuous flow of manufactured products. Statistical quality control (SQC) is the term used to describe the set of statistical tools used by quality professionals to identify quality problems in the production process as well as in the product itself. It has been defined as the monitoring of quality by application of statistical methods in all stages of production. It is used to determine the relationship between factors and to make meaningful decisions on the basis of experimental evidence. It has been used to serve (1) as a basis for improved evaluation of materials through more representative sampling techniques, and (2) as a means of achieving sharp control in certain manufacturing processes. The procedure involves proper sampling of the product, determining the quality variation of the sample and making inferences on the entire batch under investigation from the observed data. Descriptive statistics can be helpful in describing quality characteristics of a product and a process. The most important descriptive statistics are measures of central tendency such as the mean, measures of variability such as the standard deviation and range.

11.3 Advantages of Statistical Quality Control

The advantages are as follows:

- Ensures an effective check on the quality of the product.
- Minimises the losses due to defective quality products.
- Make quality conscious in the employs concerned.
- Quality can be defended before governmental organisations.
- Decide when to take corrective measures.
- Determine the efficiency of the equipment and suggests suitable replacement if any.

11.4 Causes of Variation

As per statutory requirement, the pharmaceutical products produced should be of exactly of same quality and should conform to prescribed standards. The variation in drug product quality is mainly due to two types of causes.

1. **Random and Chance Causes.** These variations are due to inevitable causes of taking out random samples and drawing inferences. These small variations are natural and inherent in the manufacturing process. These variations cannot be removed or prevented and hence called as allowable variations
2. **Assignable and Preventable Causes.** These variations are generally caused by the defects and faults in the production design and manufacturing process. The source of variation can be identified and controlled. These variations are neither natural nor inherent in the manufacturing process

11.5 Types of Statistical Quality Control

It can be divided into two broad categories.

1. **Statistical Process Control:** Statistical process control methods extend the use of descriptive statistics to monitor the quality of the product manufactured in the

process of production. It detects whether the process is going on the desired fashion. It involves inspecting a random sample of the output from a process and deciding whether the process is producing products with characteristics that fall within a predetermined range. Statistical process control answers the question of whether the process is functioning properly or not. The most commonly used statistical process control tool for monitoring the production process is control chart.

2. **Statistical Product Control:** It is concerned with the classification of raw materials or finished products into acceptable, non acceptable. It involves the inspection of already produced products, whether they are fit for release.

11.6 Control Charts

Control charts, also known as Shewhart charts or process-behavior charts is the most successful statistical process control tool, originally developed by Walter Shewhart in the early 1920s. It is used to determine whether a manufacturing of dosage form in pharmaceutical industry is in a state of statistical control or not.

Walter Shewhart identified two types of variation that prevailed among all processes. Those are chance cause variations and assignable cause variations.

1. **Natural/Allowable/Chance Cause Variations:** These variations are unavoidable. They occur despite all the preventive measures and are random in nature. These variations must be within the acceptable limits and do not exert any marked influence on the quality of product.

   Ex: Quality of raw materials, Weight variation, Friability etc.

2. **Assignable/Chaotic/Erratic/Preventive Cause Variations:** When the differences between the observable and set standards of quality are large and significant then they are known as assignable causes. Ex: Machine wear, unskilled workers.

A control chart is a graph with the adding of statistically calculated upper and lower control limits drawn above and below the process average line (center line). The charts are used to present a visual graphic demonstration of instances when a process is beginning to go out of control. The purpose of the chart is to designate this trend in order that the system may be brought back into control. They contain an average line(centerline) representing the process average or expected performance, as well as upper and lower control limits that set bounds for an acceptable area of variation from the center line. A control chart is consists of six basic elements:

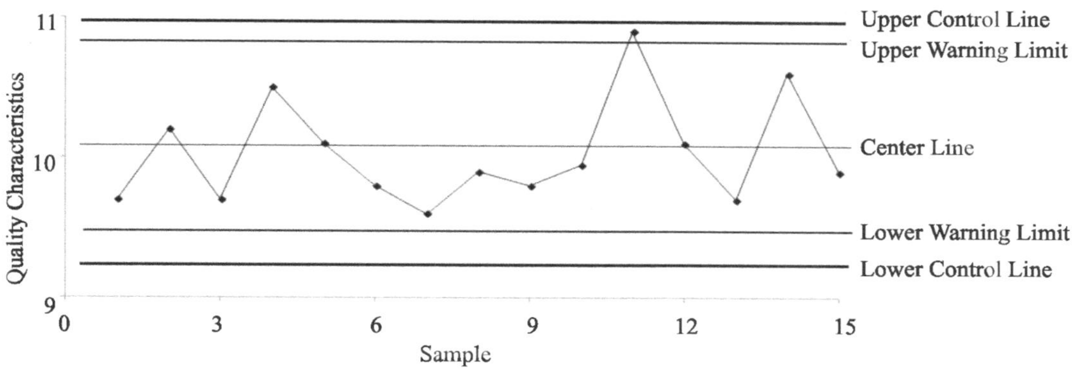

Fig. 11.1 Basic Elements of Control Chart

1. **Title**: This describes the information which is displayed.
2. **Legend**: Legend is information on how and when the data were collected.
3. **Measuring Parameters**
   - **A.** Parameter to be plotted on the x-axis (horizontal axis); e.g. Sample ID, order, sequence, run time, site, etc.
   - **B.** Parameter to be plotted on the y-axis (vertical axis); e.g. Volume of solution/ liquid filled in bottle, tablet hardness, weight of tablet.
4. **Centre Line:** Central line is drawn at the process characteristic mean, which is calculated from the data. It indicate the values of parameter which decide centrality of process i.e., arithmetic mean or any other measurement of central tendency.
5. **Control Lines**: Two dotted parallel lines represent the limits within which practically all the observed results should fall as long as the process is under normal variation (statistically control).
   - **A.** ***Upper Control Line (UCL):*** UCL is the line drawn parallel to central line from the y-axis at such a point which is considered to be an upper threshold value. This is plotted generally three times standard deviation above the central line.
   - **B.** ***Lower Control Line (LCL):*** LCL is the line drawn parallel to the central line from the y-axis at such a point which is considered to be a lower threshold value. This is plotted generally three times standard deviation below the central line. In a control chart, control line (limits) is calculated by the following formula

$$= \text{(Average Process Value)} \pm \{3 \times (\text{Standard Deviation})\}$$

The standard deviation should be due to unassigned process variation only.

6. **Data Collection Section**: The measurements or counts are recorded in the data collection section of the control chart prior to being graphed.

   Points inside the UCL and LCL are in-control-points (Fig 11.2), while points above the UCL or below the LCL lines are out-of-control points (Fig 11.3). It may have other optional features, including: Upper and lower warning limits, drawn as separate lines, typically two standard deviations above and below the center line and division into zones, with the addition of rules governing frequencies of observations in each zone.

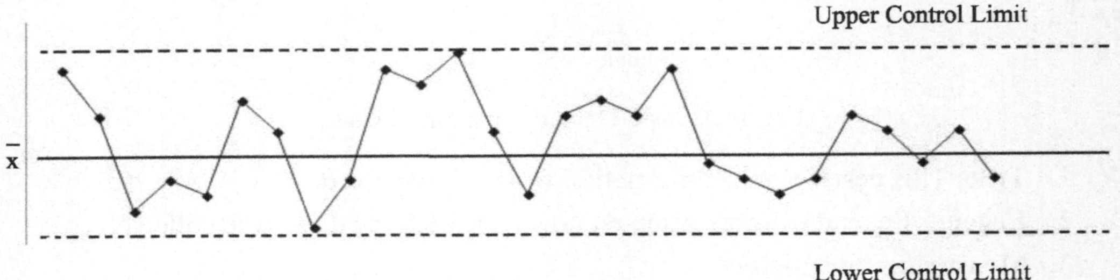

Fig. 11.2 Graphical Representation of "In Control" Process

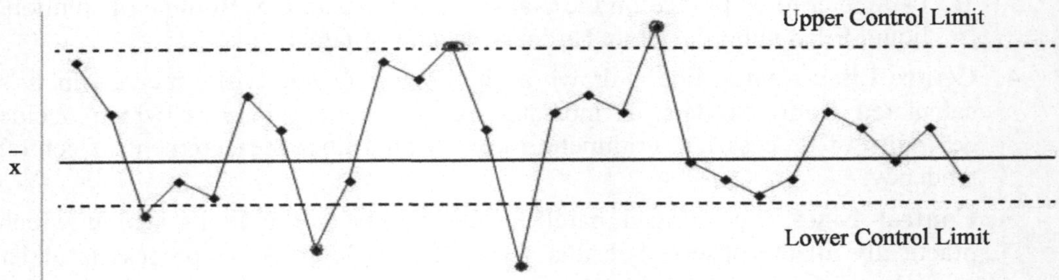

Fig. 11.3 Graphical Representation of "Out of Control" Process

Development of Control Charts

Select the quality characteristics to be controlled.

Establish the control limits basing on statistical principles (3 sigma limits)

Analyze the production process to know the cause of irregularities.

Determine how the inspection data are to be collected and recorded.

Interpretation of Control Charts

Arbitrarily, it can be declared as the process is out of control when the graph is above UCL and LCL.

Uses of Control Charts

1. It is used to monitor, control, and improve process performance over time by studying variation and its source.
2. It provides statistical ease for detecting and monitoring process variation over time.
3. It provides a tool for ongoing control of a process.
4. It differentiates special from common causes of variation in order to be a guide for local or management action.
5. It helps to improve a process to perform consistently and predictably to achieve higher quality, lower cost, and higher effective capacity.

Types of Control Charts

The different characteristics that can be measured by control charts can be divided into two groups: **variables** and **attributes.**

11.6.1 Variables Control Charts

It is used to monitor quantitative characteristics that can be measured and have a continuum of values, such as height, weight or thickness. These charts are based on a continuous distribution of measurements that can measure degrees of unacceptability.

Control charts for variables include x-Bar charts and R-charts.

X-Bar charts monitor the mean or average value of a product characteristic.

Mean (X-Bar) Charts

A mean control chart is often referred to as an X-***Bar chart.*** It is used to monitor changes in the mean of a process. The primary and foremost step in constructing of mean chart is the accurate measurement of the random sample. The sample size and frequency of sampling have to be considered. Generally not less than 4 or 5 samples are selected.

Construction of X-Bar Chart

The mean chart is constructed by taking sample number (n) on x-axis and statistics under study i.e., X-Bar on y-axis. The mean values of various samples are plotted against the coressponding sample number. The central dark line is drawn at $\bar{x}$. The Upper and lower control limits are represented as dotted lines at the computed values.

Inference of X-Bar Chart

If all the sample points fall with in the control limits, then the process is termed to be in statistical control and the variations that arise are due to chance cause that are inevitable. If one or more points fall out of control limits or when they show an unusual pattern of behaviour either towards upper control or towards lower control limit, then the process is not in statistical control. Such a situation indicates presence of assignable causes which should be identified and rectified immediately.

Control Limits for x-Bar Charts

$$\text{Upper Control Limit (UCL)} = \bar{\bar{x}} + z\sigma_{\bar{x}}$$

$$\text{Lower Control Limit (LCL)} = \bar{\bar{x}} - z\sigma_{\bar{x}}$$

$$\sigma_{\bar{x}} = \frac{\sigma}{\sqrt{n}}$$

Where $\bar{\bar{x}}$ = the average of the sample means $\left(\bar{\bar{x}} = \frac{\bar{x}_1 + \bar{x}_2 + \cdots \bar{x}_K}{K}\right)$; z = standard normal variable;

σ = standard deviation; n = sample size (number of observations per sample)

Problem **1:** A drug inspector at the pharmaceutical company has taken 25 samples with 4 observations each of the volume of bottles filled. The data was given below. The standard deviation of bottling operation was 0.14 ounces. By using this information develop the control limits (three standard deviations) for the bottling operation.

Sample Number	**Observations (Bottle Volume in Ounces)**			
	1	**2**	**3**	**4**
1	15.86	16.03	15.84	15.94
2	16.13	15.99	15.86	16.00
3	15.83	15.94	16.00	15.91
4	15.93	15.85	15.74	16.20
5	16.11	16.01	15.82	16.00
6	15.74	15.86	16.21	16.10
7	15.94	16.01	16.03	16.14
8	15.75	16.01	16.21	15.86
9	15.83	15.98	15.98	16.04
10	15.64	15.86	15.94	15.89
11	16.01	16.10	16.20	15.89
12	15.72	15.85	16.12	16.15
13	15.85	15.74	15.76	15.98
14	15.73	15.84	15.96	16.10
15	16.01	16.10	16.20	15.89
16	16.12	16.08	15.94	15.83
17	16.01	15.68	15.93	15.81
18	16.04	16.11	16.12	15.78
19	16.05	16.12	15.84	15.92
20	15.93	15.92	16.09	16.12
21	15.88	16.00	16.02	16.11
22	15.89	15.89	15.98	15.82
23	16.05	15.73	15.93	15.73
24	15.86	15.89	16.01	16.01
25	15.78	15.92	15.98	16.08

Solution: The center line of the control data is the average of the samples:

$$\bar{\bar{x}} = \frac{398.75}{25} = 15.95$$

The control limits are

$$UCL = \bar{\bar{x}} + z\sigma_{\bar{x}} = 15.95 + 3\left(\frac{0.14}{\sqrt{4}}\right) = 16.16$$

$$LCL = \bar{\bar{x}} - z\sigma_{\bar{x}} = 15.95 - 3\left(\frac{0.14}{\sqrt{4}}\right) = 15.74$$

The resulting control chart is:

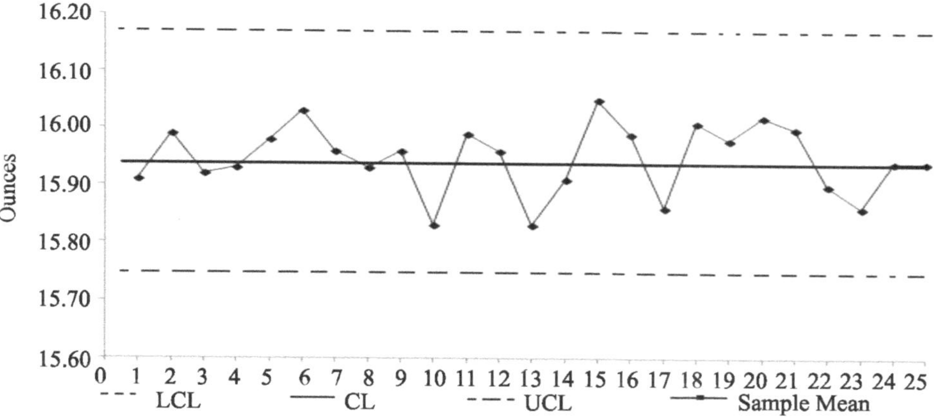

Fig. 11.4

Mean (X-Bar) Charts can also be Calculated from the Sample Range

Control limits for X-Bar charts using sample range as an estimate of variability

$$\text{Upper Control Limit (UCL)} = \bar{\bar{x}} + A_2\bar{R}$$

$$\text{Lower Control Limit (UCL)} = \bar{\bar{x}} - A_2\bar{R}$$

where $\bar{\bar{x}}$ = average of the sample mean and $\bar{R}$ = average range of the samples.

A_2 factor can be obtained from Table 11.1. Notice that A_2 is a factor that includes three standard deviations of ranges and is dependent on the sample size being considered.

Table 11.1 Factors for three sigma control limits of X-Bar and range chart

Sample Size n	Factor for X-Chart	Factor for R-Chart	
	A_2	D_3	D_4
2	1.88	0	3.27
3	1.02	0	2.57
4	0.73	0	2.28
5	0.58	0	2.11
6	0.48	0	2.00
7	0.42	0.08	1.92
8	0.37	0.14	1.86
9	0.34	0.18	1.82
10	0.31	0.22	1.78
11	0.29	0.26	1.74
12	0.27	0.28	1.72
13	0.25	0.31	1.69
14	0.24	0.33	1.67
15	0.22	0.35	1.65
16	0.21	0.36	1.64
17	0.20	0.38	1.62
18	0.19	0.39	1.61
19	0.19	0.40	1.60
20	0.18	0.41	1.59
21	0.17	0.43	1.58
22	0.17	0.43	1.57
23	0.16	0.44	1.56
24	0.16	0.45	1.55
25	0.15	0.46	1.54

***Problem* 2:** If the average range for the twenty-five samples is 0.29 ounces (Data is same as that of Problem 1) and the 25 average mean of the observations is 15.95 ounces, develop three-sigma control limits for the bottling operation.

Solution:

$$\bar{\bar{x}} = 15.95 \text{ ounces} \qquad \bar{R} = 0.29$$

The value of A_2 is obtained from Table 11.1.

For $n = 4, A_2 = 0.73.$

This leads to the following limits:

The center of the control chart = CL = 15.95

$$UCL = 15.95 + (0.73)(0.29) = 16.16$$

$$LCL = 15.95 - (0.73)(0.29) = 15.74$$

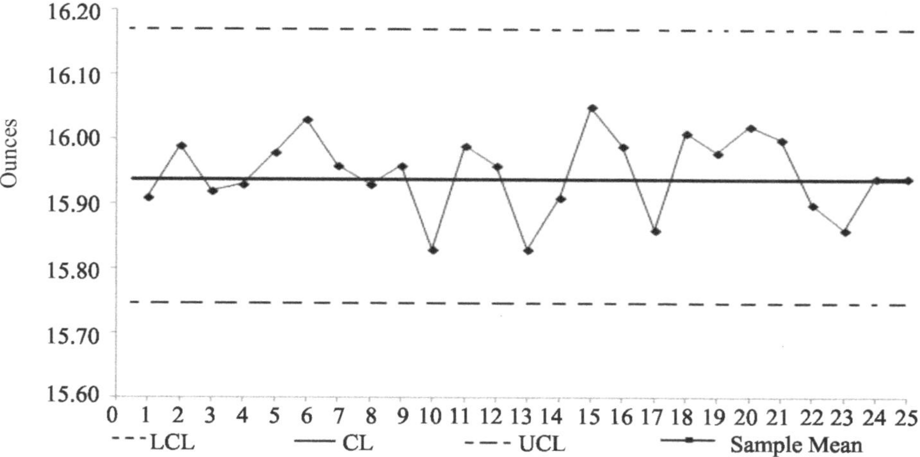

Fig. 11.5

Range(R) Charts: Range(R) charts are another type of control chart for variables. Whereas X-bar charts measure shift in the central tendency of the process, range charts monitor the dispersion or variability of the process. The method for developing and using R-charts is the same as that for X-bar charts.

Construction of R Chart: The range chart is constructed by taking sample number (n) on x-axis and range on y-axis. The range values of various samples are plotted against the coressponding sample number. The central dark line is drawn at $\overline{R}$. The upper and lower control limits are represents as dotted lines at the computed values.

Inference of R Chart: If all the sample points fall with in the control limits, then the process is termed to be in statistical control. If one or more points fall out of control limits or when they show an unusual pattern of behaviour either towards upper control or towards lower control limit, then the process is not in statistical control. Such a situation indicates presence of assignable causes which should be identified and rectified immediately.

The center line of the control chart is the average range, and the upper and lower control limits are computed as follows:

$$CL = \overline{R}$$

$$UCL = D_4\overline{R}$$

$$LCL = D_3\overline{R}$$

The values for D_4 and D_3 can be obtained from Table 11.1.

***Problem* 3:** Construct the range chart for the data shown in the Problem 1.

Solution:

$$\overline{R} = \frac{7.17}{25} = 0.29$$

$$n = 4$$

$$D_4 = 2.28$$

$$D_3 = 0$$

$$UCL = D_4\overline{R} = 2.28(0.29) = 0.6612$$

$$LCL = D_3\overline{R} = 0(0.29) = 0$$

The resulting control chart is

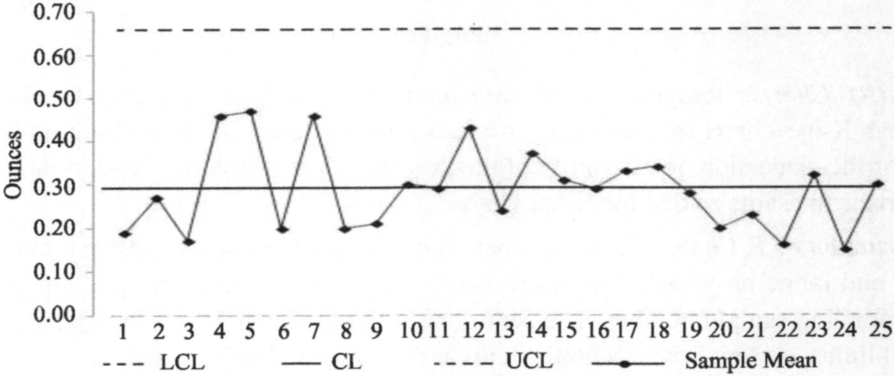

Fig. 11.6

Advantages of Variable Control Charts

1. Variable control charts are more sensitive than attribute control charts.
2. Variable control charts may alert us to quality problems before any actual "un–acceptable" condition (as detected by the attribute chart) will occur.
3. Variable control charts are leading indicators of trouble that will sound an alarm before the number of rejects increases in the production process.
4. It is used for the statistical optimization study of drug formulations in pharmaceutical industry.

11.6.2 Attribute Control Charts

It is used when measurements are qualitative, for example, accept/reject. It is used to monitor characteristics that have discrete values and can be counted. Often they can be evaluated with a simple yes or no decision. Examples include color, taste, or smell. Control charts for attributes include p-charts and c-charts. The desirable properties for applying this chart are as follows.

1. Choose the samples at random.
2. Fixed number of samples to be taken each time for inspection.
3. Each sample is evaluated for acceptance/rejection.
4. Each sample must be independent.

P-Charts: P-charts are used to measure the proportion that is defective in a sample. The computation of the centre line as well as the upper and lower control limits is similar to the computation for the other kinds of control charts. The centre line is computed as the average proportion defective in the population. This is obtained by taking a number of samples of observations at random and computing the average value of p across all samples.

Construction of P Chart

The P chart is constructed by taking sample number (n) on x-axis and static P on y-axis. The fraction defectives are plotted against the corresponding sample number. The central dark line is drawn at $\bar{P}$. The upper and lower control limits are represent as dotted lines at the computed values.

Centre line $\bar{P}$ = total number of defects/ total number of observations

Upper and lower control lines are calculated by using the following formula

$$UCL = \bar{P} + z\sigma_p$$

$$LCL = \bar{P} - z\sigma_p$$

Where z = standard normal vaiable.

$$\sigma_P = \sqrt{\frac{\bar{P}(1-\bar{P})}{n}}$$

$\bar{P}$ = the sample proportion defective

n = the sample size

Inference of P Chart

If all the sample points fall with in the control limits, then the process is termed to be in statistical control. If one or more points fall out of control limits or when they show an unusual pattern of behavior either towards upper control or towards lower control limit, then the process is not in statistical control. Such a situation indicates presence of assignable causes which should be identified and rectified immediately.

***Problem* 4:** A quality control analyst at a pharmaceutical industry has inspected number of defective tablets in twenty random samples with twenty observations each. Following are the number of defective tablets found in each sample. Construct the p-chart for the data.

Sample number	Number of defective tablets	Number of observations sampled	Fraction defective
1	3	20	0.15
2	2	20	0.10
3	1	20	0.05
4	2	20	0.10
5	1	20	0.05
6	3	20	0.15
7	3	20	0.15
8	2	20	0.10
9	1	20	0.05
10	2	20	0.10
11	3	20	0.15
12	2	20	0.10
13	2	20	0.10
14	1	20	0.05
15	1	20	0.05
16	2	20	0.10
17	4	20	0.20
18	3	20	0.15
19	1	20	0.05
20	1	20	0.05

Solution:

$$\text{CL} = \overline{\text{P}} = \frac{\text{total number of defective}}{\text{total number of observations}} = \frac{40}{400} = 0.10$$

$$\sigma_P = \sqrt{\frac{\overline{\text{P}}\left(1-\overline{\text{P}}\right)}{\text{n}}} = \sqrt{\frac{(0.10)(0.90)}{20}} = (0.67)$$

$$\text{UCL} = \overline{\text{P}} + \text{z}(\sigma_P) = 0.10 + 3(0.067) = 0.301$$

$$\text{LCL} = \overline{\text{P}} - \text{z}(\sigma_P) = 0.10 - 3(0.067) = -\ 0.1101 \rightarrow 0$$

The resulting control chart is as follows

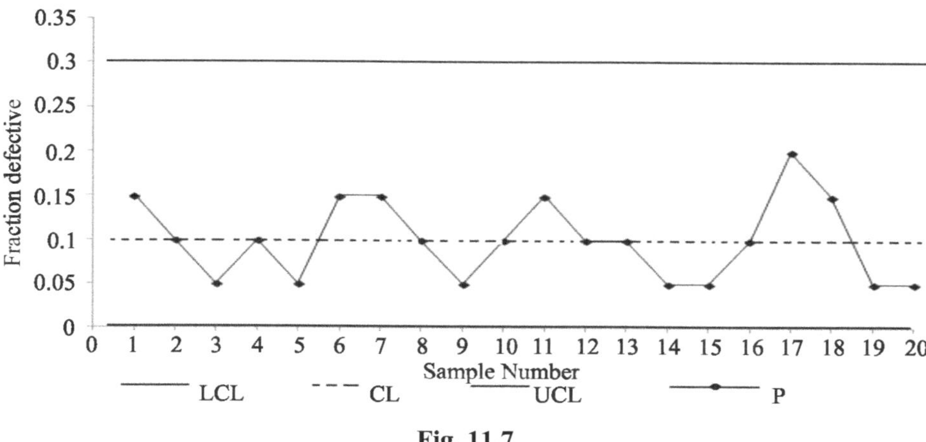

Fig. 11.7

np- Charts

A np chart shows the actual number of defectives found in each sample. If the number of items inspected on each occasion is the same, the np charts may be more convenient and meaningful than p-chart. The construction and interpretation is similar to p-chart. The actual number of defective (np) in samples of fixed size (n) is plotted and the central line is drawn at $n\bar{P}$. The following values are required to construct a np-chart.

- n = constant sample size
- The number of defectives in each sample, d = np.
- The average number of defectives per sample of a constant size $n\bar{P}$ = total number of defectives of all samples/number of samples inspected.
- UCL np and LCL np values which are computed with the following formulae

$$\text{UCL np} = n\bar{P} + 3\sqrt{n}\ \bar{P}(1 - \bar{P})$$

$$\text{LCL np} = n\bar{P} - 3\sqrt{n}\ \bar{P}(1 - \bar{P})$$

- Plot the values of $n\bar{P}$ and mark the control limits.

The np chart can be constructed for the data given in problem 4.

$$n = 20,\ \bar{P} = 0.1,\ np = 2$$

$$\text{UCL np} = 2 + 3\sqrt{2\ (0.9)} = 6.02$$

$$\text{LCL np} = 2 - 3\sqrt{2\ (0.9)} = -2.02 = 0$$

The np chart drawn with the above data is shown in the following Fig 11.8.

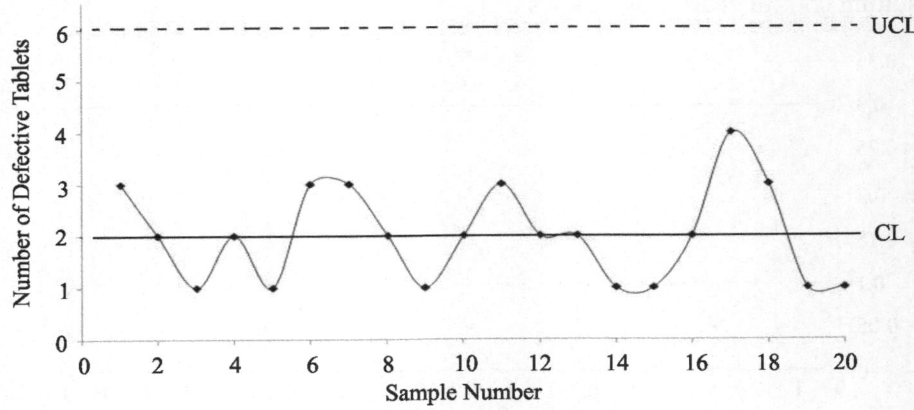

Fig. 11.8

C-Charts

C-Charts: It is used when the number of defects per individual units are counted instead of classifying the unit to be defective or non defective. In C-chart, the size of sample remains constant. For example, we can count the number of complaints from quality control department in a month, the number of bacteria on a petri dish.

Construction of C Chart: The C chart is constructed by taking sample on x-axis and statistics on y-axis. The range values of various samples are plotted against the coressponding sample number. The central dark line is drawn at $\overline{c}$. The Upper and lower control limits are represent as dotted lines at the computed values.

Inference of C Chart: If all the sample points fall with in the control limits, then the process is termed to be in statistical control. If one or more points fall out of control limits or when they show an unusual pattern of behaviour either towards upper control or towards lower control limit, then the process is not in statistical control. Such a situation indicates presence of assignable causes which should be identified and rectified immediately

The average number of defects, $\overline{c}$, is the center line of the control chart.

The upper and lower control limits are computed as follows.

$$UCL = \overline{c} + z\sqrt{\overline{c}}$$

$$LCL = \overline{c} - z\sqrt{\overline{c}}$$

Where z = standard normal variable; $\overline{c}$ = the average number of defects

***Problem* 5:** The number of weekly quality control department complaints is monitored at a pharma plant using C-chart. Complaints have been recorded over the past 20 weeks. Develop three sigma control limits using the following data.

1	2	3	4	5	6	7	8	9	10	11	12	13	14	15	16	17	18	19	20
3	2	3	1	3	3	2	1	3	1	3	4	2	1	1	1	3	2	2	3

Solution:

The average number of complaints per week is $\frac{44}{20} = 2.2$, Therefore, $\bar{c} = 2.2$

$$UCL = \bar{c} + z\sqrt{\bar{c}} = 2.2 + 3\sqrt{2.2} = 6.65$$

$$LCL = \bar{c} - z\sqrt{\bar{c}} = 2.2 - 3\sqrt{2.2} = -2.25 \rightarrow 0$$

As in the previous example, the LCL is negative and should be rounded up to zero. Following is the control chart for this example.

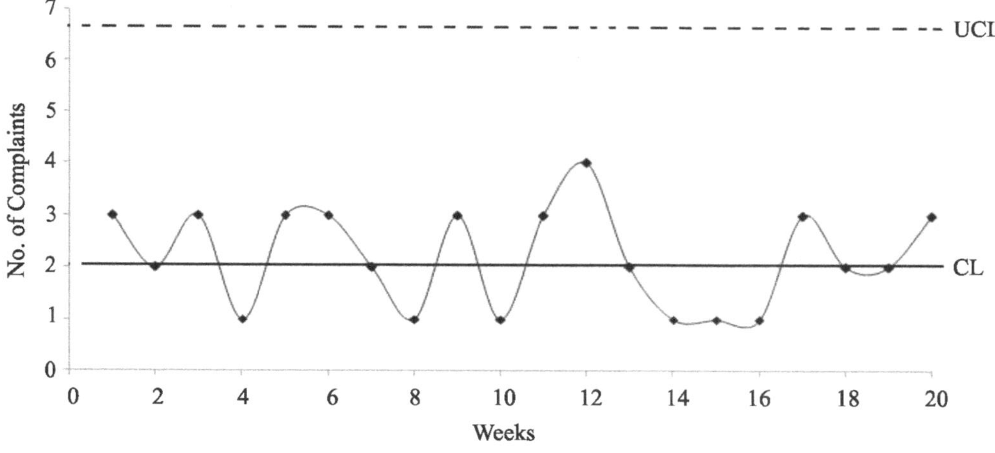

Fig. 11.9

11.7 Acceptance Sampling

Acceptance sampling is the process of randomly inspecting a sample of goods and deciding whether to accept the entire lot based on the results. Sampling may be defined as the process of removing an appropriate number of items from a population in order to make inferences on entire population. Representative of material to be sampled are drug substances, intermediate products and finished products before, during and after manufacturing and packaging operations. Acceptance sampling determines whether a batch of goods should be accepted or rejected. Acceptance sampling different from statistical process control is performed before or after the process, rather than during the process.

Acceptance sampling before the process involves sampling materials received from a supplier.

Sampling after the process involves sampling finished items that are to be shipped either to market or to a distribution center.

When the defects found in a particular sample are well below the desired value, then the lot is passed or accepted, otherwise it is rejected. Therefore, acceptance sampling helps to decide whether to accept or reject a lot without actually having to analyze the entire lot.

It is the technique of choice in the following conditions.

1. When large number of items are to be inspected in short time period.
2. When 100% inspection of the items proves to be expensive.
3. When the inspection is of destructive nature.
4. When there are high chances of errors during the inspection of large number of items, called either due to fatigue or boredom.

Acceptance sampling can be used for sampling both attributes and variables. However the former is widely used.

Sampling Plans

A good statistical sampling plan should able to pass a high percentage of acceptable batches and reject the unacceptable batches. Sampling plan is one of the key factors is acceptance sampling. It indicates the following.

- Lot size (N)
- Sample size (n)
- No. of samples to be chosen
- Acceptance or rejection criteria

There are three types of sampling plans which differ with respect to their nature and the number of samples to be taken.

The construction of a statistical sampling plan normally requires the following quality standard specifications.

1. An acceptable quality level(AQL)
2. An unacceptable quality level(UQL)
3. The risk or error (α), which is the probability of rejecting a good batch.
4. The risk or error (β), which is the probability of accepting a bad batch.

The usual approach is the determination of desirable AQL, UQL, α, β, and subsequent computation of sample size and acceptance criteria. In case of potent medicaments, the AQL and UQL are close together α, β, are very small and large sample is required.

Single Sampling Plan: In this plan one sample is picked from each lot and every item in the single random sample is analyzed. Therefore, the number of items in this random sample is actually the sample size (n) which would represent the acceptance value c i.e., the maximum number of acceptable defective in sample. If the number of defectives in a sample exceeds c, then that lot is rejected or otherwise undergoes 100% inspection. If the defective number lies below c, then the lot is accepted.

In single sampling plan, although only one sample is drawn, the sample size is relatively larger than the number of observations in double or multiple sampling. It is the method of choice when the cost of obtaining a sample is higher than its analysis.

Double Sampling Plan: In this type of plan, a second sample can be drawn from the lot when the analysis of the first sample gives inconclusive results. For instance, a lot would be accepted or rejected when the number of defectives in a first sample is lower or higher than c, respectively. However, if the result is intermediate between the two, then the second sample is drawn, analyzed and then accepted or rejected depending upon the combined results of both the samples. In double sampling; two values are given for the number of defective items in a sample.

c_1 = lower level $\qquad$ c_2 = higher level

If the number of defectives of first sample $\leq c_1$, the lot is accepted, but if it is $\geq c_1$ the lot is rejected. However if it lies in between c_1 and c_2, then second sample is drawn and analyzed. If the number of defectives of both first and second sample is less than a third value c_3, then the lot is accepted or vice-versa.

Sequential (multiple) Sampling Plan: In this plan, the sample size and two limits for each and every sample is specified. The lot is rejected when the cumulative number of defectives i.e., defectives of this sample + defectives of all previous samples exceeds the specified upper limit. The lot passes or accepted when cumulative number of defectives is less than the specified lower limit for that sample. However, if it lies in between upper and lower limits, then another sample is drawn and the process is repeated until the lot is either accepted or rejected.

Double or multiple sampling is preferred when the cost of analysis of sample is high.

CHAPTER 12

FACTORIAL DESIGNS

12.1 Introduction

Factorial designs are used to study the influence of multiple factors simultaneously. For example the dissolution rate of a tablet is influenced by the concentrations of binder, disintegrant, lubricant. So to study the influence of each individual factor and their interactions, these factorial designs are used. A factor is an assigned variable found to influence the experimental objective. It may be either qualitative or quantitative. The qualitative factors are usually represented with name. The quality of a dosage form may vary from batch to batch, laboratory to laboratory or equipment to equipment. Then the experiment may be designed to investigate the affect of these factors on quality/performance of the dosage forms. Such qualitative factors are represented as names (batch, laboratory, equipment). Some of the factors like applied compression force, concentration of excipients are designated with numerical value and they are examples for quantitative variables.

The numerical values of the factors included in the experiment are referred as level. Generally these levels are decided based on earlier experimental findings and they are usually kept at their low concentrations and high concentrations which are found to influence the experimental objective. The effect of factor is determined by observing the changes in response caused by the variations in level (s) of the factor. If the additive value of the factors effects was not observed, then it is called as interaction. The interaction may be either synergistic or antagonistic effect. For example the disintegration time and dissolution rate of a water insoluble drug may be influenced by the concentration of complexing agent and surfactant. To investigate the impact of these two factors at two levels (0% and high concentration of each independent variable), four sets of experiments (formulations) are required. This experimental design is shown below.

Table 12.1

Symbol	Formulation
1	Formulation without complexing agent and surfactant (0% concentration of each variable)
a	Formulation containing the selected ratio of complexing agent and not having surfactant
b	Formulation containing surfactant
ab	Formulation contains both complexing agent and surfactant

In the above example, the initial level of the assigned factors complexing agent and surfactant are kept as zero concentration because the formulations can be produced without these excipients. If the above formulations were subjected to dissolution studies and the observed dissolution rates are 15 mg/min in case of complexing agent and 20 mg/min in case of surfactant, then the theoretical dissolution rate expected from the formulation containing both the excipients should be 35 mg/min. The practically observed dissolution rate is more than that then it is referred as synergistic activity and the observed dissolution rate is less than the theoretical value then it is treated as antagonistic affect.

The other example for factorial design is the stability studies of a drug. The drug stability in presence of temperature, humidity and in presence of both the conditions can be studied with the design showed as follows.

Table 12.2

Symbol	Storage Condition
1	Room temperature and humidity
a	Temperature 40 °C and normal humidity conditions
b	Relative humidity of 75% RH and room temperature
ab	40 °C temperature and 75% RH.

In the above example it is not possible to keep zero level for the variable. The effect of each factor on decomposition rate constant is showed in the following figure.

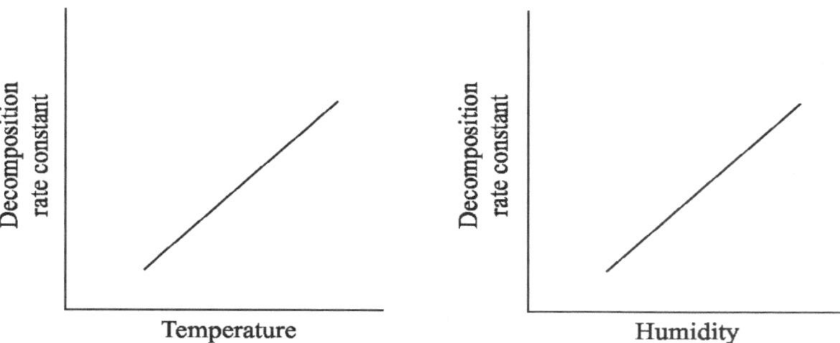

Fig. 12.1

In the above graphs, linear relationships are observed to get more precise information and more than two levels are required to obtain the effect of level on response. For example, the decomposition rate may be less influenced at low temperatures as shown in the following figure.

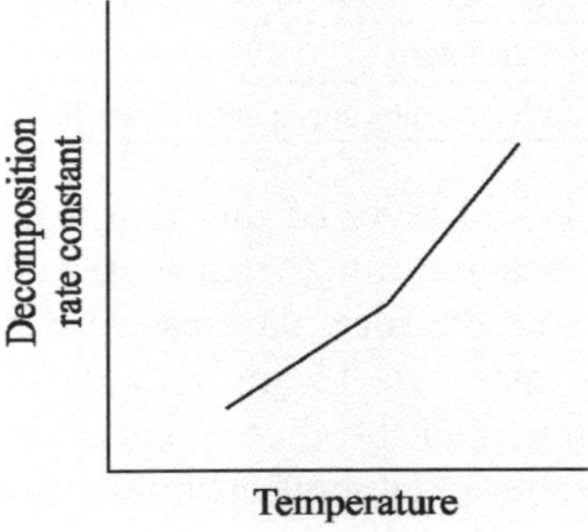

Fig. 12.2

The interaction can be observed from a formulation containing the factors binding agent and disintegrant concentrations. The experimental trial for this example is as follows.

Table 12.3

Symbol	Composition of Formulation
1	Contains low concentrations of disintegrate and binder
a	Contains high concentration of binder
b	Contains high concentration of disintegrate
ab	Contains high concentrations of binder & disintegrates

If the above formulations are subjected to dissolution studies, then the following observations are expected.

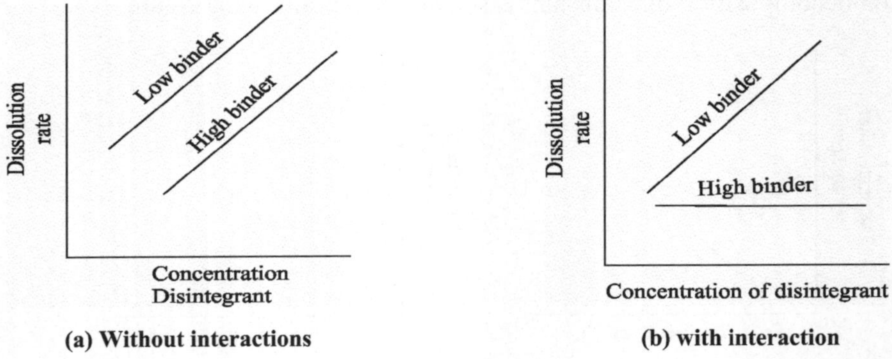

(a) Without interactions (b) with interaction

Fig. 12.3

12.2 Advantages of Factorial Design

1. Useful for detailed analysis of interactions among the factors.
2. Runs are made for all possible combinations of factors levels
3. These designs have maximum application in estimation of main effect in absence of interaction.
4. The experimental findings are effectively interpretive as the each factor effect is measured over varying levels of other factors.
5. Suitable for detection of interactions (if any) existing between the factors
6. The estimated effects are independent of effect of other factor.

A common experimental design for factorial design is the design with all input factors set at levels each. These levels are called high and low or +1 and –1. Such factorial design is called a full factorial design in two levels. If there are K factors, each at 2 levels, a full factors design has 2^K runs. If an experiment is influenced by 3 factors and each is fixed at 2 levels then the number of runs are 2^3 (8 runs). In case of 5 factors it is 32 and incase of 8 factors it is 256. When the number of factors is 5 or greater, a full factorial design requires a large number of runs and is not very efficient.

The significance of factorial design can be explained with the following example. The effect of exercise, drug on blood glucose levels was determined. Group A was under exercises group B was received drug and group C was exposed to exercise and received drug. The observed results were as follows.

Exercise decreased blood glucose levels to 5%

Drug decreased blood glucose levels to 10%

Combination of both approaches reduced blood glucose levels to 15%

The above result indicates there was no interaction between the two factors as the effects are additive. However the subjects neither performing exercise nor received drug may show a reduction in 5% blood glucose level due to fasting or other such conditions. It is happened then the conclusion is that the exercise is not having blood glucose reduction effect. Thus the precise interpretation of the result is possible only with the presence of fourth group (without exercise and drug). Inclusion of fourth group in to the experiment results complete factorial design and missing of this group may result incorrect results. Thus the interactions can be clearly calculated based on the results observed from these four groups.

12.3 Notation in Factorial Experiments

The following general notation is recommended in factorial experiments.

1. When all the factors are at low levels, then it is represented as 1.
2. If factor A is higher level and other factors are low level, it is represented as A.

3. If factors A and B are higher levels then it is represented as ab.
4. If factor A, B, C are higher levels, then it is represented as abc.

12.4 Guidelines in Factorial Design

For effective implementation of factorial experimental, the following guidelines are recommended.

1. The objectives of experiment should be clearly defined.
2. The factors influencing the selected experiment should be carefully selected.
3. The factors out of interest should be kept constant while conducting the experiment with interested variables.
4. The choice of qualitative factors should be clearly defined. For example in an analytical method, the influence of spectrophotometer make on absorbance values is supposed to be studied then the name of the spectrophotometer to be used in the experiment should be clearly specified.
5. In case of quantitative factors, 2 levels in case of linear response and 3 levels in case of curved response are preferable. In case of 2 levels, generally ¼ th and ¾ th of usual ranges are used. For example if an excipient is used in the concentration of 5-10%, then it can be divided into 4 equal parts as 6%, 7%, 8%, 9% . 6% and 8% concentrations of excipient are preferable for experiment if these concentrations factor linearly alters the response (effect).
6. To achieve unbiased results, the trials of the factorial experiments should be done randomly.
7. The residual error observed due to extraneous variables should be controlled to get the more sensitive effect contributed by the chosen factor. For example, if an experiment is designed to study the influence of polymer concentration and the composition of solvent (changes in pH or binary/ternary solvent system) then the experiment is conducted as follows.

Table 12.4

Trial	Condition
1	1% Polymer in water
a	2% Polymer in water
b	1% Polymer in water: ethanol (60 : 40)
ab	2% Polymer in water: ethanol (60: 40).

In the above experiment, the viscosity may get modified with temperature. So until unless temperature is not maintained constant, the result obtained from the above factorial design can be properly interpretive.

Working Examples for a 2^2 Factorial Design

The influence of two factors namely paddles speed and the composition of dissolution medium on the dissolution rate of drug was conducted. The experimental design and the results obtained are given in following table.

Table 12.5

Notation	Dissolution study condition	Observed released rate (mg/hr)
1	50 rpm, water	2
a	100 rpm, water	4
b	50 rpm, 6.8 pH phosphate buffer	12
ab	100 rpm, 6.8 pH phosphate buffer	18

Working Example for 2^3 Factorial Design

The effect of three factors namely concentration of binder, disintegrate and lubricant on disintegration time of an uncoated tablet was investigated with the factorial design. The experimental condition and the results are furnished below. Interpret ate the results.

Table 12.6

Notation	Concentration of excipients	Disintegration time (min)
1	1% binder, 5% disintegrate, 0.5% Lubricant	10
a	2% binder 5% disintegrant, 0.5% Lubricant	12
b	1% binder 10% disintegrant 0.5% Lubricant	6
c	1% binder, 5% disintegrate, 1% Lubricant	13
ab	2% binder, 10% disintegrate, 0.5% Lubricant	11
bc	1% binder, 10% disintegrate, 1% Lubricant	9
ac	2% binder, 5% disintegrant, 1% Lubricant	14
abc	2% binder, 10% disintegrate, 1% Lubricant	16

In the above problem, the factors and the levels are as follows.

Table 12.7

Factor	Low level (% w/w)	High level (% w/w)
A Binder	1	2
B Disintegrant	5	10
C Lubricant	0.5	1

The effects can be calculated by applying the signs (+ or –) arithmetically for each of the eight responses as shown in the following table. The factor at highest level is denoted with + sign and the lowest level is represented with – sign. The sign in the columns containing interaction terms a and b is obtained by multiplying the signs of the individual components.

For the above problem, the allotted signs for representation of various factors are shown below.

Table 12.8

Factor Combination	Level of factor in experiment			Interaction			
1	A	B	C	AB	BC	AC	ABC
1	–	–	–	+	+	+	–
a	+	–	–	–	+	–	+
b	–	+	–	–	–	+	+
c	–	–	+	+	–	–	+
ab	+	+	–	+	–	–	–
bc	–	+	+	–	+	–	–
ac	+	–	+	–	–	+	–
abc	+	+	+	+	+	+	+

12.5 Interpretation of the Factorial Design Experimental Results

The average effect is calculated by multiplying the response with the sign of each run in the column and dividing the result by 2^n, n is number of factors. For illustration, the main effect of factor A is calculated –1, where by following the second column of the above table. It is –1 + a – b + ab – c + ac – bc + abc/4

The main effect and the interaction effect of the parameters are calculated as follows. The main effect is calculated by subtracting the average results of the selected factor observed when the factor is at its low level from the results observed while the same factor is at its high level.

Main effect of A = [(a + ab + ac + abc)/4] – [(1) + b + c + bc/4]

12 + 11 + 14 + 16/4 – 10 + 6 + 13 + 9/4 = 53/4 + 38/4 = 13.5-9.5 = 4 minutes

Increasing the concentration of binder from 1% to 2% results an increase in disintegration time of 4 minutes.

Main effect of B= –1 – a + b + ab – c – ac + bc + abc = [(b + ab + bc + abc)/4] – [(1) + a + c + ac/4]

6 + 11 + 9 + 16/4 – 10 + 12 + 13 + 14/4

42/4 – 49/4

10.5 – 12.5 = –2.5 minutes

Increasing the concentration of disintegrating agent from 5% to 10% results a decrease in disintegration time of 2.5 minutes

Main effect of C= – 1 – a – b – ab + c + ac + bc + abc = [(c + ac + bc + abc)/4] – [(1) + a + b + ab/4]

13 + 14 + 9 + 16/4 – 10 + 12 + 6 + 11/4

52/4 – 39/4 = 13 – 9.75 = 3.25. Increasing the concentration of lubricant from 0.5% to 1% results an increase in disintegration time of 3.25 minutes

Interaction effect of the factors A and C

1/4 [(abc + ac – bc – c) – (ab + a – b – 1)]

1/4 [(16 + 9 – 12 – 6) – (11+ 6 – 12 – 10)] = 0.25 minutes

The disintegration time of tablet in presence of high concentration of binder and low concentration of lubricant (factor A) is increased to 4 minutes and low concentration of binder and high concentration of lubricant (factor C) is increased the disintegration time to 3.75 minutes. The disintegration time is increased to 0.25 minutes in presence of high concentrations of binder and lubricant indicating an interaction effect between the two factors. The results can be summarized as follows.

1. The effect of lubricant on disintegration time is more at low binder concentration.
2. The effect of binder on disintegration time is more at low lubricant concentration.

Interaction effect of the factors B and C

1/4 [(abc + bc – ac – c) – (ab + b – a – 1)]

1/4 [(16 + 9 – 14 – 13) – (11+ 6 – 12 – 10)] = 0.75 minutes

The disintegration time of tablet in presence of high concentration of disintegrant and low concentration of lubricant (factor B) is decreased to 2.5 minutes and low concentration of disintegrant and high concentration of lubricant (factor C) is increased the disintegration time to 3.75 minutes. The disintegration time is increased to 0.75 minutes in presence of high concentrations of disintegrant and lubricant indicating an interaction effect between the two factors.The results can be summarized as follows.

1. The effect of lubricant on disintegration time is more at low disintegrant concentration.
2. The effect of disintegrant on disintegration time is more at low lubricant concentration.

Interaction effect of the factors A and B

1/4 [(abc – bc – ac + c) – (a + b – ab – 1)]

1/4 [(16 – 9 – 14 + 13) – (12 + 6 – 11 – 10)] = 2.25 minutes

The disintegration time of tablet in presence of high concentration of binder and low concentration of disintegrant (factor A) is increased to 4 minutes and high concentration of disintegrant and low concentration of binder (factor B) decreased the disintegration time to 2.5 minutes. The disintegration time is increased to 2.25 minutes in presence of high concentrations of disintegrant and binder indicating an interaction effect between the two factors. The results can be summarized as follows

1. The effect of disintegrant on disintegration time is high at low binder concentration.
2. The effect of binder on disintegration time is more at low disintegrant concentration.

Interaction effect of the factors A, B and C

1/4 [(abc – bc – ac + c) – (ab – b – a + 1)]

1/4 [(16 – 9 – 14 + 13) – (11 – 6 – 12 + 10)] =0.75 minutes

The results can be summarized as follows

1. Incorporation of high concentration of lubricant (ABC) in to the tablet containing high concentrations of binder and disintegrant (AB) reduced the disintegration time from 2.25 minutes to 0.75 minutes and interactions are significant.
2. Incorporation of high concentration of binder (ABC) in to the tablet containing high concentrations of lubricant and disintegrant (BC) does not changed the disintegration time and no interactions are observed.
3. Incorporation of high concentration of disintegrant (ABC) in to the tablet containing high concentrations of binder and lubricant (AC) increased the disintegration time from 0.25 minutes to 0.75 minutes and interactions are significant.

Data analysis

A method devised by Yates is used to analyze the data generated from 2^n factorial experiments.

Consider the above data:

Table 12.9

Notation	Disintegration time (min)
1	10
a	12
b	6
c	13
ab	11
bc	9
ac	14
abc	16

The data are first tabulated in standard order, the data are added in pair followed by taking differences in pairs as shown in column of the table.

$$10 + 12 = 22$$
$$6 + 13 = 19$$
$$11 + 9 = 20$$
$$14 + 16 = 30$$
$$10 - 1\ 2 = -2$$
$$6 - 13 = -7$$
$$11 - 9 = 2$$
$$14 - 16 = -2$$

This addition and subtraction process is repeated sequentially on the n columns (n is number of factors, three columns for three factors) the process is repeated in column (2), operating on the results in column (1) of below table.

Finally, the process is repeated operating on column (2) to form column (3), then column (3) is divided by 2^{n-1} ($2^{n-1} = 4$ for 3 factors) to obtain average effect. The mean squares for the ANOVA are obtained by dividing the square of column (3) by 2^n

Table 12.10 Yates analysis of the factorial tableting experiment for analysis of variance.

Notation	Disintegration	(1)	(2)	(3)	Effect (3)/4	Mean square $(3)^2/8$
1	10	(22)	41	91	--	--
a	12	19	50	–9	–2.25	10.125
b	6	20	–9	–7	–1.75	6.125
c	13	30	0	9	2.25	10.125
ab	11	–2	3	–9	–2.25	10.125
bc	9	–7	–10	–9	–2.25	10.125
ac	14	2	5	13	3.25	21.125
abc	16	–2	4	1	0.25	0.125

The error estimated from the average of the

$$\text{AB, BC and ABC is } \frac{10.125 + 10.125 + 0.125}{3} = 6.7916$$

Based on the mean squares of the above table, the effects which are of interest are A, C, AB, BC and AC as their mean square values are greater than the estimated average.

12.6 Fractional Factorial Design

One of the limitations of factorial design is that the number of experiments increases exponentially with the number of factors. The fractional factorial designs examine lot of factors efficiently with few runs compared to full factorial design. This design assumes that some or all interactions are negligible. These methods are useful to screen large number of factors and to identify the important factors. These designs are also known as main effect designs or screening designs because they are able to screen out few important main effects from the many less important effects. Fractional factorial experiments can be used as fraction (1/2, 1/4, etc.) of the original experiments. These designs are preferable to overcome the cost and time consideration of the full factorial design.

Types of fractional factorial designs

1. Homogenous fractional
2. Mixed level fractional
3. Box hunter
4. Plackett Burman design
5. Taguchi design
6. Latin square design

12.6.1 Homogenous Fractional Factorial Design

These designs are useful when the experimenter wants to evaluate a large number of factors efficiently. This design assumes that interactions are negligible. It is useful to identity the important variables. All the variables should have same number of levels.

To construct 1/2 replicate to estimate the main effect of 4 factors, we equate one of the effect to the fourth factor. The following example describes this.

The drug release rate from the microcapsules formulated by employing ionic gelation technique is influenced by the processing variables such as concentration of sodium alginate, stirring speed, concentration of curing reagent and curing time. To study these factor 2^4 (16) experiments are required. If we ignore the affect of stirring speed, as it is not much significant, the half of 2^4 factorial designs requires 8 experiments and it is presented below.

Table 12.11 Fractional Factorial Design for 4 Factors at two levels

Experi -ment	**A (% w/w)**	**B (rpm)**	**C (% w/v)**	**D (mix)**	**AB = CD**	**AC = BD**	**AD = BC**	**rate (mg/ha)**
1	–	–	–	–	+	+	+	8.6
ab	+	+	–	–	+	–	–	6.4
bc	–	+	+	–	–	–	+	10.3
ac	+	–	+	–	–	+	–	5.2
ad	+	–	–	+	–	–	+	5.7
bd	–	+	–	+	–	+	–	3.1
cd	–	–	+	+	+	–	–	2.2
abcd	+	+	+	+	+	+	+	6.6

A = Concentration of sodium alginate, B = Stirring speed, C = Concentration of curing reagent, D = Curing time.

To estimate the effects we add the responses multiplied by sings in the appropriate column and divided by 4.

For Example effects AB is

$$= \frac{(8.6 + 6.4 - 10.3 - 5.2 - 5.7 - 3.1 + 2.2 + 6.6)}{4}$$

$$= \frac{-0.5}{4} = 0.125$$

$$A = \frac{-8.6 + 6.4 - 10.3 + 5.2 + 5.7 - 3.1 - 2.2 + 6.6}{4}$$

$$= \frac{-0.3}{4} = -0.075$$

$$B = \frac{-8.6 + 6.4 + 10.3 - 5.2 - 5.7 + 3.1 - 2.2 + 6.6}{4}$$

$$= \frac{-6.9}{4} = -1.725$$

$$C = \frac{-8.6 - 6.4 + 10.3 + 5.2 - 5.7 - 3.1 + 2.2 + 6.6}{4}$$

$$= \frac{0.5}{4} = 0.125$$

$$D = \frac{-8.6 - 6.4 - 10.3 - 5.2 + 5.7 + 3.1 + 2.2 + 6.6}{4}$$

$$= \frac{-12.9}{4} = -3.225$$

$$AC = \frac{8.6 - 6.4 - 10.3 + 5.2 - 5.7 + 3.1 - 2.2 + 6.6}{4}$$

$$= \frac{-1.1}{4} = -0.275$$

$$AD = \frac{8.6 - 6.4 + 10.3 - 5.2 + 5.7 - 3.1 - 2.2 + 6.6}{4}$$

$$= \frac{14.3}{4} = 3.57$$

We cannot perform tests for the significance of these parameters without an estimate of the error. The variance can be estimated from duplicate experiments, nonexistent interactions or experiments from previous studies. For example, based on the above estimated values, factors C, D, and AD are having the largest effects. To clarify the possible confounding effects, eight more experiments can be performed. The results are summarized as follows:

1. Concentration of curing reagent (factor C) has positive effect on the drug release rate from microcapsules
2. There is a strong interaction between factors A and D. (concentration of Sodium alginate A curing time D)
3. Curing time (factor D) is having negative effect on yield
4. The AB and AC interaction is considered to be small.

Fractional Factorial Design for 3 Factors at two Levels

To study the influence of 3 factors – compression force, binder concentration and the disintegrate concentration on disintegration time, 2^3 (8) factors are required. If the factor c is equated to interaction factor AB, then the experimental design become 1/2 replicate design of 3 factors.

Table 12.12

Experiment	Parameter			AC	BC	Response (min)
	A (kg)	B (% w/w)	C = AB (% w/w)			
a	+	–	–	–	+	12
b	–	+	–	+	–	10
c	–	–	+	–	–	6
abc	+	+	+	+	+	8

The above data can be treated by following the procedure explained in 12.6.1 to interpretative the experimental results.

12.6.2 Mixed Level Fractional Factorial Designs

In this level fractional factorial designs, some parameters are at two levels and the other parameter is having more than 2 levels.

Example

The experiment employed to study the influence of two parameters – binder and disintegrates at 2 levels and lubricant at 3 levels on disintegration time of a tablet.

Table 12.13

Parameter	Low level	High Level
Binder	2	5
Disintegrate	0	10
Lubricant levels	0	0.5,1

Table 12.14 Generation of a mixed design

Run	Binder	Disintegrate	Lubricant
1	2	0	0
2	5	0	0.5
3	2	10	0.5
4	5	10	1

Mix up design for the same problem but the concentration of lubricant is 4 levels. i.e. two factors at two levels and one factor at 4th level (Lubricant Concentrations) 0, 0.5,1 and 2%).

Table 12.15

Run	Binder	Disintegrate	Lubricant
1	2	0	0
2	2	0	0.5
3	2	0	1
4	2	0	2
5	5	0	0
6	5	0	0.5
7	5	0	1.0
8	5	0	2.0
9	2	10	0
10	2	10	0.5
11	2	10	1
12	2	10	2
13	5	10	0
14	5	10	0.5
15	5	10	1
16	5	10	2

12.6.3 Box, Hunter Fractional Factorial Design

These designs are expressed using the notation $l^{K\text{-}P}$, where l is the number of level of each factor investigated, k is the number of factors investigated and P describes the size of the fraction of full factorial used. $2^{5\text{-}2}$ indicates 5 factorial design with 2 levels and 1/2 fraction.

Table 12.16 The design for 3 Factors and 2 levels (2^{3-1}) Factors

Run	A	B	C
1	–1	–1	+1
2	+1	–1	–1
3	–1	+1	–1
4	+1	+1	+1

– Indicates low level

+ Indicates high level

Table 12.17 Design for four Factors and 2 levels. (2^{4-1})

Run	A	B	C	D
1	– 1	–1	–1	–1
2	+1	–1	–1	+1
3	–1	+1	–1	+1
4	+1	+1	–1	–1
5	–1	–1	+1	+1
6	+1	–1	+1	–1
7	–1	+1	+1	–1
8	+1	+1	+1	+1

12.6.4 Plackett-Burman Designs

This design is a very efficient screening design when only main effects are of interest. It assumes that all interactions are negligible compared to main effects.

Table 12.18 Example for a Design of 12 runs and 11 Factors

Run	Factors										
	A	B	C	D	E	F	G	H	I	J	K
1	+1	+1	+1	+1	+1	+1	+1	+1	+1	+1	+1
2	–1	+1	–1	+1	+1	+1	–1	–1	–1	+1	–1
3	–1	–1	+1	–1	+1	+1	+1	–1	–1	–1	+1
4	+1	–1	–1	+1	–1	+1	+1	+1	–1	–1	–1
5	–1	+1	–1	–1	+1	–1	+1	+1	+1	–1	–1
6	–1	–1	+1	–1	–1	+1	–1	+1	+1	+1	–1

Table 12.18 *Contd...*

Run	Factors										
	A	B	C	D	E	F	G	H	I	J	K
7	–1	–1	–1	+1	–1	–1	+1	–1	+1	+1	+1
8	+1	–1	–1	–1	+1	–1	–1	+1	–1	+1	+1
9	+1	+1	–1	–1	–1	+1	–1	–1	+1	–1	+1
10	+1	+1	+1	–1	–1	+1	+1	–1	–1	+1	–1
11	–1	+1	+1	+1	–1	+1	–1	+1	–1	–1	+1
12	+1	–1	+1	+1	+1	–1	–1	–1	+1	–1	–1

By using this design, the effect of 12 factors can be determined with 20 experiments.

The above design can be explained with the following problem.

The solubility of a drug is tested with six variables (A–F) and 12 mixtures are prepared with high and low levels of six variables containing constant amount of drug. The variables include particle size, complexing agent, buffer, surfactant, hydrophilic carrier and co solvent.

The samples subjected to solubility studies and the observed solubility (response y) is given in the following table.

Table 12.19

Trial	Solubility (mg/ml)
1	2.4
2	1.6
3	2.8
4	3.2
5	4.1
6	8.2
7	7.4
8	9.2
9	10.6
10	6.2
11	5.4
12	7.3

Interpretation of Results

1. The y values for each column when the variable is having positive affect was added and recorded as “sum +”.
2. y values for each column when the variable is having negative affect was added and recorded as “sum –”.

3. In the variables have no effects on measured y, then sum + equal sum –
4. The average effect of each variable x on response y =

$$\frac{\text{Difference between sum + and sum –}}{\text{number of times each variable was +}}$$

5. The effect of un assigned variables are used to calculate experimental error.

Taguchi Design

This design utilize 2, 3 or mixed level fractional factorial designs. Large screening designs are favored by this model. This design ensures good performance in the design state of product or process.

Example

Table 12.20 Fractional Factorial Design for 4 Factors at Three Levels. (3^{4-2})

	Variable			
Run	**A**	**B**	**C**	**D**
1	1	1	1	1
2	1	2	2	2
3	1	3	3	3
4	2	1	2	3
5	2	2	3	1
6	2	3	1	2
7	3	1	3	2
8	3	2	1	3
9	3	3	2	1

Working Example: To improve the dissolution rate of glimepride, four factors such as concentration of diluents, super disintegrant, lubricants and the compression force were varied at 3 levels. The formulated tablets were subjected to *in vitro* dissolution studies. The dissolution rate constants were calculated. The design and the response are furnished in the Table 12.21.

Table 12.21

Parameter	**Notation**	**Concentration**		
		Low	**Medium**	**High**
Spray dried Lactose (Diluent)	A	120	160	180
Sodium starch glycholate (super disintegrant)	B	5	8	10
Magnesium stearate (Lubricant)	C	1	1.5	2
Compression force	D	2	3	5

Table 12.22

	Parameter				
Run	**A**	**B**	**C**	**D**	**Dissolution rate constant (min^{-1})**
1	Low	Low	Low	Low	0.102
2	Low	Medium	Medium	Medium	0.108
3	Low	High	High	High	0.096
4	Medium	Low	Medium	High	0.114
5	Medium	Medium	High	Low	0.134
6	Medium	High	Low	Medium	0.126
7	High	Low	High	Medium	0.132
8	High	Medium	Low	High	0.092
9	High	High	Medium	Low	0.146

Latin Square and Related Designs

Latin square design and its related Graeco-latin square and Hyper-Graeco-Latin square designs are also useful as fractional factorial designing. In these designs there is a single factor of primary interest (treatment factor) along with other non interested factors (nuisance factors). These nuisance factors are 2, 3 and 4 for Latin square, Graeco-Latin square and Hyper-Graeco-Latin square design respectively.

Advantages

1. This design handles the case where there are several nuisance factors which cannot be combined or the objective is to keep the nuisance factors separately.
2. Small number of runs is sufficient.

Disadvantages

1. Number of levels of each nuisance factor must equal to the number of levels of treatment factor.
2. This design assumes that the interacting between the blocking variables or between treatment and blocking variable is not significant, but practically it is not possible in all cases.

Example: **Design for 3 levels Factors**

A	B	C
1	1	1
1	2	2
1	3	3
2	1	3
2	2	1
2	3	2
3	1	2
3	2	3
3	3	1

CHAPTER 13

OPTIMIZATION TECHNIQUES AND SCREENING DESIGNS

13.1 Introduction

Optimization is the process of finding the best possible values of dependent variable by changing the independent variables. Pharmaceutical formulation effectiveness, stability, safety and quality are influenced by different formulation and process variables. So there is a need to determine the influence of these variables and to set these variables at their optimum levels.

In a pharmaceutical formulation, there may be numerous important parameters due to the involvement of several components in the formulation. If all the variables are included in the design, it may be prohibitive due to the increase in number of experiments; hence, the formulator must identify the potentially dominating variables in order to limit the number of experiments. It is based on experience, for example, previously prepared products of similar kind.

Optimization by means of experimental design is helpful to reduce the experimenting time. These designs with appropriate mathematical models are useful to interprete the effects and interactions of the variables. When all responses of interest (dependent variable) and either the formulation or process variables (independent variables) are expressed in models (mathematical graphical), then the suitable composition and process satisfying all the requirements can be developed.

The general procedure consists of a preparing a series of formulations by varying the independent variables and the formulation are evaluated for its response. Based on these experimental results, one formulation is selected as optimum. If the formulation is optimized for a single attribute like dissolution then optimization is not complicated. It is complicated when the attributes are more. For example the optimization of particle size in suspension. In this example, the particle size influences the dissolution rate and physical stability of suspensions. Small particle size is preferred for physical stability of suspension and large size is desired for controlled release of drug from suspensions. Similarly the hardness and disintegration time optimization with the variables like compression force, binder concentration etc. Hardness is desired to withstand the packaging, transshipment and handling. Low disintegration time is desired for rapid response. Thus hardness as well as disintegration time are the desired attributes of the tablet dosage form. The binder at high concentration increases the hardness and disintegration time. So the concentration of binder should be optimized to obtain the tablet having the required mechanical strength and desired disintegration time.

13.2 Design of Experiments

Design of experiment is an organized method to determine the relationship between factors affecting the process (x) and the output of the process (y). It refers to the process of planning to collect the appropriate data from an experiment and interpretation of the experimental results. Statistically meaningful conclusions from the experimental data is drawn with these designs.

Objectives of Design of Experiments

The following experimental objectives are met by the statistically well designed experiment.

1. **Comparative Objective:** The primary objective in this study is to make a conclusion about one important factor in presence to other factors and to find out whether the chosen factor is significant or not.
2. **Screening Objective:** The purpose of these experiments is to select few important effects from the many less important factors. They are also known as main effect designs.
3. **Response Surface Objective:** They are designed to estimate interaction and adverse effects.
4. **Mixture Design:** This design is meant to identify a best proportion of factors that are required to maximize or minimize a response.

Phases of Design of Experiments

1. Identification of factors that may affect the outcome of the experiment.
2. Choosing an appropriate experimental design.
3. Conducting the experiments.

4. Fitting the data in suitable mathematical models and generating the plots to describing the results.
5. Drawing Conclusions and Planning for the next step.

Advantages of experimental design

- To reduce experimental error.
- To determine the important variables that needs to be controlled during the experiment.
- To measure the interaction effects of two variables.
- To allow extrapolation of the data to finalize the best possible product.

Optimization using factorial designs

The experimental results obtained from the factorial design can be fitted into an empirical polynomial equation. The resulting equation can be used to predict the response of a formulation comprised with the same formulation variables but with different levels.

Illustration: **Construction of empirical equation for 3 factors with 2 levels.**

This can be exemplified with the formulation having three variables – binder, disintegrate and lubricant at two different levels. The composition of this formulation is shown below. The response observed from this experiment is disintegration time.

Composition of aspirin tablets

Quantity (mg/tablet)

Ingredients (mg)	F_1	F_2	F_3	F_4	F_5	F_6	F_7	F_8
Aspirin	500	500	500	500	500	500	500	500
PVP (binder)	–	10			10	–	10	10
Starch (disintegrate)	–	–	50		50	50	–	50
Mg stearate (lubricant)	–	–	–	5	–	5	5	5

The levels of these factors are shown as follows

Variable	Low level	High level (mg)
X_1	0	10
X_2	0	50
X_3	0	5

X_1 indicates binder, X_2 indicates disintegrant and X_3 indicates lubricant

Results observed from the above experiment can be summarized as follows

Formulation	Factor level			Observed Disintegration line (min)
	X_1	X_2	X_3	
F_1	0	0	0	8
F_2	10	0	0	20
F_3	0	50	0	6
F_4	0	0	5	10
F_5	10	50	–	9
F_6	0	50	5	11
F_7	10	0	5	18
F_8	10	50	5	15

The polynomial equation generated for 2^3 factorial designs is follows.

$$Y = B_0 + B_1X_1 + B_2X_2 + B_3X_3 + B_{12}X_1X_2 + B_{23}X_2X_3 + B_{13}X_1X_3 + B_{13}X_1X_3 + B_{123}X_1X_2X_3$$

The experimental results obtained from the above experiment can be substituted in the above equation to get a generalized polynomial equation suitable to find out the disintegration time of aspirin tablet containing the same variables (same binder, disintegrate and lubricant) but containing the concentration within the minimum and maximum levels.

$$Y = 8 + 20X_1 + 6X_2 + 10X_3 + 9X_1X_2 + 11X_2X_3 + 18X_1X_3 + 15X_1X_2X_3$$

If such equations are available, one can predict the response theoretically as follows.

Suppose the levels of the factors as follows

Variable	Low level	High level
X_1 binder	0	5
X_2 disintegrate	0	10
X_3 lubricant	0	2

The observed empirical equation is

$$Y = 8 + 20X_1 + 6X_2 + 10X_3 + 9X_1X_2 + 11X_2X_3 + 18X_1X_2 + 15X_1X_2X_3$$

Solving for B_0 ($X_1 = X_2 = X_3 = 0$) formulation 1.

$Y = 8 = B_0$

$X_1 = 5$; X_2 and $X_3 = 0$ In case of formulation 2

$X_1 = X_3 = 0$ and $X_2 = 10$ for formulation 3

Substitution of the X_1, X_2 and X_3 values of the concerned formulations in the above polynomial equations results the predicted disintegration time. The observed values are depicted in the following table.

Formulation	**Factor level**			**predicted Disintegration line (min)**
	X_1	X_2	X_3	
F_1	0	0	0	8
F_2	5	0	0	108
F_3	0	10	0	68
F_4	0	0	2	20
F_5	5	10	–	618
F_6	0	10	2	308
F_7	5	0	2	318
F_8	5	10	2	2538

The level of the factors can be coded so that low level is represented with –1 and the high level with +1. The transformed values of the above data are furnished in the following table.

Formation	**Factor level**							
	x_1	x_2	x_3	x_1x_2	x_2x_3	$x_1x_2x_3$	**Total**	y
F_1	–1	–1	–1	+1	1	–1	+1	8
F_2	+1	–1	–1	–1	–1	+1	+1	108
F_3	–1	+1	–1	–1	1	+1	+1	68
F_4	–1	–1	+1	+1	–1	+1	+1	20
F_5	+1	+1	-1	+1	–1	–1	+1	618
F_6	–1	+1	+1	–1	–1	–1	+1	308
F_7	+1	–1	+1	–1	+1	–1	+1	318
F_8	+1	+1	+1	+1	+1	+1	+1	2538

The above method of prediction of theoretical response is not applicable, if the low level values of the factor are not zero. The first step in the solution is to code the levels of factors so that the high level of each factor is +1, and the low levels of each factor are –1. This procedure requires a transformation of each variable and their interactions are transformable with appropriate signs.

Example

A pharmaceutical dosage form was subjected to the following stability conditions and the observed shelf life is given in the following table.

Factorial Design (2^3) for the Stability Study

	Temperature (°C)	Humidity (RH)	Light (foot candles)	Shelf life(y)
1	32 °C	40	200	2.5
A	40 °C	40	200	1.6
B	32 °C	80	200	1.4
C	32 °C	40	400	2.0
Ab	40 °C	80	200	1.2
Bc	32 °C	80	400	1.8
Ac	40 °C	40	400	1.7
Abc	40 °C	80	400	1.0

The above levels can be transformed into signs as follows

Condition	X_1	X_2	X_3	X_1X_2	X_2X_3	X_1X_3	$X_1X_2X_3$	Total	y
1	–1	–1	–1	+1	+1	+1	–1	+1	2.5
a	+1	–1	–1	–1	+1	–1	+1	+1	1.6
b	–1	+1	–1	–1	–1	+1	+1	+1	1.4
c	–1	–1	+1	+1	–1	–1	+1	+1	2.0
ab	+1	+1	–1	+1	–1	–1	–1	+1	1.2
bc	–1	+1	+1	–1	+1	–1	–1	+1	1.8
ac	+1	–1	+1	–1	–1	+1	–1	+1	1.7
abc	+1	+1	+1	+1	+1	+1	+1	+1	1.0

The polynomial equation generated for this stability studies is given below.

$$Y = 2.5 + 1.6X_1 + 1.4\,X_2 + 2.0X_3 + 1.2\,X_1X_2 + 1.8\,X_2X_3 + 1.7\,X_1X_3 + 1.0\,X_1X_2X_3$$

By using above equation, it is possible to predict theoretical response at different conditions.

Example

Calculate the shelf life of the dosage forms when it is subjected to 40°C (X_1), 60% (RH) (X_2) and 200 foot candles of light. (X_3)

$$= 2.5+1.6\times40+1.4\times60+2\times200+1.2\times40\times60+1.8\times60\times200+1.7\times40\times200+1\times40\times60\times200$$

$$= 518630.5$$

13.3 Replication (Sample Size)

To test the significance of coefficients to binomial equation, an estimate of the experimental error is required. This error is best estimated with the replicating runs. The best procedure for replication consists of performing an experiment in duplicate and randomizing the order of experiment. To determine the sample size for hypothesis tests, we may use the formulae $N = 4\ (S^2 / \Delta^2)$, where N is Sample size, S = Standard deviation and Δ = Difference in response.

Example

Calculate the sample size for an experiment where the required response difference and standard deviation are 2 and 0.5 respectively.

$$N = 4\ (0.5^2 / 2^2)\ 10 = 4\left(\frac{0.25}{4}\right) \times 10 = 2.5 = 3.$$

Illustration: Optimization of a formulation containing combination of two approaches.

In order to enhance the dissolution rate of a poorly soluble drug, the formulator may use the combination of two approaches like the use of complexing agents A_1 (x_1) and surfactant B (X_2). The formulations consist of combinations containing each excipient at two levels. The concentration of excipient A is 100 mg and 200 mg and the surfactant (B) is 2 and 4 mg. The 4 formulations (full factorial design of 2^2) were subjected to dissolution studies.

Factorial design for the formulation developments study

Formulation	Exceipient Concentration (mg)		Concentration (transformed)			Response Dissolution rate (min^{-1})
	A(X_1)	**B(X_2)**	**A(X_1)**	**B(X_2)**	**AB(X_1X_2)**	
1	100	2	−1	−1	+1	0.104
2	200	2	+1	−1	−1	0.152
3	100	4	−1	+1	−1	0.128
4	200	4	+1	+1	+1	0.175

For drug B, the transformation is $\frac{Potency - 75}{25}$ (high-level is 200, low level is 100)

The response equation has the form

$$Y = B_0 + B_1\ (X_1) + B_2(X_2) + B_{12}(X_1)\ (X_2)$$

Compute B_1

Column A(X_1)	Y	X_1Y
100 (– 1)	0.104	– 0.104
200 (+1)	0.152	0.152
100 (–1)	0.128	– 0.128
200 (+1)	0.175	0.175
		0.095

$$\therefore B_1 = \text{The sum of } \frac{x_1 y}{4} = \frac{0.095}{4} = 0.02375$$

To Compute B_2

Column $B(X_2)$	Y	X_2Y
–1	0.104	–0.104
–1	0.152	–0.152
+1	0.128	0.128
+1	0.175	0.175

$$B_2 \text{ is the sum of } \frac{x_2 y}{4} = \frac{0.047}{4} = 0.01175$$

To Compute B_{12}

Column $B(X_1\ X_2)$	Y	$X_1\ X_2Y$
+1	0.104	0.104
–1	0.152	–0.152
–1	0.128	–0.128
+1	0.175	0.175
		–0.001

$$B_{12} \text{ is the sum of } \frac{x_1 x_2 y}{4} = \frac{-0.001}{4} = -0.00025$$

$$B_0 = \frac{0.104 + 0.152 + 0.128 + 0.175}{4} = 0.13975$$

$$\therefore \text{ the polynamial equaiton is } y = B_0 + B_1(X_1) + B_2(X_2) + B_{12}(X_1)(X_2)$$

$$\therefore \quad y = 0.13975 + 0.02375X_1 + 0.01175X_2 + 0.13975\ X_1X_2$$

13.4 Composite Designs to Estimate Curvature

The response predicted from the empirical equation is valid only if there is a linear relationship between the response parameter y and variable parameter X (both levels X_1 and X_2) at its low and high levels. The response calculated by substituting the X value which is between low and high value parameter relation is not curvature as shown in Fig. 13.1.

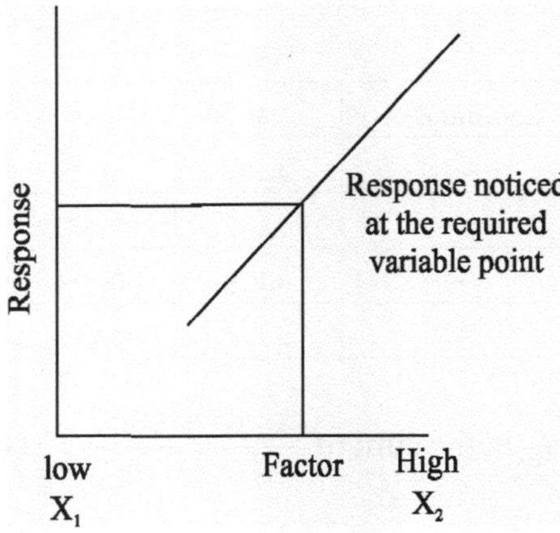

Fig. 13.1 Linear relation between variable and response factors

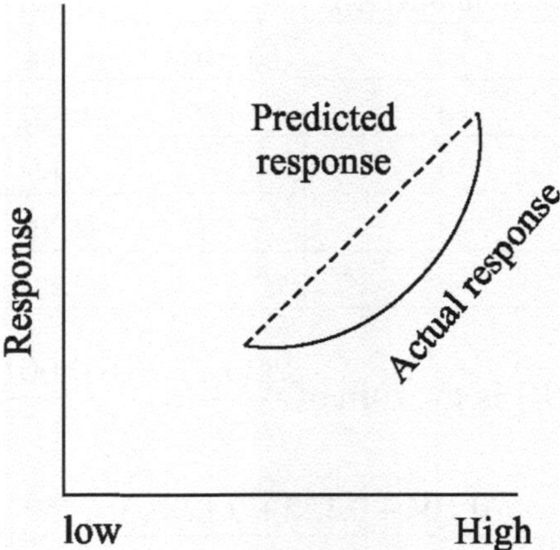

Dotted point indicates the actual and predicted response

Fig. 13.2 Curved response between the two factors.

In order to estimate curvatures, more than two levels of the factor should be included in the experimental (Low, medium and high) design. The polynomials of order 2 are required to describe the curved response and the example of such equation for a 2. Factor design is shown as follows.

$$Y = B_o + B_1X_1 + B_{11}\ X_1^2 + B_2\ X_2 + B_{22}\ X_2^2 + B_{12}\ X_1X_2 + \ldots$$

These composite designs estimates second-order terms and useful for effective estimation of curvature. This design allows the performance of an experiment in stepwise.

Example

The above design can be converted to composite design by incorporating third level i.e. center of the low and high concentrations. The resulting design is as follows.

Formulation	Excipient Concentration		True Response	Predicted response
	A (X_1)	B(X_2)		
1	100(– 1)	2(– 1)	0.104	0.1176
2	100(– 1)	3(0)	0.119	0.126073
3	100(– 1)	4(+1)	0.128	0.1274
4	150(0)	2(– 1)	0.132	0.1383
5	150(0)	3(0)	0.139	0.1467
6	150(0)	4(+1)	0.142	0.1533
7	200(+1)	2(–1)	0.152	0.1656
8	200(+1)	3(0)	0.168	0.173798
9	200(+1)	4(+1)	0.175	

Transform the above data in to coded levels as follows.

Formation	X_1	X_2	X_1X_2	$X_1^2-\frac{2}{3}$	$X_2^2-\frac{2}{3}$	True response	Predicted response
1	–1	–1	+1	1/3	1/3	0.104	
2	–1	0	0	1/3	–2/3	0.119	
3	–1	1	–1	1/3	1/3	0.128	
4	0	–1	0	–2/3	1/3	0.132	
5	0	0	0	–2/3	–2/3	0.139	
6	0	1	0	–2/3	1/3	0.142	
7	1	–1	–1	1/3	1/3	0.152	
8	1	0	0	1/3	–2/3	0.168	
9	1	1	1	1/3	1/3	0.175	

$$Y = B_0 + B_1X_1 + B_{11}\left(X_1^2 - \frac{2}{3}\right) + B_2X_2 + B_{22}\left(X_2^2 - \frac{2}{3}\right) + B_{12}X_1X_2$$

$$B_0 = \text{the average of the nine responses} = \frac{0.104 + 0.119 + \ldots\ldots\ldots\ldots + 0.175}{9}$$

$$= \frac{1.259}{9} = 0.14$$

To find B_1

Column (X_1)	Response (Y)	X_1Y
–1	0.104	–0.104
–1	0.119	–0.119
–1	0.128	–0.128
0	0.132	0
0	0.139	0
0	0.142	0
1	0.152	0.152
1	0.168	0.168
1	0.175	0.175
		0.114

$$\therefore \qquad B_1 = \frac{0.114}{6} = 0.024$$

To find B_{11}

$X_i^{12} = X_i^2 - \frac{2}{3}$	Y	$(X_i^{12})(Y)$
1/3	0.104	0.104/3
1/3	0.119	0.119/3
1/3	0.128	0.128/3
– 2/3	0.132	– 0.132 × – 2/3
– 2/3	0.139	– 0.139 × – 2/3
– 2/3	0.142	– 0.142 × 2/3
1/3	0.152	0.152/3
1/3	0.168	0.168/3
1/3	0.175	0.175/3
		$\frac{0.02}{3} = 0.0066$

The sum of square in Colum $X_i^2 - \frac{2}{3}$ is 2

$$\therefore \qquad B_{11} = \frac{0.00666666}{2} = 0.003$$

To find B_2

Colum B (X_2)	Y	B(X_2) Y
–1	0.104	– 0.104
0	0.119	0
1	0.128	0.128
–1	0.132	– 0.132
0	0.139	0
1	0.142	0.142
–1	0.152	– 0.152
0	0.168	0
1	0.175	0.175
		0.057

$$B_2 = \frac{0.057}{6} = 0.0095$$

$$B_2 = 0.0095$$

To find B_{22}

Colum $\left(X_2^2 - \frac{2}{3}\right)$	Y	$\left(X_2^2 - \frac{2}{3}\right)Y$
1/3	0.104	$\frac{(0.104)}{3}$
– 2/3	0.119	$\frac{-(0.119)2}{3}$
1/3	0.128	$\frac{0.128}{3}$
1/3	0.132	$\frac{0.132}{3}$
–2/3	0.139	$\frac{-(0.139)2}{3}$
1/3	0.142	$\frac{(0.142)1}{3}$
1/3	0.152	$\frac{(0.152)}{3}$
–2/3	0.168	$\frac{-(0.168)2}{3}$
1/3	0.175	$\frac{-0.175}{3}$
		$\frac{-0.019}{3} = -\ 0.006$

$$\therefore \qquad B_{22} = \frac{-0.0063333}{6} = -\ 0.0011$$

To find B_{12}

Colum X_1X_2	Y	$(X_1X_2)Y$
1	0.104	+0.104
0	0.119	0
–1	0.128	– 0.128
0	0.132	0
0	0.139	0
0	0.142	0
–1	0.152	– 0.152
0	0.168	0
1	0.175	0.175
		– 0.001

$$\therefore \quad B_{12} = \frac{-0.001}{4} = -0.00025$$

$$\therefore \quad Y = B_0 + B_1X_1 + B_{11}\left(X_1^2 - \frac{2}{3}\right) + B_2X_2 + B_{22}\left(X_2^2 - \frac{2}{3}\right) + B_{12}X_1X_2$$

$$= 0.1398 + 0.024X_1 + 0.0033\left(X_1^2 - \frac{2}{3}\right) + 0.0095X_2 - 0.001055\left(X_2^2 - \frac{2}{3}\right)$$

$$= -\ 0.00025\ X_1X_2$$

1	$X_1 = -1$	$X_2 = -1$	$X_1X_2 = 1$	$X^2 - \frac{2}{3} = \frac{1}{3}$	$X_2^2 - \frac{2}{3} = \frac{1}{3}$	Response value predicted
2	–1	0	0	1/3	–2/3	0.118
3	–1	1	–1	1/3	1/3	0.126
4	0	–1	0	–2/3	1/3	0.127
5	0	0	0	–2/3	–2/3	0.138
6	0	1	0	–2/3	1/3	0.147
7	0	–1	–1	1/3	1/3	0.153
8	1	0	0	1/3	–2/3	0.166
9	1	1	1	1/3	1/3	0.174

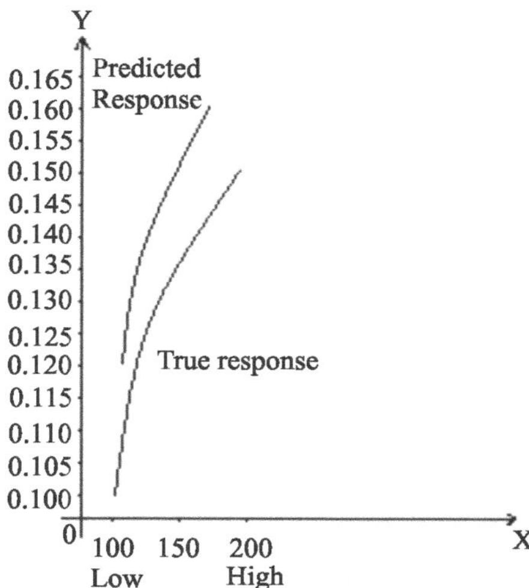

13.5 The Simplex Lattice Design

These designs are used in optimization of formulations containing the constant composition of total excipients(%), however the percentage of each individual excipient may vary. For example, if the maximum volume of the solvent for intramuscular injection is 2 ml, then the total volume of the solvent is 2 ml. The 2 ml may not be necessarily the same solvent and it may be a mixture of solvents to achieve the desired solubility, stability and safety. They are also known as mixture designs.

The following steps are involved in this design.

1. Preparation of various formulations containing the different compositions of the selected ingredients.
2. Observe the experimental response,
3. Construct the polynomial equation.

The combinations should be selected in such a way that the observed response can be used to cover the interested space i.e. within the range of composition. Extrapolation of the data outside the experimental range is not recommended for interpretation.

E.g. optimize the composition of the solvents to achieve the desired solubility.

The above problem is pursued by preparing the saturated solutions of the drug in water (100% solvent A), Alcohol (100% solvent B) and the blend containing the equal proportion of solvents A and B.

The solubility of the drug from these saturated solutions was observed and found to be 2, 10 and 12 mg/ml respectively.

The empirical equation from the above experimental data is as follows.

$$Y = B_1 (A) + B_2 (B) + B_{12} (A) (B)$$

B_1 is the response at (A) equal to 100% = 2 mg/ml

B_2 is the response at (B) equal to 100% = 10 mg/ml

B_{12} is equal to 4(response observed from a mixture containing equal proportion of solvents) – 2(sum of responses observed in presence of pure solvents A and B)

$$B_{12} = 4(12) - 2(2 + 10) = 24$$

The response equation is

$$Y = 2(A) + 10(B) + 24(A) (B)$$

The above equation is used to predict the solubility from the blend containing the 40% solvent A and 60% solvent B as follows.

Such response equations are used to predict the response at extra design points which are not included in the design but lie within the simplex space (100%A and 100%B).

The maximum solubility is calculated by applying calculus and considering the first derivative of the following equation is equal to zero.

$$Y = 2(A) + 10(1.00 - A) + 24(A) (1.00 - A)$$

$$Y = 2A + 10 (1 - A) + 24(A) (1 - A)$$

$$\frac{dY}{dA} = 2 - 10 + 24 - 28A$$

$$\frac{d^2Y}{dA^2} = -48 < 0 \qquad \therefore \text{ Y has Maximum}$$

To get Maximum and Minimum, take $\frac{dY}{dA} = 0$

$$\Rightarrow \qquad 2 - 10 + 24 - 48A = 0$$

$$16 - 48A = 0 \Rightarrow \qquad A = \frac{16}{48} = \frac{1}{3} = 0.33$$

$$Y_{Max} = 2\left(\frac{1}{3}\right) + 10\left(1 - \frac{1}{3}\right) + 24\left(\frac{1}{3}\right)\left(1 - \frac{1}{3}\right)$$

$$= \frac{2}{3} + \frac{20}{3} + \frac{16}{3}$$

$$= 38/3 = 12.67$$

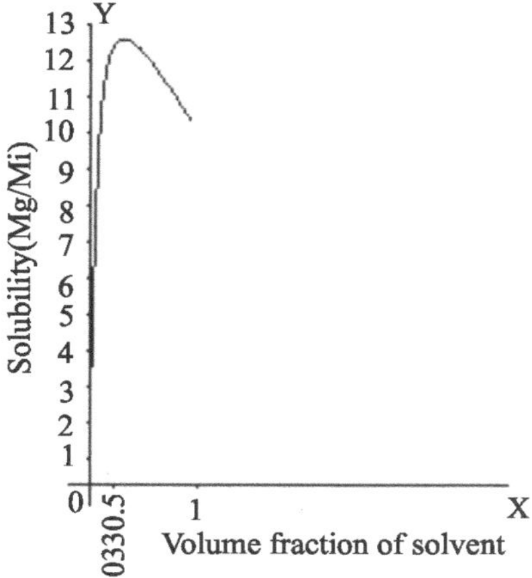

The general equation for the response based on simple design contains terms for pure components and all mixtures of components. The values observed from the empirical response equation derived from a limited number of experimental observations may not be correct due to the variability in empirical model or due to observational errors.

The simplex equation coefficients for 3 component system is calculated with the following formulae

$B_1 = Y_1$ observed response at 100% A

$B_2 = Y_2$ observed response at 100% B

$B_3 = Y_3$ observed response at 100% C

$B_{12} = 4(Y_{12}) - 2(Y_1+Y_2)$ Y_{12} response at 50-50 AB

$B_{13} = 4(Y_{13}) - 2(Y_1+Y_3)$ Y_{13} response at 50-50 AC

$B_{23} = 4(Y_{23}) - 2(Y_2+Y_3)$ Y_{23} response at 50-50 BC

$B_{123} = 27(Y_{123}) - 12(Y_{12}+Y_{13}+Y_{23}) + 3(Y_1+Y_2+Y_3)$,

where Y_{123} response at ⅓A, ⅓B and ⅓C.

In pharmaceutical formulations, the dose of the drug is fixed and the composition of remaining components may be varied and the lower limit for the varying component is more often not equal to zero. In case of tablets, a minimum amount of disintegrant/ lubricant is required. Then a design with constraints should be applied. In such cases, the composition of variable parameters is transformed to 100%.

Example: The disintegrantation time observed from the tablets containing three variables A (diluent), B (binder) and C (disintegrant) is given below. Drug dose is 50 mg and the tablet weight is 200 mg and the amount of A, B and C in the tablet is 150 mg. Calculate the response coefficients.

Transformed proportion is calculated with the following formulae

Amount used- minimum/maximum- minimum

The transformed value of A in experiment 1 is shown below.

$$139–129/139–129 = 10/10 = 1$$

The transformed value of B in experiment 1 is shown below.

$$10–10/20 – 10 = 0/10 = 0$$

The transformed value of C in experiment 1 is shown below

$$1–1/4 – 1= 0/3 = 0$$

Similarly calculate the A, B and C values for the remaining experiments

Experiment no	**Composition of components (mg/tablet)**			**Transformed proportion**			**Average disintegration time (min)**
	A	**B**	**C**	**A**	**B**	**C**	
1	139	10	1	1	0	0	20
2	129	20	1	0	1	0	15
3	139	10	11	0	0	1	18
4	134	15	1	0.5	0.5	0	16
5	134	10	6	0.5	0	0.5	19
6	129	15	6	0	0.5	0.5	17
7	132.5	13.33	4.33	0.33	0.33	0.33	21

$$B_1 = 20$$

$$B_2 = 15$$

$$B_3 = 18$$

$$B_{12} = 4(16) – 2(20+15)$$
$$= 64 – 70= – 6$$

$$B_{13} = 4(19) – 2 (20+18)$$
$$= 76-76 = 0$$

$$B_{23} = 4(17) – 2 (15+18)$$
$$= 68 – 66 = 2$$

$$B_{123} = 27(21) – 212(16 + 19 + 17) + 3(20 + 156 + 18)$$
$$= 567 – 624 + 159$$
$$= 102$$

∴ The response equation is

$$Y = B_1(A) + B_2(B) + B_3(C) + B_{12}(AB) + B_{13}(AC) + B_{23}(BC) + B_{123}(ABC)$$

∴ $$Y = 20(A) + 15(B) + 18(C) - 6AB + O\ (AC) + 2BC + 102(ABC)$$

ie $$Y = 20A + 15B + 18C - 6AB + 2BC + 102ABC$$

13.6 Sequential Optimization

In this technique, the process is optimized in a step wise fashion. Evolutionary operation (EVOP) uses factorial design and requires a large number of experiments. The procedure involved the generation of data from n + 1 experiments (n= number of independent variables). From the results, one result is eliminated (worst response) and a new experiment is performed. A decision is made on the result of the new experiment, generating new experiment and so on until the optimal response is generated.

The disintegration time (response) observed from a tablet containing three dependent variables – Diluent(A), Binder(B) and Disintegrant(C) is given below.

Experiment	Percentage composition			Disintegration time(min)
	Diluent	Binder	Disintegrant	
1	50	5	2	12
2	20	20	2	20
3	20	5	4	8
4	50	20	4	10

It is subjected to sequential simplex optimization as follows.

Eliminate the formulation (worst response, experiment 2) and set the composition for new formulation with the following formulae

$$R = p^{-} + (p^{-}\ W) \qquad \text{Step1}$$

p^{-} = average composition of the formulations other than eliminated formulation and

W = composition of the worst formulation.

Average % of diluent= 50 + 20 + 50/3= 120/3= 40

Average % of binder= 5 + 5 + 20/3= 30/3= 10

Average % of disintegrant= 2 + 4 + 4/3= 10/3= 3.33.

Therefore p^{-} = 40, 10, 3.33

W = 20, 20, 2(experiment2)

R = [40, 10, 3.33] + [(40 – 20), (10 – 20), (3.33 – 2)]

= [40, 10, and 3.33] + [20, –10, 1.33]

= (60, 0, and 4.66)

Experiment 5

The new formulation should be formulated with 60% diluents, 0% binder and 4.66% disintegrant and observe the response. If the observed disintegration time is 6 minutes i.e. better than best performance (formulation 3), then proceed for the next experiment with the following formulae

$$E = p^{-} + 2(p^{-} - W) \qquad \text{Step2}$$

Now the worst response is formulation 1 (W_2) and p^- is calculated as follows.

Average % of diluents = 60 + 20 + 50/3 = 130/3 = 43.33

Average % of binder = 0 + 5 + 20/3 = 25/3 = 8.33

Average % of disintegrant = 4.66 + 4 + 4/3 = 12.66/3 = 4.22.

$p^- = 43.33, 8.33, 4.22$

$W = 50, 5, 2$

$E = [43.33, 8.33, 4.22] + 2[(43.33 – 50), (8.33 – 5), (4.22 – 2)]$

$= [43.33, 8.33, 4.22] + 2[– 6.27, 3.33, 2.22]$

$= (43.33 – 12.54, 8.33 + 6.66, \text{ and } 4.22 + 4.44)$

$= (30.79, 14.99, \text{ and } 8.66)$

The new formulation (experiment 6) should be prepared with 30.79% diluent, 14.99% binder and 8.66% disintegrant and observe the response. If the observed response is better than the best response, then follow the same procedure.

If the observed response from step 1 is worse than the second worst response (more than 12 minutes), then evaluate the response with the following formulae

$c_r = p^{-} + 0.5(p^{-} w)$ Step3

$p^- = (50 + 20 + 50/3, 5 + 5 + 20/3, 2 + 4 + 4/3) = (40, 10, 3.33)$

$c_r = (40,10,3.33) + 0.5[(40 – 60),(10 – 0), (3.33 – 4.66)]$

$= (40, 10, 3.33) + 0.5(–20, 10, –1.33)$

$= [(40 – 10), (10 + 5), (3.33 – 0.66)]$

$= (30, 15, 2.67)$

The new formulation should be prepared with 30% diluent, 15% binder and 2.67% disintegrant and observe the response. If the observed response is better than the best response, then follow the step2 and the response observed is worse than the second worst response then follow the step3. If the response observed in any stage is worse than worst response, then follow step4.

$c_w = p^{-} 0.5(p^{-} w)$

For example, if the observed response in experiment 5 is 20 minutes i.e. worse than the worst response, then the p^- is calculated as follows.

Average % of diluents = 50 + 20 + 50/3 = 120/3 = 40

Average % of binder = 5 + 5 + 20/3 = 30/3 = 10

Average % of disintegrant = 2 + 4 + 4/3 = 10/3 = 3.33.

Therefore p^- = 40, 10, 3.33

The above average values are computed from the experiments 1, 3 and 4.

$$W = (60, 0, 4.66) = \text{experiment } 5$$

$$c_W = (40,10,3.33) - 0.5[(40 - 60),(10 - 0), (3.33 - 4.66)]$$

$$= (40, 10, 3.33) - 0.5(-20, 10, -1.33)$$

$$= [(40 + 10), (10 - 5), (3.33 + 0.66)]$$

$$= (50, 5, 3.99)$$

Now the new experiment should be started with 50% diluent, 5% binder and 3.99% disintegrant and observe the response. If the observed response is better than the best response, then follow the step 2 and the response observed is worse than the second worst response then follow the step 3. If the response observed is worse than worst response, then follow step4. This procedure should be continued until the observed response is not showing improved response from the repeated experiments.

13.7 Process Optimization by Surface Response Analysis

Response surface methodology, or RSM, is a collection of mathematical and statistical techniques that are useful for modelling and analysis in applications where a response of interest is influenced by several variables and the objective is to optimize this response. In most RSM cases, the form of the relationship between the response and the independent variables is unknown. Thus, the first step in RSM is to find a suitable approximation for the relationship between Y and the independent variables. Usually, a low-order polynomial in some region of the independent variables is employed. If the response is well modelled by a linear function of the independent variables, the approximating function is the first-order model if there is curvature in the system, then a polynomial of higher degree must be used, such as the second-order model. Many RSM problems use one or both of these approximating polynomials. Of course, it is unlikely that a polynomial model will be a reasonable approximation of the true functional relationship over the entire space of the independent variables, but for a relatively small region they usually work quite well.

Illustration: Surface response analysis was applied to optimize the concentration of SSG Magnesium sterate (MS) and Talc for a tablet dosage form and k value was taken as response. Central composite design was developed using Minitab 15 trail version and results were furnished below

Actual coded values of the experiment

S.no	Factors	Levels	
		Lower level (%w/w)	**Higher level (%w/w)**
1	Sodium Starch Glycol ate (SSG)	4	6
2	Magnesium Sterate (MS)	2	5
3	Talc	0	2

13.7.1 Central Composite Design

A better design that encompasses the advantages of factorial design or fractional factorial design and the star design, is the central composite design (CCD) developed by Box and Wilson. It is composed of

1. 2k Factorial design or 2 (k–p) Fractional factorial designs.
2. 2* k star design.

This design enables the estimation of a full second-order model. The equation for two factors is given by

Design

Factors: 03 Replicates: 01

Base runs: 20 Total runs: 20

Base blocks: 01 Total blocks: 01

Two-level factorial: Full factorial

Cube points: 08

Center points in cube: 06

Axial points: 06

Center points in axial: 00

Alpha: 1.68179

The design of central composite design is depicted in the following table.

Std. order	Run order	SSG	MS	Talc	K (hr^{-1} min^{-1})
5	1	4	2	1	0.1021
16	2	5	3.5	0.5	0.1423
1	3	4	2	0	0.1011
19	4	5	3.5	0.5	0.1082

Table *Contd...*

Std. order	Run order	SSG	MS	Talc	k (hr^{-1} min^{-1})
3	5	4	5	0	0.098
14	6	5	3.5	1.3409	0.1126
12	7	5	6.02269	0.5	0.1012
9	8	3.31821	3.5	0.5	0.1021
2	9	6	2	0	0.1423
6	10	6	2	1	0.1011
13	11	5	3.5	– 0.3409	0.0981
11	12	5	0.97731	0.5	0.1123
17	13	5	3.5	0.5	0.1321
15	14	5	3.5	0.5	0.1083
10	15	6.68179	3.5	0.5	0.1136
7	16	4	5	1	0.1261
18	17	5	3.5	0.5	0.1821
4	18	6	5	0	0.1075
8	19	6	5	1	0.1463
20	20	5	3.5	0.5	0.123

Response Surface Regression: K versus SSG, MS, TALC

The analysis was done using coded units.

Estimated Regression Coefficients for K

Term	Coef	SE Coef	T	p
Constant	0.132328	0.008929	14.82	0
SSG	0.006534	0.005924	1.13	0.296
MS	0.000925	0.005924	0.156	0.879
TALC	0.003741	0.005924	0.631	0.542
SSG*SSG	– 0.00656	0.000577	– 1.138	0.282
MS*MS	– 0.00695	0.000577	– 1.205	0.256
TALC*TALC	– 0.00744	0.000577	– 1.291	0.226
SSG*MS	– 0.00131	0.00774	– 0.17	0.869
SSG*TALC	– 0.00394	0.00774	– 0.509	0.622
MS*TALC	– 0.01339	0.00774	1.73	0.114

Analysis of Variance for K

Source	DF	Seq SS	Adj SS	AdjMS	F	P
Regression	9	0.004123	0.004123	0.000458	0.96	0.522
Linear	3	0.000786	0.000786	0.000262	0.55	0.662
SSG	1	0.000583	0.000583	0.000583	1.22	0.296

Table *Contd...*

Source	DF	Seq SS	Adj SS	AdjMS	F	P
MS	1	0.000012	0.000012	0.000012	0.02	0.879
Talc	1	0.000191	0.000191	0.000191	0.4	0.542
Square	3	0.001766	0.001766	0.000589	1.23	0.35
SSG*SSG	1	0.000406	0.00062	0.00062	1.29	0.282
MS*MS	1	0.000521	0.000696	0.000696	1.45	0.256
Talc*Talc	1	0.000799	0.000799	0.000799	1.67	0.226
Interaction	3	0.001572	0.001572	0.000524	1.09	0.396
SSG*MS	1	0.000014	0.000014	0.000014	0.03	0.869
SSG*Talc	1	0.000124	0.000124	0.000124	0.26	0.622
MS*Talc	1	0.001434	0.001434	0.001434	2.99	0.114
Residual Error	10	0.004793	0.004793	0.000479		
Lack-OF-Fit	5	0.000971	0.000971	0.000194	0.25	0.921
Pure error	5	0.003823	0.003823	0.000765		
Total	19	0.008917				

Estimated Regression Coefficients for K using data in uncoded units

Term	Coef
Constant	– 0.11929
SSG	0.079137
MS	0.017686
TALC	0.014158
SSG*SSG	– 0.00656
MS*MS	– 0.00309
TALC*TALC	– 0.02978
SSG*MS	– 8.75
SSG*TALC	– 0.00788
MS*TALC	0.1785

A polynomial equation is derived with the help of estimated regression coefficients for k of coded units is as follows

$$Y = B_0 + B_1 X_1 + B_2 X_2 + B_3 X_3 + B_{12} X_1 X_2 + B_{23} X_1 X_3 + B_{13} X_1 X_3 + B_{11} X_1^2 + B_{22} X_2^2 + B_{33} X_3^2$$

Y = Dependent variable

X_1 = independent variable SSG % concentration

X_2 = independent variable MS % concentration

X_3 = independent variable Talc % concentration

b_0 = Arithmetic mean of 4 responses

b_1, b_2, b_{12}, b_{11}, b_{22} and b_{33}= Estimated co-efficient for the corresponding factors X_1, X_2, X_3, X_1^2, $X_2^2 X_3^2$.

Therefore the equation is

$$Y = 0.132328 + 0.006534X_1 + 0.000925X_2 + 0.003741X_3 - 0.00131X_1X_2 - 0.00394X_1X_3 - 0.01339X_2X_3 - 0.00656X_1^2 - 0.00695X_2^2 - 0.00744X_3^2$$

The results can also be furnished with help of different plots such as Contour plots and the desired concentrations of the variables can be readily obtained from the following graph.

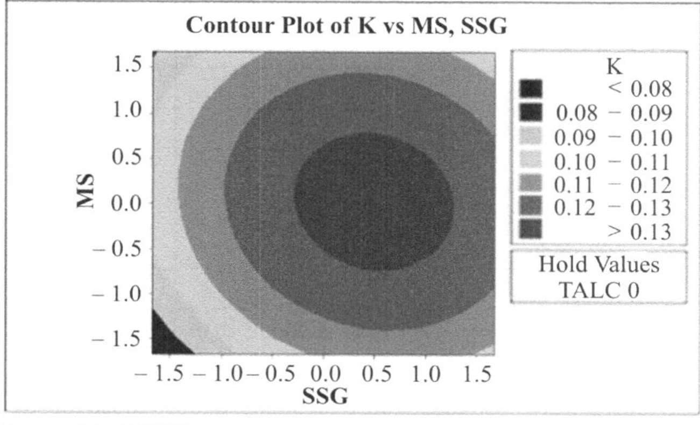

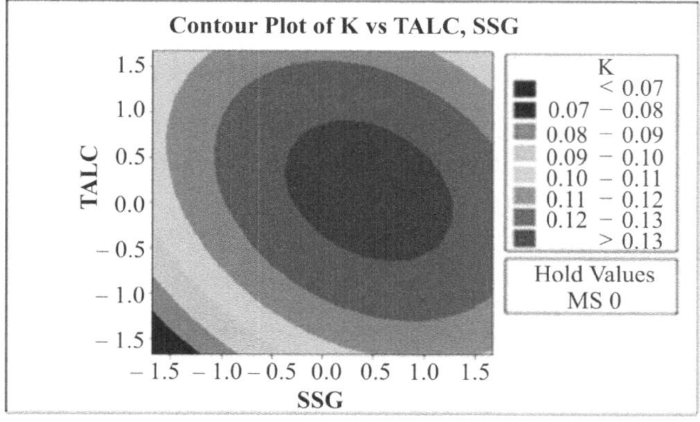

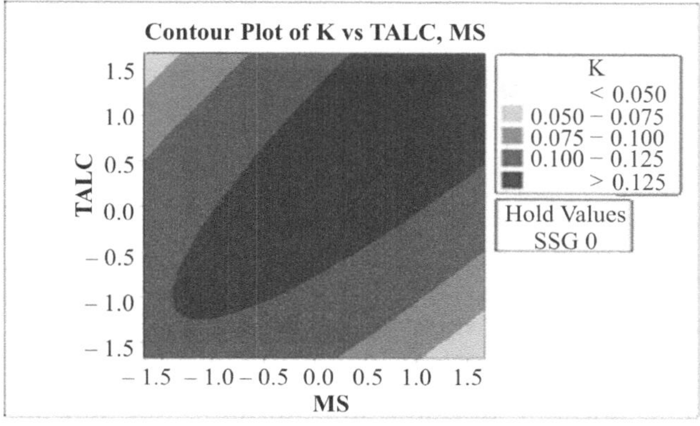

Illustration: Surface response analysis was applied to optimize the concentration of binder and disintegrating agent by conducting the experiments with the % w/w concentrations of binder and disintegrating agent generated from the protocol given by the Minitab 15 software.

Binder (% w/w)	Disintegrant (%w/w)	K value
5.00	2.00	0.102
7.50	3.50	0.094
7.50	3.50	0.094
11.04	3.50	0.068
7.50	3.50	0.094
7.50	5.62	0.108
5.00	5.00	0.124
7.50	1.37	0.072
10.00	5.00	0.089
3.96	3.50	0.108
7.50	3.50	0.094
10.00	2.00	0.064
7.5	3.50	0.094

In the above table, the concentrations of the binder and disintegrant are obtained by entering the low and high concentrations of the selected factors in the software. The result observed by analyzing the above data is furnished below.

Central Composite Design

Factors: 2 Replicates: 1

Base runs: 13 Total runs: 13

Base blocks: 1 Total blocks: 1

Two-level factorial: Full factorial

Cube points: 4

Center points in cube: 5

Axial points: 4

Center points in axial: 0

Alpha: 1.41421

Response Surface Regression: K value versus Binder, Disintegrant

The analysis was done using coded units.

Estimated Regression Coefficients for K value

term	coef	SE Cof	T	p
Constant	0.094	0.001705	55.121	0.000
Binder	– 0.0162	0.001348	– 12.013	0.000
Disintegrant	0.012239	0.001348	9.078	0.000
Binder*Binder	– 0.00156	0.001446	– 1.081	0.316
Disintegrant*Disintegrant	– 0.00056	0.001446	– 0.389	0.709
Binder*Disintegrant	0.00075	0.001907	0.393	0.706

Analysis of Variance for K value

source	DF	Seq SS	Adj SS	Adj Ms	T	P
Regression	5	0.003317	0.003317	0.000663	45.62	0.000
Linear	2	0.003297	0.003297	0.001648	113.36	0.000
Square	2	0.000018	0.000018	0.000009	0.62	0.567
Interaction	1	0.000002	0.000002	0.000002	0.15	0.706
Residual Error	7	0.000102	0.000102	0.000015		
Lack-of-Fit	3	0.000102	0.000102	0.000034		
Pure Error	4	0	0	0		
Total	12	0.00319	0.00319			

The observed P value for the both the variables is less than 0.05 and thus these results indicating that these selected two variables are significantly influencing the dissolution rate constant (k).

Estimated Regression Coefficients for K value using data in uncoded units

Term	Coef
Constant	0.102156
Binder	– 0.00342843
Disintegrant	0.00840931
Binder*Binder	– 2.50000E-04
Disintegrant*Disintegrant	– 2.50000E-04
Binder*Disintegrant	0.000200000

The optimized dissolution rate constant can be obtained from the following polynomial equation.

$$Y = b_0 + b_1 X_1 + b_2 X_2 + b_{12} X_1 X_2 + b_{11} X_1^2 + b_{22} X_2^2$$

Y = Dependent variable

X_1 = independent variable binder % concentration

X_2 = independent variable disintegrating agent % concentration

$X_1 X_2$ = Interaction term between binder and disintegrating agent

X_1^2, X_2^2 are polynomial terms

b0 = Arthmatic mean of 4 responses

b_1, b_2, b_{12}, b_{11}, b_{22} = Estimated co-efficients for the coresponding factors X_1, X_2, $X_1 X_2$, X_1^2, X_2^2

$$Y = 0.1022 - 0.00343 X_1 + 0.00841 X_2 + 0.0002 X_1 X_2 - 0.00025 X_1^2 - 0.00025 X_2^2$$

The optimised concentrations of binder and the disintegrating agent can be readily visualized from the following contour plot

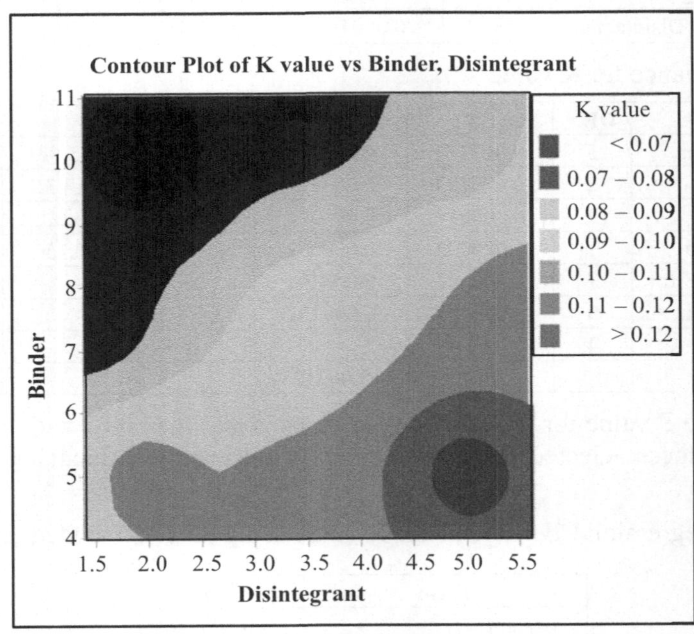

13.7.2 Formulation Optimization with Box-Bhenken Design

The Box-Behnken experimental design, developed by Box and Behnken in 1980, is a useful method for developing second-order response surface models. The Box-Behnken design is based on the construction of balanced incomplete block designs and requires at least three levels for each factor/variable. In Box-Behnken experimental design, the level of one of the factors is fixed at the center level while combinations of all levels of the other factors are applied, the level of the factor C was fixed and then, the combinations of all levels of the factors A and B were applied and subsequently, the same procedures are performed for the factors B and A, respectively.

The variables like A, B and C were chosen as the critical variables and designated as X_1, X_2 and X_3 respectively. The low, middle, and high levels of each variable were designated as –1, 0, and +1 respectively, and given in the following table. The actual design of experiments is given in Table 13.1.

S.no	Level of variable	Number of variables		
		A	B	C
1	Lower (-1)	-1	0	+1
2	Middle (0)	-1	0	+1
3	Higher (+1)	-1	0	+1

In a system involving three significant independent variables (X1, X2 and X3) the mathematical relationship of the response on these variables can be approximated by the second order polynomial equation:

$$Y = \beta_0 + \beta_1X_1 + \beta_2X_2 + \beta_3X_3 + \beta_{12}X_1X_2 + \beta_{13}X_1X_3 + \beta_{23}X_2X_3 + \beta_{11}X_1^2 + \beta_{22}X_2^2 + \beta_{33}X_3^2$$

Where, *Y* is the dependent variable; X_1, X_2 and X_3 are the levels of independent variables; β_0 is the regression coefficient at the center point; β_1, β_2 and β_3 are linear coefficients; β_{12}, β_{13} and β_{23} are quadratic coefficients. The quality of the fit of the polynomial model equation was expressed by the coefficient of determination, R_2. The reason is why 3rd and 5th harmonic currents are chosen as dependent variable is being the highest harmonic currents produced by the single phase loads. For the evaluation of results, a statistical program was used.

Table 13.1 Design matrix of box-Bhenken

Rank	Box-Bhenken experimental design		
	A	B	C
1	–1	–1	0
2	+1	–1	0
3	–1	1	0
4	+1	1	0
5	–1	0	–1
6	+1	0	–1
7	–1	0	+1
8	+1	0	+1
9	0	–1	–1
10	0	+1	–1
11	0	–1	+1
12	0	+1	+1
13	0	0	0
14	0	0	0
15	0	0	0

Coded Values

Sno	Level of variable	Number of variables		
		SSG	MS	TALC
1	Lower (-1)	0.5	0.5	0.5
2	Middle (0)	0.75	1.25	1.25
3	Higher (+1)	1	2	2

Box-Behnken Design

Factors: 03 Replicates: 01

Base runs: 15 Total runs: 15

Base blocks: 01 Total blocks 01 Centre points: 03

Table 13.2 Box-Behnken Design

Std Order	Run Order	Pt Type	Blocks	SSG	MS	TALC	K
02	01	02	01	1.00	0.50	1.25	0.2002
10	02	02	01	0.75	2.00	0.50	0.1277
08	03	02	01	1.00	1.25	2.00	0.1547
11	04	02	01	0.75	0.50	2.00	0.1994
01	05	02	01	0.50	0.50	1.25	0.1022
12	06	02	01	0.75	2.00	2.00	0.1652
07	07	02	01	0.50	1.25	2.00	0.0827
13	08	02	01	0.75	1.25	1.25	0.1054
05	09	02	01	0.50	1.25	0.50	0.1352
14	10	02	01	0.75	1.25	1.25	0.1021
09	11	02	01	0.75	0.50	0.50	0.2175
15	12	02	01	0.75	1.25	1.25	0.0993
04	13	02	01	1.00	2.00	1.25	0.1369
06	14	02	01	1.00	1.25	0.50	0.1108
03	15	02	01	0.50	2.00	1.25	0.1077

Response Surface Regression: K versus SSG, MS, Talc

The analysis was done using coded units.

Estimated Regression Coefficients for K

S.no	Term	Coefficients	SE Coef	T	P
01	Constant	0.101200	0.009809	10.317	0.000
02	SSG	0.021850	0.006007	03.638	0.015
03	MS	– 0.022725	0.006007	– 03.783	0.013
04	TALC	0.001350	0.006007	0.225	0.831
05	SSG*SSG	– 0.010525	0.008842	– 1.190	0.287
06	MS*MS	0.046075	0.008842	5.211	0.003
07	TALC*TALC	0.030175	0.008842	3.413	0.019
08	SSG*MS	– 0.017200	0.008495	– 2.025	0.099
09	SSG*TALC	0.024100	0.008495	2.837	0.036
10	MS*TALC	0.013900	0.008495	1.636	0.163

$S = 0.0169897$

$R - Sq = 94.25\%$

$R - Sq(pred) = 8.34\%$

$R - Sq(adj) = 83.91\%$

Table 13.3 Analysis of Variance for K

Source	DF	Seq SS	Adj SS	Adj MS	F	P
REGRESSION	9	0.023670	0.023670	0.002630	09.11	0.013
LINEAR	3	0.007965	0.007965	0.002655	09.20	0.018
SSG	1	0.003819	0.003819	0.003819	13.23	0.015
MS	1	0.004131	0.004131	0.004131	14.31	0.013
TALC	1	0.000015	0.000015	0..000015	00.05	0.831
SQUARE	3	0.011425	0.011425	0.003808	13.19	0.008
SSG*SSG	1	0.000952	0.000409	0.00409	1.42	0.287
MS*MS	1	0.007111	0.007838	0.007838	27.16	0.003
TALC*TALC	1	0.003362	0.003362	0.003362	11.65	0.019
INTERACTION	3	0.004279	0.004279	0.001426	04.94	0.059
SSG*MS	1	0.001183	0.001183	0.001183	04.10	0.099
SSG*TALC	1	0.002323	0.002323	0.002323	08.05	0.036
MS*TALC	1	0.000773	0.000773	0.000773	02.68	0.163
RESIDUAL ERROR	5	0.001443	0.001443	0.000289		
LACK-OF-FIT	3	0.001438	0.001438	0.000479	176.85	0.006
PURE ERROR	2	0.000005	0.000005	0.000003		
TOTAL	14	0.025113				

Table 13.4 Estimated Regression Coefficients for K using data in uncoded units

S.no	Term	Coefficients
01	Constant	0.261467
02	SSG	0.29400
03	MS	– 0.197167
04	TALC	– 0.259600
05	SSG*SSG	– 0.168400
06	MS*MS	0.0819111
07	TALC*TALC	0.0536444
08	SSG*MS	– 0.091733
09	SSG*TALC	0.128533
10	MS*TALC	0.0247111

Equation

A polynomial equation is derived with the help of estimated regression coefficients for k of coded units is as follows

$$Y = B_0 + B_1X_1 + B_2X_2 + B_3X_3 + B_{12}X_1X_2 + B_{23}X_1X_3 + B_{13}X_1X_3 + B_{11}X_1^2 + B_{22}X_2^2 + B_{33}X_3^2$$

Y = Dependent variable

X_1 = independent variable SSG % concentration

X_2 = independent variable MS % concentration

X_3 = independent variable Talc % concentration

b_0 = Arithmetic mean of 4 responses

b_1, b_2, b_{12}, b_{11}, b_{22} and b_{33} = Estimated co-efficient for the corresponding factors X_1 X_2, X_3, X_1^2, $X_2^2X_3^2$

Therefore the equation is

$$Y = 0.101200 + 0.021850X_1 – 0.022725X_2 + 0.001350X_3$$

$$0.010525X_1^2 + 0.046075X_2^2 + 0.030175X_3^2 – 0.017200X_1X_2 + 0.024100X_1X_3 + 0.013900X_2 X_3$$

The results can also be furnished in the form of graphs.

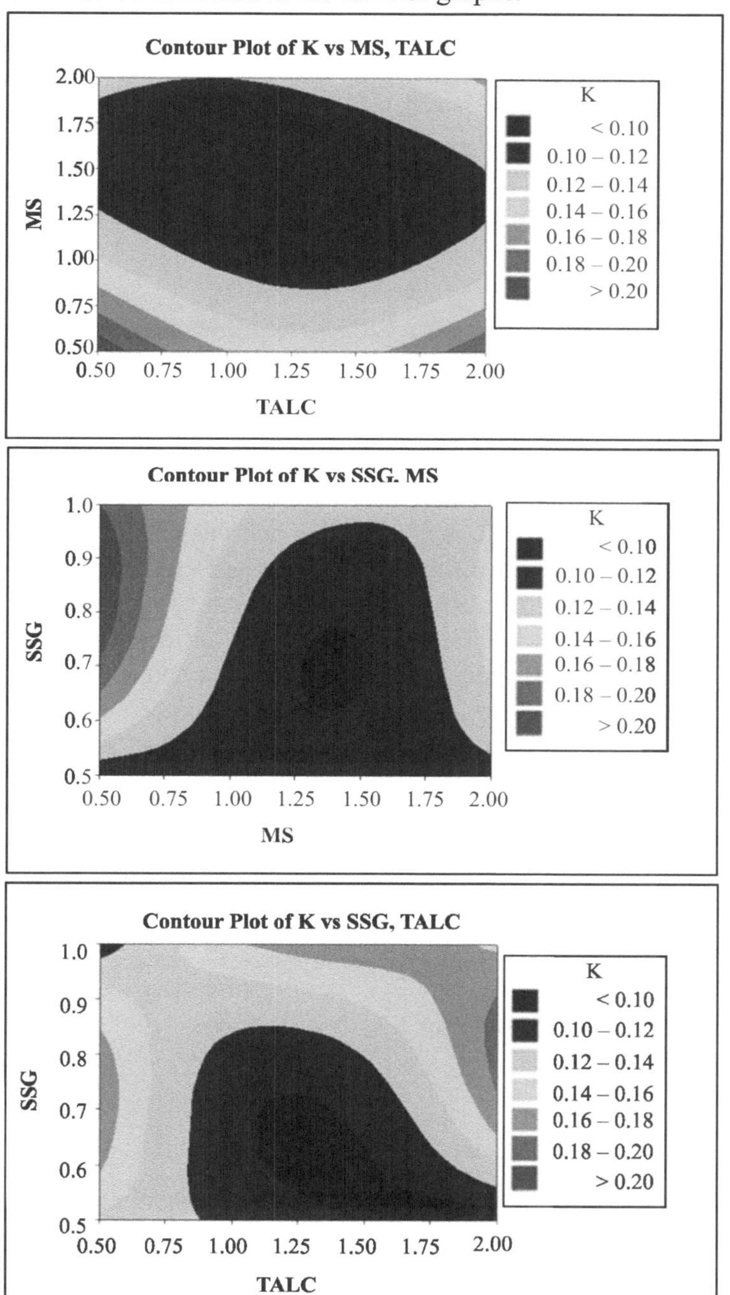

The above plots are useful to select the required concentration of varibles.

CHAPTER 14

DESIGN OF EXPERIMENTS

Design of experiment in a proper or well organized manner is essential for maximum interpretation of the experimental results. It is important in all scientific disciplines, but it is more essential in clinical trials. Good clinical protocol design and data analysis is required by the pharmaceutical manufactures and regulatory authorities. Statistical expertise is helpful in planning, design, execution and analysis of clinical investigations in order to ensure the validity of estimates of safety and efficiency obtained from these studies. In all biological experimentation, the design should be planned so that differences in treatment do not coincide with factors that could influence the outcome such as differences in age, weight, sex, sequence of administration and so forth. This is known as confounding in statistical sciences. Animals/patients/volunteers should be assigned to doses or treatments at random, taking advantage of the availability of optimal experimental designs. The distribution of variables such as age, sex, severity of disease etc may be controlled by stratification.

14.1 Principles of Experimental Design

To study the objectives of experimental pathway, the following factors are important.

- Absence of bias
- Absence of systematic error
- Adequate precision
- Choice of variable
- Simplicity and symmetry

Randomization, replication and local control are the basics of experimental design.

Randomization is a process involving an equal chance to each object for their participation in an experiment. In an experiment of four treatments i.e., placebo, drug A, drug B, drug C, on a group of 60 volunteers, each volunteer has an equal chance of getting into any of the four treatments. Randomization is a scientific procedure to confirm that the experiment is free from bias. The samples can be chosen in a randomized manner from random number table to eliminate bias in the experiment.

Replication is the repetition of an experimental observation to increase the confidence on experimental results. To get error free results, the total number of volunteers required to make the necessary replications is increased. Increased size of the sample minimizes the standard and sample error and the confidence in the result increases. The increase in replication increases the accuracy of experimental results but it is decided by other factors such as time, economy and ethical considerations.

Local control involves the controlling of heterogeneity of a population involving the experiment. In case of human volunteers, the heterogeneity may be due to the variations in age, weight, sex, disease, etc. The heterogeneity is controlled by dividing the population into blocks such a way that each block is homogeneous and heterogeneity is distributed in between blocks.

The choice of design depends primarily on the variability in observations. The major source of variability arises from the differences between subjects. The criteria for choosing an appropriate design is whether the design can identify, estimate and isolate inter-subject variability in data analysis. The designs often considered for bioavailability/ bio equivalence studies include the complete randomized designs (parallel design), the randomized block designs, the cross over designs, the balanced incomplete block design and the Latin square design.

14.2 Selection of Design

The factors to be considered in selecting the suitable design is given below

- Number or formulations to be compared.
- Characteristics of the drug and its disposition.
- Study objectives.
- Availability of subjects.
- Inter and intra subject variability.
- Duration of study and the number of dosing periods allowed.
- Cost considerations.
- Dropout rates.

14.3 Parallel Design

In a parallel design, two are more drugs effects are compared with each other. The drugs are randomly assigned to different patients. Each patient is allowed to take a single drug.

An equal number of patients are assigned to equal group.

Two group parallel designs is shown below

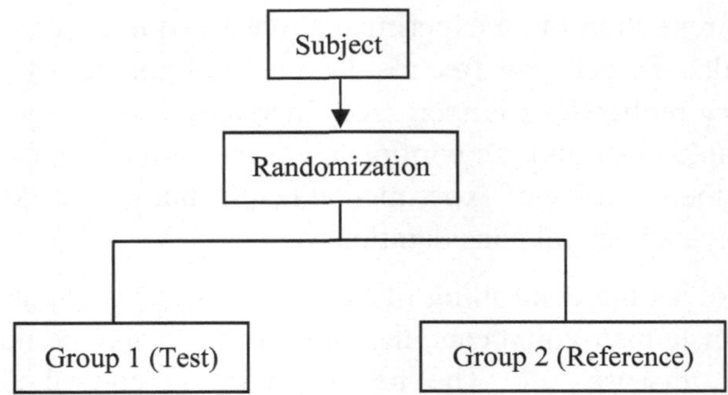

For phase II and phase III clinical trials, the parallel design is most frequently used. However it may not appropriate design for BA/BE studies due to intra and inter subject variability.

The parallel design is not widely used for bio availability studies owing the in capability of identifying and removing the inter subject variability from the comparison between formulations.

In following occasions, parallel design is more appropriate.

- FDA requires a three - arm parallel design (test, reference and vehicle control)
- The drugs having very long half life.
- In case of studies conducted on very ill patients, as studies can be completed quickly.

If 40 subjects are included in the design, then 20 subjects are allotted to each group based on randomization. To go for randomization, the animals are assigned numbers and then selected by following random number table.

Example: Two formulations containing the drug diclofenac may differ in their *in vivo* performance. The differences in their pharmacokinetic parameters may be determined as follows.

40 volunteers are selected for these studies and divided in to two groups. The observed AUC values are showed in the Table 14.1.

Table 14.1

Person number	AUC Value	
	Formulation 1	Formulation 2
1	140	-
2	-	162
3	-	165
4	-	168
5	-	164
6	136	-
7	138	-
8	142	-
9	-	166
10	-	167
11	-	169
12	144	-
13	143	-
14	142	-
15	147	-
16	-	170
17	-	172
18	145	-
19	-	167
20	144	-
21	143	-
22	144	-
23	-	165
24	-	167
25	-	170
26	142	-
27	149	-
28	146	-
29	-	162
30	-	165

Table 14.1 *Contd...*

Person number	AUC Value	
	Formulation 1	Formulation 2
31	-	164
32	143	-
33	-	166
34	148	-
35	146	-
36	142	-
37	141	-
38	-	168
39	-	170
40	-	172

In the above problem, the 40 persons are divided into two independent groups so that each group containing 20 persons. The first group was administered with the formulation 1 and the next group received the second formulation. The participants are allotted to each group based on randomization. The plasma concentrations of the formulations were determined and the corresponding AUC values were calculated and depicted in the above table. The differences in between the mean AUC values noticed with these two formulations are determined statistically by applying t-test.

Null Hypothesis: There is no significance difference between the mean AUC values observed from formulation 1 and formulation 2.

Table 14.2 AUC Values

S.No.	Formulation 1 (x_1)	Formulation 2 (x_2)
1	140	162
2	136	165
3	138	168
4	142	164
5	144	166
6	143	167
7	142	169
8	147	170
9	145	172
10	144	167
11	143	165
12	144	167
13	142	170

Table 14.2 *Contd...*

S.No.	Formulation 1 (x_1)	Formulation 2 (x_2)
14	149	162
15	146	165
16	143	164
17	148	166
18	146	168
19	142	170
20	141	172
	$\Sigma x_1 = 2865$	$\Sigma x_2 = 3339$

$$\therefore \quad \overline{x_1} = \frac{\Sigma x_1}{20} = \frac{2865}{20} = 143.25$$

and $$\overline{x_2} = \frac{\Sigma x_2}{20} = \frac{3339}{20} = 166.95$$

Standard derivation of value of AUC for formulation 1 is

$$s_1 = \sqrt{\frac{\Sigma x_1^2 - \frac{(\Sigma x_1)^2}{n_1}}{n_1 - 1}} = \sqrt{\frac{410603 - \frac{8208225}{20}}{19}} = \sqrt{\frac{410603 - 410411.25}{19}}$$

$$= \sqrt{\frac{191.75}{19}} = \sqrt{10.09210526} = 3.1768$$

Standard deviation of AUC values for formulation 2 is

$$s_2 = \sqrt{\frac{\Sigma x_2^2 - \frac{(\Sigma x_2)^2}{n_2}}{n_2 - 1}} = \sqrt{\frac{557611 - \frac{11148921}{20}}{19}} = \sqrt{\frac{557611 - 557446.05}{19}} = \sqrt{\frac{164.95}{19}}$$

$$= \sqrt{8.6815} = 2.946$$

Now test statistic

$$t = \frac{\bar{X}_1 - \bar{X}_2}{S_p\sqrt{\frac{1}{N_1} + \frac{1}{N_2}}} = \frac{|2865 - 3339|}{3.063\sqrt{\frac{1}{20} + \frac{1}{20}}} = \frac{474}{(3.063)(0.316227766)}$$

$$= \frac{474}{0.969} = 489.16$$

Calculated t value = 489.16

Tabled t value at the significant level 0.05 with degrees of freedom $n_1 + n_2 - 2 = 38$ is 2.02

∴ Calculated t > tabled t

∴ There is a significance difference between the AUC values observed from two formulations

Example : Three formulations containing the drug diclofenac may differ in their *in vivo* performance. The differences in their pharmacokinetic parameters may be determined as follows. 45 volunteers are selected for these studies and divided in to three groups. The observed c_{max} values are showed in the following table.

Table 14.3

Participant number	C_{max} values		
	Formulation 1	**Formulation 2**	**Formulation 3**
1	4.5	-	-
2	-	-	7.1
3	4.3	-	-
4	-	-	6.9
5	-	-	6.8
6	-	5.8	-
7	-	6.0	-
8	-	5.7	-
9	-	5.9	-
10	4.6	-	-
11	-	5.5	-
12	4.2	-	-
13	4.6	-	-
14	4.7	-	-
15	4.5	-	-
16	4.3	-	-
17	4.4	-	-
18	4.6	-	-
19	-	6.1	-
20	-	-	6.4
21	-	-	7.0
22	4.5	-	-
23	-	-	7.2
24	-	-	6.8
25	-	5.8	-
26	-	6.1	-
27	4.4	-	-
28	-	6.2	-
29	-	5.7	-
30	-	-	6.5
31	4.2	-	-
32	-	5.9	-

Table 14.3 *Contd...*

Participant number	C_{max} values		
	Formulation 1	Formulation 2	Formulation 3
33	-	-	6.7
35	-	-	6.6
36	4.3	-	-
37	-	5.8	-
39	-	6.3	-
40	-	-	6.8
41	-	6.2	-
42	-	-	7.0
43	-	-	7.2
44	4.5	-	-
45	-	6.4	-

In the above problem, the 45 persons are divided in to three independent groups so that each group containing 15 persons. The first group was administered with the formulation 1, second group received formulation 2 and the next group received the third formulation. The participants are allotted to each group based on randomization. The plasma concentrations of the formulations were determined and the corresponding c_{max} values were calculated and depicted in the above table. The differences in between the mean c_{max} values noticed with these three formulations are determined statistically by applying ANOVA.

Table 14.4 C_{max} values

S.No.	Formulation 1	Formulation of 2	Formulation of 3
1	4.5	5.8	7.1
2	4.3	6.0	6.9
3	4.6	5.7	6.8
4	4.2	5.9	6.4
5	4.6	5.5	7.0
6	4.7	6.1	7.2
7	4.5	5.8	6.8
8	4.3	6.1	6.5
9	4.4	6.2	6.7
10	4.6	5.7	6.8
11	4.5	5.9	6.6
12	4.4	5.8	6.9
13	4.2	6.3	6.8
14	4.3	6.2	7.0
15	4.5	6.4	7.2
Total	$\Sigma x_{c_1} = 66.6$	$\Sigma x_{c_2} = 89.4$	$\Sigma x_{c_3} = 102.7$

Null Hypothesis: There is no significance difference between the C_{max} values of 3 formulations

$$\Sigma x^2 = (4.5)^2 + (4.3)^2 + \dots\dots\dots + (7.2)^2 = 1533.69$$

$$\frac{\Sigma x_c^2}{n_c} = \frac{(\Sigma x_{c_1})^2}{n_1} + \frac{(\Sigma x_{c_2})^2}{n_2} + \frac{(\Sigma x_{c_3})^2}{n_3}$$

$$= \frac{(66.6)^2}{15} + \frac{(89.4)^2}{15} + \frac{(102.7)^2}{15}$$

$$= \frac{(66.6)^2 + (89.4)^2 + (102.7)^2}{15}$$

$$= \frac{22975.21}{15} = 1531.68$$

and $$\frac{(\Sigma x)^2}{N} = \frac{(66.6 + 89.4 + 102.7)^2}{45} = \frac{(258.7)^2}{45} = \frac{66925.69}{45}$$

$$= 1487.24$$

Total sum of squares $= \text{SST} = \Sigma x^2 - \frac{(\Sigma x)^2}{N}$

$$= 1533.69 - 1487.24$$

$$= 46.45$$

Between sum of squares

$$\text{SSC} = \left(\frac{(\Sigma x_{c_1})^2}{^nC_1} + \frac{(\Sigma x_{c_2})^2}{^nC_2} + \frac{(\Sigma x_{c_3})^2}{^nC_3}\right) - \frac{(\Sigma x)^2}{N}$$

$$= 1531.680 - 1487.237$$

$$= 44.44$$

Error sum of squares SSE = SST – SSC

$$= 46.45 - 44.44$$

$$= 2.01$$

Table 14.5 ANOVA Table

Source of variation	D.f	Sum of squares (S.S)	Mean squares	F
Between groups	C – 1 = 3 – 1 = 2	SSC = 44.44	$\frac{SSC}{C-1}$ =22.22= x	$\frac{x}{y}$ =132.711
Error	N – C = 15 –3=12	SSE=2.01	$\frac{SSE}{N-C}$ =0.167= y	
Total	N – 1 = 15 – 1=14	46.452444		

Calculated F value = 132.71

Tabled F value at the level of significance 0.05 with degrees of freedom n_1 = 2, and n_2 = 12 is 3.89

∴ Calculated F values > table F value

∴ There is a significance difference between the three formulations

14.4 Randomized Block Design

In the randomized block design (RBD), or, more correctly, the randomized complete block design, the experimental units are first classified in to groups or blocks that are as needs alike as possible. The treatments are then assigned to the experimental units (subjects/ patients/ volunteers) within blocks in such a way that each treatment occurs the same number of times, usually once, within each block. The object is to make variation from subject to subject as small as possible within the blocks while maximizing the variation among blocks. The subjects are blocked based on identifiable sources of variation. The subjects are blocked based on differences in age, weight, sex etc.

Advantages and Disadvantages

The following advantages are observed from this design.

- It can remove one source of variation from experimental error and thus increase precision.
- By placing blocks under different conditions, the scope of the experiment can be broadened.
- Any number of treatments and any number of blocks may be used.
- Statistical analysis of the results is fairly simple.

The following disadvantages are noticed.

- Missing data can cause some difficulty in the analysis.
- Assignment of treatment to the wrong block can lead to the problem in the analysis.

- Less efficient in presence of more than one source of unwanted variation
- RBD is less efficient than CRD, if the experimental units are uniform.

In BA/BE studies, the performance of a dosage form may vary due to the differences in age of the subject or due to the differences in formulation factors of the dosage form. Such studies can be conducted with RBD design. The subjects should be categorized based on their age into different blocks. The subjects confined to each block are randomly divided into different groups depending on the number of formulations to be tested. For example to study the influence of age and formulation on bioavailability, the subjects can be categorized based on their age. Based on age, the subjects can be categorized as age group 20-30, 31-40, 41-50, 51-60. A total number of 60 volunteers can be selected and then divided based on their age in such a way that each group contains 15 volunteers. Each subject in this group of 15 numbers is allotted a number starting from 1 to 15. From 15 subjects, 5 subjects should be selected randomly and placed in an independent group to receive a formulation separately. The patient numbers 4, 2, 9, 7 and 11 may receive formulation 1, the subjects 8, 5, 1, 15 and 10 may receive the second formulation and the remaining subjects 3, 6, 14, 13 and 12 receive the third formulation. These subjects are chosen based on random tables. Then their plasma concentrations are evaluated and the mean plasma concentrations from each block corresponding to the selected age is calculated and tabulated as follows. The tabulated data can be treated statistically to detect the differences existing in between age of the patient and the formulation.

Table 14.6

Age group	**AUC Values**		
	R	**T_1**	**T_2**
20-30	140	150	160
31-40	132	145	154
41-50	128	140	149
51-60	120	136	142

Where R indicates reference product and T_1, T_2 indicates test products 1 and 2 respectively.

The above data is subjected to two-way ANOVA in order to determine the statistical differences existing in between rows (age) and in between the columns (formulations).

Null Hypothesis: There is no significance difference between the rows (age) and columns (formulations)

Table 14.7

AUC values				
Age group	**R**	$\mathbf{T_1}$	$\mathbf{T_2}$	
20-30	140	150	160	$R_1 = 450$
31-40	132	145	154	$R_2 = 431$
41-50	128	140	149	$R_3 = 417$
51-60	128	136	142	$R_4 = 406$
	$C_1 = 528$	$C_1 = 571$	$C_3 = 605$	1704

$$\Sigma x^2 = (140)^2 + (132)^2 + (128)^2 + \ldots\ldots\ldots\ldots\ldots\ldots + (142)^2$$

$$= 243094$$

$$\Sigma x = 1704$$

$$\frac{(\Sigma x)^2}{N} = \frac{(1704)^2}{12} = 241968$$

$\therefore$ Total Sum of squares SST = 243094 – 241968

$$= 1126$$

Column sum of squares SSC $= \dfrac{C_1^2 + C_2^2 + C_3^2}{4} - \dfrac{(\Sigma x)^2}{N}$

$$= \frac{(528)^2 + (571)^2 + (605)^2}{4} - 241968$$

$$= \frac{970850}{4} - 241968$$

$$= 242712.5 - 241968 = 744.5$$

Row Sum of Squares SSR $= \dfrac{R_1^2 + R_2^2 + R_3^2 + R_4^2}{3} - \dfrac{(\Sigma x)^2}{N}$

$$= \frac{(450)^2 + (431)^2 + (417)^2 + (406)^2}{3} - 241968$$

$$= \frac{726986}{3} - 241968$$

$$= 242328.6667\text{-}241968$$

$$= 360.67$$

∴ Error sum of squares SSE =

Total sum of squares – (column sum of squares + row sum of squares)

= 1126 – (744.5+360.67)

= 1126-1105.17

= 20.83

Table 14.8 ANOVA Table

Source of variance	D.f	S.S	M.S	F
Between Columns	c –1 = 3 – 1= 2	SSC 744.5	$\frac{SSC}{c-1} = \frac{744.5}{2} = 372.5 = x$	$\frac{x}{z} = 107.35$
Between Rows	r – 1 = 4 – 1 = 3	SSR 360.67	$\frac{SSR}{r-1} = \frac{36067}{3} = 12022 = y$	$\frac{y}{z} = 34.65$
Error	(c – 1) (r – 1) = 6	SSE 20.83	$\frac{SSE}{(c-1)(r-1)} = \frac{20.83}{6} = 3.47 = z$	
Total	cr – 1 = 12 –1 = 11	1126		

F values for Formulation

Calculated F value = 100.075

Tabled F value at the level of significance 0.05 with degrees of freedom $n_1 = 2$, $n = 6$ is. 5.14

∴ Calculated F > Tabled F

∴ Null hypothesis Rejected

∴ There is a significance difference between formulations

F values for Age

Calculated f value = 32.30

Tabled f value at the level of significance 0.05 and degrees of freedom for $n_1 = 3$, $n_2 = 6$ is 4.76

∴ Calculated F > tabled F

∴ Null hypothesis Rejected

∴ There is significance difference between age factors

14.5 Crossover Designs

A crossover design is a modified, randomized block design in which each block receives more than one formulation of a drug at different periods. A block may be a subject or a group of subjects. Subject in each block receive a different sequence of formulations. A crossover design is called as complete crossover design if each sequence contains each formulation. Crossover design is referred as g*p crossover design if there are g sequences of formulations administered at p different periods.

In this design, each subject takes each treatment on different occasions. To conduct BA/BE studies on two or more formulations of the same drug, usually crossover designs are designed. Each person/patient acts as his/her own control. These designs are efficient in reduction of inter individual variations. Crossover designs are not appropriate in case of carryover effects or other such anticipated interactions. The crossover design is more economical (cost effective) with good precision. The number of subjects is reduced in crossover designs. The time required to complete crossover study is more compared to parallel design due to extra testing period. The crossover designs are not useful in case of higher intra subject variations. Missing data pose a more serious problem in case of crossover designs and such chances of missing data is more in crossover design due to the collection of data in more number of occasions compared to single occasion of parallel design. Further patient compliance is less in crossover designs.

For BA/BE studies, the cross over design is favored by several regulatory agencies due to the following advantages

- Each subject serves as his / her own control
- It removes the inter subject variability
- It provides the best un biased estimate for the difference between formulations.

Crossover Design for two Formulations

The most commonly used statistical design for comparing average bio-availability between two formulations of a drug probably is a two sequence, two periods cross over design (2 × 2 cross over design). In 2 × 2 standard cross over design, each subject is randomly assigned to either sequence reference-standard or the sequence standard-reference at two dosing periods. The dosing periods are separated by a wash out period. An example of 2 × 2 designs is shown below

Table 14.9

<table>
<tr><th></th><th></th><th></th><th colspan="3">Period</th></tr>
<tr><td rowspan="2">subjects</td><td rowspan="2">Randomization</td><td>Sequence 1</td><td>Reference</td><td rowspan="2">Washout period</td><td>Test</td></tr>
<tr><td>Sequence 2</td><td>Test</td><td>reference</td></tr>
</table>

The number of formulations may not be equal to the number of dosing periods. In 2 × 3 cross over design, two formulations are fed in a sequence with three dosing periods and this design is given below.

Table 14.10

<table>
<tr><th></th><th></th><th></th><th colspan="5">Period</th></tr>
<tr><td rowspan="2">subjects</td><td rowspan="2">Randomization</td><td>Sequence 1</td><td>Test</td><td rowspan="2">Washout period</td><td>Reference</td><td rowspan="2">Washout period</td><td>Test</td></tr>
<tr><td>Sequence 2</td><td>Reference</td><td>Test</td><td>Reference</td></tr>
</table>

To increase the precision, to get a un biased estimate for the direct drug effect, for precise estimation of carry over effects and for estimation of intra subject variability, higher order cross over designs are used. In higher order cross over design, either the number of periods or the number of sequences is greater than the number of formulations to be compared. Such designs (A-C) are furnished below.

Table 14.11 Design A

Sequence	Dosing period	
	1	2
1	T	T
2	R	R
3	R	T
4	T	R

Table 14.12 Design B

Sequence	Dosing period		
	1	2	3
1	T	R	R
2	R	T	T

Table 14.13 Design C

Sequence	Dosing period			
	1	2	3	4
1	T	T	R	R
2	R	R	T	T
3	T	R	R	T
4	R	T	T	R

The following assumptions are involved in the above designs:

- The number of items in each column are same
- The variations in each column are same. $\sigma_1^2 = \sigma_2^2 = \sigma_3^2$
- There is no direct drug by carry over interactions.

Design A is formed by adding two more sequences to the standard 2 × 2 crossover designs. This design is used to estimate the carryover effect and intra subject variability. Design B is the class of the crossover design with two sequences, three periods and two formulations. This design is obtained by adding an additional dosing period to the standard 2 × 2 crossover designs. Design C is a class of crossover design with four sequences, four periods and two formulations. In this design, the variances associated with direct drug effects and carryover effects are reduced.

Crossover Design for three Formulations

The crossover design suitable for study of BA/BE studies containing three formulations one reference product (R) and two test products (T_1, T_2) is shown below.

Table 14.14

Sequence	Dosing period		
	1	2	3
1	R	T_1	T_2
2	T_1	T_2	R
3	T_2	R	T_1
4	R	T_2	T_1
5	T_1	R	T_2
6	T_2	T_1	R

Cross Over Design for Four Formulations

The crossover design suitable for study of BA/BE studies containing four formulations one reference product (R) and three test products (T_1, T_2, T_3) is shown below.

Table 14.15

Sequence	Dosing period			
	1	**2**	**3**	**4**
1	R	T_1	T_2	T_3
2	T_1	R	T_3	T_2
3	T_2	T_3	R	T_1
4	T_3	T_2	T_1	R
5	R	T_3	T_1	T_2
6	T_1	T_2	R	T_3
7	T_2	T_1	T_3	R
8	T_3	R	T_2	T_1
9	R	T_2	T_3	T_1
10	T_1	T_3	T_2	R
11	T_2	R	T_1	T_3
12	T_3	T_1	R	T_2

Example: Patients are involved in a bioavailability study and they are divided into two groups. The observed C_{max} values are furnished in the following table.

Table 14.16

Group	Subject number	Dosing period	
	1	1	2
1	1	78	86
1	4	80	91
1	5	77	93
1	7	82	88
1	10	81	90
2	2	88	77
	3	92	79
	6	90	83
	8	87	82
	9	89	80

Table 14.17 The above table can also be presented as follows

Subject no	C_{max} values		Order of administration
	R	T_1	2
1	78	86	R-T_1
2	77	88	T_1-R
3	79	92	T_1-R
4	80	91	R-T_1
5	77	93	R-T_1
6	83	90	T_1-R
7	82	88	R-T_1
8	82	87	T_1-R
9	82	87	T_1-R
10	81	90	R- T_1

Table 14.18 The above data is treated statistically as follows

Subject No.	C_{max} values		
	R	T_1	Order of administration
1	78	86	R-T_1
2	77	88	T_1-R
3	79	92	T_1-R
4	80	91	R-T_1
5	77	93	R-T_1
6	83	90	T_1-R
7	82	88	R-T_1
8	82	87	T_1-R
9	82	87	T_1- R
10	81	90	R-T_1
	$\Sigma x_1 = 801$	$\Sigma x_2 = 892$	

$$\text{Average } \overline{x_1} = \frac{\Sigma x_1}{10} = \frac{801}{10} = 80.1$$

$$\text{Average } \overline{x_2} = \frac{\Sigma x_2}{10} = \frac{892}{10} = 89.2$$

$$\Sigma x_i^2 = x_1^2 + x_2^2 + \dots\dots\dots\dots\dots + x_{10}^2 = 78^2 + 77^2 + \dots\dots(90)^2 = 143821$$

$$\text{C.F} = \frac{(\Sigma x)^2}{N} = \frac{(801+892)^2}{20} = \frac{(1693)^2}{20} = \frac{2866249}{20} = 143312.45$$

$$\text{Total sum of squares} = \Sigma x^2 - \frac{(\Sigma x)^2}{N} = 143821 - 143312.45 = 508.55$$

formulation sum of squares

$$= \frac{\Sigma C_i^2}{10} - C.F$$

$$= \frac{C_1^2 + C_2^2}{10} - C.F = \frac{(801)^2 + (892)^2}{10} - C.F$$

$$= \frac{641601 + 795664}{10} - 143312.45$$

$$= 414.05$$

Subject sum of squares

$$= \frac{\Sigma R_i^2}{2} - \text{C.F}$$

$$= \frac{(78+86)^2 + (77+88)^2 + \dots\dots\dots + (81+90)^2}{2} - (C.F)$$

$$= \frac{\left((164)^2 + (165)^2 + (171)^2 + (171)^2 + (170)^2 + (173)^2 + (170)^2 + (169)^2 + (169)^2 + (171)^2\right)}{2} - \text{C.F}$$

$$= \frac{286695}{2} - 143312.45 = 143347.5 - 143312.45 = 35.05$$

$$\text{Order sum of squares} = \frac{\Sigma O_i^2}{10} - \text{C.F}$$

$$= \frac{(842)^2 + (851)^2}{10} - \text{C.F}$$

$$= \frac{708964 + 724201}{10} - 143312.45$$

$$= \frac{1433165}{10} - 143312.45$$

$$= 143316.5 - 143312.45$$

$$= 4.05$$

Where O_1 is sum of result of first visit and O_2 is the sum of results for second visit and error sum of squares = 508.55 – (35.05 + 414.05 + 4.05)

Table 14.19 ANOVA Table

Source of variance	**D.f**	**S.S**	**M.S**	**F**
Subjects	r – 1 = 10 – 1 = 9	35.05	$\frac{35.05}{9} = 3.89$	$\frac{3.89}{6.925} = 0.5617$
Formulations	c – 1 = 2 – 1 = 1	414.05	$\frac{414.05}{1} = 414.09$	$\frac{414.09}{6.925} = 59.7963$
Order	2 – 1 = 1	4.05	4.05	$\frac{4.05}{6.925} = 0.5848$
Error	8	55.4	6.925	
Total	N – 1 = 19	508.55		

F values for subjects

Calculated F value = 0.5617

Tabled F value at the level of significance 0.05

With degree of freedom 9, 8

$F_{(9, 8)} = 3.39$

Calculated F value < tabled F value

∴ There is no significance difference between subjects.

F values for formulation:

Calculated f value = 59.7963

Tabled F value = F(1, 8) = 5.32

Calculated F value > tabled f value

So there is a significance difference between formulations

F values for order of administration

Calculated F value = 0.5848

Tabled F value = F(1,8) = 5.32

∴ Calculated F value < tabled F value

So, there is no significance difference between the orders of administration

14.6 Balanced Incomplete Block Design (BIBD)

A BIBD design is an incomplete block design in which any two treatments appear together an equal number of times. In this design, every treatment is not present in every block. When all treatment comparisons are equally important, the treatment combinations used in each block should be selected in a balanced manner so that any pair of treatments occurs together the same number of times as any other pair.

In certain experiments using RBD design, we may not be able to run all the treatment combinations in each block. This problem occurs because of shortage of experimental apparatus or facilities or physical size of the block. Randomized incomplete block designs are used for this type of problem.

Crossover designs are not appropriate for BA/BE studies containing more than three formulations due to the following reasons.

- More time consumption due to the washout period in between treatment periods.
- Not desirable to draw many blood samples for each subject.
- Increase in number of dropouts due to too many dosing periods

Incomplete block designs may be used for bioavailability studies when more than 2 formulations are to be compared. For example, if three formulations are included in a study, a full crossover would require three visits. In BIBD design, each subject takes only 2 of the three possible formulations (R, T_1 and T_2) in a symmetric pattern. Six subjects taking the order R – T_1, R – T_2, T_1 – T_2, T_1 – R, T_2 – R, T_2 – T_1.

Table 14.20 The BIBD design for four formulations with two dosing periods is shown below

Sequence	Dosing period	
	1	2
1	R	T_1
2	T_1	T_2
3	T_2	T_3
4	T_3	R
5	R	T_2
6	T_1	T_3
7	T_3	T_1
8	T_2	R
9	R	T_3
10	T_3	T_2
11	T_2	T_1
12	T_1	R

Table 14.21 The BIBD design for four formulations with three dosing periods is shown below

Sequence	Dosing period		
	1	2	3
1	T_1	T_2	T_3
2	T_2	T_3	R
3	T_3	R	T_1
4	R	T_1	T_2

E.g. three formulations R, T_1 and T_2 are administered to six subjects in BIBD design as follows

Table 14.22

Subject no	Dosing period	
	1	2
1	R	T_1
2	R	T_2
3	T_1	T_2
4	T_1	R
5	T_2	R
6	T_2	T_1

The time required to attain the maximum plasma concentration (T_{max}) with these three formulations is depicted in the following table. Treat the data statistically to detect the differences in T_{max} values observed due to the differences in subjects, formulations and order of administration.

Table 14.23

Subject no	T_{max} Values			Order of administration
	R	T_1	T_2	
1	4.5	3.8	-	R-T_1
2	4.7	-	3.4	R-T_2
3	-	3.6	3.2	T_1-T_2
4	4.6	3.8	-	T_1-R
5	4.7	-	3.4	T_2-R
6	-	3.7	3.3	T_2- T_1

The variations in the T_{max} values may be due to three different sources such as subjects, formulation and the order of administration. These variations are identified by treating the data with three-way ANOVA.

Table 14.24

Subject Number	T_{max} values				
	R	T_1	T_2	Row Total	Order of Administration
1	4.5	3.8	-	8.3	$R\text{-}T_1$
2	4.7	-	3.4	8.1	$R\text{-}T_2$
3	-	3.6	3.2	6.8	$T_1\text{-}T_2$
4	4.6	3.8	-	8.4	$T_1\text{-}R$
5	4.7	-	3.4	8.1	$T_2\text{-}R$
6	-	3.7	3.3	7	$T_2\text{-}T_1$
Column Total	18.5	14.9	13.3	46.7	

$$\Sigma x^2 = (4.5)^2+(4.7)^2 + (4.6)^2 + \dots\dots\dots + (3.3)^2 = 187.5$$

$$\text{C.F} = \frac{(\Sigma x)^2}{n} = \frac{(46.7)^2}{12} = \frac{2180.89}{12} = 181.7408333$$

$$\text{Total sum of squares} = \Sigma x^2 - \frac{(\Sigma x)^2}{n}$$

$$= 187.5 - 181.7408333$$

$$= 5.759166667$$

Formulation sum of squares

$$= \frac{(18.5)^2 + (14.9)^2 + (13.3)^2}{4} - \text{C.F}$$

$$= \frac{741.15}{4} - 181.7408333$$

$$= 185.2875 - 181.7408333 = 3.5466667$$

Subject sum of squares =

$$= \frac{(8.3)^2 + (8.1)^2 + (6.8)^2 + (8.4)^2 + (8.1)^2 + (7)^2}{2} - \text{C.F}$$

$$= \frac{(18.5)^2 + (14.9)^2 + (13.3)^2}{4} - 181.7408333$$

$$= 182.955 - 181.7408333 = 1.21416667$$

Order of administration sum of squares

Subject no	T_{max} value	
1	4.5	3.8
2	4.7	3.4
3	3.6	3.2
4	3.8	4.6
5	3.4	4.7
6	3.3	3.7
Total	23.3	23.4

Order of administration sum of squares = square of the sum of the first column + square of the sum of the second column/number of subjects

$$= \frac{(4.5+4.7+3.6+3.8+3.4+3.3)^2 + (3.8+3.4+3.2+4.6+4.7+3.7)^2}{6} - \text{C.F}$$

$$= \frac{(23.3)^2 + (23.4)^2}{6} - 181.7408333$$

$$= 181.7416667 - 181.7408333$$

$$= 0.000833367$$

∴ Error sum of squares = (Total sum of squares) – (Formulation sum of squares + subject sum of squares + order of administration sum of squares)

$$= 5.759166667 - (3.5466667 + 1.21416667 + 0.000833367)$$

$$= 5.759166667 - (4.761666737)$$

$$= 0.999749993$$

Table 14.25 ANOVA Table

Source of variance	**D.f**	**S.S**	**M.S**	**F value**
Between formulation	c – 1 = 3 – 1 = 2	3.5466667	$\frac{3.5466667}{2}$ = 1.7733335	$\frac{1.7733335}{0.249937}$ = 7.0951219
Between Subjects	r – 1 = 6 – 1 = 5	1.21416667	$\frac{1.21416667}{5}$ = 0.242833334	$\frac{0.242833337}{0.249937}$ = 0.97144991
Between order of administration	2 – 1 = 1	0.000833367	0.000833367	$\frac{0.000833367}{0.249937}$ = 0.003334308
Error	4	0.999749993	$\frac{0.999749993}{4}$ = 0.249937	
Total	n – 1 = 12			

F value for Formulation

Calculated F value = 7.0951219

Tabled F value at the level of significance 0.05 with degrees of freedom (2, 4) = $F_{(2,4)}$ = 6.94

Calculated F > tabled F

∴ There is a significance difference between formulations

F values for subjects

Calculated F value = 0.9714

Tabled F value at the of significance 0.05

With degrees of freedom (5, 4) = $F_{5,4}$ = 6.26

∴ Calculated F value < tabled F value

∴ There is no significance difference between subjects

F values for order of administration

Calculated F value = 0.0033343

Tabled F value at the level of significance 0.05

With degrees of freedom 1, 4 = $F_{1,4}$ = 7.71

∴ Calculated F value < tabled F value

∴ There is no significance difference between order of administration

14.7 Latin Square Designs

A Latin square design is an n × n array filled with n different symbols, each occurring exactly once in each row and exactly once in column. It is shown as follows.

A B C
C A B
B C A

A Latin square design is said to be reduced/ normalized/ standard if both its first row and its first column are in their natural order. Such design is showed as follows.

A B C
B C A
C A B

Silent features of Latin square design

- Able to control variation in two directions.
- Treatments are arranged in rows and columns.

- Each row contains every treatment.
- Each column contains every treatment.
- The most common sizes are 5 × 5 to 8 × 8

Advantages of the Latin square design

- Possible to control variations in two directions.
- Increase efficiency compared to randomized complete block designs

Disadvantages of the Latin square design

- The number of treatments must equal to number of replicates.
- The experimental error is likely to increase with the size of the square.
- Small squares have very few degrees of freedom for experimental error.
- Not possible to evaluate interactions between
  1. Rows and columns
  2. Rows and treatments
  3. Columns and treatments

The pharmacokinetic parameters are influenced by the parameters such as age and formulation factors. To conduct BA/BE studies on four different formulations and to evaluate the influence of age (21-60) the subjects can be blocked or divided into different groups/rows based on the age of the subject. The number of blocks should be equal to the number of formulations to be tested in contrast to RBD design. For example, to compare the AUC values of four formulations on different subjects, the subjects are blocked as four groups such a way that group A contains the group of subjects having age between 21-30, group B contains the group of subjects having age between 31-40, group C contains the group of subjects having age between 41-50 and the group D contains the group of subjects having age between 51-60. In this design each group receives each formulation in contrast to RBD design. In the first dosing period, group 1-4 patients receive formulations A-D respectively. In second dosing period, group 1-4 patients receive formulations B, C, D, A respectively. In third dosing period, group 1-4 patients receive formulations C, D, A, B respectively. In last dosing period, group 1-4 patients receive formulations D, A, B, C respectively. In between two dosing periods, washout period is maintained. Generally each group contains 4-8 subjects. Thus in Latin square designs each subject receives each formulation and intra and inter subject variations are minimized. This design is more time consuming and expensive. These designs are also not useful in case of carry over effects and interactions.

E.g. the AUC values observed from four groups administered with four formulations by following Latin square design is furnished below. Treat the data statistically.

Table 14.26

Group	Formulation				Order of administration
	1	2	3	4	
1	124	140	152	161	A-B-C-D
2	110	132	138	152	B-C-D-A
3	94	123	129	138	C-D-A-B
4	86	115	118	127	D-A-B-C

The variations in the AUC values may be due to three different sources such as groups (rows), formulation (columns) and the order of administration. These variations are identified by treating the data with three-way ANOVA.

Solution

Table 14.27

Group	Formulation					
	A	B	C	D	Row Total	Order of Administration
1	124	140	152	161	577	A-B-C-D
2	110	132	138	152	532	B-C-D-A
3	94	123	129	138	484	C-D-A-B
4	86	115	118	127	446	D-A-B-C
Column Total	414	510	537	578	2039	

$$\Sigma x^2 = (124)^2 + (110)^2 + (94)^2 + \ldots\ldots\ldots\ldots + (127)^2 = 265997$$

$$\text{C.F.} = \frac{(\Sigma x)^2}{N} = \frac{(2039)^2}{16} = \frac{4157521}{16} = 259845.0625$$

Total sum of squares = $\Sigma x^2 - \text{C.F} = 265997 - 259845.0625 = 6151.9375$

Column (formulation) sum of squares

$$= \frac{C_1^2 + C_2^2 + C_3^2 + C_4^2}{4} - \text{C.F}$$

$$= \frac{(414)^2 + (510)^2 + (537)^2 + (578)^2}{4} - 259845.0625$$

$$= \frac{1053949}{4} - 259845.0625$$

$$= 263487.25 - 259845.0625$$

$$= 3642.1875$$

Row (Groups) sum of squares

$$= \frac{R_1^2 + R_2^2 + R_3^2 + R_4^2}{4} - \text{C.F}$$

$$= \frac{(577)^2 + (532)^2 + (484)^2 + (446)^2}{4} - 259845.0625$$

$$= \frac{1049125}{4} - 259845.0625$$

$$= 262281.25 - 259845.0625 = 2436.225$$

Order of administration sum of squares:

$$= \frac{(512)^2 + (502)^2 + (513)^2 + (512)^2}{4} - 259845.0625$$

$$= \frac{1039461}{4} - 259845.0625$$

$$= 259865.25 - 259845.0625$$

$$= 20.1875$$

Error sum of squares = Total sum of squares – (Column sum of squares + Row sum of squares + order of administration sum of squares)

$$= 6151.9375 - (3642.1875 + 2436.225 + 20.1875)$$

$$= 6151.9375 - 6098.6 = 53.3375$$

Table 14.28 ANOVA Table

Source of variation	**D.f**	**S.S**	**M.S**	**F value**
Between Formulation (Columns)	c – 1 = 4 – 1 = 3	3642.187	$\frac{3642.187}{3} =$ 1214.395	$\frac{1214.395}{8.8895833} =$ 136.60876
Between Groups (rows)	r – 1 = 4 – 1 = 3	2436.225	812.075	91.35136068
Between order of administration	0 – 1 = 4 – 1 = 3	20.187	6.729	0.756953363
Error	6	53.3375	8.889583333	
Total	N – 1 = 15	6151.9375		

F values for Formulation

The calculated F value = 136.60876

Tabled F value at the level of Significance 0.05 with degrees of freedom 3, 6 = $F_{(3, 6)}$ = 4.76

∴ Calculated F value > Tabled F value

∴ There is a significance difference between formulation

F values for groups

Calculated F value = 91.35130068

Tabled F value at the level of significance 0.05

With degrees of freedom 3, 6 = $F_{3,6}$ = 4.76

Calculated F value > tabled F value

∴ There is a significance difference between groups

F values for order of administration

Calculated F value = 0.756953363

Tabled F value at the level of significance 0.05 with degrees of freedom 3,6 = F_{36} = 4.76

∴ Calculated F value < tabled F value

∴ There is no significance difference between the order of administration

14.8 Carryover Effects in Crossover Designs

The carryover or residual effect is observed when the response on the second period is dependent on the response in the first period. It occurs when the interval between administrations of dosage forms A&B is not sufficient to rid the body of the drug when the dosage form A is given first. It results high plasma concentrations for formulation B when it is subsequently administered with formulation A. More complex designs with suitable number of replications are required to analyze the data and such designs are not easily accommodated in clinical studies.

The test for carryover or sequence effects is performed as follows.

- Compute the sum of squares due to carryover effects by comparing the results of the groups which are differ in their order of administration. In absence of sequence effects, the average results of the groups should be equal. The sum of squares is calculated with the following formulae.

$$(\varepsilon \text{ Group } 1)^2/N_1 + (\varepsilon \text{ Group } 2)^2/N_2 - CT$$

- Calculate the sum of squares between subjects within each group with the following formulae.

$$\frac{1}{2}\,\varepsilon(\text{subject total})^2 - (CT)_1 - (CT)_2$$

- Test the sequence effects by comparing the sequence mean square to the within group mean square.

E.g. The C_{max} values for the bioequivalence study comparing drugs A and B is given below.

Table 14.29

Group 1 (Treatment A first, B second)				**Group 2 (Treatment B first, A second)**			
Subject	**A**	**B**	**Total**	**Subject**	**A**	**B**	**Total**
1	6	15	21	2	8	12	20
3	7	17	24	5	9	14	23
4	5	18	23	6	7	15	22
7	7	15	22	8	8	13	21
10	6	17	23	9	8	15	23
11	7	16	23	12	9	14	23
Total	38	98	136	Total	49	83	132

Solution:

Table 14.30

Group 1 Treatment A first B second				**Group 2 Treatment B first, A second**			
Subject	**A**	**B**	**Total**	**Subject**	**A**	**B**	**Total**
1	6	15	21	2	8	12	20
3	7	17	24	5	9	14	23
4	5	18	23	6	7	15	22
7	7	15	22	8	8	13	21
10	6	17	23	9	8	15	23
11	7	16	23	12	9	14	23
Total	38	98	136	Total	49	83	132

Compute the sum of squares due to carry over effects by comparing the results for group 1 to group 2. It can be determined that in absence of sequence effects the average results for group1 (A first B second) is expected to be equal to the average result for group 1 (B first A second)

The sum of squares is calculated as

$$\frac{(\Sigma \text{ group } 1)^2}{N_1} + \frac{(\Sigma \text{ group } 2)^2}{N_2} - \text{C.F}$$

$$= \frac{(136)^2}{12} + \frac{(132)^2}{12} - \frac{(136+132)^2}{24}$$

$$= 2993.33 - 2992.66 = 0.66$$

The proper error term to test the sequence effect is the within group (sequence) mean square, represented by sum of squares between subjects with in groups (sequence).

The sum of squares is, calculated as follows

$$\frac{1}{2}\Sigma\,(\text{subject total})^2 - (\text{C.F})_1 - (\text{C.F})_2$$

where $(\text{C.F})_1$, $(\text{C.F})_2$) are correction factors for group 1 and group 2 respectively

$$\frac{1}{2}\begin{pmatrix}(21)^2+(24)^2+(23)^2+(22)^2+(23)^2+(23)^2+(20)^2+(23)^2 \\ +(22)^2+(21)^2+(23)^2+(23)^2\end{pmatrix}$$

$$= \frac{6000}{2} - \frac{(136)^2}{12} - \frac{(132)^2}{12}$$

$$= \frac{1}{2}\,[6000] - 1541.333333 - 1452$$

$$= 3000 - 2993.33$$

$$= 6.67$$

Here sequence some of square d.f = 1

and this within group (or subject with in square) sum of squares has 10 d.f from each group the mean square is $\frac{6.666667}{10} = 0.67$

ANOVA Table for the cross over Bioequivalence study (AUC) without sequence effect

$$\text{Correction factor} = \frac{(\Sigma x)^2}{N} = \frac{(136+132)^2}{24} = \frac{71824}{24} = 2992.67$$

$$\text{Total sum of squares} = \Sigma x^2 - \frac{(\Sigma x)^2}{N}$$

$$= 3410 - 2992.67$$

$$= 417.33$$

Subjects sum of squares

$$\frac{1}{2}\begin{pmatrix}(21)^2 + (24)^2 + (23)^2 + (22)^2 + (23)^2 + (23)^2 + (20)^2 + \\ (23)^2 + (22)^2 + (21)^2 + (23)^2 + (23)^2\end{pmatrix} - 2992.6667$$

$$= 3000 - 2992.67 = 7.33$$

Period sum of squares

$$\frac{(38+83)^2}{12} + \frac{(78+49)^2}{12} - 2992.66667$$

$$= 7.333333\ \frac{(121)^2 + (147)^2}{12} - 2992.66667$$

$$= 3020.833333 - 2992.66667$$

$$= 28.167$$

Treatment sum of squares

$$\frac{(38+49)^2}{12} + \frac{(98+83)^2}{12} - 2992.66667$$

$$= \frac{(87)^2 + (181)^2}{12} - 2992.66667$$

$$= 3360.83 - 2992.67$$

$$= 368.167$$

Error sum of squares

Total sum of squares – [subject sum of squares + Period sum of squares + Treatment sum of squares]

$$= 417.33333 - [7.33333 + 28.16666333 + 368.1666633]$$

$$= 417.33 - 403.67$$

$$= 13.66$$

Table 14.31 ANOVA Table

Source of variance	D.f	S.S	M.S	F
Subject	11	7.33	0.66	
Period	1	28.16	28.16	$F_{(1,10)} = 20.61$
Treatments	1	368.16	368.16	
Error	10	13.66	1.36	
Total	23	417.33		

Now test sequence effect by comparing the square mean square to the within group mean square (F test)

$$F_{1,10} = \frac{0.6666666}{0.6666667}$$

$$F_{1,10} = \frac{0.6666666}{0.6666667} = 0.99$$

Calculated F value = 0.99

Tabled F value at the level of significance 0.05 with degrees of freedom 1,10 is 4.84

∴ Calculated F value < tabled F value

If the sequence (carryover) effect is not significant, one would proceed with the usual analysis and interpretation as shown above.

If the sequence (carryover) effect is significant, the usual analysis is not valid. The recommended analyses uses only the first period results and treat the data with one-way ANOVA as follows.

Table 14.32

Subject	A	Subject	B
1	6	2	12
3	7	5	14
4	5	6	15
7	7	8	13
10	6	9	15
11	7	12	14

Solution:

Table 14.33

Subject	A	Subject	B	Total
1	6	2	12	18
3	7	5	14	21
4	5	6	15	20
7	7	8	13	20
10	6	9	15	21
11	7	12	14	21

$$\text{C.F} = \frac{(38+83)^2}{12} = \frac{14641}{12} = 1220.083333$$

$$\Sigma x^2 = 6^2 + 7^2 + \ldots\ldots\ldots\ldots\ldots + (15)^2 + (14)^2 = 1399$$

$$\text{Total sum of squares} = \Sigma x^2 - \frac{(\Sigma x)^2}{N}$$

$$= 1399 - 1220.083333$$

$$= 178.92$$

Between treatment sum of squares =

$$= \frac{(38)^2 + (83)^2}{6} - 1220.083333$$

$$= \frac{8333}{6} - 1220.083 = 1388.83 - 1220.083 = 168.75$$

Table 14.34 ANOVA Table

Source of variance	D.f	S.S	M.S	F
Between treatment	1	168.750	168.250	$\frac{168.125}{1.016} = 165.60$
Within treatments	10	10.1666666	1.016	

Calculate F = 165.60

Tabled F value at level of significance 0.05 with degrees of freedom 1,10 = 4.96

$\therefore$ Calculate F > tabled F

The analysis leads to the conclusion of significant treatment difference.

14.9 ANOVA in Crossover Designs

The usual ANOVA in crossover designs separates the total sum of squares in to 4 components; subjects, periods, treatments and error. In the evaluation of BE analysis, the null hypothesis at 5% level of significance was simply that the products were equal. Such hypothesis had the obvious problem with regard to the power of the test products that showed nearly the same average results with very small variance could show a significant difference, which may not be of clinical significance. Analysis of the log transformed values is currently required by the FDA. The critical parameters are AUC and $C_{max.}$

Table 14.35 Treat with one-way ANOVA after log transformation of the C_{max} values

Subject	A	Subject	B
1	6	2	12
3	7	5	14
4	5	6	15
7	7	8	13
10	6	9	15
11	7	12	14

Solution:

Table 14.36

Subject	A	Log A	Subject	B	Log B
1	6	0.77815125	2	12	1.079181246
3	7	0.84509804	5	14	1.146128036
4	5	0.698970004	6	15	1.176091259
7	7	0.84509804	8	13	1.113943352
10	6	0.77815125	9	15	1.176091259
11	7	0.84509804	12	14	1.146128036
		4.790566625			6.837563188

Correction factor =

$$C.F = \frac{(\Sigma x)^2}{N} = \frac{(11.62812981)^2}{12} = 11.27$$

$$\Sigma x^2 = (0.77815125)^2 + (0.84509804)^2 + \ldots\ldots\ldots + (1.146128036)^2 = 11.64$$

Total sum of squares =

$$\Sigma x^2 - \frac{(\Sigma x)^2}{N} = 11.64 - 11.27 = 0.37$$

Between treatment sum of squares

$$= \frac{(4.790566625)^2 + (6.3375631)^2}{6} - C.F$$

$$= \frac{22.94952859 + 46.75226915}{6} - \text{C.F}$$

$$= \frac{69.70179774}{6} - 11.27$$

$$= 11.62 - 11.27$$

$$= 0.35$$

With in treatment sum of squares = Total sum of squares – between treatment sum of squares

$$= 0.37 - 0.35$$

$$= 0.02$$

Table 14.37 ANOVA Table

Source of variance	S.S	D.f	M.S	F
Between treatment	0.35	1	0.35	$\frac{0.35}{0.0024} = 143.66$
Within treatments	0.024	10	0.0024	
Total	0.37	11		

∴ F calculate value = 143.66

Tabled F value at the level of significant 0.05 with degrees of freedom 1, 10

$$= F_{1,10} = 4.96$$

∴ Calculate F value > tabled F value

∴ The analysis leads to the conclusion of significant treatment differences.

14.10 Sample Size in BA/BE Studies

The approximate sample size needed for bioavailability / bioequivalence studies can be calculated with the following equation.

$$N = 2(t_{a.}2N - 2 + t_{\beta .}2N - 2)^2 \left[\frac{CV}{(V - \delta)}\right]^2$$

N = the number of subjects per sequence.

t = appropriate value from the t distribution

α = the significance level (usually 0.1)

1– β= the power (usually 0.8)

Cv = the coefficient of variation.

V= bioequivalence limit (ln1.25 = 0.223)

δ = difference between products (for 5% difference δ = [ln (1.05) = 0.0488]

If the initial approximation of sample size is 20 per sequence (a total of 40 subjects)

In the above case, the coefficient of variation is assumed to be 30% , as the drug product is highly variable if the error variance shows a coefficient of variation of 30% or greater. The number of subjects increases with the coefficient of variation and the confidence interval (usually 90%) or the significance level (0.1). If the % cv is increased to 60,

Then $N = 2[(1.69 + 0.85)^2 (0.6/ 0.22 - 0.48)]^2 = 12.90 (5.09)$

$= 65.66 \approx 66$

14.11 Replicate Designs for BE Studies

Replicate designs may be defined as designs with more than 2 periods where products are given on more than one occasion. The subjects are administered one or both products on more than one occasion. Replicate studies can isolate the within subject variance of each product separately as well as potential product – by – subject interactions. The FDA recommends the submission of studies with replicate designs be analyzed for average bioequivalence. FDA also recommends using only two sequence designs with three /four period designs because the interaction. Variability estimate, subject x formulation, will be otherwise controlled. Two sequence designs for three and four period studies are shown below.

Three-Period Design:

Sequence	**Period**		
	1	2	3
1.	A	B	B
2.	B	A	A

Four-Period Design:

Sequence	**Period**			
	1	2	3	4
1.	A	B	B	A
2.	B	A	A	B

With replicate designs, carry over effects, within subject variances and subject x formulation interactions can be estimated, uncontrolled with other effects. The analysis of studies with replicate designs would be very difficult without access to a computer program.

CHAPTER 15

VALIDATION

15.1 Introduction

The term validation has in recent years become familiar in connection with pharmaceutical process. Validation of a process is the demonstration that controlling the critical steps of a process results in products of repeatable attributes such as content uniformity or causes reproducible events like sterilization. A validated process is a systematic, documented program that provides a high degree of assurance that a specific process will consistently produce a product meeting its predetermined specifications and quality attributes. The validation performed should be auditable by an appropriate responsible individual.

Validation can be categorized as either prospective or retrospective. The former is applied to new drug entities or formulations in anticipation of the product requirement and expected performance and the later is applied for the validation of the existing product. Retrospective validation would be applicable for a product that has been on the market for which adequate data is available for evaluation. It consists of evaluation of product characteristics such as content uniformity, hardness, disintegration time, viscosity etc over time. Generally 20 lots are used as lower limit. The most simple and direct way of evaluating and displaying these characteristics is via control charts. Always, prospective validation should be performed for a new product during its initial development. Usually first three production batches are evaluated in detail to demonstrate consistency and accuracy. In prospective validation, the measured attributes reflect the critical characteristics of the process.

Few objectives of Process validation are furnished below.

- To confirm that the observed values from the finished product are within limits.
- To show that the average results observed from the different regions of the bulk are same.
- To claim that the average potency values observed from the bulk and the finished product are same.
- To indicate that the batches are under control based on quality control charts.

15.2 Sampling in Process Validation

The method of sampling, sample size and number of samples to be taken are important considerations in process validation. Random sampling is advised, but it is not possible to take samples randomly during production. The samples should be representative of the material being tested. For example, in validation of mixing process, the samples should be collected from different locations of the blender based on the geometry of the blender. The number and size of the samples to be taken depend on the purpose of the study. Few large size samples are desired for testing drug content/assay. Many small size samples are necessary for validation study on content uniformity. Actually there are no rules for the number of samples to be taken for validation testing. The number of samples to be taken is also dependent on the product and the process. A product having a history of good performance will need less number of samples compared to a product that showed problems.

15.3 Example for Retrospective Validation

The quality control data observed from a tablet dosage form is furnished below. The potency of the raw material during in process was examined by collecting the sample from the top, middle and the lower regions of the blender. The finished tablet dosage form is tested in triplicate.

Table 15.1

Batch	In-process bulk material (%)				Finished product (%)			
	Top	Middle	Bottom	Average	1	2	3	Average
1	98	100	106	101.3333	96	99	102	99
2	100	103	108	103.6666	97	101	103	100.3333
3	104	102	99	101.6666	102	99	97	99.3333
4	106	108	104	106	102	103	102	102.3333
Average	102	103.25	104.25	103.166625	99.25	100.5	101	100.249975
S.D	3.6514	3.4034	3.8622		3.2015	1.9148	2.7080	

The above data should address the following issues

- Whether all the batches are within the in-house specifications

  The in-house specifications may vary from product- product. If the specifications are in between 90-110%, then all the batches passed the requirements based on average values observed from the bulk and the finished dosage form.

- Whether the mean potency values observed from the different regions of the bulk and from different batches are same.

The above question is solved by treating the data statistically with two-way ANOVA

A two-way analysis of variance is used to test for equality of drug content means observed from top, middle and bottom regions of the bulk container

Table 15.2

Batch	Top	Middle Bottom	Bottom	Total
1	98	100	106	304
2	100	103	108	311
3	104	102	99	305
4	106	108	104	318
Total	408	413	417	1238

$$\Sigma x^2 = (98)^2 + (100)^2 + \ldots\ldots\ldots\ldots + (104)^2 = 127850$$

$$(\Sigma x) = 1238 ; \frac{(\Sigma X)^2}{N} = \frac{(1238)^2}{12} = 127720.3333 = \text{C.F}$$

Total sum of squares = $\Sigma x^2 - \frac{(\Sigma x)^2}{n} = 129.67$

Batch Sum of Squares

$$\frac{(304)^2}{3} + \frac{(311)^2}{3} + \frac{(305)^2}{3} + \frac{(318)^2}{3} - \text{C.F}$$

$$= \frac{383286}{3} - \text{C.F}$$

$$= 127762 - 127720.3333$$

$$= 41.67$$

Top-Middle-Bottom Sum of Squares

$$= \frac{(408)^2}{4} + \frac{(413)^2}{4} + \frac{(417)^2}{4} - \text{C.F}$$

$$= \frac{510922}{4} - 127720.3333$$

$$= 127730.5 - 127720.3333 = 10.17$$

Error Sum of Squares

(Total sum of squares) – (Batch sum of squares + Top-Middle-Bottom Sum of Squares)

$$= 129.67 - (41.67 + 10.17)$$
$$= 129.666667 - 51.83334 = 77.83336$$
$$= 129.67 - 51.84 = 77.83$$

Table 15.3 ANOVA Table for Top-Mid-Bottom of Bulk

Source of Variation	D.F	S S	M.S	F
Batches	4 – 1 = 3	41.67	$\frac{41.67}{3} = 13.89$	___
Top-middle-Bottom	3 – 1 = 2	10.17	$\frac{10.17}{2} = 5.08$	0.39
Error	6	77.83	77.833/6 =12.97	
Total	11	129.66		

∴ Calculated F value = 0.39

Tabled F value at the level of significance 0.05 with degrees of freedom 2, 6 $= F_{(2,6)} = 5.14$

Calculated F value < tabled F value

∴ There is no significance difference in process between

Top-middle –Bottom of the bulk Container

Whether the homogeneity observed from the bulk and finished dosage form is same.

There is a possibility in loss of potency from processing of bulk to dosage form. To confirm such possible losses, the data should be treated statistically with paired t-test.

Paired t test comparing the average of the bulk and Finished Product

Average of bulk $=103.166625 = \bar{x}_1$

Average of finished bulk = $100.2499975 = \bar{x}_2$

Table 15.4

Product bulk(A)	Finished Product (B)	$\delta = A - B$
101.3333 103.6666 101.6666 106	99 100.3333 99.3333 102.3333	2.3333 3.3333 2 3.6667
$\bar{x}_1 = 103.166625$	$\bar{x}_2 = 100.249975$	$\bar{\delta} = \frac{11.3333}{4} = 2.833325$

$$S_\alpha = \sqrt{\frac{0.250025 + 0.249967501 + 0.694443055 + 0.69450139}{3}}$$

$$= \sqrt{\frac{1.888936946}{3}} = \sqrt{0.629645648} = 0.793502$$

$$t = \frac{\bar{\delta}}{S_\alpha \sqrt{N}} = \frac{2.833325}{0.793502 \times \sqrt{4}} = 1.785329464$$

∴ Calculated t value =1.78

Tabled value at the level of significance 0.05 with degrees of freedom n – 1 = 3 is 3.18

∴ Calculated t < tabled t.

∴ There is no significance difference between average results of bulk and finished Product.

Whether the batches are in control based on control charts

The above problem is solved by constructing the control charts separately for the bulk and finished dosage form separately.

Control chart for data within batch variation to construct limit = Control limits are

$$= x \pm 3 \times \frac{S}{\sqrt{3}}$$

$$s = \sqrt{\frac{10.1667 + 77.83336}{2 + 6}} = \sqrt{11.0000075}$$

$$= 3.32$$

$$\text{Control limits} = 103.1666625 \pm 3 \times \frac{3.316625921}{\sqrt{3}}$$

$$= 103.1666625 \pm 5.744564605$$

Upper control limit = 103.1666625 + 5.7444564605

= 108.91

Lower control limit = 103.1666625 – 5.7444564605

= 97.42

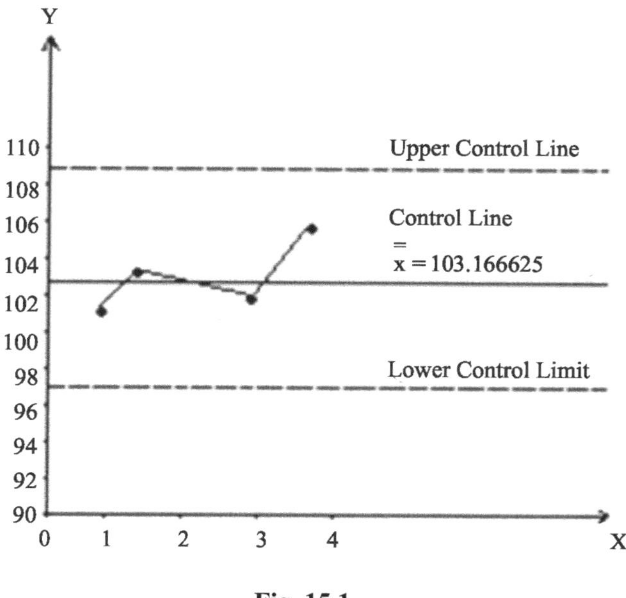

Fig. 15.1

∴ All the values lie between the control limits.

∴ The process is under control

A moving control chart may be appropriate for this data and this chart is constructed for the average of the three bulk assays for 4 batches, using a control chart with a moving average of size 3.

Table 15.5 Compilation for the moving Average Control Chart

Moving average(N = 3)	**Moving range**
$\frac{101.3333+103.6666+101.6666}{3}=\frac{306.6665}{3}$ =102.2221667	2
$\frac{103.6666+101.6666+106}{3}=\frac{311.3332}{3}$ =103.77773333	4.3334
Average = 102.99995	3.1667

For sample size 3, A factor for $\overline{X}$ chart =1.02

The control limits are

$$102.99 \pm (1.02)(3.1667)$$

$$= 102.99 \pm 3.230034$$

Upper control limit = 102.99 + 3.23 = 106.23

Lower control limit = 102.99 – 3.23 = 99.77

Moving control chart for above data

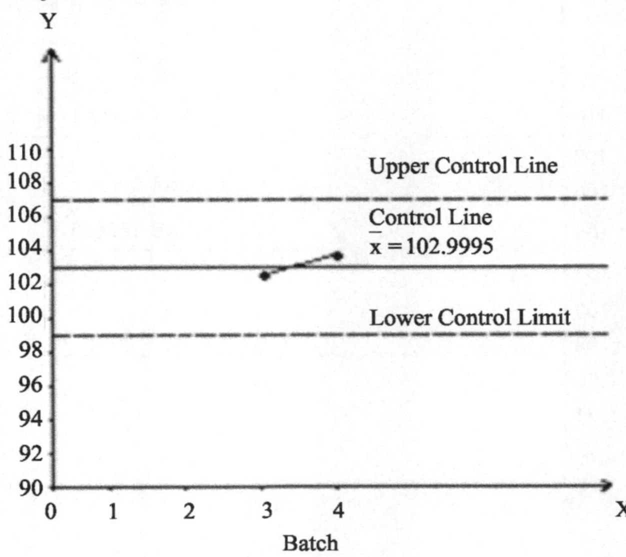

Fig. 15.2

To construct range chart

Sample size n = 3

Factors for Range Chart

D_L For lower limit = 0

D_U For upper limit = 2.57

Table 15.6

Top	middle	Bottom	Range
98	100	106	106 – 98 = 8
100	103	108	108 – 100 = 8
104	102	99	104 – 99 = 5
106	108	104	108 – 104 = 4
			$\overline{R} = \frac{8+8+5+4}{4} = \frac{25}{4} = 6.25$

Upper control limit = 2.57 × 6.25 = 16.0625

Lower control limits = 2.57 × 0 = 0

Range chart

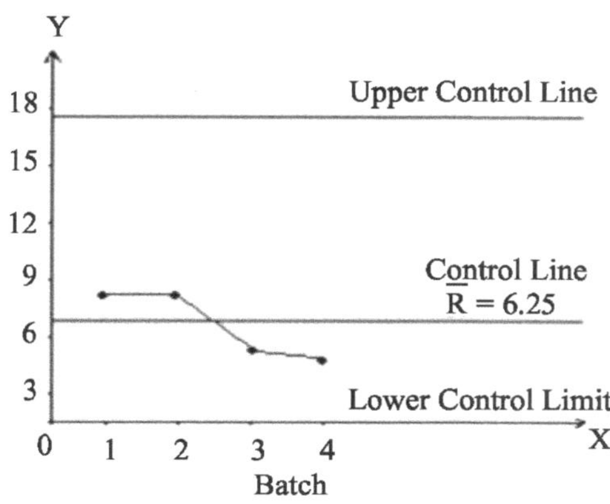

Fig. 15.3

15.4 Example for Prospective Validation

Generally prospective validation is desired during development of a new dosage form or during the reformulation of the existing formulation. Under such circumstances it is difficult to generate data from many batches due to the availability of limited batches. So the recommended procedure is to collect and analyze the data from at least three batches. Prior to design of validation, the critical steps influencing the quality of the product should be identified. The number of experiments is dependent on the number of critical factors. In such cases the main important factors should be identified based on experience or by conducting the studies with factorial design.

In case of tablet production, the quality of the tablet is influenced by mixing time and the conditions observed during compression. The uniformity or homogeneity during blending should be analyzed by collecting the samples from different regions of the blender at regular intervals of time. Such data (percent drug content) is furnished in the following table.

Table 15.7

Sample no	Mixing time(5min)				Mixing time(10min)			
	Top	Middle	Bottom	Average	1	2	3	Average
1	96	102	105	101	102	104	105	103.666
2	99	104	102	101.66	104	103	104	103.666
3	103	105	98	102	105	100	102	102.333
4	105	104	102	103.66	104	102	103	103
Average	100.75	103.75	101.75		103.75	102.25	103.5	
s.d	4.031	1.2583	2.87		1.2583	1.707	1.290	

Such data is analyzed statistically with ANOVA as shown below. At each time interval, the mixing efficiency is varied from different regions and different samples collected from the same region. Hence the data is treated statistically with ANOVA.

Solution:

Table 15.8 Mixing time (5 min)

Sample no	Top	middle	Bottom	Average
1	96	102	105	101
2	99	104	102	101.66
3	103	105	98	102
4	105	104	102	103.66
Total	403	415	407	

$$\Sigma x^2 = (96)^2+(99)^2+............+(102)^2=125149.$$

$$= \frac{(\Sigma x)^2}{N} = \frac{(403+415+407)^2}{12} = \frac{(1225)^2}{12} = 125052.083$$

$$\text{Total sum of squares} = \Sigma x^2 - \frac{(\Sigma x)^2}{N}$$

$$= 96.91$$

$$\text{Between sum of Squares} = \frac{(403)^2+(415)^2+(407)^2}{4} - \text{C.F}$$

$$= \frac{500283}{4} - \text{C.F}$$

$$= 125070.75 - 125052.0833$$

$$= 18.67$$

With in sum of Squares = Total sum of squares – Between sum of Squares

$$= 96.91 - 18.67$$

$$= 78.24$$

Table 15.9 ANOVA Table (For mixing time 5 min)

Source of variation	D.f	S. S	M.S	F
Between	2	18.67	9.34	1.07
with in	9	78.24	8.69	
Total	**11**	**96.91**		

Table 15.10 ANOVA Table (Mixing time 10 mints)

S. No	1	2	3
1	102	104	105
2	104	103	104
3	105	100	102
4	104	102	103
Total	**415**	**409**	**414**

$$\Sigma x^2 = 127744.$$

$$C.F = \frac{(\Sigma x)^2}{N} = \frac{(415 + 409 + 414)^2}{12}$$

$$\text{Total Sum of Squares} = \Sigma x^2 - \frac{(\Sigma x)^2}{N}$$

$$= 23.67$$

$$\text{Between sum of squares} = \frac{(415)^2 + (409)^2 + (414)^2}{4}$$

$$= \frac{510902}{4} - C.F$$

$$= 127725.5 - 127720.33$$

$$= 5.17$$

With in sum of squares = Total sum of Squares – Between sum of squares

$$= 23.67 - 5.17$$

$$= 18.5$$

Table 15.11 ANOVA Table

Source of variation	D.f	S. S	M.S	F
Between	2	5.17	2.58	1.25
with in	9	18.5	2.06	
Total	**11**			

Table F value at the level of significance 0.05 with degrees of freedom 2, 9 = $F_{2,9}$ = 4.10.

$\therefore$ F ratios for between sampling locations mean square are not significant.

($\because$ Calculated F< tabled F)

This suggests that drug is dispersed uniformly to all locations after 5 minutes and 10 minutes of mixing. However the within MS is significantly large in 5 minute mixing time compared to the 10 minute mixing time.

Illustration: During compression, the die fill and the compression force are varied and hence the tablet properties such as hardness, thickness, weight, drug content and disintegration time are changed. So the tablets are collected at the beginning, middle and the ending of the compression. In a compression cycle of 3 hours, the tablets were collected at the time interval 0.5-1(beginning of compression), 1.5-2.0 (middle of compression) and 2-5-3.0 (ending of compression) and analyzed for their drug content. The observed values are given in the following table. The variations noticed within the tablets collected at same stage of compression and different stages of compression are detected by treating the data statistically with ANOVA.

Table 15.12

S. No	Drug content (%)			
	Starting	**Middle**	**Ending**	**Average**
1	98	100	106	101.33
2	100	103	108	103.66
3	104	102	99	101.66
4	106	108	104	106
5	102	99	105	102
6	98	102	103	101
7	96	99	104	99.66
8	100	96	98	98
9	99	101	103	101
10	103	104	106	104.33
Average	100.6	101.4	103.6	
s.d	3.098	3.2727	3.09838	

Table 15.13

S.no	Starting	Middle	Ending	Total
1	98	100	106	304
2	100	103	108	311
3	104	102	99	305
4	106	108	104	318
5	102	99	105	306
6	98	102	103	303
7	96	99	104	299
8	100	96	98	294
9	99	101	103	303
10	103	104	106	313
Total	1006	1014	1036	

$$\Sigma x^2 = (98)^2 + (100)^2 + \ldots\ldots\ldots\ldots\ldots + (106)^2$$

$$= 101\,290 + 1029\,16 + 1074\,16$$

$$= 311622$$

$$\text{C.F} = \frac{(\Sigma x)^2}{N} = \frac{(1006+1014+1036)^2}{30} = \frac{9339136}{30}$$

$$= \ 311304.5333.$$

$$\text{Total sum of squares} = \Sigma x^2 - \frac{(\Sigma x)^2}{N}$$

$$= 311622 - 311304.5333$$

$$= 317.4666667$$

$$\text{Starting –middle – end sum of squares} = \frac{(1006)^2 + (1014)^2 + (1036)^2}{10} - C.F$$

$$= \frac{3113528}{10} - 311304.53333$$

$$= 311352.8 - 311304.53333$$

$$= 48.26667$$

$$\text{Batches Sum of Squares} = \frac{(304)^2 + (311)^2 + (305)^2 + \ldots\ldots\ldots(313)^2}{3}$$

$$= \frac{934346}{3} - C.F$$

$$= 311448.6667 - 311304.5333$$

$$= 144.1333367$$

Error Sum of Squares

Total sum of squares = {Starting middle – ending sum of squares + Batch sum of squares}

$$= 317.47 - (48.27 + 144.13)$$

$$= 317.47 - 192.40$$

$$= 125.06$$

Table 15.14 ANOVA Table

Source	D.F	S.S	M.S	F
Batches	9	144.13	16.015	___
Starting-middle-bottom	2	48.27	24.13	3.47
Error	18	125.06	6.95	___
Total	29	317.47		

∴ Calculated F value = 3.47

Tabled F value at the level of significance 0.05 with degrees of

freedom 2,19 = F_{219} = 3.52

∴ Calculated F value < tabled F value.

∴ There is no significance of difference in the process of means bulk material in starting, middle& bottom

15.5 Analytical Method Validation

Analytical method validation is just one type of validation required during drug development and manufacturing. A validation study is performed on an analytical method to ensure that validate results are always obtained. The purpose of validation is to show that the process involved in the development and manufacturing of drugs, such as production cleaning and analytical drug can be performed in an effective and reproducible manner.

Statistical analysis of the data produced during a method validation study is required to demonstrate the validity of the method. The use of mean, standard deviation and confidence intervals in the method validation is furnished below. Statistical input is essential for validation procedures. For example, experimental design and data analysis are integral parts of assay and process validation.

Mean

It is the total sum of all the values in the data set divided by the number of values in the data set. It is calculated with the following formulae

$$\overline{X} = \frac{1}{n}\Sigma(x_i)$$

The single determination of the amount of active pharmaceutical ingredient will provide an estimation of true value but the mean of the results obtained from several determinations is normally used to provide a better estimate for the true value. The calculated sample mean will be nearer to the true value, when the number of determinations is increased.

Standard Deviation (S)

It is measure of the spread of the values in the data set and calculated by measuring the difference between mean and the individual value in the dataset. It is determined using the equation

$$S = \sqrt{\frac{\Sigma\left(X - \overline{X}\right)^2}{(n-1)}}$$

The smaller the number of data, higher is the variability of the sample standard deviation. As a result of this, it is recommended to calculate the standard deviation when the $n > 3$. Relative standard deviation should be less than the permitted value to confirm the uniformity.

Confidence Intervals

They are used to indicate the relativity of an estimate. It produces limits around the experimentally determined value of the mean within which the true value lies with a given degree of probability usually 95%.

Statistical Significance

A statistically significant difference means that there is statistical evidence that a difference means that there is a difference exists which is unlikely to have occurred as chance. It is tested with ANOVA or t-test.

Comparative Studies

In the following cases, the comparative studies could be performed during method validation

- Specificity and accuracy, to compare the results obtained using the method being validated against a second well characterized method.
- Precision to compare the results obtained during intermediate precision and reproducibility studies.
- Robustness to compare the effects due to the factors under evaluation.
- Stability of solutes to compare the results obtained from solution which has been stored at specified conditions and compared against the original results.
- Sensitivity of the analytical method is determined with the parameters limit of detection (LOD), Limit of quantification (LOQ). They are calculated with the following formulas

$$\text{LOD} = 3.3 \times \frac{\text{standard deviation of Y intercept}}{\text{Slope of calibration curve}}$$

$$LOQ = 10 \times \frac{\text{standard deviation of Y intercept}}{\text{Slope of calibration curve}}$$

Table 15.15 The application of statistics in precision is shown in the following table

Name of the sample	Sample area	%Assay on the label claim
Precision-1	69875689	99.5
Precision-2	70696785	100.7
Precision-3	70817890	100.9
Precision-4	71235912	101.5
Precision-5	70908597	101.0
Precision-6	71612231	102.0
Mean		100.9
Std.Dev		0.8
%RSD		0.8

Table 15.16

Name of the Sample	Sample area	%Assay on label claim
Sample-1	71254836	101.6
Sample-2	70369845	100.3
Sample-3	69998742	99.8
Sample-4	70854600	101.0
Sample-5	71548753	102.0
Sample-6	70826473	101.0
Mean		101.0
Std.Dev		0.8
%RSD		0.8

The observed **%** RSD value is very less in tables 15.15-16 indicating that the method is having good precision. The linear relationship existing between concentration and the response can be determined with correlation and regression analysis. From the data given in table 15.17, the correlation coefficient and regression equation is calculated and shown in Fig 15.4.

Table 15.17 calibration curve data

Linearity of Detector response		
S. no	**Concentration Level**	**Area**
1	25%	13381329
2	50%	29359866
3	100%	66414478
4	150%	105742405
5	200%	144017712
	Correlation Coefficient	0.9997
	Intercept	–7211259.1
	Slope	752327.8

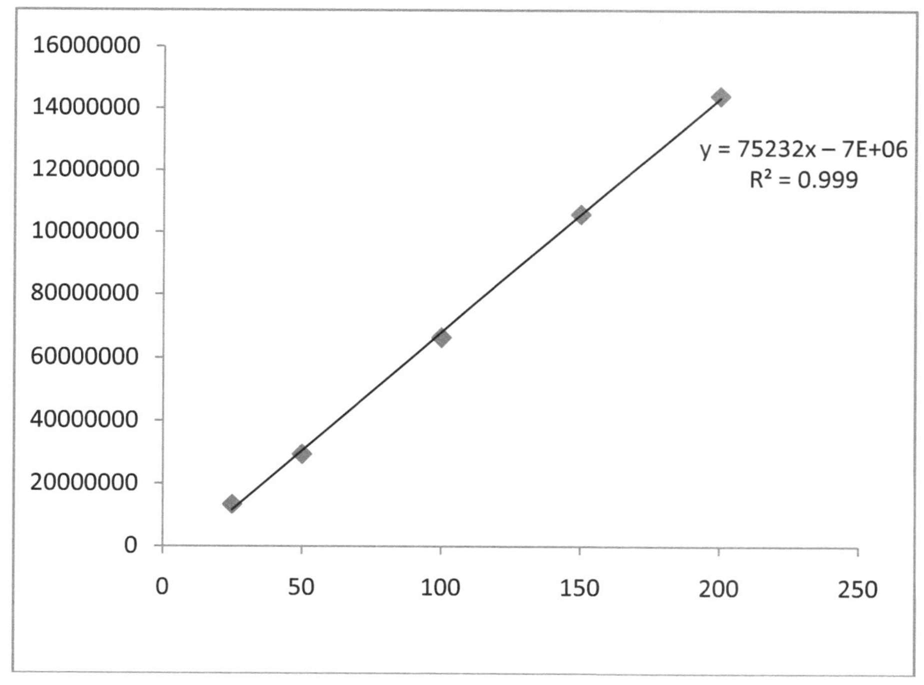

Fig. 15.4

Determination of LOD & LOQ: these values are calculated as follows

Values of Y intercept

7211259.1

7189224.2

7211359.1

Standard deviation of Y intercept is 10411.04

Slope of calibration curve is 752327.8

$$\text{LOD} = 3.3 \times \frac{1041104}{752927.8} = 0.045$$

$$\text{LOQ} = 10 \times \frac{1041104}{752227.0} = 0.138$$

$$\text{LOD} = 3.3 \times \frac{1041104}{752927.8} = 0.045$$

Application of Statistics for calculation of Accuracy of the analytical method is shown in the following table

Table 15.18

%level	Amount of API Spiked in (μg/ml)	Peak Area	Amount Found in (μg/ml)	%Recovery	Mean %Recovery	Std.Dev	%RSD
50	6	609893	6.023	100.38	100.44	0.118	0.117
	6	610720	6.032	100.58			
	6	609823	6.022	100.37			
100	8	785207	7.962	99.529	100.4	0.735	0.732
	8	794624	8.0624	100.781			
	8	784906	7.959	99.487			
150	10	1056267	10.226	102.26	102.83	0.501	0.487
	10	1064472	10.305	103.05			
	10	1065892	10.319	103.19			

The low RSD values in the above method indicate good accuracy.

The application statistical inference test (t test) is shown in the following table

Table 15.19

S.no	%Recovery	
	UV	HPLC
1.	98.56	100.02
2.	99.22	99.82
3.	95.99	101.23
4.	100.1	102.44
5.	95.66	98.96
6.	98.22	100.23

Solution:

Table 15.20

S. no	% RECOVERY	
	U V (X_1)	**HPLC (X_2)**
1	98.56	100.02
2	99.22	99.82
3	95.99	101.23
4	100.1	102.44
5	95.66	98.96
6	98.22	100.23
TOTAL	587.75	602.7

$$\bar{x}_1 = \frac{587.75}{6} = 97.958333,\ \bar{x}_2 = \frac{602.7}{6} = 100.45$$

$$S_1^2 = \frac{\Sigma(x_1 - \bar{x}_1)^2}{n_1 - 1} = \frac{(98.56 - 97.95833)^2 + \ldots\ldots\ldots + (98.22 - 97.9583)^2}{5}$$

$$= \frac{(0.60166)^2 + (1.26166)^2 + (-1.96833)^2 + (2.141666)^2 + (-2.29833)^2 + (0.26166)^2}{5}$$

$$= 15.765623715 = 3.15312474$$

$S_2^2 = (100.02 - 100.45)^2 + (99.82 - 100.45)^2 + (101.23 - 100.45)^2 + \ldots\ldots\ldots + (100.23 - 100.45)$

$$= \frac{(-0.43)^2 + (-0.63)^2 + (0.78)^2 + (1.99)^2 + (-1.49)^2 + (-0.22)^2}{5}$$

$$= \frac{7.4188}{5} = 1.48376$$

Test statistic $t = \dfrac{|\bar{x}_1 - \bar{x}_2|}{S_p\sqrt{\dfrac{1}{n_1} + \dfrac{1}{n_2}}}$

Where $$S_P = \sqrt{\frac{(n_1 - 1)S_1^2 + (n_2 - 1)S_2^2}{n_1 + n_2 - 2}}$$

$$5 \times = \sqrt{\frac{5 \times 3.15312474 - 5 \times 1.48376}{6+6-2}}$$

$$= \sqrt{\frac{5(3.15312474 - 1.48376)}{10}} = \sqrt{\frac{1.66936474}{2}}$$

$$= \sqrt{0.83468237}$$

$$= 0.913609528$$

$$\therefore \quad t = \frac{|97.958333 - 100.45|}{(0.913609528)\sqrt{\frac{1}{6} + \frac{1}{6}}}$$

$$= \frac{2.491667}{(0.913609528) \times \left(\frac{1}{\sqrt{3}}\right)}$$

$$= \frac{2.491667 \times \sqrt{3}}{0.913609528} = \frac{4.31569384}{.913609528}$$

$$= 4.72$$

$\therefore$ Calculated value = 4.72

Tabled value at the level of significance 0.05 with degrees of freedom $n_1 + n_2 - 2 = 12 - 2 = 10$ is 2.23.

$\therefore$ Calculated value > t tabled value.

$\therefore$ There is a significance difference between the mean of % recovery observed from two methods UV, HPLC.

INDEX

D

E

F

G

H

I

K

L

M

N

O

P

Q

R

Printed at Repro India Ltd.

Table of Contents

For Resource files, visit:

https://autocadforbeginners.weebly.com

For Technical Support, contact us at:

online.books999@gmail.com

AutoCAD 2024 For Beginners

CADFolks

Introduction

CAD is an abbreviation for Computer-Aided Design. It is the process used to design and draft components on your computer. This process includes creating designs and drawings of the product or system. AutoCAD is a CAD software package developed and marketed by Autodesk Inc. It can be used to create two-dimensional (2D) and three-dimensional (3D) models of products. These models can be transferred to other computer programs for further analysis and testing. Also, you can convert these computer models into numerical data. This numerical data can be used in manufacturing equipment such as machining centers, lathes, mills, or rapid prototyping machines to manufacture the product.

AutoCAD is one of the first CAD software packages. It was introduced in the year 1982. Since that time, it has become the industry leader among all CAD products. It is the most widely used CAD software. Other systems utilize the commands and concepts introduced by AutoCAD. As a student, learning AutoCAD provides you with a more significant advantage as compared to any other CAD software.

Scope of this Book

The ***AutoCAD 2024 for Beginners*** book provides a learn-by-doing approach for users to learn AutoCAD. It is written for students and engineers who are interested to learn AutoCAD 2024 for creating designs and drawing of components or anyone who communicates through technical drawings as part of their work. The topics covered in this book are as follows:

- Chapter 1, "Introduction to AutoCAD 2024", gives an introduction to AutoCAD. The user interface and terminology are discussed in this chapter.
- Chapter 2, "Drawing Basics," explores the essential drawing tools in AutoCAD. You will create simple drawings using the drawing tools.
- Chapter 3, "Drawing Aids," explores the drawing settings that will assist you in creating drawings.
- Chapter 4, "Editing Tools," covers the tools required to modify drawing objects or create new objects using the existing ones.
- Chapter 5, "Multi View Drawings," teaches you to create multi-view drawings standard projection techniques.
- Chapter 6, "Dimensions and Annotations," teaches you to apply dimensions and annotations to a drawing.

- Chapter 7, "Parametric Tools," teaches you to create parametric drawings. Parametric drawings are created by using the logical operations and parameters that control the shape and size of a drawing.

- Chapter 8, "Section Views," teaches you to create section views of a component. A section view is the inside view of a component when it is sliced.

- Chapter 9, "Blocks, Attributes, and Xrefs," teaches you to create Blocks, Attributes, and Xrefs. Blocks are a group of objects in a drawing that can be reused. Attributes are notes or values related to an object. Xrefs are drawing files attached to another drawing.

- Chapter 10, "Layouts and Annotative Objects," teaches you to create layouts and annotative objects. Layouts are the digital counterparts of physical drawing sheets. Annotative objects are dimensions, and notes, which change their sizes to drawing scale.

- Chapter 11, "Templates and Plotting," teaches you to create drawing templates and plot drawings.

- Chapter 12, "3D Modeling Basics," explores the necessary tools to create 3D models.

- Chapter 13, "Solid Editing Tools," covers the tools required to edit solid models and create new objects by using the existing ones.

- Chapter 14, "Creating Architectural Drawings," introduces you to architectural design in AutoCAD. You will design floor plan, staircase, elevations, roof plan, and roof and wall details.

Chapter 1: Introduction to AutoCAD 2024

In this chapter, you will learn about:

- **AutoCAD user interface**
- **Customizing user interface**
- **Important AutoCAD commands**

Introduction

AutoCAD is a legendary software in the world of Computer Aided Designing (CAD). It has completed 41 years by 2023. If you are a new user of this software, then the time you spend on learning this software will be a wise investment. If you have used previous versions of AutoCAD, you will be able to learn the new enhancements. I welcome you to learn AutoCAD using this book through step-by-step examples to learn various commands and techniques.

System requirements

The following are system requirements for running AutoCAD smoothly on your system.

- 64-bit Microsoft® Windows® 11 and Windows 10 version 1809 or above.
- CPU Type:
    - Basic: 2.5 to 2.9 gigahertz (GHz)
    - Recommended: 3+ GHz processor (base), 4+ GHz (turbo)
- 8 GB of RAM (32 GB Recommended).
- Resolution 1920 x 1080 or higher recommended with True Color.
- Resolutions up to 3840 x 2160 supported on Windows 10, 64 bit systems (with capable display card) for High Resolution & 4K Displays.
- 10 GB of free space for installation.
- Google Chrome Browser.
- .NET Framework Version 4.8 or later

Starting AutoCAD 2024

To start **AutoCAD 2024**, double-click the **AutoCAD 2024** icon on your Desktop (or) click **Start > All apps > AutoCAD 2024 > AutoCAD 2024**.

AutoCAD user interface

When you double-click the AutoCAD 2024 icon on the desktop, the AutoCAD 2024 Start screen appears. On the Start Screen, click the **New** button to open a new drawing file. The drawing file consists of a graphics window, ribbon, menu bar, toolbars, command line, and other screen components, depending on the workspace that you have selected.

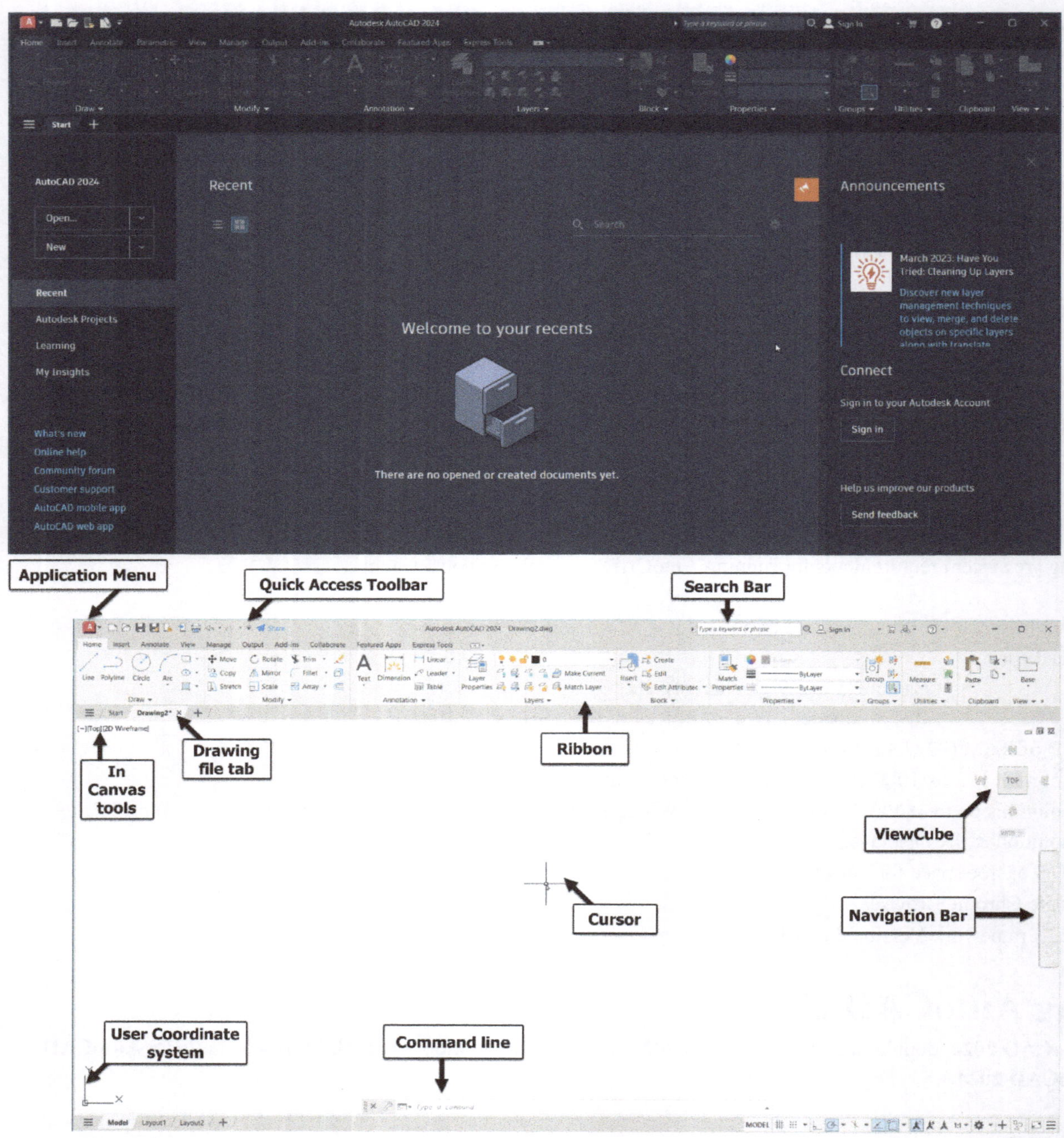
Recent
Announcements
Welcome to your recents
There are no opened or created documents yet.
Connect
Sign in to your Autodesk Account
Sign in
Help us improve our products
Send feedback
AutoCAD 2024
Open...
New
Recent
Autodesk Projects
Learning
My Insights
What's new
Online help
Community forum
Customer support
AutoCAD mobile app
AutoCAD web app
Application Menu
Quick Access Toolbar
Search Bar
Ribbon
Drawing file tab
In Canvas tools
ViewCube
Cursor
Navigation Bar
User Coordinate system
Command line

Changing the Color theme

AutoCAD 2024 is available in two different color themes: **Dark** and **Light**. You can change the Color theme by using the **Options** dialog. Click the right mouse button and select **Options** from the shortcut menu. On the **Options** dialog, click the **Display** tab and select an option from the **Color theme** drop-down.

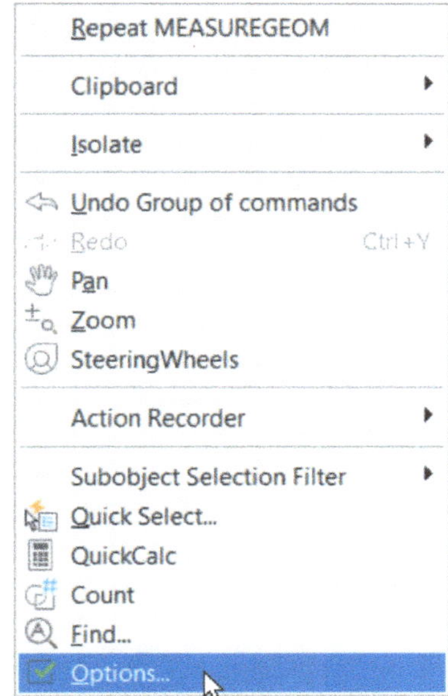

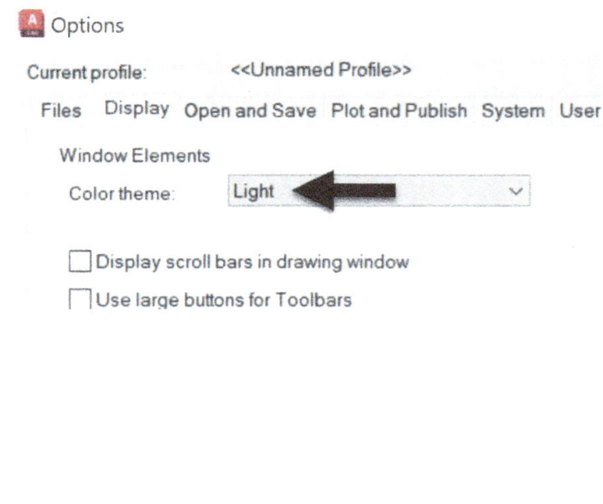

Workspaces in AutoCAD

There are three workspaces available in AutoCAD: **Drafting & Annotation**, **3D Basics**, and **3D Modeling**. By default, the **Drafting & Annotation** workspace is activated. You can create 2D drawings in this workspace. You can also enable other workspaces by using the **Workspace** drop-down on the top-left corner or the **Workspace Switching** menu on the lower-right corner of the window.

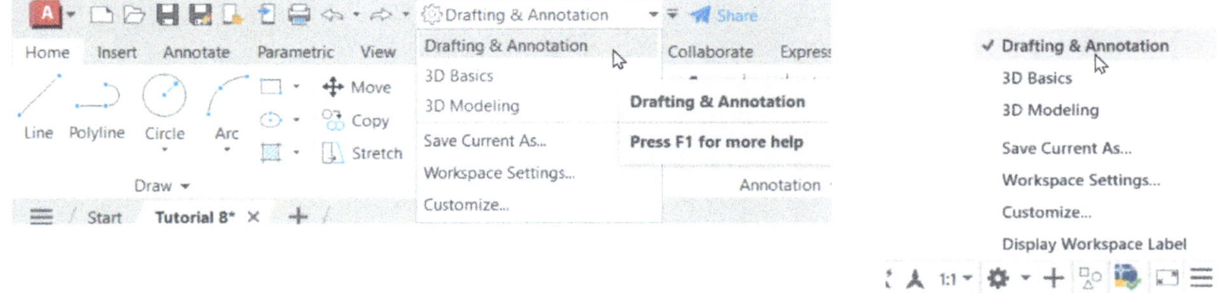

Tip: *If the **Workspace** drop-down is not displayed at the top left corner, then click the down arrow next to Quick Access Toolbar. Next, select **Workspace** from the drop-down; the **Workspace** drop-down will be visible on the Quick Access Toolbar.*

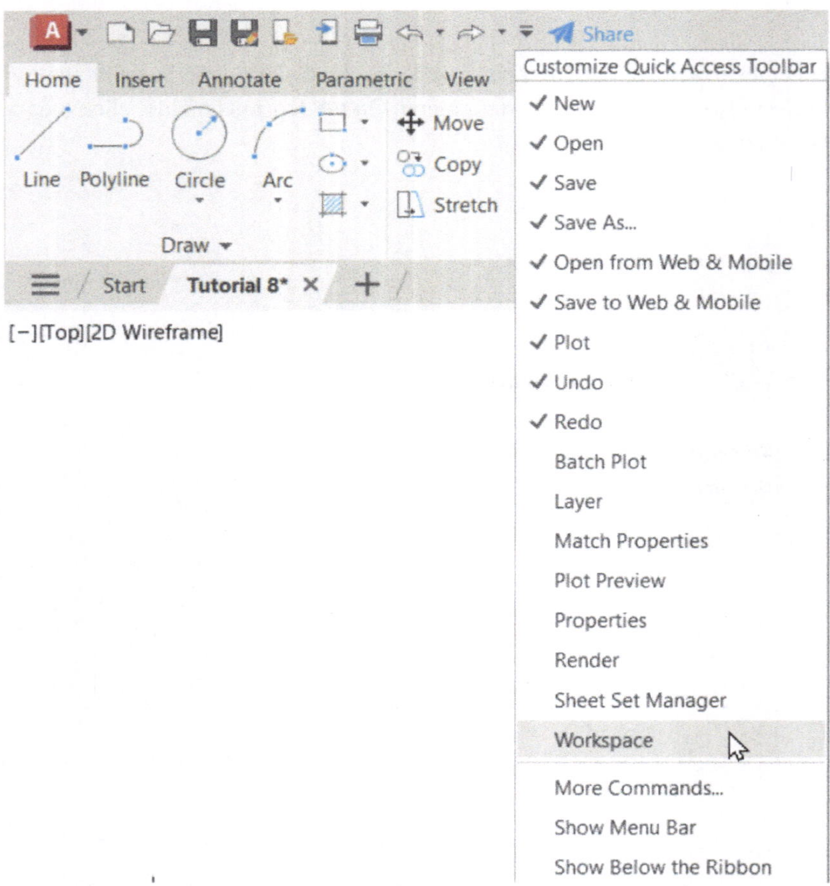

Drafting & Annotation Workspace
This workspace has all the tools to create a 2D drawing. It has a ribbon located at the top of the screen. The ribbon is arranged in a hierarchy of tabs, panels, and tools. Panels such as **Draw**, **Modify**, and **Layers** consist of tools that are grouped based on their usage. Panels, in turn, are grouped into various tabs. For example, the panels, such as **Draw**, **Modify**, and **Layers**, are located in the **Home** tab.

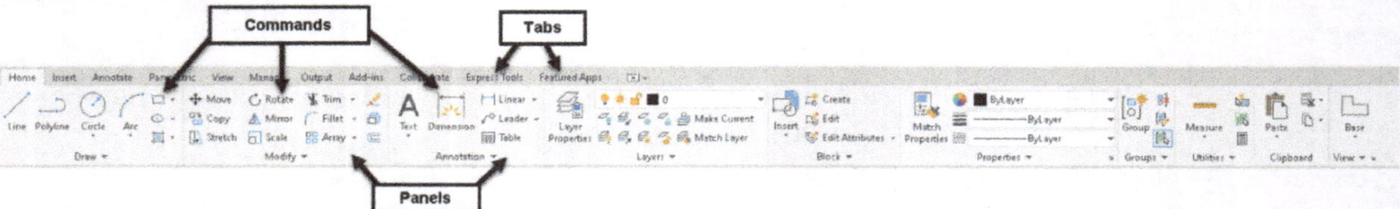

3D Basics and 3D Modeling Workspaces
These workspaces are used to create 3D models. You will learn more about these workspaces in Chapter 12. The other components of the user Interface are discussed next.

Application Menu

The **Application Menu** appears when you click on the icon located at the top left corner of the window. The **Application Menu** consists of a list of options such as **New**, **Open**, **Save**, and **Print** on the left side. Place the pointer on any one of the options on the left side; a list of additional options is displayed on the right side. You can see a list of recently opened documents or a list of currently opened documents by clicking the **Recent Documents** and **Open Documents** buttons, respectively. The Search Bar is used to search for any command. You can type any keyword in the search bar and find a list of commands related to it. In addition to that, there are two buttons, **Options** and **Exit Autodesk AutoCAD 2024** located at the bottom of the **Application Menu**. You can use the **Options** button to open the **Options** dialog. In this dialog, you can specify various settings related to display, printing, and user preferences.

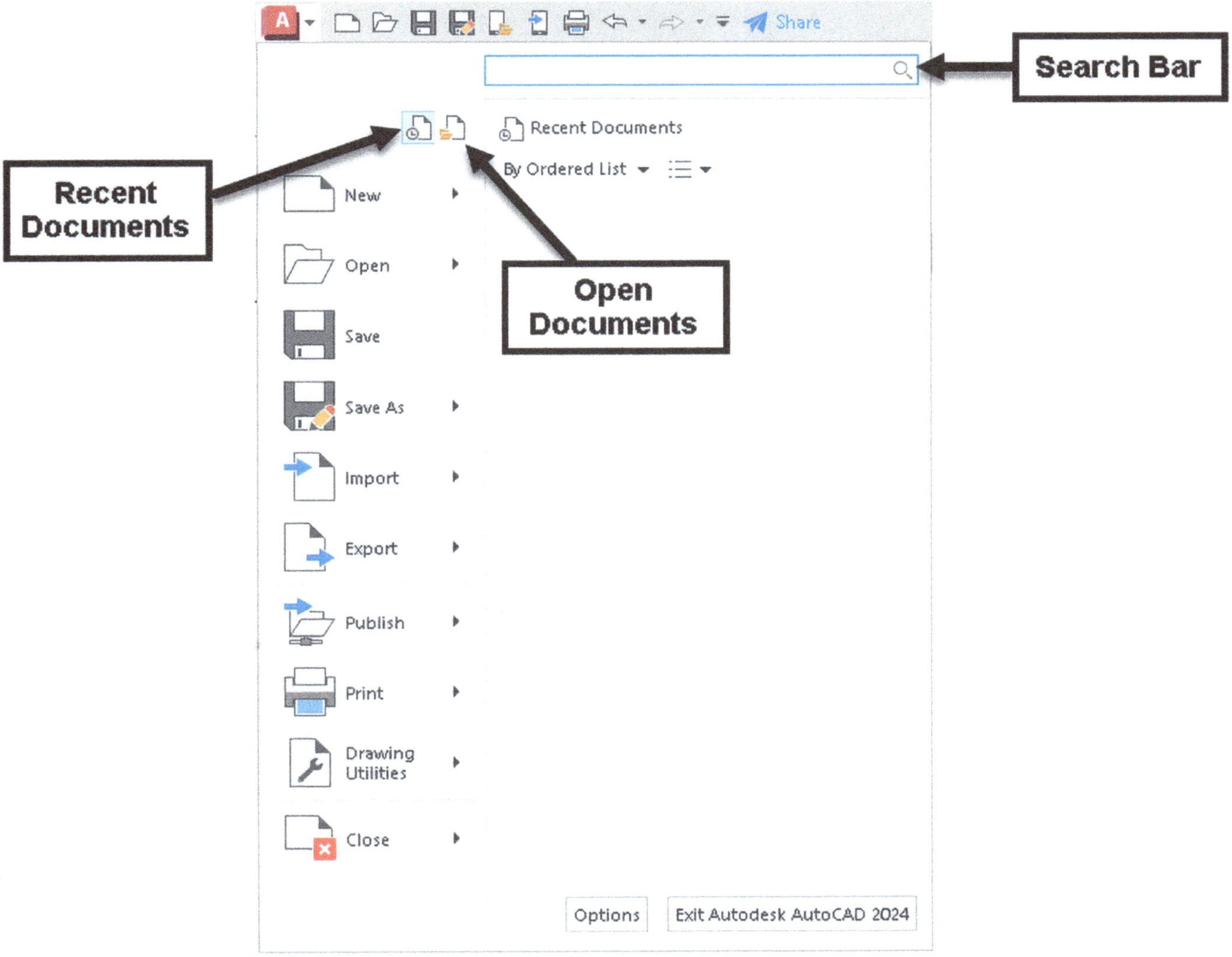

Quick Access Toolbar

The Quick Access Toolbar is located at the top left corner of the window and helps you to access commands quickly. It consists of commonly used commands such as **New**, **Save**, **Open**, **Save As**, **Open from Web & Mobile**, and **Save to Web & Mobile**.

File tabs

The File tabs are located below the ribbon. You can switch between different drawing files by using the file tabs. Also, you can open a new file by using the + button, easily.

You can use the **File tab menu** located on the left side of the Drawing File tabs to switch between the file tabs, open, save, or close a drawing file tab.

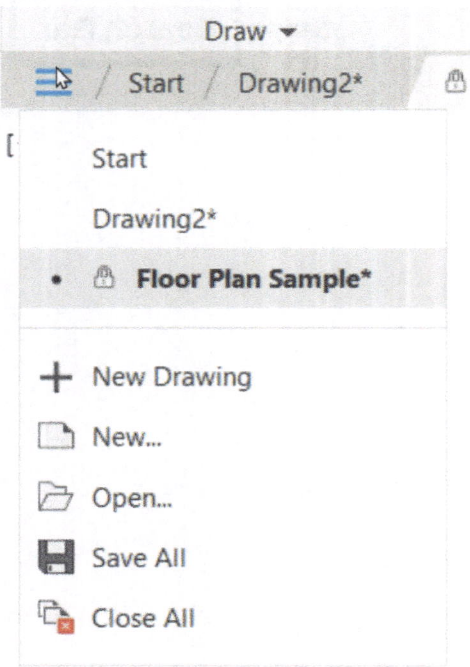

Graphics Window

The Graphics window is the blank space located below the file tabs. You can draw objects and create 3D graphics in the graphics window. The top left corner of the graphics window has **In-Canvas Controls**. Using these controls, you can set the orientation and display style of the model.

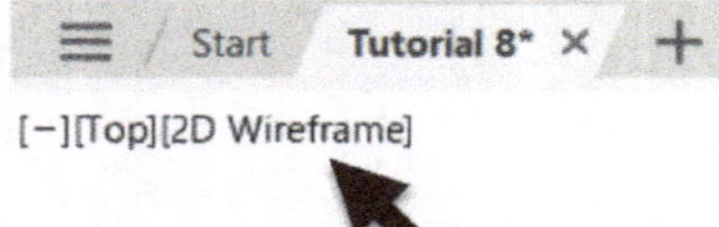

ViewCube

The ViewCube allows you to navigate in the 3D Modeling and 2D drafting environments. Using the ViewCube, you can set the orientation of the model. For example, you can select the top face of the ViewCube to set the orientation to Top. You can click the corner points to set the view to Isometric.

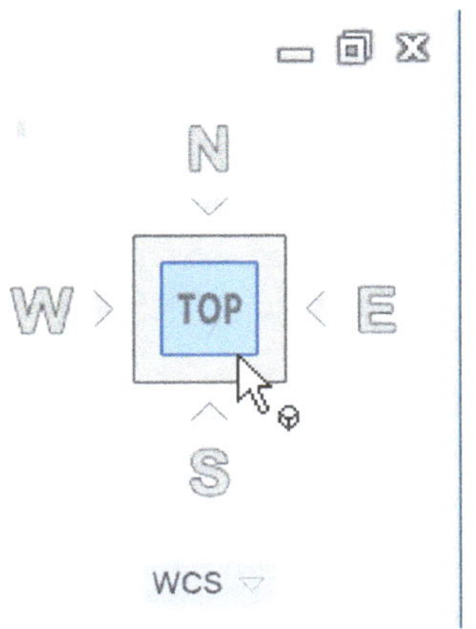

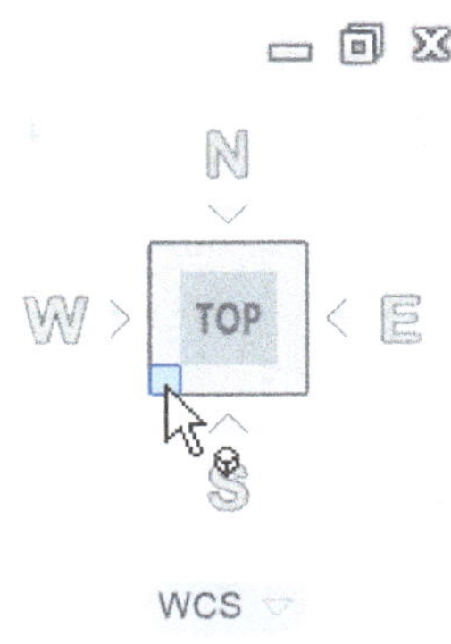

Navigation Bar

The Navigation Bar contains navigation tools such as **Steering wheel**, **Pan**, **Zoom**, **Orbit**, and **Show Motion**.

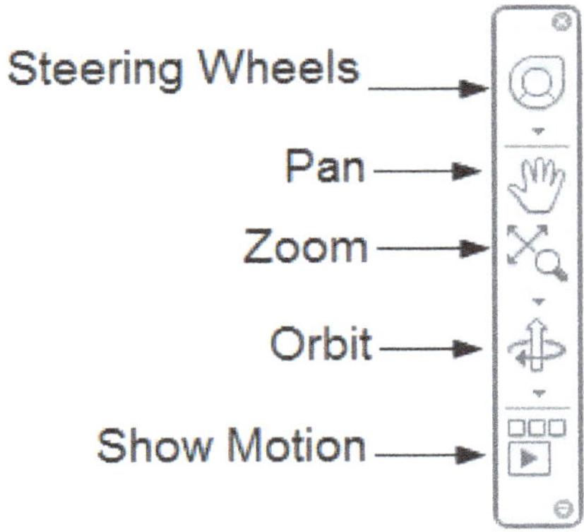

Command line

The command line is located below the graphics window. It is effortless to execute a command using the command line. You can just type the first letter of a command, and it lists all the commands starting with that letter. It helps you to activate commands very quickly and increases your productivity.

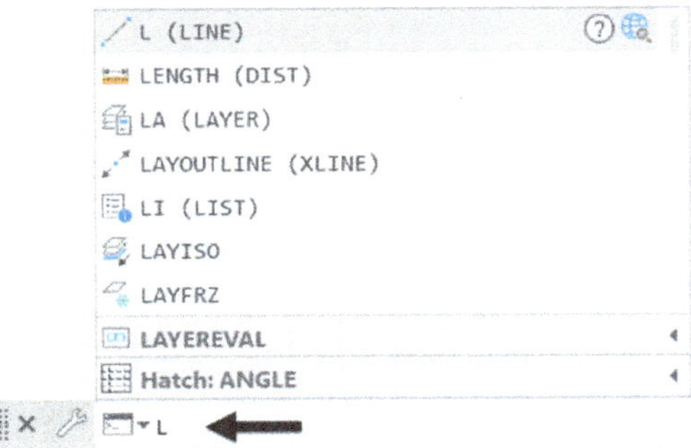

Also, the command line shows the current state of the drawing. It shows various prompts while working with any command. These prompts are a series of steps needed to execute a command successfully. For example, when you activate the LINE command, the command line displays a prompt, "Specify the first point." You need to click in the graphics window to specify the first point of the line. After defining the first point, the prompt, "Specify next point or [Undo]:" appears. Now, you need to determine the next end of the line. It is recommended that you should always have a look at the command line to know the next step while executing a command.

Status Bar

Status Bar is located at the bottom of the AutoCAD window. It contains many buttons which help you to create a drawing very easily. You can turn ON or OFF these buttons just by clicking on them. Some buttons are hidden by default. You can display more buttons on the status bar by clicking the **Customization** button at the bottom right corner and selecting the options from the menu. The buttons available on the status bar are briefly discussed in the following section.

Button	Description
6505.7547, 178.4229, 0.0000 **Coordinates**	It displays the drawing coordinates when you move the pointer in the graphics window. You can turn OFF this button by clicking on it. If this button is not displayed, you can show it by using the **Customization** menu.
Infer Constraints	This icon automatically creates constraints when you draw objects in the graphics window. Constraints are logical operations that control the shape of a drawing. You can turn it ON or OFF by clicking on it.
Snap mode (F9)	The Snap mode aligns pointer only with the Grid points. When you turn ON this button, the pointer will be able to select only the Grid points.
GRIDMODE (F7)	It turns the Grid display ON or OFF. You can set the spacing between the grid lines by clicking the down arrow next to the Snap Mode button and selecting the **Snap Settings** option. You can use the grid lines along with the Snap Mode to draw objects quickly and accurately.
ORTHOMODE (F8)	It turns the Ortho Mode ON or OFF. When the Ortho Mode is ON, only horizontal or vertical lines can be drawn.

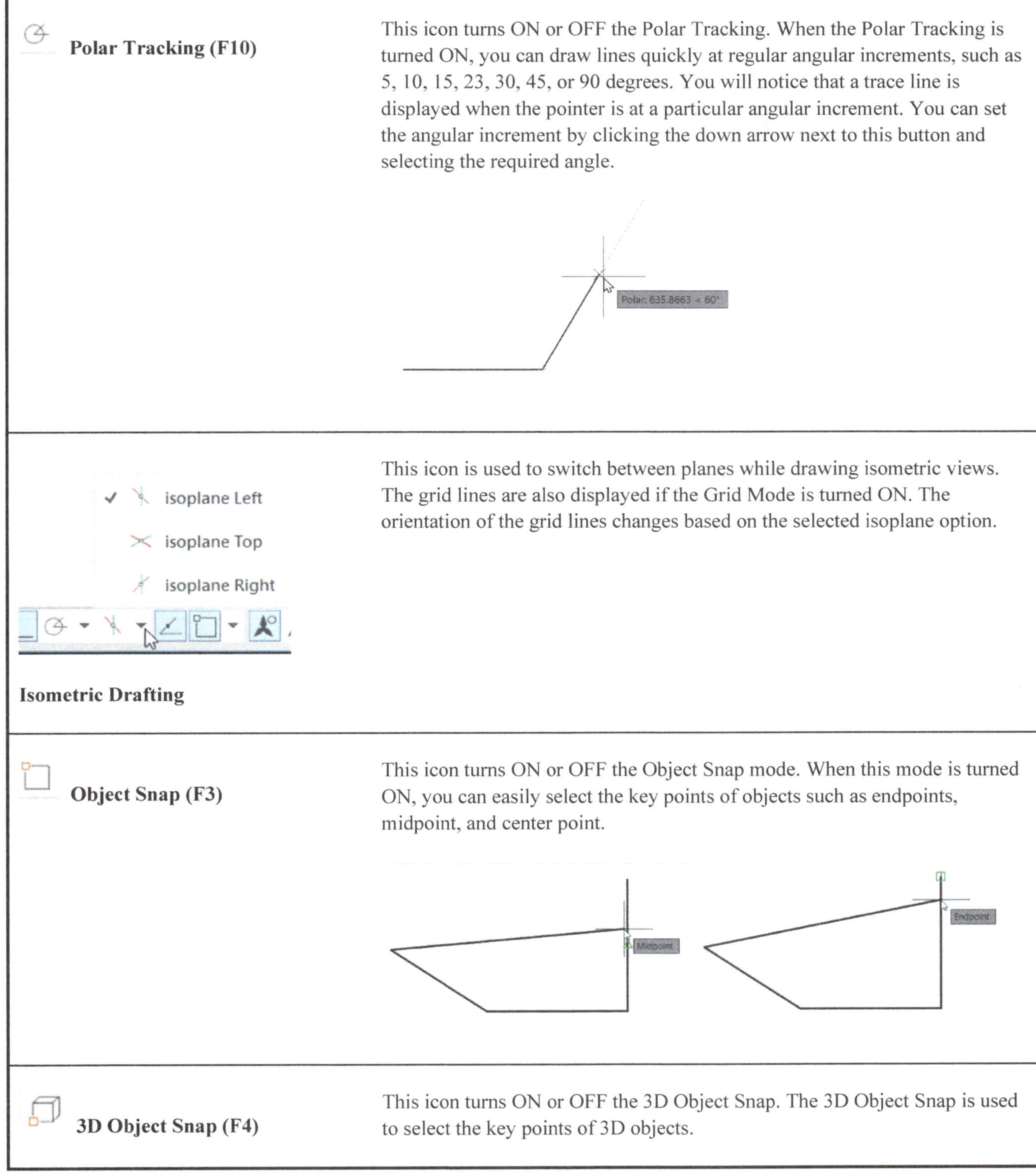

Polar Tracking (F10)	This icon turns ON or OFF the Polar Tracking. When the Polar Tracking is turned ON, you can draw lines quickly at regular angular increments, such as 5, 10, 15, 23, 30, 45, or 90 degrees. You will notice that a trace line is displayed when the pointer is at a particular angular increment. You can set the angular increment by clicking the down arrow next to this button and selecting the required angle.
isoplane Left isoplane Top isoplane Right **Isometric Drafting**	This icon is used to switch between planes while drawing isometric views. The grid lines are also displayed if the Grid Mode is turned ON. The orientation of the grid lines changes based on the selected isoplane option.
Object Snap (F3)	This icon turns ON or OFF the Object Snap mode. When this mode is turned ON, you can easily select the key points of objects such as endpoints, midpoint, and center point.
3D Object Snap (F4)	This icon turns ON or OFF the 3D Object Snap. The 3D Object Snap is used to select the key points of 3D objects.

<table>
<tr>
<td> Object Snap Tracking (F11)</td>
<td>This icon is used to turn ON or OFF the Object Snap Tracking mode. When this mode is turned ON, you can easily select points by using the trace lines originating from the key points.
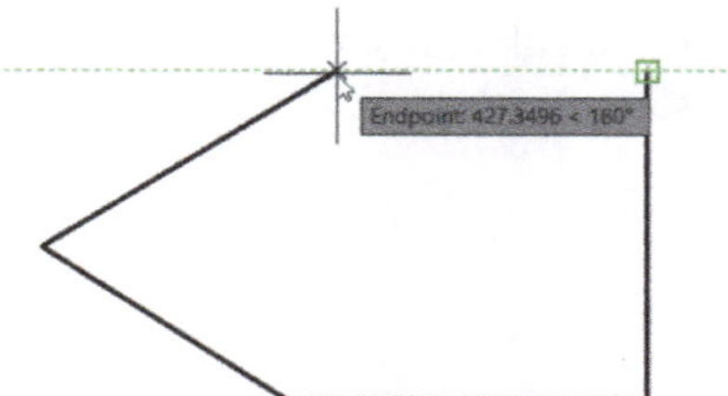
</td>
</tr>
<tr>
<td>Dynamic UCS (F6)</td>
<td>This icon turns ON/OFF the Dynamic UCS. When the Dynamic UCS is turned ON, you can draw and create objects on any face of a 3D Model, dynamically.
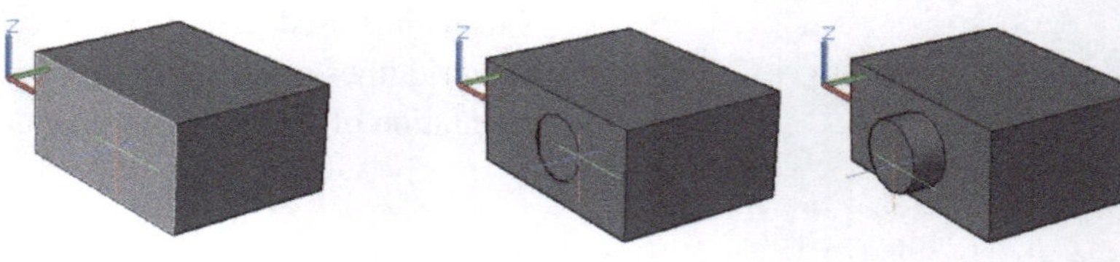</td>
</tr>
<tr>
<td> Dynamic Input (F12)</td>
<td>This icon turns ON or OFF the Dynamic Input mode. When this mode is turned ON, a dynamic input box is attached to the pointer along with a prompt. You can directly enter a value in the dynamic input box. You can use Dynamic Input in place of the command line.
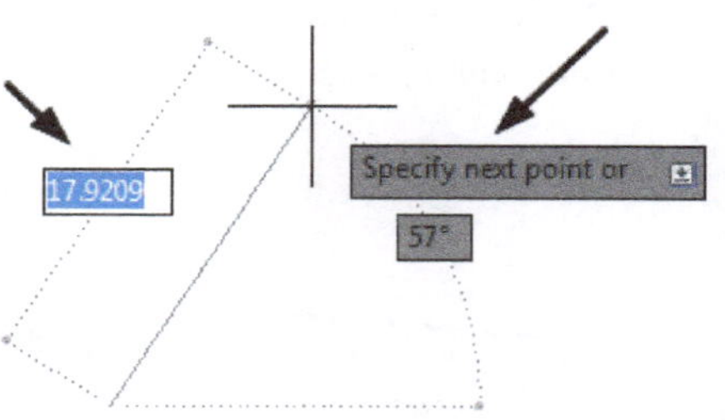
</td>
</tr>
<tr>
<td> Show/Hide Lineweight</td>
<td>This icon turns ON or OFF the lineweight. Line weight is the thickness of objects. You can set the thickness of objects by specifying the lineweight. If the Lineweight is turned OFF, the objects are displayed with the default thickness.</td>
</tr>
</table>

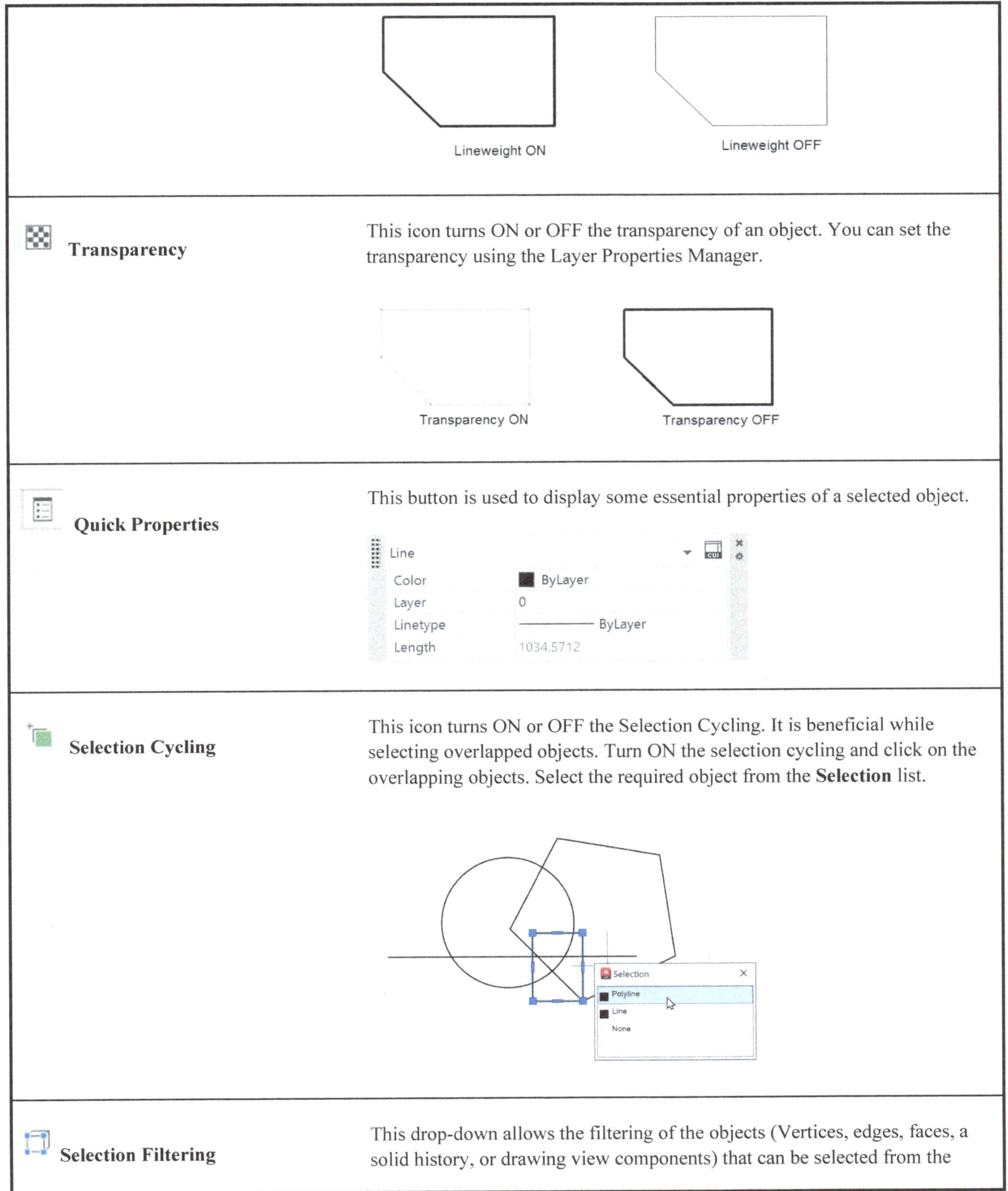

	Lineweight ON Lineweight OFF
Transparency	This icon turns ON or OFF the transparency of an object. You can set the transparency using the Layer Properties Manager. Transparency ON Transparency OFF
Quick Properties	This button is used to display some essential properties of a selected object. Line Color ByLayer Layer 0 Linetype ByLayer Length 1034.5712
Selection Cycling	This icon turns ON or OFF the Selection Cycling. It is beneficial while selecting overlapped objects. Turn ON the selection cycling and click on the overlapping objects. Select the required object from the **Selection** list. Selection Polyline Line None
Selection Filtering	This drop-down allows the filtering of the objects (Vertices, edges, faces, a solid history, or drawing view components) that can be selected from the

	graphics window. Select an option from the drop-down and click on the **Selection Filter** icon to activate the selected filters.
Annotation Monitor	This icon turns the Annotation Monitor ON or OFF. The Annotation Monitor checks whether the annotations are attached to their respective objects. When an annotation is not connected to any object, it displays an error message. 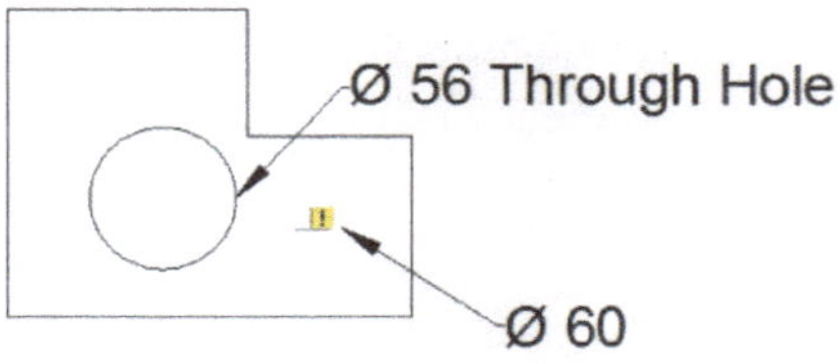
MODEL **Model or Paper Space**	This icon switches between the Model space and Paper space. Model space is used to create drawings, and Paper space is used to print drawings.
1:1 **Annotation Scale**	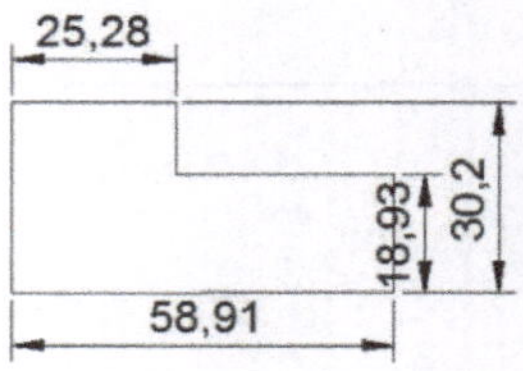 This icon controls the size of annotative objects. Annotative objects are dimensions, texts, notes, and other objects which can be sized as per the drawing scale. 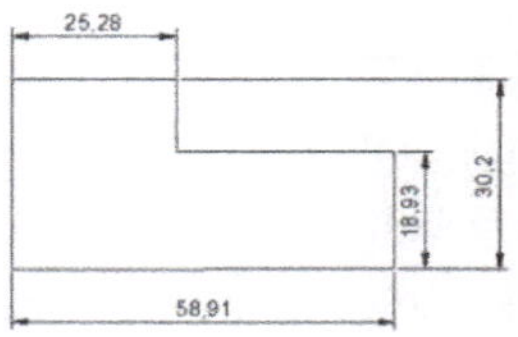
Annotation Visibility	This icon displays annotative objects that are not created in the current scale.
AutoScale	This icon resizes the annotative objects as per the new drawing scale.
Workspace Switching	This icon changes the workspace.

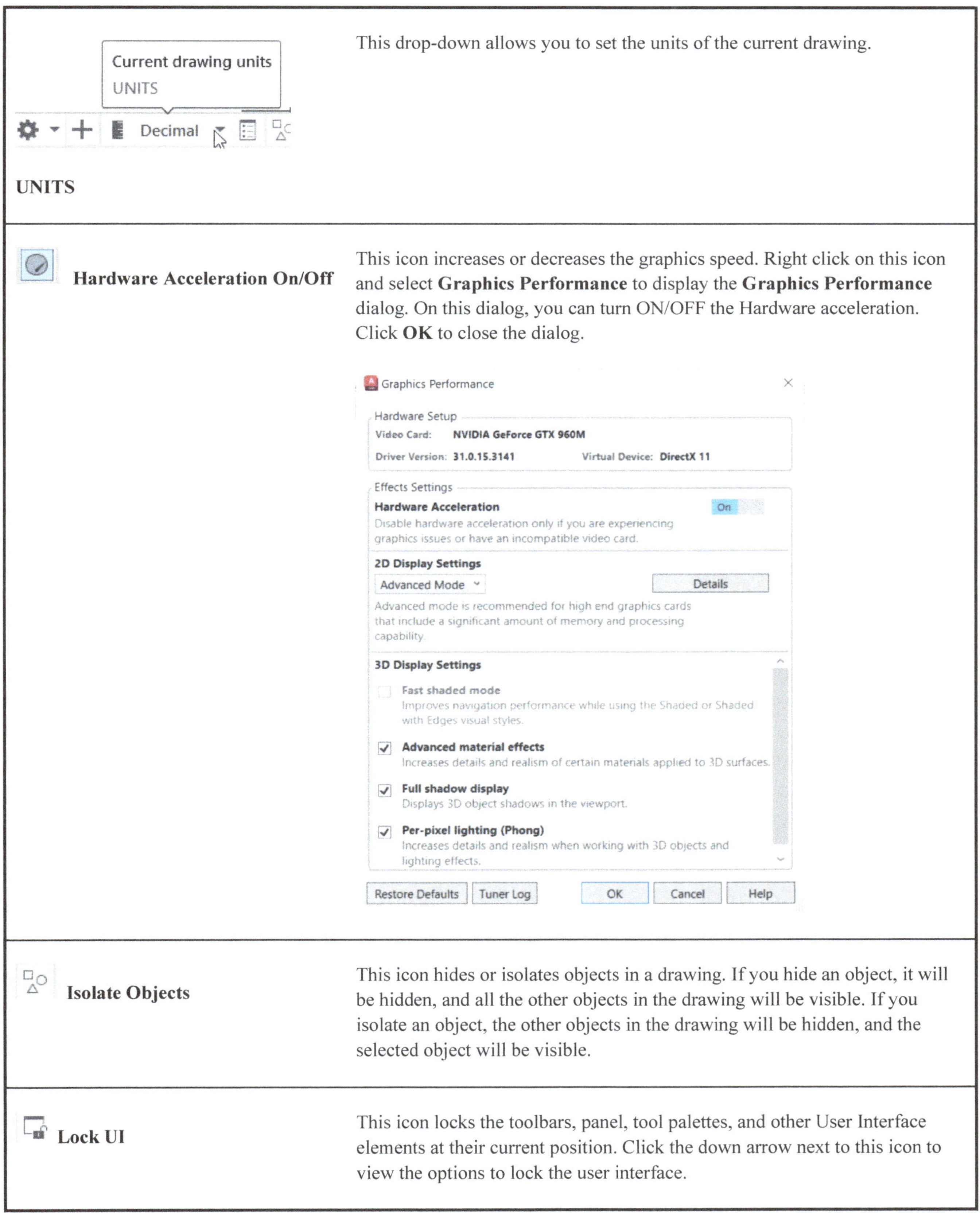

UNITS	This drop-down allows you to set the units of the current drawing.
Hardware Acceleration On/Off	This icon increases or decreases the graphics speed. Right click on this icon and select **Graphics Performance** to display the **Graphics Performance** dialog. On this dialog, you can turn ON/OFF the Hardware acceleration. Click **OK** to close the dialog.
Isolate Objects	This icon hides or isolates objects in a drawing. If you hide an object, it will be hidden, and all the other objects in the drawing will be visible. If you isolate an object, the other objects in the drawing will be hidden, and the selected object will be visible.
Lock UI	This icon locks the toolbars, panel, tool palettes, and other User Interface elements at their current position. Click the down arrow next to this icon to view the options to lock the user interface.

System Variables

System variables control the behavior of various functions and commands in AutoCAD. Usually, system variables have two or more values. You can control a system variable value from the command line. For example, the MIRRTEXT system variable controls the direction of text when you mirror it. The 0 value retains the text direction when you mirror it. Whereas, the 1 value reverses the text direction when you mirror it.

In AutoCAD, you can also control the system variables by using the **System Variable Monitor** dialog. Type SYSVARMONITOR in the command line and press Enter to open this dialog. A list of system variables, which are monitored by default, appears on the dialog. You can know the function of a system variable by clicking the **Help** icon located next to it. You can change a system variable value in the **Preferred** column of the dialog. The **Status** column shows a yellow triangle if you have changed the default value of a system variable. The **Enable balloon notification** option shows a balloon on the status bar if you have changed any system variable value. You can click the **Reset All** button to restore the default values of system variables.

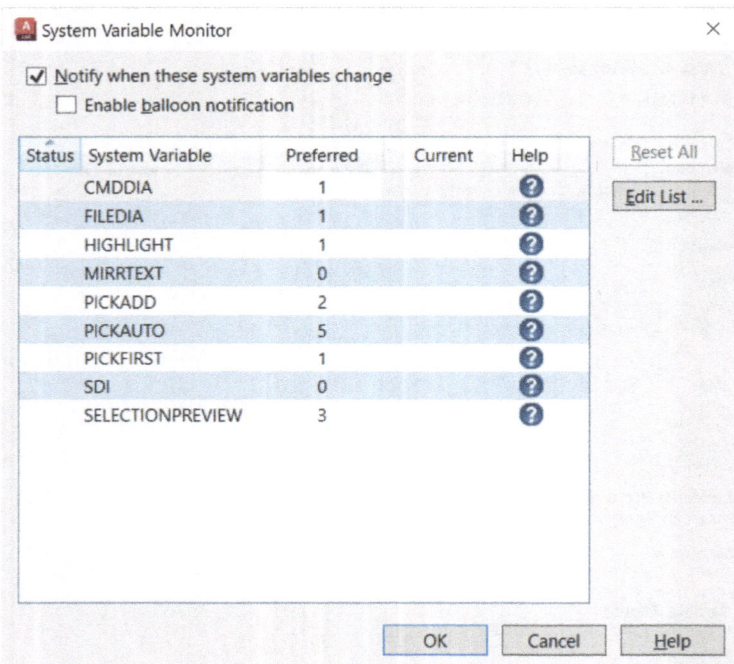

You can monitor more system variables by clicking the **Edit List** button. Next, on the **Edit System Variable List** dialog, select a system variable from the **Available system variables** list, and click the **Add** (>>) button. You can also remove system variables from the Monitored system variables list by selecting them and clicking the **Remove** (<<) button. Click **OK** on both the dialogs after changing the values.

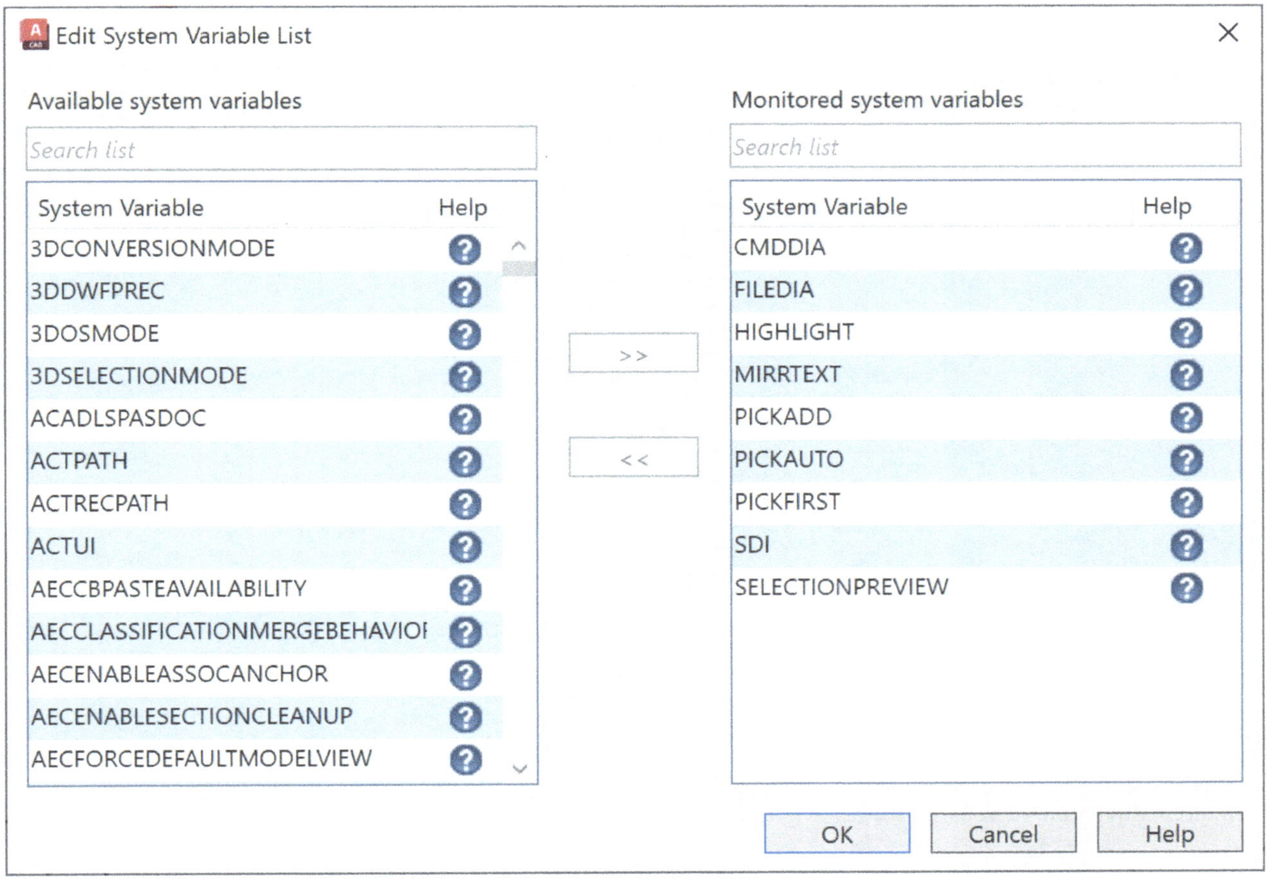

Edit System Variable List

Available system variables

System Variable	Help
3DCONVERSIONMODE	?
3DDWFPREC	?
3DOSMODE	?
3DSELECTIONMODE	?
ACADLSPASDOC	?
ACTPATH	?
ACTRECPATH	?
ACTUI	?
AECCBPASTEAVAILABILITY	?
AECCLASSIFICATIONMERGEBEHAVIOI	?
AECENABLEASSOCANCHOR	?
AECENABLESECTIONCLEANUP	?
AECFORCEDEFAULTMODELVIEW	?

Monitored system variables

System Variable	Help
CMDDIA	?
FILEDIA	?
HIGHLIGHT	?
MIRRTEXT	?
PICKADD	?
PICKAUTO	?
PICKFIRST	?
SDI	?
SELECTIONPREVIEW	?

OK Cancel Help

In AutoCAD, the **System Variable Monitor** icon appears on the Status bar when you change the value of any one of the system variables. Right click on this icon to display a menu. The options on this menu are **Configure System Variable Monitor**, **Reset System Variables**, and **Display Notification**. The **Configure System Variable Monitor** option displays the **System Variable Monitor** dialog, whereas the **Reset System Variables** option resets the system variables to default values. The **Display Notification** option displays a balloon when there is a change in the value of any system variable.

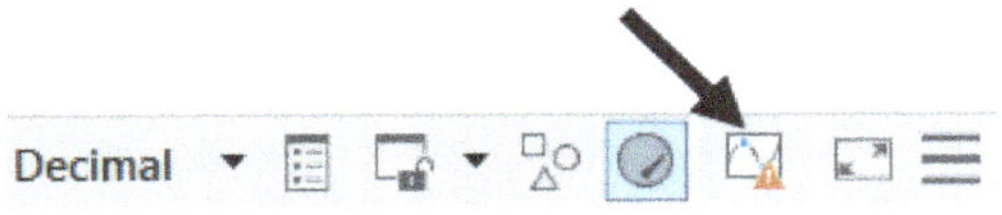

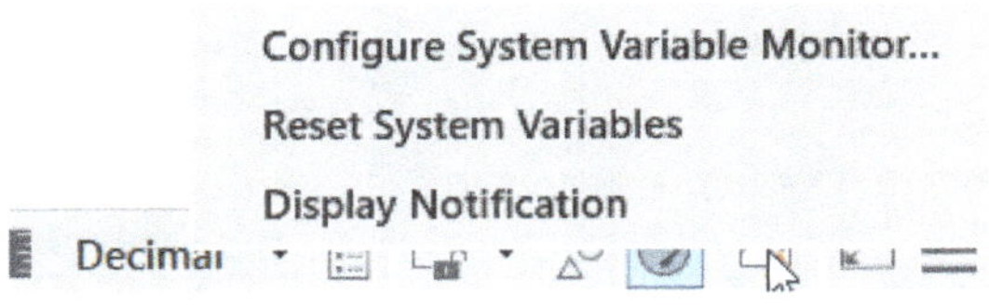

Menu Bar

Menu Bar is not shown by default. However, you can view the Menu Bar by clicking on the down-arrow located on the right side of the Quick Access Toolbar and selecting the **Show Menu Bar** option. The Menu Bar is located at the top of the window just below the title bar. It contains various menus such as File, Edit, View, Insert, Format, Tools, Draw, Dimensions, Modify, and so on. Clicking on any of the words on the Menu Bar displays a menu. The menu contains various tools and options. There are also sub-options available on the list. These sub-options are displayed if you click on an option with an arrow. If you click on an option with (…), a dialog will appear.

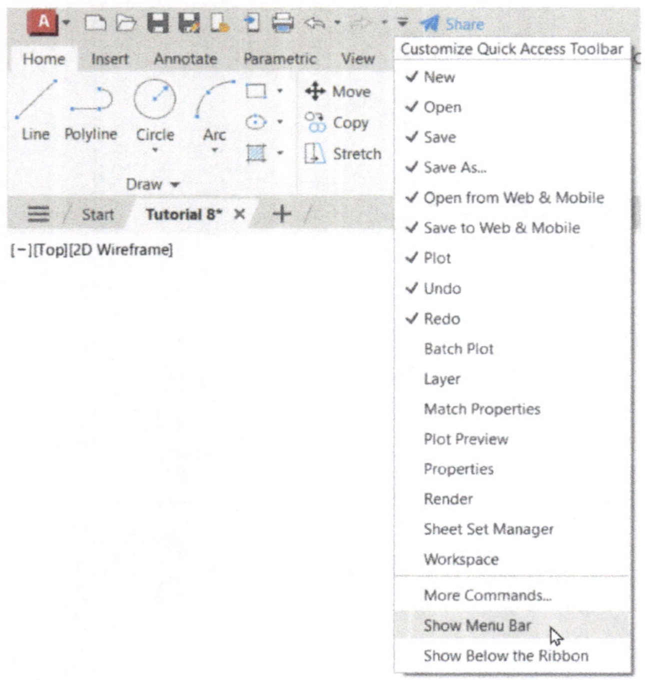

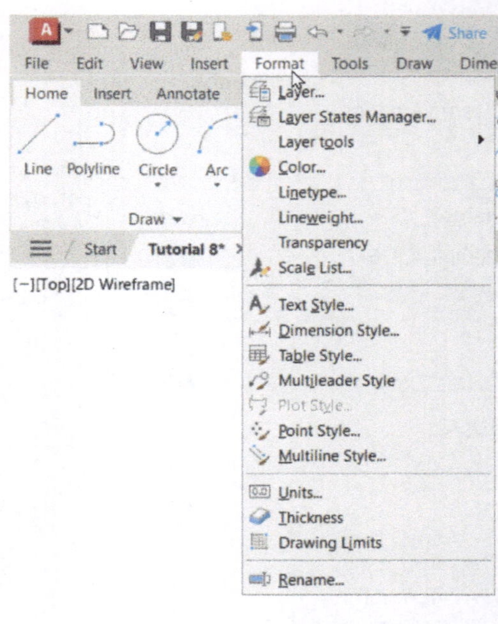

Changing the display of the Ribbon

You can change the presentation of the ribbon by clicking the arrow button located at the top of it. The ribbon can be displayed in three different modes, as shown below.

Minimized to Panel Buttons

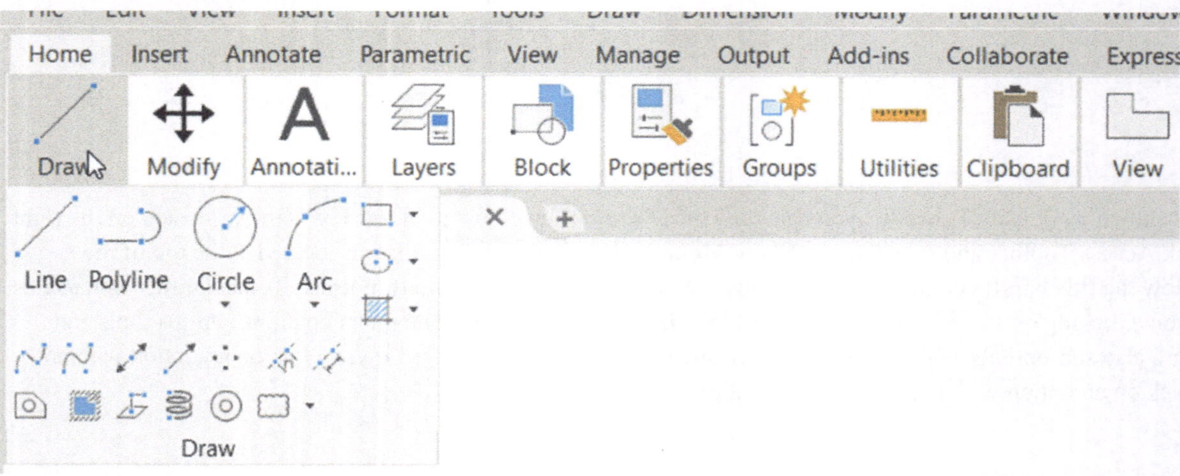

Minimized to Panel Titles

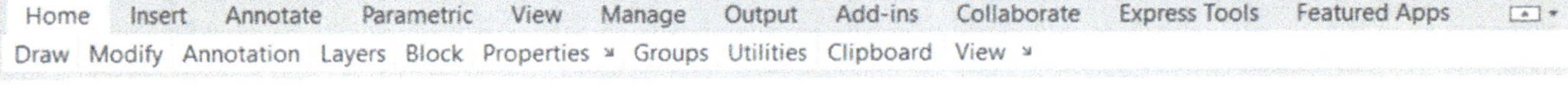

Minimized to Tabs

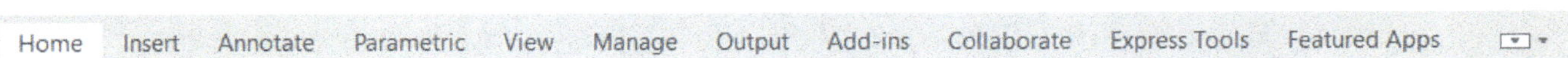

You can use the GALLERYVIEW system variable to hide or show galleries on the ribbon. Enter the GALLERYVIEW system variable in the command line and set the system variable value. The system variable value 1 displays a gallery for dimension styles, blocks, table styles, and mleader styles. The value 0 hides the gallery view.

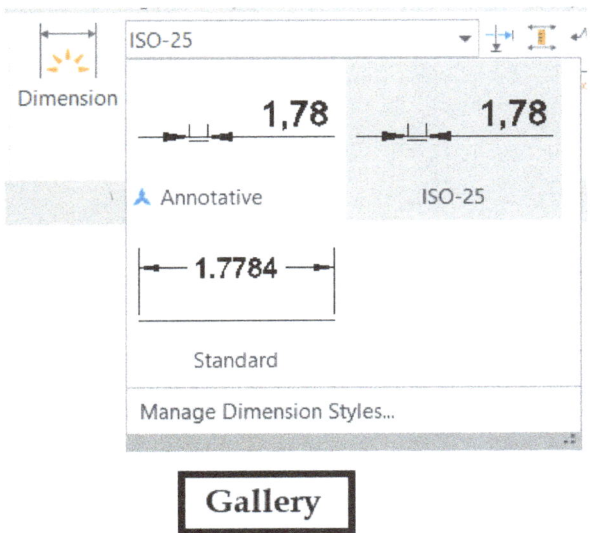

Gallery

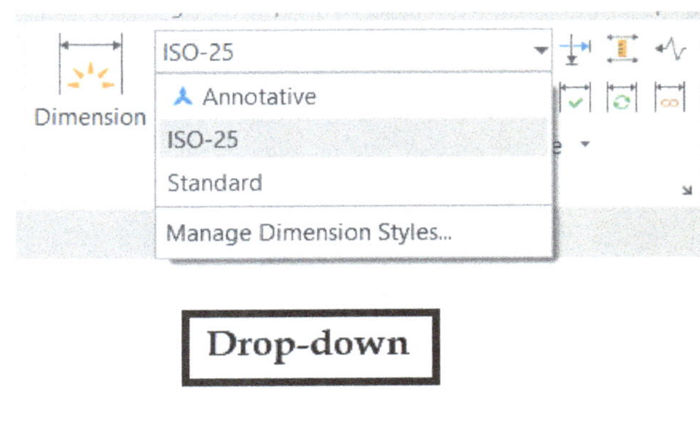

Drop-down

Dialogs and Palettes

Dialogs and Palettes are part of the AutoCAD user interface. Using a dialog or a palette, you can easily specify many settings and options at a time. Examples of dialogs and palettes are as shown below.

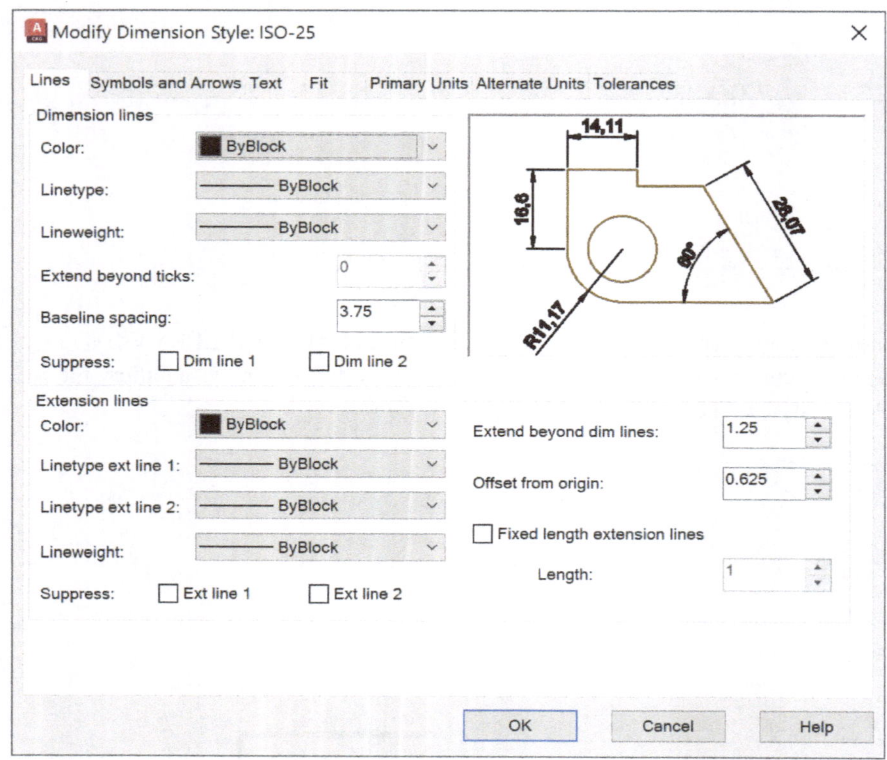

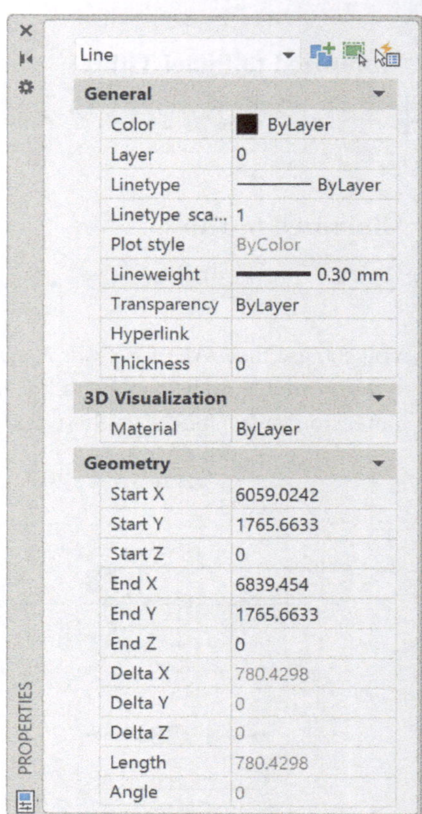

Tool Palettes

Tool Palettes provide you with another way of selecting tools and placing objects. You can display Tool Palettes by clicking **View > Palettes > Tool Palettes** on the ribbon. A Tool Palette is similar to a palette except that it has many palettes grouped in the form of tabs. You can select tools from the Tool Palettes as well as drag and place objects (blocks) into the drawing. You can also create a new Tool Palette and add frequently used tools and objects to it.

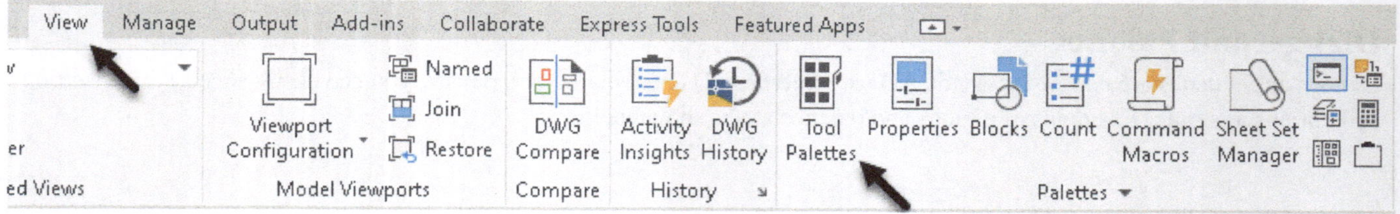

Shortcut Menus

Shortcut Menus appear when you right-click in the graphics window. AutoCAD provides various shortcut menus to help you access tools and options very easily and quickly. There are multiple types of shortcut menus available in AutoCAD. Some of them are discussed next.

Right-click Menu

This shortcut menu appears whenever you right-click in the graphics window without activating any command or selecting an object.

Select and Right-click menu

This shortcut menu appears when you select an object from the graphics window and right-click. It consists of editing and selection options.

Command Mode shortcut menu

This shortcut menu appears when you activate a command and right-click. It shows options depending upon the active command. The shortcut menu below shows the options related to the RECTANGLE command.

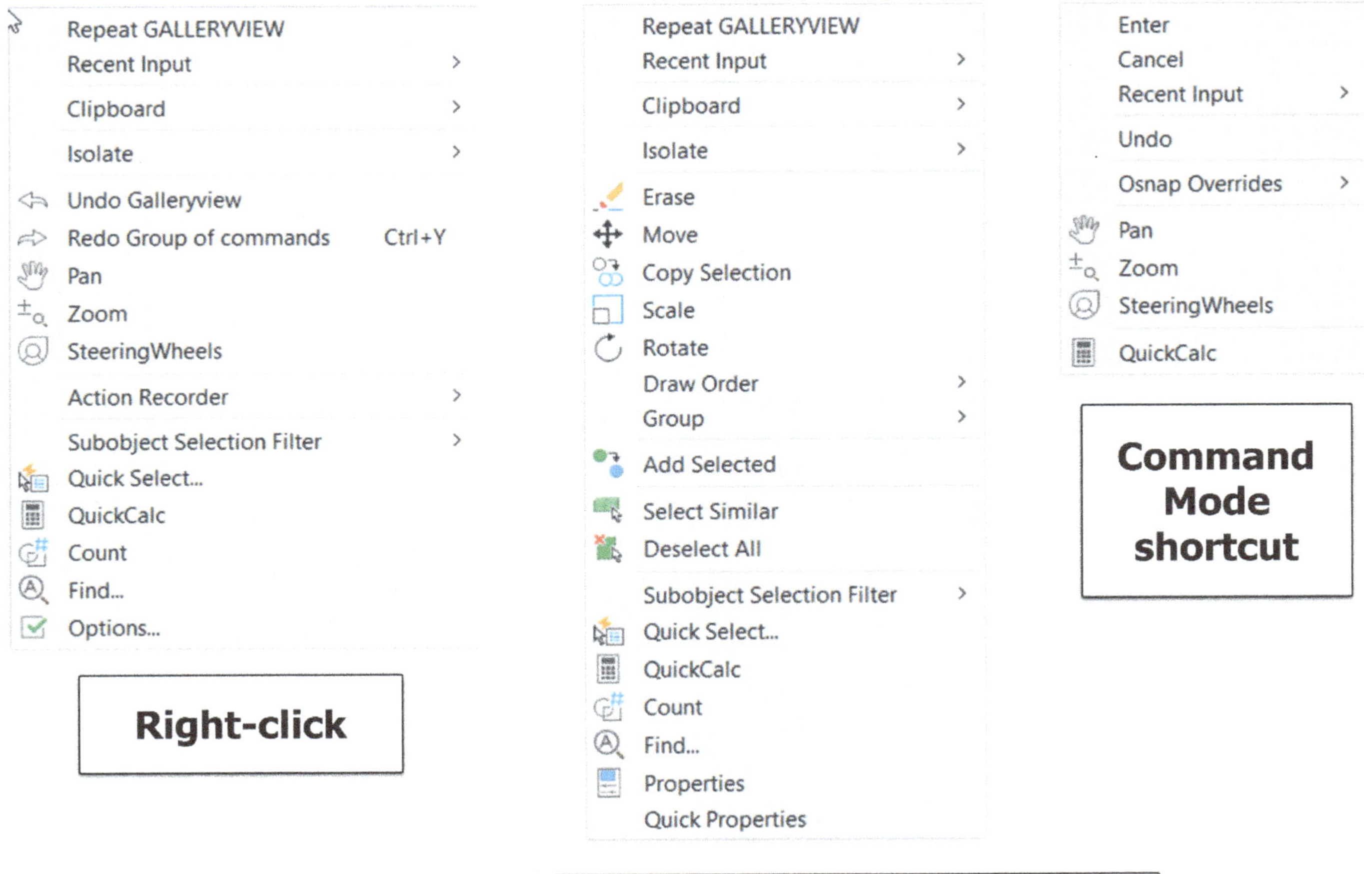

Grip shortcut menu

This shortcut menu is displayed when you select a grip of an object, move the pointer and right-click. It shows various operations that can be performed using grips.

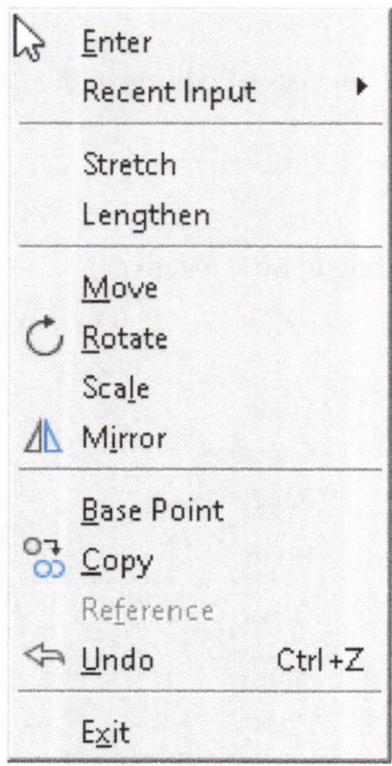

Selection Window

A selection window is used to select multiple elements of a drawing. You can select various elements by using two types of selection windows. The first type is a rectangular selection window. You can create this type of selection window by defining its two diagonal corners. When you set the first corner of the selection window on the left and second corner on the right side, the elements which fall entirely under the selection window will be selected.

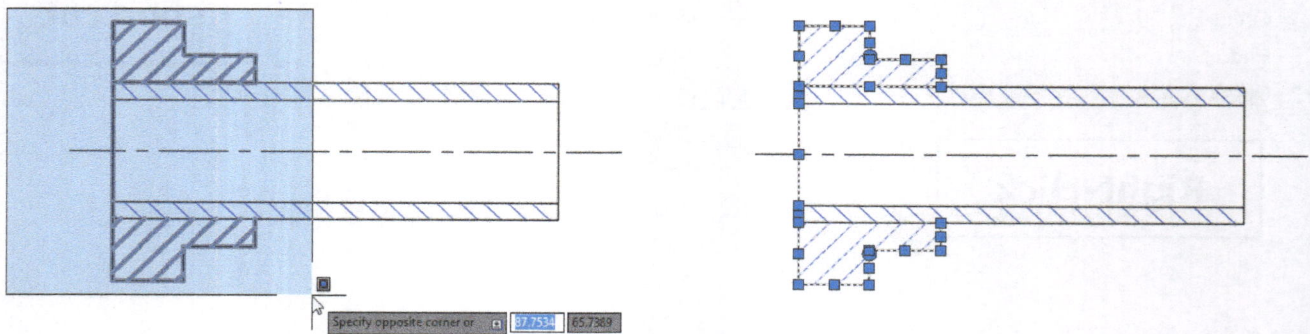

However, if you define the first corner on the right side and the second corner on the left side, the elements, which fall entirely or partially under the selection window, will be selected.

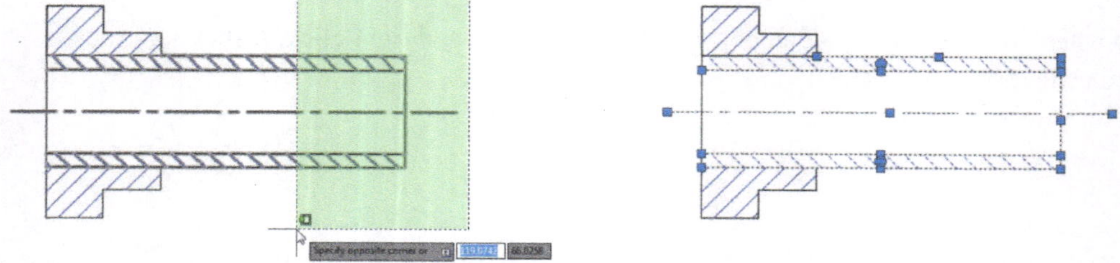

The second type of selection window is Lasso. Lasso is an irregular shape created by holding the left mouse button and dragging the pointer across the elements to select. If you drag the pointer from left to right, the elements falling entirely under the lasso will be selected.

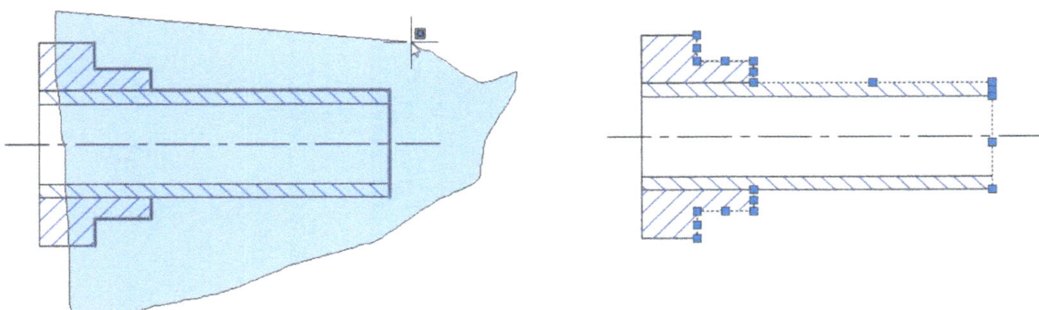

If you drag the pointer from right to left, the elements which fall wholly or partially under the lasso will be selected.

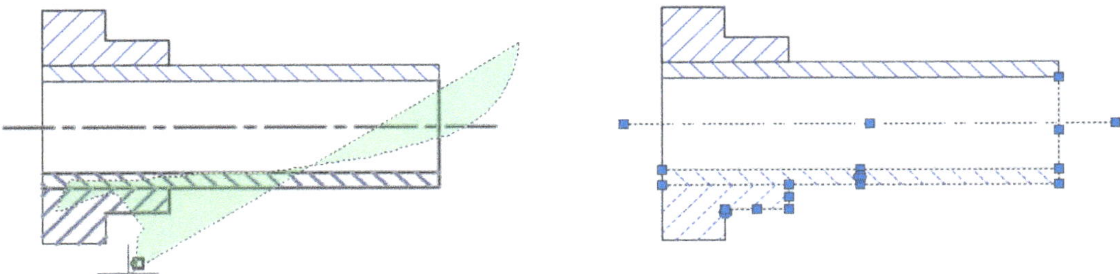

In AutoCAD, you can specify the first corner of the selection window at one portion of a large drawing. Next, zoom and pan to the rest of the drawing, and then specify the second corner of the selection window. By doing so, you can select the portion of the drawing, which is currently not visible on the screen.

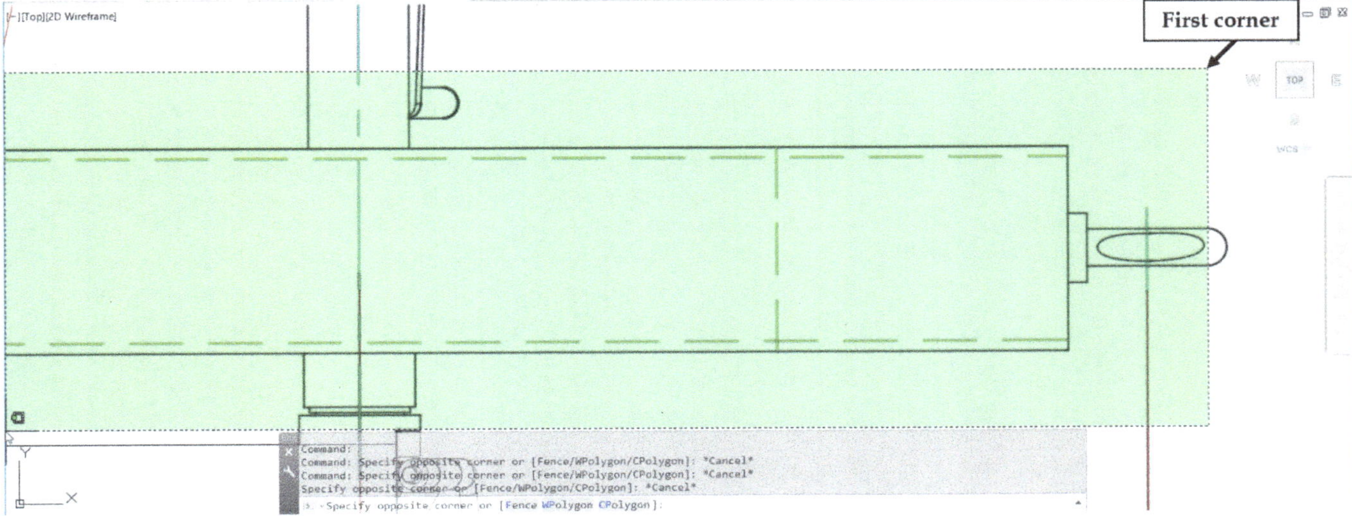

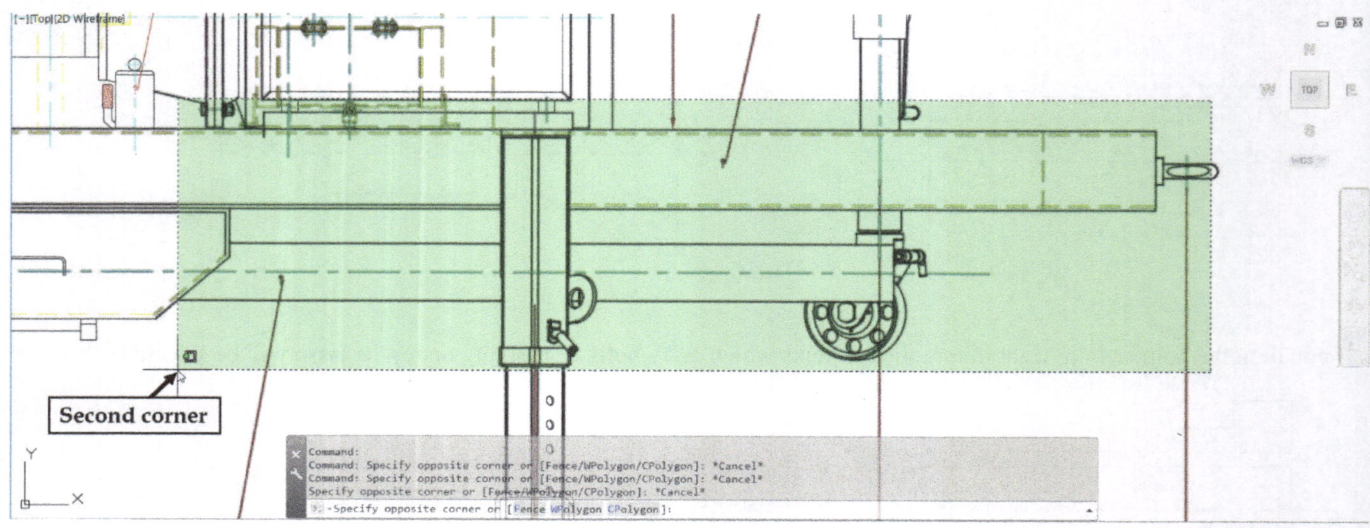

Starting a new drawing

You can start an AutoCAD document by using the **New** drop-down available on the Start screen or by using the **Select template** dialog.

New drop-down on the Initial Screen

To start a new drawing, select the New option from the New drop-down available at the left side of the **Start** screen. You can also select the acadiso.dwt template from the **New** drop-down to start a new drawing in the ISO format.

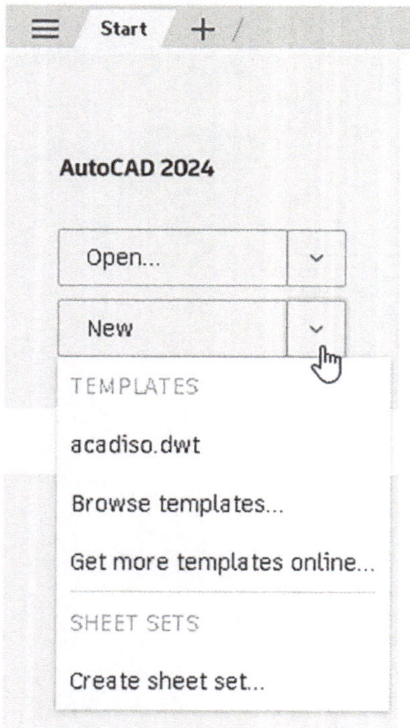

The Select Template dialog

- To start a new drawing using the **Select Template** dialog, click the **New** button on the Quick Access Toolbar (or) Application Menu. You can also select the **Browse Templates** option from the **New** drop-down available on the **Start** screen.

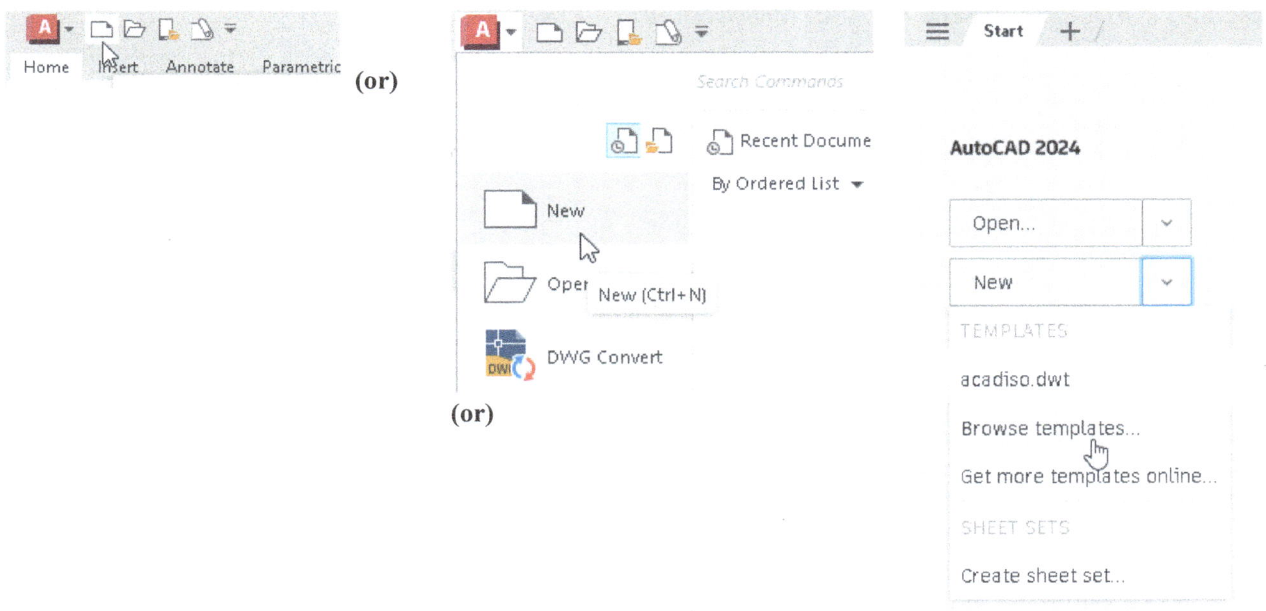

On the **Select Template** dialog, select the **acad.dwt** (inch units) or **acadiso.dwt** (metric units) template for creating a 2D drawing. Select the **acad3D.dwt** or **acadiso3D.dwt** template for creating 3D models.

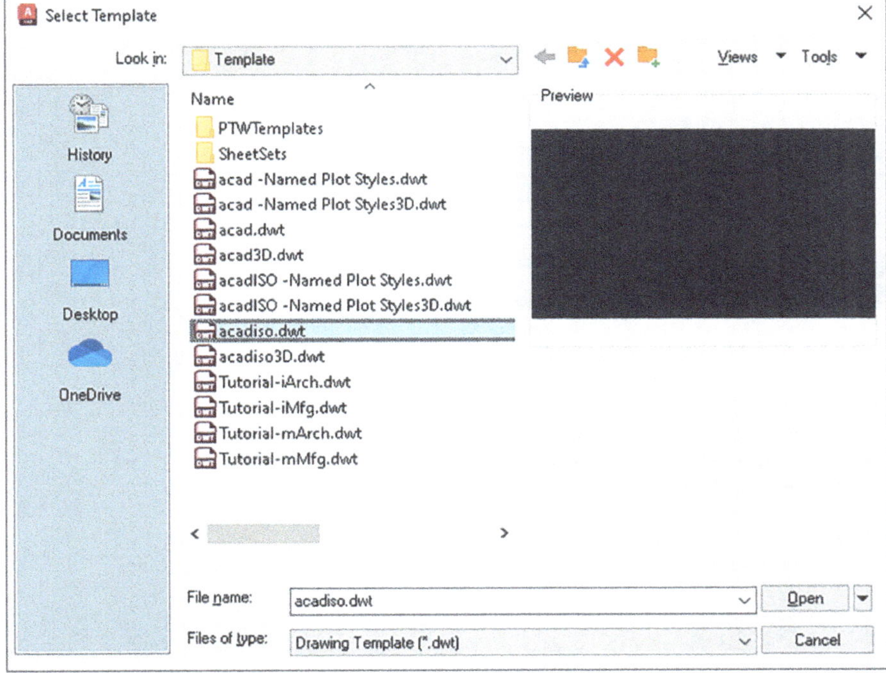

Saving a new drawing to Web & Mobile

You can save a drawing to AutoCAD Web & Mobile using the Save to AutoCAD Web & Mobile application. This application helps you view and make changes to the drawing from AutoCAD Web and AutoCAD Mobile application. The drawings edited in AutoCAD Web and AutoCAD Mobile can be opened in AutoCAD.

- Click the **Save to AutoCAD Web & Mobile** icon on the Quick Access Toolbar; the **Install Save to AutoCAD Web & Mobile** dialog appears.

- Check the **I have read and agree to the App Store End Use License Agreement** option.
- Click the **Install** button; the **Save to AutoCAD Web & Mobile** application is installed in the background.
- On the **Save in AutoCAD Web & Mobile** dialog, type in the **File name** box and then click the **Save** button.

To open the file saved in AutoCAD Web and Mobiles applications, click the **Open from Web & Mobile** icon on the Quick Access Toolbar. Next, select the files from the **Open from Web & Mobile** dialog, and then click **Open**.

Help

Press F1 or type a keyword in the search bar located at the top right corner of the window to get help for any topic. On the Autodesk AutoCAD 2024 –Help window, click the **Find** option next to the topic; an animated arrow appears on the window showing the tool location.

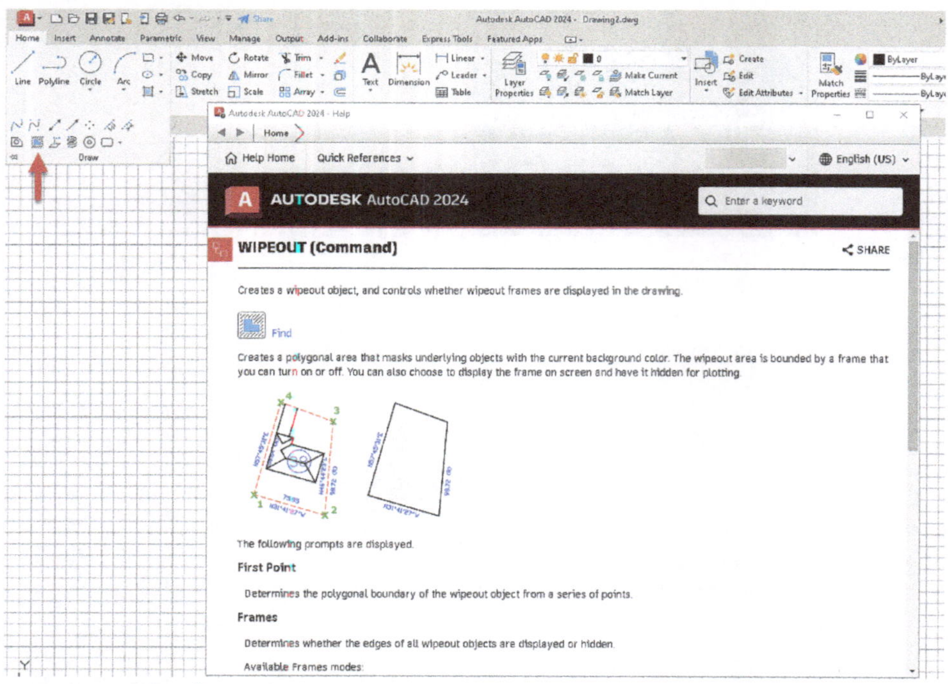

Command List

Various commands in AutoCAD are given in the table below:

Command	Alias	Description
APPLOAD		This command activates the **Load/Unload Applications** dialog.
ADCENTER	**DC**	Opens the **DESIGNCENTER** palette.
ALIGN	**AL**	It is used to align objects with other objects.
ARC	**A**	It is used to create an arc.
AREA		This command displays the area of a selected closed object.
ARRAY	**AR**	Creates Rectangular, Path, or Polar 2D arrays.
ASE		Displays the **DBCONNECT MANAGER** palette.
ATTDEF	**ATT**	It displays the **Attribute Definition** dialog.
ATTEDIT	**ATE**	It is used to edit Attributes.
AUDIT		It is used to check and fix errors.
AUTOCONSTRAIN		It is used to apply constraints automatically.
AUTOPUBLISH		It is used to create a DWF file.
BACTION	**AC**	It is used to add an action to a dynamic block. This command is available in Block Editor.
BLOCK		It is used to create a block.

BMAKE	**B**	It is used to create a block.
BMPOUT		It is used to create a Raster image out of the drawing.
BOUNDARY	**BO**	It is used to create a hatch boundary.
BREAK	**BR**	It is used to break an object.
CAL		It is used to calculate mathematical expressions.
CHAMFER	**CHA**	It is used to create chamfers.
CHPROP		This command changes the properties of a selected object.
CIRCLE	**C**	It is used to create a circle.
COLOR	**COL**	This command displays the **Select Color** dialog.
COPYTOLAYER		It is used to copy objects from one layer to another.
COPY	**CO**	It is used to copy objects inside a drawing.
COPYCLIP		It is used to copy objects from one drawing to another.
CUSTOMIZE		It is used to customize the tool palettes.
DDEDIT	**ED**	It is used to edit a note or annotation.
DIMSTYLE	**D**	It is used to create or modify a dimension style.
DDMODIFY		Displays the Properties palette.

DELCONSTRAINT		It is used to delete constraints.
OSNAP	**OS**	It is used to set the **Object Snap** settings.
DDPTYPE		It is used to set the point style and size.
VIEW	**V**	It is used to save views by names.
DGNEXPORT		It is used to export the drawing to Microstation (DGN) format.
DGNIMPORT		It is used to import a Microstation (DGN) format file.
DIMCONSTRAINT	**DCON**	It is used to apply dimensional constraints to objects.
DIMLINEAR	**DLI**	It is used to create a linear dimension.
DIMALIGNED	**DAL**	It is used to create an aligned dimension.
DIMARC	**DAR**	It is used to dimension the arc length.
DIMRADIUS	**DIMRAD**	It is used to create a radial dimension.
DIMJOGGED	**JOG**	It is used to create a jogged dimension.
DIMDIAMETER	**DIMDIA**	It is used to create a diameter dimension.
DIMANGULAR	**DAN**	It is used to create an angular dimension.
DIMORDINATE	**DOR**	It is used to create an ordinate dimension.
DIMCONTINUE	**DIMCONT**	It is used to create continuous dimensions from an existing one.

DIMBASELINE	**DIMBASE**	It is used to create baseline dimensions.
DIMINSPECT		It is used to create an inspection dimension.
-DIMSTYLE		Update a dimension according to the current dimension style.
DIMSPACE		It is used to adjust the space between dimensions.
DIMBREAK		It is used to break the extension line of a dimension when it intersects with another dimension.
DIMOVERRIDE		It is used to override the system variables of a selected dimension.
DIMCENTER		It is used to create a center mark of a circle.
DIMEDIT	**DIMED**	It is used to edit a dimension.
DIMTEDIT	**DIMTED**	It is used to edit the dimension text.
DIMDISASSOCIATE		This command disassociates a dimension from the object.
DIST	**DI**	It is used to measure the distance between two points.
DISTANTLIGHT		It is used to create a distant light.
DIVIDE	**DIV**	Places evenly spaced objects on a line segment
DONUT	**DO**	It is used to create a donut.
DVIEW		It is used to get the aerial view of a drawing.
DXBIN		It is used to open a DXB file.

DXFIN		It is used to open a DXF file.
DXFOUT		It is used to save a file in the DXF format.
ELLIPSE	**EL**	It is used to create an ellipse.
ERASE	**E**	It is used to erase objects.
EXIT		It is used to close AutoCAD.
EXPLODE	**X**	It is used to explode or ungroup objects.
EXPLORER		This command displays the Windows Explorer.
EXPORT	**EXP**	It is used to export data.
EXTEND	**EX**	It is used to extend an object up to another.
FILLET	**F**	It is used to create a fillet at the corner.
FILTER		It is used to set object selection filters.
GEOMCONSTRAINT	**GCON**	It is used to apply geometric constraints.
GRADIENT		It is used to apply the gradient to a closed area.
GROUP	**G**	It is used to group objects.
HATCH	**H**	It is used to apply hatch to a closed area.
HATCHEDIT	**HE**	It is used to edit hatch.

HELP		Display the Help window.
HIDE	**HI**	This command changes the Visual Style to Hidden.
ID		This command displays the coordinate values of a selected point.
IMAGE, IMAGEATTACH	**IM**	It is used to attach an Image reference.
IMAGEADJUST	**IAD**	It is used to adjust images.
IMAGECLIP		It is used to crop an image.
IMPORT	**IMP**	It is used to import other forms of CAD data.
INSERT	**I**	It is used to insert a block.
INSERTOBJ		It is used to insert an object into the drawing.
ISOPLANE	**CTRL+E**	It is used to set the current isometric plane.
JOIN	**J**	It is used to join the endpoints of two linear or curved objects.
LAYCUR		The Layer of the selected objects will be made current.
LAYER	**LA**	It is used to create a new layer and modify its properties.
LAYFRZ		It is used to freeze the layer of a selected object.
LAYISO		This command isolates the layer of a selected object.

LAYOUT		It is used to modify layouts.
LAYOFF		It is used to turn off the layer of a selected object.
LAYON		It is used to turn ON all the layers.
LAYOUTWIZARD		This command displays the Create Layout dialog.
LENGTHEN	**LEN**	It is used to increase the length of an object.
LIMITS		It is used to set the drawing limits.
LIMMAX		It is used to set the maximum limit of a drawing.
LINE	**L**	It is used to create a line.
LINETYPE	**LT**	It is used to set the line type.
LIST	**LI**	This command lists the properties of a selected object in the text window.
LOAD		This command imports the shapes that can be used by the SHAPE command.
LTSCALE	**LTS**	It is used to set the linetype scale.
MEASURE	**ME**	It is used to place points or blocks at regular intervals on an object.
MENU		It is used to load a customization file.
MENULOAD		It is used to load or unload a customizable file.

MIRROR	**MI**	It is used to create a mirror image of an object.
MLEDIT		It is used to edit a multiline.
MLINE	**ML**	It is used to create multiple parallel lines.
MLSTYLE		It is used to create and modify a multiline style.
MOVE	**M**	It is used to move selected objects.
MSLIDE		It is used to create a slide out of a drawing.
MSPACE	**MS**	I used to switch from paper space to model space.
MSTRETCH		It is used to stretch multiple objects at a time.
MTEXT	**MT or T**	It is used to write text in multiple lines.
MVIEW	**MV**	It is used to create and modify viewports.
MVSETUP		It is used to set drawing specifications for printing purposes.
NEW	**CTRL+N**	It is used to open a new file.
NOTEPAD		It is used to edit a file in Notepad.
OFFSET	**O**	This command creates a parallel copy of a selected object at a specified distance.
OOPS		It is used to undo the ERASE command.
OPEN		It is used to open an existing file.

OPTIONS	**OP**	It is used to set various options related to the drawing.
ORTHO		This command turns ON/OFF the Ortho Mode.
OSNAP	**OS**	I am used to the **Object Snap** settings.
PAGESETUP		It is used to specify the printing properties of a layout.
PAN	**P**	It is used to drag a drawing to view its different portions.
PARAMETER	**PAR**	It is used to assign expressions to a dimensional constraint.
PBRUSH		Opens the **Windows Paint** application.
PEDIT	**PE**	It is used to edit polylines.
PLINE	**PL**	It is used to create a polyline. A polyline is a single object which can have continuous lines and arcs.
PLOT	**CTRL+P**	It is used to plot a drawing.
POINT	**PO**	It is used to place a point in the drawing.
POLYGON	**POL**	It is used to create a polygon.
PREVIEW	**PRE**	It is used to preview the plotted drawing.
PROPERTIES	**PR**	Displays the **Properties** palette.
PSOUT		It is used to create a postscript file.
PURGE	**PU**	It is used to remove unwanted data from the drawing.

QDIM		It is used to create a quick dimension.
QSAVE		It is used to save the current drawing.
QUICKCALC	**QC**	This command displays the QuickCalc calculator.
QUIT		It is used to close the current drawing session.
RAY		It is used to create a line that starts from a selected point and extends up to infinity.
RECOVER		It is used to repair and open the damaged files.
RECOVERALL		It is used to repair a damaged file along with the attached external references.
RECTANG		It is used to create a polyline rectangle.
REDEFINE		It is used to restore an AutoCAD command, which has been overridden.
REDRAW	**R**	Refreshes the current viewport.
UNDEFINE		It is used to override an existing command with a new one.
REDO		It is used to cancel the previous UNDO command.
REDRAWALL	**RA**	This command refreshes all the viewports in a drawing.
REGEN	**RE**	This command regenerates the current viewport of a drawing.
REGENALL	**REA**	This command regenerates all the viewports of a drawing.

REGION	**REG**	This command converts the area enclosed by objects into a region.
RENAME	**REN**	It is used to rename blocks, viewports, and dimension styles.
REVCLOUD		It is used to highlight a portion of drawing by creating a cloud around it.
RIBBON		This command displays the ribbon.
RIBBONCLOSE		This command hides the ribbon.
SAVE	**CTRL+S**	This command saves the currently opened drawing.
SAVEAS		This command saves the drawing with another name and location.
SAVEIMG		It is used to save a rendered output file.
SCALE	**SC**	It is used to increase or decrease the size of a drawing.
SCRIPT	**SCR**	It is used to load a script file. A script is used to run various commands sequentially.
SETVAR	**SET**	It is used to list or change a system variable.
SHAPE		It is used to insert a shape into a drawing.
SHELL		It is used to enter MS-DOS commands.
SKETCH		It is used to draw freehand sketches.
SOLID	**SO**	It is used to create filled triangles or quadrilaterals.

SPELL	**SP**	It is used to check the spelling of a text.
SPLINE	**SPL**	It is used to create a spline (curved object).
SPLINEDIT	**SPE**	It is used to edit a spline.
STATUS		It is used to display the details of a drawing, such as limits, model space usage, and layers.
STRETCH	**S**	It is used to stretch objects.
STYLE	**ST**	It is used to create or modify the text style.
TABLET		Allows using a tablet for creating drawings.
TBCONFIG		It is used to customize the user interface.
TEXT		It is used to enter text in the drawing.
THICKNESS	**TH**	It is used to set a thickness value to 2D objects.
TOLERANCE		It is used to apply geometric tolerances to the drawing.
TOOLBAR	**TO**	It is used to customize toolbars.
TRIM	**TR**	It is used to trim unwanted portions of an object.
UCS		It is used to specify the location of the user coordinate system.
UNDO	**CTRL+Z (or) U**	It is used to undo the last operation.
UNITS	**UN**	Set the units of the drawing

VIEW		It is used to save and restore the model space, layout, and preset views.
VPLAYER		It is used to control layer visibility in paper space.
VPORTS		It is used to create multiple viewports in model space of paper space.
VSLIDE		It is used to show an image slide file.
WBLOCK	**W**	It is used to convert a block into a drawing.
WMFIN		It is used to import a Windows Metafile. This file contains the drawing and image data. But only the drawing data is imported.
WIPEOUT		It is used to wipe out a portion of the drawing.
WMFOPTS		It is used to specify the options for importing a Windows Metafile.
WMFOUT		It is used to save objects as Windows Metafile.
XATTACH	**XA**	It is used to attach a drawing as an external reference.
XLINE	**XL**	It is used to create construction lines. Construction lines extend to infinity and help in drawing objects.
XREF	**XR**	It is used to attach a drawing as an external reference.
ZOOM	**Z**	It is used to Zoom in or out of a drawing.

3D Commands

Command	Shortcut	Description
3DARRAY	**3A**	It is used to create three-dimensional arrays of an object.
3DALIGN	**3AL**	It is used to align 3D objects.
3DFACE	**3F**	It is used to create a three or four-sided 3D surface.
3DMESH		It is used to create a freeform 3D mesh.
3DCORBIT		It is used to rotate a view in the 3D space with continuous motion.
3DDISTANCE		It is used to control the distance.
3DEDITBAR		It is used to add and edit control vertices on a NURBS surface or spline.
3DFLY		It is used to view the 3D model as if you are flying through.
3DFORBIT		It is used to rotate a view in 3D space freely.
3DMOVE	**3M**	It is used to move the objects in 3D space.
3DORBIT	**3DO**	It is used to rotate the view constrained along the horizontal or vertical axis.
3DORBITCTR		It is used to set the center for rotating the view in 3D space.
3DPAN		It is used to pan the 3D models horizontally or vertically. This is used when working in perspective view.
3DPOLY	**3P**	It is used to create a 3D polyline.

3DPRINT	**3DP**	It is used to print the model in 3D (plastic prototype).
3DROTATE		It is used to rotate 3D objects in 3D space.
3DSCALE	**3S**	It is used to increase or decrease the size of a 3D object along the X, Y, Z directions.
3DSIN		It is used to import a 3ds Max file.
3DDWF		Export the 3D model to a 3D DWF file.
3DWALK		It is used to view the 3D model as if you are walking through it.
ANIPATH		It is used to create an animation when you are navigating through the model.
BOX		It is used to create a 3D box.
CONE		It is used to create a 3D cone.
CONVERTOLDLIGHTS		It is used to convert lights created in previous releases to the current format.
CONVERTOLDMATERIALS		It is used to convert old materials to the new format
CONVTONURBS		It is used to convert a surface to NURBS. You can easily edit a NURBS by using control vertices displayed on it.
CONVTOSOLID		It is used to convert 3D meshes, polylines, and circles to 3D solids.
CONVTOSURFACE		It is used to convert objects to surfaces.

CVADD		It is used to add control vertices to a NURBS surface or spline.
CVREMOVE		It is used to remove control vertices from a NURBS surface or spline.
CVHIDE		It is used to hide the control vertices of a NURBS surface or splines,
CVSHOW		It is used to display the control vertices of a NURBS surface or splines.
CVREBUILD		It is used to rebuild the control vertices of a NURBS surface.
CYLINDER		It is used to create a Cylinder.
EDGESURF		It is used to create a mesh surface from four adjacent edges.
EXTRUDE	**EXT**	It is used to extrude a closed region or polyline.
FILLETEDGE		It is used to blend an edge of a 3D object.
FLATSHOT		It is used to create a 2D representation of a 3D model.
FREEPOINT		It is used to create point light that emits light in all directions.
FREESPOT		It is used to create a spotlight without any target.
HELIX		It is used to create a helical or spiral curve.
INTERFERE		It is used to create a 3D solid at the interference point of the various solid objects.
INTERSECT	**IN**	It is used to create a 3D solid at the intersection portion of solid.

LIGHT	It is used to create a light.
LIGHTLIST	This command displays the lights available in the current 3D model.
LOFT	It is used to create a 3D solid or surface between various cross-sections.
MATERIALS	This command displays the Material Browser.
MATERIALASSIGN	It is used to assign a material to the model.
MATERIALMAP	It is used to control the texture.
MATERIALATTACH	It is used to associate materials with layers.
MESH	It is used to create 3D mesh objects.
MESHREFINE	It is used to refine the mesh of 3D mesh objects.
MESHSMOOTH	It is used to increase the smoothness of mesh objects.
MIRROR3D	It is used to mirror 3D objects in 3D space.
OFFSETEDGE	It is used to create a parallel copy of an edge at a specified distance.
PFACE	It is used to create a 3D Polyface mesh by specifying vertices.
PLAN	This command displays the top view of the 3D model.
PLANESURF	It is used to create a planar surface.

POINTLIGHT		It is used to create point light that emits light in all directions.
PRESSPULL		It is used to extrude or subtract material.
PYRAMID		It is used to create a pyramid.
-RENDER		It is used to specify settings for rendering.
RENDERCROP		It is used to render a rectangular portion of a 3D model.
RENDERENVIRONMENT		It is used to control the visual properties of the rendered image.
RENDEREXPOSURE		It is used to control the lighting of a rendered image.
RENDERONLINE		It is used to render an image in Autodesk 360 (cloud).
RENDERPRESETS		It is used to specify preset values for rendering an image.
RENDERWIN		Displays the render window.
REVOLVE	**REV**	It is used to create a revolved solid.
REVSURF		It is used to create a revolved surface.
RMAT		This command displays the Material Browser.
RPREF	**RPR**	It is used to specify advanced render settings.
SECTION	**SEC**	It is used to create a section plane in a 3D model.
SLICE	**SL**	It is used to slice a 3D model.

SOLPROF		Create a profile from a 3D model in the paper space.
SOLIDEDIT		It is used to edit faces and edges of a 3D solid.
SPACETRANS		It is used to calculate equivalent model space and paper space distance.
SPHERE		It is used to create a sphere.
SPOTLIGHT		It is used to create a spotlight that emits light like a torch.
STLOUT		It is used to export a file to STL format.
SUNPROPERTIES		Displays the Sun properties palette.
SURFBLEND	**BLENDSRF**	It is used to create a continuous blend surface between two surfaces.
SURFEXTEND		It is used to lengthen a surface up to another surface.
SURFEXTRACTCURVE		It is used to create Isoline curves on a surface, solid, or a face in U and V directions.
SURFFILLET		It is used to create a surface fillet between two surfaces.
SURFOFFSET		It is used to create a parallel surface at a specified distance.
SURFNETWORK		It is used to create a surface from various curves in U and V directions.
SURFPATCH		It is used to create a surface using the edges, forming a closed loop.

SURFSCULPT		It is used to create a closed surface by trimming and combining the surfaces that form a region together.
SURFTRIM		It is used to trim portions of a surface at intersections with other surfaces.
SURFUNTRIM		It is used to untrim the trimmed surface.
SWEEP		It is used to create a 3D solid or surface by sweeping a profile along a path.
TABSURF		It is used to create a mesh from a line or curve swept along a straight path
TORUS	**TOR**	It is used to create a torus.
UNION	**UNI**	It is used to combine various solids into one.
VISUALSTYLES		It is used to create and modify visual styles.
VPOINT		It is used to set the viewing direction of the 3D model.
WEDGE	**WE**	It is used to create a wedge shape.
XEDGES		It is used to create a 3D wireframe from a 3D solid.

Chapter 2: Drawing Basics

In this chapter, you will learn to do the following:

- **Draw lines, rectangles, circles, ellipses, arcs, polygons, and polylines**
- **Use the Erase, Undo and Redo tools**
- **Draw entities using the absolute coordinate points**
- **Draw objects using the relative coordinate points**
- **Draw objects using the tracking method**

Drawing Basics

This chapter teaches you to create simple drawings. You will create these drawings using the essential drawing tools. These tools include **Line**, **Circle**, **Polyline**, and **Rectangle**, and they are available in the **Draw** panel of the ribbon, as shown below. You can also activate these tools by typing them in the command line.

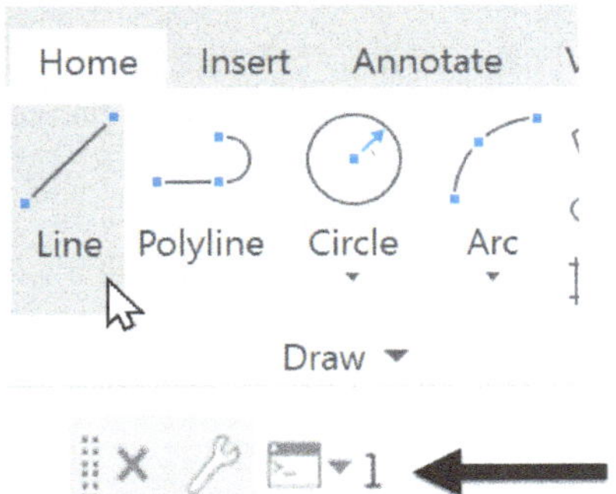

Drawing Lines

You can draw a line by specifying its start point and endpoint using the **Line** tool. However, there are various methods to specify the start and endpoint of a line. These methods are explained in the following examples.

Example 1 (using the Absolute Coordinate System)

In this example, you will create lines by specifying points in the absolute coordinate system. In this system, you specify the points with respect to the origin (0, 0). A point will be defined by entering its X and Y coordinates separated by a comma, as shown in the figure below.

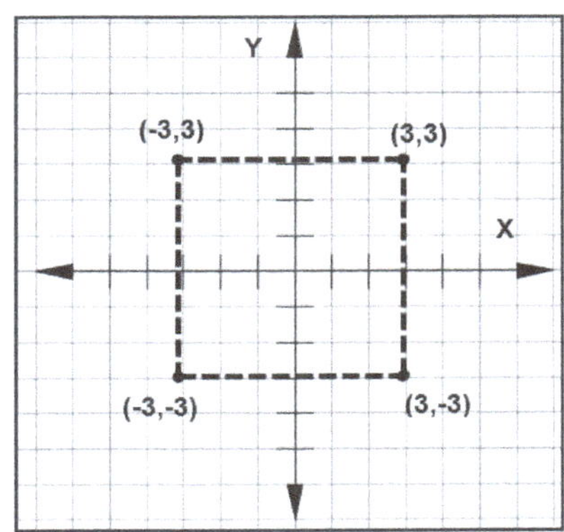

- Start AutoCAD 2024 by clicking the **AutoCAD 2024** icon on your desktop.
- On the Start screen, click **New > Templates > acadiso.dwt**. This starts a new drawing using the ISO template.

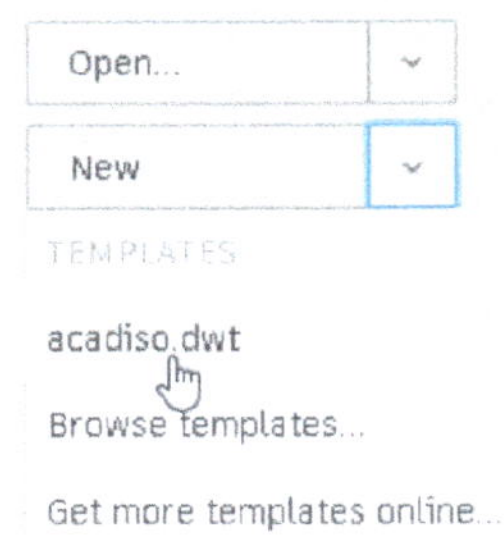

- Click **Zoom > Zoom All** on the **Navigation Bar**; the entire area in the graphics window will be displayed.

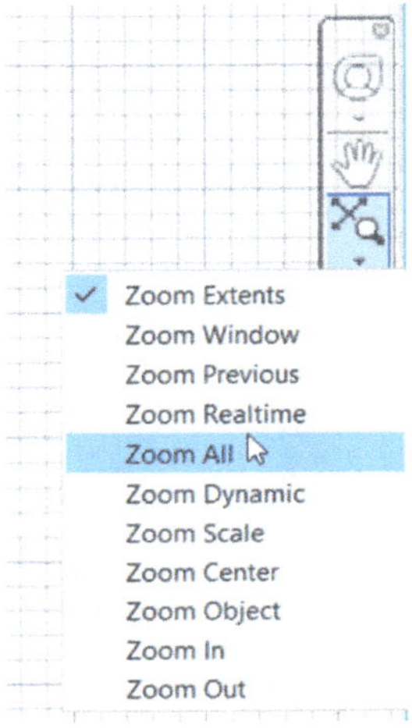

- Turn OFF the **GRIDMODE** by pressing the F7 key.

- Click the **Customization** button on the status bar, and then select **Dynamic Input** from the flyout. This displays the **Dynamic Input** icon on the status bar.

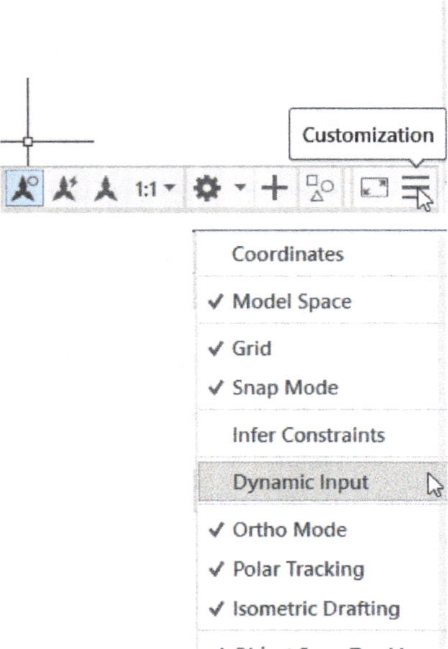

- Turn OFF the **Dynamic Input** icon. You will learn about **Dynamic Input** later in this chapter.

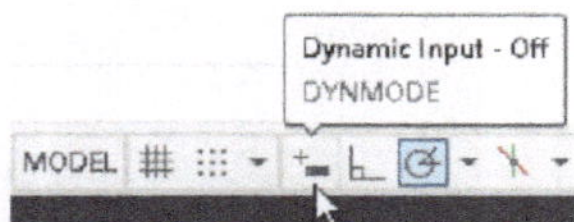

- To draw a line, click **Home > Draw > Line** on the ribbon, or enter **LINE** or **L** in the command line.
- Type **50, 50**, and press ENTER.
- Type **150, 50**, and press ENTER.
- Type **150,100** and press ENTER.
- Type **50,100** and press ENTER.
- Select the **Close** option from the command line. It creates a rectangle, as shown below.

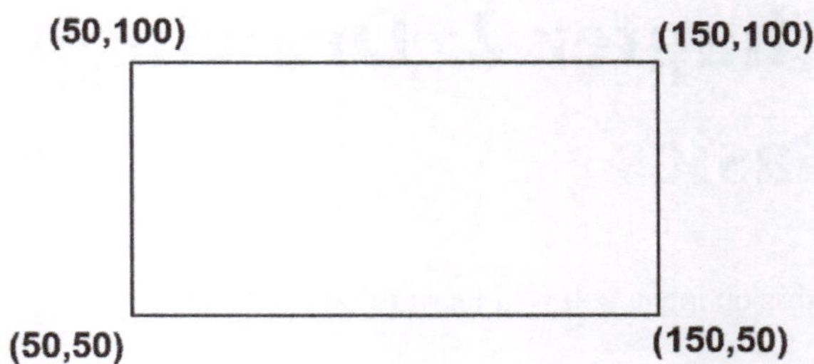

- Click **Save** on the **Quick Access Toolbar**.

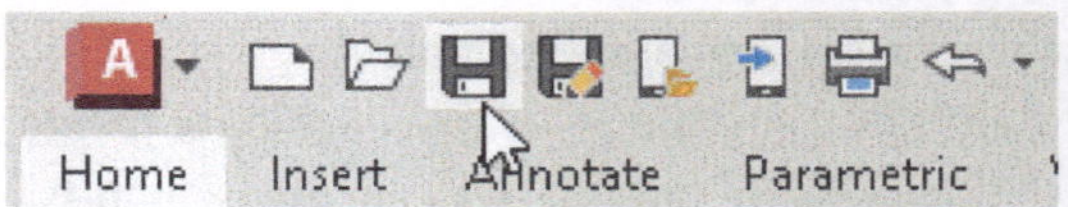

- Browse to a location on your computer.
- Type **Line-example1.dwg** in the **File name** box.
- Click **Save**.
- Close the file by closing the file tab, as shown.

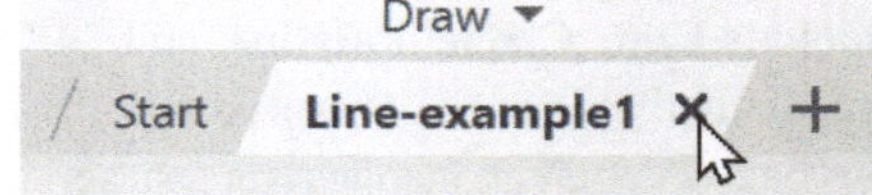

Example 2 (using the Relative Coordinate system)

In this example, you will draw lines by defining its endpoints in the relative coordinate system. In the relative coordinate system, you define the location of a point with respect to the previous point. For this purpose, the symbol, '@' is used before the point coordinates. This symbol means that the coordinate values are defined in relation to the previous point.

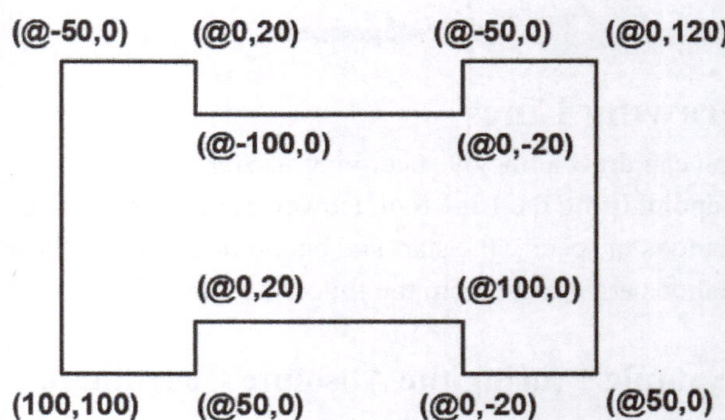

- Click **New** on the **Quick Access Toolbar**.

- Select the **acadISO-Named Plot Styles** template.

- Click **Open**.
- Type-in **Z** in the command line
- Press ENTER to activate the **ZOOM** command.
- Click the **All** option in the command line. It displays the entire area in the graphics window.
- Turn OFF the **Grid** icon on the status bar.

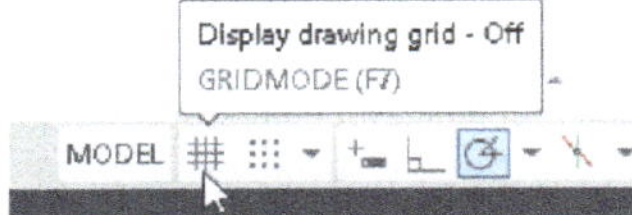

- Turn OFF the **Dynamic Input** mode, if active.
- Click **Home > Draw > Line** on the ribbon, or enter **LINE** or **L** in the command line.
- Type **100,100** and press ENTER. It defines the first point of the line.
- Type **@50,0** and press ENTER.
- Type **@0,20** and press ENTER.
- Type **@100,0** and press ENTER.
- Type **@0,-20**, and press ENTER.
- Type **@50,0** and press ENTER.
- Type **@0,120** and press ENTER.
- Type **@-50,0** and press ENTER.
- Type **@0,-20**, and press ENTER.
- Type **@-100,0** and press ENTER.
- Type **@0,20** and press ENTER.
- Type **@-50,0** and press ENTER.
- Select the **Close** option from the command line.
- Save the file as **Line-example2.dwg**.
- Close the file.

Example 3 (using the Polar Coordinate system)

In the polar coordinate system, you define the location of a point by entering two values: distance from the previous point and angle from the zero degrees. You enter the distance value along with the @ symbol and angle value with the < symbol. You have to make a note that AutoCAD measures the angle in the anti-clockwise direction.

Drawing Task

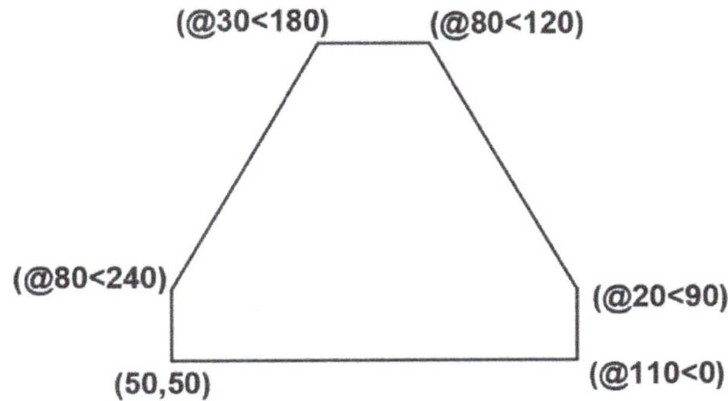

- Open a new file using the **acadISO-Named Plot Styles.dwt** template.
- Click **Zoom > Zoom All** on the **Navigation Bar**.
- Turn OFF the **Grid** icon on the status bar.
- Turn OFF the **Dynamic Input** mode, if active.
- Click **Home > Draw > Line** on the ribbon, or enter **LINE** or **L** in the command line.
- Type **50,50** and press the **Enter** key.
- Type **@110<0** and press ENTER.
- Type **@20<90** and press ENTER.
- Type **@80<120** and press ENTER.
- Type **@30<180** and press ENTER.
- Type **@80<240** and press ENTER.
- Select the **Close** option from the command line.
- Save the file as **Line-example3.dwg**.
- Close the file.

Example 4 (using the Direct Input method)

In the direct input method, you draw a line by entering its distance and angle values. You use the **Dynamic Input** mode in this method.

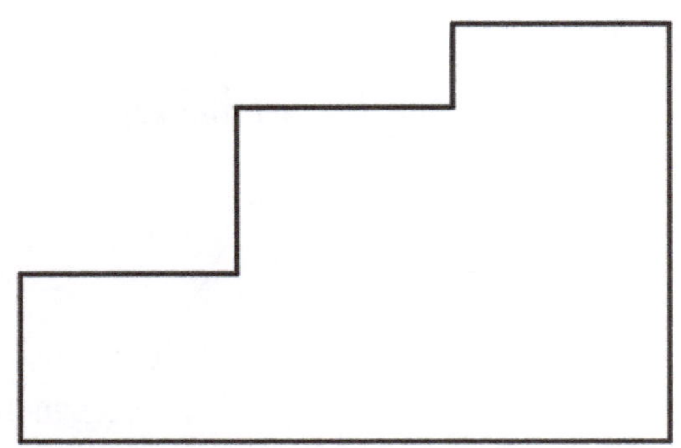

- Open a new file using the **acadISO-Named Plot Styles.dwt** template.
- Turn OFF the **Grid** and **Snap Mode** icons on the Status Bar.
- Click **Zoom > Zoom All** on the **Navigation Bar**.
- Activate the **Dynamic Input** icon on the Status Bar.

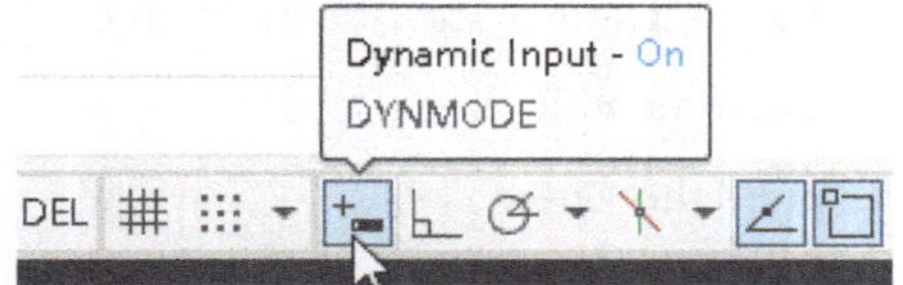

- Click **Home > Draw > Line** on the ribbon, or enter **LINE** or **L** in the command line.
- Define the first point of the line by typing 50,50 and pressing ENTER.
- Move the pointer horizontally toward right and type in 150 in the length box.
- Press the TAB key and type 0 as the angle. Next, press ENTER.

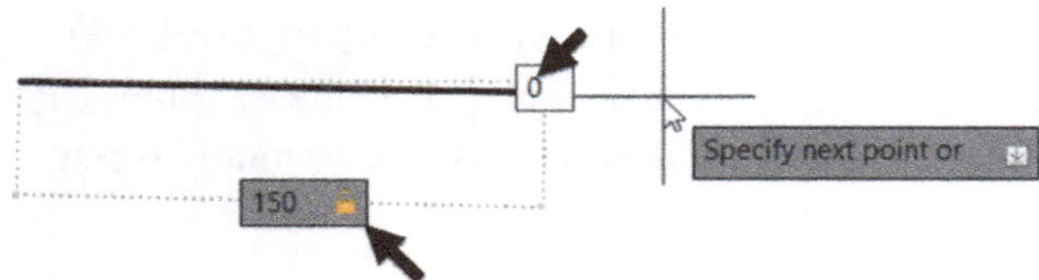

- Move the pointer vertically upwards and type-in 100 as length.
- Press the TAB key and type 90 as the angle — next, press ENTER.
- Move the pointer horizontally toward left and type 50.
- Press the TAB key and type 180 as the angle — next, press ENTER.
- Move the pointer vertically downwards and type 20.
- Press the TAB key and type 90 as the angle — next, press ENTER.
- Move the pointer horizontally toward left and type 50.
- Press the TAB key and type 180 as the angle — next, press ENTER.
- Move the pointer vertically downwards and type 40.
- Press the TAB key and type 90 as the angle — next, press ENTER.
- Move the pointer horizontally toward left and type 50.
- Press the TAB key and type 180 as the angle — next, press ENTER.
- Click the **Close** option in the command line.
- Save and close the file.

Erasing, Undoing and Redoing

- Draw the sketch shown below using the **Line** tool. You can use the Direct Input method to create this sketch. Do not dimension the drawing.

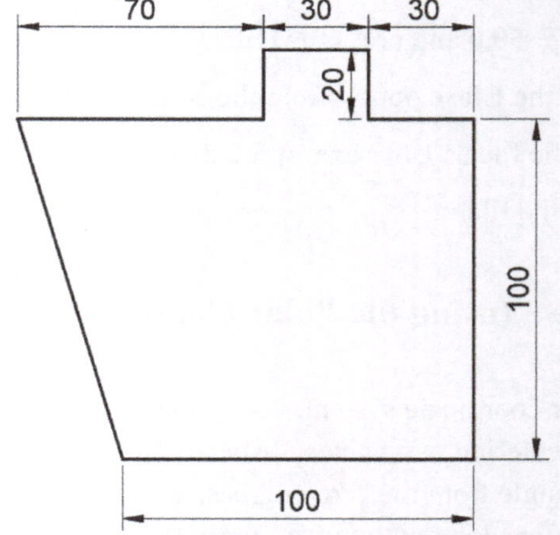

- Click **Home > Modify > Erase** on the ribbon or Enter **ERASE** or **E** in the command line.

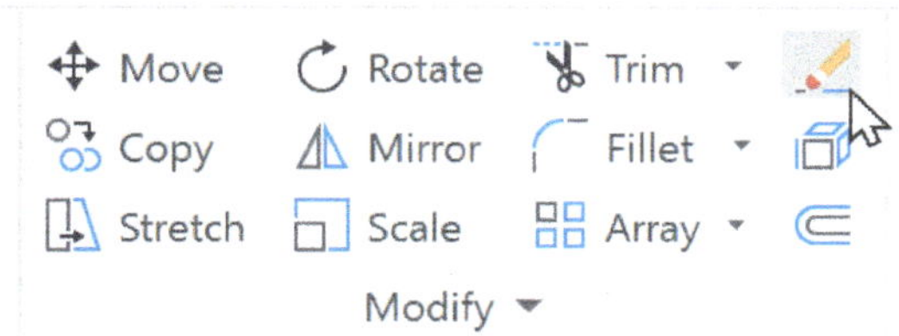

- Select the lines shown below and press ENTER. It erases the lines.

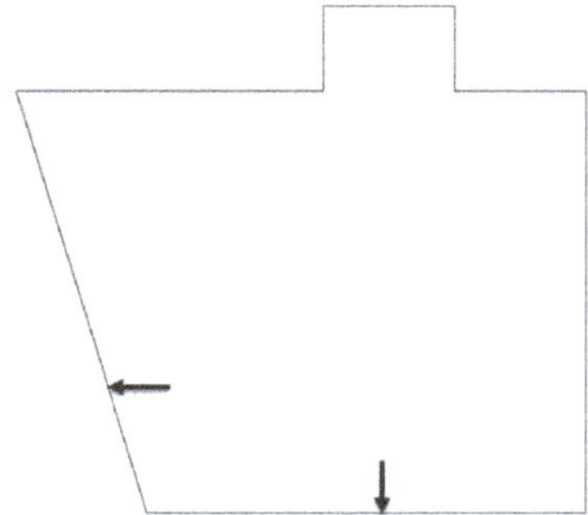

- Click the **Undo** button on the **Quick Access Toolbar**. This action restores the lines.

- Click the **Redo** button on the **Quick Access Toolbar**. This action erases the lines again.

- Type **E** in the command line and press the SPACEBAR; the **ERASE** command will be activated.
- Drag a selection lasso as shown below and press ENTER; the entities will be erased. (Refer to the Selection Window section in Chapter 1 to learn about the selection lasso)

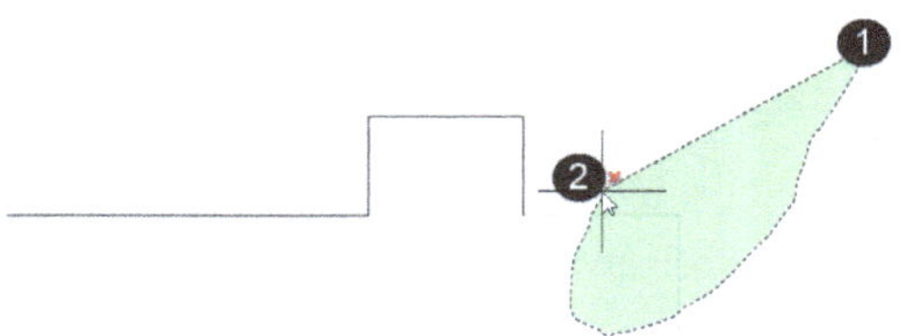

Drawing Circles

The tools in the **Circle** drop-down on the **Draw** panel can be used to draw circles. You can also type-in the **CIRCLE** command in the command line and create circles. There are various methods to create circles. These methods are explained in the following examples.

Example 1(Center, Radius)

In this example, you will create a circle by specifying its center and radius value.

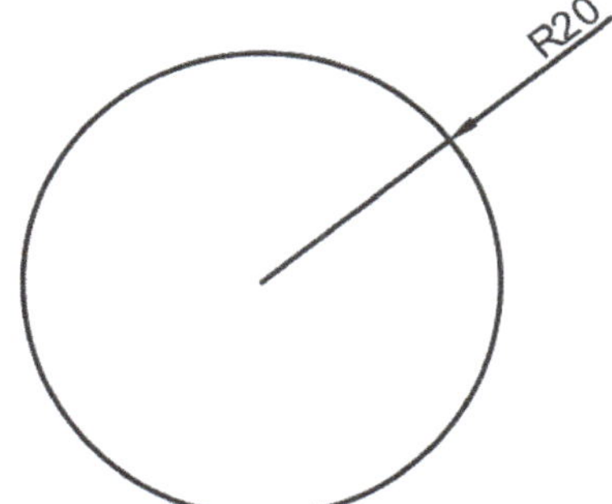

- Click **Home > Draw > Circle** drop-down **> Center, Radius** on the ribbon.
- Click in the graphics window.
- Type 20 as the radius and press ENTER.

Example 2(Center, Diameter)

In this example, you will create a circle by specifying its center and diameter value.

- Click **Home > Draw > Circle > Center, Diameter**  on the ribbon. The message, "Specify center point for circle or [3P/2P/Ttr (tan tan radius)]:" appears in the command line.

- Pick a point in the graphics window, which is approximately horizontal to the previous circle.

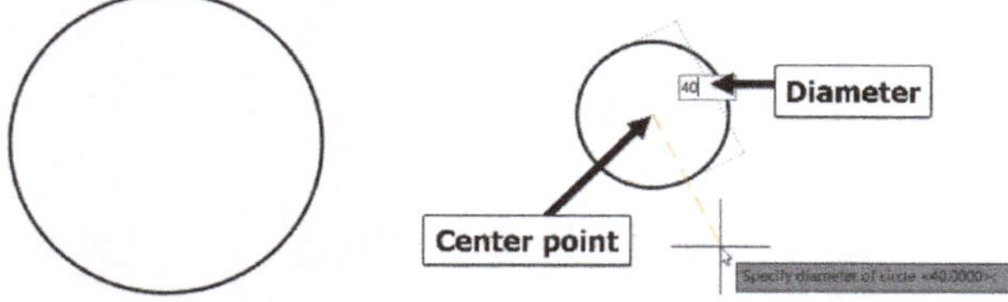

- Type 40 as the diameter and press ENTER; the circle will be created.

Example 3(2-Point)

In this example, you will create a circle by specifying two points. The first point is to specify the location of the circle, and the second defines the diameter.

- Click the down arrow next to the **Object Snap** icon on the status bar. A flyout appears. The options in this flyout are called Object Snaps. You will learn about these Object Snaps later in Chapter 3.
- Activate the **Center** option, if it is not already active.
- Now, you will create a circle by selecting the center points of the previous circles.
- Click **Home > Draw > Circle > 2-Point** on the ribbon. The message, "Specify the first endpoint of circle's diameter:" appears in the command line.
- Select the center point of the left side circle; the message, "Specify the second endpoint of circle's diameter:" appears in the command line.
- Select the center point of the right-side circle; the circle will be created, as shown below.

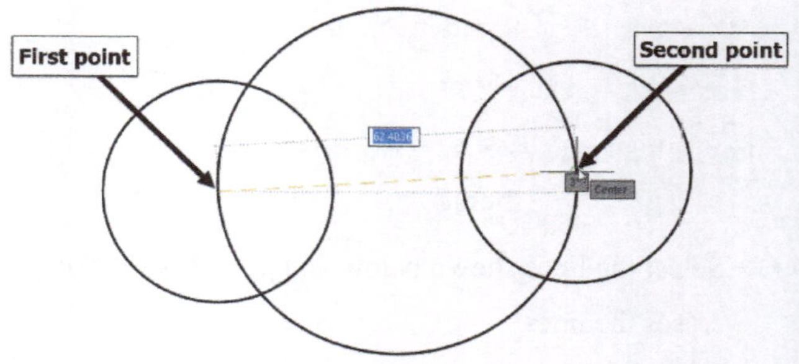

Example 4(3-Point)

In this example, you will create a circle by specifying three points. The circle will pass through these three points.

- Open a new file.
- Use the **Line** tool and create the drawing shown in the figure below. The coordinate points are also given in the figure.

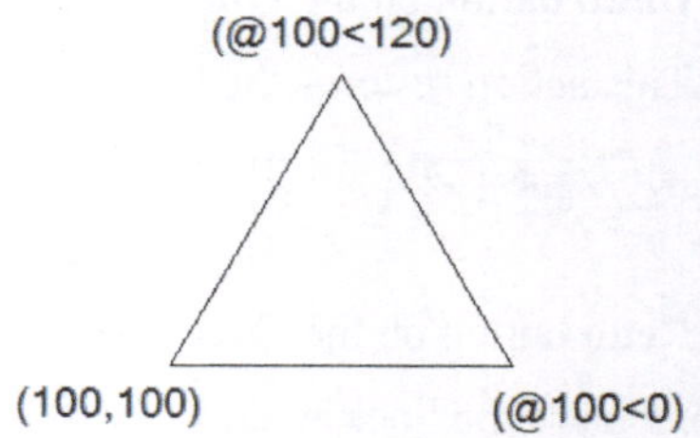

- Click **Home > Draw > Circle > 3-Point** on the ribbon.
- Select the three vertices of the triangle; a circle will be created, passing through the selected points.

Example 5 (Tan, Tan, Radius)

In this example, you will create a circle by selecting two objects and then specifying the radius of the circle. A circle is created tangent to the selected objects.

- Click **Home > Utilities > Measure > Radius** on the ribbon. The message, "Select arc or circle: "appears in the command line.
- Select the circle passing through the three vertices of the triangle; the radius and diameter values of the circle will be displayed above the command line.

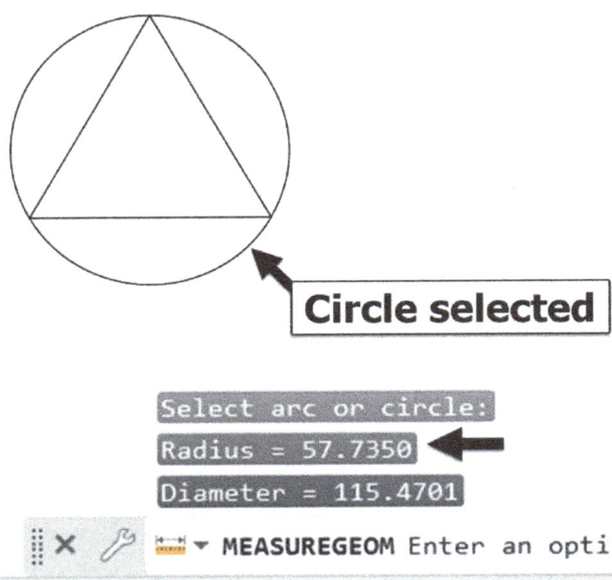

- Select the **eXit** option from the command line.
- Click **Home > Utilities > Quick Calculator** on the ribbon; the **Quick Calculator** appears.

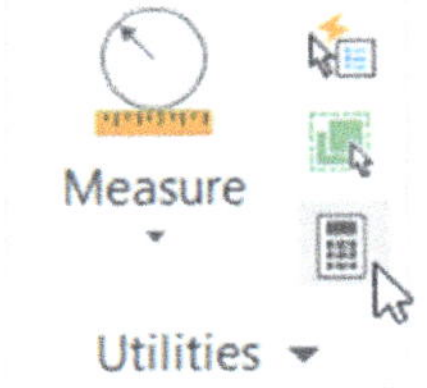

- Type-in **57.7350** in the **Quick Calculator**.

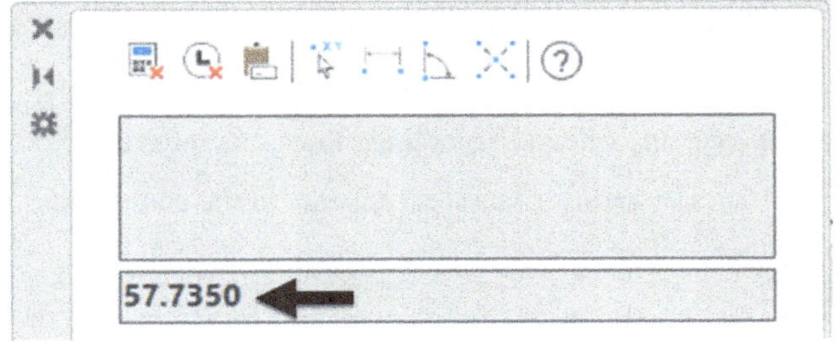

- Click the / button and then the **2** button on the **Number Pad**.
- Click the = button; the value **28.8675** is displayed in the value box.

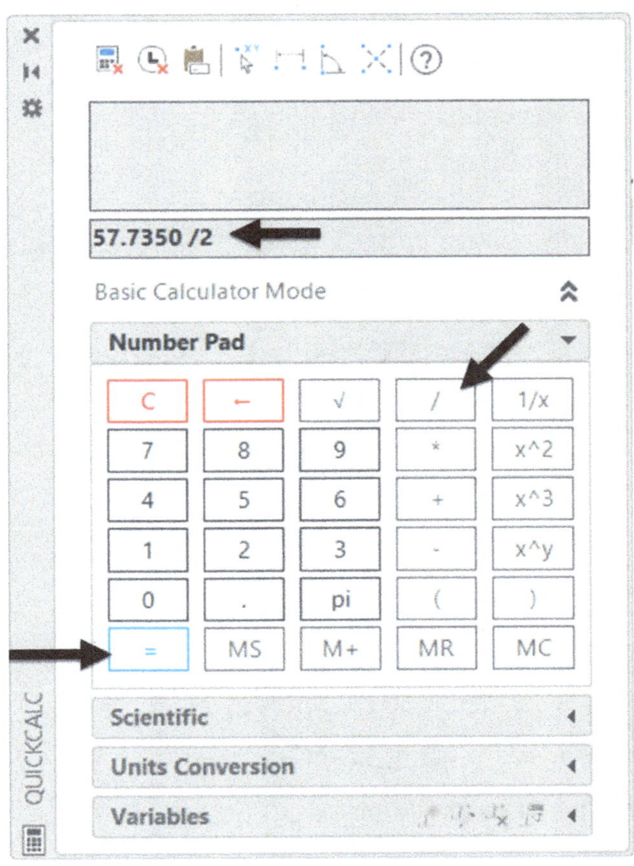

- Click **Home > Draw > Circle > Tan, Tan, Radius** on the ribbon; the message, "Specify point on the object for first tangent of circle:" appears in the command line.
- Select the horizontal line of the triangle; the message,

"Specify point on object for second tangent of circle:" appears in the command line.

- Select any one of the inclined lines; the message, "Specify radius of circle," appears in the command line.
- Click the **Paste value to command line** button on the **Quick Calculator**; the value **28.8675** will be pasted in the command line.

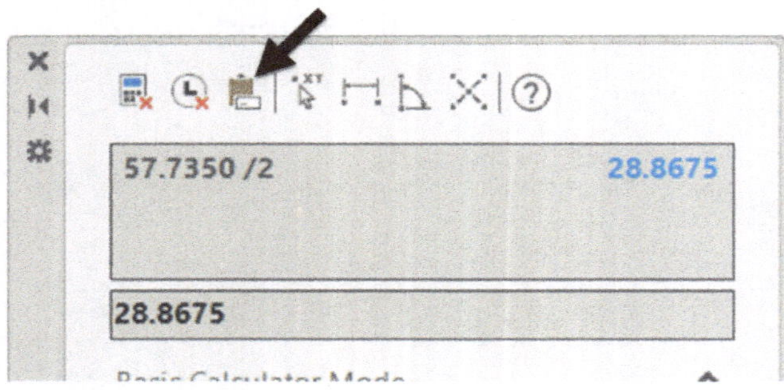

- Press ENTER to specify the radius; the circle will be created, touching all three sides of the triangle.

- Save and close the file.

Example 6 (Tan, Tan, Tan)

In this example, you will create a circle by selecting three objects to which it will be tangent.

- Click the **Open** button on the **Quick Access Toolbar**; the **Select File** dialog appears.

- Browse to the location of the **Line-example3.dwg** file and double-click on it; the file will be opened.
- Click **Home > Draw > Circle > Tan, Tan, Tan** on the ribbon.
- Select the bottom horizontal line of the drawing.
- Select the two inclined lines. This creates a circle tangent to the selected lines.

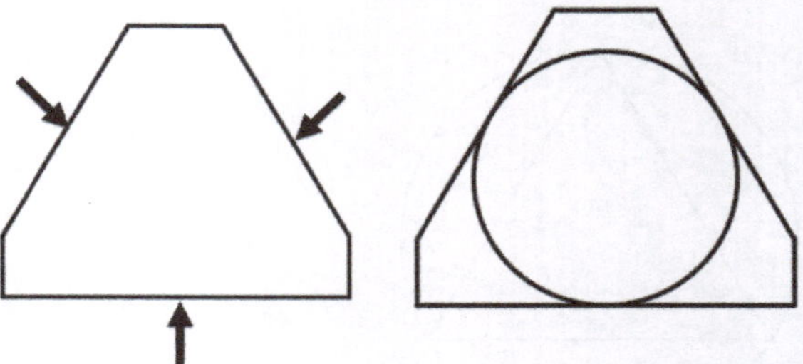

- Save and close the file.

Drawing Arcs

An arc is a portion of a circle. The total angle of an arc will always be less than 360 degrees, whereas the total angle of a circle is 360 degrees. AutoCAD provides you with eleven ways to draw an arc. You can draw arcs in different ways by using the tools available in the **Arcs** drop-down of the **Draw** panel. The usage of these tools will depend on your requirement. Some methods to create arcs are explained in the following examples.

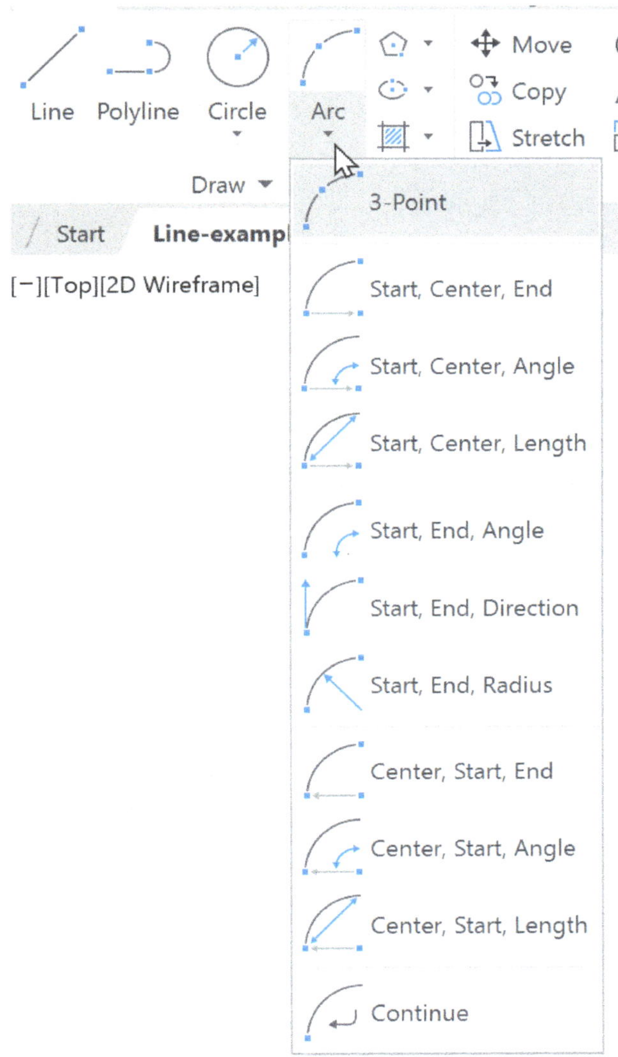

Example 1 (3-Point)

In this example, you will create an arc by specifying three points. The arc will pass through these points.

- Open the **Line-example1.dwg** file.
- Expand the **Draw** panel in the **Home** tab and select the **Multiple Points** tool.

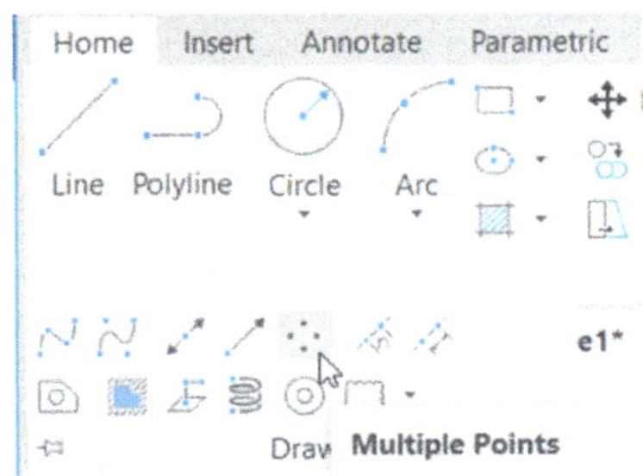

- Type 100,120 in the command line and press ENTER. It places a point above the rectangle.

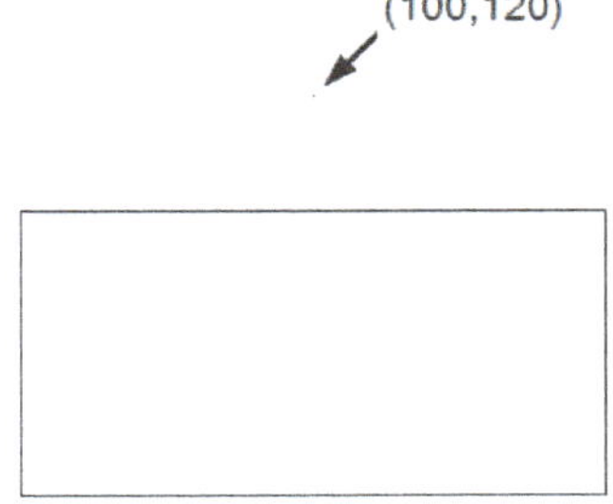

- Press ESC.
- Click the down arrow next to the **Object Snap** icon on the status bar, and then select the **Node** option from the menu.
- Click **Home > Draw > Arc > 3-Point** on the ribbon. The message, "Specify start point of arc or [Center]:" appears in the command line.
- Select the top left corner of the rectangle.
- Select the point located above the rectangle.
- Select the top right corner of the rectangle; the three-point arc will be created.

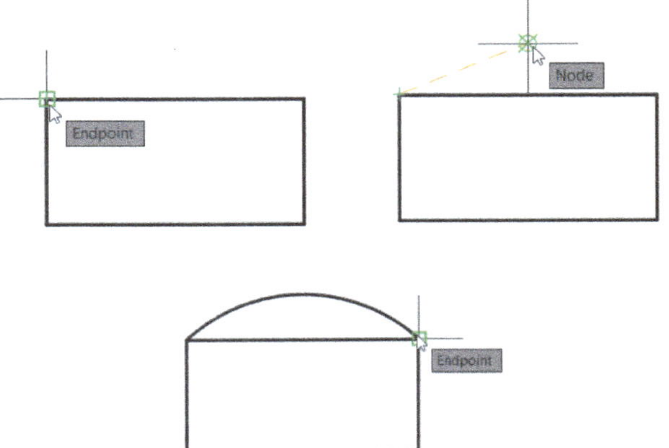

Example 2 (Start, Center, End)

In this example, you will draw an arc by specifying its start, center, and endpoints. The first two points define the radius of the arc, and the third point defines its included angle.

- Click **Home > Draw > Arc > Start, Center, End** on the ribbon. The message, "Specify start point of arc or [Center]:" appears in the command line.

- Pick an arbitrary point in the graphics window to define the start point of an arc. The message, "Specify center point of arc:" appears.
- Pick a point to define the radius of the circle. You can also type in the radius value and press ENTER; the message, "Specify end point of arc or [Angle/chord Length]:" appears.
  You will notice that, as you move the pointer, the included angle of the arc changes. The included angle of the arc is measured in the counter-clockwise direction. Press and hold the Ctrl key if you want to reverse the direction.

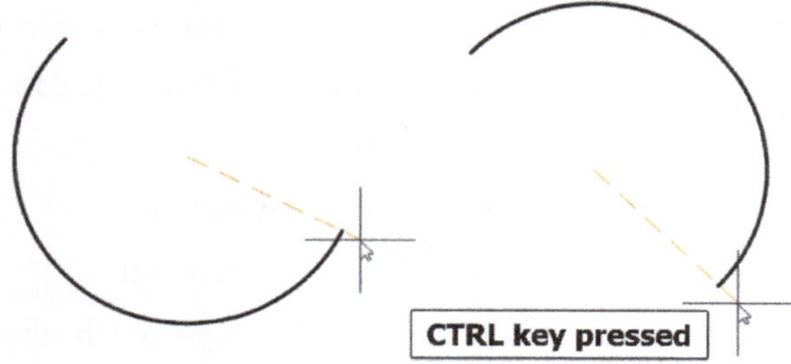

- Pick a point to define the included angle of the arc. You can also type the angle value and press ENTER.

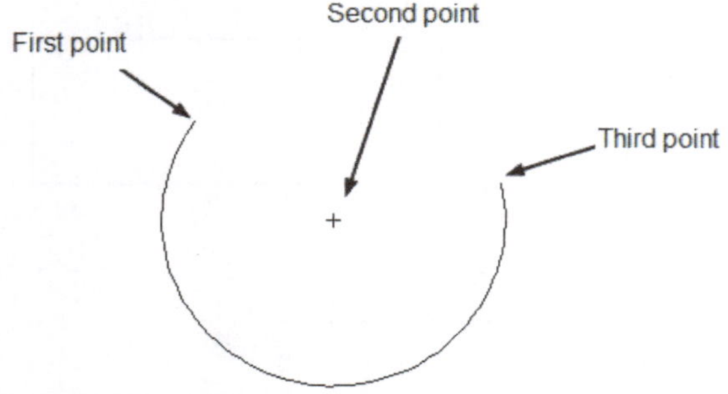

Example 3 (Start, End, Direction)

- Use the **Line** tool and create the drawing shown in the figure below. The dimensions are also given in the figure. (Use any one of the procedures given in the **Drawing Lines** section)

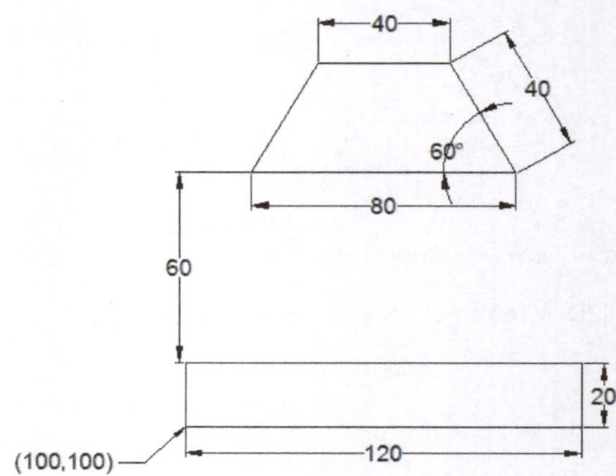

- Click **Home > Draw > Arc > Start, End, Direction** on the ribbon.
- Select the start and end points of the arc, as shown in the figure.

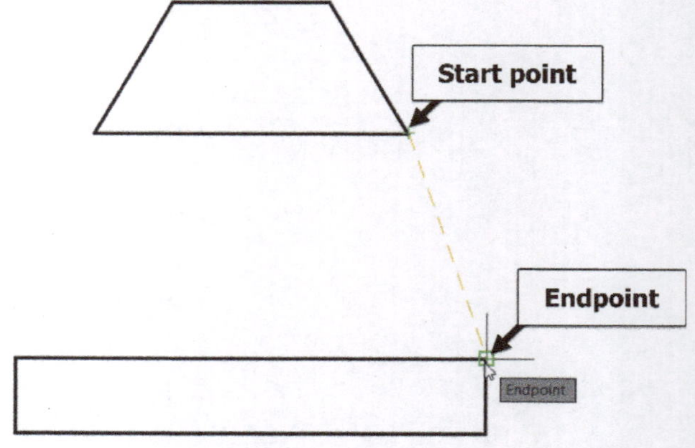

- Activate the **ORTHOMODE** icon on the status bar.
- Move the pointer vertically downward and click to specify the direction.

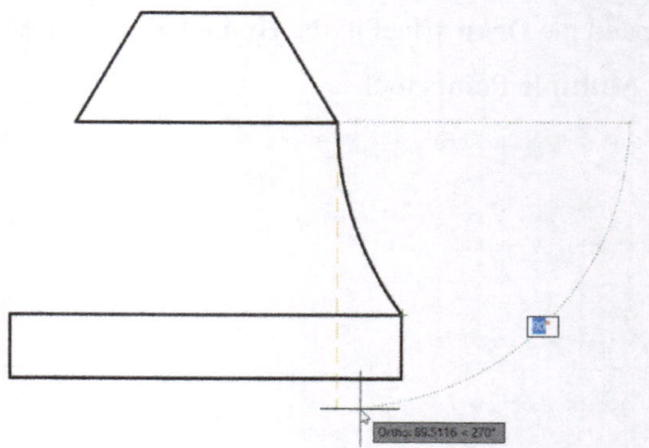

- Likewise, create another arc.

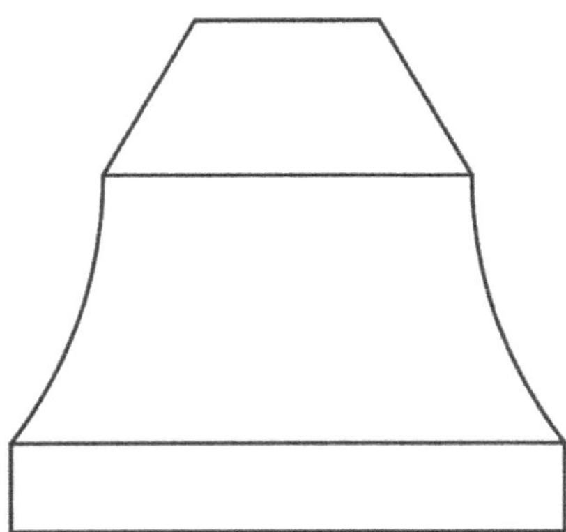

Drawing Polylines

A Polyline is a single object that consists of line segments and arcs. It is more versatile than a line, as you can assign a width to it. In the following example, you will create a closed polyline.

Example 1

- Activate the **Ortho Mode** on the Status Bar.
- Click **Home > Draw > Polyline** on the ribbon or enter **PLINE** or **PL** in the command line; the message, "Specify start point:" appears in the command line.
- Select an arbitrary point in the graphics window.
- Move the pointer horizontally toward right and type 100 — next, press ENTER.
- Select the **Arc** option from the command line.
- Move the pointer vertically upward and type **50** — next, press ENTER.
- Select the **Line** option from the command line.
- Move the pointer horizontally toward left and type **50** — next, press ENTER.
- Move the pointer vertically upward and type **50** — next, press ENTER.
- Select the **Arc** option from the command line.
- Move the pointer horizontally toward left and type **50** — next, press ENTER.
- Select the **CLose** option from the command line.

Now, when you click on a line segment from the sketch, the whole sketch will be selected. It is because the polyline created is a single object.

Drawing Rectangles

A rectangle is a four-sided single object. You can create a rectangle by just specifying its two diagonal corners. However, there are various methods to create a rectangle. These methods are explained in the following examples.

Example 1

In this example, you will create a rectangle by specifying its corner points.

- Open a new file.
- Click **Home > Draw > Rectangle** on the ribbon, or enter **RECTANG** or **REC** in the command line; the message, "Specify first corner point or [**Chamfer Elevation Fillet Thickness Width**]:" appears in the command line.
- Pick an arbitrary point in the graphics window; the message "Specify other corner point or [**Area Dimensions Rotation**]:" appears in the command line.
- Move the pointer diagonally toward the right and click to create a rectangle.

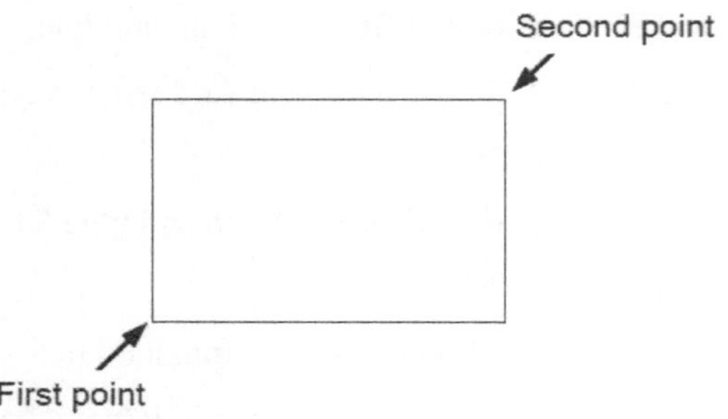

Example 2

In this example, you will create a rectangle by specifying its length and width.

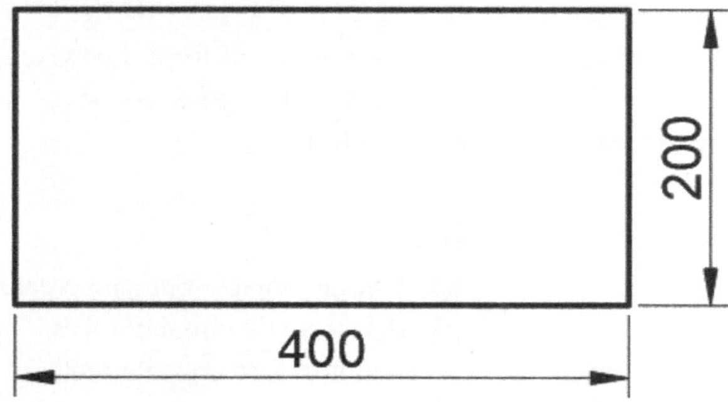

- Click **Home > Draw > Rectangle** on the ribbon, or enter **RECTANG** or **REC** in the command line.
- Specify the first corner of the rectangle by picking an arbitrary point in the graphics window.
- Follow the prompt sequence given next:

  Specify other corner point or [Area Dimensions Rotation]: Select the **Dimensions** option from the command line

  Specify length for rectangles: Type **400** and press ENTER.

  Specify width for rectangles: Type **200** and press ENTER.

  Specify other corner point or [Area Dimensions Rotation]: Move the pointer upward and click to create the rectangle.

Example 3

In this example, you will create a rectangle by specifying its area and width.

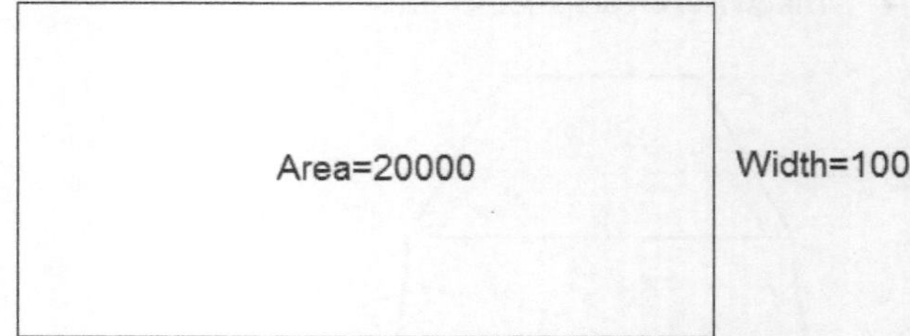

- Click **Home > Draw > Rectangle** on the ribbon, or enter **RECTANG** or **REC** in the command line.
- Specify the first corner of the rectangle by picking an arbitrary point.
- Follow the prompt sequence given next:

  Specify other corner point or [Area Dimensions Rotation]: Select the **Area** option from the command line

  Enter area of rectangle in current units: Type **20000** and press ENTER.

  Calculate rectangle dimensions based on [Length Width] <Length>: Select the **Width** option from the command line.

  Enter rectangle width: Type **100** and press ENTER; the length will be calculated automatically.

Example 4

In this example, you will create a rectangle with chamfered corners.

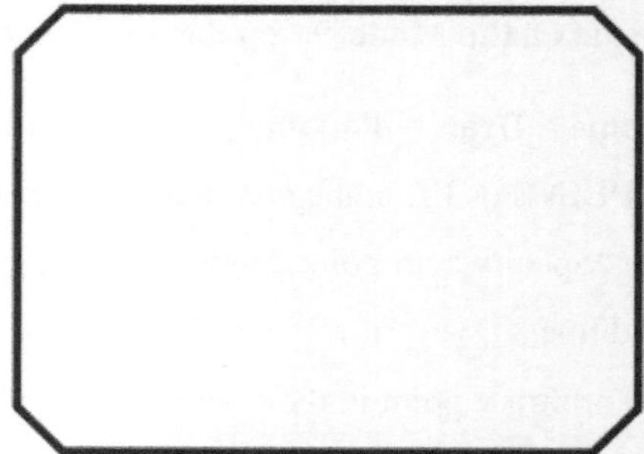

- Click **Home > Draw > Rectangle** on the ribbon, or enter **RECTANG** or **REC** in the command line.
- Follow the prompt sequence given next:

Specify first corner point or [Chamfer Elevation Fillet Thickness Width]: Select the **Chamfer** option from the command line.

Specify first chamfer distance for rectangles: Type **20** and press ENTER.

Specify second chamfer distance for rectangles: Type **20** and press ENTER.

Specify first corner point or [Chamfer Elevation Fillet Thickness Width]: Click at an arbitrary point in the graphics window to specify the first corner.

Specify other corner point or [Area Dimensions Rotation]: Move the pointer diagonally toward the right and click to specify the second corner.

Example 5

In this example, you will create a rectangle with rounded corners.

- Click **Home > Draw > Rectangle** on the ribbon, or enter **RECTANG** or **REC** in the command line.
- Follow the prompt sequence given next:

Specify first corner point or [Chamfer Elevation Fillet Thickness Width]: Select the **Fillet** option from the command line.

Specify fillet radius for rectangles: Type **50** and press ENTER.

Specify first corner point or [Chamfer Elevation Fillet Thickness Width]: Click at an arbitrary point in the graphics window to specify the first corner.

Specify other corner point or [Area Dimensions Rotation]: Move the pointer diagonally toward the right and click to specify the second corner.

Example 6

In this example, you will create an inclined rectangle.

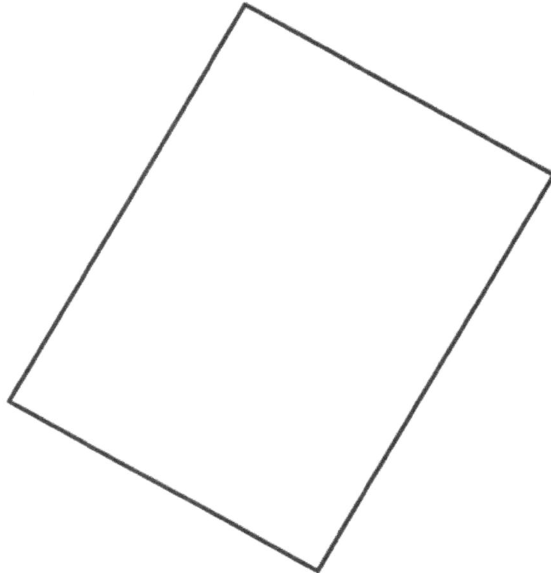

- Click **Home > Draw > Rectangle** on the ribbon, or enter **RECTANG** or **REC** in the command line.
- Select the **Fillet** option from the command line.
- Type 0 and press Enter.
- Specify the first corner of the rectangle by picking an arbitrary point.
- Follow the prompt sequence given next:

Specify other corner point or [Area/Dimensions/Rotation]: Select the **Rotation** option from the command line.

Specify rotation angle or [Pick points]: Type **60** and press ENTER.

Specify other corner point or [Area/Dimensions/Rotation]: Select the **Dimensions** option from the command line.

Specify length for rectangles: Type 400 and press ENTER.

Specify width for rectangles: Type 300 and press ENTER.

- Move the pointer toward the right and click to position the rectangle.

Drawing Polygons

A Polygon is a single object having many sides ranging from 3 to 1024. In AutoCAD, you can create regular polygons having sides with equal length. There are two methods to create a polygon. These methods are explained in the following examples.

Example 1

In this example, you will create a polygon by specifying the number of sides and then determining the length of one side.

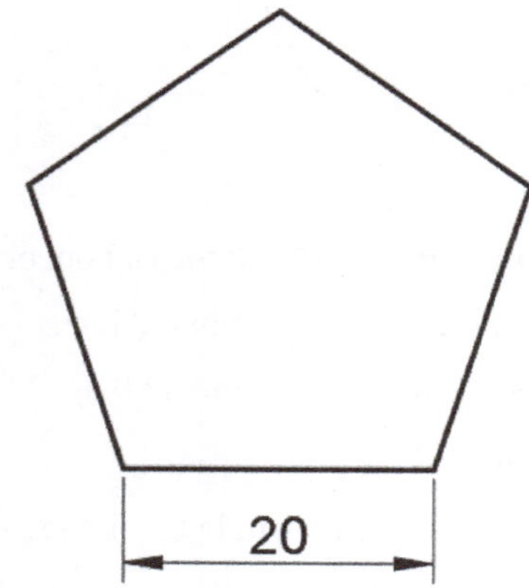

- Click **Home > Draw > Polygon** on the ribbon.

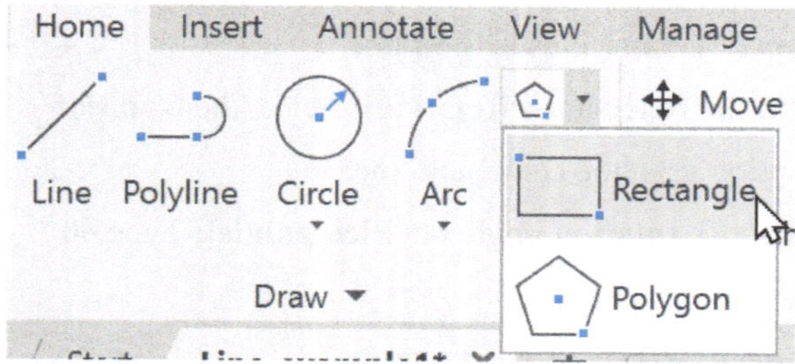

- Follow the prompt sequence given next.

Enter number of sides <4>: Type **5** and press **ENTER.**

Specify center of polygon or [Edge]: Select the **Edge** option from the command line.

Specify first endpoint of edge: Select an arbitrary point.

Specify second endpoint of edge: Move the pointer horizontally toward the right. Next, type **20**, and press **ENTER**.

Example 2

In this example, you will create a polygon by specifying the number of sides and drawing an imaginary circle (inscribed circle). The polygon will be created with its corners located on the imaginary circle. You can also create a polygon with the circumscribed circle. A circumscribed circle is an imaginary circle which is tangent to all the sides of a polygon.

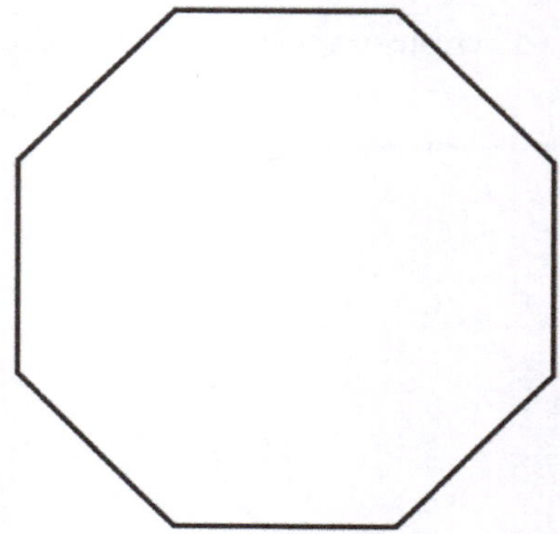

- Type **POL** in the command line and press ENTER; the **POLYGON** command will be activated.
- Follow the prompt sequence given next:

Enter number of sides <5>: Type **8** and press **ENTER.**

Specify center of polygon or [Edge]: Select an arbitrary point

Enter an option [Inscribed in circle Circumscribed about circle] <C>: Select the **Inscribed in circle** option from the command line.

Specify radius of circle: Type **20** and press ENTER; a polygon will be created with its corners touching the imaginary circle.

Drawing Splines

Splines are non-uniform curves, which are used to create irregular shapes. In AutoCAD, you can create splines by using two methods: **Spline Fit** and **Spline CV**. These methods are explained in the following examples:

Example 1: (Spline Fit)

In this example, you will create a spline using the **Spline Fit** method. In this method, you need to specify various points in the graphics window. The spline will be created, passing through the specified points.

- Start a new drawing file.
- Use the **Line** tool and create a sketch similar to the one shown below.

- Expand the **Draw** panel in the **Home** tab and select the **Spline Fit** button; the message, "Specify first point or [**Method Knots Object**]:" appears in the command line.

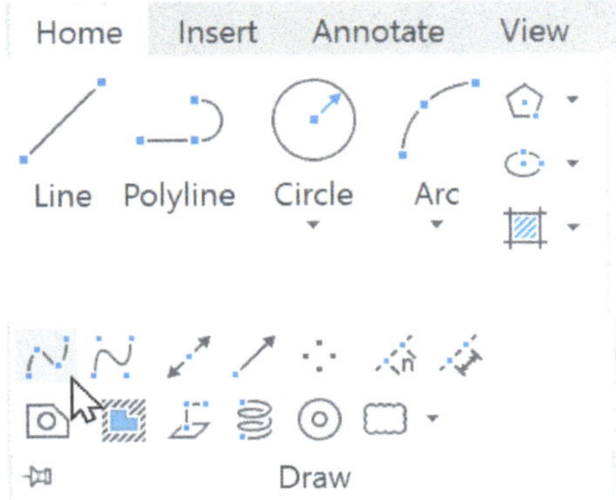

- Select the lower-left corner of the sketch; the message, "Enter next point or [**start Tangency toLerance**]:" appears in the command line.

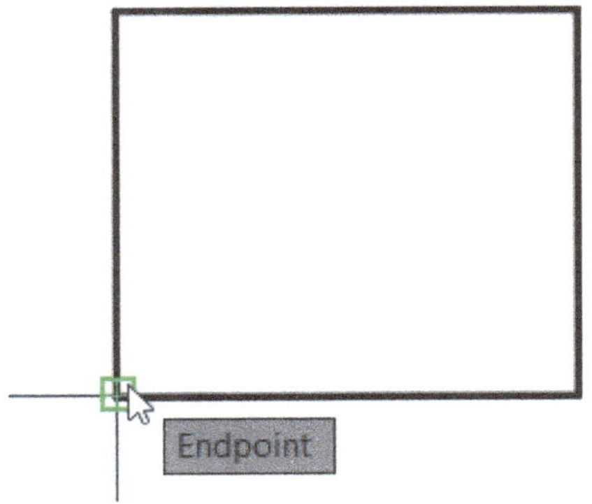

- Select the top-left corner point of the sketch.

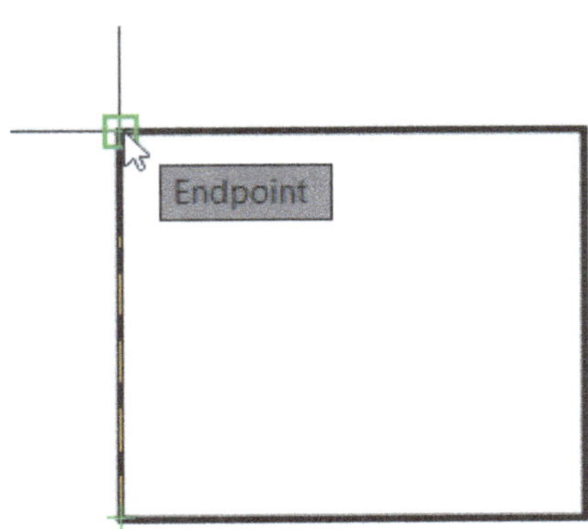

- Similarly, select the top-right and lower-right corners; a spline will be attached to the pointer.
- Press ENTER.

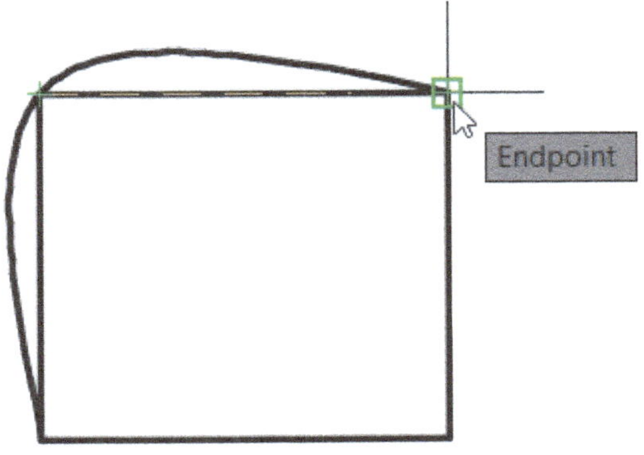

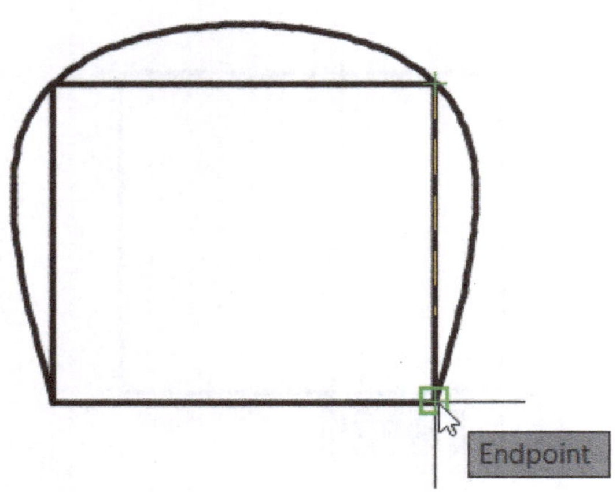

Example 2: (Spline CV)

In this example, you will create a spline by using the **Spline CV** method. In this method, you will specify various points called control vertices. As you specify the control vertices, imaginary lines are created connecting them. The spline will be drawn tangent to these lines.

- Expand the **Draw** panel in the **Home** tab and select the **Spline CV** button.

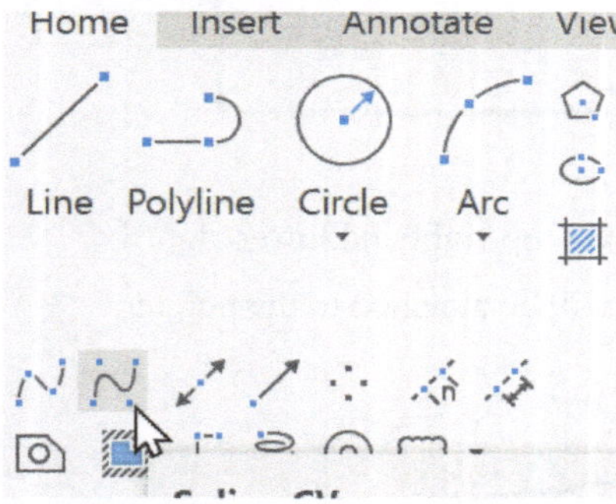

- Select the four corners of the sketch in the same sequence as in the earlier example.

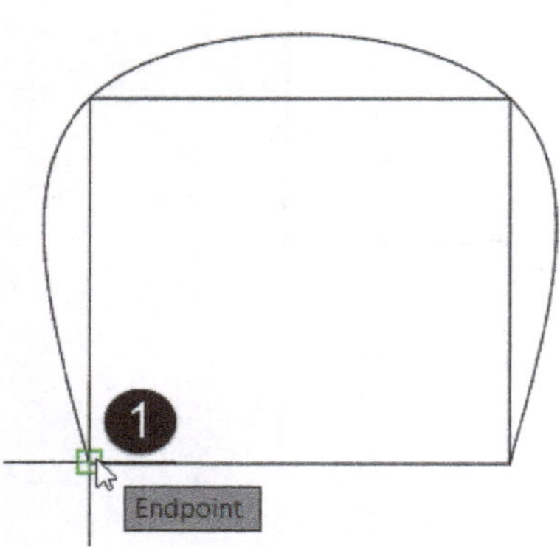

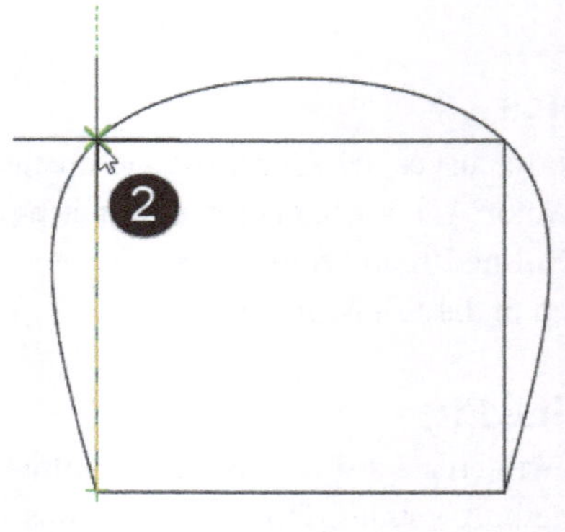

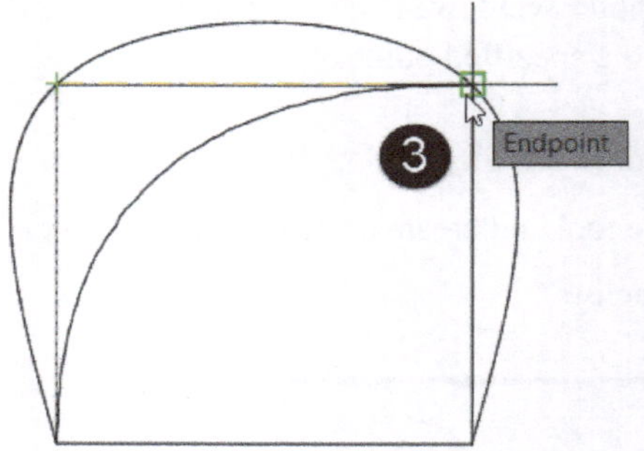

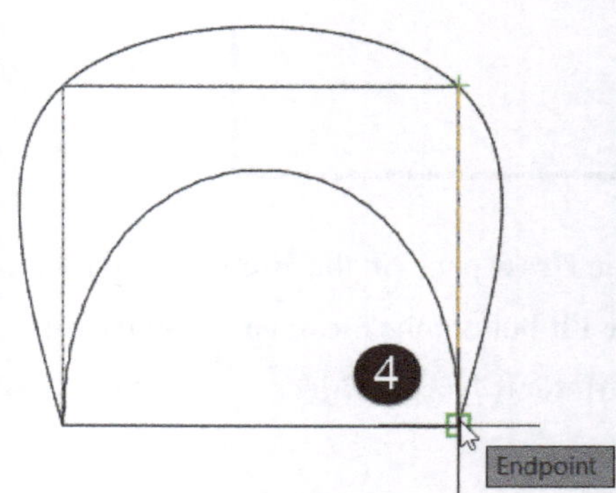

- Press ENTER; a spline with control vertices will be created.

Example 3:

- Create a polyline, as shown.

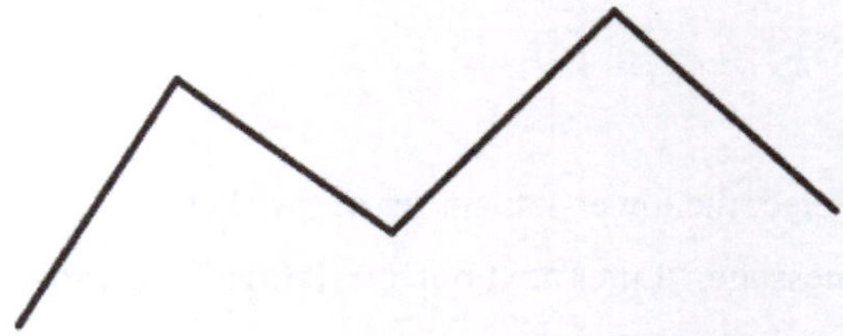

- Activate the **Spline CV** command.
- Select **Object** from the command line.
- Select the polyline and press Enter; the polyline is converted into a spline.

Drawing Ellipses

Ellipses are also non-uniform curves, but they have a regular shape. They are actually splines created in a proper closed shape. In AutoCAD, you can draw an ellipse in three different ways by using the tools available in the **Ellipse** drop-down of the **Draw** panel. The three different ways to draw ellipses are explained in the following examples.

Example 1 (Center)

In this example, you will draw an ellipse by specifying three points. The first point defines the center of the ellipse. The second and third points define the two axes of the ellipse.

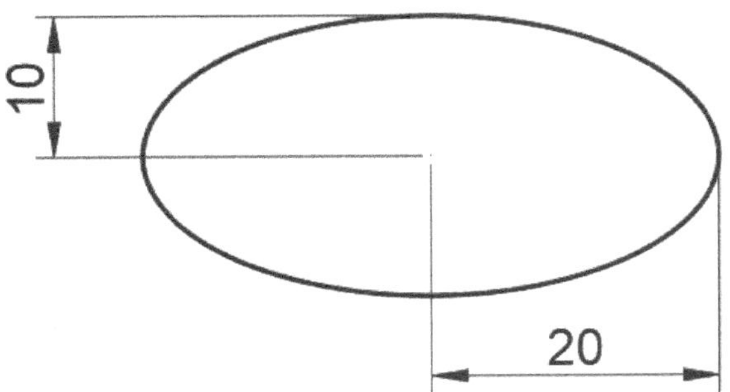

- Click **Home > Draw > Ellipse > Center** on the ribbon; the message, "Specify center of ellipse:" appears in the command line.
- Select an arbitrary point in the graphics window; the message, "Specify endpoint of axis:" appears in the command line.
- Move the pointer horizontally and type 20. Next, press ENTER; the message, "Specify distance to other axis or [Rotation]:" appears in the command line.
- Type 10 and press ENTER; the ellipse will be created.

Example 2 (Axis, End)

In this example, you will draw an ellipse by specifying three points. The first two points define the location and length of the first axis. The third point defines the second axis of the ellipse.

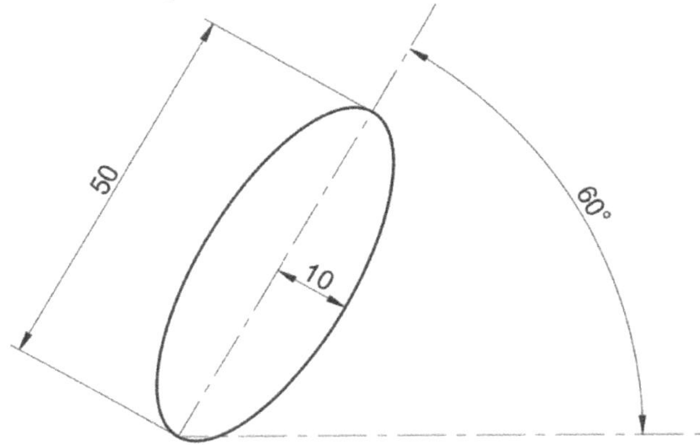

- Activate the **Dynamic Input** icon on the status bar, if it is not active.
- Click **Home > Draw > Ellipse > Axis, End** on the ribbon.
- Select an arbitrary point to specify an axis endpoint.
- Type **50** as the length of the first axis and press TAB.
- Type **120** as angle and press ENTER.
- Type **10** as the radius of the second axis and press ENTER; the ellipse will be created inclined at a 60-degree angle.

Example 3 (Elliptical Arc)

In this example, you will draw an elliptical arc. To draw an elliptical arc, first, you need to define the location and length of the first axis. Next, set the radius of the second axis; an ellipse will be displayed. Next, you need to set the start angle of the elliptical arc. The start angle can be any angle between 0 and 360. After defining the start angle, you need to specify the end angle of the elliptical arc.

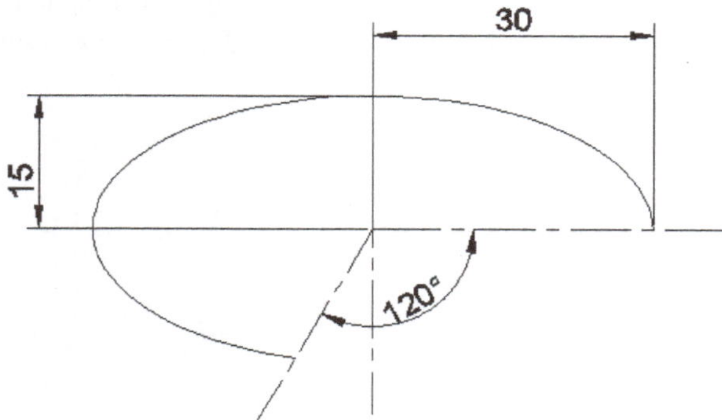

- Turn on the **ORTHOMODE** on the Status bar.

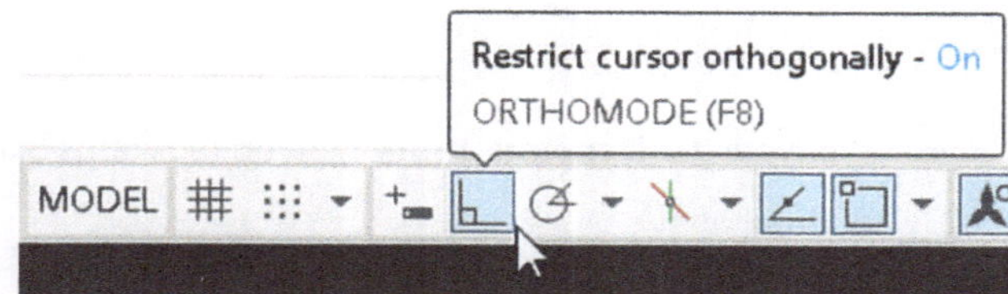

- Click **Home > Draw > Ellipse > Elliptical Arc** on the ribbon.
- Select an arbitrary point to specify an axis endpoint.
- Move the pointer horizontally toward left and type **60**. Next, press ENTER to specify the axis length.
- Move the pointer upward and type 15. Next, press ENTER to specify the length of another axis.
- Type **0** and press ENTER to specify the start angle.
- Type **240** and press ENTER to specify the end angle.

Solid

The **Solid** command creates a two-dimensional triangle or quadrilateral. The area inside the triangle or quadrilateral is filled with a solid color. Type SOLID in the command line and press ENTER. Next, specify the first point. Next, move the pointer and specify the second point. Move the pointer and select the third point; you are prompted to specify the fourth point. Now, you can press ENTER to create a solid triangle or specify the fourth point to create a solid quadrilateral.

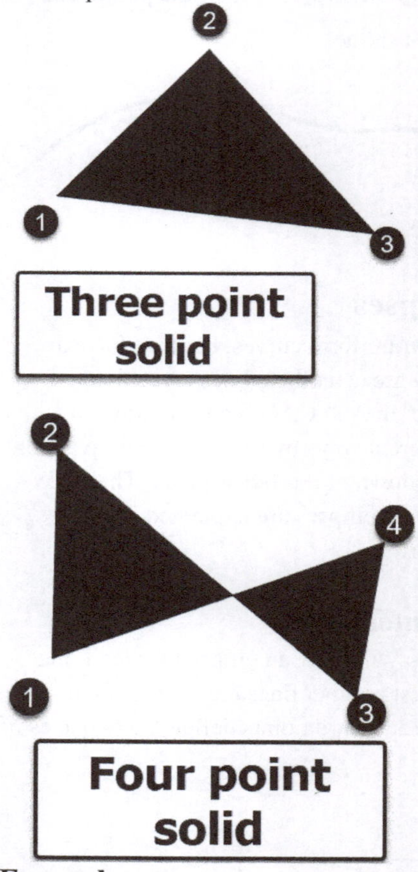

Example

- Download and open the SOLID_example.dwg file.
- Type CIRCLE in the command line and press ENTER.
- Select the corner point, as shown.
- Move the pointer outward, type 1 and press ENTER; a circle is created.

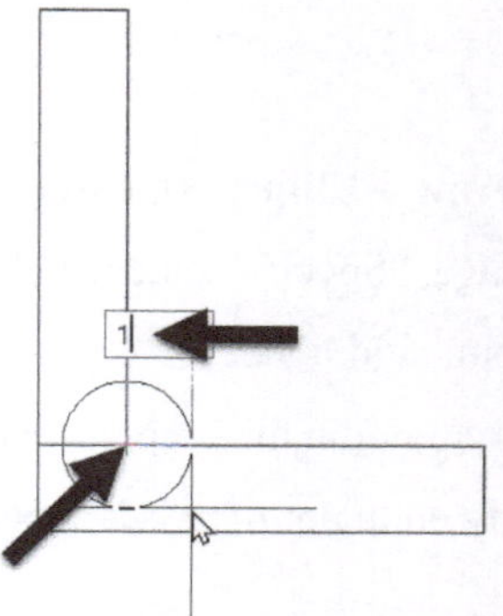

- Type SOLID in the command line and press ENTER.
- Select the left quadrant point of the circle.

- Select the top quadrant point of the circle.
- Select the center point of the circle and press ENTER twice.

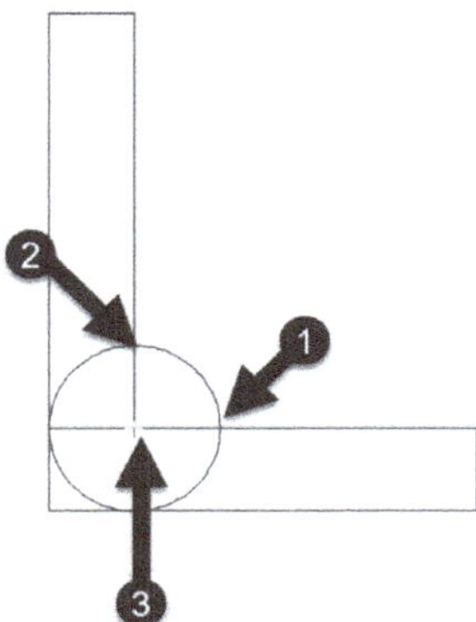

- Select the circle and press DELETE.

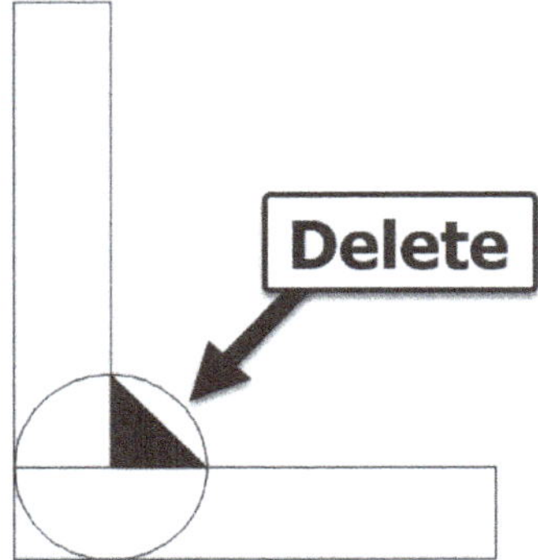

Donut

The **DONUT** command is used to create a closed ring or filled circle with a hollow center. This command is useful in creating designs that require shapes with circular openings or rings. To use the DONUT command, follow these steps:

- Download and open the Donut_example.dwg file.
- Type DONUT into the command line (or) click **Home** tab > **Draw** panel > **Donut** on the ribbon.

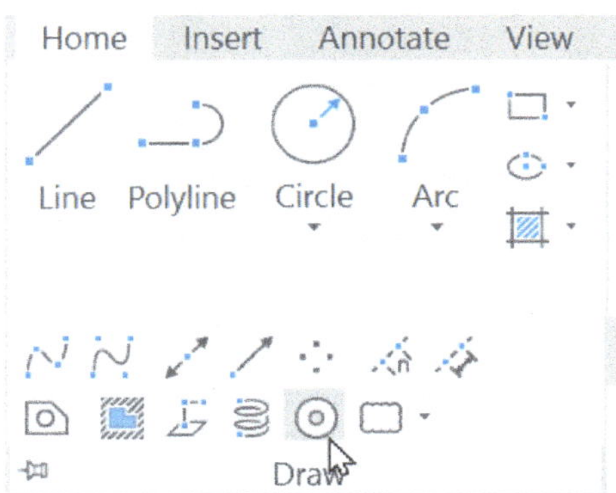

- Specify the diameter of the inside edge of the donut by either selecting two points or typing in the desired value.

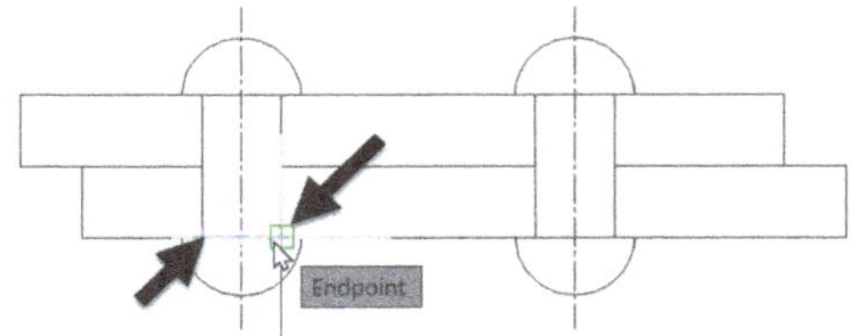

Two points selected to define the inner diameter

- Specify the diameter of the outside edge of the donut by either selecting an existing point or typing in the desired value. The outside diameter must be larger than the inside diameter.

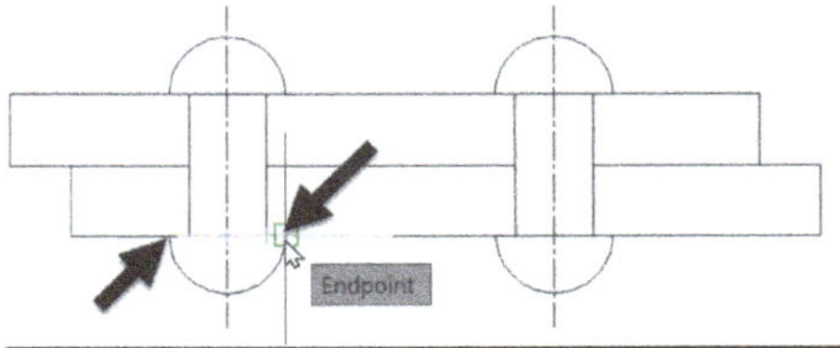

Two points selected to define the outside diameter

- Specify the center point of the donut by either selecting an existing point or typing in the X, Y, and Z coordinates.

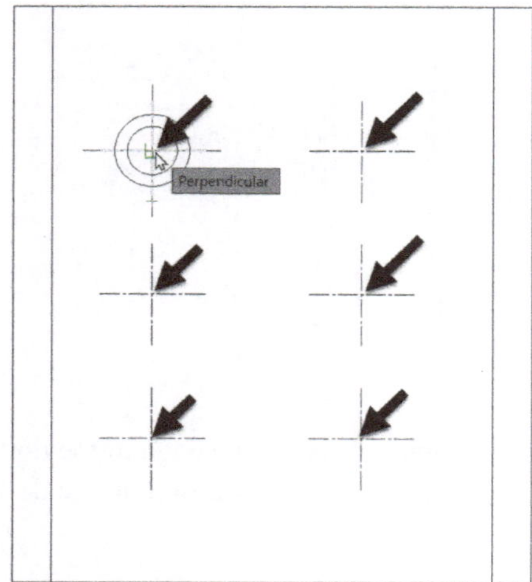

- Press "Enter" to complete the command.

The MLINE Command

The **MLINE** command is used to create complex lines and shapes that have multiple segments with different properties such as width, style, and color. This command allows you to create parallel lines, zigzag lines, and other types of shapes with ease. To use the MLINE command, simply type **MLINE** in the command line. Once the command is activated, you can specify the start point of the first segment. Next, specify the endpoint of the first segment. Then, specify the endpoint of the second segment, and so on, until all the required multiline segments have been established.

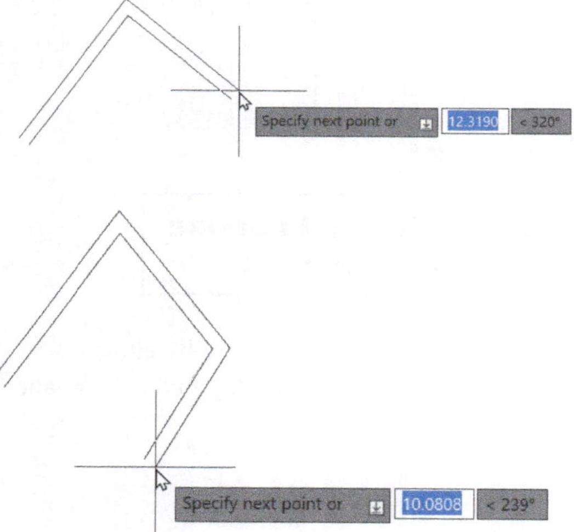

To generate a closed shape, choose the **Close** option from the command line. Alternatively, press ENTER to finalize the creation of the multiline segments.

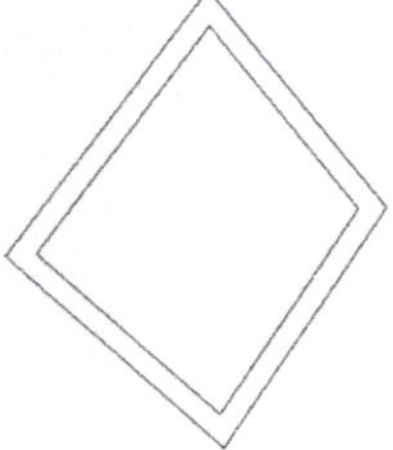

The command line displays various options when you activate the **MLINE** command. these options are discussed next.

Style

Allows you to select the style of the line segments. You can choose from the default STANDARD style or any other line style that has been created. You can create you own multiline style using the **MLSTYLE** command. To do this, type MLSTYLE in the command line and press ENTER. In the **Multiline Style** dialog box that appears, click on the **New** button to create a new multiline style.

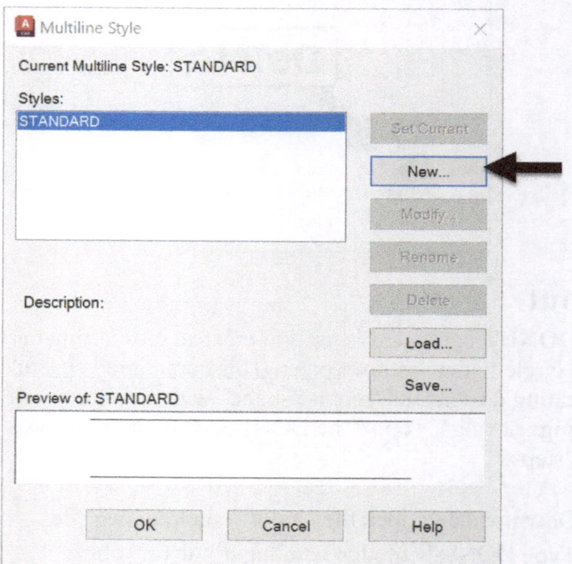

Enter a name for the new multiline style in the **New Style name** box of the **Create New Multiline Style** dialog. Next, click the **Continue** button.

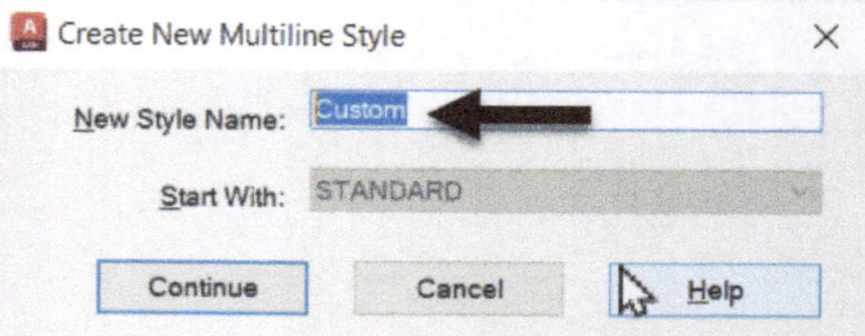

On the **New Multiline Style** dialog, in the **Elements** section, click on the **Add** button to add a new element to the multiline style. Next, select the newly added element and specify a value in the **Offset** box located below the **Elements** section. Next, specify the **Color** and **Linetype** of the newly added element using the **Color** drop-down and **Linetype** button, respectively.

Use the options in the **Caps** section to add caps to the start and end points of the multiline. You can add a Line, Outer arc, or Inner arcs at the start and endpoints of the multline.

To specify a color to be filled inside the area enclosed by the elements of the multiline, utilize the **Fill color** drop-down. If you wish to see lines at the points where two segments are joined, select the **Display joints** option. Once you have defined all the elements, click **OK** to save the new multiline style.

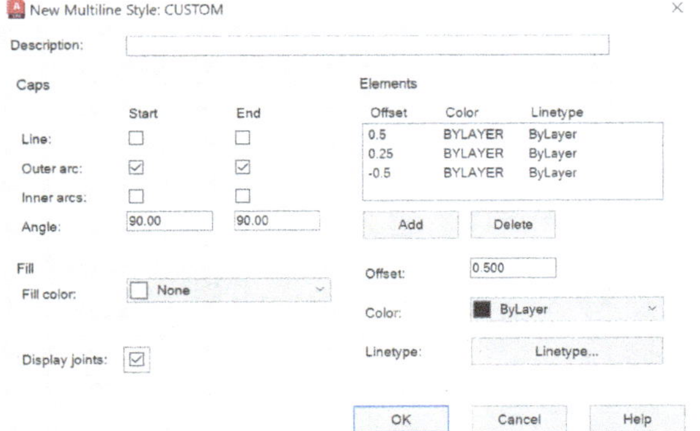

Click **Set Current** on the **Multiline Style** dialog and click to set the newly create multiline style as current. Your newly created multiline style can now be used in the current drawing or saved for use in future drawings.

Justification

The way in which the multiline is drawn between specified points is dependent on the chosen justification. The available justification options include **Top**, **Zero**, or **Bottom**.

When **Top** is selected, the top of the line elements are aligned with the chosen points during the drawing process.

On the other hand, selecting **Zero** will result in the center of the line elements being used as the chosen point.

Selecting **Bottom** will align the bottom line element between the specified points.

The **Justification** methods are designed for drawing in the positive X direction, from left to right. This means that when using the Bottom justification, the bottom of the multiline will be positioned between specified points when drawn from left to right. However, if you draw the multiline from right to left, the multiline will be flipped upside down.

Scale

Specifies the scale factor for the multiline. This option is useful if you need to change size of the multiline. The scale factor that you input will directly affect the width of the defined multiline, either increasing or decreasing it proportionately. It helps you change the size of the multiline without creating a new multiline style. For instance, a scale factor of 0.5 would produce a multiple line pattern that is half of the width defined using the MLSTYLE command. On the other hand, a scale factor of 1.5 would create a pattern that is one and a half times the width, while a scale factor of 4.0 would result in a pattern four times wider than the multiline width defined using the MLSTYLE command.

Note that you also have the option to input negative scale values, which will reverse the order of the multline. Alternatively, using a value of 0 will collapse the multiline pattern into a single line.

Positive Scale value

Negative Scale value

Scale value =0

Exercises

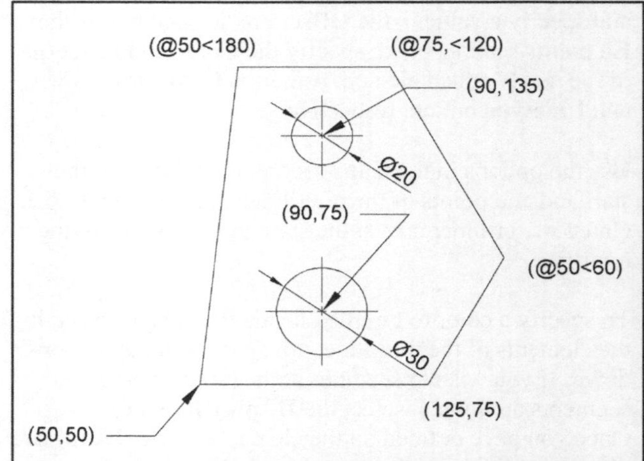

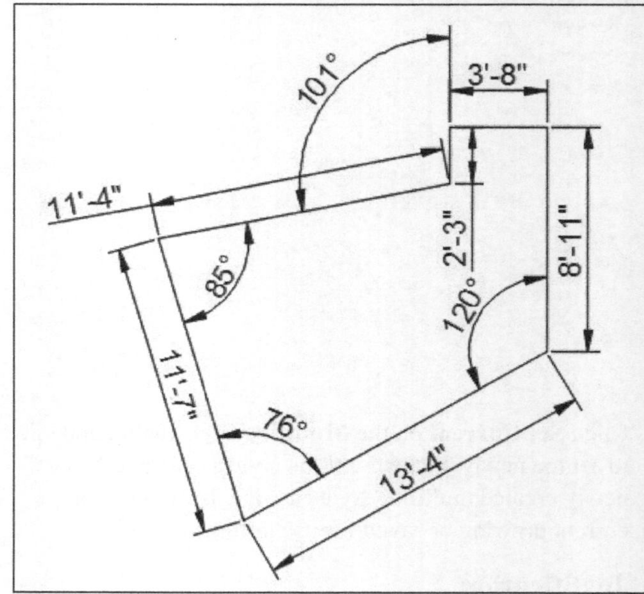

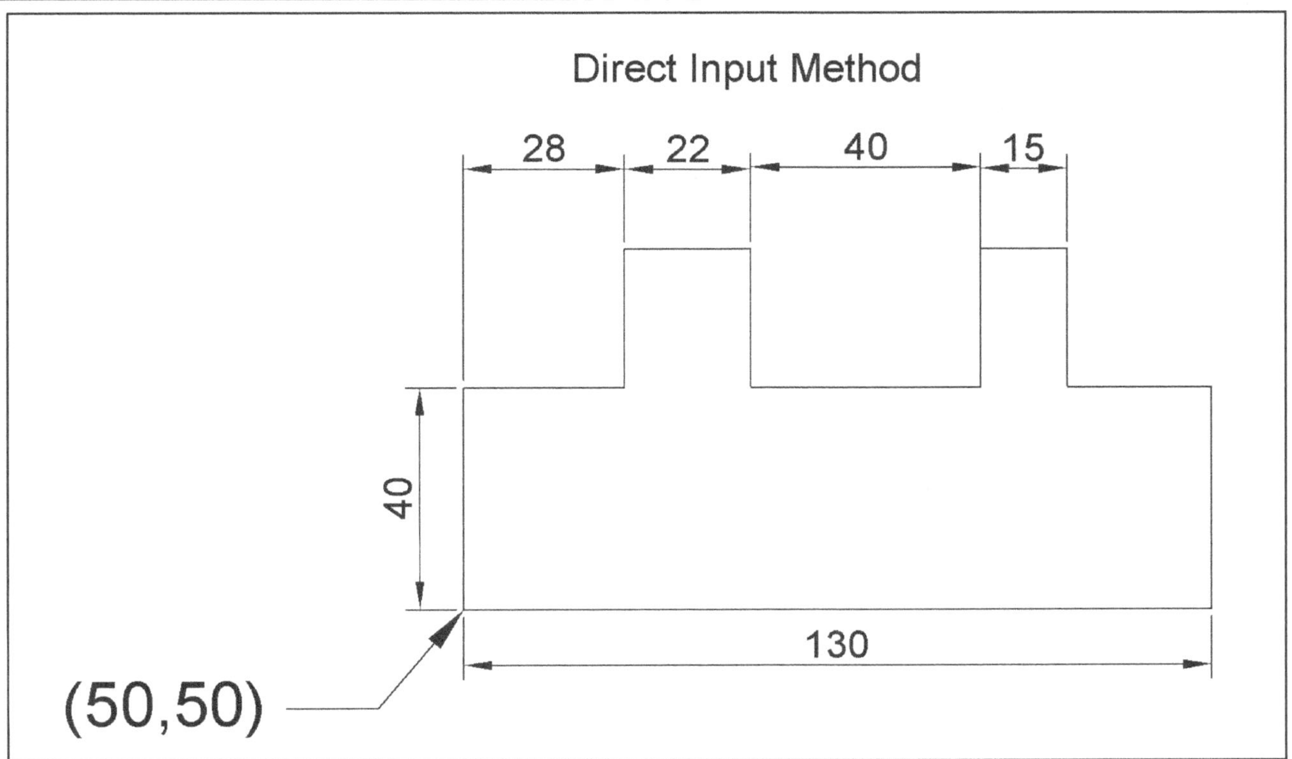
Direct Input Method
28
22
40
15
40
130
(50,50)

Chapter 3: Drawing Aids

In this chapter, you will learn to do the following:

- **Use Grid and Snap**
- **Use Ortho Mode and Polar Tracking**
- **Use Object Snaps and Object Snap Tracking**
- **Create Layers and assign properties to it**
- **Zoom and Pan drawings**

Drawing Aids

This chapter teaches you to define the drawing settings, which will assist you in creating a drawing in AutoCAD quickly. Most drawing settings can be turned on or off from the status bar. You can also access additional drawing settings by right-clicking on the button located on the status bar.

Setting Grid and Snap

Grid is the primary drawing setting. It makes the graphics window appear like a graph paper. You can turn ON the grid display by clicking the **Grid** icon on the status bar or just pressing **F7** on the keyboard.

Snap is used for drawing objects by using the intersection points of the grid lines. When you turn the **Snap Mode** ON, you will be able to select only grid points. In the following example, you will learn to set the grid and snap settings.

Example:

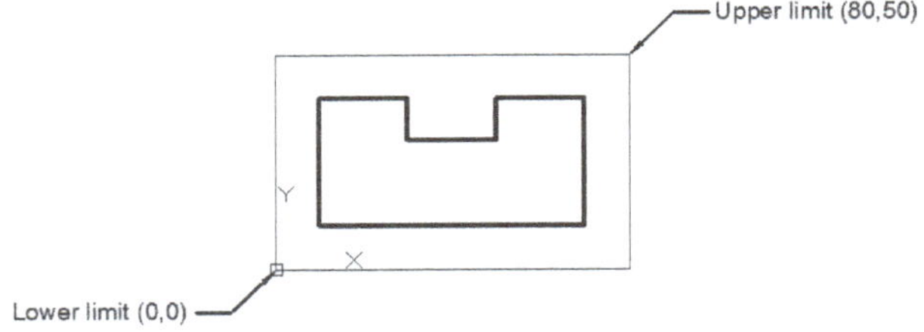

- Click **Application Menu > New**; the **Select Template** dialog appears.
- Select the **acadISO-Named Plot Styles** template. Click **Open**.
- On the Status bar, click the down arrow next to the **Snap Mode** icon and select **Snap Settings**. The **Drafting Settings** dialog appears.
- Click the **Snap and Grid** tab on the dialog.
- Set **Grid X spacing** to **10** and press the TAB key; the **Grid Y spacing** is updated with the same value.
- Set **Major line every** to **10**.

- Select the **Snap On** checkbox.
- Make sure that **Snap X spacing** and **Snap Y spacing** is set to **10**.

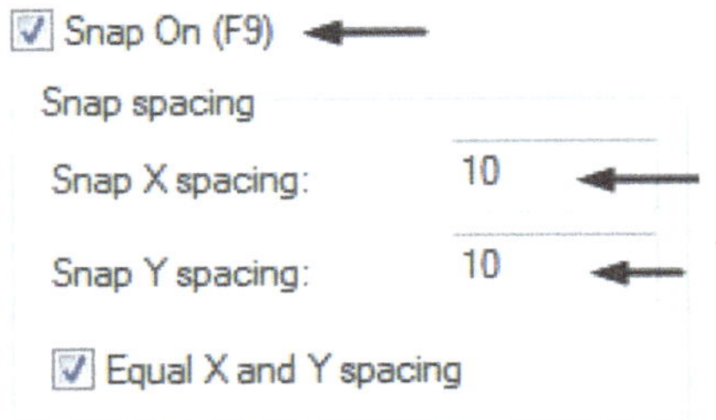

- Make sure that the **Grid snap** option is selected in the **Snap type** group.

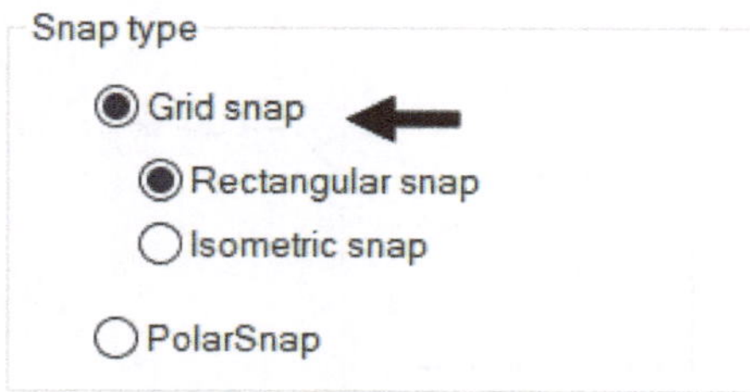

- Select the **Rectangular snap** option under **Grid snap**.
- Click **OK** on the dialog.
- Activate the **Grid** icon on the Status Bar.

Setting the Limits of a drawing

You can set the limits of a drawing by defining its lower-left and top-right corners. By setting Limits of a drawing, you will determine the size of the drawing area. In AutoCAD, limits are set to some default values. However, you can redefine the limits to change the drawing area as per your requirement.

- Type **Limits** at the command line and press ENTER.

- Type 0,0 and press ENTER to define the lower limit. Now, you need to define the upper limit.
- Type 80,50 and press the ENTER key.
- On the Navigate Bar, click **Zoom > Zoom All**; the graphics window will be zoomed to the limits.

Setting the Lineweight

Line weight is the thickness of the objects that you draw. In AutoCAD, there is a default lineweight assigned to objects. However, you can set a new lineweight. The method to set the lineweight is explained below.

- On the Status bar, click the **Customization** option, and then select **LineWeight** from the flyout. This shows the **LineWeight** icon on the status bar.
- Activate the **Show/Hide Lineweight** icon located on the status bar.
- Right click on the **Show/Hide Lineweight** icon, and then select **Lineweight Settings**. The **Lineweight Settings** dialog appears.
- On the **Lineweight Settings** dialog, select **0.40** mm from the **Default** drop-down.

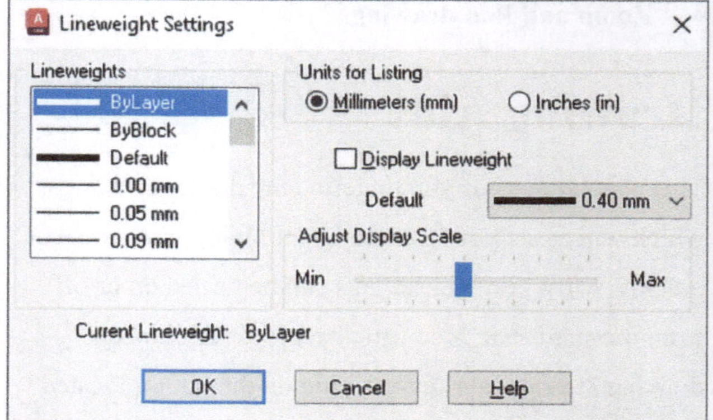

- Click **OK**.
- Type **L** in the command line and press ENTER.
- Type 10,10 and press ENTER to define the first point.
- Move the pointer horizontally toward the right and click on the sixth grid point from the first point.
- Move the pointer vertically upwards and select the third grid point from the second point.
- Move the pointer horizontally toward the left and select the second grid point from the previous point.
- Move the pointer vertically downwards and select the grid point next to the previous point.
- Move the pointer horizontally toward the left and select the second grid point from the previous point.
- Move the pointer vertically upwards and select the grid point next to the previous point.

- Move the pointer horizontally toward the left and select the second grid point from the previous point.
- Right-click and select **Close**.

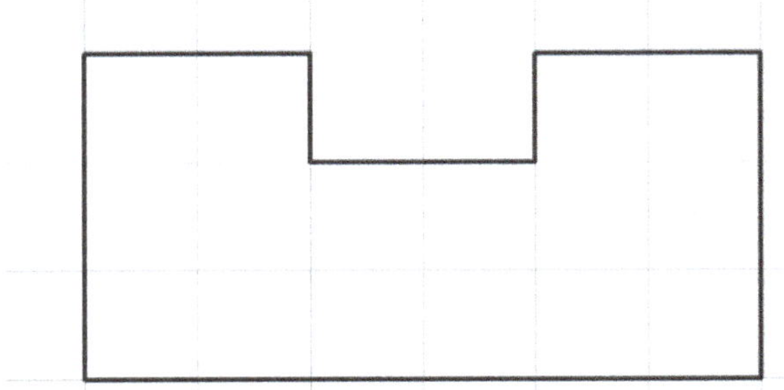

- Save and close the file.

Using Ortho mode and Polar Tracking

Ortho mode is used to draw orthogonal (horizontal or vertical) lines. Polar Tracking is used to constrain the lines to angular increments. In the following example, you will create a drawing with the help of Ortho Mode and Polar Tracking.

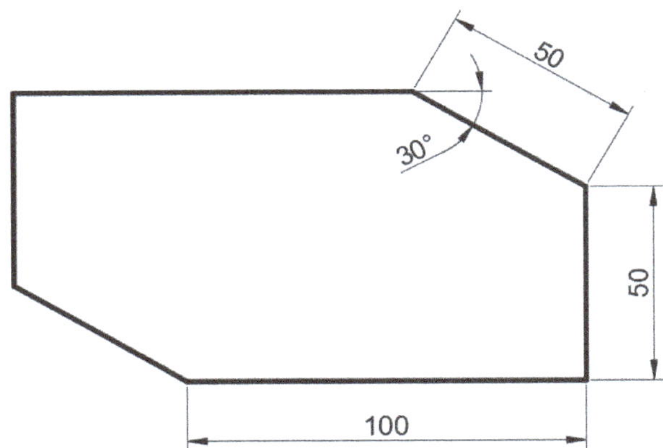

- Open a new AutoCAD file.
- Deactivate the **Grid Mode** and **Snap Mode** icons on the status bar.
- Click the **Ortho Mode** icon on the status bar.
- Click **Zoom All** on the **Navigation Bar**.
- Click the **Line** button on the **Draw** panel.
- Select an arbitrary point to define the starting point.
- Move the pointer toward the right, type 100, and press ENTER; you will notice that a horizontal line is created.
- Move the pointer upwards, type 50, and press ENTER; you will notice that a vertical line is created.
- Click the **Polar Tracking** icon on the status bar.
- Click the down arrow next to the **Polar tracking** icon and select **30** from the menu.

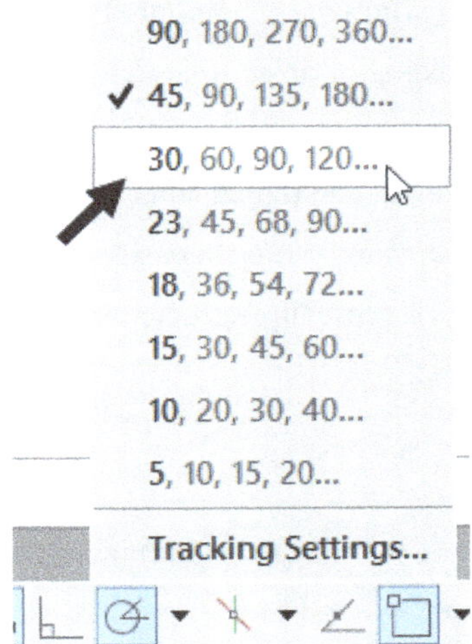

You will notice a track line at 30-degree increments when you rotate the pointer.

- Move the pointer and stop when the tooltip displays <150 angle value.

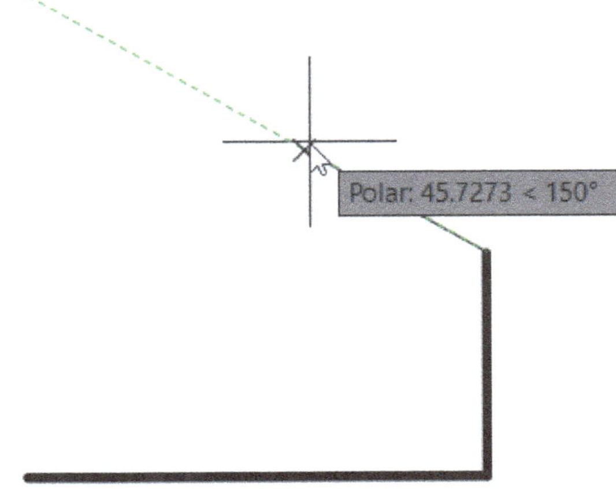

- Type 50 and press ENTER when the tooltip displays <150°.
- Move the pointer toward left.

- Type 100 and press ENTER when the tooltip displays <180°.
- Move the pointer vertically downward.
- Type 50 and press ENTER when the tooltip displays <270 °.
- Right-click and select **Close**.

Using Layers

Layers are like a group of transparent sheets that are combined into a complete drawing. The figure below displays a drawing consisting of object lines and dimension lines. In this example, the object lines are created on the 'Object' layer, and dimensions are created on the layer called 'Dimension.' You can easily turn-off the 'Dimension' layer for a more unobstructed view of the object lines.

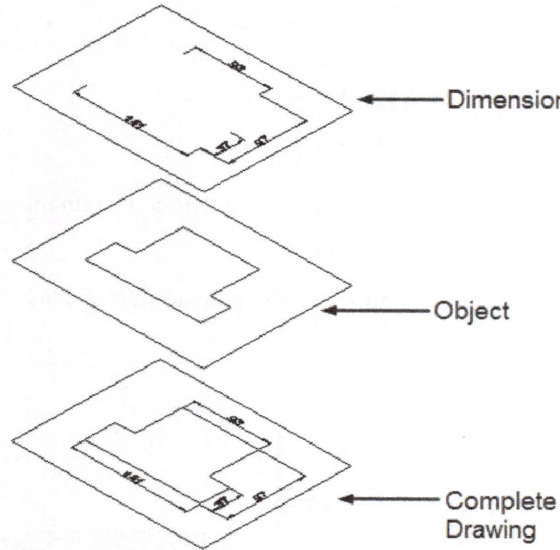

Layer Properties Manager

The **Layer Properties Manager** is used to create and manage layers. To open **Layer Properties Manager**, click **Home > Layers > Layer Properties** on the ribbon or enter **LA** in the command line.

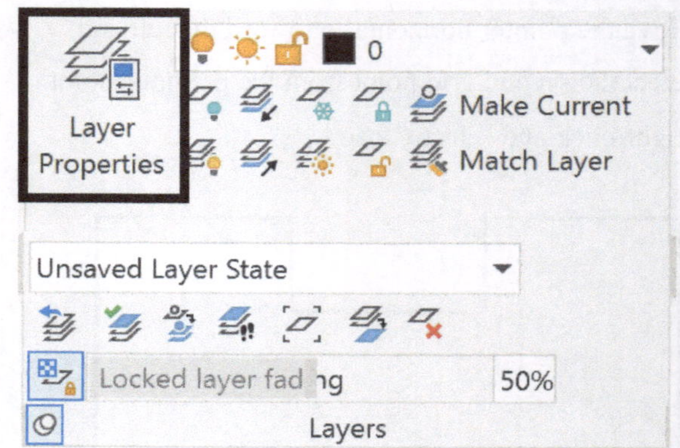

The components of the **Layer Properties Manager** are shown below. The **Tree View** section is used for displaying layer filters, group, or state information. The **List View** section is the main body of the **Layer Properties Manager**. It lists the individual layers that currently exist in the drawing.

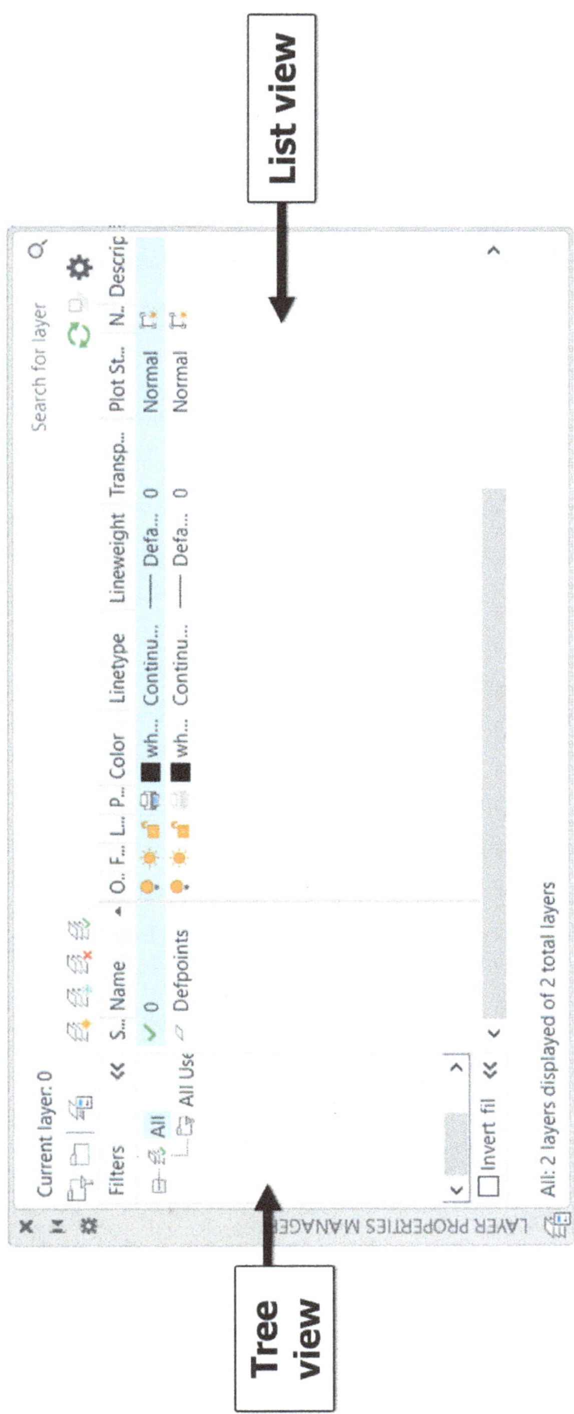

The **List View** section contains various properties. You can set layer properties and perform multiple operations in the **List View** section. A brief explanation of each layer property is given below.

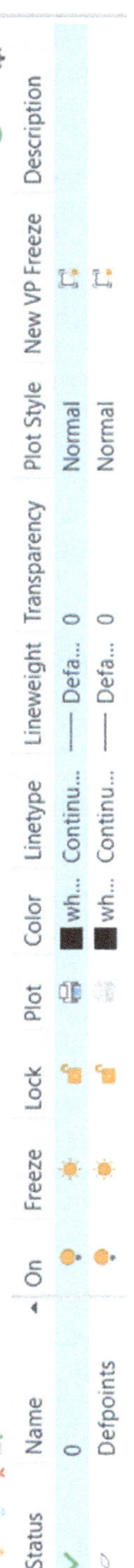

Status –Shows a green check when a layer is set to current.

Name - Shows the name of the layer.

On – It is used to turn on/off the visibility of a layer. When a layer is turned on, it shows a yellow light-bulb. When you turn off a layer, it shows a grey light-bulb.

Freeze/Thaw – It is used to freeze the objects of a layer so that they cannot be modified. Also, the visibility of the object is turned off.

Lock/Unlock- It is used to lock the layer so that the objects on it cannot be modified.

Color – It is used to assign a color to the layer.

Linetype – It is used to assign a linetype to the layer.

Lineweight – It is used to define the lineweight (thickness) of objects on the layer.

Transparency – It is used to define the transparency of the layer. You set a transparency level from 0 to 90 for all objects on a layer.

Plot Style – It is used to override the settings such as color, linetype, and lineweight while plotting a drawing.

Plot – It is used to control which layer will be plotted.

New VP Freeze – It is used to create and freeze a layer in any new viewport.

Description – It is used to enter a detailed description of the layer.

Creating a New Layer

You can create a new layer by using any one of the following methods:

1. Click the **New Layer** button on the **Layer Properties Manager**; a new layer with the name '**Layer1**' appears in **Name** field. Next, enter the name of the layer in the **Name** field.

2. Right-click in the **Name** field and select **New Layer** from the shortcut menu.
3. Select an existing layer, and then type ENTER or comma (,).

Making a layer current

If you want to draw objects on a particular layer, then you have to make it current. You can make a layer current using the methods listed below.

1. Select the layer from the List view and click the **Set Current** button on the **Layer Properties Manager**.

2. Double-click on the **Name** field of the layer.
3. Right-click on the layer and select **Set current**.
4. Select the layer from the **Layer** drop-down of the **Layer** panel.

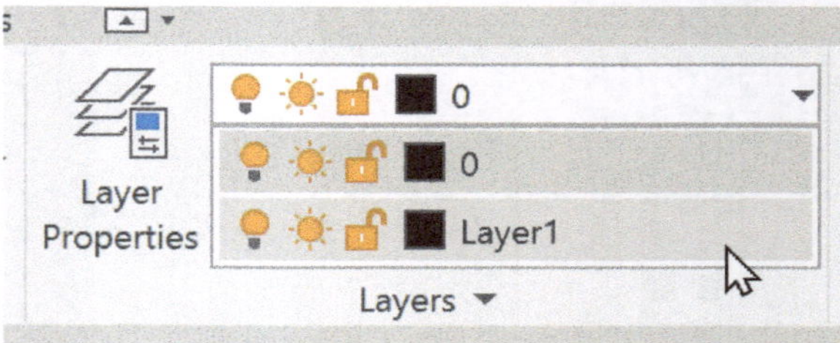

You can also display the **Layer** drop-down on the Quick Access Toolbar. To do this, click the down arrow next to the Quick Access Toolbar and select **Layer** from the menu.

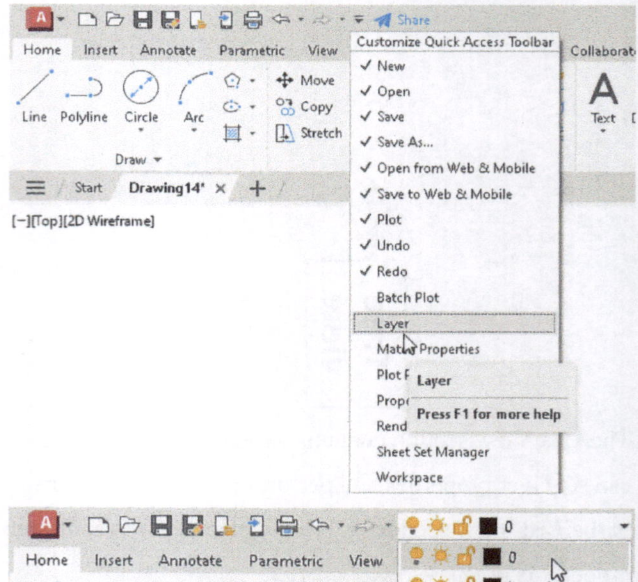

5. Click the **Make Current** button on the **Layers** panel. Next, select an object from the graphics window; the layer related to the selected object will become current.

Deleting a Layer

You can delete a layer by using any one of the following methods:

1. Click the **Delete Layer** button or press ALT+D.

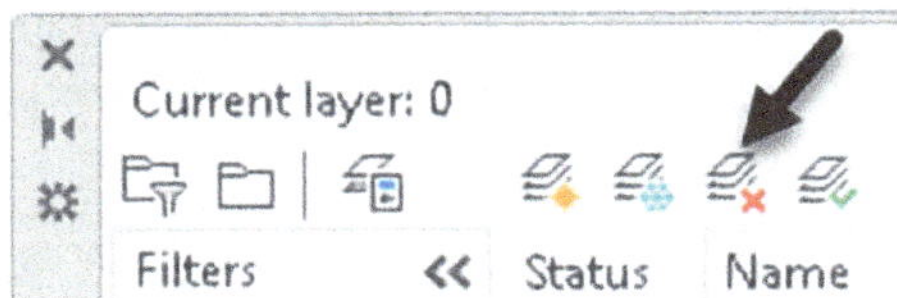

2. Right-click in the **Name** field and select **Delete Layer** from the shortcut menu.

You will learn more about layers in later chapters. You can find an example related to layers in the **Offset** tool section of chapter 4.

Using Object Snaps

Object Snaps are essential settings that improve your performance and accuracy while creating a drawing. They allow you to select key points of objects while creating a drawing. You can activate the required Object Snap by using the **Object snap** shortcut menu. Press and hold the SHIFT key and right-click to display this shortcut menu. Note that the object snaps can be used only when a drawing command is active.

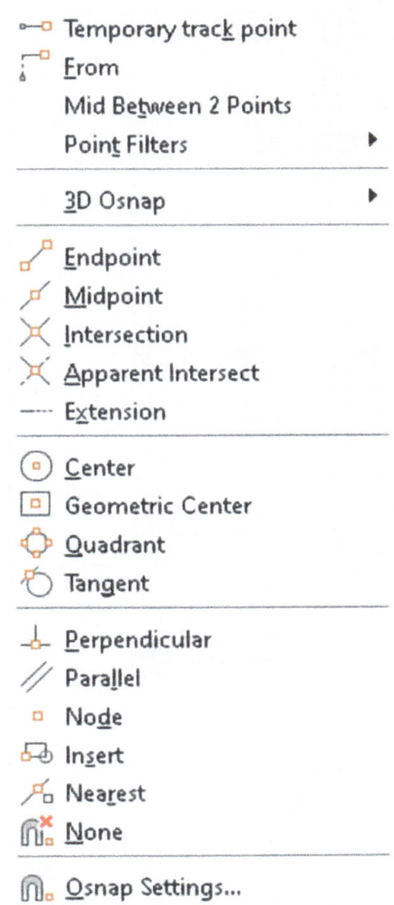

The functions of various Object Snaps are explained next.

Endpoint: Snaps to the endpoints of lines and arcs.

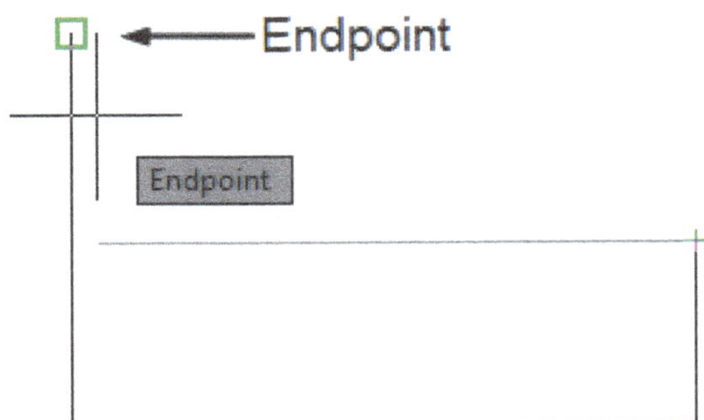

Midpoint: Snaps to the midpoint of a line.

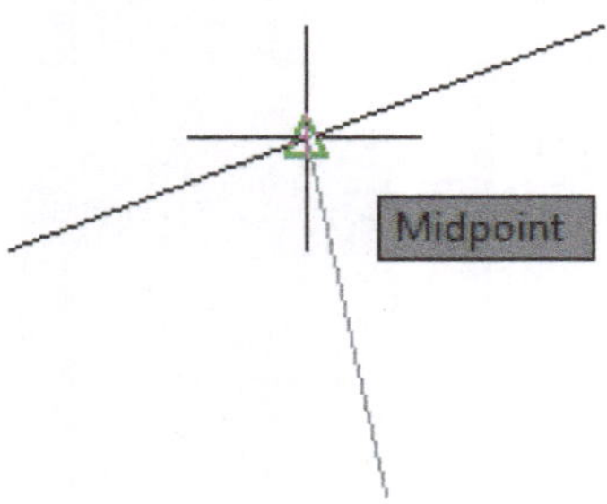

Intersection: Snaps to the intersections of objects.

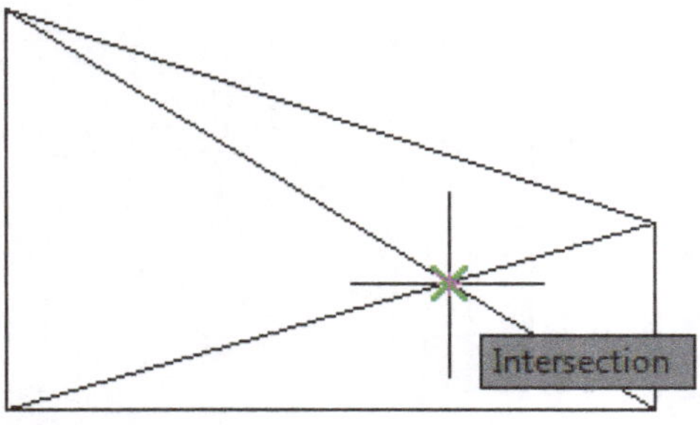

Apparent Intersection: Snaps to the projected intersection of two objects in 3D space.

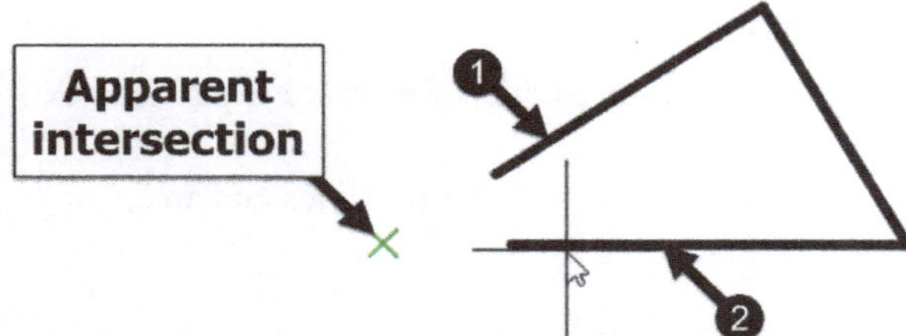

Extension: Creates a temporary extension line when the pointer passes through the endpoints of a line or an arc. You can pick points along the temporary extension lines.

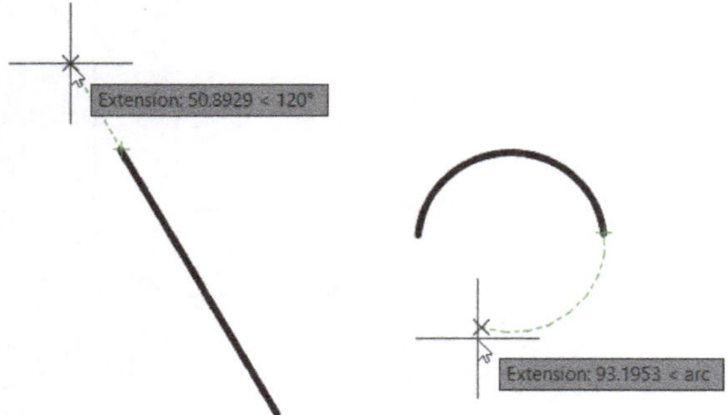

Center: Snaps to the centers of circles and arcs.

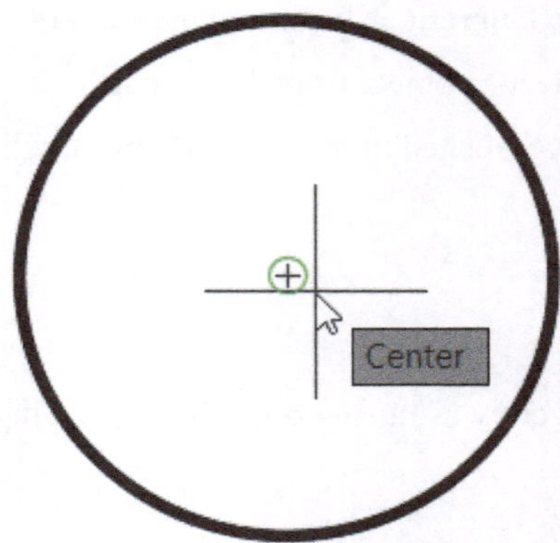

Geometric Center: Snaps to the center point of a closed geometry created by a single object such as polyline, rectangle, or polygon.

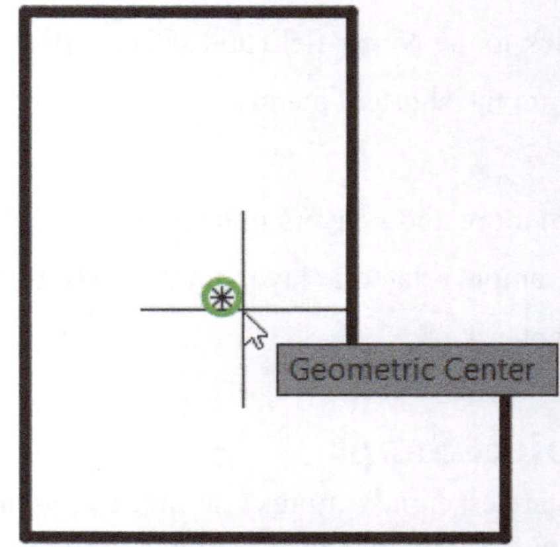

Quadrant: Snaps to four key points located on a circle.

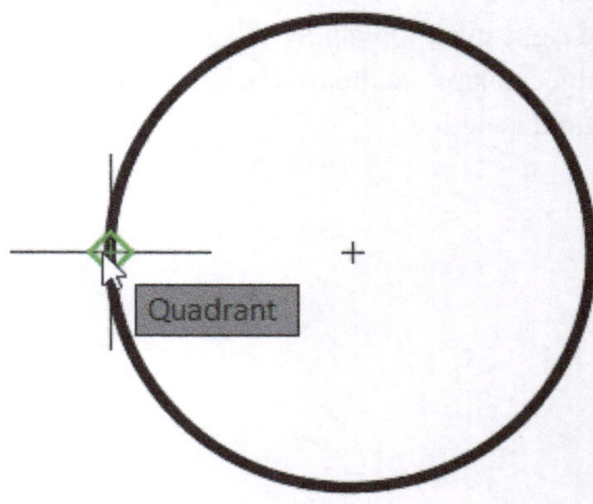

Tangent: Snaps to the tangent points of arcs and circles.

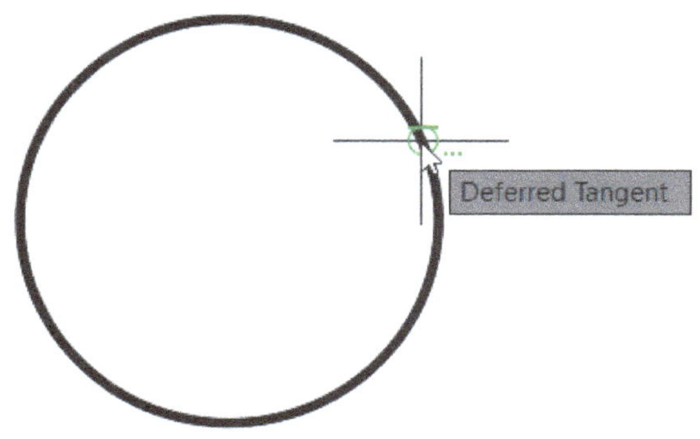

Perpendicular: Snaps to a perpendicular location on an object.

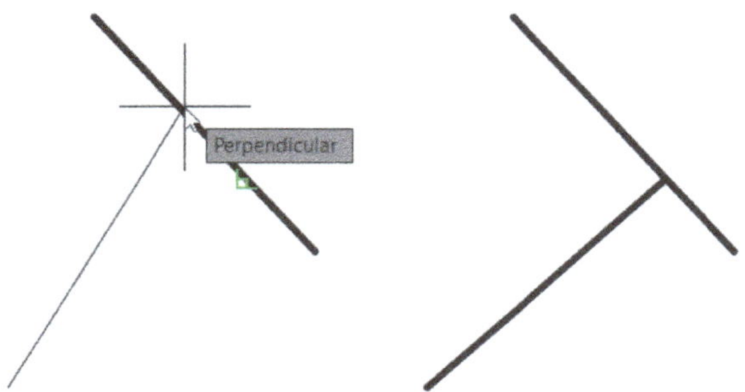

Parallel: It is used to draw an object parallel to another object. Select this option and place the pointer on the first line. Next, move the pointer and notice the track line parallel to the first object. Move along the track line and click to create a line parallel to the first line.

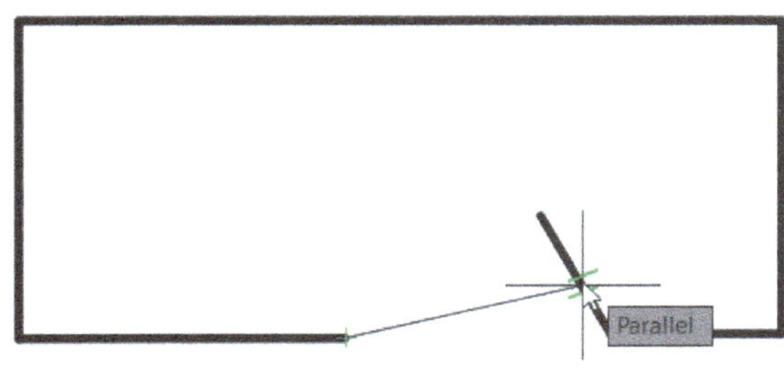

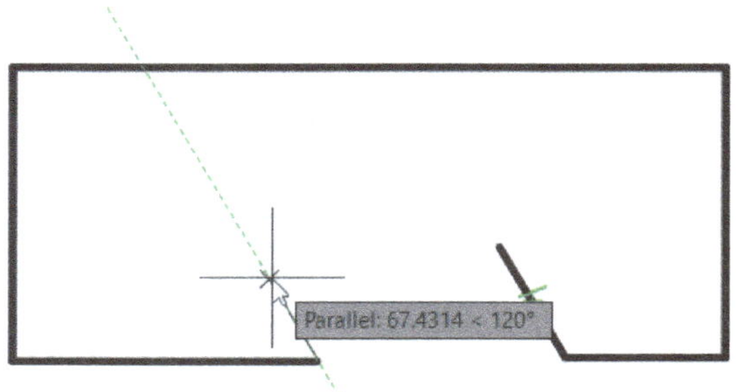

Node: Snaps to points of dimension` lines, text objects, and dimension text.

Insert: Snaps to the insertion point of blocks, shapes, and text.

Nearest: Snaps to the nearest point found along with any object.

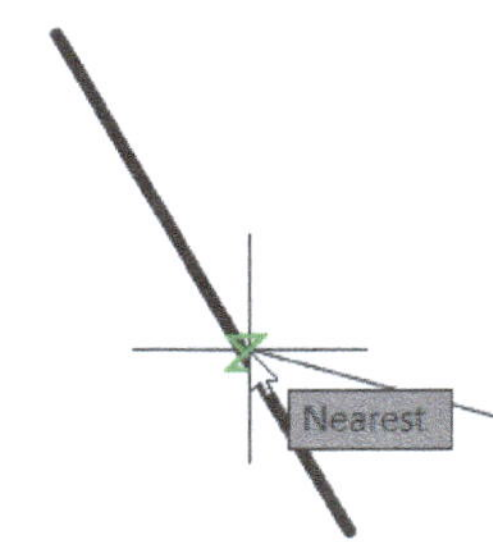

None: Deactivates the Object Snap.

Temporary Track Point: It is used to locate a point by using trace lines from a reference point.

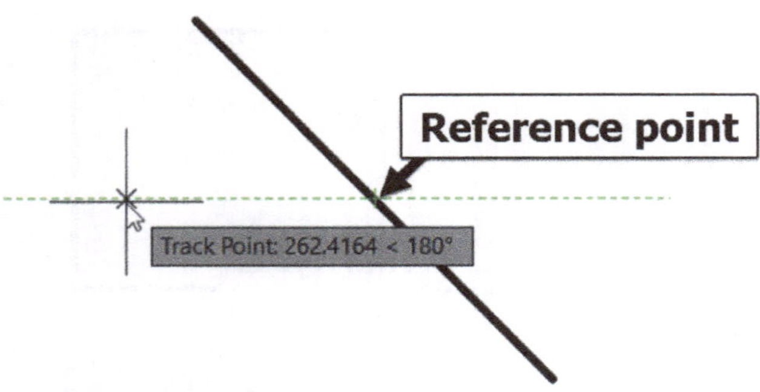

From: Locates a point at a specified distance and direction from a selected reference point.

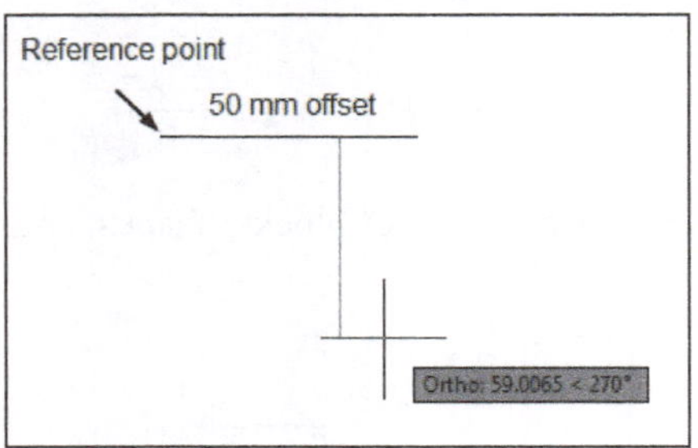

Midpoint Between 2 Points: Snaps to the middle point of two selected points.

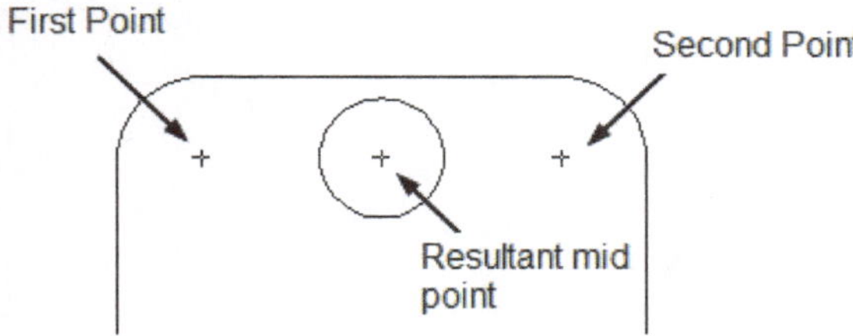

Running Object Snaps

Previously, you have learned to select Object Snaps from the shortcut menu. However, you can make Object Snap modes available continuously instead of picking them every time. You can do this by using the **Running Object Snaps**. To use the Running Object Snaps, click the down arrow next to the **Object Snaps** button on the status bar and select the required object snap from the menu.

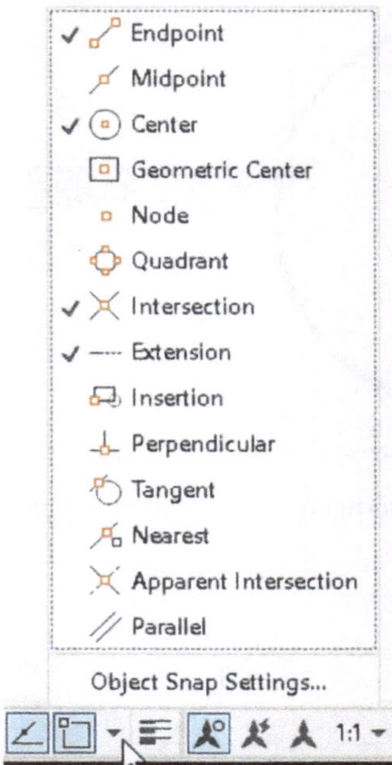

You can also select the **Object Snap Settings** option from the menu to open the **Drafting Settings** dialog. In this dialog, you can choose the required Object Snaps by selecting checkboxes.

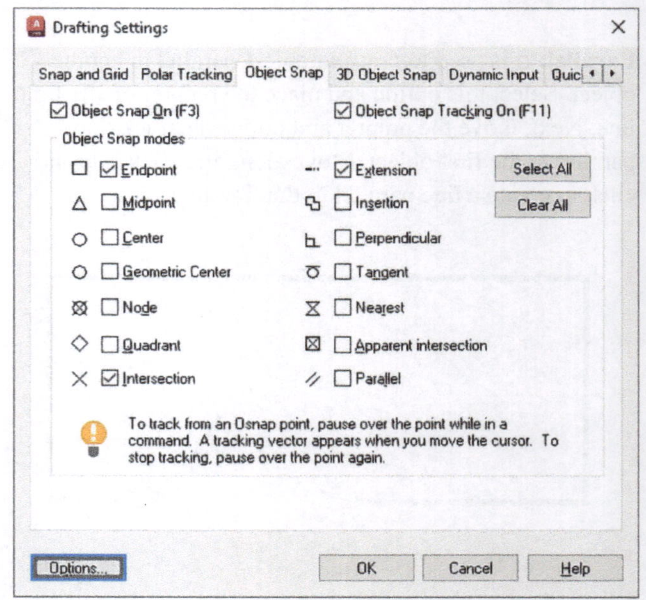

Cycling through Object Snaps

After setting the Running Object Snap settings, AutoCAD displays object snaps depending on the shape of the object. However, you can cycle through the object snaps by pressing the TAB key. In the following example, you will learn to cycle through different object snaps.

Example:

- Click the down arrow next to the **Object Snap** button and select the **Object Snap Settings** option; the **Drafting Settings** dialog appears.
- Click the **Select All** button and click the **OK** button.
- Draw the objects, as shown below.

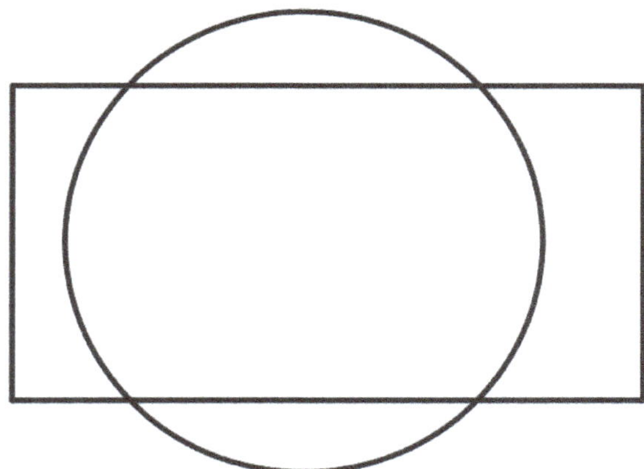

- Click the **Circle** button on the **Draw** panel.
- Place the pointer on the drawing. Press the TAB key; you will notice that the object snaps change.

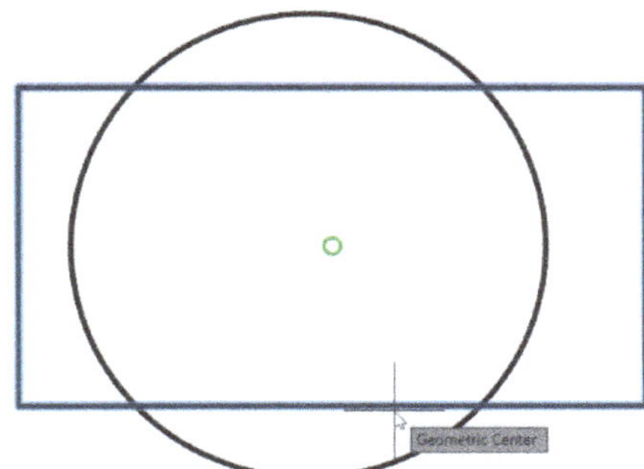

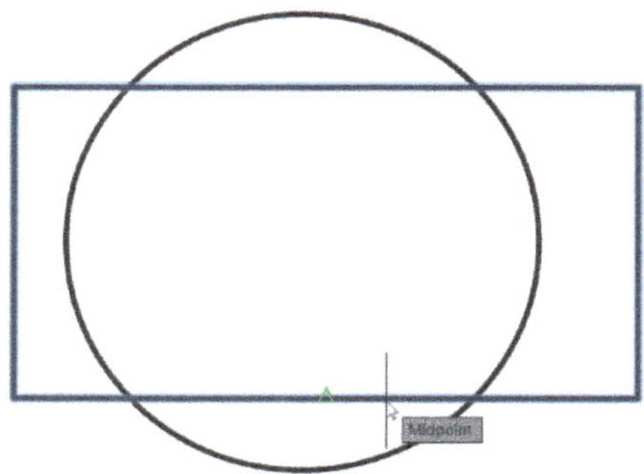

- Click when the **Center** snap is displayed and draw a circle.

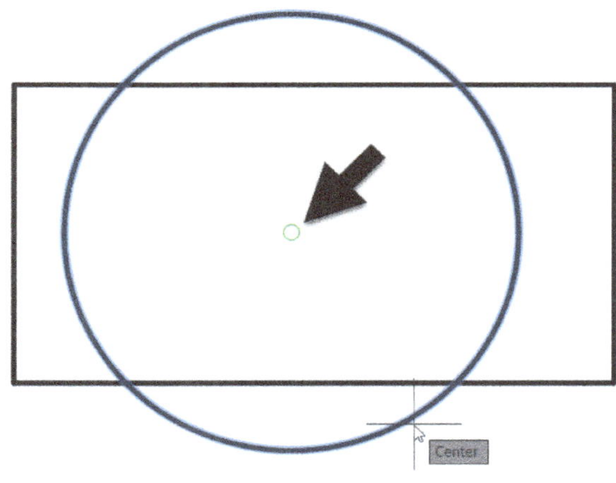

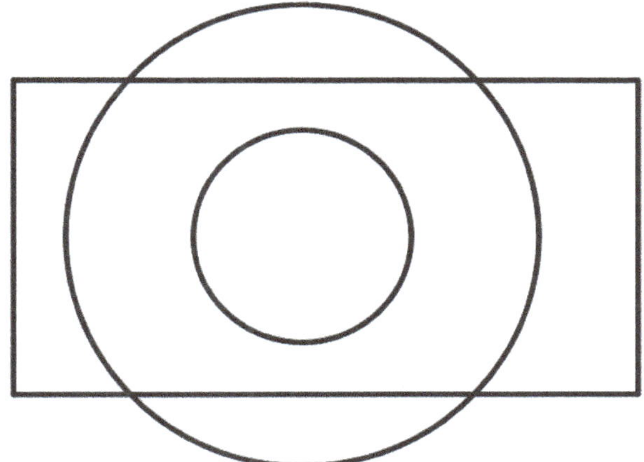

Using Object Snap Tracking

Object Snap tracking is the movement of the pointer along the trace lines originating from the key points of objects. Object Snap Tracking works only when the **Object Snap** mode is turned on. In the following example, you will learn to use Object Snap Tracking for creating objects.

Example:

- Select the **Object Snap Tracking** button from the Status bar.

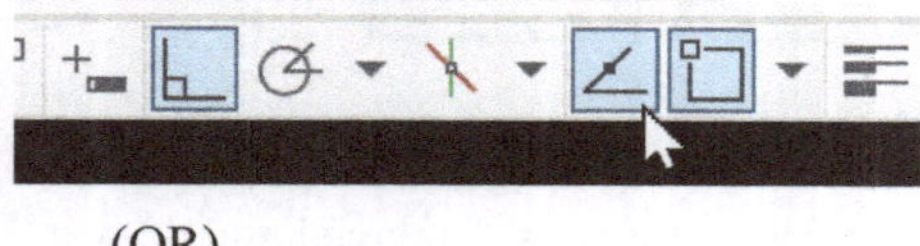

(OR)

- Open the **Drafting Settings** dialog and click the **Object Snap** tab.
- Select the **Object Snap Tracking On** checkbox.

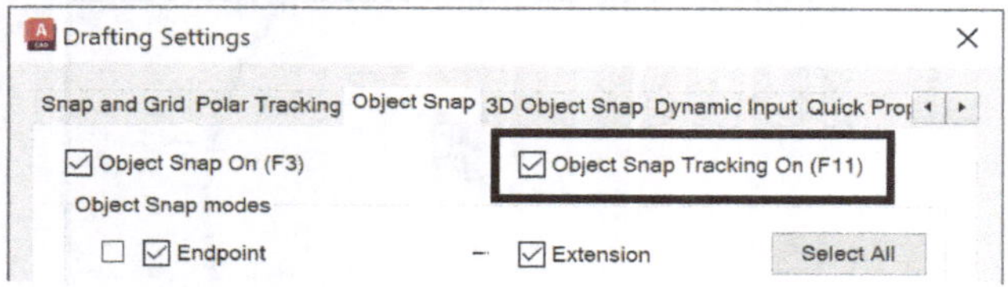

- Click **OK**.
- Use the **Line** tool and draw the objects, as shown below.

- Press the ENTER key twice to start drawing lines from the last point.
- Move the pointer and place it on the endpoint of the lower horizontal line.

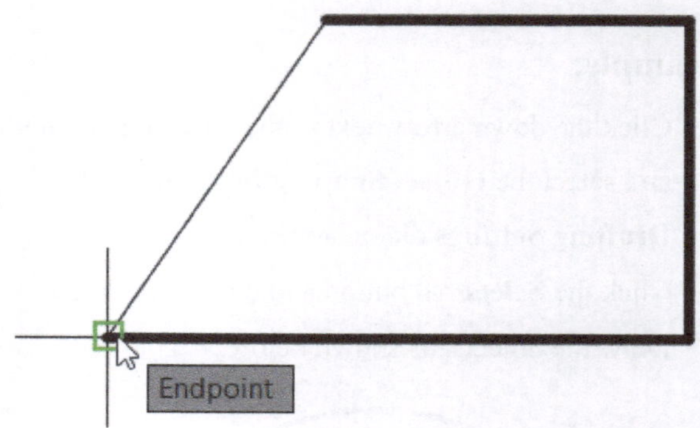

- Move the pointer vertically upward; you will notice the trace line, as shown below.

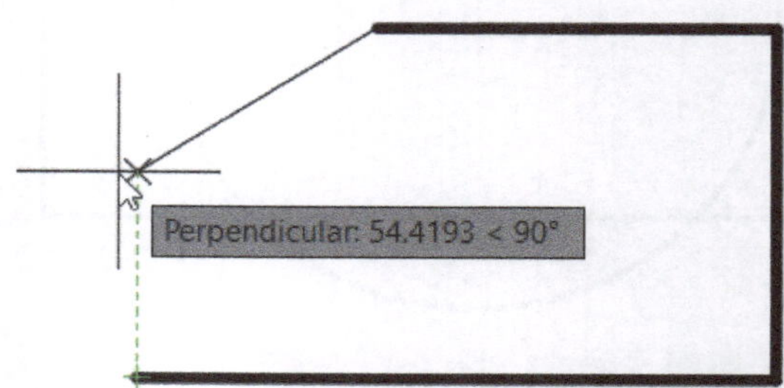

- Click on the trace line to create an inclined line.
- Snap the pointer to the endpoint of the lower horizontal line and click.

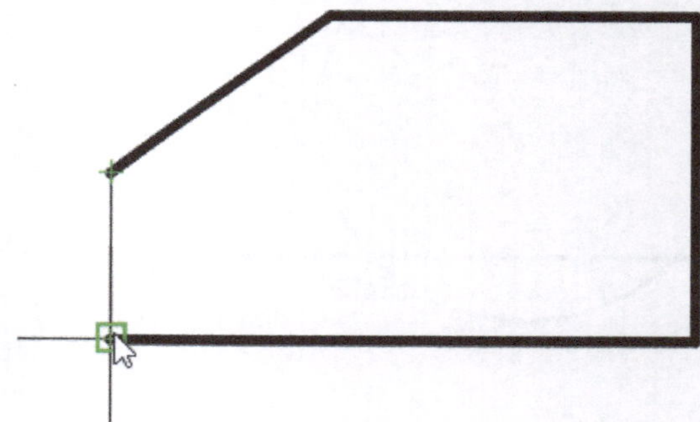

- Right-click and select **Enter**.
- Click the **Circle** button on the **Draw** panel of the ribbon.
- Place the pointer over the lower endpoint of the inclined line and move horizontally; you will notice that a trace line is displayed.

- Place the pointer on the midpoint of the lower horizontal line; a vertical trace line is displayed from the midpoint of the horizontal line, as shown below.

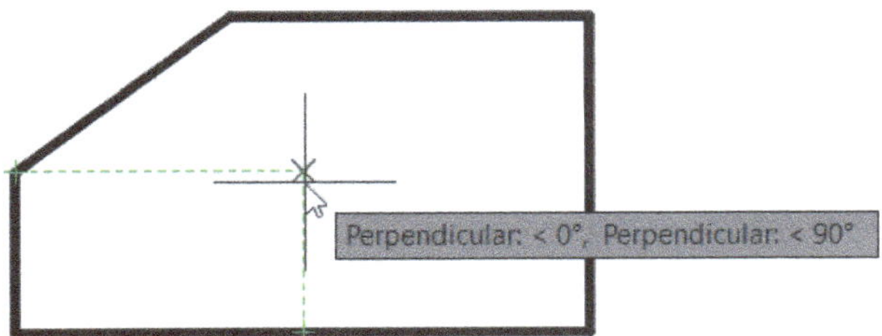

- Click at the point where the horizontal and vertical trace lines intersect. Next, create a circle, as shown below.

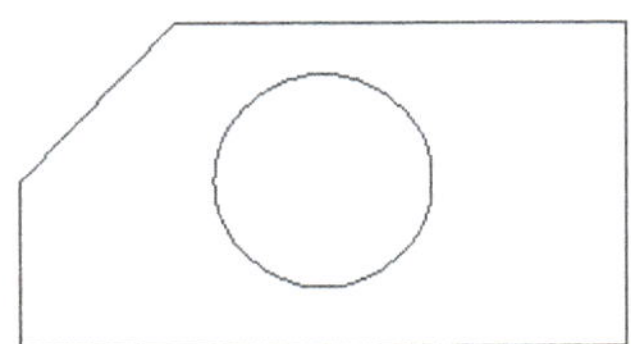

Linetype gap selection

AutoCAD makes it easy to select line types such as centerlines, dashed-dotted lines, hidden, and phantom. Earlier, it was difficult to select these linetypes by clicking in the gaps. Now, you can select them by clicking on the gaps.

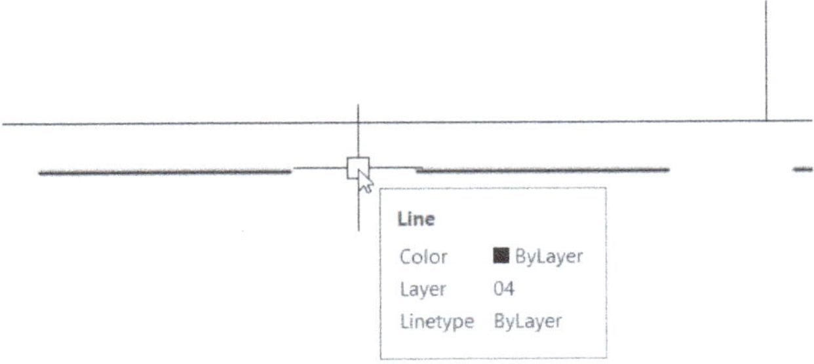

You can also snap to the line at the gaps.

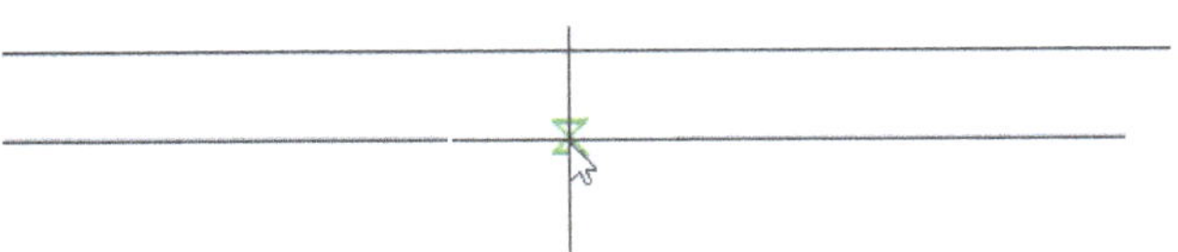

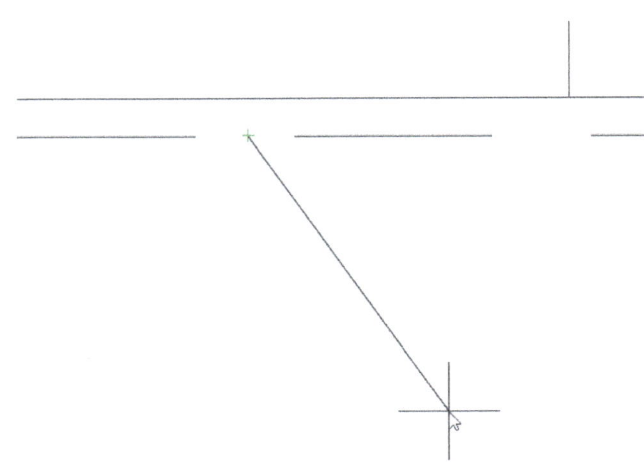

The LTGAPSELECTION system variable, when set to 1, helps you to select the line by clicking in the gaps. You can turn OFF this feature by setting the LTGAPSELECTION system variable to 0.

Using Zoom tools

Using the zoom tools, you can magnify or reduce a drawing. You can use these tools to view the minute details of a very complicated drawing. The Zoom tools can be accessed from the Navigation Bar, Command line, and Menu Bar.

Navigation Bar

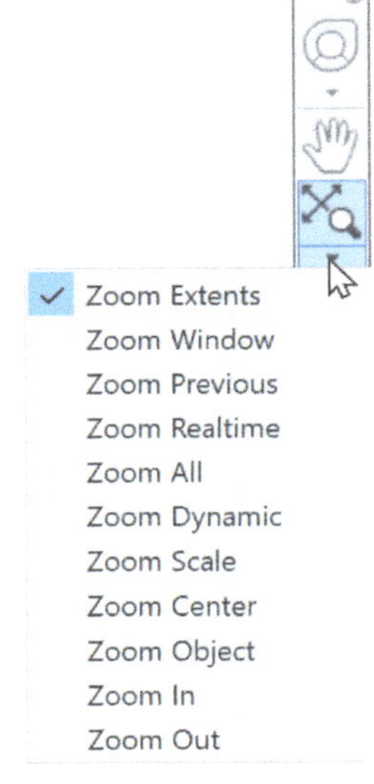

Menu Bar

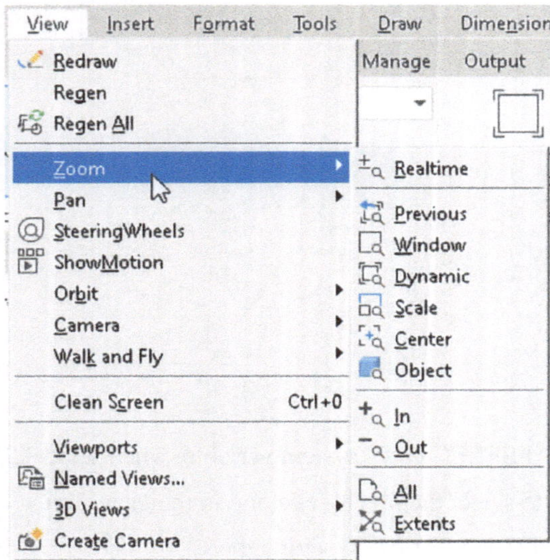

Command line

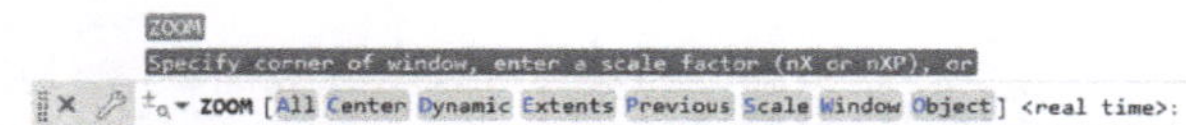

Zooming with the Mouse Wheel

Zooming using the mouse wheel is one of the easiest methods.

- Roll the mouse wheel forward to zoom into a drawing.
- Roll the mouse wheel backward to zoom out of the drawing.
- Press the mouse wheel and drag the mouse to pan the drawing.

Using Zoom Extents

Using the **Zoom Extents** tool, you can zoom to the extents of the largest object in a drawing.

- Click **Zoom Extents** on the Navigation Bar.
- You can also double-click on the mouse wheel to zoom to extents.

Using Zoom-Window

Using the **Zoom-Window** tool, you can define the area to be zoomed by selecting two points representing a rectangle.

- Click **Zoom > Zoom Window** on the Navigation Bar.
- Specify the first point of the zoom window, as shown.
- Move the pointer diagonally toward the right, and then specify the second point, as shown.
  The area inside the window will be zoomed.

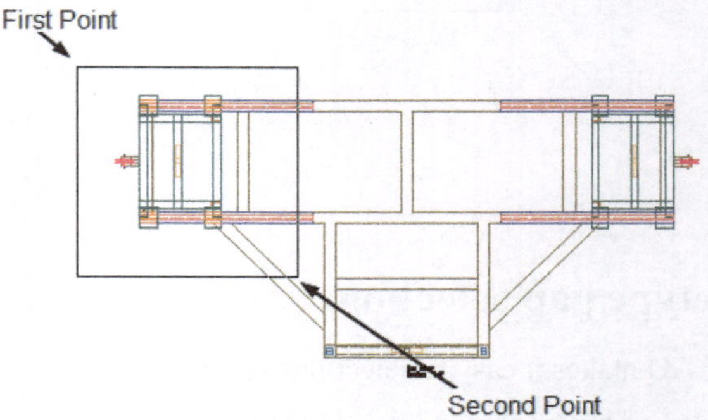

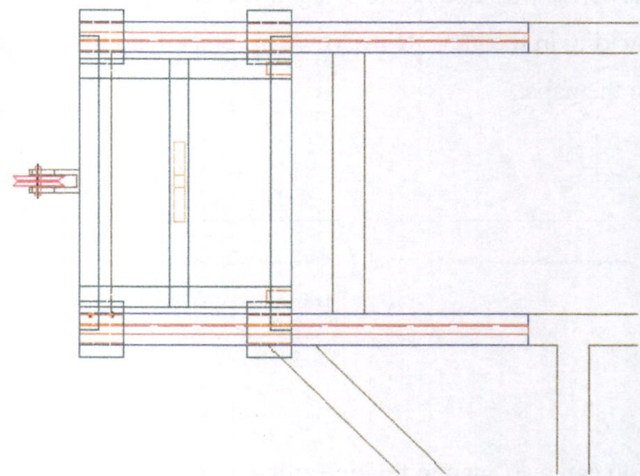

Using Zoom-Previous

After magnifying a small area of the drawing, you can use the **Zoom-Previous** tool to return to the previous display.

- Click **Zoom > Zoom Previous** on the Navigation Bar.

Using Zoom-Realtime

Using the **Zoom-Realtime** tool, you can zoom in or zoom out of a drawing dynamically.

- Click **Zoom > Zoom Realtime** on the **Navigation Bar**; the pointer is changed to a magnifying glass with plus and minus symbols.
- Press and hold the left mouse button and drag the mouse forward to zoom into the drawing.
- Drag the mouse backward to zoom out of the drawing.

Using Zoom-All

The **Zoom All** tool is used to adjust the drawing space to the limits set by using the LIMITS command.

- Click **Zoom > Zoom All** on the **Navigation Bar**; the drawing will be zoomed to its limits.

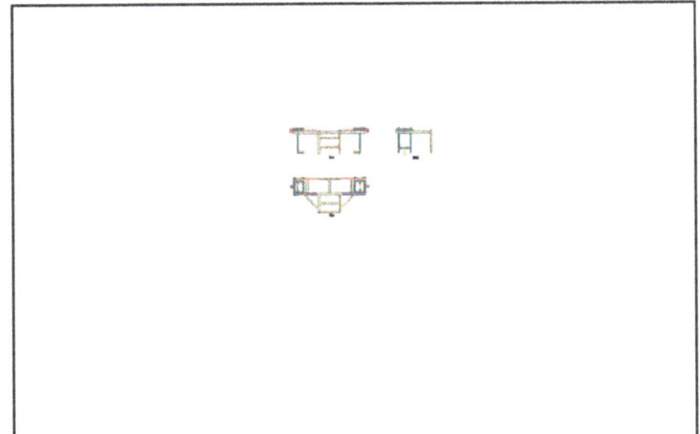

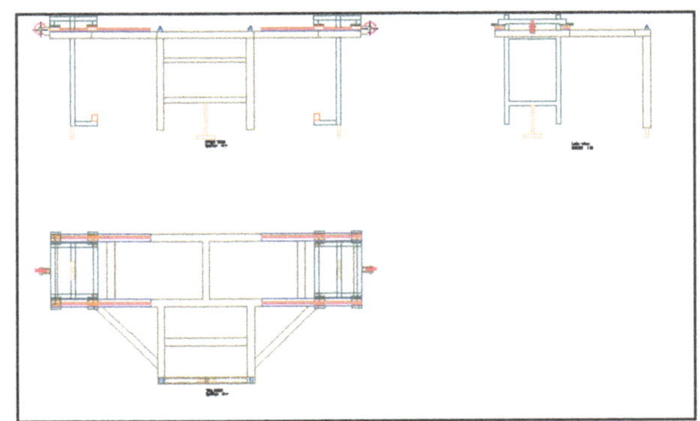

Using Zoom Dynamic

With the **Zoom Dynamic** tool, you can zoom to a particular portion of a drawing by using a viewing box.

- Click **Zoom Dynamic** on the **Navigation Bar**; the drawing will be zoomed to its limits. Also, a viewing box is attached to the pointer.

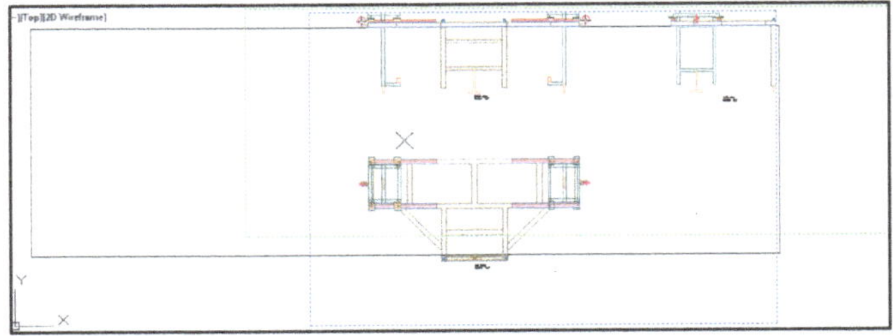

- Click and drag the pointer to define the size of the viewing box.
- Left-click and move the pointer to the area to be zoomed.

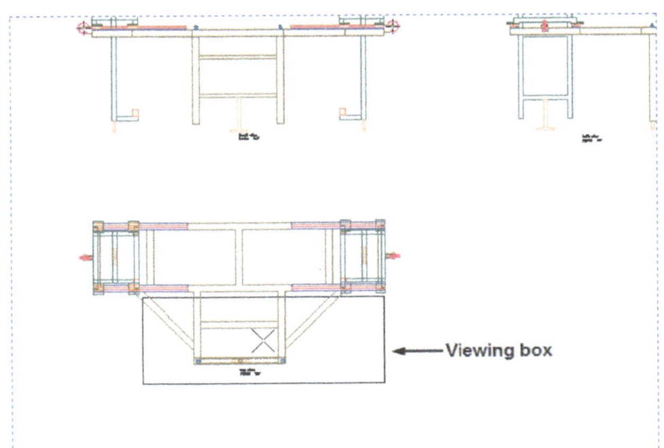

- Click the right mouse button. The area covered by the viewing box is magnified.

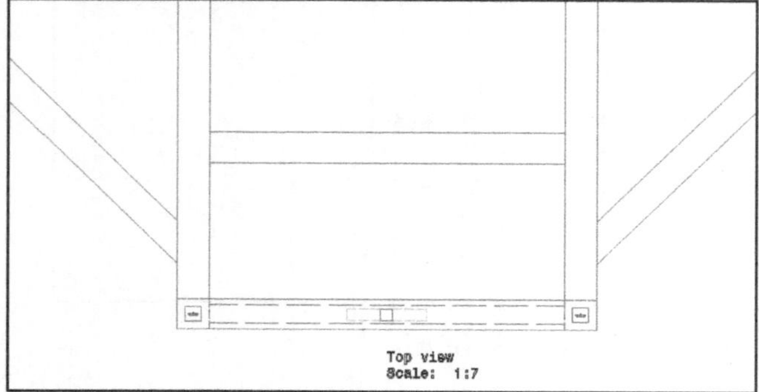

Using Zoom-Scale

Using the **Zoom-Scale** tool, you can zoom in or zoom out of a drawing by entering zoom scale factors directly from your keyboard.

- Click the **Zoom > Zoom Scale** on the Navigation Bar. The message, "**Enter a scale factor (nX or nXP),**" appears in the command line.
- Enter the scale factor 0.25 to scale the drawing to 25% of the full view.
- Enter the scale factor 0.25X to scale the drawing to 25% of the current view.
- Enter the scale factor 0.25XP to scale the drawing to 25% of the paper space.

Using Zoom-Center

Using the **Zoom Center** tool, you can zoom to an area of the drawing based on a center point and magnification value.

- Click **Zoom > Zoom Center** on the Navigation Bar; the message, "**Specify Center point**," appears in the command line.
- Select a point in the drawing to which you want to zoom in; the message, "**Enter magnification or height**," appears in the command line.

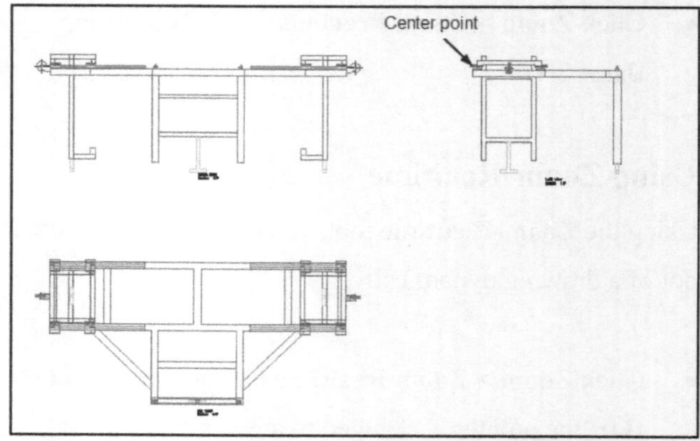

- Enter 10X in the command line to magnify the location of the point by ten times.

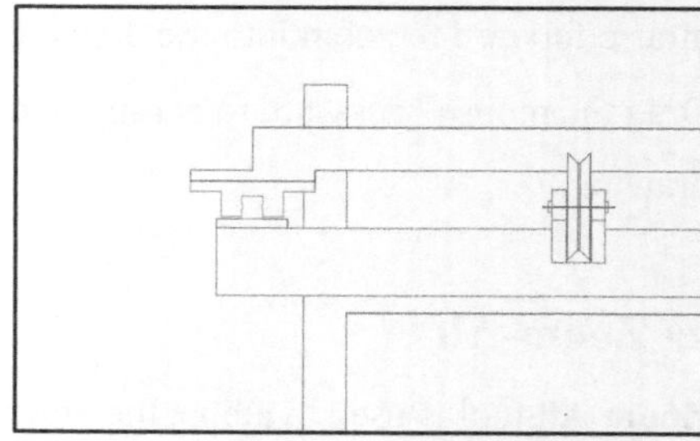

Using Zoom-Object

Using the **Zoom Object** tool, you can magnify a portion of the drawing by selecting one or more objects.

- Click **Zoom > Zoom Object** on the Navigation Bar.
- Select one or more objects from the drawing and press ENTER; the objects will be magnified.

Using Zoom-In

Using the **Zoom In** tool, you can magnify the drawing by a scale factor of 2.

- Click **Zoom > Zoom-In** on the Navigation Bar; the drawing is magnified to double.

Using Zoom-Out

The **Zoom-out** tool is used to de-magnify the display screen by a scale factor of 0.5.

Panning Drawings

After zooming into a drawing, you may want to view an area that is outside the current display. You can do this by using the **Pan** tool.

- Click **Pan** on the Navigation Bar.

- Press and hold the left mouse button and drag the mouse; a new area of the drawing, which is outside the current view, is displayed.

Exercises

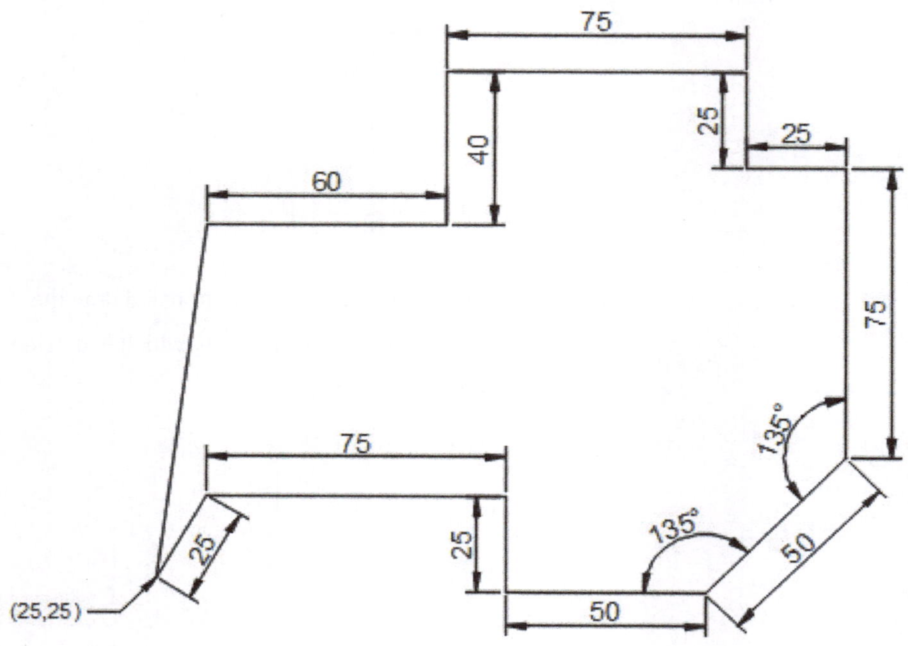
75
25
40
25
60
75
135°
75
135°
25
25
50
50
(25,25)

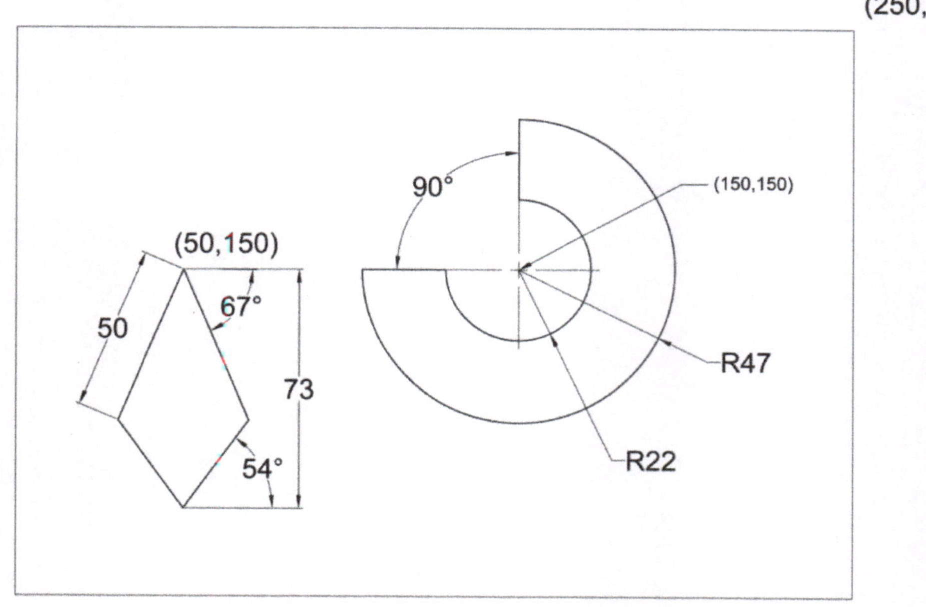
(250,175)
90°
(150,150)
(50,150)
67°
50
73
R47
54°
R22
(0,0)

Chapter 4: Editing Tools

In this chapter, you will learn the following tools:

- The **Move** tool
- The **Copy** tool
- The **Rotate** tool
- The **Scale** tool
- The **Trim** tool
- The **Extend** tool
- The **Fillet** tool
- The **Chamfer** tool
- The **Mirror** tool
- The **Explode** tool
- The **Stretch** tool
- The **Polar Array** tool
- The **Offset** tool
- The **Path Array** tool
- The **Rectangular Array** tool

Editing Tools

In previous chapters, you have learned to create some simple drawings using the basic drawing tools. However, to create complex drawings, you may need to perform various editing operations. The tools to perform the editing operations are available in the **Modify** panel on the **Home** ribbon. You can click the down arrow on this panel to find more editing tools. Using these editing tools, you can modify existing objects or use existing objects to create new or similar objects.

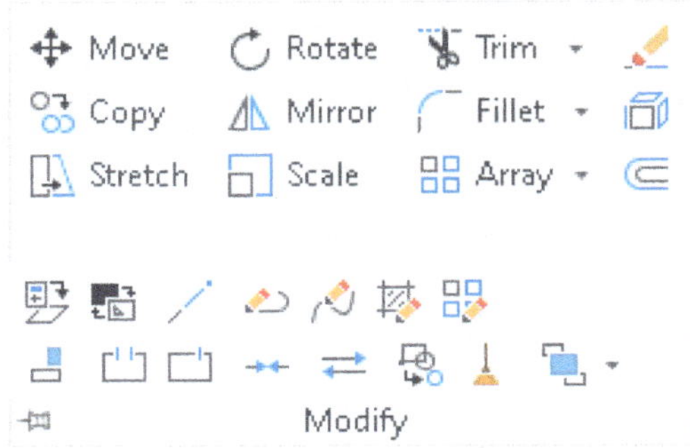

The Move tool

The **Move** tool moves a selected object(s) from one location to a new location without changing its orientation. To move objects, you must activate this tool and select the objects from the graphics window. After selecting the objects, you must define the 'base point' and the 'destination point.'

Example:

- Create the drawing, as shown below.

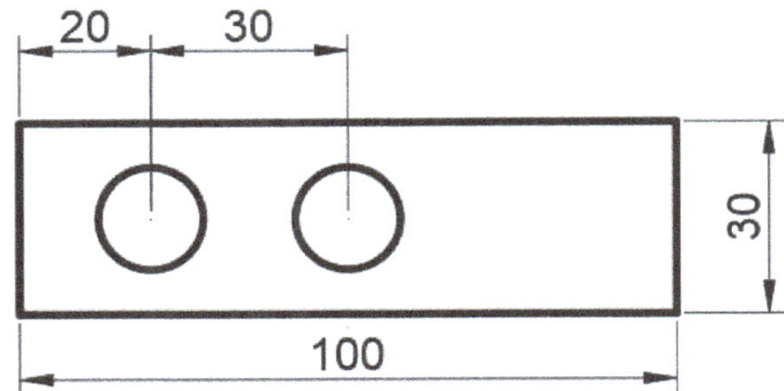

- Click **Home > Modify > Move** on the ribbon or enter **M** in the command line.
- Click on the circle located at the right side, and then right-click to accept the selection.

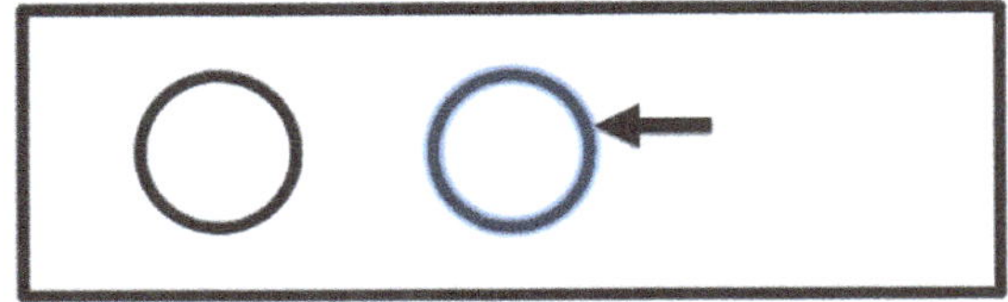

- Select the center of the circle as the base point.

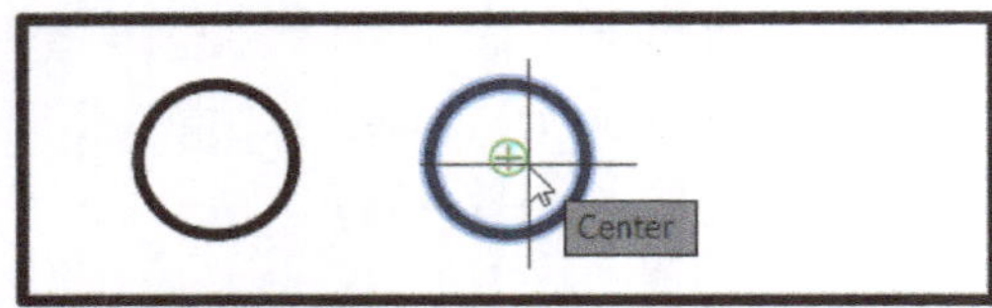

- Make sure that the **Ortho Mode** is activated.
- Move the pointer toward the right, type 30, and then press Enter. It moves the circle to the new location.

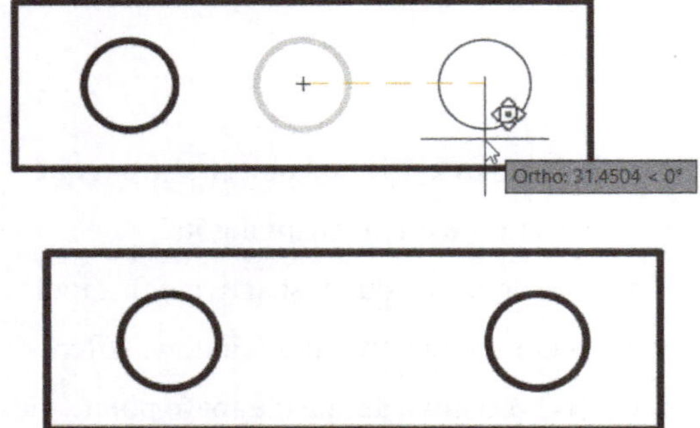

The Copy tool

The **Copy** tool is used to copy objects and place them at a required location. This tool is similar to the **Move** tool, except that the object will remain at its original position, and a copy of it will be placed at the new location.

Example:

- Draw two circles of 40 mm and 70 mm radii, respectively.

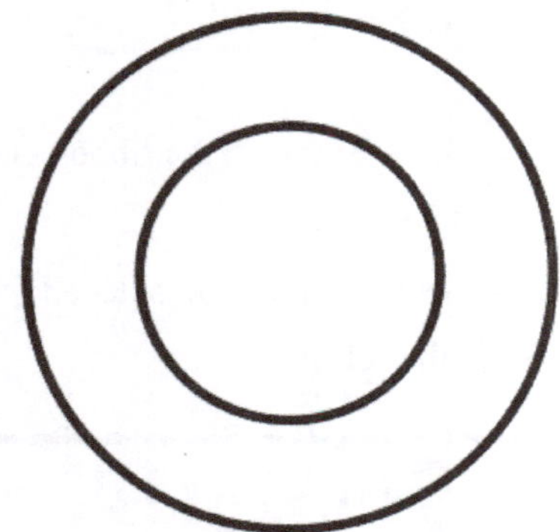

- Click **Home > Modify > Copy** on the ribbon or enter **CO** in the command line.
- Select the two circles and then right-click to accept the selection.
- Select the center of the circle as the base point.

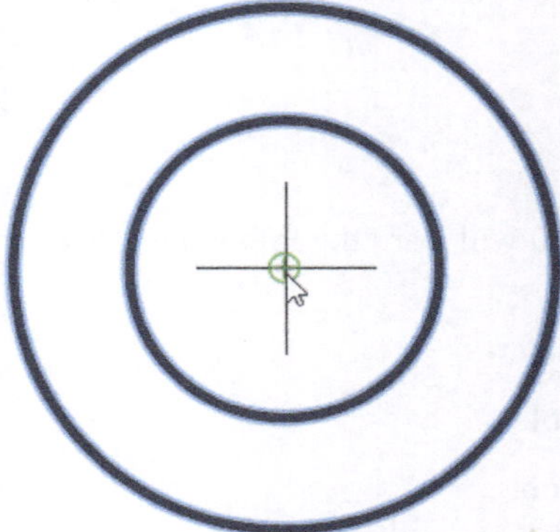

- Make sure that the **Ortho Mode** is active.
- Move the pointer toward the right.
- Type 200 and press ENTER. This action creates a copy of the circles at the new location

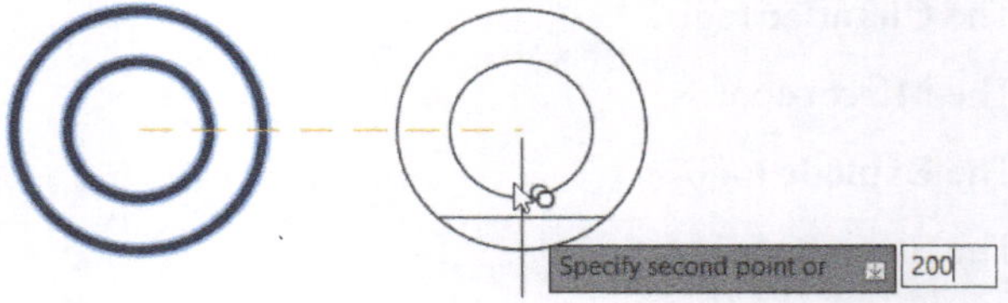

- Select **Exit** from the command line.

The Rotate tool

The **Rotate** tool rotates an object or a group of objects about a base point. Activate this tool and select the objects from the graphics window. After selecting objects, you must define the 'base point' and the angle of rotation. It rotates the object(s) about the base point.

- Click **Home > Modify > Rotate** on the ribbon or enter **RO** in the command line.

- Select the circles as shown below, and then right-click to accept.

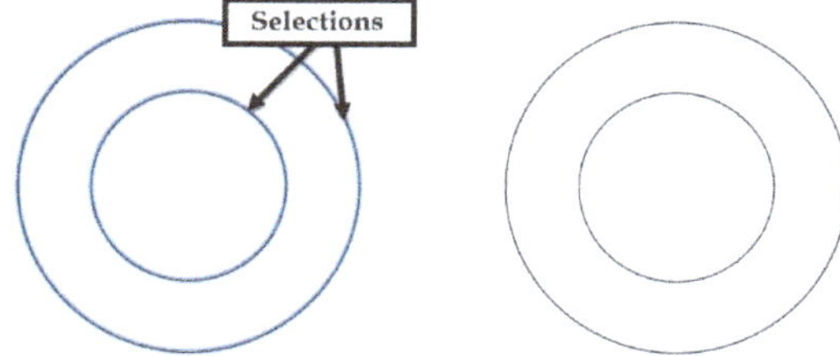

- Select the center of the other circle as the base point.

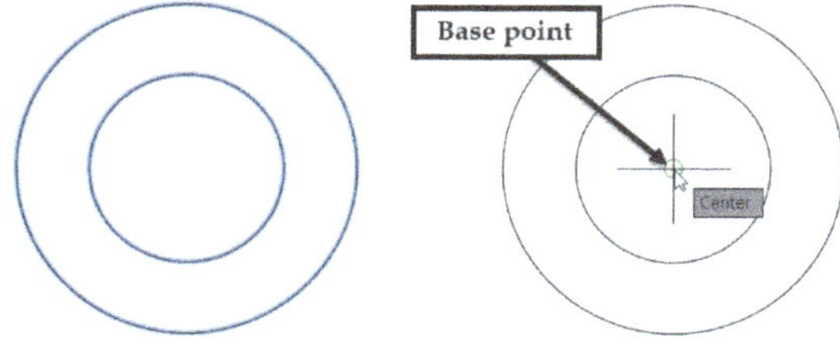

- Select the **Copy** option from the command line.
- Type -90 as the rotation angle and press ENTER; the selected objects are rotated by 90 degrees.

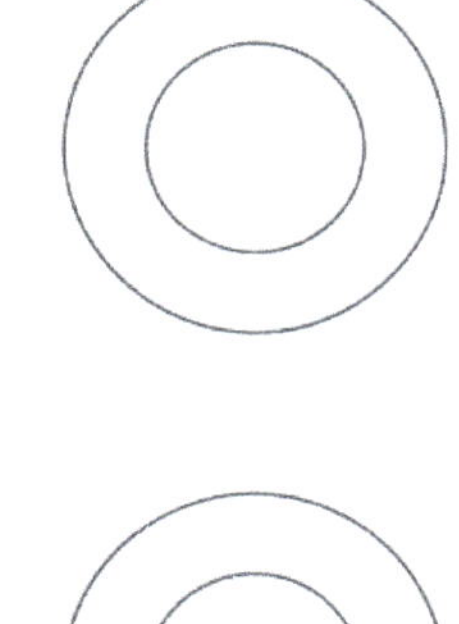

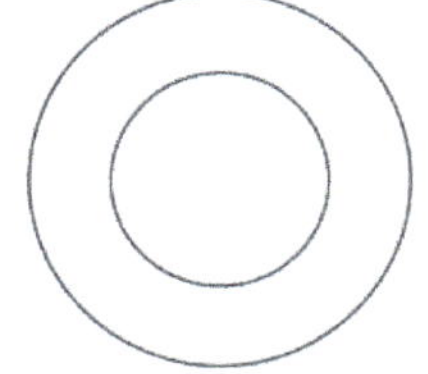

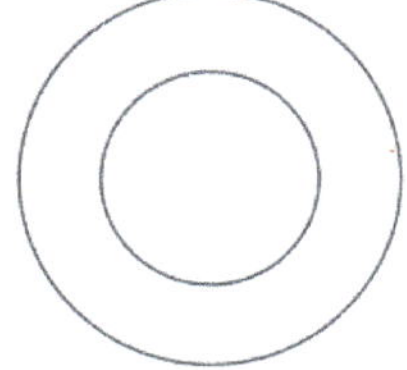

The Scale tool

The **Scale** tool changes the size of objects. It reduces or enlarges the size without changing the shape of an object.

- Click **Home > Modify > Scale** on the ribbon or enter **SC** in the command line.
- Select the circles as shown below and right-click to accept the selection.

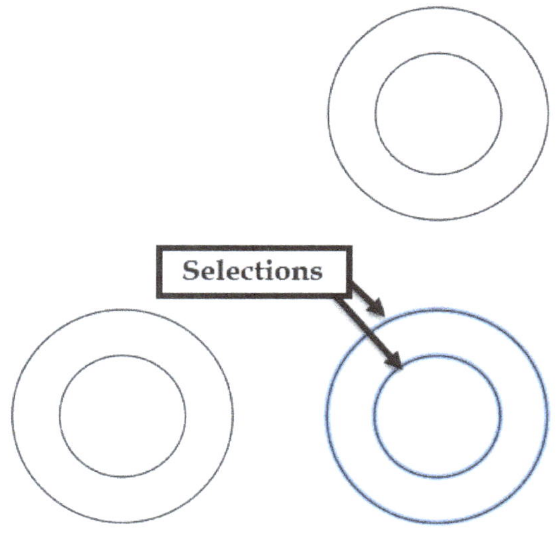

- Select the center point of the selected circles as the base point.
- Type 0.8 as the scale factor and press ENTER.

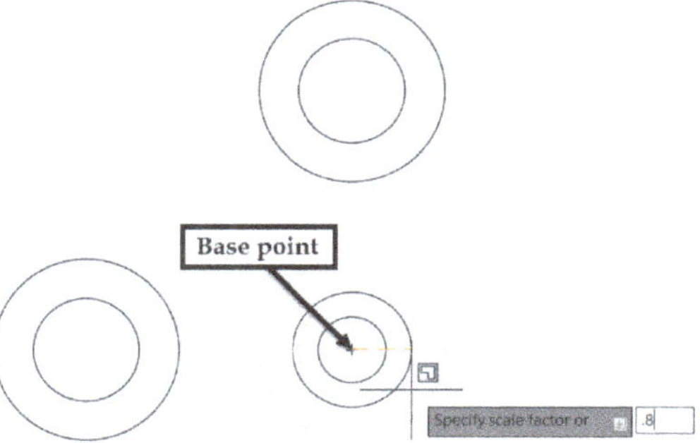

- Likewise, scale the circles located at the top to 0.7.

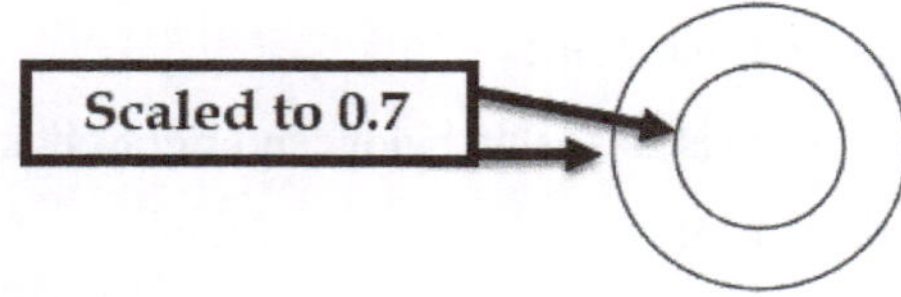

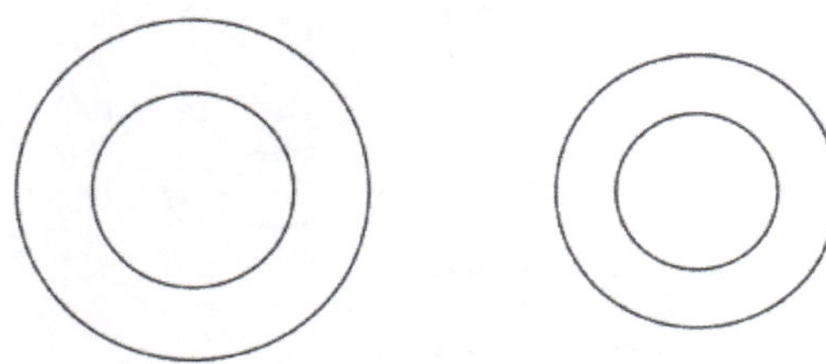

- Click **Home > Draw > Circle > Tan, Tan, Radius** on the ribbon.
- Select the two circles shown below to define the tangent points.

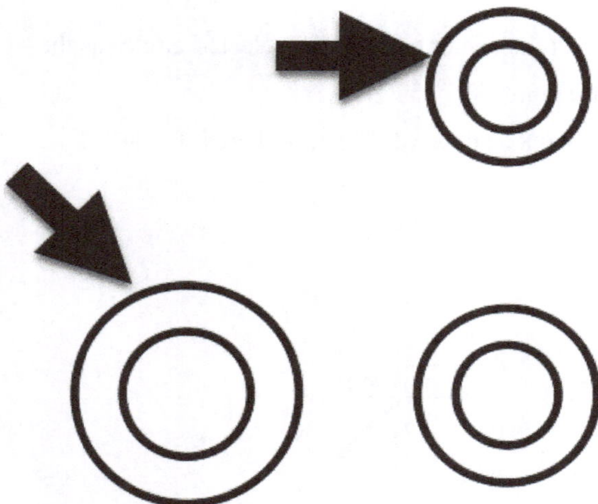

- Type 150 as the radius of the circle and press ENTER.

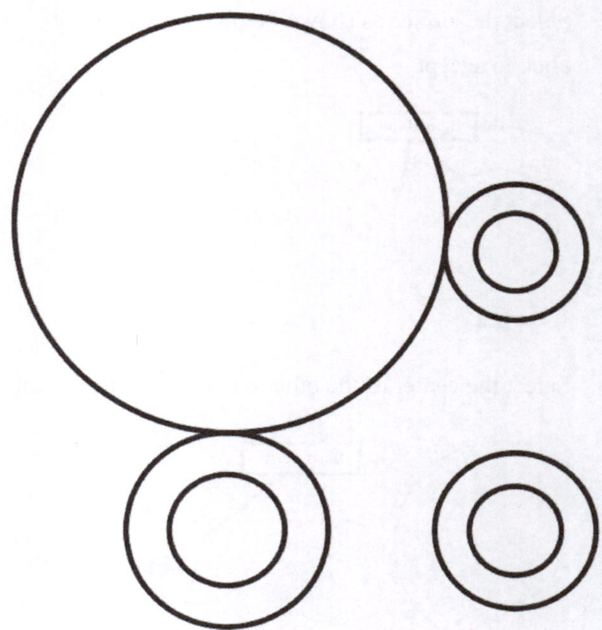

- Likewise, create other circles of radius 100 and 120.

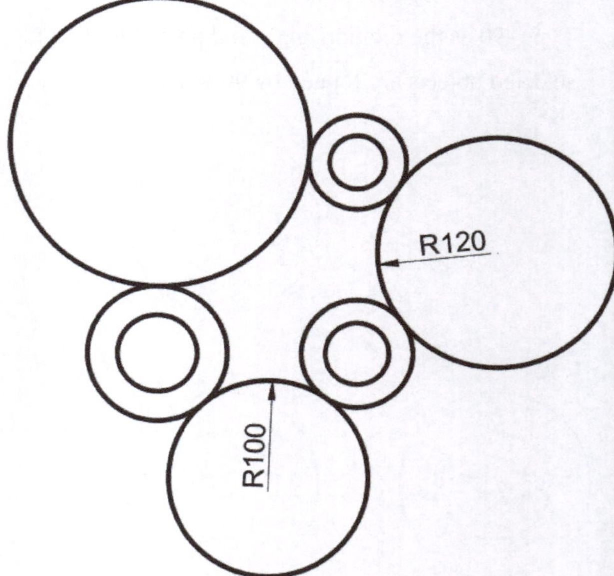

The Cut with Base Point Tool

The **Cut With Base Point** tool allows you to cut and paste the selected objects using a base point. You can paste the cut objects within a drawing or in another drawing.

- Click **Home > Clipboard > Cut** drop-down **> Cut With Base Point** on the ribbon.
- Select the base point of the selection.

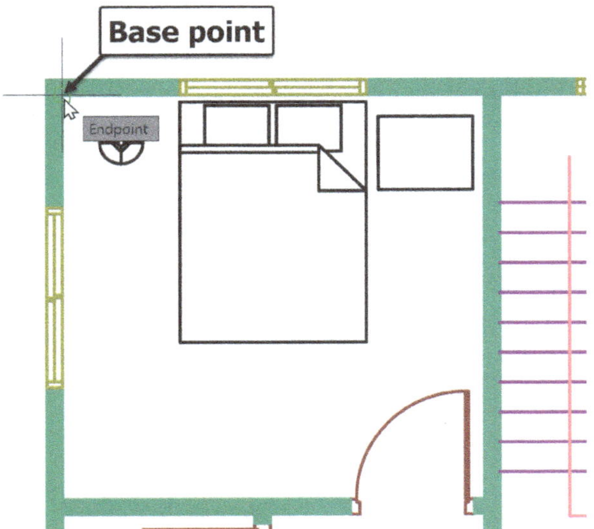

- Create a selection window across all the objects tob be moved. Next, press ENTER.

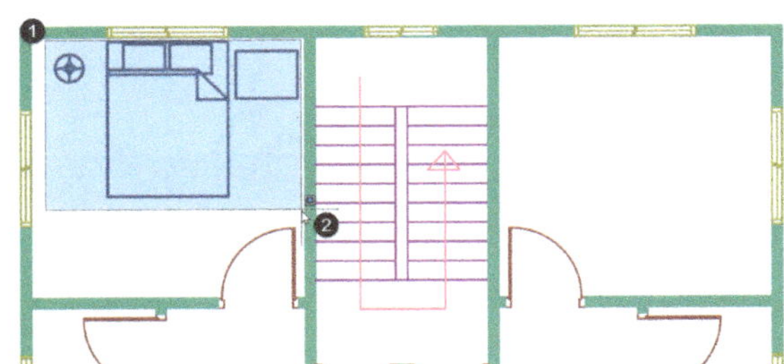

- Click **Home > Clipboard > Paste** on the ribbon.
- Select the destination point.

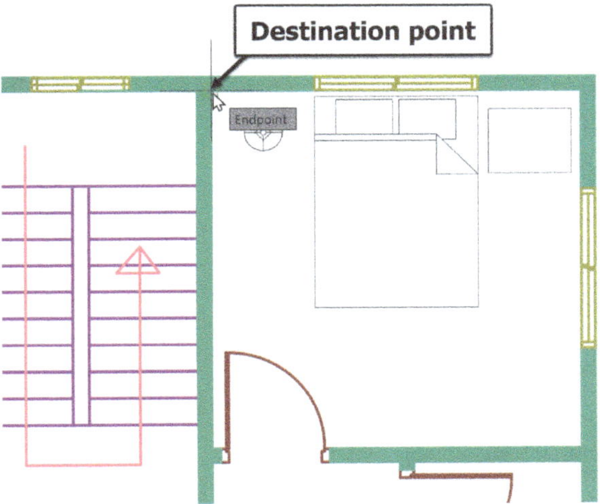

The Trim tool

When an object intersects with another object, you can remove its unwanted portion by using the **Trim** tool. To trim an object, you must first activate the **Trim** tool, and then select the elements to be trimmed. You can also select the elements to trim by dragging the cursor across them.

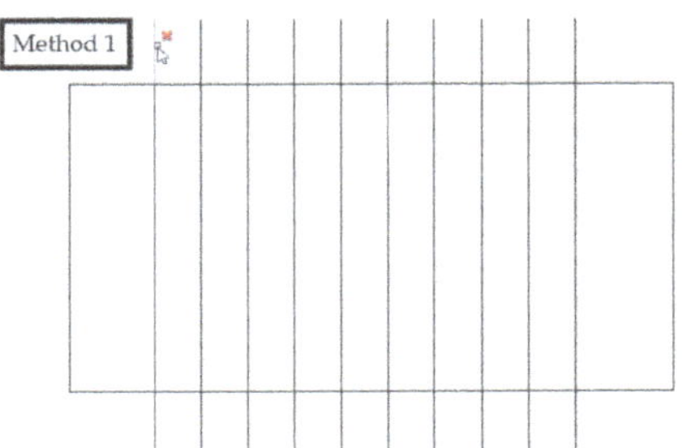

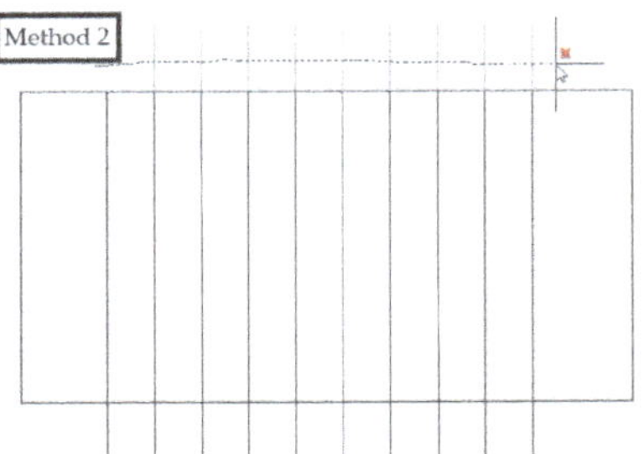

Click and move the pointer across the elements to trimmed. Next, click again to select them.

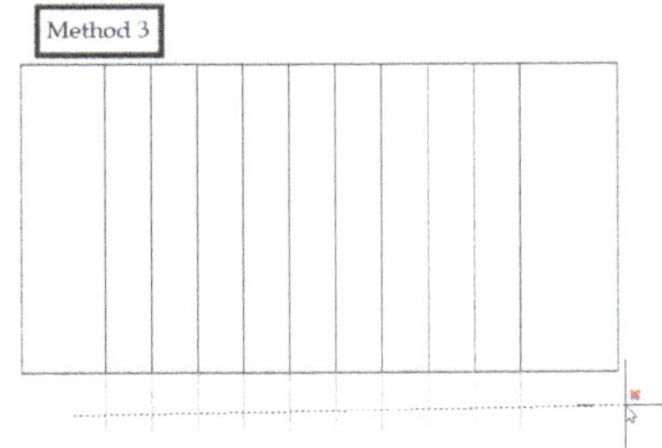

However, if you want to trim the elements using an intersecting object, then select **cuTting edges** option from the command line and select the cutting edges (intersecting objects). Next, press ENTER and select the portions to be removed.

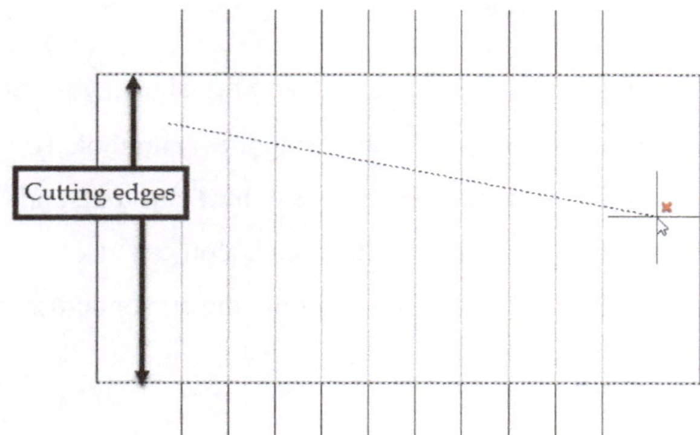

- Click **Home > Modify > Trim** on the ribbon or enter **TR** in the command line.

  Now, you must select the objects to be trimmed.
- Select the large circles one by one; the circles will be trimmed.

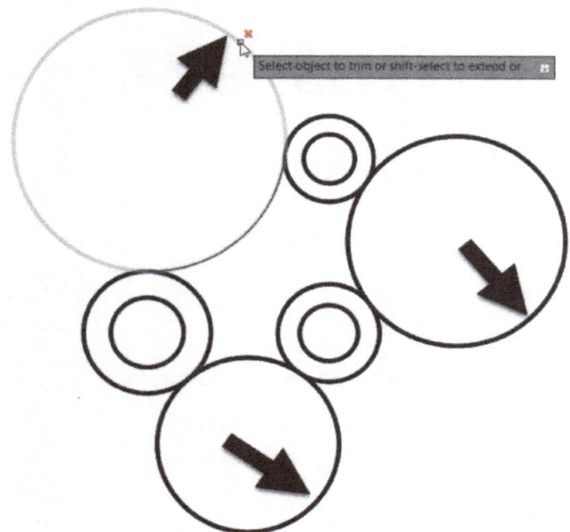

- Likewise, trim the other circles, as shown below.

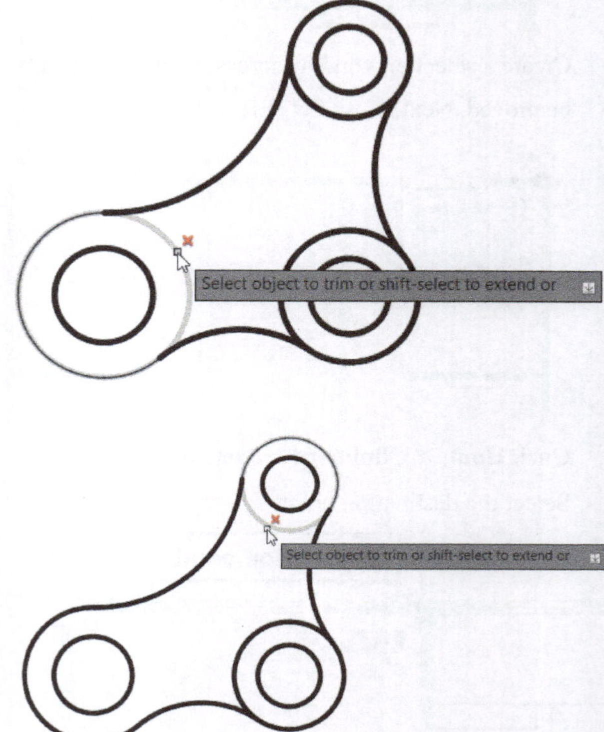

- Save and close the drawing.

The Extend tool

The **Extend** tool is similar to the **Trim** tool, but its use is the opposite of it. This tool is used to extend lines, arcs, and other open entities to connect to other objects.

- Start a new drawing.
- Create a sketch, as shown below, using the **Line** tool.

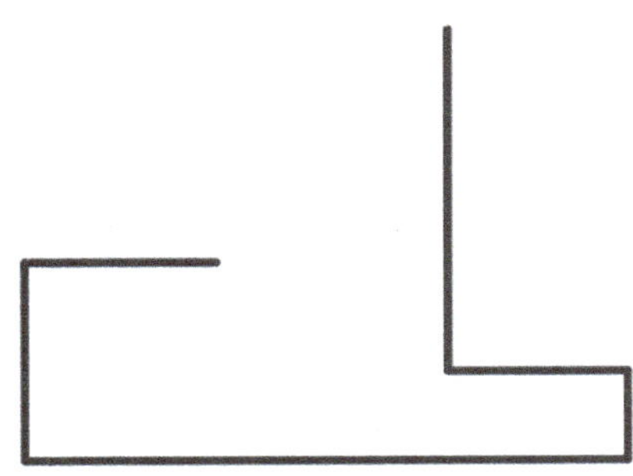

- Click **Home > Modify > Trim > Extend** on the ribbon or enter **EX** in the command line.
- Select the horizontal open line. It will extend the line up to the next element.

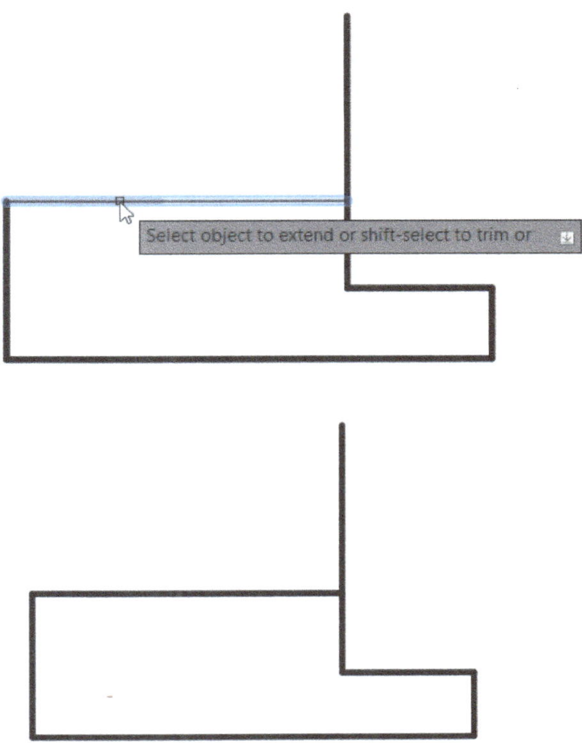

The Fillet tool

The **Fillet** tool converts the sharp corners into round corners. You must define the radius and select the objects forming a corner. The following figure shows some examples of rounding the corners.

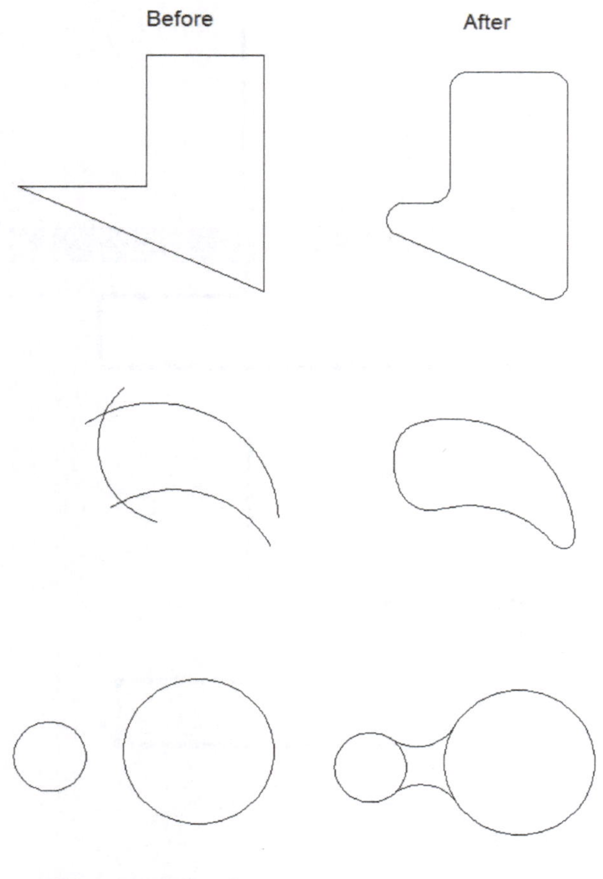

- Start a new drawing.
- Type **Limmax** in the command line and press ENTER.
- Set the maximum limit to 100,100 and press ENTER.
- Click **Zoom All** on the Navigation Bar.
- Click **Home > Draw > Polyline** on the ribbon.
- Define the start point as 20, 50.
- Draw the lines, as shown below.

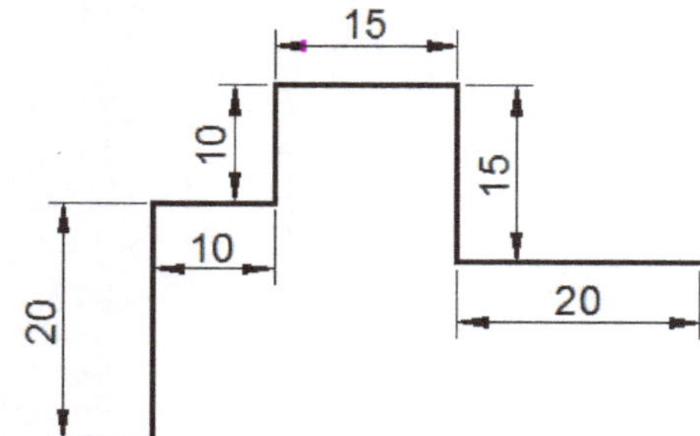

- Right-click and select **Enter**.
- Click **Home > Modify > Fillet** on the ribbon or enter **F** in the command line.
- Select the **Radius** option from the command line.
- Type **5** and press ENTER.
- Select the vertical and horizontal lines, as shown below.

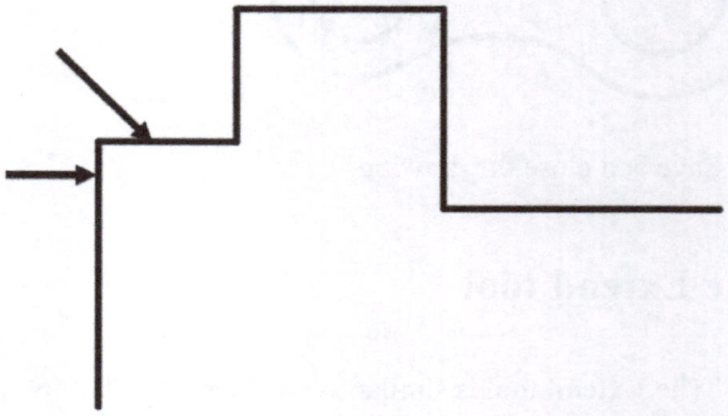

- Notice that a fillet is created.

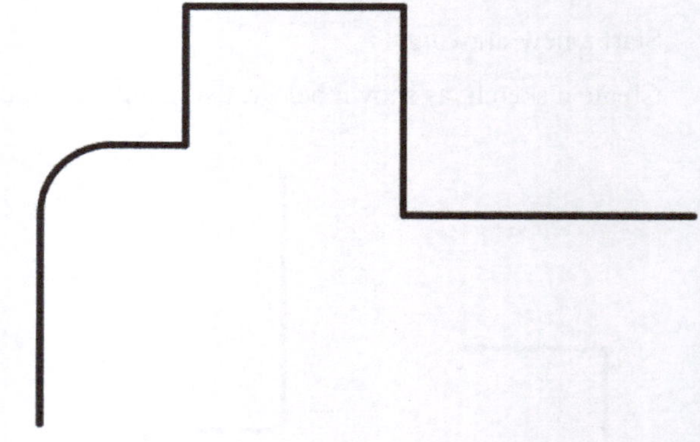

The Chamfer tool

The **Chamfer** tool replaces the sharp corners with an angled line. This tool is similar to the **Fillet** tool, except that an angled line is placed at the corners instead of rounds.

- Click **Home > Modify > Fillet > Chamfer** on the ribbon or enter **CHA** in the command line.
- Follow the prompt sequence given next:

Select first line or [Undo Polyline Distance Angle Trim mEthod Multiple]: Select the **Distance** option from the command line.

Define first chamfer distance <0.0000>: Enter **8** as the first chamfer distance and press ENTER.

Define second chamfer distance <8.0000>: Press ENTER to accept 8 as the second chamfer distance.

Select first line or [Undo Polyline Distance Angle Trim mEthod Multiple]: Select the vertical line on the right-side.

Select second line or shift-select to apply corner or [Distance Angle Method]: Select the horizontal line connected to the vertical line.

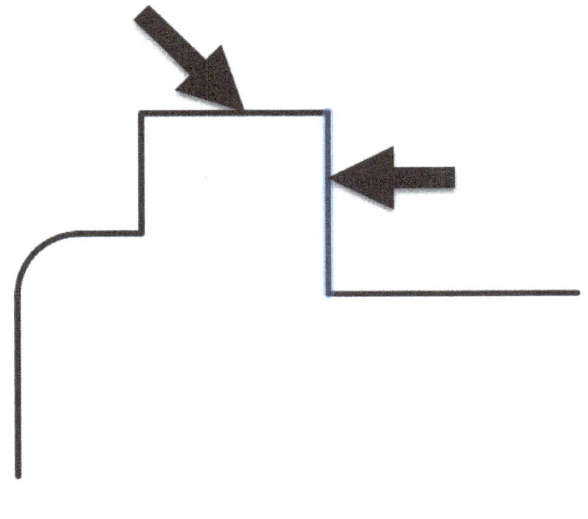

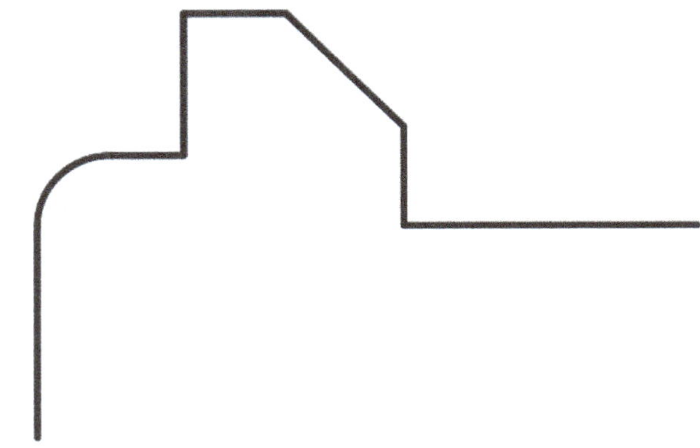

The Mirror tool

The **Mirror** tool creates a mirror image of objects. You can create symmetrical drawings using this tool. Activate this tool and select the objects to mirror, and then define the 'mirror line' about which the objects will be mirrored. You can define the mirror line by either creating a line or selecting an existing line.

- Click **Home > Modify > Mirror** on the ribbon or enter **MI** in the command line.
- Select the drawing by clicking on it, and then press Enter.

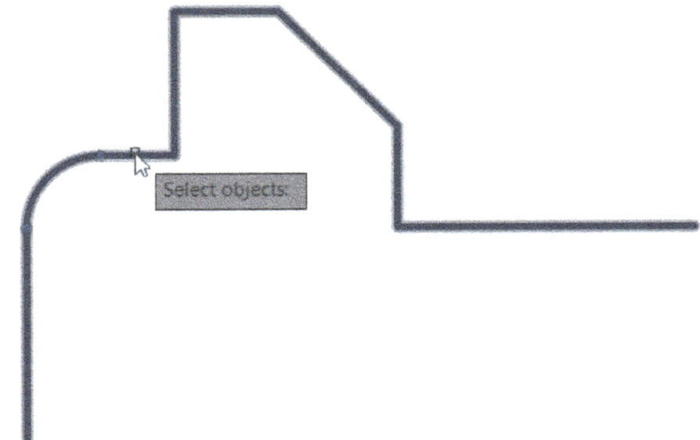

- Select the first point of the mirror line, as shown below.

- Make sure that the **Ortho Mode** on the status bar is active.
- Move the pointer toward the right and click.

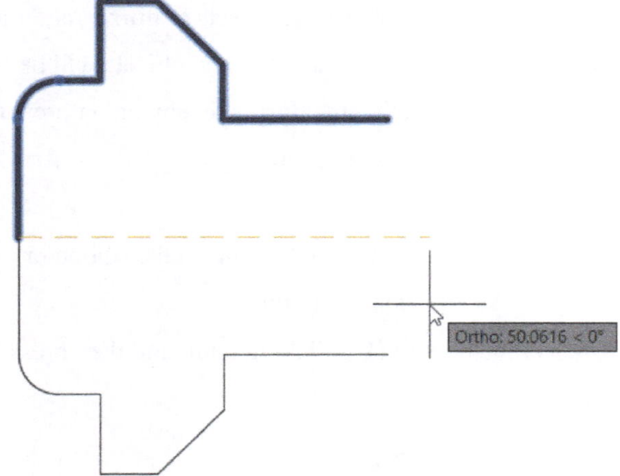

- Select the **No** option from the command line to retain the source objects.

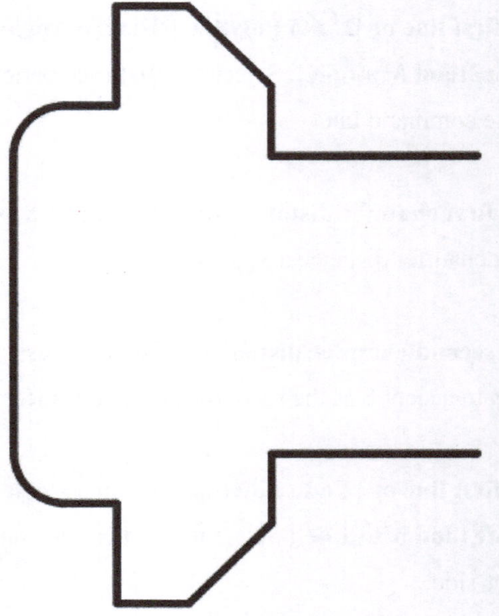

- Click **Home > Draw > Arc > Start, End, Direction** on the ribbon.
- Select the start point of the arc, as shown.
- Select the endpoint of the arc, as shown.

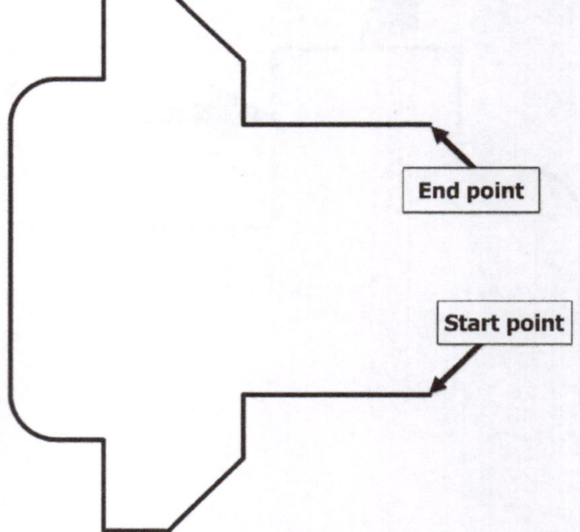

- Make sure that the **Ortho Mode** is active.
- Move the pointer toward the right and click.

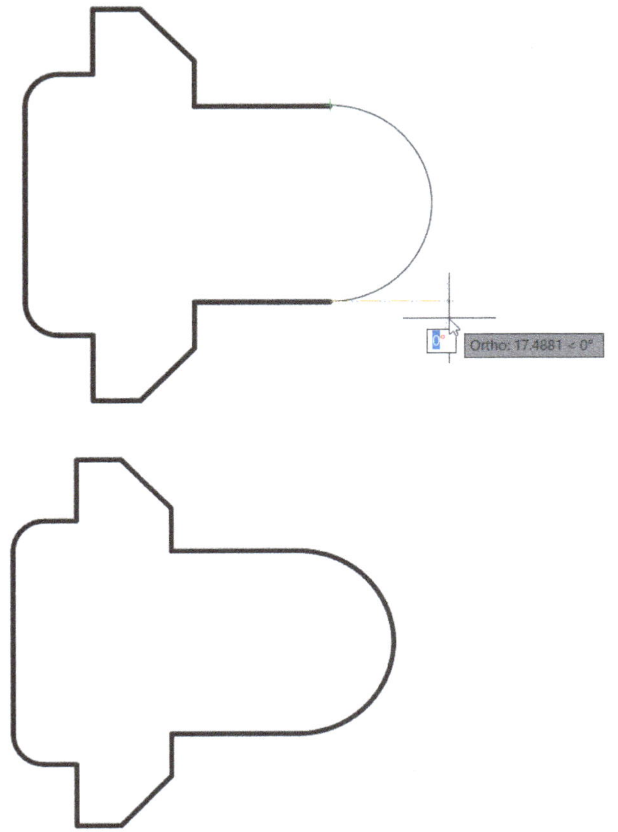

The Explode tool

The **Explode** tool explodes a group of objects into individual objects. For example, when you create a drawing using the **Polyline** tool, it acts as a single object. You can explode a polyline or rectangle or any group of objects using the **Explode** tool.

- Click on the portion of the drawing created using the **Polyline** tool; you will notice that the complete polyline is selected as a single object.

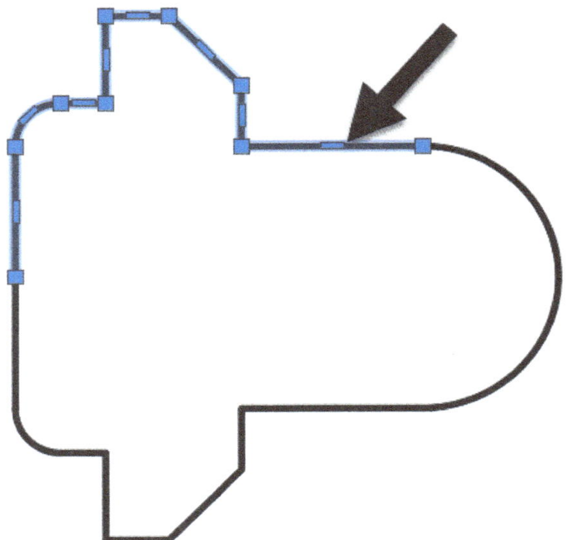

- Click **Home > Modify > Explode** on the ribbon or enter **X** in the command line.
- Select the polylines from the drawing.

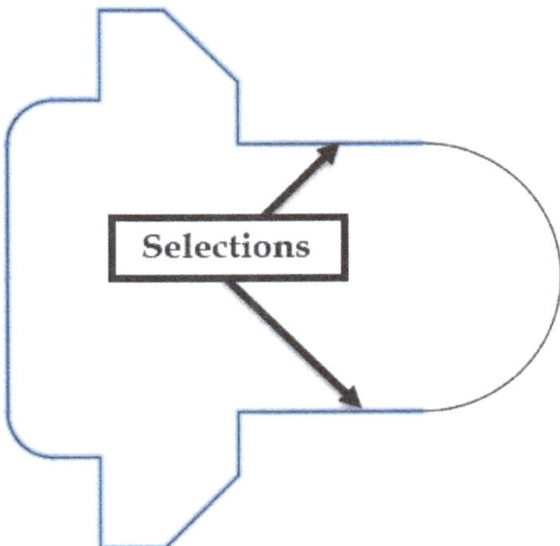

- Press ENTER; the polyline is exploded into individual objects.

Now, you can select the individual objects of the polyline.

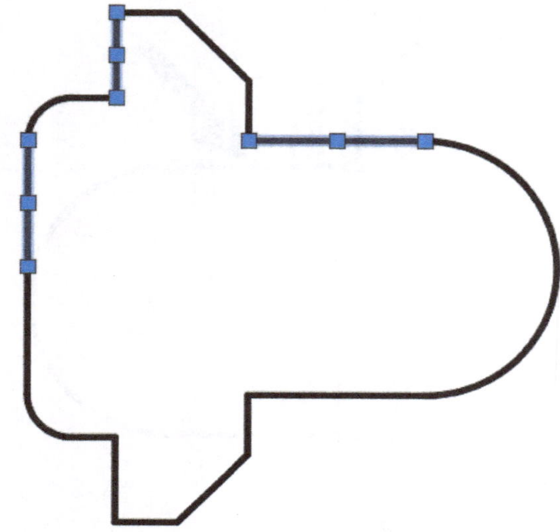

The Stretch tool

The **Stretch** tool lengthens or shortens drawings or parts of drawings. Note that you cannot stretch circles using this tool. Also, you must select the portion of the drawing to be stretched by dragging a window.

- Click **Home > Modify > Stretch** on the ribbon or enter **STRETCH** in the command line.
- Create a crossing window to select the objects of the drawing.

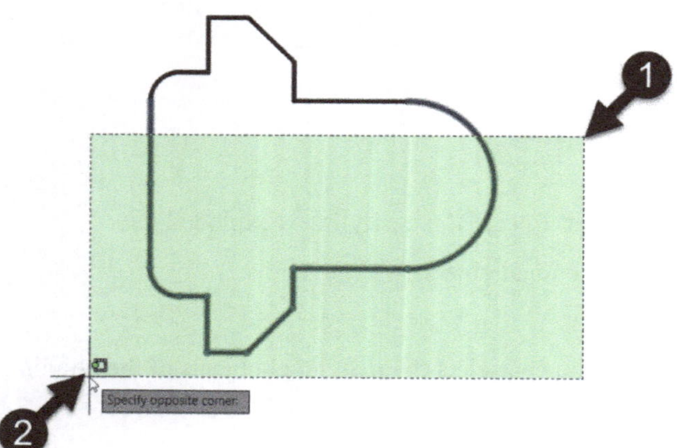

- Press ENTER (or) right-click to accept the selection.
- Select the base point, as shown below.

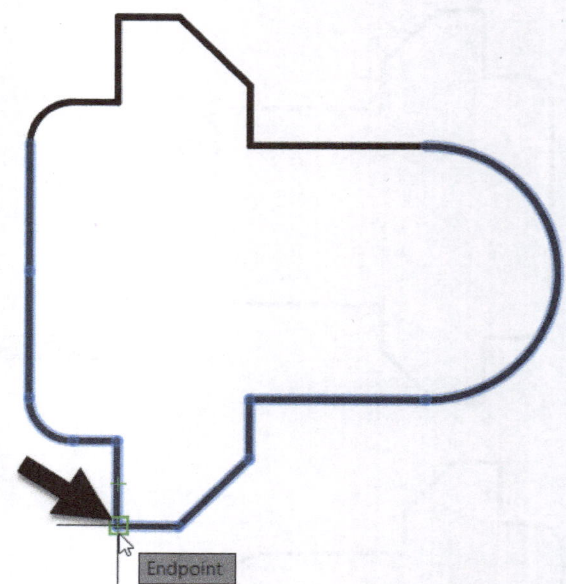

- Move the pointer downward and click to stretch the drawing.

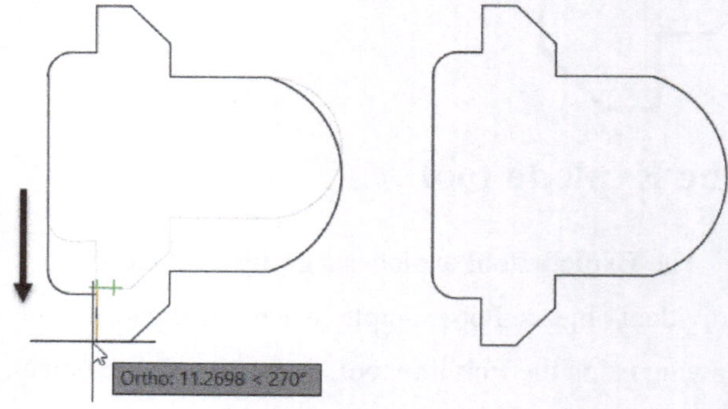

- Save and close the file.

The Polar Array tool

The **Polar Array** tool creates an arrangement of objects around a point in a circular form. The following example shows you to create a polar array.

- Create two concentric circles of 140 and 50 diameters.

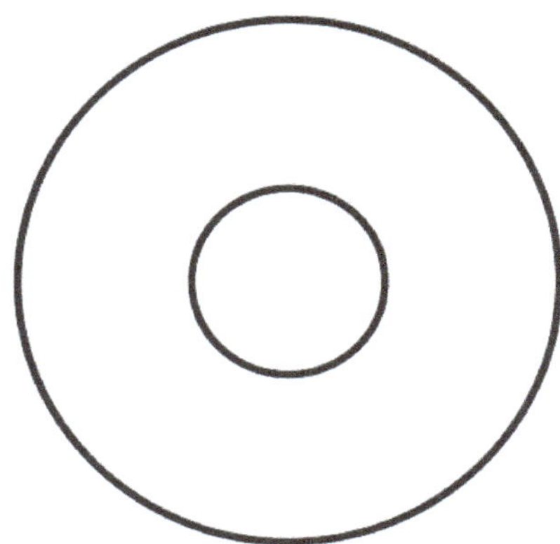

- Type **C** in the command line and press ENTER.
- Press and hold the Shift key, right-click, and select **Quadrant** from the shortcut menu.
- Select the quadrant point of the circle, as shown below.

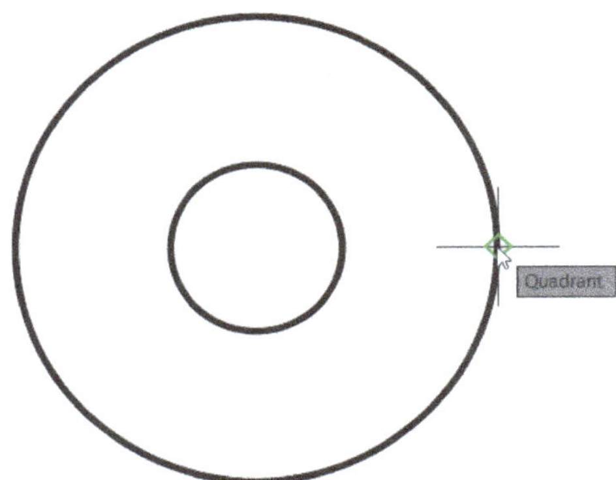

- Type 30 as radius and press ENTER.

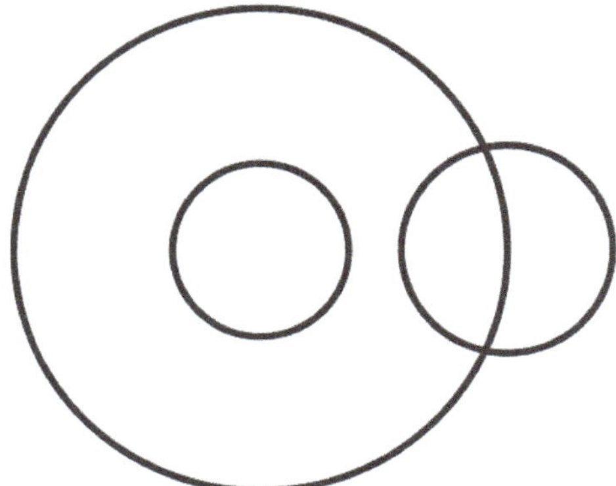

- Click **Home > Modify > Trim** on the ribbon.
- Select the circle on the quadrant as the object to be trimmed.

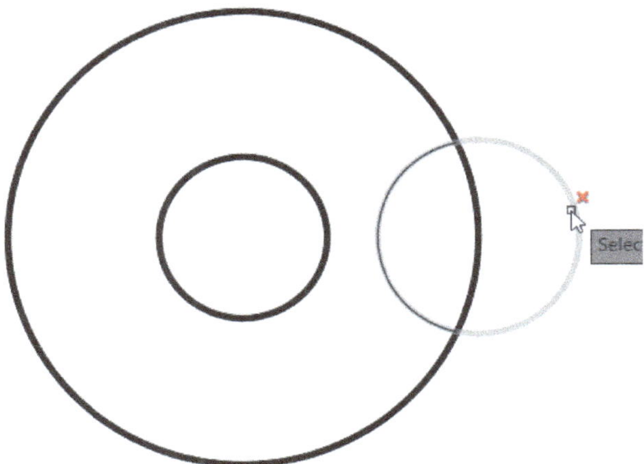

- Press Enter.
- Click **Home > Modify > Array > Polar Array** on the ribbon or **ARRAYPOLAR** in the command line.
- Select the arc created after trimming the circle. Next, right-click to accept the selection.

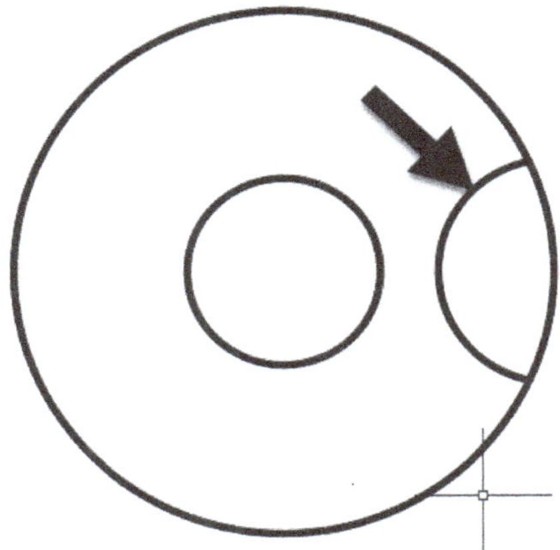

- Make sure that **Object Snap** is activated.
- Select the center of the large circle as the center of the array; the **Array Creation** tab appears in the ribbon.
- In the **Items** panel of the **Array Creation** tab, set the **Items** value to 4.

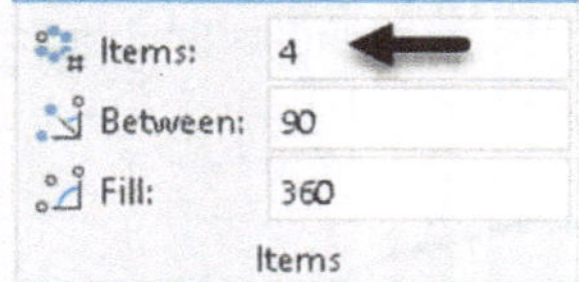

Notice that the **Rotate Items** button is active in the **Properties** panel of the **Array Creation** tab. It rotates the objects of the polar array. If you deactivate this button, the polar array is created without rotating the objects, as shown in the figure.

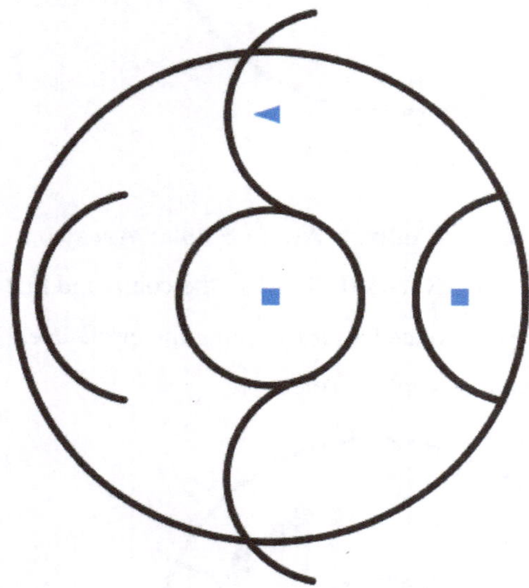

In addition, the **Associative** button is active by default. It ensures that you can edit the array after creating it.

- Make sure that the **Associative** and the **Rotate Items** buttons are active. Next, click the **Close Array** button on the ribbon.

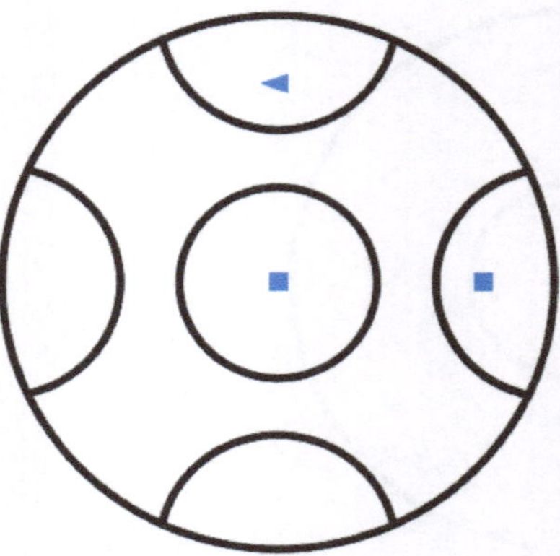

- Click the **Trim** button on the **Modify** panel.
- Trim the unwanted portions, as shown below.

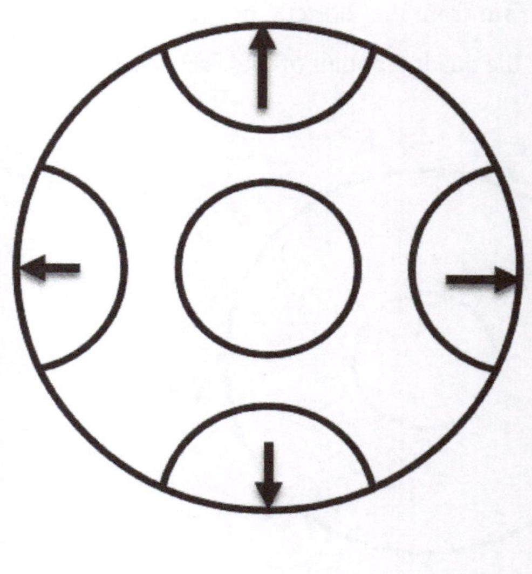

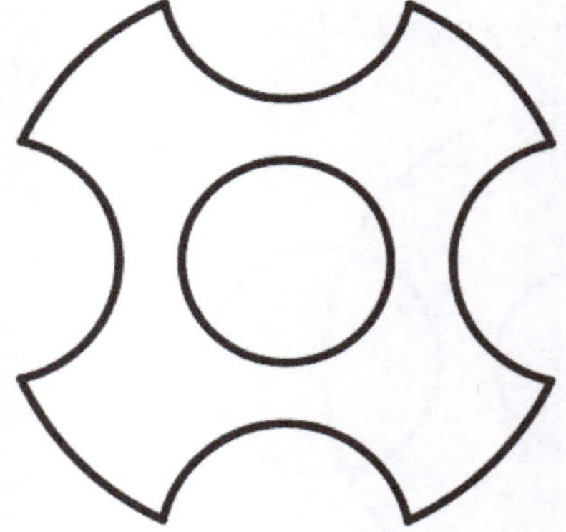

The Offset tool

The **Offset** tool creates parallel copies of lines, polylines, circles, and arcs. To create a parallel copy of an object, first, you must define the offset distance and then select the object. Next, you must define the side in which the parallel copy will be placed.

- Create the drawing shown below using the **Polyline** tool. Do not add dimensions.

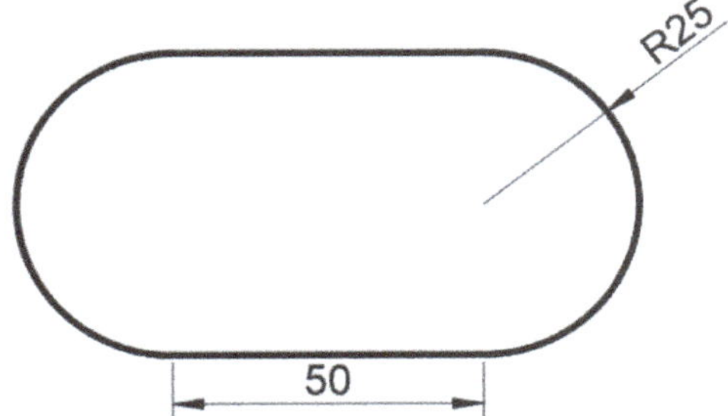

- Click **Home > Modify > Offset** on the ribbon or enter **O** in the command line.
- Type **20** as the offset distance and press ENTER.
- Select the polyline loop.
- Click outside the loop to create a parallel copy.

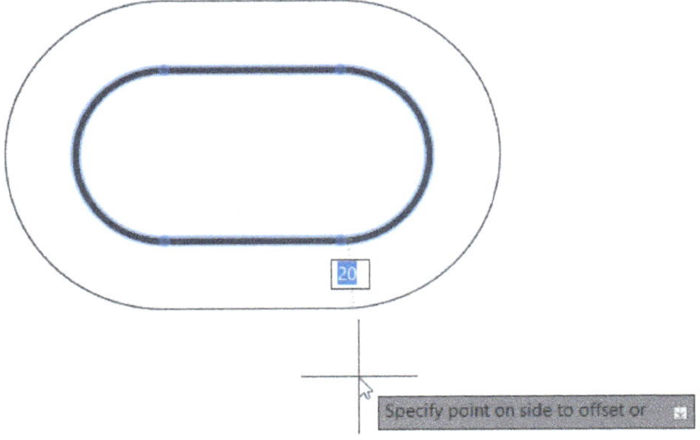

- Select **Exit** from the command line.
- Click **Home > Layers > Layer Properties** on the ribbon (or) type **LA** in the command line; the **Layer Properties Manager** appears.
- Click the **New layer** button on the **Layer Properties Manager**. Enter **Centerline** in the **Name** field.

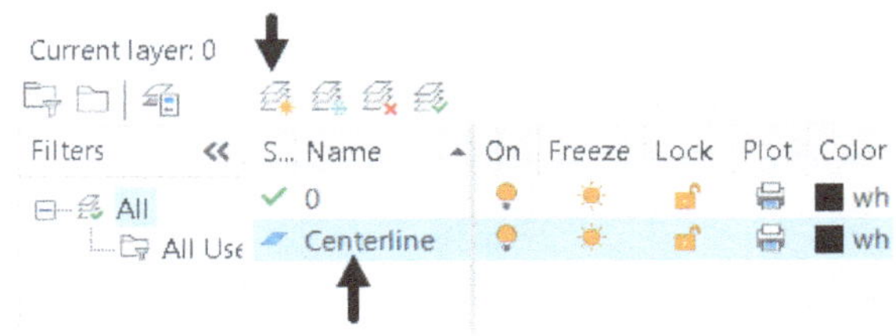

- Click the **Set current** icon. It activates the new layer.
- Click in the **Linetype** field of the current layer; the **Select Linetype** dialog appears.

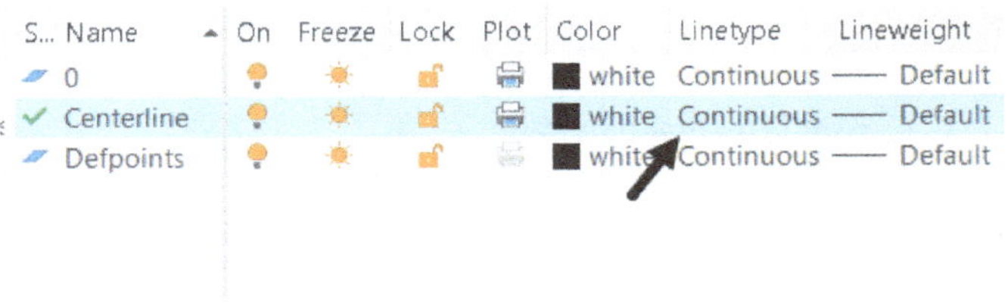

- On the **Select Linetype** dialog, click the **Load** button; the **Load or Reload Linetypes** dialog appears.

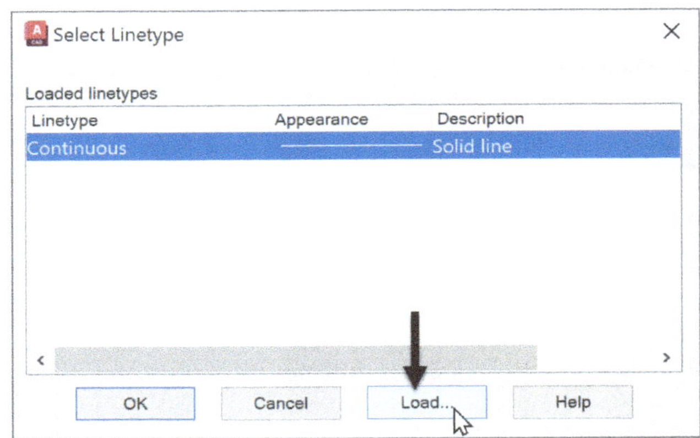

- Select the **CENTER2** Linetype from this dialog. Click **OK**. It adds the linetype to the **Select Linetype** dialog.

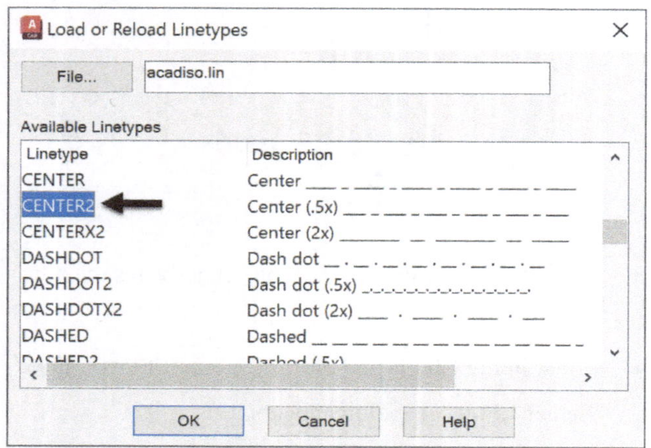

- Select the **CENTER2** linetype from the **Select Linetype** dialog and click **OK**.
- Close the **Layer Properties Manager**.
- Click the **Offset** button on the **Modify** panel.
- Select the **Layer** option from the command line.
- Select the **Current** option from the command line; this ensures that the offset entity will be created with the currently active layer properties. If you select the **Source** option, the offset entity will be created with the properties of the source object.
- Type **10** as the offset distance and press ENTER.
- Select the outer loop of the drawing.
- Move the pointer inwards and click to create the offset entity.
- Select **Exit** from the command line.

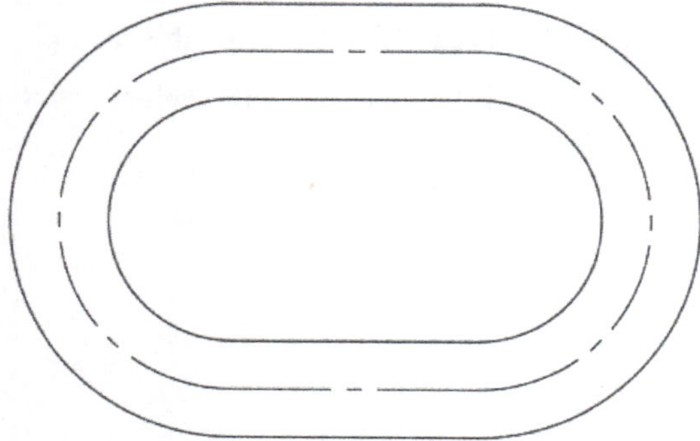

- Click on the **Layer** drop-down on the **Layer** panel of the ribbon.
- Select the **0** layer from the drop-down.

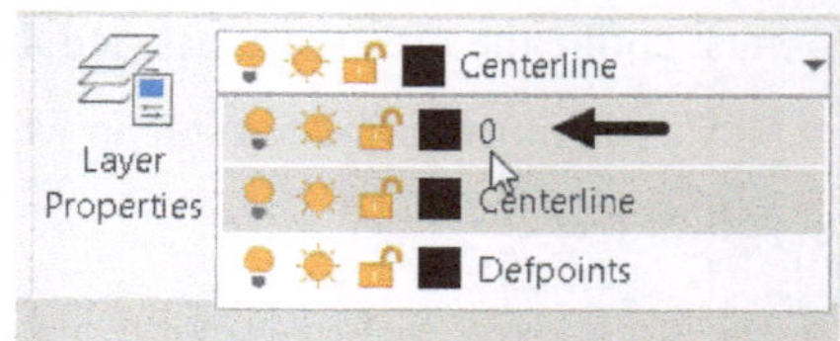

- Create a circle of 6 mm in radius, as shown.

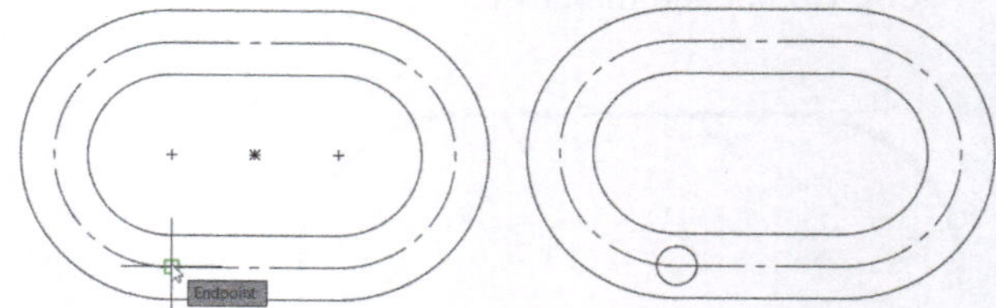

The Path Array tool

The **Path Array** tool creates an array of objects along a path (line, polyline, circle, helix, and spline).

- Click **Home > Modify > Array > Path Array** on the ribbon or enter **ARRAYPATH** in the command line.
- Select the circle and right-click.
- Select the centerline as the path; the preview of the path array appears.

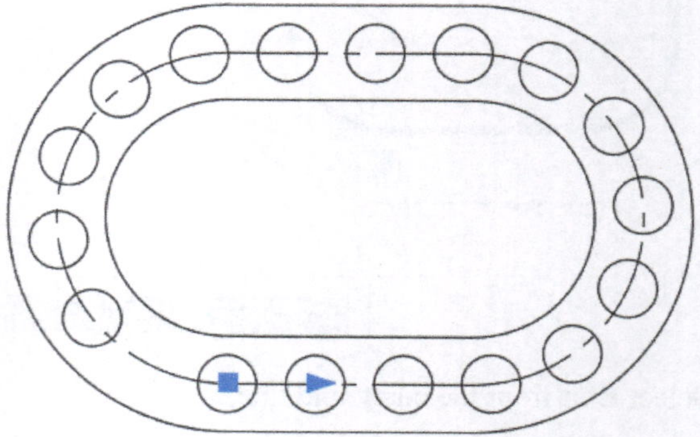

- On the **Array Creation** tab, click **Measure > Divide** on the **Properties** panel. Now, you must enter the number of items in the path array.

If you select the **Measure** method, you must enter the distance between the items in the path array.

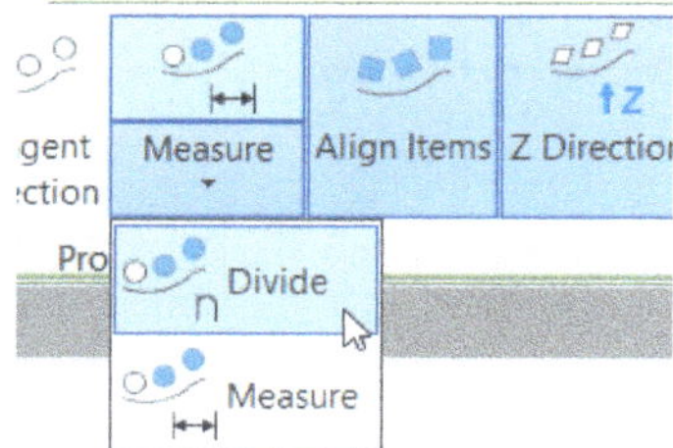

- Set the **Items** count to 12.

Notice that the **Align Items** button is active by default. As a result, the items are aligned with the path. If you deactivate this button, the items will not be aligned with the path.

With the Align Items button active

Align Items button Inactive

- Click the **Close Array** button.

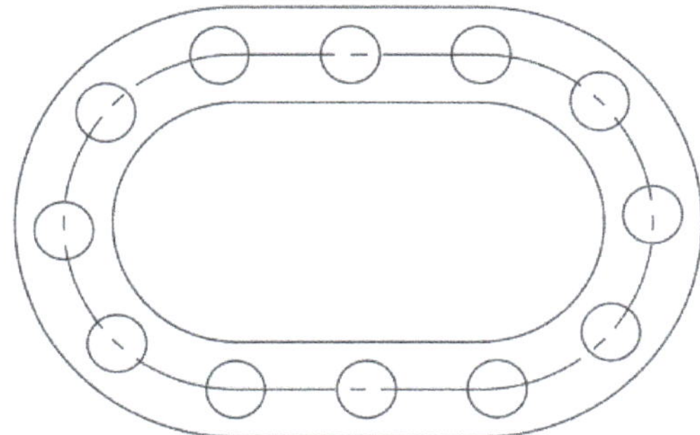

- Save and close the file.

The Rectangular Array tool

The **Rectangular Array** tool creates an array of objects along with the X and Y directions.

- Open a new AutoCAD file and draw the sketch shown below. Do not add dimensions. (refer to the **Drawing Rectangles** and **Drawing Circles** section in Chapter 2 to know the procedure to draw the rectangle and circle)

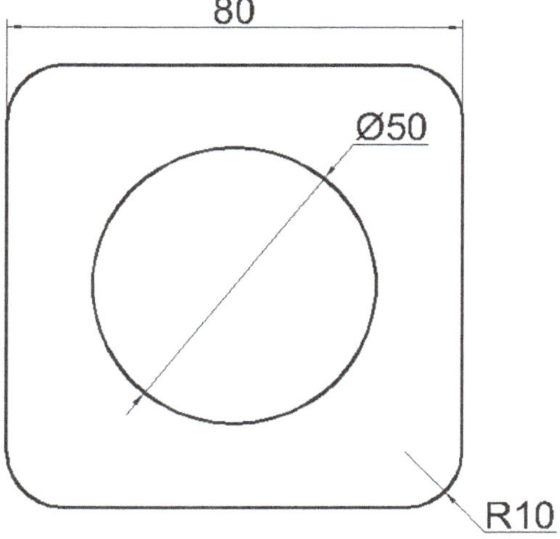

- Click **Home > Draw > Circle** drop-down > **Center, Radius** on the ribbon.
- Select the center point of the lower-left fillet.

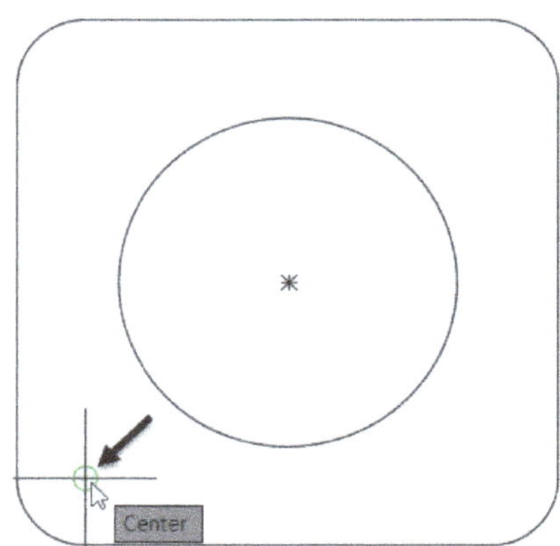

- Type 5 and press Enter.

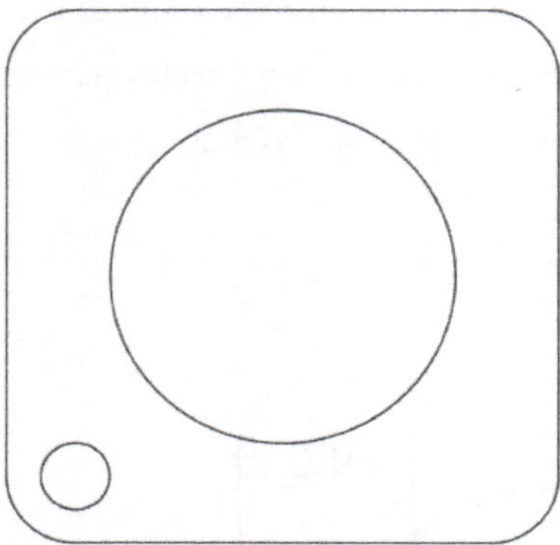

- Click **Home > Modify > Array > Rectangular Array** on the ribbon or enter **ARRAYRECT** in the command line.
- Select the small circle and right-click; a rectangular array with default values appears.

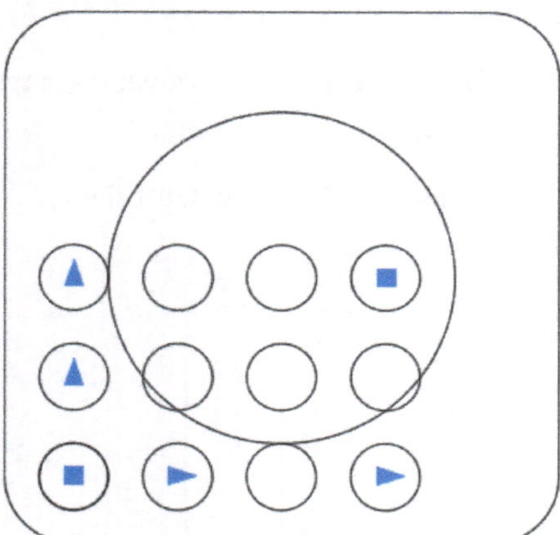

Also, the **Array Creation** tab appears.

- Set the **Columns** count to 2.
- Set the **Rows** count to 2.
- Set the **Between** value in the **Columns** panel to 60.
- Set the **Between** value in the **Rows** panel to 60.

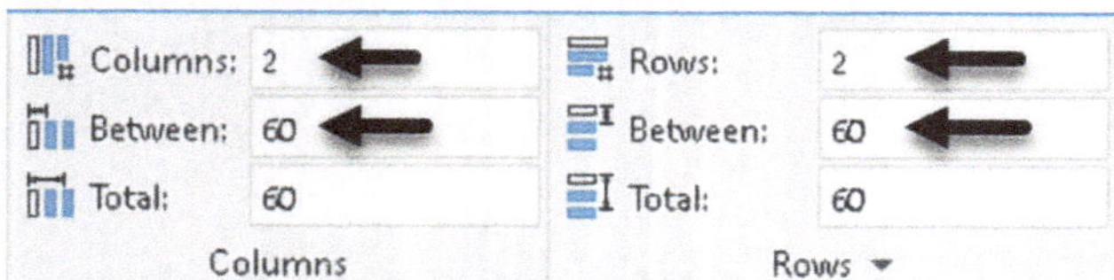

- Click **Close Array** on the ribbon.

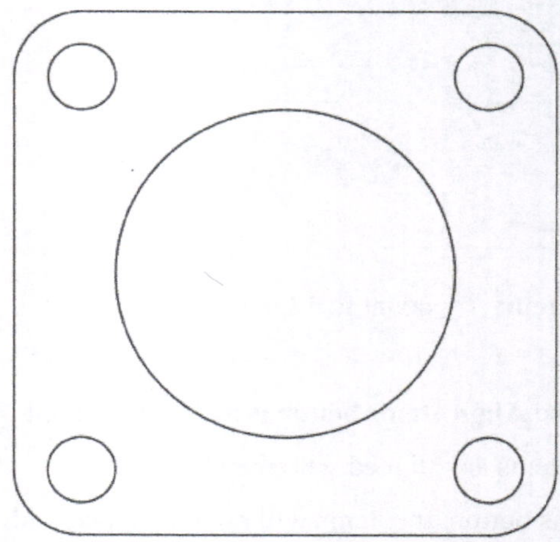

- On the ribbon, click **Home** tab > **Utilities** > **Measure** Drop-down > **Quick**.

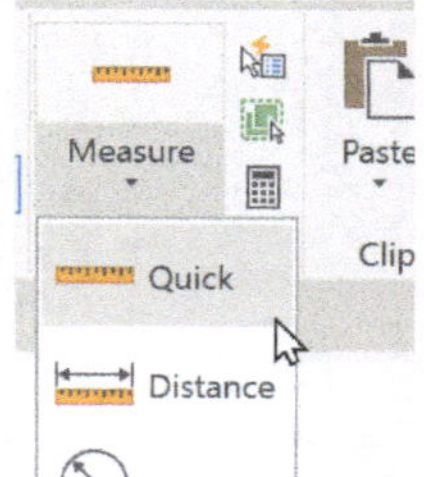

- Place the cursor inside the circle located at the center; the radius of the circle is displayed.

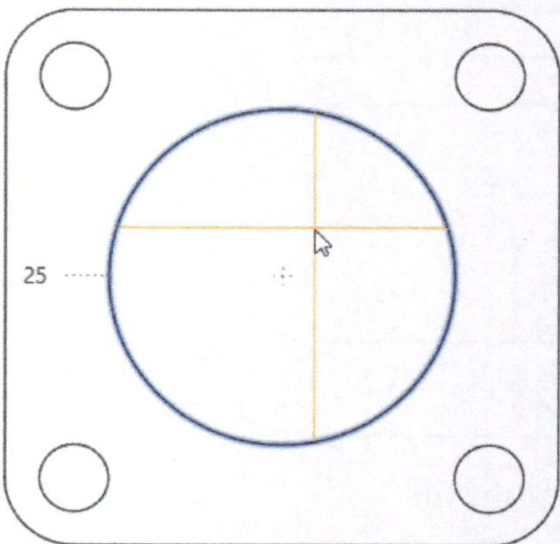

- Place the cursor between the two circles located at the bottom; the measurements of the objects located in four directions of the cursor are displayed.

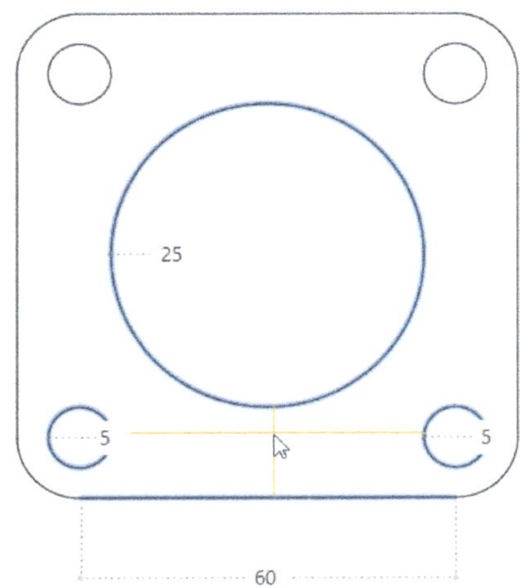

Editing Using Grips

When you select objects from the graphics window, small squares appear on them. These squares are called grips. You can use these grips to stretch, move, rotate, scale and mirror objects, change properties, and perform other editing operations. Grips displayed on selecting different objects are shown below.

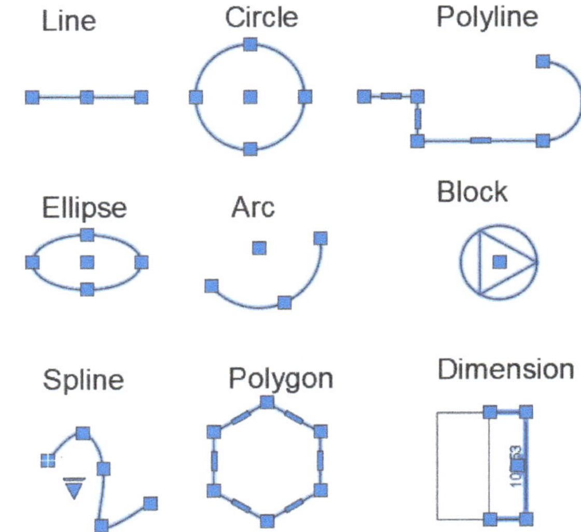

The following table gives you the details of the editing operations that can be performed when you select and drag grips.

Object	Grip	Editing Operation
Circle	Grip on circumference	**Scale**: Select any one of the grips on the circumference and move the pointer to scale a circle.
	Center point grip	**Move**: Select the center grip of the circle and move the pointer.

		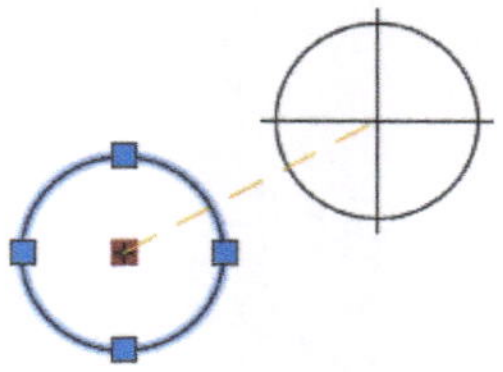
Arc	Grip on circumference	**Stretch**: Select the grip on the circumference and move the pointer. 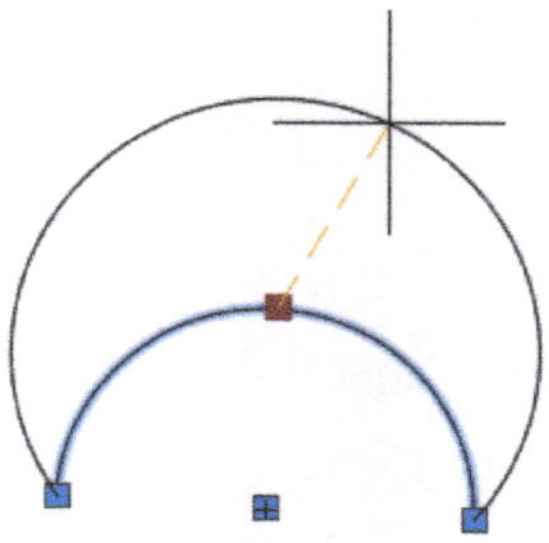
	Center point grip	**Move**: Select the center grip of the arc and move the pointer. 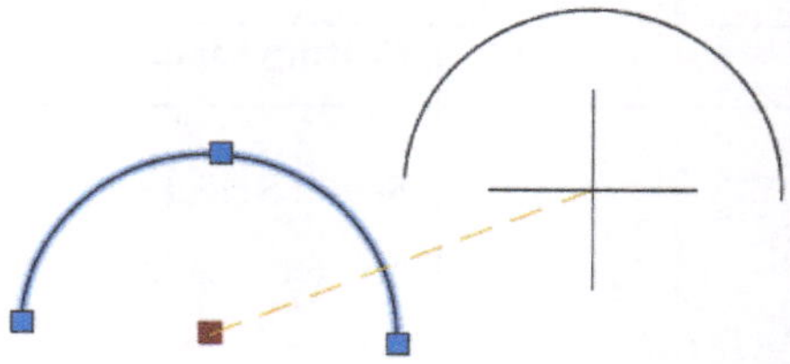
Line	Midpoint Grip	**Move**: Select the Midpoint grip and move the pointer. 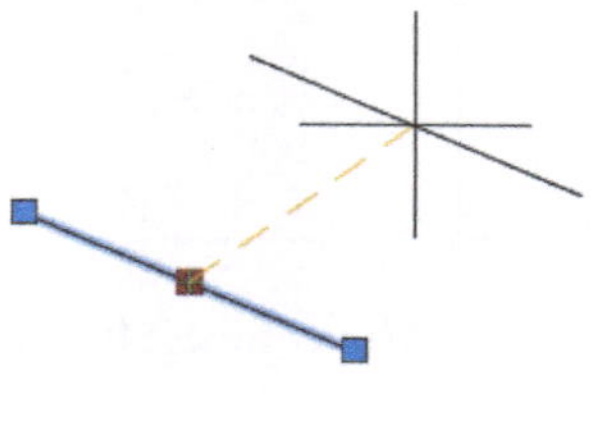
	Endpoint Grip	**Stretch/Lengthen**: Select an endpoint grip and move the pointer.

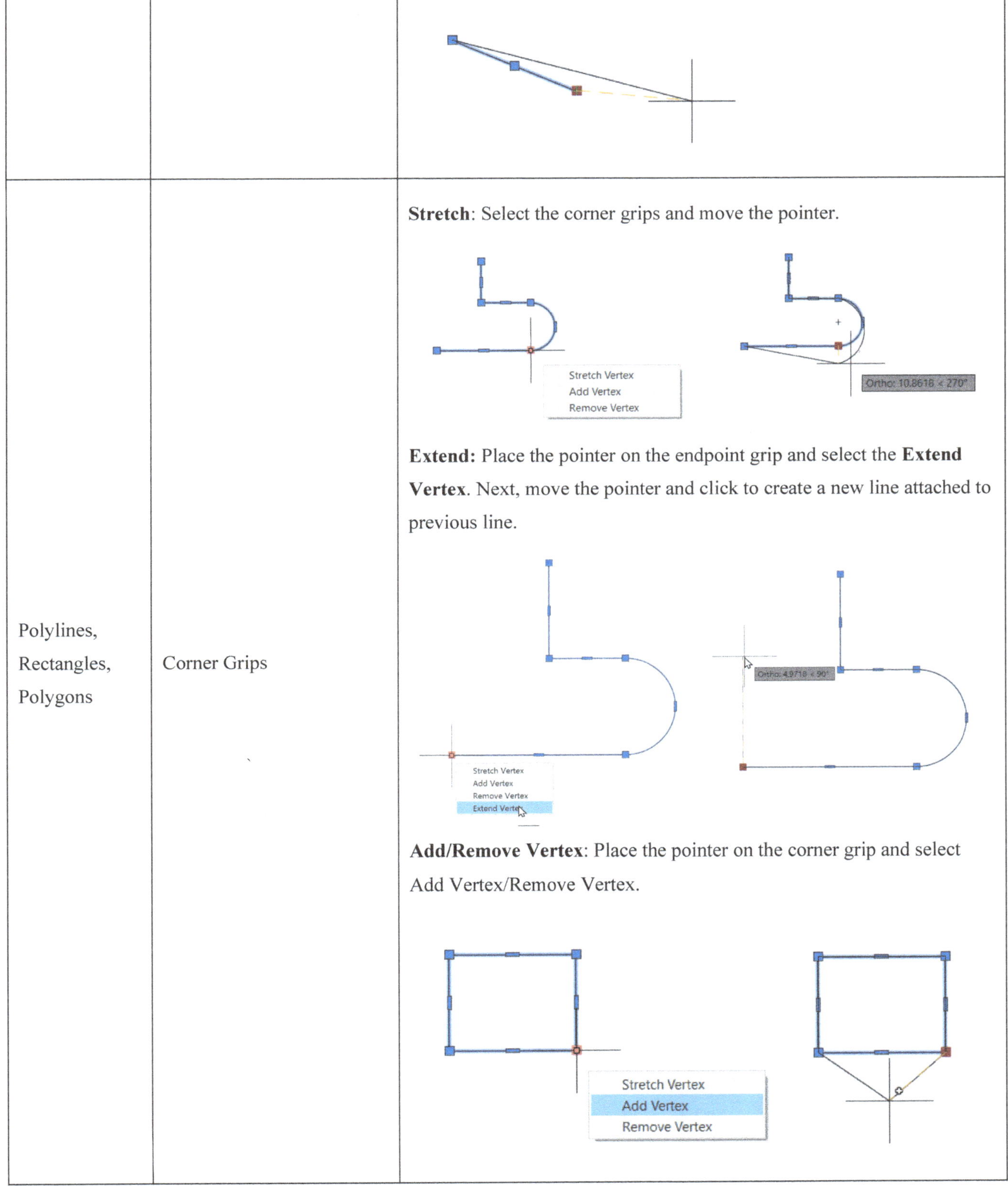

Polylines, Rectangles, Polygons	Corner Grips	**Stretch**: Select the corner grips and move the pointer. **Extend:** Place the pointer on the endpoint grip and select the **Extend Vertex**. Next, move the pointer and click to create a new line attached to previous line. **Add/Remove Vertex**: Place the pointer on the corner grip and select Add Vertex/Remove Vertex.

	Midpoint Grips	**Convert to Arc**: Place the pointer on the midpoint grip and select **Convert to Arc**. 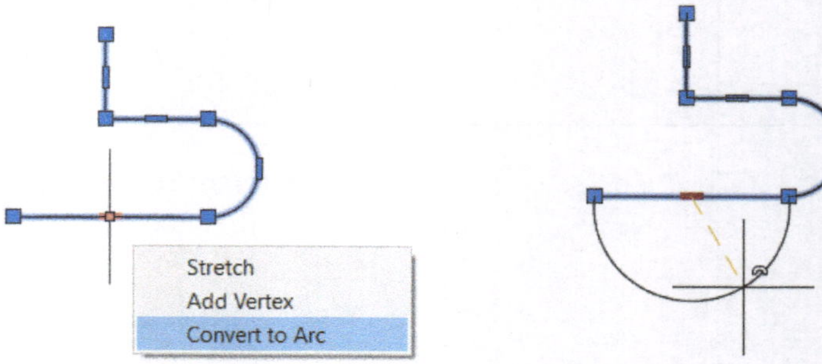 **Convert to Line**: Place the pointer on the midpoint grip of a polyline arc and select **Convert to Line**. 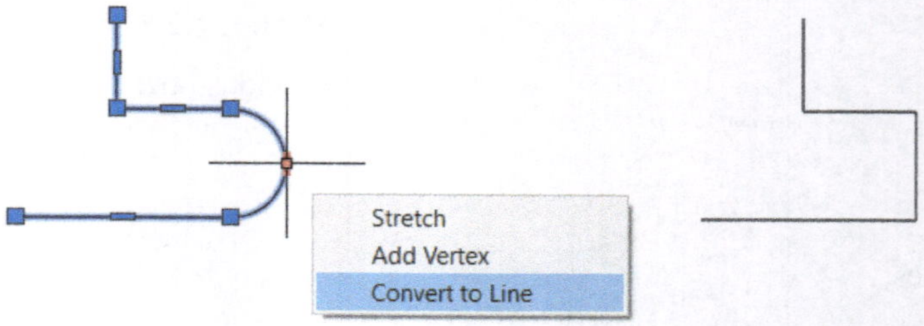
Ellipse	Center Grip	**Move**: Select the center grip and move the pointer. 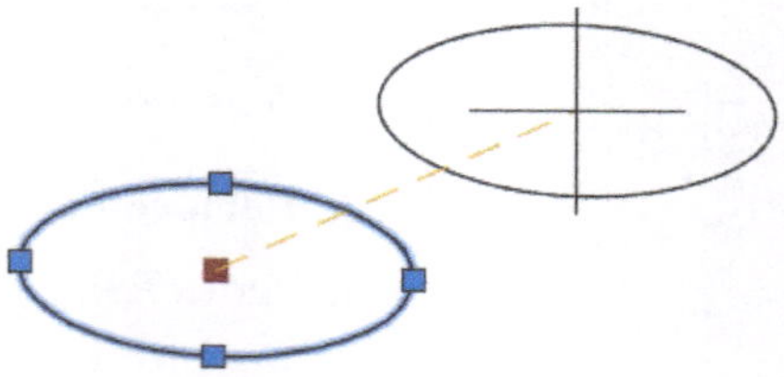
	Grips on circumference	**Stretch**: Select a grip on the circumference and move the pointer. 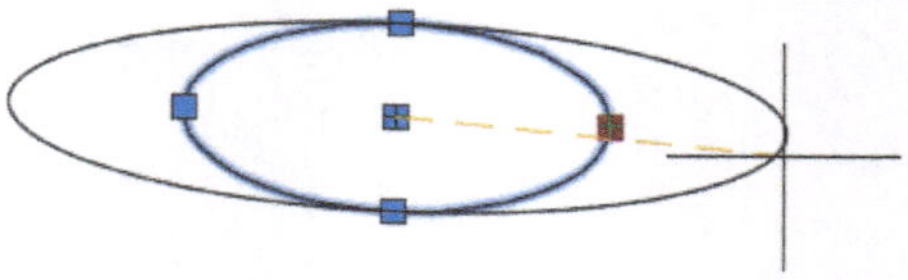

Spline	Fit Points	**Stretch**: Select a grip on the spline and move the pointer. 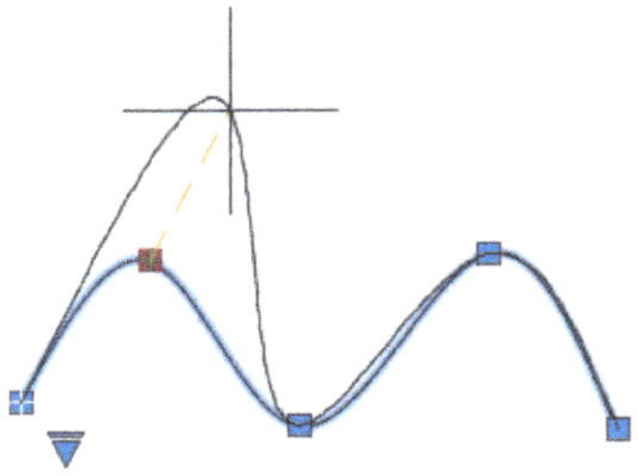 **Add/Remove Fit Point**: Place the pointer on a fit point and select **Add Fit Point** or **Remove Fit Point**. 
	Control Vertices	**Stretch Vertices**: Select the control vertices of a CV spline and move the pointer. 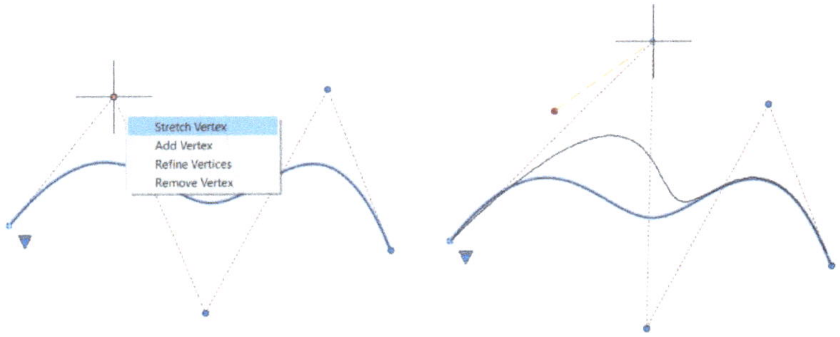 **Add/Remove Vertex:** Place the pointer on a control vertex and select **Add Vertex** or **Remove Vertex**. 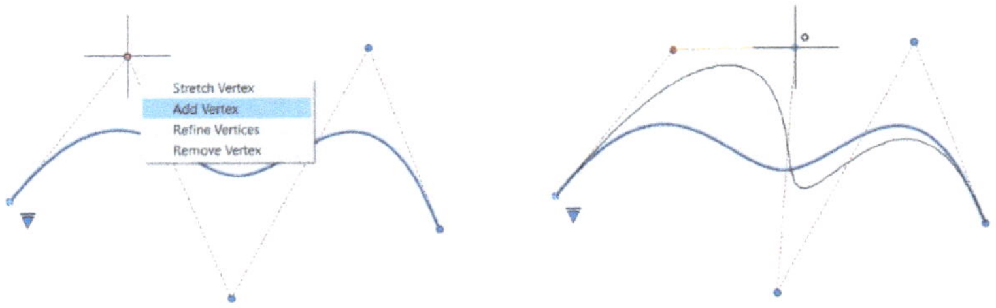 **Refine Vertices:** Place the pointer on a control vertex and select **Refine Vertices**.

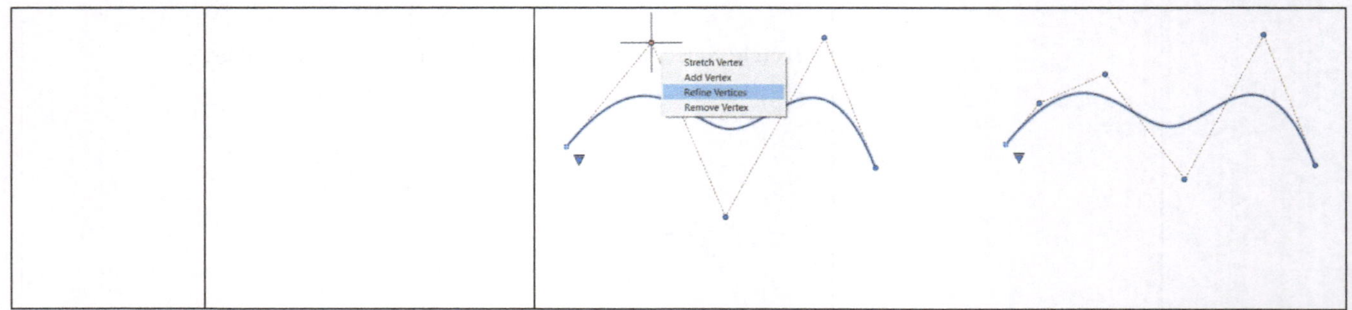

Modifying Rectangular Arrays

You can use grips to edit rectangular arrays dynamically. Various array editing operations using grips are given next.

Moving a Rectangular array

- Create a rectangular array, as shown below.

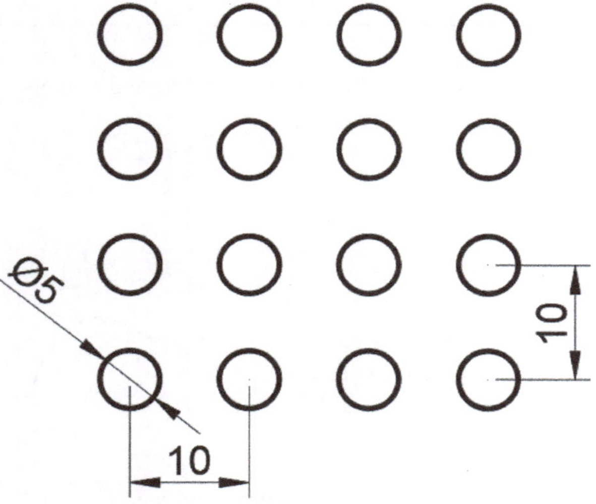

- Select the array; you will notice that grips are displayed on it.
- Select the grip located at the lower-left corner and move the array, as shown below.

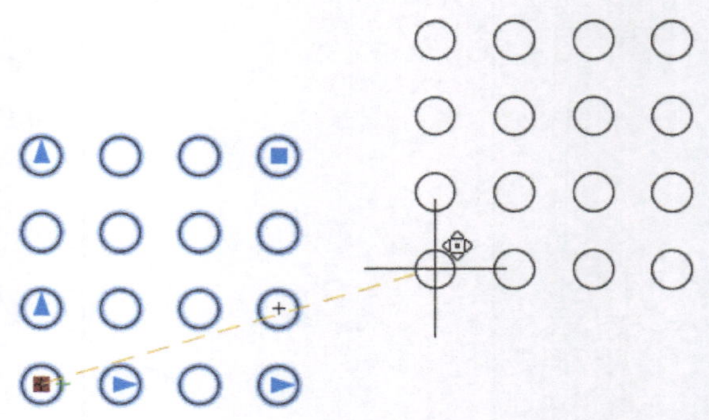

Adding/Removing Level to a Rectangular array

- Place the pointer on the lower left grip of the rectangular array; a shortcut menu appears.
- Select **Level Count** from the shortcut menu; the message, "**Specify number of levels**," appears in the command line.
- Type 3 and press ENTER.
- Click the **Home** button near the ViewCube to view the levels.

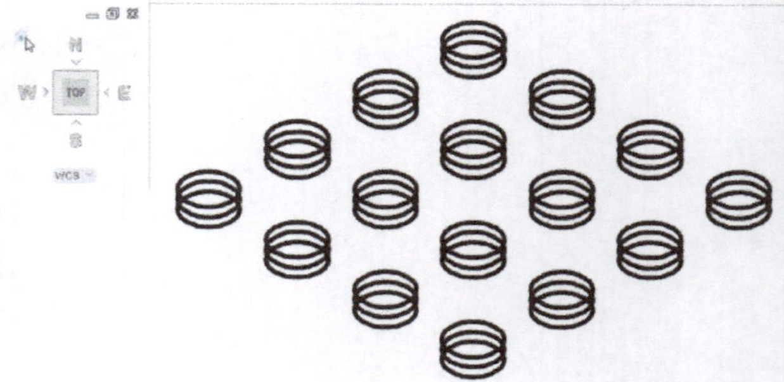

- Change the view to Top view by using the In-Canvas controls.

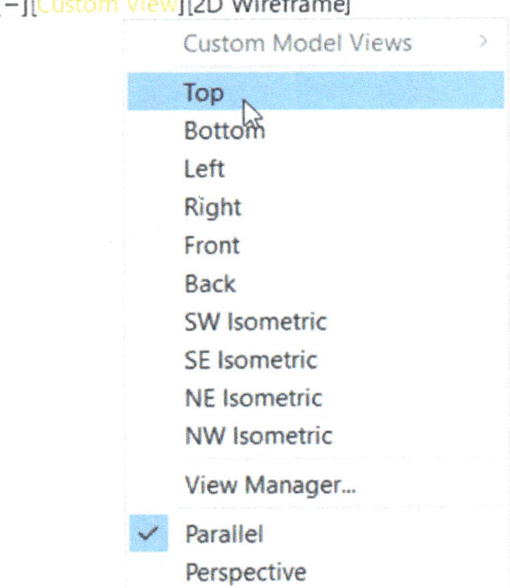

Changing the Column and Row Count

- To change the column and row count, place the pointer on the top right corner grip; a shortcut menu appears.
- Select **Row and Column Count** from the shortcut menu; the message, "**Specify number of rows and columns**," appears in the command line.
- Type **5** in the command line and press ENTER; the number of rows and columns is changed to 5.
- If you only want to change the column count, place the pointer on the lower right corner grip of the array.

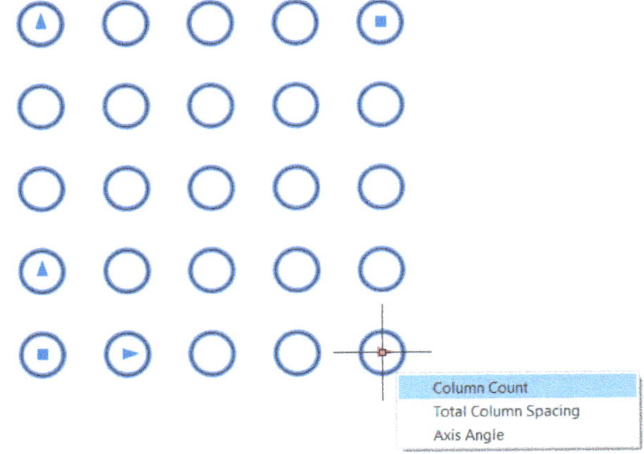

- Select **Colum Count** from the shortcut menu.
- Next, enter the number of columns or drag the pointer and click.

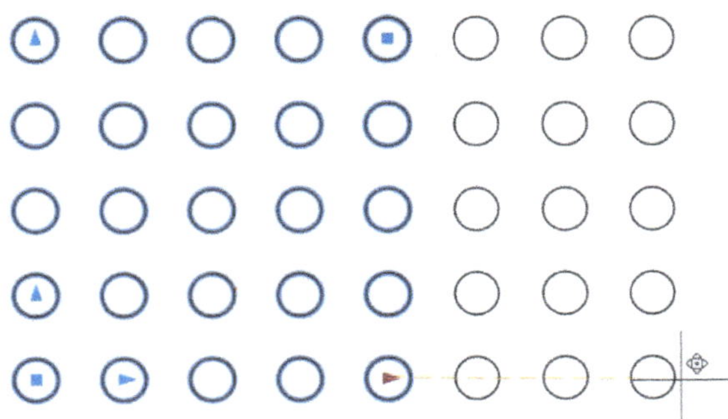

- To change the row count only, click the top left corner grip and drag the pointer. You can also enter the row count in the command line.

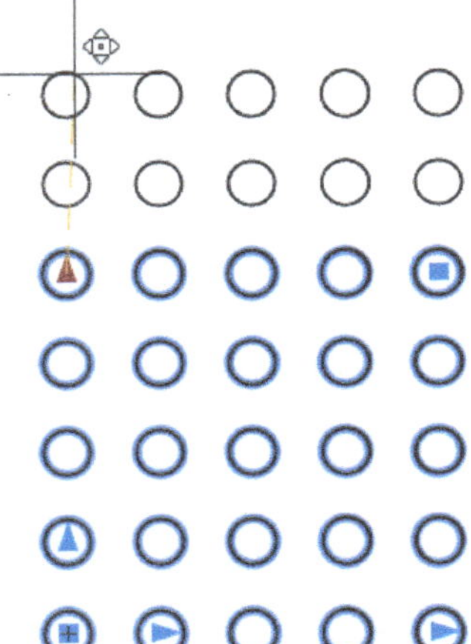

Changing the Column and Row Spacing

- To change the total column and row spacing, place the pointer on the top right corner grip and select **Total Row and Column Spacing** from the shortcut menu.

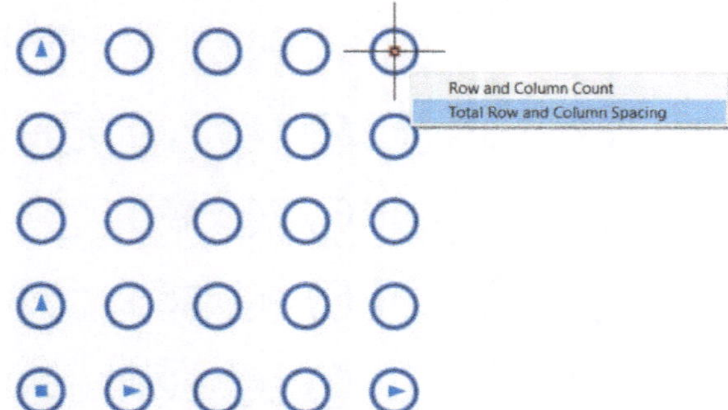

- Type 80 in the command line; the spacing between the columns and rows is adjusted to fit the total length.

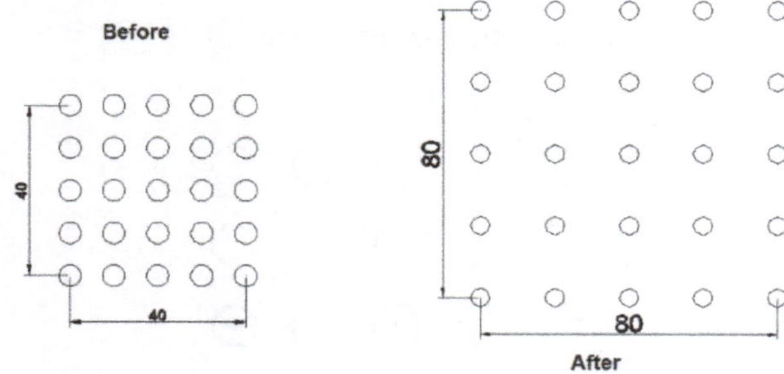

- To change the total column spacing only, place the pointer on the lower right corner grip and select **Total Column Spacing** from the shortcut menu.

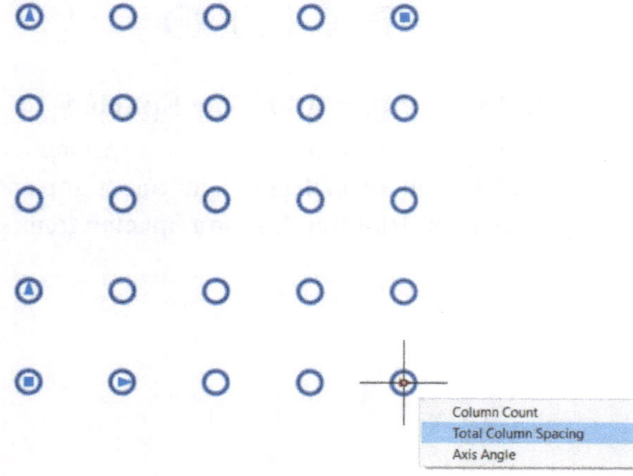

- Next, enter the total column distance or drag the pointer and click.

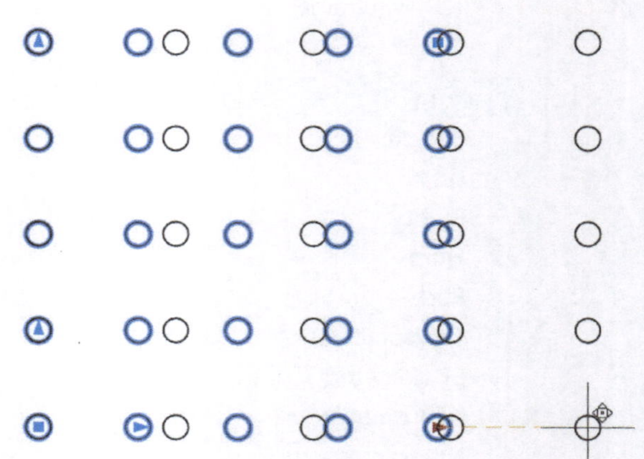

- If you want to change the distance between the individual columns, click the second column grip and drag the pointer.

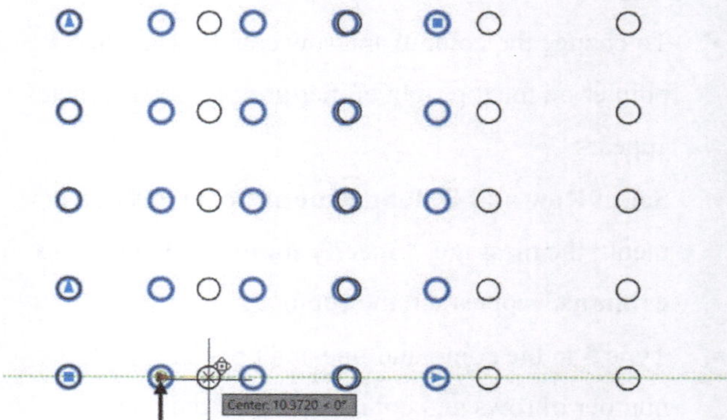

- You can also enter the distance in the command line.
- Likewise, you can change the total row spacing and distance between the individual rows by using the grips shown below.

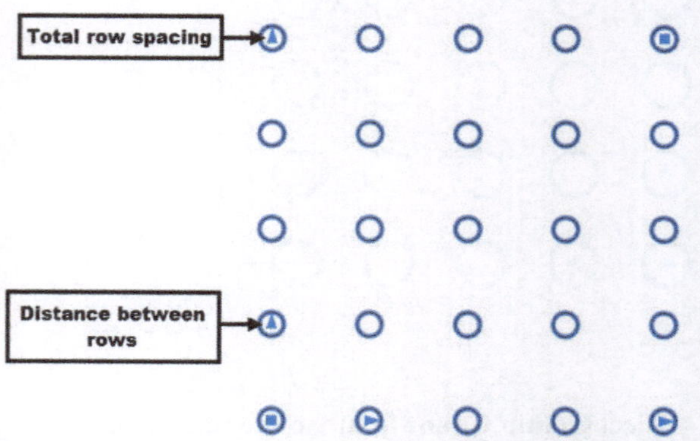

Changing the Axis Angle of the Rectangular Array

- To change the Axis angle of the rows, place the pointer on the lower right corner grip and select the **Axis angle** option from the shortcut menu.
- Type the angle and press ENTER. Note that the angle is calculated from the first column of the array. For example, if you enter 60 as the axis angle, the rows will be inclined by 60 degrees from the first column.

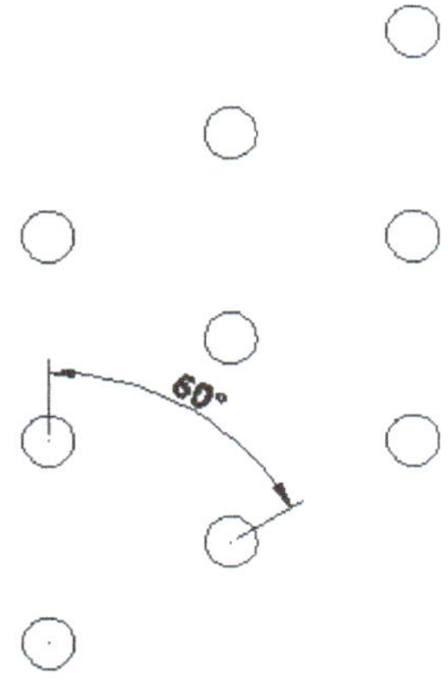

- Likewise, you can change the axis angle of the columns by using the top left corner grip.

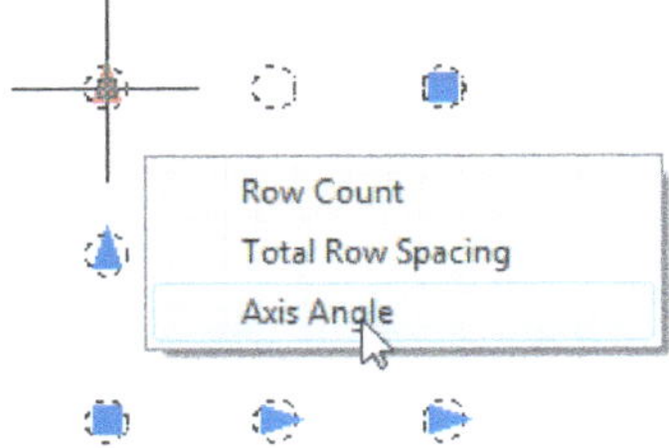

Editing the Source Item of the Rectangular Array

- Create a rectangular array, as shown below.

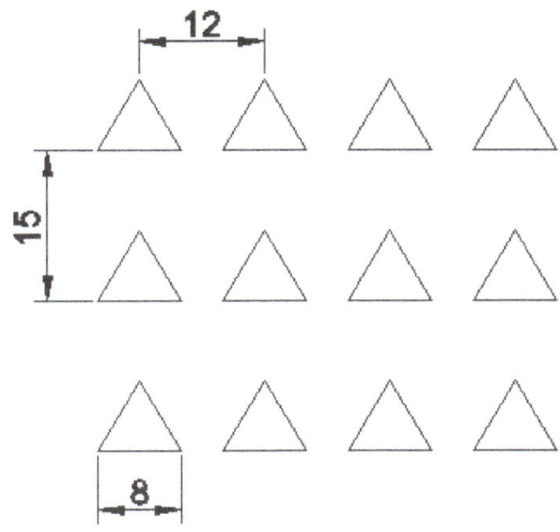

- Click **Close Array** on the **Array Creation** tab.
- Select the rectangular array; the **Array** tab appears in the ribbon.
- Click the **Edit Source** button on the **Options** panel; the message, "**Select item in array**" message appears in the command line.
- Select the lower left triangle of the rectangular array; the **Array Editing State** message box appears.

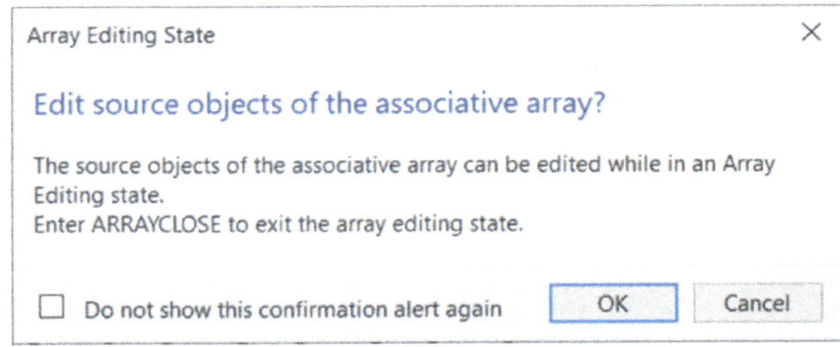

- Click **OK**; the array editing state is activated.
- Draw a circle and trim the unwanted portion, as shown below.

- Click **Save Changes** on the **Edit Array** panel of the **Home** tab of the ribbon.

Modifying Polar Arrays

Similar to editing rectangular arrays, you can also edit a polar array by using grips. Various array editing operations using grips are given next.

Changing the Radius of a Polar array

- Create the polar array, as shown in the figure.

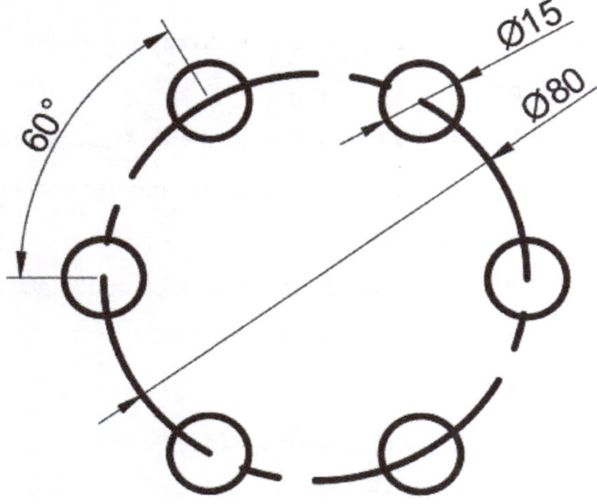

- Select the polar array; grips will be displayed on it.
- Place the pointer on the base grip, as shown in the figure.
- Select **Stretch Radius** from the shortcut menu.

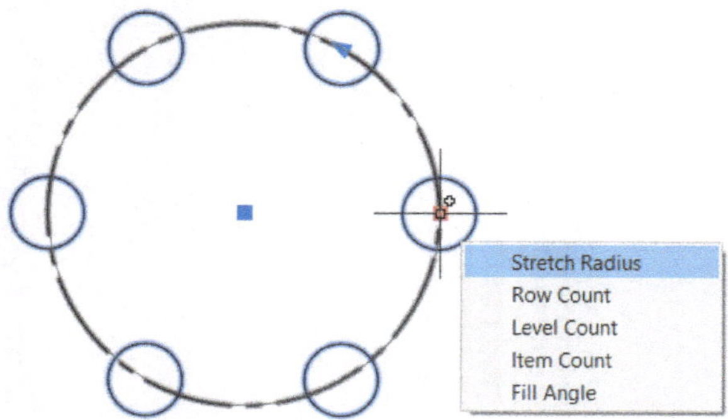

- Move the pointer outward or inward and click. You can also enter a new radius value of the polar array.

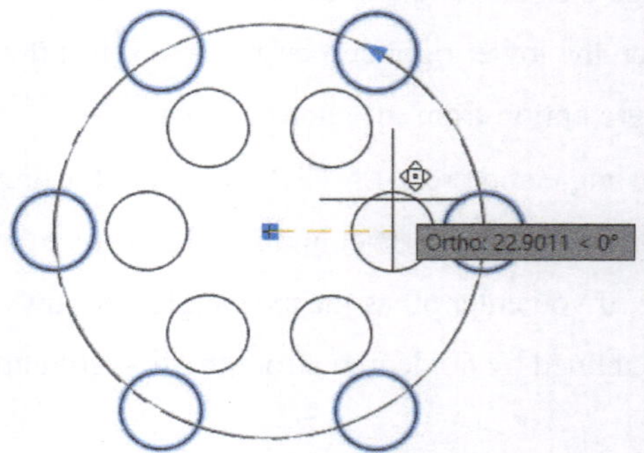

Changing the Row Count of a Polar array

- Place the pointer on the base grip of the array and select **Row Count** from the shortcut menu.
- Move the pointer outward and click. You can also enter the number of rows in the command line.

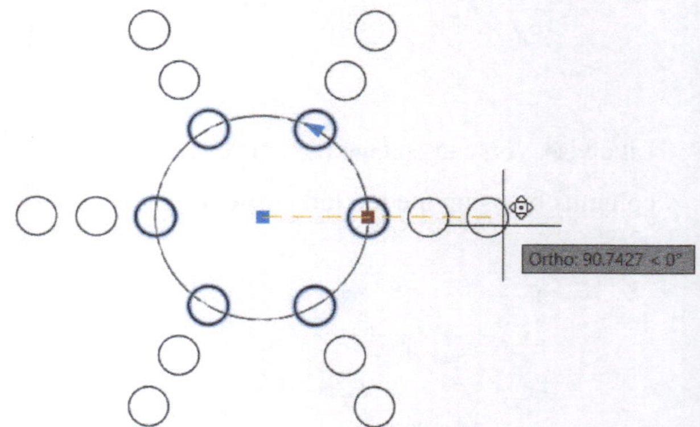

- You can again change the **Row Count** by using the last row grip.

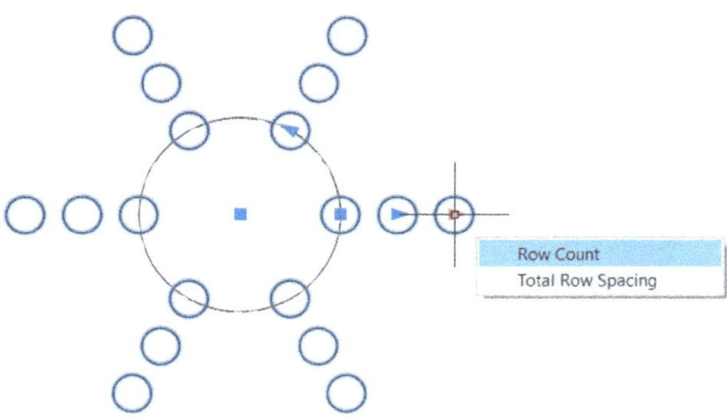

Changing the Row Spacing

- To change the total row spacing, place the pointer on the last row grip and select **Total Row Spacing**.

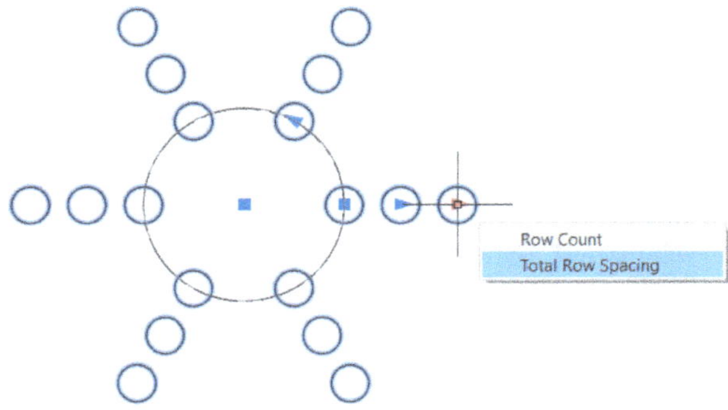

- Next, move the pointer and click. You can also enter the total row spacing value in the command line.
- To change the distance between the individual rows, click the second-row grip and move the pointer outward. You can also enter the distance in the command line.

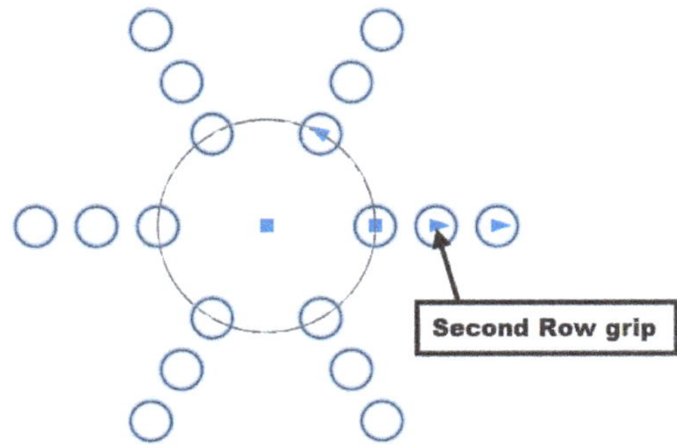

Changing the Angle between the Items

- To change the angle between the items, click the second radial grip, and enter the new angle value.

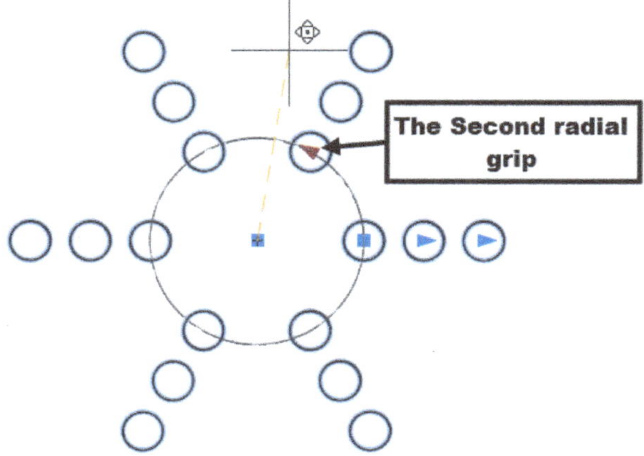

Changing the Fill angle of the array

- The default fill angle of a polar array is 360 degrees. To change the fill angle, place the pointer on the base grip and select **Fill Angle** from the shortcut menu.

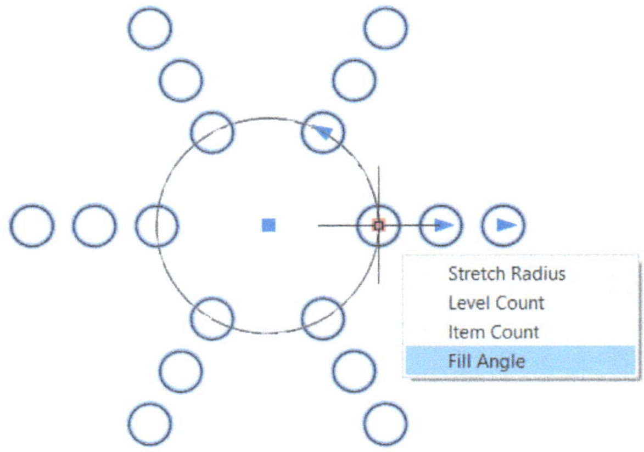

- Enter a new value for the fill angle or drag the pointer and click.

Changing the Item count of a Polar array

- Select the polar array and enter a new item count in the **Items** box of the **Array** ribbon.

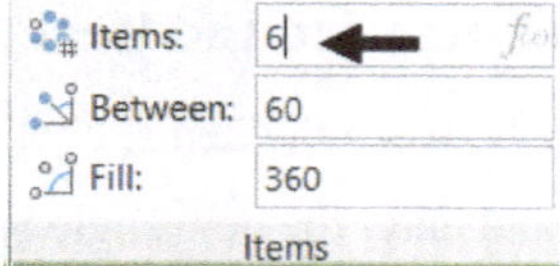

- Click **Close Array** on the ribbon.

Revision Clouds

Revision clouds are used to highlight the areas in a drawing. You can create revision clouds using three different tools.

Example 1:

- Start a new drawing using the acadISO template.
- On the ribbon, click **Annotate > Markup > Revision Cloud > Rectangular** (or) click **Home > Draw > Revision Cloud > Rectangular**.
- Select **Arc Length** from the command line.
- Type 3 and press Enter to specify the approximate arc length.
- Specify the first and second corners of the revision cloud. You can also select the **Object** option and select an object from the graphics window. The selected object will be converted into a revision cloud.

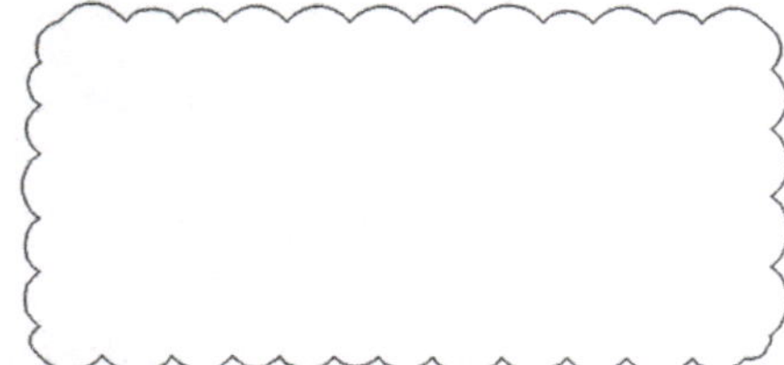

- Select the revision cloud and notice the grips. You can use the midpoint grip to stretch or add new vertices to the revision cloud.

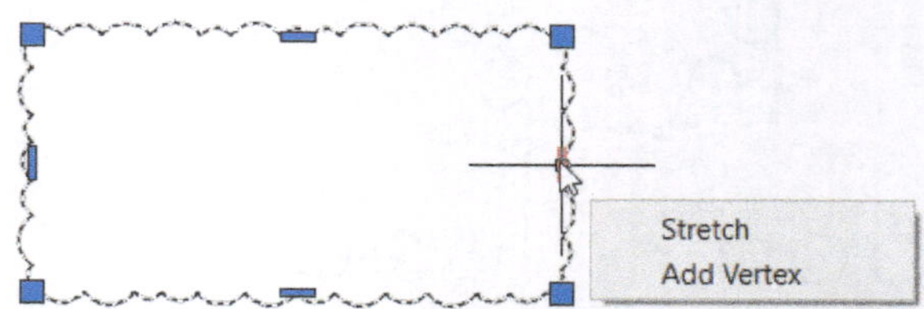

You can use the corner point grip to stretch, add, or remove vertices.

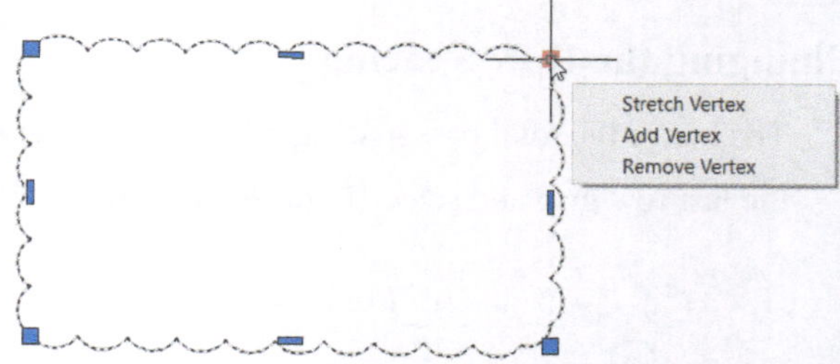

- Select the revision cloud, right-click, and select **Properties**. Notice that object type is displayed at **Revcloud** in the drop-down displayed at the top of the **Properties** palette.
- Scroll down to the **Misc** section of the **Properties** palette.
- Click in the **Arc Length** box, type **5**, and press ENTER. The arc length is changed.

Example 2:

- On the ribbon, click **Annotate > Markup > Revision Cloud > Polygonal** (or) click **Home > Draw > Revision Cloud > Polygonal.**
- Select **Style** from the command line.

- Select **Calligraphy** from the command line.
- Specify the corners of the revision cloud and press Enter.

Example 3:

- On the ribbon, click **Annotate > Markup > Revision Cloud > Freehand** (or) click **Home > Draw > Revision Cloud > Freehand**.
- Specify the start point of the revision cloud.
- Move the pointer around the area to be highlighted.
- Move the pointer onto the start point to close the cloud.

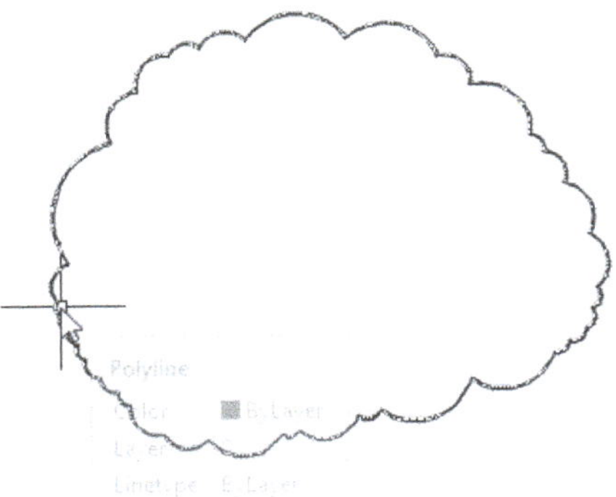

PEDIT

The **PEDIT** command is a powerful tool for modifying polylines. Its various options allow you to perform a wide range of edits, from simple vertex modifications to complex curve conversions.

On the ribbon, click **Home** tab > **Modify** panel > **Edit Polyline** (or) type **PEDIT** command in the command line and press ENTER.

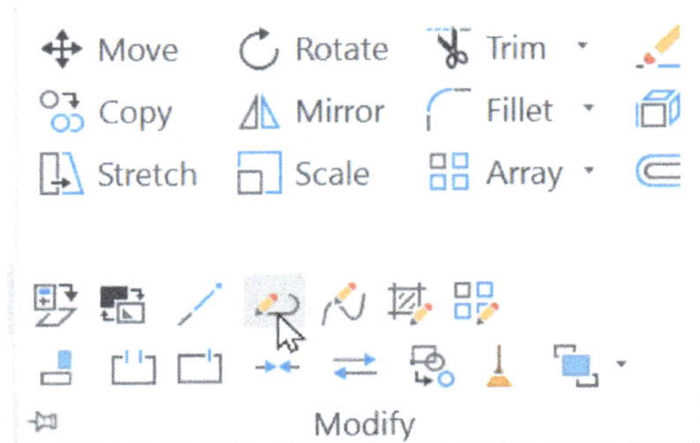

Select a polyline to edit (or) select the **Multiple** option from the command line. With the **Multiple** option, it's possible to edit several polylines at once. The poylines that have been selected don't have to be related or connected in any way, and can be completely separate objects. After selecting the polyline(s), you will see a series of options that allow you to perform different types of edits on the polyline or spline. Here is a detailed explanation of each option:

Close

This option allows you to close an open polyline or spline by connecting the end points. If the polyline or spline is already closed, this option will have no effect.

Open

If the **Close** option was used before, selecting the **Open** option will eliminate the closing segment of the polyline.

Join

This option allows you to join two or more open polylines, lines, arcs, or splines into a single entity.

Example 1

- Open a new drawing file and create a horizontal and inclined lines.

- Click **Home > Modify > Edit Polyline** on the ribbon, or enter **PEDIT** in the command line.
- Select the **Multiple** option from the command line.
- Select the two lines and press the **Enter** key.
- Select the **Yes** option from the command line to convert the lines into a polyline.
- Select the **Join** option from the command line.
- Select the **Jointype** option from the command line.
- Select the **Extend** option from the command line. By selecting this option, AutoCAD will join the chosen polylines by either extending or trimming the segments to the closest endpoints. However, you can also select the **Add** option. It allows you to insert a straight segment connecting the closest endpoints.
- Type-in a value in the command line and press ENTER (or) select two points from the graphics window to specify the Fuzz distance. The endpoints of the line segments will be only if they fall within specified distance.

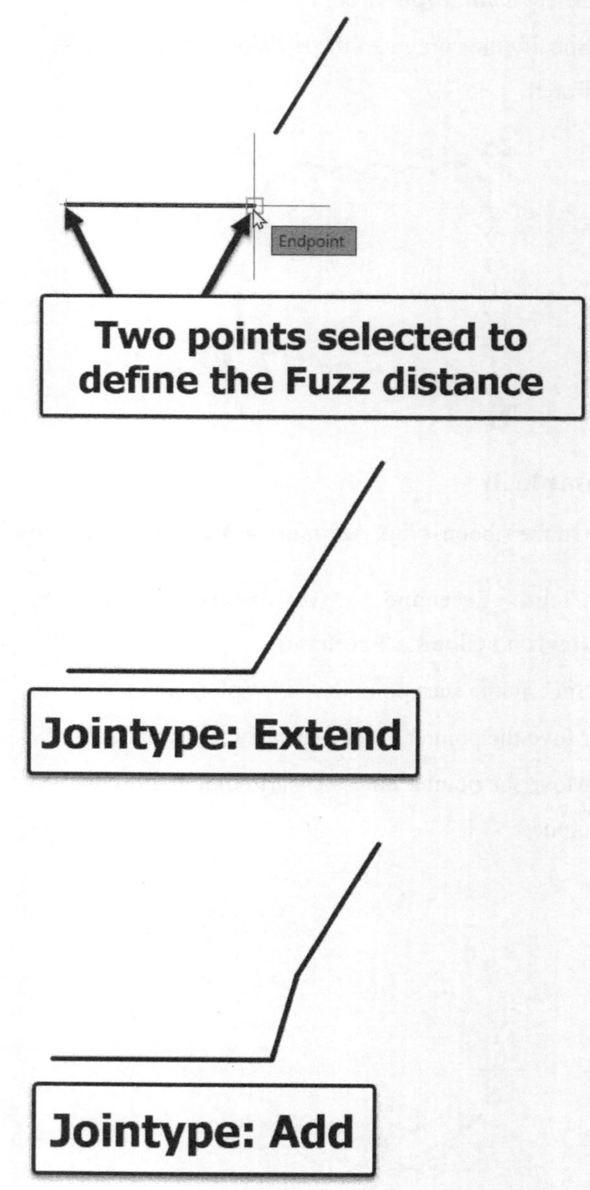

- Press ESC and close the file.

Example 2

- Open a new drawing file and create an arc and four lines, as shown.

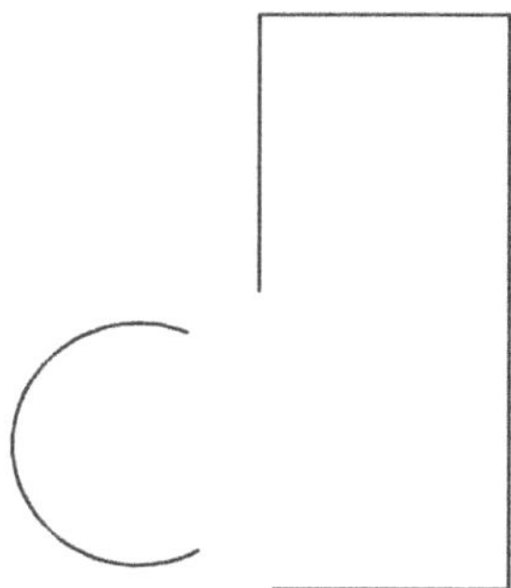

- Click **Home > Modify > Edit Polyline** e.
- Select the **Multiple** option from the command line.
- Create a selection window across all the objects of the drawing and press the **Enter** key.
- Select the **Yes** option from the command line to convert the lines into a polyline.
- Select the **Join** option from the command line.
- Select the **Jointype** option from the command line.
- Select the **Both** option from the command line. By selecting this option, you can join the specified polylines by either extending or trimming them wherever possible.
- Select the end points of the horizontal line, as shown; the fuzz distance is defined. Also, the endpoint that fall within the fuzz distance are joined together.

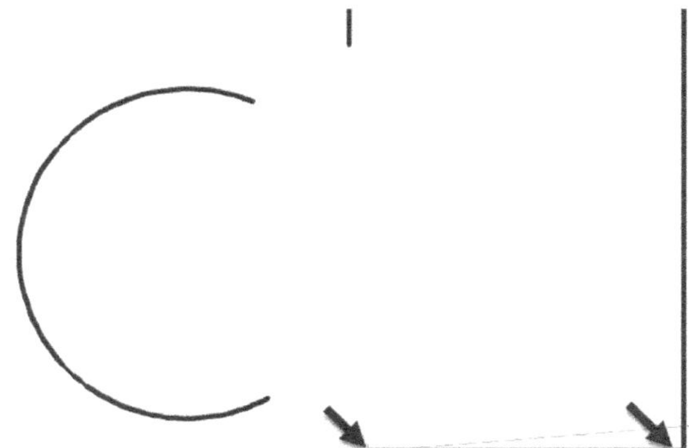

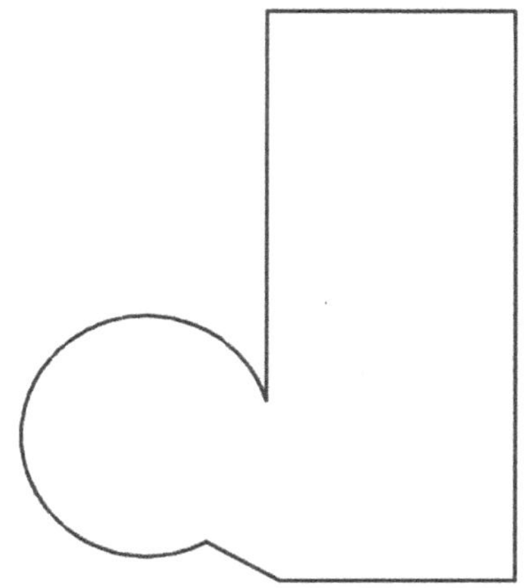

Note that when the two line segments to be joined are near almost parallel lines, a straight segment will be added between the closest endpoints.

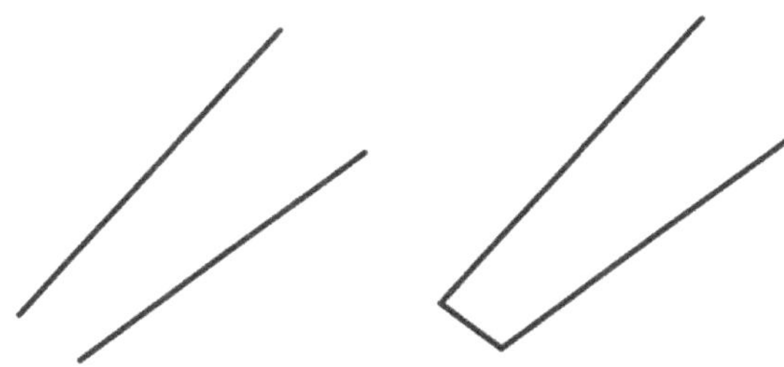

Width

This option allows you to change the width of the polyline by assigning a new consistent width. Select this option and type-in a new width value. Press ENTER to apply the new width.

Edit Vertex

This option allows you to move, add, or delete vertices of the polyline or spline. After selecting this option from the command line, a group of sub-options will appear in the command line.

Next

After selecting the **Edit Vertex** option, a marker in the shape of an X is positioned at the initial endpoint. You can use the Next and Previous options to move the marker along the polyline to the desired vertex in sequence.

Break

By selecting this option, you can create a gap between the marked vertex and another vertex. Using the **Next** or **Previous** option, you can select another vertex.

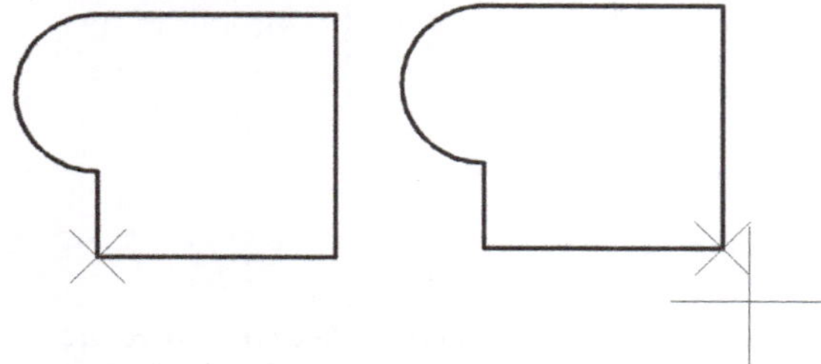

Next, select the **Go** option to create a between two select vertices.

Insert

With the help of the **Insert** option, you have the ability to add a fresh vertex at any point beyond the vertex labeled with the X marker. Select this option and specify the location of the new vertex to inserted.

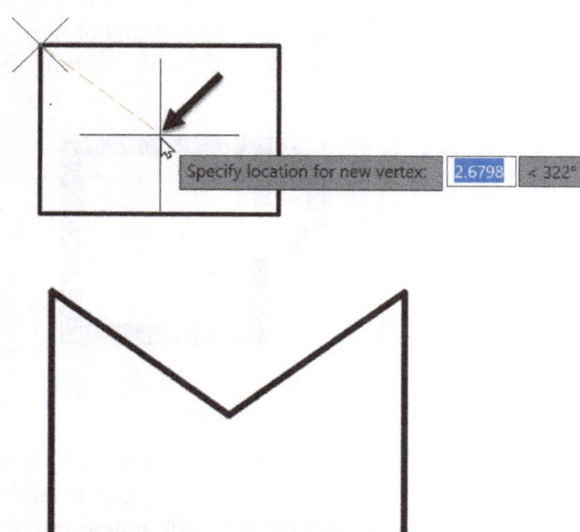

Move

This option allows you move the X marked vertex to a new location. Select this option and specify a location in the graphic window; the X marked vertex is moved to the specified location.

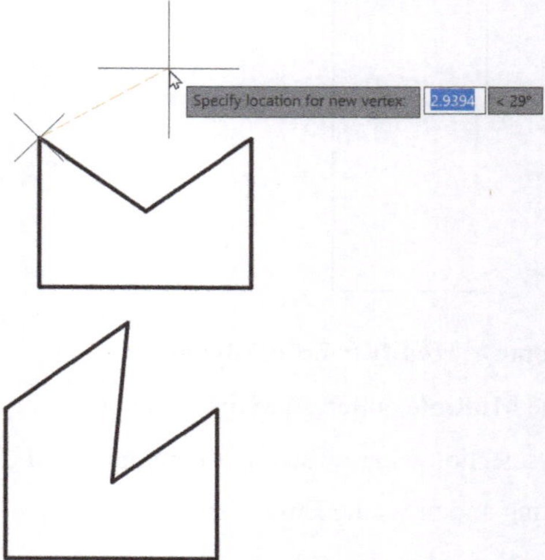

Regen

This is used to regenerate the current viewport to view the changes made to the polyline.

Straighten

This option straightens the Zig-Zag line segments between the X marked vertex and another vertex. You can select another vertex by using the **Next** or **Previous** options.

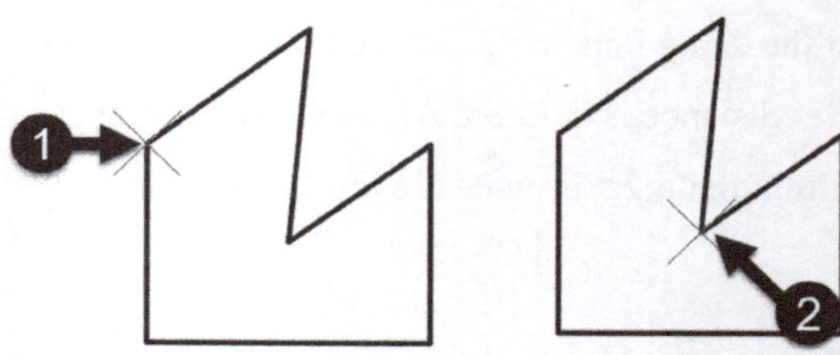

Select the **Go** option after select the two points.

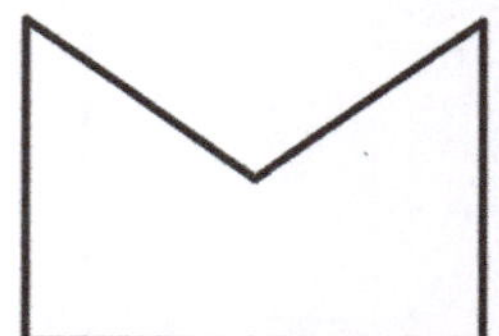

Tangent

With the **Tangent** option, you can specify the tangency direction selected vertex. This option will be useful when a polyline is converted to a curve.

Width

By using this option, you can modify the width of the polyline segment next to the X marked vertex. Select this option and specify the start width of the line segment. Next, specify the end width of the line segment.

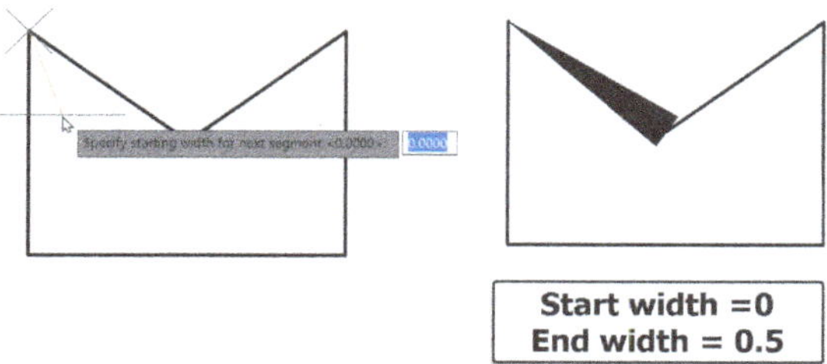

eXit

Select this option to exit the **Edit Vertex** mode.

Note that the **Edit Vertex** option can be quite lengthy, tedious, and difficult if you are a beginner. You can explode the polyline, perform any necessary modifications using the editing tools, and then join the lines and arcs into a single polyline again.

Fit

If you choose this option, the polyline you have will be transformed from straight line segments into arcs. The resulting curve will pass through all the vertices in the polyline. This means that for every pair of vertices in the polyline, there will be two arcs that form the curve.

Note that if the original polyline has steep angles, the resulting curve can be quite unexpected.

Spline

This option can be used to convert a polyline into a B-spline (Bezier spline) curve. A spline is a smooth curve that passes only through the end vertices of the polyline. By using this option, you will be able to create a more precise curve.

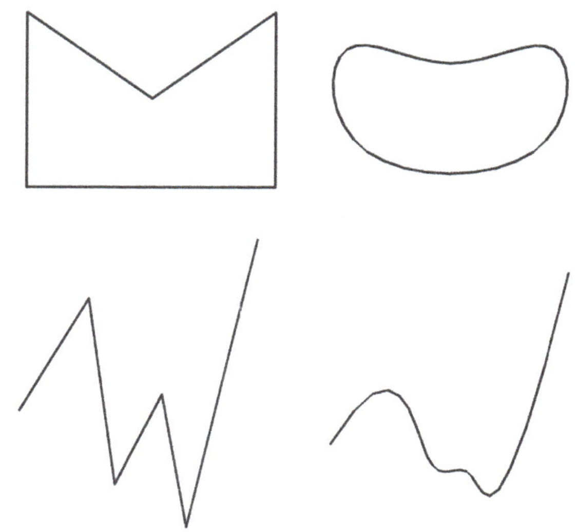

Decurve

When you select this option, it converts the curved lines created by **Spline** or **Fit** options into the original polyline.

Ltype gen

This option determines how non-continuous lines are drawn. Select this option, and then select the **Off** option from the command line; it generated non-continuous lines with dashes that start and stop at each vertex. This means that each line segment will begin and end with a full dashed segment. If you select the **ON** option; it generates a non-continuous line in a consistent pattern, regardless of the vertices. This means that some vertices may have a space instead of a dash.

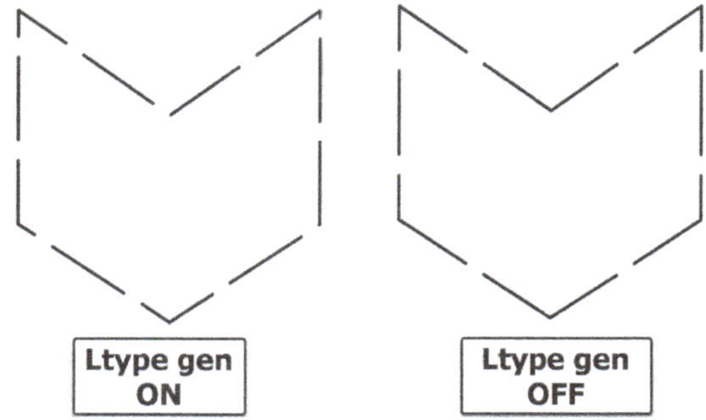

Reverse

Select this option to change the order of the first and last points of a polyline.

Note that the PLINEREVERSEWIDTHS system variable controls direction in which a polyline's width is displayed. When the value of this variable is set to 0, the polyline's width will not be reversed. However, if the value is changed to 1, the polyline's width will be reversed.

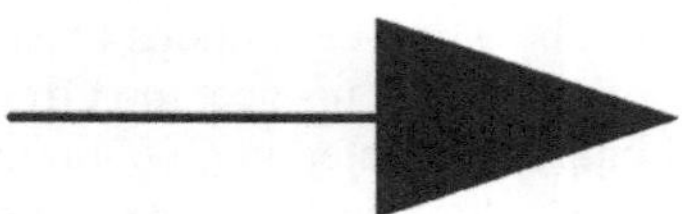

PLINEREVERSEWIDTHS set to 0

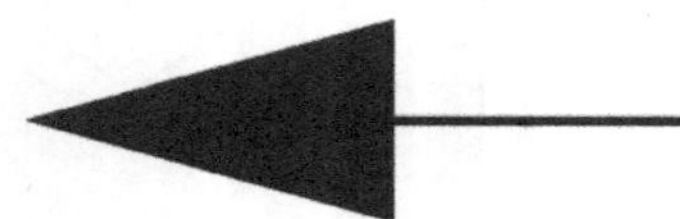

PLINEREVERSEWIDTHS set to 1

Undo

If you make a mistake when using **Pedit** command, you can use the **Undo** option to reverse the most recent operation you performed.

Example 1

In this example, you will create the drawing shown in the figure.

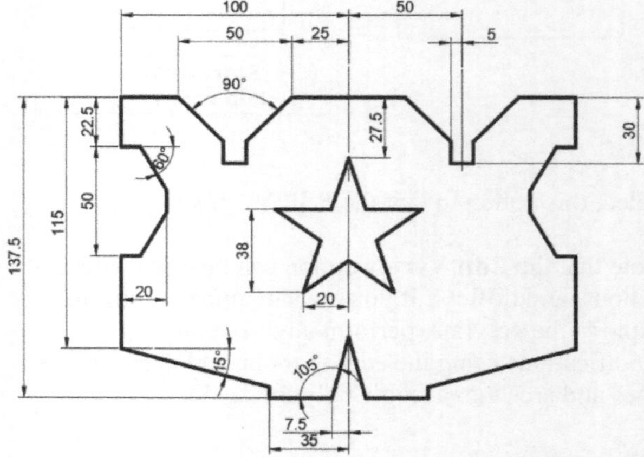

- Start a new drawing file using the acadISO-Named Plot Styles.dwt.
- Type LIMMAX in the command line and press ENTER.
- Type 200,200 and press ENTER.
- Click **Zoom** drop-down > **Zoom All** on the Navigation bar.
- Click the **Line** tool on the **Draw** panel of the **Home** ribbon tab.
- Pick a point in the graphics area.
- Move the pointer toward the right.
- Type 35 and press ENTER.
- Move the pointer upward.
- Type 22.5 and press ENTER.
- Move the pointer toward the right.
- Type 65 and press ENTER.
- Move the pointer upward.
- Type 42.5 and press ENTER.
- Move the pointer toward left.
- Type 20 and press ENTER.
- Move the pointer upward.
- Type 50 and press ENTER.
- Move the pointer toward the right.
- Type 20 and press ENTER.
- Move the pointer upward.
- Type 22.5 and press ENTER.

- Move the pointer toward left.
- Type 45 and press ENTER.
- Move the pointer downward.
- Type 30 and press ENTER.
- Move the pointer toward left.
- Type 10 and press ENTER.
- Move the pointer upward.
- Type 30 and press ENTER.
- Move the pointer toward left.
- Type 45 and press ENTER.
- Move the pointer downward and select the start point of the drawing.

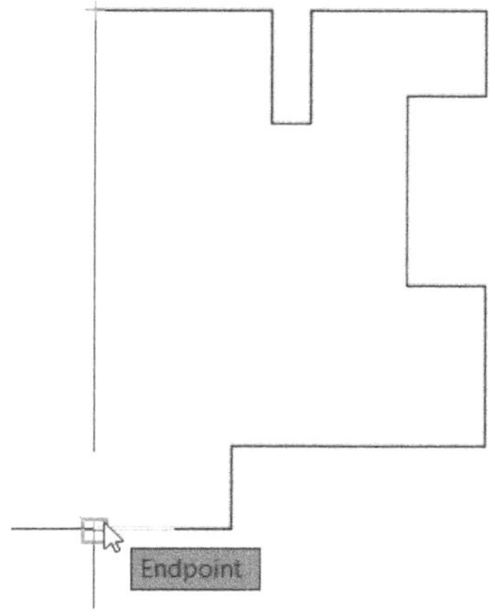

- Click **Fillet** drop-down > **Chamfer** on the **Modify** panel of the **Home** ribbon tab.
- Select **Angle** from the command line.
- Type 20 as the chamfer distance.
- Type 45 as the chamfer angle.
- Select the horizontal and vertical lines, as shown in the figure.

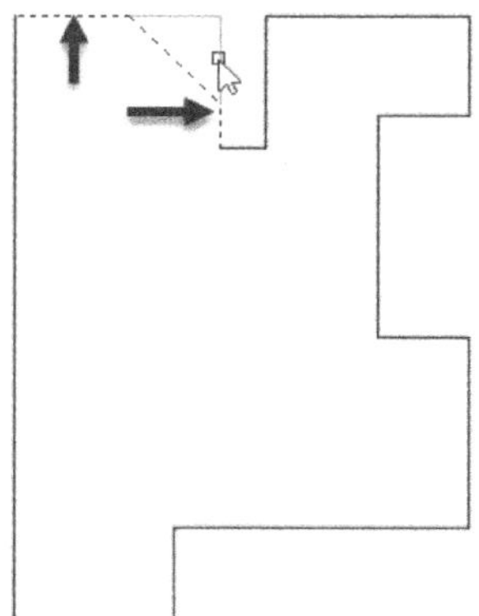

- Press ENTER to activate the **Chamfer** tool.
- Select the horizontal and vertical lines, as shown.

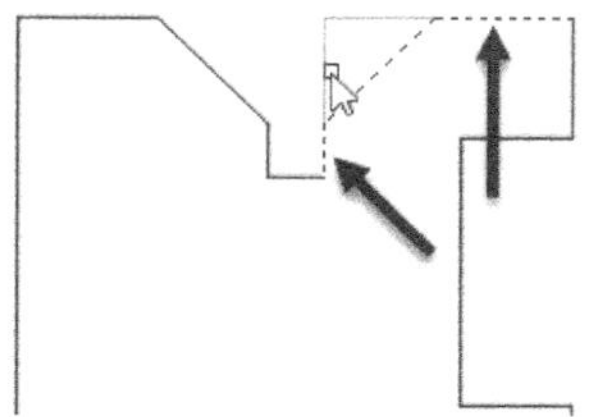

- Press ENTER to activate the **Chamfer** tool.
- Select **Angle** from the command line.
- Type 20 as the chamfer distance.
- Type 30 as the chamfer angle.
- Select the vertical and horizontal lines, as shown in the figure.

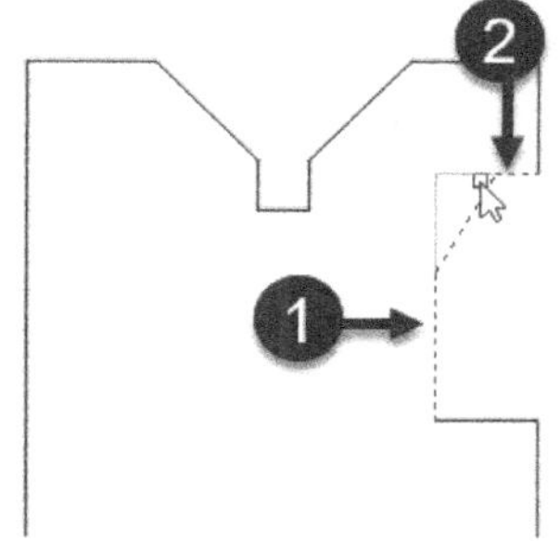

- Press ENTER to activate the **Chamfer** tool.
- Select the vertical and horizontal lines, as shown in the figure.

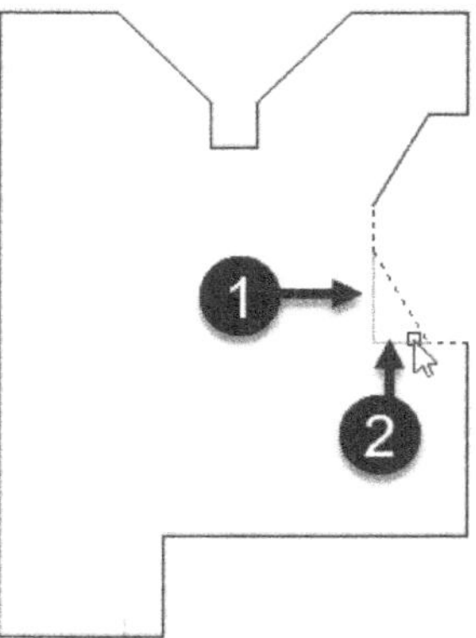

- Press ENTER to activate the **Chamfer** tool.
- Select **Angle** from the command line.
- Type 65 as the chamfer distance.
- Type 15 as the chamfer angle.
- Select the horizontal and vertical lines, as shown in the figure.

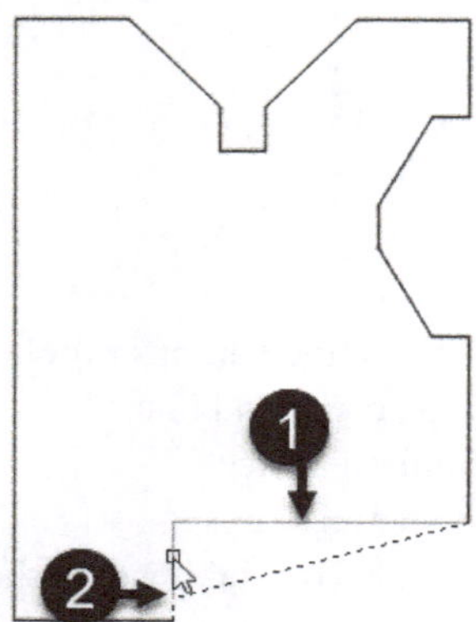

- Press ENTER to activate the **Chamfer** tool.
- Select **Angle** from the command line.
- Type 7.5 as the chamfer distance.
- Type 75 as the chamfer angle.
- Select the horizontal and vertical lines, as shown in the figure.

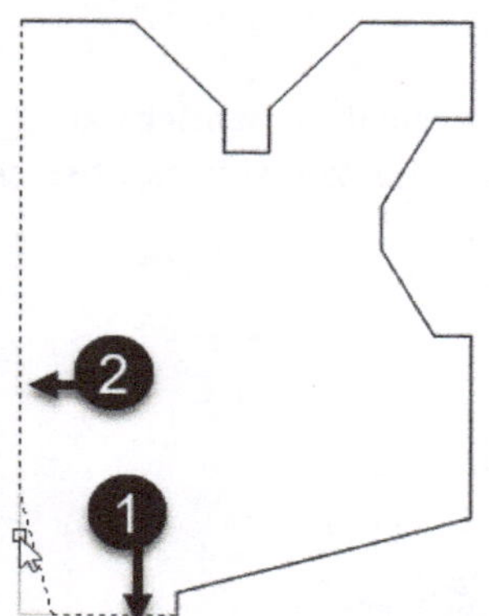

- Select the left vertical line and press Delete.

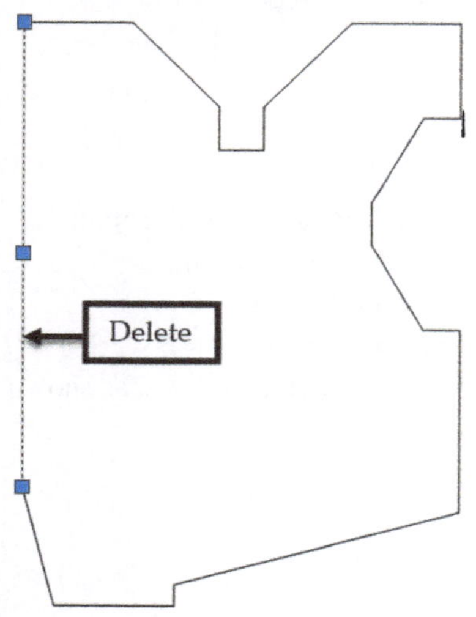

- Create a selection window across all the objects.

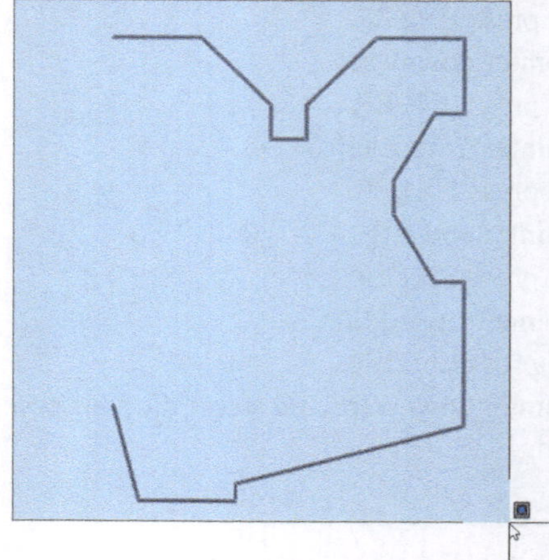

- Click **Mirror** on the **Modify** panel of the **Home** ribbon tab.
- Select the points, as shown. The mirror line is defined.

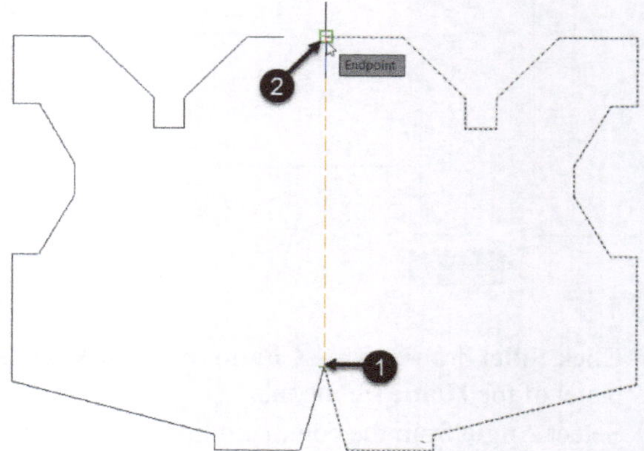

- Select **No** from the command line. The selected objects are mirrored about the mirror line.
- Click **Rectangle** drop-down > **Polygon** on the **Draw** panel of the **Home** ribbon tab.
- Type 5 and press ENTER.
- Select the **Edge** option from the command line.
- Click in the empty area.
- Move the pointer toward the right.

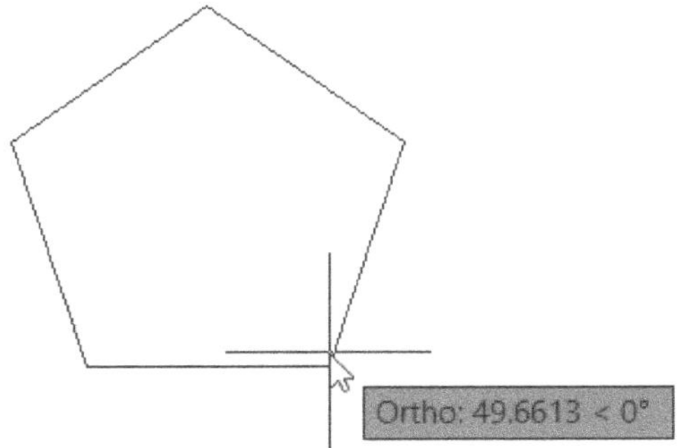

- Type **40** and press ENTER.
- Select the polygon and click the **Move** icon on the **Modify** panel of the **Home** ribbon tab.
- Select the vertex point of the polygon, as shown.

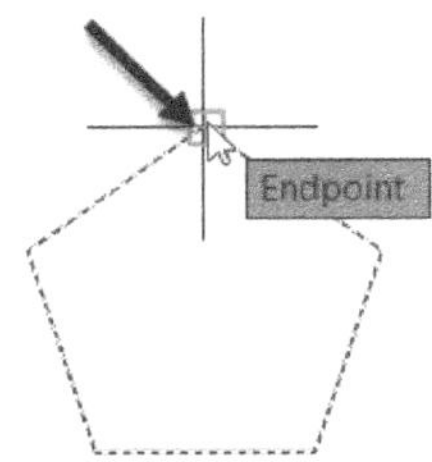

- Move the pointer and select the endpoint of the line, as shown.

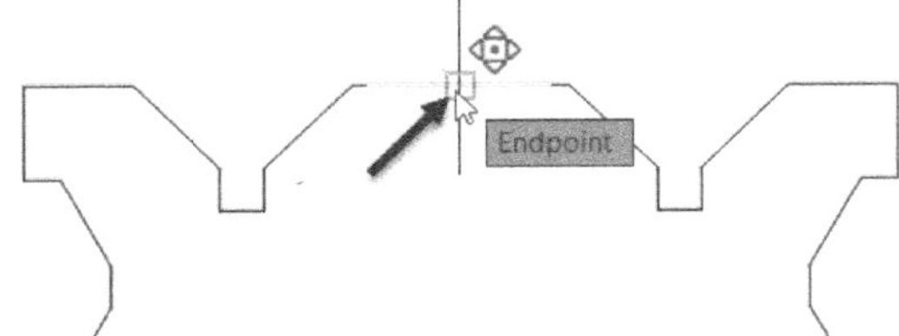

- Select the polygon and click the **Move** icon on the **Modify** panel of the **Home** ribbon tab.

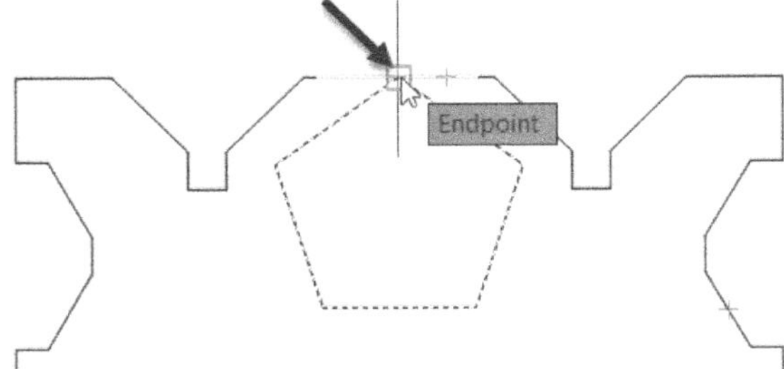

- Select the vertex point of the polygon, as shown.
- Move the pointer downward.
- Type 27.5 and press ENTER.

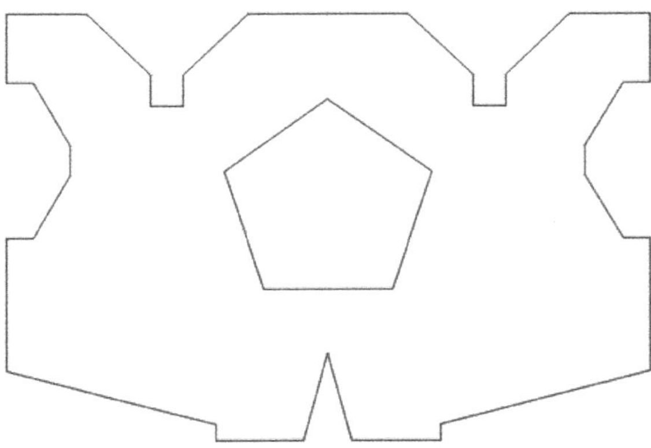

- Click the **Line** icon on the **Draw** panel of the **Home** ribbon tab.
- Deactivate the ORTHOMODE icon on the status bar.
- Select the vertices of the polygon, as shown.

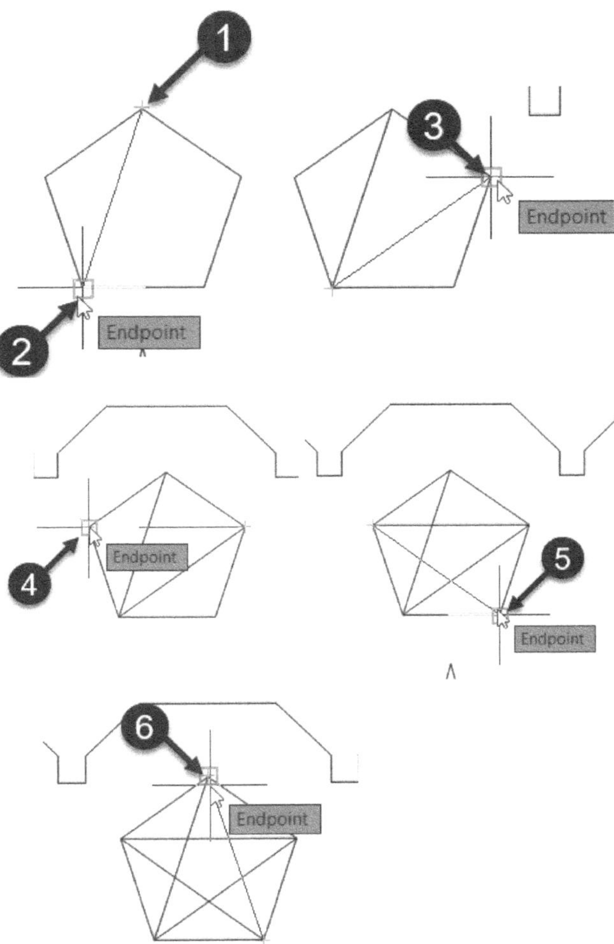

- Press ESC.
- Select the polygon and press Delete.

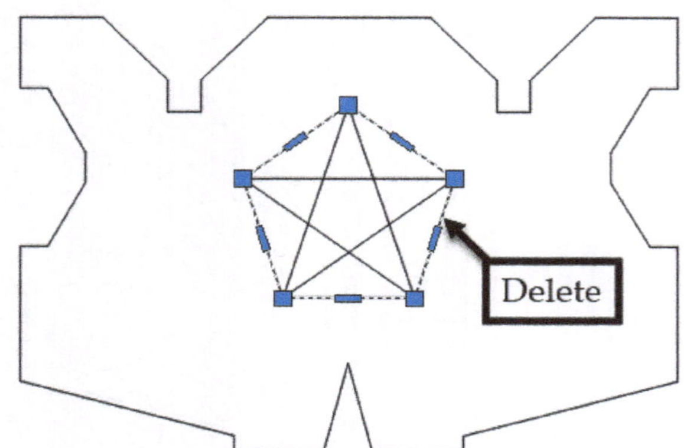

- Click the **Trim** icon on the **Modify** panel of the **Home** ribbon tab.
- Select the inner lines of the star, as shown.

- Save and close the drawing file.

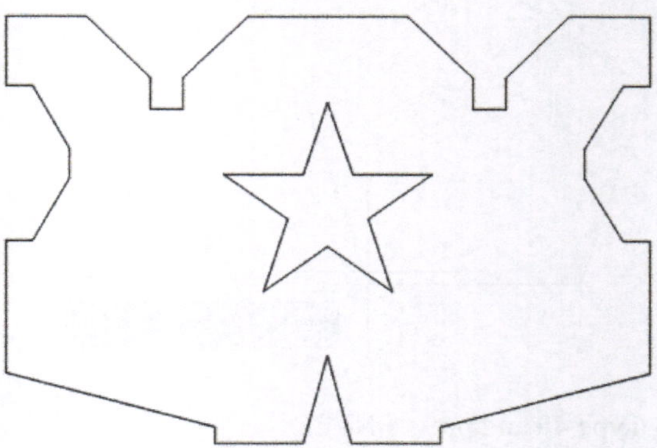

Exercises

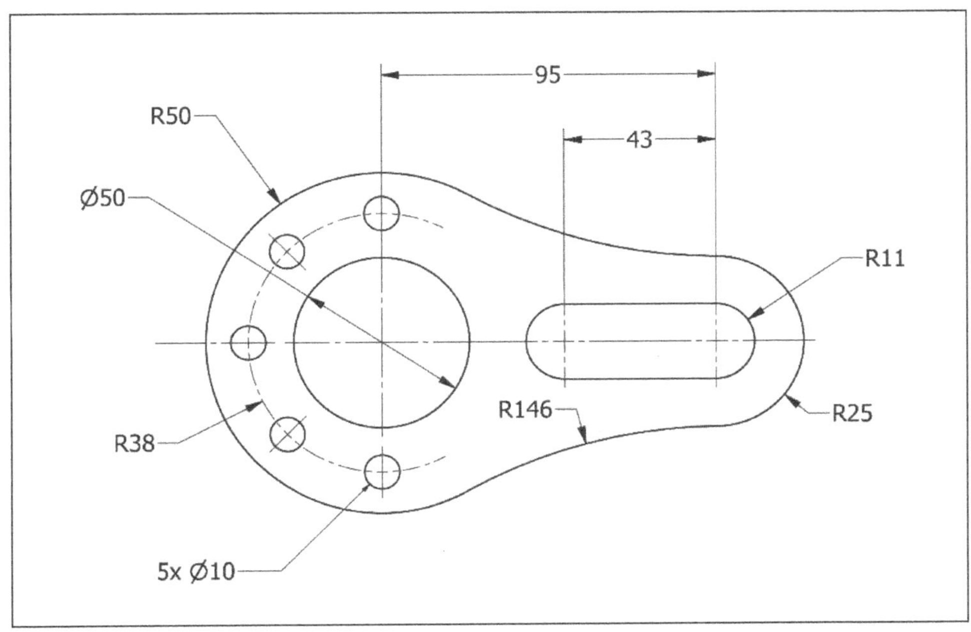

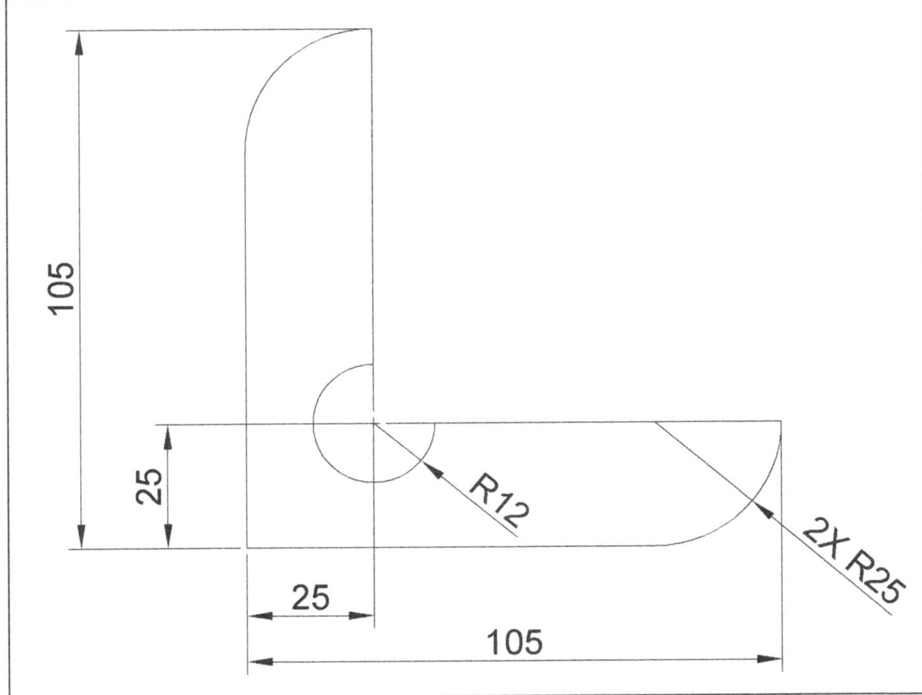

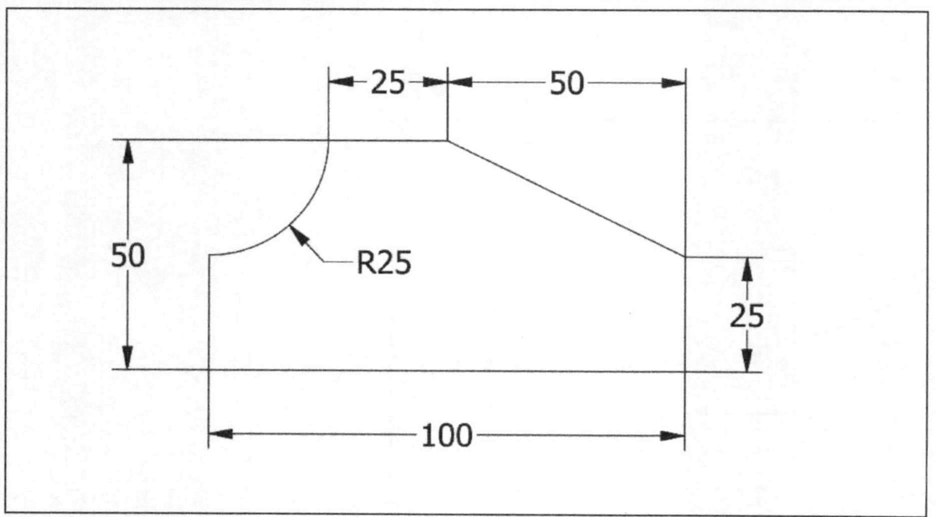
25
50
50
R25
25
100

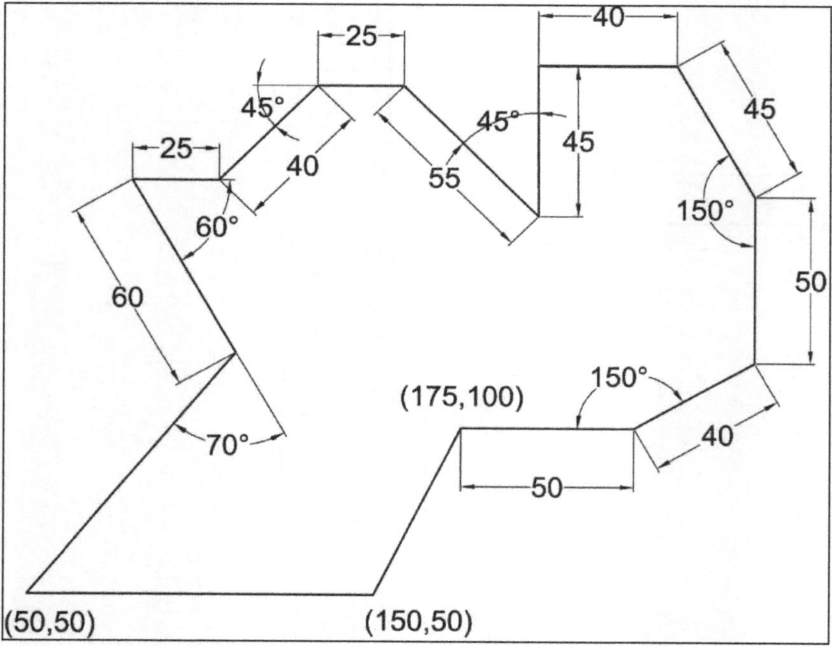
25
40
45°
25
40
45°
45
55
45
60°
150°
60
50
150°
(175,100)
70°
40
50
(50,50)
(150,50)

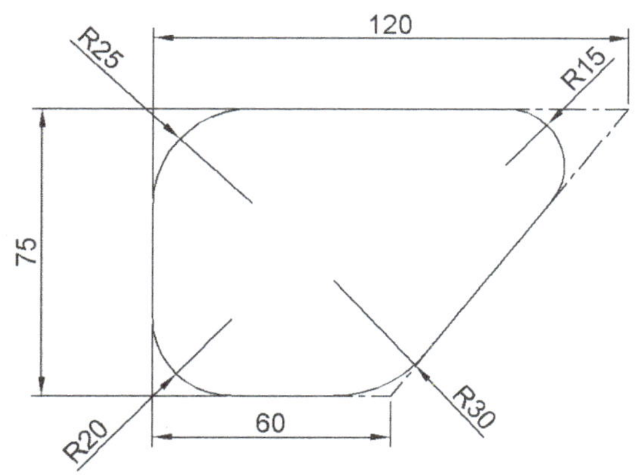
120
R25
R15
75
R20
60
R30

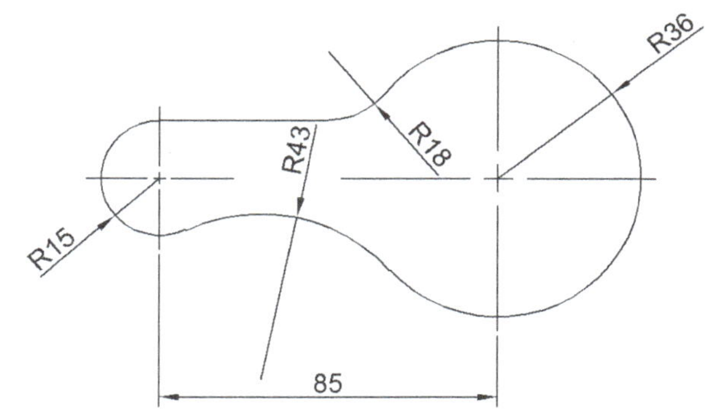
R36
R18
R43
R15
85

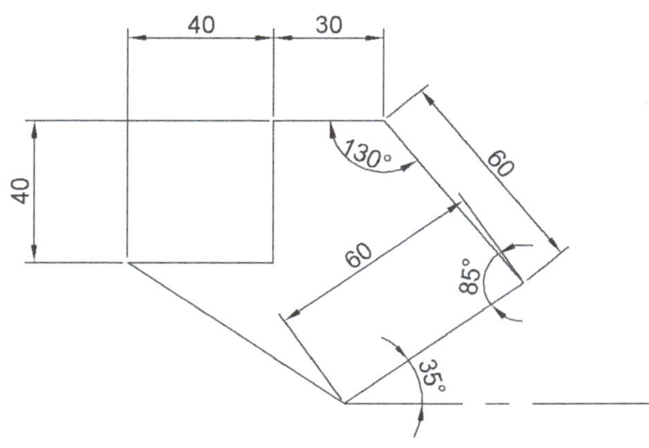
40
30
40
130°
60
60
85°
35°

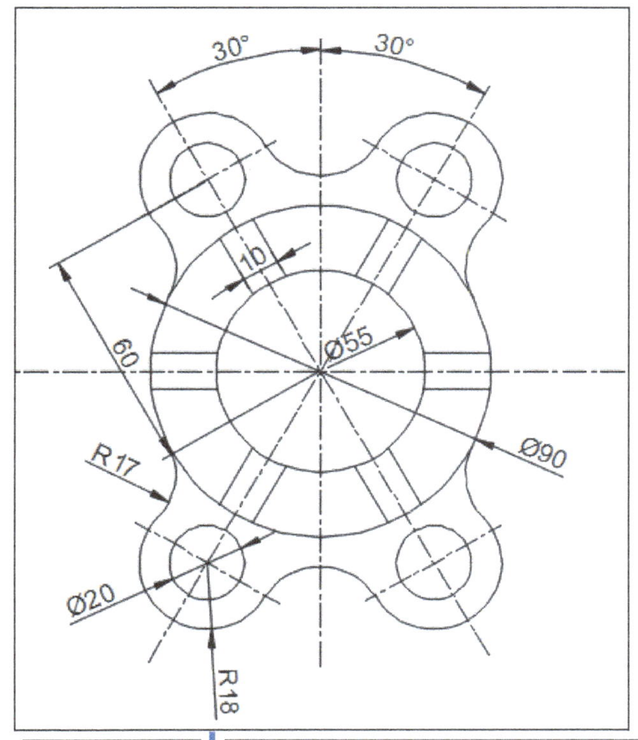
30°
30°
10
60
Ø55
R17
Ø90
Ø20
R18

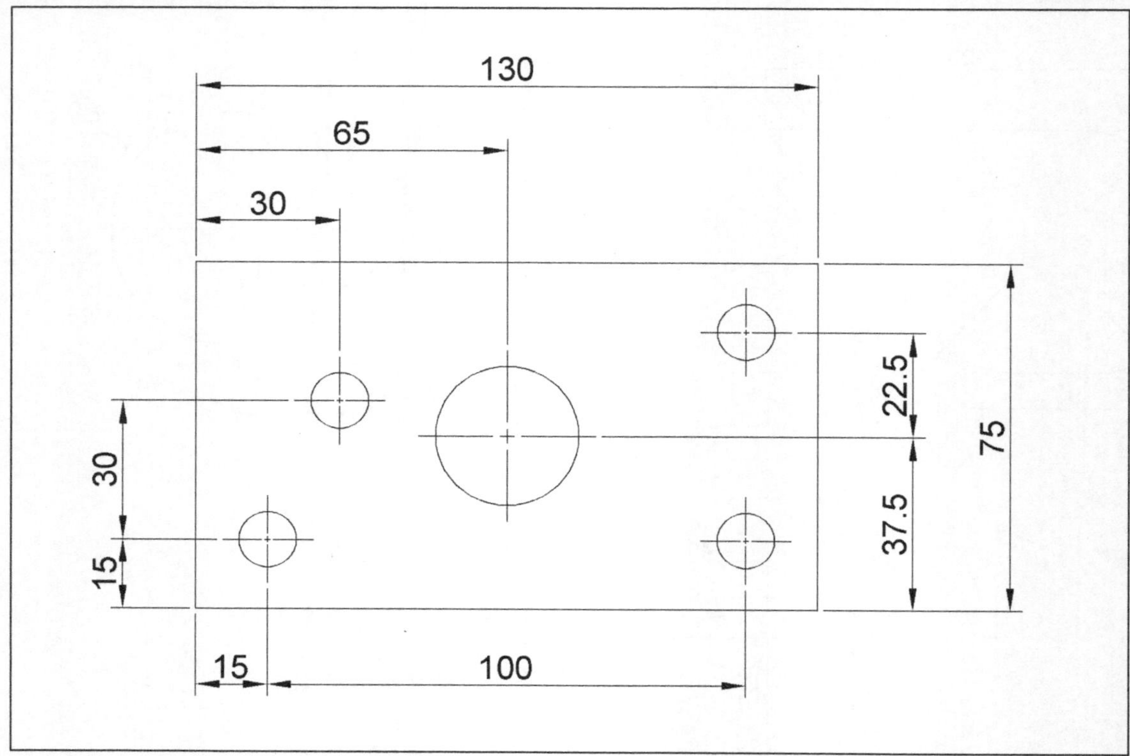
130
65
30
22.5
75
30
37.5
15
15
100

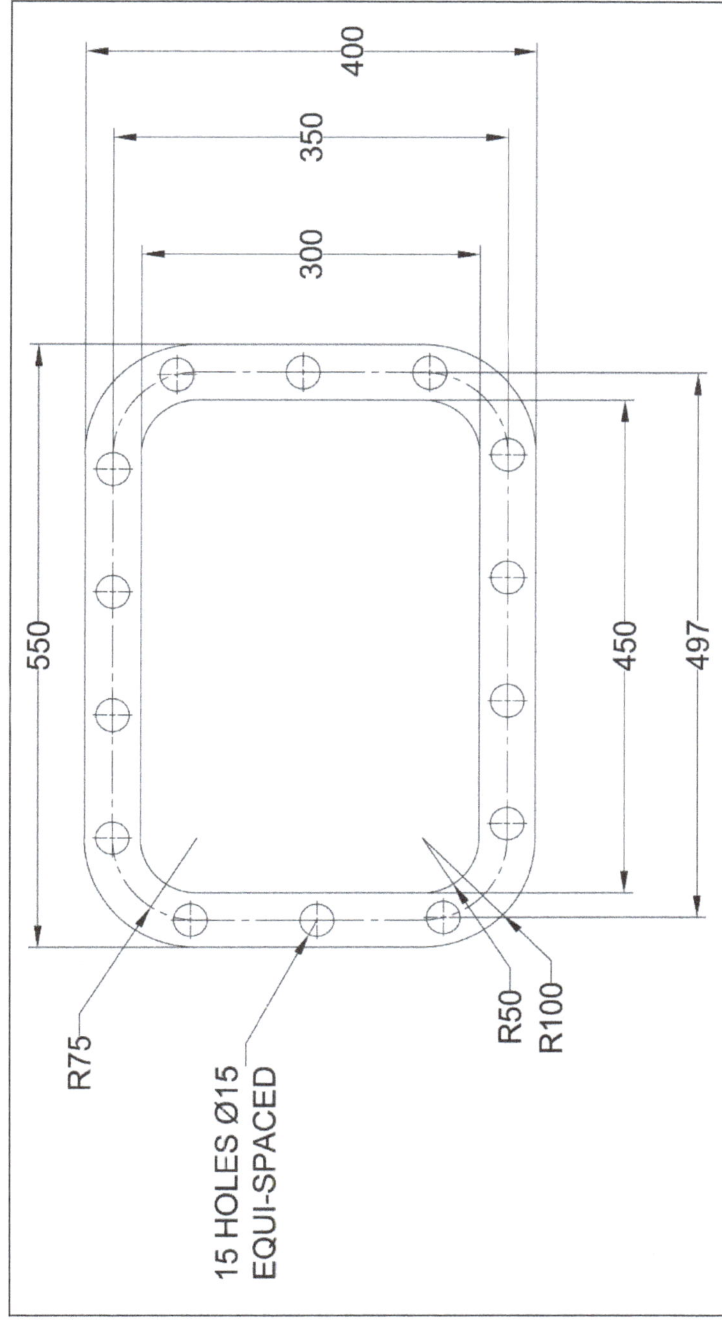
400
350
300
550
450
497
R75
15 HOLES Ø15
EQUI-SPACED
R50
R100

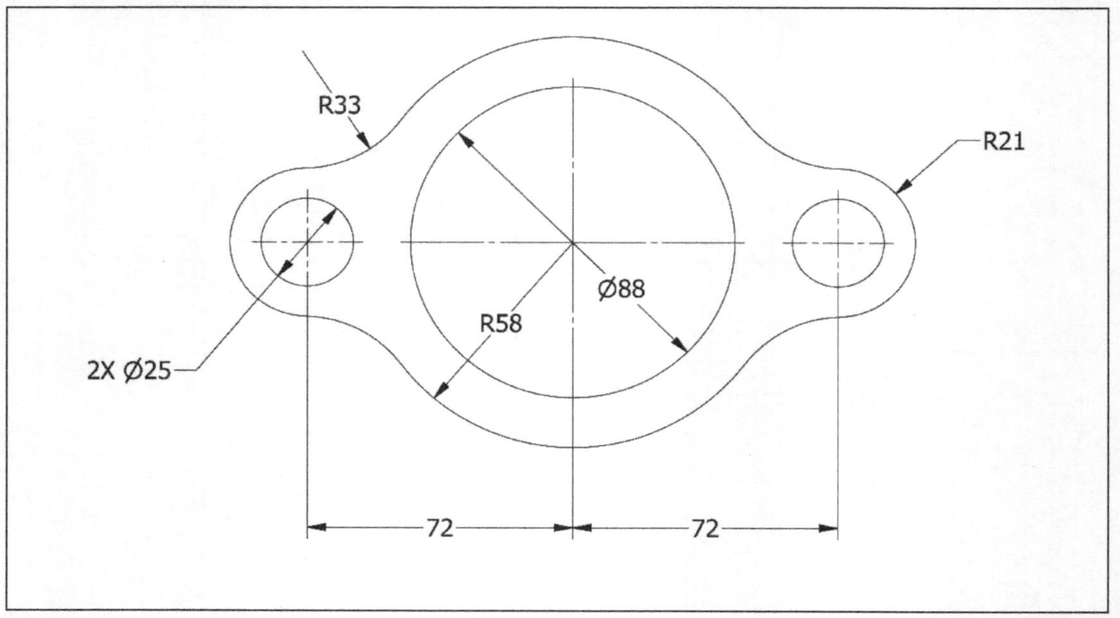
R33
R21
Ø88
R58
2X Ø25
72
72

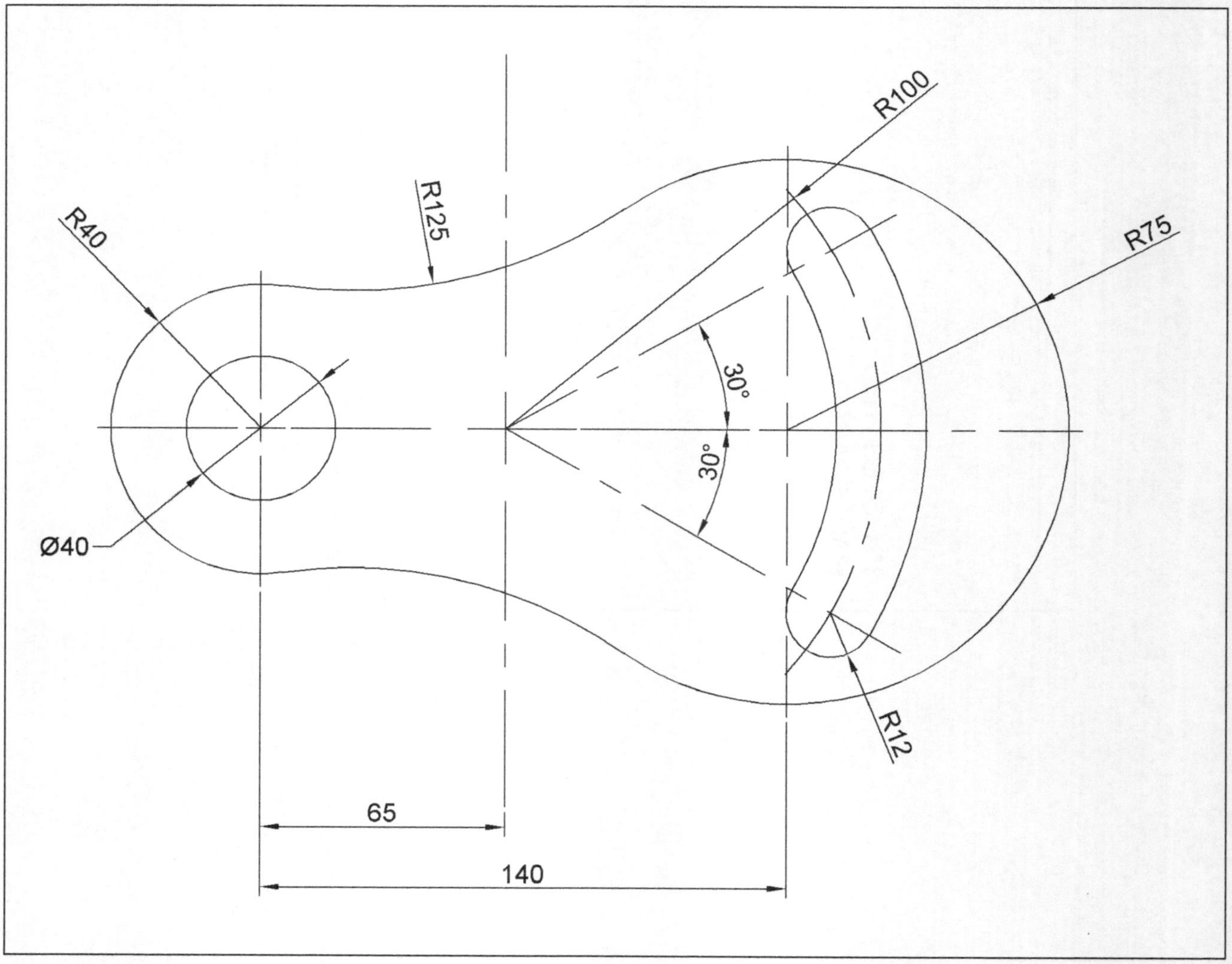

R100
R125
R40
R75
30°
30°
Ø40
R12
65
140

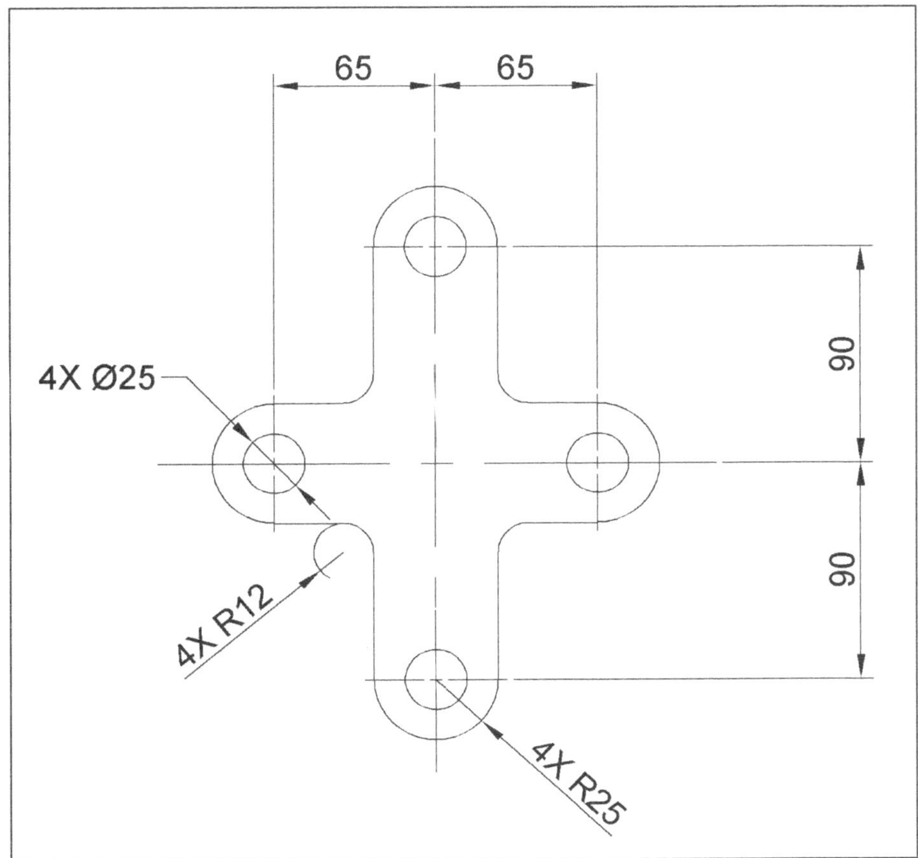
65
65
4X Ø25
90
90
4X R12
4X R25

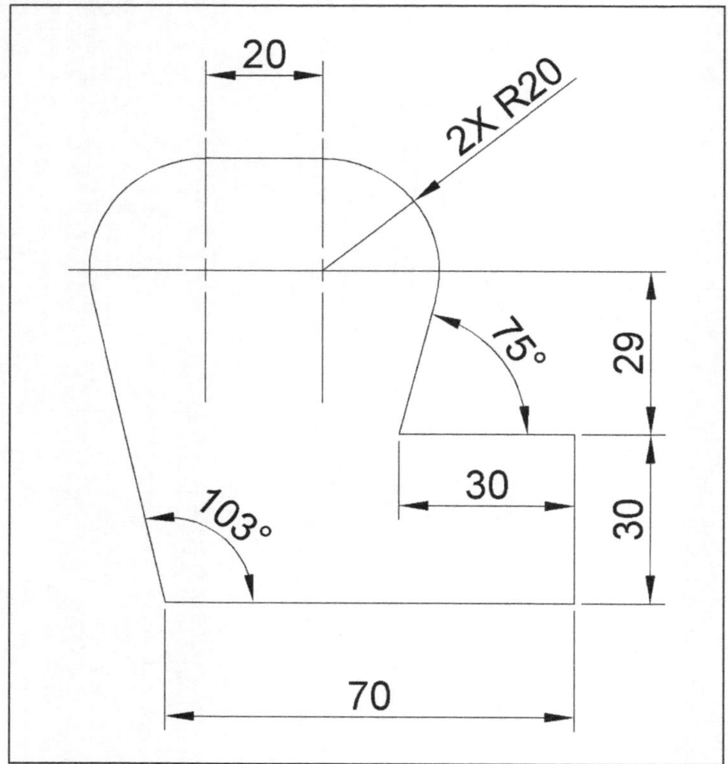
20
2X R20
75°
29
30
103°
30
70

Chapter 5: Multi View Drawings

In this chapter, you will learn to create:

- **Orthographic Views**
- **Auxiliary Views**
- **Named Views**

Multi-view Drawings

To manufacture a component, you must create its engineering drawing. The engineering drawing consists of various views of the object, showing its true shape and size so that it can be dimensioned. It can be achieved by creating the orthographic views of the object. In the first section of this chapter, you will learn to create orthographic views of an object. The second section introduces you to auxiliary views. The auxiliary views clearly describe the features of a component, which are located on an inclined plane or surface.

Creating Orthographic Views

Orthographic Views are standard representations of an object on a sheet. These views are created by projecting an object onto three different planes (top, front, and side planes). You can project an object by using two different methods: **First Angle Projection** and **Third Angle Projection**. The following figure shows the orthographic views that will be created when an object is projected using the **First Angle Projection** method.

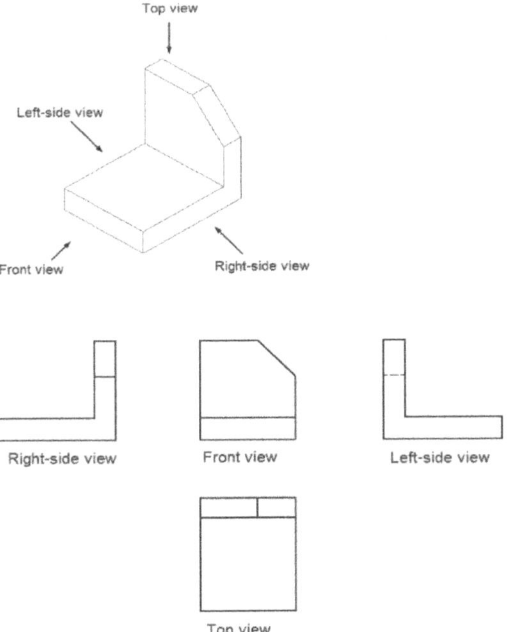

The following figure shows the orthographic views that will be created when an object is projected using the **Third Angle Projection** method.

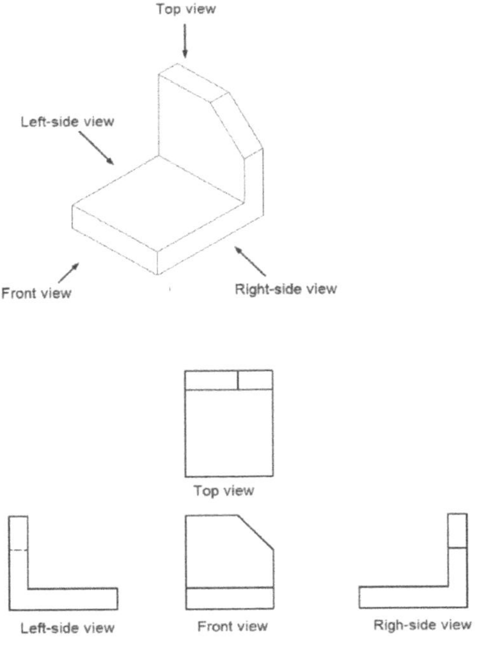

Example:

In this example, you will create the orthographic views of the part shown below. The views will be created by using the **Third Angle Projection** method.

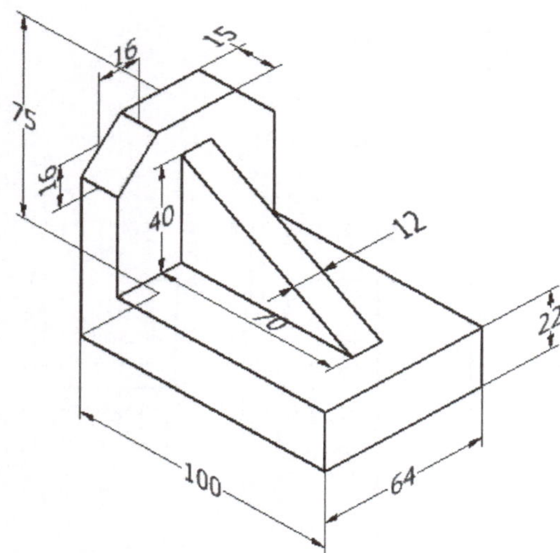

- Open a new drawing using the **acadISO – Named Plot Styles.dwt** template.
- Click the **Layer Properties** button on the **Layers** panel; the **Layer Properties Manager** appears.
- Click the **New Layer** button on the **Layer Properties Manager** to create new layers.
- Create two new layers with the following properties.

Layer Name	Lineweight	Linetype
Construction	0.00 mm	Continuous
Object	0.30 mm	Continuous

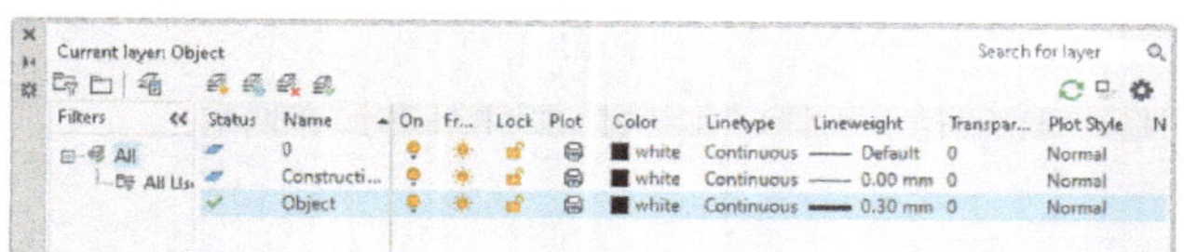

- Right-click on the **Construction** layer and select **Set current**.
- Close the **Layer Properties Manager**.
- Activate the **Ortho Mode** icon on the status bar.
- Click **Zoom > Zoom All** on the Navigation Bar. Next, you need to draw construction lines. They are used as references to create actual drawings. You will create these construction lines on the **Construction** layer so that you can hide them when required.
- Click **Home > Draw > Construction** line on the ribbon or enter **XLINE** in the command line.

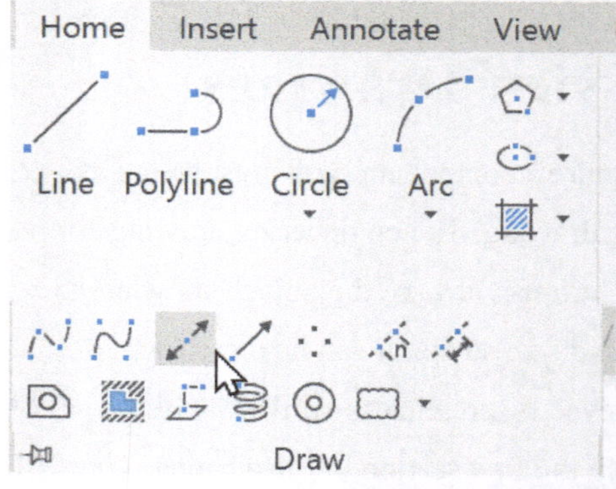

- Click anywhere in the lower-left corner of the graphics window.
- Move the pointer upward and click to create a vertical construction line.
- Move the pointer toward the right and click to create a horizontal construction line.

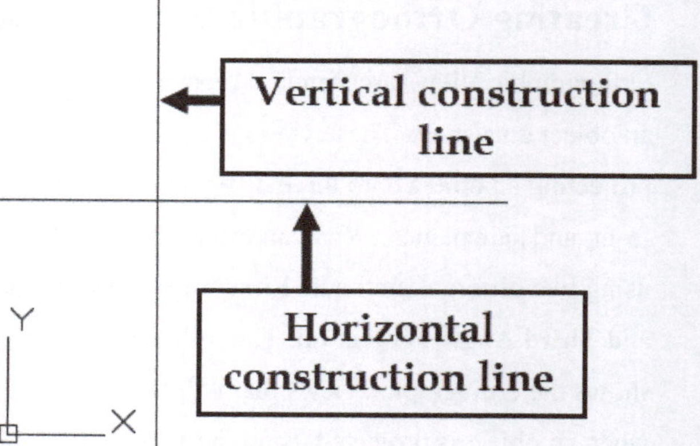

- Press ENTER to exit the tool.
- Click the **Offset** button on the **Modify** panel.

- Type 100 as the offset distance and press ENTER.
- Select the vertical construction line.
- Move the pointer toward the right and click to create an offset line.

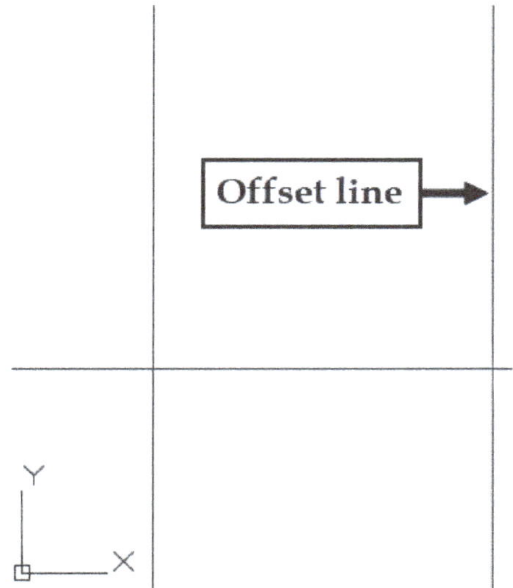

- Right-click and select **Enter** to exit the **Offset** tool.
- Press the SPACEBAR on the keyboard to start the **Offset** tool again.
- Type 75 as the offset distance and press ENTER.
- Select the horizontal construction line.
- Move the pointer above and click to create the offset line.

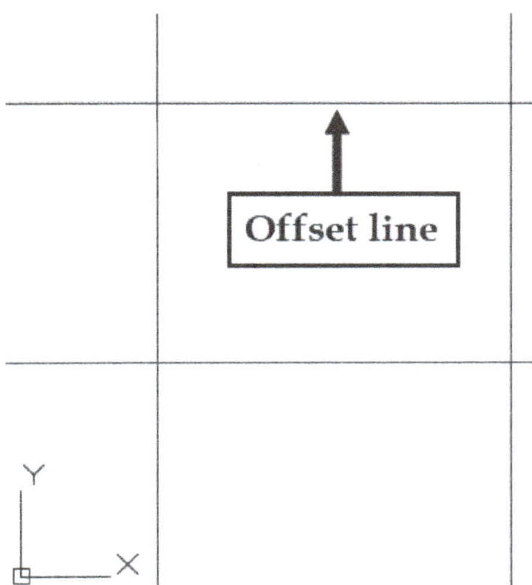

- Press ENTER to exit the **Offset** tool.
- Likewise, create other offset lines, as shown below. The offset dimensions are displayed in the image. Do not add dimensions to the lines.

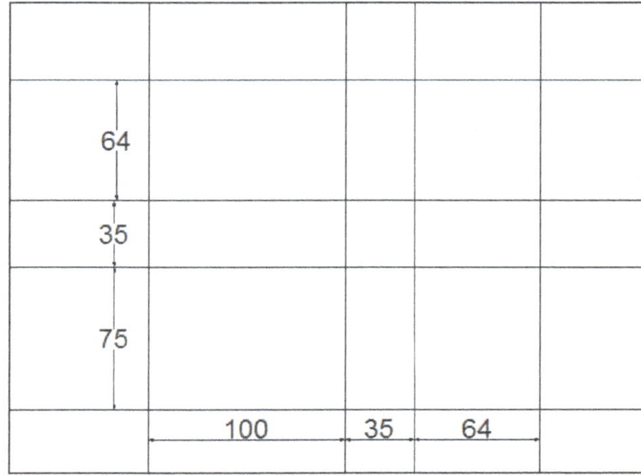

- Activate the **Object** layer.

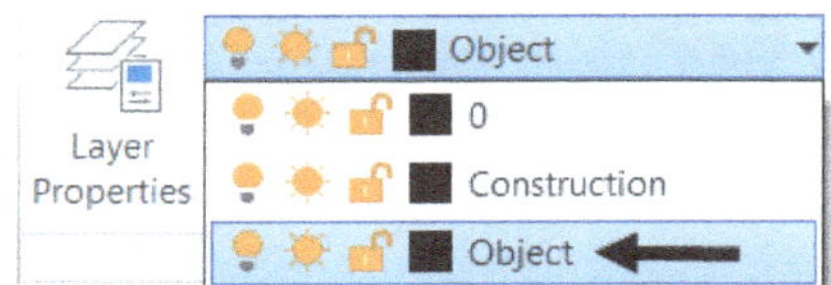

Now, you must create the object lines.

- Activate the **Show/Hide Lineweight** button on the status bar.
- Click the **Line** button on the **Draw** panel.
- Select the intersection points of the construction lines, as shown.

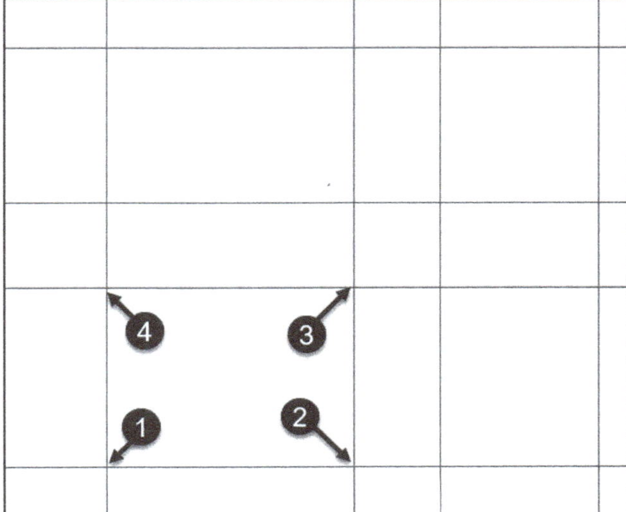

- Select the **Close** option from the command line to create the outline of the front view.

- Likewise, create the outlines of the top and side views.

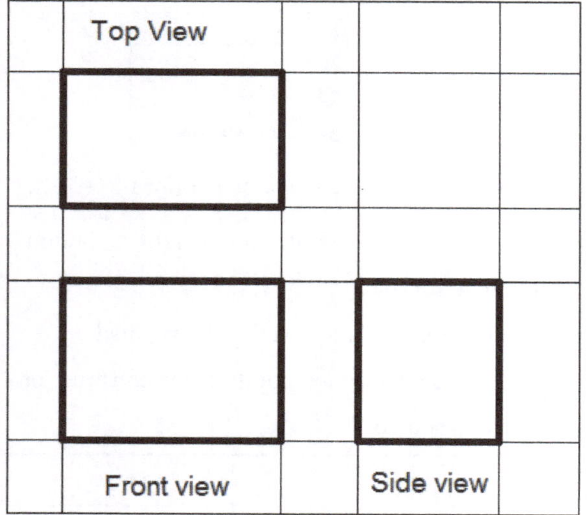

Next, you must turn off the **Construction** layer.

- Click on the **Layer** drop-down in the **Layers** panel.
- Click the light-bulb of the **Construction** layer; the layer will be turned off.

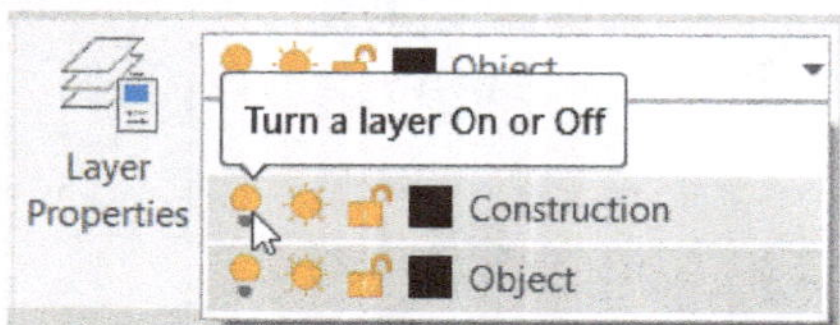

- Use the **Offset** tool and create two parallel lines on the front view, as shown below.

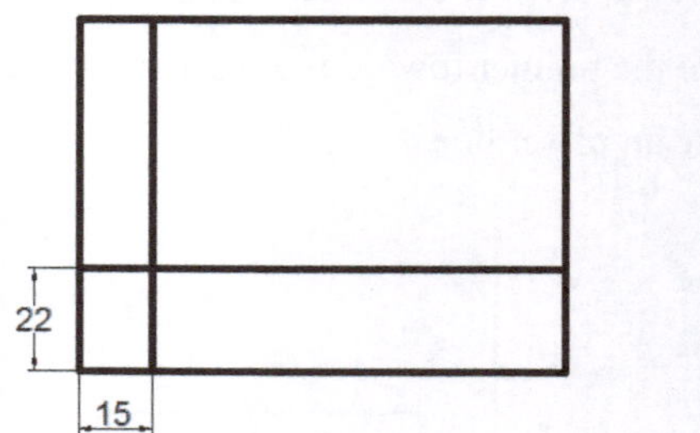

- Use the **Trim** tool and trim the unwanted lines of the front view, as shown below.

- Use the **Offset** tool to create the parallel line, as shown below.

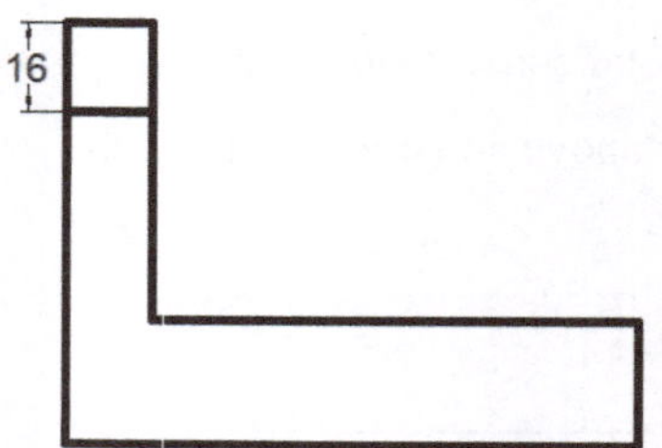

- Use the **Offset** tool and create offset lines in the Top view, as shown below.

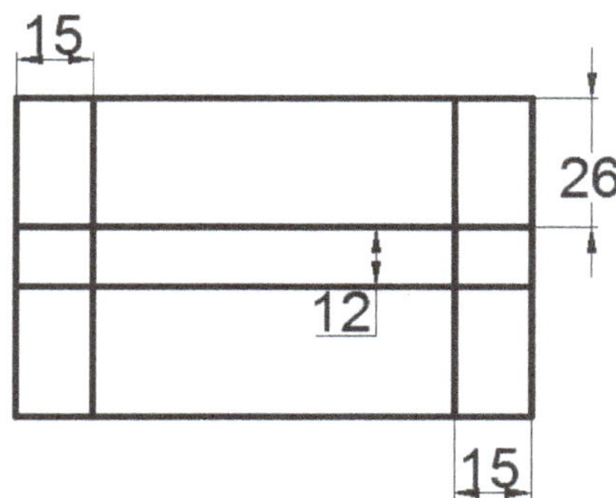

- Use the **Trim** tool and trim unwanted objects.

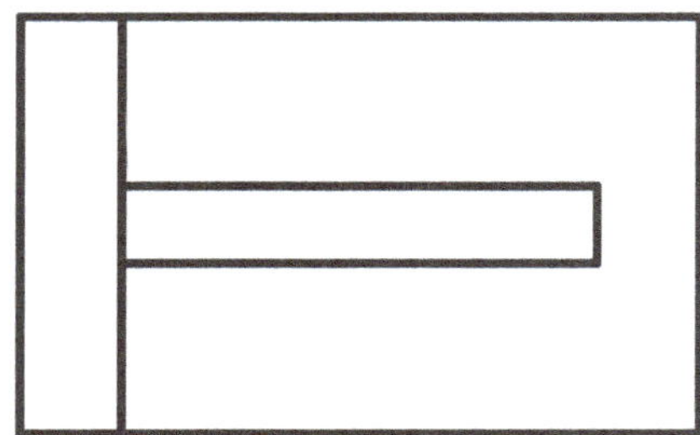

- Create other offset lines and trim the unwanted portions, as shown below.

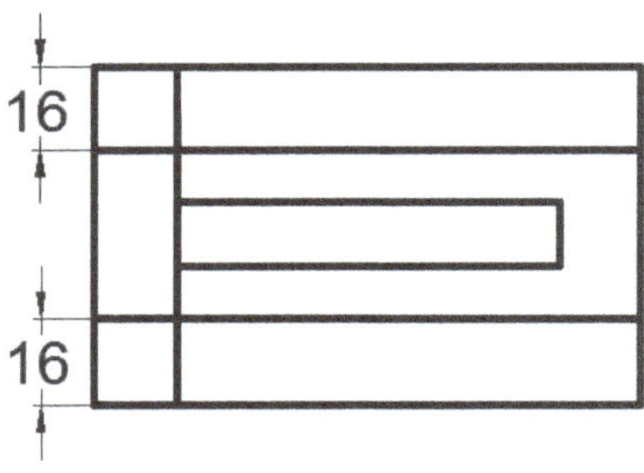

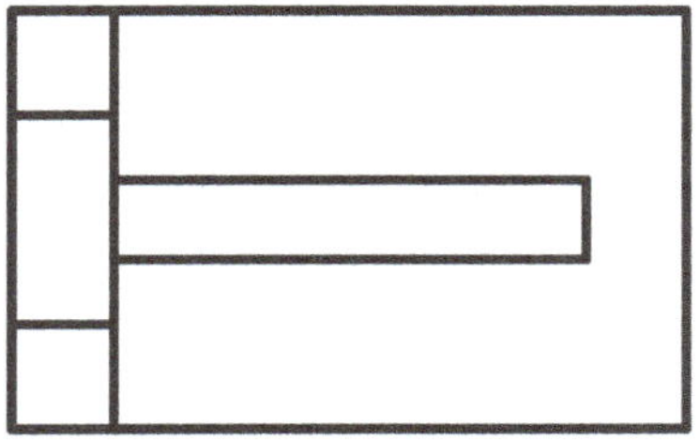

- Deactivate the **Ortho Mode** icon on the status bar.
- Click the **Line** button on the **Draw** panel.
- Press and hold the SHIFT key and right-click. Select the **From** option.

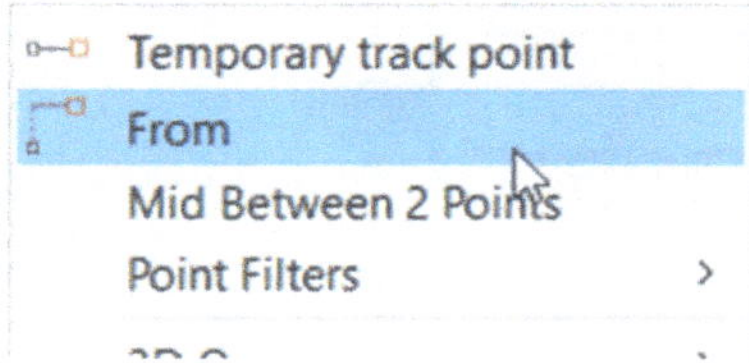

- Select the endpoint of the line in the front view, as shown below.

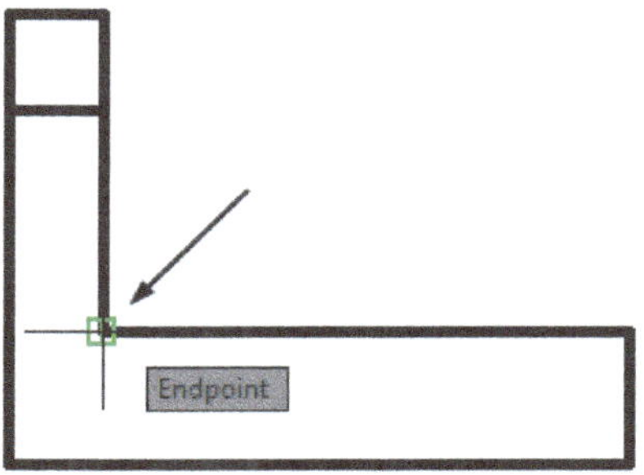

- Move the pointer on the vertical line and enter **40** in the command line; the first point of the line is specified at a point 40 mm away from the endpoint. Also, a rubber band line will be attached to the pointer.

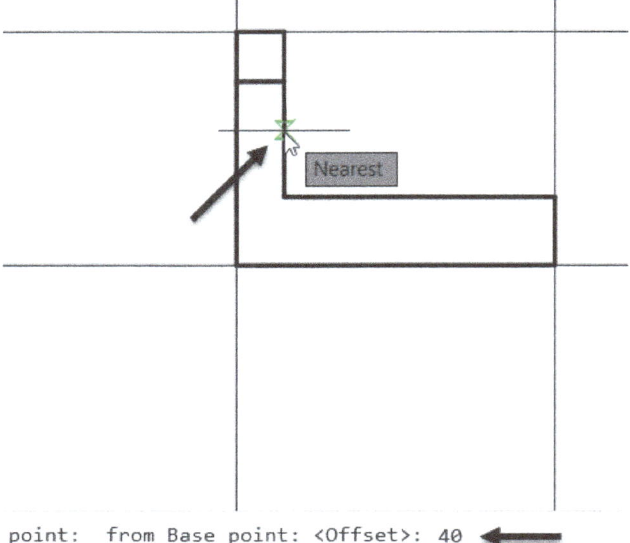


```
point: _from Base point: <Offset>: 40
```

- Move the pointer onto the endpoint on the top view, as shown below.

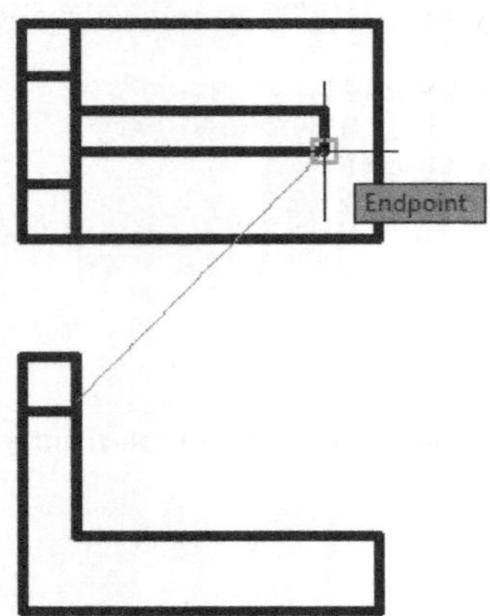

- Move the pointer vertically downward; you will notice the track lines.
- Move the pointer near the horizontal line of the front view and click at the intersection point, as shown below. Press ENTER to exit the tool.

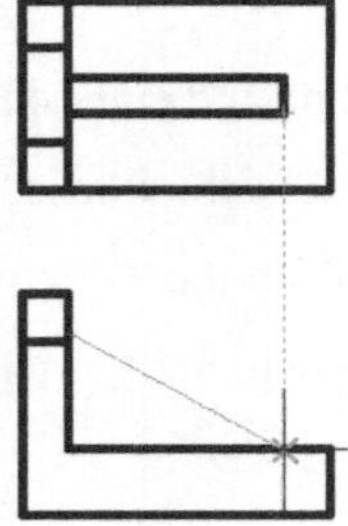

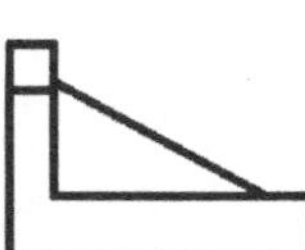

Next, you must create the right-side view. To do this, you must draw a 45- degree miter line and project the measurements of the top view onto the side view.

- Click on the **Layer** drop-down in the **Layers** panel.
- Click the light-bulb icon of the **Construction** layer; the **Construction** layer is turned on.
- Select the **Construction** layer from the **Layer** drop-down to set it as the current layer.
- Draw an inclined line by connecting the intersection points of the construction lines, as shown below.

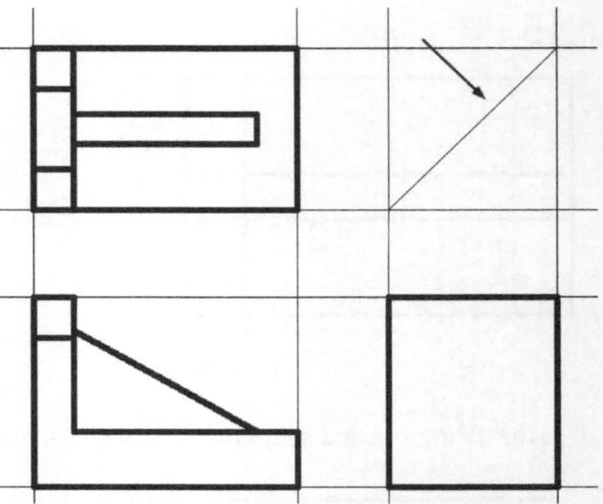

- Click the **Construction Line** button on the **Draw** panel.
- Select the **Hor** option from the command line and select the points on the top and front views, as shown below.

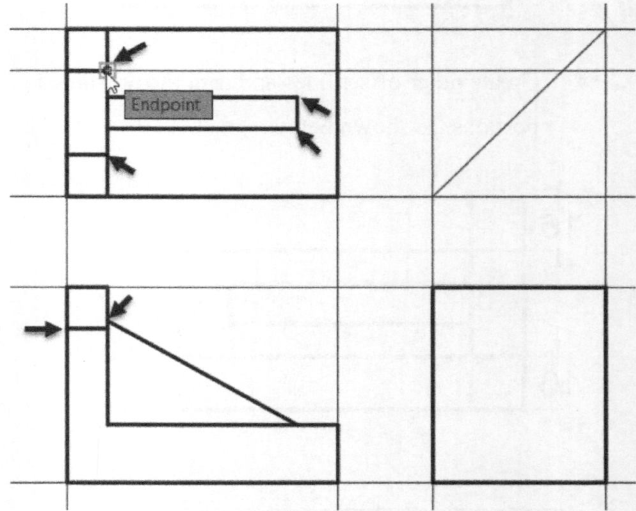

The projection lines are created, as shown below.

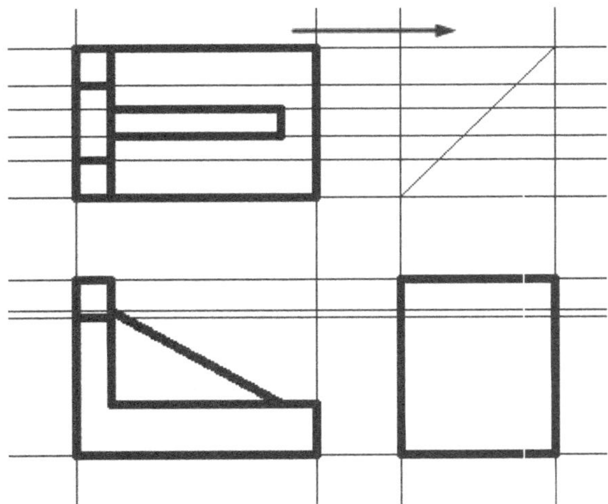

- Right-click to exit the **Construction Line** tool.
- Press ENTER and select the **Ver** option from the command line.
- Create vertical projection lines, as shown below.

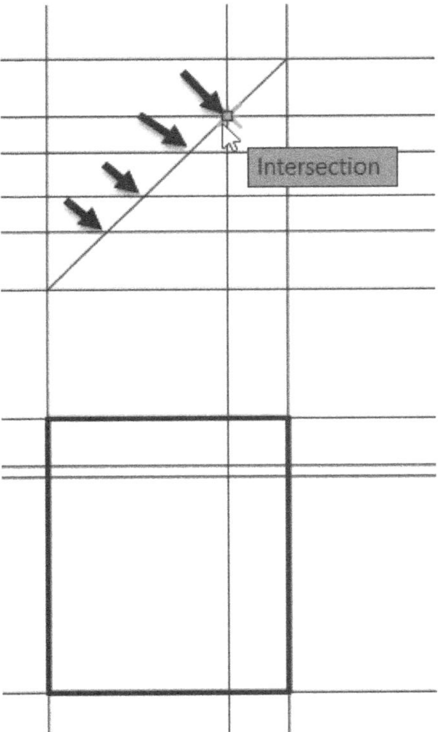

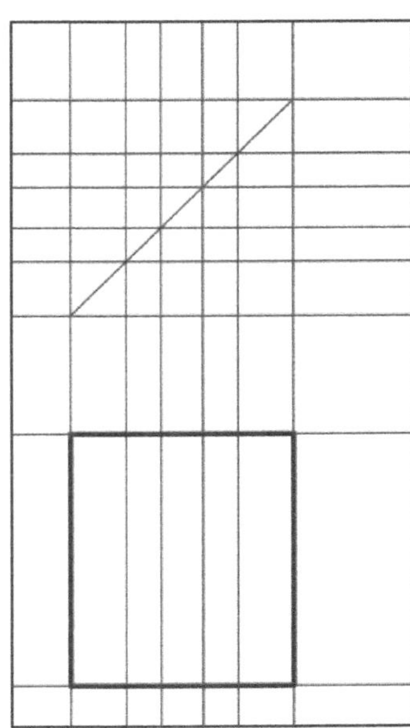

- Use the **Trim** tool and trim the extended portions of the construction lines.

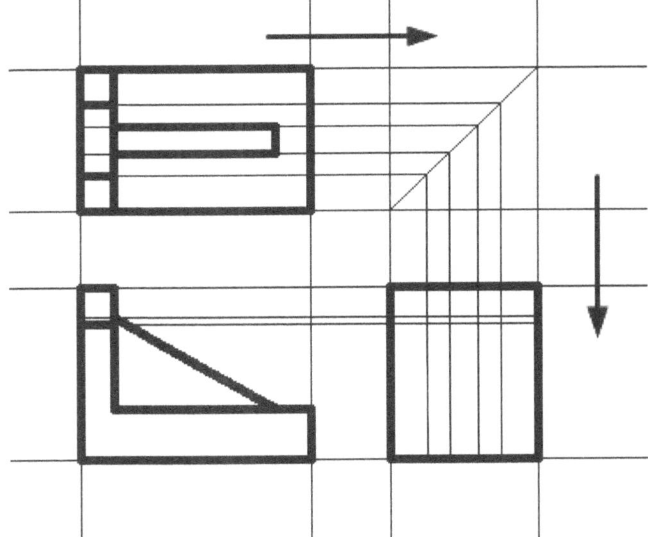

- Set the **Object** layer as current.
- Click the **Offset** button on the **Modify** panel.
- Select the **Through** option from the command line.
- Select the lower horizontal line of the side view.

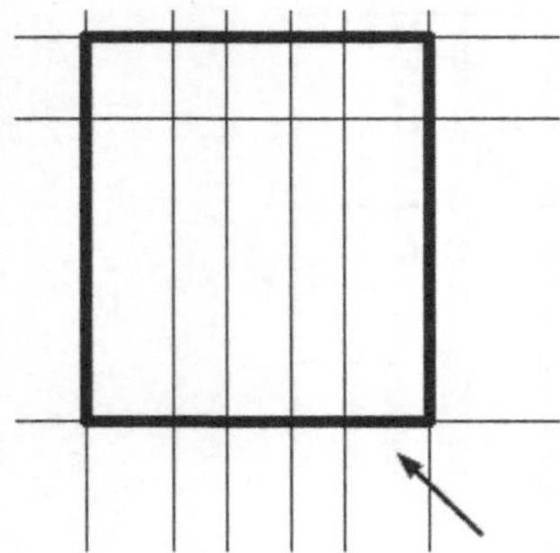

- Select the endpoint on the front view, as shown below.

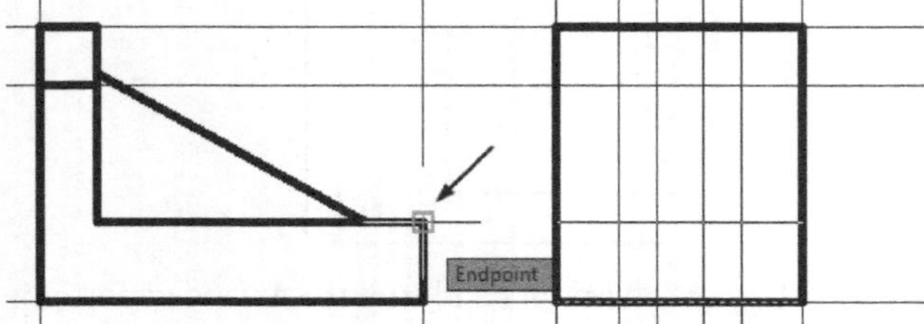

- Click **Exit** in the command line.
- Use the **Line** tool and create the objects in the side view, as shown below.

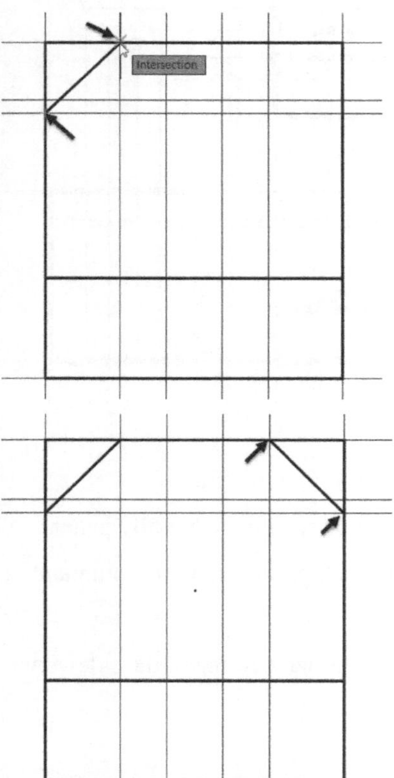

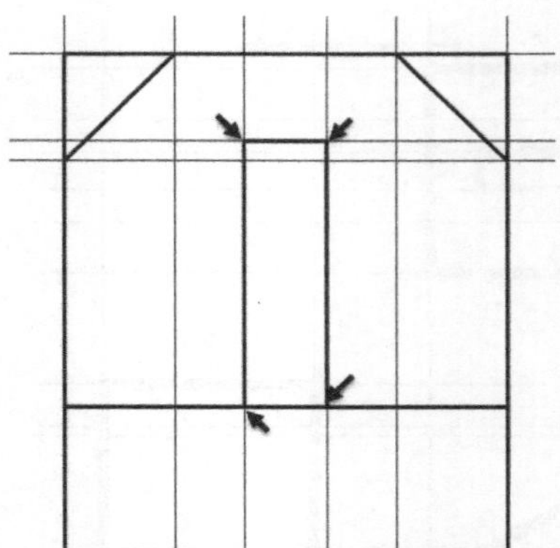

- Turn off the **Construction** layer by clicking on the light-bulb of the **Construction** layer.
- Trim the unwanted portions on the right-side view.

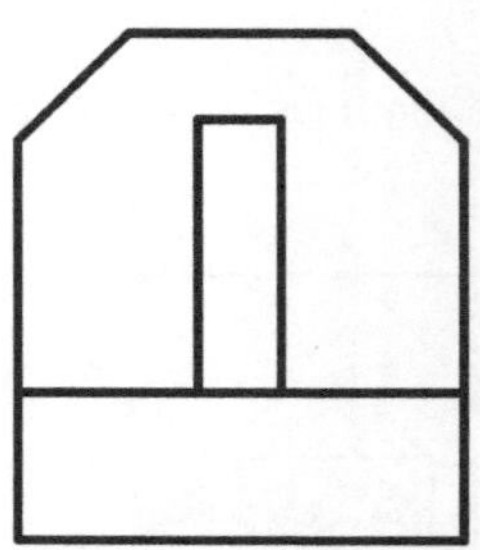

The drawing after creating all the views is shown below.

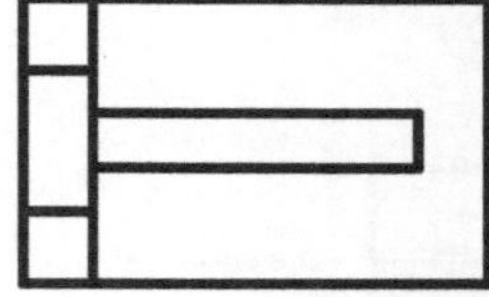

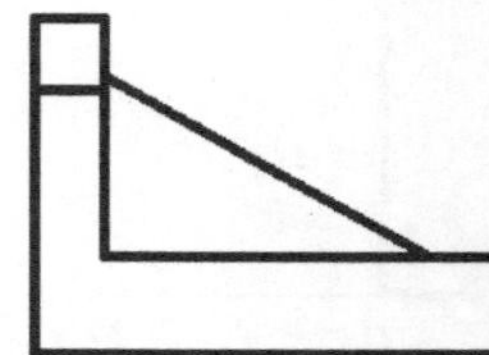

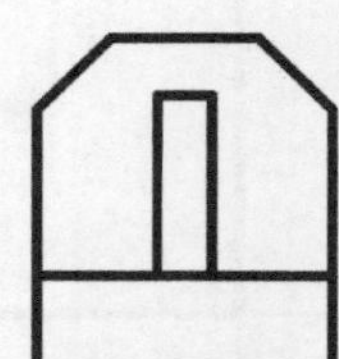

- Save the file as **ortho_views.dwg**. Close the file.

Creating Auxiliary Views

Most of the components are represented by using orthographic views (front, top and side views). But many components have features located on inclined faces. You cannot get the true shape and size for these features by using the orthographic views. To see an accurate size and shape of the inclined features, you must create an auxiliary view. An auxiliary view is created by projecting the component onto a plane other than horizontal, front, or side planes. The following figure shows a component with an inclined face. When you create orthographic views of the component, you will not be able to get the true shape of the hole on the inclined face.

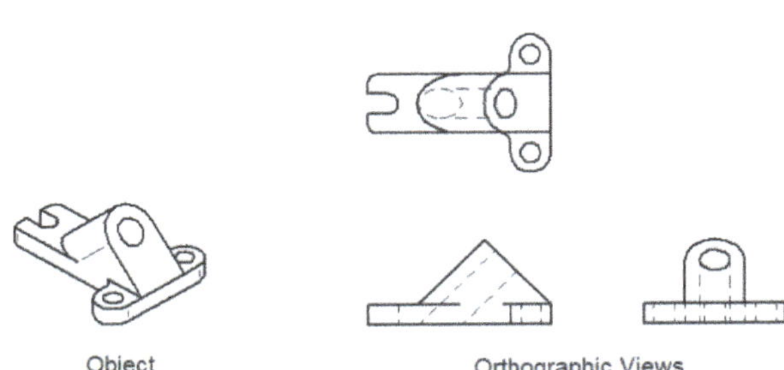

To get the actual shape of the hole, you must create an auxiliary view of the object, as shown below.

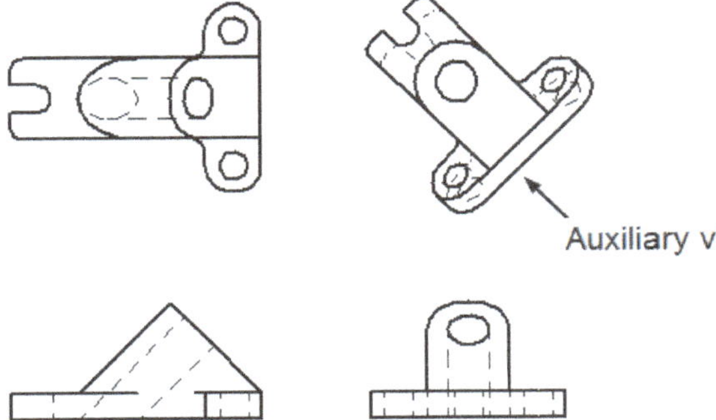

Example:

In this example, you will create an auxiliary view of the object shown below.

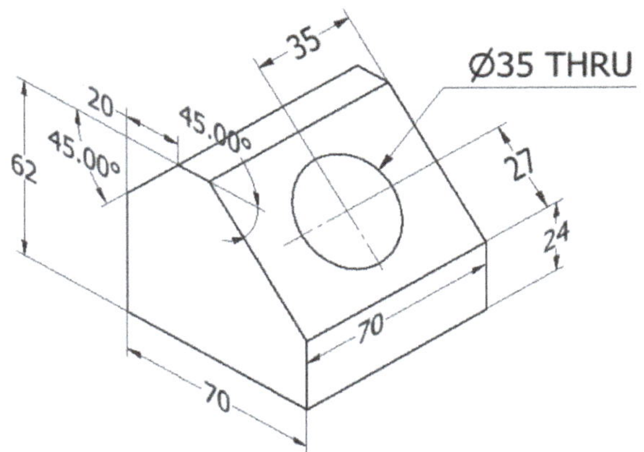

- Open a new AutoCAD file.
- Create four new layers with the following properties.

Layer Name	Lineweight	Linetype
Construction	0.00 mm	Continuous
Object	0.50 mm	Continuous
Hidden	0.30 mm	HIDDEN
Centerline	0.30 mm	CENTER

- Select the **Construction** layer from the **Layer** drop-down in the **Layers** panel.
- Create a rectangle at the lower-left corner of the graphics window, as shown in the figure.

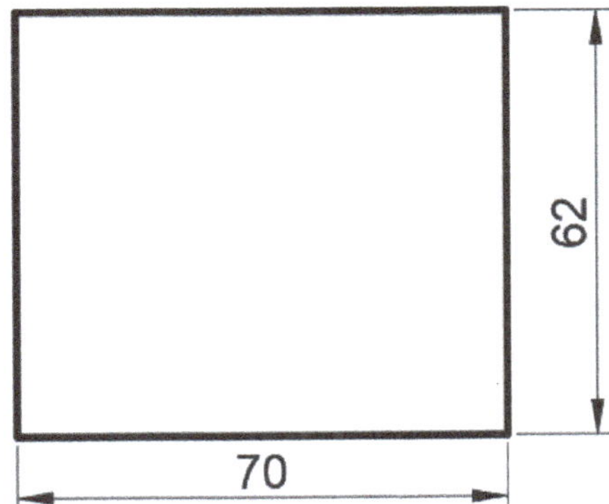

- Select the rectangle and click the **Copy** button on the **Modify** panel.

- Select the lower-left corner of the rectangle as the base point.
- Make sure that the **Ortho mode** is activated.
- Move the pointer upward and type **25** in the command line — next, press ENTER.
- Press ESC to exit the **Copy** tool.

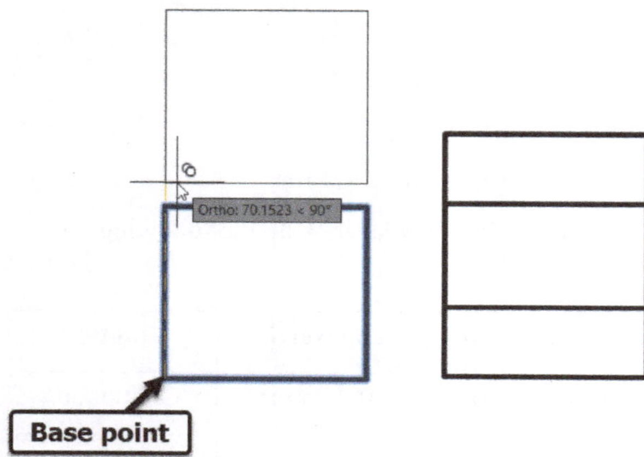

- Click the **Rotate** button on the **Modify** panel and select the copied rectangle. Press ENTER to accept.
- Select the lower right corner of the copied rectangle as the base point.
- Type 45 as the angle and press ENTER.

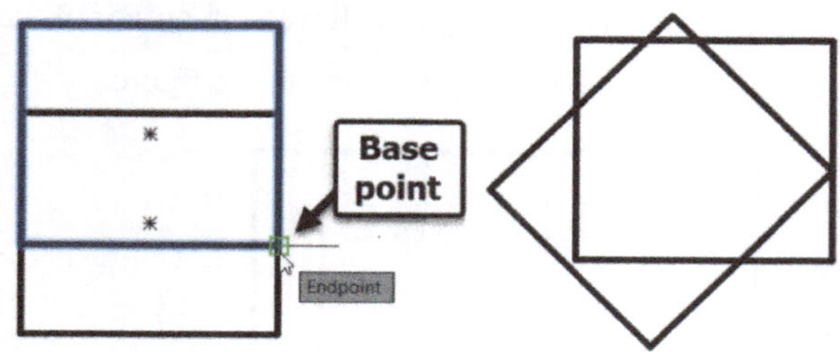

- Click the **Object Snap Tracking** icon on the Status Bar.
- Click the down arrow next to the **Object Snap** icon, and then select **Endpoint** from the list.
- Activate the **Rectangle** command.
- Place the pointer on the top left corner of the existing rectangle.
- Move the pointer vertically upward, and then notice a vertical tracking line from the top left corner of the rectangle.
- Move the pointer along the tracking line up to an approximate distance of 60 mm.

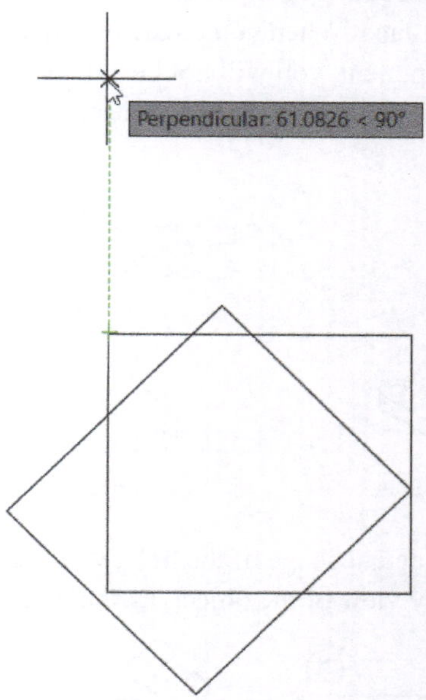

- Click to specify the first corner of the rectangle.
- Select **Dimensions** from the command line.
- Type 70 and press Enter to specify the length of the rectangle.
- Again, type 70 and press Enter to specify the width of the rectangle.
- Move the pointer up and click to position the rectangle.

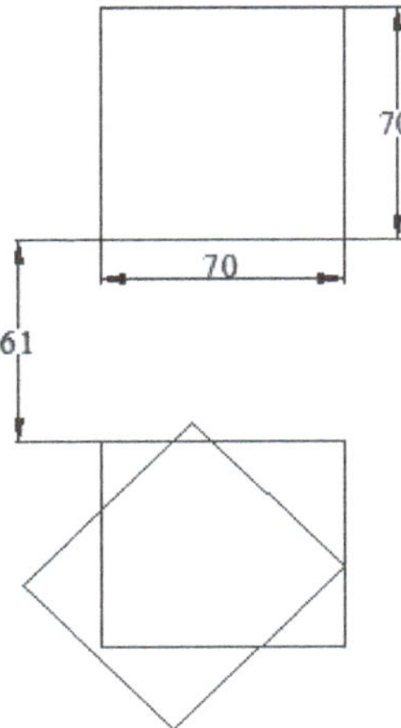

The rectangle located at the top is considered as top view and the below one as the front view.

- Click the **Explode** button on the **Modify** panel and select the newly created rectangle. Next, right-click to explode the rectangle.
- Activate the **Offset** tool and select the **Through** option from the command line.
- Select the left vertical line of the top rectangle.
- Select any one of the through points, as shown; the selected vertical line is offset through the selected point.
- Again, select the left vertical line.
- Move the pointer, and then select the remaining through point.

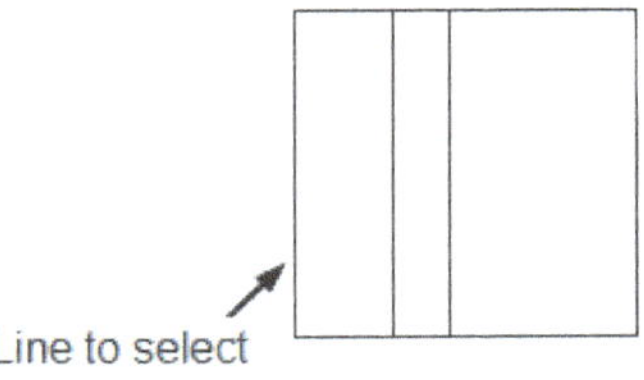

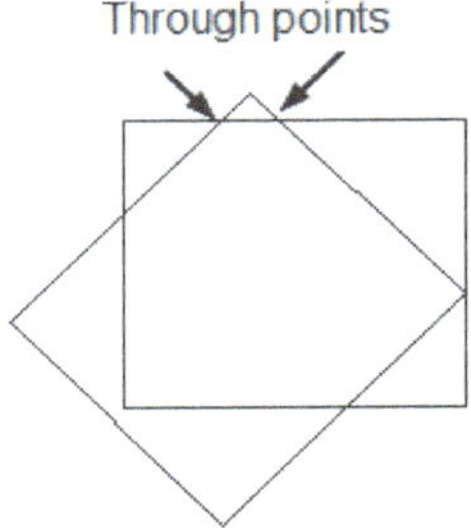

- Press Esc to deactivate the **Offset** tool.
- Select the **Object** layer from the **Layer** drop-down in the **Layers** panel.
- Activate the **Show/Hide Lineweight** button on the status bar.
- Activate the **Line** tool and select the intersection points on the front view, as shown.

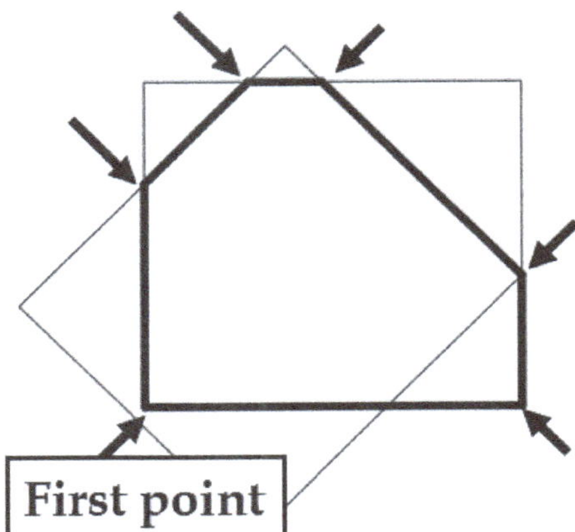

- Likewise, create the object lines in the top view, as shown below.

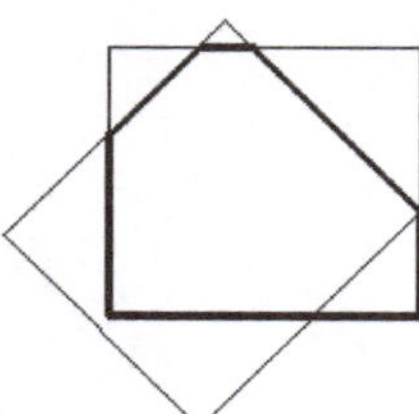

- Select the **Construction** layer from the **Layers** panel.
- Click the **Construction Line** button on the **Draw** panel.
- Select the **Offset** option from the command line. Next, select the **Through** option.
- Select the inclined line on the front view. Next, select the intersection point, as shown below.

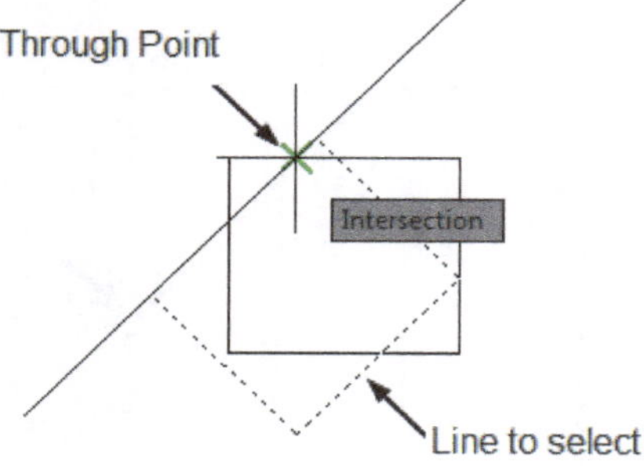

- Likewise, create other construction lines, as shown below.

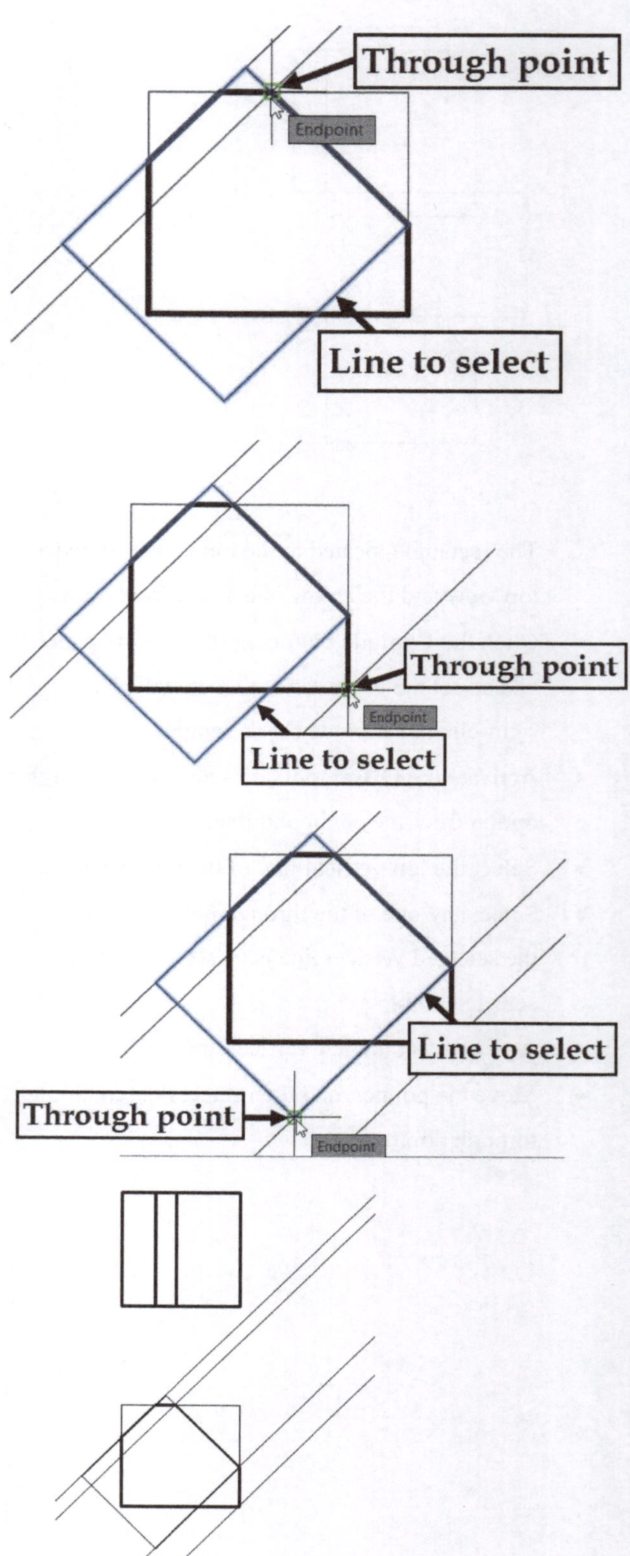

- Press Esc to deactivate the **Construction Line** command.

- Activate the **Construction Line** command and select **Offset** from the command line.
- Type 80 and press ENTER.
- Select the inclined line of the front view, as shown.
- Move the pointer toward the right and click to create the construction line.

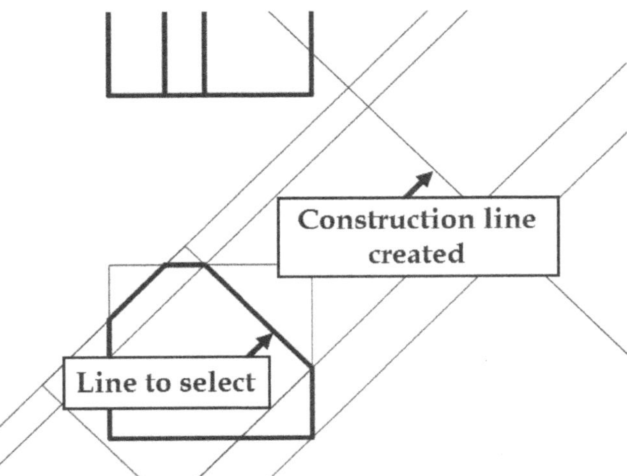

- Create other construction lines, as shown. The offset dimensions are given in the figure.

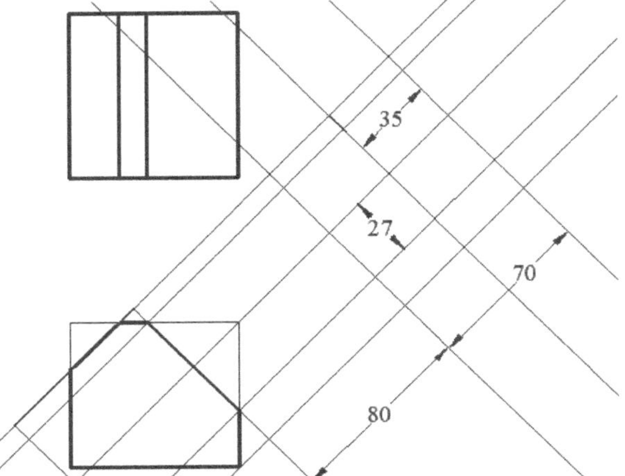

- Set the **Object** layer as the current layer. Next, create the object lines using the intersection points between the construction lines.
- Use the **Circle** tool and create a circle of 35 mm in diameter.

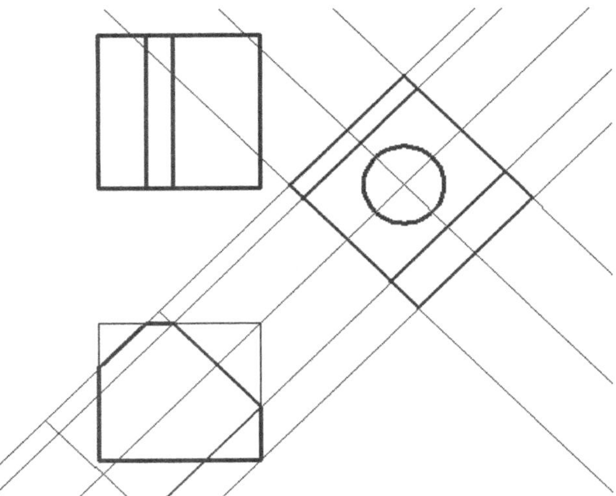

- Set the **Construction** layer as the current layer.
- Create projection lines from the circle.

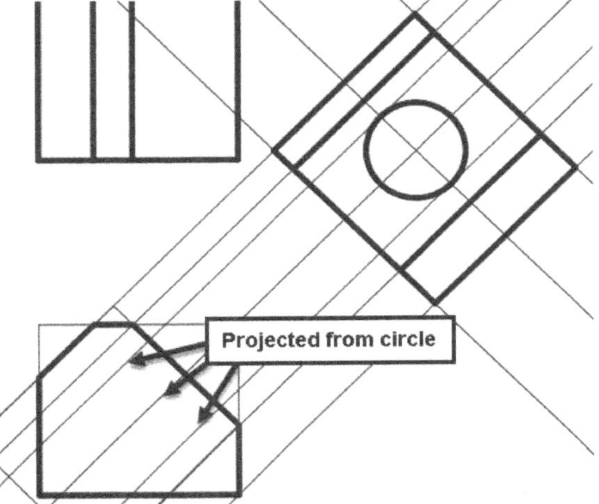

- Set the **Hidden** layer as the current layer,
- Create the hidden lines, as shown.

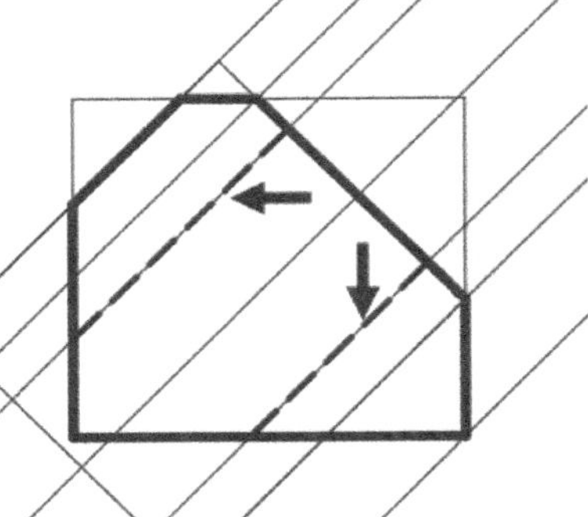

- Set the **Centerline** layer as the current layer.
- Activate the **Line** tool.
- Create the centerlines, as shown.

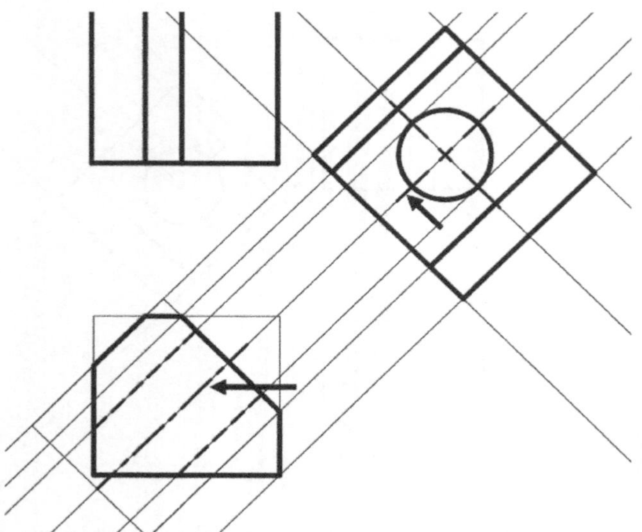

- Set the **Construction** layer as the current layer.
- On the ribbon, click **Home > Draw > Ray** .
- Select the intersection point of the centerline and object line, as shown.
- Move the pointer upward and click to create a ray.

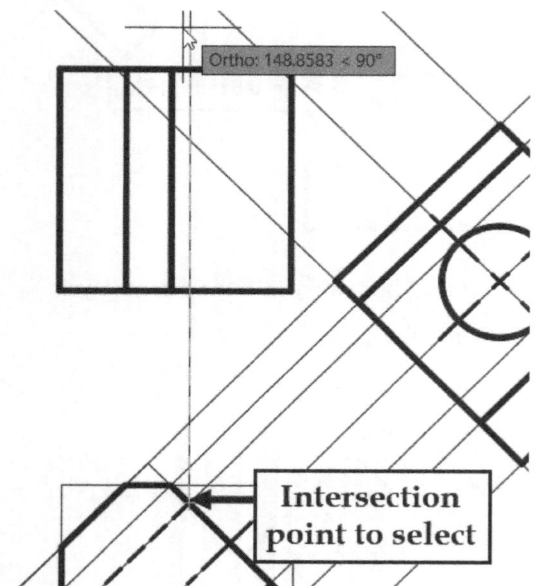

- Press ENTER twice.
- Likewise, create two more rays, as shown.

- Create a horizontal construction line passing through the midpoint of the top view, as shown.

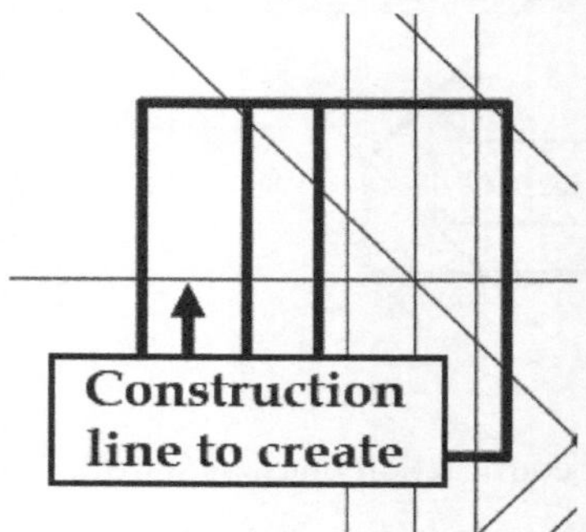

- Set the **Object** layer as the current layer,
- On the ribbon, click **Home > Draw > Ellipse drop-down > Axis, End** .
- Specify the first and second points, as shown.

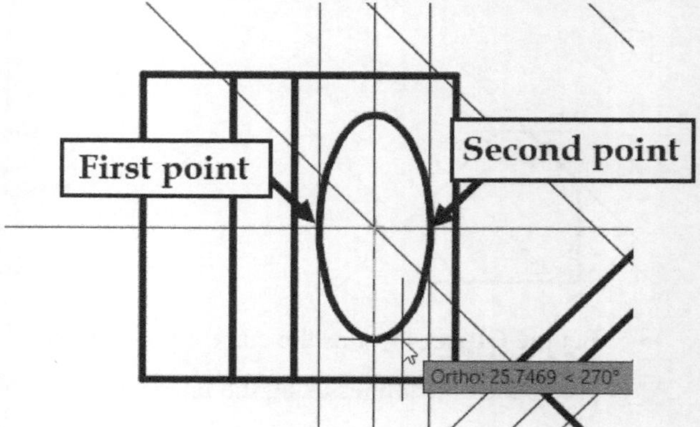

- Move the pointer downward, type-in 17.5, and then press ENTER.

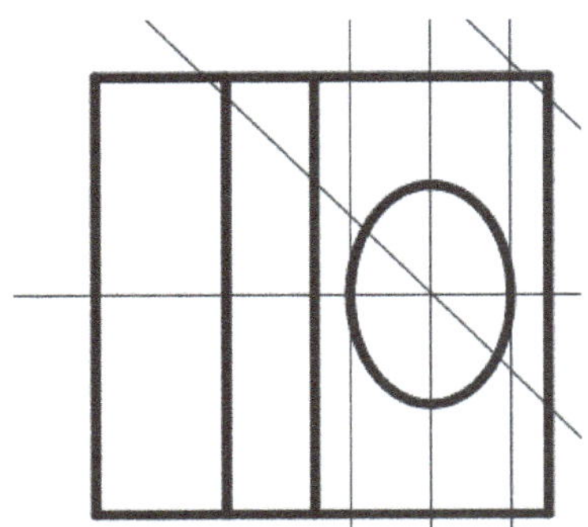

- Set the **Centerline** layer as a current layer,
- Create the remaining centerlines.
- The drawing after hiding the **Construction** layer is shown next.
- Save the file as auxiliary_views.dwg.

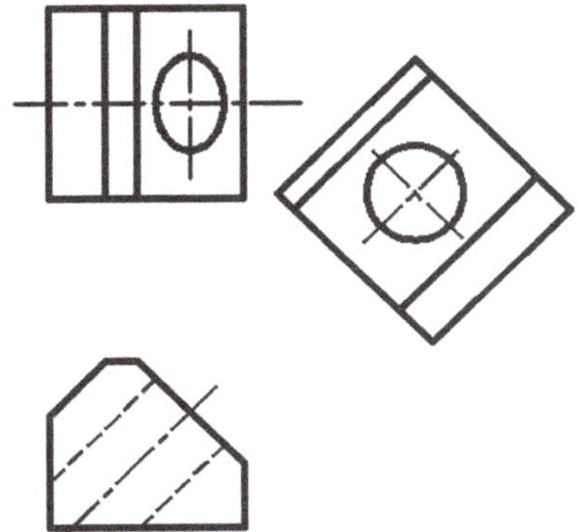

Creating Named views

While working with a drawing, you may need to perform numerous zoom and pan operations to view key portions of a drawing. Instead of doing this, you can save these portions with a name. Then, restore the named view and start working on them.

- Open the **ortho_views.dwg** file (The drawing file created in the Orthographic Views section of this chapter).
- Click the **View** tab on the ribbon.
- To create a named view, click **View > Named Views > View Manager** on the ribbon; the **View Manager** dialog appears.

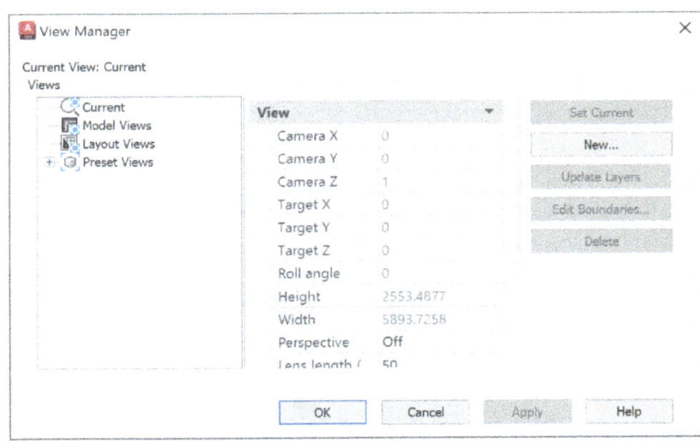

- Click the **New** button on the **View Manager** dialog; the **New View/Shot Properties** dialog appears.
- Select the **Define Window** option from the **Boundary** section of the **New View/Shot Properties** dialog.
- Create a window on the front view, as shown below.

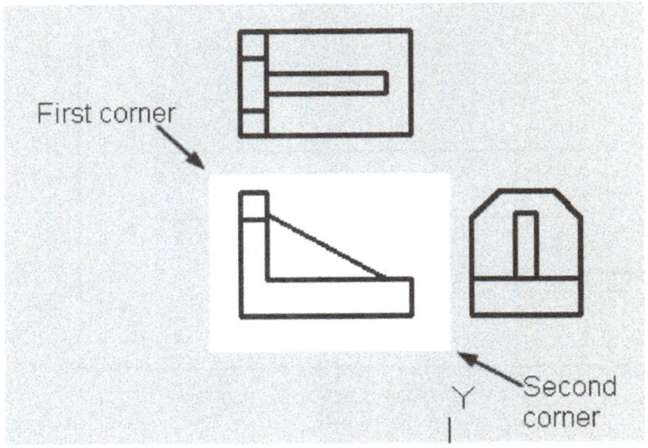

- Press ENTER to accept.
- Enter **Front** in the **View name** box.

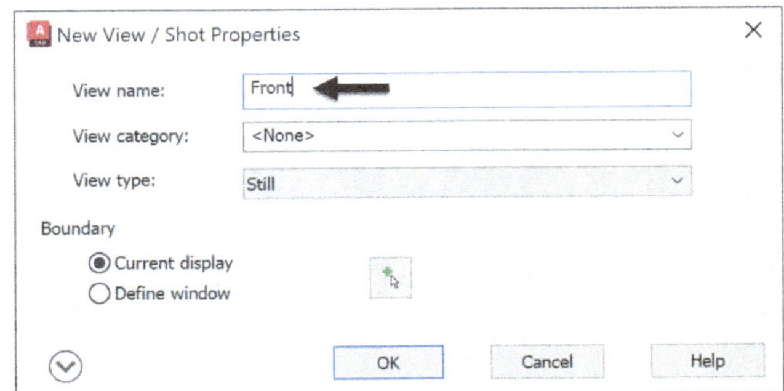

- Click **OK** on the **New View/Shot Properties** dialog.

- Likewise, create the named views for the top and right views of the drawing.

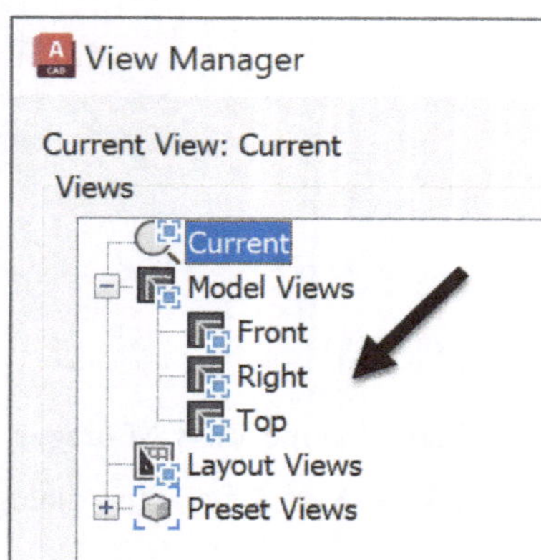

- To set the **Top** view to current, select it from the **Views** tree and click the **Set Current** button on the dialog. Next, click **OK** on the **View Manager** dialog; the **Top** view will be zoomed and fitted to the screen.

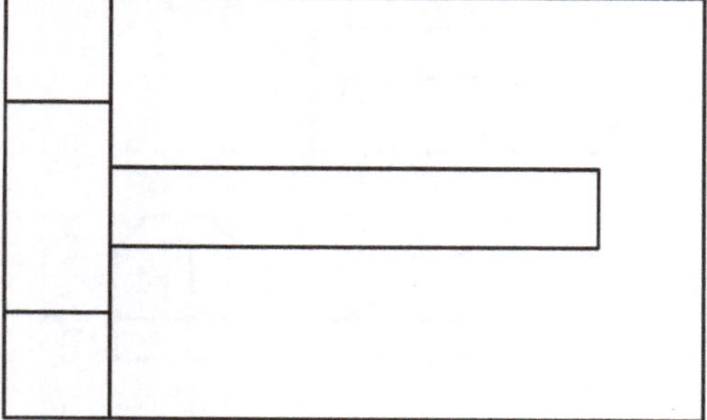

- Save and close the file.

Exercises

Exercise 1

Create the orthographic views of the object shown below.

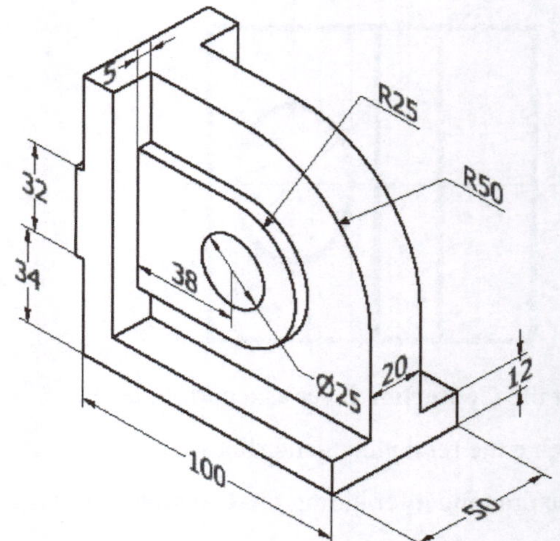

Exercise 2

Create the orthographic views of the object shown below.

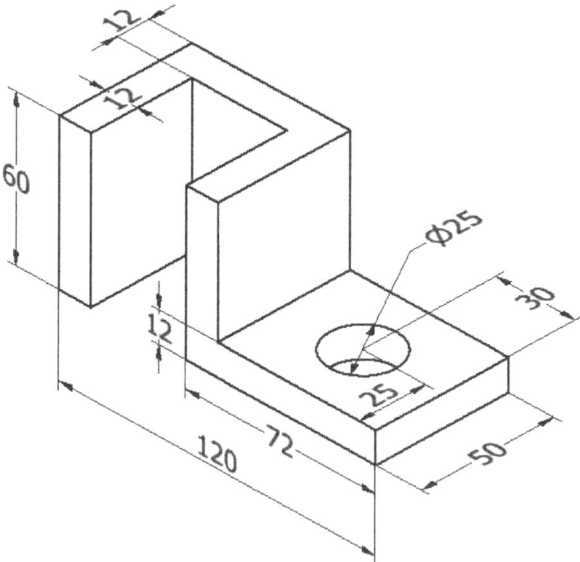

Exercise 3

Create the orthographic and auxiliary views of the object shown below.

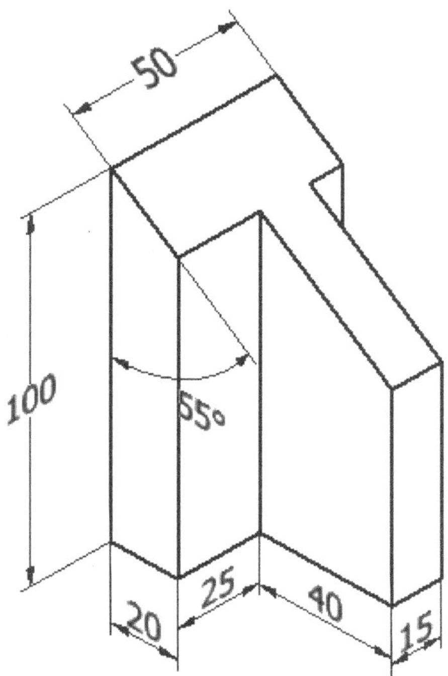

Exercise 4

Create the orthographic and auxiliary views of the object shown below.

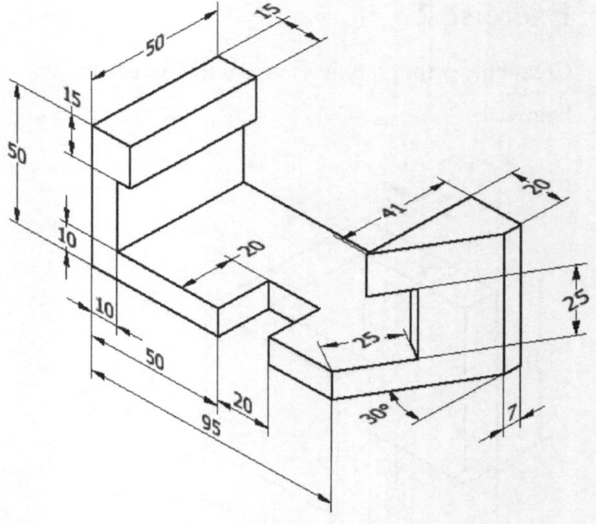

Chapter 6: Dimensions and Annotations

In this chapter, you will learn to do the following:

- **Create Dimensions**
- **Create Dimension Style**
- **Add Leaders**
- **Create Centerlines**
- **Add Dimensional Tolerances**
- **Add Geometric Tolerances**
- **Edit Dimensions**

Dimensioning

In previous chapters, you have learned to draw shapes of various objects and create drawings. However, while creating a drawing, you also need to provide the size information. You can provide the size information by adding dimensions to the drawings. In this chapter, you will learn how to create various types of dimensions. You will also learn about some standard ways and best practices of dimensioning.

Creating Dimensions

In AutoCAD, there are many tools available for creating dimensions. You can access these tools from the Ribbon, Command line, and Menu Bar.

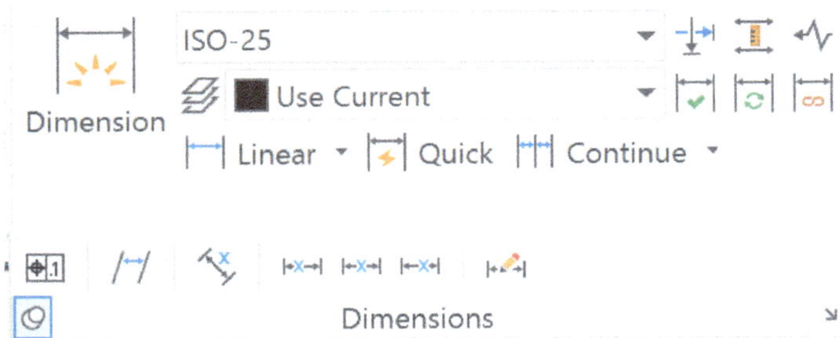

The following table gives you the functions of various dimensioning tools.

Tool	Shortcut	Function
Dimension	DIM	This tool creates a dimension based on the selected geometry.

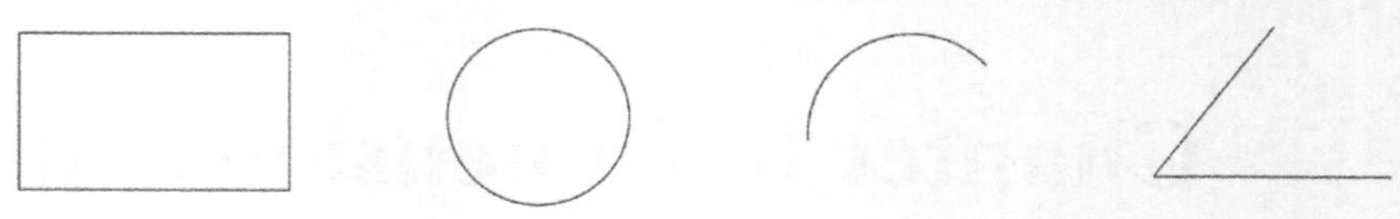

- Create a rectangle, circle, arc, and two intersecting lines, as shown in the previous figure.
- Click **Annotate > Dimensions > Dimension** on the ribbon.
- Select a line, move the pointer, and click to create the linear dimension.

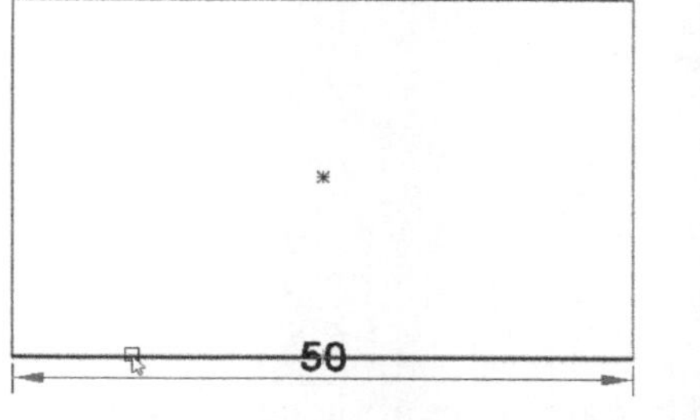

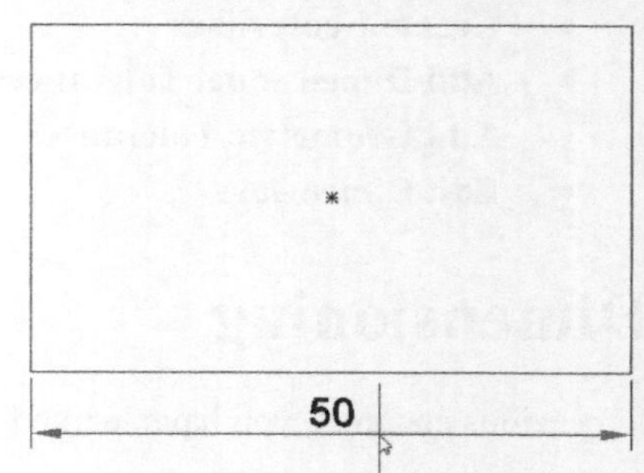

- Select a circle, move the pointer, and click to position the diameter dimension.

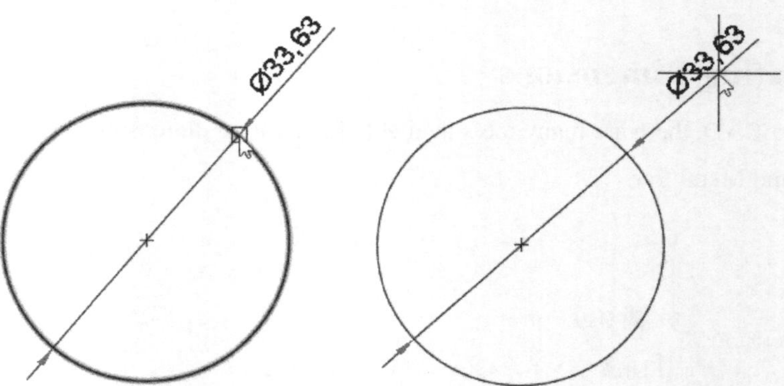

- Select an arc, move the pointer, and click to position the radial dimension.

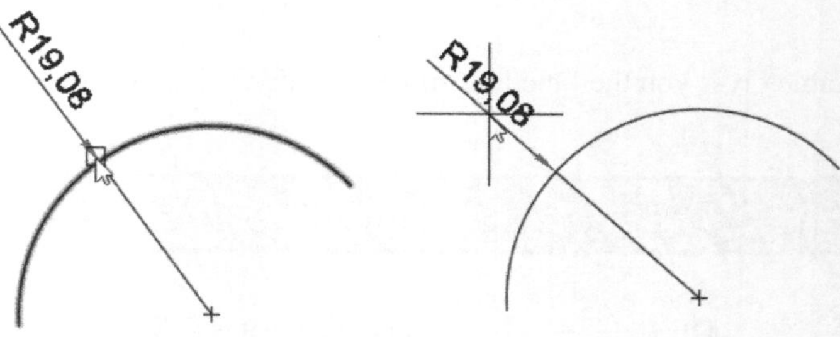

- Place the pointer on the arc, type L, and press Enter. Select the arc, move the pointer, and click to position the arc length dimension.

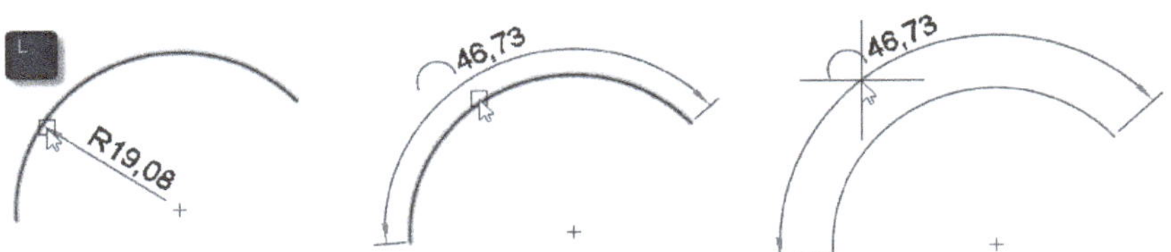

- Place the pointer on the arc, type A, and press Enter. Select the arc, move the pointer, and click to position the angle of the arc.

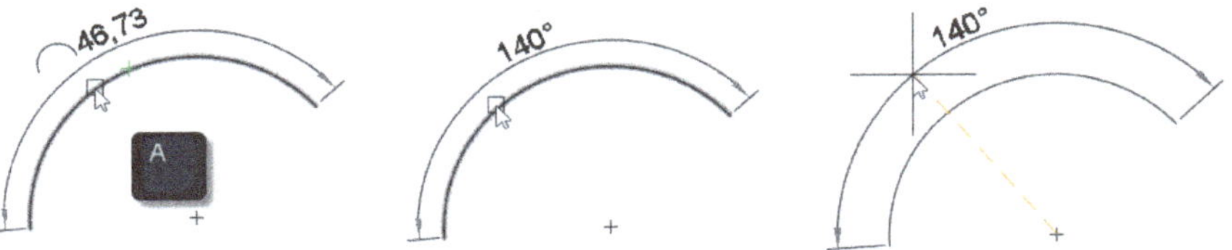

- Select two non-parallel lines and position the angular dimension between them.

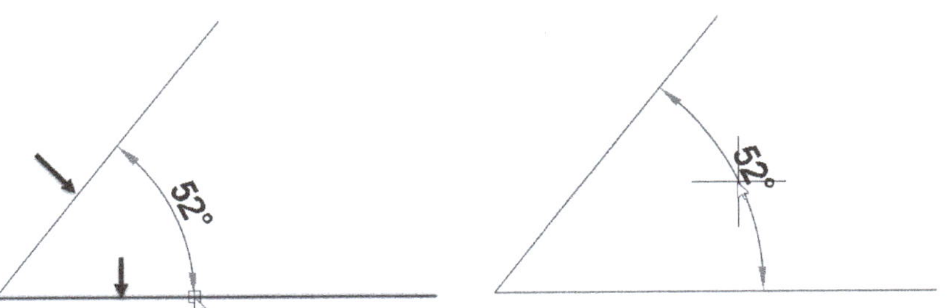

Likewise, you can create other types of dimensions using the **Dimension** tool.

Linear

DLI

This tool creates horizontal and vertical dimensions.

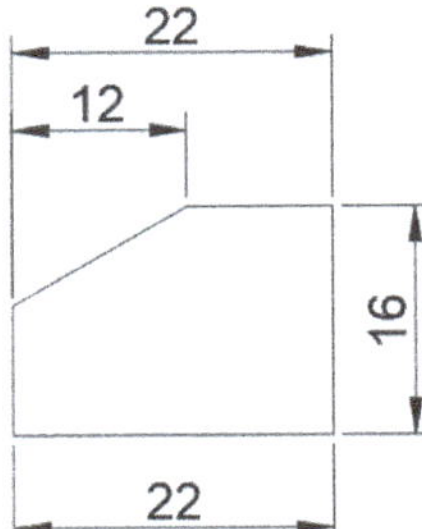

- Click **Annotate > Dimensions > Dimension** drop-down > **Linear** on the ribbon.

<table>
<tr>
<td></td>
<td></td>
<td>

- Select the first and second points of the dimension.
- Move the pointer in the horizontal direction to create a vertical dimension (or) move in the vertical direction to create a horizontal dimension.

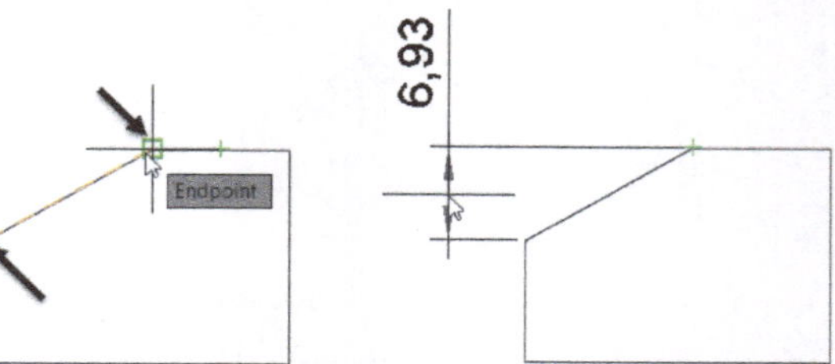

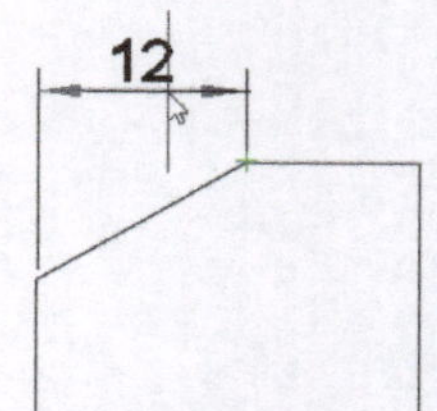

- Click to position the dimension.

</td>
</tr>
<tr>
<td>

Aligned

</td>
<td>DAL</td>
<td>

This tool creates a linear dimension parallel to the object.

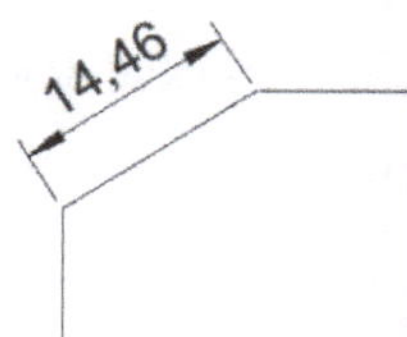

- Click **Annotate > Dimensions > Dimension** drop-down **> Aligned** on the ribbon.
- Select the first and second points of the dimension line.

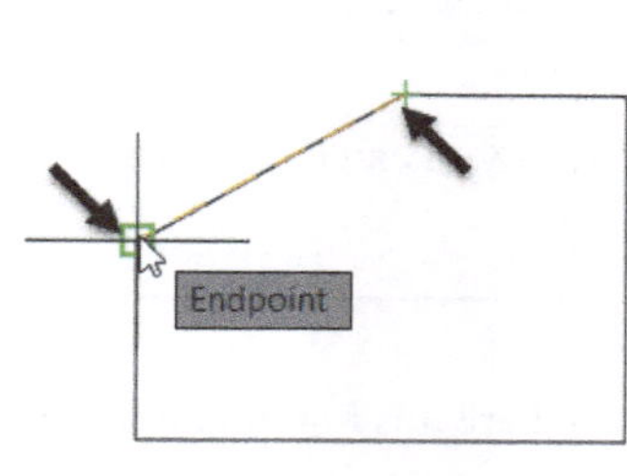

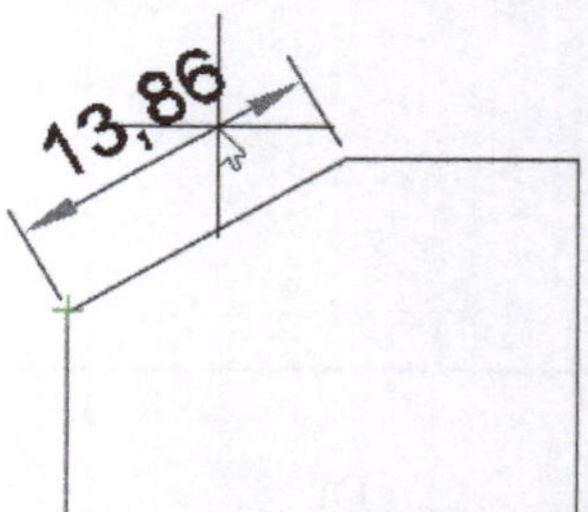

(or) press Enter and select the line.

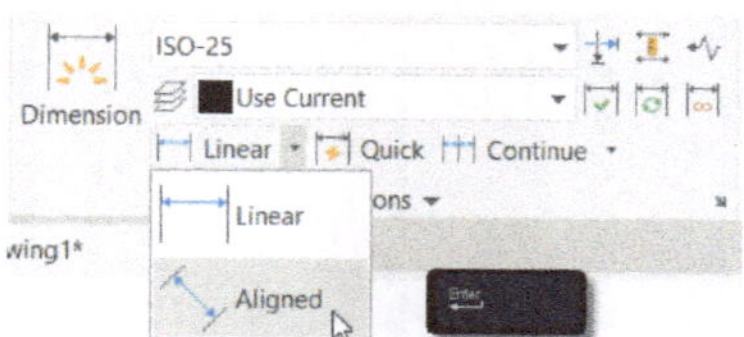

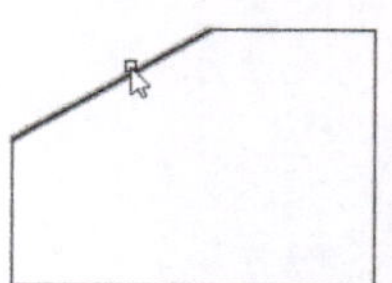
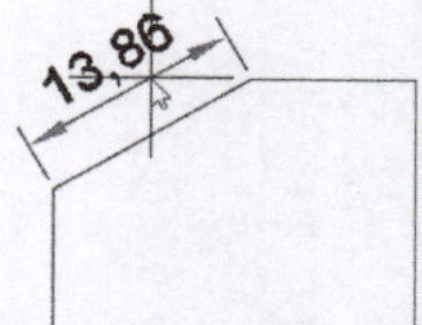

- Move the pointer and click to position the dimension.

</td>
</tr>
</table>

Arc Length	DAR	It dimensions the total or partial length of an arc.  • Click **Annotate > Dimensions > Dimension** drop-down **> Arc Length** on the ribbon. • Select an arc from the drawing. • If you want to dimension only a partial length of an arc, select **Partial** option from the command line. Next, select the two points on the arc. • Move the pointer and click to position the dimension.
Continue	DCO	It creates a linear dimension from the second extension line of the previous dimension. 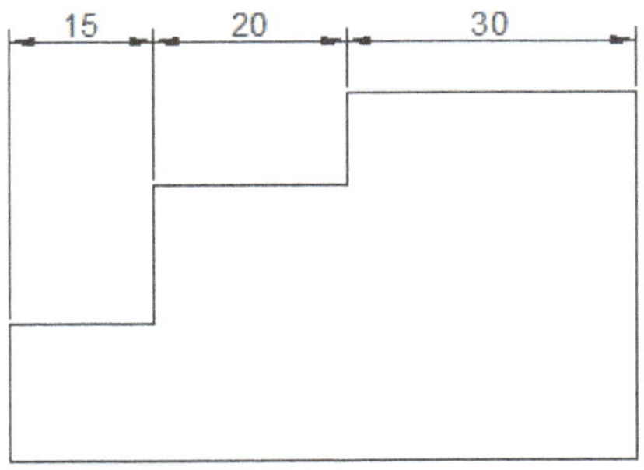 • Create a linear dimension by selecting the first and second points. 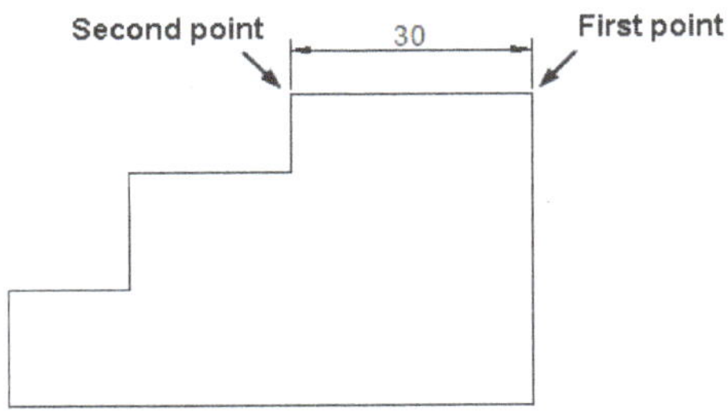 Second point 30 First point • Click **Annotate > Dimensions > Continue** on the ribbon; a chain dimension is attached to the pointer.

<table>
<tr>
<td></td>
<td></td>
<td>
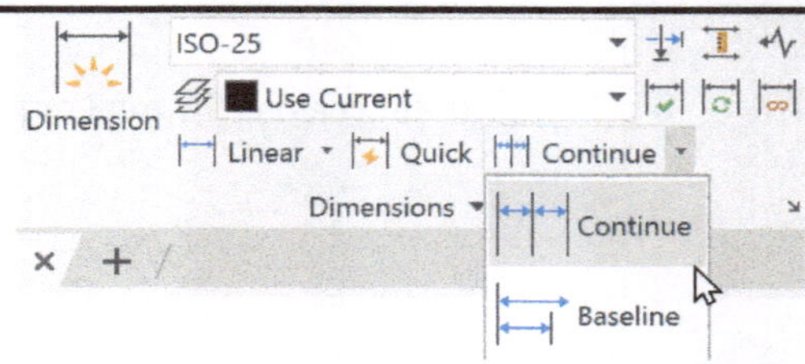

<ul><li>Select the third and fourth points of the chain dimension.</li></ul>

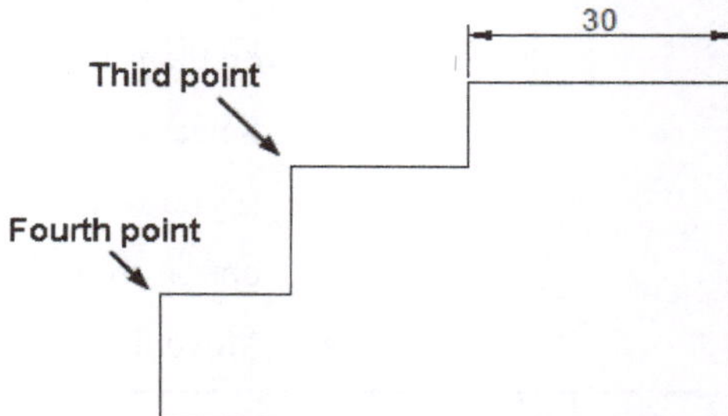

<ul><li>Position the chain dimension. Next, right-click and select Enter.</li></ul>
</td>
</tr>
<tr>
<td>Baseline</td>
<td>DBA</td>
<td>
It creates dimensions by using the previously created dimension, as shown below.

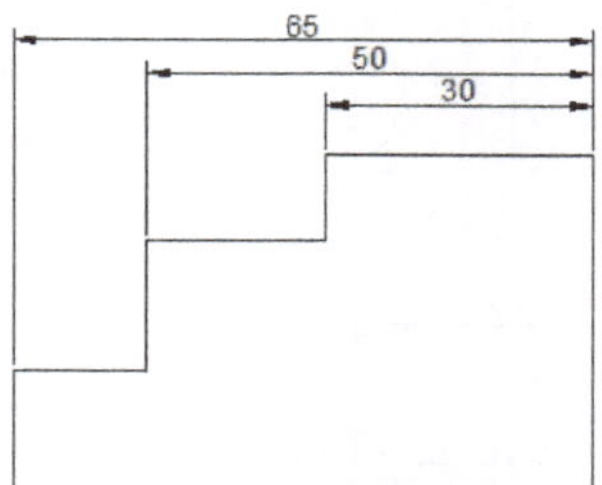

<ul><li>Create a linear dimension by selecting the first and second points.</li></ul>

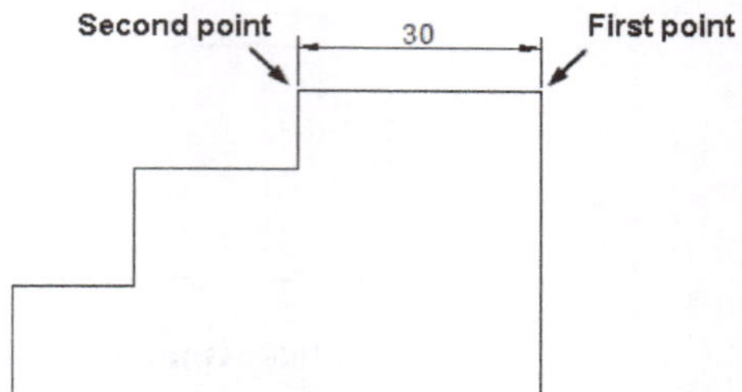

<ul>
<li>Click Annotate > Dimensions > Continue > Baseline on the ribbon.</li>
<li>Select the third and fourth points of the baseline dimension. Next, right-click and select Enter.</li>
</ul>
</td>
</tr>
</table>

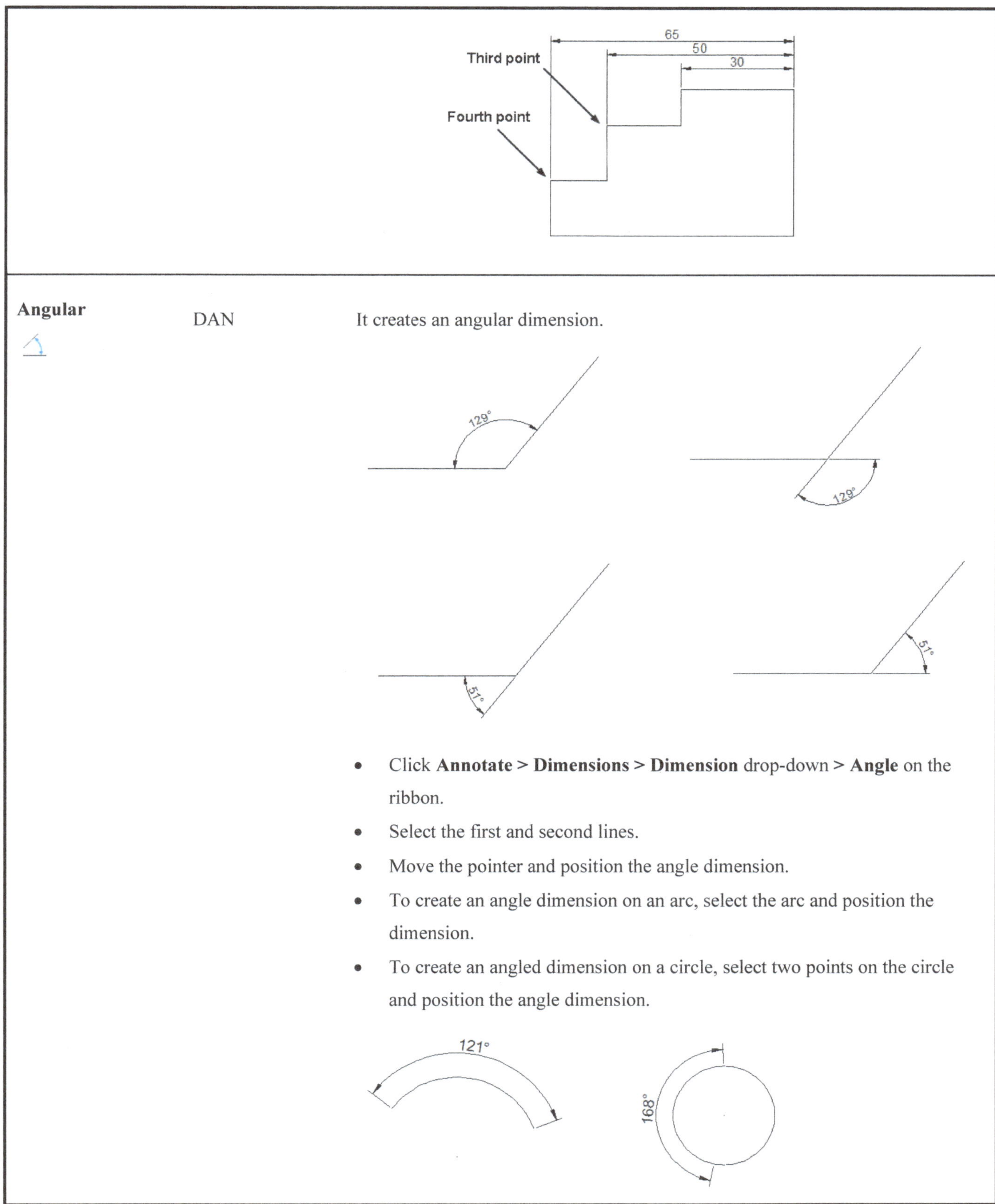

Angular	DAN	It creates an angular dimension.

- Click **Annotate > Dimensions > Dimension** drop-down **> Angle** on the ribbon.
- Select the first and second lines.
- Move the pointer and position the angle dimension.
- To create an angle dimension on an arc, select the arc and position the dimension.
- To create an angled dimension on a circle, select two points on the circle and position the angle dimension.

Diameter	DIA	It adds a diameter dimension to a circle or an arc. 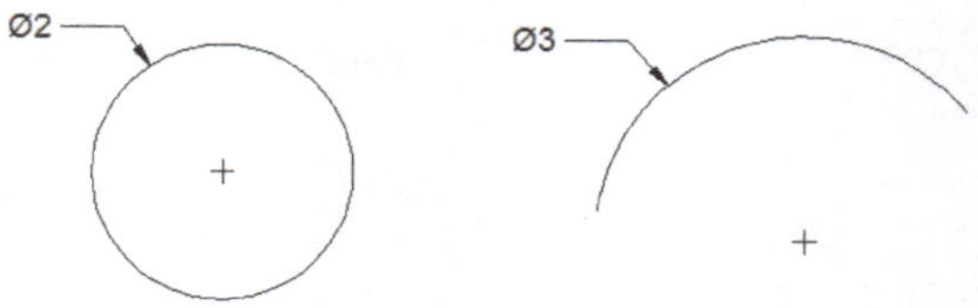 Ø2 Ø3 • Click **Annotate > Dimensions > Dimension** drop-down> **Diameter** on the ribbon. • Select a circle or an arc and position the dimension.
Radius	DRA	It adds a radial dimension to an arc or circle. 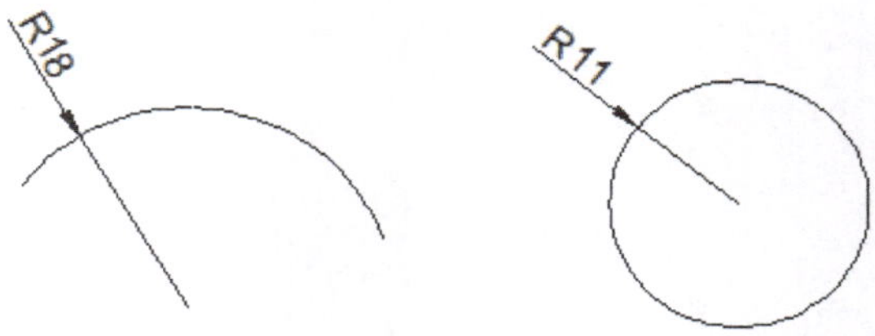 R18 R11
Jogged	DJO	It creates jogged dimensions. A jogged dimension is created when it is not possible to show the center of an arc or circle. 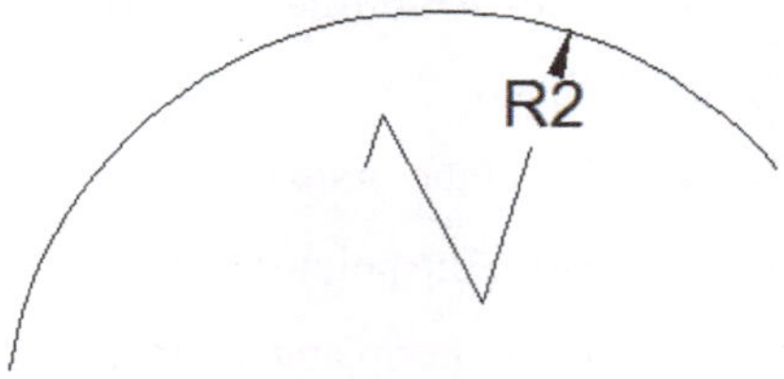 R2 • Click **Annotate > Dimensions > Dimension** drop-down **> Jogged** on the ribbon. • Select an arc or circle. • Select a new center point override. • Locate the dimension and the jog location.

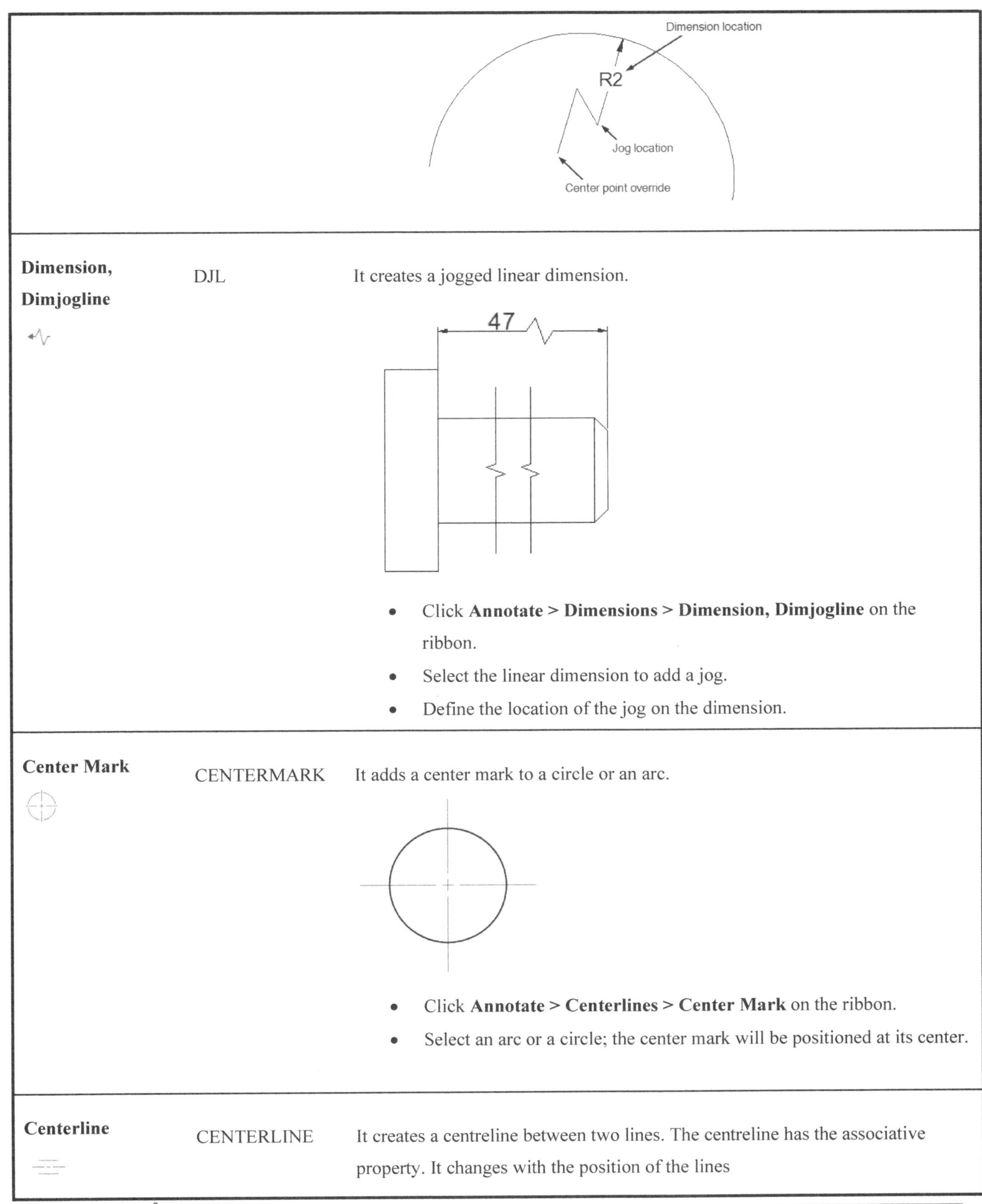

Dimension, Dimjogline	DJL	It creates a jogged linear dimension. • Click **Annotate > Dimensions > Dimension, Dimjogline** on the ribbon. • Select the linear dimension to add a jog. • Define the location of the jog on the dimension.
Center Mark	CENTERMARK	It adds a center mark to a circle or an arc. • Click **Annotate > Centerlines > Center Mark** on the ribbon. • Select an arc or a circle; the center mark will be positioned at its center.
Centerline	CENTERLINE	It creates a centreline between two lines. The centreline has the associative property. It changes with the position of the lines

<table>
<tr>
<td></td>
<td></td>
<td>

- Click **Annotate > Centerlines > Centerline** on the ribbon.
- Select two lines that are parallel or non-parallel to each other; a centreline is created between them.

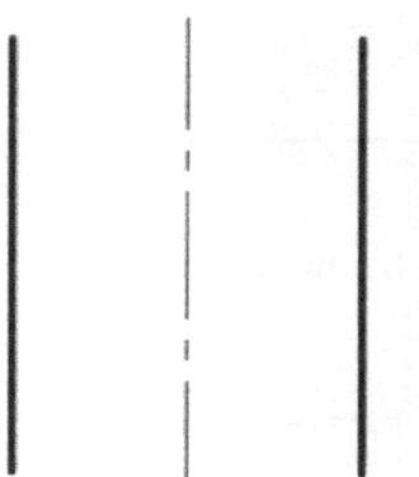
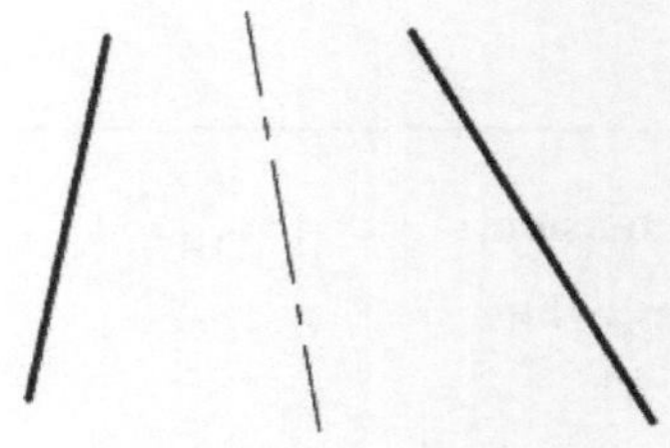

- Change the position of the lines; the centreline also changes.

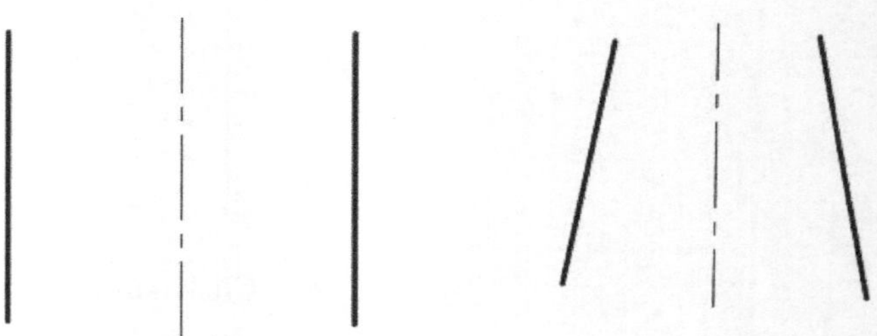

</td>
</tr>
<tr>
<td>Ordinate</td>
<td>DOR</td>
<td>

It creates ordinate dimensions based on the current position of the User Coordinate System (UCS).

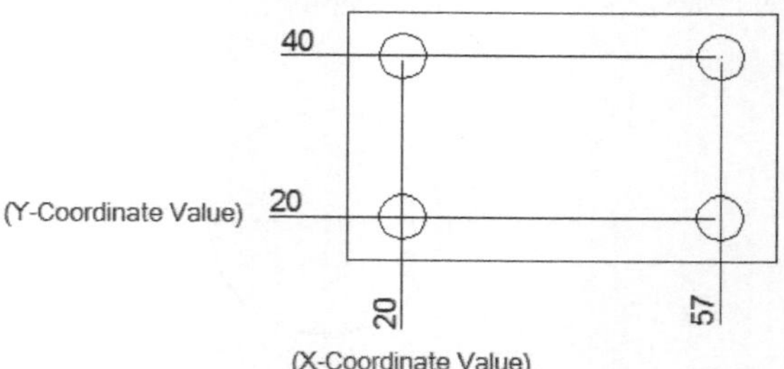

- Click **Annotate > Dimensions > Dimension** drop-down **> Ordinate** on the ribbon.
- Select the point of the object.
- Move the pointer in the vertical direction and click to position the X-Coordinate value.
- Select the point of the object.

</td>
</tr>
</table>

		• Move the pointer in the horizontal direction and click to position the Y-Coordinate value.
Quick	QDIM	It dimensions one or more objects at the same time. 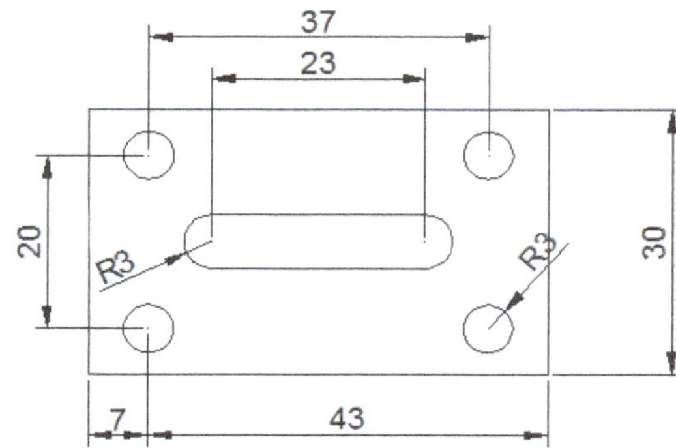 • Click **Annotate > Dimensions > Quick** on the ribbon. • Select one or more objects from a drawing. • Right-click and position the dimensions.
Adjust Space 	DIMSPACE	In the following figure, the drawing on the left side has congested dimensions, whereas the right side drawing has dimensions with ample space between them. You can use the **Adjust Space** tool to adjust the space between the dimensions. 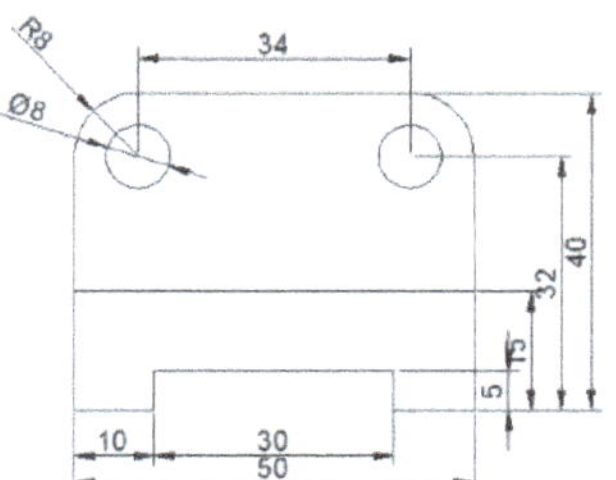 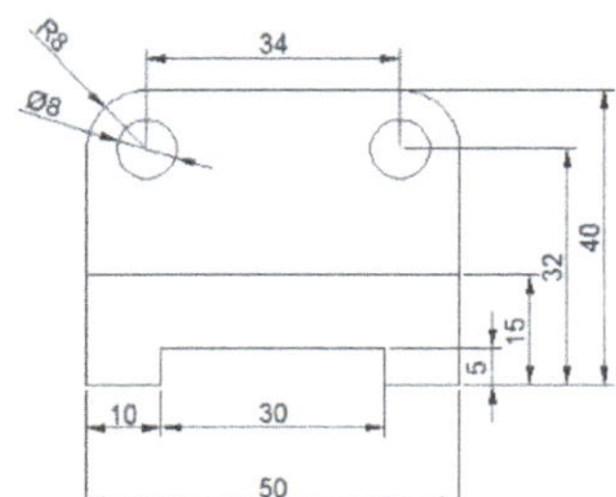 • Click **Annotate > Dimensions > Adjust Space** on the ribbon. • Select the base dimension from which the other dimensions are to be adjusted. • Select the dimensions to adjust. • Right-click to accept. • Enter the value of the spacing between the dimensions or select the **Auto** option; the dimensions will be adjusted with respect to the base dimension.

Break	DIMBREAK	It adds breaks to a dimension, extension, and leader lines. 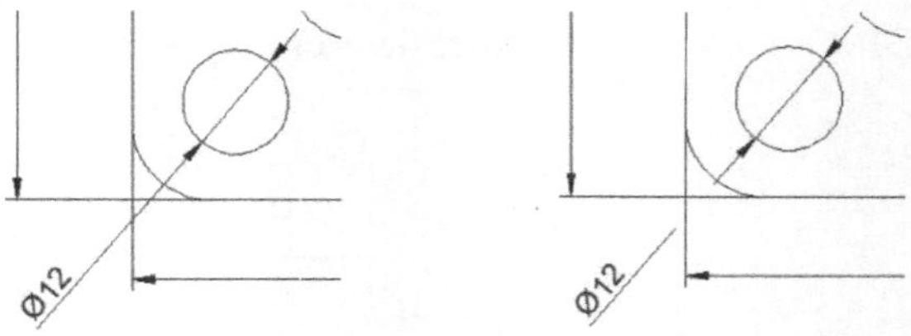 Ø12 Ø12 • Click **Annotate > Dimensions > Break** on the ribbon. • Select the dimension to add a break. • Select the dimension or object intersecting the dimension selected in the previous step. It breaks the dimension by the intersecting object. • Right-click to exit the tool.
Inspect	DIMINSPECT	It creates an inspection dimension. The inspection dimension describes how frequently the dimension should be checked during the inspection process to ensure the quality of the component. 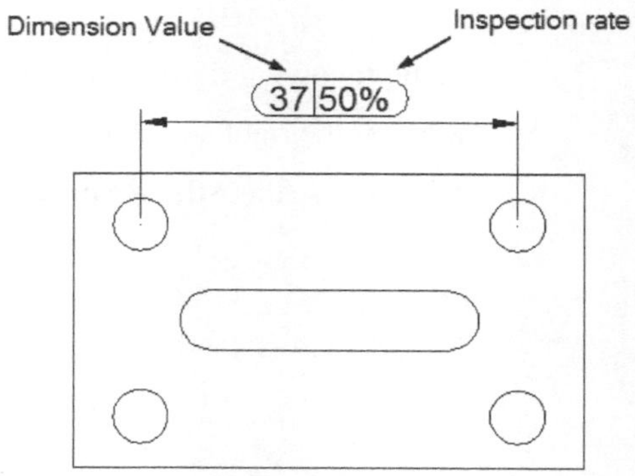 Dimension Value Inspection rate 37 50% • Click **Annotate > Dimensions > Inspect** on the ribbon; the **Inspection Dimension** dialog appears.

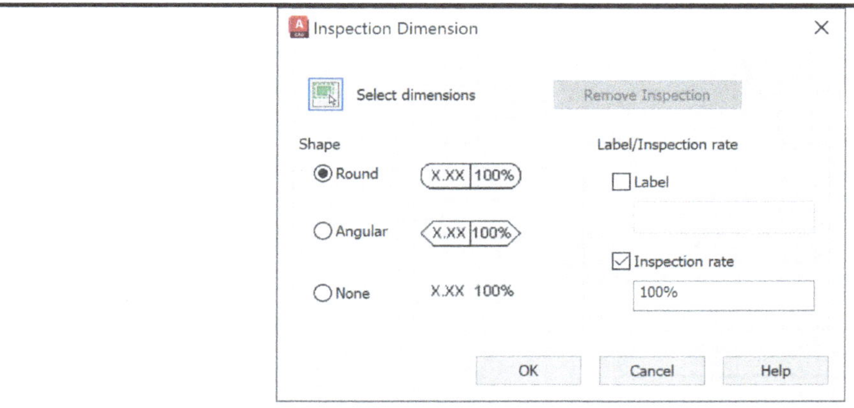

- Click the **Select dimensions** button on the dialog and select the dimension to apply the inspection rate.
- Right-click to accept.
- Select the shape of the inspection from the **Shape** section.
- Enter the **Inspection rate**. 100% means that the value will be checked every time during the inspection process. 50% means half the time.
- If required, select the **Label** checkbox and enter the inspection label.
- Click **OK**.

Example:

In this example, you will create the drawing, as shown in the figure, and add dimensions to it.

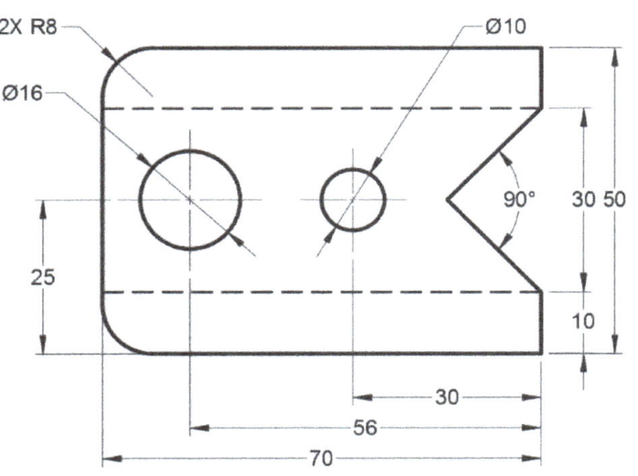

- Create four new layers with the following settings.

Layer	Lineweight	Linetype
Construction	0.00 mm	Continuous
Object	0.50 mm	Continuous
Hidden	0.30 mm	HIDDEN2
Dimensions	0.30 mm	Continuous

- Type LIMMAX and press ENTER.
- Type 100, 100, and press ENTER to set the maximum limit of the drawing.
- Click **Zoom All** on the **Navigation Bar**.
- Create the drawing on the **Object** and **Hidden** layers.

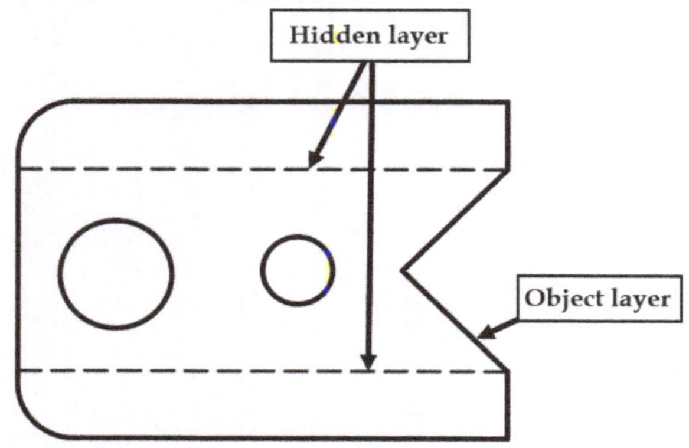

- Select the **Dimensions** layer from the **Layer** drop-down in the **Dimensions** panel.

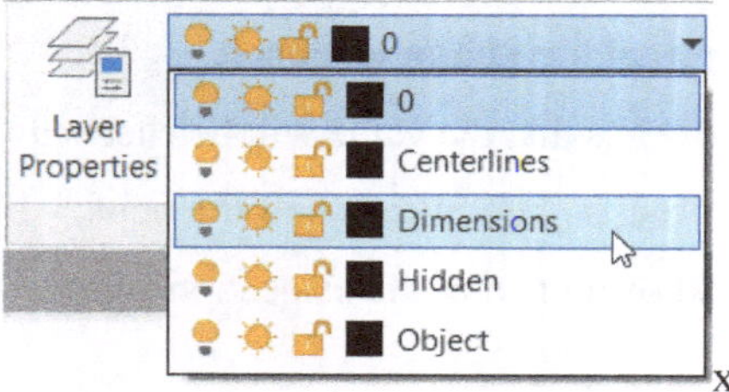

Creating a Dimension Style

The appearance of the dimensions depends on the dimension style that you use. You can create a new dimension style using the **Dimension Style Manager** dialog. In this dialog, you can specify various settings related to the appearance and behaviour of dimensions. The following example helps you to create a dimension style.

- Expand the **Annotation** panel on the **Home** ribbon tab and click **Dimension Style**.

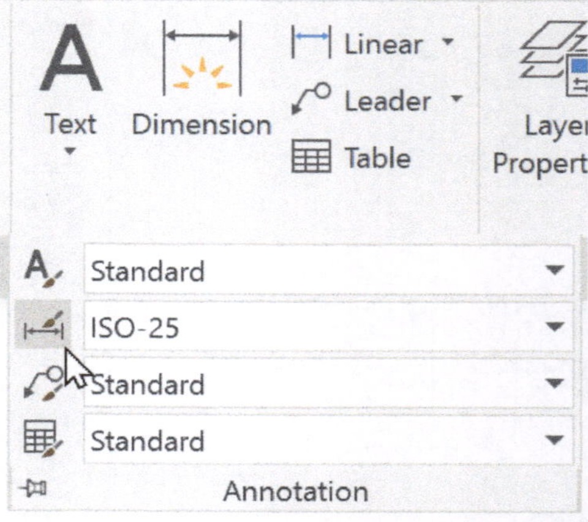

The **Dimension Style Manager** dialog appears.

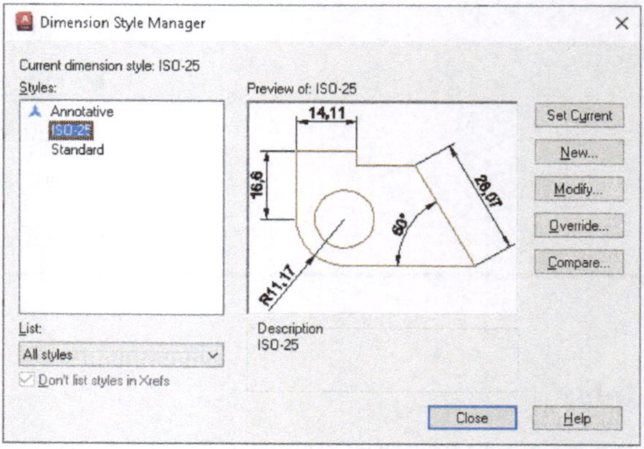

The basic nomenclature of dimensions is given below.

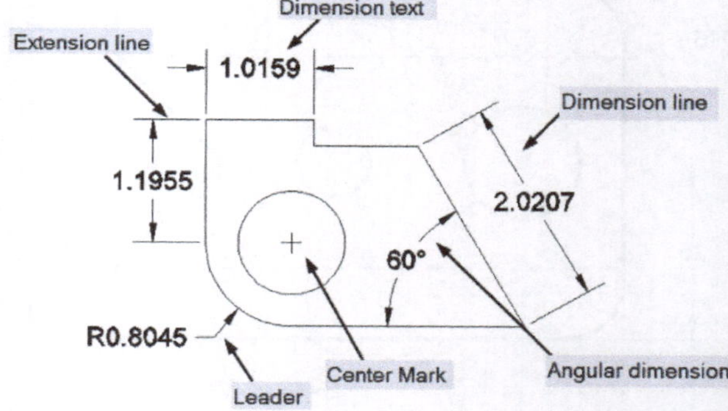

By default, the **ISO-25** or the **Standard** dimension style is active. If the default dimension style does not suit the dimensioning requirement, you can create a new dimension style and modify the nomenclature of the dimensions.

- To create a new dimension style, click the **New** button on the **Dimension Style Manager** dialog; the **Create New Dimension Style** dialog appears.
- In the **Create New Dimension Style** dialog, enter **Mechanical** in the **New Style Name**.
- Select **ISO-25** from the **Start With** drop-down and click **Continue**.

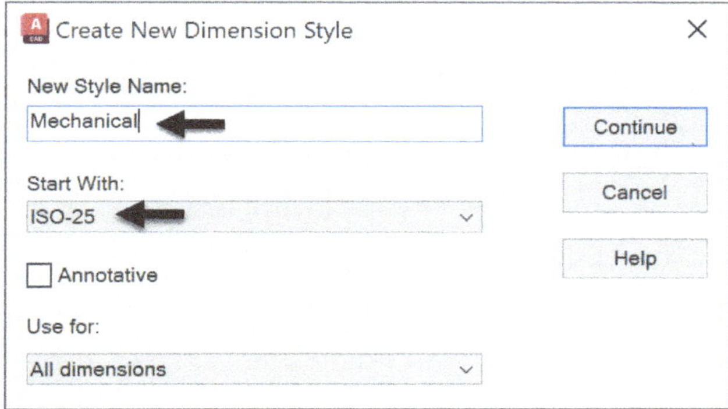

- In the **New Dimension Style** dialog, click the **Primary Units** tab.
- Ensure that the **Unit Format** is set to **Decimal**.
- Set **Precision** to **0**.
- Select **Decimal separator** > **'.'(Period)**.

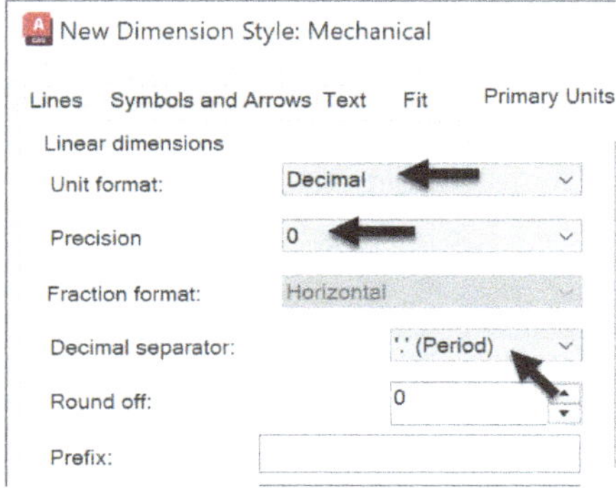

Study the other options in the **Primary Units** tab. Most of them are self-explanatory.

- Click the **Text** tab.
- Ensure that the **Text height** is set **2.5**.
- In the **Text placement** section, set the **Vertical** and **Horizontal** values to **Centered**.
- Select **Text alignment** > **Horizontal**.

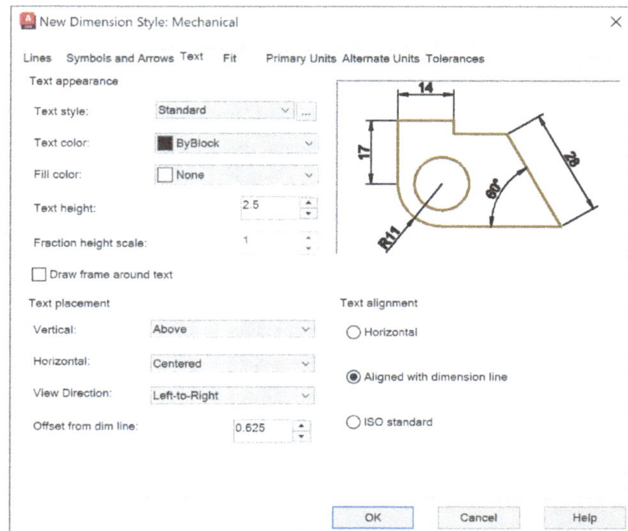

Study the other options in the **Text** tab. These options let you change the appearance of the dimension text.

- Click the **Lines** tab on the dialog.
- In this tab, notice the two options in the **Extension lines** section: **Extend beyond dim lines** and **Offset from origin.**

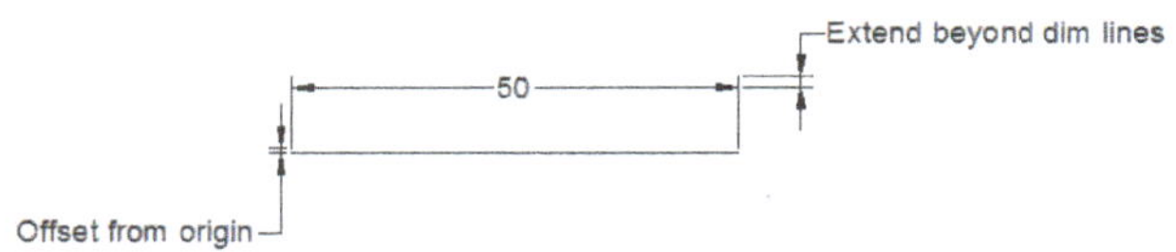

- Set **Extend beyond dim lines** and **Offset from origin** to **1.25**.
- Set the **Baseline spacing** in the **Dimension lines** section to **5**.

Study the different options in this tab. The options in this tab are used to change the appearance and behavior of the dimension lines and extension lines.

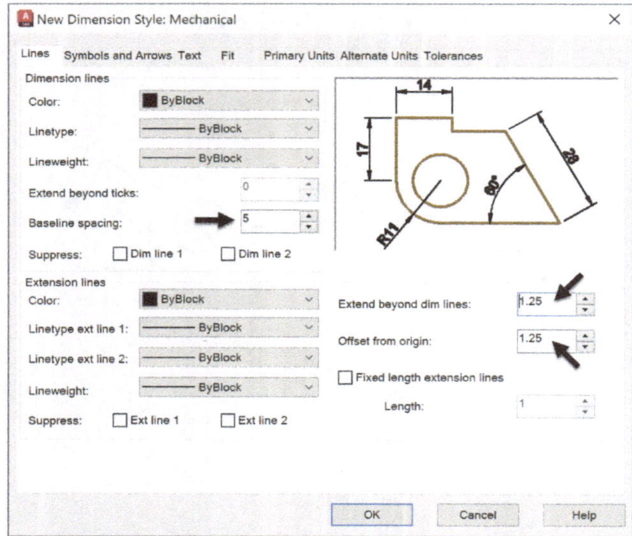

- Click the **Symbols and Arrows** tab, and then set **Arrow size** and **Center Marks** to 3.
- Select the **Line** option in the **Center marks** section.

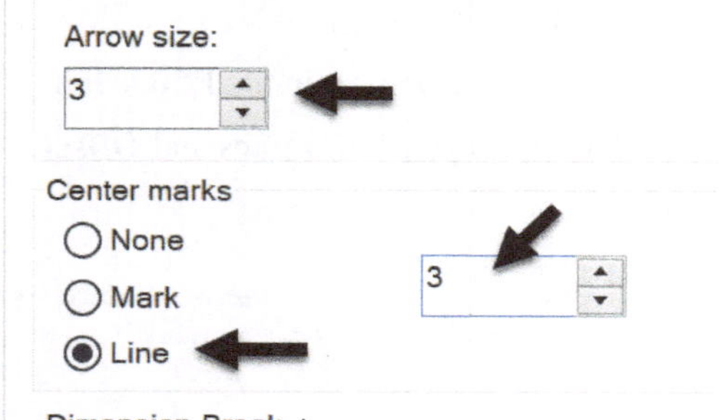

Notice the different options in this tab. The options in this tab are used to change the appearance of the arrows and symbols. Also, you can set the appearance of the center marks and centerlines of circles and arcs.

- Click **OK** to accept the settings.
- Click **Set Current** on the **Dimension Style Manager** dialog; the **Mechanical** dimension style will be set as current.
- Click **Close** to close the dialog.
- On the **Annotate** tab of the ribbon, click **Centerlines > Center Mark** .
- Select the circles from the drawing to apply the center mark to them.

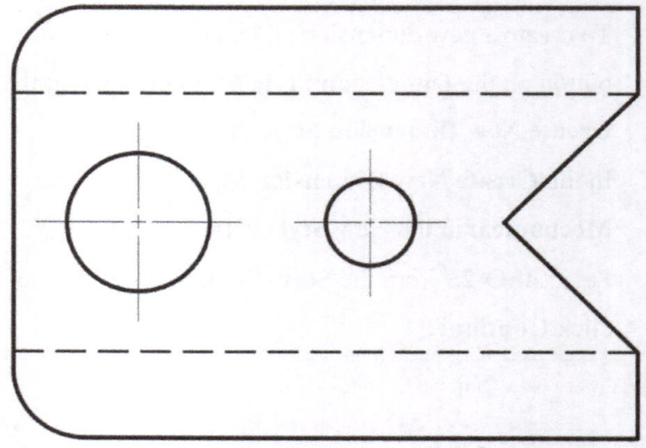

- On the ribbon, click **Annotate > Dimensions > Dimension**  .
- Make sure that the **Object Snap** icon is turned on the status bar.
- Select the lower right corner of the drawing.
- Select the endpoint of the center mark of the small circle; the dimension is attached to the pointer.
- Move the pointer vertically downwards and position the dimension, as shown below.

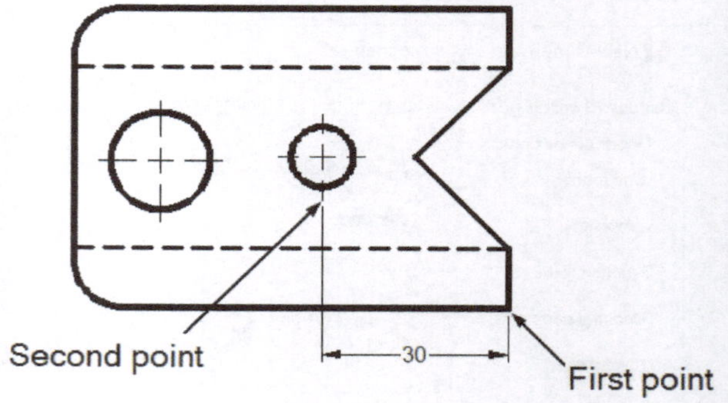

- Select the **Baseline** option from the command line.
- Select the right extension line of the linear dimension; a dimension is attached to the pointer.
- Select the endpoint of the center mark of the large circle; another dimension is attached to the pointer.
- Select the lower-left corner of the drawing.

- Press ENTER twice.

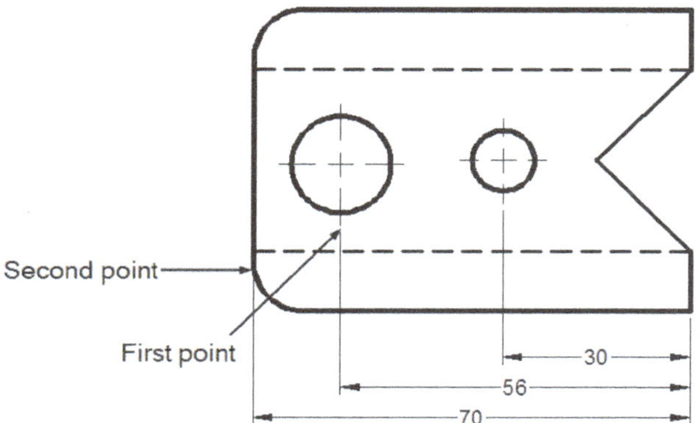

- Select the **Angular** option from the command line.
- Select the two angled lines of the drawing and position the angle dimension.

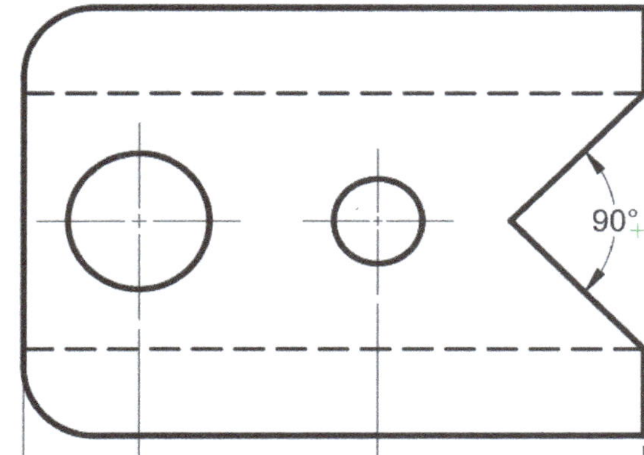

- Select the large circle and position the diameter dimension.
- Likewise, select the small circle and position the dimension.
- Press ENTER twice.
- Select the fillet located at the top left corner.
- Right-click and select **Radius**; the radial dimension is attached to the pointer.
- Select **Mtext** from the command line and type **2X** and press SPACEBAR.
- Click in the graphics window to update the dimension text.
- Next, position the radial dimension approximately at 45 degrees.
- Likewise, apply the other dimensions, as shown.
- Save and close the drawing.

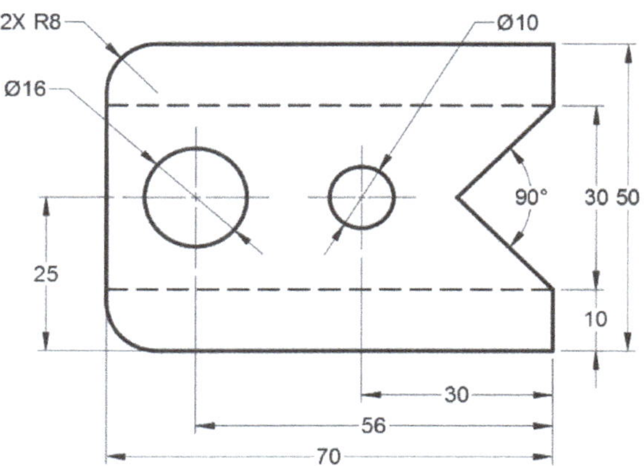

Adding Leaders

A leader is a thin solid line terminating with an arrowhead at one end and a dimension, note, or symbol at the other end. In the following example, you will learn to create a leader style and then create a leader.

Example 1:

- Draw a square of 24 mm side length.
- Create a circle of 10.11 mm diameter at the center of the square.

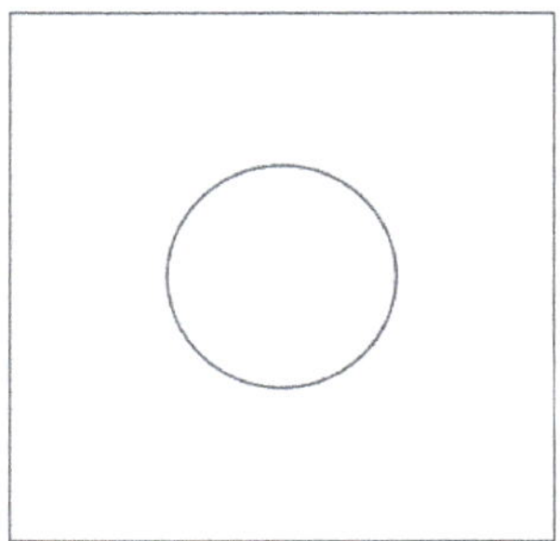

- Click **Home > Draw > Arc > Center, Start, Angle**  on the ribbon.
- Select the center point of the circle.
- Move the pointer horizontally toward the right.
- Type 6 as the radius and press ENTER.
- Move the pointer vertically downward and click.

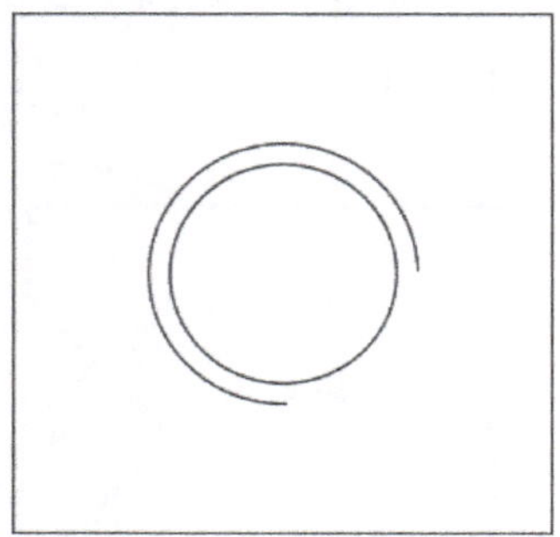

- On the **Home** tab of the ribbon, expand **Annotation** panel and click the **Multileader Style** icon; the **Multileader Style Manager** appears.

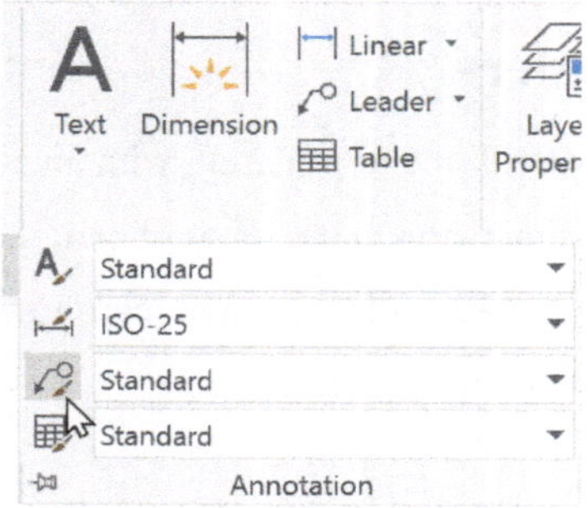

- In the **Multileader Style Manager** dialog, click the **New** button; the **Create New Multileader Style** dialog appears.

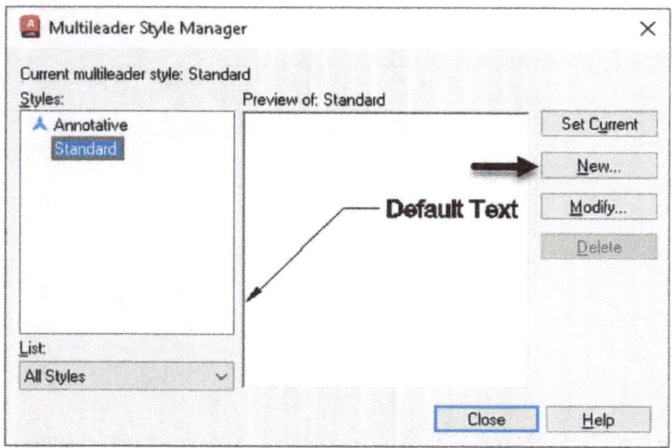

- In the **Create New Multileader Style** dialog, enter **Hole callout** in the **New style name** box and select **Standard** from the **Start with** drop-down.

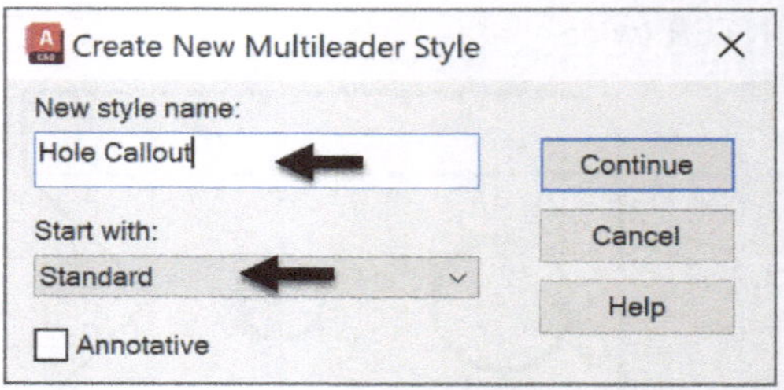

- Click **Continue**; the **Modify Multileader Style** dialog appears.
- Click the **Leader Format** tab and set the **Arrowhead Size** to **2.5**.

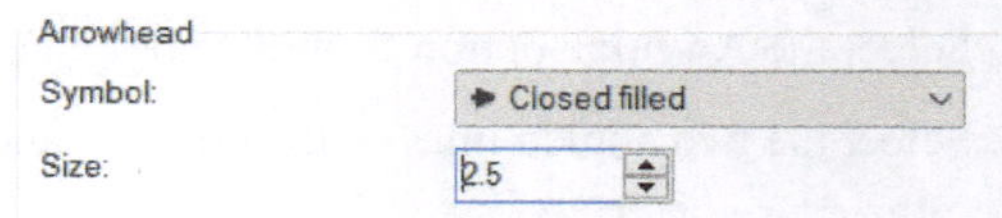

Also, notice the other options in this tab. They are used to set the appearance of the multileader lines and the arrowhead.

- Click the **Leader Structure** tab and set the **Landing distance** to **5**.

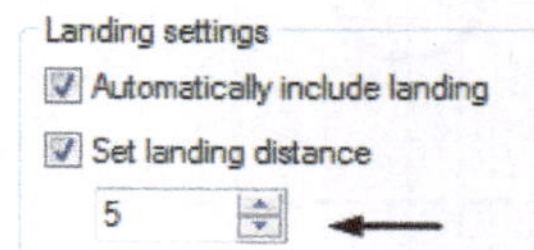

- Click the **Content** tab and set the **Text height** to **2.5**.

The other options in this tab are used to define the appearance of the text or block that will be attached at the end of the leader line.

- Click **OK** on the **Modify Multileader Style** dialog.
- Click **Set Current** on the **Multileader Style Manager** dialog.
- Click **Close** to close the dialog.
- Click **Annotate > Leaders > Multileader** on the ribbon.

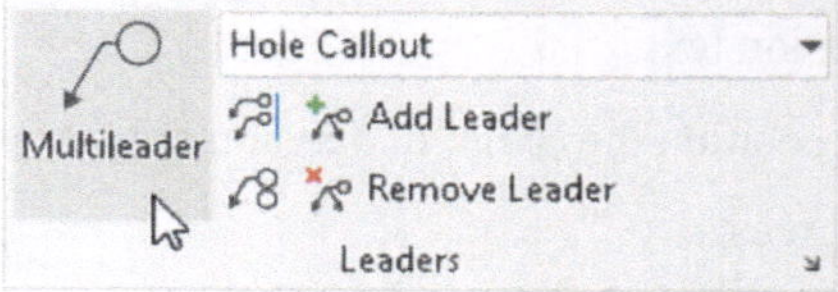

- Click the down arrow next to the **Polar tracking** button on the status bar and select **45** from the menu.
- Activate the **Polar tracking** button on the status bar.
- Select a point in the first quadrant of the arc.
- Move the pointer in the top-right direction and click to create the leader.

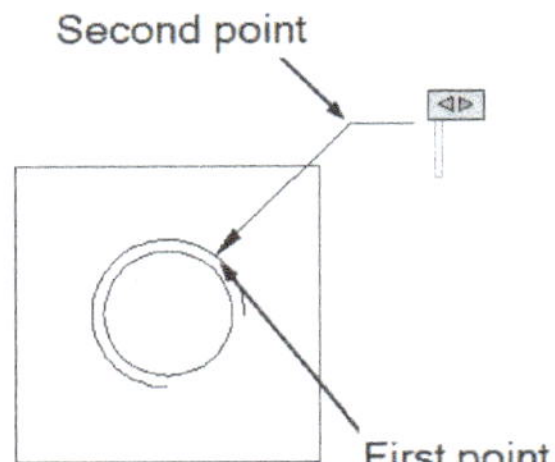

- Type **M12x1.75 – 6H 16** in the text editor. Next, you must insert the depth symbol before 16.
- Position the pointer before 16 and click the **Symbol** button on the **Insert** panel of the **Text Editor** ribbon; a menu appears.

- Click **Other** on the menu; the **Character Map** dialog appears.

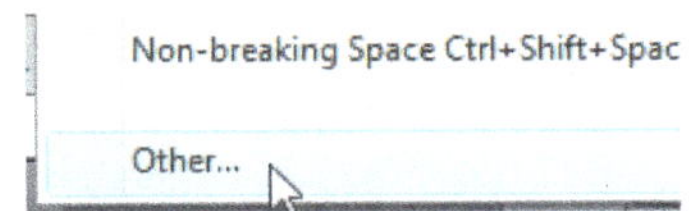

- In the **Character Map** dialog, select **GDT** from the **Font** drop-down.
- Select the Depth symbol from the fonts table.

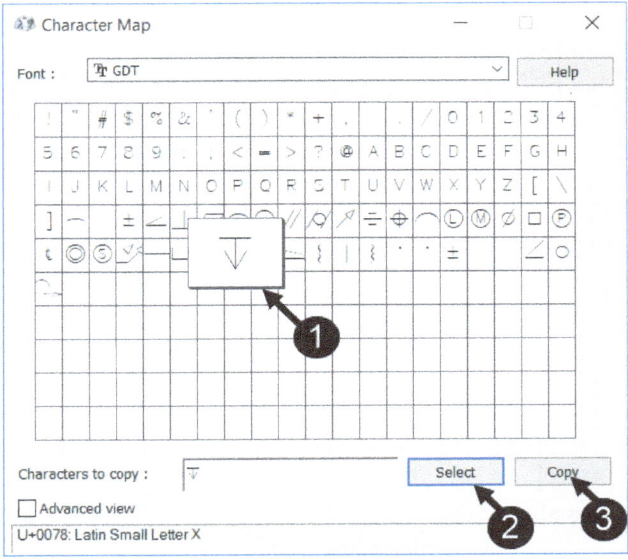

- Click **Select** and **Copy** buttons.
- Close the **Character Map** dialog.
- Right-click and select **Paste**; the depth symbol is pasted in the text editor.
- Adjust the spacing so that the complete text is in one line.
- Click in the graphics window.

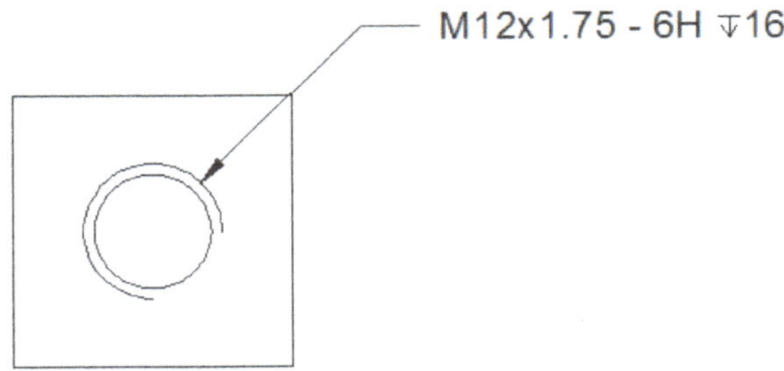

Adding Dimensional Tolerances

During the manufacturing process, the accuracy of a part is an important factor. However, it is impossible to manufacture a part with the exact dimensions. Therefore, while applying dimensions to a drawing, we provide some dimensional tolerances, which lie within acceptable limits. The following example shows you to add dimension tolerances in AutoCAD.

Example:

- Create the drawing, as shown below. Do not add dimensions to it.

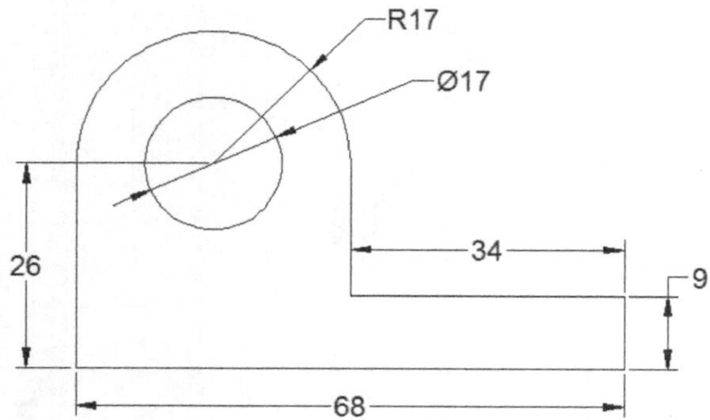

- Create a new dimension style with the name **Tolerances**.
- In the **New Dimension Styles** dialog, click the **Tolerances** tab.
- In the **Tolerances** tab, set the **Method** as **Deviation**.
- Set **Precision** as **0.00**.
- Set the **Upper Value** and **Lower Value** to **0.05**.
- Set the **Vertical position** as **Middle**.
- Specify the following settings in the **Primary Units**, **Text**, and **Symbols and Arrows** tab:

  The **Primary Units** tab:
  Unit format: Decimal
  Precision: 0.00
  Decimal Separator: '.' Period

  The **Text** tab:
  Text Height: 2.5
  Text placement:
  Vertical: Centered
  Horizontal: Centered
  Text alignment: Horizontal

  The **Symbols and Arrows** tab:
  Arrow Size: 2.5
  Center Marks: Line

- Click **OK** on the **New Dimension Styles** dialog.
- Click **Set Current** and **Close** on the **Dimension Style Manager** dialog.
- Apply dimensions to the drawing.

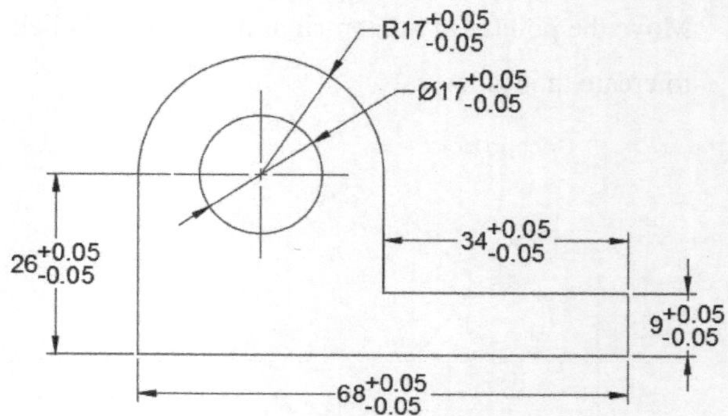

Geometric Dimensioning and Tolerancing

Earlier, you have learned how to apply tolerance to the size (dimensions) of a component. However, the dimensional tolerances are not sufficient for manufacturing a component. You must give tolerance values to its shape, orientation, and position as well. The following figure shows a note which is used to explain the tolerance value given to the shape of the object.

Note: The vertical face should not taper over 0.08 from the horizontal face

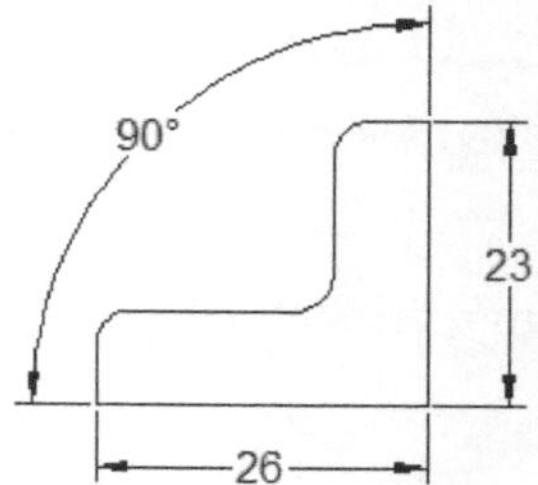

Providing a note in a drawing may be confusing. To avoid this, we use Geometric Dimensioning and Tolerancing (GD&T) symbols to specify the tolerance values to shape, orientation, and position of a component. The following

figure shows the same example represented by using the GD&T symbols. In this figure, the vertical face to which the tolerance frame is connected must be within two parallel planes 0.08 apart and perpendicular to the datum reference (horizontal plane).

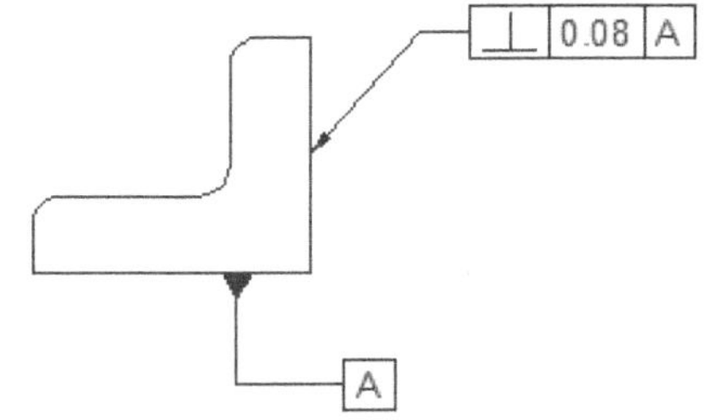

The Geometric Tolerancing symbols that can be used to interpret the geometric conditions are given in the table below.

Purpose		**Symbol**
To represent the shape of a single feature.	Straightness	⏤
	Flatness	⏥
	Cylindricity	⌭
	Circularity	○
	Profile of a surface	⌓
	Profile of a line	⌒
To represent the orientation of a feature with respect to another feature.	Parallelism	∥
	Perpendicularity	⊥
	Angularity	∠
	Position	⌖

To represent the position of a feature with respect to another feature.	Concentricity and coaxiality	
	Run-out	
	Total Run-out	
	Symmetry	

Example 1:

In this example, you will apply geometric tolerances to the drawing shown below.

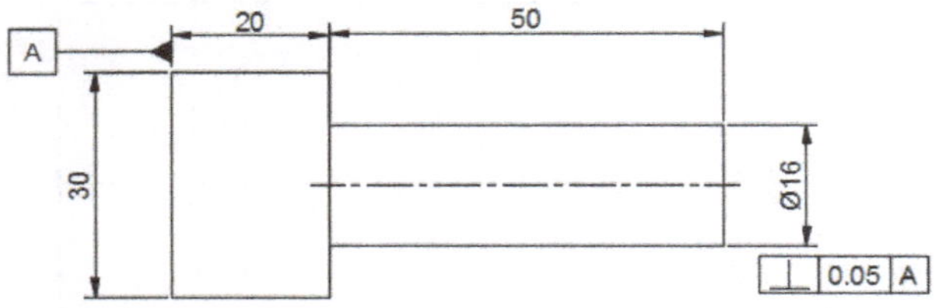

- Create the drawing, as shown below.

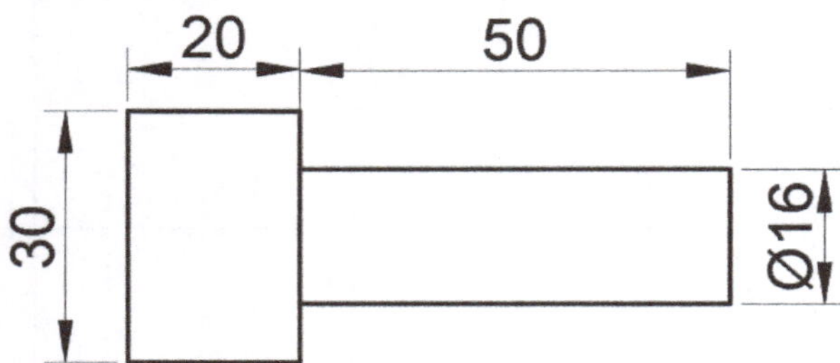

- Click **Annotate > Dimensions > Tolerance** on the ribbon; the **Geometric Tolerance** dialog appears.

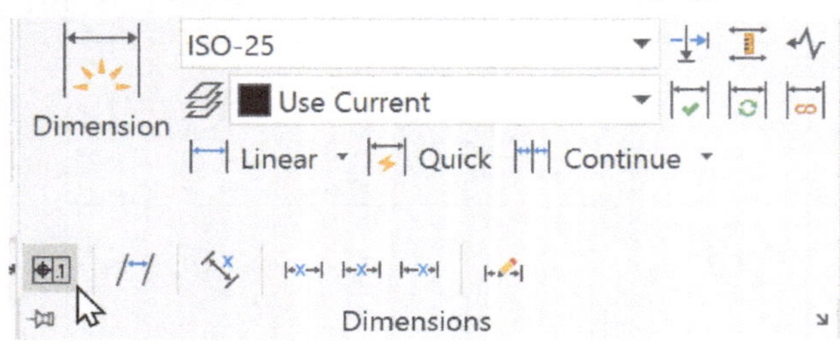

- In the **Geometric Tolerance** dialog, click the upper box of the **Sym** group. The **Symbol** dialog appears.

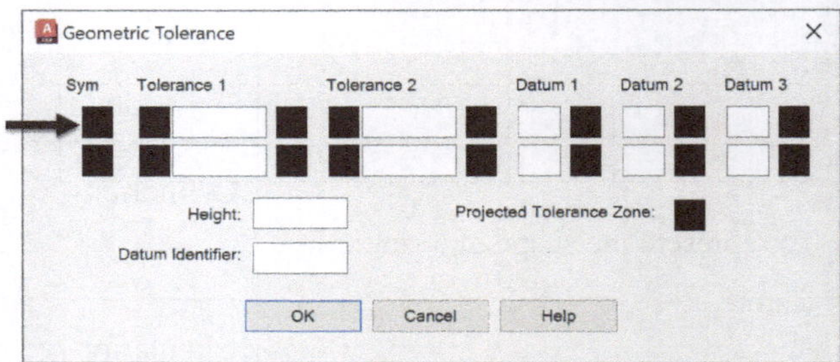

- In the **Symbol** dialog, click the **Perpendicularity** symbol. The symbol appears in the **Sym** group.

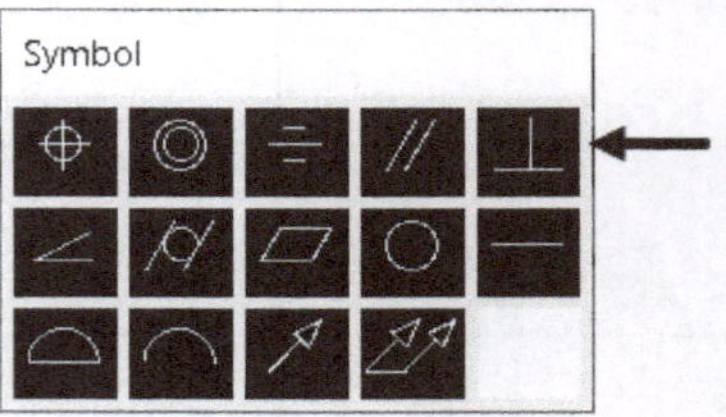

- Click in the top left box in the **Tolerance 1** group. The diameter symbol appears in the box.
- Enter **.05** in the box next to the diameter symbol.

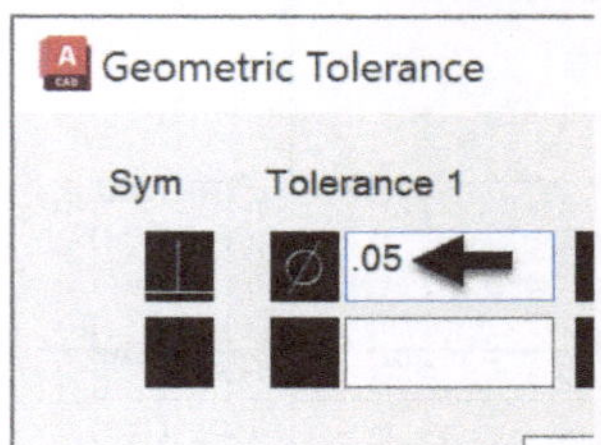

- Enter **A** in the upper box of the **Datum 1** group.

- Click **OK** and position the **Feature Control frame,** as shown below.

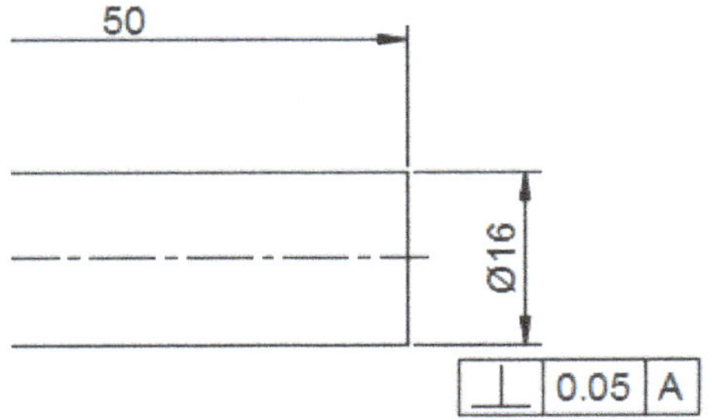

Next, you must add the datum reference.

- On the **Home** ribbon tab, expand the **Annotation** panel and click **Multileader Style** .
- Click the **New** button.
- On the **Create New Multileader Style** dialog, type **Tolerance** in the **New Style name** box, and click **Continue**.
- Click the **Leader Format** tab and select **Arrowhead > Symbol > Datum triangle filled**.
- Set the **Size** to 2.5.
- On the **Leader Structure** tab, set **Maximum leader points** to **2**.
- Click the **Content** tab and select **Multileader type > Block**.
- Select **Source block > Box**.
- Set the **Scale** value to 0.75.
- Click **OK**, **Set Current**, and **Close**.
- On the **Home** ribbon tab, click **Annotation > Leader**

.

- Specify the first and second points of the datum reference, as shown.
- On the **Edit Attributes** dialog, type **A** in the **Enter tag number** box.
- Click **OK**.

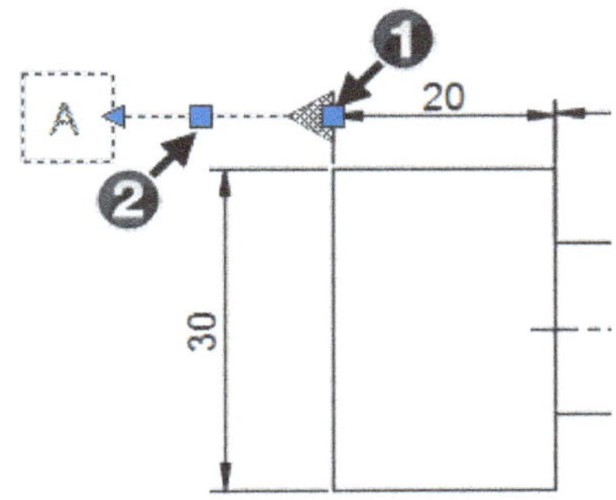

Editing Dimensions by Stretching

In AutoCAD, the dimensions are associated with the drawing. If you modify a drawing, the dimensions will be modified automatically. In the following example, you will stretch the drawing to modify the dimensions.

Example:

- Create the drawing, as shown below, and apply dimensions to it.

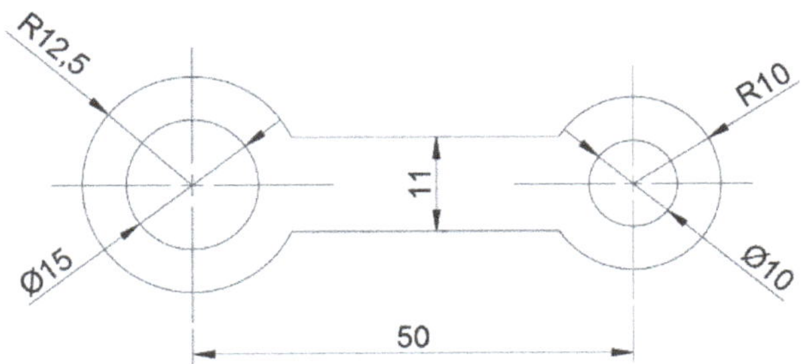

- Click **Home > Modify > Stretch** on the ribbon.
- Drag a window and select the right-side circles and the horizontal lines.

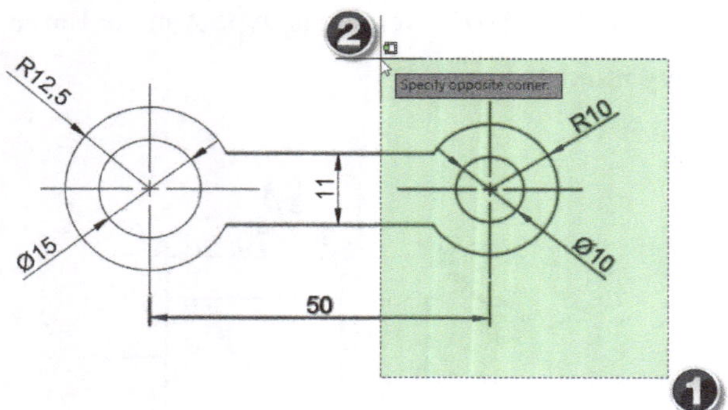

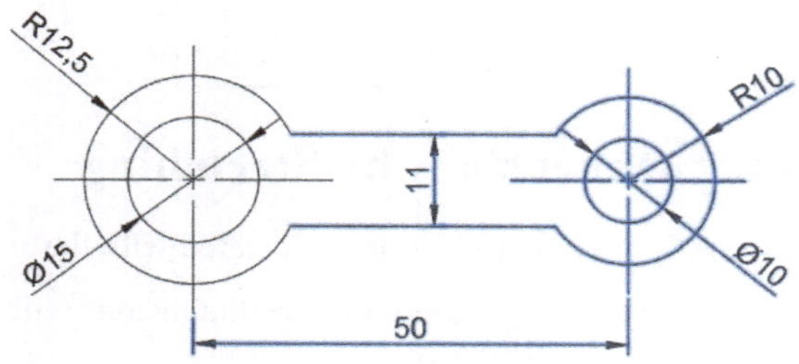

- Right-click and select the center point of the right-side circles.
- Move the pointer to stretch the drawing; you will notice that the horizontal dimension also changes.
- Type **30** and press ENTER; the horizontal dimension is updated to 80.

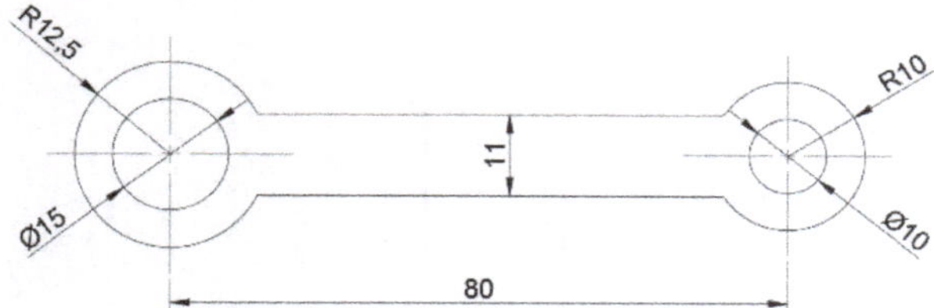

Modifying Dimensions by Trimming and Extending

In earlier chapters, you have learned to modify drawings by trimming and extending objects. In the same way, you can modify dimensions by trimming and extending. The following example shows you to modify dimensions by this method.

Example:

- Create a drawing, as shown below, and add dimensions to it.

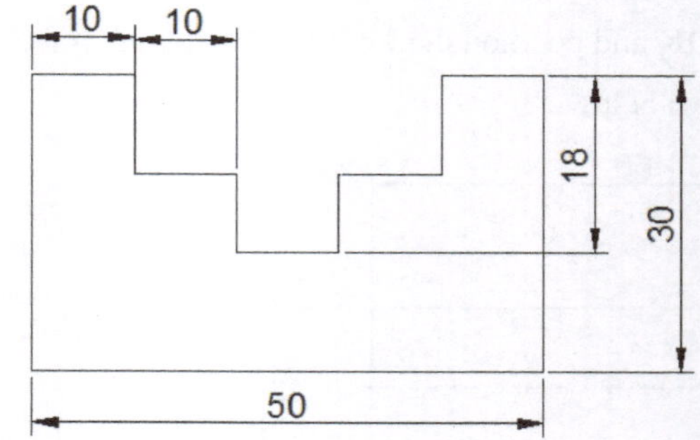

- Click **Home > Modify > Trim** on the ribbon.
- Select the **cuTting edges** option from the command line and press ENTER.
- Select the horizontal edge, as shown in the figure, to define the cutting edge.
- Right-click to accept.

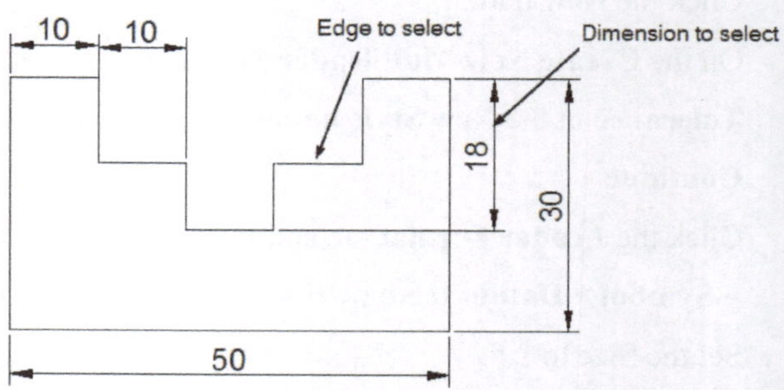

- Select the vertical dimension with the value 18. It trims the dimension up to the selected edge.

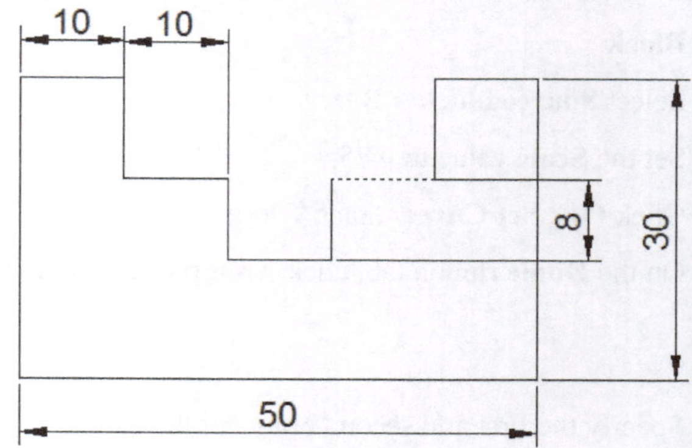

- Press ESC.

- Click **Home > Modify > Trim > Extend** on the ribbon.
- Select **Boundary edges** option from the command line and press ENTER.
- Select the vertical edge as the boundary, as shown below. Next, right-click to accept.

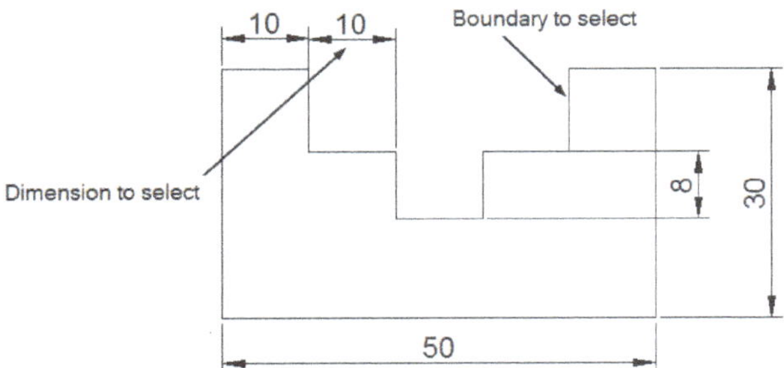

- Select the horizontal dimension with the value 10. It will extend the dimension up to the selected boundary.
- Press Esc.

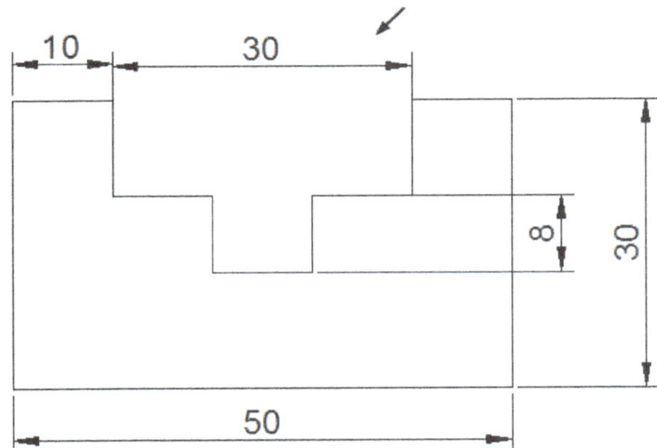

Using the DIMEDIT command

The **DIMEDIT** command can be used to modify dimensions. Using this command, you can add text to a dimension, rotate the dimension text and extension lines, or reset the position of the dimension text.

Example 1: (Adding Text to the dimension)

- Type **DED** in the command line and press ENTER.
- Select the **New** option from the command line; a text box appears.
- Enter **TYP** in the text box and press the SPACEBAR.

- Left-click and select the dimension with value 10.

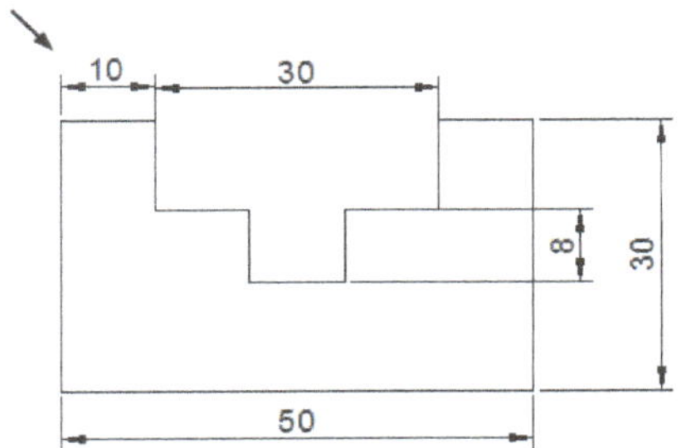

- Press ENTER; the dimension text will be changed.

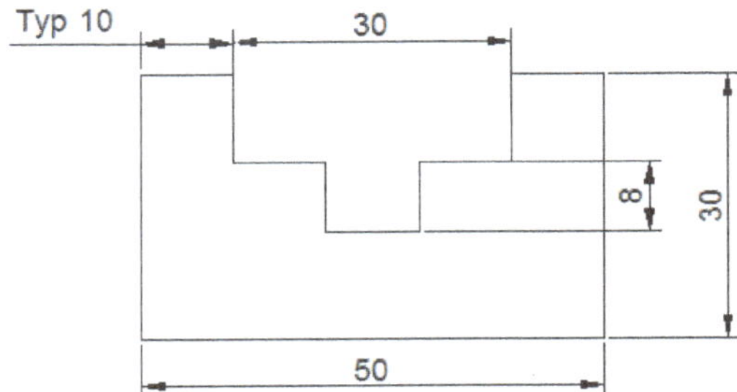

Example 2: (Rotating the dimension text)

- Enter **DED** in the command line and select the **Rotate** option; the message, "Specify angle for dimension text," appears in the command line.
- Type **30** and press ENTER.
- Select the dimension with the value 50 and right-click. The angle of the dimension text is changed to 30 degrees. Note that the angle is measured from the horizontal axis (X-axis).

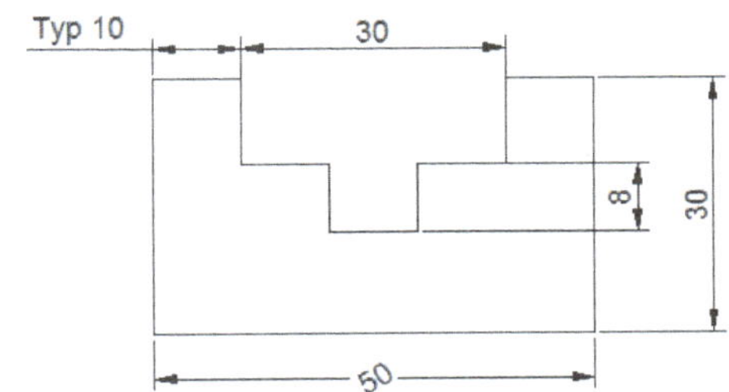

Using the Update tool

The **Update** tool is used to update a dimension with the currently active dimension style. For example, if you have created a new dimension style, you can apply it to an already existing dimension using the **Update** tool. The following example shows you to update a dimension.

- Type **D** in the command line and press ENTER; the **Dimension Style Manager** dialog appears.
- In the **Dimension Style Manager** dialog, select **Standard** from the **Styles** list and click **Modify**.
- In the **Modify Dimension Style** dialog, click the **Text** tab, and then set the **Text height** to **2.5**.
- Click the **Text Style** button; the **Text Style** dialog appears.

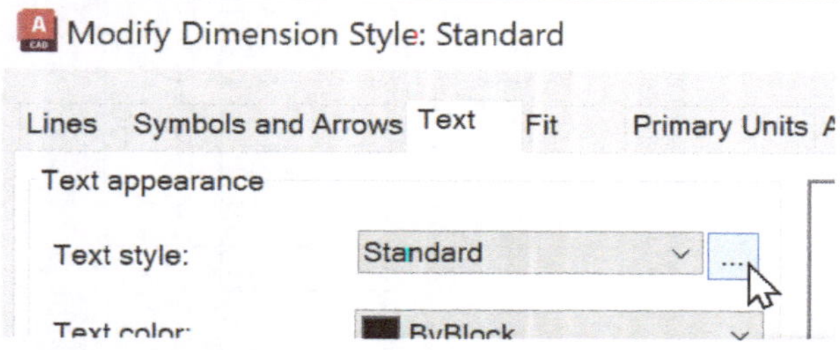

- In the **Text Style** dialog, change the **Font Style** to **Italic**.
- Click **Apply** and close the **Text Style** dialog.
- Click **OK** on the **Modify Dimension Style** dialog.
- Click **Close** on the **Dimension Style Manager** dialog.
- In the **Dimensions** panel, set the dimension style to **Standard**.

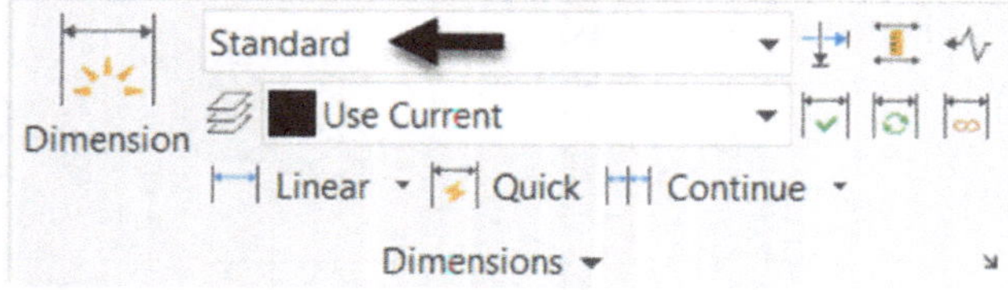

- Click the **Update** button on the **Dimensions** panel.

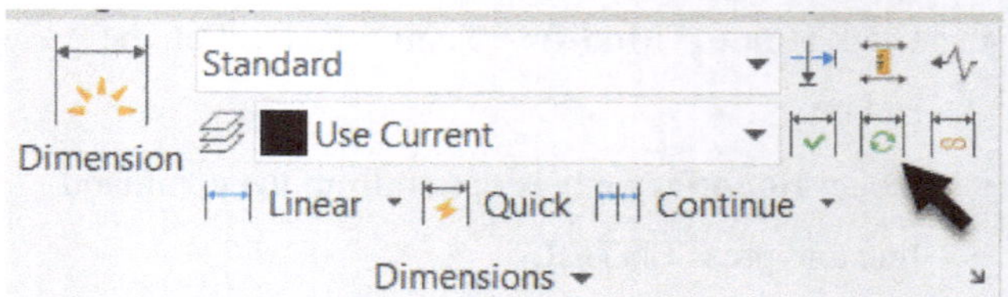

- Select the horizontal dimension with the value 30. Next, right-click; the dimension will be updated with the current dimension style.

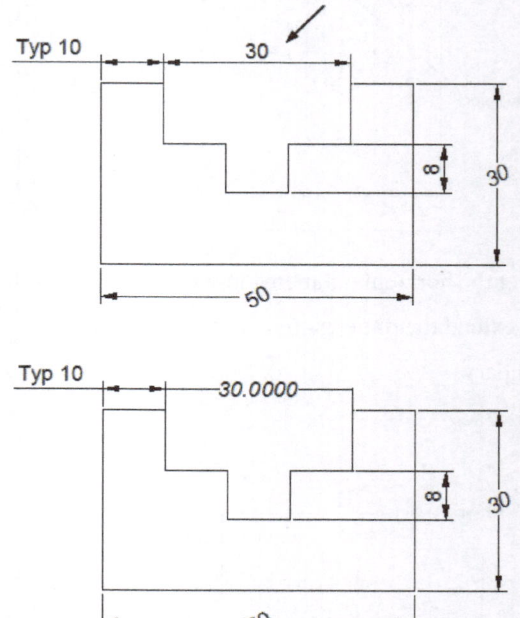

Using the Oblique tool

The **Oblique** tool is used to incline the extension lines of a dimension. This tool is very useful while dimensioning the isometric drawings. It can also be used in 2D drawings when the dimensions overlap with each other.

Example:

In this example, you will create an isometric drawing and add dimensions to it. Next, you will use the **Oblique** tool to change the angle of the dimension lines.

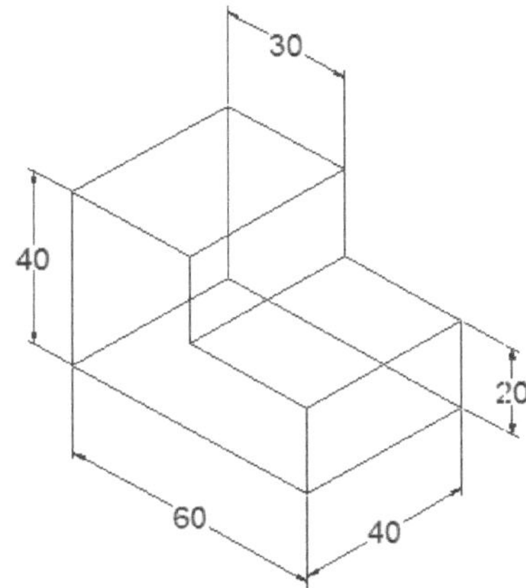

- Type-in **DS** in the command line, and then press Enter.
- On the **Drafting Settings** dialog, click the **Snap and Grid** tab.
- In the **Drafting Settings** dialog, set **Snap type** to **Isometric snap** and click **OK**.

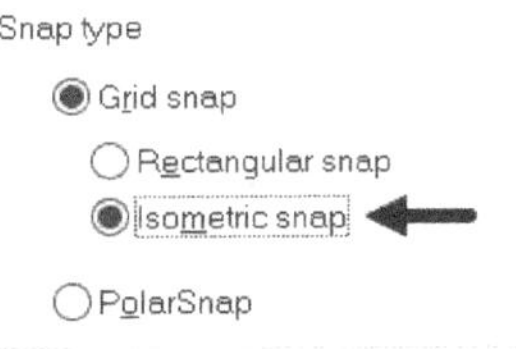

- Turn on the **Snap Mode** and the **Ortho Mode**. Also, turn on the **Dynamic Input**.
- Click **Zoom All** on the Navigation Bar.
- Type **L** in the command line and press ENTER.
- Click at a random point and move the pointer vertically.
- Type 40 in the command line and press ENTER; a vertical line will be created.
- Move the pointer toward the right; you will notice that an inclined line is attached to the pointer.

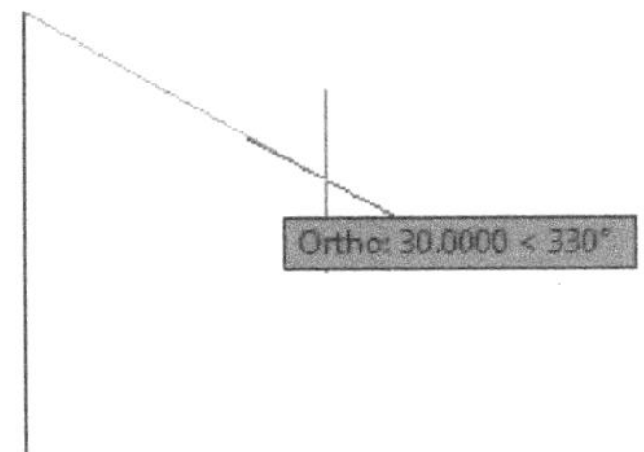

- Type 30 and press ENTER; an inclined line is drawn.
- Move the pointer downward.
- Type 20 and press ENTER.

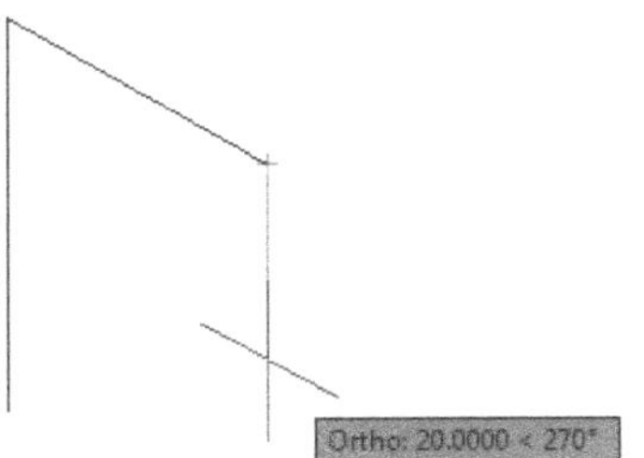

- Move the pointer toward the right.
- Type 30 and press ENTER.
- Move the pointer downward, type 20, and then press ENTER.
- Move the pointer toward the left and click on the start point of the sketch.
- Right-click select **Enter**.

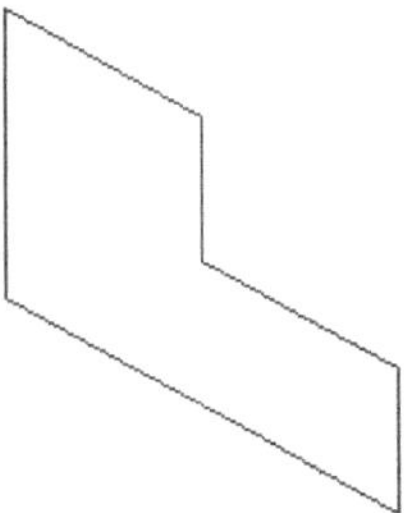

- Turn off the **Ortho Mode**.
- Click the down-arrow next to the **Polar Tracking (F10)** icon on the Status bar, and then select **30** from the flyout.
- Create a selection window and select all the objects of the sketch.
- Right-click and select **Copy-Selection** from the shortcut menu.
- Select the lower-left corner point as the base point.

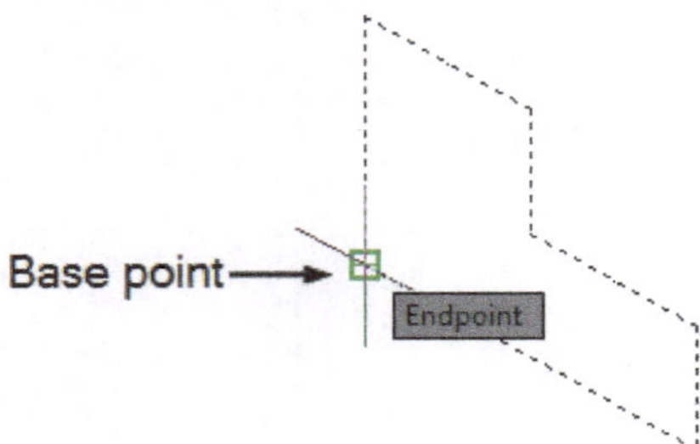

- Move the pointer in the direction perpendicular to the drawing; the track line is displayed.
- Move the pointer along the track line.

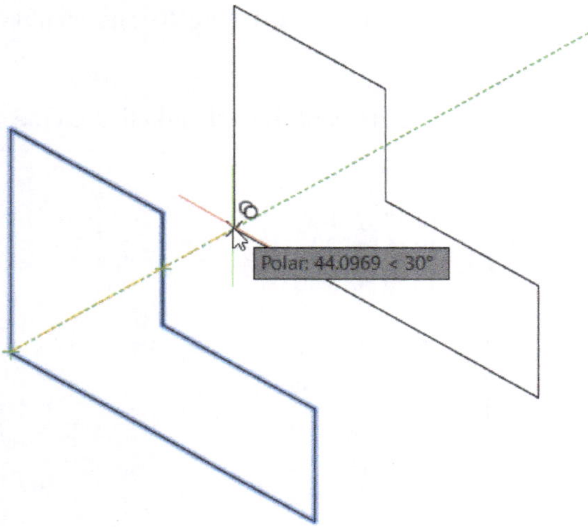

- Type 40 and press ENTER.
- Right-click and select **Enter**.

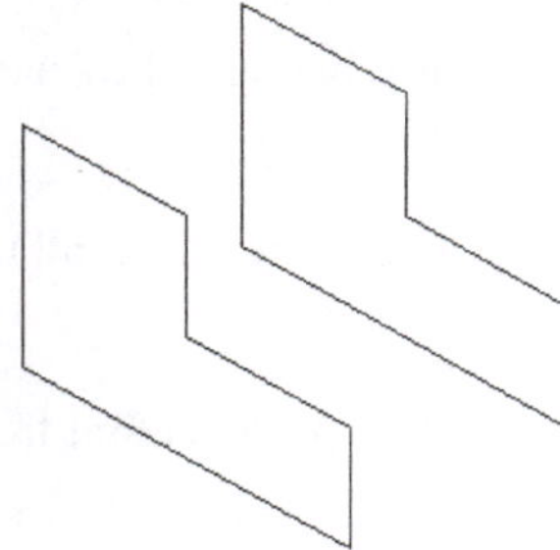

- Use the **Line** tool and connect the endpoints of the two sketches.

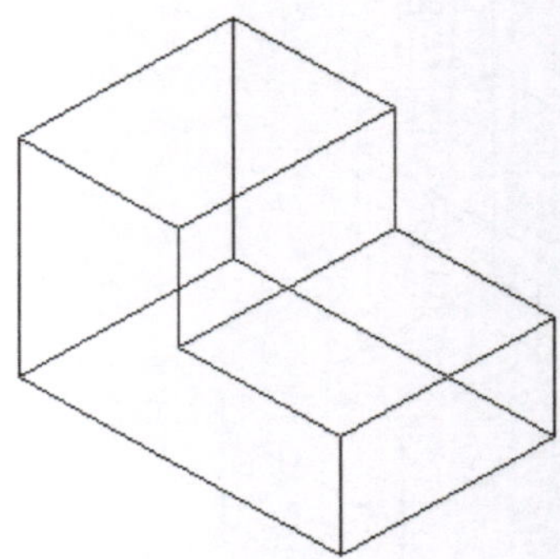

- Deactivate the ISODRAFT and the Snap Mode icons on the Status bar.
- Use the dimensioning tools and apply dimensions to the sketch.

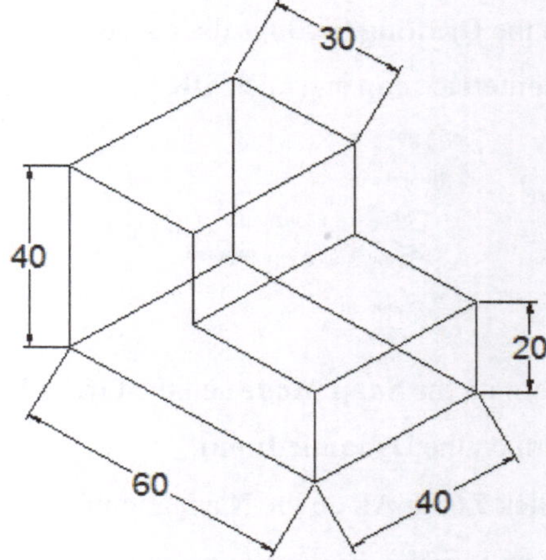

- Expand the **Dimensions** panel on the **Annotate** ribbon tab and click the **Oblique** button.

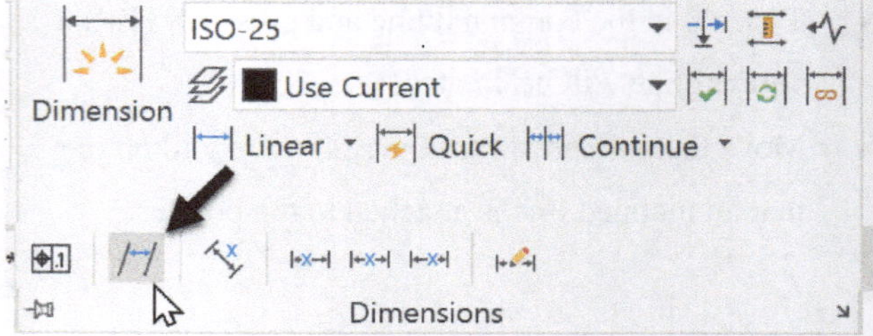

- Select the vertical dimensions and right-click to accept; the message, "**Enter obliquing angle**," appears in the command line.

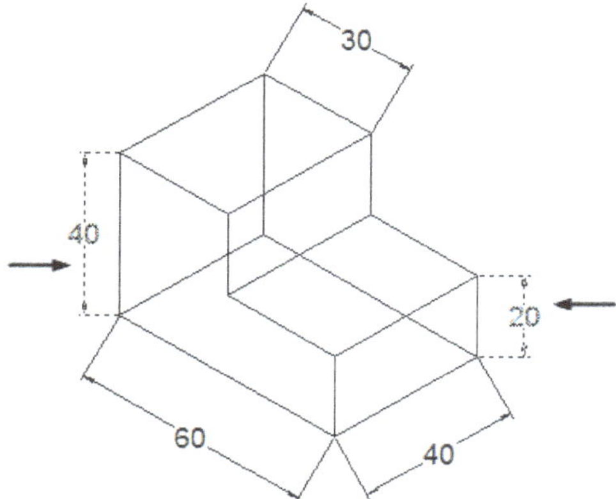

- Type 150 as the oblique angle and press ENTER; the dimensions are oblique, as shown below.

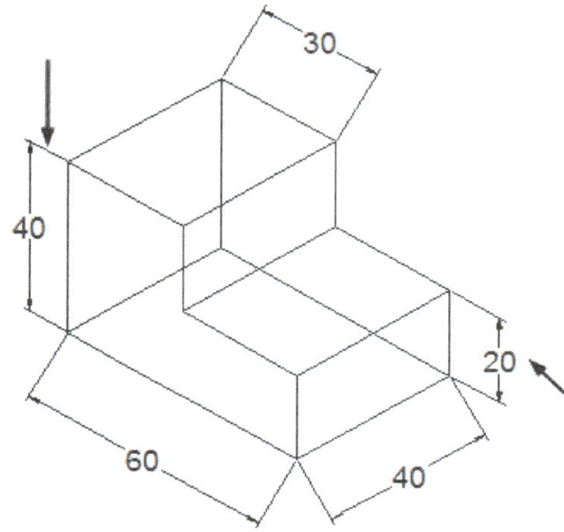

- Again, click the **Oblique** tool on the **Dimensions** panel and select the aligned dimensions. Next, right-click to accept.

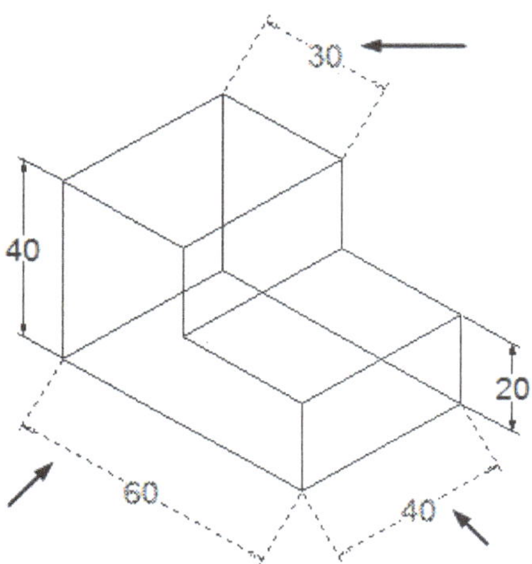

- Type 90 as the oblique angle and press ENTER; the dimensions will be oblique, as shown below.

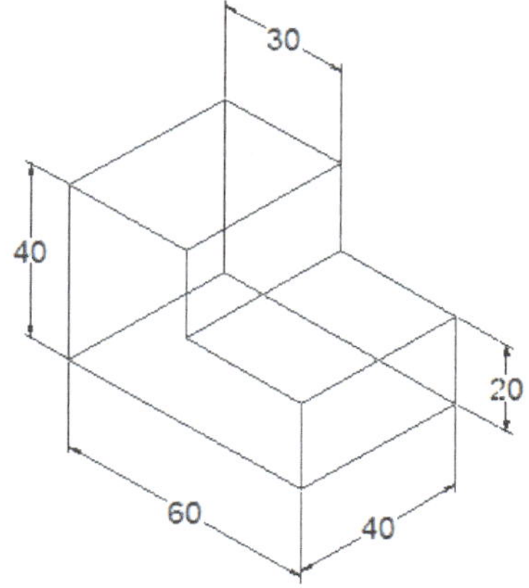

Editing Dimensions using Grips

In Chapter 4, you have learned to edit objects using grips. In the same way, you can edit dimensions using grips. The editing operations using grips are discussed next.

Example 1: (Stretching the Dimension)

- Select the dimension to display grips on it.
- Select the endpoint grip of the dimension.
- Next, move the pointer and select a new point; the dimension value will be updated automatically.

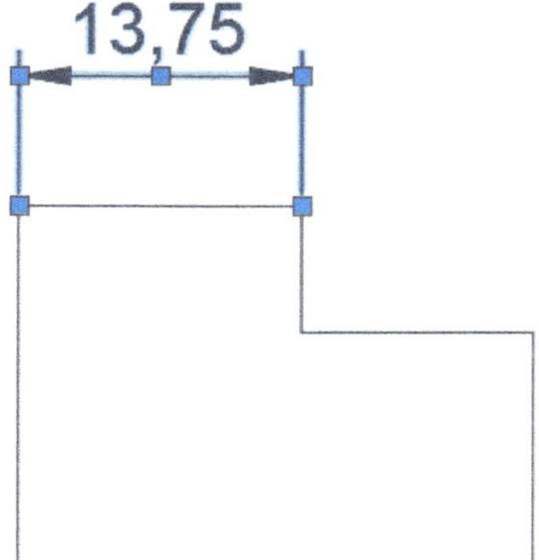

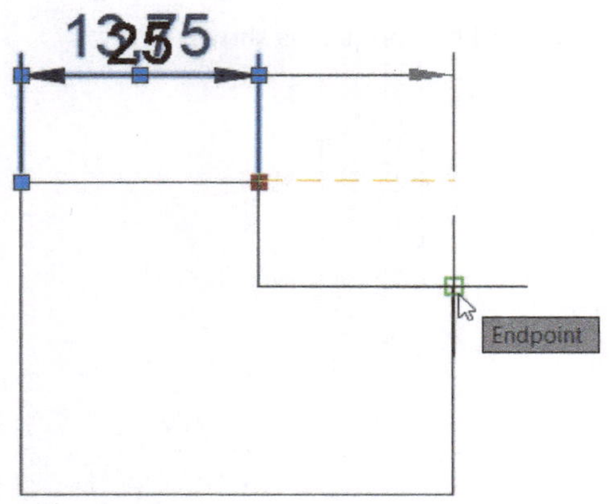

- You can also stretch the angular or radial dimensions.

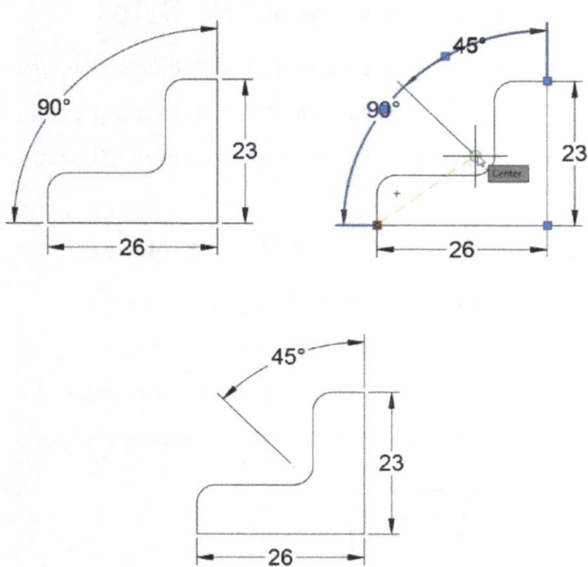

Example 2: (Moving the Dimension)

- To move a linear dimension, select the middle grip and move the pointer.

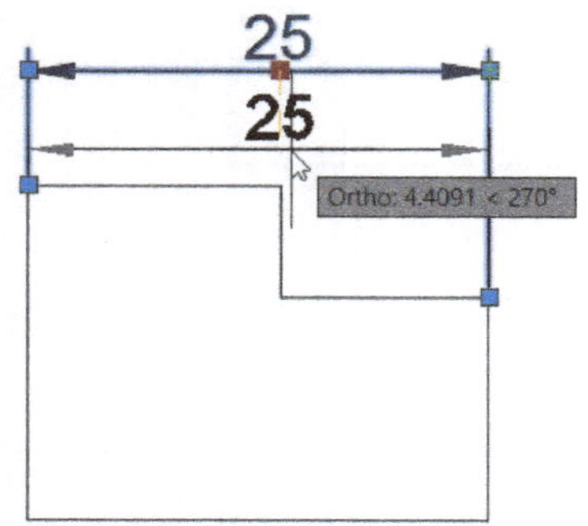

- Likewise, you can move the angular and radial dimensions.

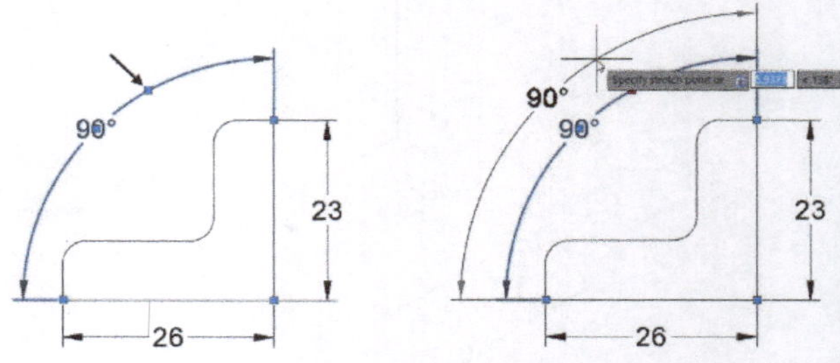

Example 3: (Modifying the Dimension text)

- Select the dimension and position the pointer on the middle grip; a shortcut menu appears as shown below.

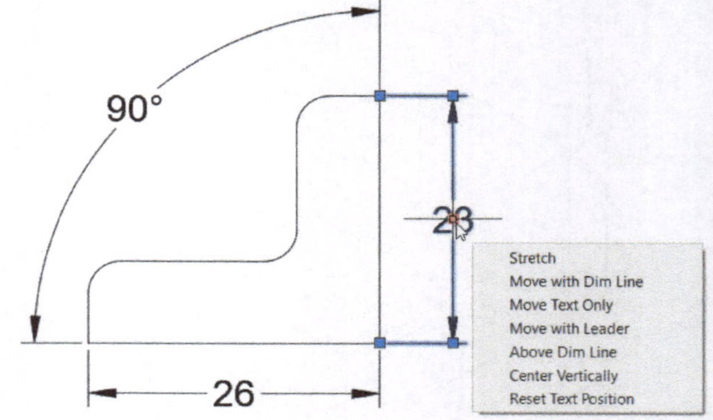

The options in the menu are self-explanatory. You can perform the required operation by selecting the corresponding option.

- Likewise, position the pointer on the endpoint of the dimension line and select the required option from the menu.

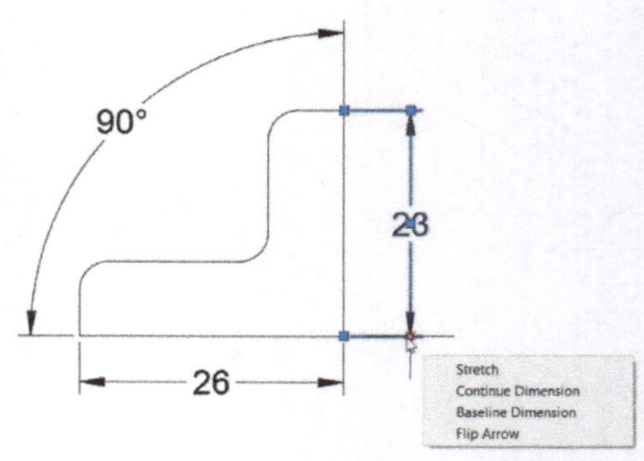

Modifying Dimensions using the Properties palette

Using the **Properties** palette, you can modify the dimensional properties such as text, arrow size, precision, linetype, and lineweight. The **Properties** palette comes in handy when you want to modify the properties of a particular dimension only.

Example:

- Create the drawing shown in the figure and apply dimensions to it.

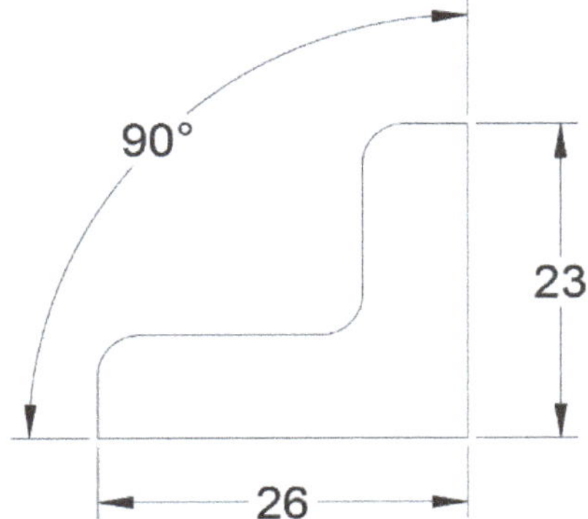

- Select the vertical dimension and right-click.
- Select **Properties** from the shortcut menu; the **Properties** palette appears.
- In the **Properties** palette, under the **Lines & Arrows** section, set the **Arrow size** to **2**.

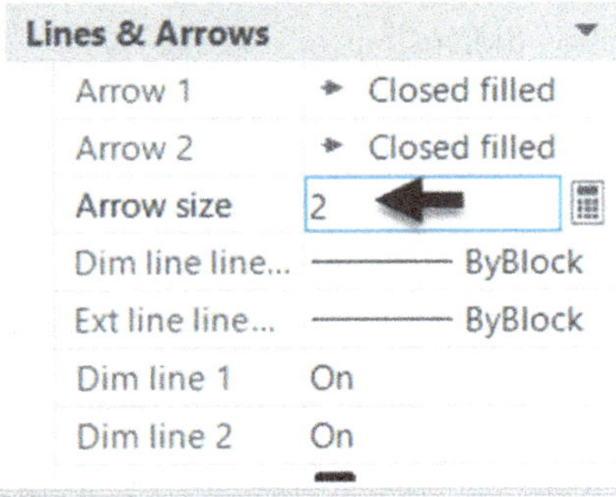

- Under the **General** section, set **Color** to **Blue**.

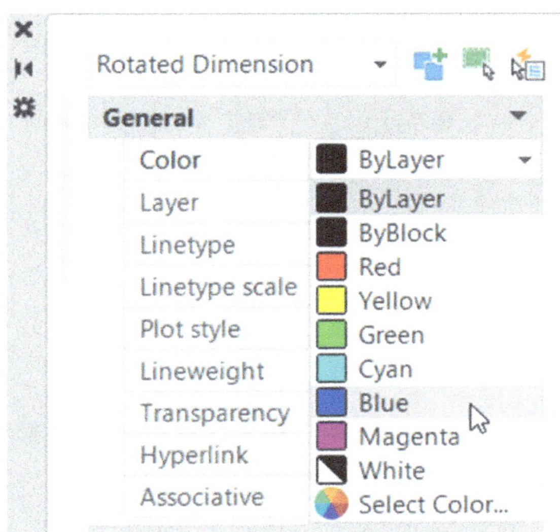

- In the **Properties** palette, under the **Lines & Arrows** section, set the **Ext line offset** value to **1.25**.
- Scroll down to the **Text** section and set **Text height** to **2**.

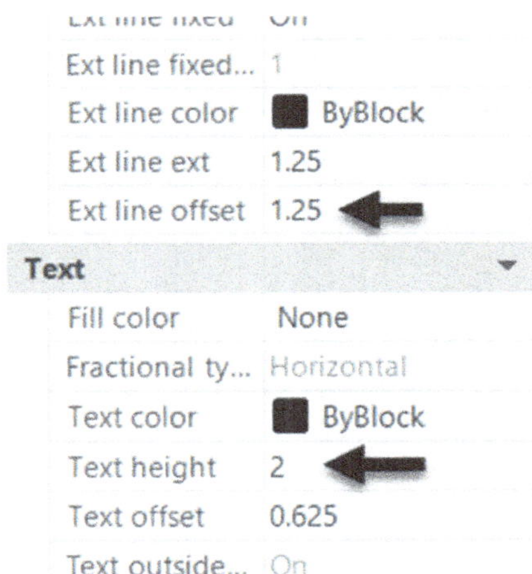

- Close the **Properties** palette; you will notice that the properties of the dimension are updated as per the changes made.

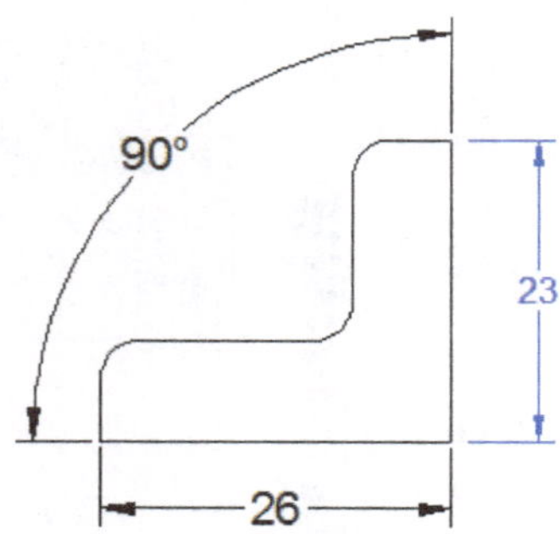

Matching Properties of Dimensions or Objects

In the previous section, you have learned to change the properties of a dimension. Now, you can apply these properties to other dimensions by using the **Match Properties** tool.

- Click **Home > Properties > Match Properties** on the ribbon or type **MA** and press ENTER; the message, "Select source object," appears in the command line.

- Select the vertical dimension from the drawing; the message, "Select destination object(s) or [Settings]:" appears in the command line.

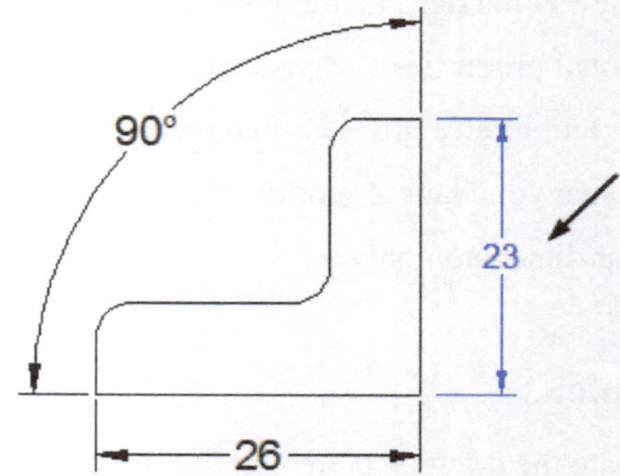

- Select the **Settings** option from the command line; the **Property Settings** dialog appears.

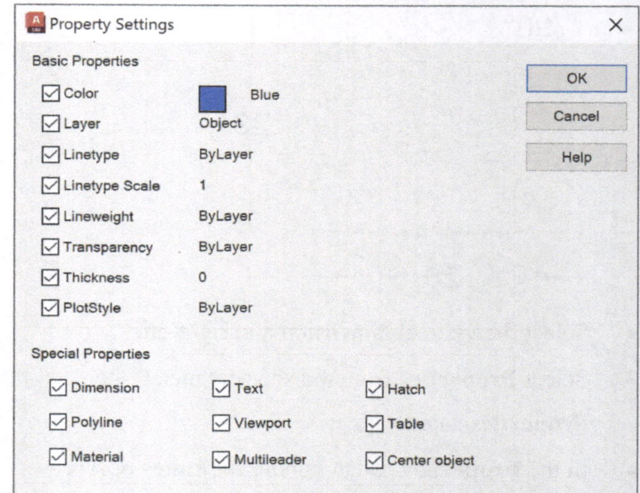

In this dialog, you can select the settings that can be applied to the destination dimensions or objects. By default, all the options are selected in this dialog.

- Click **OK** on the **Property Settings** dialog. Next, you must select the destination objects.
- Select the other dimensions from the drawing; the properties of the source dimension are applied to other dimensions.

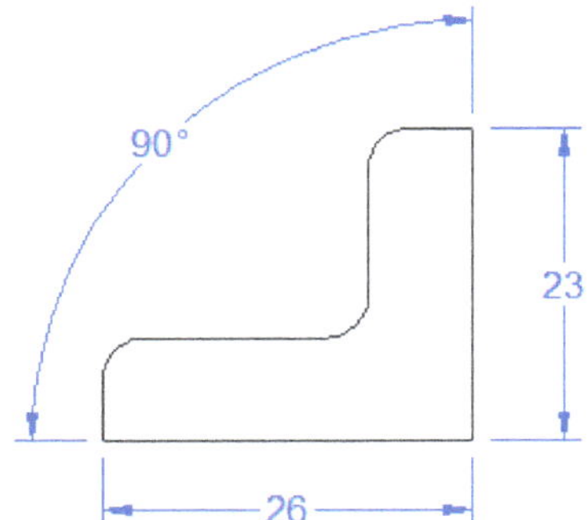

- Right-click and select **Enter**.

Exercises

Exercise 1

Create the drawing shown below and create hole callouts for different types of holes. Assume missing dimensions.

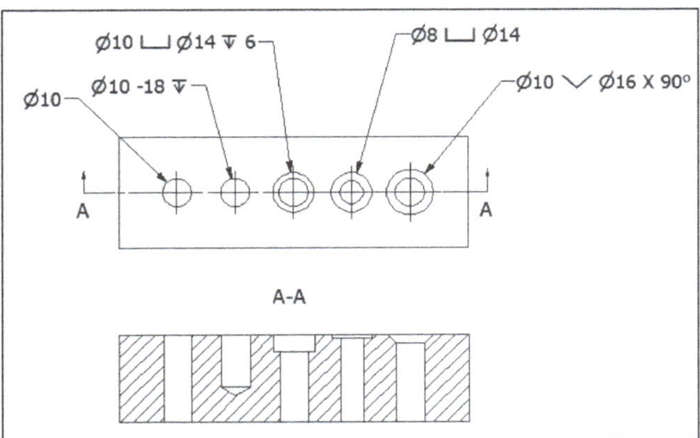

Exercise 2

Create the following drawings and apply dimensions and annotations. The Grid Spacing X= 10 and Grid Spacing Y=10.

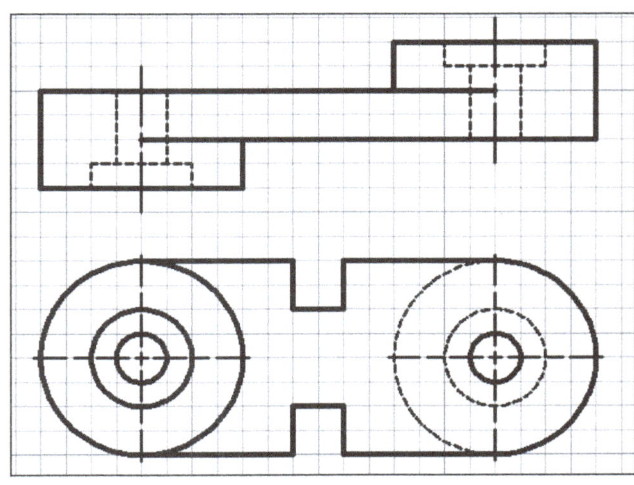

Exercise 3

Create the drawing shown below. The Grid spacing is 10 mm. After creating the drawing, apply dimensional tolerances to it. The tolerance specifications are given below.

Method: Limits
Precision: 0.00
Upper Value: 0.05
Lower Value: 0.05

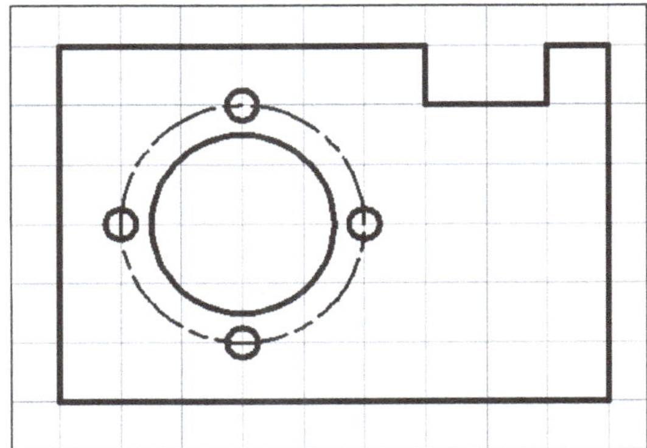

Exercise 4

Create the drawing shown below.

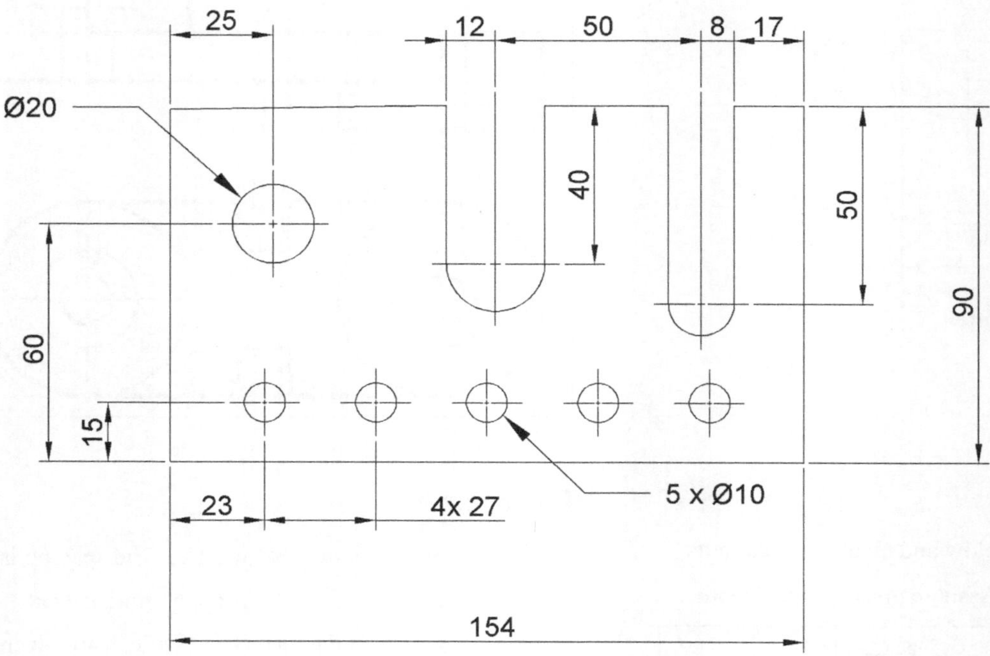

Exercise 5

Create the drawing shown below.

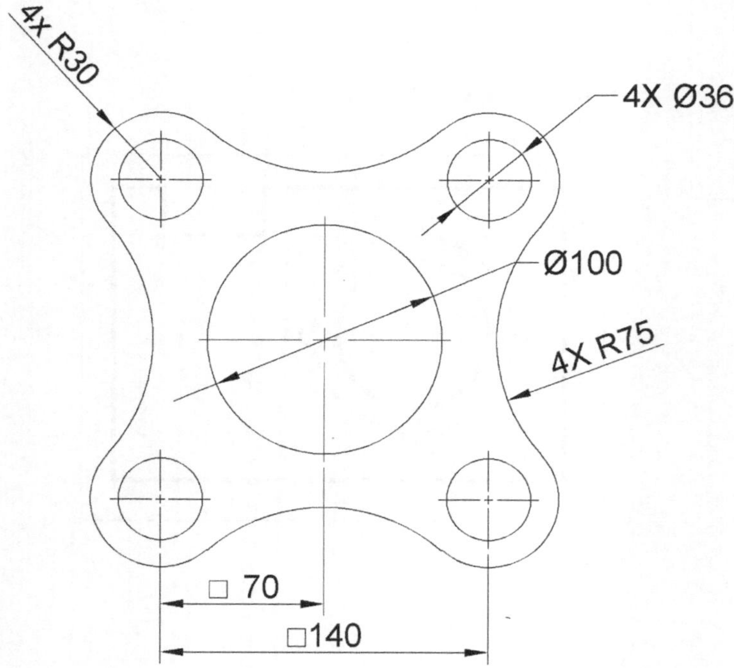

Chapter 7: Parametric Tools

In this chapter, you will learn to do the following:

- **Apply Geometric and Dimensional Constraints**
- **Create Equations using the Parameter Manager**
- **Create Inferred Constraints**

Parametric Tools

Parametric tools are one of the main advancements in Computer-Aided Design. Using the parametric tools, you can define the shape and size of a drawing by applying relations and dimensions between the objects. You can also use equations in place of dimensions. Changing one parameter of an equation would change the entire shape and size of the drawing. It makes it easy to modify the design.

The parametric tools can be accessed from the Ribbon, Command line, and Menu Bar.

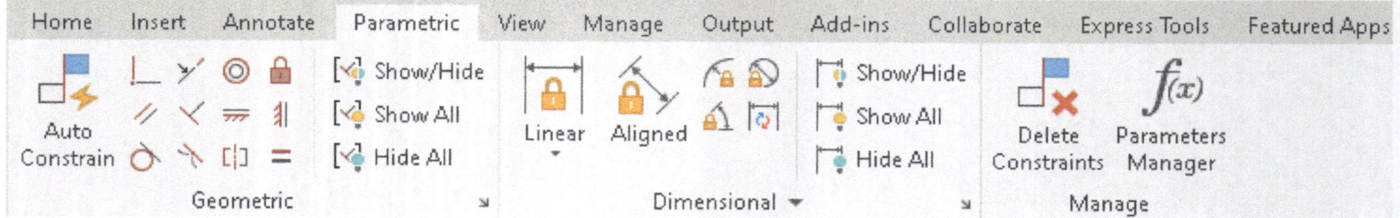

Geometric Constraints

Geometric Constraints are used to control the shape of a drawing by applying geometric relationships between the objects. For example, you can apply the **Tangent** constraint to make a line tangent to a circle. You can use the **Equal** constraint to make two lines equal in length.

The following table shows various geometric constraints and their functions.

Constraint	Function
 **Coincident**	It is used to constrain a point to lie on another point or an object. • Click **Parametric > Geometric > Coincident** on the ribbon. • Select a point on a line or arc. • Select a point on another object; the two points will coincide with each other.  First point Second point
 **Collinear**	It is used to constrain a line along another line. The lines are not required to touch each other. Second object First object • Click **Parametric > Geometric > Collinear** on the ribbon. • Select the first line and the second line; the second line will be made collinear with the first line.
Concentric	It is used to make the center points of arcs, circles, or ellipses coincident. 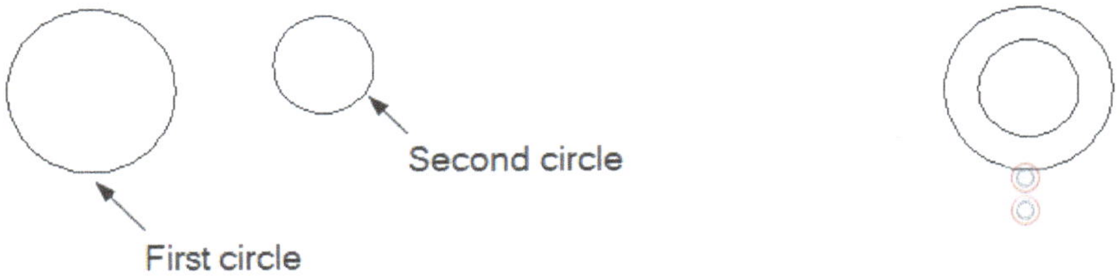 Second circle First circle • Click **Parametric > Geometric > Concentric** on the ribbon. • Select a circle or arc from the drawing. • Select another circle or arc; the second circle will be concentric with the first circle.

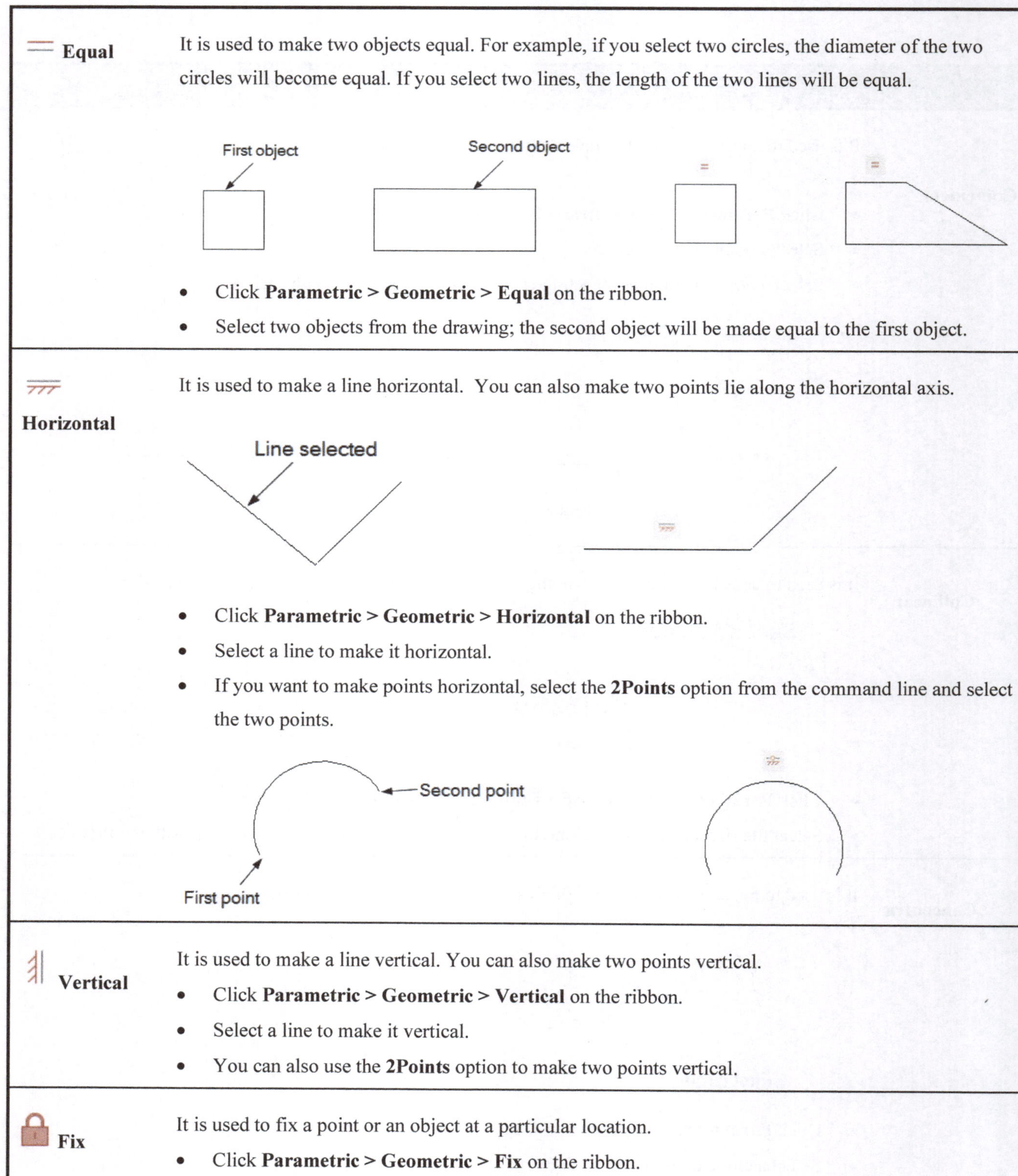

Equal	It is used to make two objects equal. For example, if you select two circles, the diameter of the two circles will become equal. If you select two lines, the length of the two lines will be equal. • Click **Parametric > Geometric > Equal** on the ribbon. • Select two objects from the drawing; the second object will be made equal to the first object.
Horizontal	It is used to make a line horizontal. You can also make two points lie along the horizontal axis. • Click **Parametric > Geometric > Horizontal** on the ribbon. • Select a line to make it horizontal. • If you want to make points horizontal, select the **2Points** option from the command line and select the two points.
Vertical	It is used to make a line vertical. You can also make two points vertical. • Click **Parametric > Geometric > Vertical** on the ribbon. • Select a line to make it vertical. • You can also use the **2Points** option to make two points vertical.
Fix	It is used to fix a point or an object at a particular location. • Click **Parametric > Geometric > Fix** on the ribbon. • Select a point to make it fixed at its location.

	  • You can also use the **Object** option to select objects from the drawing.
 **Perpendicular**	It is used to make two lines perpendicular to each other. 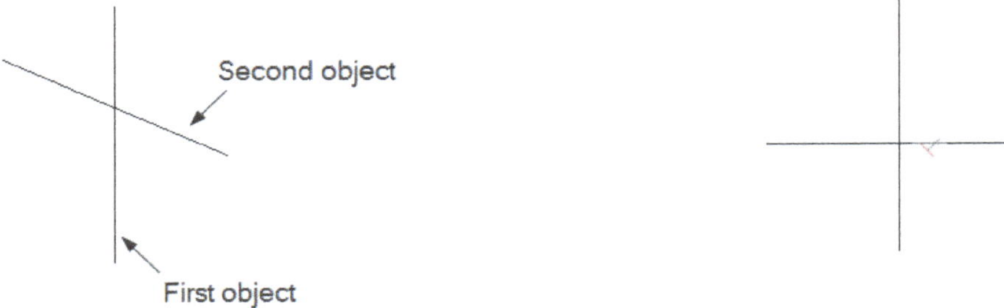 • Click **Parametric > Geometric > Perpendicular** on the ribbon. • Select two lines from the drawing; the second line is made perpendicular to the first line.
Smooth	It is used to make a spline continuous with another spline or arc. 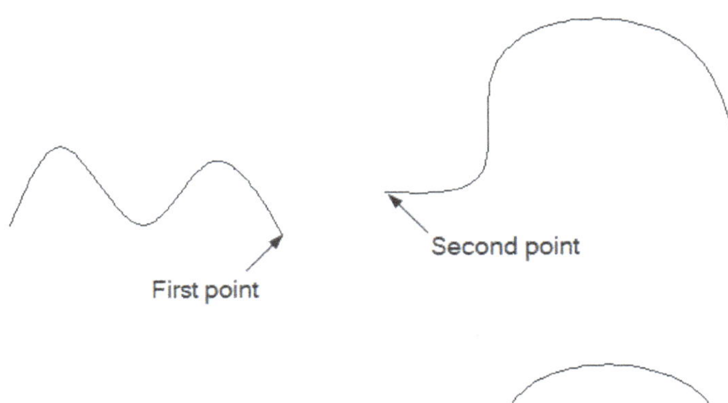 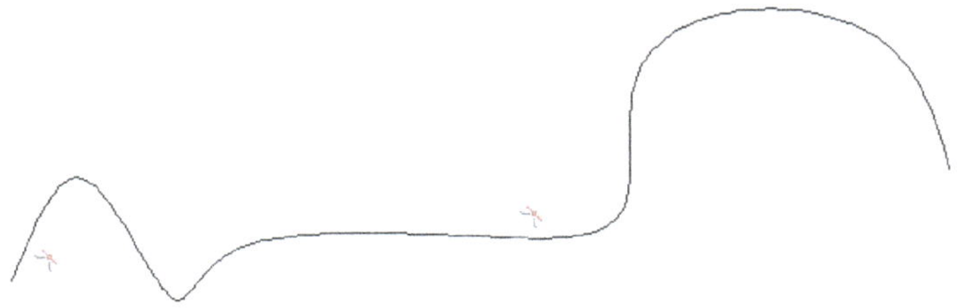 • Click **Parametric > Geometric > Smooth** on the ribbon. • Select a spline curve. • Select another spline or arc; the first curve will become continuous with the second curve.
 **Parallel**	It is used to make two lines parallel to each other.

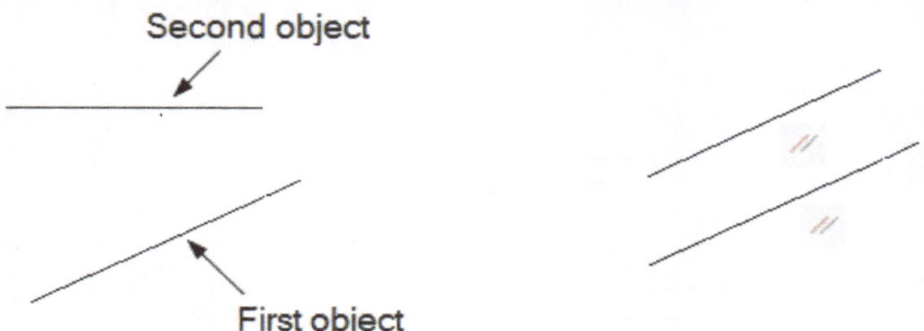

- Click **Parametric > Geometric > Parallel** on the ribbon.
- Select two lines from the drawing; the second line is made parallel to the first line.

Symmetric

It is used to make two objects symmetric about a line. The objects will have the same size, position, and orientation about a line.

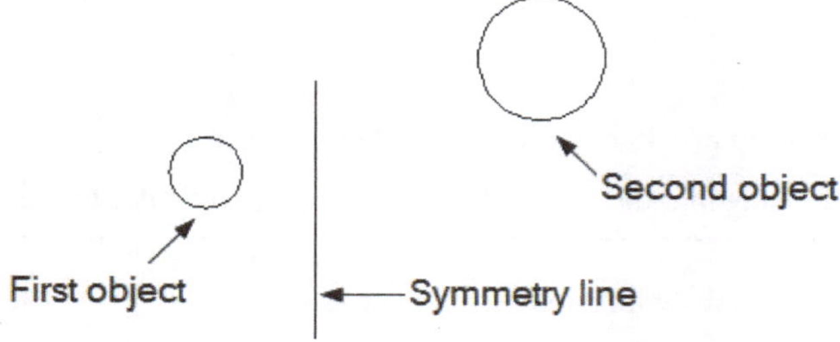

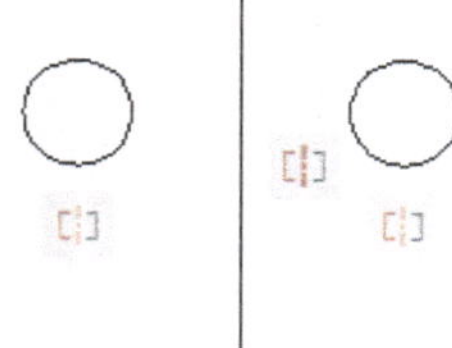

- Click **Parametric > Geometric > Symmetric** on the ribbon.
- Select two objects from the drawing.
- Select the symmetry line; the objects will be made symmetric about the selected line.
- You can also use the **2Points** option to make two points symmetric about a line.

Tangent

It is used to make an arc, circle, or line tangent to another arc or circle.

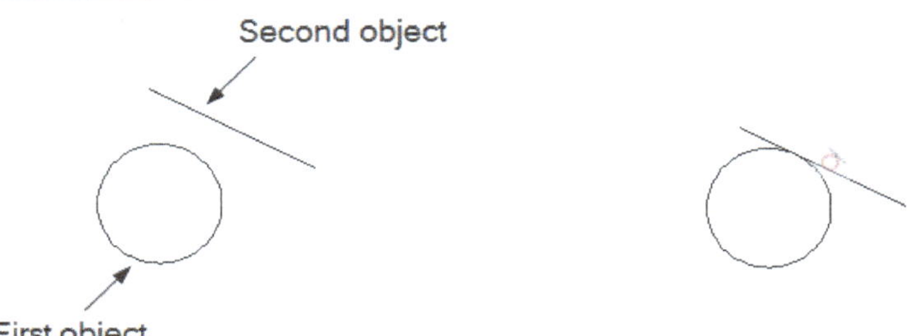

- Click **Parametric > Geometric > Tangent** on the ribbon.
- Select a circle, arc, or line.
- Select another circle, arc, or line; the second object will be tangent to the first object.

**Auto Constrain**

The **Auto Constrain** tool is used to apply constraints to the objects automatically.

- Click **Parametric > Geometric > Auto Constrain** on the ribbon.
- Select the **Settings** option from the command line; the **Constraint Settings** dialog appears.

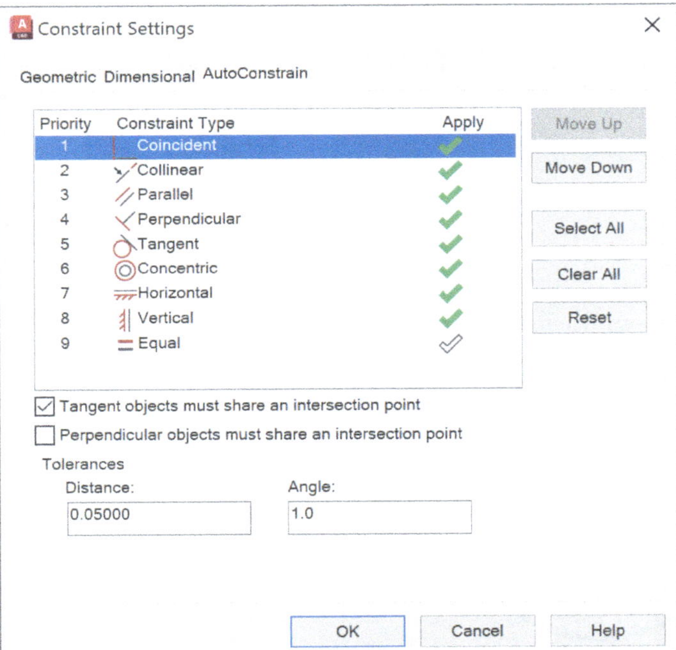

- In this dialog, select the constraints that you want to apply. You can also select the **Tangent objects must share an intersection point**, and **Perpendicular objects must share an intersection point** option.
- Click **OK**.
- Select multiple objects by clicking on them or by dragging a selection window.

- Right-click and select **Enter**; geometric constraints are applied to the objects based on their geometric condition.

Example:

In this example, you will create the following drawing by using the drawing and parametric tools.

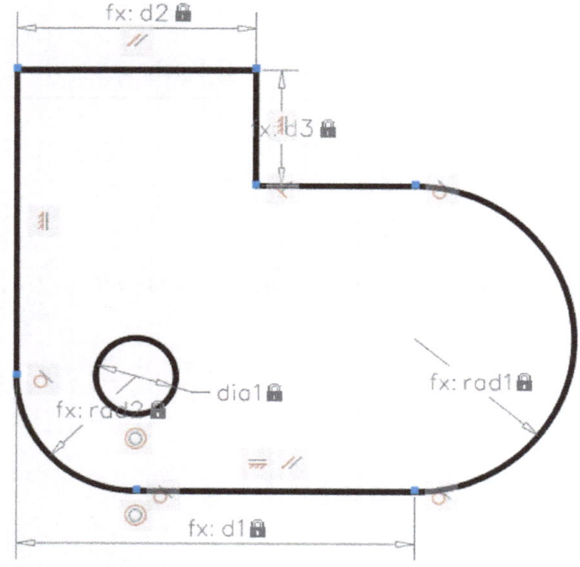

- Open a new AutoCAD file.
- Create two circles and a line, as shown in the figure.

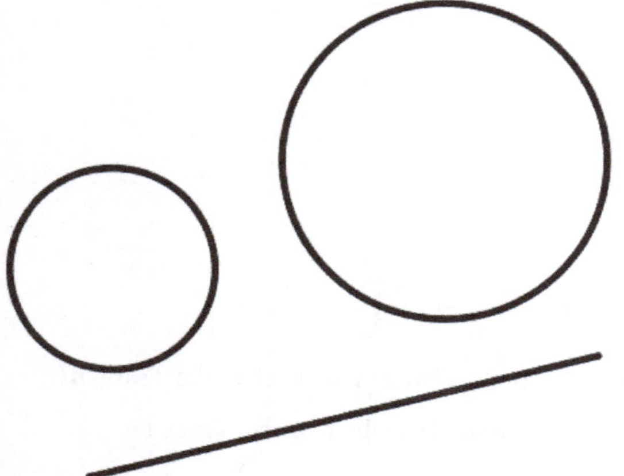

- Click **Parametric > Geometric > Horizontal** on the ribbon.
- Select the line to make it horizontal.

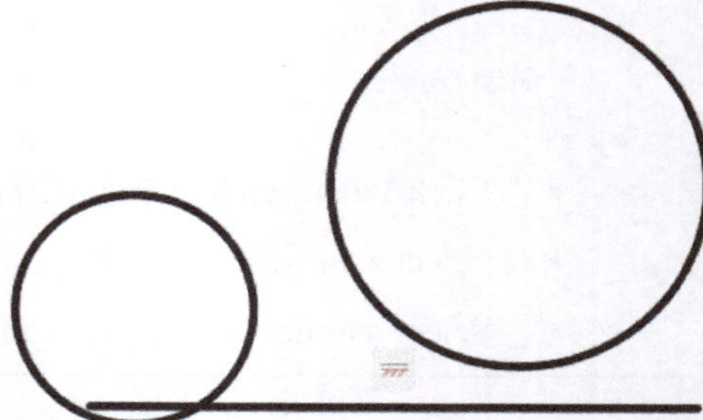

- Press the SPACEBAR and select the **2Points** option from the command line.
- Select the large circle and the small circle; the center points of the two circles will be horizontal.

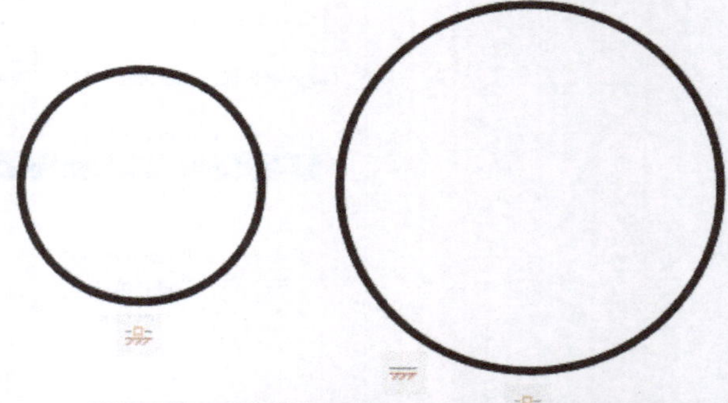

- Create four lines, as shown below.

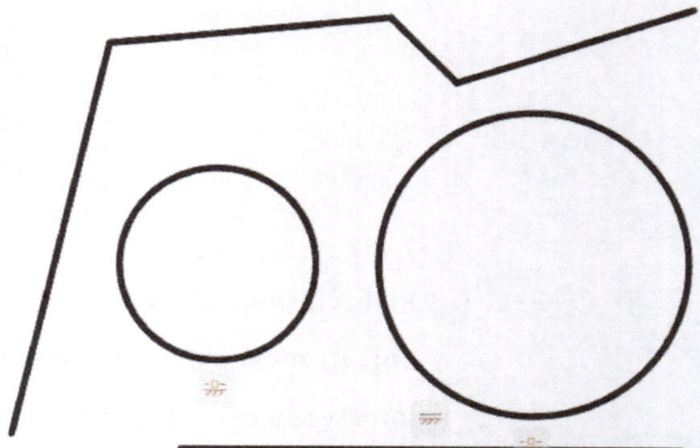

- Click the **Coincident** button on the **Geometric** panel and select the two endpoints of the lines as shown below; the endpoints will be made coincident.

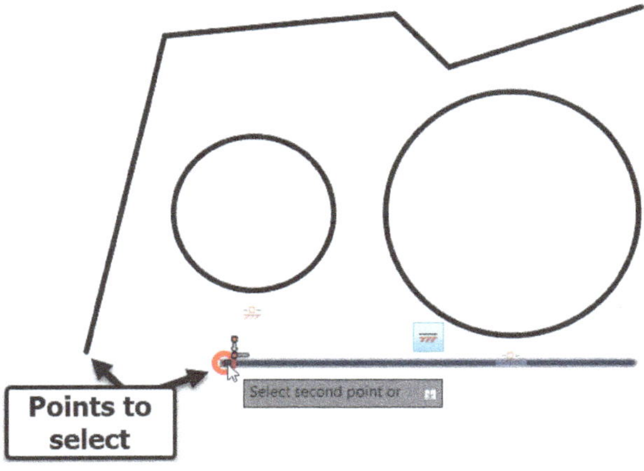

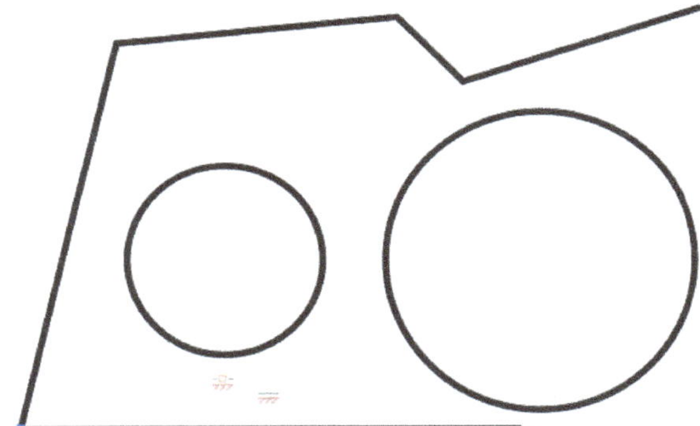

- Click the **Auto Constrain** button on the **Geometric** panel and select the four lines, as shown below.

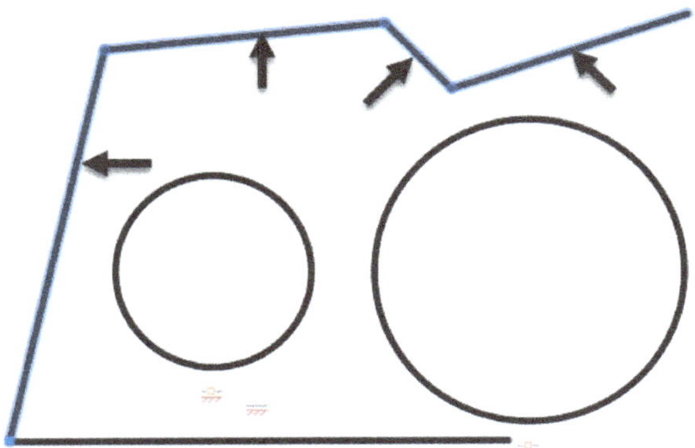

- Right-click and select **Enter**; constraints are applied to the selected objects, automatically.

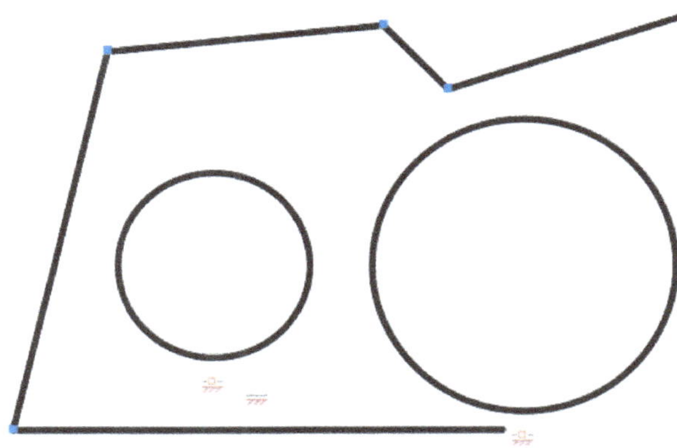

- Click the **Vertical** button on the **Geometric** panel and select the line as shown below; the line will become vertical.

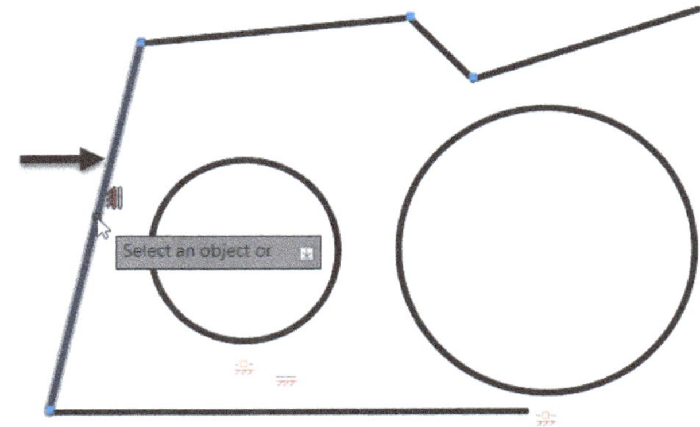

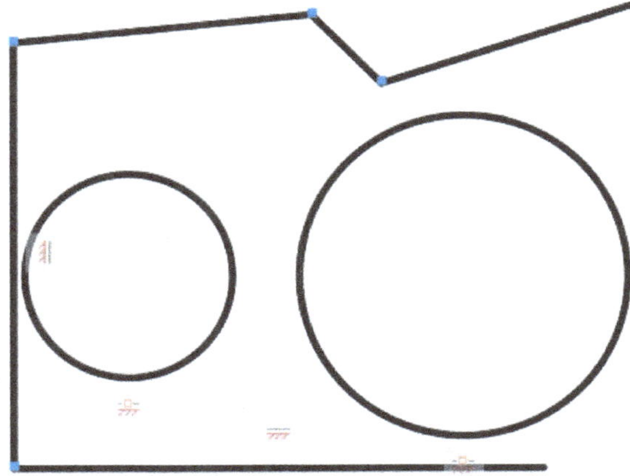

- Use the **Parallel** tool and make the two lines parallel to each other, as shown below.

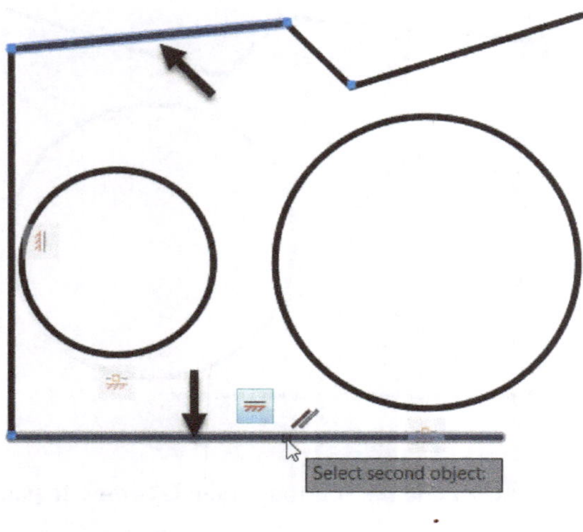

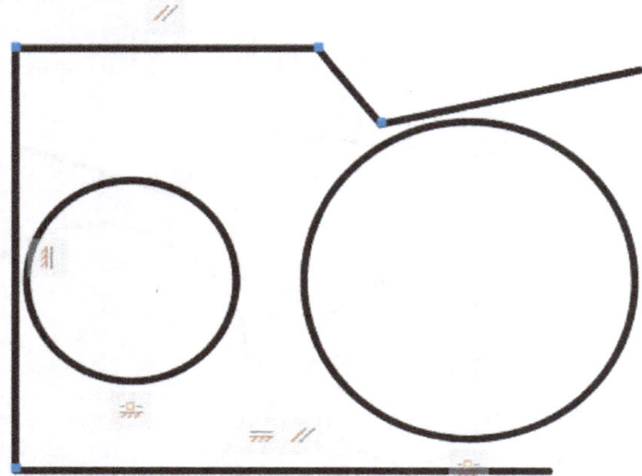

- Use the **Tangent** tool and make the two lines tangent to the large circle, as shown below.

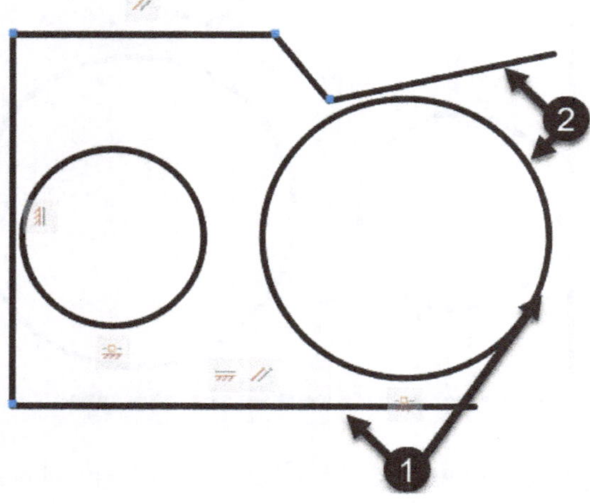

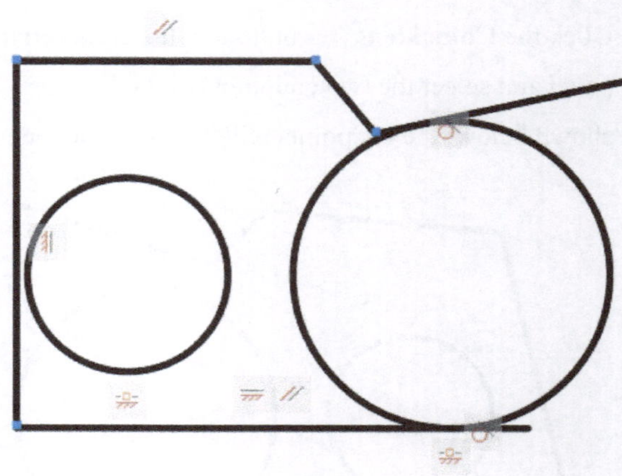

- Click the **Coincident** button on the **Geometric** panel.
- Select the **Object** option from the command line and select the large circle.
- Select the endpoint of the lower horizontal line to make it coincident with the circle.

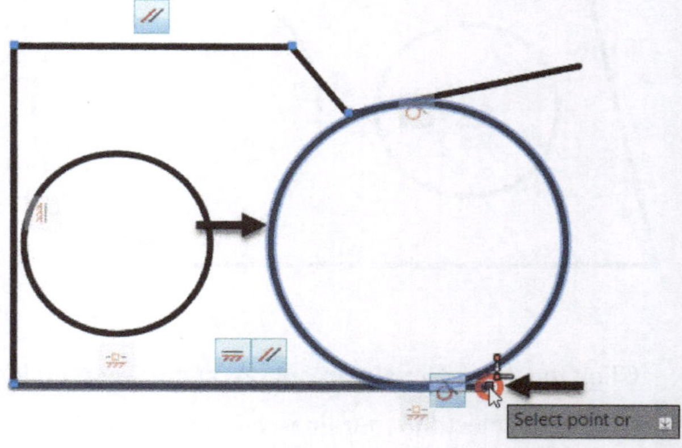

- Likewise, apply the **Coincident** constraint between the large circle and the upper horizontal line.

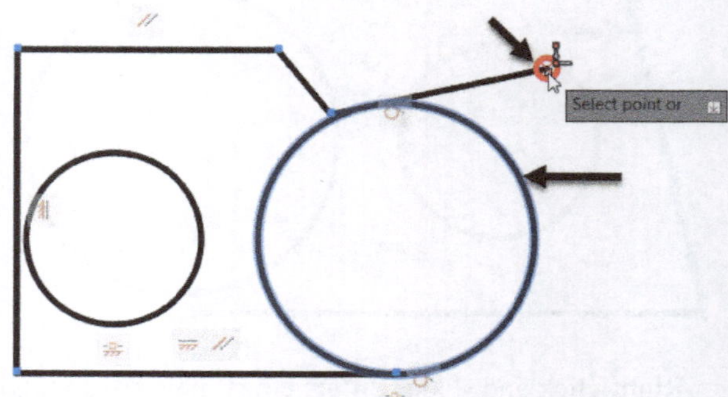

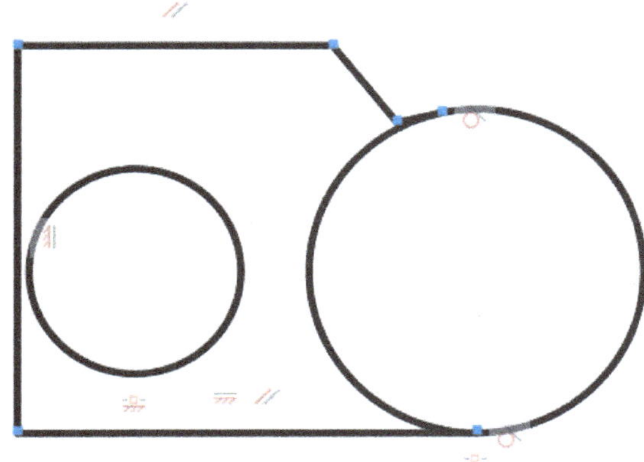

- Use the **Trim** tool and trim the unwanted portion of the circle.

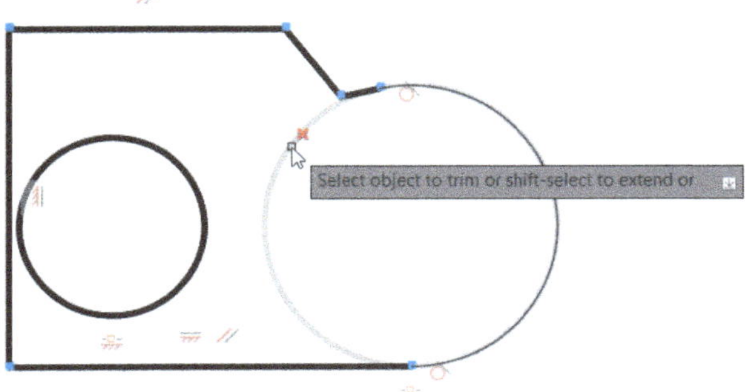

Follow the next three steps if the **Tangent** and **Coincident** constraints are deleted.

- Click the **Auto Constrain** button on the **Geometric** panel.
- Drag a window around the arc and lines, as shown.
- Right-click and select **Enter**; the **Tangent** and **Coincident** constraints are applied between the arc and the horizontal lines.

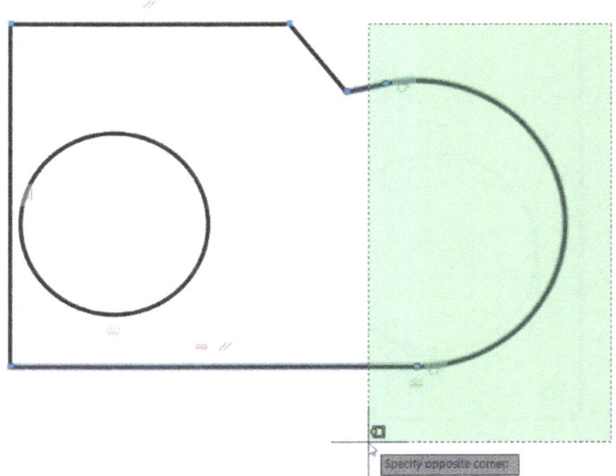

- Click the **Perpendicular** button on the **Geometric** panel.
- Select the two lines, as shown.

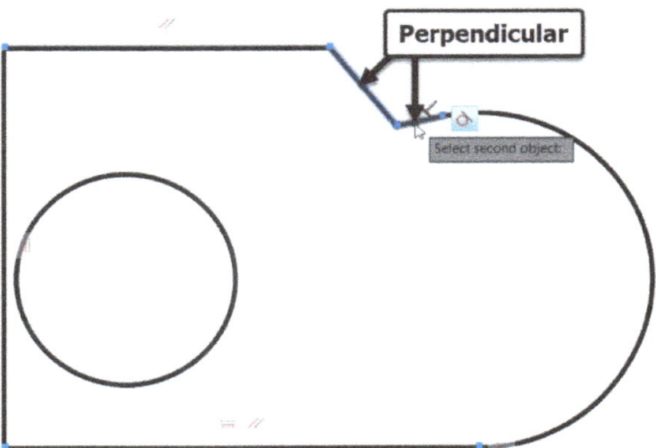

- Click the **Vertical** button on the **Geometric** panel.
- Select the line, as shown.

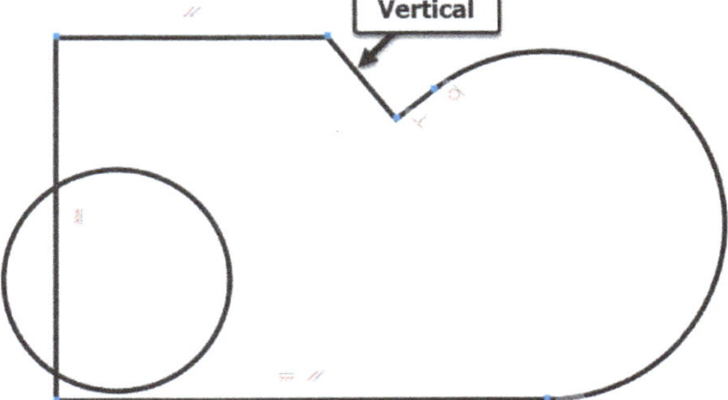

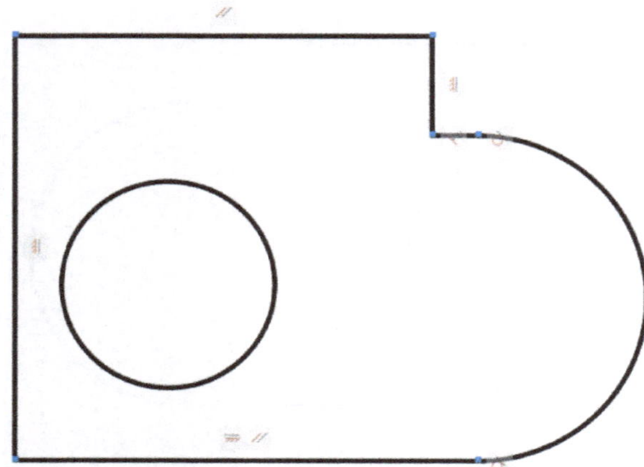

Dimensional Constraints

Dimensional constraints are applied to a drawing after applying the Geometric constraints. They are used to control the size and position of the objects in a drawing. You can apply the dimensional constraints using the tools available in the **Dimensional** panel of the **Parametric** ribbon.

- Click the inclined arrow on the **Dimensional** panel; the **Constraint Settings** dialog appears.

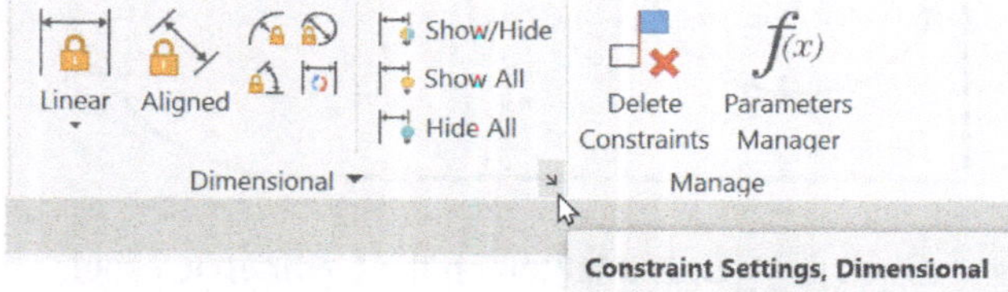

- On the **Constraint Settings** dialog, set **Dimension name format** to **Name**.

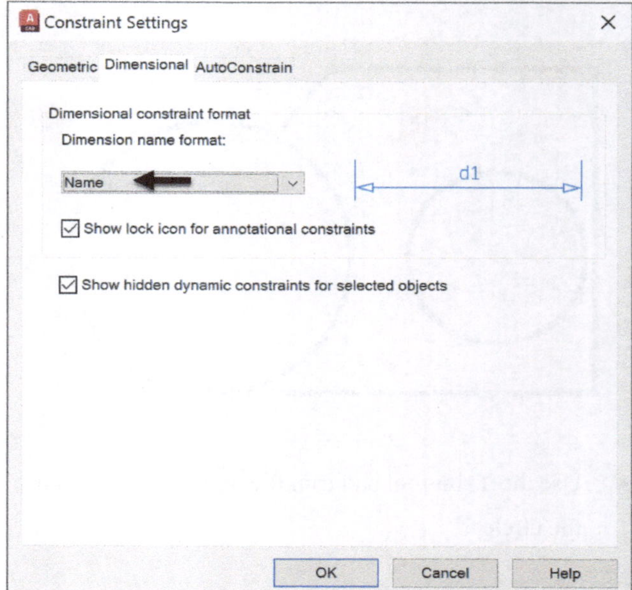

- Click the **OK** button.
- Click **Parametric > Dimensional > Linear** on the ribbon.

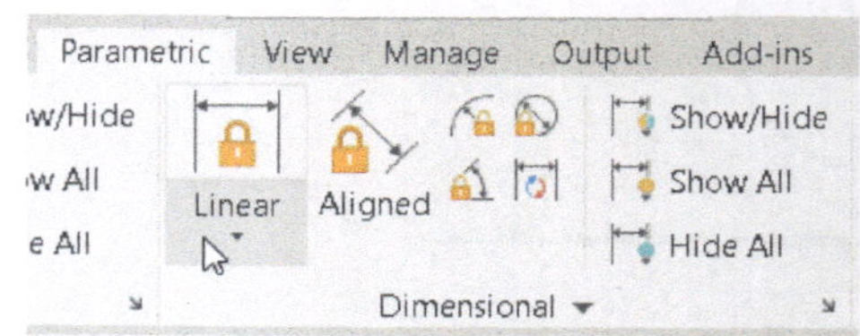

- Select the two endpoints of the lower horizontal line; the dimensional constraint is attached to the pointer.

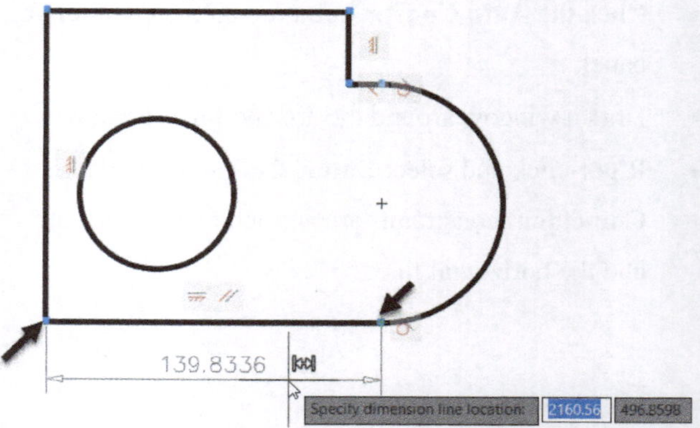

- Place the dimension constraint and left-click.

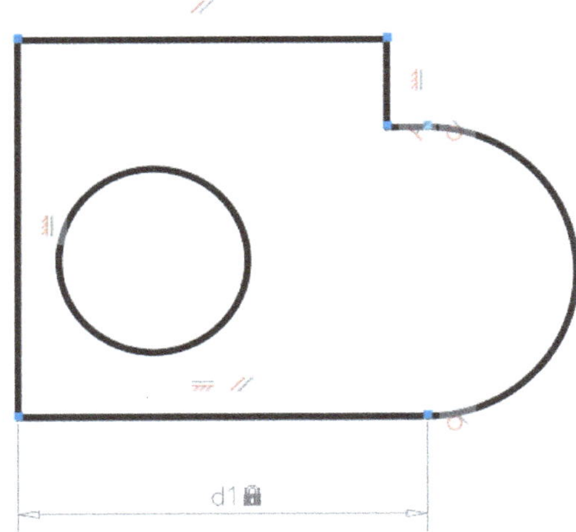

- Similarly, apply linear dimensions to other lines, as shown below.

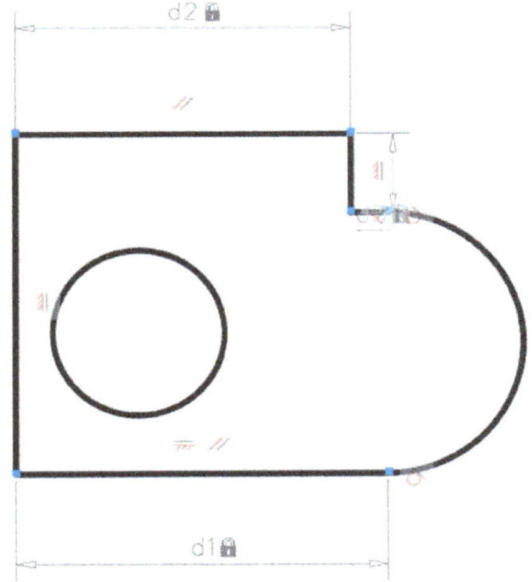

You will notice that when you try to apply the dimensional constraint to the horizontal line connected to the arc, the **Dimensional Constraints** message box appears. It shows that the dimension will over-constrain the geometry. In an over-constrained geometry, there are conflicting dimensions or relations or both. Click the **Cancel** button on the **Dimensional Constraints** message box.

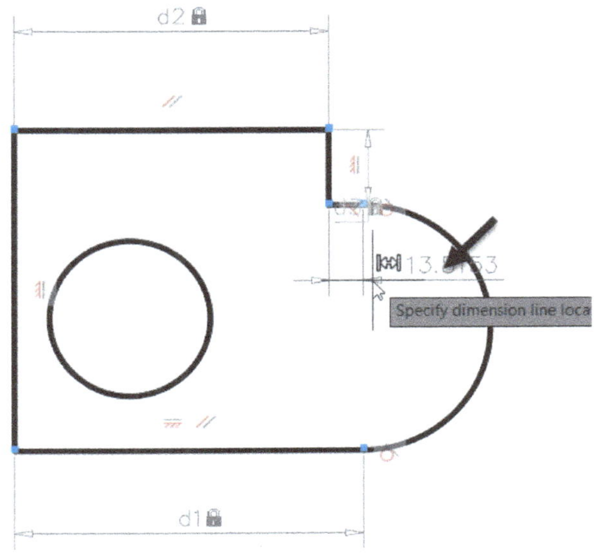

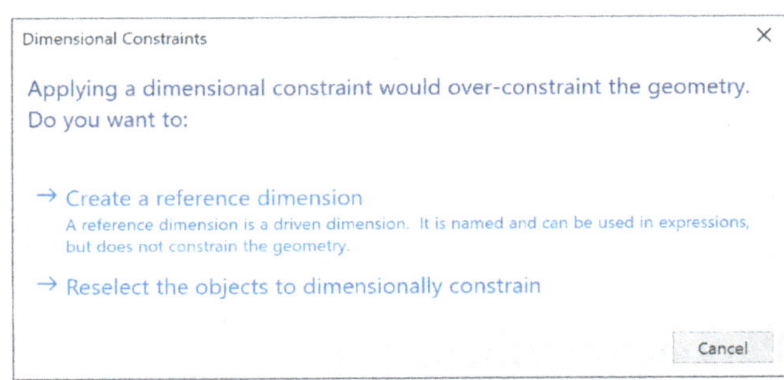

- Click the **Diameter** button on the **Dimensional** panel and apply the diameter dimension to the circle located on the left side.

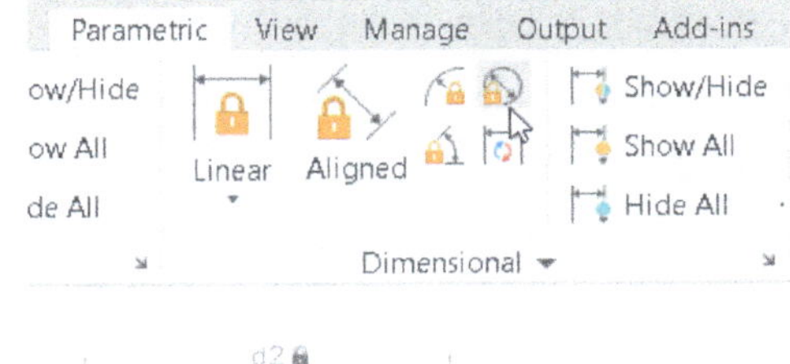

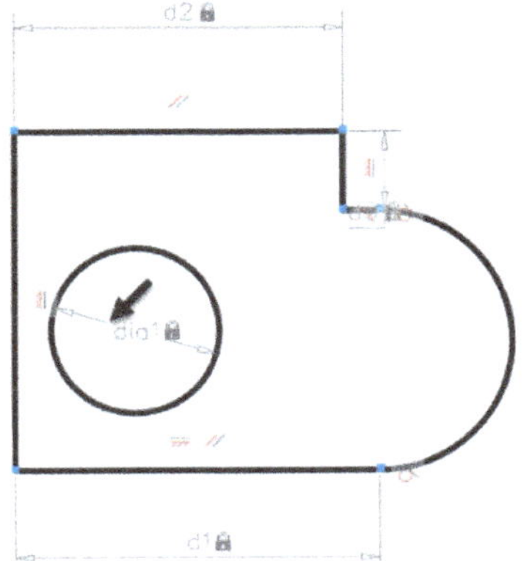

- Click the **Radius** button on the **Dimensional** panel and apply the radial dimension to the arc.

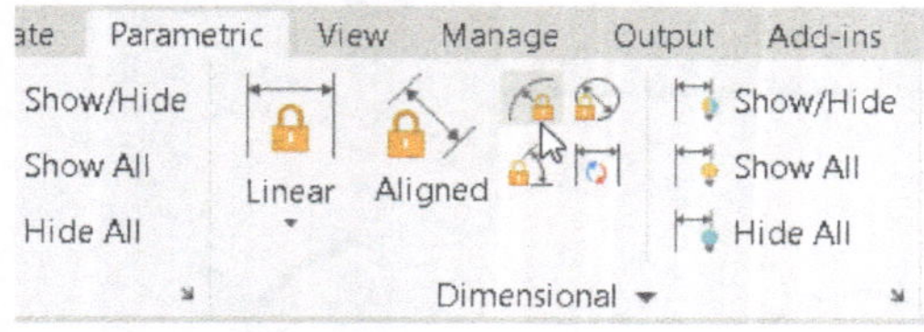

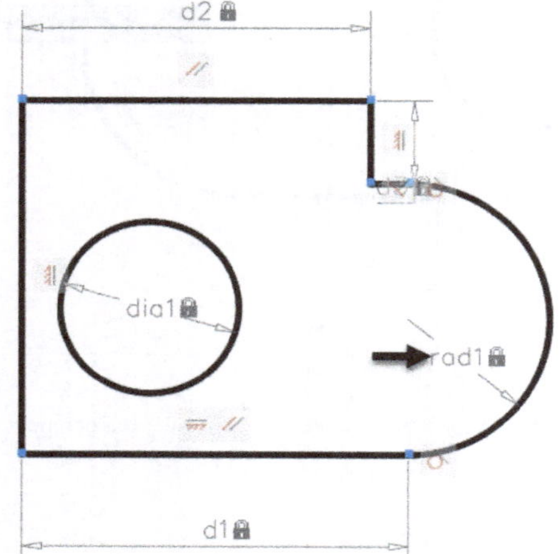

Creating equations using the Parameters Manager

Equations are relations between the dimensional constraints. Look at the drawing given below. In this drawing, all the dimensions are controlled by the diameter of the hole. In AutoCAD, you can create this type of relations between dimensions very easily using the **Parameters Manager** palette.

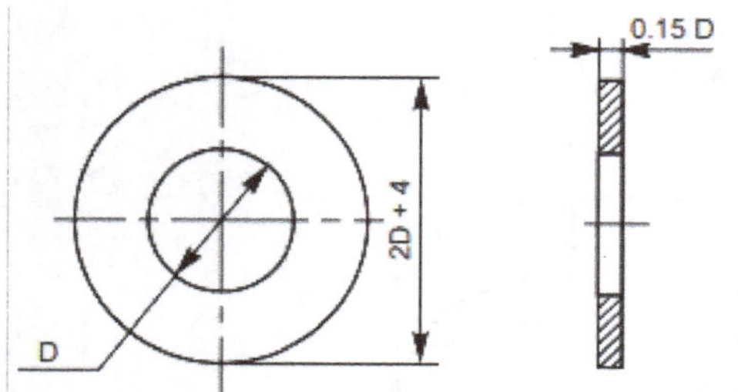

- Click the **Parameters Manager** button on the **Manage** panel of the **Parameters** tab; the **Parameters Manager** palette appears.

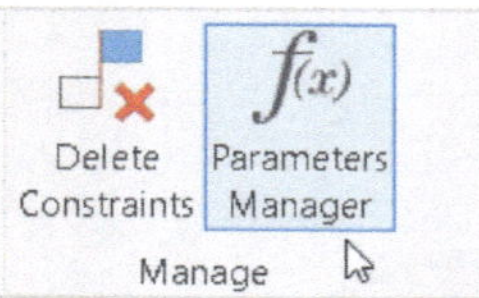

- Double-click in the box next to the **dia1** and enter **50**.

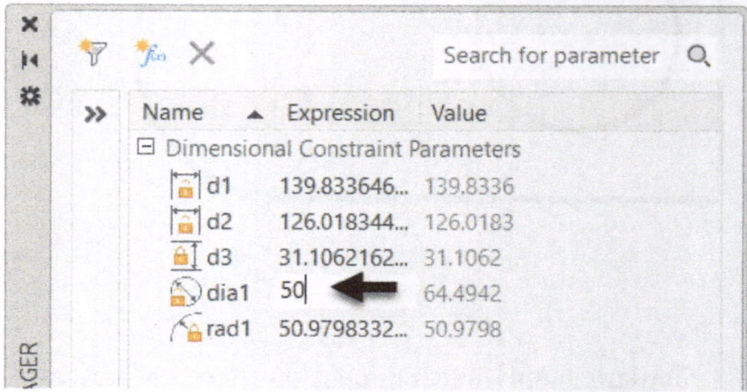

Name	Expression	Value
Dimensional Constraint Parameters		
d1	139.833646...	139.8336
d2	126.018344...	126.0183
d3	31.1062162...	31.1062
dia1	50	64.4942
rad1	50.9798332...	50.9798

- Likewise, change the values of the other dimensions, as shown below.

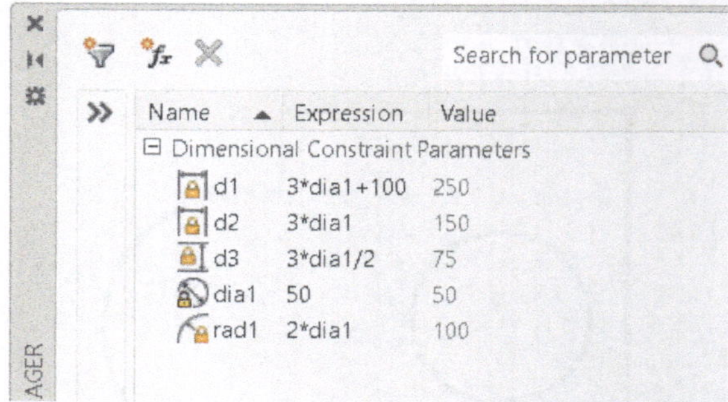

Name	Expression	Value
Dimensional Constraint Parameters		
d1	3*dia1+100	250
d2	3*dia1	150
d3	3*dia1/2	75
dia1	50	50
rad1	2*dia1	100

You will notice that the circle is placed outside the loop.

- Click **Zoom All** on the **Navigation Bar** to view the circle.

Creating Inferred Constraints

The **Infer Constraints** button helps you to create constraints automatically. With this button active on the status bar, you can automatically create constraints while drawing a sketch.

- On the status bar, click the **Customization** button and select **Infer Constraints** from the flyout. It adds the **Infer Constraints** button to the status bar.
- Activate the **Infer Constraints** button on the status bar.

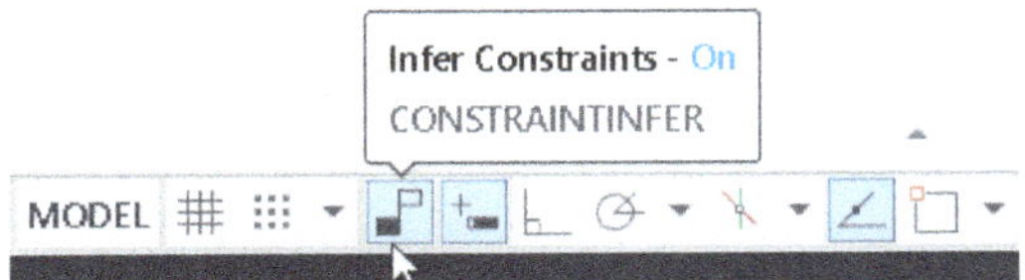

- Click the **Fillet** button on the **Modify** panel of the **Home** ribbon.
- Select the **Radius** option from the command line and enter **50** as the radius.
- Create a fillet at the lower-left corner of the sketch.

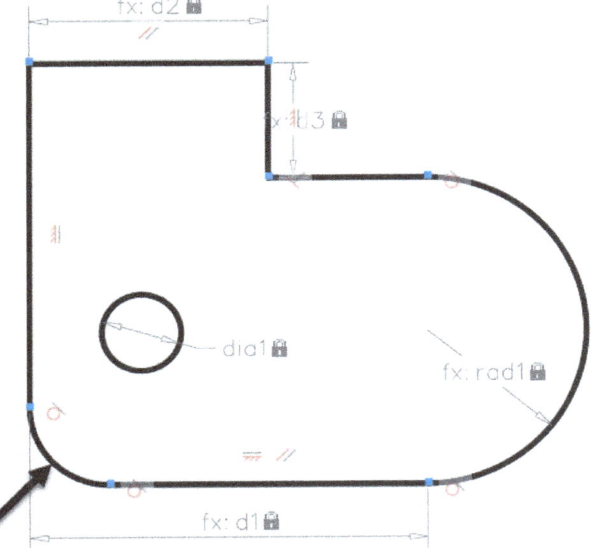

You will notice that the **Tangent** and **Coincident** constraints are applied automatically.

- Click the **Concentric** button on the **Geometric** panel.
- Select the circle located outside the loop and the fillet; they both will be concentric.

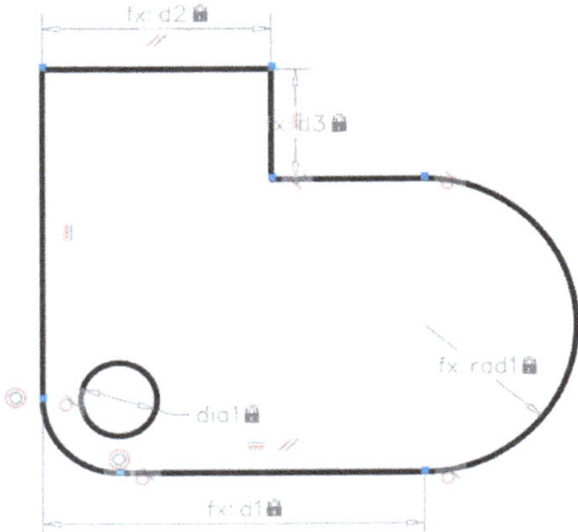

- Use the **Radius** tool from the **Dimensional** panel and apply the dimensional radius constraint to the fillet.
- Open the **Parameters Manager** palette and modify the **rad2** value to **3/2*dia1**.

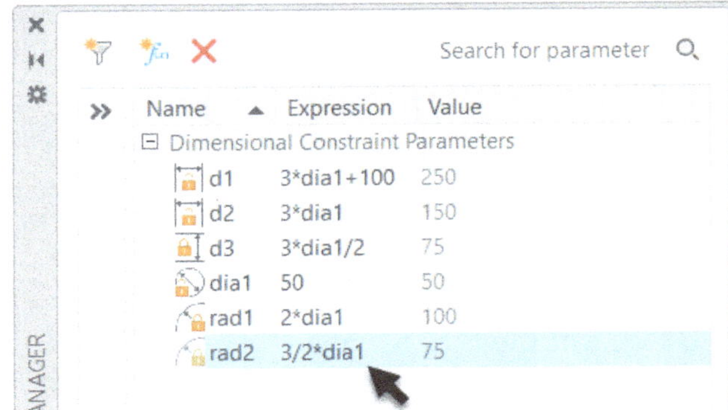

- To hide all the **Geometric Constraints**, click the **Hide All** button on the **Geometric** panel.
- Similarly, click **Hide All** on the **Dimensional** panel to hide all the dimensional constraints.

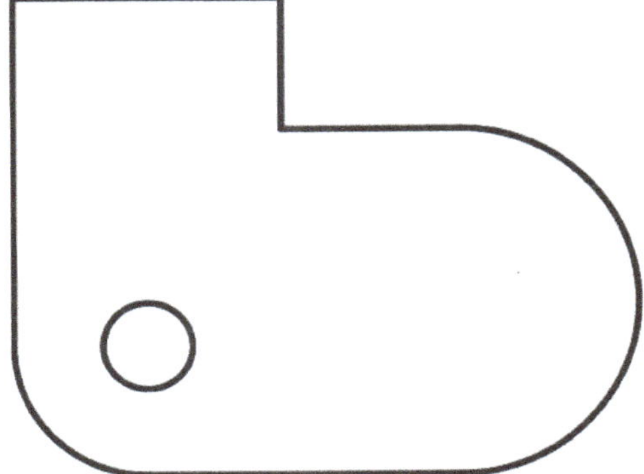

- To modify the size of the drawing, change the value of **dia1** in the **Parameters Manager** window; you will notice that all the values will be changed automatically.
- Save and close the file.

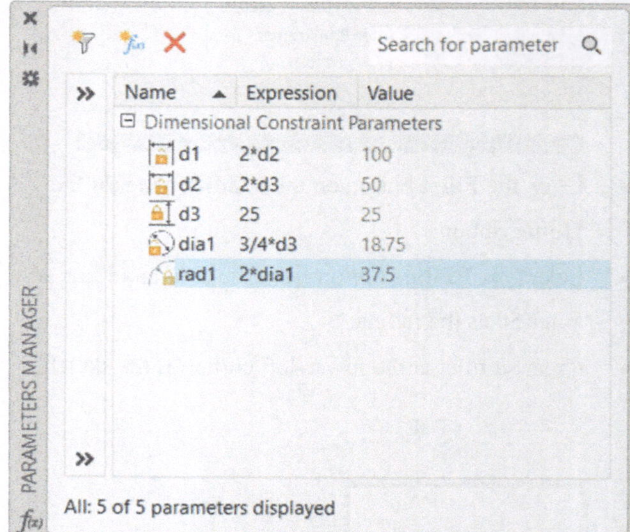

Exercises

Exercise 1

In this exercise, you need to create the drawing shown in the figure and apply geometric and dimensional constraints to it.

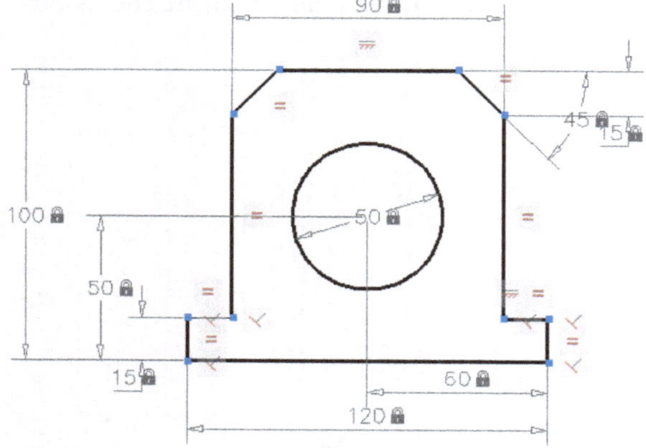

Exercise 2

In this exercise, you need to create the drawing as shown below and apply geometric and dimensional constraints to it. Also, create relations between dimensions in the **Parameters Manager**.

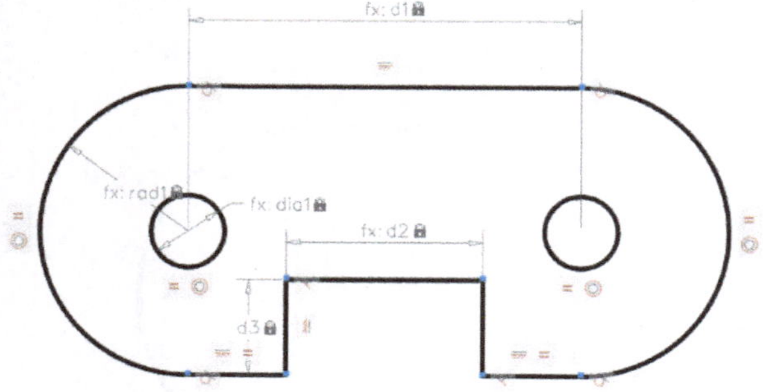

Chapter 8: Section Views

In this chapter, you will learn to:

- **Create Section Views**
- **Set Hatch Properties**
- **Use Island Detection tools**
- **Create text in Hatching**
- **Edit Hatching**

Section Views

In this chapter, you will learn to create section views. You can create section views to display the interior portion of a component that cannot be shown clearly using hidden lines. It can be done by cutting the component using an imaginary plane. In a section view, section lines, or cross-hatch lines are added to indicate the surfaces that are cut by the imaginary cutting plane. In AutoCAD, you can add these section lines or cross-hatch lines using the **Hatch** tool.

The Hatch tool

The **Hatch** tool is used to generate hatch lines by clicking inside a closed area. When you click inside a closed area, a temporarily closed boundary will be created using the PLINE command. The closed boundary will be filled with hatch lines, and then it will be deleted.

Example 1:

In this example, you will apply hatch lines to the drawing, as shown in the figure below.

- Open a new AutoCAD file.
- Create four layers with the following properties.

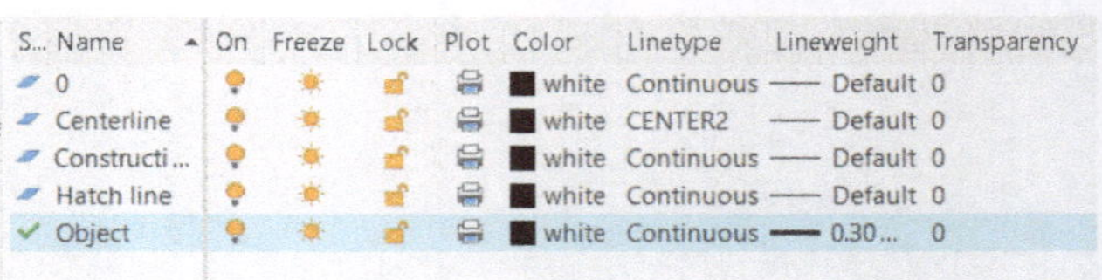

S...	Name	On	Freeze	Lock	Plot	Color	Linetype	Lineweight	Transparency
	0					white	Continuous	Default	0
	Centerline					white	CENTER2	Default	0
	Constructi...					white	Continuous	Default	0
	Hatch line					white	Continuous	Default	0
✓	Object					white	Continuous	0.30...	0

- Create the drawing, as shown below. Do not apply dimensions.

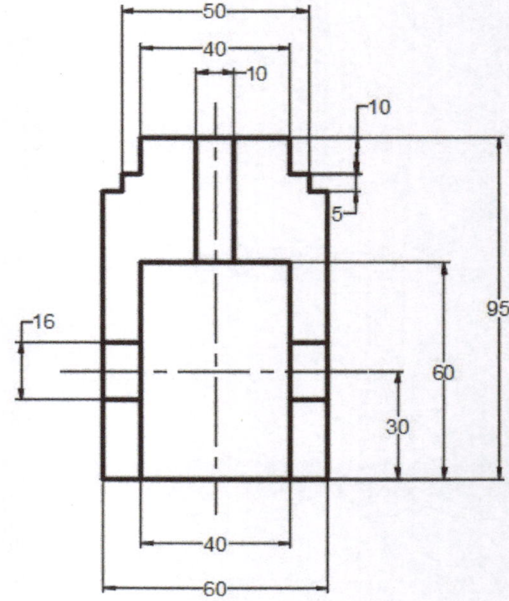

- Select the **Hatch lines** layer from the **Layer** drop-down of the **Layers** panel.
- Click **Home > Draw > Hatch** on the ribbon or enter **H** in the command line; the **Hatch Creation** tab appears in the ribbon.

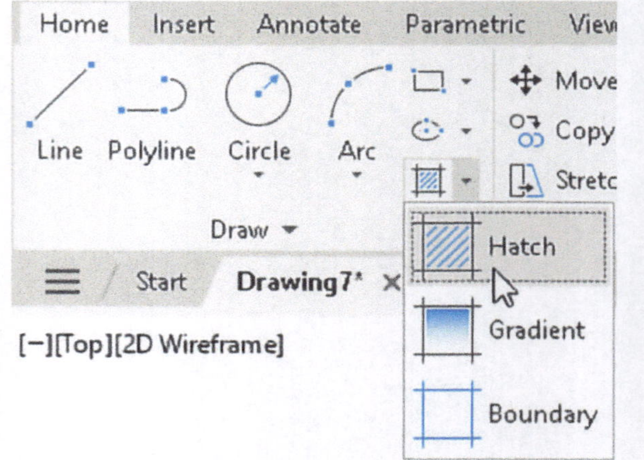

- Select **ANSI31** from the **Pattern** panel of the **Hatch Creation** ribbon tab.

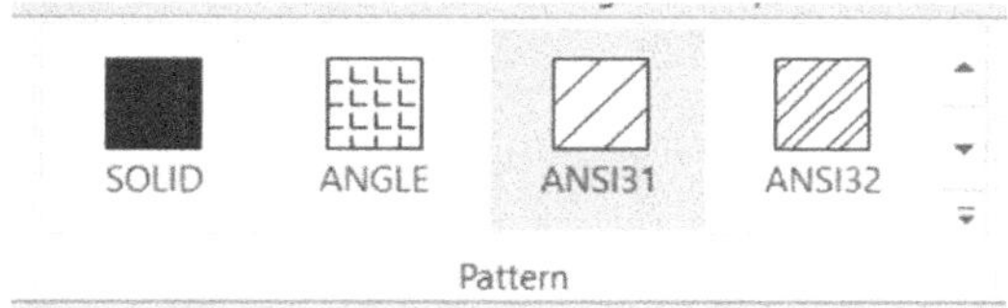

- Click on the four regions of the drawing, as shown below.

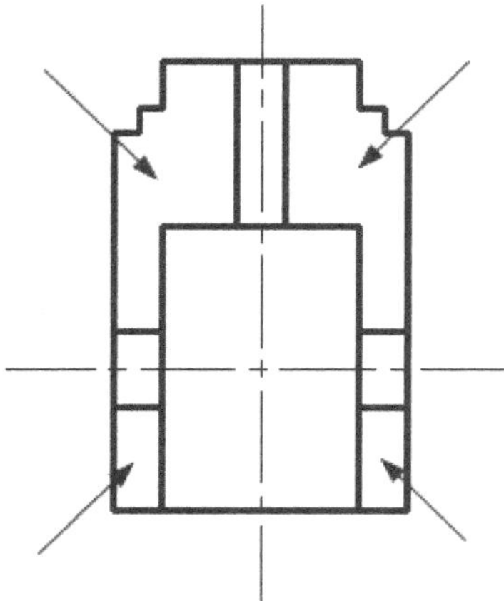

- Click the **Close Hatch Creation** button on the ribbon.

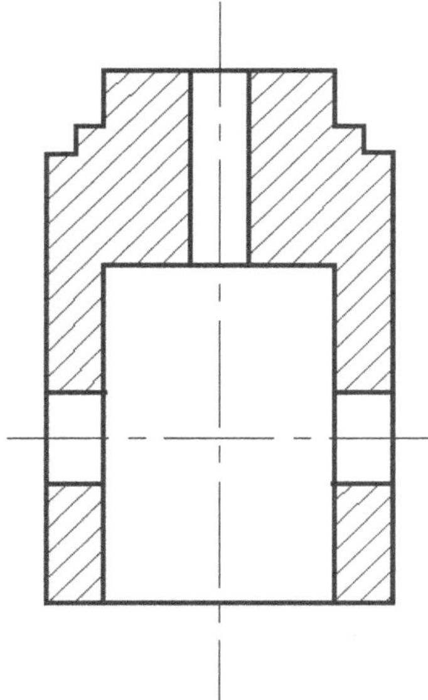

Example 2:

In this example, you will create the front and section views of a crank.

- Create five layers with the following settings:

Layer	Lineweight	Linetype
Construction	0.00 mm	Continuous
Object	0.30 mm	Continuous
Centerline	0.00 mm	CENTER
Hatch lines	0.00 mm	Continuous
Cutting Plane	0.30 mm	PHANTOM

- Activate the **Construction** layer and create construction lines, as shown.

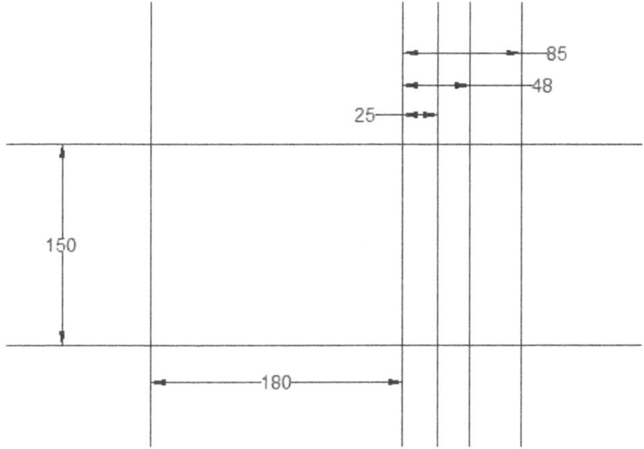

- Set the **Object** layer as current and create circles, as shown below.

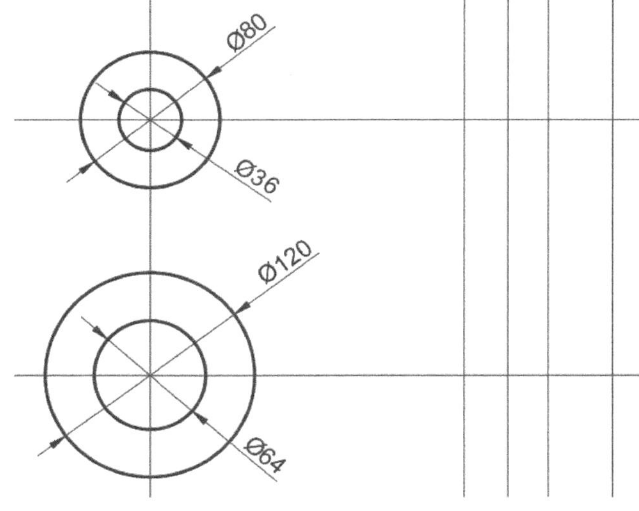

- Switch to the **Construction** layer and create construction lines, as shown.

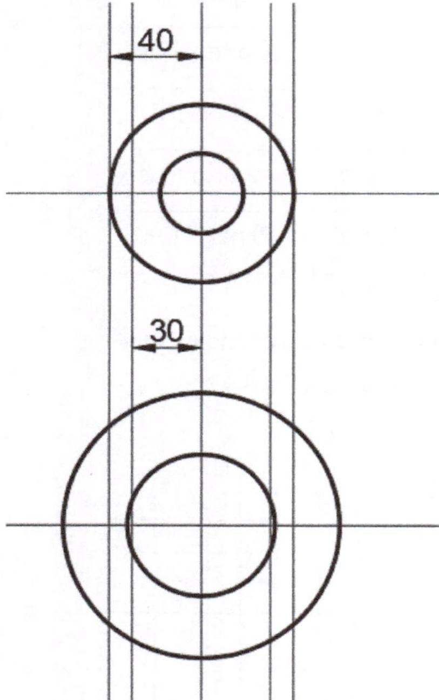

- Switch to the **Object** layer and create two lines, as shown.

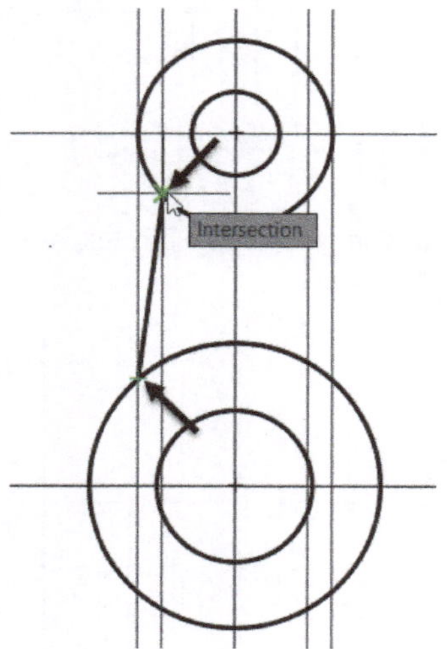

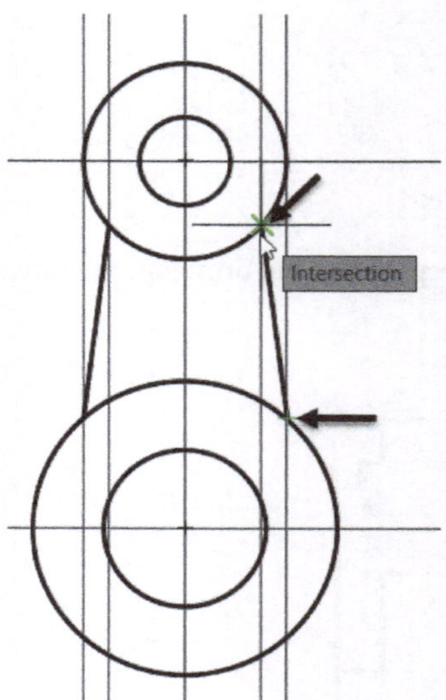

- Create the fillets at the corners, as shown.

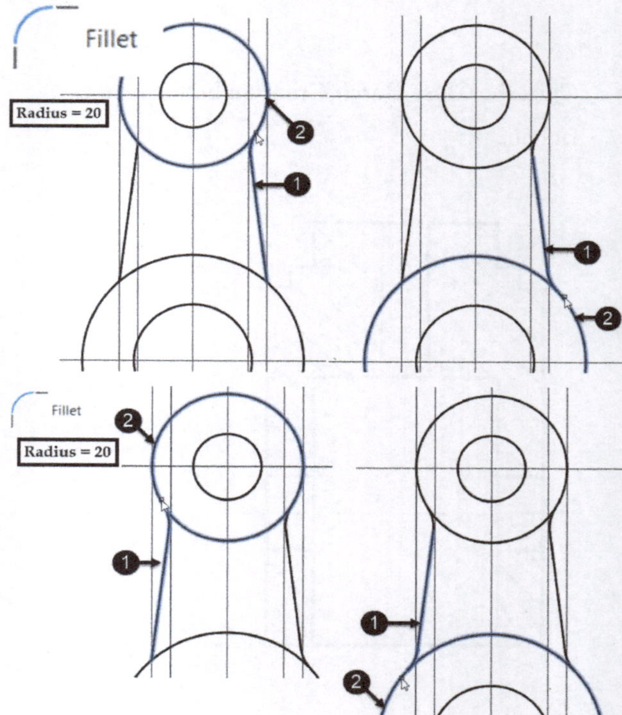

- Create a 16X38 rectangle, as shown.

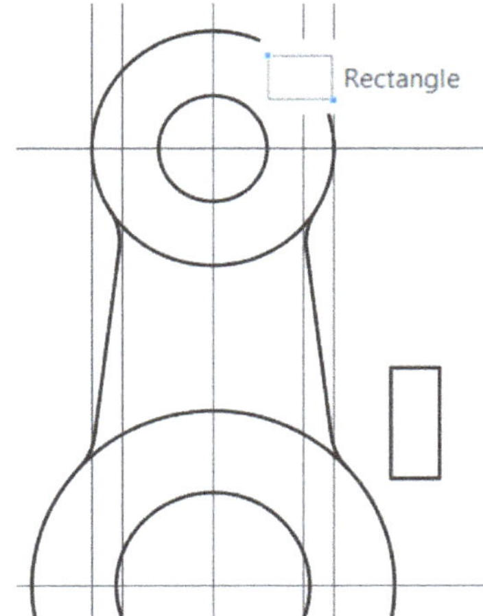

- Move the rectangle and place it at the centerpoint of the bottom circle, as shown.

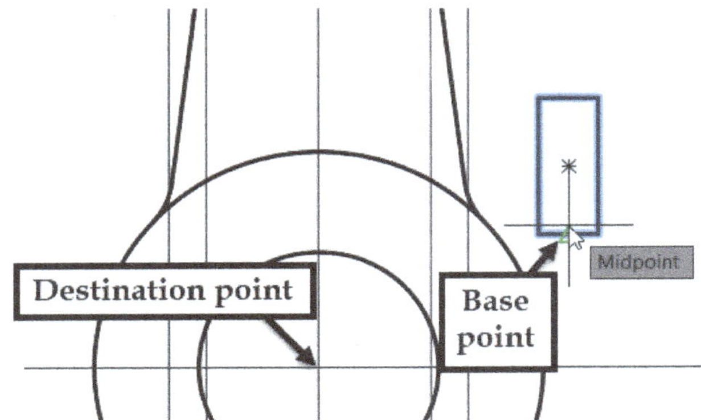

- Trim the unwanted entities of the rectangle, as shown.

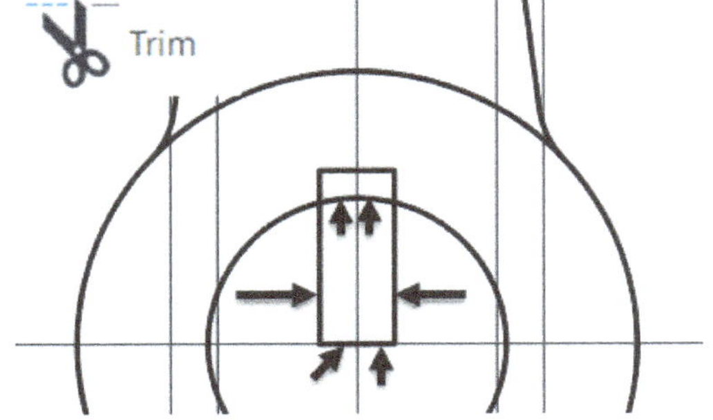

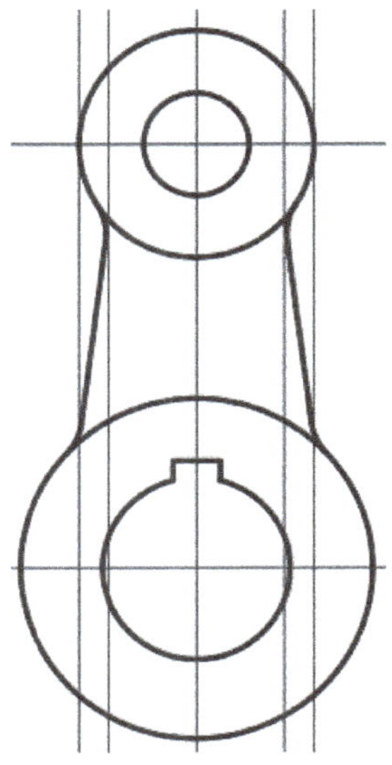

- On your own, create the objects of the section view, as shown below. (For any help, refer to the **Multi view Drawings** section of Chapter 5)

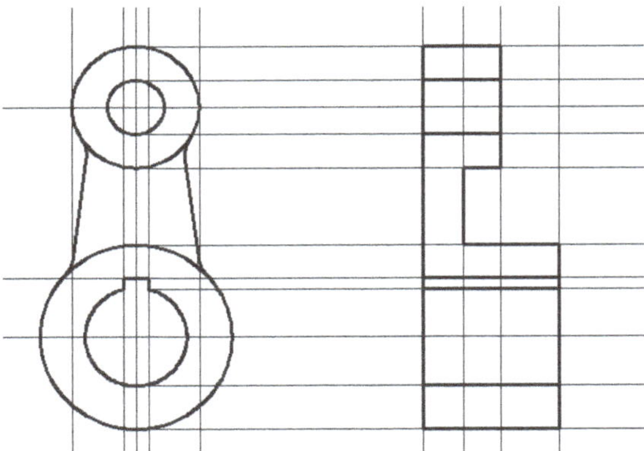

- Hide the **Construction** layer.
- Set the **Centerlines** layer as current and create center marks and centrelines.

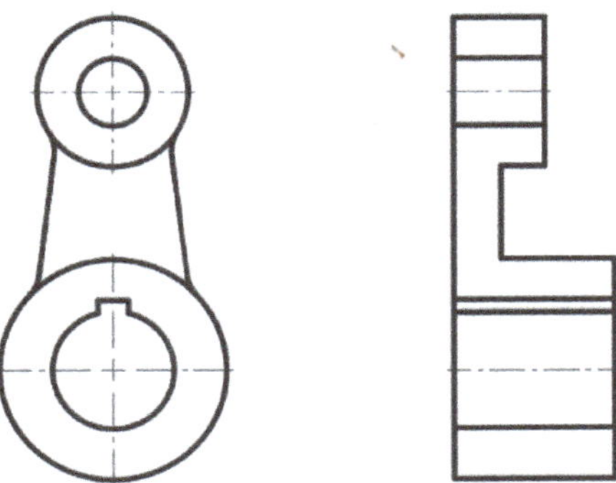

- Set the **Cutting Plane** layer as current.

- Activate the **Ortho Mode** icon on the Status bar, if not already active.
- Click the **Polyline** button on the **Draw** panel and pick a point below the front view, as shown.

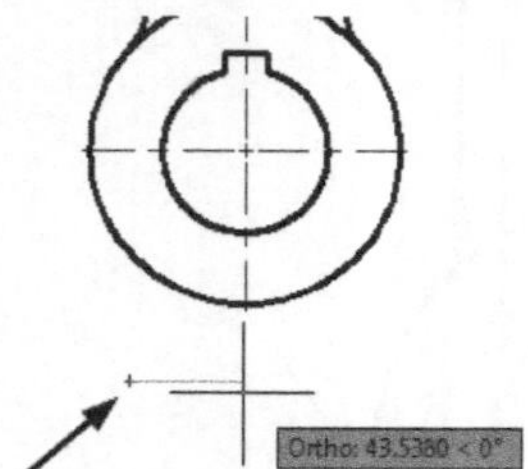

- Select the **Width** option from the command line.
- Type 0 as the starting width and press ENTER.
- Type 10 as the ending width and press ENTER.
- Move the pointer horizontally toward the right and enter 20.
- Again, select the **Width** option from the command line.
- Set the starting and ending width to 0.
- Move the pointer horizontally and click when trace lines are displayed, as shown below.

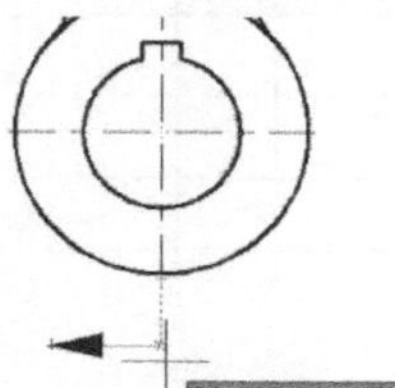

- Move the pointer vertically up and click.
- Move the pointer to the endpoint of the lower horizontal line.

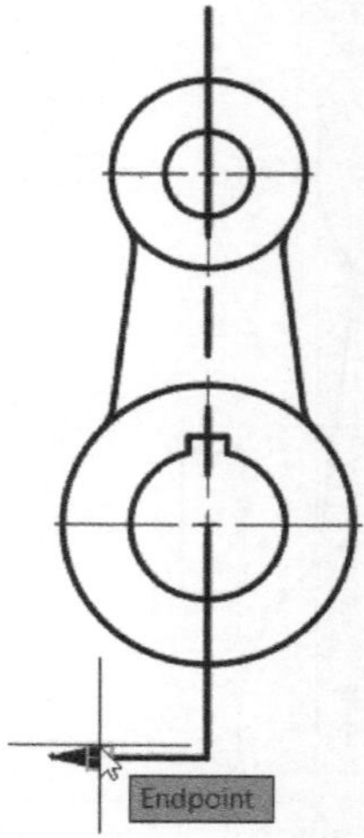

- Move the pointer upward.
- Move the pointer toward the left and click when trace lines are displayed from the endpoint of the lower horizontal line.

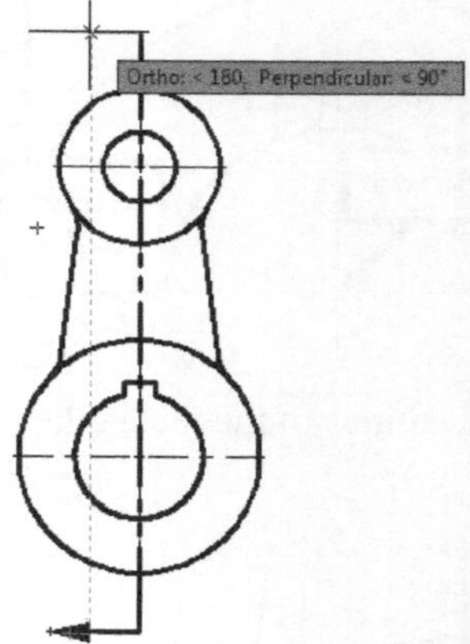

- Click the **Width** option in the command line.
- Type 10 and press Enter.
- Type 0 and press Enter.
- Move the pointer toward the left, type 20, and then press Enter.

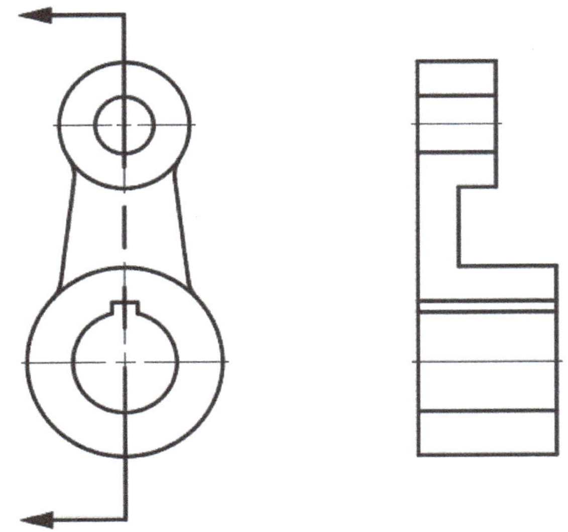

- Press Esc to deactivate the **Polyline** tool.
- Activate the **Hatch lines** layer.
- Type **H** in the command line and press ENTER.
- Select the **seTtings** option from the command line; the **Hatch and Gradient** dialog appears.
- Click in the **Swatch** box under the **Type and pattern** group; the **Hatch Pattern Palette** dialog appears.

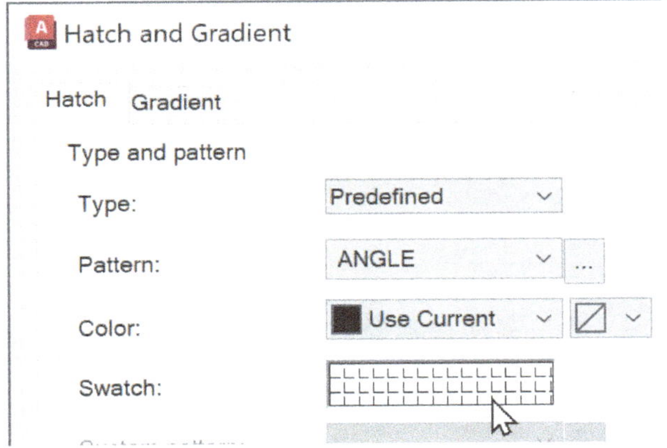

- Click the **ANSI** tab, select **ANSI31** from the dialog, and then click **OK**.
- Set the **Scale** value to **2**.

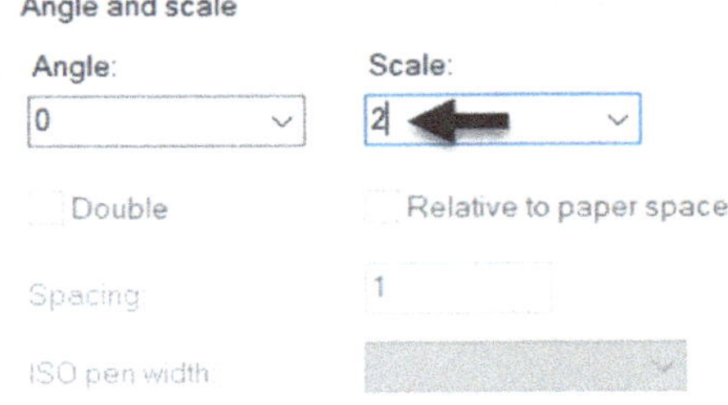

- Click the **Add Pick Points** button from the **Boundaries** group and click in Region 1, Region 2, and Region 3.

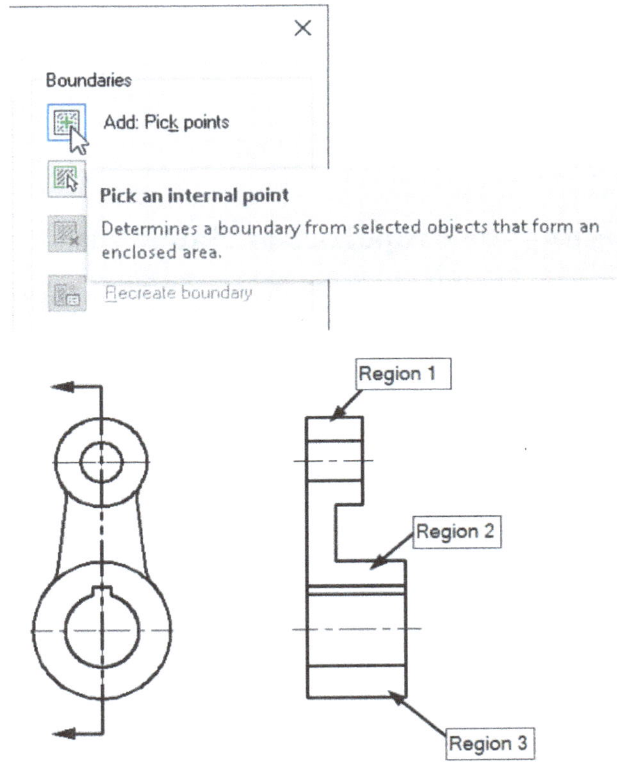

- Press ENTER to create hatch lines.

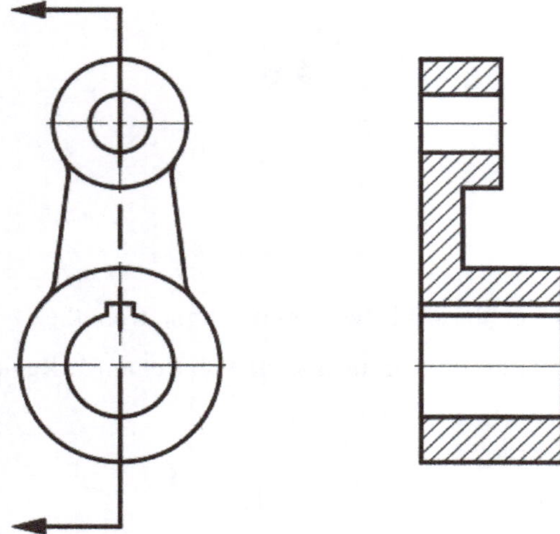

- Save the drawing as **Crank.dwg** and close.

Setting the Properties of Hatch lines

You can set the properties of the hatch lines such as angle, scale, and transparency in the **Properties** panel of the **Hatch Creation** ribbon.

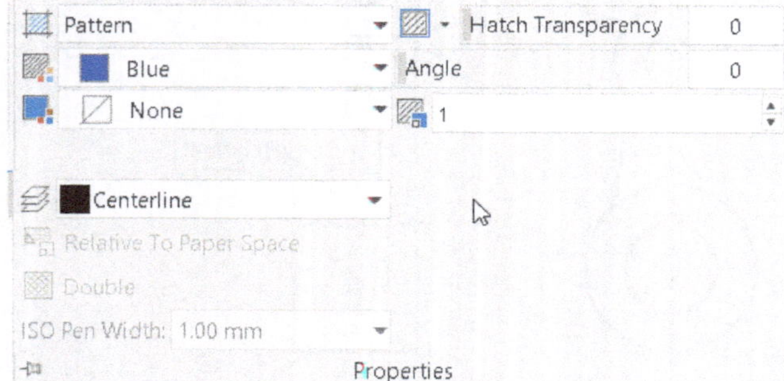

Example:

- Create four layers with the following settings.

Layer	Lineweight	Linetype
Construction	0.00 mm	Continuous
Object	0.30 mm	Continuous
Centerline	0.00 mm	CENTER2
Hatch lines	0.00 mm	Continuous

- Create the following drawing in different layers. Do not apply dimensions.

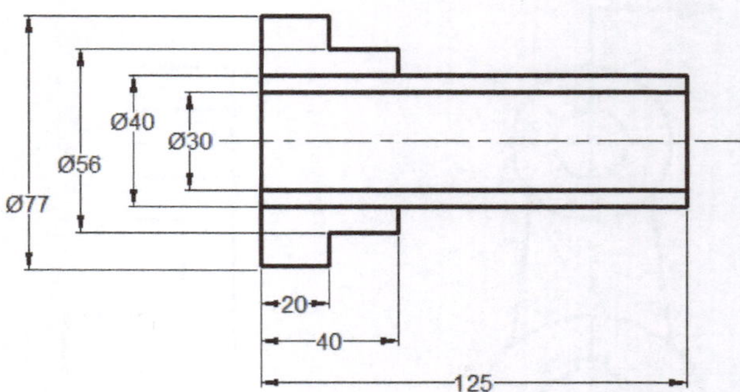

- Type **H** and press ENTER; the **Hatch Creation** tab appears in the ribbon.
- Select the **Pattern** option from the **Hatch Type** drop-down in the **Properties** panel.

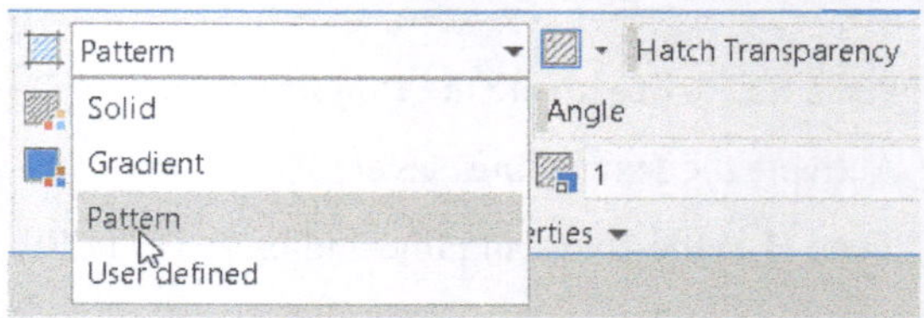

You can also select a different hatch type, such as Solid, Gradient, and User-defined.

- Select **ANSI31** from the **Pattern** panel.
- Select **Blue** from the **Hatch Color** drop-down.

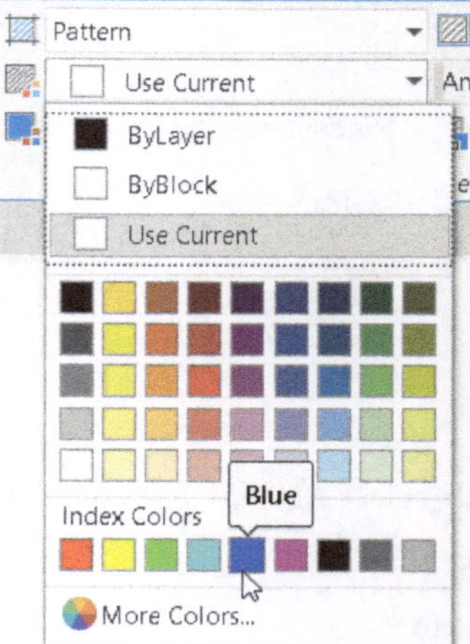

- Expand the **Properties** panel and set the **Hatch Layer Override** to **Hatch lines**.

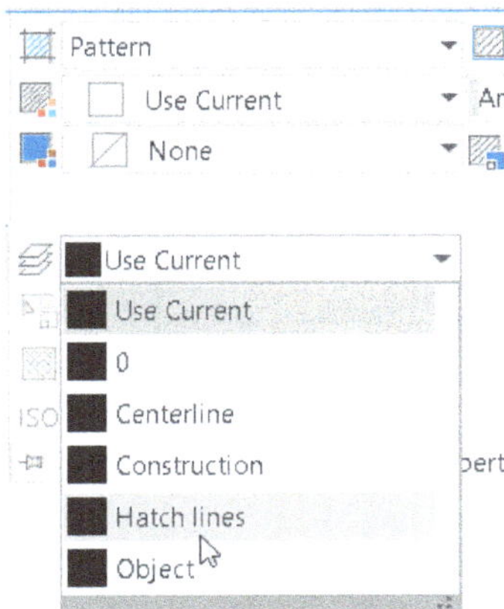

- Click the **Pick Points** button from the **Boundaries** panel.

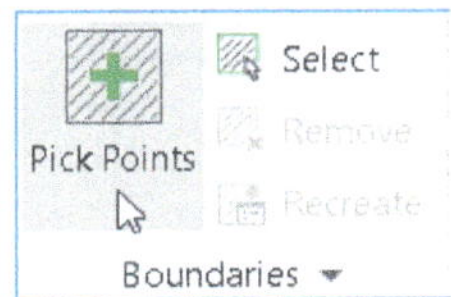

- Pick points in the outer areas of the drawing, as shown below.

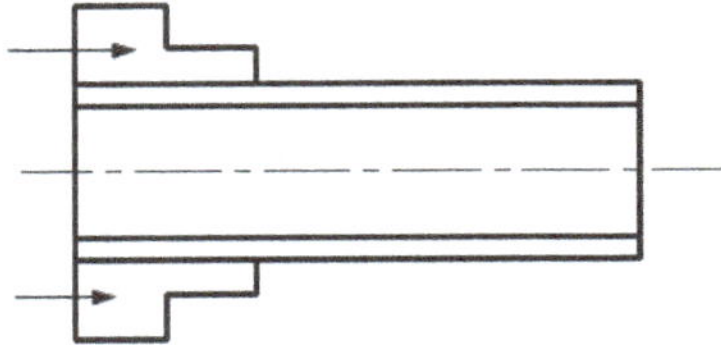

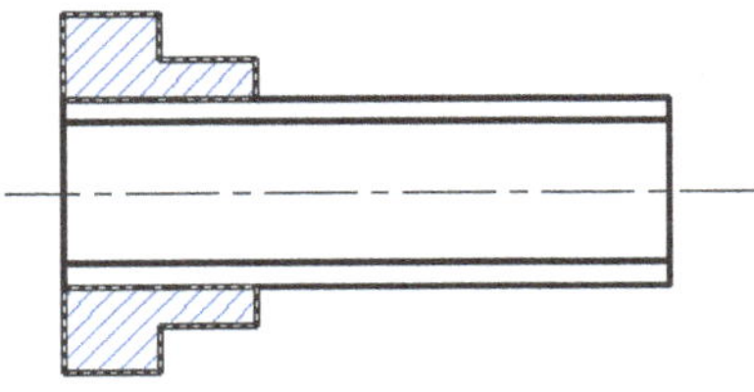

- Adjust the **Hatch Pattern Scale** to **1.5**; you will notice that the distance between the hatch lines changes.

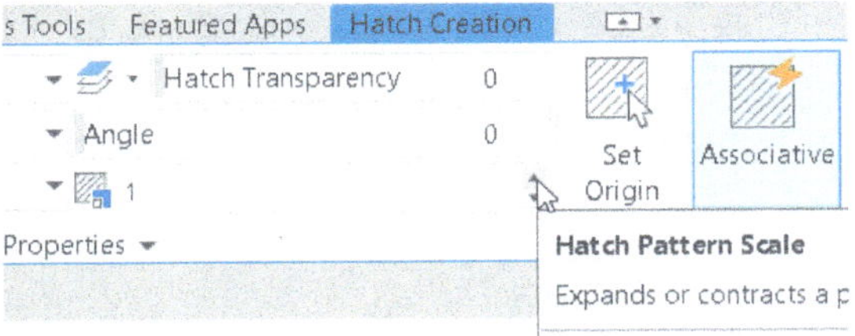

- Click the **Close Hatch Creation** button on the **Close** panel.

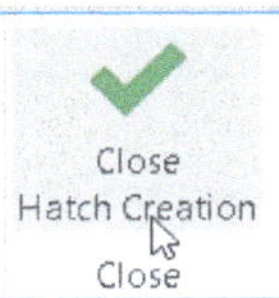

- Press the SPACEBAR to activate the **HATCH** command again.
- Change the **Hatch Angle** value to **90** in the **Properties** panel.

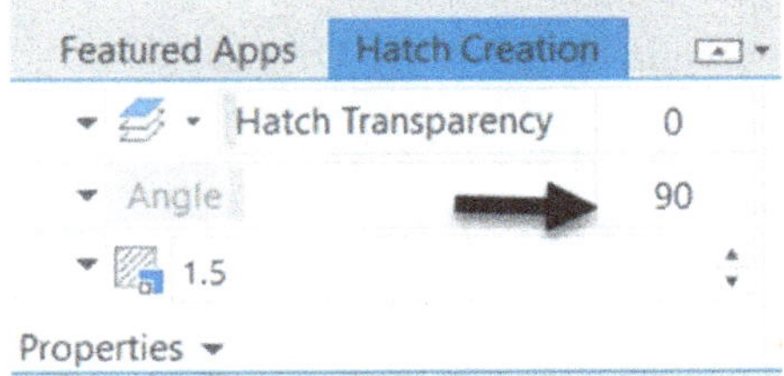

- Pick points in the area, as shown below.

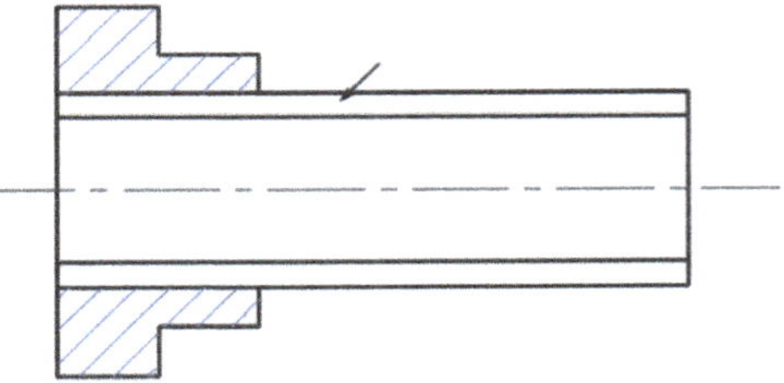

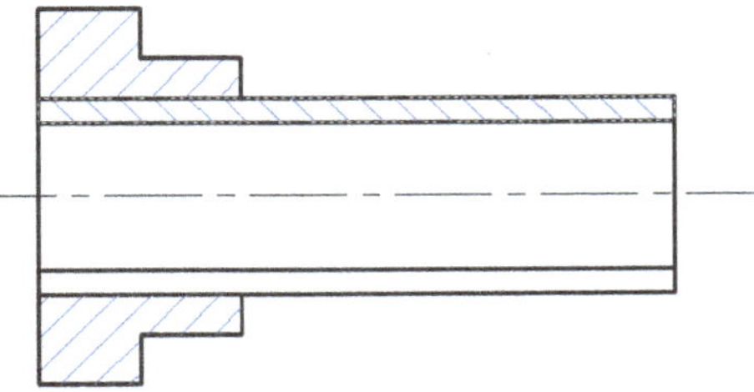

On zooming into the hatch lines, you may notice that they are not aligned properly. It is because the **Use Current Origin** button activated in the **Origin** panel. As a result, the origin of the drawing will act as the origin of the hatch pattern. However, you can change the origin of the hatch pattern.

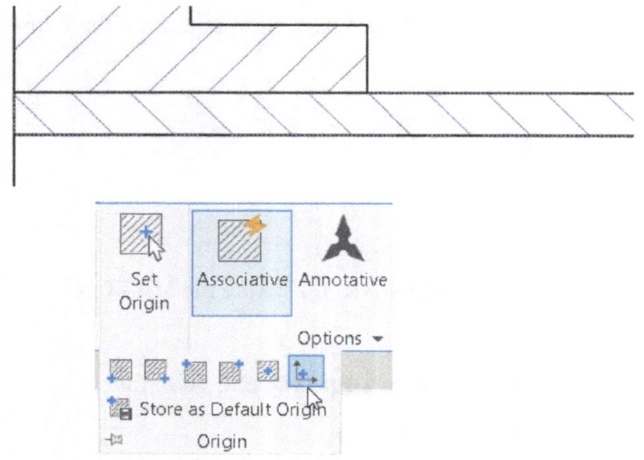

- Click **Set Origin** button on the **Origin** panel.
- Set the origin point, as shown below.

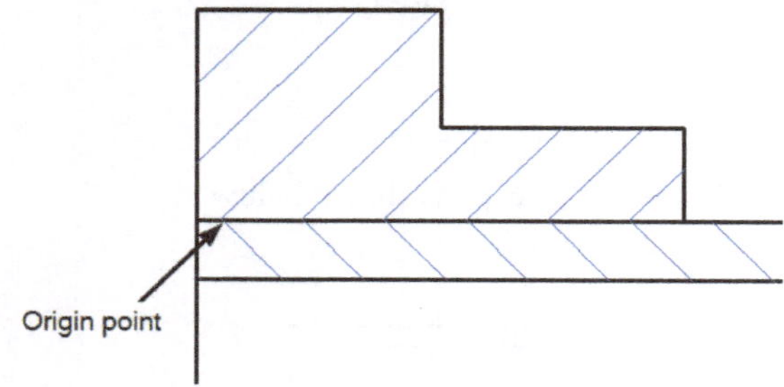

- Click **Close Hatch Creation**.
- Activate the **Hatch** tool and click **Match Properties > Use source hatch origin** on the **Options** panel.

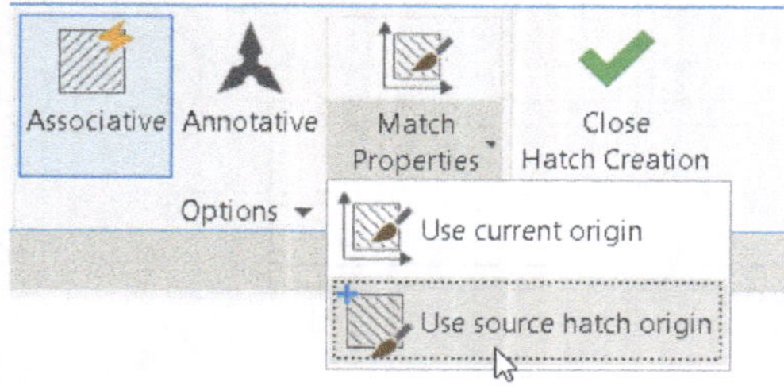

The **Match Properties** tools are used to create new hatch lines by using the properties of an existing one. The **Use source hatch origin** tool will create new hatching using the origin of the source.

- Select the source hatching, as shown in the figure.
- Pick a point in the empty area, as shown below.

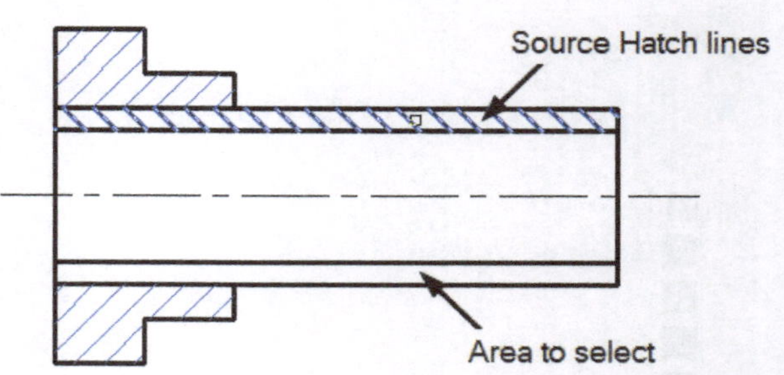

New hatch lines are created using the properties and origin of the source hatching.

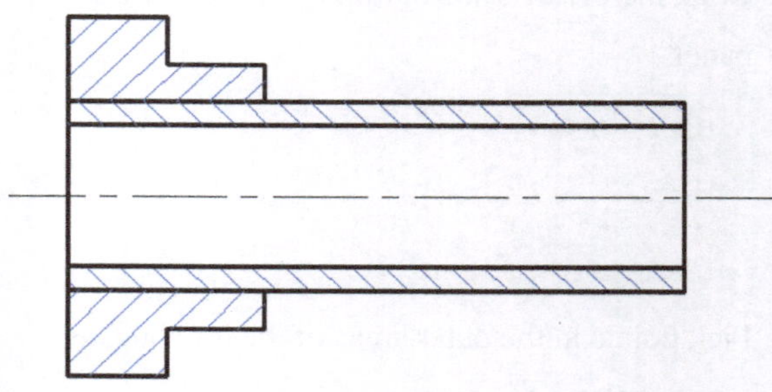

- Save and close the file.

Island Detection tools

While creating hatch lines, the island detection tools help you to detect the internal areas of a drawing.

Example:

- Create the drawing, as shown below. Do not apply dimensions.

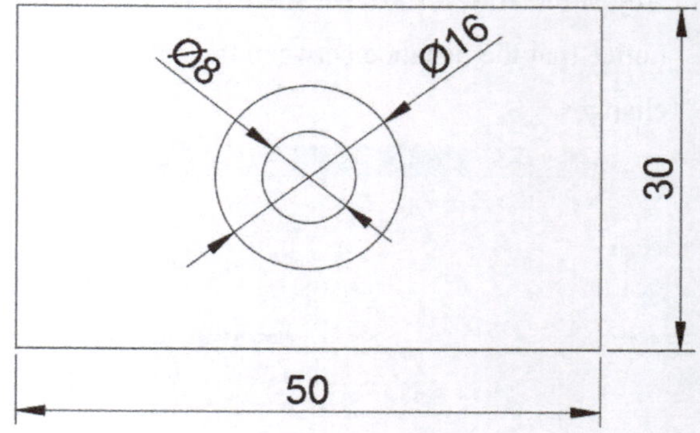

- Click **Home > Draw > Hatch** on the ribbon.

- Select **ANSI31** from the **Pattern** panel.
- Expand the **Options** panel and select the **Normal Island Detection** tool.

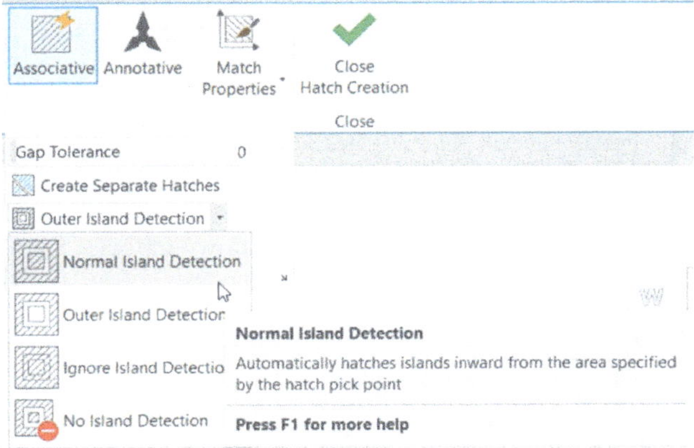

- Pick a point in the area outside the large circle; you will notice that the area inside the small circle is detected automatically. Also, hatch lines are created inside the small circle.

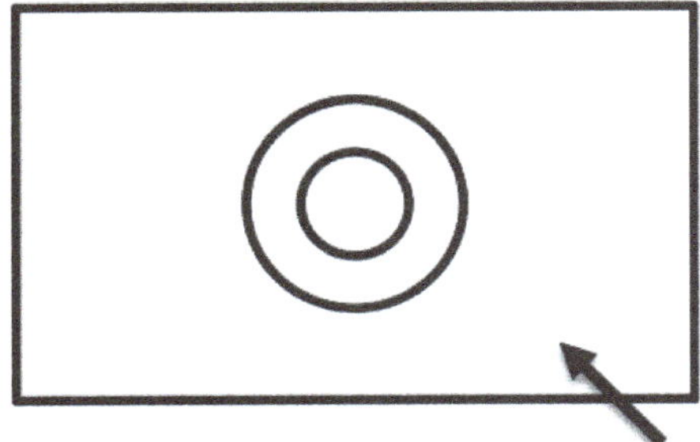

- Press **Enter**.

- Click **Undo** on the **Quick Access Toolbar**.

- Activate the **Hatch** tool and select **ANSI31** from the **Pattern** panel.
- Expand the **Options** panel and select the **Outer Island Detection** option.

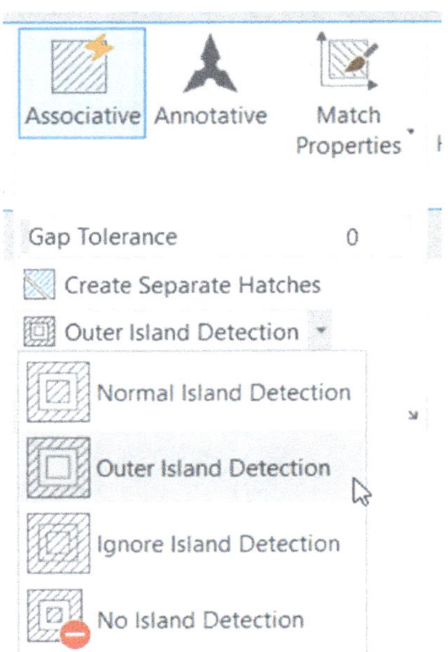

- Pick a point in the area outside the large circle and press ENTER; you will notice that hatch lines are created only outside the large circle. The **Outer Island Detection** tool will enable you to create hatch lines only in the outermost level of the drawing.

- Repeat the process using the **Ignore Island Detection** tool. You will notice that the internal loops are ignored while creating the hatch lines.

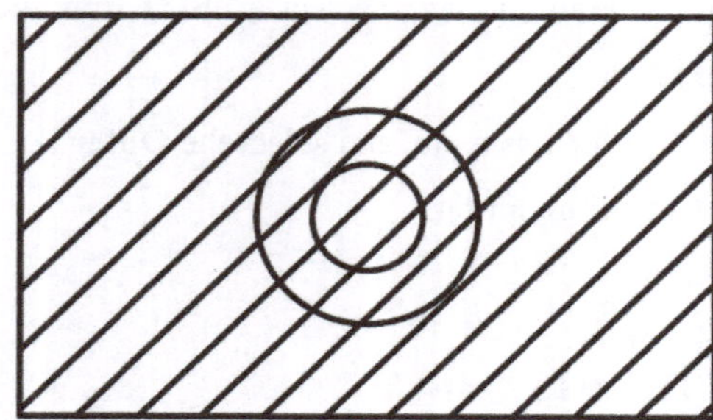

Text in Hatching

You can create hatching without passing through the text and dimensions.

- Create a drawing, as shown in the figure.

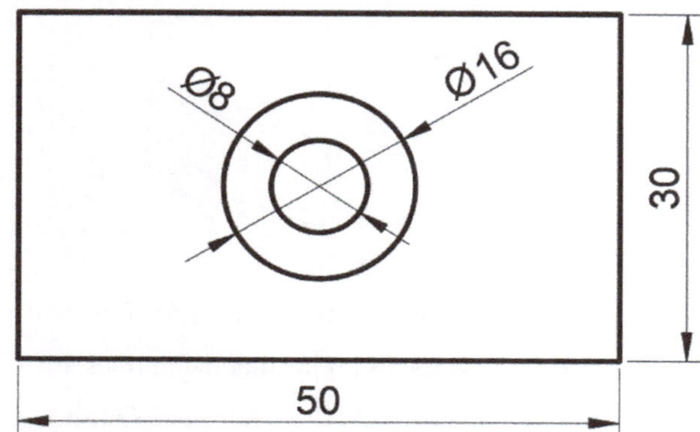

- Click **Home > Annotation > Multiline Text** on the ribbon.

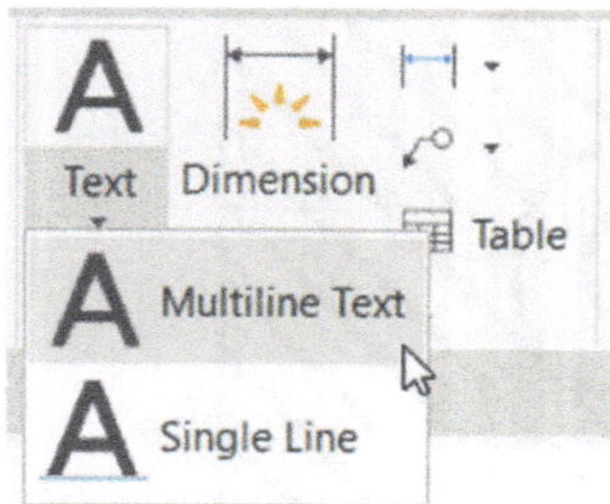

- Specify the first and second corners of the text editor, as shown below.

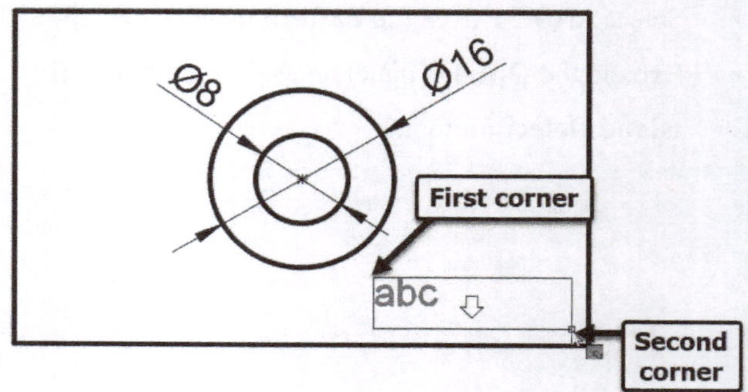

- Select **Arial** from the **Font** drop-down of the **Formatting** panel.

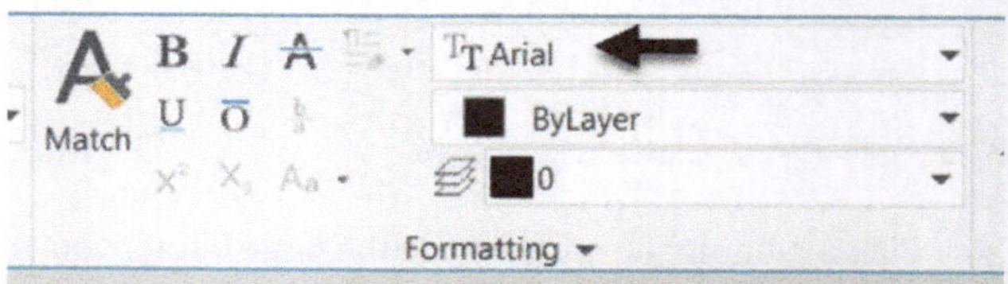

- Ensure that **Text Height** is set to **2.5**.

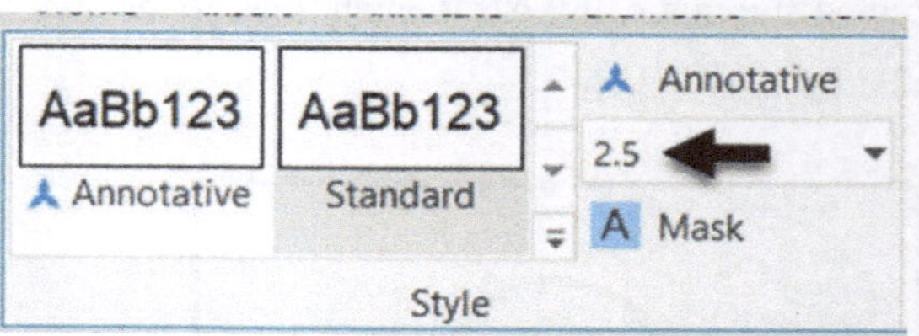

- Type **AutoCAD** in the text editor. Left-click in the space of the graphics window.

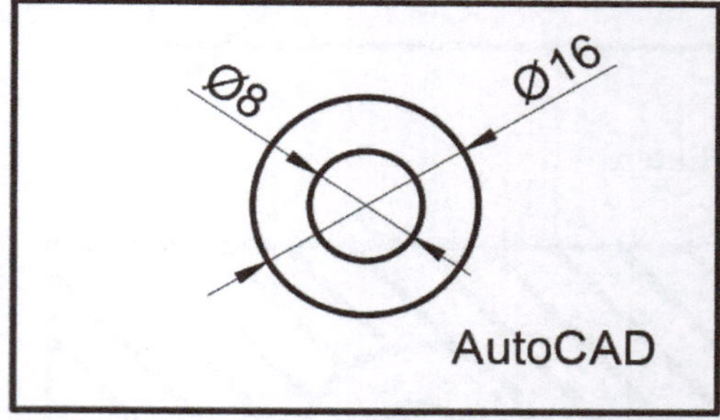

- Activate the **Hatch** tool and select the **Normal Island Detection** option from the **Options** panel.
- Pick a point in the area covered by the outside boundary and press ENTER; hatch lines are created. You will notice that hatch lines do not pass through the text and dimension.

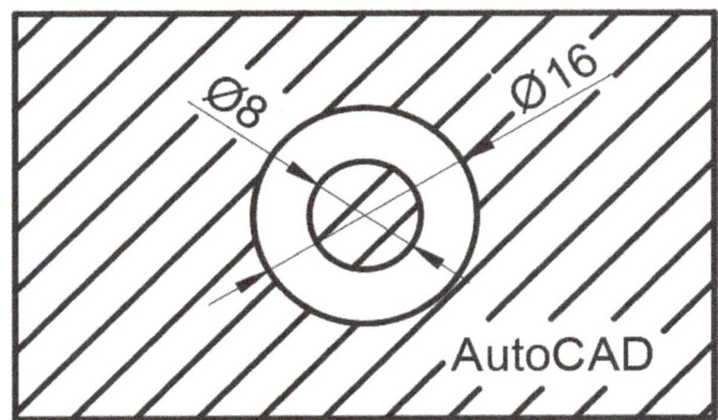

Editing Hatch lines

You can edit a hatch by using the **Edit Hatch** tool or simply selecting the hatch.

- To edit a hatch using the **Edit Hatch** tool, expand the **Modify** panel of the **Home** ribbon and select the **Edit Hatch** tool.

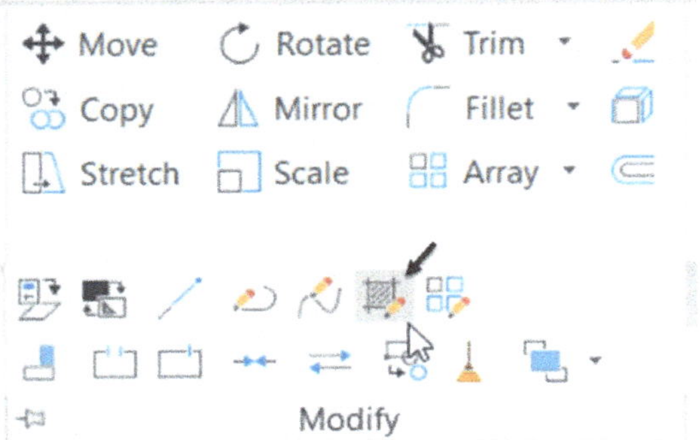

- Select the hatch from the drawing, the **Hatch Edit** dialog appears. The options in this dialog are the same as that available in the **Hatch Creation** ribbon. Expand this dialog by clicking the **More Options** button located at the bottom right corner.

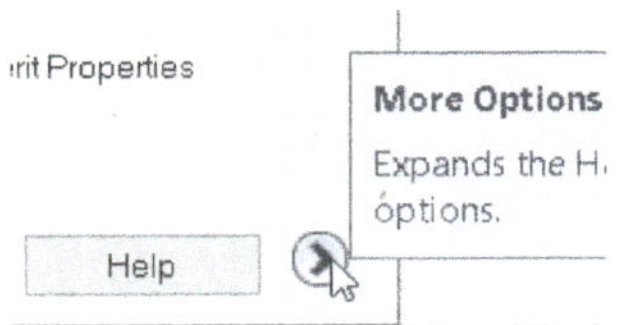

The expanded dialog will display more options, as shown below. The options in this dialog are the same as that available in the **Hatch Creation** tab.

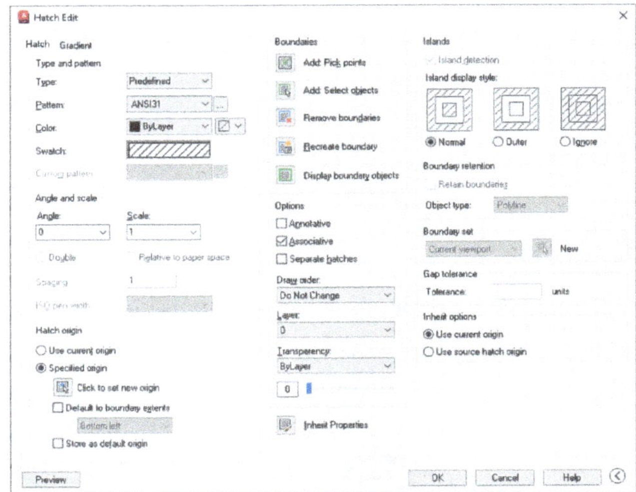

- Edit the options in the **Hatch Edit** dialog and click the **OK** button; the hatch pattern will be modified.

Exercises

Exercise 1

Create the half-section view of the object shown below.

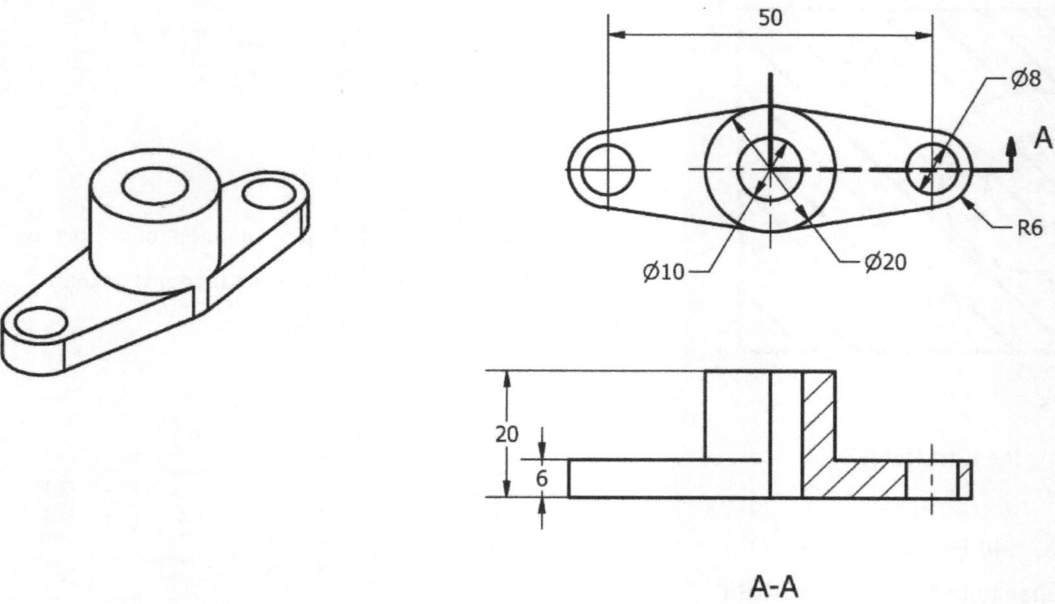

Exercise 2

In this exercise, the top, front, and right-side views of an object are given. Replace the front view with a section view. The section plane is given in the top view.

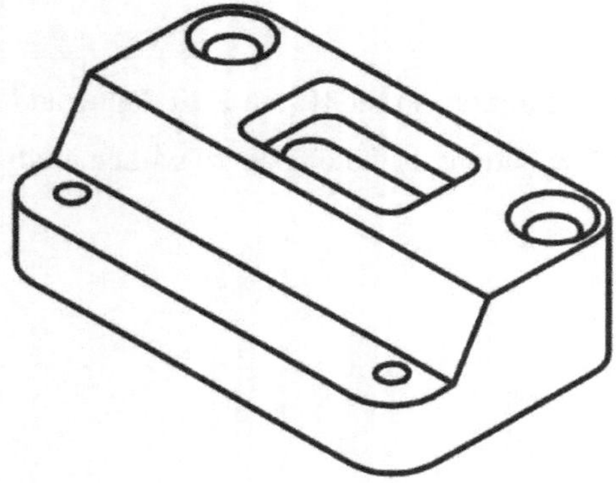

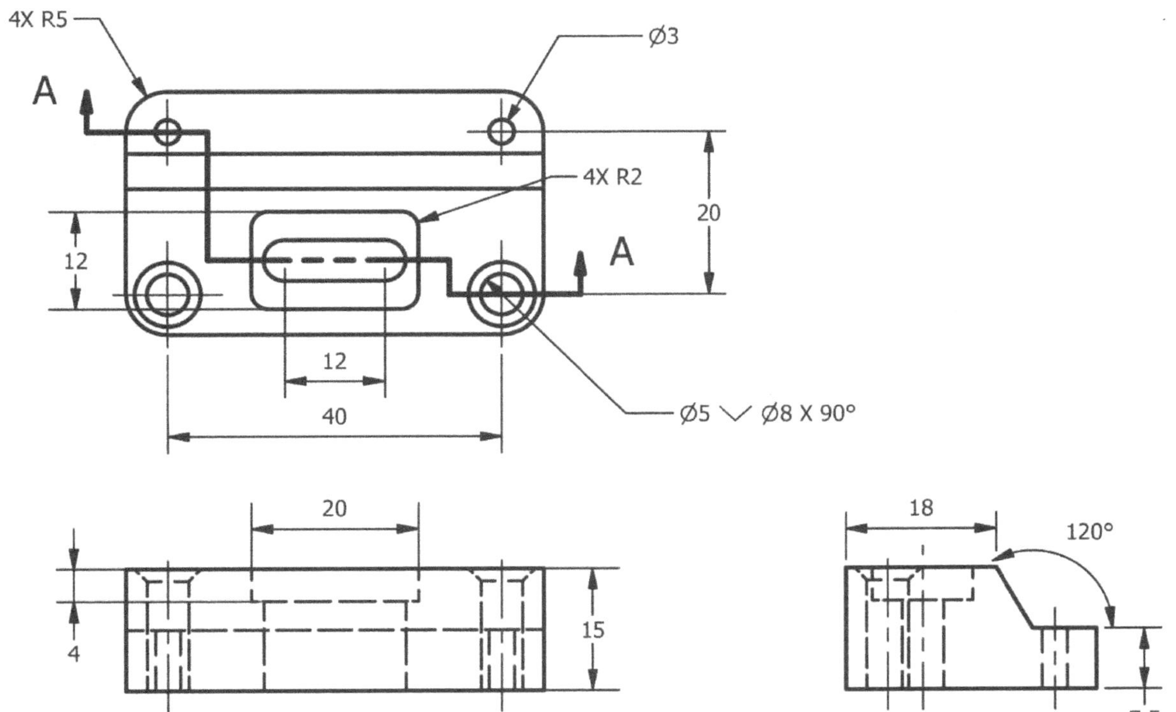
4X R5
Ø3
A
4X R2
20
12
A
12
40
Ø5 ⌵ Ø8 X 90°
20
4
15
18
120°
7.5

Chapter 9: Blocks, Attributes, and Xrefs

In this chapter, you will learn to do the following:

- **Create and insert Blocks**
- **Create Annotative Blocks**
- **Explode and purge Blocks**
- **Use the Divide tool**
- **Use the DesignCenter and Tool Palettes to insert Blocks**
- **Insert Multiple Blocks**
- **Edit Blocks**
- **Create Blocks using the Write Block tool**
- **Define and insert Attributes**
- **Work with Xrefs**

Introduction

In this chapter, you will learn to create and insert Blocks and Attributes in a drawing. You will also learn to attach external references to a drawing. The first part of this chapter deals with Blocks. A Block is a group of objects combined and saved together. You can later insert it in drawings. The second part of this chapter deals with Attributes. An Attribute is an intelligent text attached to a block. It can be any information related to the block such as description, part name, and value, and so on. The third part of the chapter deals with the Xrefs (external references). External references are drawing files, images, PDF files attached to a drawing.

Creating Blocks

To create a block, first, you need to create shapes using the drawing tools and use the BLOCK command to convert all the objects into a single object. The following example shows the procedure to create a block.

Example 1

- Create the drawing, as shown below. Do not apply dimensions. Assume the missing dimensions.

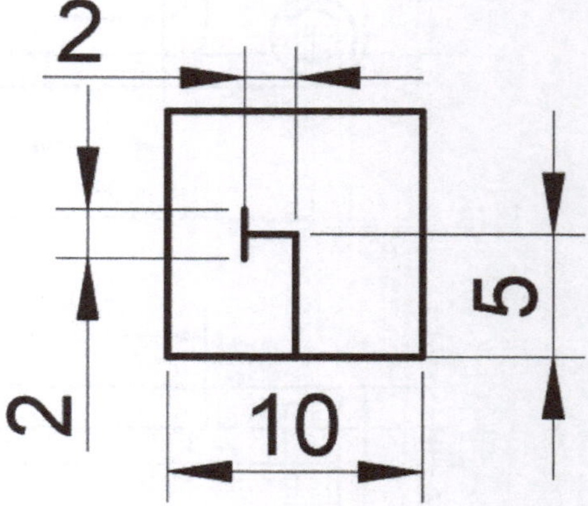

- Click **Insert > Block Definition > Create Block** on the ribbon; the **Block Definition** dialog appears.

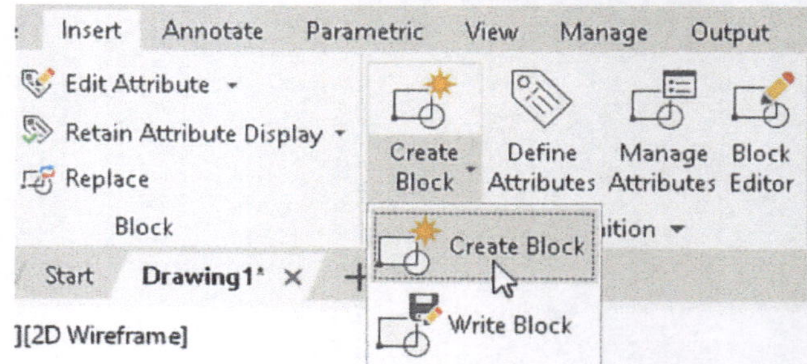

- Enter **Target** in the **Name** field.

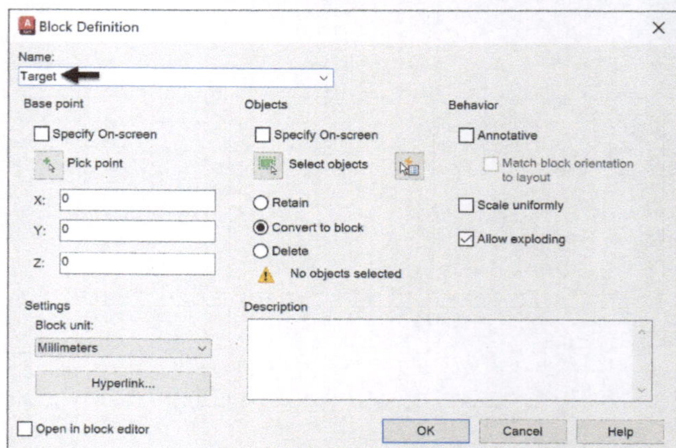

- Click the **Select Objects** button on the dialog. Drag a window and select all the objects of the drawing.

- Right-click to accept; the dialog appears again.
  You can choose to retain or delete the objects after defining the block. The **Retain** option under the **Objects** section retains the objects in the graphics window after defining the block. The **Convert to Block** option deletes the objects and displays the block in place of them. The **Delete** option completely deletes the objects from the graphics window.
- Select the **Delete** option under the **Objects** section.
- Click the **Pick point** button on the dialog.
- Select the midpoint of the left vertical line. The selected point will be the insertion point when you insert this block into a drawing.

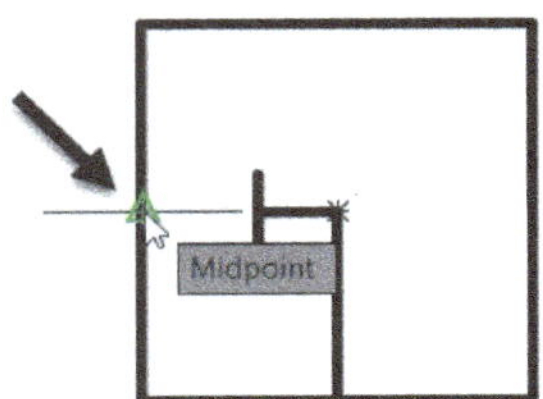

You can also add a description to the block in the **Description** box. In addition to that, you can set the behavior of the block such as scalability, annotative, and explode ability using the options in the **Behavior** section. The options in the **Settings** area can be used to set the units of the block and link a website or other files with the block.

- Uncheck the **Scale uniformly** option (for this example).
- Uncheck the **Open in block editor** option, if selected.
- Click **OK** on the dialog; the block will be created and saved in the database.

Inserting Blocks

After creating a block, you can insert it at the desired location inside the drawing using the INSERT command.

The procedures to insert blocks are explained in the following examples.

Example 1

- Click **Insert > Block > Insert > Target** on the ribbon.

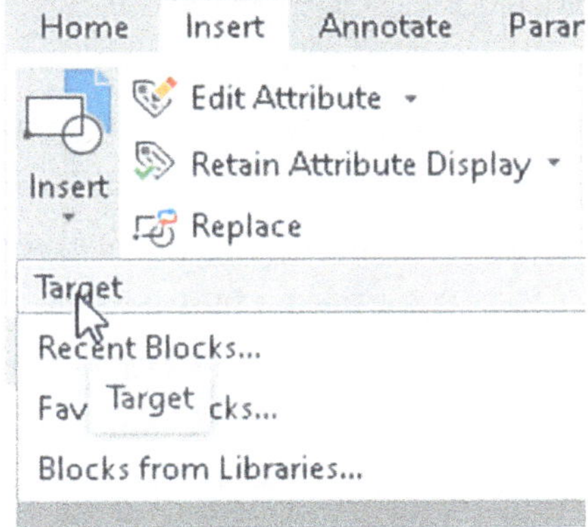

- Pick a point in the graphics window to place the block.

Example 2 (Scaling the block)

- On the ribbon, click **Insert > Block > Insert > Recent Blocks**; the **BLOCKS** palette appears.

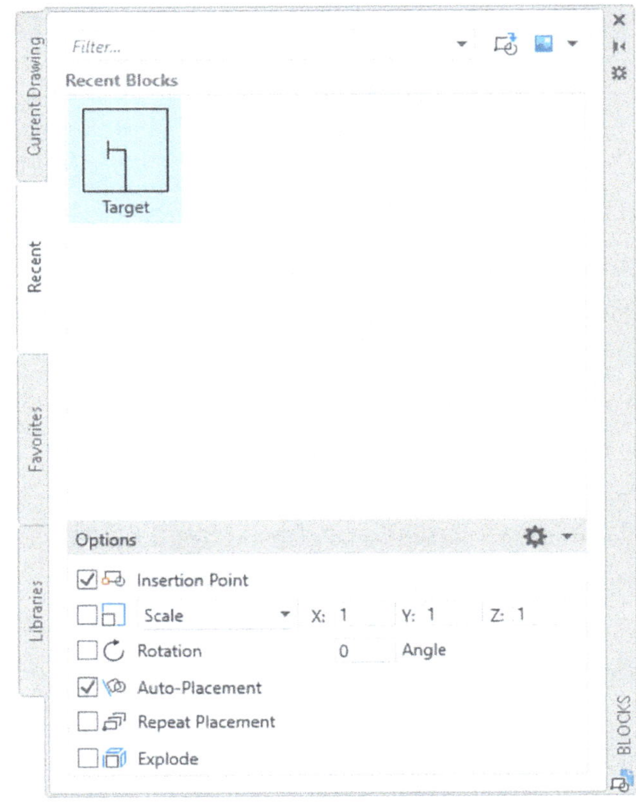

The BLOCKS palette can be used to access a large number of blocks. There are four tabs on this palette: **Current Drawing**, **Recent**, **Favorites**, and **Libraries**. The **Current Drawing** tab displays the blocks available in the current drawing. The **Recent** tab displays the recently used blocks. The **Favorites** tab displays your favorite blocks. You can add a block to this tab by right-clicking on it and selecting the **Copy to Favorites** option. The **Libraries** tab displays the blocks available in the selected drawing files. To do this, click the **Libraries** tab, and then click the **Open Block Libraries** button. Next, browse to the required location and select the drawing file; the blocks in the drawing file are displayed on the palette. Note that you can also insert an entire drawing file as a block.

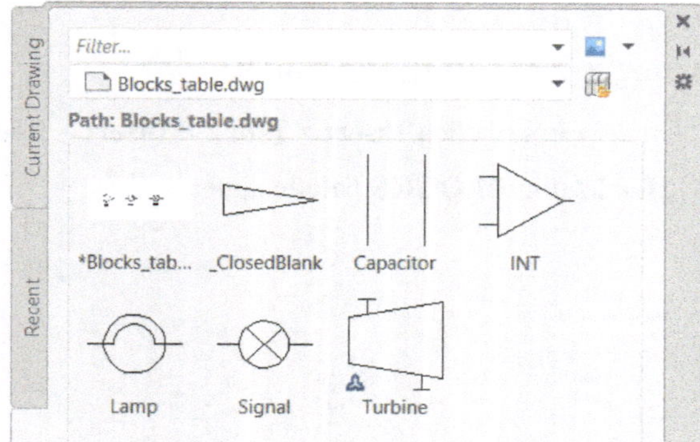

If you want to select another drawing file, then click the **Browse block libraries** icon next to the **File** drop-down. Next, select the required drawing file.

You can use the options in the **Scale** drop-down available in the **Options** section to scale the block. The **Uniform Scale** option can be used to scale the block uniformly. You can select the **Scale** option to specify the scale factor separately in the X, Y, and Z boxes. If you select the checkbox next to the **Scale** drop-down, the block can be scaled dynamically in the graphics window.

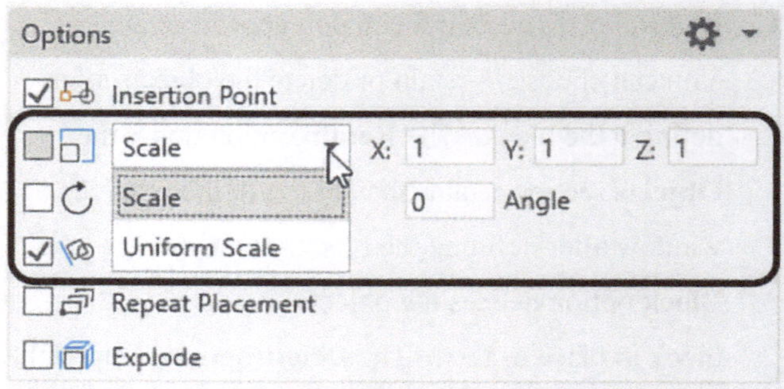

- Make sure that the **Insertion Point** option is checked.
- Select the checkbox next to the **Scale** drop-down.
- Select the **Scale** option from the **Scale** drop-down.
- Select the **Target** block from the **Recent Blocks** section; the block is attached to the pointer.
- Pick a point in the graphics window; the message, "Enter X scale factor, specify opposite corner, or [Corner/XYZ]:" appears in the command line. Also, as you move the pointer, the block automatically scales.
- Type 3 and press ENTER; the message, "Enter Y scale factor <use X scale factor>:" appears.
- Type 2 as the Y scale factor and press ENTER; the block will be scaled, as shown below.

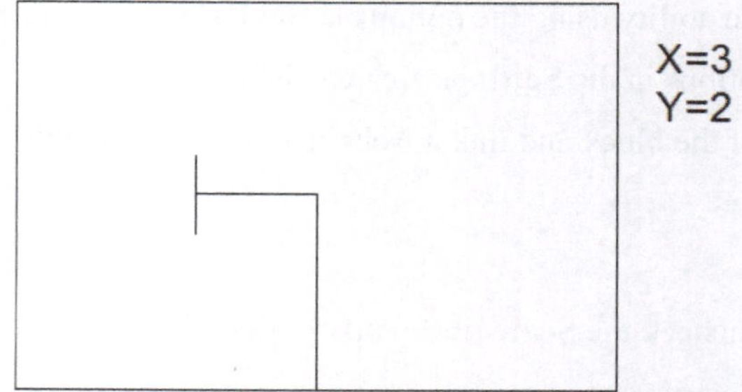

Example 3 (Rotating the block)

- Click **Home > Block > Insert > Recent Blocks** on the ribbon; the **BLOCKS** palette appears.
- Clear the checkbox next to the **Scale** drop-down.
- Select the **Uniform scale** option from the **Scale** drop-down.

The **Rotation** option can be used to rotate the block. You can enter the rotation angle in the **Angle** box. You can dynamically rotate the block by selecting the checkbox next to the **Rotation** option.

- Select the checkbox next to the **Rotation** option.
- Select the **Target** block from the **Recent Blocks** section; the block is attached to the pointer.
- Pick a point in the graphics window; the message, "Specify rotation angle <0>:" appears in the command line. As you rotate the pointer, the block also rotates. You can dynamically rotate the block and pick a point to orient the block at an angle or type a value and press ENTER to specify the angle.
- Type **45** and press **ENTER**; the block will be rotated by **45** degrees.

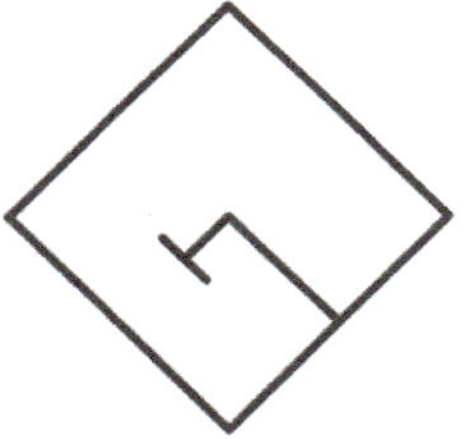

- Save and close the drawing file.

Example 4 (Using the Auto-Placement option)

The Auto-Placement option provides you recommendations for Insertion Point, taking into account previous instances of the block within the drawing.

- Create the drawing, as shown below.
- Click **Insert > Block Definition > Create Block** on the ribbon; the **Block Definition** dialog appears.
- Create a selection window across the elements of the drawing.
- Enter **Auto_placement** in the **Name** field.
- Click the Pick Point button in the Base point section and select the bottom-right corner of the drawing, shown.

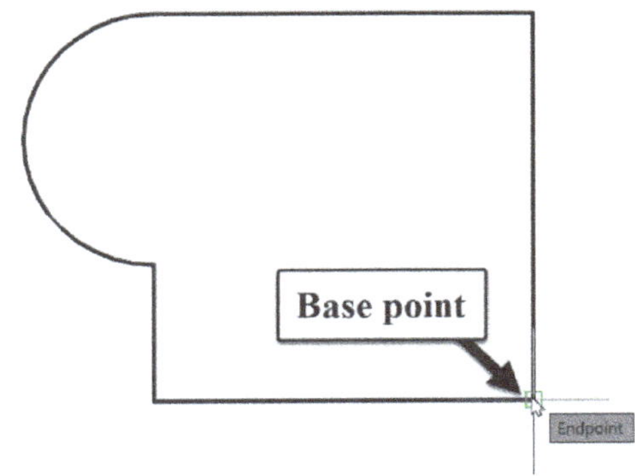

- Leave the other default settings and click **OK**.
- Create a square.
- Create a rectangular array of the square, as shown.

- Click **Home > Block > Insert > Recent Blocks** on the ribbon; the **BLOCKS** palette appears.
- On the BLOCKS palette, check the **Auto-placement** option.
- Select the **Auto_placement** block from **Recent Blocks** section.
- Select the lower-right corner of the square, as shown.

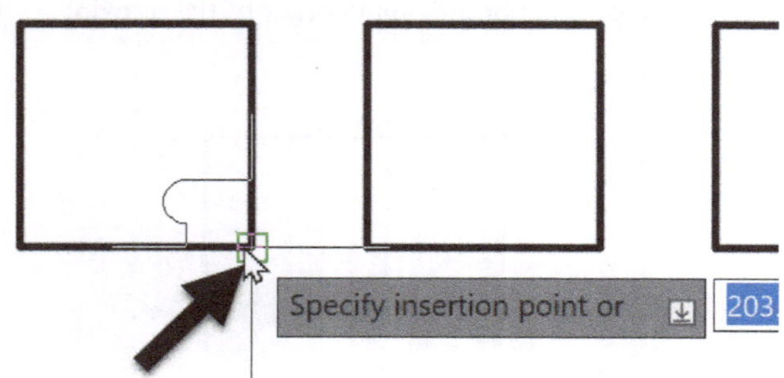

- Deactivate the **Object Snap** icon on the Status bar.
- Select the Auto_placement block from the BLOCKS palette.
- Move the pointer near to the placement point of the previous block placement; the block attached to the pointer is snaped to the previous placement point.
- Click to place the block.

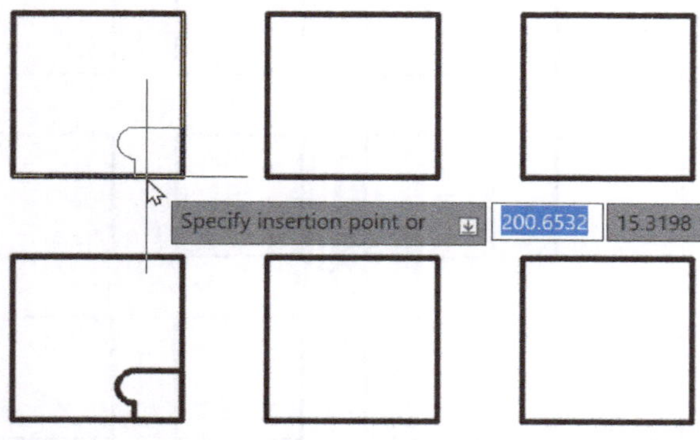

- Likewise, place the other blocks using the **Auto-placement** option.

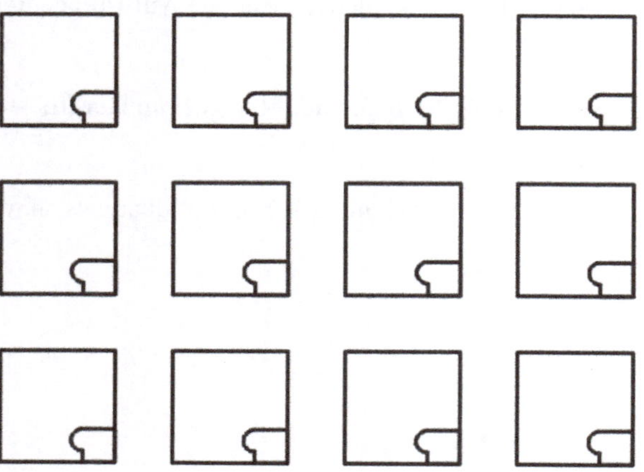

Replacing Blocks

The **Replace Block** command allows you replace a block with another block inside a drawing. It retains the original block's insertion point, scale, and rotation.

- Download the Replace_block.dwg and open it.

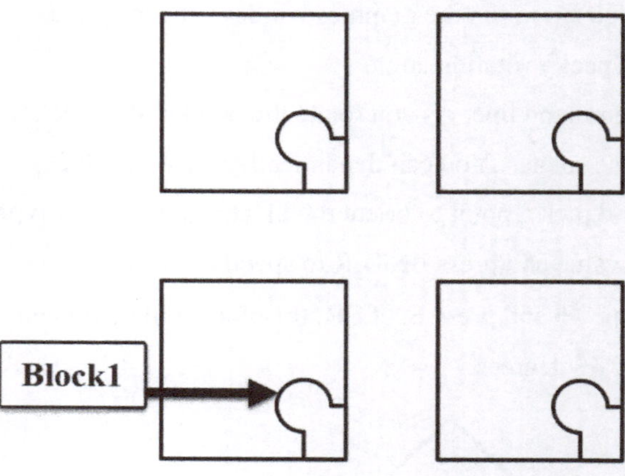

- Type BLOCKREPLACE on the command line and press ENTER (or) click **Express Tools > Blocks > Replace Block with Another Block** on the ribbon.

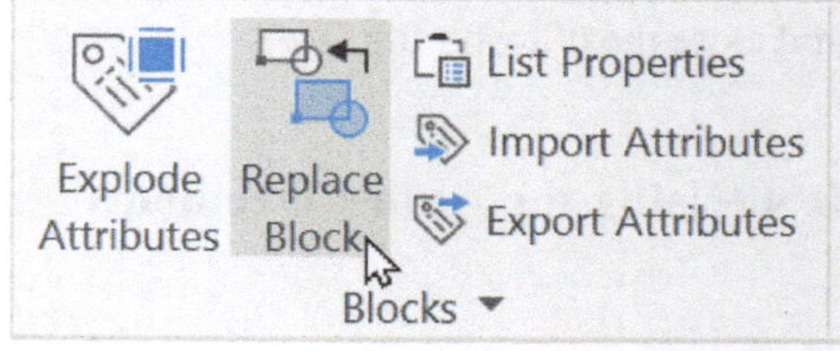

- In the **BLOCKREPLACE** dialog box, select the block you want to replace from the list of blocks in the drawing. You can also click the Pick button and select the block to be replaced directly from the graphics window.
- Click **OK**.
- Select the block you want to replace it with from the list of blocks available in the drawing.
- Click **OK**; the command line displays "Purge unreferenced items when finished? <Y>:"

- Type **N** and press ENTER. The original block is replaced with the new one.

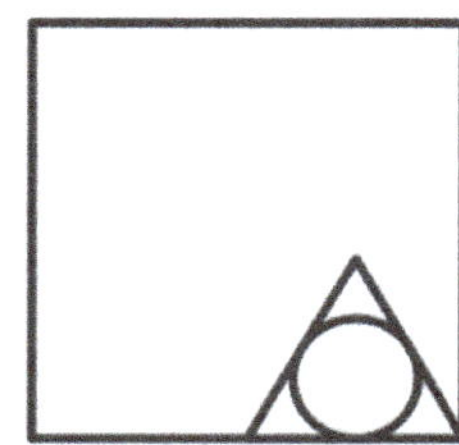

Redefining Blocks

AutoCAD allows you to redefine an already-created block.

- Download the Redefining Blocks.dwg file from the companion website.
- Open the downloaded drawing file.

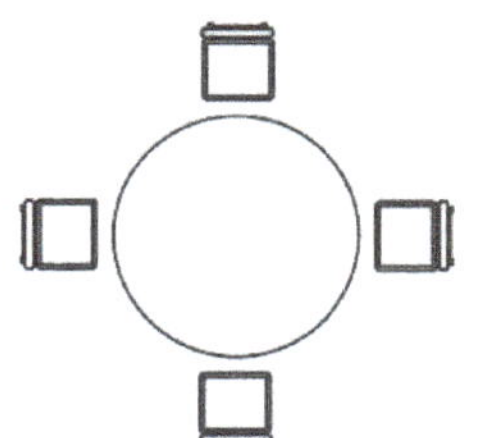
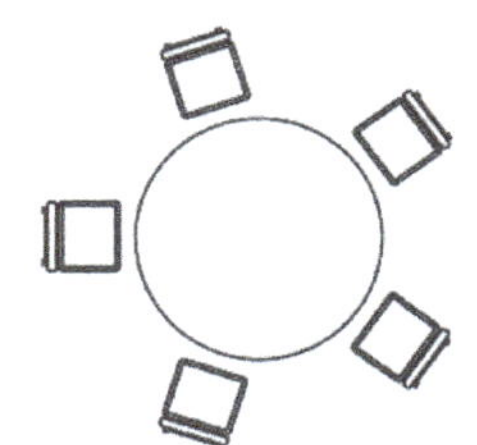

- On the ribbon, click **Insert > Block Definition > Create Block**.
- Click the **Select Objects** button on the **Block Definition** dialog.
- Create a selection window across the objects located on the left side, as shown.

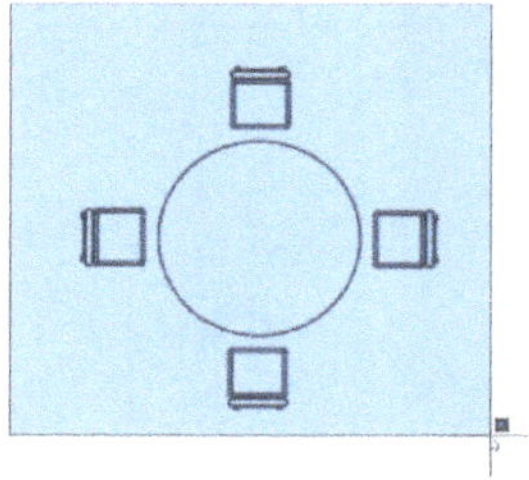
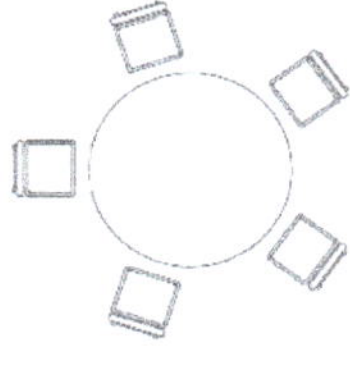

- Press Enter to accept the selection.
- On the **Block Definition** dialog, click the **Pick Point** button in the **Base point** section.
- Select the center point of the circle to define the base point of the block.

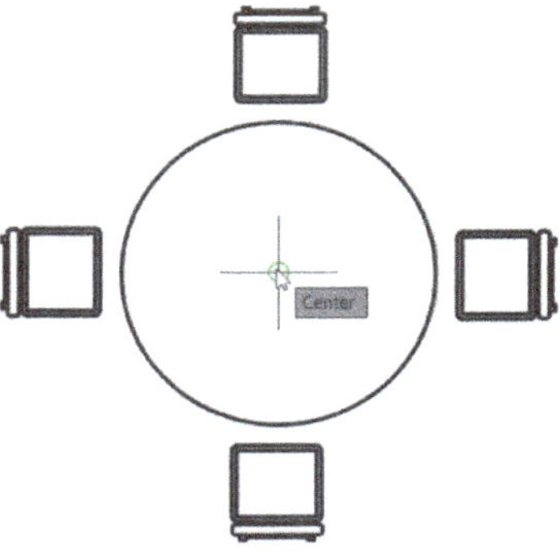

- Type **Dining Table** in the **Name** box.
- Select the **Retain** option in the **Objects** section.
- Click **OK** on the **Block Definition** dialog; the block is created.
- Click **Insert > Block > Insert > Dining Table**.
- Click in the graphics window to insert the block.

Now, you need to redefine the Dining Table block with the objects located on the right side.

- On the ribbon, click **Insert > Block Definition > Create Block**.
- Click on the down-arrow located next to the **Name** box.
- Select **Dining Table** from the list.
- Click the **Select Objects** button on the **Block Definition** dialog.

- Create a selection across the objects located on the right side.

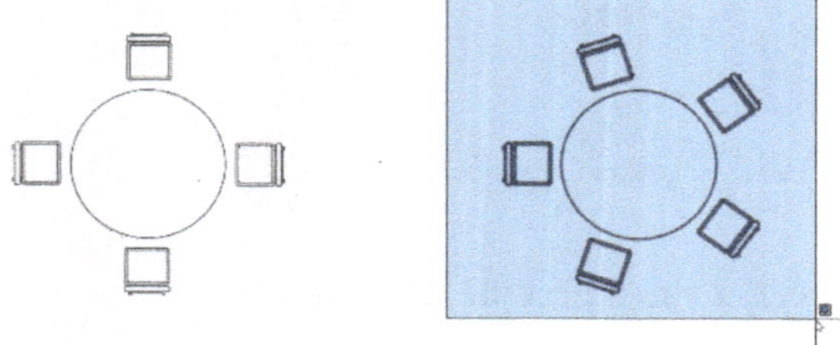

- Press Enter to accept the selection.
- On the **Block Definition** dialog, click the **Pick Point** button in the **Base point** section.
- Select the center point of the circle to define the base point of the block.

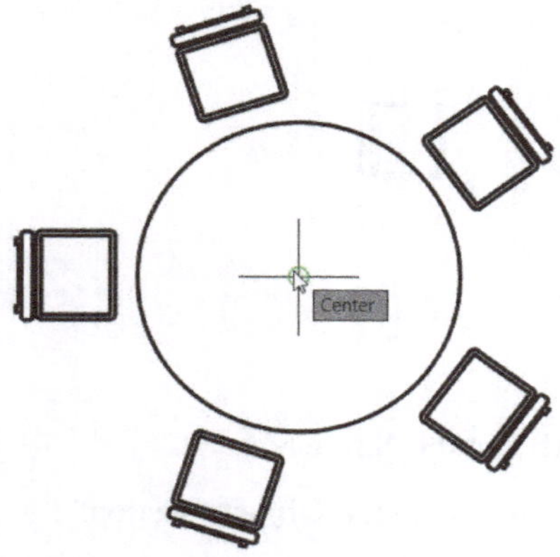

- Click **OK** on the **Block Definition** dialog; the **Block – Redefine Block** message box appears.
- Click the **Redefine Block** option; the block is redefined and updated in the graphics window.

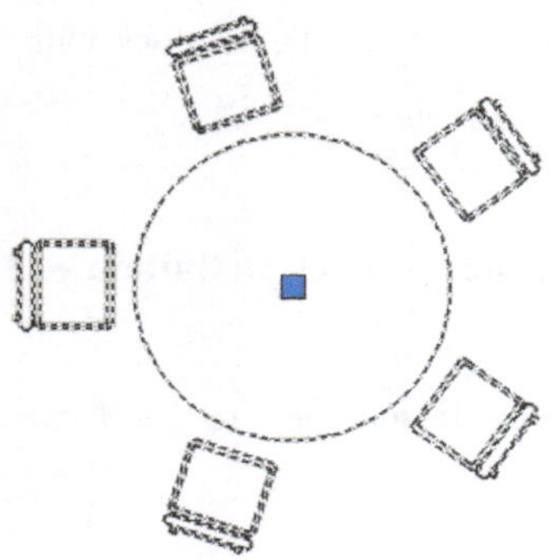

- Close the Drawing file without saving.

Creating Annotative Blocks

Annotative blocks possess annotative properties. They will be scaled automatically depending upon the scale of the drawing sheet. The procedure to create and insert annotative blocks is explained in the following example.

Example:

- Create the drawing shown in the figure. Assume the missing dimensions.

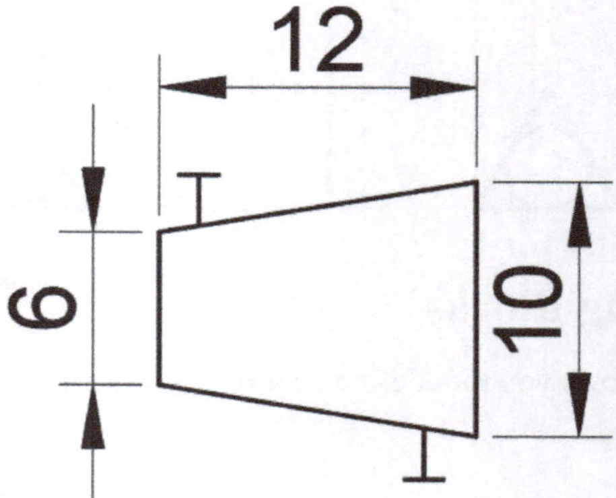

- Click **Insert > Block Definition > Create Block** on the ribbon; the **Block Definition** dialog appears.
- Enter **Turbine Driver** in the **Name** field.
- Click the **Select Objects** button on the dialog. Create a window and select all the objects of the drawing. Right-click to accept the selection.
- Select the **Delete** option under the **Objects** section.
- Click the **Pick Point** button and select the midpoint of the left vertical line.

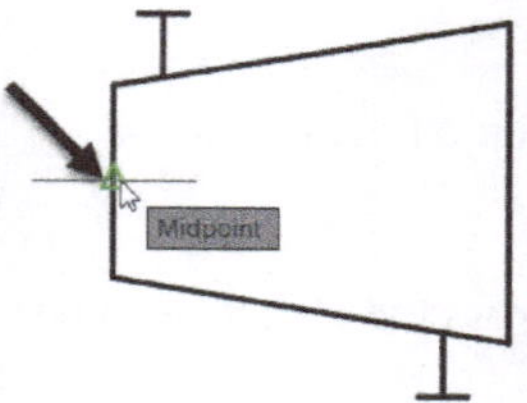

- Check the **Annotative** option under the **Behavior** section. Click the **OK** button on the dialog.
- Activate the **Automatically Add Scale to Annotative**

Objects button located on the right side of the Status Bar.

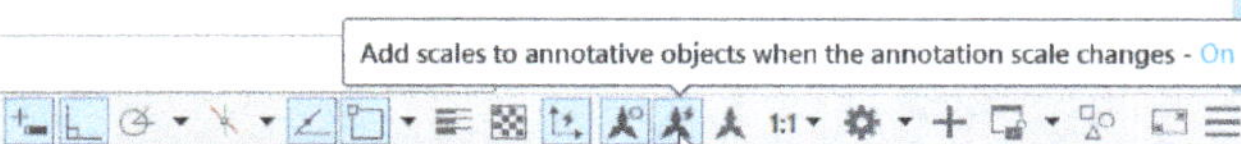

- Set the **Annotation Scale** to **1:10**.

- Click **Insert > Block > Insert > Turbine Driver**.
- Pick a point in the graphics window; the block will be inserted with the scale factor 1:10.
- Click **Zoom All** on the **Navigation Bar** to view the block.
- Change the **Annotation Scale** to **1:2**; you will notice that the block is automatically scaled to **1:2**.

Exploding Blocks

When you insert a block into a drawing, it will be considered as a single object, even though it consists of numerous individual objects. At many times, you may require to break a block into its parts. Use the **Explode** tool to break a block into its individual objects.

- To explode a block, click **Home > Modify > Explode** on the ribbon or type EXPLODE in the command line and press ENTER.
- Select the block and press ENTER; the block will be broken into individual objects. You can select the individual objects by clicking on them. (Refer to the Explode Tool section of Chapter 4).

Using the Purge tool

You can remove the unused blocks and other unwanted drawing items from the database using the **Purge** tool. This tool also allows you to find non-purgeable items from the current drawing.

- To activate the **Purge** tool, click **Manage > Cleanup > Purge** on the ribbon; the **Purge** dialog appears.

- Click the **Find Non-Purgeable Items** button on the dialog to view the items that cannot be purged. The items that are currently displayed in the drawing cannot be purged. Also, the items such as styles, layers, and linetypes that are currently used in the drawing cannot be purged. You can expand the node under the All items tree to view the individual items that cannot be purged. Next, select the item to view the possible reasons not to purge in the **Possible reasons** section.

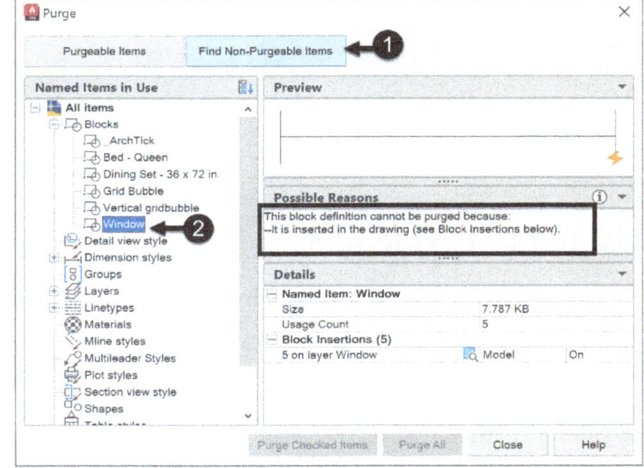

- Click the **Purgeable Items** button to view the items in the drawing that can be purged.
- To remove unwanted blocks from the database, expand the **Blocks** tree, and select the blocks.

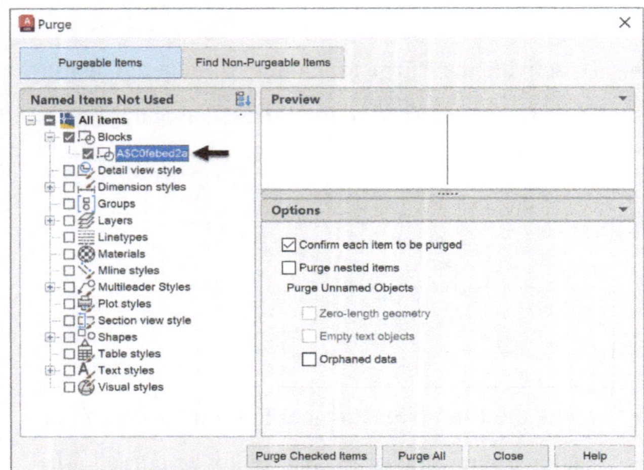

- Click the **Purge Checked Items** button on the dialog; the **Purge – Confirm Purge** message box appears.
- Click **Purge this item** to delete the item from the database.

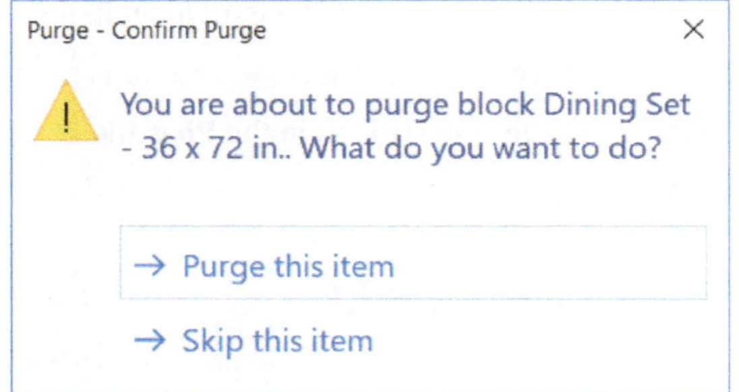

- Click **Close** on the **Purge** dialog.

Using the Divide tool

The **Divide** tool is used to place many instances of an object equally spaced on a line segment. You can also place blocks on a line segment. The following example shows you to divide a line using the **Divide** tool.

Example:

- Create the object, as shown in the figure.

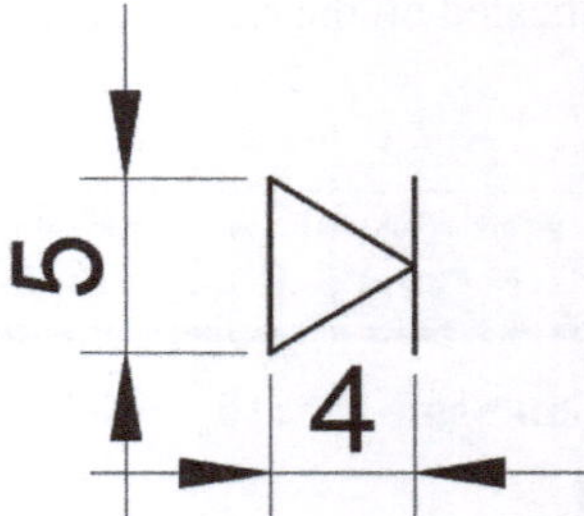

- Create a block with the name **Diode**. Specify the midpoint of the left vertical line as the base point.

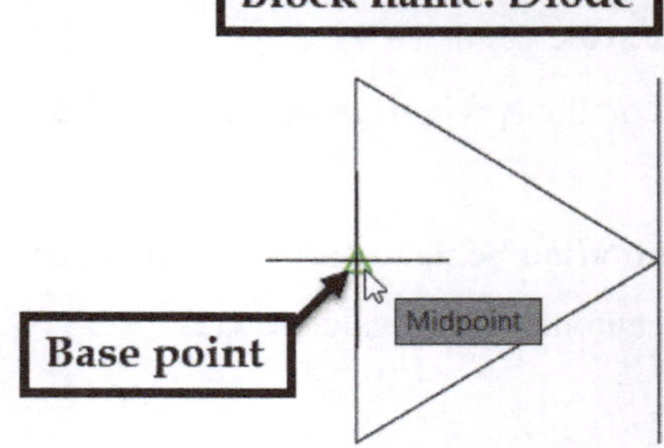

- Create a line of 50 mm length and 45 degrees inclination.

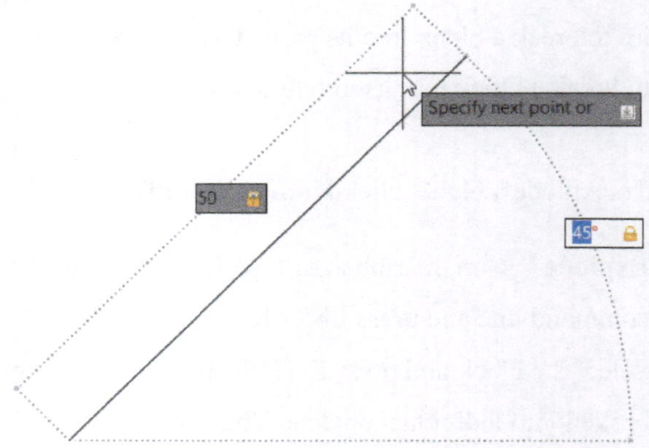

- Expand the **Draw** panel in the **Home** tab and click **Divide**.

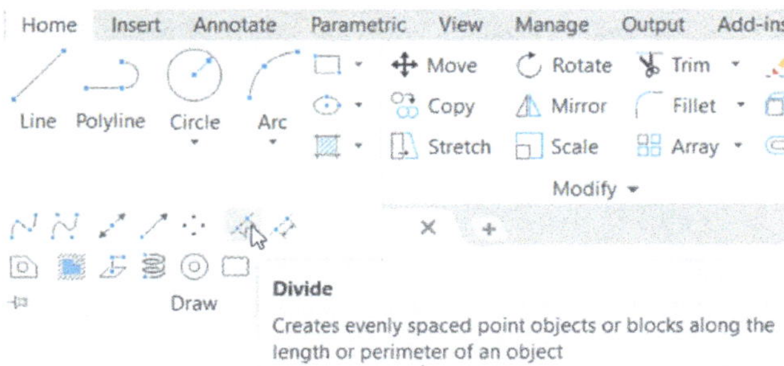

- Select the line segment; the message, "Enter the number of segments or [Block]:" appears.
- Select the **Block** option from the command line; the message, "Enter name of block to insert," appears.
- Type **Diode** and press ENTER; the message, "Align block with object? [Yes/No] <Y>:" appears.
- Select the **Yes** option; the message, "Enter the number of segments:" appears.
- Type **5** and press ENTER; the line segment will be divided into five segments, and four instances of blocks will be placed.

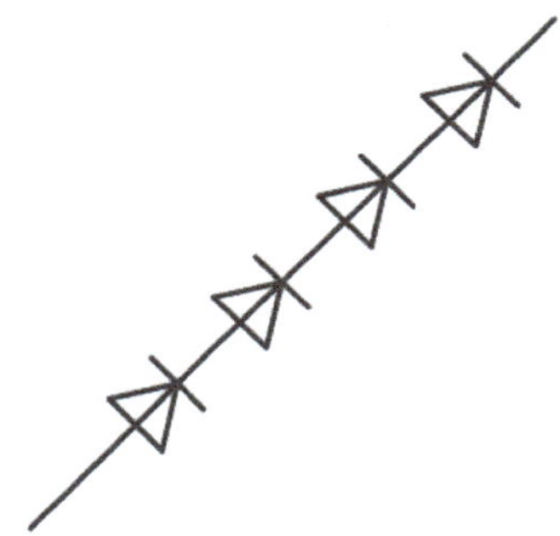

- Trim the unwanted portions, as shown below.

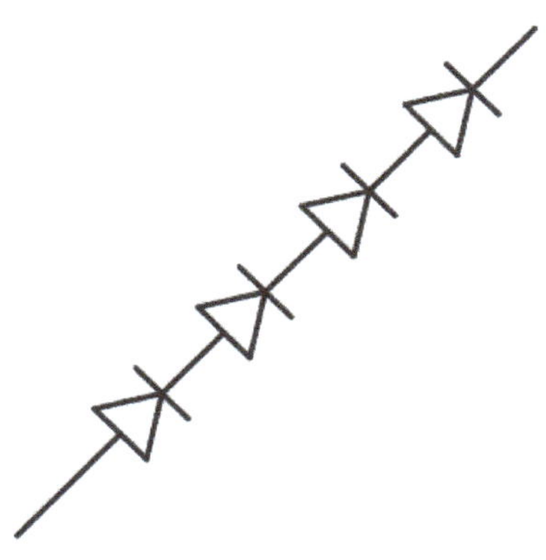

Renaming Blocks

You can rename blocks at any time. The procedure to rename blocks is discussed next.

- On the Quick Access Toolbar, click the down arrow located on the right side.
- Select **Show Menu Bar** from the drop-down.
- On the Menu bar, click **Format > Rename** or type **RENAME** in the command line and press ENTER; the **Rename** dialog appears.
- In the **Rename** dialog, select **Blocks** from the **Named Objects** list.
- Select the block to be named from the **Items** list and enter a new name in the **Rename To** box.

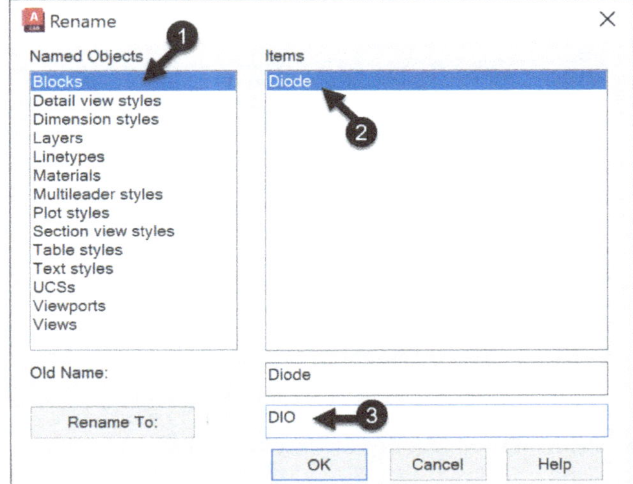

- Click **OK**; the block will be renamed.

Inserting Blocks in a Table

You can insert blocks in a table and fit inside the table cells. Note that you cannot insert Annotative blocks in a table. The following example shows you to insert a block in a table.

Example:

- Create three blocks, as shown below. You can also download them from the companion website.

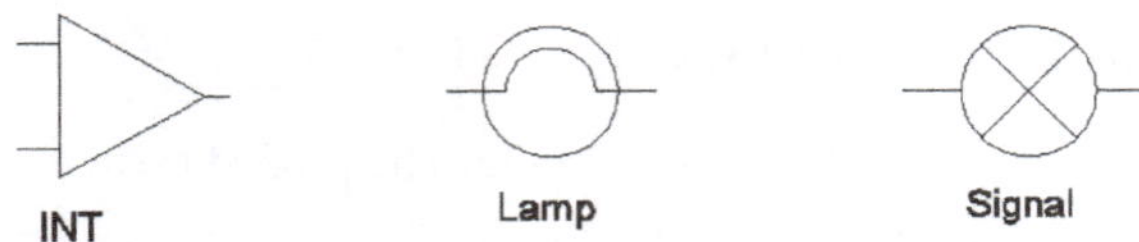

- On the ribbon, click **Annotate > Tables > Table**.

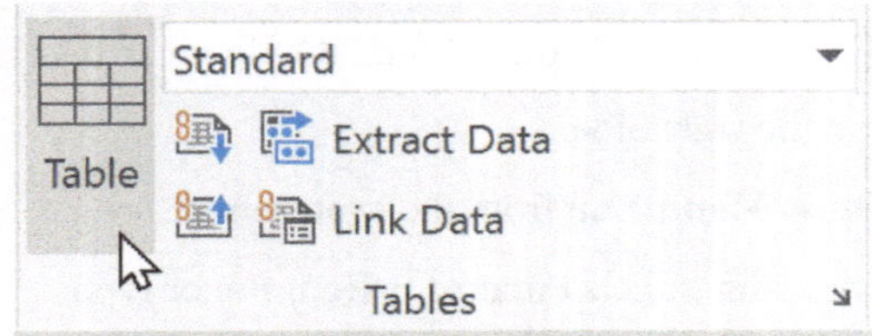

- On the **Insert table** dialog, specify the values, as shown below.

  Columns:2

  Column width: 60

  Data rows:3

  Row height:2

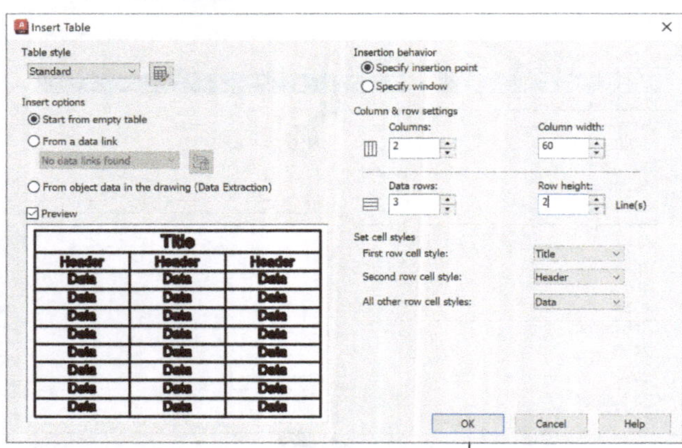

- Click **OK** and define the insertion point of the table.
- Type-in text in the table cells (double-click in the cells and type), as shown below.

Electronic/Electrical Symbols	
Symbol	Name
	INT
	Lamp
	Signal

- Select the first cell in the **Symbol** row and right-click.
- Select **Insert > Block** from the shortcut menu; the **Insert a Block in a Table Cell** dialog appears.

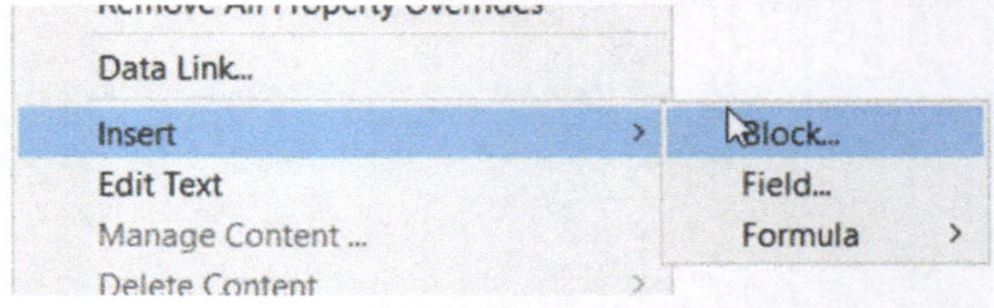

- In the **Insert a Block in a Table Cell** dialog, select **INT** from the **Name** drop-down.
- Set **Overall cell alignment** to **Middle Center**.

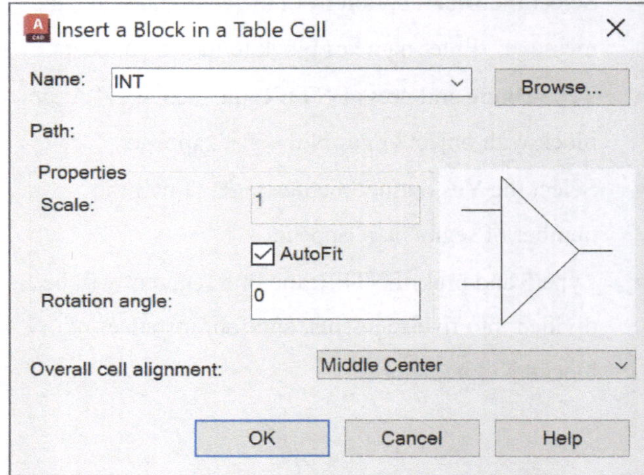

- Click **OK**; the **INT** symbol will be placed in the selected cell.

Electronic/Electrical Symbols	
Symbol	Name
	INT
	Lamp
	Signal

- Likewise, insert the other symbols in the corresponding cells.

Electronic/Electrical Symbols	
Symbol	Name
	INT
	Lamp
	Signal

Using the DesignCenter

DesignCenter is one of the additional means by which you can effectively insert blocks and drawings. Using the DesignCenter, you can insert blocks created in one drawing into another drawing. You can display the DesignCenter by clicking **View > Palettes > DesignCenter** on the ribbon (or) entering **DC** in the command line. The following example shows you to insert blocks using the DesignCenter.

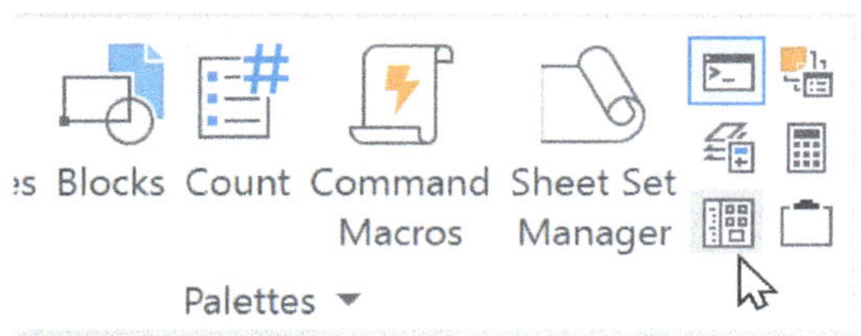

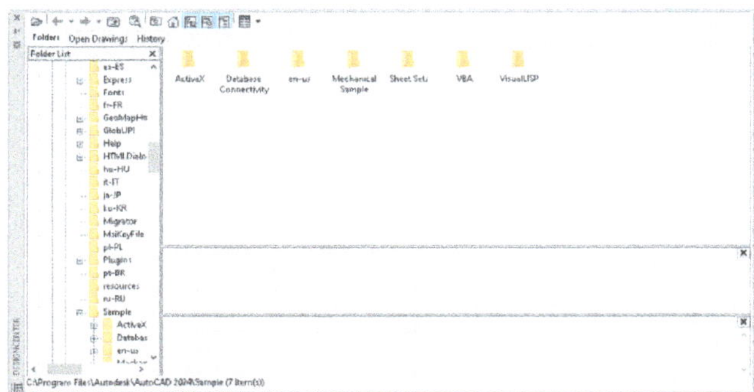

Example:

- Open a new drawing file.
- Create the following symbols and convert them into blocks. You can also download them from the companion website.

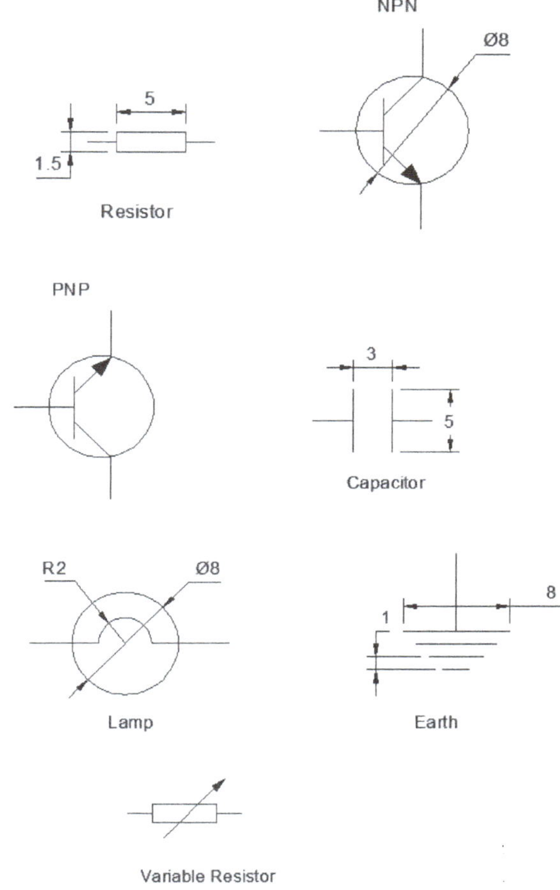

- Save the file as **Electronic Symbols.dwg**. Close the file.
- Open a new drawing file.
- Set the maximum limit of the drawing to 100,100.
- Click **Zoom All** on the Navigation Bar.
- Click **View > Palettes > DesignCenter** on the ribbon; the **DesignCenter** palette appears.
- In the **DesignCenter** palette, browse to the location of the **Electronic Symbols.dwg** file using the **Folder List**.
- Select the file and double-click on the **Blocks** icon; all the blocks present in the file will be displayed.
- Drag and place the blocks in the graphics window.

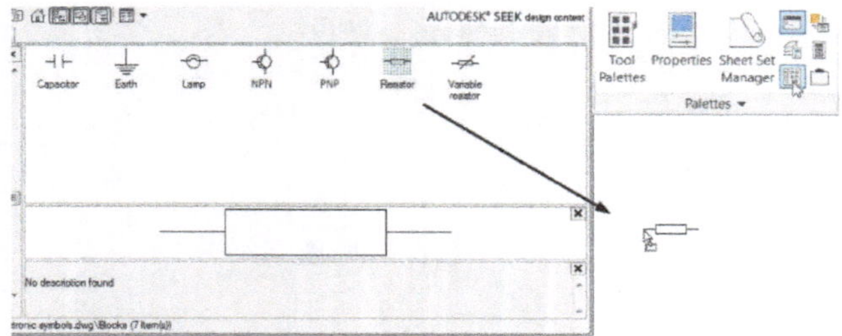

You can also insert blocks by activating the **Insert** dialog. To do this, right-click on the block in the Design Center, and then select **Insert Block**; the **Insert** dialog appears. Click **OK** on the **Insert** dialog; the selected block is attached to the pointer. Next, click in the graphics window to insert the block.

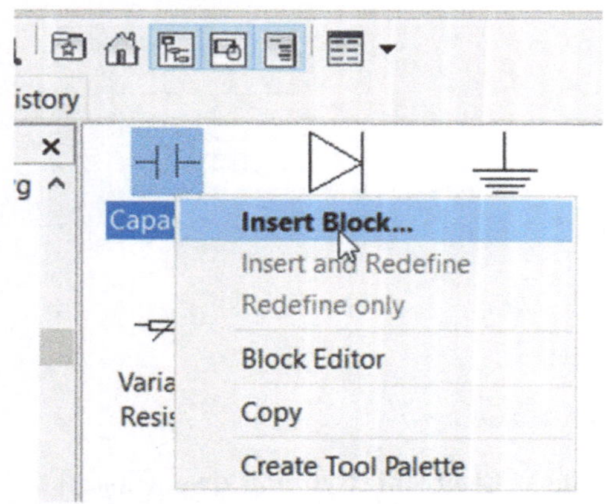

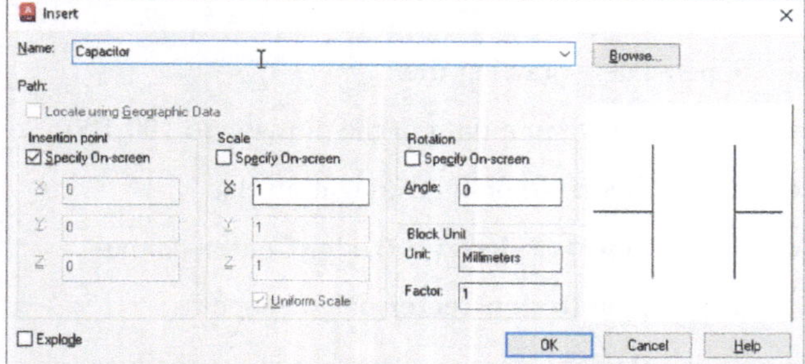

- Use the **Move** and **Rotate** tools and arrange the blocks, as shown below.

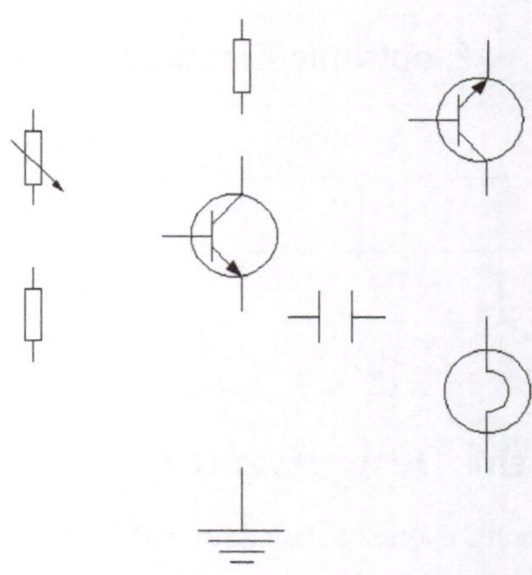

- Close the **DesignCenter** palette.

- Use the **Line** tool and complete the drawing, as shown below.

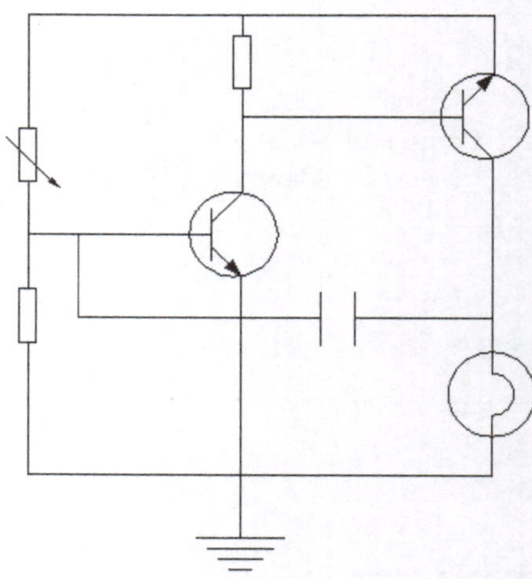

Using Tool Palettes

You can arrange blocks, dimensions, hatch patterns, and other frequently used tools in Tool Palettes. Similar to the **DesignCenter** palette, you can drag and place various features from Tool Palettes into the drawing. You can display the Tool Palettes by clicking **View > Palettes > Tool Palettes** on the ribbon or entering **TOOLPALETTES** in the command line.

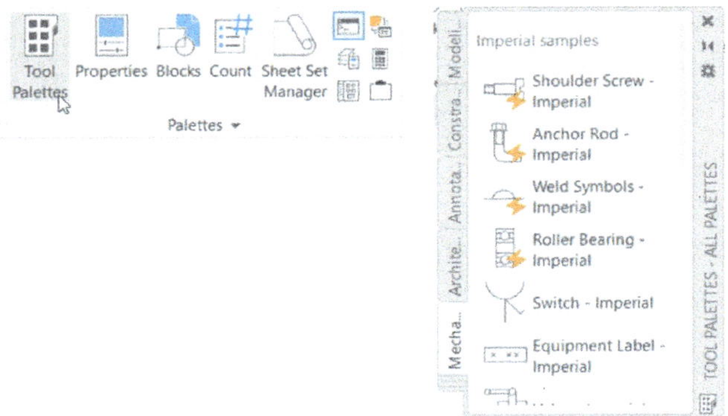

There are many palettes arranged in the Tool Palettes window. You can display more palettes by clicking the lower-left corner of the Tool Palettes and selecting the required palettes.

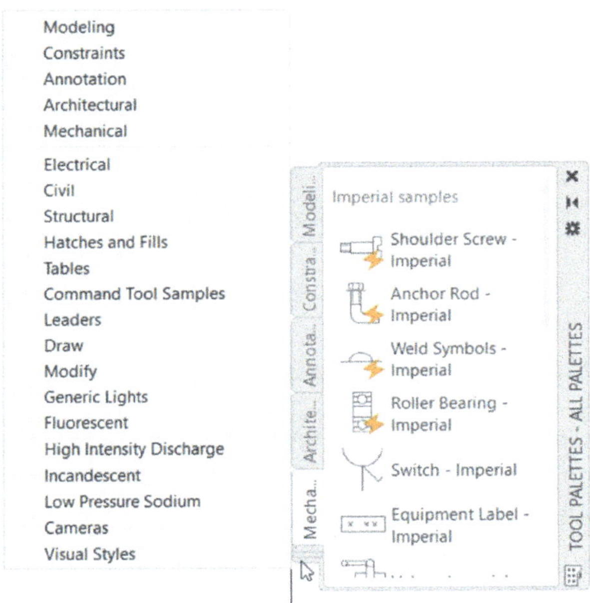

There are many blocks available in the **Architectural**, **Mechanical**, **Electrical**, **Civil**, and **Structural** palettes. You can drag and place blocks from these palettes. You can also right-click on a block and perform various operations using the shortcut menu displayed, as shown in the figure.

Creating a New Tool Palette

- Right-click on the Tool Palette and select **New Palette** from the shortcut menu; a new palette is added to Tool Palettes.
- Enter **Electronic Symbols** as the name.

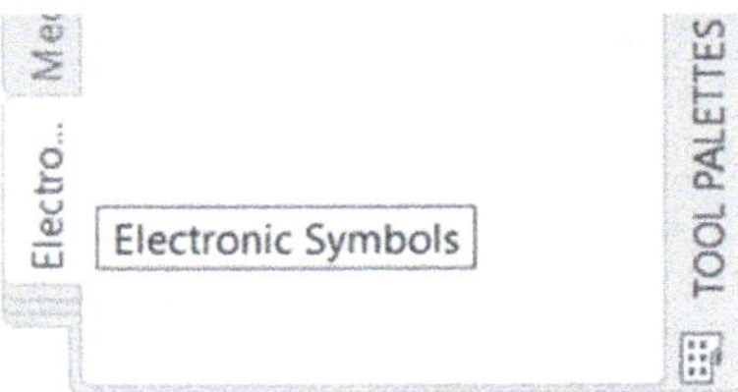

You can also create a new tool palette using the **Customize** dialog.

- Right-click on Tool Palettes and select **Customize Palettes**; the **Customize** dialog appears.
- In the **Customize** dialog, right-click in the **Palettes** list and select **New Palette**.

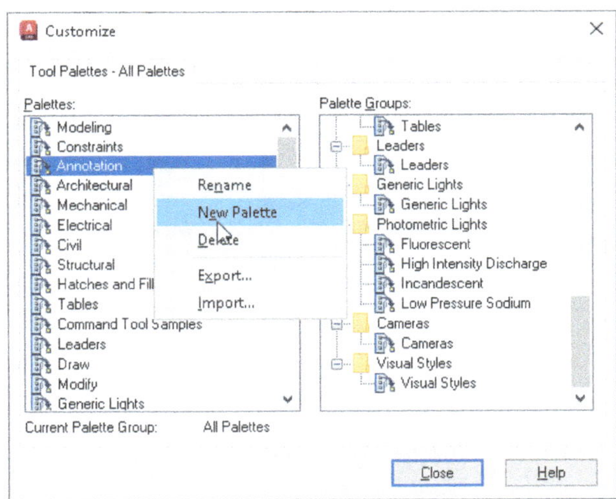

- Enter the name of the palette and click the **Close** button.

Adding Blocks to a Tool Palette

- Open the **DesignCenter** palette and select the **Electronic Symbols.dwg** file from the **Folders** list; the blocks available in the selected file appear.

- Drag the blocks from the **DesignCenter** and place them in the Tool Palette.

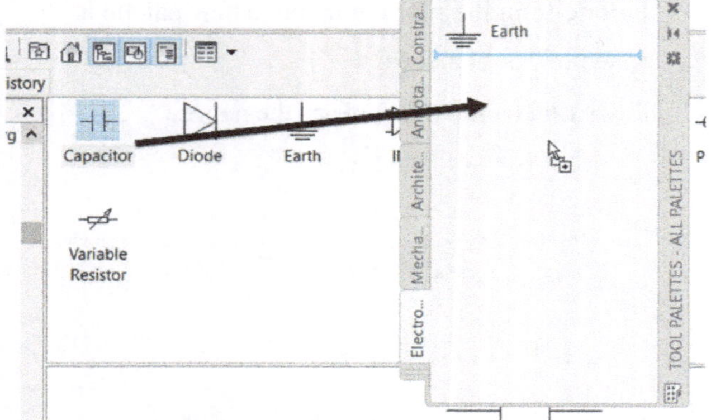

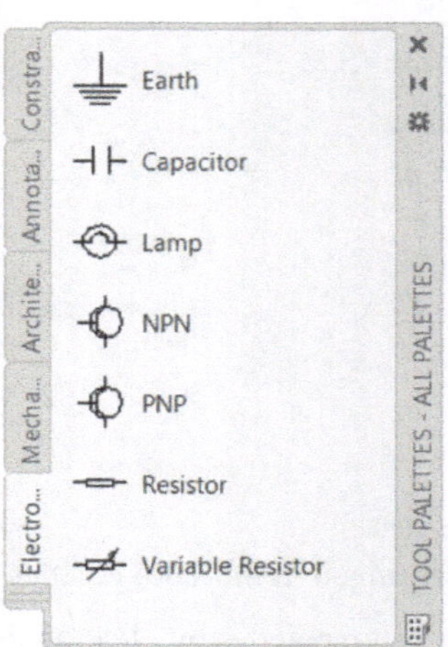

In the Tool Palette, you can group blocks depending on their function.

- Right-click on the **Tool Palette** and select **Add Separator**; a separator will be added.
- Right-click and select **Add Text**. Enter the name of the group.

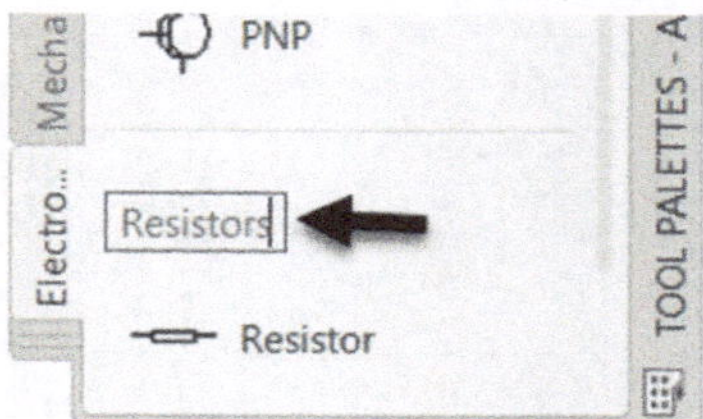

You can also create a new tool palette from a drawing consisting of blocks.

- In the **DesignCenter** palette, select the **Electronic symbols.dwg** file from **Folder List**.
- Select the **Blocks** option.
- Right-click and select **Create Tool Palette**; a new palette will be created from the drawing file.

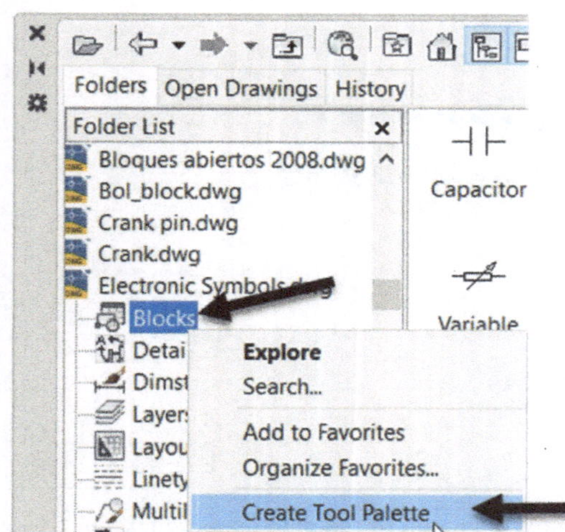

Inserting Multiple Blocks

You can insert multiple instances of a block at a time by using the **MINSERT** command. This command is similar to the **ARRAY** command. The following example explains the procedure to insert multiple blocks at a time.

Example:

- Create two blocks, as shown below.

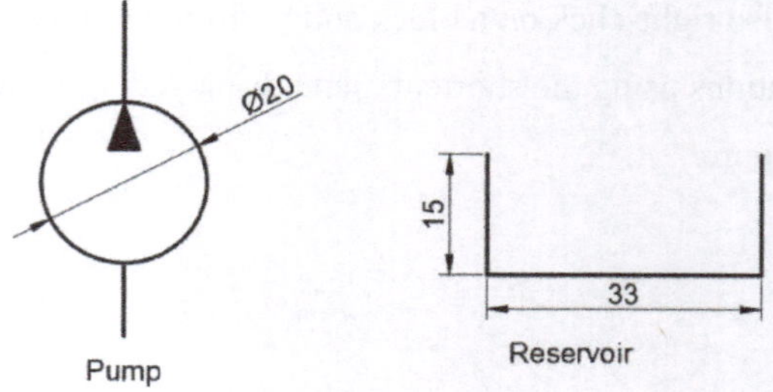

- Type **MINSERT** in the command line and press ENTER; the message, "Enter block name or [?]:" appears.
- Type **Pump** and press ENTER; the Pump is attached to the pointer.
- Pick a point in the graphics window.
- Enter 1 as the scale factor.
- Enter 0 as the rotation angle; the message, "Enter number of rows (---) <1>:" appears.
- Enter 1 as the row value; the message, "Enter number of columns (|||) <1>:" appears.
- Enter 4 as the column value; the message, "Specify distance between columns (|||):" appears.
- Type 60 and press ENTER; the pumps will be inserted as shown below.

- Likewise, insert the reservoirs and create lines, as shown below.

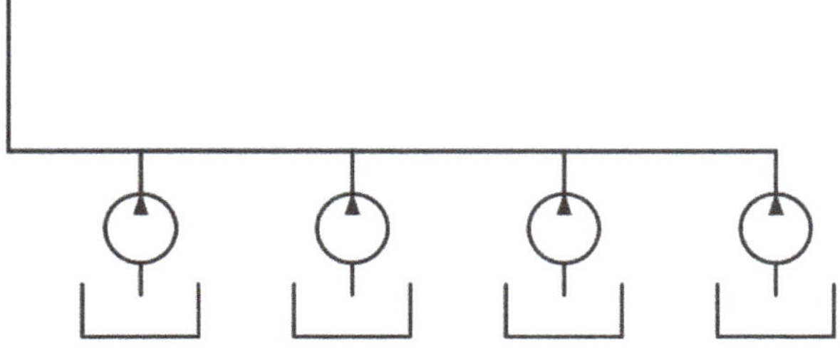

Editing Blocks

During the design process, you may need to edit blocks. You can easily edit a block using the **Block Editor** window. As you edit a block, all the instances of it will be automatically updated. The procedure to edit a block is discussed next.

- Click **Insert > Block Definition > Block Editor** on the ribbon; the **Edit Block Definition** dialog appears.

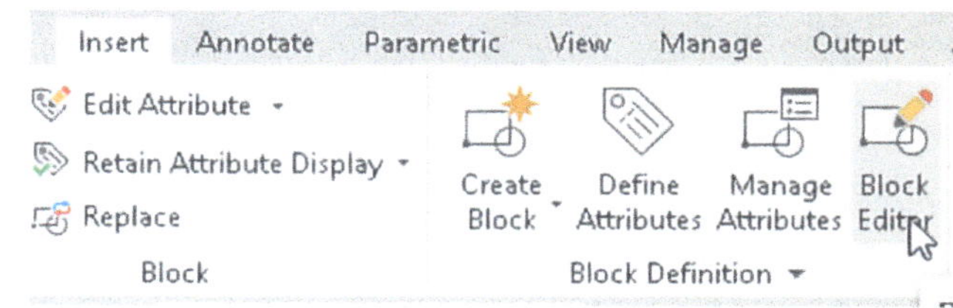

- In the **Edit Block Definition** dialog, select **Pump** from the list and click **OK**; the **Block Editor** window appears.

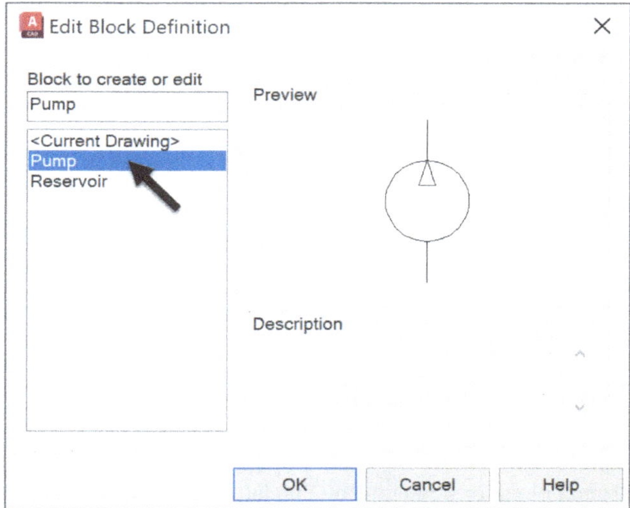

- Click **Home > Draw > Polyline** on the ribbon and draw a polyline, as shown below.

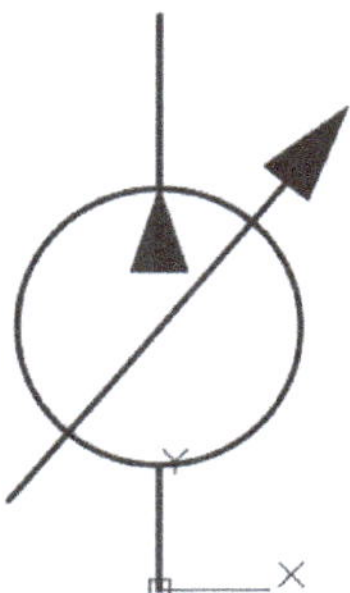

- Click **Close Block Editor** on the **Close** panel.

- In the **Block – Changes Not Saved** dialog, click **Save the changes to Pump**.

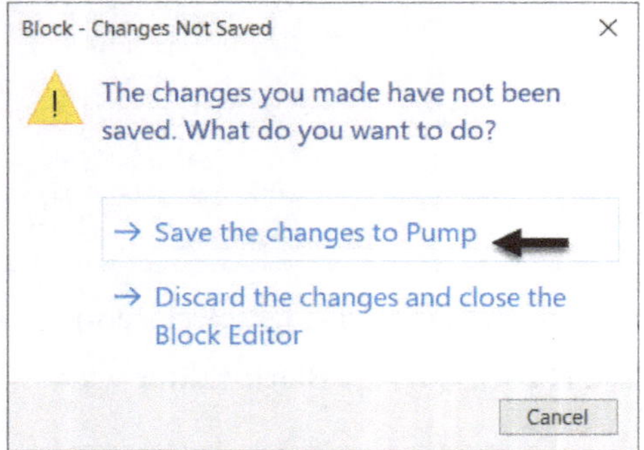

All the instances of the block will be updated automatically.

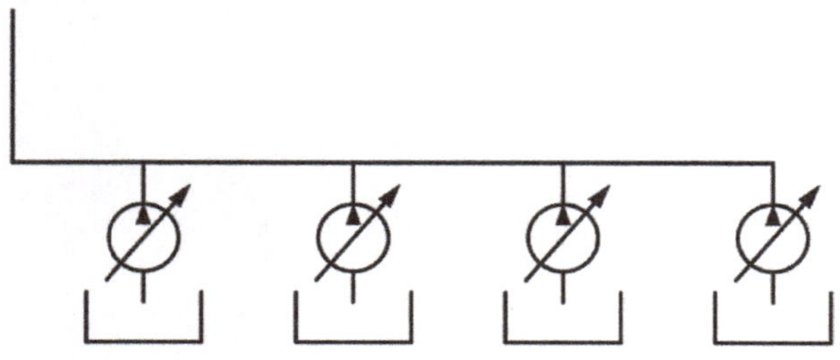

Using the Write Block tool

Using the **Write Block** tool, you can create a drawing file from a block or objects. You can later insert this drawing file as a block into another drawing. The procedure to create a drawing file using blocks is discussed in the following example.

Example:

- Start a new drawing file and create two blocks, as shown below. You can also download them from the companion website.

- Insert the blocks and create the drawing, as shown below.

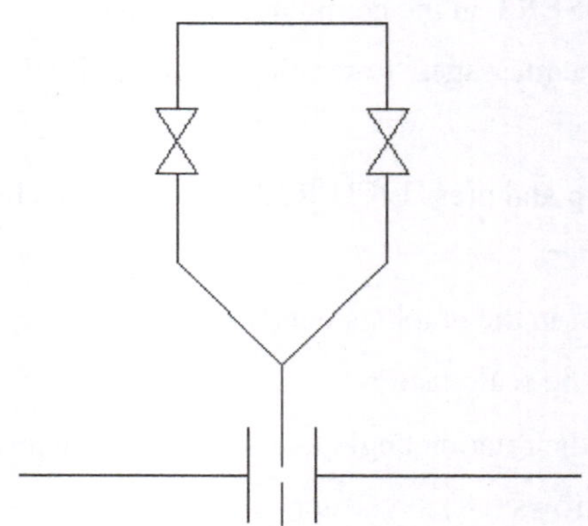

- Expand the **Block Definition** panel and select the **Set Base point** button.

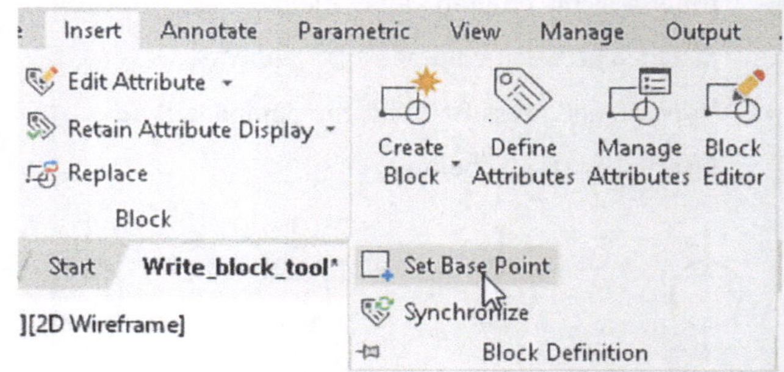

- Select the endpoint of the lower horizontal line, as shown.

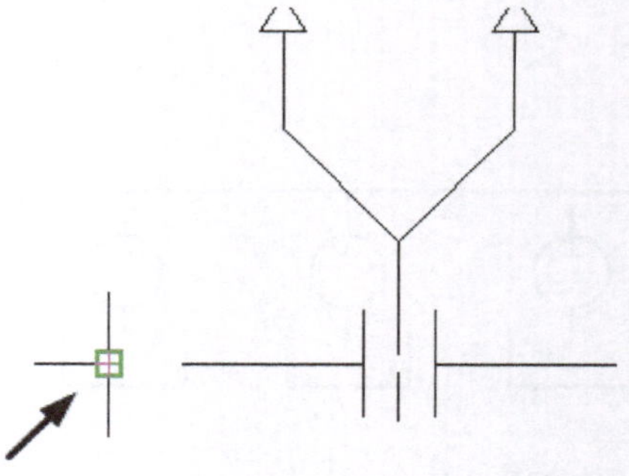

- Click **Insert > Block Definition > Write Block** on the ribbon; the **Write Block** dialog appears.

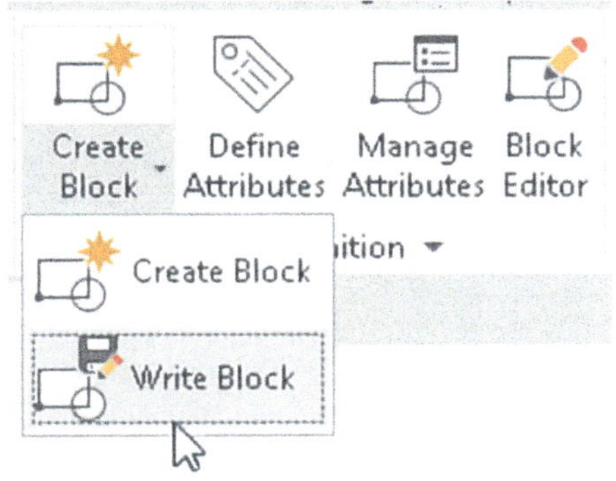

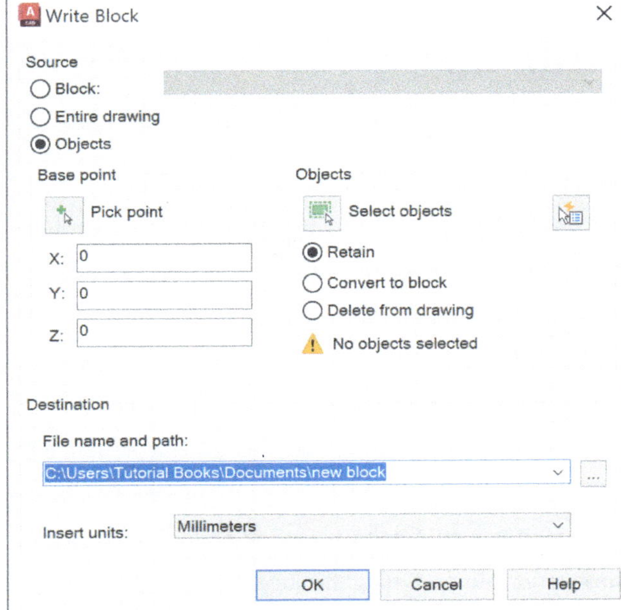

In the **Write Block** dialog, you can select three different types of sources (Block, entire drawing, or objects) to create a block. If you select the **Block** option, you can select blocks present in the drawing from the drop-down.

- Select the **Entire drawing** option.
- Specify the location of the file and name it as **Tap-in line**.

- Click the **OK** button.
- Close the drawing file.
- Open a new drawing file, and then type **I** in the command line and press ENTER; the **BLOCKS** palette appears.
- Click the **Recent** tab and select the **Tap-in Line** file from the **Recent Blocks** section.

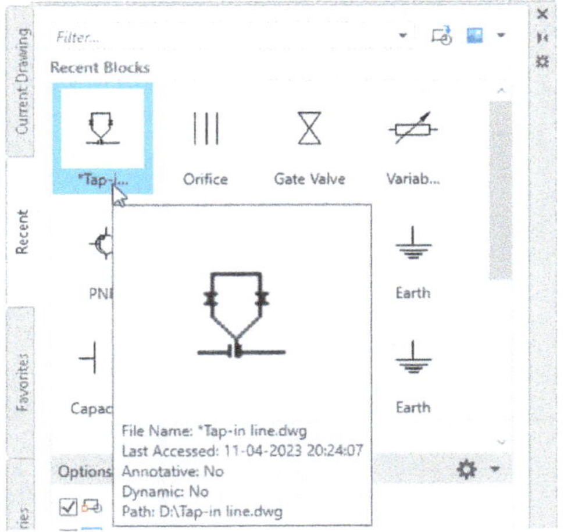

- Pick a point in the graphics window to insert the block.

Defining Attributes

An attribute is a line of text attached to a block. It may contain any type of information related to a block. For example, the following image shows a Compressor symbol with an equipment tag. The procedure to create an attribute is discussed in the following example.

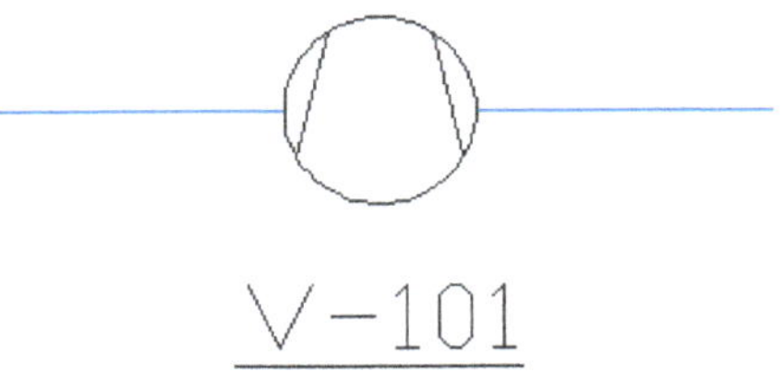

Example 1:

- Open a new drawing file.
- Create the symbols, as shown below.

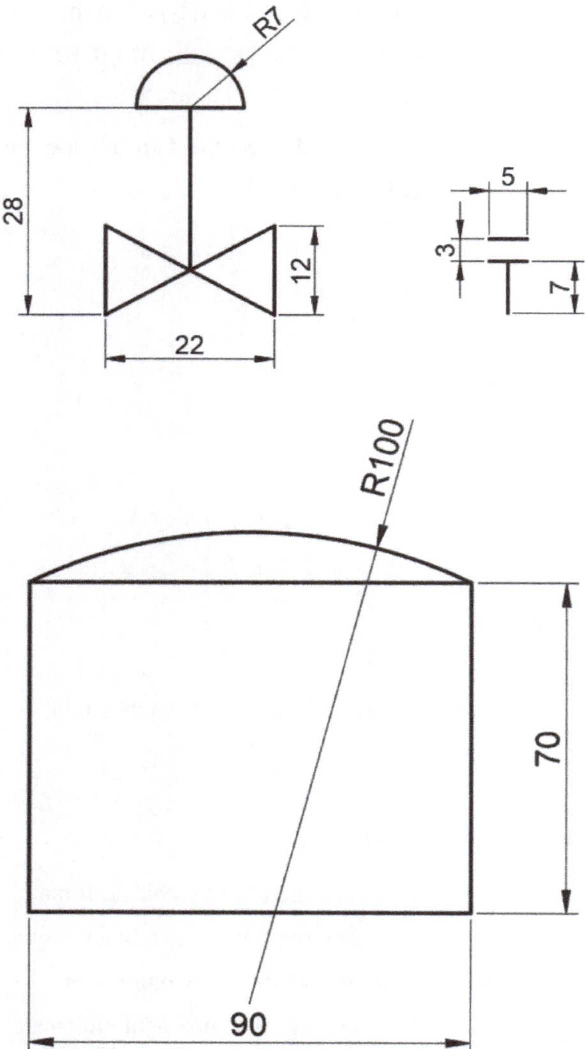

- Click **Insert > Block Definition > Define Attributes** on the ribbon; the **Attribute Definition** dialog appears.

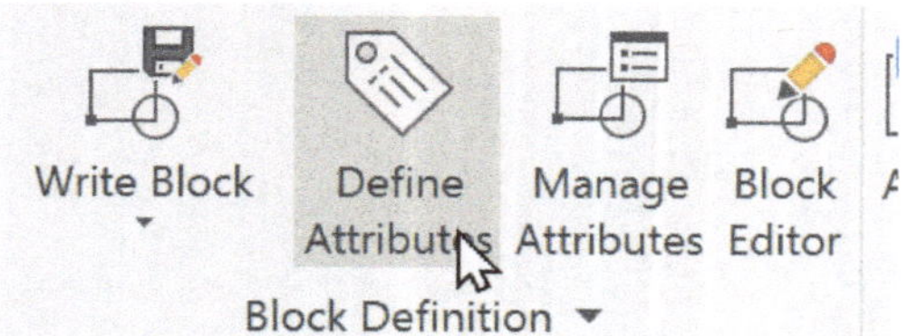

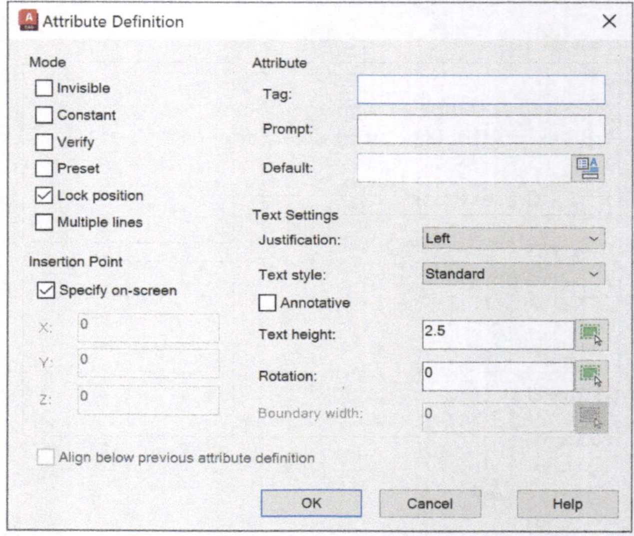

The options in the **Mode** group of the **Attribute Definition** dialog define the display mode of the attribute. If you check the **Invisible** option, the attribute will be invisible. The **Constant** option makes the value of the attribute constant. You cannot change the value. The **Verify** option prompts you to verify after you enter a value. The **Preset** option can be used to set a predefined value for the attribute. The **Lock position** option fixes the position of the attribute to a selected point. The **Multiple lines** option allows typing the attribute value in single or multiple lines.

- Ensure that the **Lock position** option is selected.

The options in the **Attribute** group define the values of the attribute. The **Tag** box is used to enter the label of the attribute. For example, if you want to create an attribute called RESISTANCE, you must type **Resistance** in the **Tag** box. The **Prompt** box defines the prompt message that appears after placing the block. The **Default** box defines the default value of the attribute.

- In the **Attribute Definition** dialog, enter **Valvetag** in the **Tag** box.

The **Text Settings** options define the display properties of the text, such as style, and height. Observe the other options in this dialog. Most of them are self-explanatory.

- Enter **5** in the **Text height** box.
- Set the **Justification** to **Middle** and click **OK.**
- Specify the location of the attribute, as shown below.

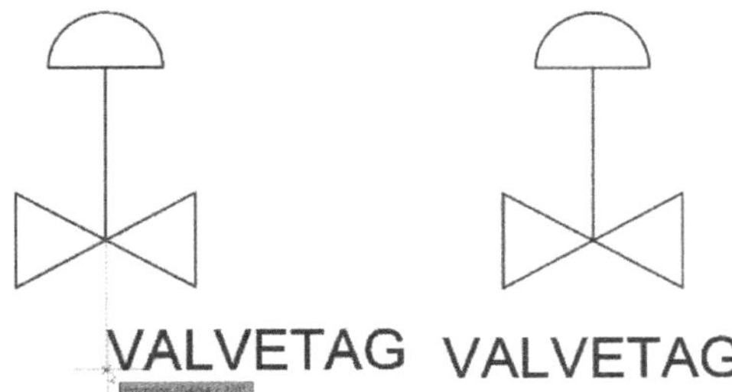

- Click the **Create Block** button on the **Block Definition** panel; the **Block Definition** dialog appears.
- On the dialog, click the **Select objects** button.
- Drag a window and select the control valve symbol and attribute — next, press ENTER.
- Select the **Delete** option from the **Objects** group.
- Click the **Pick Point** button under the **Base point** group and select the point, as shown below.

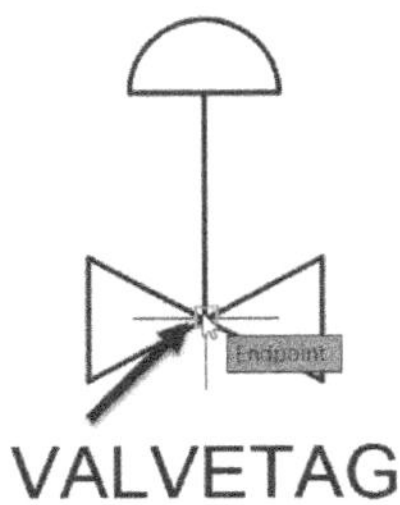

- Enter **Control Valve** in the **Name** box and click **OK**.
- Likewise, create the **Equipmenttag** attribute and place it inside the tank symbol.
- Create a block and name it as **Tank**.

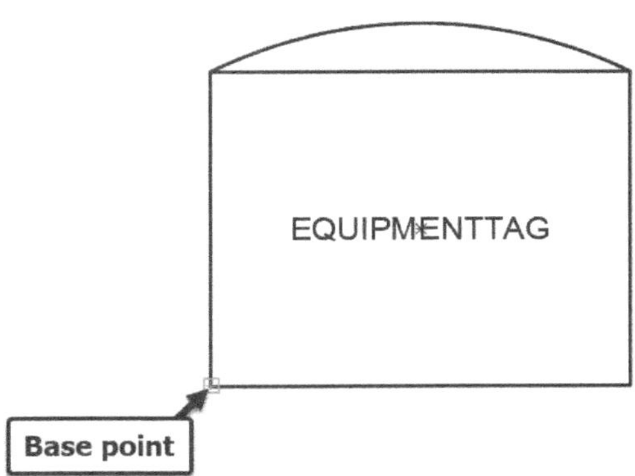

- Also, create a block of the nozzle symbol and name it **Nozzle**.

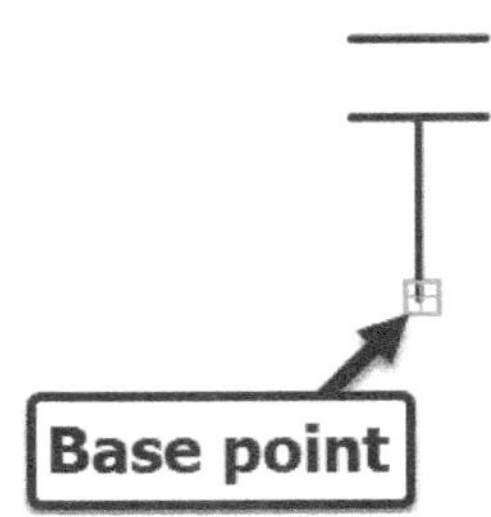

Inserting Attributed Blocks

You can use the INSERT command to insert the attributed blocks into a drawing. The procedure to insert attributed blocks is discussed next.

- On the ribbon, click **Insert > Block > Insert > Tank**.
- Click in the graphics window to define the insertion point. The **Edit Attributes** dialog appears.
- Enter **TK-001** in the **EQUIPMENTTAG** field and click **OK**; the block will be placed along with the attribute.

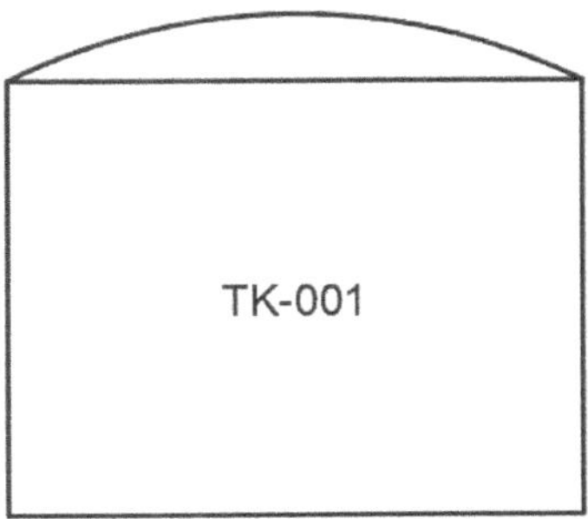

- Likewise, place the control valves, as shown below.

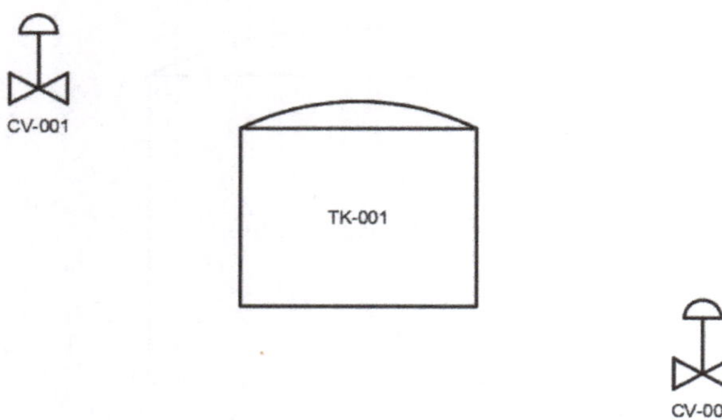

- Place and rotate the nozzles on the tank, as shown below.

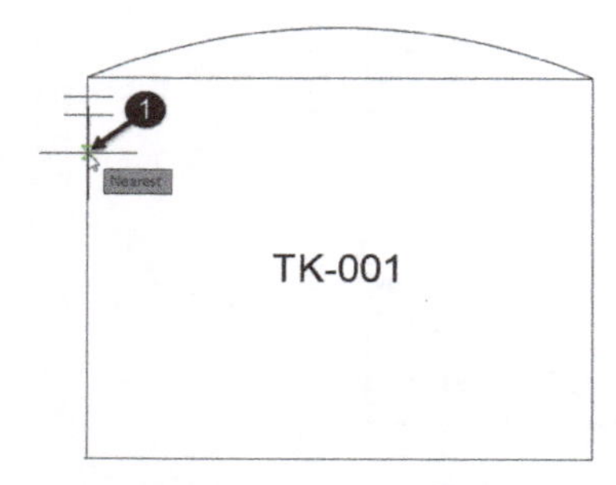

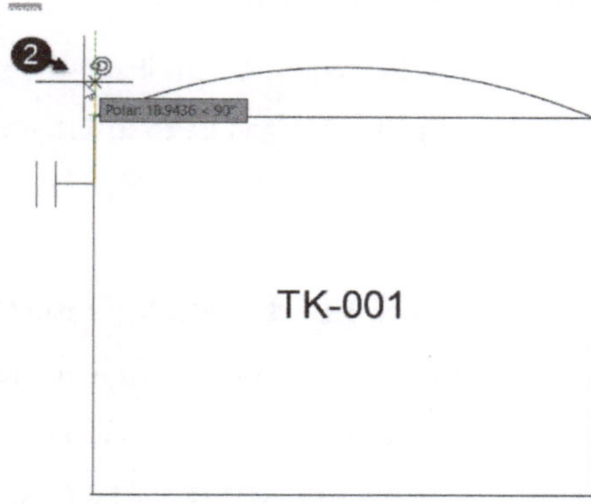

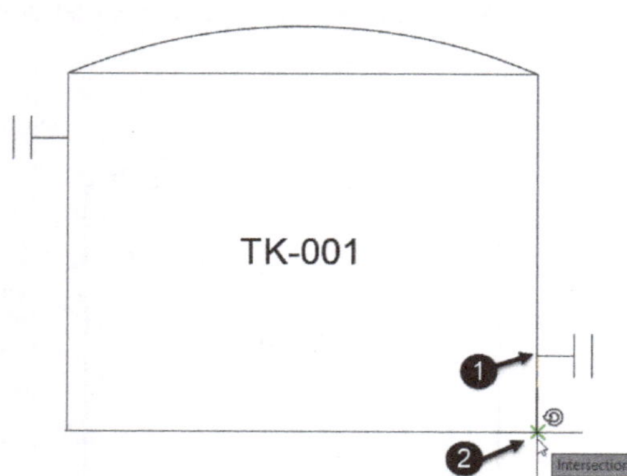

- Use the **Polyline** tool and connect the control valves and tank.

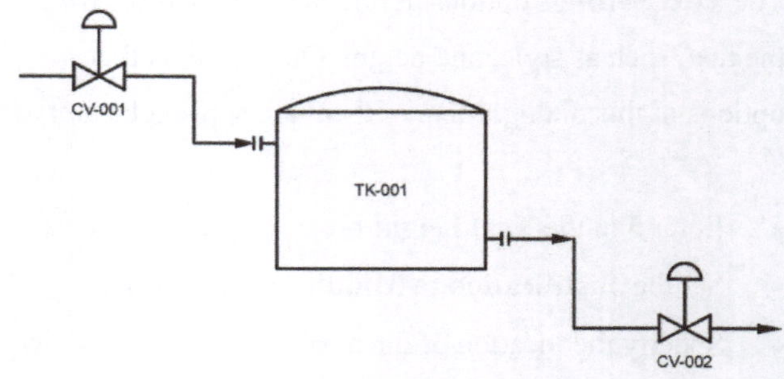

Working with External references

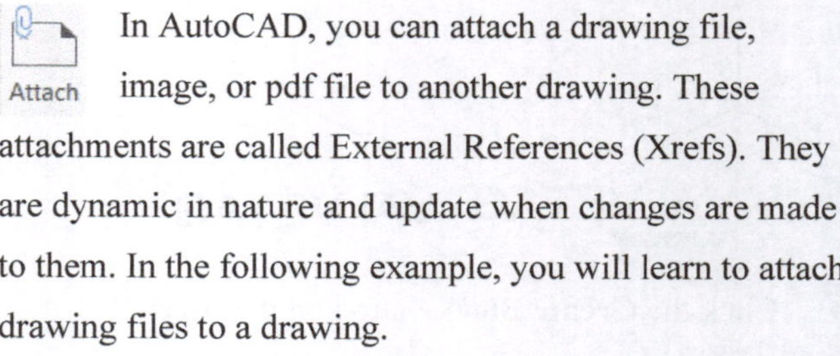

In AutoCAD, you can attach a drawing file, image, or pdf file to another drawing. These attachments are called External References (Xrefs). They are dynamic in nature and update when changes are made to them. In the following example, you will learn to attach drawing files to a drawing.

Example 1:

- Create the drawing shown below.

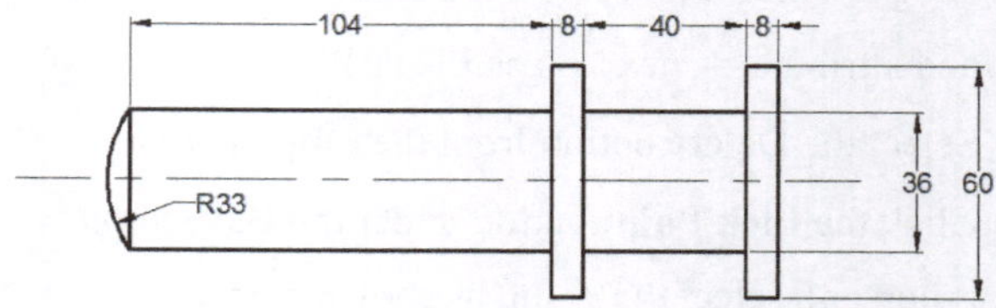

- Type **BASE** in the command line and press ENTER.
- Select the midpoint of the vertical line as the base point, as shown.

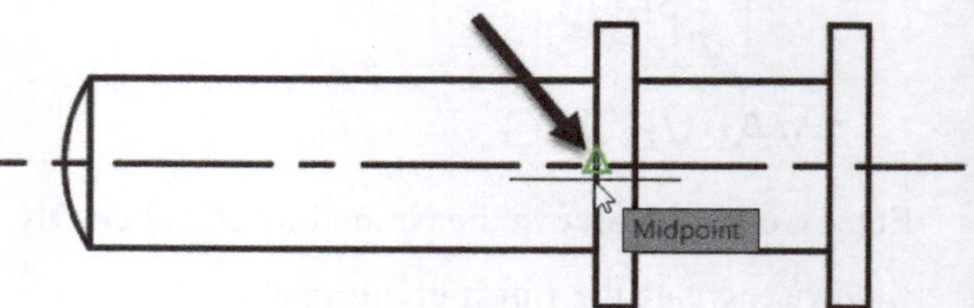

- Save the drawing as **Crank pin.dwg**
- Create another drawing as shown below (For help, refer to the **Multi View Drawings** section in Chapter 5).

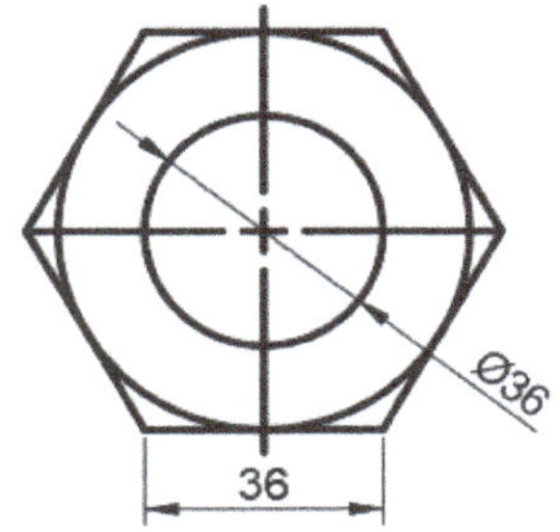

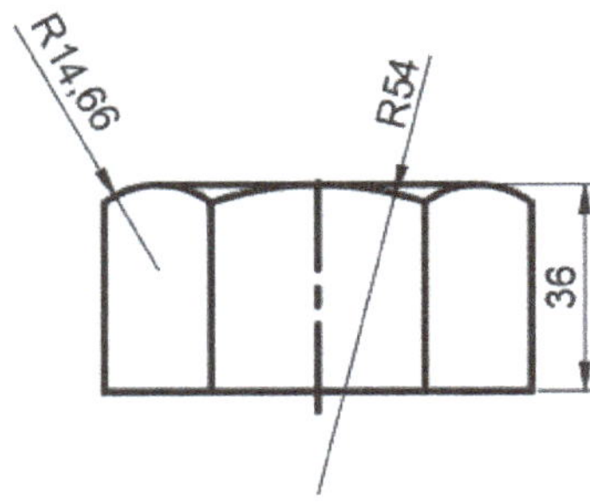

- Use the **Set Base Point** tool and specify the base point, as shown below.

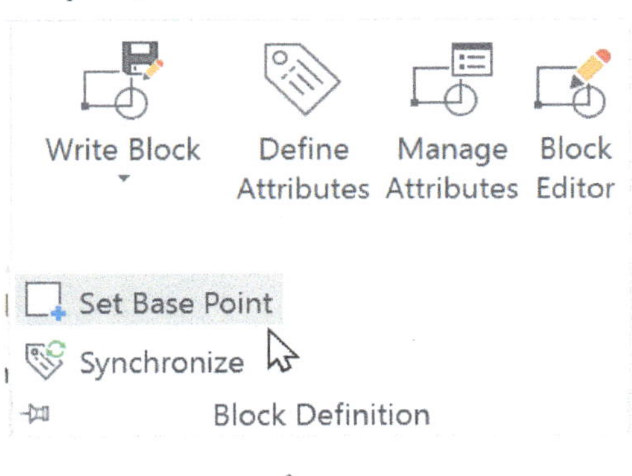

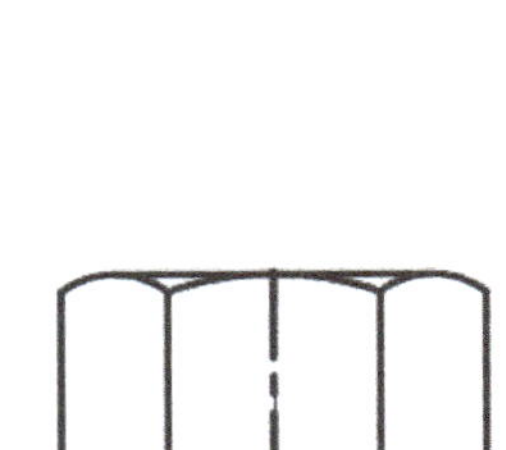

- Save the drawing as **Nut.dwg** and close it.
- Open the **Crank.dwg** file created in Chapter 8.
- Click **Insert > Reference > Attach** on the ribbon; the **Select Reference file** dialog appears.

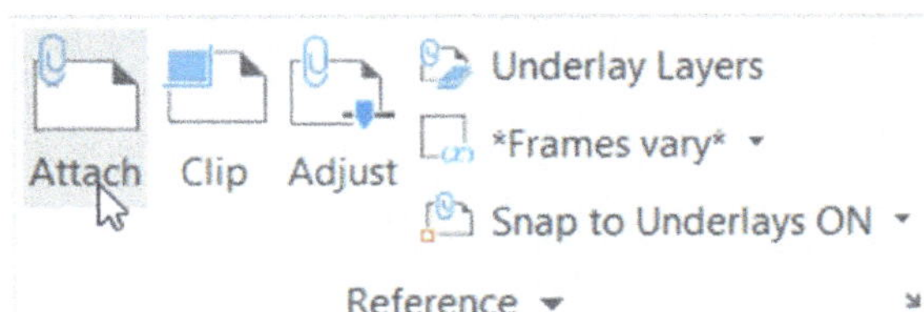

- Browse to the location of the **Crankpin.dwg** and double-click on it; the **Attach External Reference** dialog appears.

Some of the options available in this dialog are similar to that in the **Insert** dialog. The options include the insertion point, scale, and rotation angle of the external reference.

- Accept the default settings in this dialog and click **OK**; the crankpin will be attached to the pointer.
- Select the point on the hatched view, as shown below.

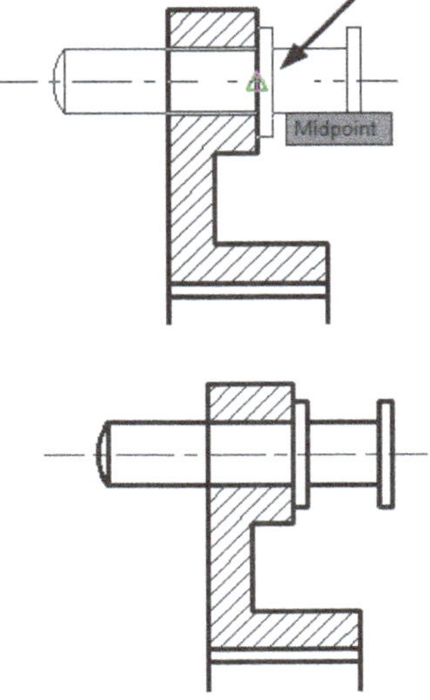

- On the ribbon, click **View** tab > **Palettes** panel > **External References Palette**.

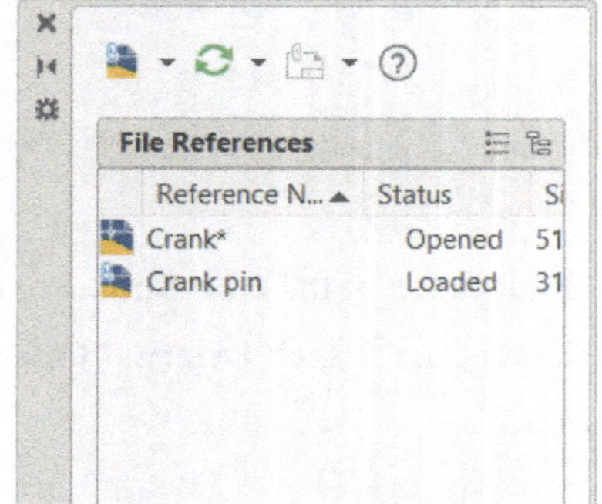

- In the **External References** palette, open the **Attach** drop-down and select the **Attach DWG** option; the **Select References file** dialog appears.

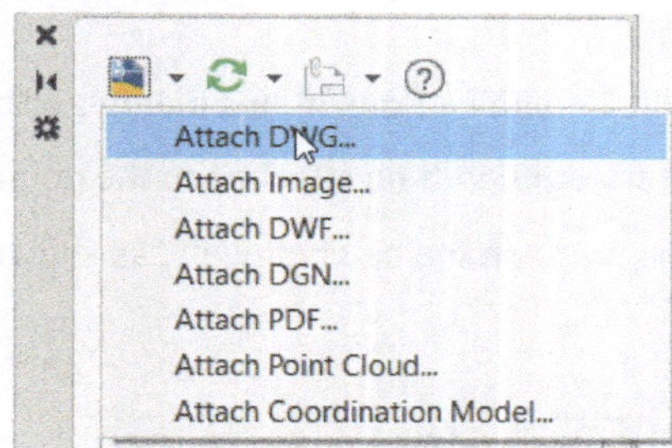

- Browse to the location of the **Nut.dwg** and double-click on it; the **Attach External Reference** dialog appears.
- In the **Attach External Reference** dialog, enter **90** in the **Angle** box under the **Rotation** group and click **OK**.
- Select the insertion point on the section view, as shown below.

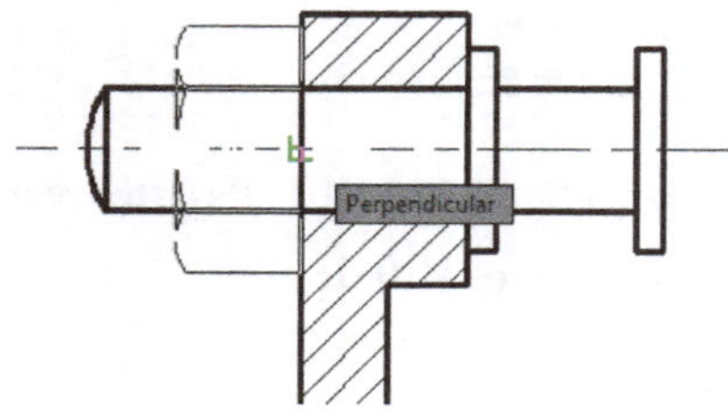

Fading an Xref

You can change the fading of Xref by using the Xref fading slider available in the expanded **Reference** panel.

- Expand the **Reference** panel of the **Insert** ribbon and use the **Xref fading** slider to adjust the fading.

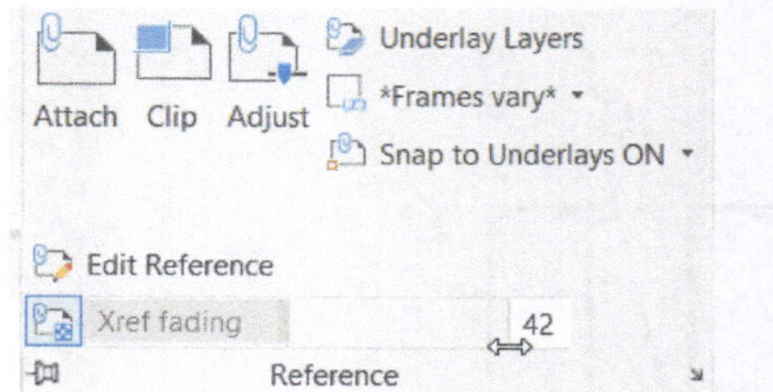

Clipping External References

You can hide the unwanted portion of an external reference by using the **Clip** tool.

- Click **Insert > Reference > Clip** on the ribbon; the message, "Select Object to clip," appears in the command line.

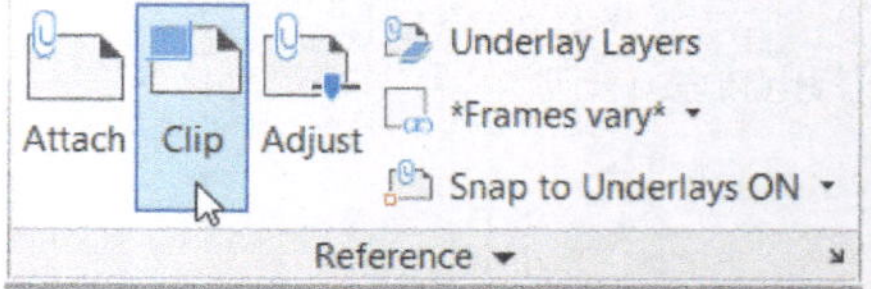

- Select the **Nut.dwg** from the graphics window.
- Select the **New boundary** option from the command line.
- Select the **Rectangular** option from the command line.
- Draw a rectangle as shown below; only the front view of the nut is visible, and the top view is hidden. Also, the clipping frame is visible.

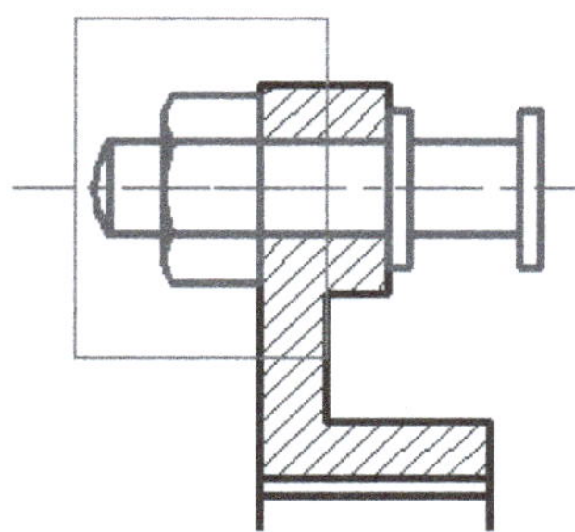

- To hide the clipping frame, type XCLIPFRAME in the command line
- Type 0 and press ENTER.
- You can also hide the frame by clicking **Modify > Object > External reference > Frame** on the Menu Bar.

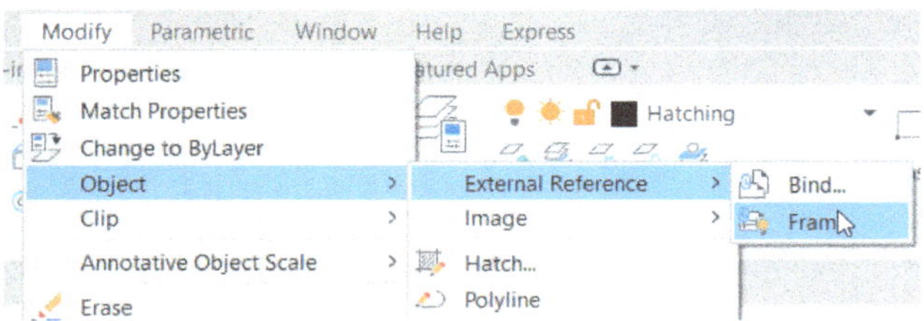

- Attach another instance of the **Nut.dwg** file.
- Use the **Rotate** and **Move** tools to position the top view, as shown below.

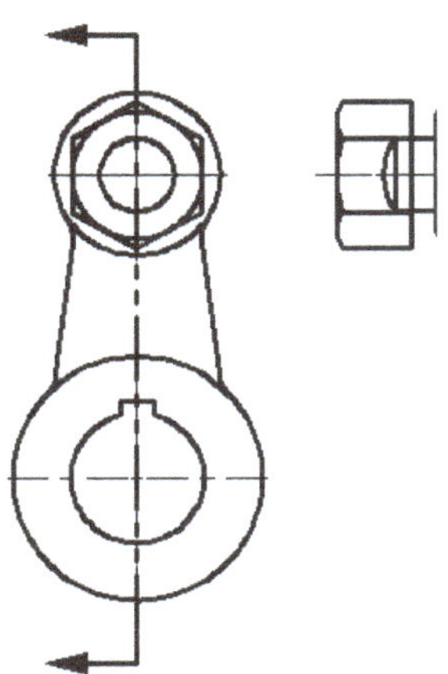

- Use the **Clip** tool and clip the Xref.

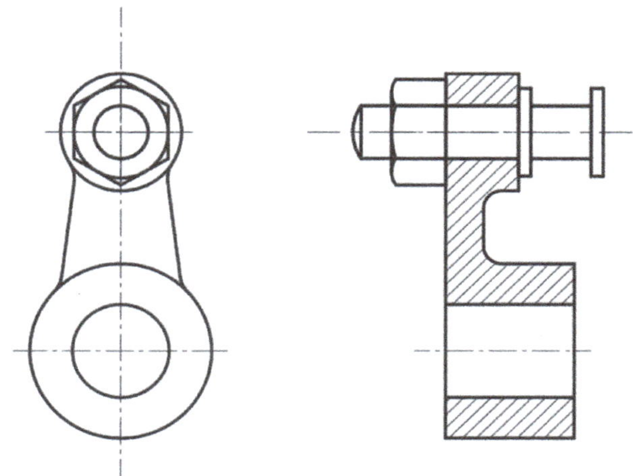

Editing the External References

AutoCAD allows you to edit the external references in the file to which they are attached. You can also edit them by opening their drawing file. The procedure to edit an external reference is discussed next.

- To edit an external reference, expand the **Reference** panel and click the **Edit Reference** button.

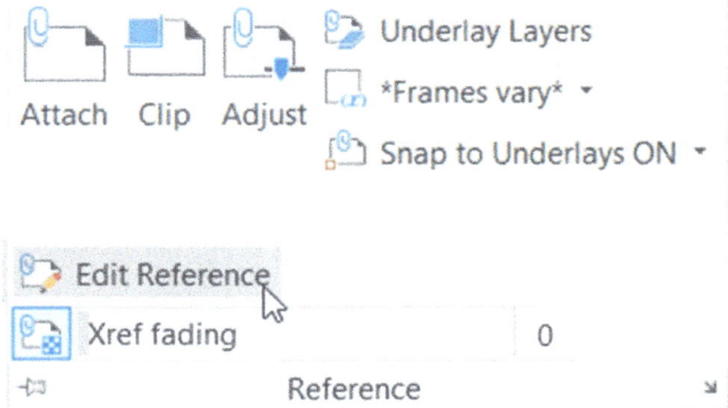

- Select **Nut** from the drawing; the **Reference Edit** dialog appears
- Click **OK** to get into the reference editing mode.
- In the drawing, the centerlines of the nut are overlapping on the centerlines of the crank. Delete the centerlines and center marks of the nut.
- Click **Save Changes** on the **Edit Reference** panel of the **Insert** tab; the **AutoCAD** message box appears.

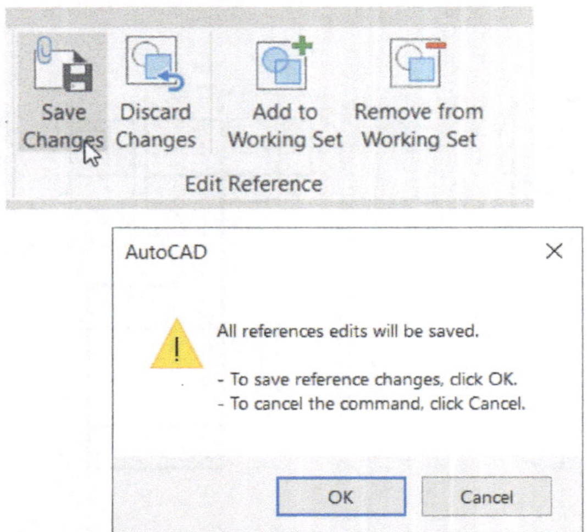

- Click **OK**.

Adding Balloons

- Click **Annotate > Leaders > Multileader Style Manager** (inclined arrow) button on the ribbon; the **Multileader Style Manager** dialog appears.

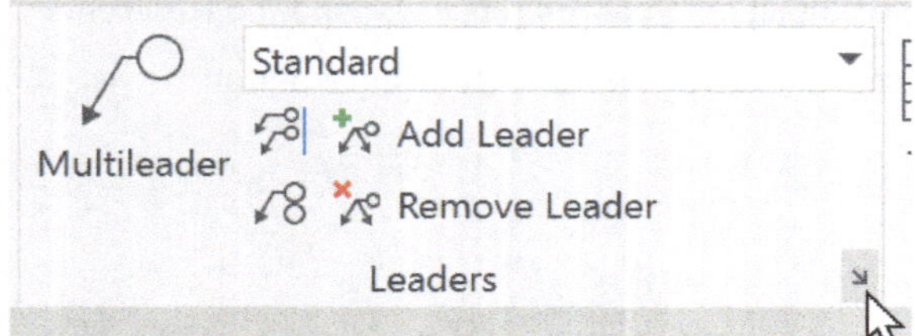

- Click the **New** button on the dialog.
- In the **Create New Multileader Style** dialog, enter **Balloon Callout** in the **New Style name** box and click **Continue**.

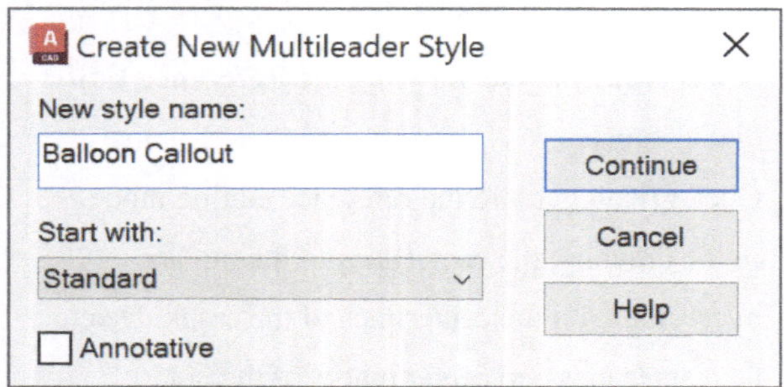

- In the **Modify Multileader Style** dialog, click the **Content** tab, and set the **Multileader type** to **Block**.
- Under **Block Options**, set the **Source block** to **Circle**.
- Set the **Scale** to **3**.

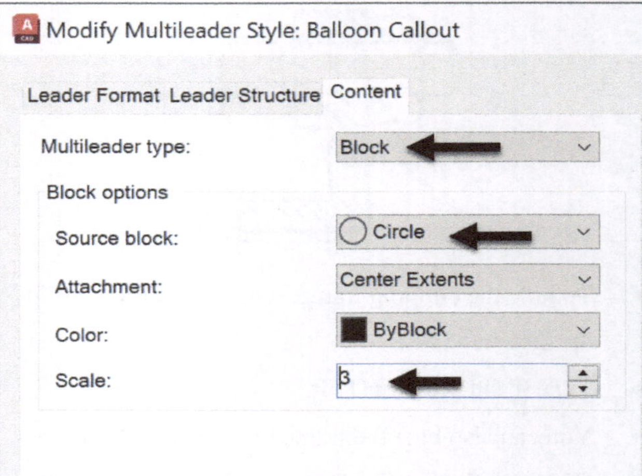

- Click the **Leader Format** tab and set the Arrowhead **Size** to 8.

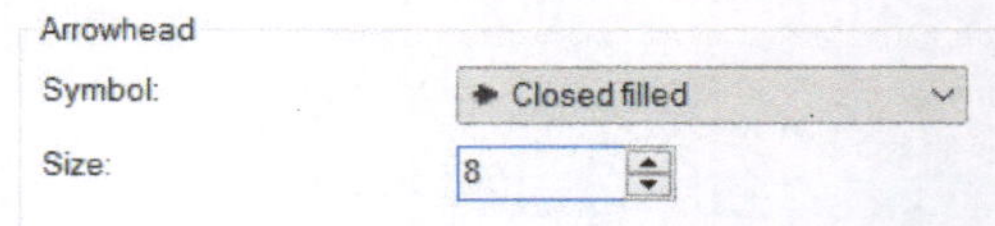

- Click **OK** and set the **Balloon Callout** style as current.
- Click **Close**.
- Click **Annotate > Leaders > Multileader** on the ribbon.
- Click the down arrow next to the **Polar Tracking** icon on the status bar and select **45** from the menu. Activate the **Polar Tracking**.
- Select a point on the section view of the crank.
- Move the pointer along the polar trace lines and click; the **Edit Attributes** dialog appears.

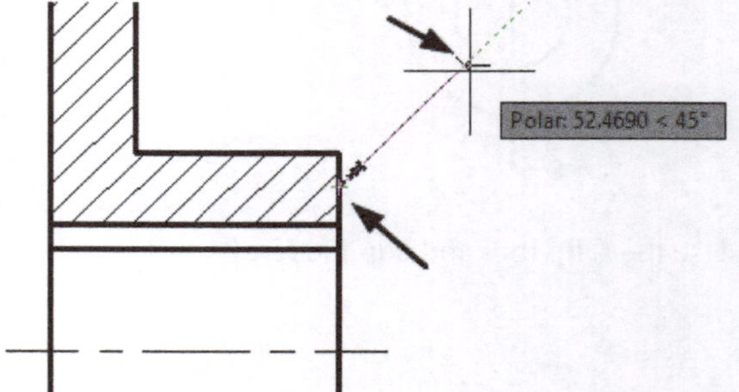

- In the **Edit Attributes** dialog, enter 1 in the **Enter tag number** field.

- Click **OK**; the balloon will be created.

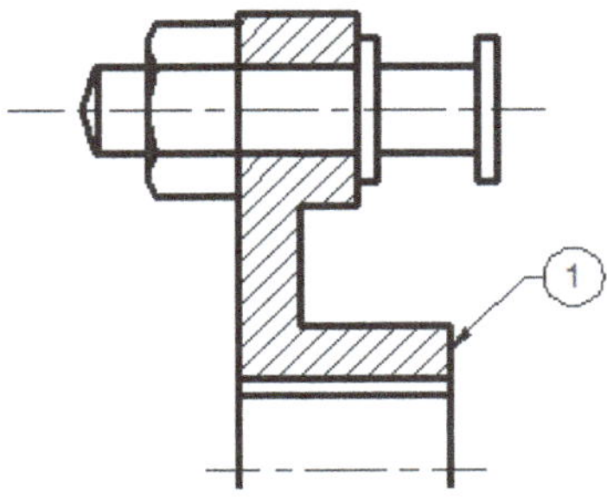

- Likewise, create other balloons.

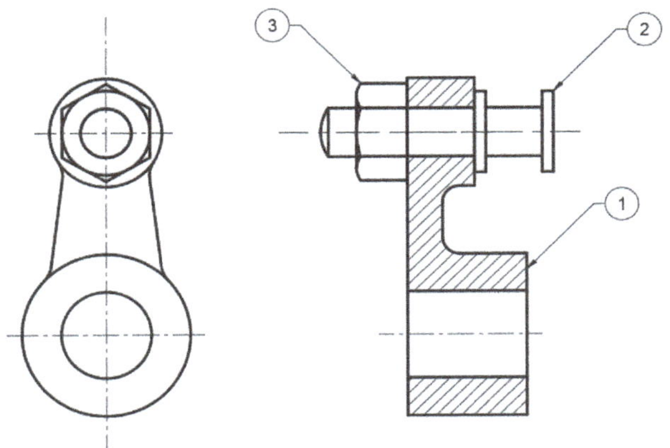

Creating the Part List

- Click **Annotate > Tables > Table Style** on the ribbon; the **Table Style** dialog appears.

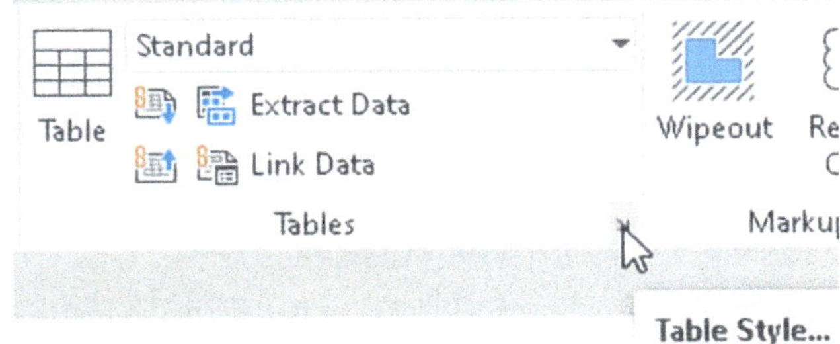

- In the **Table Style** dialog, click the **New** button; the **Create New Table Style** dialog appears,
- In the **Create New Table Style** dialog, enter **Part List** in the **Name** box. Click **Continue**; the **New Table Style** dialog appears.
- Click the **Text** tab and set the **Text height** to **10**.

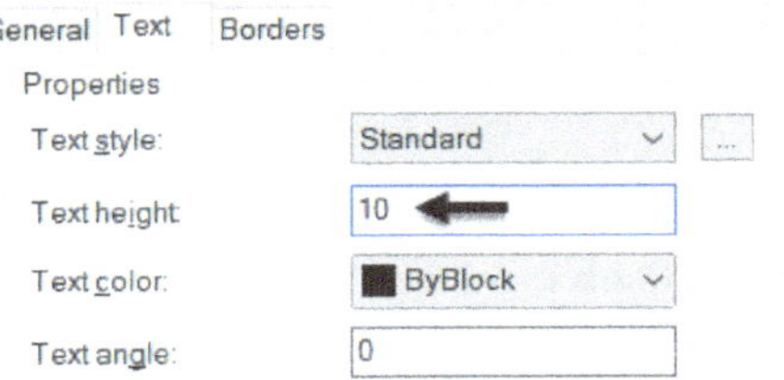

- Select the **Header** option from the **Cell Styles** drop-down and set the **Text height** to **10**.

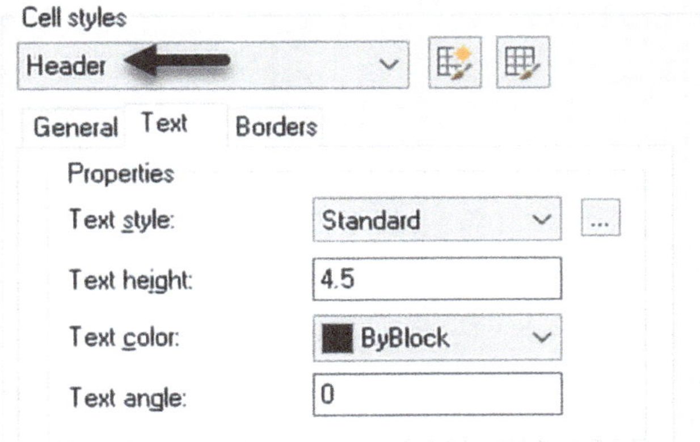

- Click **OK**.
- In the **Table Style** dialog, select the **Part List** style and click **Set current**.
- Close the dialog.
- Click the **Table** button on the **Tables** panel; the **Insert table** dialog appears.

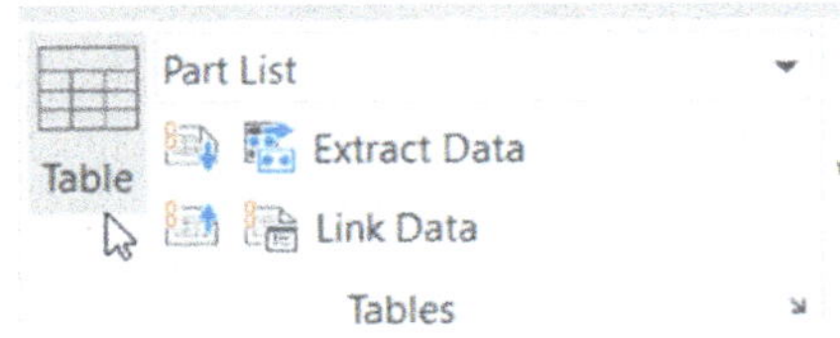

- Ensure that the **Table style** is set to **Part List**.

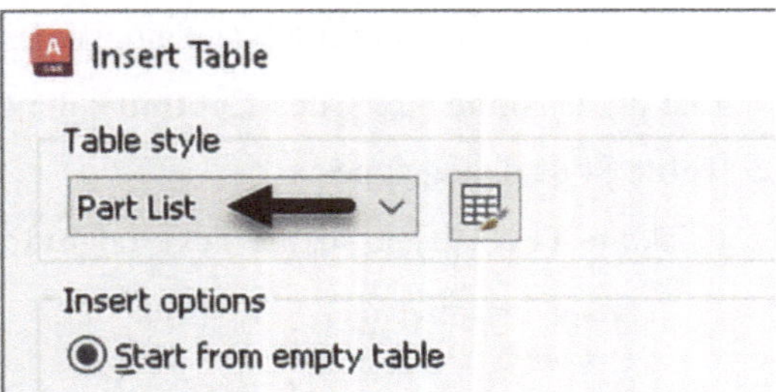

- Under the **Set cell Styles** group, set the **First row cell style** to **Header**.
- Set the **Second row cell style** and **All other row cell styles** to **Data**.

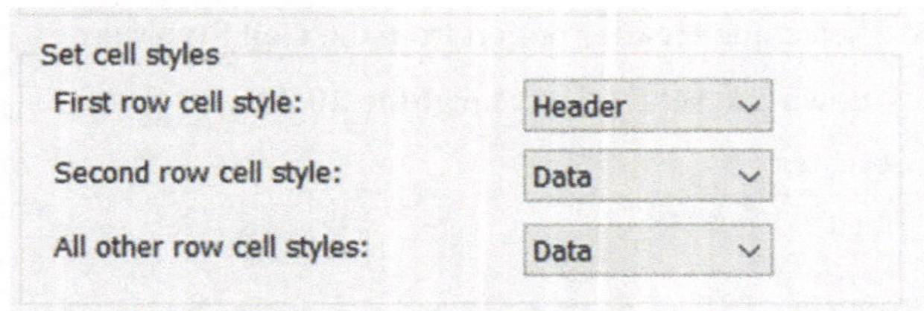

- Set the number of **Columns** to **4** and **Column width** to **65**.
- Set the **Data rows** and **Row height** to 1 and 1, respectively.
- Click **OK** and place the table at the lower right corner of the graphics window.
- Enter **PART No.**, **NAME**, **MATERIAL**, **QTY** in the first row of the **Part list** table. Use the **TAB** key to navigate between the cells.

Part No.	Name	Material	Qty

- Click the **Close Text Editor** button on the ribbon.

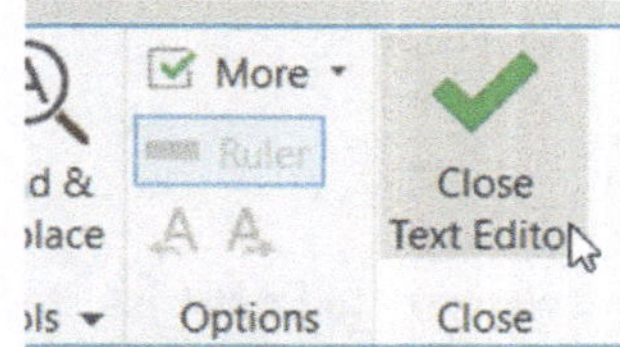

- Click on any one of the edges of the table; you will notice that grips are displayed on it. You can edit the table using these grips.

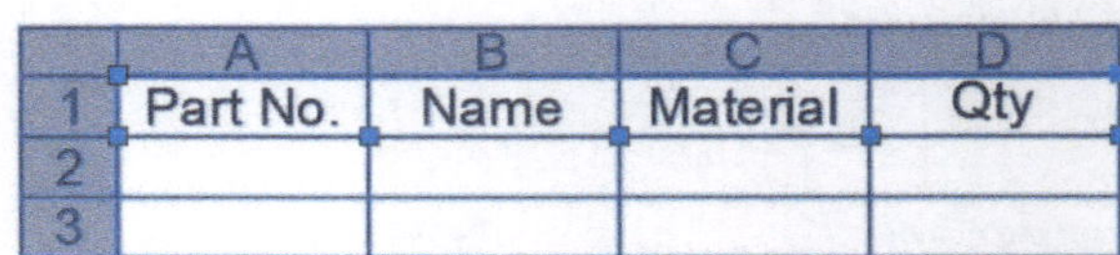

- Click and drag the square grip below the MATERIAL cell; the width of the cell will be changed.

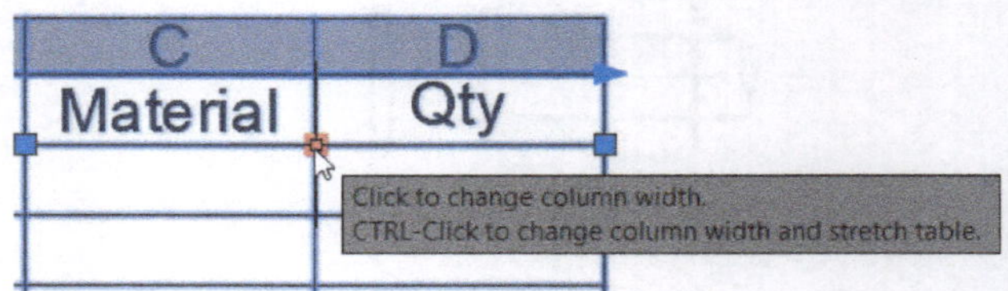

- Click and drag the triangular grip located at the bottom left corner of the table; the height of the rows will be increased uniformly.

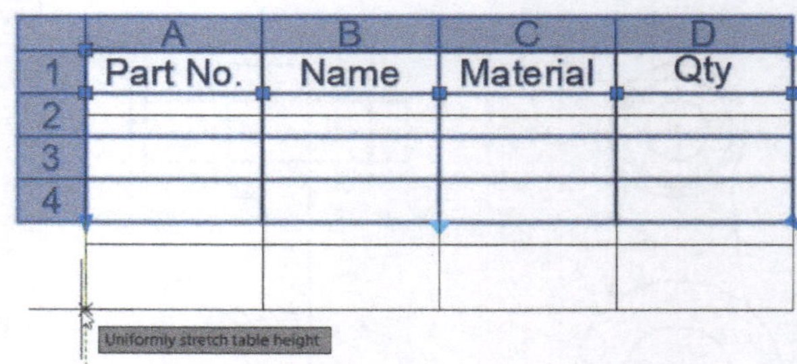

- Click on the second cell of the first column; the **Table Cell** ribbon appears. You can use this ribbon to modify the properties of the table cell.
- Click the **Insert Below** button on the **Rows** panel; a new row will be added to the cell.

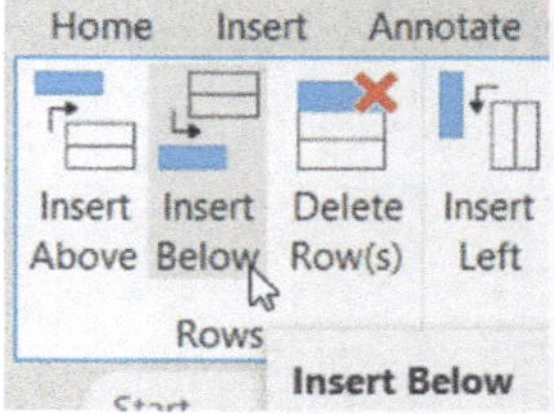

- Click in the top left corner cell of the table.
- Press and hold the SHIFT key and click in the lower right corner of the table; all the cells in the table will be selected.

	A	B	C	D
1	PART No.	NAME	MATERIAL	QTY
2				
3				
4				

- In the **Table Cell** ribbon, click **Cell Styles > Alignment** drop-down **> Middle Center**; the data in all the cells will appear in the middle center of the cells.

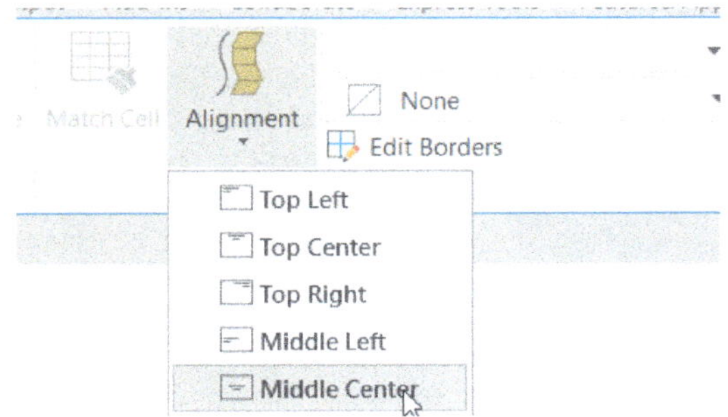

- Double-click in the cell below the **PART No.**; the text editor will be activated.
- Enter the following data in the cells. Use the TAB key to navigate between the cells.

PART No.	NAME	MATERIAL	QTY
1	Crank	Forged Steel	1
2	Crank pin	45C	1
3	Nut	MS	1

Exercise

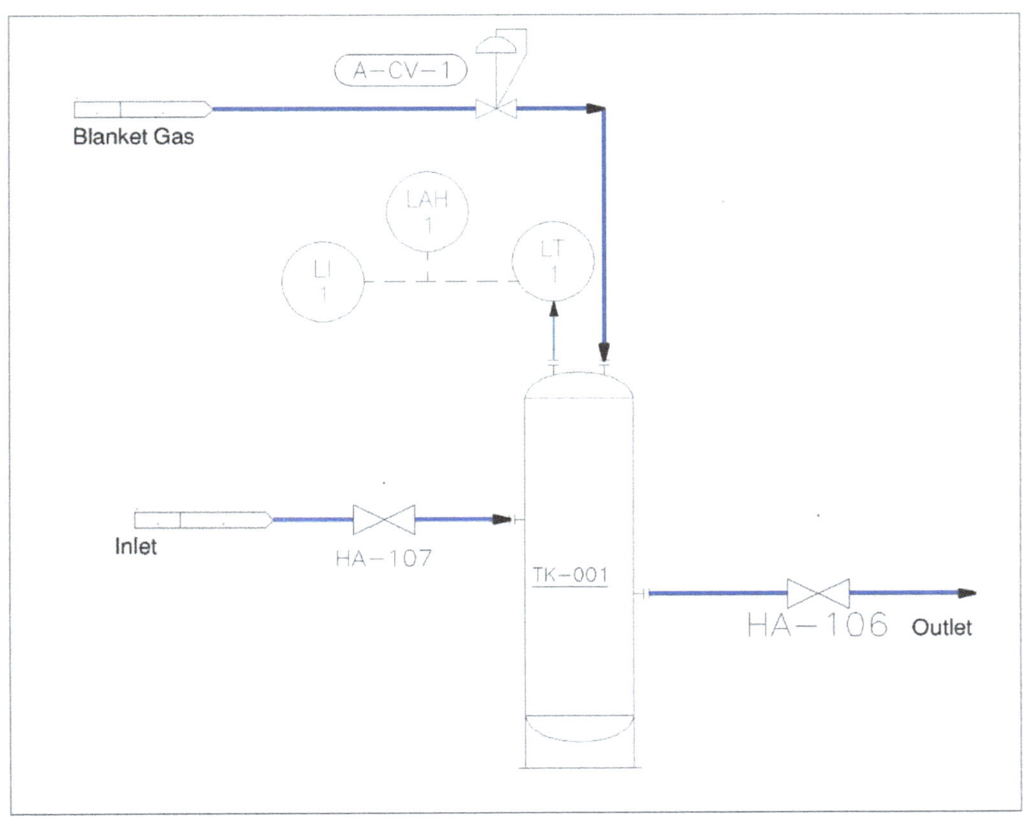

Chapter 10: Layouts & Annotative Objects

In this chapter, you will learn to do the following:

- **Create Layouts**
- **Specify the Paper space settings**
- **Create Viewports in Paper space**
- **Change Layer properties in Viewports**
- **Create a Title Block on the layout**
- **Use Annotative objects in Viewports**

Drawing Layouts

There are two workspaces in AutoCAD: The Model space and the Paper space. In the Model space, you create 2D drawings and 3D models. You can even plot drawings from the model space. However, it is difficult to plot drawings at a scale or if a drawing consists of multiple views arranged at different scales. For this purpose, we use Layouts or paper space. In Layouts or paper space, you can work on notes and annotations and perform the plotting or publishing operations. In Layouts, you can arrange a single view or multiple views of a drawing or multiple drawing by using Viewports. These viewports display drawings at specific scales on layouts. They are mainly rectangular in shape, but you can also create circular and polygonal viewports. In this chapter, you will learn about viewports and various annotative objects.

Working with Layouts

Layouts represent the conventional drawing sheet. They are created to plot a drawing on a paper or in electronic form. A drawing can have multiple layouts to print in different sheet formats. By default, there are two layouts available: Layout 1 and Layout 2. You can also create new layouts by clicking the plus (+) symbol next to the layout. Next, select **New layout** from the shortcut menu. In the following example, you will create two layouts, one representing the ISO A1 (841 X 594) sheet and another representing the ISO A4 (210 X 297) sheet.

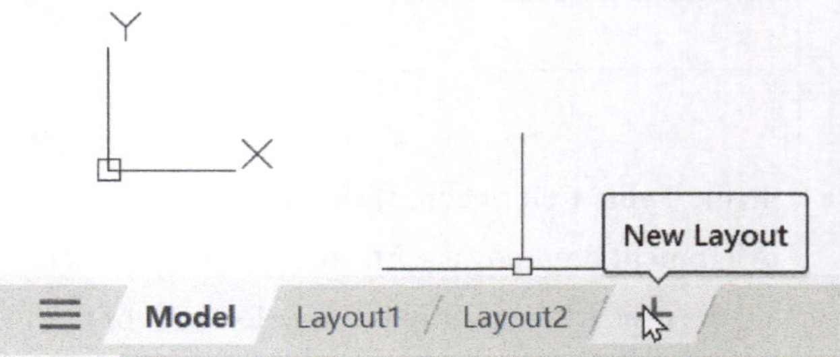

You can use the **Layout tab menu** located on the left side of the **Model** tab to create a new layout, define the page settings of the model tab or layouts, select all layouts, publish the selected layouts, and dock the model and layout tabs above the status bar or inline with it.

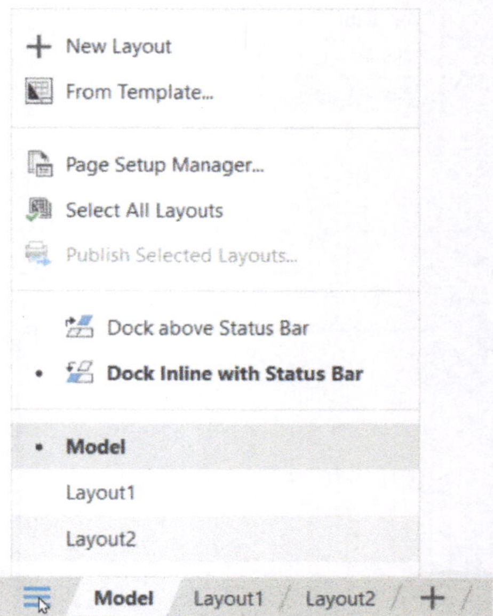

Example:

- Open a new drawing file.
- Create layers with the following settings:

Layer	Linetype	Lineweight
Construction	Continuous	0.25mm
Object Lines	Continuous	0.6 mm

Hidden Lines	Hidden	0.3 mm
Center Lines	CENTER	0.25 mm
Dimensions	Continuous	0.25 mm
Title Block	Continuous	1.2 mm
Viewport	Continuous	0.25 mm

- Create the drawing, as shown next. Do not add dimensions.

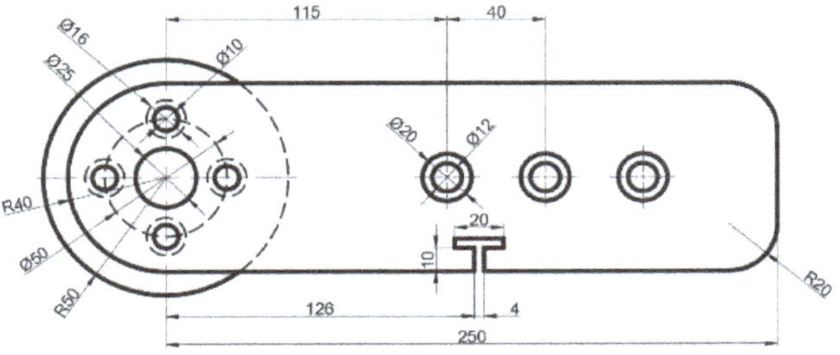

- Click the **Layout 1** tab at the bottom of the graphics window.

You will notice that a white paper is displayed with the viewport created automatically. The components of a layout are shown in the figure below.

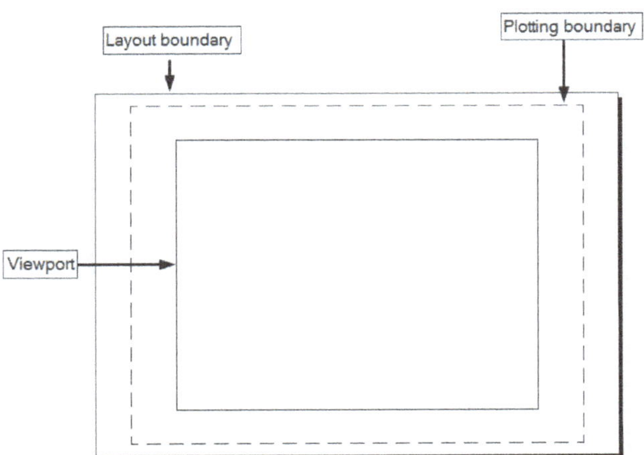

- Click **Output > Plot > Page Setup Manager** on the ribbon; the **Page Setup Manager** dialog appears.

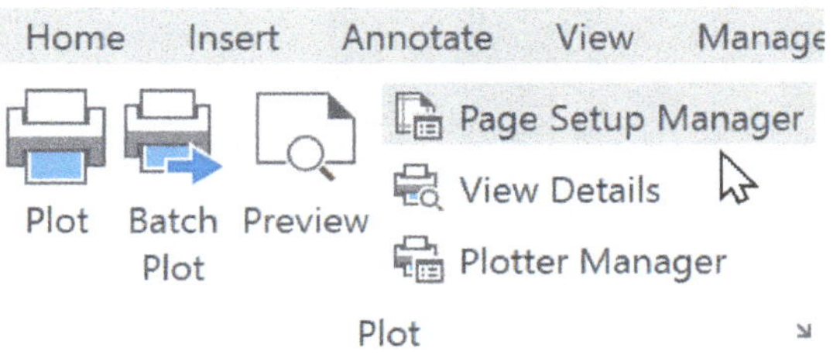

- In the **Page Setup Manager** dialog, click the **Modify** button; the **Page Setup –Layout1** dialog appears.

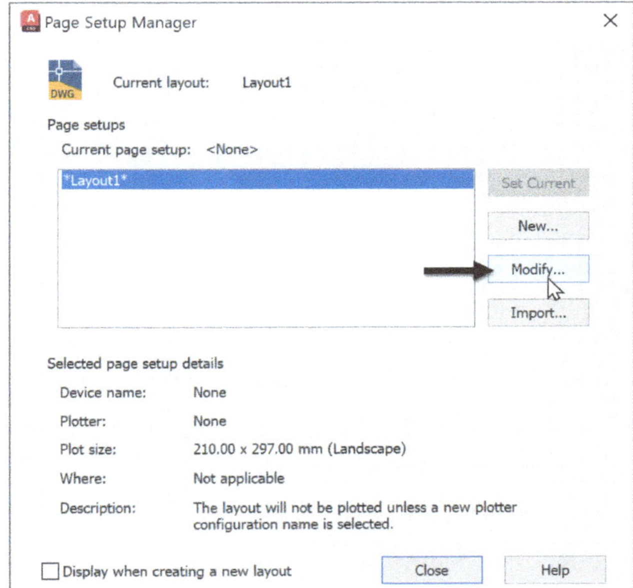

- In the **Page Setup** dialog, select **DWG to PDF.pc3** from the **Name** drop-down under the **Printer/Plotter** group.
- Set the **Plot Style table** to **acad.stb**.
- Set the **Paper size** to **ISO A1 (841.00 x 594.00 MM)**. Set the **Plot scale** to **1:1**.

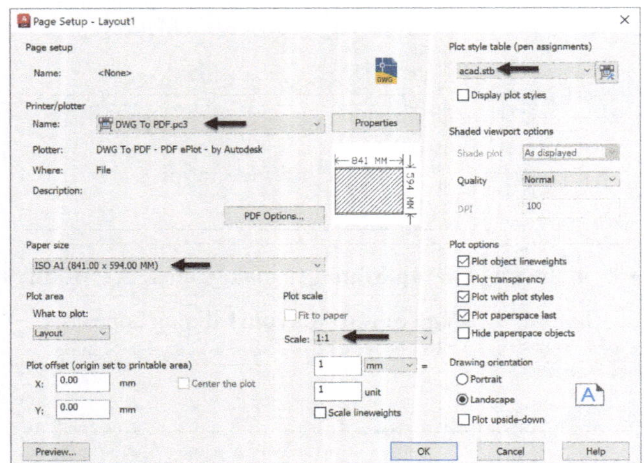

- Click **OK**, and then click **Close** on the **Page Setup Manager** dialog.
- Click the **Layout2** tab below the graphics window.
- Double-click on the **Layout1** tab and enter **ISO A1**; the **Layout1** is renamed.
- Similarly, rename the **Layout2** to **ISO A4**.
- Click **Layout > Layout > Page Setup** on the ribbon; the **Page Setup Manager** dialog appears.
- Select **ISO A4** from the list.
- Click the **Modify** button on the dialog.
- In the **Page Setup** dialog, select the **DWG to PDF.pc3** plotter and select **acad.stb** from the **Plot style table** drop-down.
- Set the **Paper size** to **ISO A4 (210 x 297 MM)** and **Scale** to**1:1.**
- Set **Drawing Orientation** as **Portrait** and click **OK**; you will notice that the size of the Layout is changed to A4 size.
- Close the **Page Setup Manager** dialog.

Creating Viewports in the Paper space

The viewports that exist in the paper space are called floating viewports. It is because you can position them anywhere in the layout and modify their shape and size with respect to the layout.

Creating a Viewport in the ISO A4 layout

- Open the **ISO A4** layout, if not already open.
- Select the default viewport that exists in the **ISO A4** layout.
- Press the DELETE key; the viewport will be deleted.
- Click **Layout > Layout Viewports > Rectangular** on the ribbon.

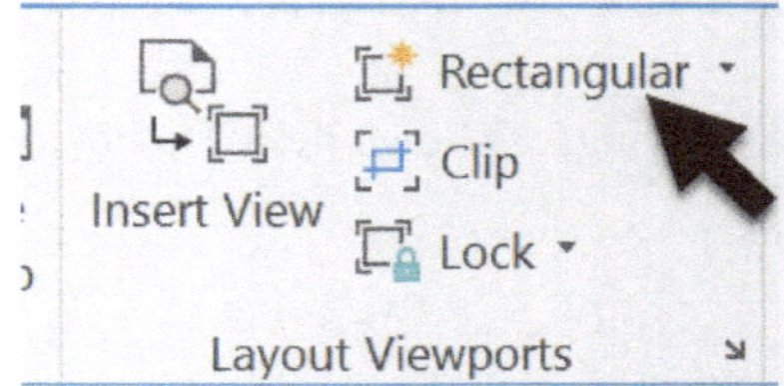

- Create the rectangular viewport by picking the first and second corner points, as shown in the figure.

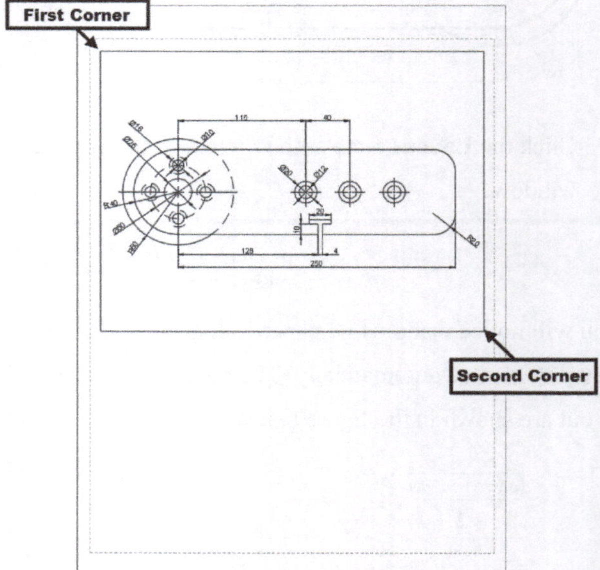

- Click the **PAPER** button on the status bar; the model space inside the viewport will be activated. Also, the viewport frame will become thicker when you are in model space.

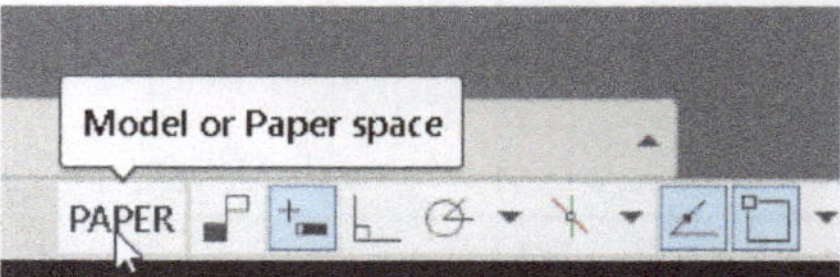

- Click the **Viewport Scale** button and select **1:2** from the menu; the drawing will be zoomed out.

- Use the **Pan** tool and position the drawing in the center of the viewport.
- After fitting the drawing inside the viewport, you can lock the position by clicking the **Lock/Unlock Viewport** button on the status bar.

After locking the viewport, you cannot change the scale or position of the drawing.

- Click the **MODEL** button on the status bar to switch back to paper space.

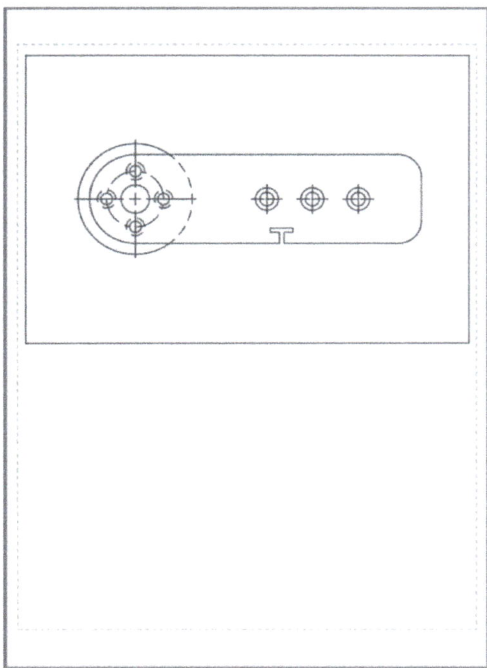

Creating Viewports in the ISO A1 layout

- Click the **ISO A1** tab below the graphics window.
- Select the viewport frame and modify the viewport using the grips, as shown below.

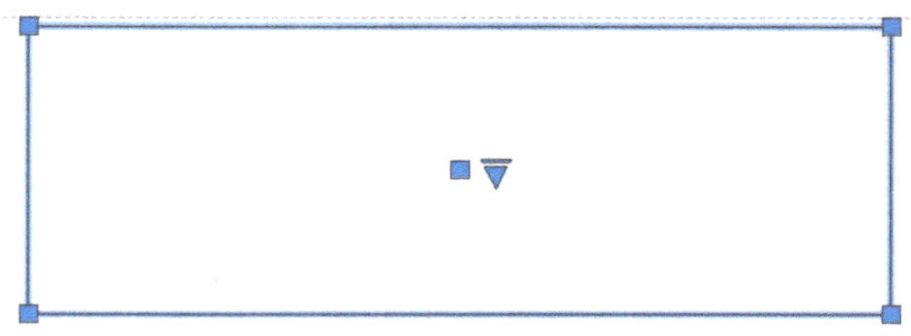

- Double-click inside the viewport to switch to the model space.
- Use the **Zoom** and **Pan** tools and drag the drawing to the center of the viewport.
- Click the **Viewport Scale** button and select **2:1** from the menu.
- Use the **Pan** tool and position the drawing, as shown in the figure.
- Click the **Lock/Unlock Viewport** button on the status bar.
- Double-click outside the viewport to switch to the paper space.
- Use the **Circle** tool and create a 180 mm diameter circle on the layout, as shown below.

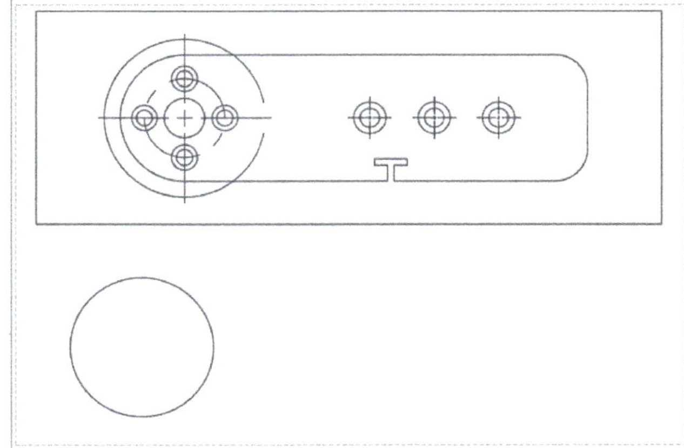

- Click **Layout > Layout Viewports > Viewport drop-down > Object** on the ribbon.

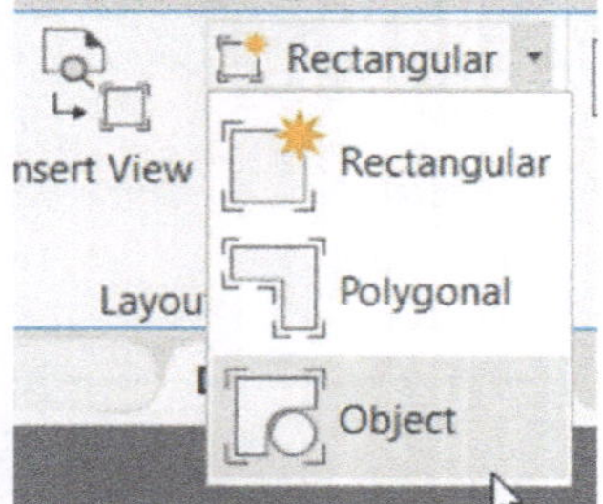

- Select the circle from the layout; it will be converted into a viewport.

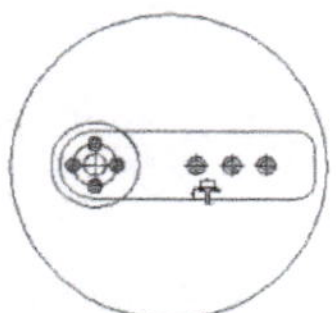

- Double-click in the circular viewport to switch to the model space.
- Click the **Viewport Scale** button on the status bar and select **4:1** from the menu; the drawing will be zoomed in to its center.
- Use the **Pan** tool and adjust the drawing, as shown below.

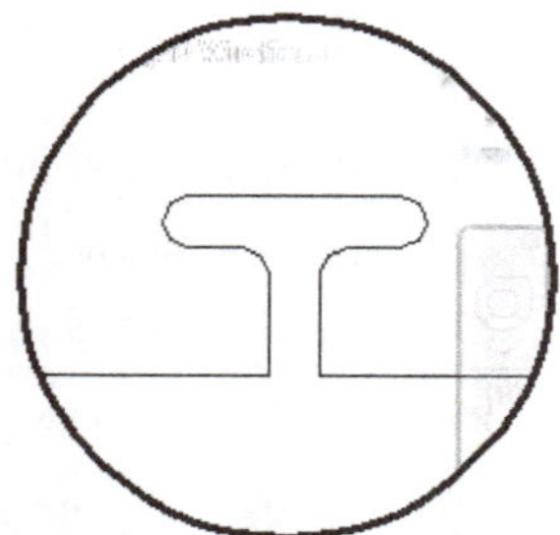

- Click the **Lock** button on the **Layout Viewports** panel.

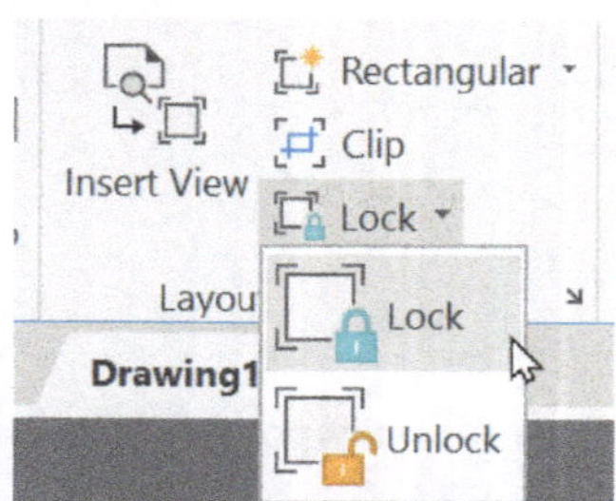

- Select the circular viewport and press ENTER; the drawing inside the viewport will be locked. Now, you cannot zoom or pan the drawing.

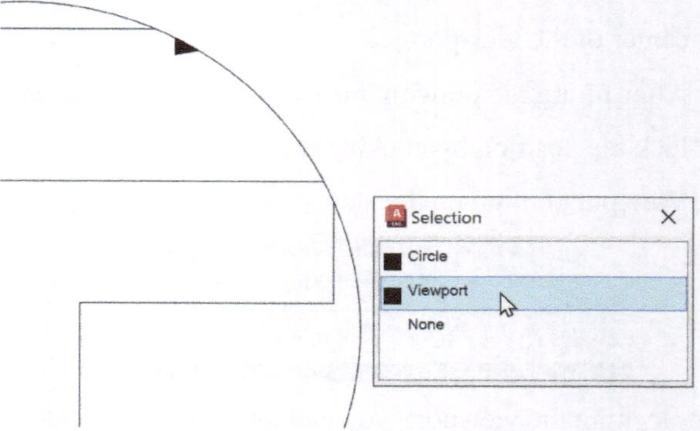

- Click **Output > Plot > Preview** on the ribbon; the plot preview will be displayed. You will notice that the viewport frames are also displayed in the preview.

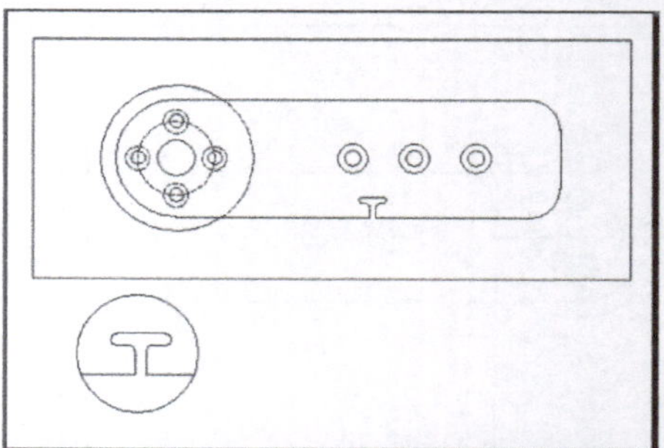

- Press ESC to close the preview window.

To hide viewport frames while plotting a drawing, follow the steps given below.

- Type **LA** in the command line to open the **Layer Properties Manager**.
- In the **Layer Properties Manager**, create a new layer called **Hide Viewports** and make it current.
- Deactivate the plotter symbol under the **Plot** column of the **Hide Viewports** layer; the object on this layer will not be plotted.
- Close the **Layer Properties Manager**.

- Click the **Home** tab on the ribbon and expand the **Layers** panel.
- Click the **Change to Current Layer** button on the **Layers** panel.

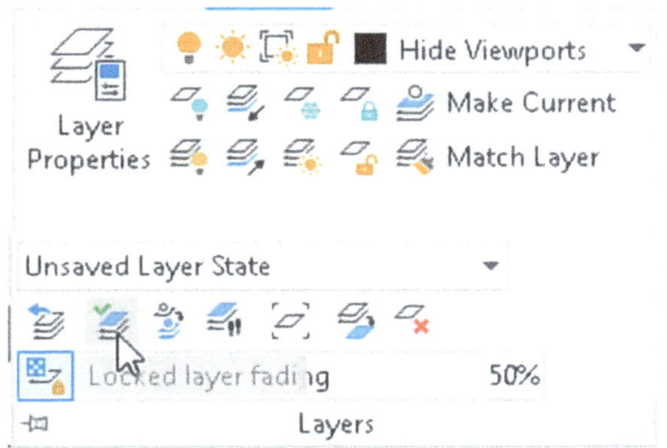

- Select the viewports in the **ISO A1** layout and press ENTER; the viewport frames will become unplottable. To check this, click the **Preview** button on the **Plot** panel of the **Output** ribbon tab; the plot preview will be displayed as shown below.

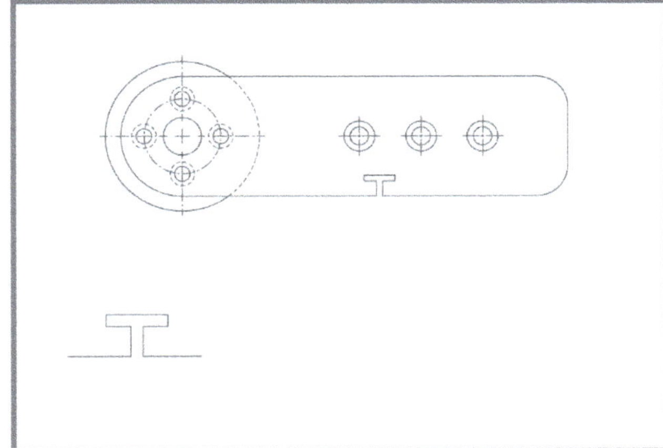

- Close the preview window.

Changing the Layer Properties in Viewports

The layer properties in viewports are not related to the layer properties in model space. You can change the layer properties in viewports without any effect in the model space.

- Double-click inside the larger viewport to activate the model space.
- Type **LA** in the command line to open the **Layer Properties Manager**.
- In the **Layer Properties Manager**, click the icon in the **VP Freeze** column of the **Hidden lines** layer; the hidden lines will disappear in the viewport, as shown below.

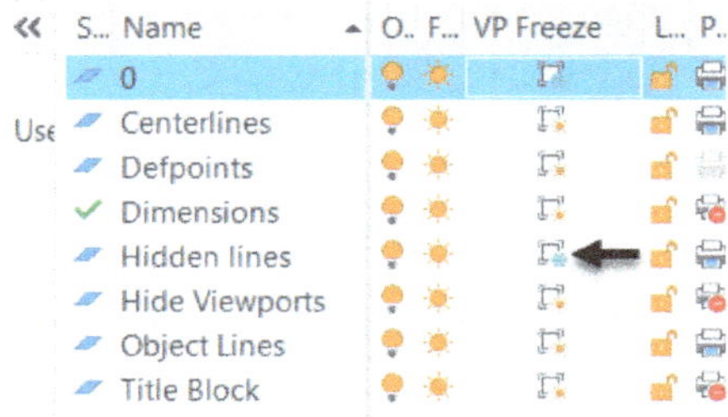

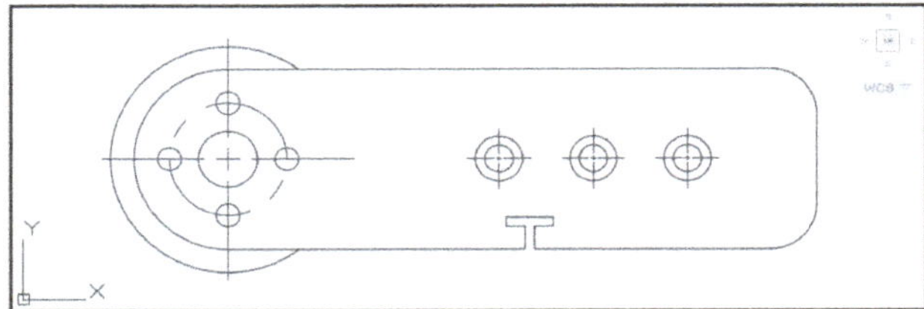

- Double-click outside the viewport to switch to paper space.
- Click the **Model** tab below the graphics window; you will notice that the hidden lines are retained in the model space.

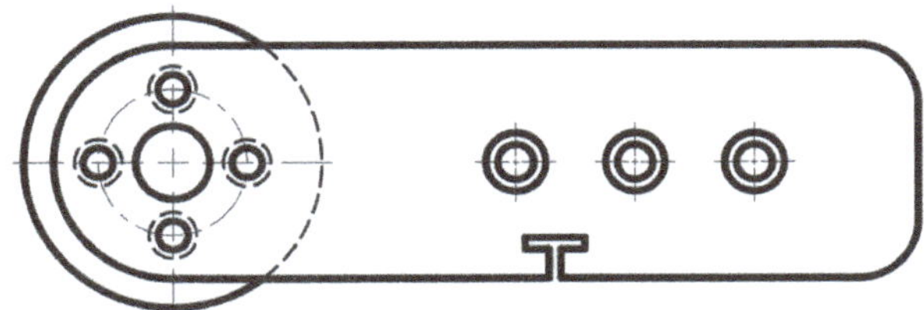

Creating the Title Block on the Layout

You can draw objects on layouts to create a title block, borders, and viewports. However, it is not recommended to draw the actual drawing on layouts. You can also create dimensions on layouts.

Example1:

- Click the **ISO A1** layout tab.
- Set the **Title Block** layer as current.
- Click the **Rectangle** button on the **Draw** panel.
- Pick a point at the lower right corner of the layout.
- Select the **Dimensions** option from the command line.
- Specify the length of the rectangle as **820** and width as **550**.
- Click in the upper area of the layout; a rectangular border will be created.
- Create a title block at the lower right corner, as shown below (**Use the Line** and **Multiline Text** tools).

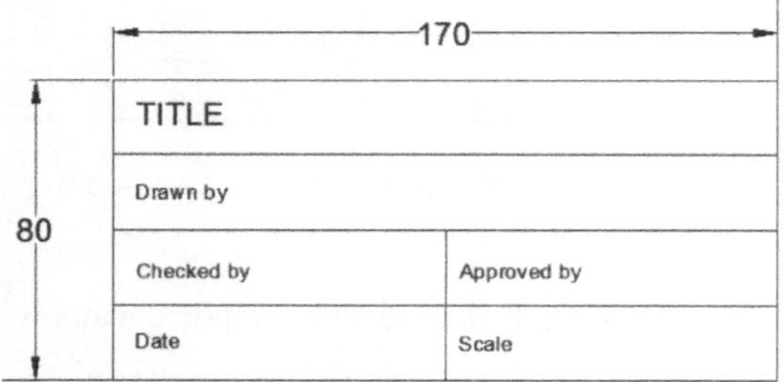

- Create attributes and place them inside the title, as shown below (refer to *Chapter 9: Blocks, Attributes, and Xrefs* to learn how to create attributes).

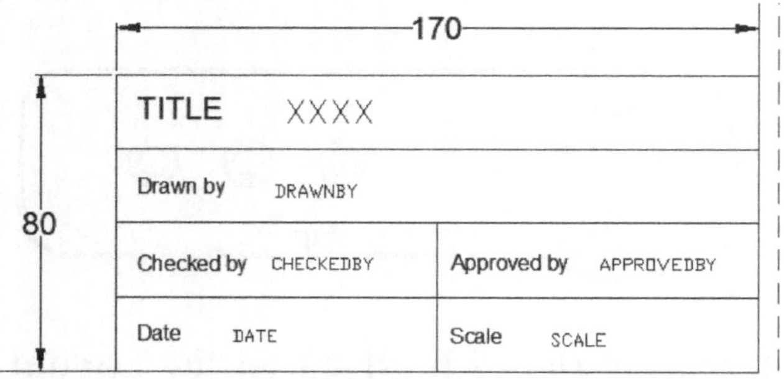

- Use the **Create Block** tool and convert it into a block.
- Use the **Insert** tool and insert it at the lower right corner of the layout.
- Save the drawing file as **Viewports-Example.dwg**.

Working with Annotative Dimensions

In AutoCAD, you create drawings at their actual size. However, when you scale a drawing to fit inside a viewport, the size of the dimensions will not be appropriately scaled. For example, in the following figure, the first viewport is scaled to 1:2, and the second viewport is scaled to 1:1. The dimensions in the first viewport appear much smaller.

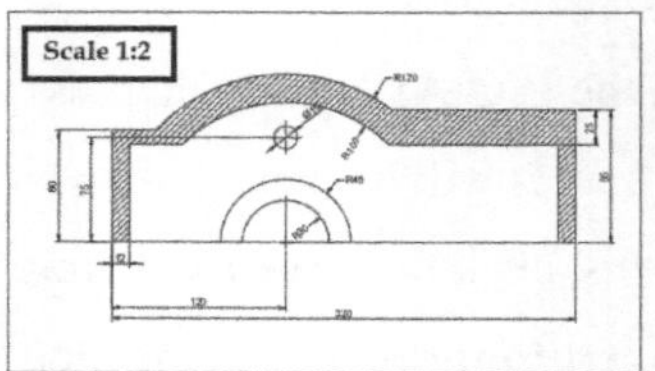

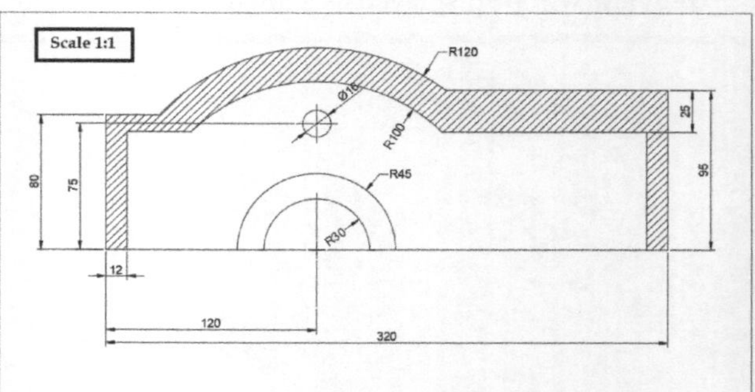

You can fix this problem by applying the Annotative property to dimensions.

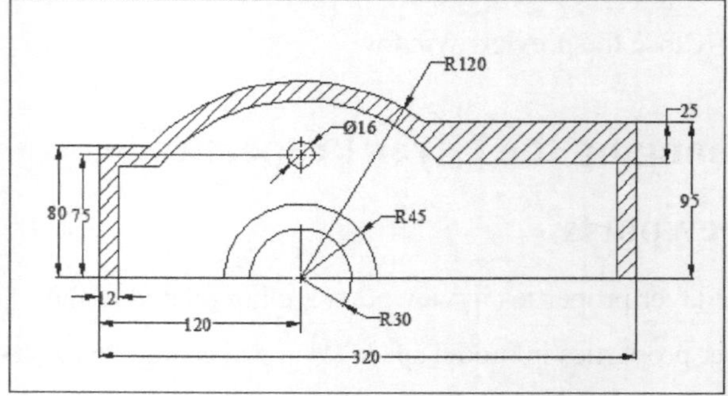

- Open the **Viewports-Example.dwg**, if not already opened.
- Set the **Dimensions** layer as current.
- Type **D** in the command line and press ENTER.
- In the **Dimension Style Manager**, click the **New** button.
- In the **Create New Dimension Style** dialog, enter **New style name** as **Dim_Anno** and select the **Annotative** checkbox. Click **Continue**.

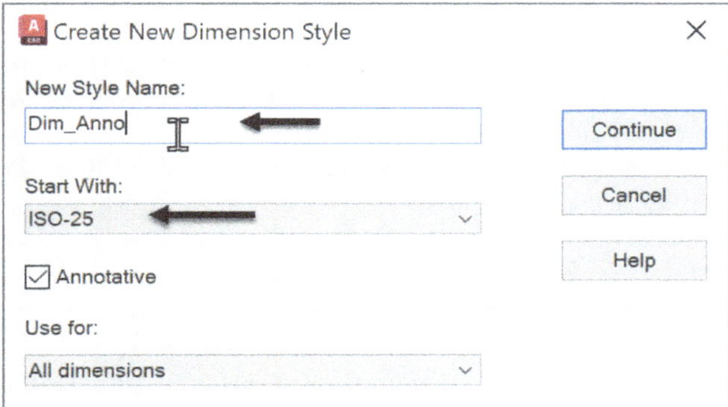

- Set the following settings in the **New Dimension Style** dialog.

  Lines tab: Offset from origin 1.25
  Symbols and Arrows tab: Arrow size 2.5, Center Marks-Line.
  Text tab: Text height – 2.5, Text placement - Vertical-Centered, Text alignment - Horizontal
  Primary Units tab: Units Format – Decimal, Precision – 0, Decimal separator – '.' period

- In the **Fit** tab, ensure that the **Annotative** checkbox is selected.

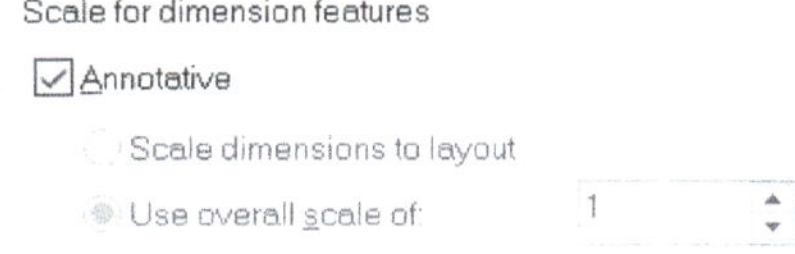

- Click **OK** on the **New Dimension Style** dialog; you will notice that the **Dim_Anno** style is listed in the **Dimension Style Manager**. Also, the annotation symbol is displayed next to it. It indicates that all dimensions created using this style will have annotative property. Click on the **Close** button.

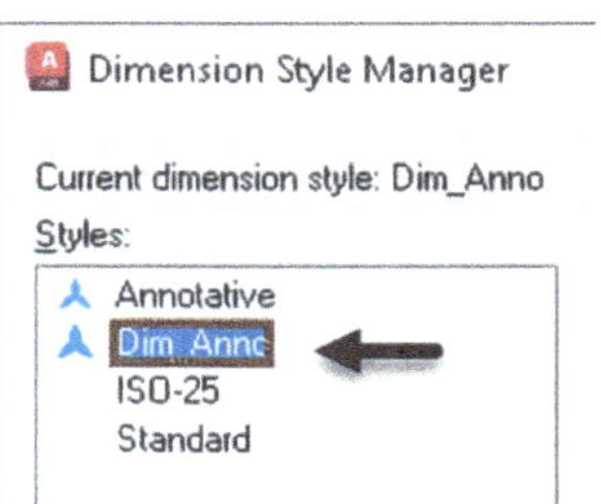

- Activate the **Dimension** tool and set the Annotation Scale to **1:1**. Click **OK**.
- Create a linear dimension, as shown below.

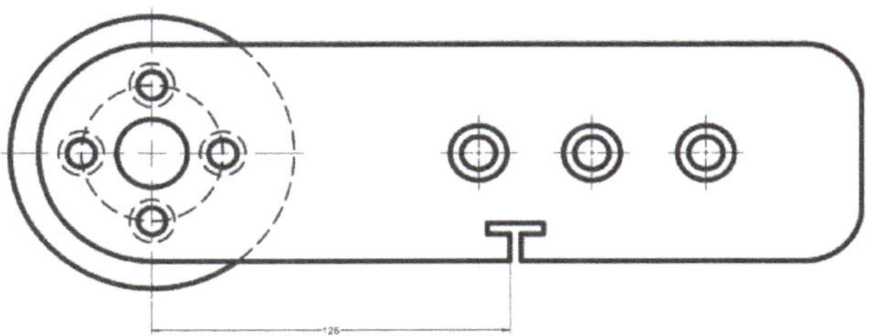

- Activate **Automatically add scales to annotative objects when the annotation scale changes** on the status bar.

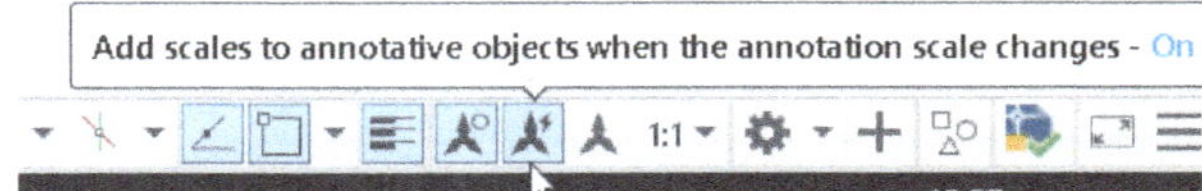

- Set the **Annotation Scale** to **1:2**; the size of the dimension will automatically increase by two times.

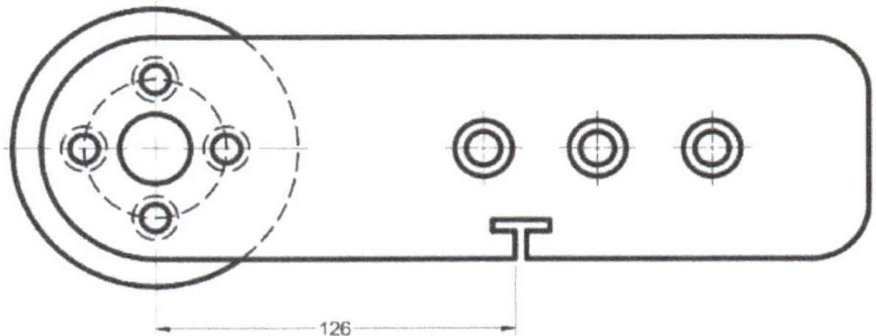

Example 2:

- Ensure that the **Annotation Scale** is set to **1:2** and create another linear dimension, as shown in the figure.

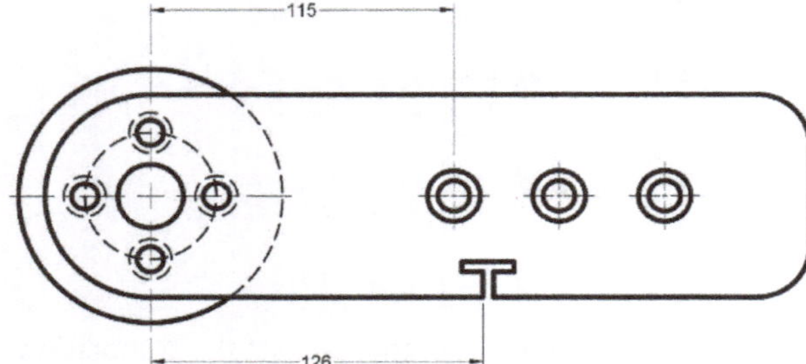

- Click the **ISO A4** layout in which the viewport scale is set to 1:2; you will notice that the dimensions are scaled with respect to the viewport.

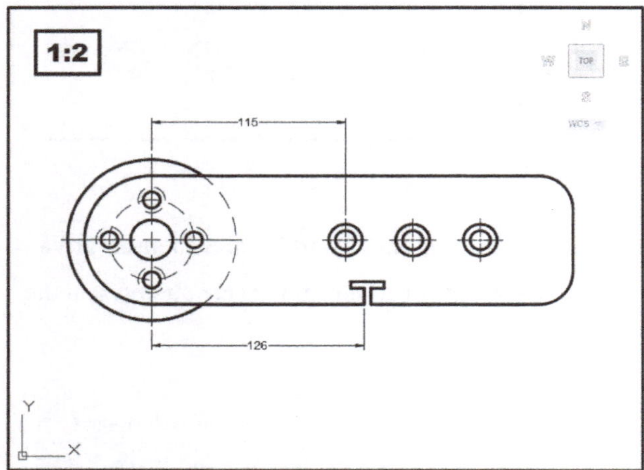

- Click the **ISO A1** layout; you will notice that the dimensions are not displayed in the 2:1 viewport. To display dimensions in the 2:1 viewport, you need to add 2:1 scale to dimensions.

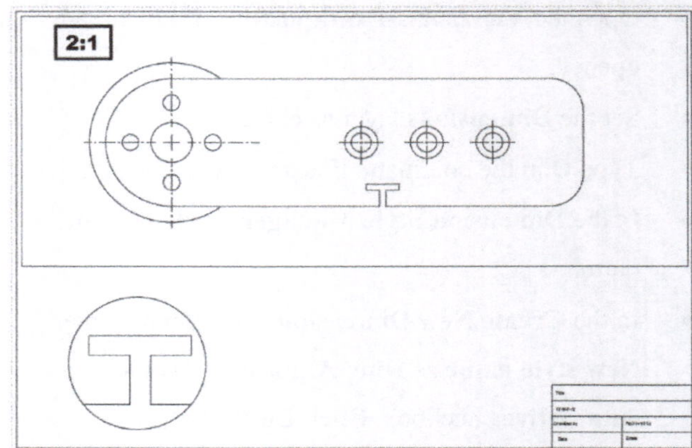

- Click the **Model** tab below the graphics window to switch to the model space.
- Click **Annotate > Annotation Scaling > Add/Delete Scales** on the ribbon.

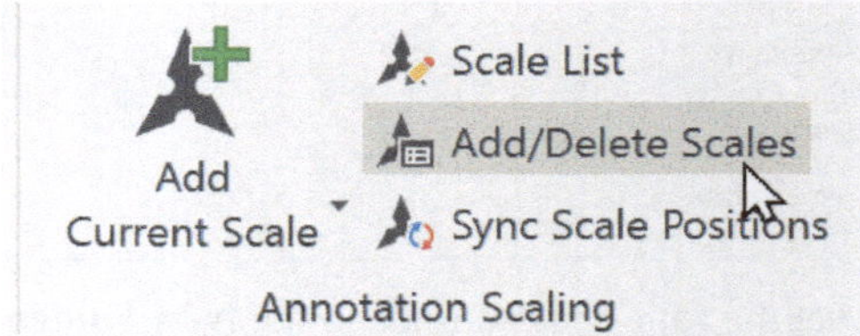

- Select the dimensions from the graphics window and right-click; the **Annotation Object Scale** dialog appears. In this dialog, the **Object Scale List** shows the scales applied to the selected dimensions. You need to add a 2:1 scale to the dimensions so that they will be visible in the 2:1 viewport.

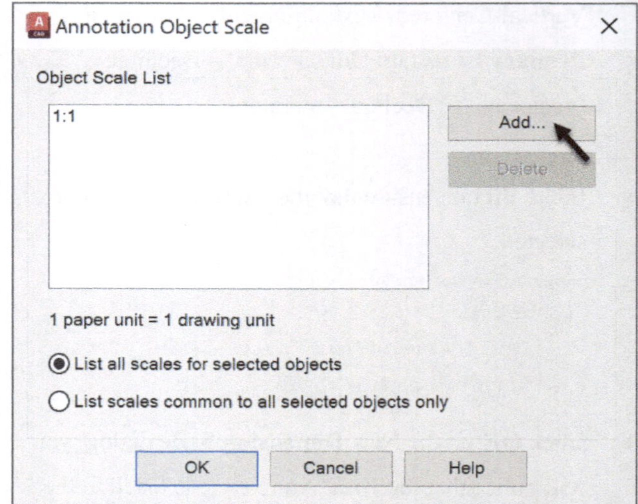

- To add a new scale to the dimensions, click the **Add**

button; the **Add Scales to Object** dialog appears.

- Select the **2:1** scale from the list and click **OK**; the scale will be added to the **Object Scale list**.
- Click **OK** on the **Annotation Object Scale** dialog.
- Click the **ISO A1** layout; the dimensions are displayed in both 2:1 and 1:2 viewports.

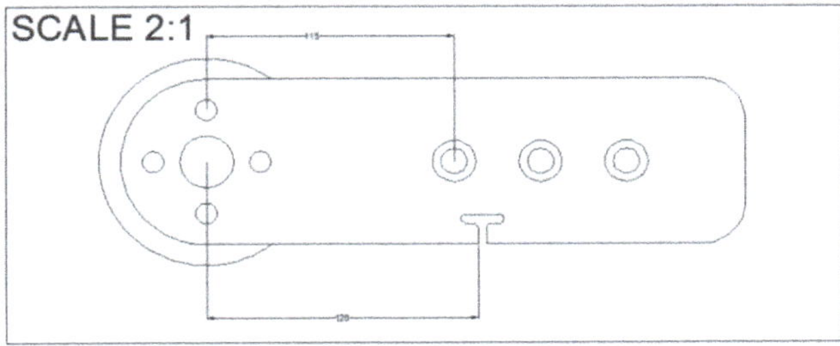

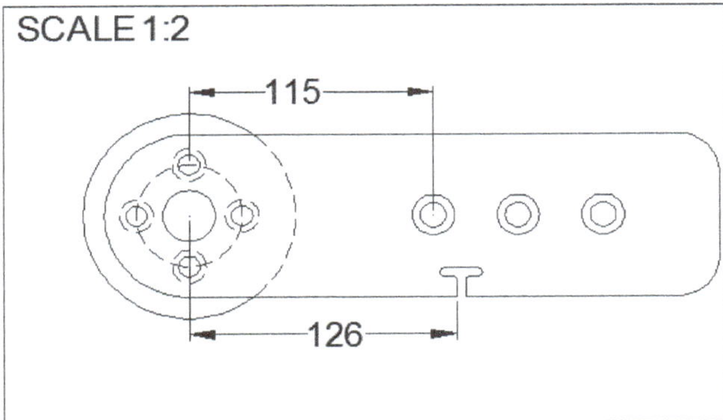

- Similarly, create other dimensions, as shown below. Add 2:1 and 1:2 scales to dimensions and check the drawing in two different layouts.

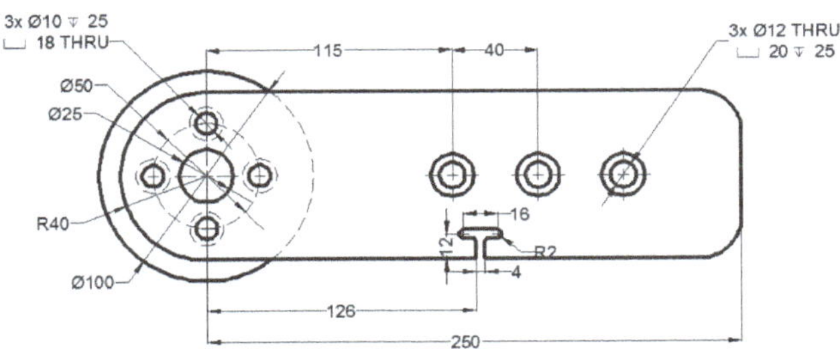

Scaling Hatches relative to Viewports

While working in layouts, you may also need to scale the hatch with respect to the viewport scale. The following figure shows a drawing in two different viewports 1:2 and 1:1. The hatch in the left viewport is smaller than that in the right-side viewport. You can correct this problem by using the **Relative to Paper Space** option.

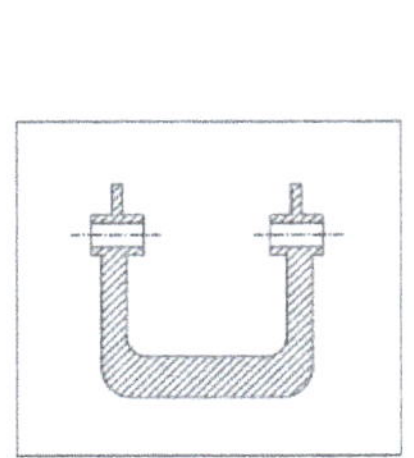

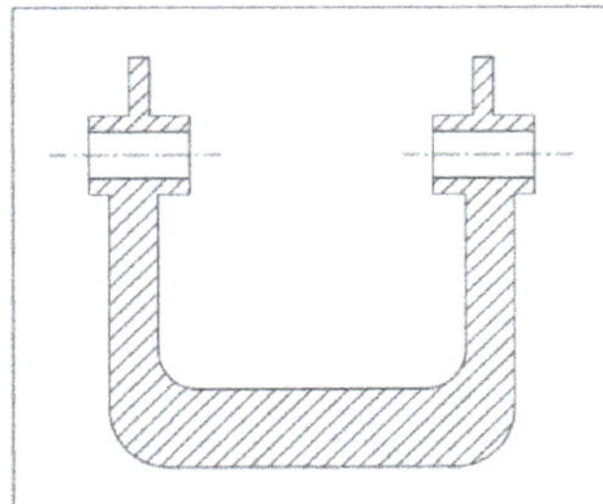

- Double-click inside a viewport; the model space will be activated.
- Select the hatch patterns from the drawing; the **Hatch Editor** tab appears.
- In the **Hatch Editor** tab, expand the **Properties** panel and select the **Relative to Paper Space** button.

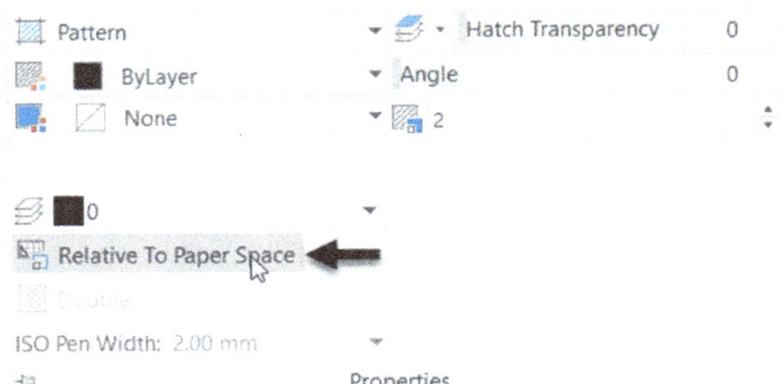

- Click the **Close Hatch Editor** button; you will notice that the hatch will be scaled with respect to the viewport scale. Double-click outside the viewport to switch to the paper space.

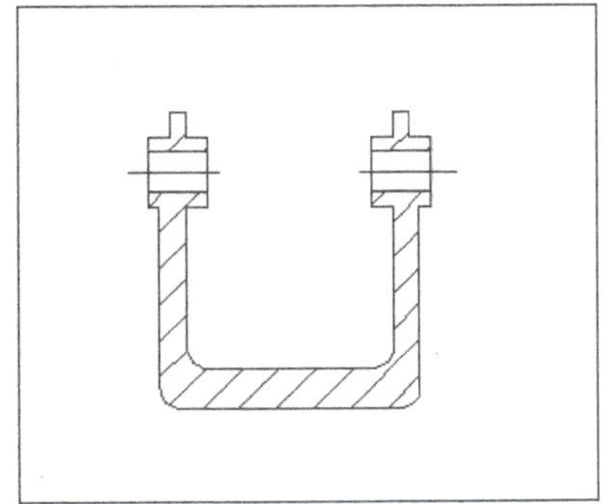

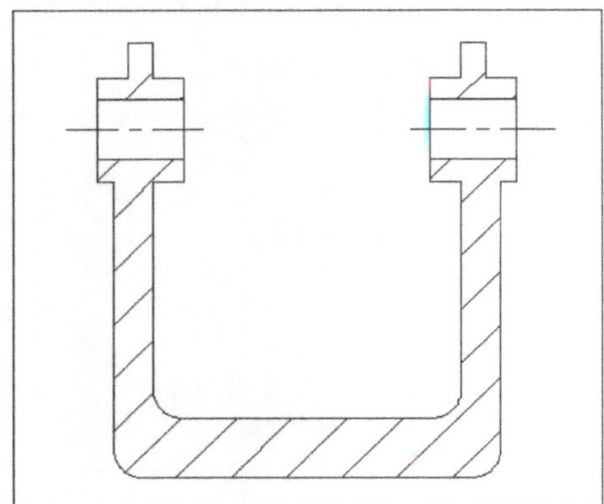

Working with Annotative Text

Annotative property can also be assigned to text. The annotative text will be scaled with respect to the viewport scale.

- Open the **Viewports-Example.dwg**, if not already opened.
- Click **Annotate > Text > Text Style** on the ribbon; the **Text Style** dialog appears.

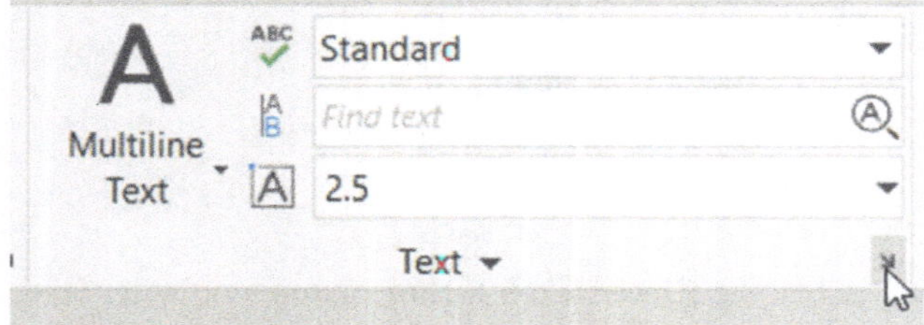

- Click the **New** button on the **Text Style** dialog; the **New Text Style** dialog appears.
- Enter **Text_Anno** as the **Style name** and click **OK**.

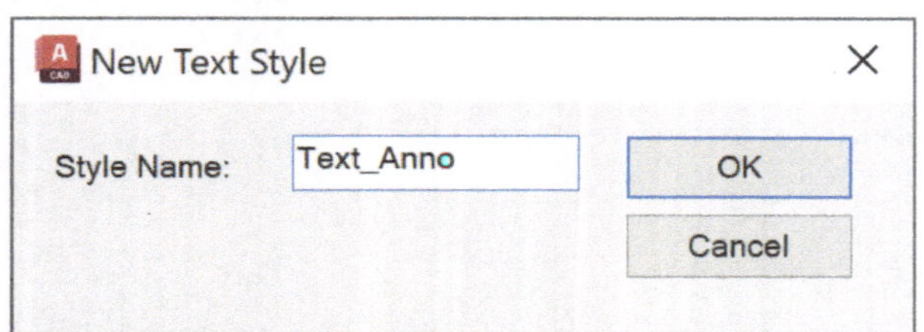

- Select the **Text_Anno** style from the **Styles** list.
- Set **Font Name** to **Arial** and select the **Annotative** checkbox.
- Set **Paper Text Height** to **2.5** and **Width Factor** to **1**.
- Click **Apply** and **Close**.
- Select **1:1** from the **Viewport Scale** menu at the status bar.
- Click **Annotate > Text > Multiline Text** on the ribbon.
- Specify the first corner of the text editor by picking an arbitrary point.
- Select the **Justify** option from the command line; the command line displays:

- Select the **MC** option from the command line.
- Move the pointer toward the right and specify the second corner of the text editor.
- Type **All dimensions are in mm** and click the **Close Text Editor** button on the **Close** panel.
- Move the text and place it at the bottom left corner of the drawing, as shown below. You can also add a frame to the text. Right-click on it and select **Properties**. On the **Properties** palette, under the **Text** section, set **Text Frame** to **Yes**.

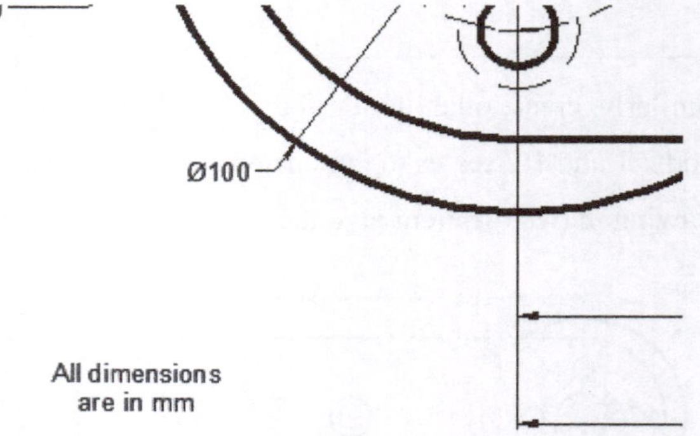

- View the drawing in the **ISO A4** layout; you will notice that the text is not displayed. It is because the text is set to a 1:2 scale.
- On the status bar, click the **Show annotation objects** button.

The text is visible in the ISO A4 layout.

- Save the drawing as **Layout Example.dwg** and close.

Exercises

Exercise 1

Create the drawing, as shown below. After creating the drawing, perform the following tasks:

- Create a layout of the A3 size and then create a viewport.
- Set the viewport scale to 1:2.
- Set the scale of the dimensions and hatch lines with respect to the viewport.

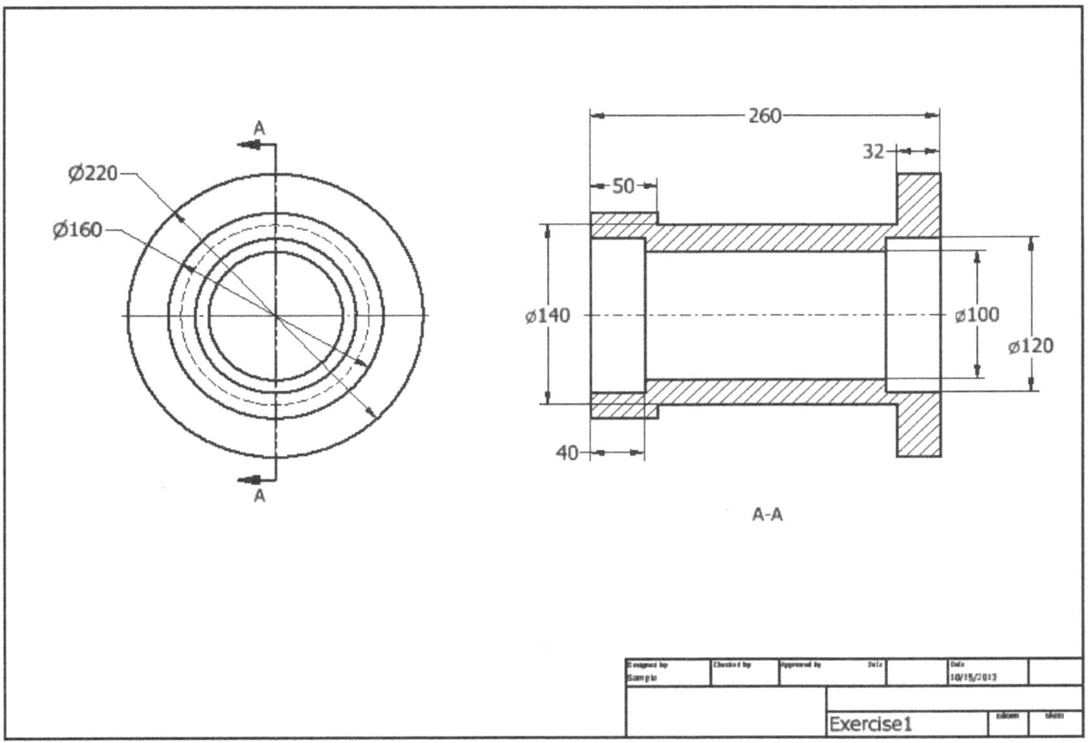

Chapter 11: Templates and Plotting

In this chapter, you will learn to do the following:

- **Configure Plotters**
- **Create Plot Style Tables**
- **Use Plot styles**
- **Create Templates**
- **Plot/Print the drawing**

Plotting Drawings

Plotting is the process of producing a physical copy of the drawing using a printer or plotter. The printer may be directly connected to an AutoCAD workstation or on the network of workstations. Although the process of plotting is straightforward, it is essential to know how to establish communication between AutoCAD and the plotter. In this chapter, you will learn to connect a plotter with AutoCAD, define plotting style, and produce professional prints of drawings. You will also learn to print and publish drawings in digital format.

Configuring Plotters

It is assumed that you have connected the plotter to your workstation and installed the drivers related to it. Even after doing so, you need to set a connection between the plotter and AutoCAD. You can establish this connection by using the Add-plotter wizard. The following example explains the procedure to connect a plotter to AutoCAD.

Example:

- Start AutoCAD 2024 and open a new drawing file.
- Click **Application Menu > Print > Manage Plotters** or type PLOTTERMANAGER in the command line; the **Plotters** folder will be opened, as shown below. All the configured plotters are displayed in this folder.

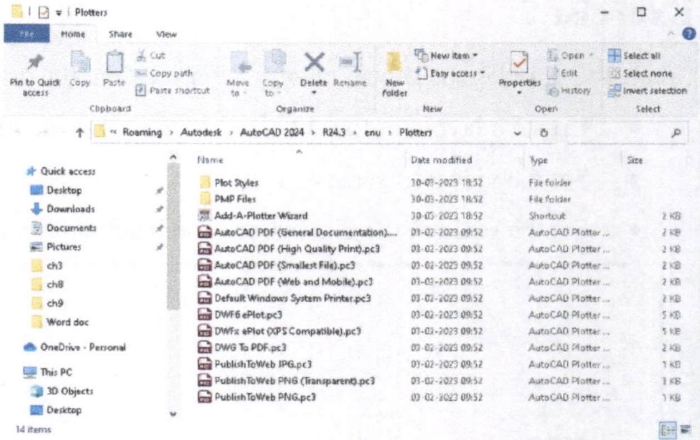

- In the **Plotters** folder, double-click on the **Add-A-Plotter Wizard** icon; the **Add Plotter – Introduction** page appears.
- Click the **Next** button; the **Add Plotter – Begin** page appears. On this page, three options allow you to set up a plotter: **My Computer**, **Network Plotter Server**, and **System Printer**. These options are explained in the dialog itself, as shown below.

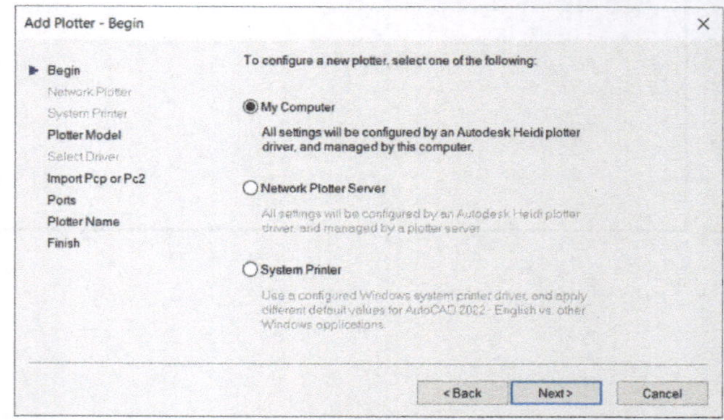

- Select the **System Printer** option and click **Next**; the **System Printer** page appears. A list of printers installed on your workstation is displayed.

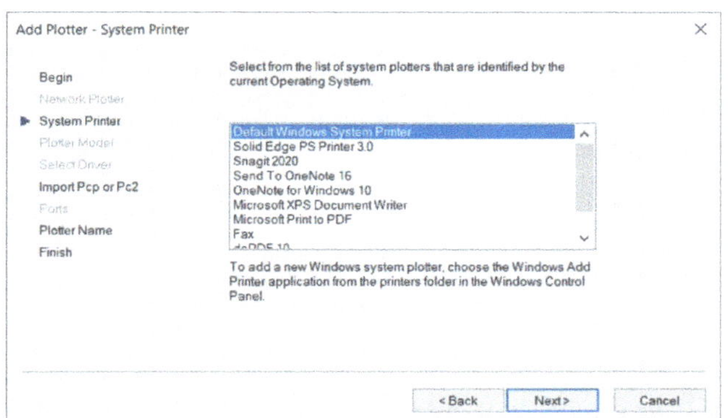

- From the list, select the required printer (**Adobe PDF** in this case) and click **Next**; the **Import** page appears.
- Click the **Next** button; the **Plotter Name** page appears.
- Type name of the plotter in the **Plotter Name** box and click **Next**; the **Finish** page appears. You can edit the configuration of the plotter by using the **Edit Plotter Configuration** button.

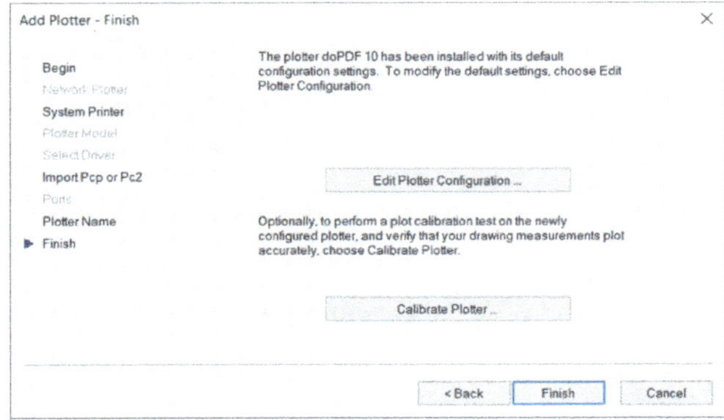

If you click the **Edit Plotter Configuration** button, the **Plotter Configuration Editor** dialog appears. In this dialog, you can modify the default settings of the plotter. The **Calibrate Plotter** button is used to test the plotter.

- Click the **Finish** button; a new plotter will be added to the **Plotters** folder.

Creating Plot Style Tables

Plot styles determine the final look of the plotted drawing. They are used to override the layer properties such as color, linetype, and lineweight when the drawing is printed. After configuring a plotter, you need to create a plot style. There are two types of plot styles: **Color-dependent** and **Named** plot style. **Color-dependent** plot styles are assigned based on the object color, whereas the **Named** plot styles are assigned based on layer or by an object.

- On the **Application Menu**, click **Print > Manage Plot styles** or type STYLESMANAGER in the command line; the **Plot Styles** folder appears.

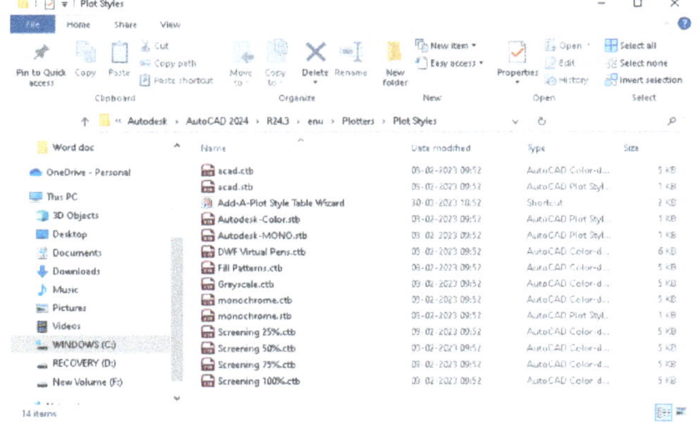

- Double-click on the **Add-A-Plot Style Table Wizard** icon; the **Add Plot Style Table** dialog appears. Read the information on this dialog and click **Next**.
- Select the **Start from Scratch** option and click **Next**.
- Select the **Named Plot Style Table** option and click **Next**.
- Enter **Sample** in the **File name** box and click **Next**; the **Finish** page appears.
- Click the **Plot Style Table Editor** button; the **Plot Style Table Editor** dialog appears.
- Click the **Add Style** button available at the bottom left of the dialog; a new style named **Style 1** is added.
- Enter **PS1** in the **Name** box.
- Select **Black** from the **Color** drop-down.
- Set the **Screening** value to 70. The screening factor will fade objects in the printed output. A 20%

screening factor will result in more fading of objects than a 50% screening factor.

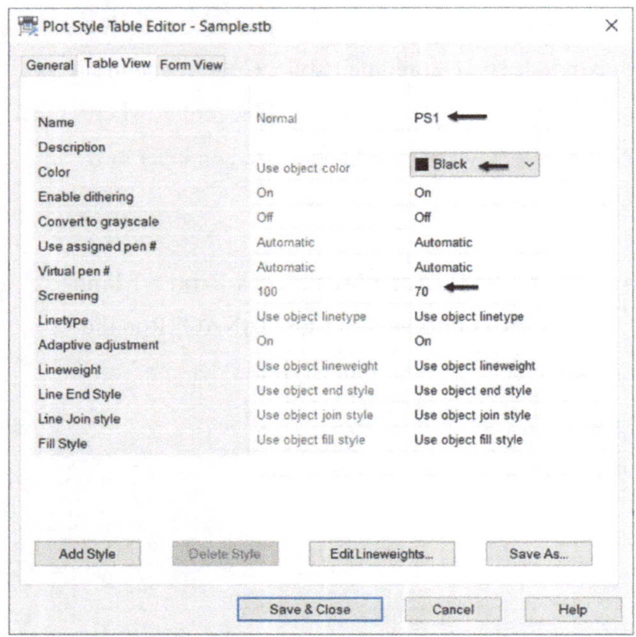

- Click **Save & Close** on the **Plot Style Table Editor** dialog.
- Click **Finish** to close the **Add Plot Style Table** dialog; the **Sample** plot style will be added to the **Plot Styles** folder.

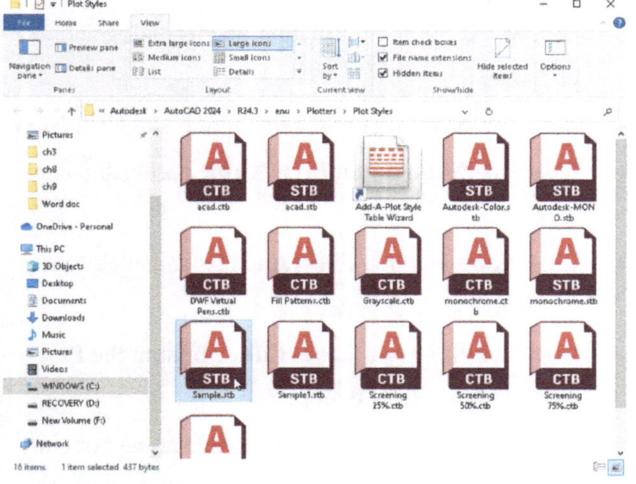

Using Plot Styles

In AutoCAD, the Color-Dependent Plot style is used by default. To use the newly created plot style, you need to specify a setting in the **Options** dialog.

- Right-click in the graphics window and select **Options**; the **Options** dialog appears.
- Select the **Plot and Publish** tab in the **Options** dialog and click the **Plot Style Table Settings** button; the **Plots Style Table settings** dialog appears.

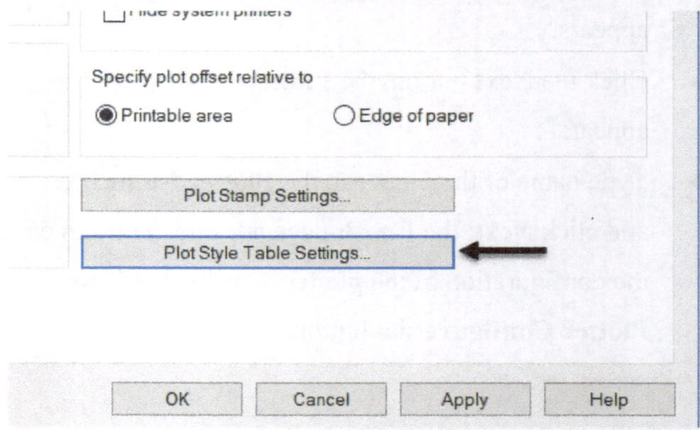

- Select the **Use named plot styles** option from the dialog.
- Select **Sample.stb** from the **Default plot style table** drop-down.
- Select **PS1** from the **Default plot style for layer 0** drop-down.
- Set the **Default plot style for objects** to **ByLayer**.

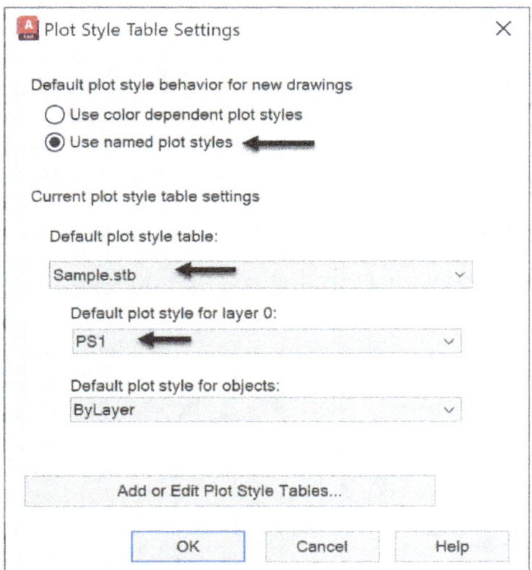

- Click **OK** twice to close both the dialogs.
- Close the drawing file by clicking the **Close** button located at the top-right corner.

- Click **NO** on the **AutoCAD** alert message.
- Click the **New** button on the **Quick Access Toolbar**; the **Select Template** dialog appears.
- Select **Open > Open with no Template – Metric** from the bottom right corner of the dialog; a drawing file will be opened.

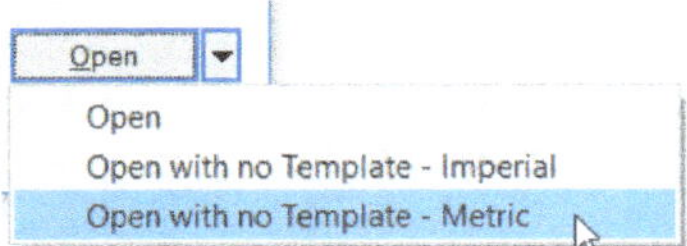

- Open the **Layers Properties Manager** and create the layers shown in the table below:

Layer	Linetype	Lineweight	Plot Style
Construction	Continuous	0.25	PS1
Object	Continuous	0.7 mm	PS1
Hidden Lines	Hidden	0.3 mm	PS1
Center Lines	CENTER	0.25 mm	PS1
Dimensions	Continuous	0.25 mm	PS1
Section Lines	Continuous	0.5 mm	PS1
Cutting Plane	Phantom	0.6mm	PS1
Title Block	Continuous	1mm	PS1
Viewport	Continuous	0.25 mm	PS1
Text	Continuous	0.25	PS1
Title block text	Continuous	0.25	PS1

- Click the **Layout 1** tab to activate the paper space.
- Click **Output > Plot > Page setup Manager** on the ribbon; the **Page Setup Manager** dialog appears.
- Click **Modify** on the **Page Setup** Manager; the **Page Setup** dialog appears.
- Under the **Printer/plotter** group, select the plotter that you have configured to your workstation.
- Set the **Paper Size** to **A3** and **Drawing orientation** to **Landscape**.
- Click **OK** and **Close** to exit both the dialogs.
- Draw a title block in the paper space, as shown below.

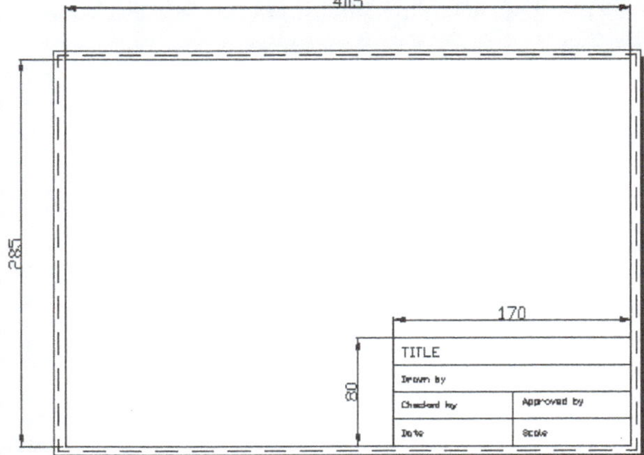

- Create a viewport inside the title block (refer to the

Creating Viewports in the Paper space section discussed earlier in this chapter).

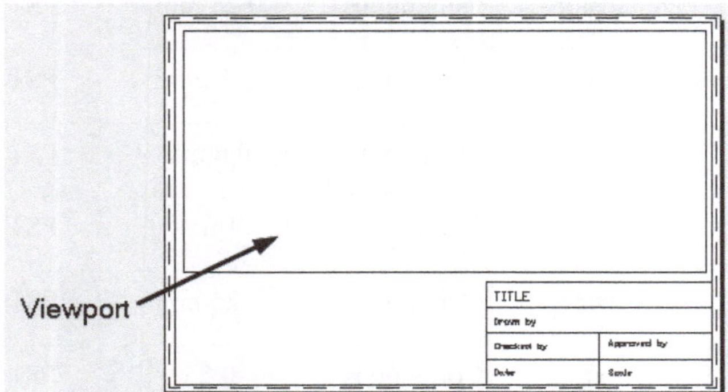

Creating Templates

After specifying the required settings in a drawing file, you can save those settings for future use. You can do so by creating a template. Template files have settings such as units, limits, and layers already created, which will increase your productivity. In previous sections, you have configured various settings, such as layers, colors, linetypes, and plotting settings. Now, you will create a template file containing all of these settings and the title block that you have created.

- On the **Quick Access toolbar**, click the **Save** button; the **Save Drawing As** dialog appears.
- In the **Save Drawing As** dialog, set **Files of type** to **AutoCAD Drawing Template (*.dwt).**
- Enter **ISOA3** in the **File name** box and click **Save**.
- In the **Template Options** dialog, enter **ISO-A3 Horizontal layout with title block** in the **Description** box.
- Click **OK** to close the dialog and save the template file.

Plotting/Printing the drawing

- Click the **New** button.
- Double-click on **ISOA3**. A new drawing will be started with the selected template.
- Open the **Layer Properties Manager**; you will notice that the layers saved in the template file are loaded automatically.
- Close the **Layer Properties Manager**.
- Create a drawing, as shown below. You can also download the drawing from the companion website.

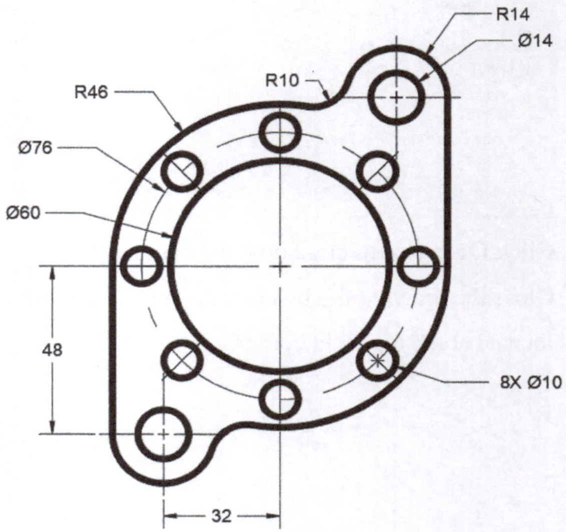

- Click the **Layout 1** tab to activate the paper space.
- Double-click inside the viewport to activate the model space.
- Set the **Viewport Scale** to 1:1 on the status bar.
- Use the **Pan** tool and position the drawing at the center of the viewport.
- Double-click outside the viewport to activate the paper space.
- Hide the viewport frame by freezing the **Viewport** layer.

- Click the **Plot** button on the **Quick Access Toolbar**; the **Plot** dialog appears.

- Make sure that the options in this dialog are the same as that you specified while creating the template.
- Click the **Preview** button located at the bottom left corner; the preview window appears.
- Click the **Zoom Original** button to fit the drawing to the window.

- Examine the print preview for the desired output and click the **Plot** button; the drawing will be plotted.

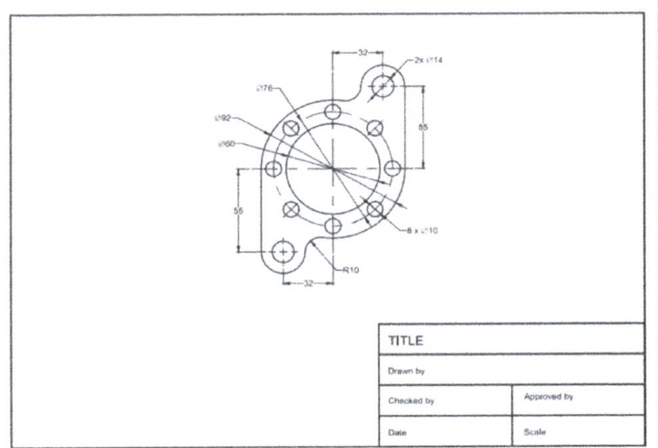

- Save and close the drawing file.

Exporting to PDF

The PDF and DWF files are one of the commonly used file formats to exchange drawings between designers and clients. AutoCAD makes it easy to export the drawing to the PDF or DWF formats.

- On the ribbon, click **Output > Export to DWF/PDF > Export to PDF Options**.

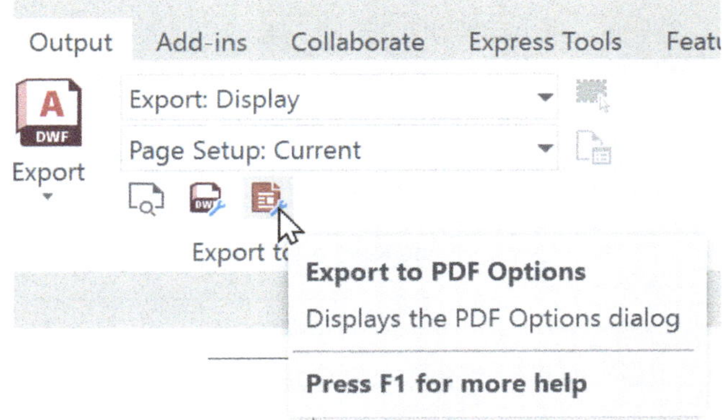

- On the **Export to PDF Options** dialog, set the Vector quality, Raster image quality, line merge control, and Data options.

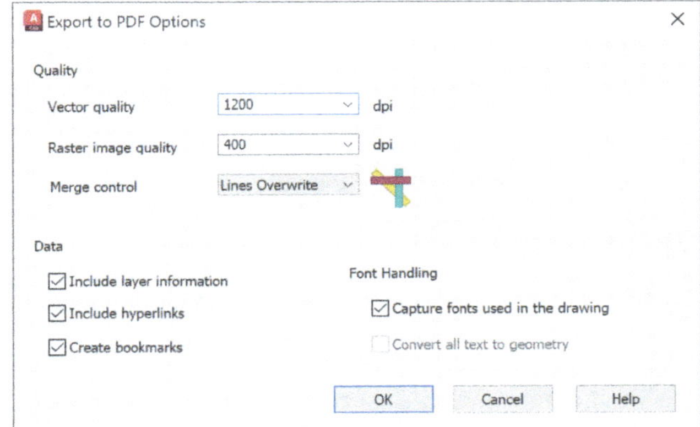

The options in the **Data** section help you to include layer information, hyperlinks, and create bookmarks. Also, you can capture fonts used in the drawing.

- Click **OK**.
- On the ribbon, click **Output > Export DWF/PDF > Export > PDF** .
- On the **Save As PDF** dialog, select a **PDF Preset** to specify the PDF quality.

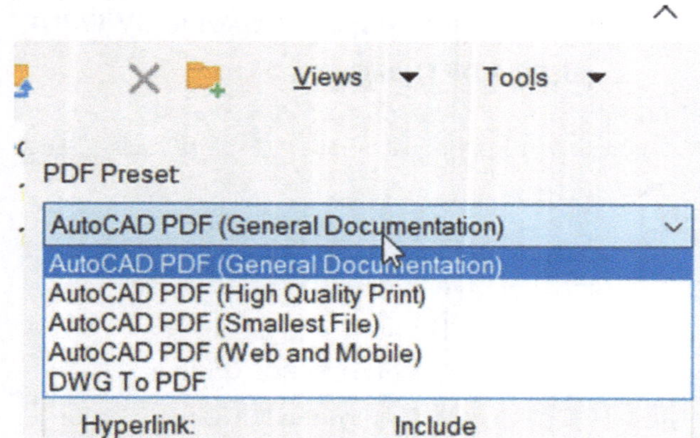

- Examine the options in the **Output Controls** section, Export, and Page Setup.
- Specify the location of the PDF file and click **Save**. A bubble appears on the status bar after exporting the PDF file.
- Open the PDF file in a PDF viewer and notice the layers, bookmarks, hyperlinks in the drawing. Also, you can find any text in the drawing using the text search option in the PDF viewer.

Importing a PDF

AutoCAD allows you to import a PDF into a drawing file.

- On the ribbon, click **Insert > Import > PDF Import** .
- On the **Select PDF File** dialog, browse to the location of the PDF file.
- Select the PDF file, and a preview appears in the **Preview** area of the **Select PDF File** dialog.
- Click **Open**; the **Import PDF** dialog appears.
- On the **Import PDF** dialog, specify the **Scale** factor.
- Under the **PDF data to import** section, check the **TrueType text** option.
- Under the **Layers** section, check the **Use PDF Layers** option.

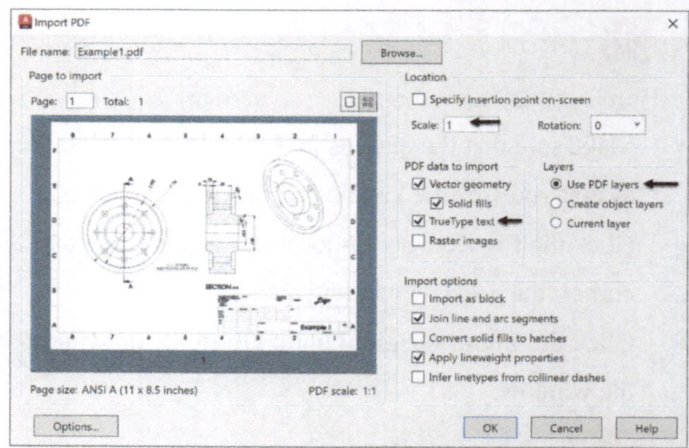

- Click **OK** to insert the PDF into the drawing.

Combining the Text of the Imported PDF

After importing the PDF into an AutoCAD drawing, the text of the PDF is converted into single-line text objects. However, you can combine them into a multi-line text using the **Combine Text** command.

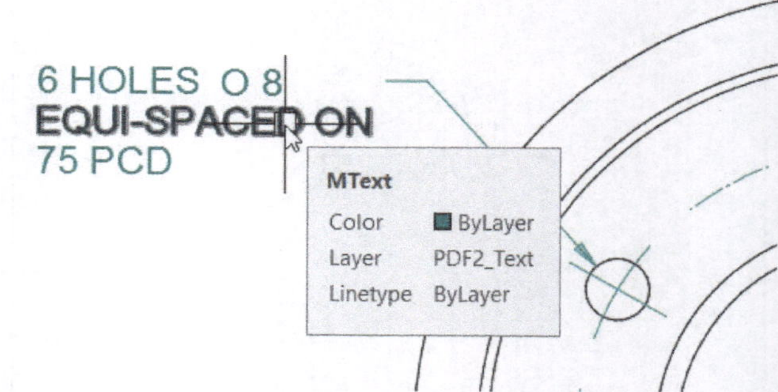

- On the ribbon, click **Insert > Import > Combine Text** .
- Select the fragments of the text to combine them.
- Press Enter.

- Click on the text and notice that the text is converted into a multi-line text.

Publishing a 2D Drawing to a Browser

AutoCAD allows you to publish a drawing to a web browser so that you can view it without any application installed on your device. To publish drawing to a web browser, follow the steps given next:

- Click **Application Menu > Publish > Share View**.

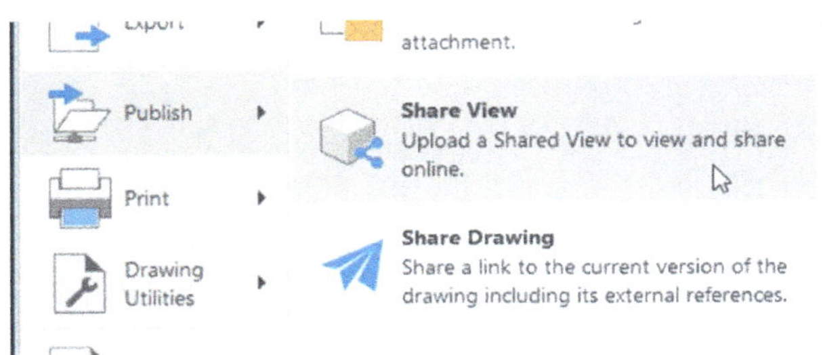

The **Autodesk Sign In** dialog appears if you have not signed in to your Autodesk Account.

- Enter your Autodesk ID and password, and then click **Sign in**.

The **Share View** dialog appears.

- Enter the name of the drawing in the **Name** box.
- Select the appropriate option from the **View to share** section.

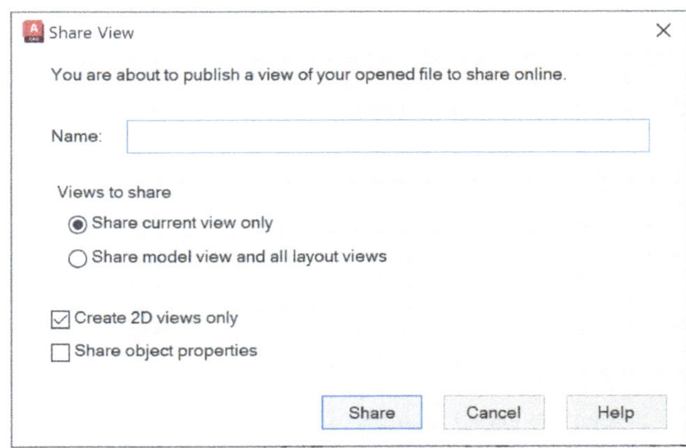

- Click **Share** and **Proceed** to share the views online.
- To view the shared drawing, click the **View in Browser line** displayed at the bottom right corner of the application window; the drawing view will open in the internet browser.

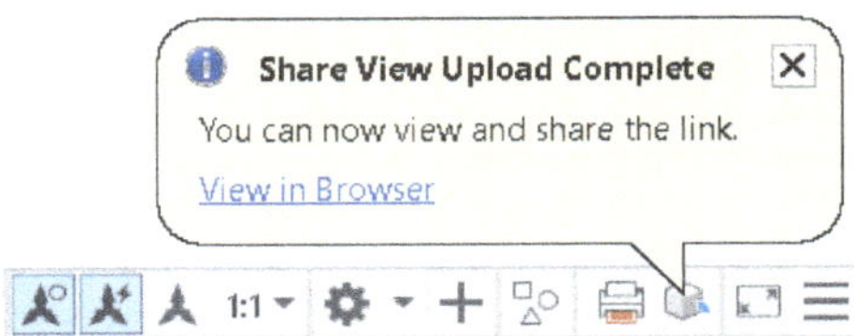

- To add comments to the shared views, click **Collaborate > Share > Shared Views** on the ribbon.
- On the **Shared Views** palette, click the **Refresh** icon to display the shared views.
- Select the shared view from the palette and click **Add Comment**.

Share Drawings

The **Share** command helps you to create a link to copy the drawing. You can view and edit the share drawing in the AutoCAD web app. The shared drawing has all the external references and image attachments. Click the **Share** icon on the Quick Access Toolbar. Next, select the **View only** or **Edit and save copy** option. Click the **Copy Link** button to copy the drawing.

Compare Drawings

The **DWG Compare** command compares two revisions of a drawing or two different drawings.

- Download the DWG_compare1 and DWG_compare2 drawing files from the companion website.
- Open the DWG_compare1 drawing.

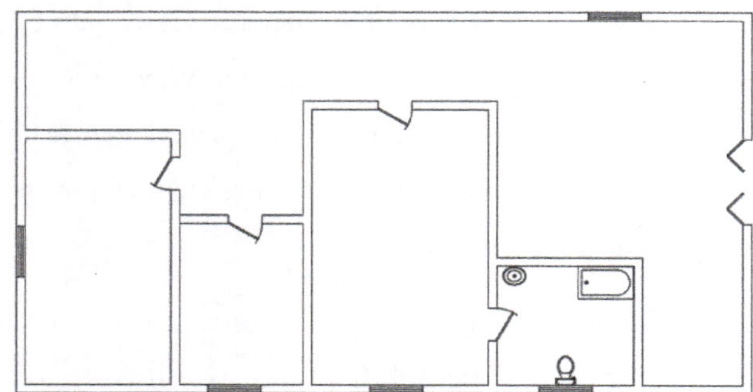

- On the ribbon, click the **Collaborate > Compare > DWG Compare** .
- Browse to the location of the DWG_compare2 file and double click on it; The differences between the two drawings are highlighted by a revision cloud. Also, the DWG Compare toolbar is displayed below the ribbon.

DWG Compare

You can click the **On or off** icon to turn ON or OFF the comparisons.

Use the **Next** and **Previous** icons to zoom to different results.

On the **DWG Compare** toolbar, click the **Settings** drop-down to specify the color settings, revision cloud settings, and objects to be filtered.

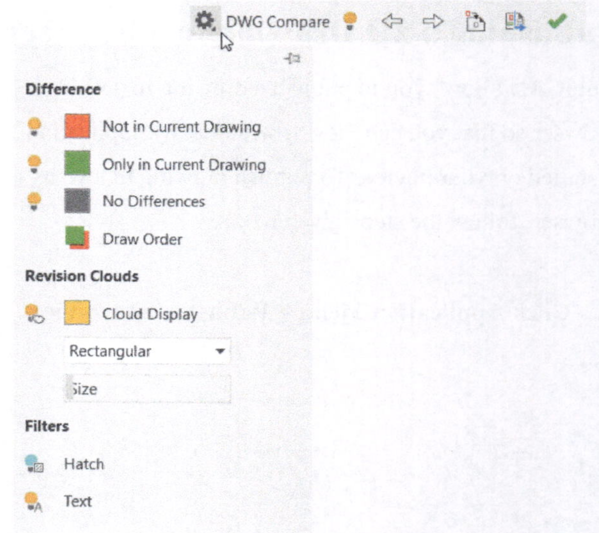

- Select the **Polygonal** option from the drop-down in the **Revision Clouds** section.
- Click and drag the **Margin** dragger to change the margin between the revision cloud and highlighted objects.

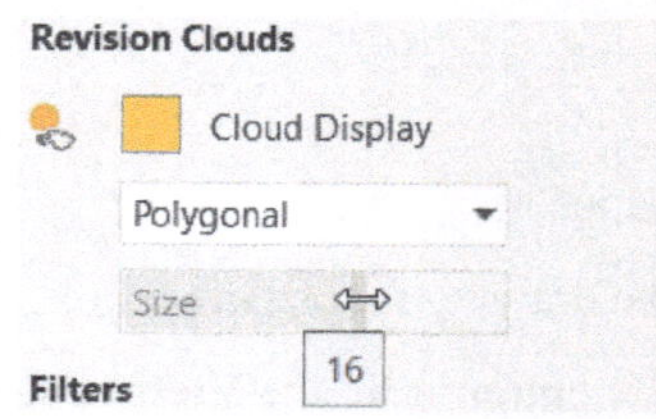

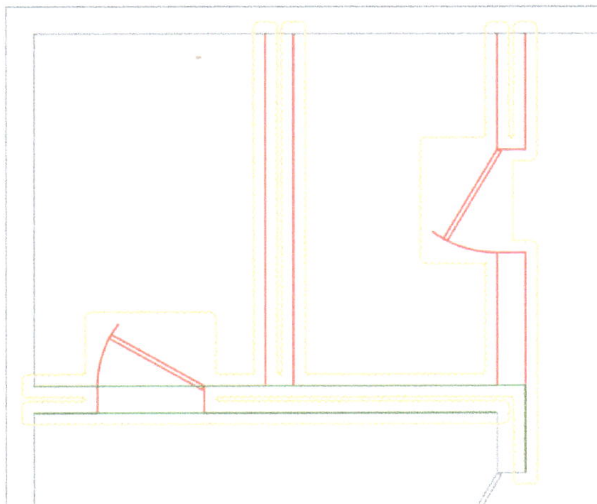

- Click the **Import Objects** icon on the **DWG Compare** toolbar.

- Select the objects highlighted in red color.

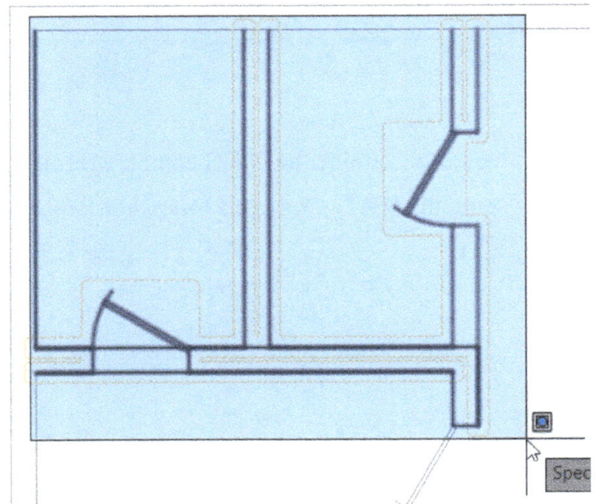

- Press Enter; the selected objects are imported into the current drawing.

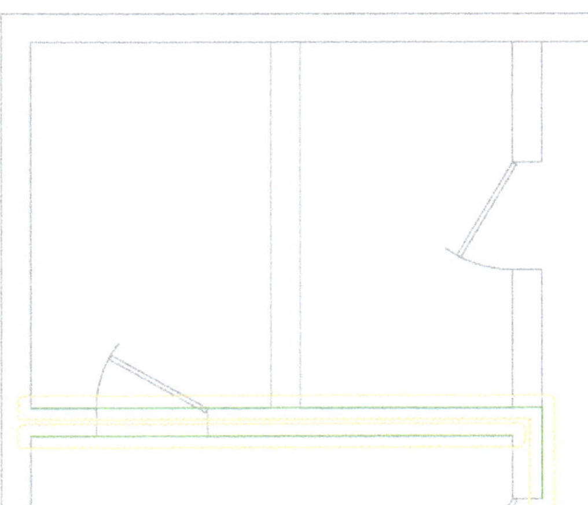

Notice that there is a revision cloud still displayed around the objects highlighted in green. These objects exist only in the current drawing.

- Select the objects highlighted in green color and press Delete on your keyboard; the revision clouds disappear.

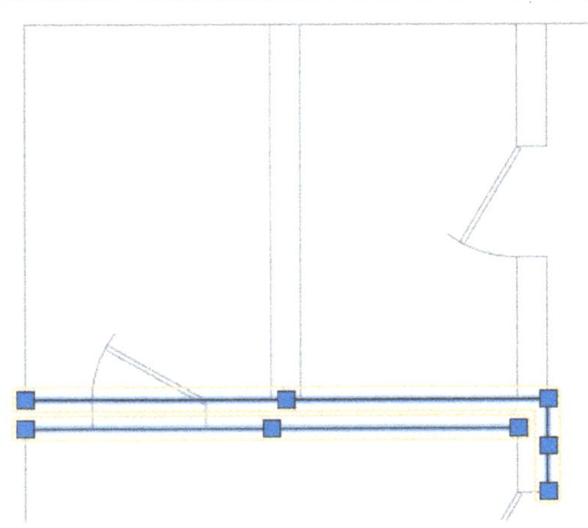

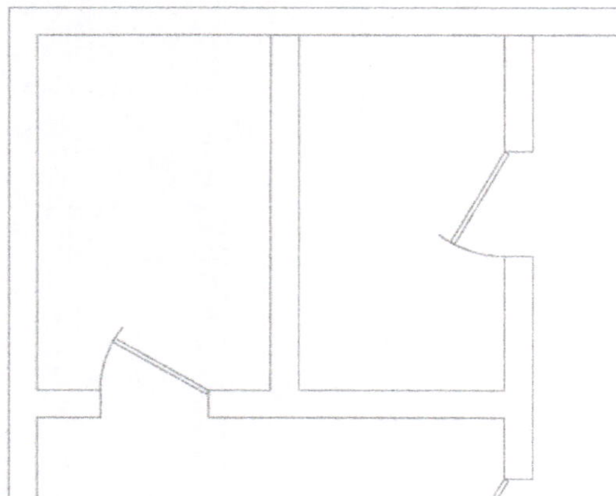

- Click the **Exit Compare** icon on the **DWG Compare** toolbar.
- Close the DWG_compare1 file without saving.

Exporting the Compared results to a new drawing

- Open the DWG_compare1 drawing.

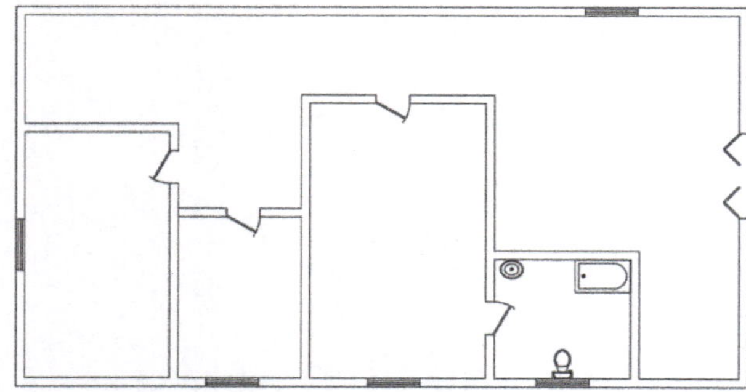

- On the ribbon, click **Collaborate > Compare > DWG Compare** .
- Browse to the location of the DWG_compare2 file and double click on it; The differences between the two drawings are highlighted by a revision cloud. Also, the DWG Compare toolbar is displayed below the ribbon.
- Click the **Export Snapshot** icon on the **DWG Compare** toolbar.

- Specify the location of the export file and click **Save**; the exported file is opened in another tab.
- Click **Continue** on the **Compare – Export a Comparison Snapshot** message box.

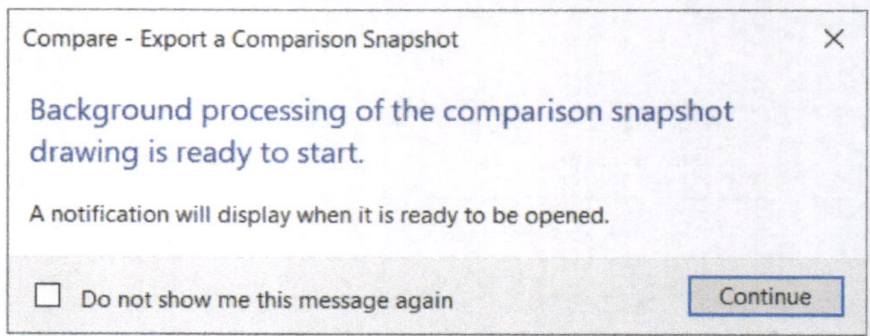

- Click **Open Drawing** on the **Snapshot Processing Complete** bubble.

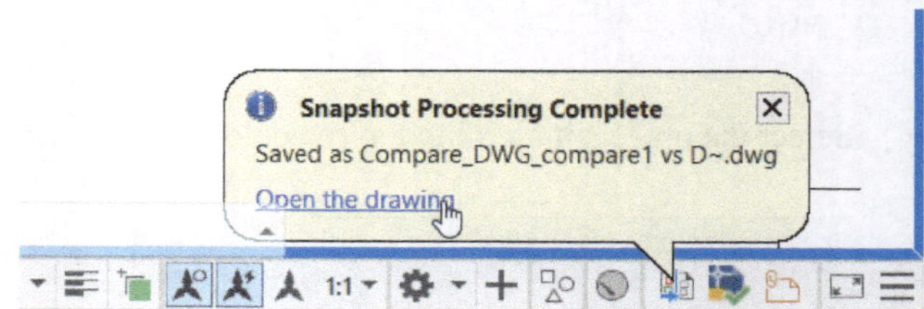

You can examine the DWG compare results using the **DWG Compare Snapshot** toolbar.

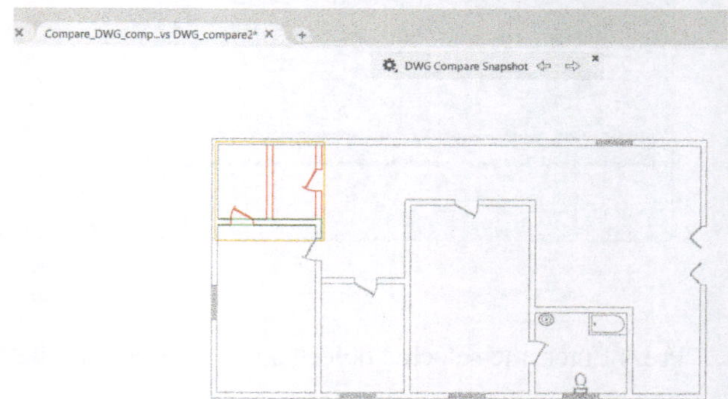

- Close all the files.

Exercise

Create and plot the drawing, as shown in the figure.

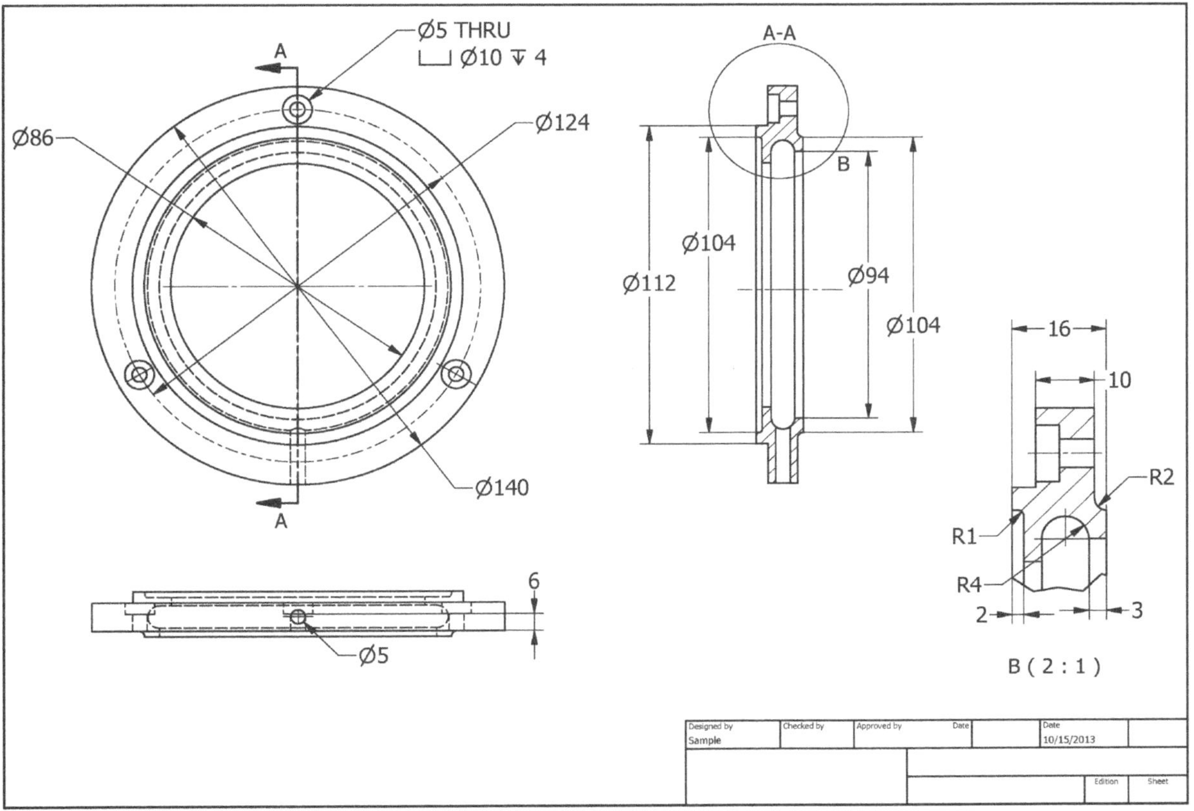

Chapter 12: 3D Modeling Basics

In this chapter, you will learn to do the following:

- **Create boxes, cylinders, wedges, cones, pyramids, spheres, and torus**
- **Create User Coordinate Systems**
- **Work with Dynamic UCS**
- **Change the View Style of objects**
- **Create Viewports in model space**
- **Create walls using the Polysolid tool**
- **Change the view orientation**
- **Create extruded, revolved, swept, lofted, and press-pulled objects**
- **Perform Boolean operations**
- **Align objects**
- **Create spiral and helical curves**

Introduction

In AutoCAD, you can create three types of 3D models: surfaces, solids, and meshes. Solids are used to create 3D models of engineering components and assemblies, and surfaces are used to create complex shapes such as plastic parts and meshes that are used for games and movies. Solids are three-dimensional models of actual objects that possess physical properties such as mass properties, the center of gravity, surface area, and moments of inertia. Surfaces are construction features without any thickness. They do not possess any physical properties. Meshes are similar to solids without mass and volume properties. In this chapter, you will learn the basics of 3D modeling, such as creating, navigating, and visualizing solid models.

3D Modeling Workspaces in AutoCAD

In AutoCAD, there are separate workspaces created to work on 3D models. In these workspaces, the tools are organized into ribbon tabs, menus, toolbars, and palettes to perform a specific task in 3D modeling. You can activate these workspaces by using the **Workspace** drop-down located on the **Quick Access Toolbar**, or by using the **Workspace Switching** menu on the status bar. You can also start an AutoCAD session directly in the 3D Modeling workspace using the **acad3D.dwt**, **acadiso3D.dwt**, **acad -Named Plot Styles3D**, or **acadISO-Named Plot Styles3D** templates.

Tip: *If the **Workspace** drop-down is not displayed at the top left corner, then click the down arrow next to the Quick Access Toolbar. Next, select **Workspace** from the drop-down; the **Workspace** drop-down will be visible on the Quick Access Toolbar.*

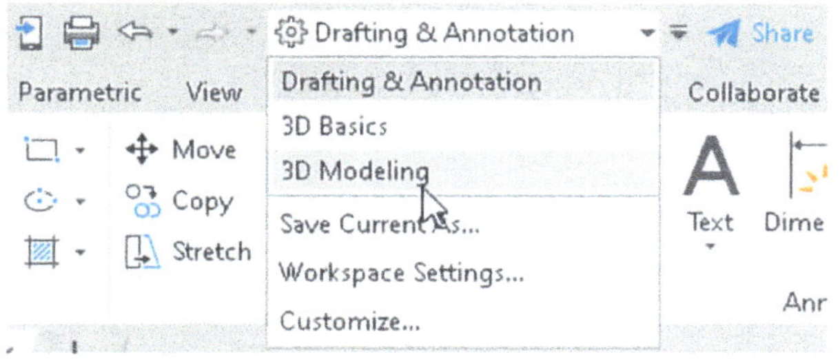

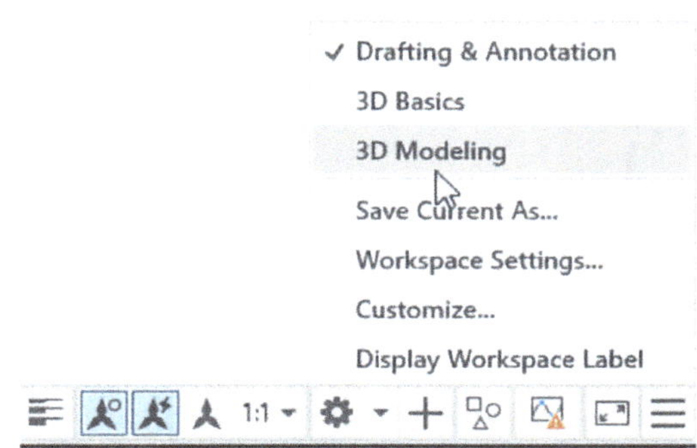

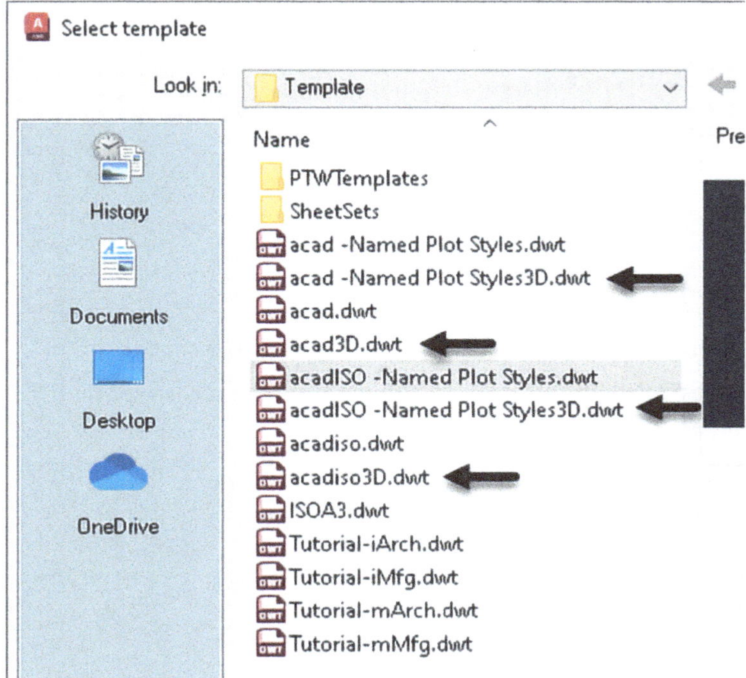

There are two workspaces of 3D modeling: **3D Basics** and **3D Modeling**. The **3D Basics** workspace has commonly used tools, whereas the **3D Modeling** workspace includes all the tools required for creating 3D models.

The 3D Modeling Workspace

Activating the **3D Modeling** workspace either by using the template or from the **Workspace** drop-down displays the screen as shown below. It contains the ribbon and tools related to 3D modeling. By default, the **Home** tab is activated in the ribbon. From this tab, you can access the tools for creating and editing solids and meshes, modifying the model display, working with coordinate systems, and sectioning 3D models.

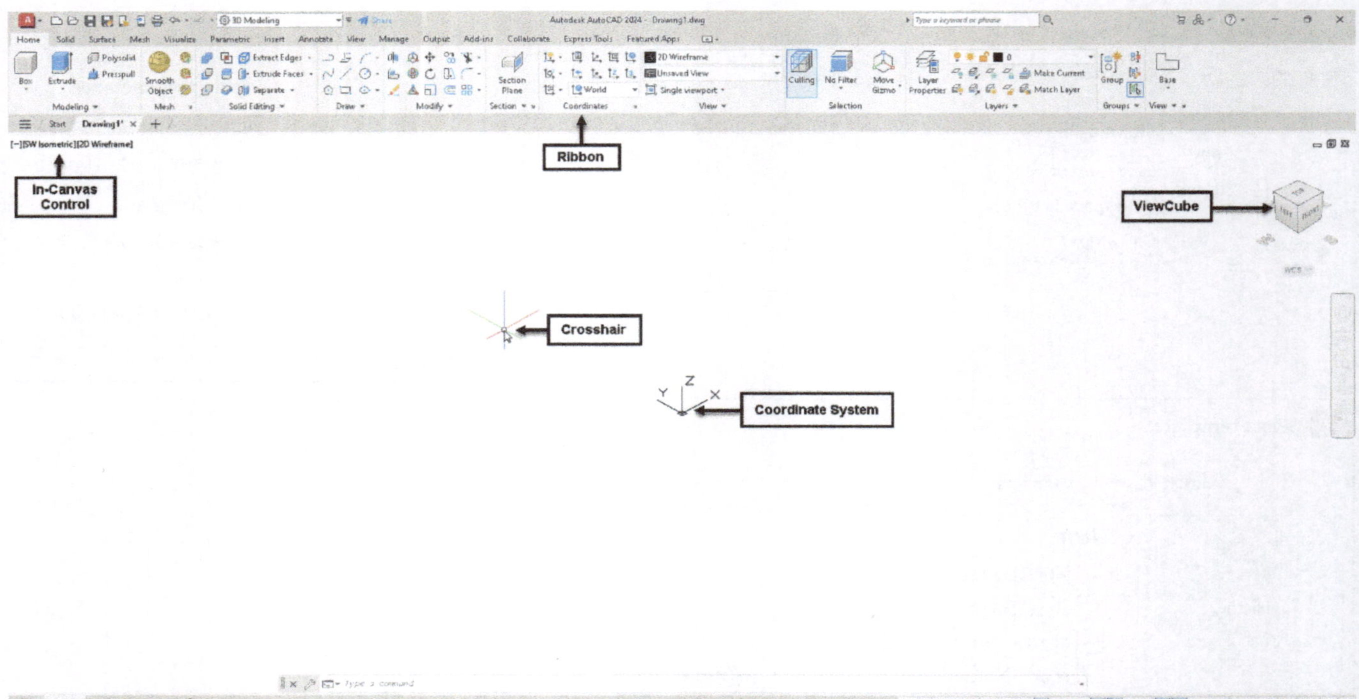

There are some additional tabs, such as **Solid**, **Surface**, **Mesh**, and **Render**. The **Solid** tab contains tools to create solid models; the **Surface** and **Mesh** tabs are used to create surface models and complex shapes; the **Visualize** tab is used for creating realistic images of solid and surface models.

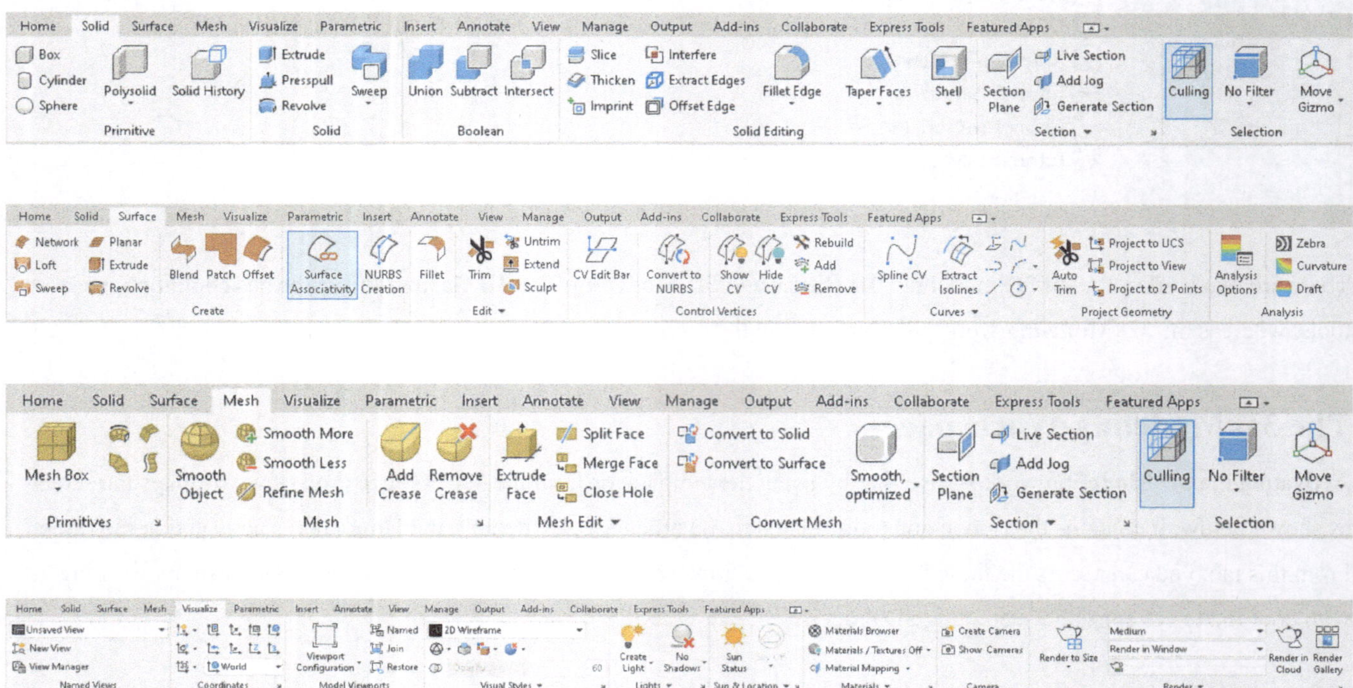

The **ViewCube** can be used to modify the view of the model quickly and easily. It is located at the top right corner of the graphics window. Using the **ViewCube**, you can switch between the standard and isometric views, rotate the model, switch to the **Home** view of the model, and create a new user coordinate system. You can also change the way the ViewCube functions by using the **ViewCube Settings** dialog. Right-click on the ViewCube, and then select the **ViewCube Settings** option; the **ViewCube Settings** dialog will be opened.

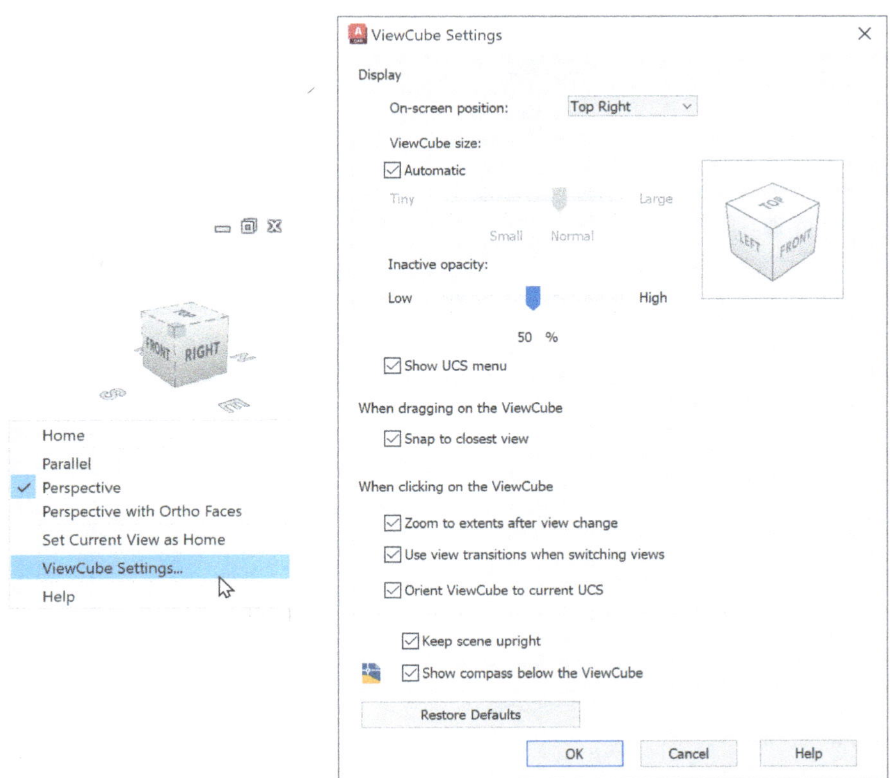

You can also modify the model view by using the In-canvas controls. In addition to that, you can also change the view style of the model and control the display of other tools in the graphics window using the In-canvas controls.

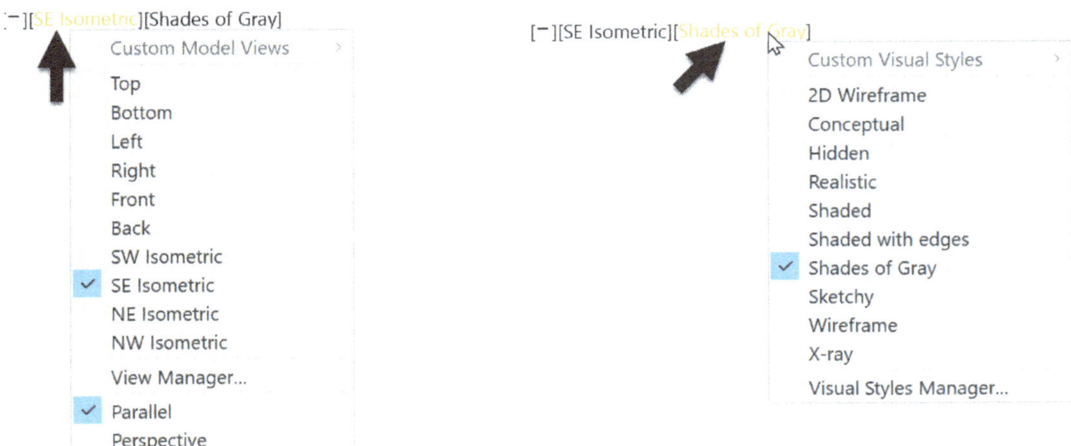

Now, you will create 3D models using the tools available in AutoCAD.

The Box tool

The **Box** tool is used to create boxes having six rectangular or square faces. It is the most commonly used tool, as many 3D objects are made of boxes.

- Click the **AutoCAD 2024** icon on your desktop.
- On the Quick Access Toolbar, click the **New** icon.
- On the **Select Template** dialog, click **acadiso3D**, and then click **Open**. A new file will be started in the **3D Modeling** workspace.
- Click **Home > Modeling > Box** on the ribbon or type **BOX** in the command line; the message, "Specify the first corner," appears in the command line.

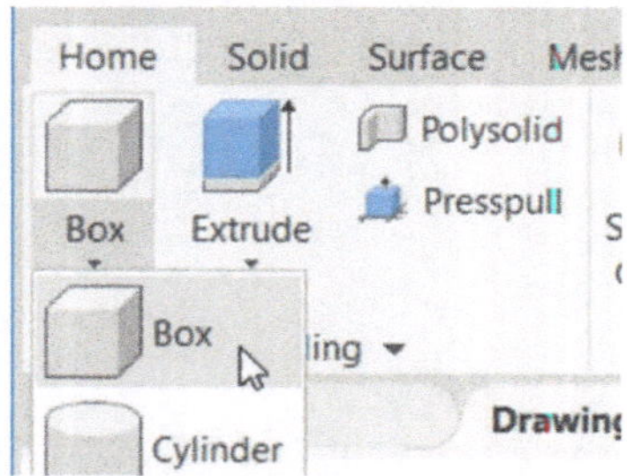

- Pick an arbitrary point in the graphics window; the message, "Specify the other corner," appears in the command line.
- Ensure that the **Dynamic Input** icon is active on the status bar. You will notice the two value boxes to specify the length and width of the box.
- Type 100 in the length box and press the TAB key.
- Type 70 in the width box and press ENTER.
- Move the pointer upward, type 60 as height, and press ENTER; the box will be created as shown.
- Click **Zoom > Zoom All** on the Navigation Bar.
- On the In-canvas controls, click **View Style Controls > Shades of Grey**.
- Right click on the **Home** icon above the ViewCube and select **Parallel**.

Creating the User Coordinate System

User Coordinate Systems assist you while creating 3D models. They are used to create construction planes on which you can add additional features to the model. Various methods to create User coordinate systems are discussed next.

Example1:

- On the status bar, click the **Customization** button and select **Dynamic UCS** from the menu. Also, select **3D Object Snap** from the menu.
- Deactivate the **Dynamic UCS** icon on the status bar. You will learn about this option later in this chapter.

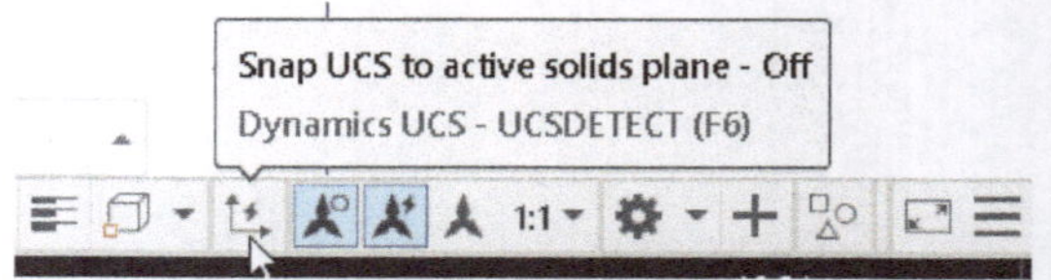

- Click **Home > Coordinates > UCS** on the ribbon; the UCS is attached to the pointer, and the message, "Specify the origin of UCS," appears.

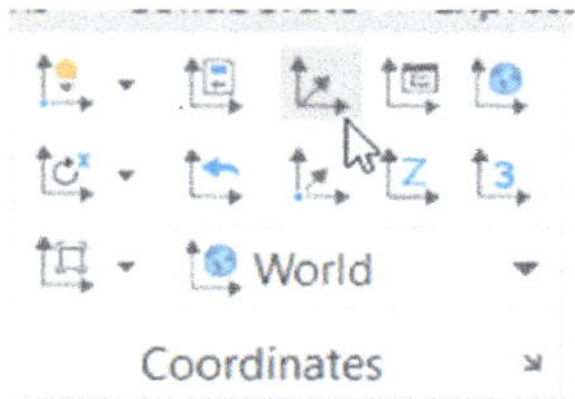

- Activate the **3D Object Snap** icon on the status bar.

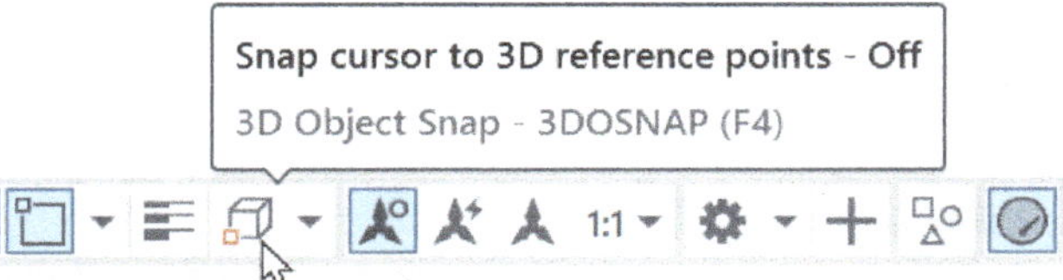

- Select the vertex point on the top left corner of the box as shown below; the message, "Specify point on X-axis or <accept>:" appears in the command line.

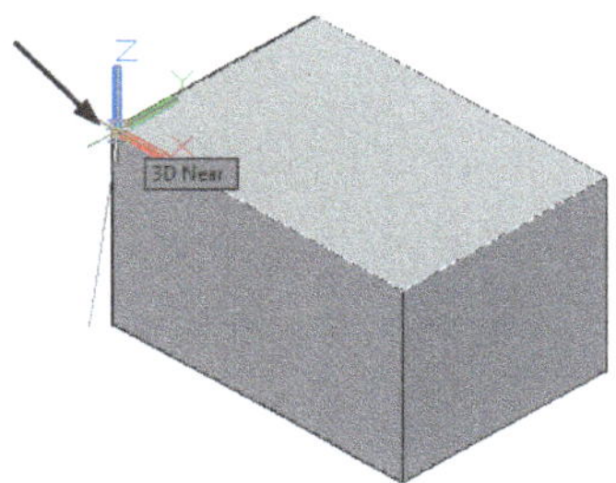

- Press ENTER to accept the orientation of the UCS, as shown below.

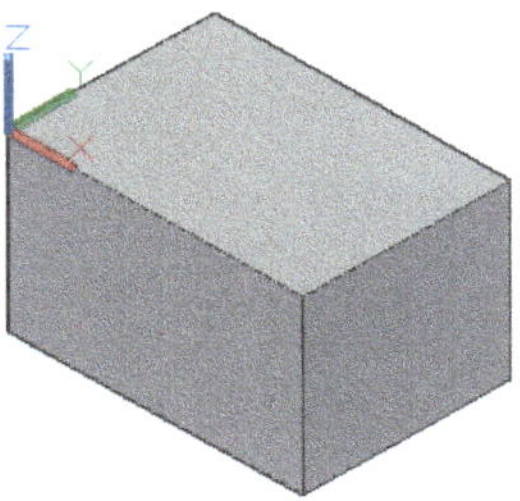

Creating a Wedge

When you slice a box diagonally, it results in a wedge. A wedge has five faces, three rectangular and two triangular.

- Click **Home > Modeling > Primitives** drop-down > **Wedge** on the ribbon or type **WE** in the command line and press ENTER; the message, "Specify first corner or [Center]" appears in the command line.
- Select the endpoint of the top face of the box, as shown in the figure; the message, "Specify other corner or [Cube Length]:" appears in the command line.
- On the Status bar, click the down arrow next to the **3D Object Snap** icon, and select **Midpoint on edge**.
- Select the midpoint of the front edge of the box, as shown below.

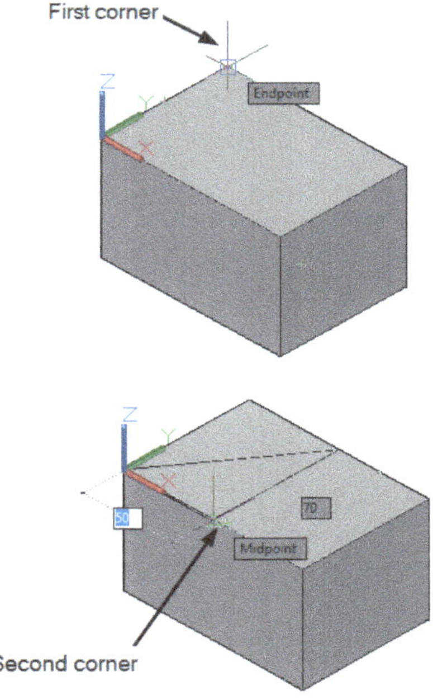

- Move the pointer upward and enter 40 as the height; the wedge will be created, as shown below.

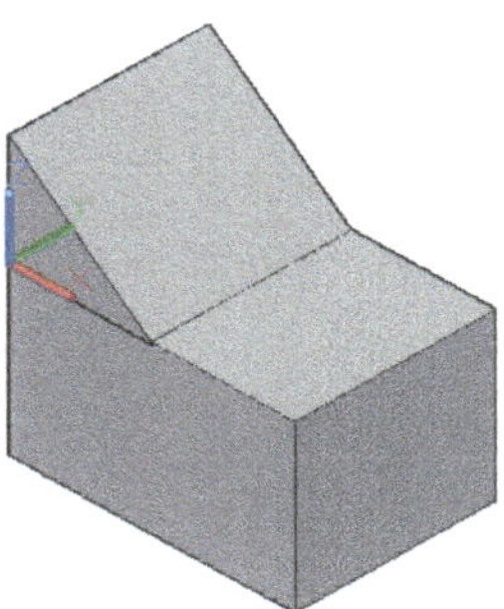

Example2: (Creating UCS by selecting 3-points)

You can create a UCS by selecting three points. The first point will be the origin of the UCS, the second point will define the X-axis, and the third point defines the Y-axis.

- Click **Home > Coordinates > 3 Point** on the ribbon; the UCS is attached to the pointer and the message, "Specify new origin point <0,0,0>:" appears.

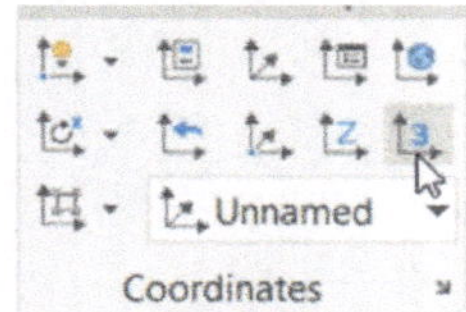

- Select the lower endpoint of the wedge, as shown in the figure.

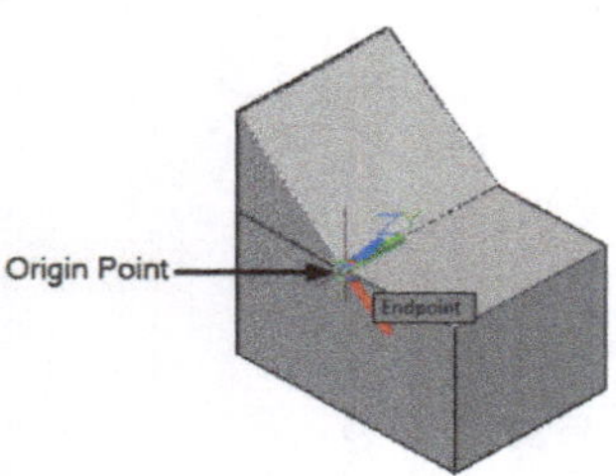

- Move the pointer toward the right.
- Select the other endpoint of the bottom edge of the wedge, as shown in the figure.

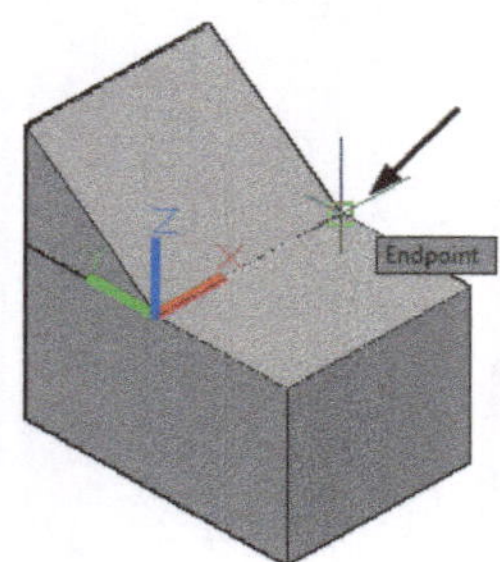

- Move the pointer along the diagonal edge of the wedge
- Select the endpoint on the top edge, as shown. The UCS will be created and aligned to the inclined face of the wedge.

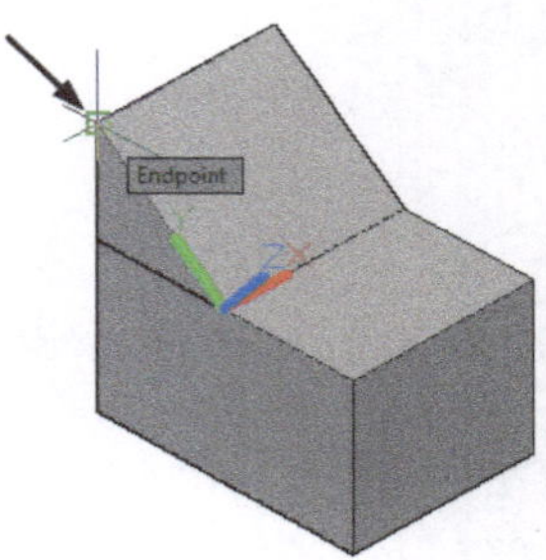

Creating a Cylinder

Cylinders are commonly used features after boxes. In AutoCAD, you can create cylinders easily by using the **Cylinder** tool. You can create a circular or elliptical cylinder by using this tool.

- Click **Home > View > View Styles** drop-down > **Wireframe** on the ribbon; the view style of the model will be changed to the wireframe style.

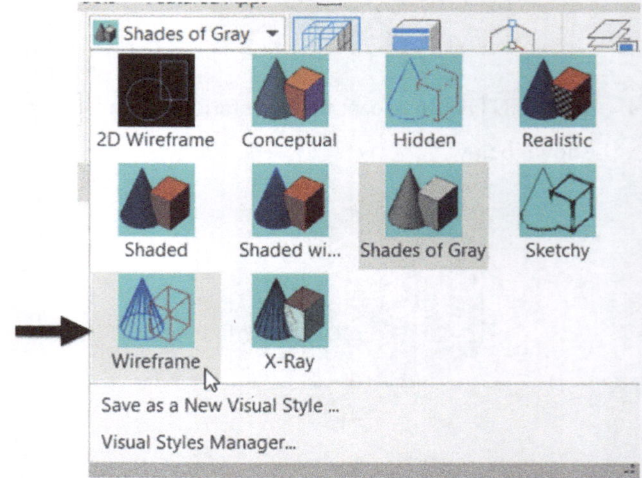

- On the Status bar, click the down arrow next to the **3D Object Snap** icon and select the **Center of face** option, if not already selected.
- Click **Home > Modeling > Primitives drop-down > Cylinder** on the ribbon or type **CYL** in the command line.
- Snap to the center point of the inclined face, and then click to select it; the center point of the cylinder is

specified on the inclined face of the wedge.

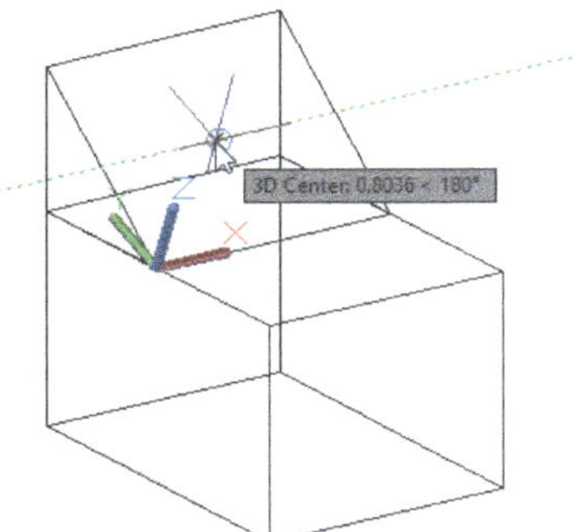

- Type 20 as the base radius and press ENTER.
- Move the pointer upward; you will notice that the pointer moves along the Z-axis of the UCS.
- Type 25 as height and press ENTER; the cylinder will be created as shown below.

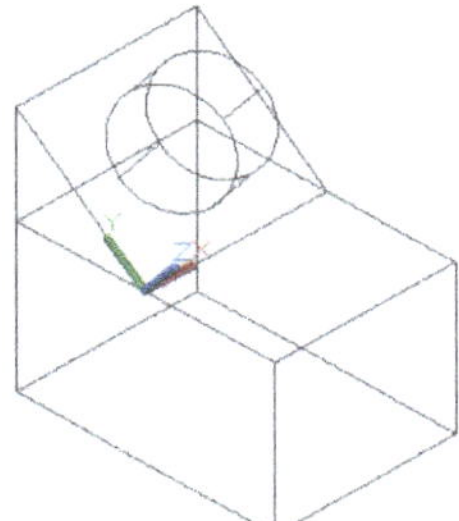

Example 3: (Returning to the previous position of the UCS)

- Click **Home > Coordinates > UCS, Previous** on the ribbon; the UCS will return to its previous position.

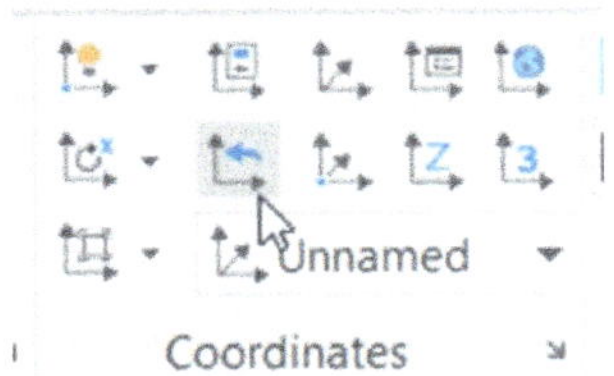

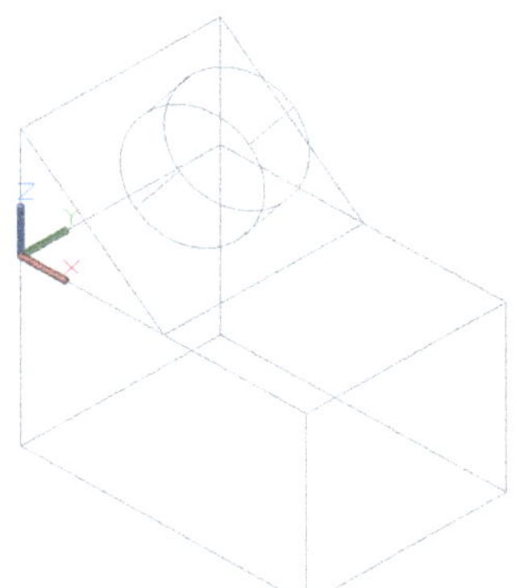

Example 4: (Creating a UCS by specifying its origin)

- Click **Home > Coordinates > Origin** on the ribbon; the UCS will be attached to the pointer.

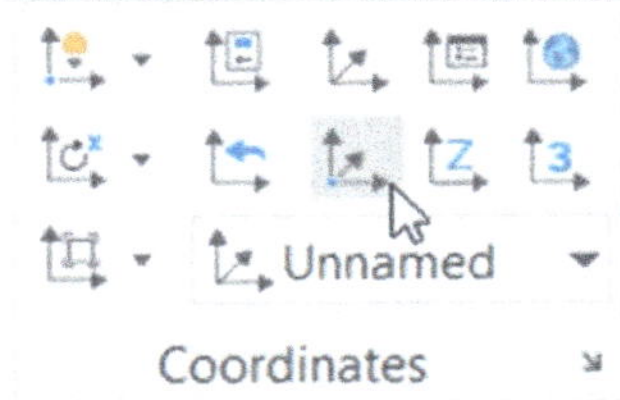

- Select the lower-left corner point of the box; the UCS will be placed at that point. Note that the orientation will not change.

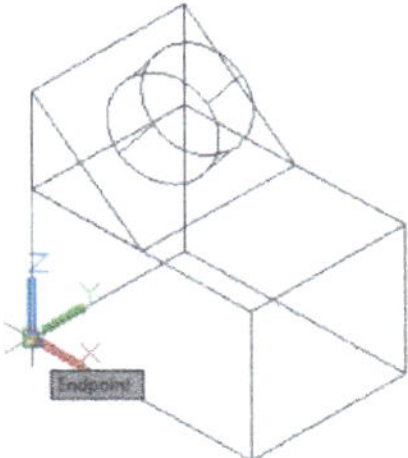

Example 5: (Rotating the UCS about X, Y, and Z axes)

You can rotate a UCS about X, Y, or Z axes by using the drop-down available in the **Coordinates** panel, as shown below.

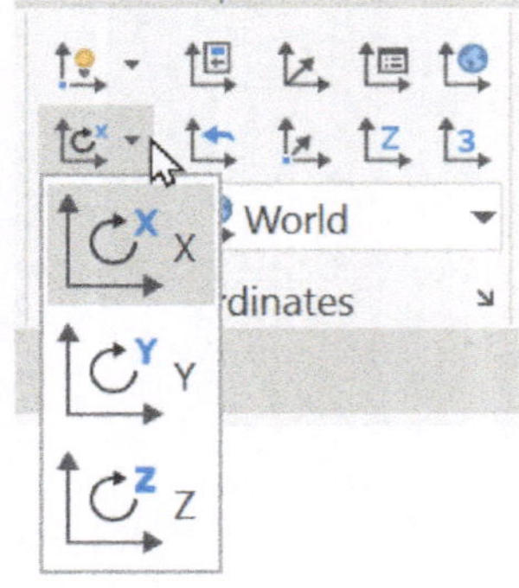

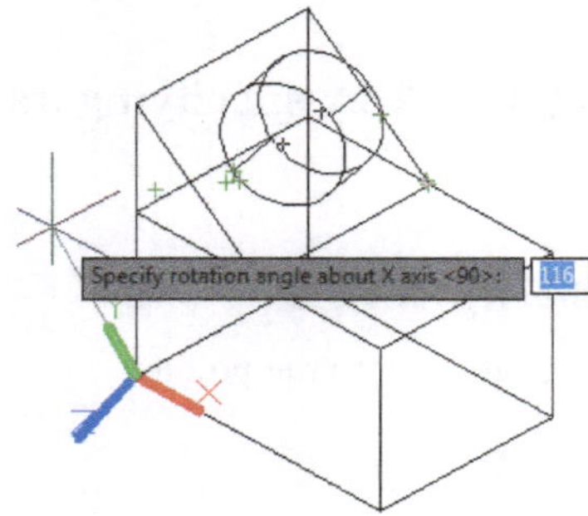

- Click the **X** option from the drop-down shown in the above figure; the message, "Specify rotation angle about X axis <90>:" appears in the command line. Also, a rubber band line originating from the Y axis is attached to the pointer.
- Rotate the pointer and pick a point to specify the rotation angle. You can also type in the rotation angle in the dynamic input or command line.
- Similarly, you can rotate the UCS about the Y and Z axes using the respective options from the drop-down.

Example 6: (Creating the UCS by specifying the Z-axis)

Using the **Z-Axis Vector** tool, you can create a UCS by specifying its Z-axis.

- Click **Home > Coordinates > Z-Axis Vector** on the ribbon.
- Select the bottom right endpoint as the origin; the message, "Specify point on positive portion of Z-axis:" appears in the command line. Also, a rubber band line originating from the Z-axis is attached to the pointer. Now, as you move the pointer, you will notice that the Z-axis also moves.

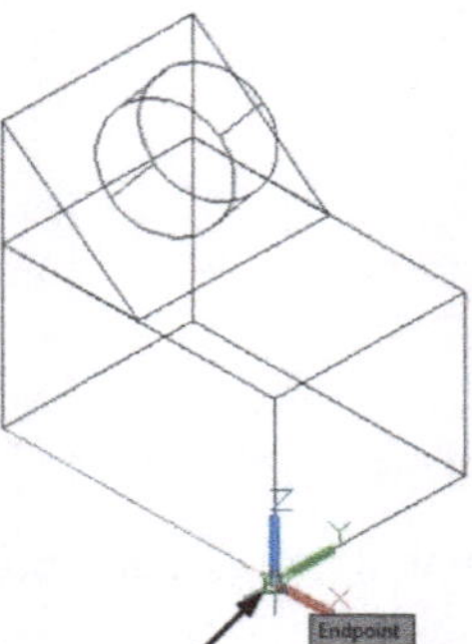

- Move the pointer and select the left endpoint of the bottom edge, as shown below; the Z-axis will be aligned to the bottom edge.

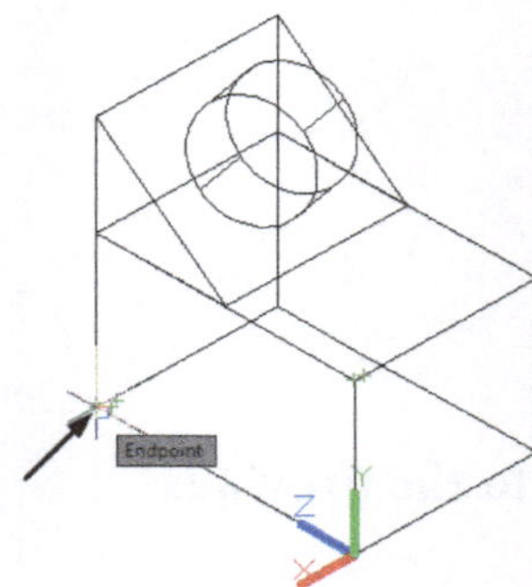

Example 7: (Creating UCS parallel to the screen)

Using the **View** tool in the **Coordinates** panel, you can create a UCS which is parallel to the screen.

- Click **Home > Coordinates > View** on the ribbon; the XY plane of the UCS will become parallel to the screen. The UCS origin will not change. This option is useful if you want to use the current view and add a title block or any other annotation.

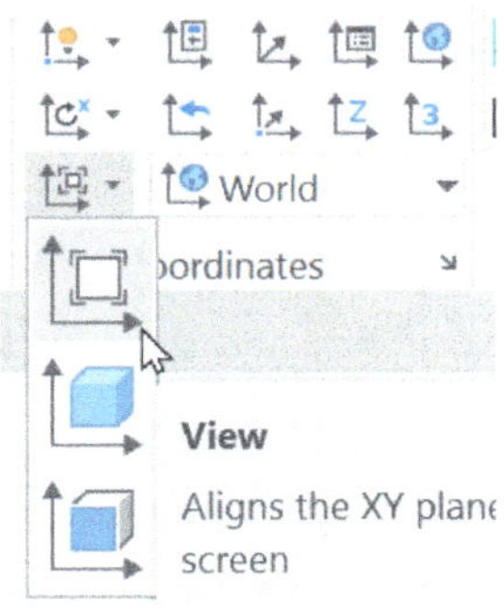

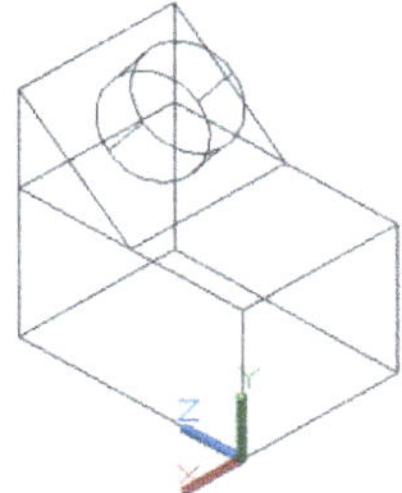

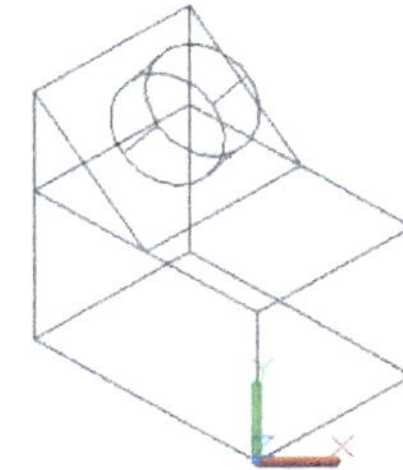

Example 8: (Creating UCS aligned to an object)

You can create a UCS aligned to an object. The origin of the UCS will be aligned to the nearest endpoint of the object.

- Click **Home > Coordinates > View > Object** on the ribbon; the message, "Select object to align UCS:" appears in the command line.
- Select the cylindrical object from the model; the UCS will be aligned to it.

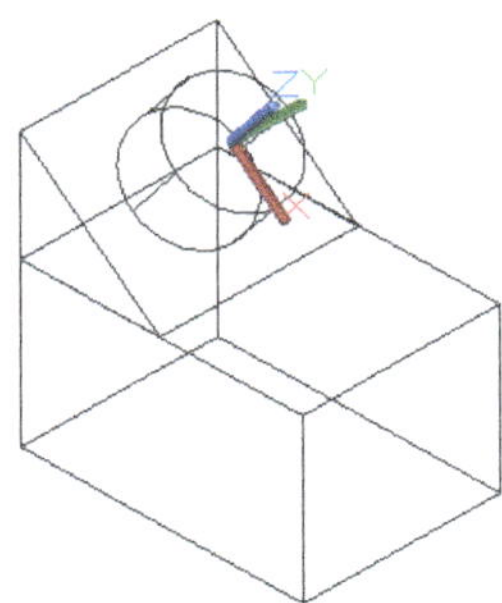

Example 9: (Creating UCS aligned to a face)

You can align a UCS to a planar or curved face of a model using the **Face** tool.

- Click **Home > Coordinates >View** drop-down > **Face** on the ribbon; the message, "Select face of solid, surface, or mesh:" appears in the command line.
- Move the pointer over the faces of the model; you will notice that the UCS is displayed on the faces.

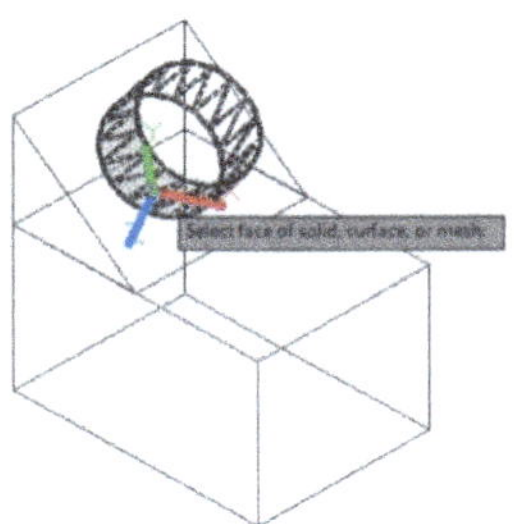

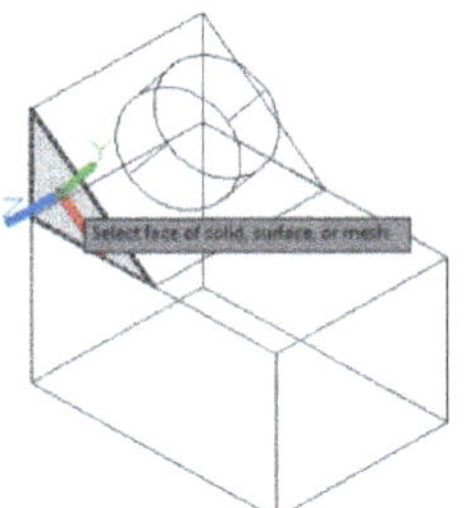

- Select the top face of the box; the message, "Enter an option [Next Xflip Yflip] <accept>:" appears in the command line.
  If you select the **Next** option, the adjacent face will be highlighted. The **Xflip** option is used to rotate the UCS 180 degrees about the X axis. The **Yflip** option is used to rotate the UCS 180 degrees about the Y axis.
- Press ENTER to accept; the UCS will be aligned to the selected face.

Using Dynamic User Coordinate System

In the previous section, you have learned to create various types of static user coordinate systems. They are active until you define another user coordinate system. You can also create dynamic user coordinate systems. A Dynamic User Coordinate System is a temporary UCS that appears automatically when you place your pointer over the face of a 3D solid object. Note that the Dynamic User Coordinate system appears only when you use tools that create objects directly (For example, drawing tools and primitive tools). To create a Dynamic UCS, you need to activate the **Dynamic UCS** option on the status bar.

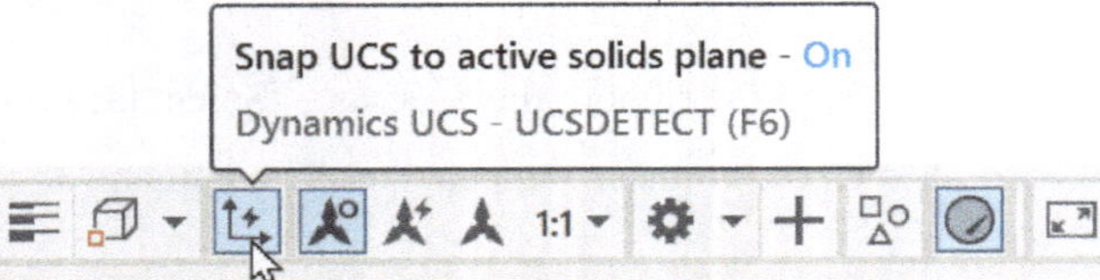

- Click the **Cylinder** button on the **Modeling** panel.
- Ensure that the **Dynamic UCS** button is active on the status bar.
- Move the pointer over the faces of the model; they will be highlighted.
- Click on the front face of the box and create the cylinder, as shown below.

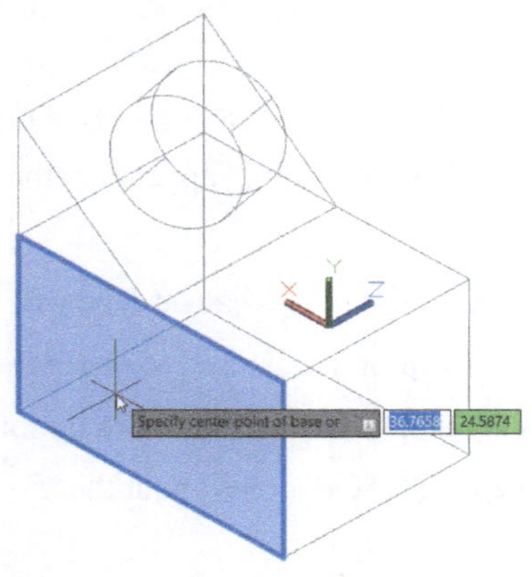

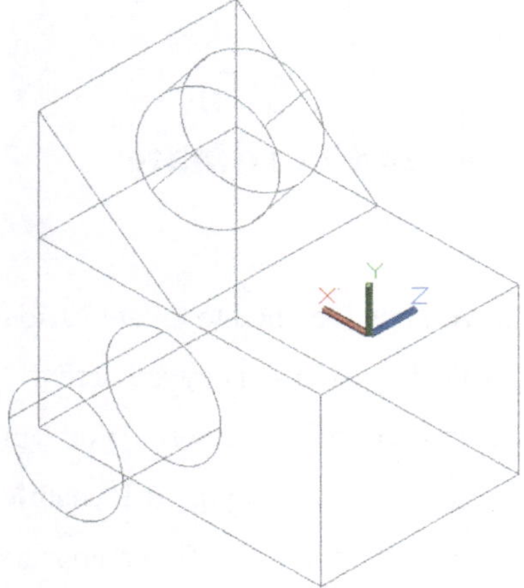

Model Space Viewports for 3D Modeling

While creating 3D models, it is useful to have a look at your model from several different orientations at the same time. For this purpose, you need to create different viewports in model space. You can create multiple viewports in model space using the **Viewport Configuration** drop-down available in the **Model Viewports** panel of the **Visualize** tab. It can also be done by using the **Viewports** dialog. To load this dialog, click **Visualize > Model Viewports > Named**; the **Viewports** dialog appears. In the dialog, select the **New Viewports** tab and then select **Four: Equal** from the **Standard viewports** list. Next, select **3D** from the **Setup** drop-down. Click the **OK** button; four tiled viewports are displayed on the screen. You can notice that each viewport has a different view and a different UCS. Click inside any viewport to activate it and perform any operation. To return to the single viewport, click the **Restore Viewports** button on the **Model Viewports** panel; the currently active viewport will fill the screen area.

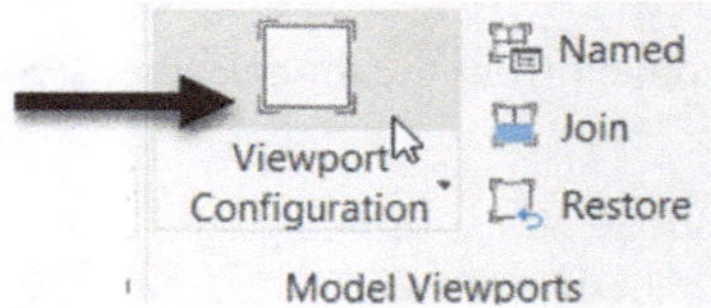

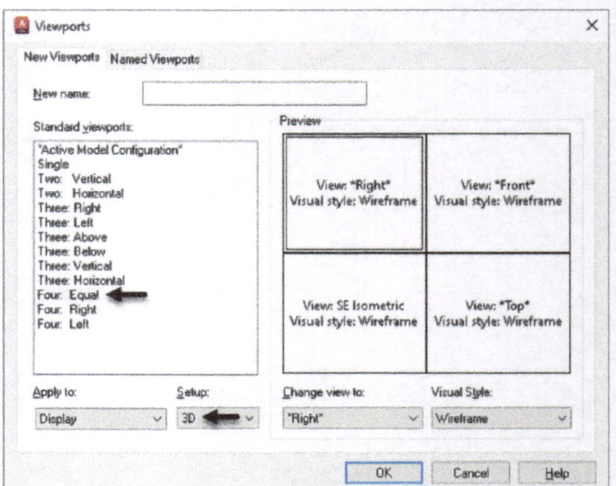

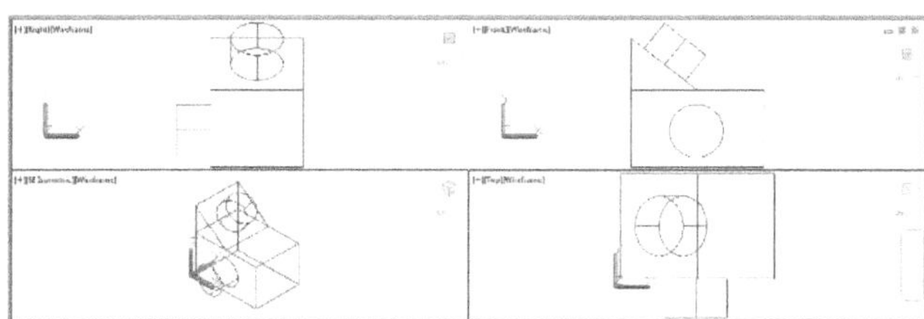

Creating Other Primitive Shapes

In AutoCAD, there is a set of tools to create basic geometric shapes. In earlier sections, you have learned to create boxes, wedges, and cylinders. Now, you will learn to create other primitive shapes.

Creating Cones

Creating a cone is similar to creating a cylinder. It has a similar shape compared to a cylinder, but it is tapered on one side.

Example 1:

- To create a cone, click **Home > Modeling > Primitives drop-down > Cone** on the ribbon; the message, "Specify center point of base or [3P/2P/Ttr/Elliptical]:" appears in the command line.
- Pick an arbitrary point from the graphics window; the message, "Specify base radius or [Diameter]:" appears.
- Type a radius value in the command line and press ENTER. You can also select the **Diameter** option to specify the diameter of the base.
- Move the pointer in the vertical direction and pick a point to specify the height of the cone. You can also type in the height value in the command line and press ENTER; the cone will be created.

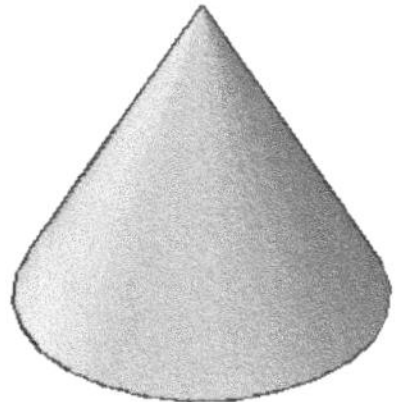

Example 2:

- Type **CONE** in the command line and press ENTER.
- Select the **Elliptical** option from the command line; the message, "Specify endpoint of first axis or [Center]:" appears in the command line.
- Pick a point to specify the endpoint of the first axis.
- Move the pointer and click to specify the other endpoint of the first axis. You can also type in the length of the first axis and press ENTER; the message, "Specify endpoint of second axis:" appears.
- Pick a point or type-in the radius value to specify the second axis.
- Move the pointer upward and pick a point to specify the height. You can also enter the value of height in the command line or the **Dynamic Input** box.

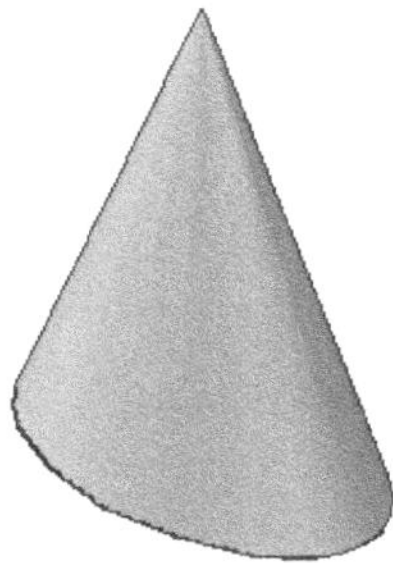

Example 3:

- Click **Solid > Primitive > Primitive drop-down > Cone** on the ribbon.
- Select the center point and specify the base radius as 20; the message, "Specify height or [2Point/Axis endpoint/Top radius]," appears in the command line.
- Select the **Top radius** option from the command line; the message, "Specify top radius:" appears.

- Type 10 as the top radius value and press ENTER.
- Move the pointer upward and enter 40 as the height.

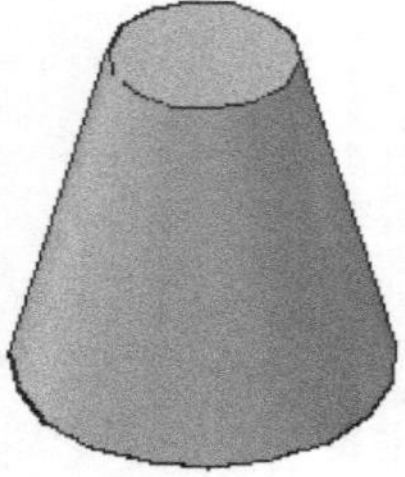

Creating a Sphere

- Click **Home > Modeling > Primitives drop-down > Sphere** on the ribbon.
- Specify the center point of the sphere.
- Move the pointer outward and enter the radius value. You can also select the **Diameter** option to specify the diameter of the sphere.

Creating a Torus

Torus is a donut-shaped solid primitive. To create a torus, you need to specify the center of the torus, radius or diameter of the torus, and radius or diameter of the tube.

- Click **Home > Modeling > Primitives drop-down >Torus** on the ribbon or type **TOR** in the command line and press ENTER.
- Specify the center point of the torus.
- Move the pointer outward and enter the radius of the torus. You can also select the **Diameter** option to specify the diameter of the torus.
- Type the tube radius and press ENTER; the torus will be created.

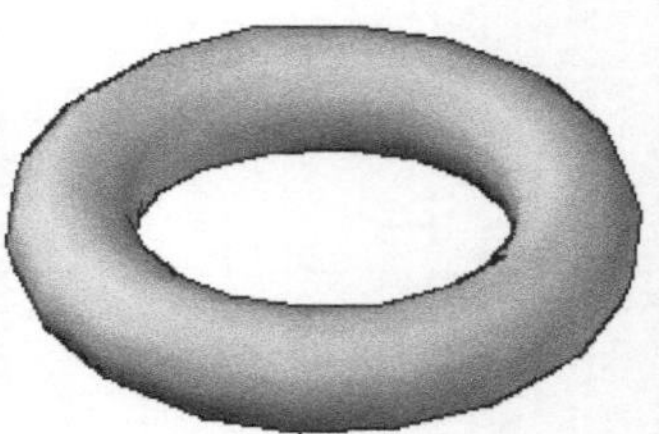

Creating a Pyramid

Pyramids are similar to cones except that the base of the pyramid is not circular in shape.

- To create a pyramid, click **Home > Modeling > Primitives drop-down > Pyramid** on the ribbon or type **PYR** in the command line and press ENTER.
- Specify the center point of the base. The base of the pyramid is a polygon. The method to create a polygon is already discussed in Chapter 2.
- After creating the base, move the pointer in a vertical direction and pick a point to specify the height of the pyramid. You can also type the value of the height and press ENTER; the pyramid will be created.

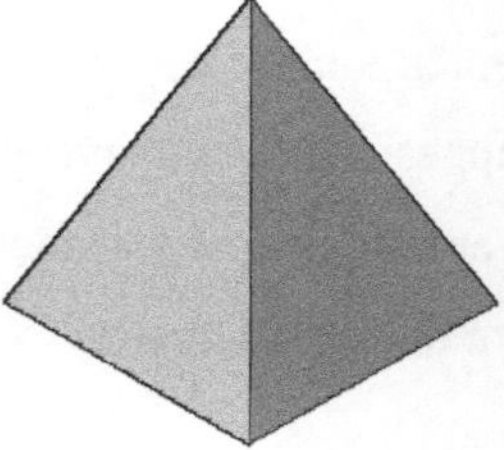

The other options displayed in the command line while creating the pyramid are the same as in the **Cone** tool.

Using the Polysolid tool

The **Polysolid** tool is used to create a 3D wall. It can also be used to convert a line, polyline, arc, or a circle to a wall. The **Polysolid** tool is similar to the **Polyline** tool except that you create a rectangular-shaped wall that has a pre-defined height and width.

- Click **Home > Modeling > Polysolid** on the ribbon; the message, “Specify start point or [Object/Height/Width/Justify] <Object>:” appears in the command line.
- Activate the **Ortho Mode** on the status bar.
- Pick an arbitrary point in the graphics window and move the pointer in the X-direction.
- Type 200 in the command line and press ENTER; a 3D wall of 200 length is created.
- Select the **Arc** option from the command line and move the pointer in the Y-direction.
- Type 100 as the arc diameter and press ENTER.
- Select the **Line** option from the command line and move the pointer in the –X-direction.
- Type 100, and press ENTER.
- Move the pointer in the Y-direction and enter 150 as the wall length.
- Move the pointer in –X-direction and enter 100 as the wall length.
- Select the **Close** option from the command line; the wall will be closed.

Using the Extrude tool

The **Extrude** tool is used to add a third dimension (height) to an existing 2D shape. If you extrude a closed shape such as circle and closed polylines, a solid is created. If you extrude an open sketch such as lines and arcs, a surface is created.

Example 1:

- Start a new AutoCAD file in the **3D Modeling** workspace.
- Click **Home > View > 3D Navigation > Front** on the ribbon; the front view will become parallel to the screen.

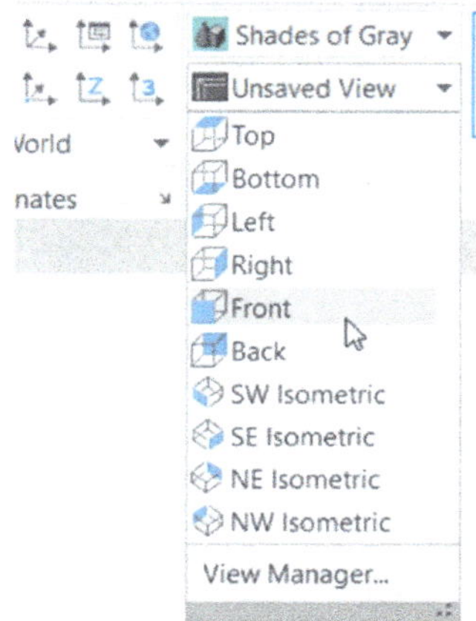

- Click **Home > Draw > Polyline** on the ribbon and create the sketch, as shown below.

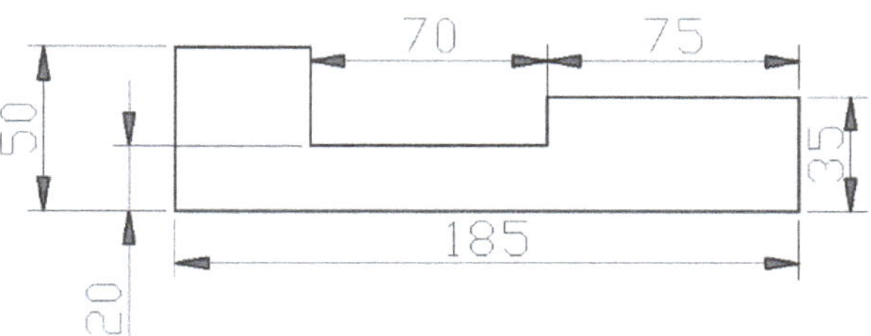

- Select **SE Isometric** from the **In-canvas controls**; the view orientation will be changed to Southeast Isometric.

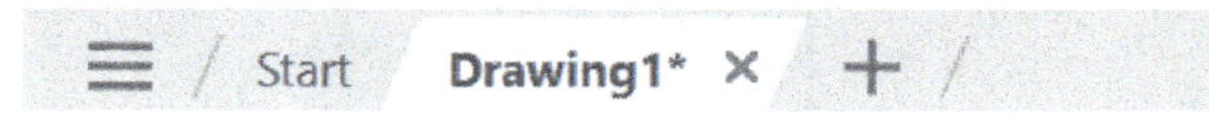

[−][SE Isometric][Shades of Gray]

- Click **Home > Modeling > Extrude.**

- Select the polyline sketch and press Enter.
- Move the pointer toward the right.
- Type **100** in the command line or **Dynamic Input** box and press ENTER; the polyline sketch will be extruded.

Example 2:

- Open a new AutoCAD file in the **3D Modeling** Workspace.
- Click **Home > View > 3D Navigation > Top** on the ribbon; the view will become parallel to the screen.

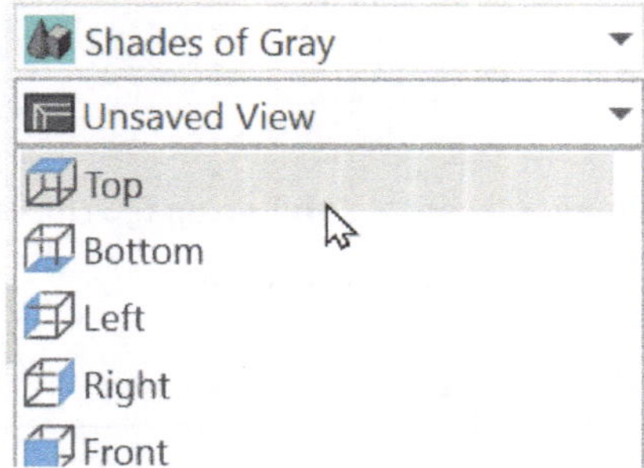

- Click **Home > Draw > Line** on the ribbon and create the sketch, as shown below.

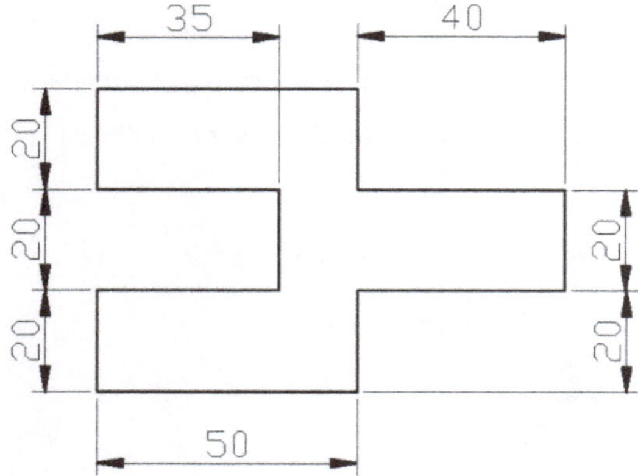

- Click **Home > View > 3D Navigation > SE Isometric** on the ribbon; the view orientation will be changed to southeast Isometric.
- Expand the **Draw** panel of the **Home** tab and click the **Region** button.

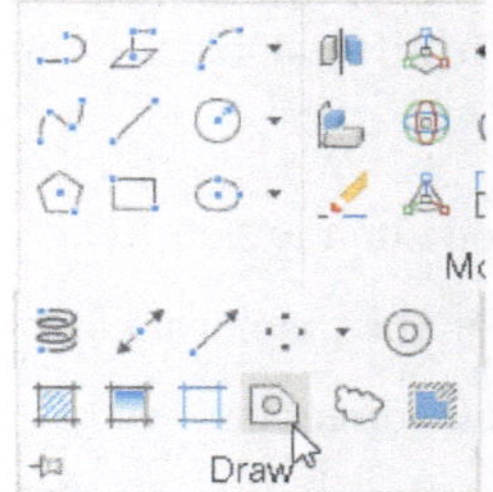

- Press and hold the left mouse button. Next, drag a window across all the objects of the sketch.
- Press ENTER; the sketch will be converted into a region. Now, you can extrude the region to create a solid. If you try to extrude the lines without creating a region, it will result in a surface.

- Click **Solid > Solid > Extrude** on the ribbon. 
- Select the region created from the sketch, and then press ENTER; the message, “Specify height of extrusion or [Direction Path Taper angle Expression]:” appears in the command line.
- Select the **Taper angle** option from the command line.
- Type 5 as the taper angle and press ENTER.
- Move the pointer upward, type 40 in the command line and press ENTER; the extruded solid will be created with a taper.

Example 3:

- Type **EXT** in the command line and press ENTER.
- Press and hold the CTRL key and select the top face of the model.
- Press ENTER and move the pointer upward.
- Type 25 as the extrusion height and press ENTER; the extruded solid will be created.

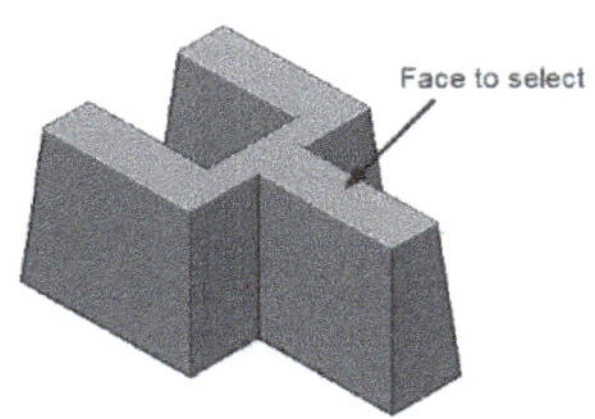

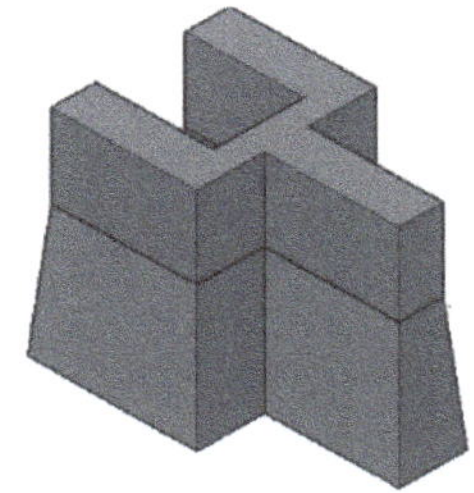

Using the Revolve tool

The **Revolve** tool is used to revolve an open or closed 2D sketch about a selected axis. If you revolve a closed profile such as a polyline sketch, polygon, circle, or a sketch region, a solid object is created. An open profile results in a surface. The sketch is deleted after revolving it. If you want to retain the sketch, you need to set the **DELOBJ** system variable to 0.

- Open a new file in the **3D Modeling** workspace.
- Set the view orientation to front and create the sketch using the **Line** tool. Do not add dimensions.

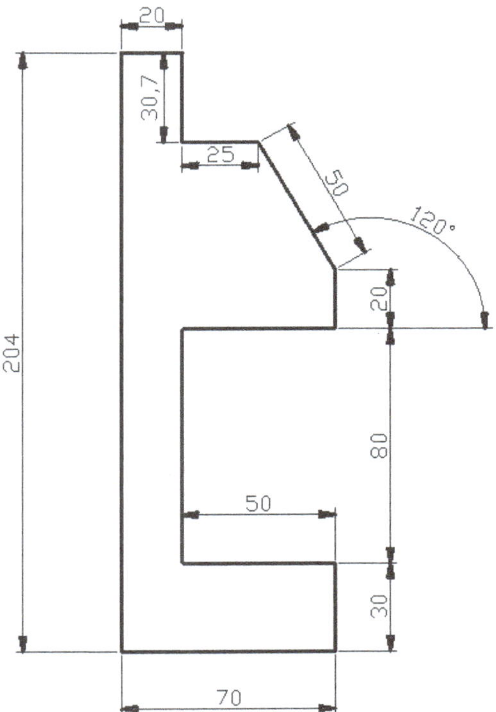

- Convert the sketch into the region using the **Region** tool.

- Create a vertical line at a distance of 10 mm from the left vertical edge of the region.

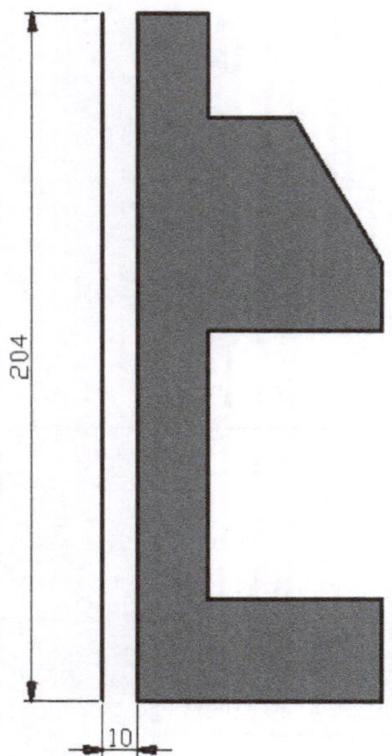

- Click **Solid > Solid > Revolve** on the ribbon or type REV in the command line.
- Select the sketch region and press ENTER; the message, “Specify axis start point or define axis by [Object/X/Y/Z] <Object>:” appears in the command line.
- Select the **Object** option from the command line and select the vertical line created at an offset; the message, “Specify angle of revolution or [STart angle/Reverse/EXpression] <360>:” appears.

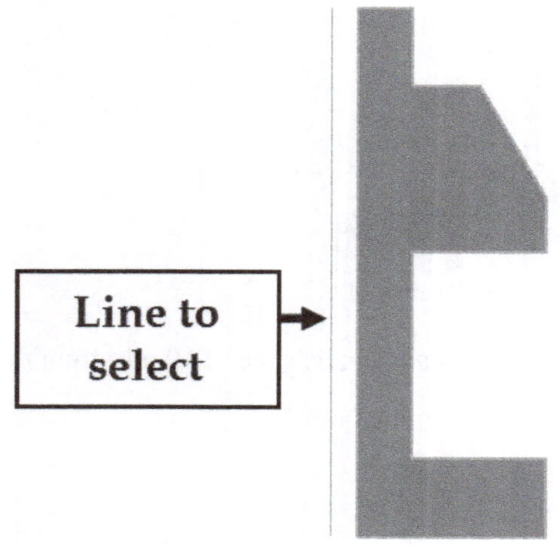

- Press ENTER to specify 360 as the revolution angle.

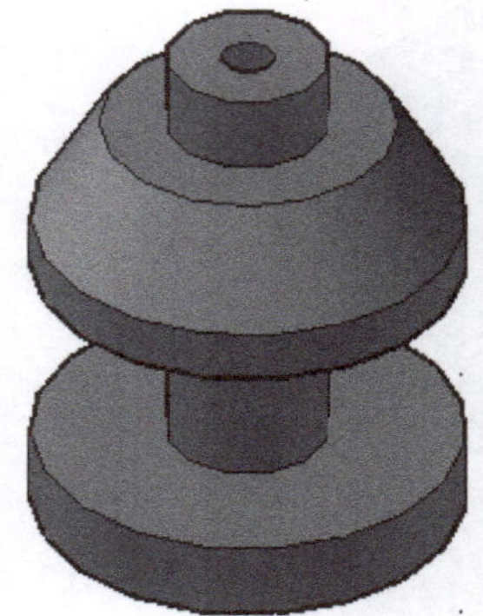

Using the Sweep tool

The **Sweep** tool is used to create a new solid or surface by sweeping a closed or open planar profile along an open or closed 2D or 3D path. The procedure to create a solid by using the **Sweep** tool is discussed next.

Example:

- Open a new file in the **3D Modeling** workspace.
- Set the view orientation to **Front**.

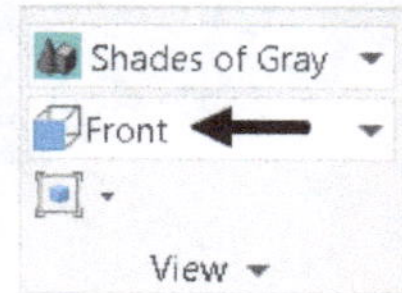

- Activate the **Polyline** tool.

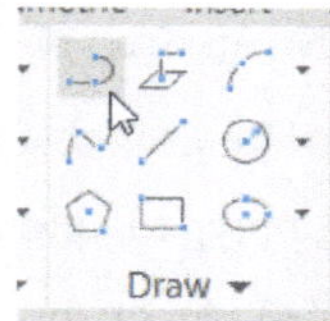

- Deactivate the **ORTHOMODE** icon on the Status bar, if already active.
- Turn ON the Dynamic Input icon on the Status bar.
- Specify the start point of the polyline.
- Move the pointer vertically downward.
- Type 38 and press the TAB key.
- Type 270, and press ENTER.

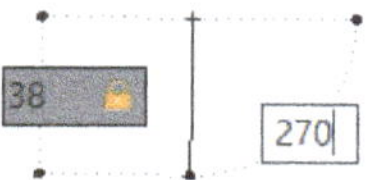

- Move the pointer toward the bottom-left corner.
- Type 83 and press the TAB.
- Type 229, and press ENTER.

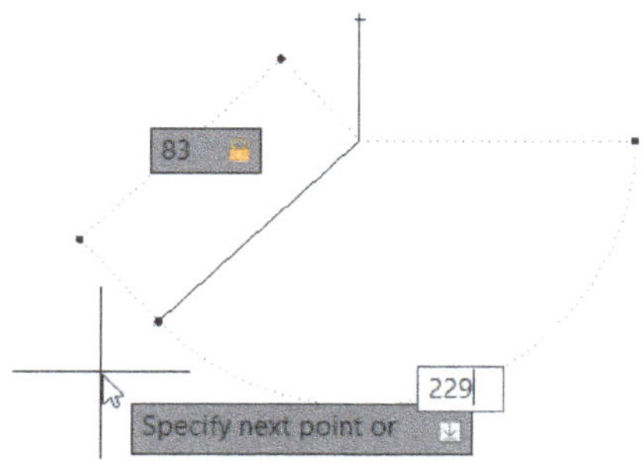

- Move the pointer downward.
- Type 66 and press the TAB key.
- Type 270, and press ENTER.

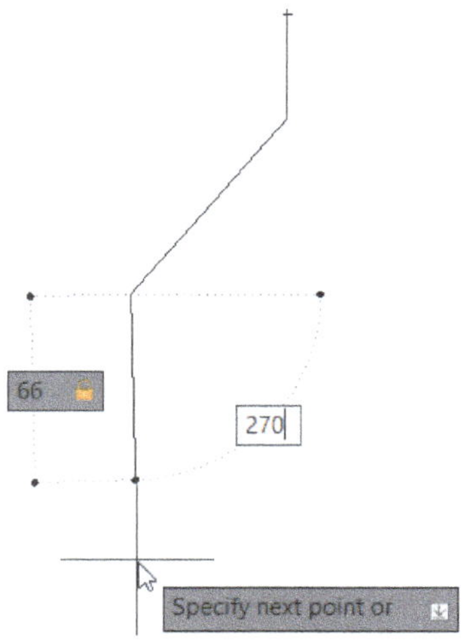

- Select the **Arc** option from the command line.
- Move the pointer toward right.
- Type 108 and press TAB.
- Type 0 and press ENTER.

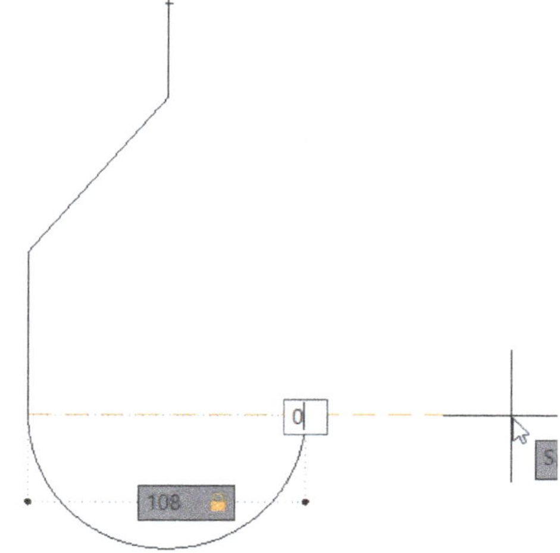

- Press Esc to deactivate the **Polyline** tool.
- Use the **Fillet** tool and apply fillets of 25 mm radius.

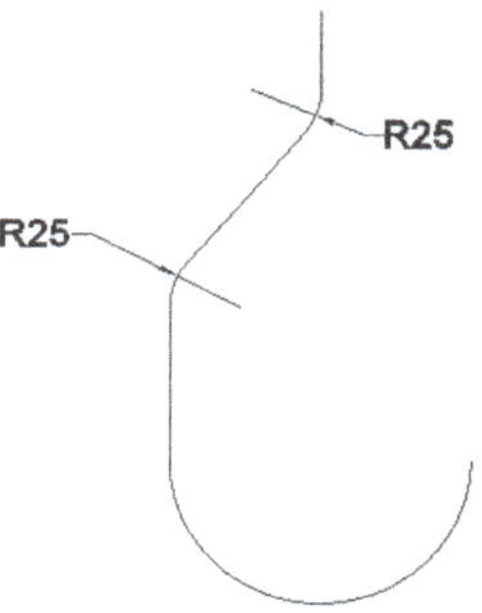

- Change the view orientation to SE Isometric.
- Click **Home > Coordinates > Z-Axis Vector** on the ribbon.
- Select the endpoint of the top vertical line as the UCS origin.
- Move the pointer downward and select the endpoint of the vertical line; the Z-axis is aligned to the vertical line.

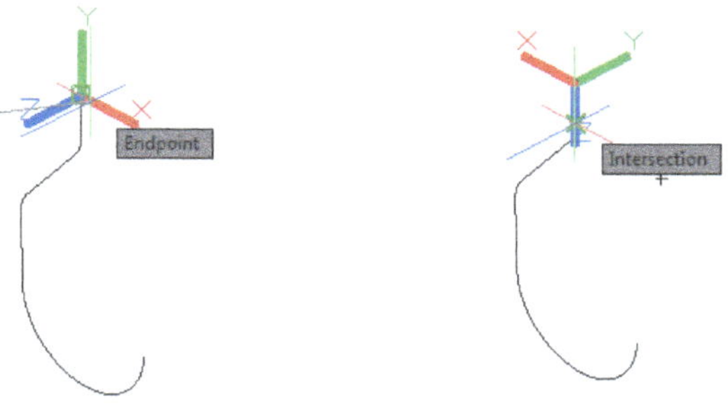

- Click the **Circle** button on the **Draw** panel.
- Select the endpoint of the vertical line to specify the center point of the circle. Specify 5 mm as the radius of the circle.
- Click the **UCS, World** button on the **Coordinates** panel; the User Coordinate System will be set to World Coordinate System (0,0,0).

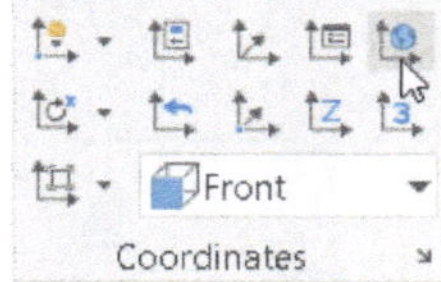

- Click **Home > Modeling > Solids** drop-down > **Sweep** on the ribbon.

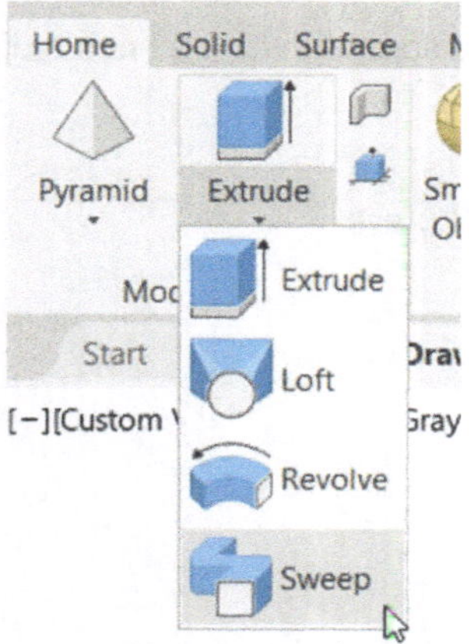

- Select the circle as the profile and press ENTER; the message, "Select sweep path or [Alignment/Base point/Scale/Twist]:" appears in the command line.

The **Alignment** option aligns the profile perpendicular to the direction of the sweep path. By default, the profile is aligned to the path.

The **Base point** specifies the base point of the profile. By default, the center of the profile is used as the base point. You can select any other point on the profile to define the base point.

The **Scale** option scales the profile along the path.

The **Twist** option twists the profile as it is swept along the length of the path.

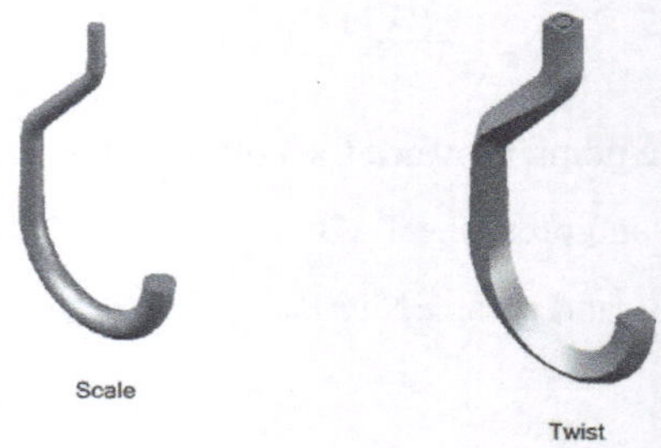

- Select the path to create the swept solid object.

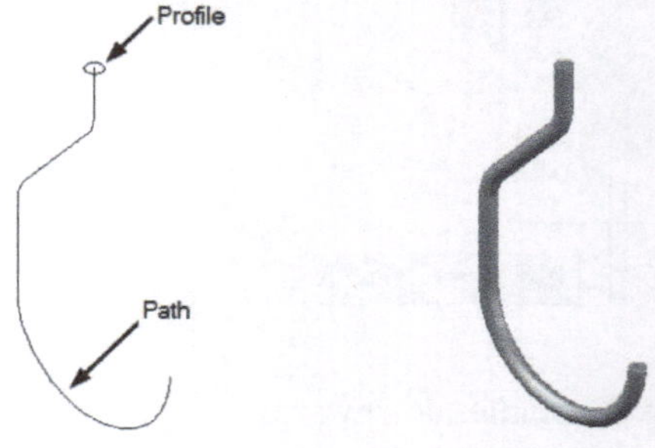

Using the Loft tool

Using the **Loft** tool, you can create a solid or surface by selecting a series of cross-sections. The selected cross-sections will define the shape of the lofted solid.

Example 1:

- Deactivate the **Dynamic Input** icon on the status bar.
- Type CIRCLE in the command line and press ENTER.
- Type 0,0,0 in the command line and press ENTER.
- Type 25 as the radius value and press ENTER.
- Type CIRCLE in the command line.
- Type 0,0,70 in the command line and press ENTER.
- Type 50 as the radius value and press ENTER.
- Type CIRCLE in the command line.
- Type 0,0,140 in the command line and press ENTER.
- Type 25 as the radius value and press ENTER.

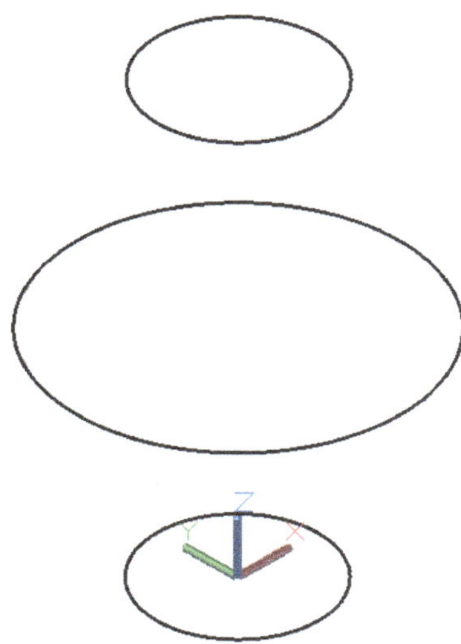

- Click **Home > Modeling > Solids drop-down > Loft** on the ribbon (or) type **LOFT** in the command line and press ENTER.

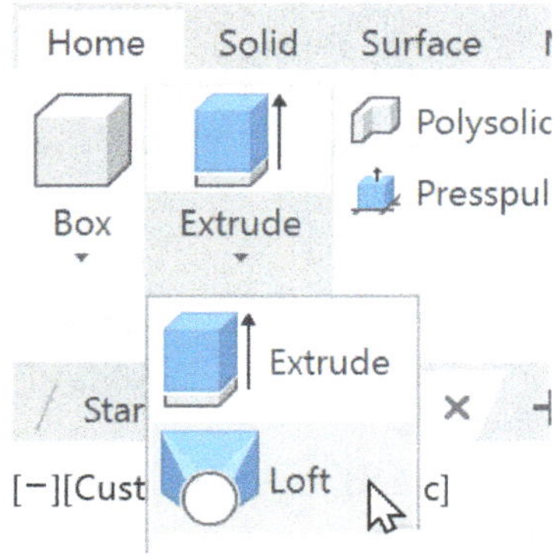

- Select the first, second, and third cross-sections one by one; the preview of the lofted solid appears.
- Press ENTER to accept the selection; the message, "Enter an option [Guides/Path/Cross sections only/Settings] <Cross sections only>:" appears in the command line.
- Select the **Settings** option from the command line; the **Loft Settings** dialog appears. In this dialog, the **Smooth Fit** option creates a smooth connection between the cross-sections. If you select the **Ruled** option, the lofted solid or surface has sharp edges.

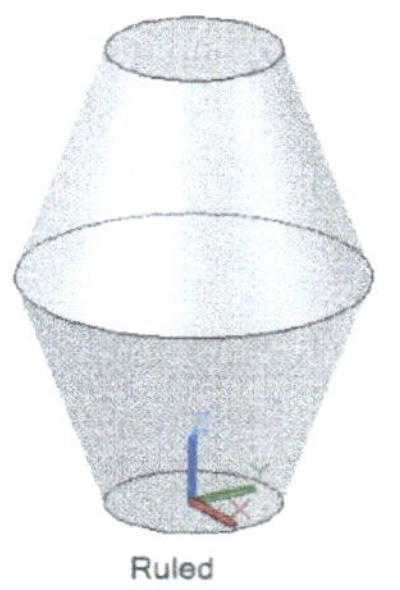

Ruled

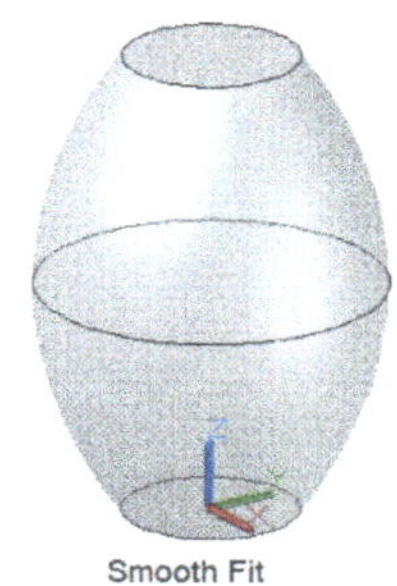

Smooth Fit

The **Normal to** option creates a solid or surface normal to the cross-section. You can select the loft solid or surface to be normal to **All cross sections**, or **Start Cross Section** or **End Cross Section** or **Start and End Cross Sections**.

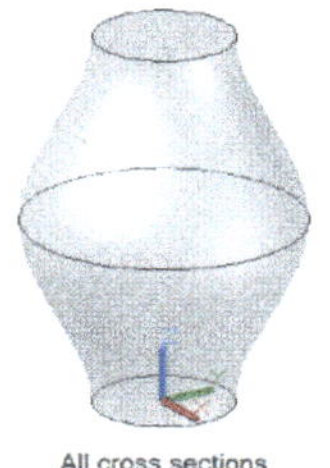

All cross sections

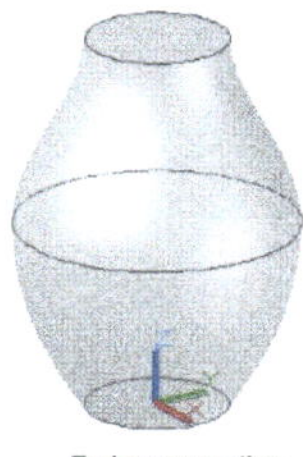

End cross section

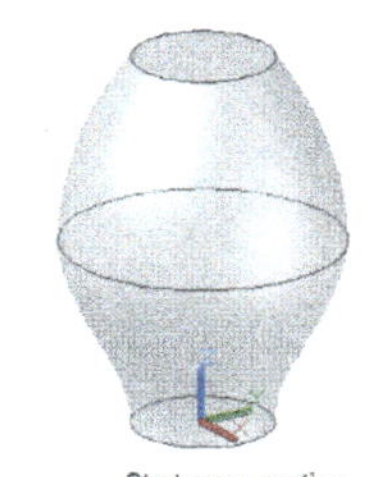

Start cross section

The **Draft angles** option defines the draft angle and magnitude at start and end cross-sections. The draft angle is the beginning direction of the loft surface. If you set the draft angle to 90 degrees, the loft surface starts vertically from the cross-section, and the 0-draft angle starts loft surface horizontally. The Magnitude is the relative distance up to which the loft surface will follow the draft angle before it bends.

The **Close surface or solid** option connects the start and end section of the lofted object.

- Select the **Normal to** option and select **All cross sections**. Click **OK**; the loft solid will be created, as shown below.

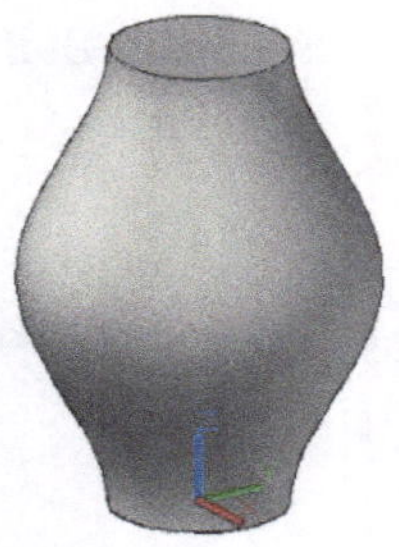

Using the Presspull tool

The **Presspull** tool is used to create and modify solid models with greater ease and speed. It can be used to accomplish two types of operations: extruding closed 2D shapes and add or remove material from a solid object based on whether you "pull" or "push" the extrusion.

- Start a new file.
- Create two layers called **Sketch** and **Solid**. Make the **Sketch** layer as current.
- Set the view orientation to **Right** and draw the sketch, as shown below.

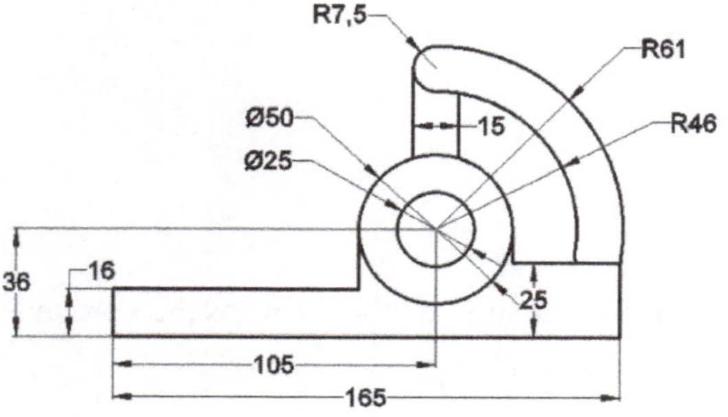

- Change the view orientation to SE Isometric.
- Ensure that the **Dynamic Input** icon is activated on the status bar.
- Set the **Solid** layer are current.
- Click **Home > Modeling > Presspull** on the ribbon.
- Click inside the bottom region of the sketch and move the pointer backward. Type 60 in the Dynamic input box and press ENTER; the extruded feature will be created.

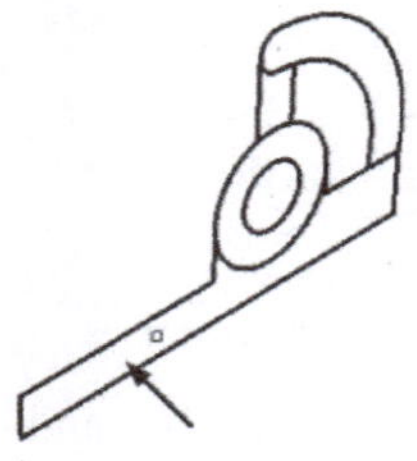

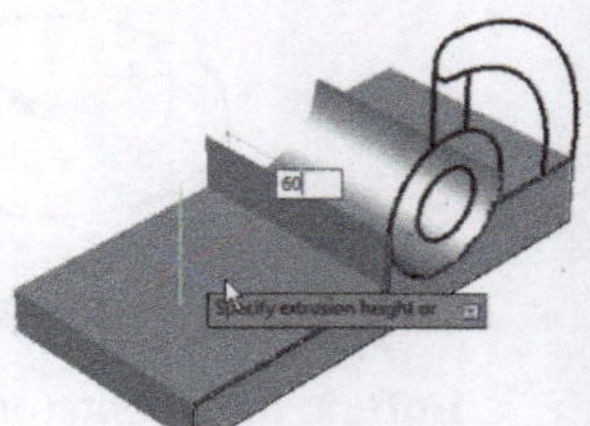

- Click on the region enclosed by the larger circle and extrude it up to 64 mm distance.

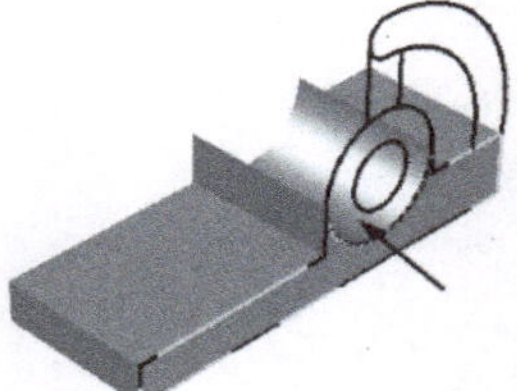

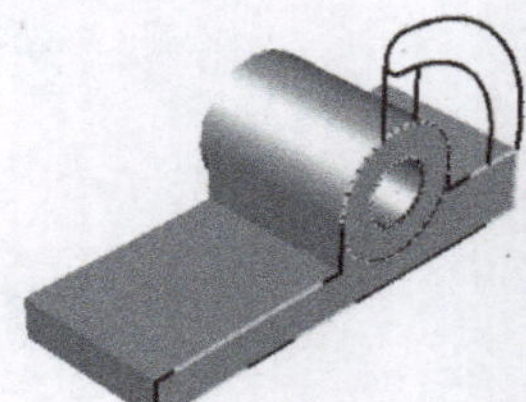

- Press and hold the CTRL key and select the front face of the cylindrical object. Move the pointer forward. Type 4 in the dynamic input box and press ENTER.

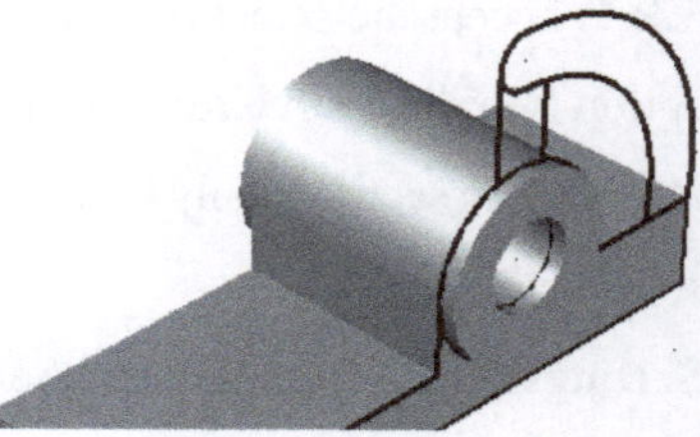

- Click in the curved slot region and move the pointer backward; the message, "Specify extrusion height or [Multiple]:" appears in the command line.
- Select the **Multiple** option from the command line and click in the region enclosed by the two vertical

lines.

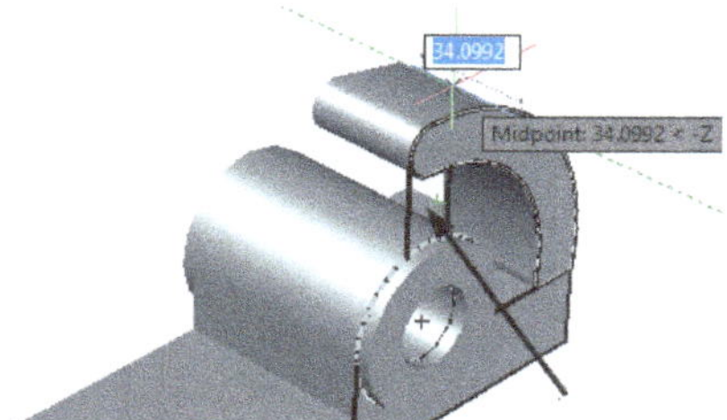

- Right-click and move the pointer backward. Type **12** in the dynamic input box and press ENTER.

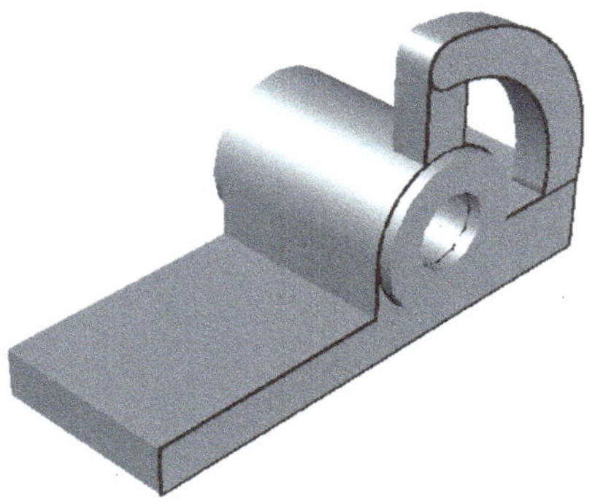

- Activate the **Presspull** tool.
- Press and hold the CTRL key and select the front face of the slot and move the pointer forward — type 4 in the dynamic input box and press ENTER.
- Set the **Sketch** layer as current.
- Activate the **Dynamic UCS** icon on the status bar.
- Click the **Circle** button on the **Draw** panel of the **Home** ribbon tab.
- Press and hold the SHIFT key. Right-click and select the **Center** option from the shortcut menu.
- Select the center point of the slot end cap and create a circle of 4 mm radius.

- Click **Home > Modify > Array** drop-down > **Path Array** on the ribbon.
- Select the circle created in the previous step and press ENTER.
- Select the arc as the path curve; the preview of the path array is displayed.

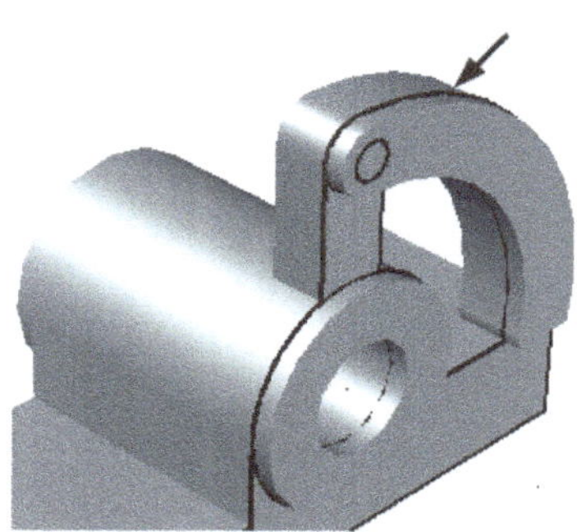

- In the **Array Creation** tab, set the **Between** value to 25 in the **Items** panel; the item count is automatically adjusted.
- Click the **Close Array** button on the ribbon; the polar array is created.

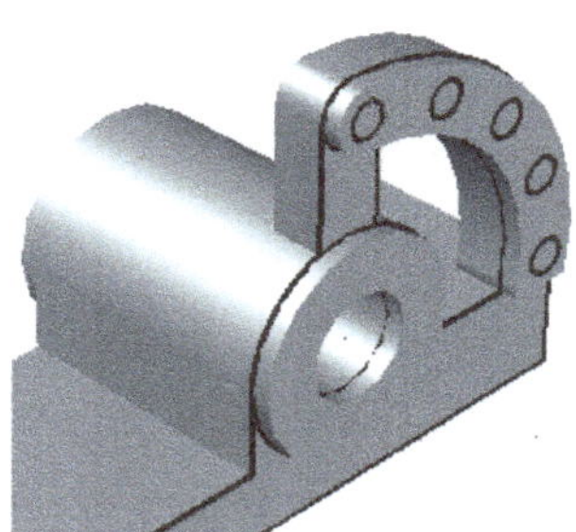

- Set the **Solid** layer as current.
- Activate the **Presspull** command.
- Click in any one of the circles and select the **Multiple** option from the command line.
- Click inside the rest of the circles of the polar array. Right-click to accept.
- Move the pointer backward and click; the holes will be created, as shown in the figure.
- Turn Off the **Sketch** layer; the sketches will be hidden.

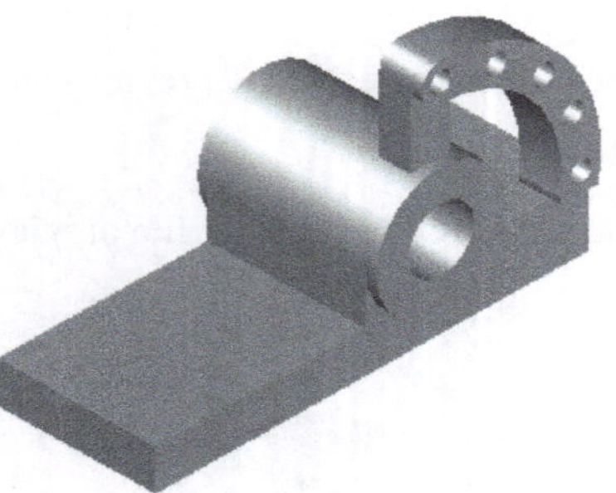

- Click the **Orbit** button on the **Navigation Bar**.
- Press and drag the left mouse button to rotate the model.

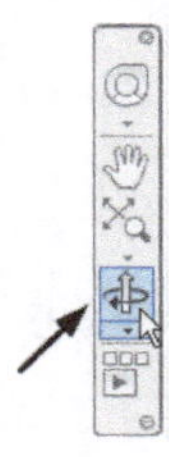

Performing the Boolean Operations

Boolean operations are performed to add two or more solids together, subtract a single solid or group of solids from another, or form a common portion when two solids are combined. You must have at least two solids to perform a Boolean operation. There are three tools available to perform Boolean Operations- **Union Subtract** and **Intersect**. These tools are discussed next.

The Union tool

The **Union** tool joins two or more solids together into a single solid. For example, when you try to select the complete model, its individual objects are selected. But, after performing the Union operation, all the solid objects are combined and act as one object.

- To perform the Union operation, click **Solid > Boolean > Union** on the ribbon.

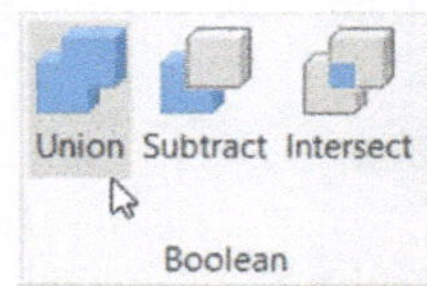

- Click the left mouse button and create a selection window across the model; all the objects of the model will be selected.
- Press ENTER; all the solid objects of the model will be combined.
  Now, when you select an individual object, the complete model will be selected.

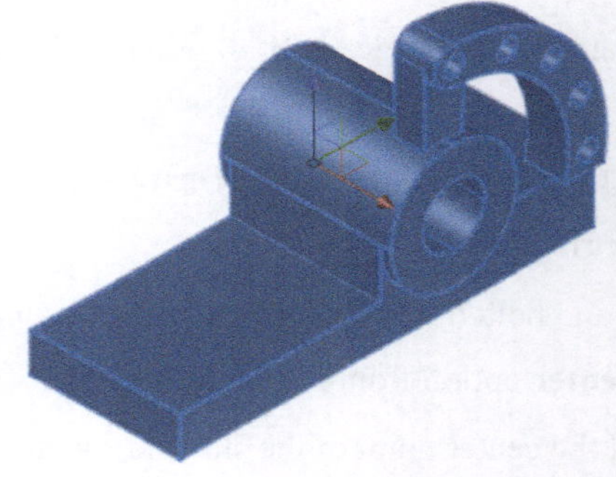

The Subtract tool

The **Subtract** tool is used to subtract one or more solid objects from another object.

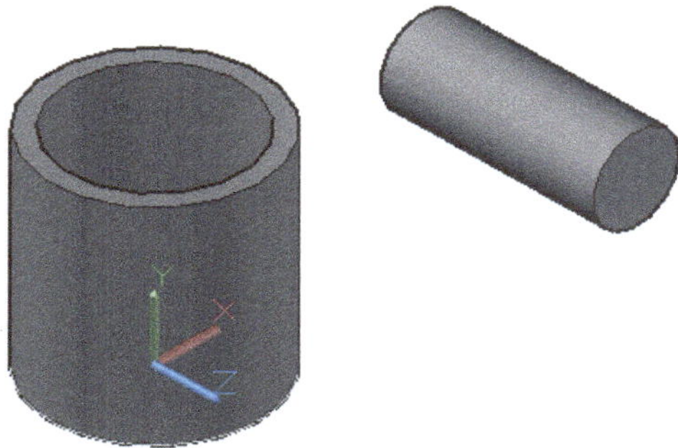

- Open a new 3D drawing file.
- Change the view orientation to Top.

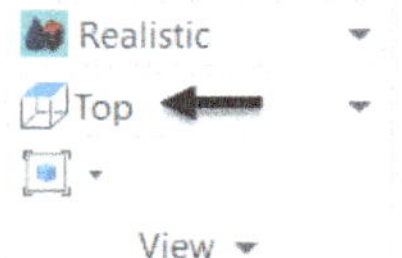

- Create two concentric circles of 240 and 200 mm in diameter.

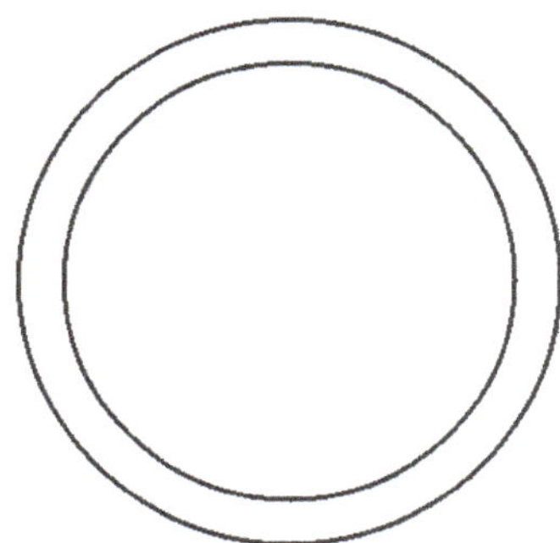

- Change the view orientation to SE Isometric.

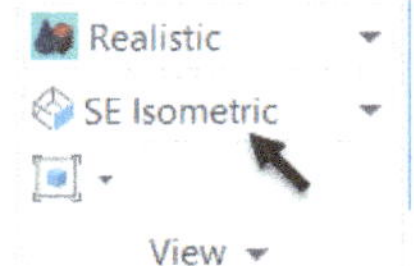

- Activate the **Presspull** tool.
- Click inside the region between the large and small circles.

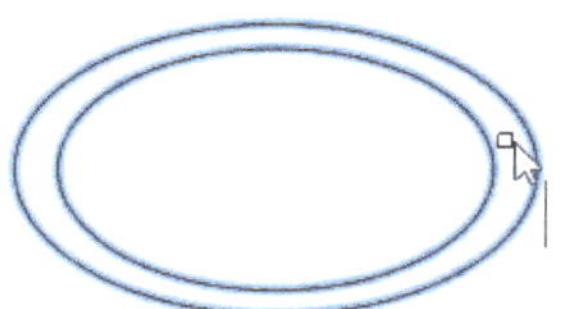

- Move the pointer upward, type 250, and then press Enter.

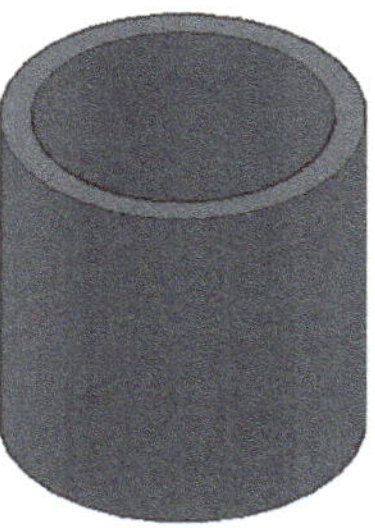

- Press Esc to deactivate the **Presspull** tool.
- Set the view orientation to **Right**.

- Click **Home > Modeling > Primitive** drop-down > **Cylinder** on the ribbon.
- Select an arbitrary point in the graphics window.
- Move the pointer outward, type 50, and then press Enter.
- Type 240 and press ENTER.
- Change the view orientation to **SE Isometric**.

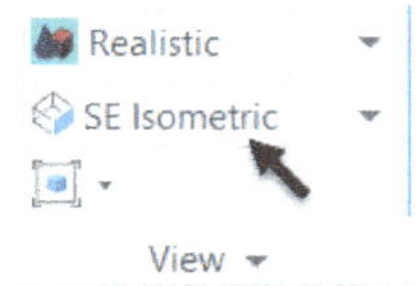

- Expand the **Modify** panel and click the **Align** button.

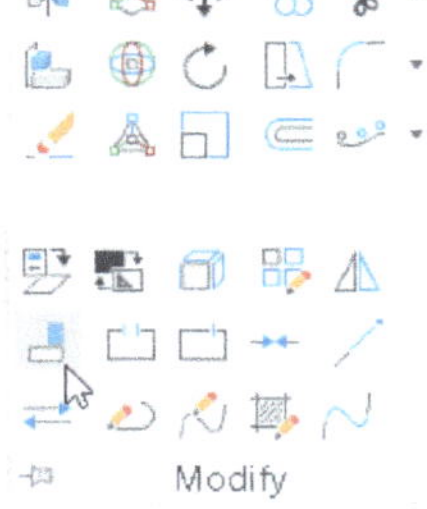

- Select the horizontal cylinder and press ENTER; the message, "Specify first source point:" appears in the command line.
- Press and hold the SHIFT key. Right-click and select the **Center** option.
- Select the center point of the front face of the horizontal cylinder; the message, "Specify first destination point:" appears in the command line.

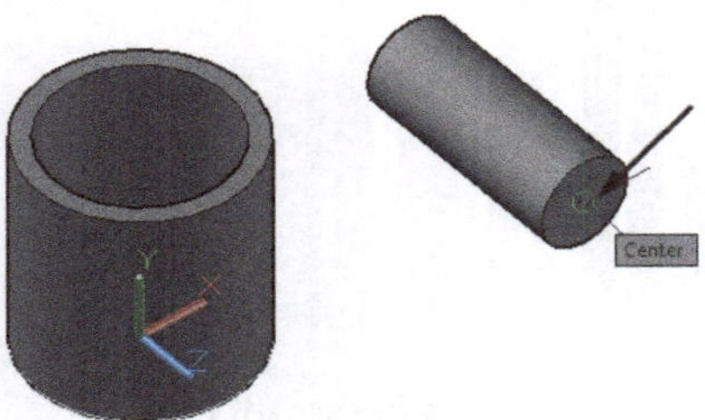

- Press and hold the SHIFT key. Right-click and select the **Quadrant** option.
- Select the quadrant point of the outer circle on the top face of the hollow cylinder.

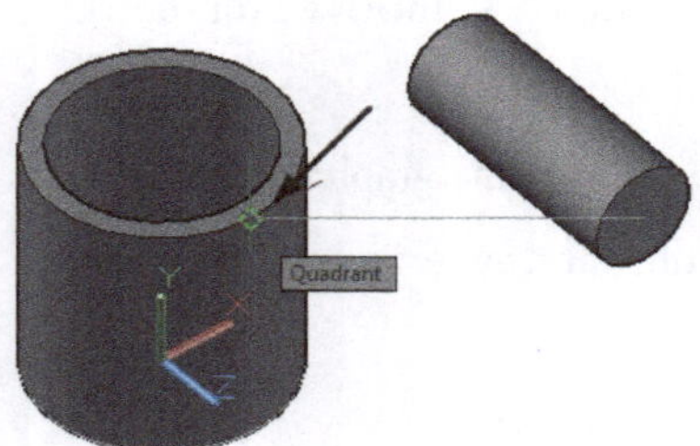

- Press ENTER; the horizontal cylinder will be aligned with the hollow cylinder.

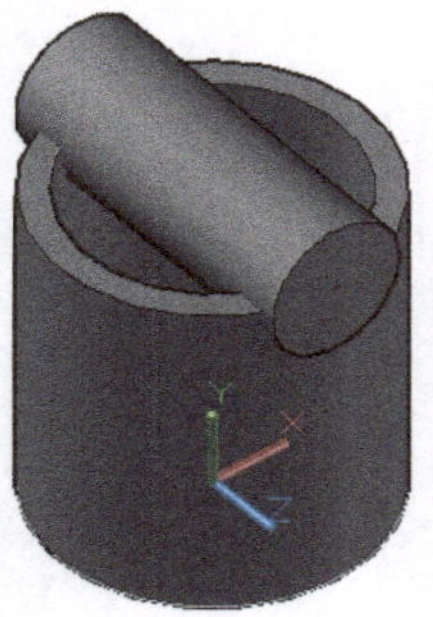

- Click **Solid > Boolean > Subtract** on the ribbon; the message, "Select objects," appears in the command line.
- Select the hollow cylinder and press ENTER; the message, "Select objects," appears in the command line.
- Select the horizontal cylinder and press ENTER; it will be subtracted from the hollow cylinder, as shown next.

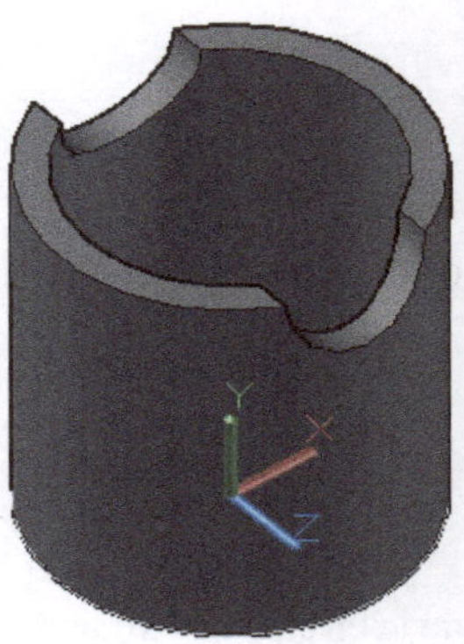

The Intersect tool

The **Intersect** tool is used to create a composite solid by finding common volume shared by the selected objects.

- Start a new file.
- Set the view orientation to **Front** and create the sketch, as shown below.

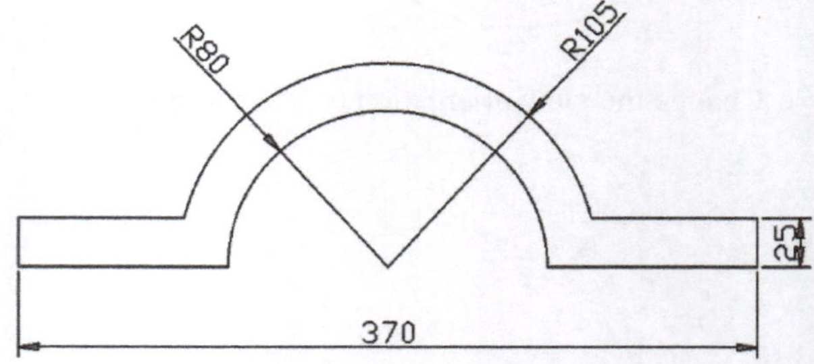

- Change the view orientation to **SE Isometric**.

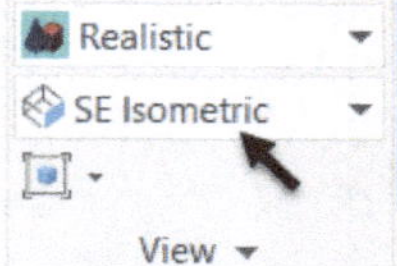

- Use the **Presspull** tool and extrude the sketch up to 150 mm distance.

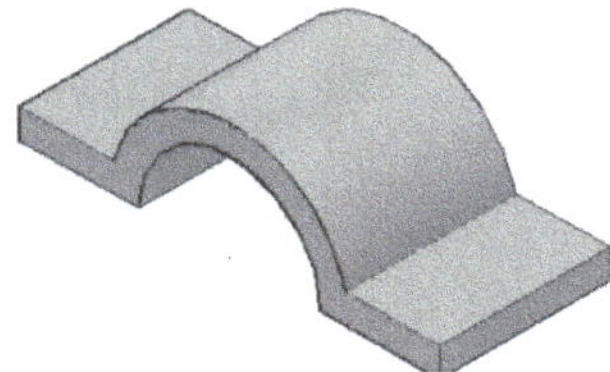

- Set the view orientation to Top.

- Activate the **Circle** tool and create three circles, as shown.

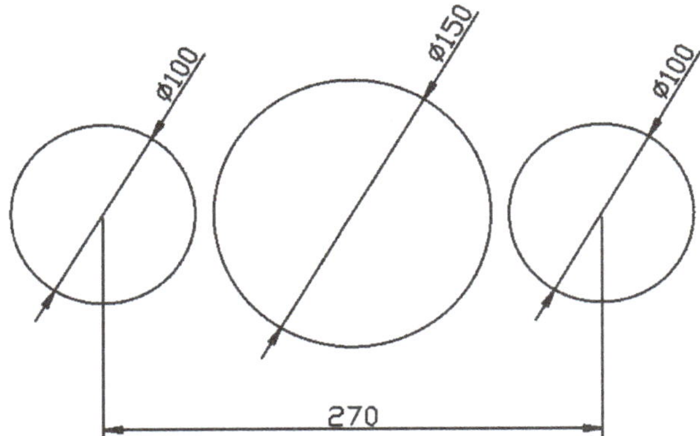

- Click **Home > Draw > Line** on the ribbon.
- Press and hold the SHIFT key, and then right-click.
- Select the **Tangent** option.
- Select the small circle by clicking at the location, as shown.

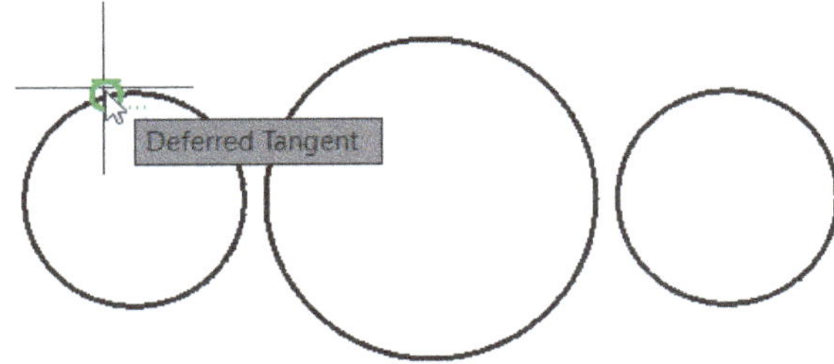

- Press and hold the SHIFT key, and then right-click.
- Select the **Tangent** option.
- Select the large circle by clicking at the location, as shown.

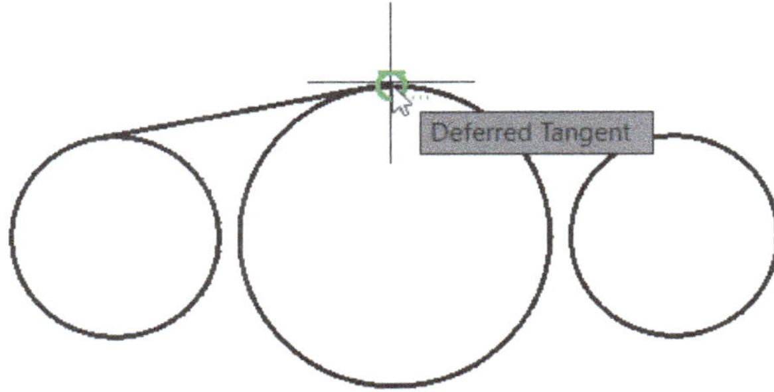

- Right-click and select **Enter**.
- Likewise, create three more lines tangent to the circles, as shown.

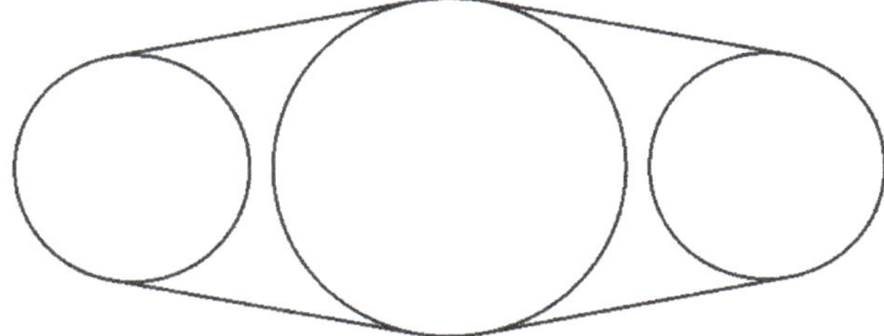

- Click **Home > Modify > Trim** on the ribbon.
- Select the entities to trim, as shown.

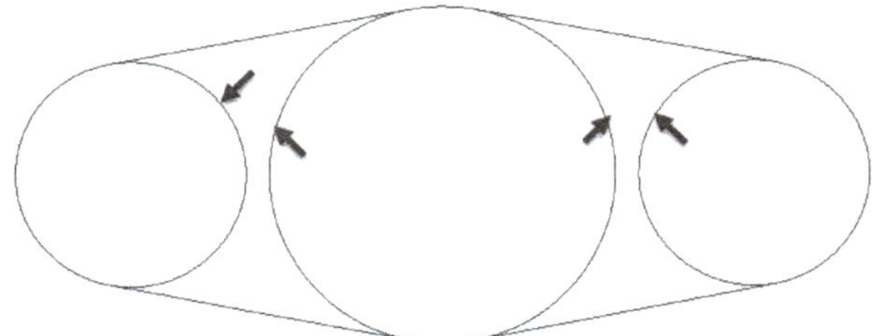

- Press Esc.

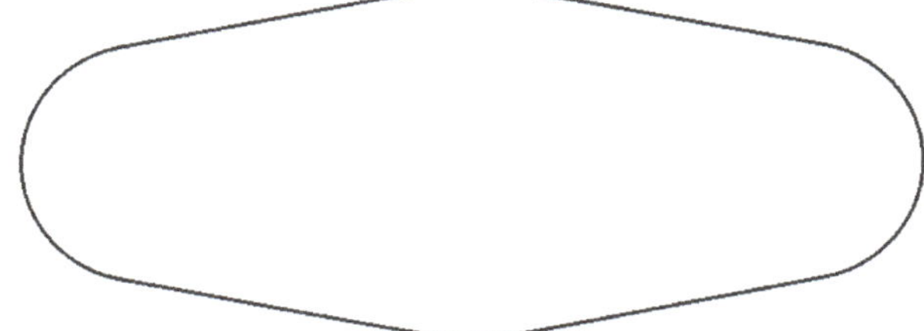

- Change the view orientation to SE Isometric.

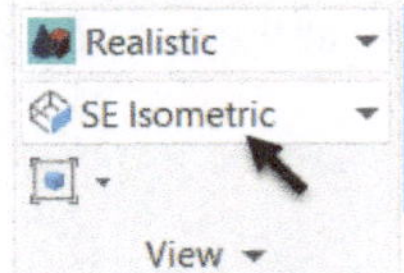

- Use the **Presspull** tool and extrude the sketch up to 200 mm height, as shown below.

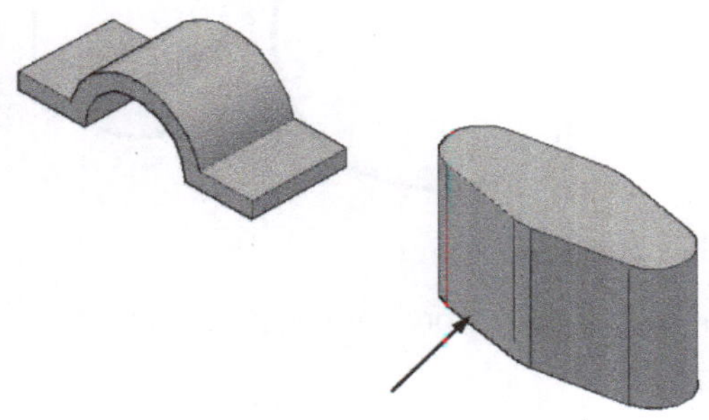

- Change the View style to **Wireframe**.

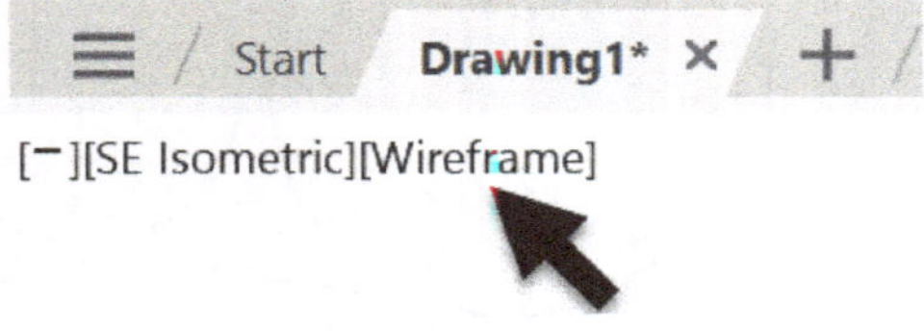

- Deactivate the **3D Object Snap** option on the status bar.
- Type **DS** in the command line and press ENTER; the **Drafting Settings** dialog appears.
- Click the **Object Snap** tab and **Clear All** the **Object Snap** modes.
- Now, select the **Quadrant** and **Midpoint** options and click **OK**.

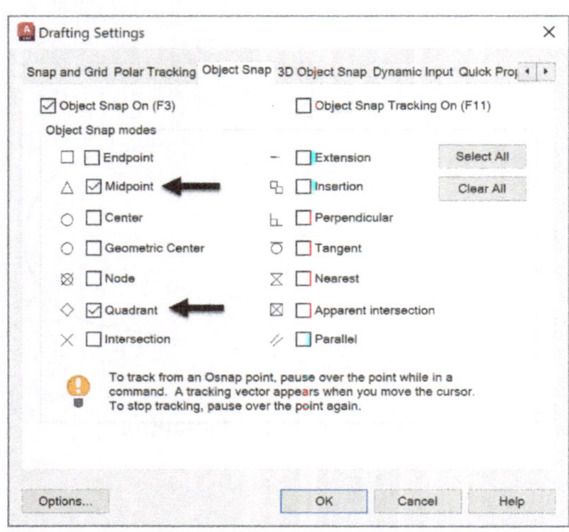

- Type **AL** in the command line and press ENTER.
- Select the second extrusion and press ENTER; the message, "Specify first source point:" appears.
- Select the point on the source object as shown below; the message, "Specify first destination point:" appears.
- Select the point on the destination object as shown below; the message, "Specify second source point:" appears.

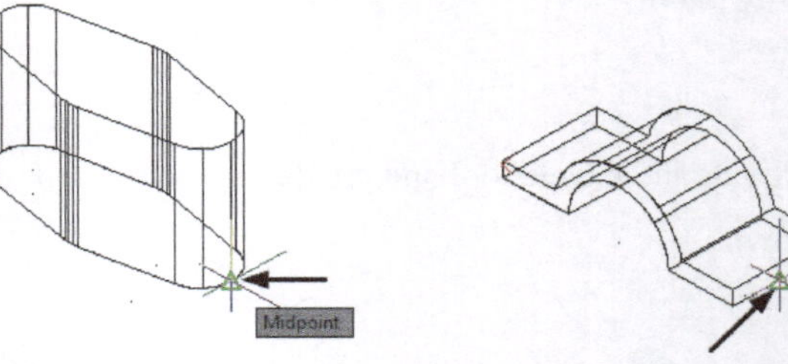

- Select another point on the source object, as shown below; the message, "Specify second destination point:" appears.
- Select another point on the destination object, as shown below; the message, "Specify third source point or <continue>:" appears.

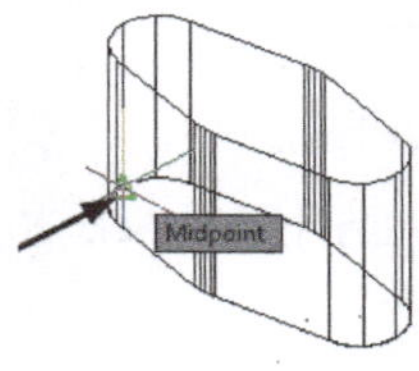

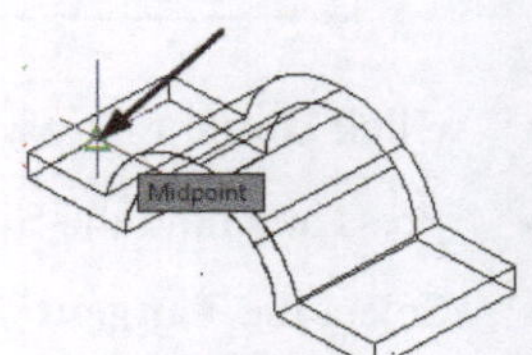

- Press ENTER to continue; the message, "Scale objects based on alignment points? [Yes/No] <N>:" appears.
- Select the **NO** option; the two objects will be aligned.
- Change the **View style** to **Shades of Gray**.

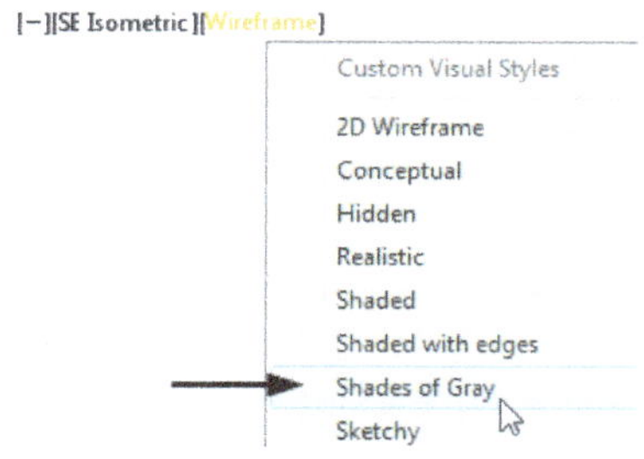

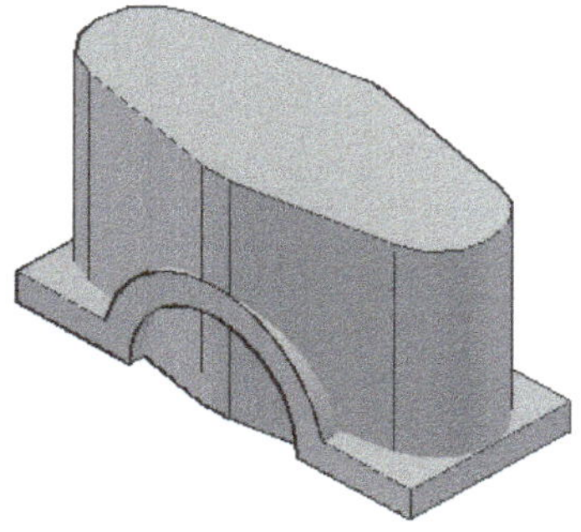

- Click **Solid > Boolean > Intersect** on the ribbon.
- Select the two objects and press ENTER; the intersection object will be created as shown below.

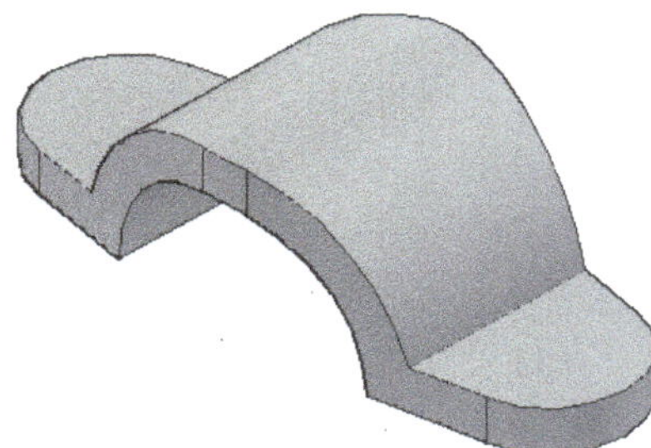

Using the Helix tool

The **Helix** tool is used to create a spiral or helix object. You can use this helix object as a path for a swept solid object.

Example 1:

- Start a new file.
- Expand the **Draw** panel in the **Home** tab and click the **Helix** button.

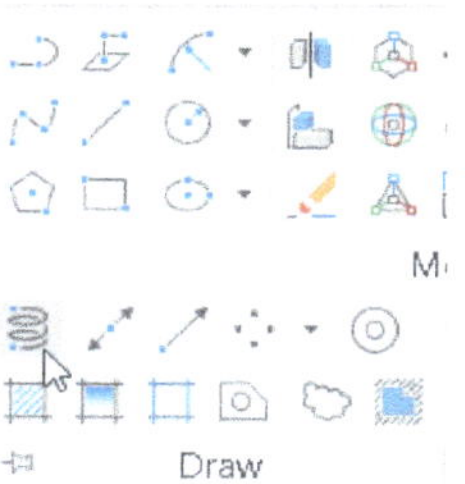

- Type 0,0 as the center point of the helix and press ENTER; the message, “Specify base radius or [Diameter]:” appears in the command line.
- Type 50 and press ENTER; the message, “Specify top radius or [Diameter] <50.0000>:” appears.
- Type 0 and press ENTER; the message, “Specify helix height or [Axis endpoint/Turns/turn Height/tWist] <1.0000>:” appears.
- Select the **Turns** option from the command line.
- Type 8 as the number of turns and press ENTER; the message, “Specify helix height or [Axis endpoint/Turns/turn Height/tWist] <1.0000>:” appears.
- Type 0 as the height and press ENTER; the spiral curve will be created as shown in the figure.

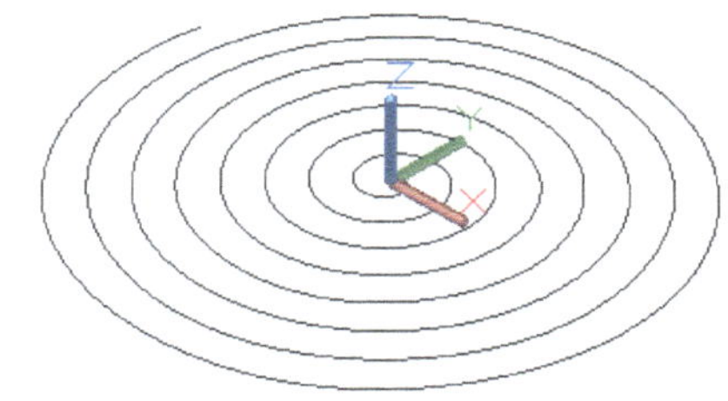

Example 2:

- Start a new file.

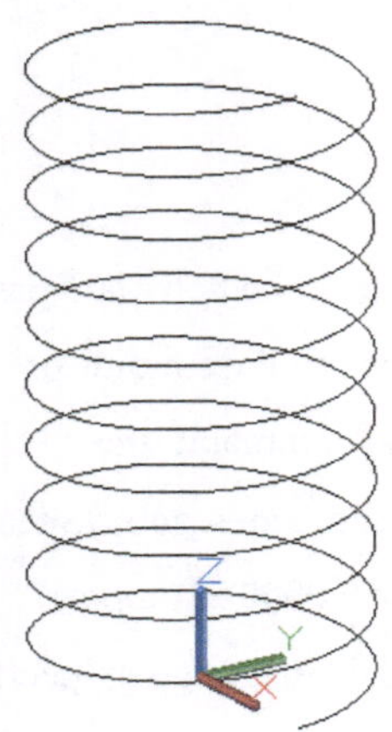

- Type HELIX in the command line and press ENTER.
- Type 0, 0 as the center point of the helix.
- Type 50 as the base radius and press ENTER.
- Press ENTER to accept 50 as the top radius.
- Select the **turn Height** option from the command line.
- Type 20 as the turn height (pitch) and press ENTER.
- Type 200 as the total height of the helix and press ENTER; the helix will be created as shown in the figure.

Exercises

Create 3D models using the drawing views and dimensions.

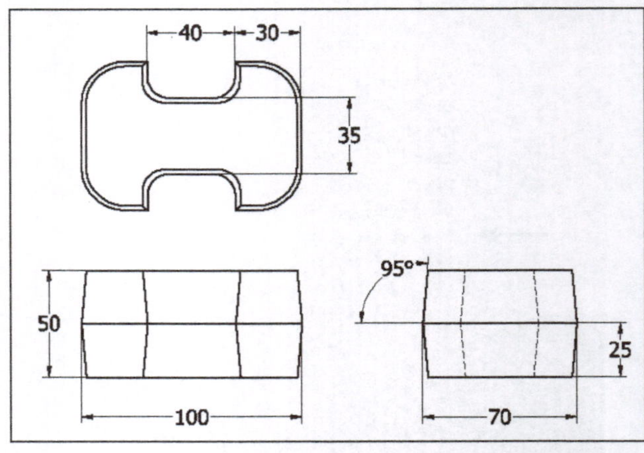

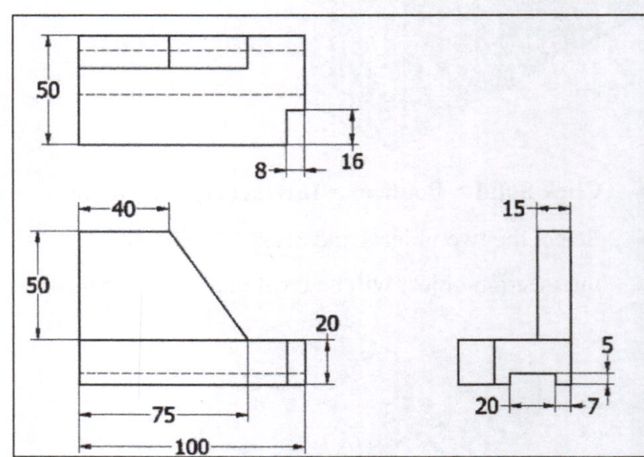

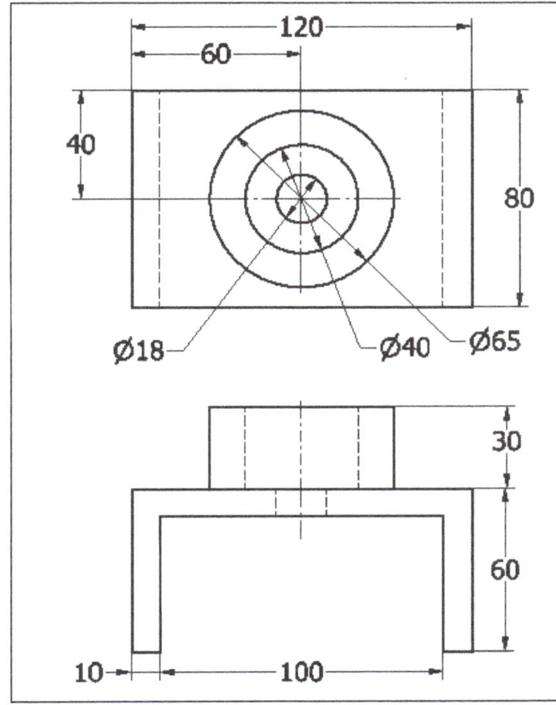
120
60
40
80
Ø18
Ø40
Ø65
30
60
10
100

Chapter 13: Solid Editing & generating 2D views

In this chapter, you will learn to do the following:

- **Move objects in 3D space**
- **Create 3D Arrays**
- **Mirror objects in 3D space**
- **Fillet edges**
- **Taper faces of a solid object**
- **Offset faces**
- **Rotate objects**
- **Create 3D Polylines**
- **Shell objects**
- **Chamfer edges**
- **Create Live sections**
- **Generate 2D views of a 3D model**
- **Create section and detailed views**

Introduction

In the previous chapter, you have learned to create simple solid objects. Now, you will learn to use the solid editing tools to create complex models. You will also learn to generate orthographic views of 3D models.

Using the Move tool

The **Move** tool that you used in 2D drawings can also be used in 3D modeling. You can change the position of an object using the **Move** tool. The application of this tool in 3D modeling is discussed next.

Example:

- Start a new file in the **3D Modeling** workspace.
- Select **Front** from the **Restore View** drop-down of the **View** panel.
- Create the sketch on the front view and presspull it up to 100 mm distance.

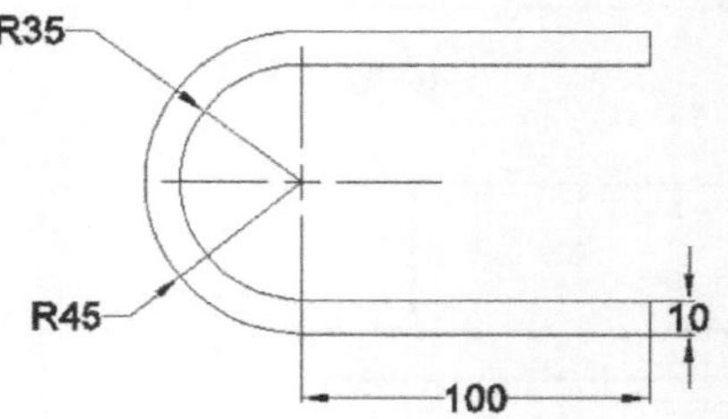

- Select **Top** from the **Restore View** drop-down of the **View** panel.
- Create a cylinder of 20 mm diameter and 20 mm in height.
- Change the view orientation to SE Isometric.

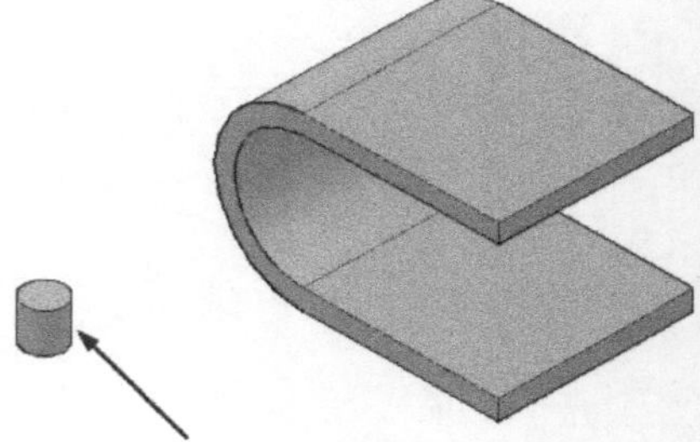

- Type **M** in the command line and press ENTER; the **Move** tool is activated.
- Select the cylinder and press ENTER.
- Select the center point of the cylinder to define the base point.

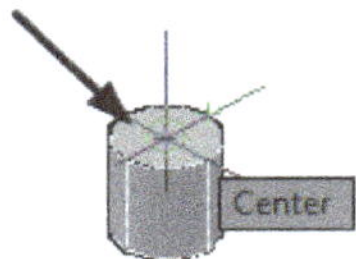

- Select the endpoint of the base object, as shown; the cylinder will be aligned to it.

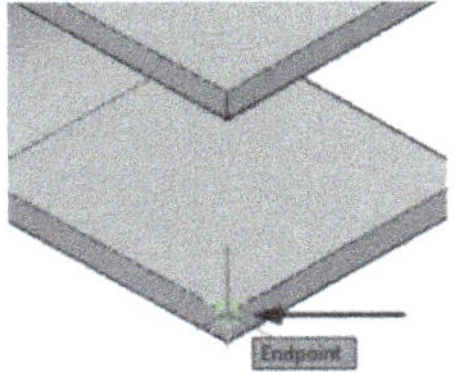

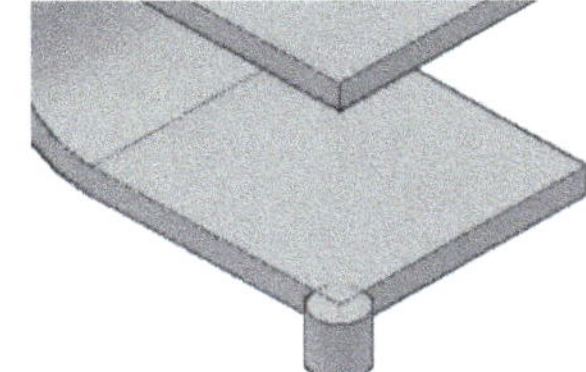

Using the 3D Move tool

The **3D Move** tool is similar to the **Move** tool. You can use this tool to move objects in 3D space. By default, the **3D Move** tool is activated, and the **Move gizmo** is displayed when you select an object. You can use the **Move gizmo** to move the object along a particular axis.

- Select the cylinder to display the **Move gizmo** tool.
- Select the X-axis (Red arrow) of the gizmo and move the pointer backward.

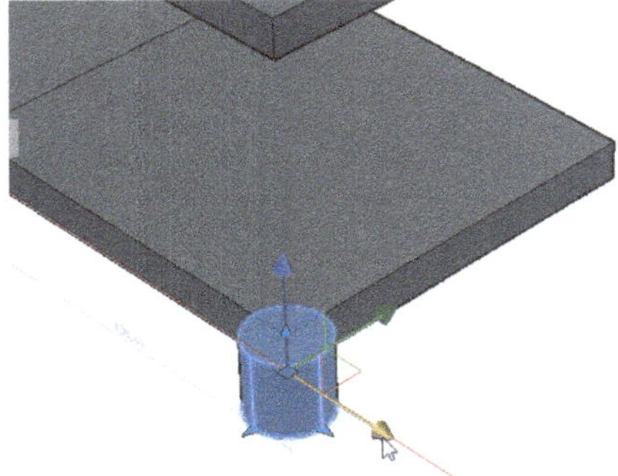

- Type 20 and press ENTER; the cylinder will be moved through a 20 mm distance along the X-axis.

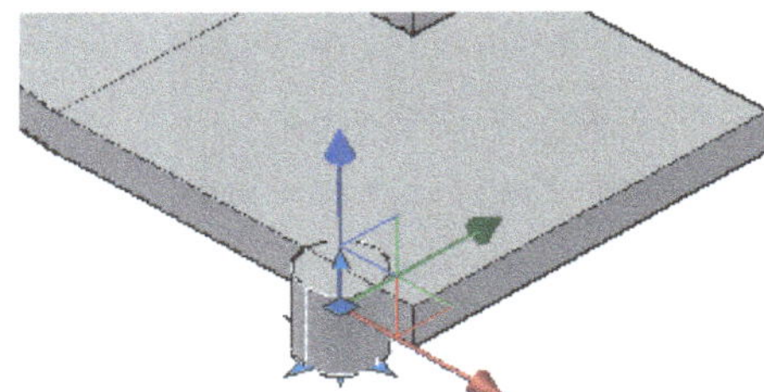

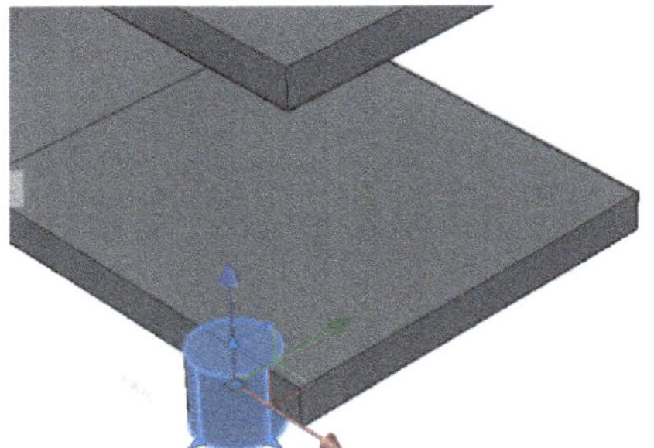

- Select the Y-axis (Green arrow) of the gizmo and move the pointer toward the right.
- Type 20 and press ENTER; the cylinder will be moved, as shown below.

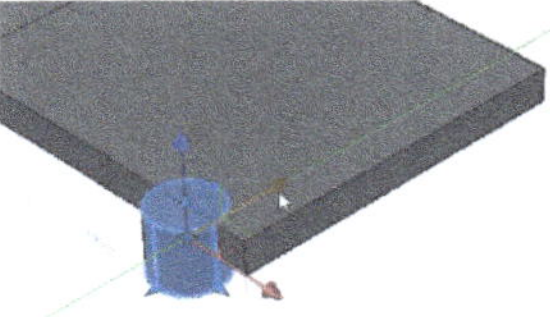

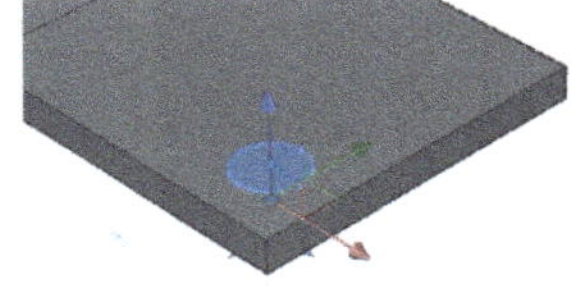

Using the Array tool

The **Array** tool is used to create Rectangular, path, and polar arrays. You can create a rectangular array by specifying the item count and distance along the X, Y, and Z directions.

Example 1 (Rectangular Array)

- Type **ARRAY** in the command line and press ENTER.
- Select the cylinder from the model and press ENTER; the message, "Enter array type [Rectangular Path POlar] <Path>:" appears in the command line.
- Select the **Rectangular** option from the command line.

- On the **Array Creation** tab, type 2 in the **Columns**, **Rows**, and **Levels** boxes, respectively.
- Type -40, 60, 80 in the **Between** boxes of the **Columns**, **Rows**, and **Levels** panels, respectively.
- Deactivate the **Associative** icon on the **Array Creation** tab.
- Click **Close Array** on the **Array Creation** tab.

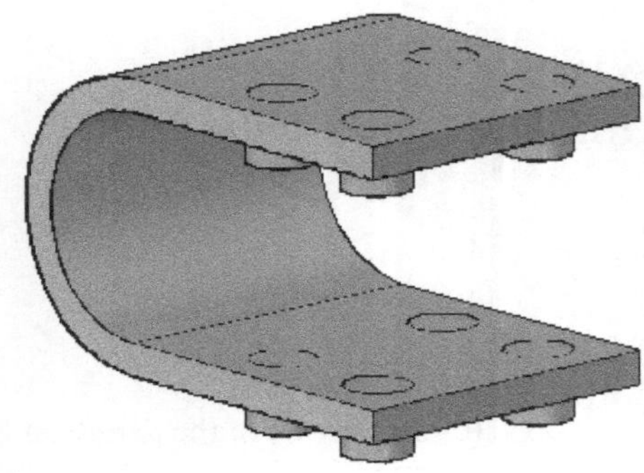

- Type **SU** in the command line and press ENTER; the **Subtract** tool will be activated.
- Select the base object and press ENTER; the message, "Select solids, surfaces, and regions to subtract" appears.
- Select all the cylinders and press ENTER; holes will be created on the model.

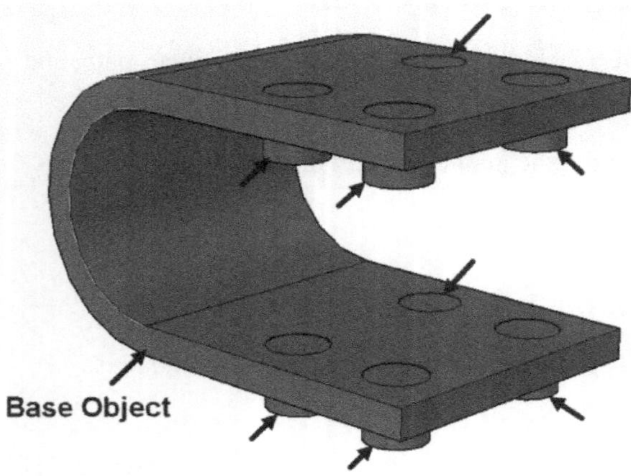

Using the 3D Align tool

The **3D Align** tool aligns one solid with another. It translates and rotates the object to align with the destination object. You need to select three points on the source object and destination object to align them together. An example of the **3D Align** tool is given next.

Example 2:

- Start AutoCAD in the **3D Modeling** workspace.
- Select **Front** from the **Restore View** drop-down in the **View** panel. Create a solid object, as shown below. The extrusion distance is 40 mm.

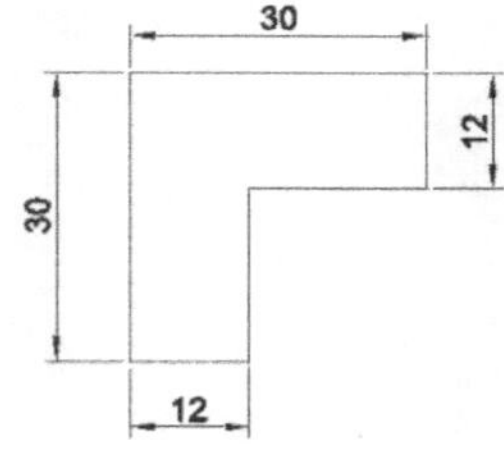

- Select **Front** from the **Restore View** drop-down in the **View** panel. Draw the sketch as shown below and extrude it up to 12 mm using the **Presspull** tool.
- Change the view orientation to SE Isometric.

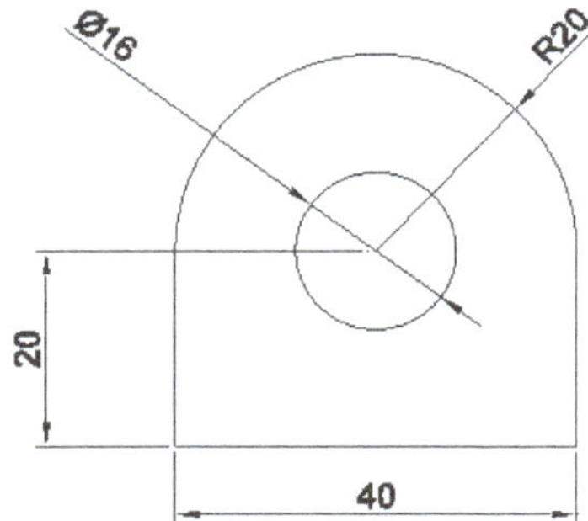

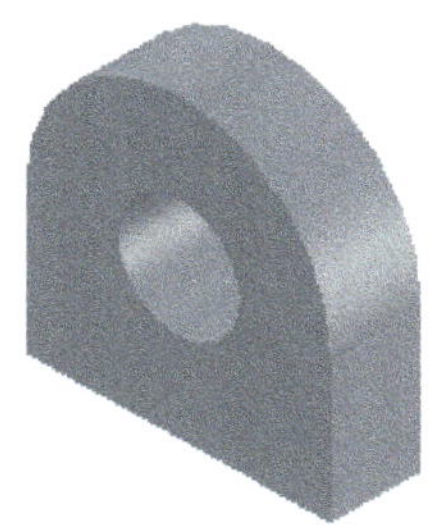

- Deactivate the **3D Object Snap** button on the status bar.
- Type **DS** in the command line and press ENTER; the **Drafting Settings** dialog appears.
- Click the **Object Snap** tab and **Clear All** the **Object Snap** modes.
- Now, select the **Endpoint** option and click **OK**.

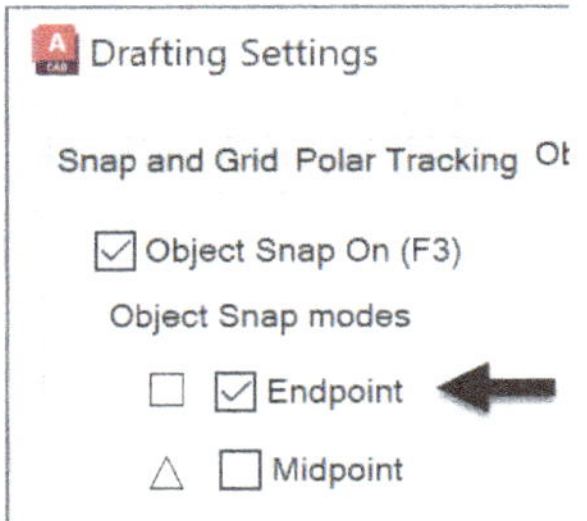

- Click **Home > Modify > 3D Align** on the ribbon.

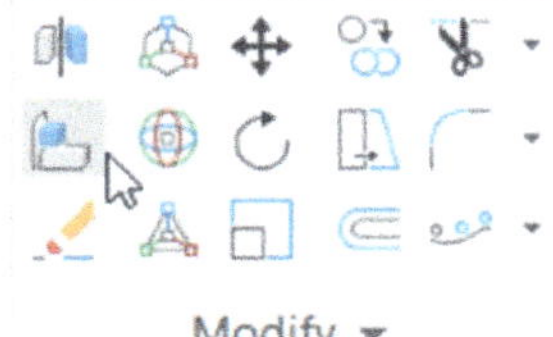

- Select the second solid object from the graphics window and press ENTER; the message, "Specify base point or [Copy]:" appears in the command line.

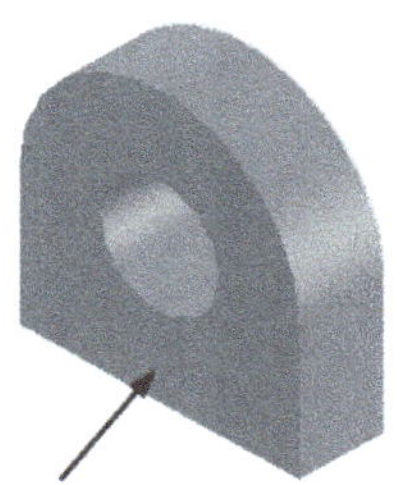

- Select the **Copy** option from the command line.
- Select three endpoints on the source object, as shown below.

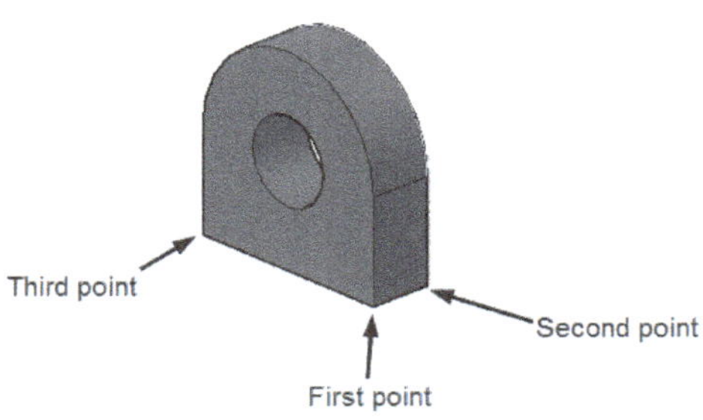

- Select three endpoints on the destination object as shown below; a copy of the source object will be aligned to the destination object.

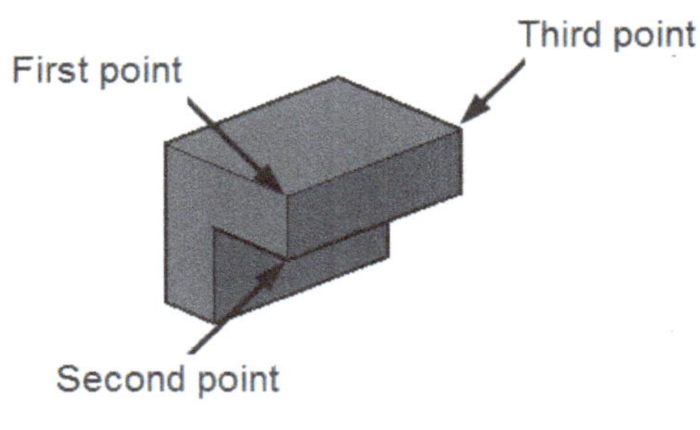

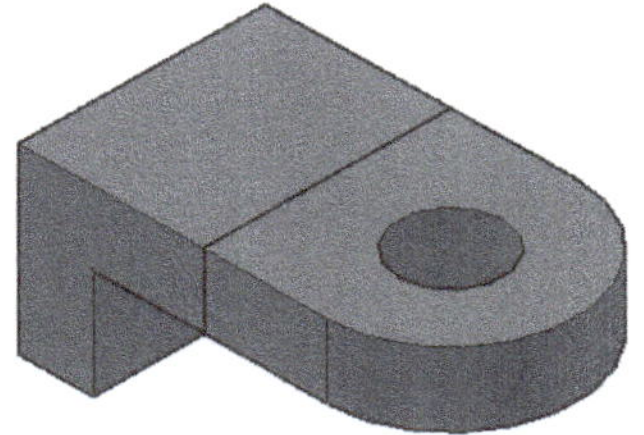

- Activate the **Ortho Mode** button on the status bar.
- Type **3DALIGN** in the command line and press ENTER.
- Select the second solid object. Press ENTER to accept the selection.

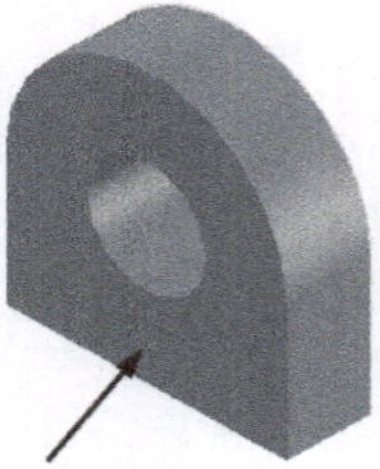

- Select the base point on the object as shown in the figure; the message, "Specify second point or [Continue] <C>:" appears in the command line.

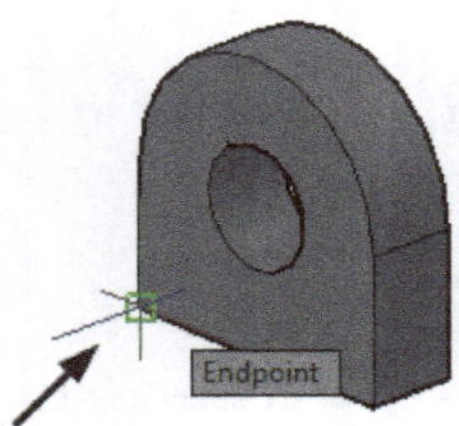

- Select the **Continue** option from the command line; the message, "Specify first destination point:" appears in the command line.
- Select the endpoint on the destination object, as shown below.

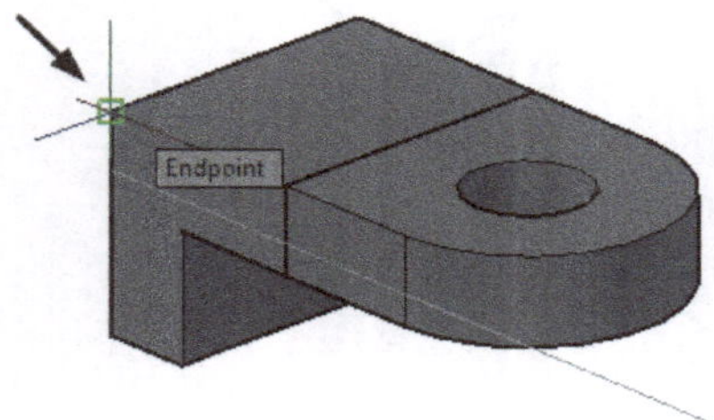

- Move the pointer along the X-direction and select the endpoint as shown in the figure; the message, "Specify third destination point or [eXit] <X>:" appears in the command line.

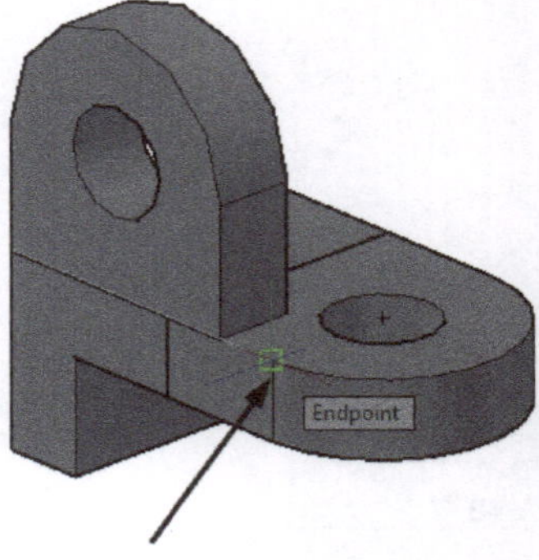

- Select the **eXit** option from the command line; the source object will be aligned, as shown below.

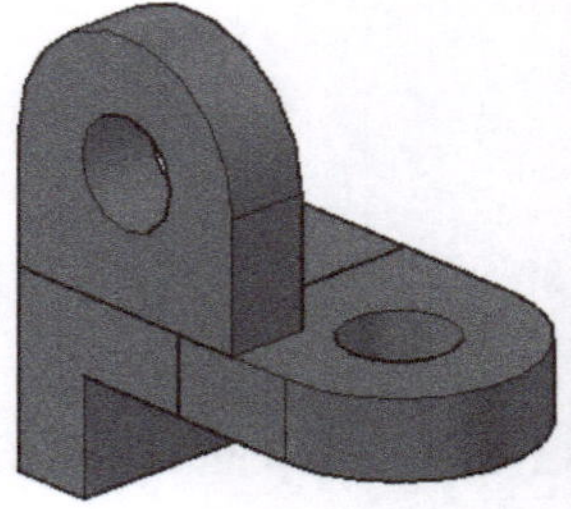

Using the 3D Mirror tool

The **3D Mirror** tool is similar to the **Mirror** tool. Using the **Mirror** tool, you can create a mirrored replica of an object in a 2D drawing. The objects are mirrored about an axis lying on a plane. But, with the **3D Mirror** tool, you need to define a plane about which the object will be mirrored. The **3D Mirror** tool provides many options to define the mirror plane.

- Click **Home > Modify > 3D Mirror** on the ribbon.

- Select the object to be mirrored from the model and press ENTER; the message, "Specify first point of mirror plane (3 points)," appears above the command line.

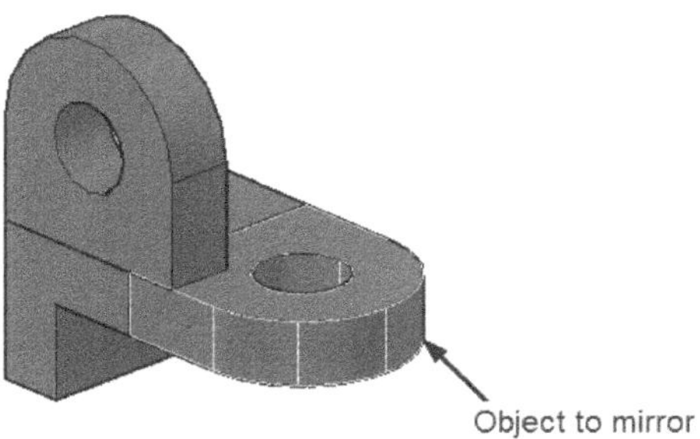

The **3points** option is selected by default to create the mirror plane. You need to specify three points to create a mirror plane. The mirror plane will pass through the selected points.

- Select the first and second points of the mirror plane, as shown below.

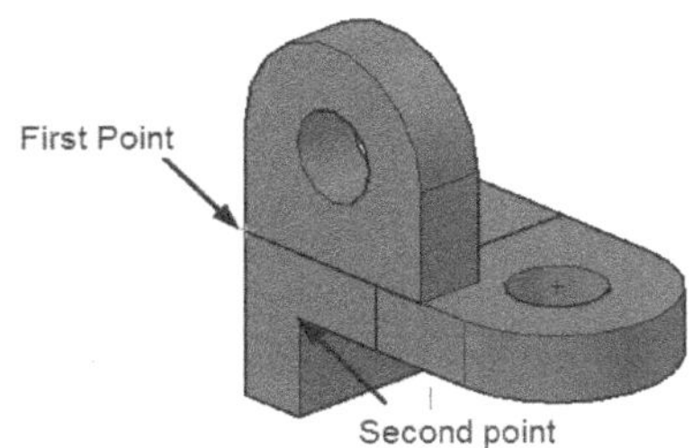

- Click the **Orbit** button on the **Navigation Bar** and rotate the model, as shown below.

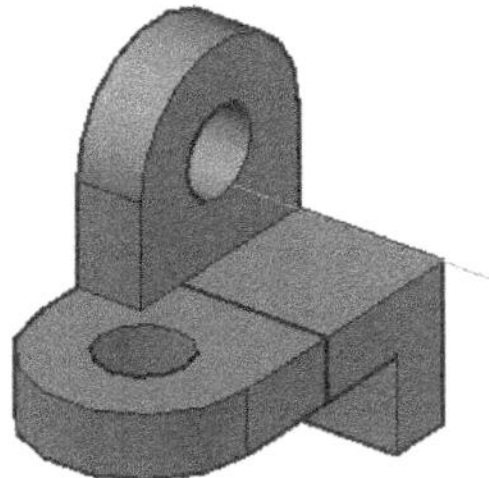

- Right-click and select **Exit** from the shortcut menu.
- Select the third point to define the mirror plane; the message, "Delete source objects? [Yes No] <N>:" appears in the command line.

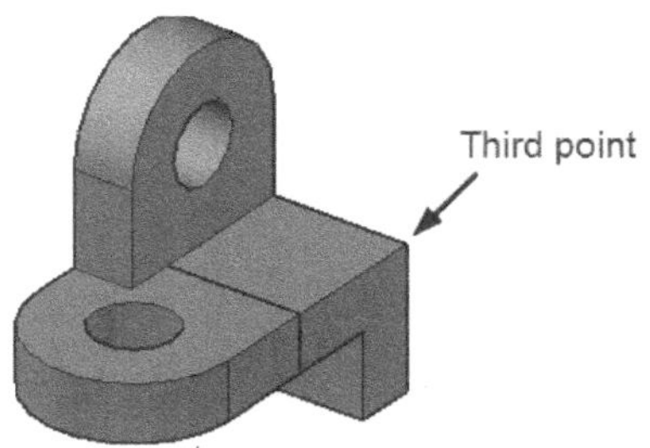

- Select the **No** option from the command line; the object will be mirrored.

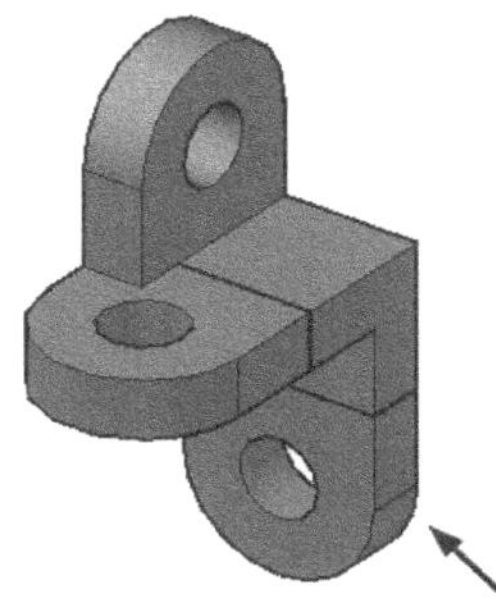

- Click the down arrow next to the **Object Snap** icon on the status bar and select **Center**.
- Type **3DMIRROR** in the command line and press ENTER.
- Select the object to be the mirror from the model and press ENTER.

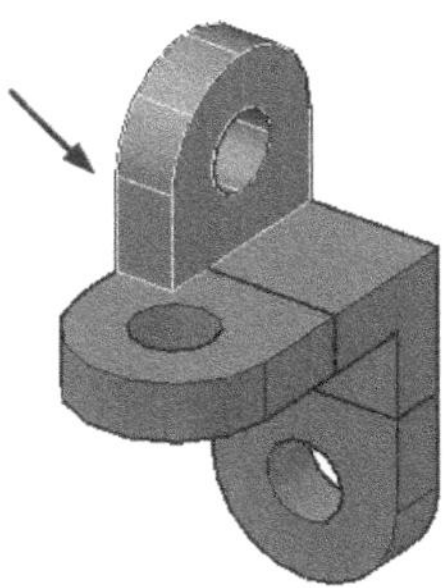

- Select the **XY** option from the command line; the message, "Specify point on XY plane <0,0,0>:" appears in the command line.

  The **XY** option creates a plane parallel to the XY plane. You need to specify a point at which the plane parallel to the XY plane will be created.

- Select the center point of the horizontal hole to define the mirror plane; the message, "Delete source objects? [Yes No] <N>:" appears in the command line.

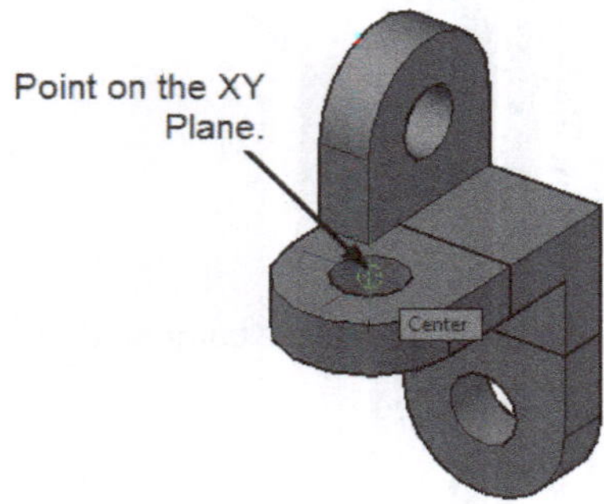

- Select the **No** option; the object will be mirrored, as shown below.

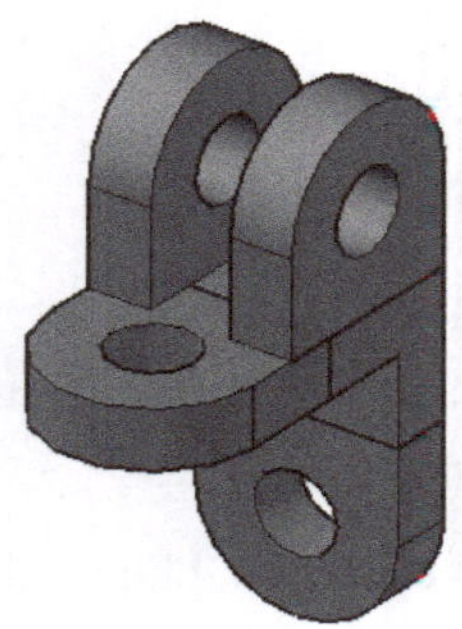

- Click the **Union** button on the **Solid Editing** panel and select all the objects of the model.
- Press ENTER; the objects will be combined into a single object.
- Change the view to SE Isometric.

Using the Fillet Edge tool

The **Fillet Edge** tool is used to create rounds (convex corners) or fillets (concave corners) on solid objects, just as in 2D drawings.

- Click **Solid > Solid Editing > Fillet Edge** on the ribbon (or) type **FILLETEDGE** in the command line and press ENTER; the message, "Select an edge or [Chain Loop Radius]:"

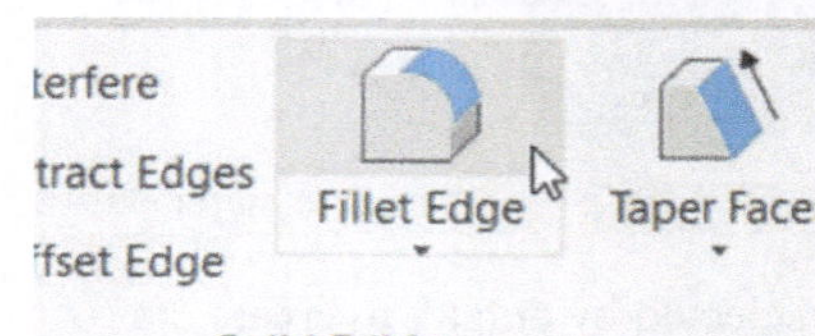

- Select the **Chain** option from the command line; the message, "Select an edge chain or [Edge Radius]:" appears.
- Select the edges from the model, as shown below; you will notice that a chain of edges is selected.

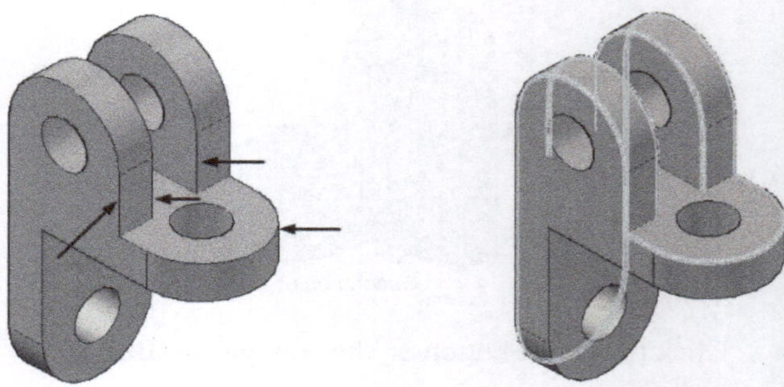

- Select the **Radius** option from the command line; the message, "Enter fillet radius or [Expression] <1.0000>:" appears.
- Type 2 in the command line and press ENTER.
- Press ENTER; the message, "Press Enter to accept the fillet or [Radius]:" appears. Also, a grip displayed on the fillets. You can use this grip to change the fillet radius dynamically.

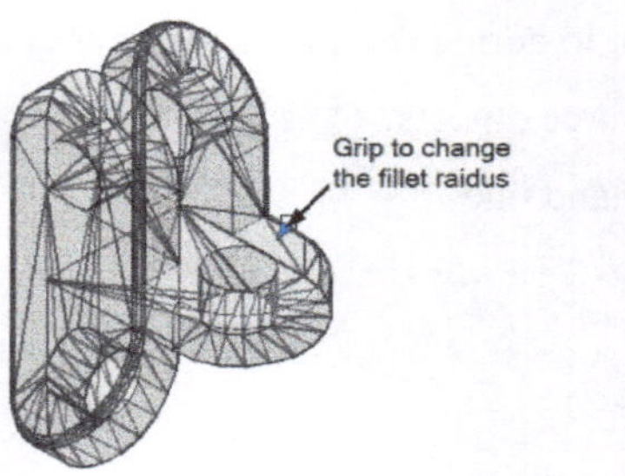

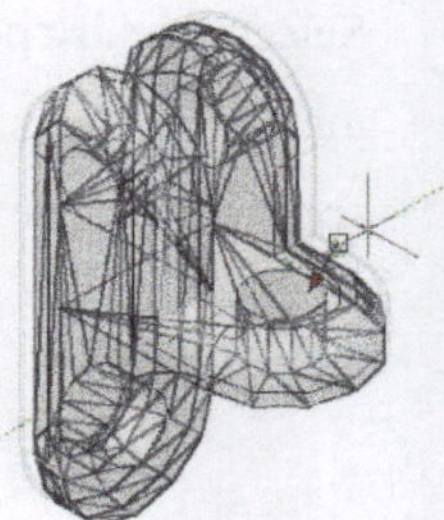

- Press ENTER to create rounds, as shown in the figure.

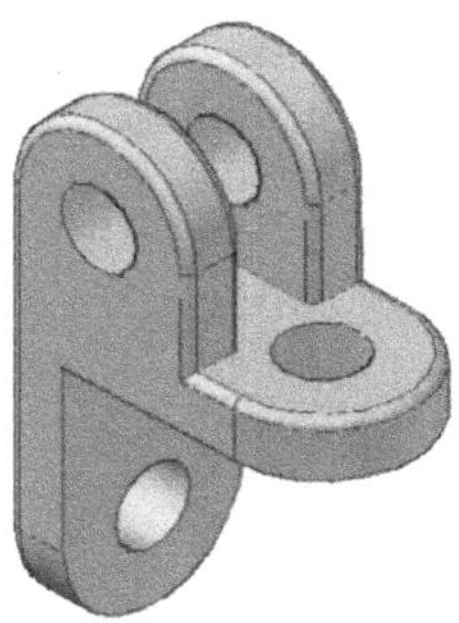

- Click the **Fillet Edge** button on the **Solid Editing** panel.
- Select the **Loop** option from the command line.
- Select the edge from the model, as shown in the figure; the edges on the front face of the model are highlighted. Also, the message, "Enter an option [Accept Next] <Accept>:" appears in the command line.

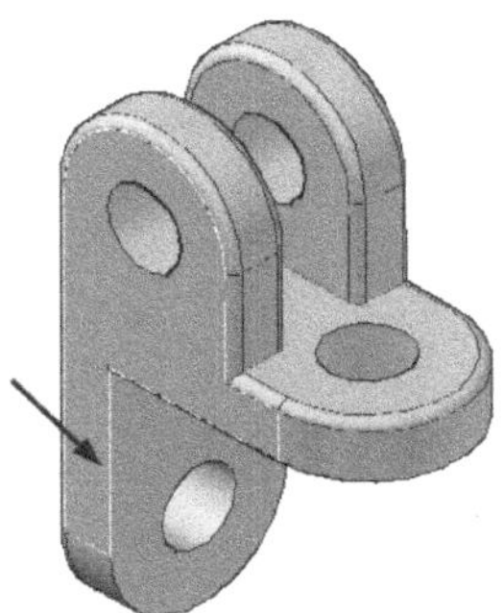

- Select the **Next** option from the command line; the edges on the side face will be highlighted.
- Select the **Accept** button; rounds and fillets are displayed on the side face.

- Likewise, select the round edge, as shown in the figure, and select the **Next** option from the command line; the edges on the bottom face of the model will be highlighted.
- Click the **Orbit** button on the **Navigation Bar** and rotate the model.
- Select the **Accept** option to view rounds and fillets on the bottom face.
- Select the **Radius** option and type 2. Press ENTER to accept.
- Press ENTER twice to create rounds, as shown in the figure.

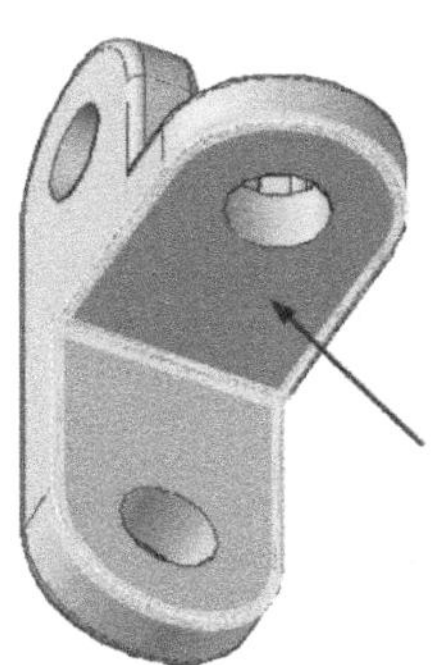

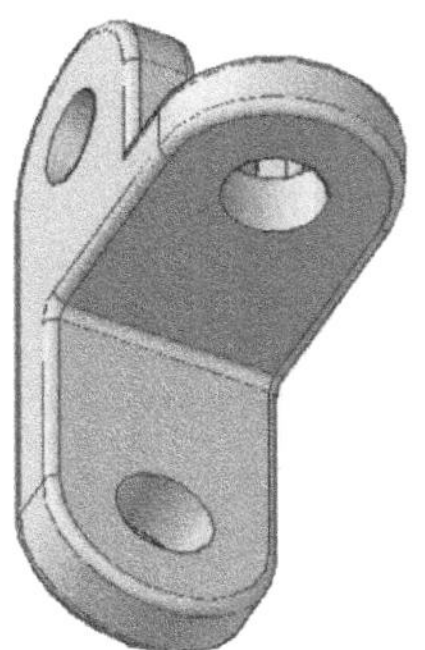

- Similarly, create fillets on the remaining edges by using the **Chain** option.

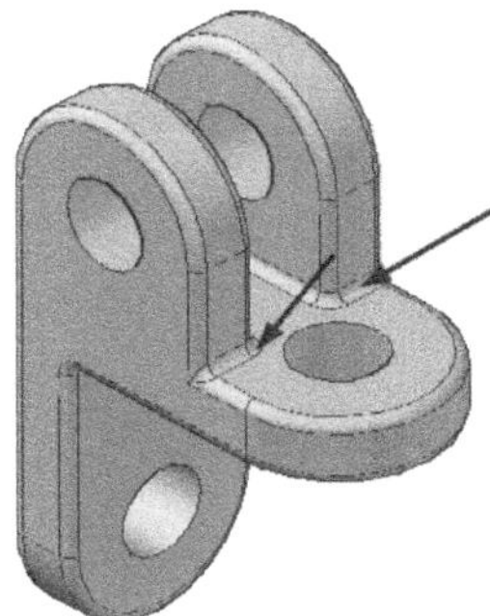

- Save and close the file.

Using the Taper Faces tool

The **Taper Faces** tool is used to taper faces. You can use this tool to change the angle of planar or curved faces.

Example 3:

In this example, you will create a cylinder and taper the outer face.

- Start a new AutoCAD file.
- Set the **Workspace** to **3D Modeling**.
- Select **Front** from the **Restore View** drop-down on the **Views** panel.
- Create a hollow cylinder with an inner and outer diameter as 140 and 150, respectively. The cylinder height is 50.

- Click **Solid > Solid Editing > Taper Faces** on the ribbon.

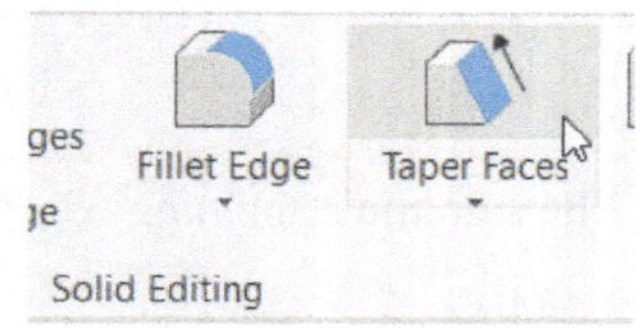

- Select the outer cylindrical face and press ENTER to accept; the message, "Specify the base point:" appears in the command line.
- Press and hold the SHIFT key and right-click to display the shortcut menu. Select the **Center** option from the shortcut menu.
- Move the pointer over the circular edge on the front face; the center point of the circular edge will be highlighted.
- Select the center point of the circular edge.
- Move the pointer along the Z-direction in the direction, as shown in the figure.
- Click to specify the axis of the taper; the message, "Specify the taper angle:" appears in the command line.

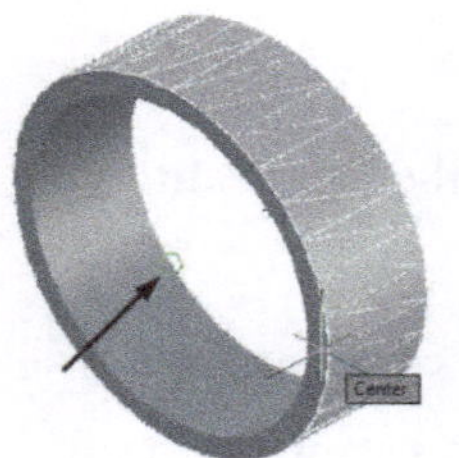

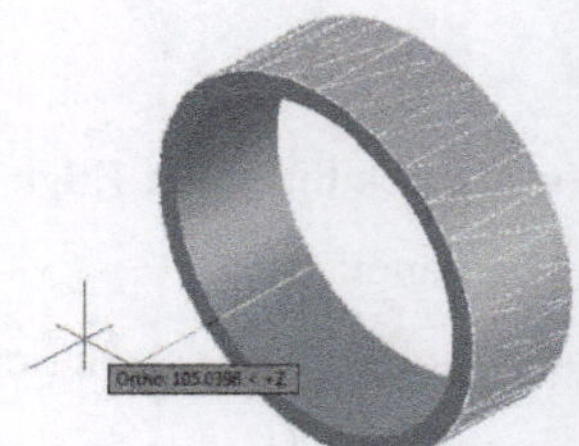

- Type 10 as the taper angle and press ENTER; the outer cylindrical face will be tapered as shown in the figure.

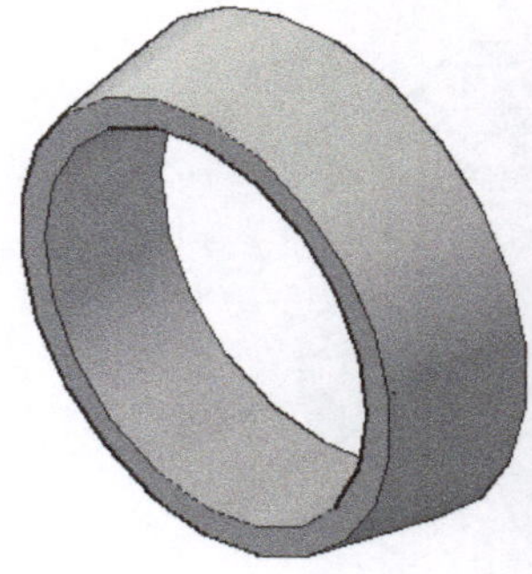

- Press Esc

Using the Offset Faces tool

The **Offset Faces** tool is used to move the faces of a 3D object in the perpendicular direction.

- On the ribbon, click **Solid > Selection > Selection Filter** drop-down > **Face**.

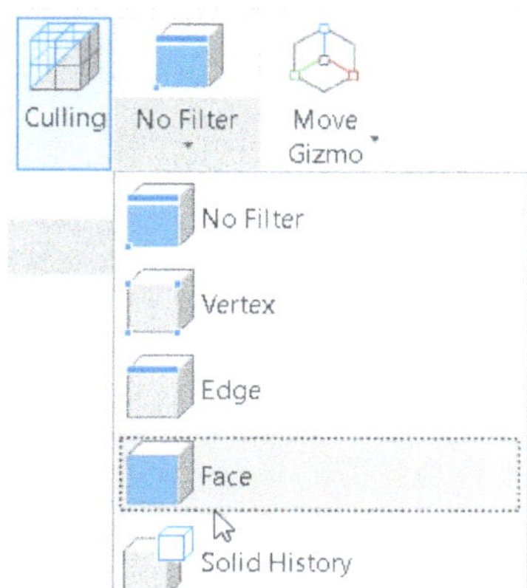

- Click **Solid > Solid Editing > Faces** drop-down > **Offset Faces** on the ribbon.

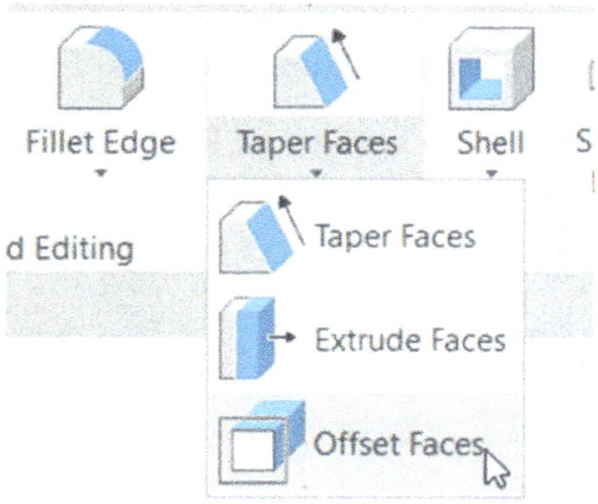

- Select the front face of the model and press ENTER; the message, "Specify the offset distance:" appears in the command line.

Note: By mistake, if you have selected the side faces, then click the Remove option in the command line. Next, click on the side faces, and then press ENTER to remove them from the selection.

- Type -20 in the command line and press ENTER; the face will be offset.

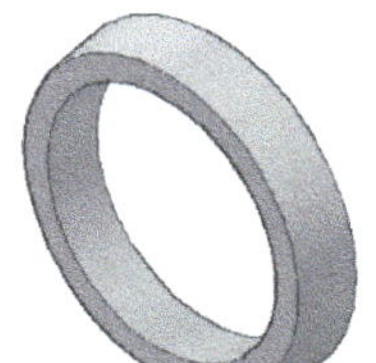

- Press Esc.

- On the ribbon, click **Solid > Selection > Selection Filter** drop-down > **No Filter**.
- Create a cylinder of 40 mm diameter and 30 mm length at the center of the hollow cylinder.

- Create a truncated cone with the following dimensions.

  Base radius: 8 mm

  Top radius: 5 mm

  Height: 65mm

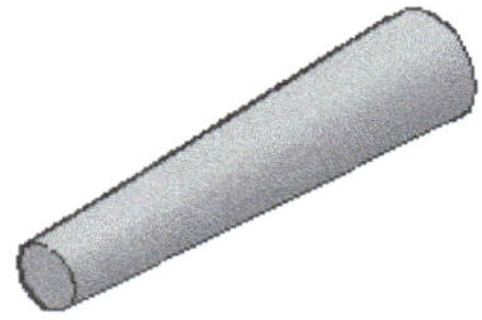

Using the 3D Rotate tool

The **3D Rotate** tool is used to rotate objects about an axis. You can define the axis of rotation by using the **Rotate Gizmo** tool. The **Rotate Gizmo** tool will be displayed when you activate the **3D Rotate** tool and select an object.

- Click **Home > Modify > 3D Rotate** on the ribbon.

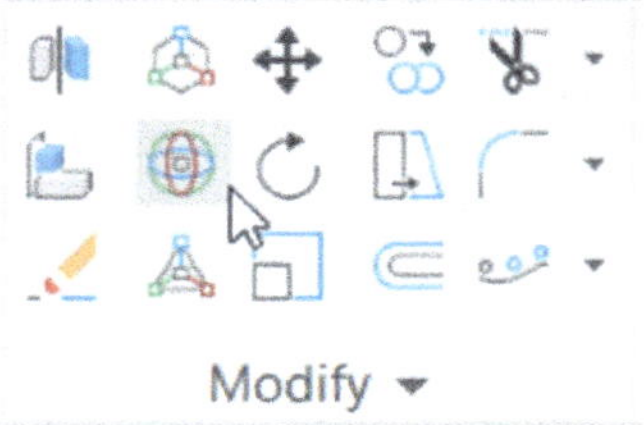

- Select the truncated cone and press ENTER; the **Rotate Gizmo** tool will be displayed.

- Select the center point of the front face as the base point; the **Rotate Gizmo** tool will be moved to the selected point.

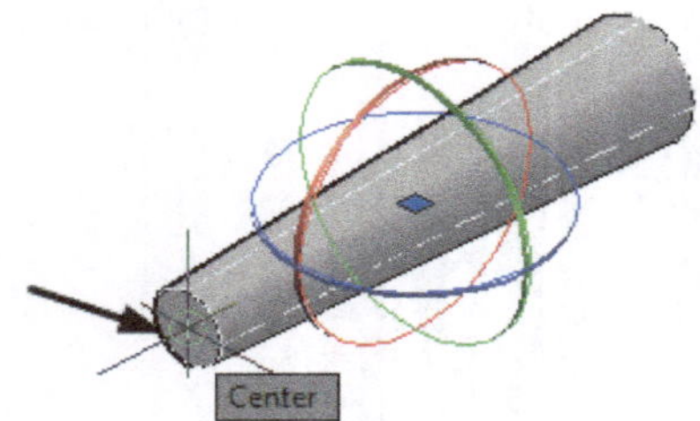

- Select the Z axis (Blue ring) of the **Rotate Gizmo**; an axis line is displayed along the Z-axis.
- Type **270** as the rotation angle and press ENTER; the cone will be rotated by 270 degrees.

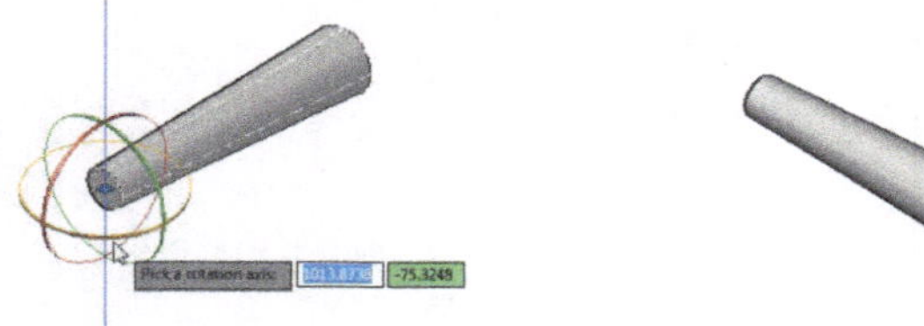

- On the status bar, click the down arrow next to the **Object Snap** icon, and then select the **Quadrant** option, if not already selected.
- Click the **Move** button on the **Modify** panel and select the cone. Press ENTER to accept.
- Select the base point and the destination point, as shown below; the cone will be placed at the destination point.

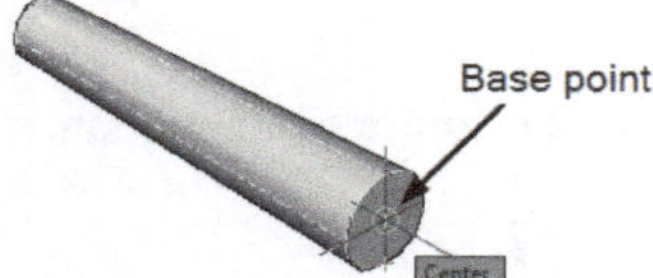

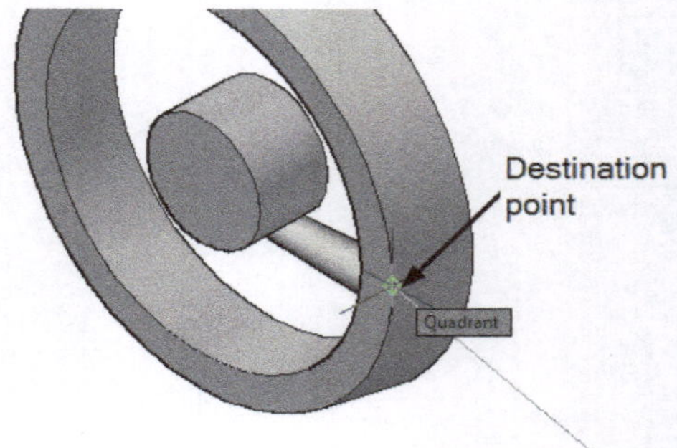

- Select the cone; the **Move Gizmo** tool will be displayed on it.
- Select the Y-axis (Green arrow) of the **Move Gizmo** tool and move the pointer toward the right.
- Type 22 in the command line and press ENTER; the cone will be moved through 22 mm.

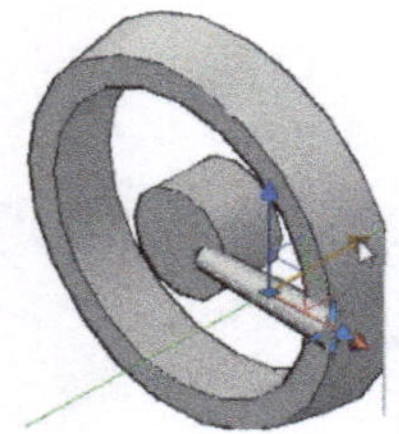

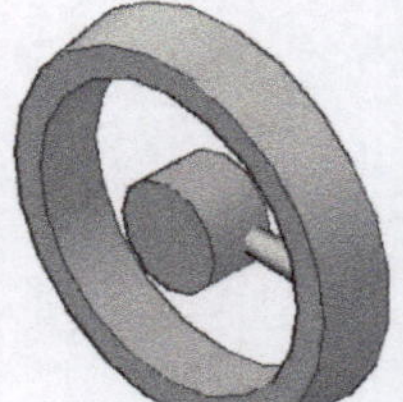

Using the 3D Polyline tool

The **3D Polyline** tool is similar to the **Polyline** tool, except that you can create a polyline by specifying coordinate points in three dimensions. Also, you can only create straight lines using this tool.

- Change the **Visual Style** of the model to **Wireframe**.
- Click **Home > Draw > 3D Polyline** on the ribbon.

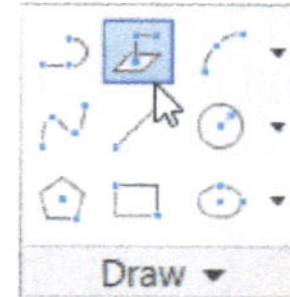

- Select the center point on the front face of the cylindrical object.
- Move the pointer toward the right and select the center point on the back face of the cylindrical object.

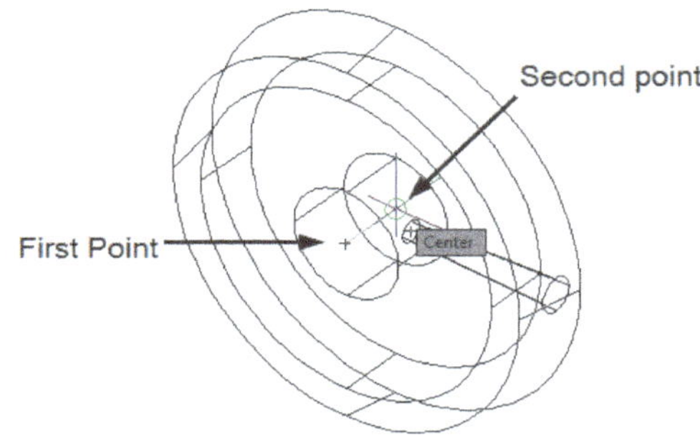

- Press ENTER; the 3D polyline will be created.

Creating a 3D Polar Array

You can create a 3D polar array by using the **Polar** option of the **3DARRAY** command. This option is similar to the **2D Polar Array** tool. The only difference between these two tools is that you need to specify an axis of rotation in 3D polar array, whereas in 2D Polar array, you need to specify an axis point. The axis of rotation in a 3D polar array can be specified by selecting two points. It allows you to create a 3D polar array about an axis in the 3D workspace.

- Type **3A** in the command line and press ENTER.
- Select the truncated cone from the model and press ENTER.
- Select the **Polar** option from the command line; the message, "Enter the number of items in the array:" appears.
- Type 6 in the command line and press ENTER; the message, "Specify the angle to fill (+=ccw, -=cw) <360>:" appears in the command line.

  Type + and press ENTER to create the polar array in counter-clockwise direction and type – to create it in the clockwise direction.
- Press ENTER to accept 360 as the fill angle; the message, "Rotate arrayed objects? [Yes No] <Y>:" appears in the command line.
- Select the **Yes** option from the command line; the message, "Specify center point of array:" appears.
- Select the first and second points of the axis, as shown in the figure; the polar array will be created.

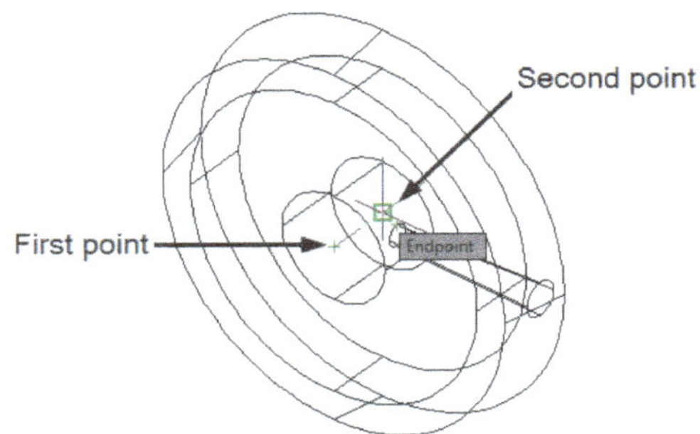

- Change the **Visual Style** to **Shades of Grey**.
- Perform the **Union** operation to combine all the objects.

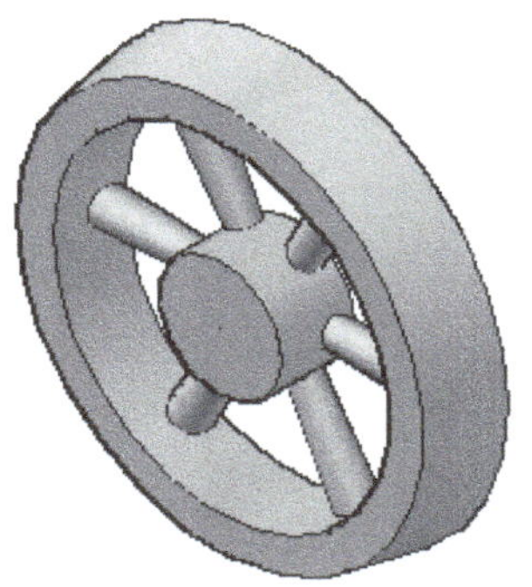

Using the Shell tool

The **Shell** tool converts a solid object into a thin-walled hollow object. You need to first select the object to be shelled, and then select the faces to be removed and enter the thickness of the walls.

- Click **Solid > Solid Editing > Shell** on the ribbon.
- Select the solid model; the message, "Remove faces or [Undo Add ALL]:" appears.
- Select the front face of the cylindrical object.
- Click the **Orbit** icon on the Navigation Bar.
- Press and hold the left mouse button and drag; the model will be rotated.

- Right-click and select **Exit**.
- Select the back face of the cylindrical object.

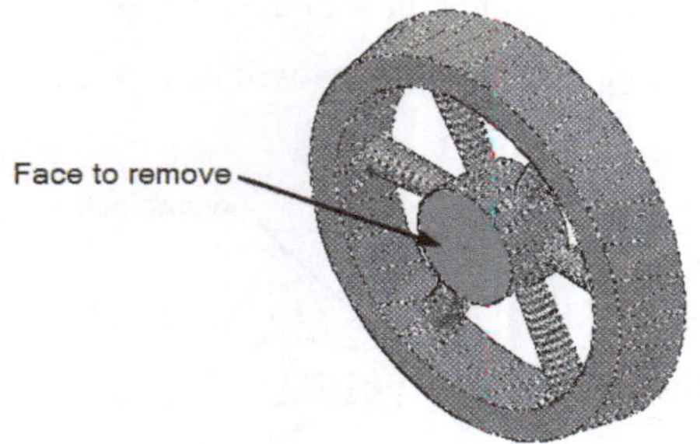

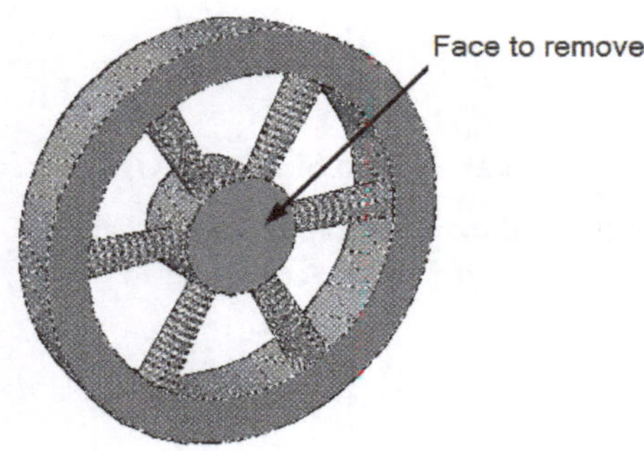

- Press ENTER; the message, "Enter the shell offset distance:" appears.
- Type 10 in the command line and press ENTER; the cylindrical object will be shelled.
- Select the **eXit** option from the command line.

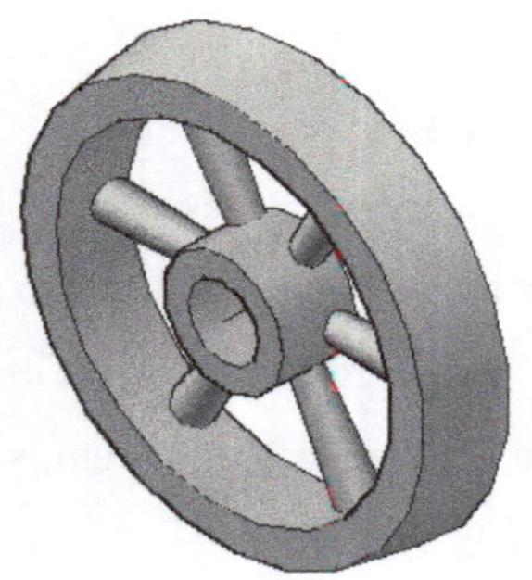

Using the Chamfer Edge tool

The **Chamfer Edge** tool is used to bevel sharp edges of a solid object. When you chamfer an edge, a wedge is created automatically, and the Boolean operation is performed to subtract it from the solid object.

- Click **Solid > Solid Editing > Chamfer Edge** on the ribbon.

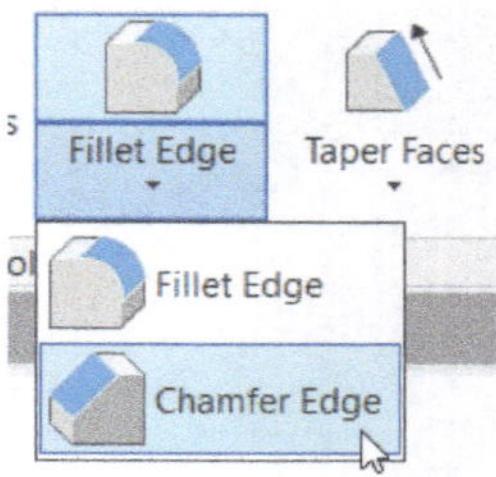

- Select the outer circular edge of the cylindrical object.
- Select the **Distance** option from the command line; the message, "Specify Distance1 or [Expression] <1.0000>:" appears.
- Type 4 in the command line and press ENTER; you will notice that the preview of the chamfer changes. Also, the message, "Specify Distance2 or [Expression] <1.0000>:" appears in the command line.
- Type 2 in the command line and press ENTER.
- Press ENTER twice to create the chamfer, as shown in the figure.

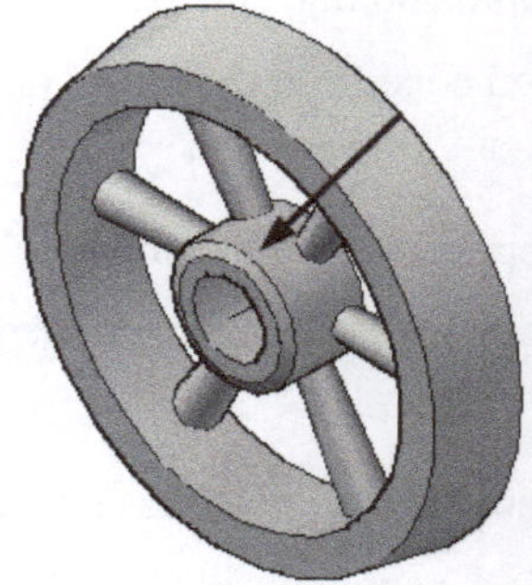

Using the Section Plane tool

The **Section Plane** tool creates a translucent cutting plane passing through a solid object to show the inside portion of it. This tool is handy when the inside portion of the solid is not visible. You can move this cutting plane dynamically to view the inside portion at different locations of the solid.

- To create a section plane, click **Solid > Section > Section Plane** on the ribbon; the message, "Select face or any point to locate section line or [Draw section Orthographic]:" appears.
- Select the **Orthographic** option from the command line.
- Select **Right**.

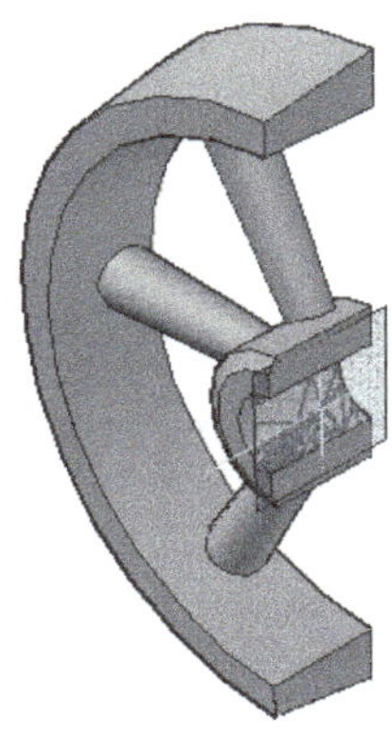

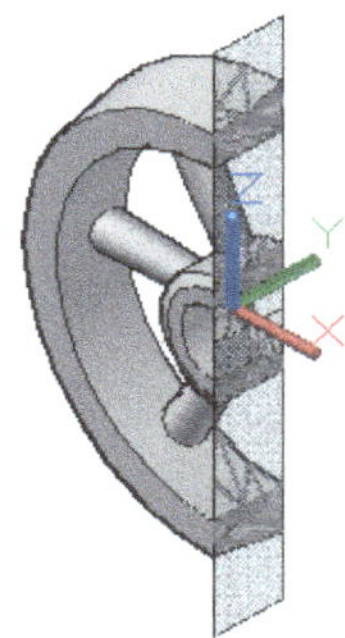

- Save the file as **Example 3**.

Using the Live Section tool

The **Live Section** tool is used to make a side of the section plane invisible or visible. When you create a section plane by selecting a plane, one side of the section plane will be invisible automatically. However, when you create a section plane by selecting points, you need to use the **Live Section** tool to make one side invisible. Click **Solid > Section > Live Section** on the ribbon. Next, select the section plane; one side of the section plane will be hidden or unhidden, as shown in the figure.

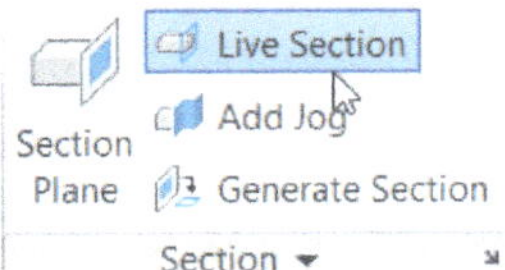

Creating Drawing Views

In Chapter 5, you have learned to create multi-view drawings using the standard projection techniques. Now, you will learn to generate views of a 3D model automatically.

Setting the Drafting Standard

Before you start generating the drawing views of the 3D model, you need to specify the drafting standard. It defines the way the views will be generated. To specify the drafting standard, click **Home > View > Drafting Standard (inclined arrow)** on the ribbon; the **Drafting Standard** dialog appears.

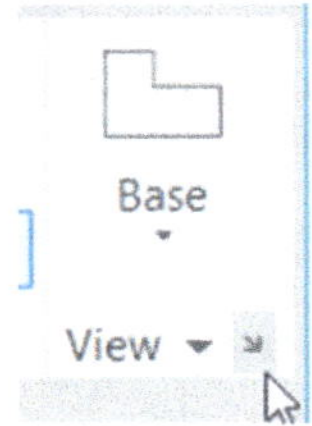

- In the **Drafting Standard** dialog, set the **Projection type** to **Third angle.**

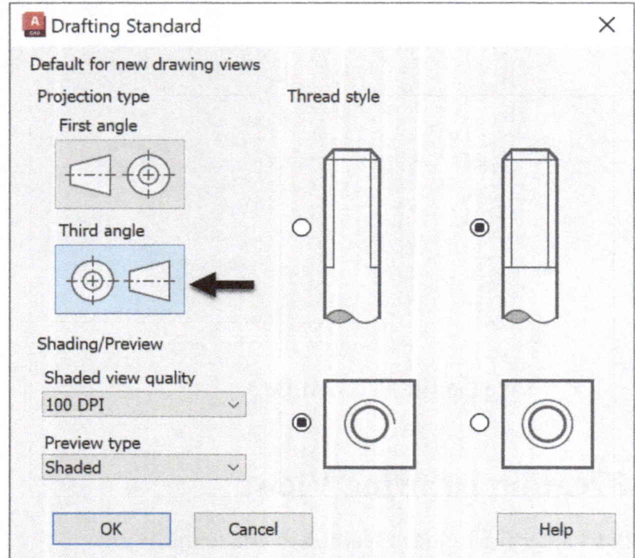

- Examine the other options in the dialog, as they are self-explanatory.
- Click the **OK** button.

Creating a Base View

The base view will be the first view of the drawing. It can be any view (front, top, right, left, bottom, or back view) of the model. But commonly, the front or top views of the model are generated first.

- Open the **Example 3.dwg,** if it is not already opened.
- To generate the base view of the model, click **Home > View > Base > From Model Space** on the ribbon; the message, "Select objects or [Entire model] <Entire model>:" appears in the command line.

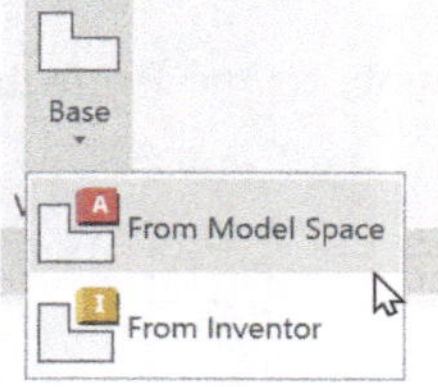

- Select the **Entire model** option from the command line; the model in the model space will be selected and the message, "Enter new or existing layout name to make current or [?] <Layout1>:" appears in the command line.
- Press ENTER to select **Layout 1**; the base view will be attached to the pointer and the message, "Specify location of base view or [Type sElect Orientation Hidden lines Scale Visibility] <Type>:" appears in the command line. Also, the **Drawing View Creation** tab appears in the ribbon.
- Specify the location of the view in the paper space, as shown below.

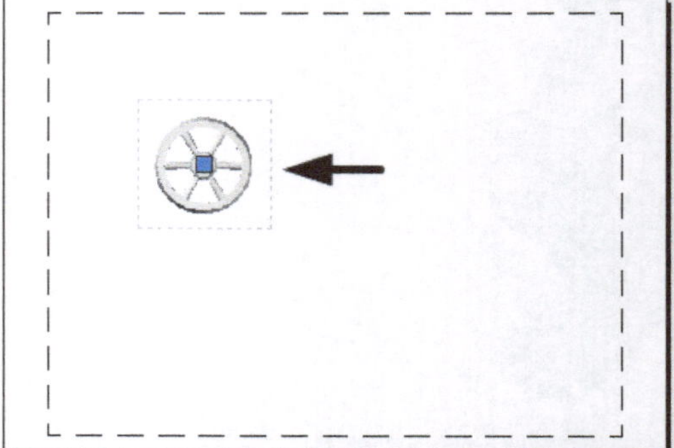

- In the **Drawing View Creation** tab, set the **Orientation** to **Front**.
- Select the **Visible Lines** option from the **Hidden Lines** drop-down.

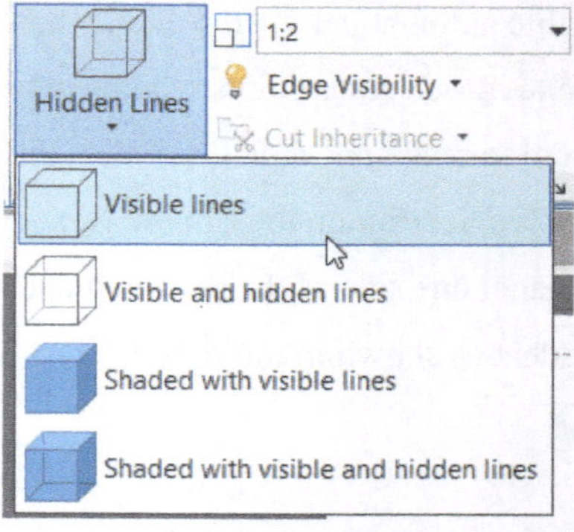

- Set the **Scale** in the **Appearance** panel to **1:4**.
- Click the **OK** button on the **Create** panel to create the base view; a projected view will be attached to the pointer, and you will be asked to specify its location. You will learn to create projected views in the next section.
- Press ENTER to exit the command.

Creating a Projected View

A projected view can be created from an existing view. It can be an orthographic view or isometric view generated by projecting from a base view or any other existing view.

- To create a projected view, click **Layout > Create View > Projected** on the ribbon, and then select the base view from **Layout 1**; the projected view will be attached to the pointer.

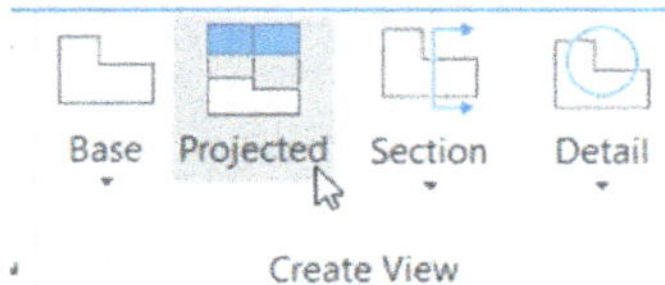

- Move the pointer downward and specify the location of the projected view, as shown below.

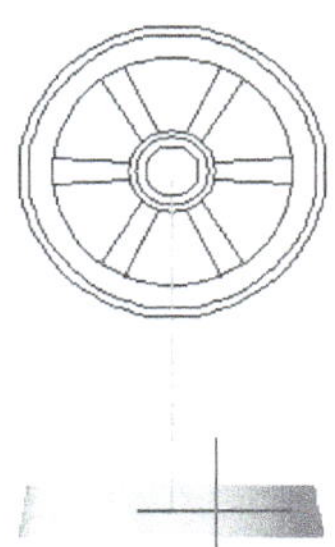

- Move the pointer diagonally toward the top-right corner and place the isometric view, as shown below.

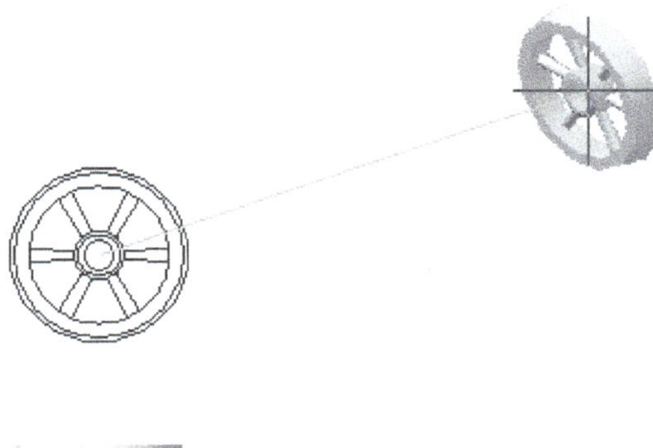

- Select the **eXit** option from the command line to exit the command.

Creating Section Views

In Chapter 8, you have learned to create section views manually. Now, you will learn to generate section views automatically from a 3D model. You can create different types of section views using the tools available in the **Section** drop-down in the **Create Views** panel.

Creating the Section View Style

Section View Style defines the display of the section view and the cutting plane. To create a section view style, click **Layout > Styles and Standards > Section View Style** on the ribbon; the **Section View Style Manager** dialog appears. Click the **New** button in the **Section View Style Manager** dialog; the **Create New Section View Style** dialog appears. Type **Example** in the **New Style Name** box and click **Continue; the New Section View Style** dialog appears. In this dialog, click the **Cutting Plane** tab and select the **Show cutting plane lines** option.

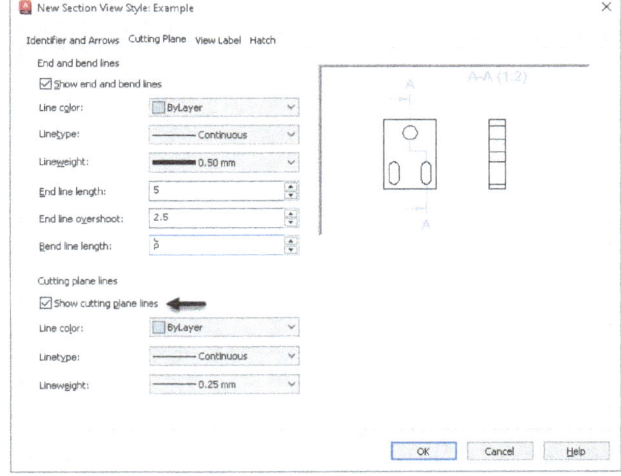

Click the **Hatch** tab and set the **Hatch Scale** to **0.5** and click **OK**. Click the **Set current** button on the **Section View Style Manager** dialog and click **Close**.

Creating a Full Section View

To create a full section view, click **Layout > Create Views > Section > Full** on the ribbon. Next, select the base view from the layout. After selecting the base view, you need to specify the start and endpoints of the cutting plane. Select the start point of the cutting plane, as shown below. Move the pointer vertically upward and specify the endpoint of the cutting plane.

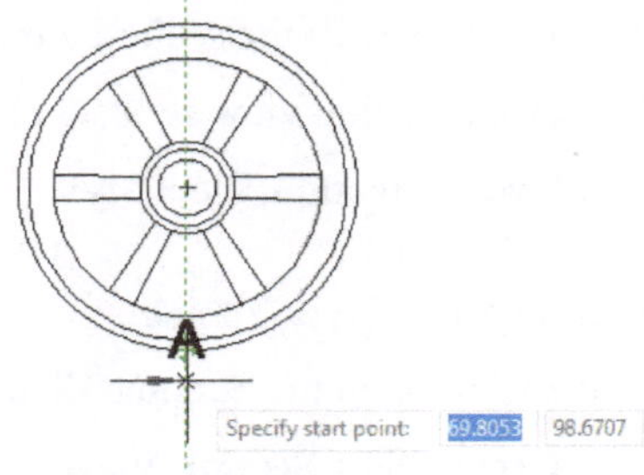

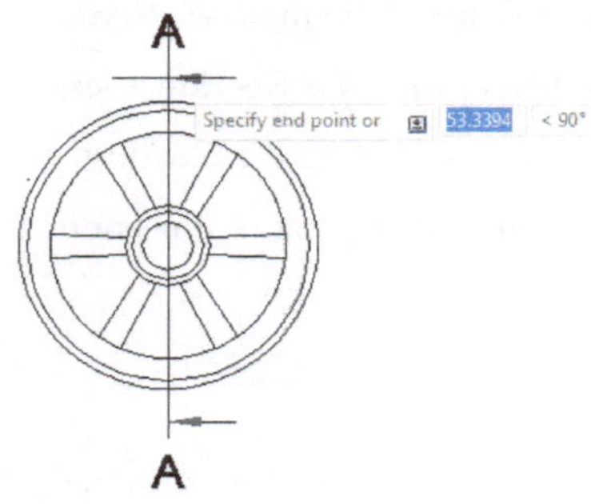

Move the pointer toward the right and click to specify the location of the section view. Select the **eXit** option to create the section view.

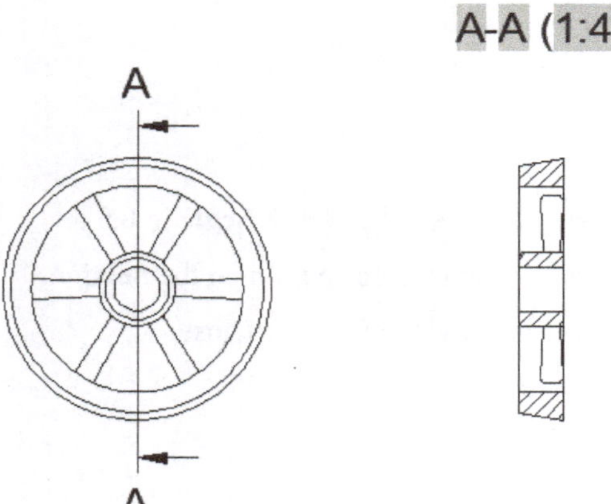

Creating a Detailed View

A detailed view is created to enlarge and view small portions of a drawing.

- To create a detailed view, click **Layout > Create Views > Detail > Circular** on the ribbon; the message, "Select parent view," appears in the command line.

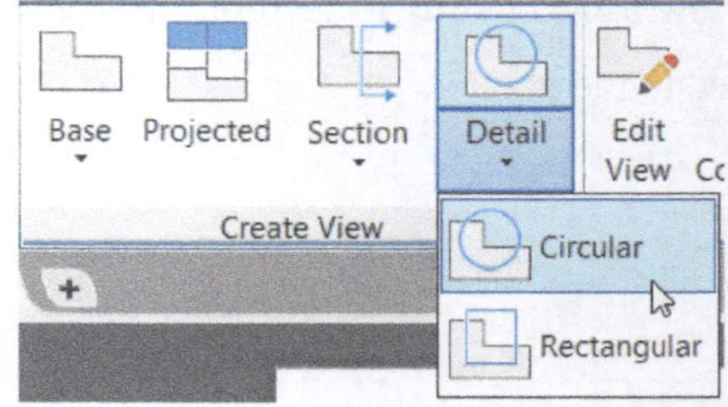

- Select the section view from the layout; the message, "Specify center point or [Hidden lines Scale Visibility Boundary model Edge Annotation] <Boundary>:" appears in the command line.
- Select a point on the section view, as shown below; the message, "Specify size of boundary or [Rectangular Undo]:" appears in the command line.
- Draw a circle similar to the one shown below.

- Next, place the detailed view on the lower right corner of the layout.
- Select the **Smooth with border** button on the **Model Edge** panel of the **Detail View Creation** tab.

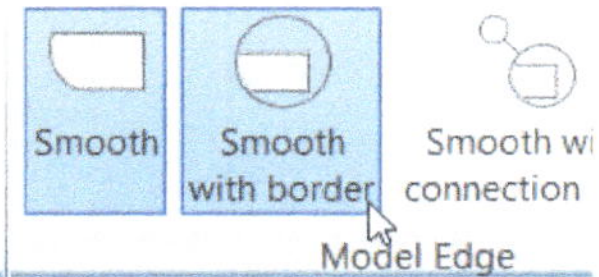

- Press Enter; the detail view will be created.

Exercises

Create 3D models using the drawing views and dimensions.

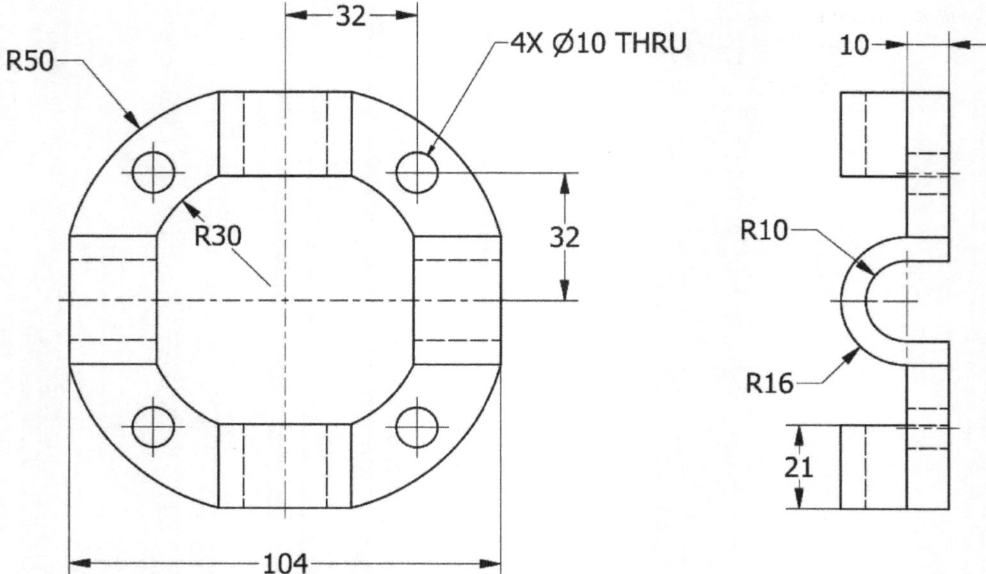

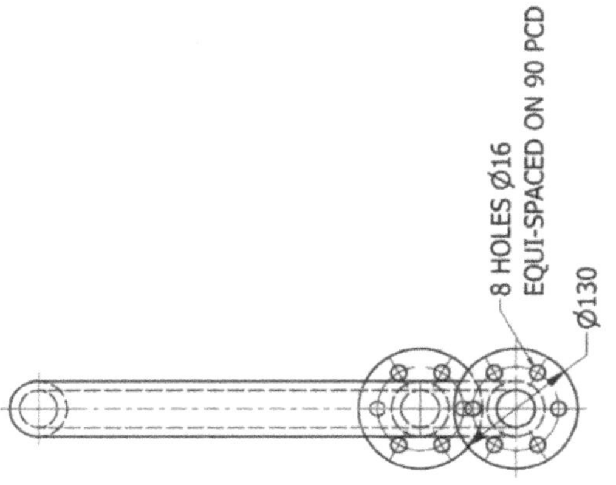
8 HOLES Ø16
EQUI-SPACED ON 90 PCD
Ø130

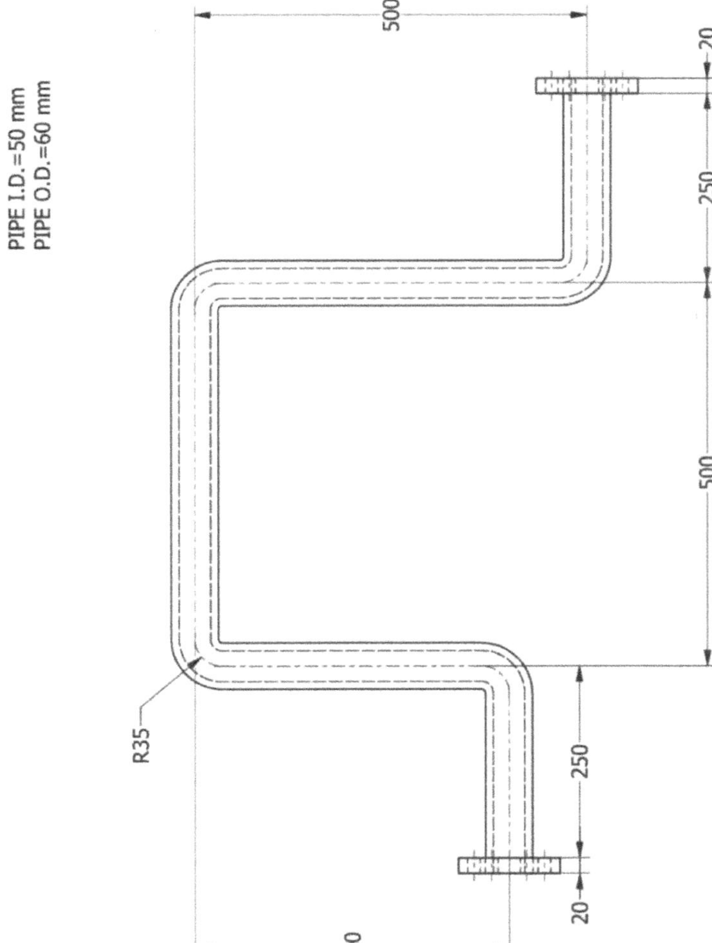
PIPE I.D.=50 mm
PIPE O.D.=60 mm
500
20
250
500
R35
250
20
400

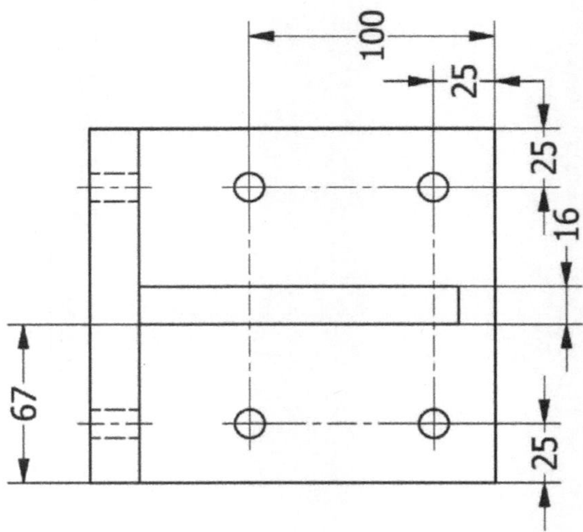
100
25
25
16
67
25

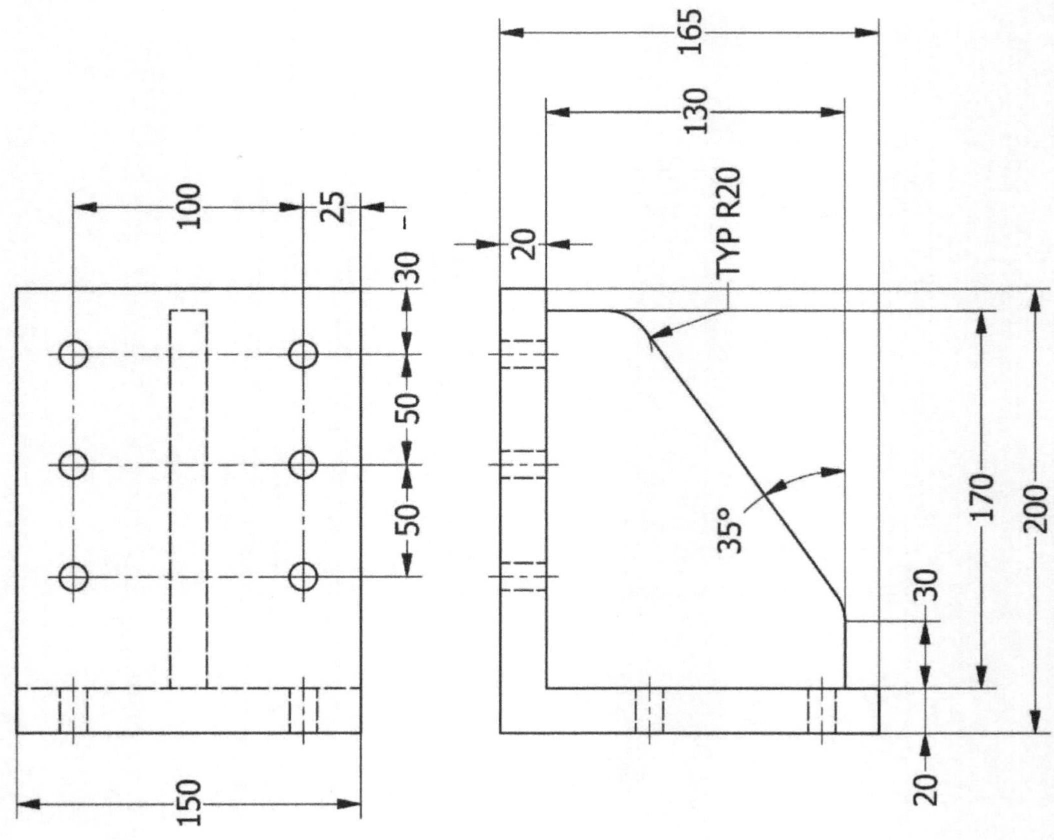
165
130
100
25
20
TYP R20
30
50
50
35°
30
170
200
20
150

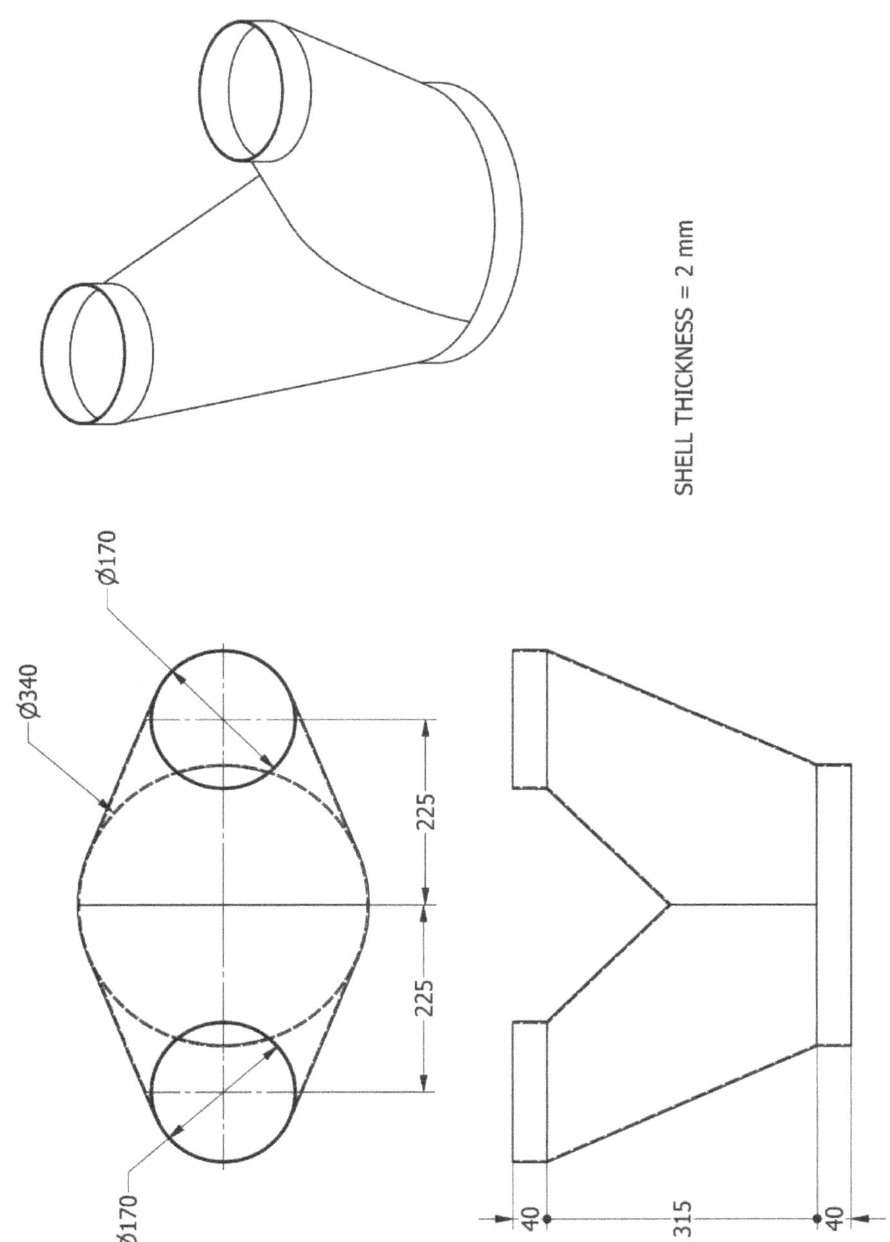
SHELL THICKNESS = 2 mm
Ø170
Ø340
225
225
Ø170
40
315
40

Chapter 14: Creating Architectural Drawings

In this chapter, you will learn to do the following:

- **Defining Settings for Architectural Drawings**
- **Creating Inner Walls**
- **Creating Openings and Doors**
- **Creating Kitchen Fixtures**
- **Creating Bathroom Fixtures**
- **Adding Furniture using Blocks**
- **Adding Windows**
- **Arranging Objects of the drawing in Layers**
- **Creating Grid Lines**
- **Adding Dimensions**
- **Create Stair Details**
- **Elevation View**
- **Roof Plan**
- **Wall and Roof details**

Example 1

In this example, you will learn to create an architectural drawing.

Creating Outer Walls

- Start **AutoCAD 2024** and click **New > Browse Templates**.

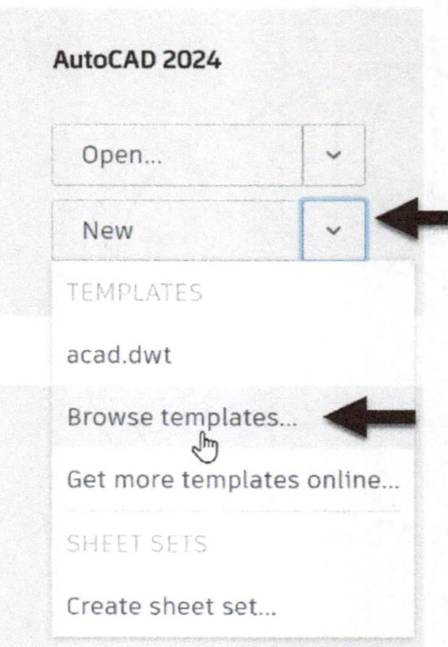

- Select the **acad** template from the **Select Template** dialog and click the **Open** button.
- Set the **Workspace** to **Drafting & Annotation**.

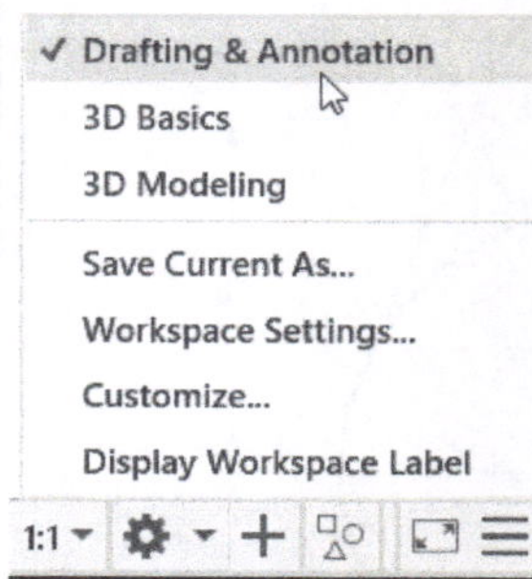

- Type **UN** in the command line and press Enter.
- On the **Drawing Units** dialog, select **Type > Architectural**.
- Select **Precision > 0-01/16**.
- Set the **Insertion Scale** to **Inches** and click **OK**.
- Type **LIMITS** in the command line and press Enter.
- Press Enter to accept 0, 0 as the lower limit.
- Type 100', 80' in the command line, and press Enter. The program sets the upper limit of the drawing.
- Turn OFF the grid.

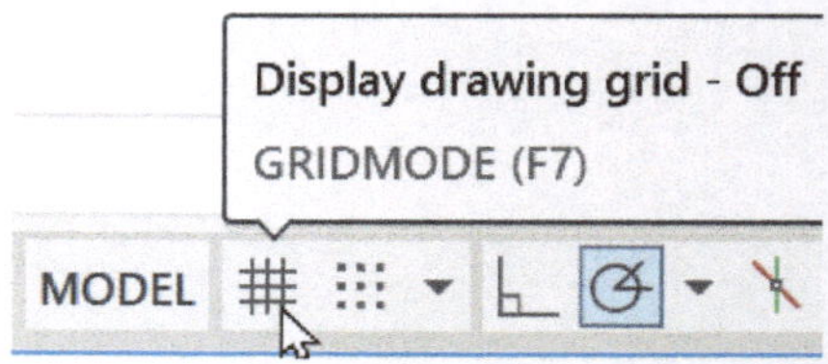

- Select **Zoom > Zoom All** on the Navigation Bar.
- On the Status bar, turn ON the **Ortho Mode** icon.

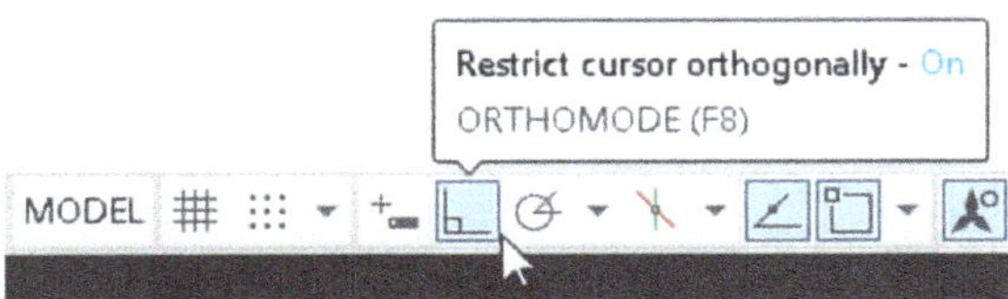

- On the ribbon, click **Home > Draw > Line**, and then select an arbitrary point. It defines the start point of the line.
- Move the pointer toward right horizontally and type-in 412 — press Enter.
- Move the pointer vertically and type-in 338 — press Enter.
- Move the pointer onto the starting point of the drawing, and then move it upwards. You will notice that a dotted line appears.

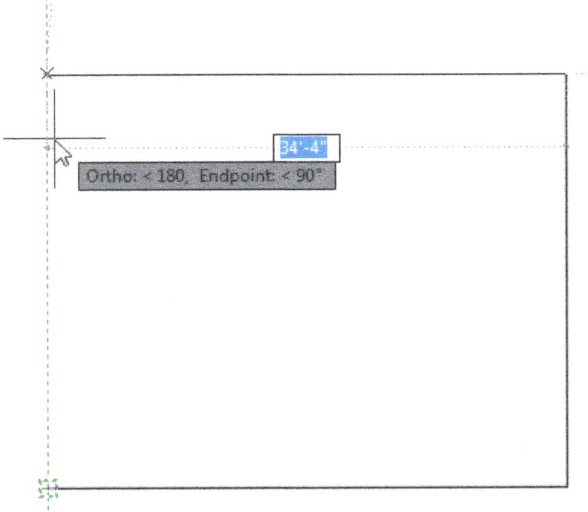

- Click to create a horizontal line. You will notice that the two horizontal lines are of the same length.
- Click the right mouse button and select **Close**.

- On the ribbon, click **Home > Modify > Offset**.
- Type-in 6 as offset distance and press Enter.
- Select the left vertical line of the drawing.
- Move the pointer inside the drawing and click to create an offset line.
- Likewise, offset the other lines, as shown below.

- On the Navigation Bar, click **Zoom > Zoom Window**.

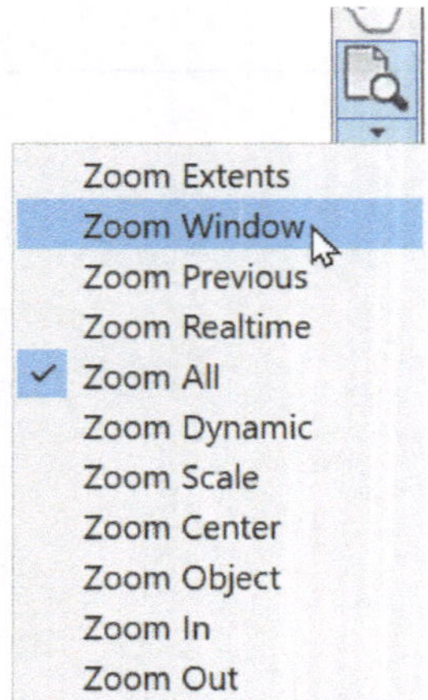

- Create a window on the top left corner of the drawing. The corner portion will be zoomed in.

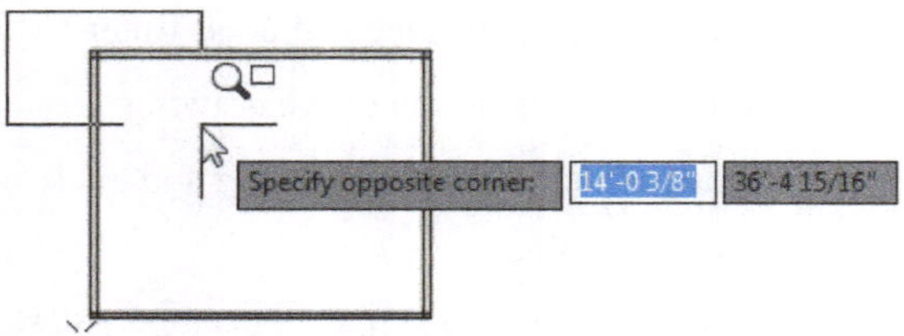

- On the ribbon, click **Home > Modify > Fillet** .
- Select the **Radius** option on the command line, and then type in 0 — press Enter.
- Select the inner offset lines, as shown below.

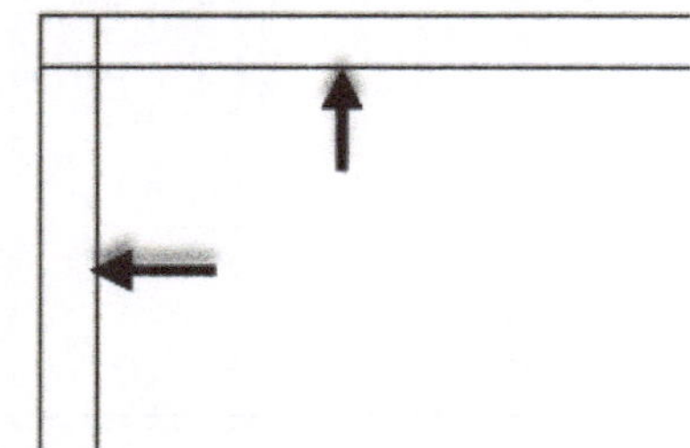

- Likewise, fillet the other corners.

- Save the drawing. Make sure that you save the drawing after each section.

Creating Inner Walls

- Activate the **Offset** command and type in **130** in the command line, and then press Enter.
- Select the inner line of the right side wall and click inside the drawing.
- Press Esc.

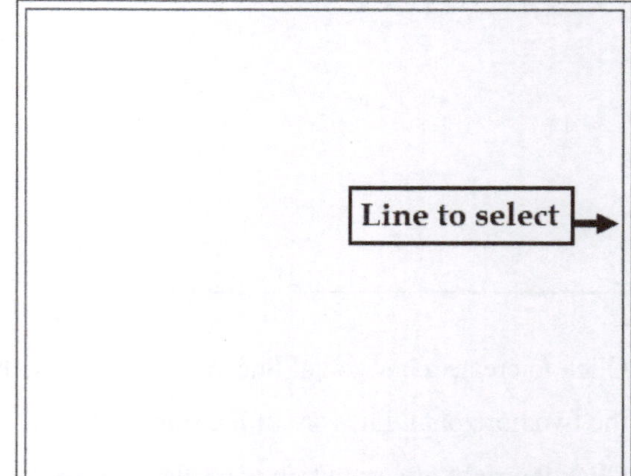

- Select the new offset line. You will notice that three grips are displayed on the line.
- Click the right mouse button and select **Copy Selection**.

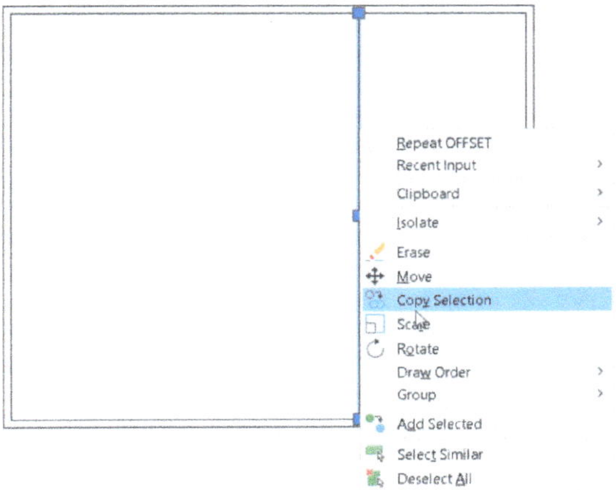

- Select the endpoint of the selected line as a base point.
- Move the pointer toward the left and type in 4, and then press Enter. A new line is created, and another line is attached to the pointer.
- Move the pointer toward the left and type in 118, and then press Enter.

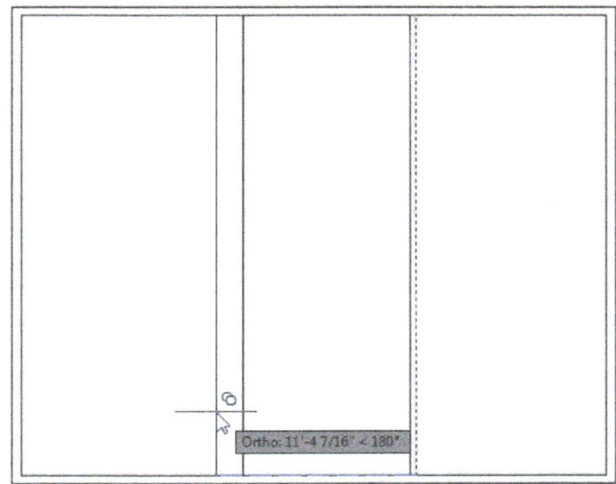

- Move the pointer toward the left and type in 122, and then press Enter.
- Press Esc to come out of the **Copy** command.

- Likewise, create horizontal offset lines, as shown below.

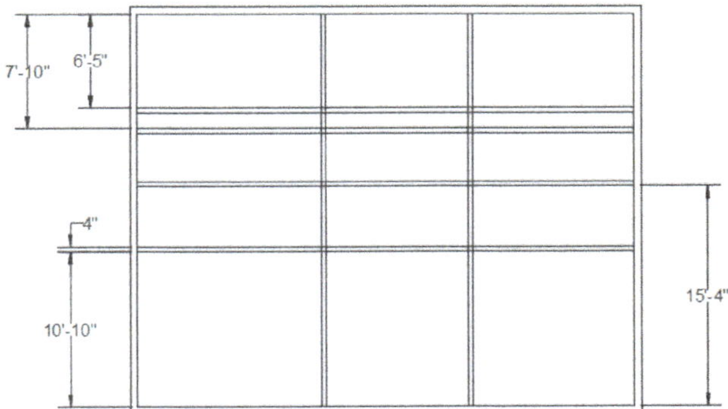

- On the ribbon, click **Home > Modify > Trim**.

- Click on the lines at the locations shown below.

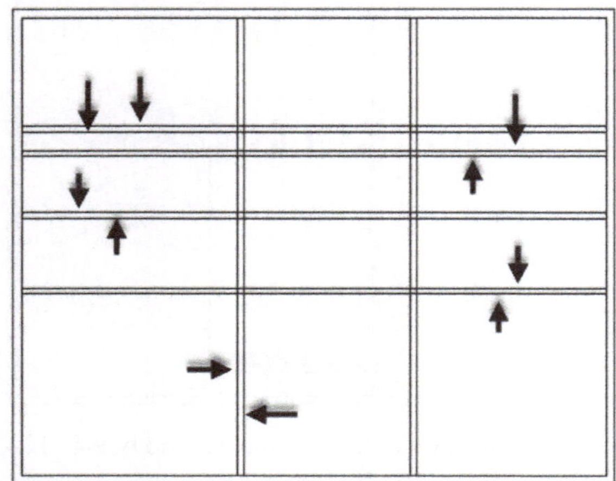

- Specify the fence points across the horizontal lines, as shown below.

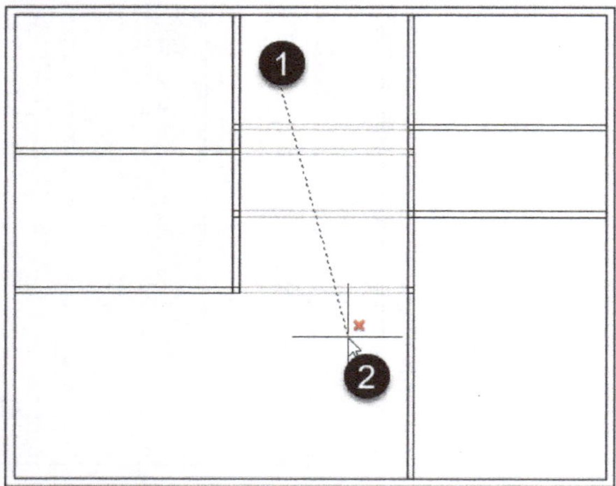

- Select the **Crossing** option from the command line and create the selection window, as shown.

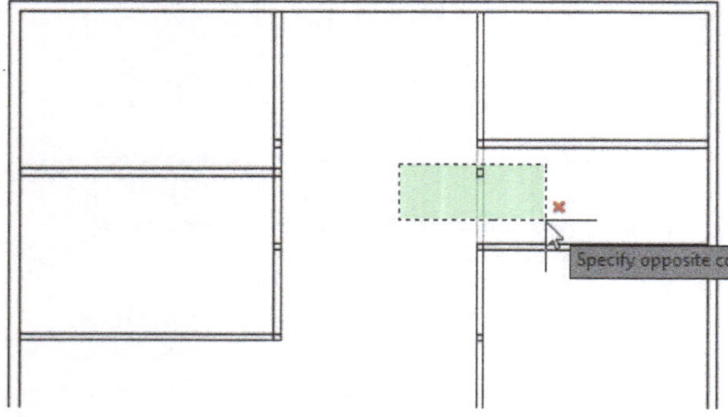

- Select the **Crossing** option from the command line and create the selection window, as shown.

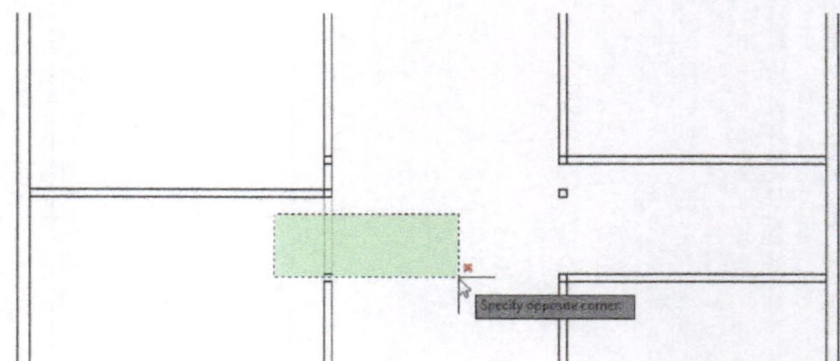

- Zoom to the top portion of the drawing by placing the pointer in the top portion and rotating the mouse scroll in the forward direction.
- Select the portion of the horizontal line that lies between the lines of the inner wall. The selected portions will be trimmed.

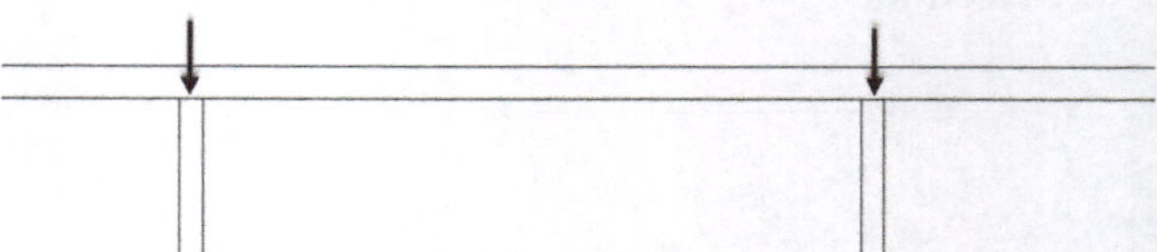

- Press and hold the mouse scroll wheel and drag downwards until the lower portion of the drawing is visible.
- Trim the unwanted portion, as shown below.

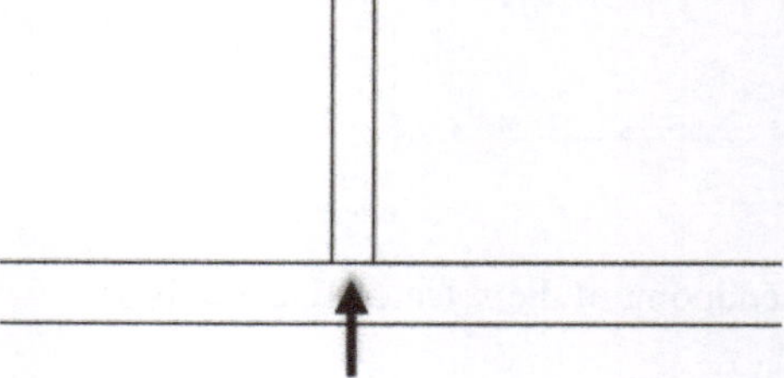

- Trim the unwanted portions, as shown below.

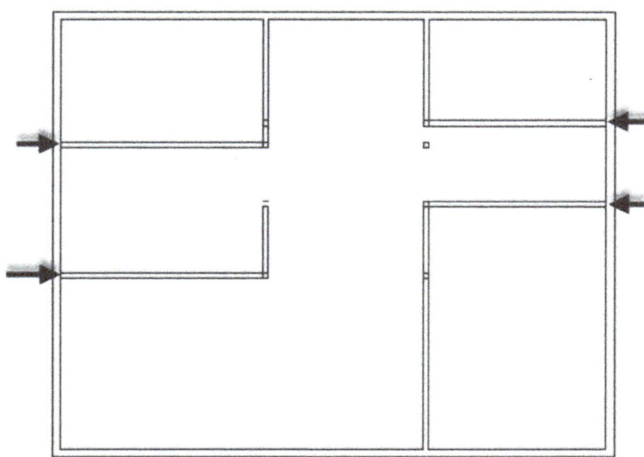

- Trim the unwanted portions at the corners.

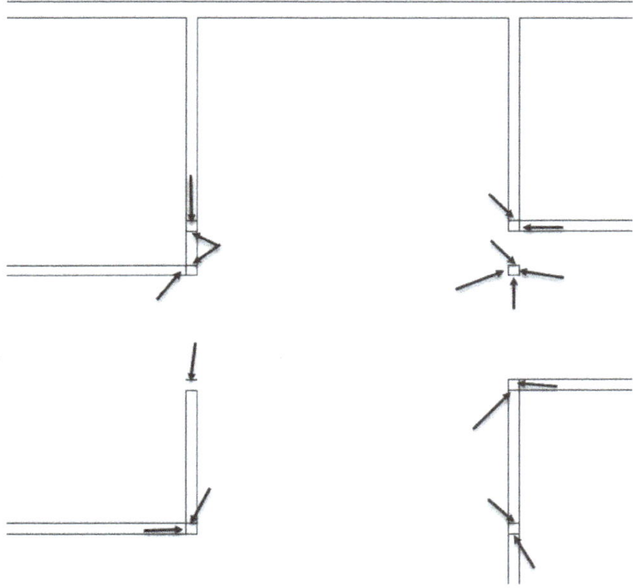

- Press Esc to deactivate the **Trim** command.

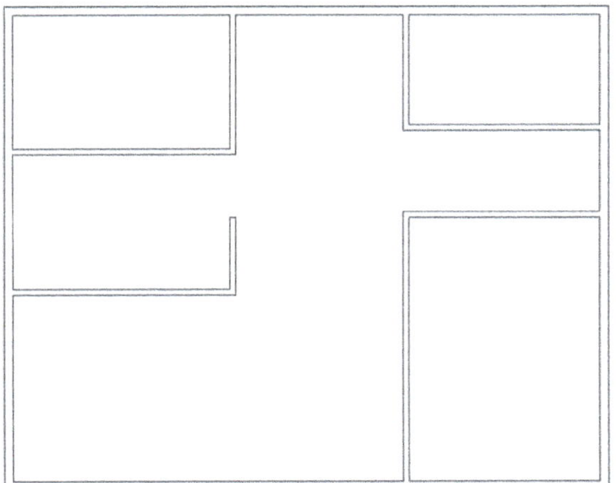

Creating Openings and Doors

- Activate the **Line** command and select the corner of the inner wall, as shown below.
- Move the pointer downward and select the other corner point.

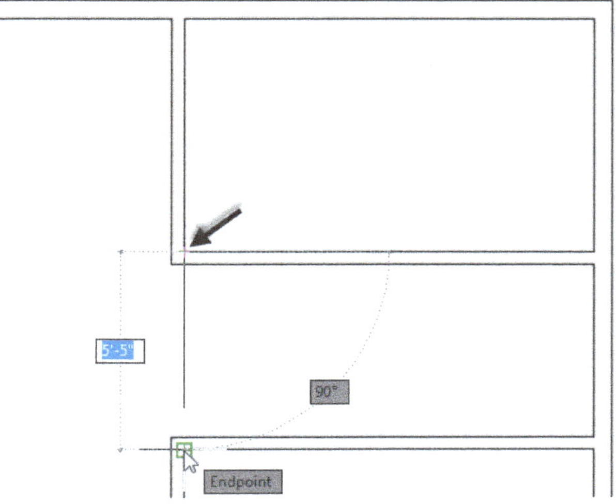

- Deactivate the **Line** command and select the new line.
- Select the middle point of the new line and move the pointer toward the right.
- Type-in 6 and press Enter.

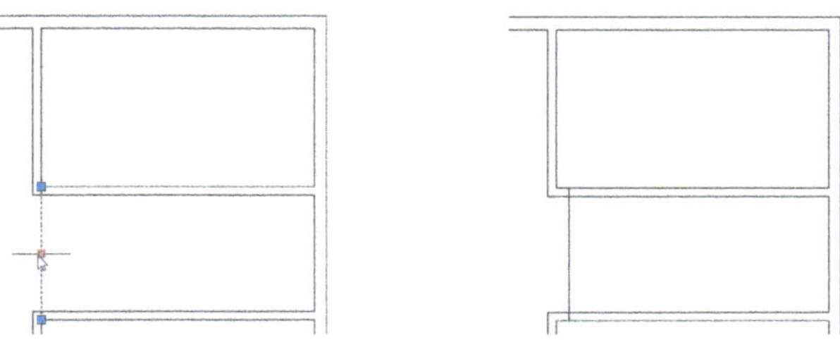

- Activate the **Offset** command and specify 32 as the offset distance.
- Select the new line and move rightwards, and then click.

- Activate the **Trim** command and trim the unwanted portions.

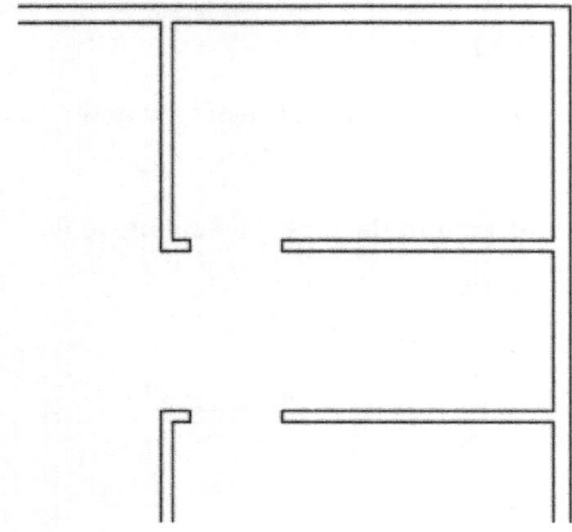

- Likewise, create other openings, as shown below (use the method described in the earlier step).

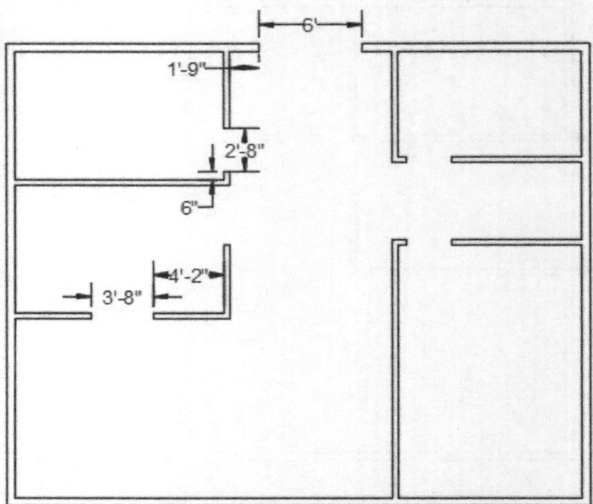

- On the ribbon, click **Home > Draw > Rectangle**.
- Select the endpoint of the opening, as shown below.

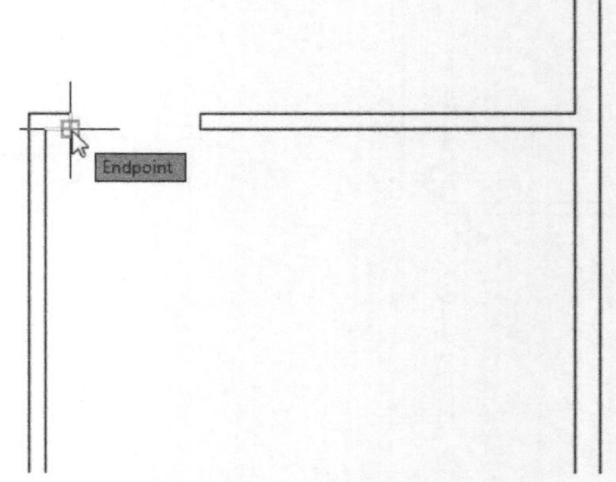

- Select **Dimensions** from the command line.
- Type-in 1 and press Enter. It defines the length of the rectangle.
- Type-in 32 and press Enter. It defines the width of the rectangle.
- Move the pointer down and click to create the door. Now, you need to create the door swing.

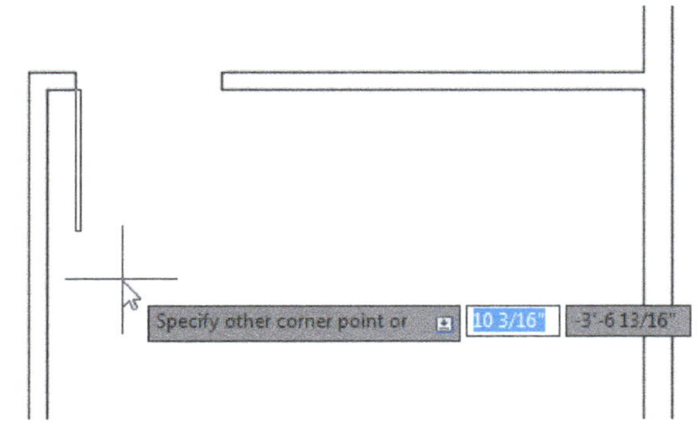

- On the ribbon, click **Home > Draw > Arc** drop-down **> Start Center End**.

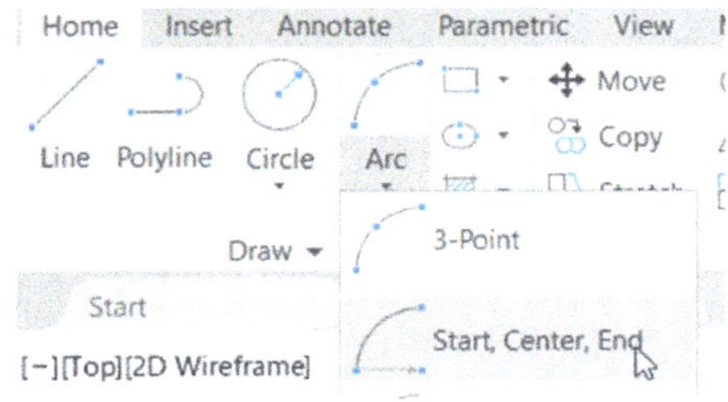

- Select the start, center, and end of the arc in the sequence shown below.

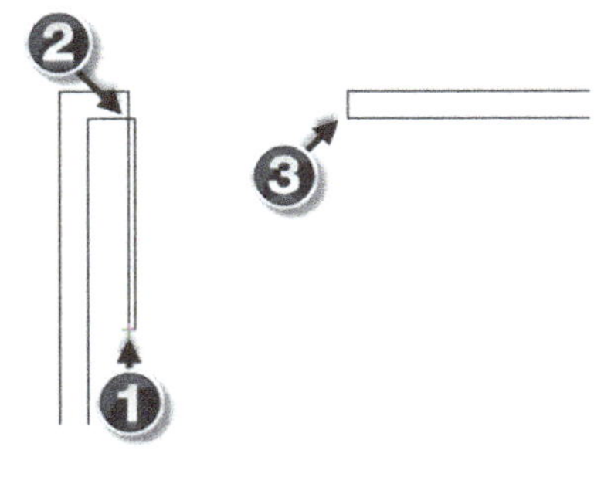

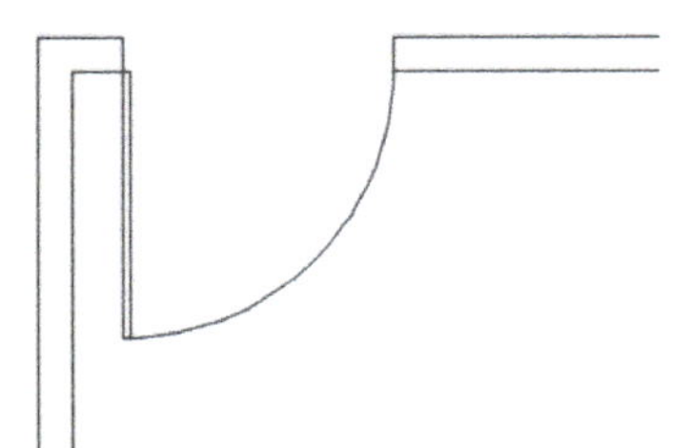

- Select the door and door swing.
- Click the right mouse button and select **Copy Selection**.
- Select the corner point of the rectangle as the base point.

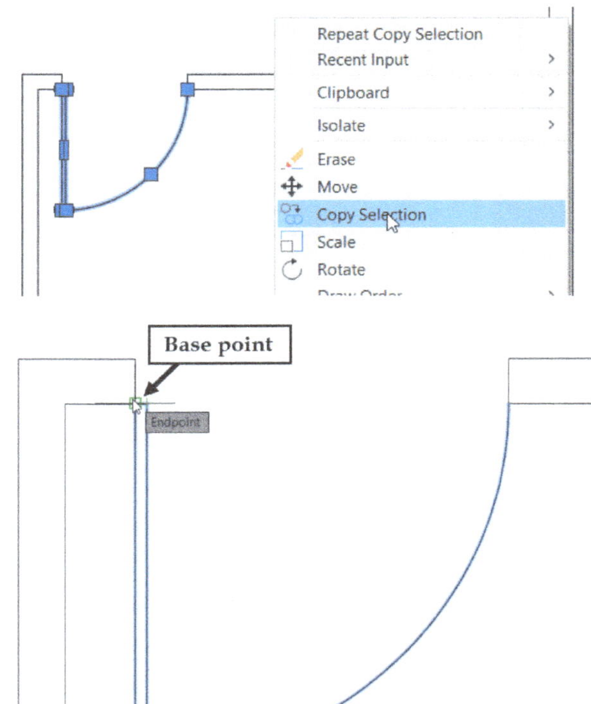

- Select the corner points of openings, as shown below.

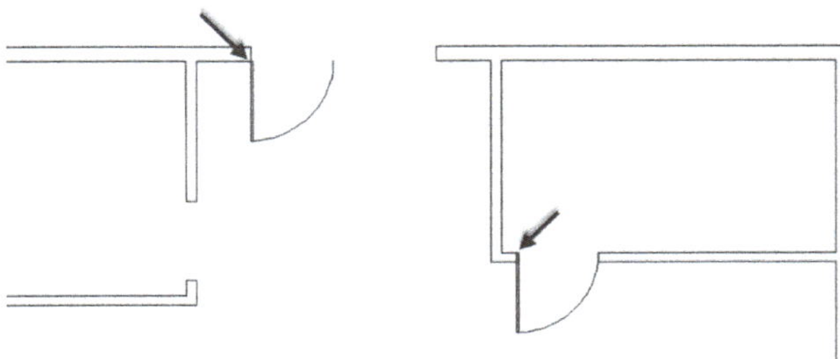

- Press Esc to deactivate the **Copy** command.
- Click the down arrow next to the **Object Snap** icon on the status bar.
- Make sure that the **Midpoint** option is selected.
- On the ribbon, click **Home > Modify > Mirror**, and then select the door and swing of the bathroom, as shown. Press Enter to accept the selection.

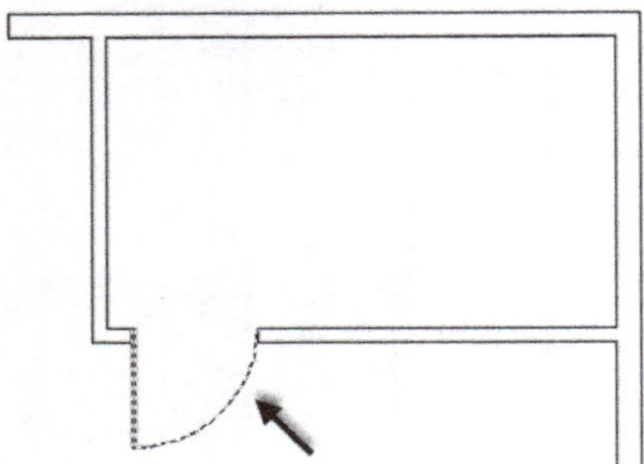

- Define the mirror line by selecting the points, as shown below.

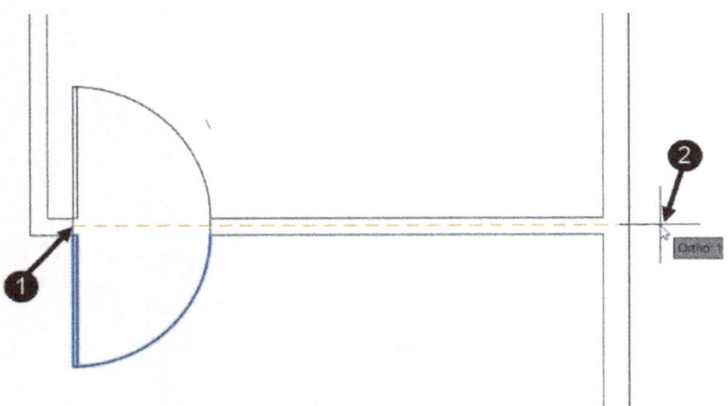

- Select **Yes** from the command line. It deletes the original object.
- On the ribbon, click **Home > Modify > Scale**, and then select the door & swing at the main entrance — press Enter.
- Select the base point, as shown below.

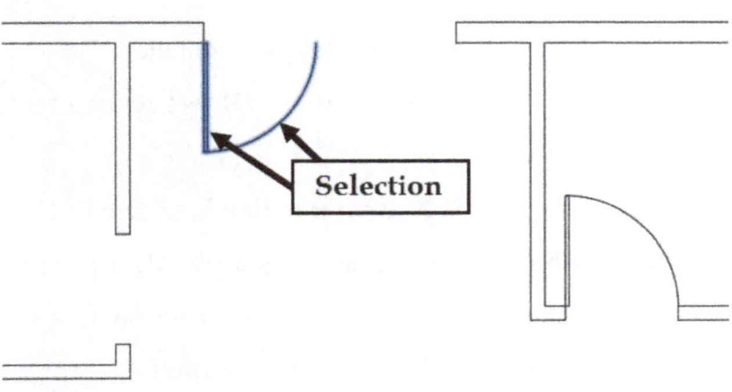

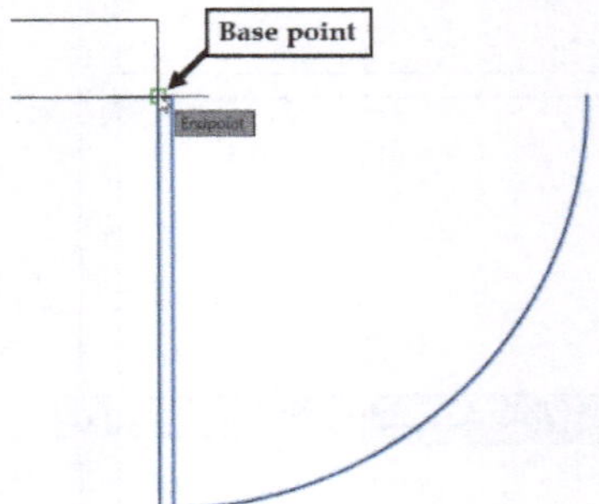

- Select the **Reference** option from the command line.
- Select the two endpoints, as shown below. It defines the reference length of the objects. Now, you need to define the length up to which you want to scale the objects.
- Type-in 36 and press Enter. The objects will be scaled.

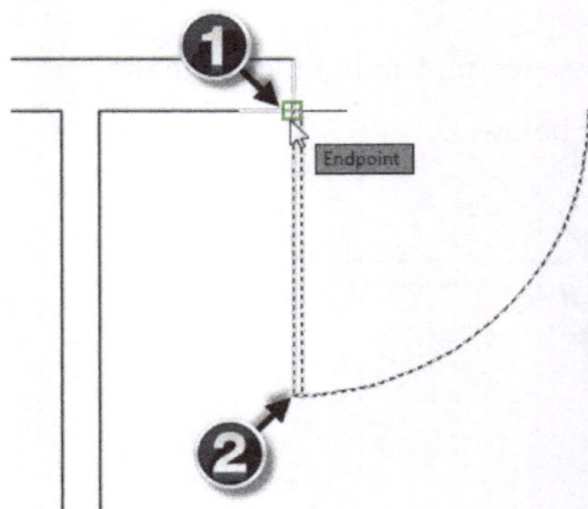

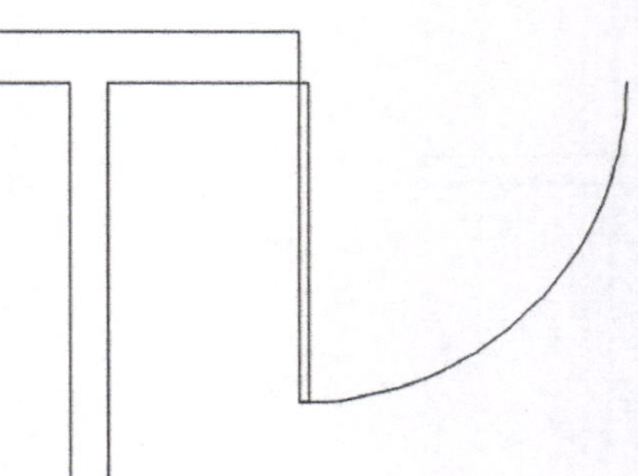

- Activate the **Mirror** command and select the door & swing at the entrance. Press Enter to accept the selection.
- Define the mirror line by selecting the points, as shown below.

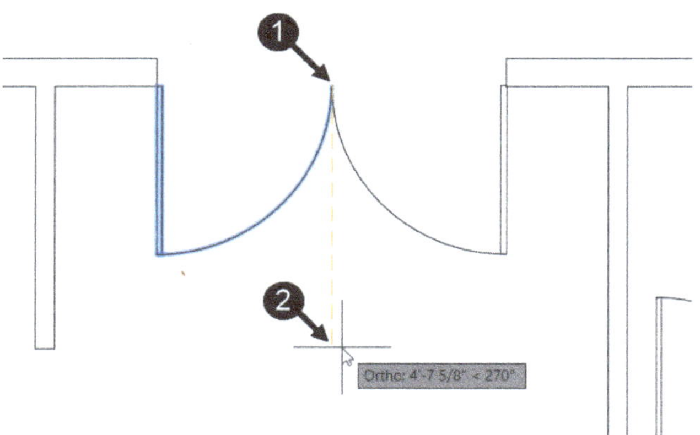

- Select **No** from the command line. It keeps the original object.
- Copy the door & swing of the bathroom and place it at the opening, as shown below.
- Press Esc.

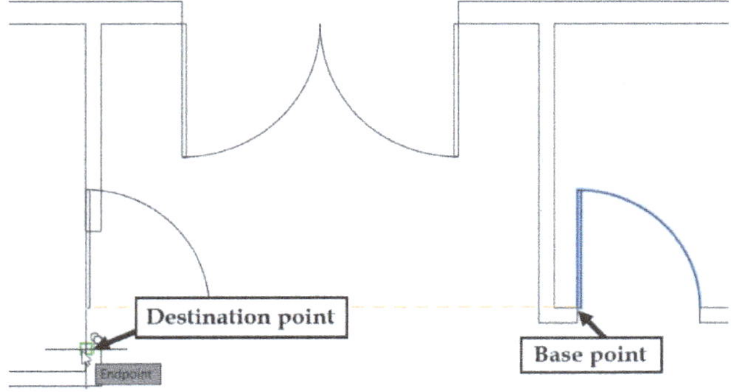

- On the ribbon, click **Home > Modify > Rotate**, and then select the copied object. Next, press Enter.
- Select the base point, as shown.

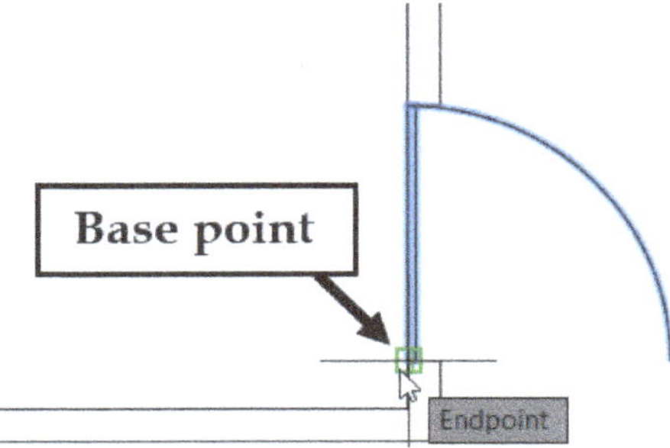

- Move the pointer vertically upward, and then click.

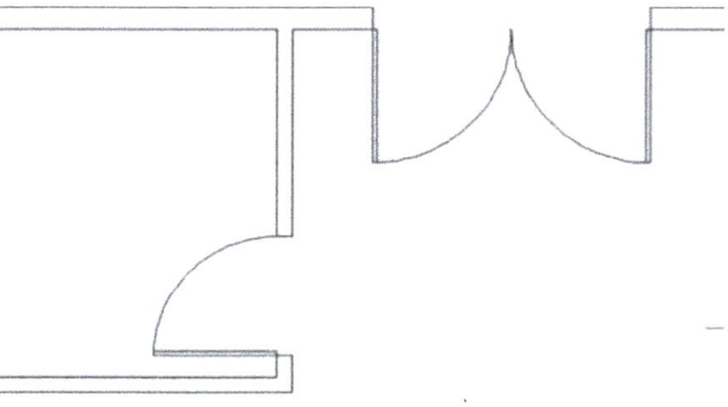

- Create an opening on the rear side of the plan, as shown below.

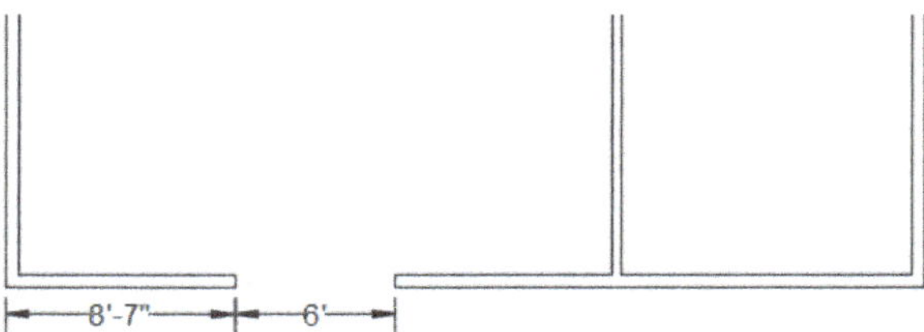

Now, you will create a sliding door in the opening.

- Activate the **Rectangle** command and select the corner point of the opening, as shown below.

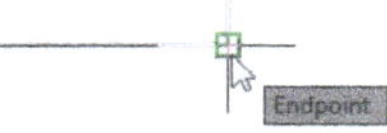

- Select the **Dimensions** option from the command line.
- Specify 37 and 2 as the length and width of the rectangle, respectively.
- Move the pointer upward and click to create the rectangle.

- Type **M** in the command line and press Enter. Select the rectangle, and then press Enter.
- Select its lower-left corner point to define the base point. Move the pointer upward and type-in 1 in the command line, and then press Enter.

- On the ribbon, click **Home > Modify > Explode**, and select the rectangle. Press Enter to explode the rectangle.
- Activate the **Offset** command and specify 2 as the offset distance.
- Offset the left and right vertical lines of the rectangle.

- Press **Esc** to deactivate the **Offset** command.
- Activate the **Line** command and select the midpoints of the offset lines. It creates a line connecting the offset lines. It creates one part of the sliding door.

- Press **Esc** to deactivate the **Line** command.
- Type-in **CO** in the command line and press Enter.
- Drag a selection window covering all the elements of the sliding door. Press Enter.

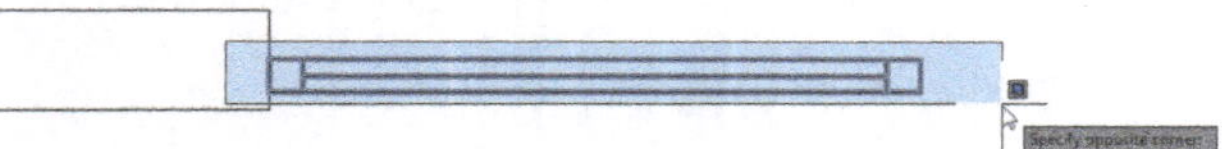

- Select the lower-left corner of the sliding door as the base point.
- Move the pointer and select the endpoint of the offset line, as shown.

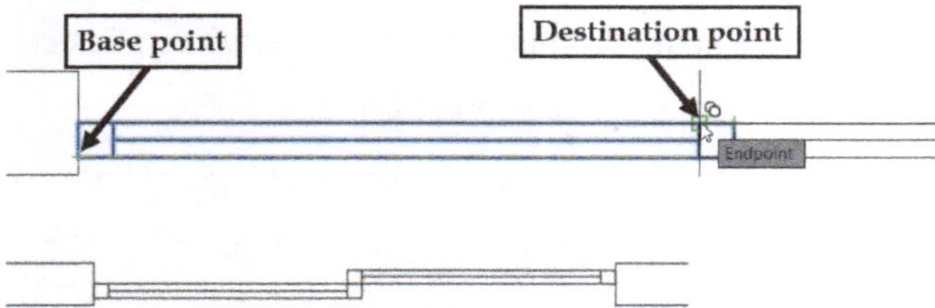

- Press **Esc** to deactivate the **Copy** command.

Now, you need to draw thresholds on the door openings.

- Zoom to the front door area using the **Zoom Window** tool.

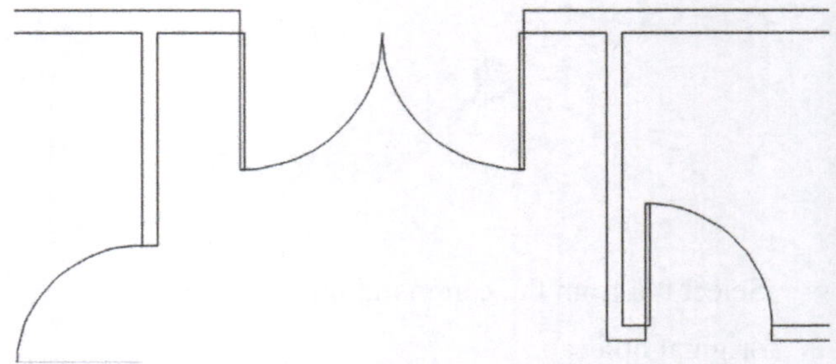

- On the status bar, click the down arrow next to the **Object Snap** button and make sure that **Endpoint, Nearest,** and **Perpendicular** options are selected.

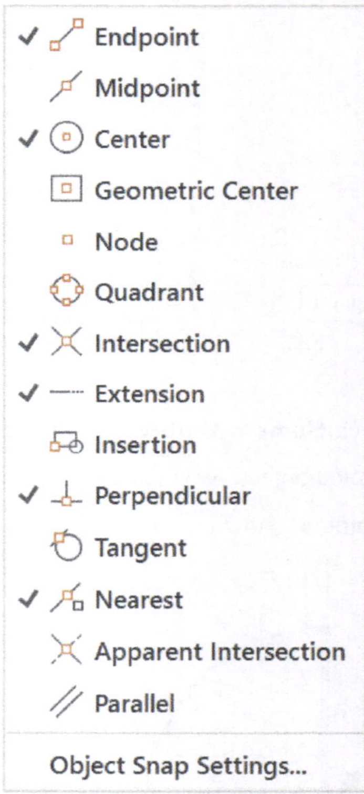

- Type-in **L** in the command line and press Enter.
- Press and hold the **Shift** key and click the right mouse button.
- Select **From** from the shortcut menu and click the endpoint of the door opening, as shown below.

- Move the pointer on the horizontal line and type in 3, and then press Enter. It defines the start point of the line at 3 distance from the endpoint.

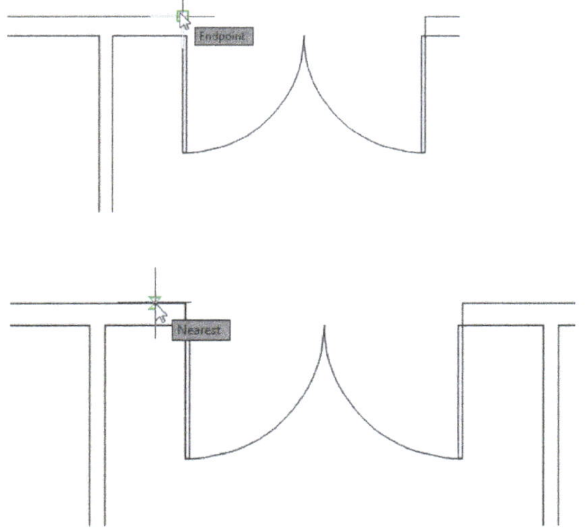

- Move the pointer up and type-in 2, and then press Enter.
- Move the pointer toward the right and type in 78, and then press Enter.
- Move the pointer downward and type in 2, and then press Enter. It creates a threshold.

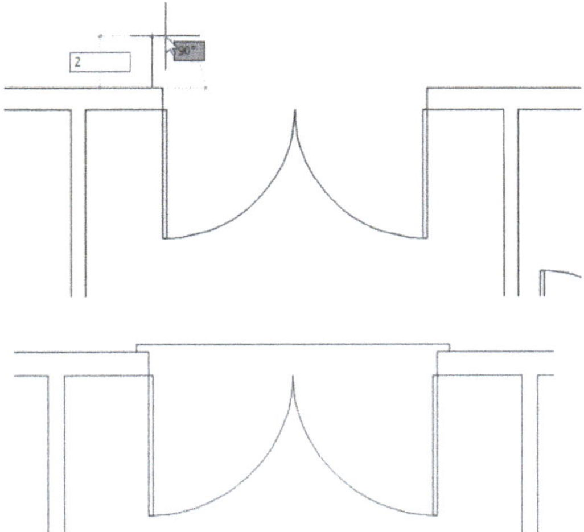

- Press Esc to deactivate the **Line** command.
- Likewise, create a threshold on the sliding glass door.

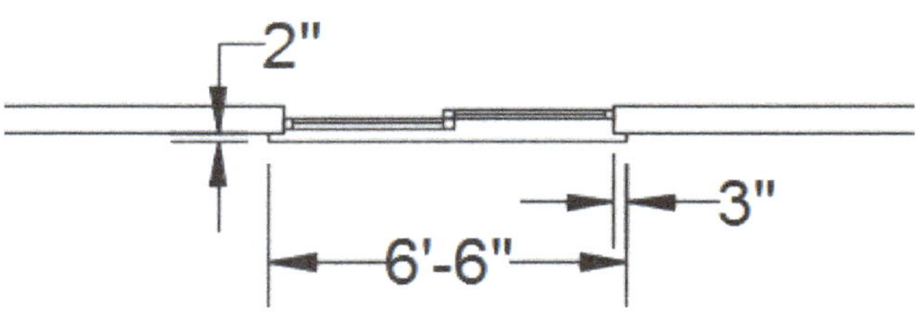

Creating Kitchen Fixtures

- Zoom to the kitchen area by using the **Zoom Window** tool.

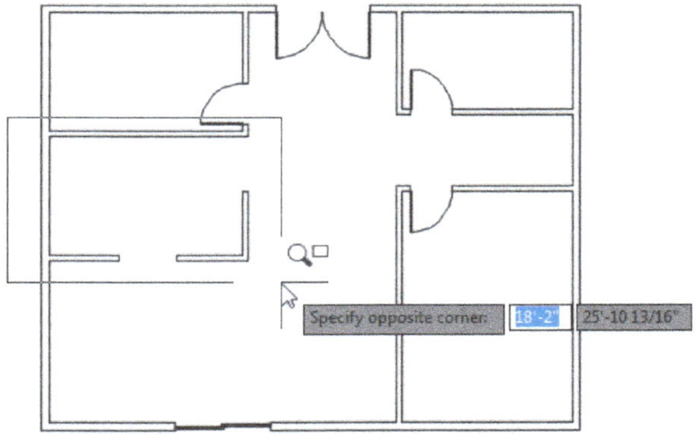

- Activate the **Offset** command and specify 26 as the offset distance.
- Offset the lines shown below.

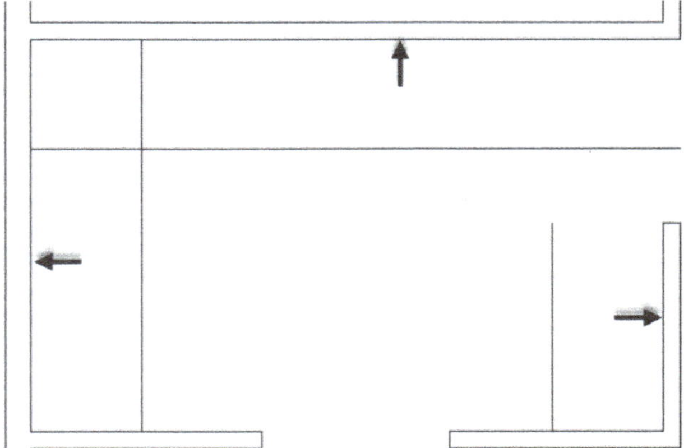

- Trim the unwanted portions.

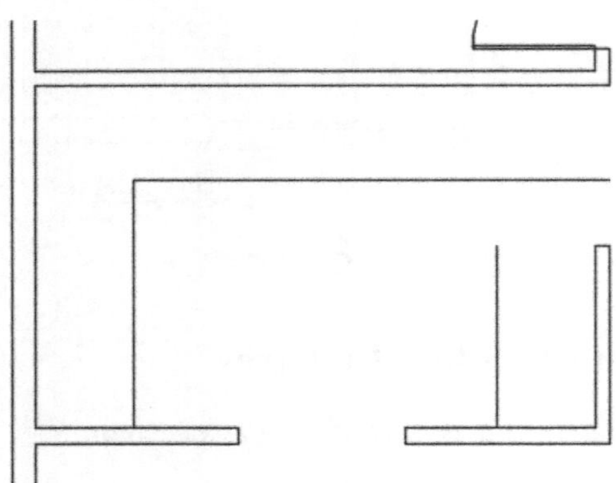

- Create another offset line at 54 distance, and then trim the unwanted elements.

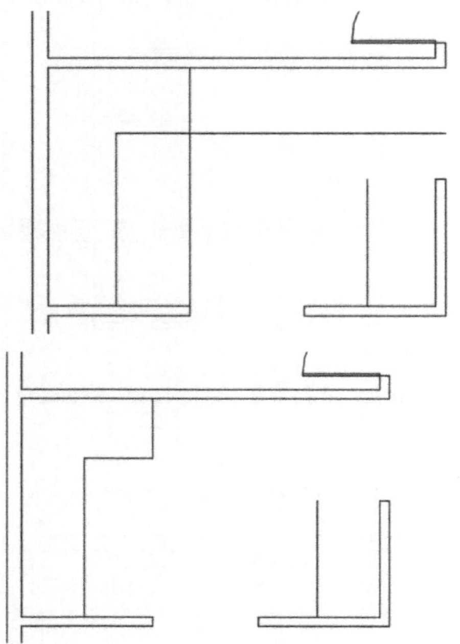

- Create another line, as shown below.

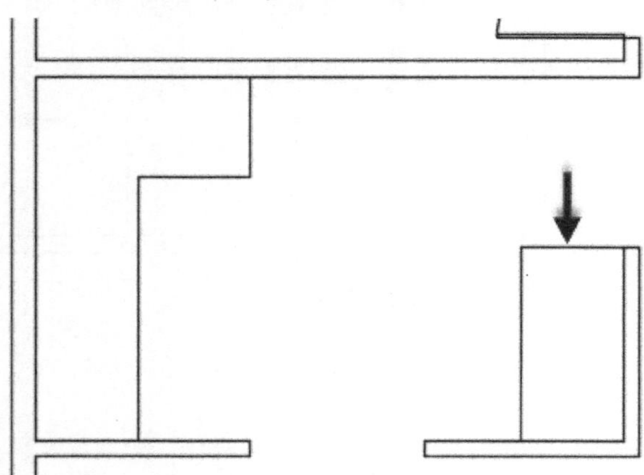

Now, you have finished drawing the counters. You need to draw a refrigerator, stove and sink.

- Type-in **REC** in the command line and press Enter. It activates the **Rectangle** command.
- Select the corner point of the counter.
- Select the **Dimensions** option from the command line.
- Specify 28 as length and width of the rectangle. Move the pointer toward the right and click to create the rectangle.

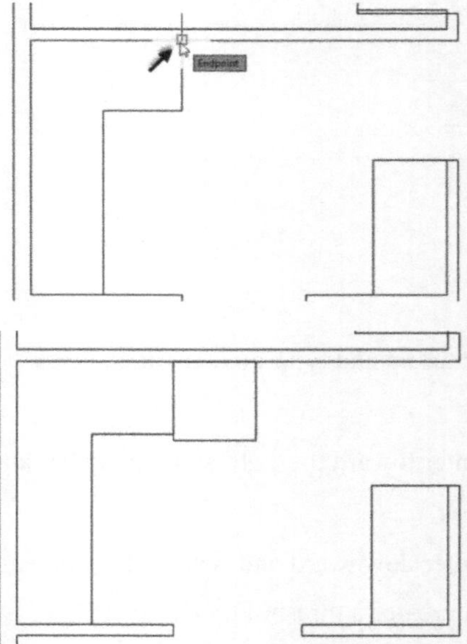

- Use the **Move** command to move the rectangle 2 inches rightwards and downwards.

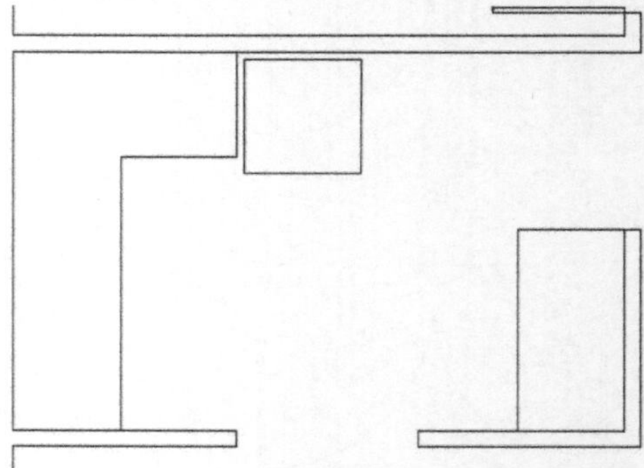

- Create the outline of the stove using the **Offset** and **Trim** commands.

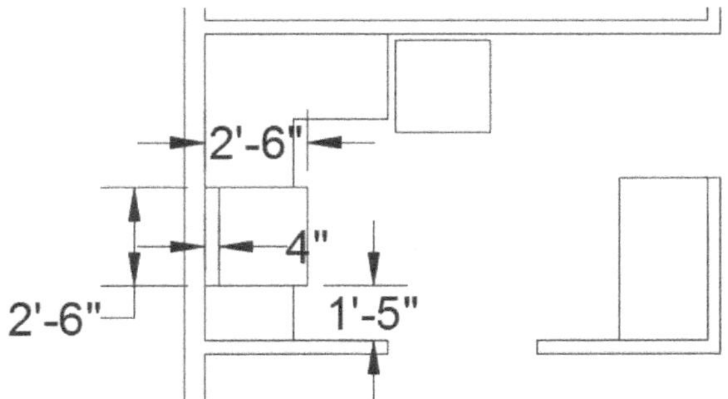

Now, you need to create the sink.

- Use the **Offset** command and create offset lines, as shown below.

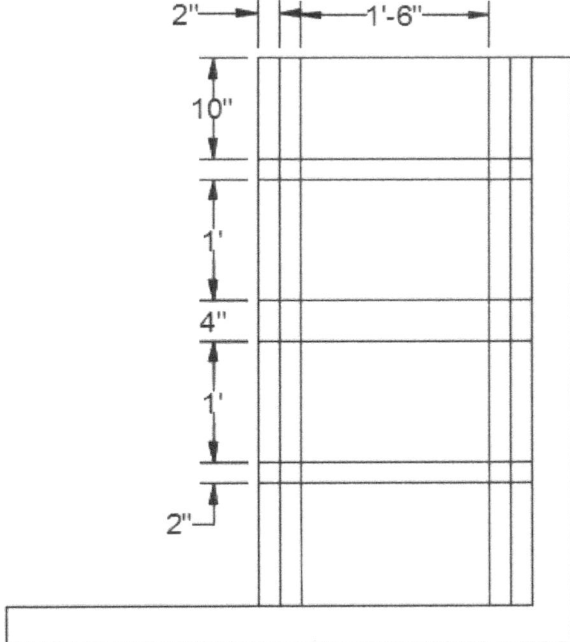

- Trim the unwanted elements, as shown below.

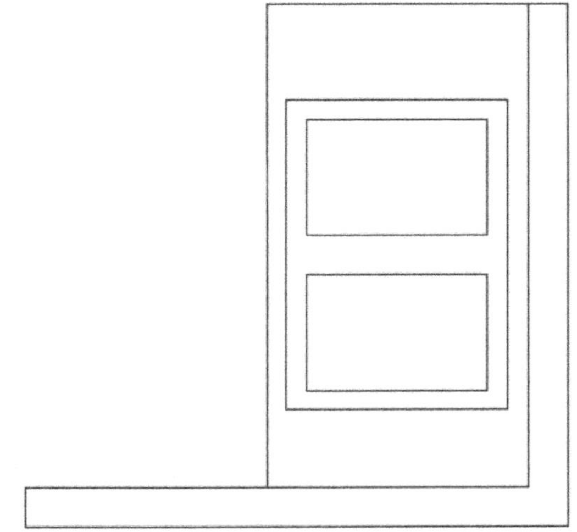

- Fillet the corners, as shown below.

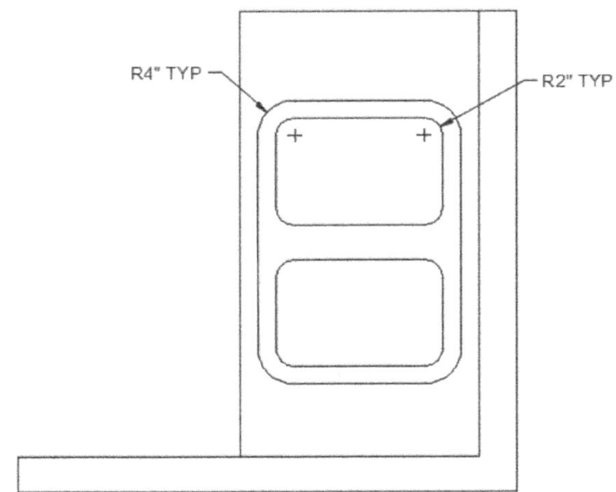

- Activate the **Circle** command and hover the pointer on the midpoints of the sink edges and move, as shown below.

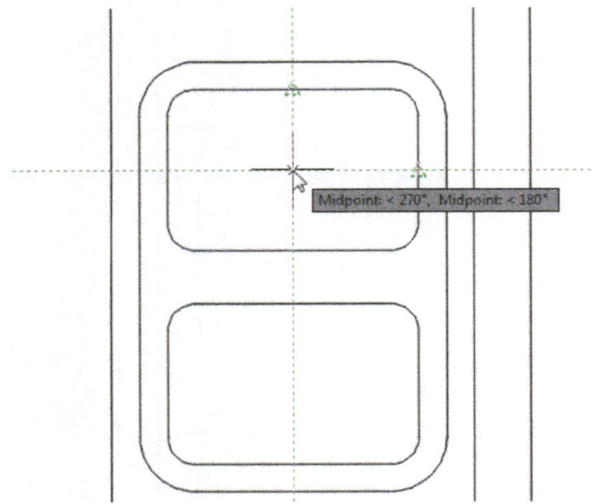

- Create the circles at the intersection points of the trace lines.

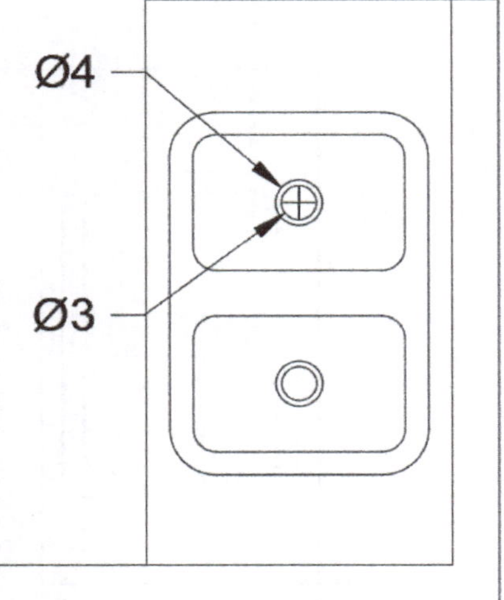

Creating Bathroom Fixtures

- Zoom into the bathroom area and create offset lines, as shown below.

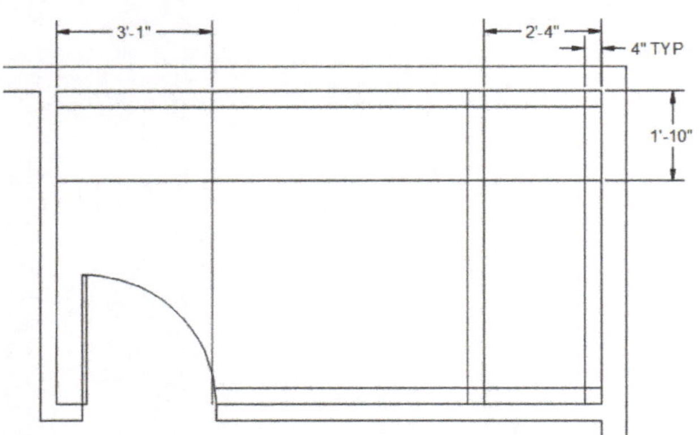

- Trim the unwanted elements, as shown below.

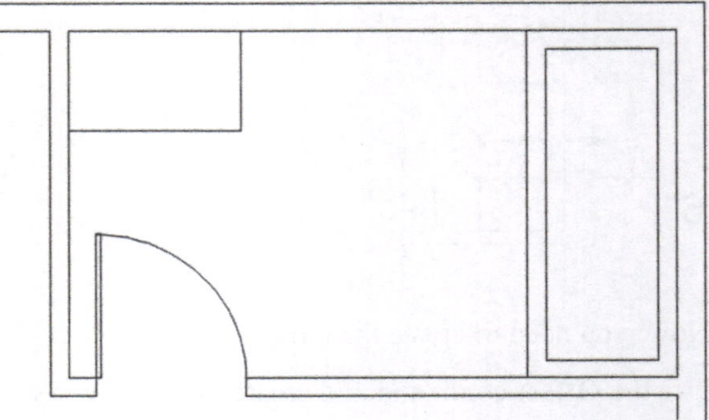

- Fillet the corners, as shown below. The fillet radius is 4.

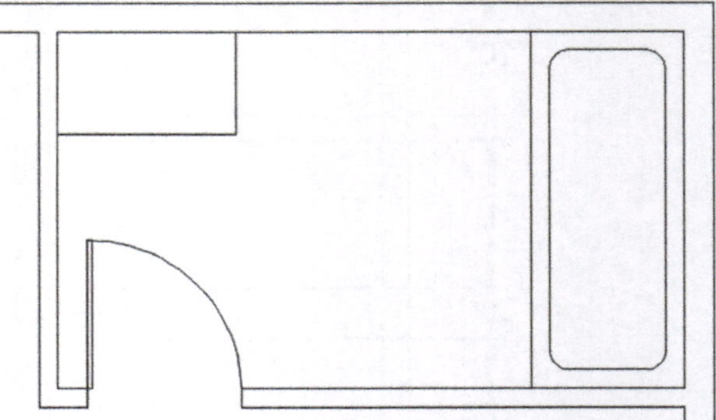

- On the ribbon, click **Home > Draw > Ellipse drop-down > Center**.

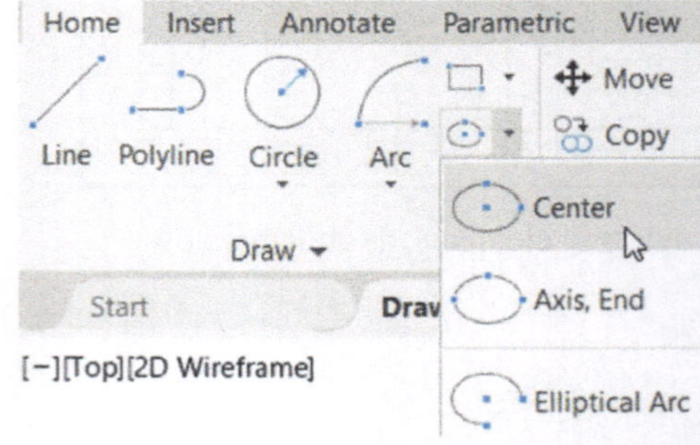

- Hover the pointer on the midpoints of the vertical and horizontal lines, as shown below.

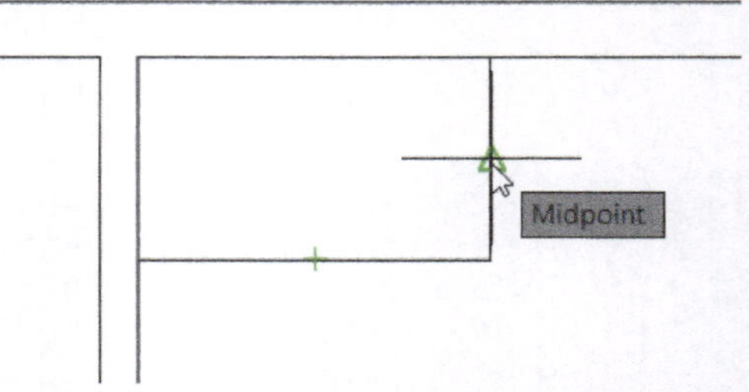

- Move the pointer and click at the intersection point of the trace lines.

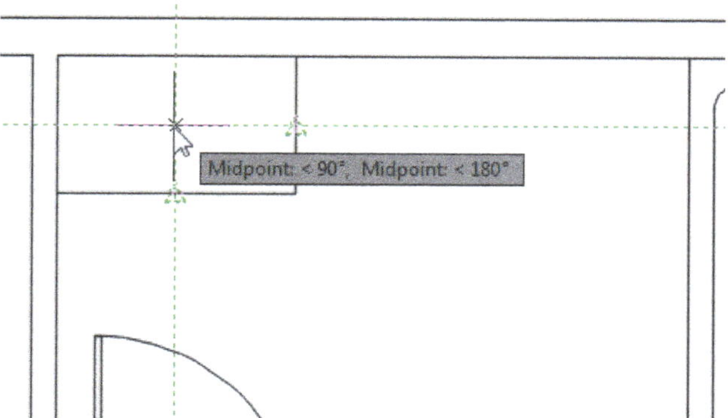

- Move the pointer toward the right and type in 10, and then press Enter. It defines the major radius of the ellipse.
- Move the pointer downward and type in 5, and then press Enter. It defines the minor radius of the ellipse.

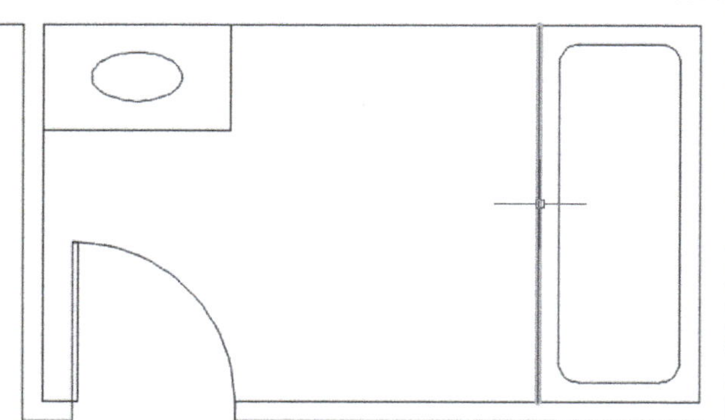

- Likewise, create another ellipse of 11 major radius and 7 minor radius.

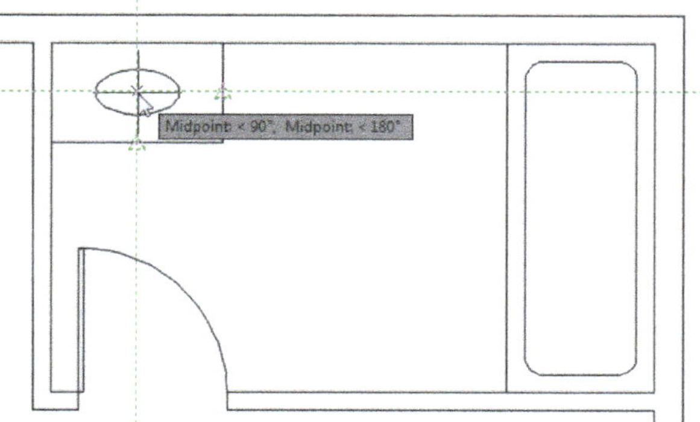

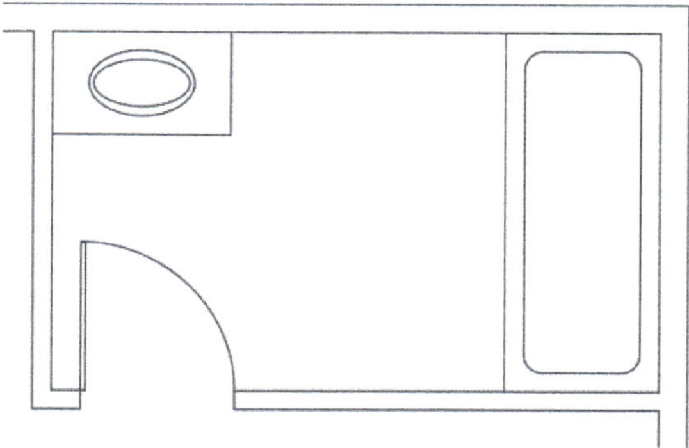

- Select the outer ellipse, and then click on the center point of the ellipse.
- Move the pointer up and type-in 1, and then press Enter. The outer ellipse moves up.

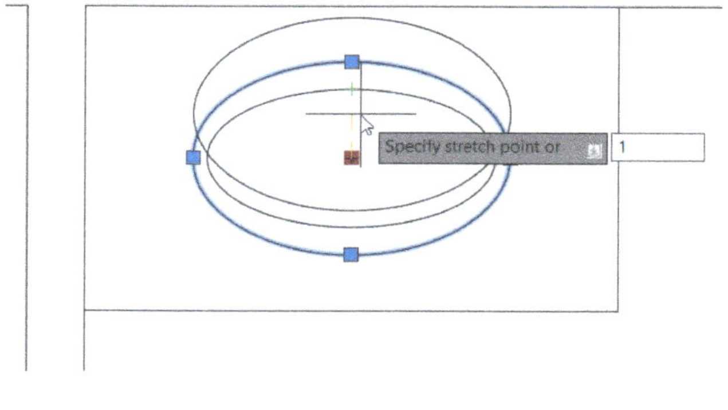

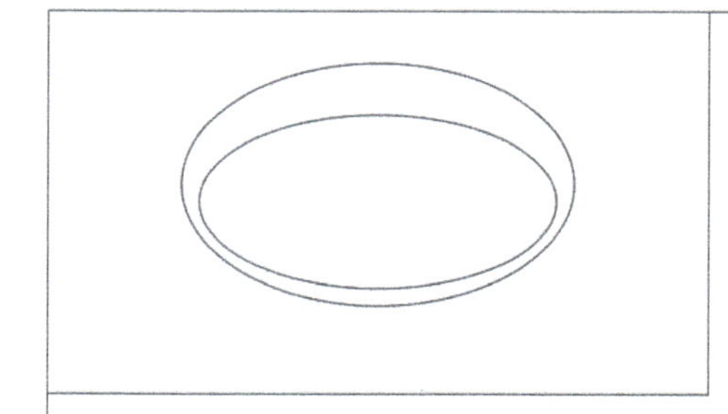

- Activate the **Rectangle** command and create a 22 x 9 rectangle, as shown below.

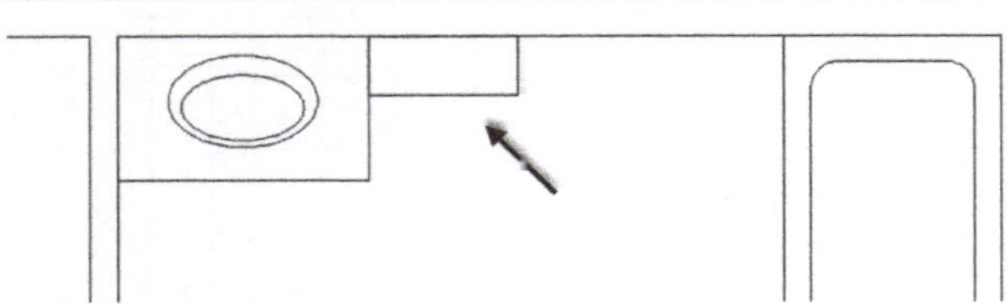

- Move the rectangle up tc 19.5 rightwards and 1 downwards.

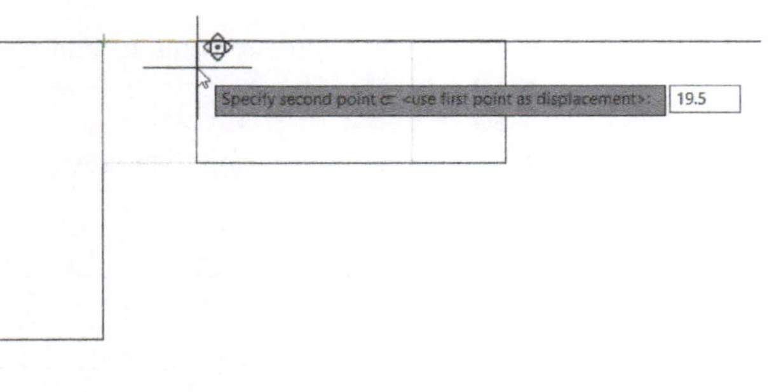

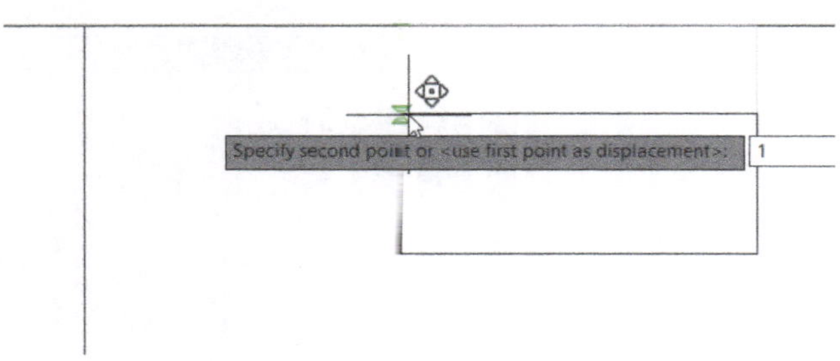

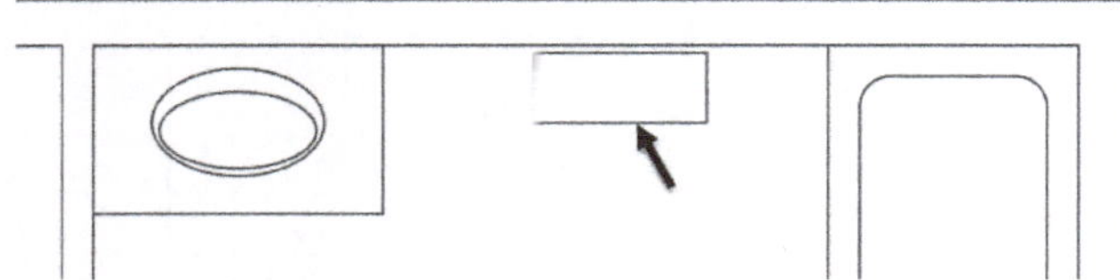

- On the ribbon, click **Home > Draw > Ellipse** drop-down **> Axis, End**.

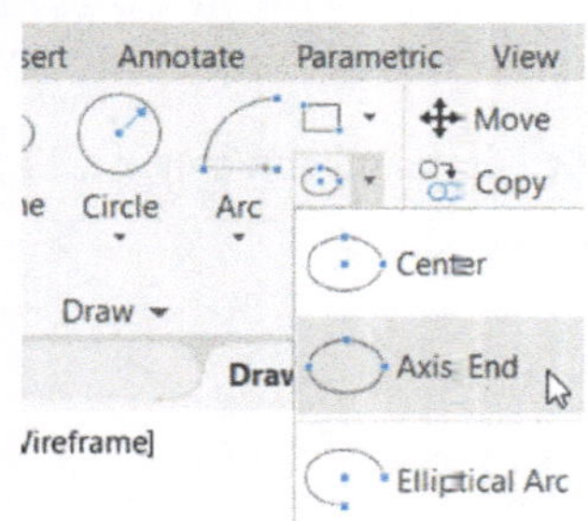

- Select the midpoint of the lower horizontal line of the rectangle.
- Move the pointer downward and type in 18, and then press Enter.
- Type-in 6 as the minor radius and press Enter.

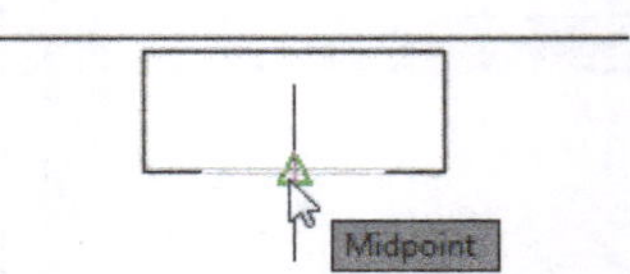

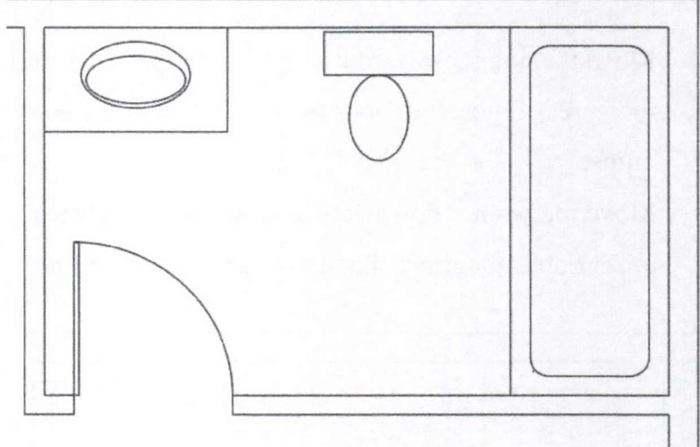

Adding Furniture using Blocks

- Type **DC** and press Enter; the **Design Center** palette appears.
- On the **Design Center** palette, click the **Home** button; the folder containing all the samples is opened.

- In the **Design Center** palette, expand the **Sample** folder and go to **en-us > Design Center > Home – Space Planner.dwg**.

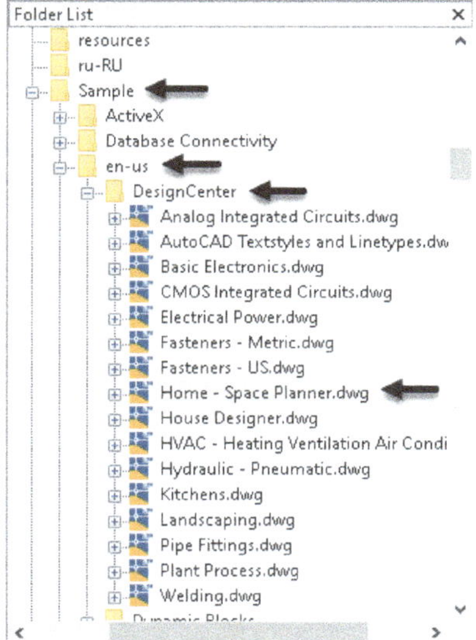

- Double-click the **Blocks** icon. It displays all the blocks available in the selected drawing file.

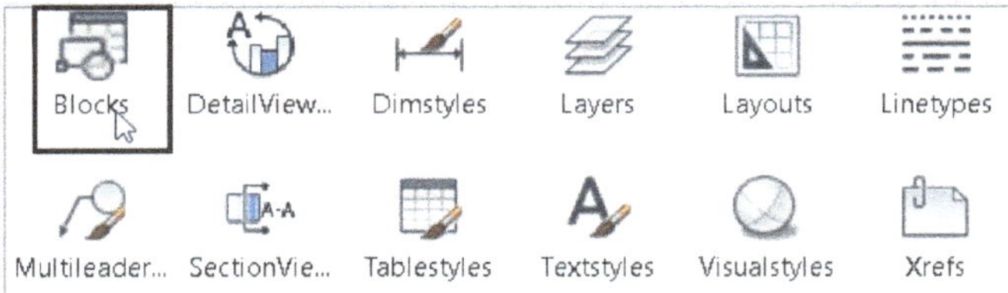

- Click and drag the highlighted blocks into the graphics window.

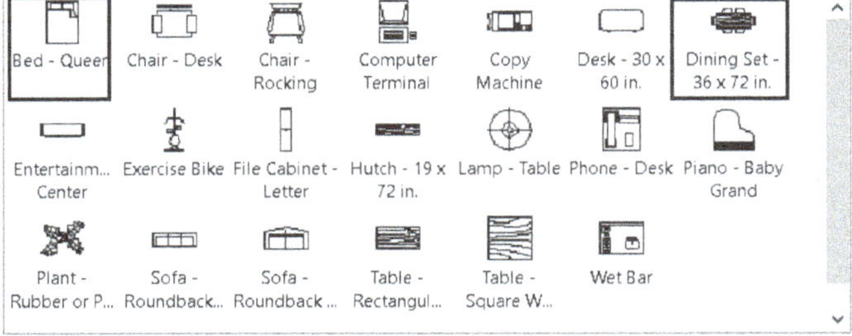

- Close the **Design Center** palette.
- Select the Dining set block, and then click on the point located at its center.
- Move the block and place it at the location shown below.

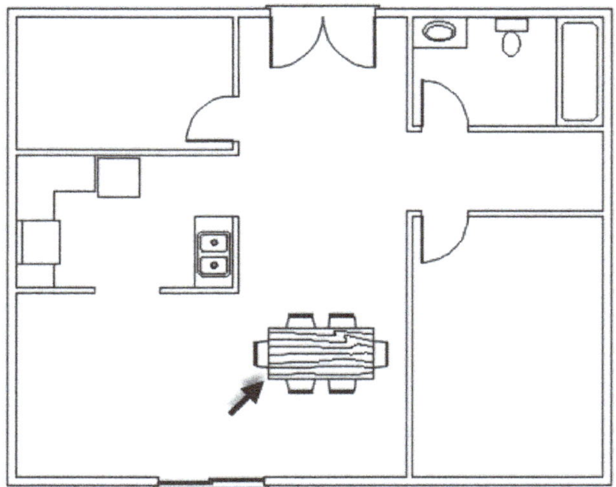

- Activate the **Rectangle** command and select the corner point of the bedroom, as shown.

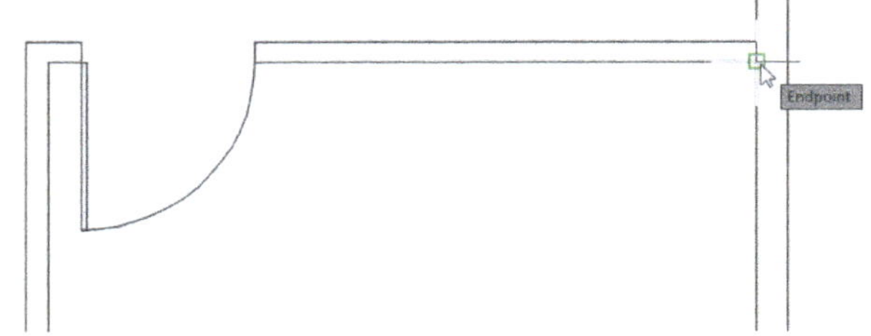

- Select the **Dimensions** option from the command line, and then specify 86 and 27.5 as the length and width of the rectangle, respectively.
- Move the pointer downward and click to create the rectangle.

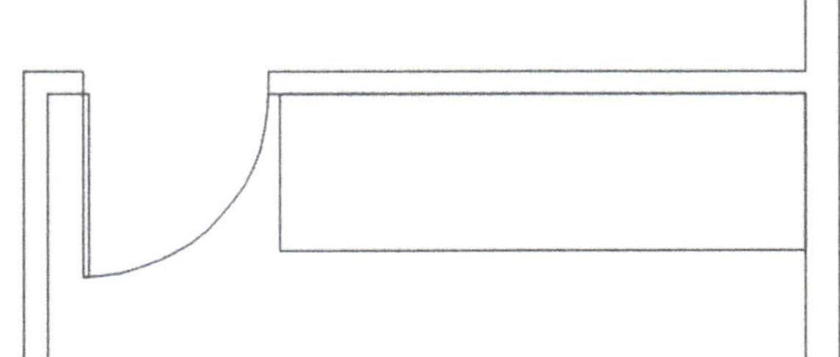

- Create another rectangle by selecting the corner points, as shown below.

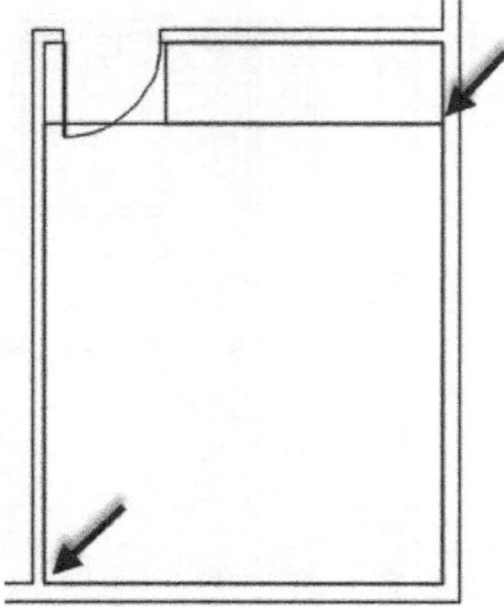

- Offset the rectangle by a distance of 47.5 inwards.
- Delete the original rectangle.

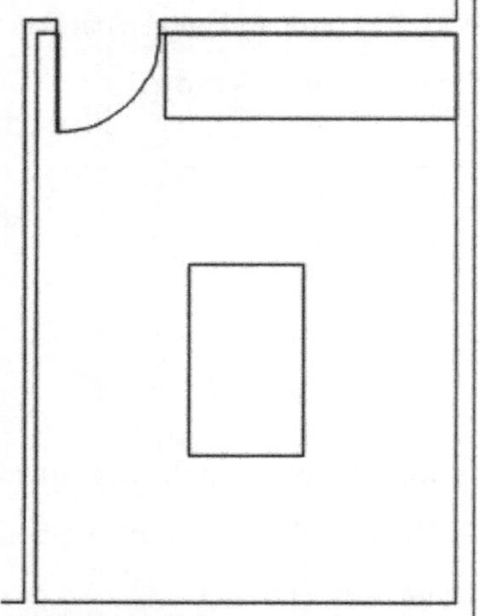

- Rotate the bed by **90** degrees.
- Activate the **Move** command and select the bed. Press Enter to accept the selection.
- Select the top left corner of the bed to define the base point.
- Select the top left corner of the offset rectangle to define the destination point.

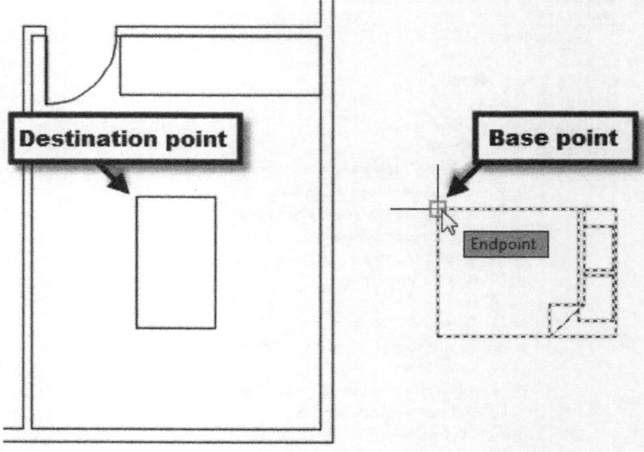

- Delete the offset rectangle.

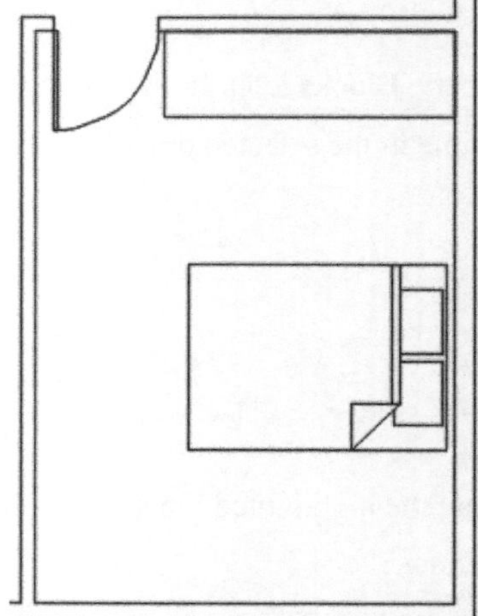

Adding Windows

- In the empty space, create the window using the **Line** command, as shown below.

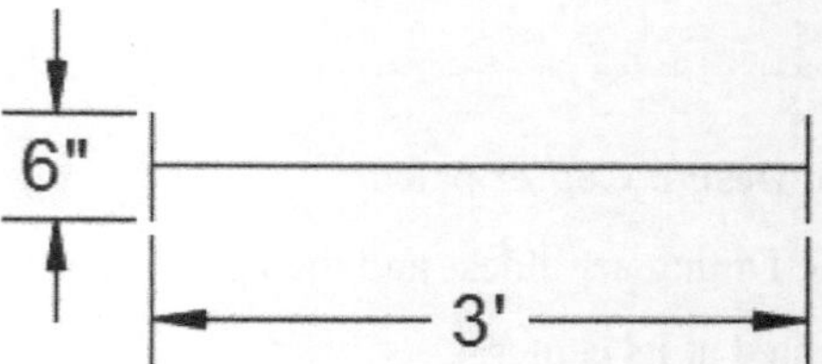

- On the ribbon, click **Insert > Block Definition > Create Block**.
- On the **Block Definition** dialog, type-in **Window** in the **Name** box.
- Click the **Select Objects** button and select all the

elements of the window by dragging a selection window.

- Press **Enter**.
- Click the **Pick point** button and select the lower-left corner of the window.

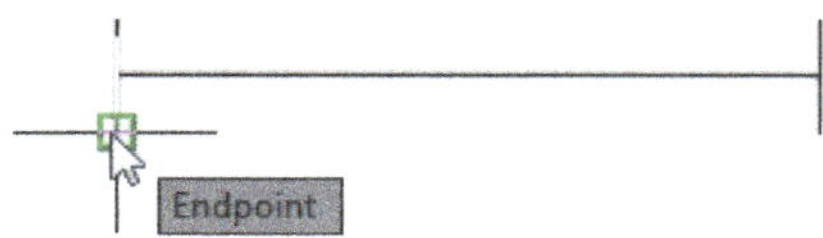

- On the dialog, check the **Open in block editor** option and click **OK**. It creates the block and opens it in the **Block Editor**.
- On the ribbon, click **Block Editor** tab > **Action Parameters** > **Parameters** drop-down > **Linear**.

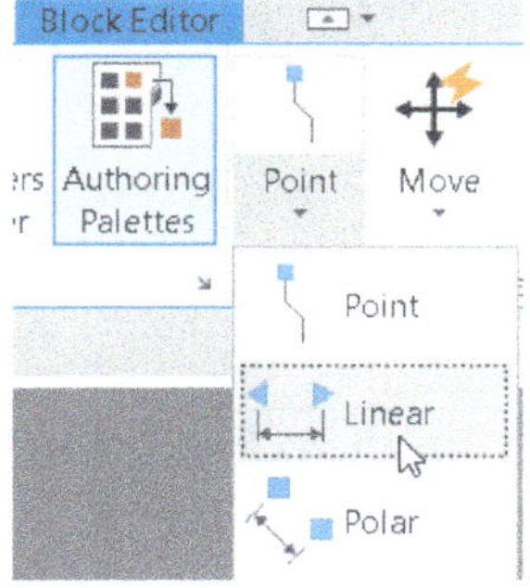

- Click the endpoints of the horizontal line.
- Move the pointer downward and click to define the parameter location.

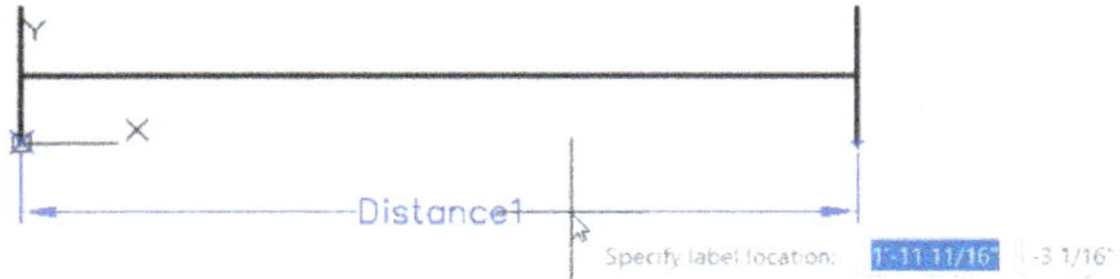

- Press Enter to specify 2 as the number of the grips to be displayed when you select the parameter.
- On the ribbon, click **Block Editor** tab > **Action Parameters** > **Actions** drop-down > **Stretch**.

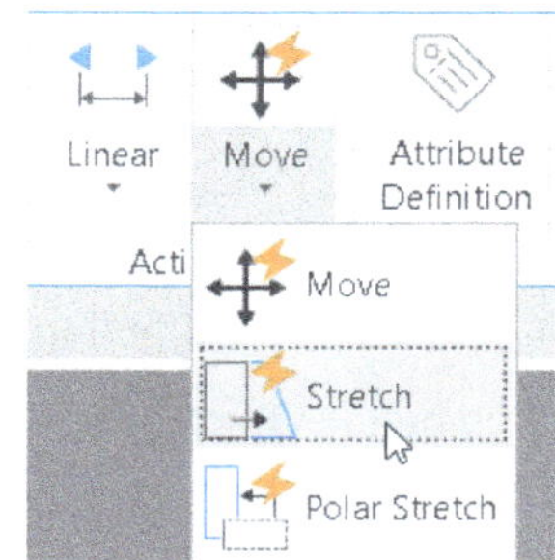

- Select the **Distance1** parameter.
- Select the right endpoint of the horizontal line. It defines the point that can be used to stretch the block.

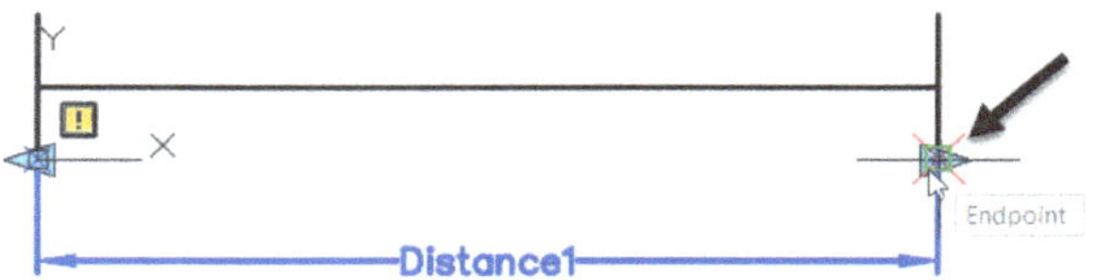

- Create a window around the selected endpoint.

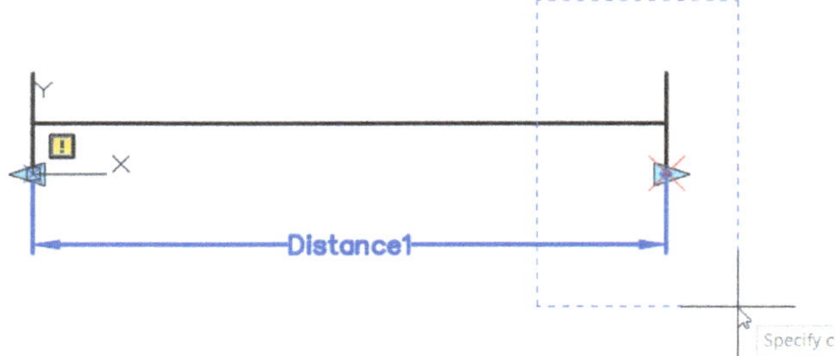

- Select the horizontal and right vertical lines, and then press Enter. It defines the elements that can be stretched.

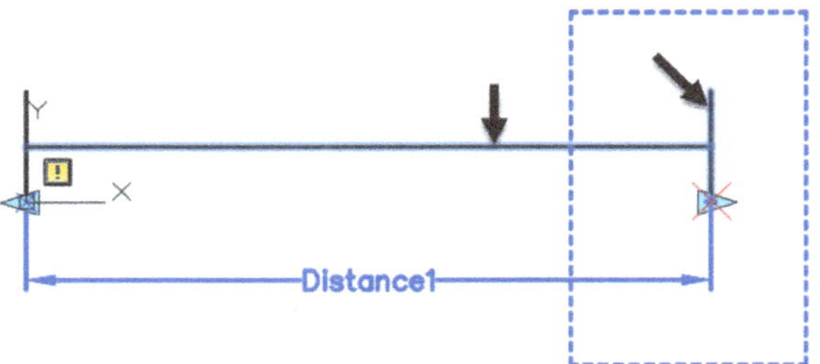

- On the **Block Editor** ribbon tab, click **Open/Save > Test Block**. The **Test Block Window** appears.
- Select the block and click the arrow grip. Drag the pointer to stretch the block.

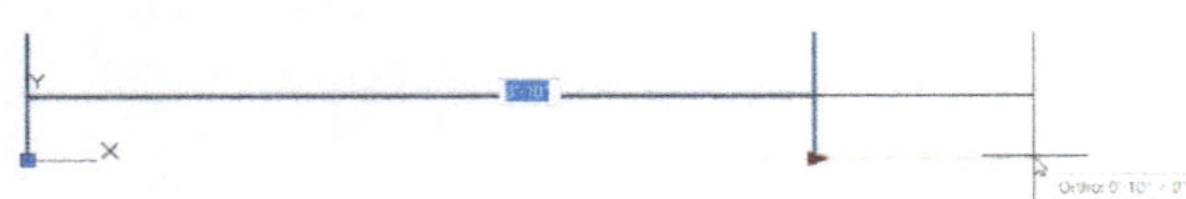

- On the ribbon, click the **Close Test Block** button.
- Click the **Save Block** button on the **Open/Save** panel.
- Click the **Close Block Editor** button on the **Close** panel. It closes the **Block Editor** window. Now, you need to place the windows.
- On the ribbon, click **Insert > Block > Insert > Window**.

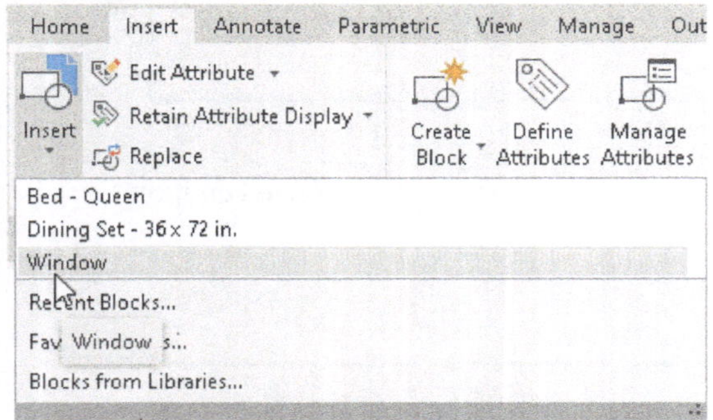

- Press and hold the Shift key and right click.
- Select **From**.
- Select the lower right corner of the bedroom.
- Move the pointer on the horizontal wall and type in 95, and then press Enter. The **Window** block will be placed at the specified location.

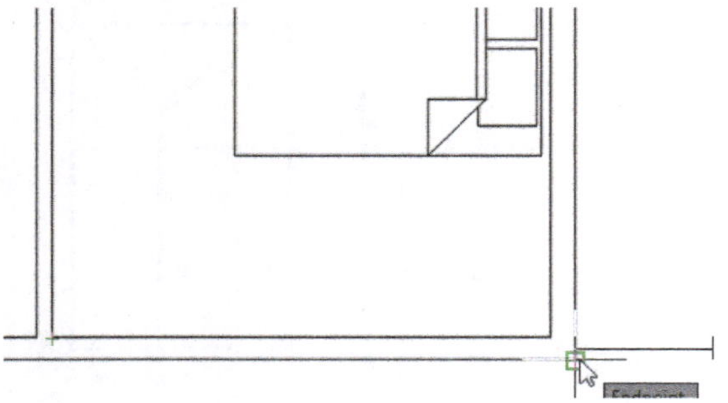

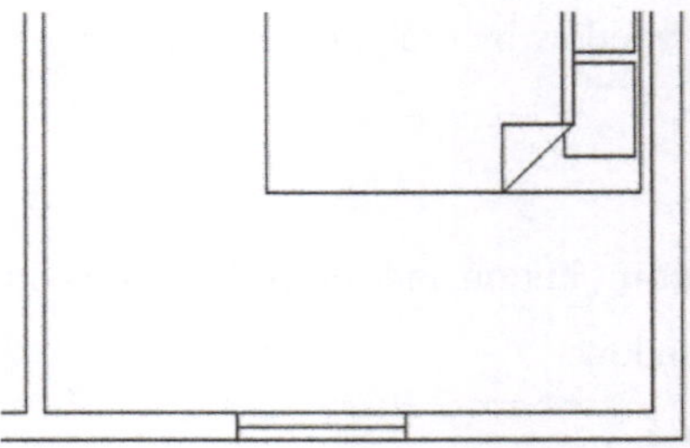

- Activate the **Dynamic Input** icon on the Status bar.
- Select the **Window** block and drag the arrow grip.
- Type-in **54** and press Enter. It changes the window size to 54.

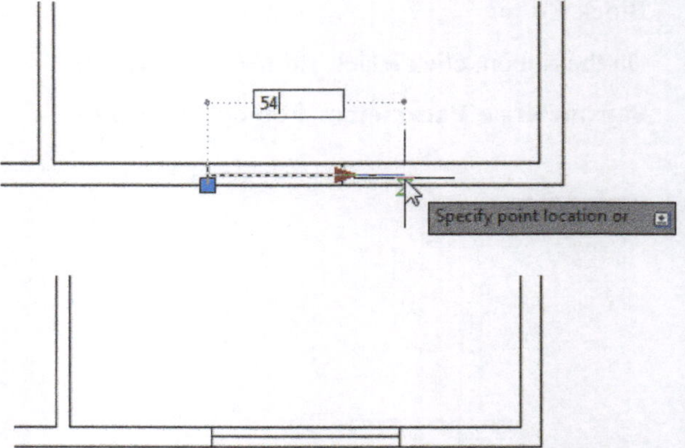

- On the ribbon, click **Insert > Block > Insert > Window**.
- Select the **Rotate** option from the command line.
- Type-in **90** and press Enter.
- Place the **Window** block on the kitchen wall, as shown below.

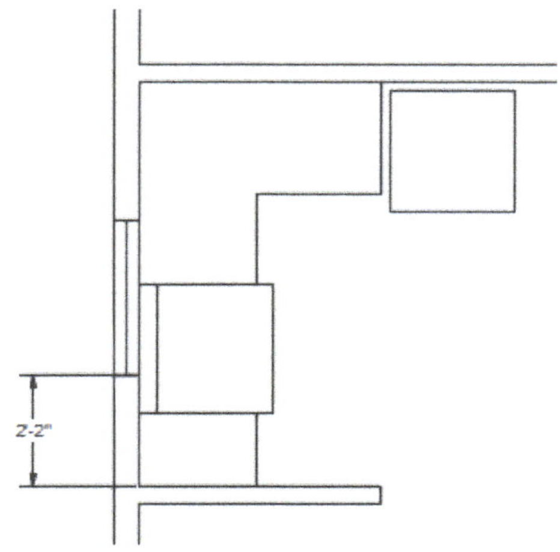

- Likewise, place the window blocks, as shown below.

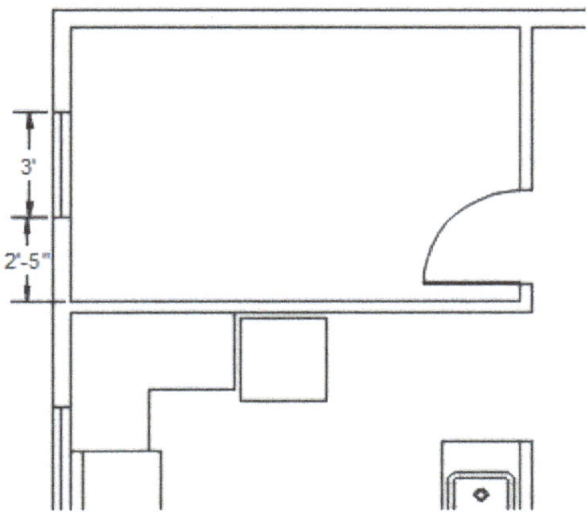

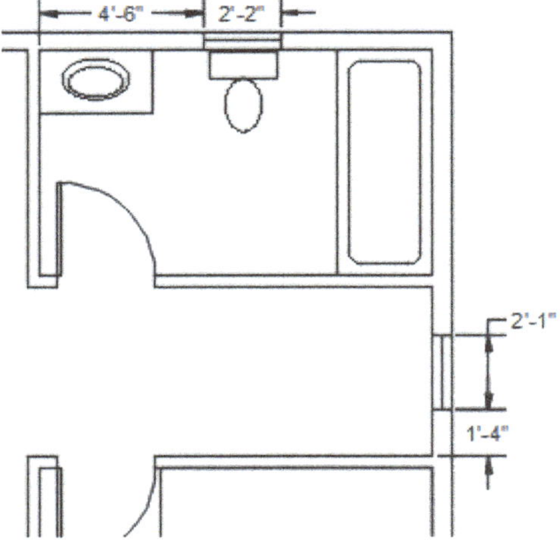

Arranging Objects of the drawing in Layers

- On the ribbon, click the **Home > Layers > Layer Properties**. It displays the Layer Properties Manager.
- On the **Layer Properties Manager** palette, click the **New Layer** button.

- Type **Wall** in the layer **Name** box and press Enter.
- Create another layer, and then type-in **Door** — press Enter.
  - Likewise, create other layers and define the layer properties, as shown below. Refer to Chapter 3 to learn more about layers.

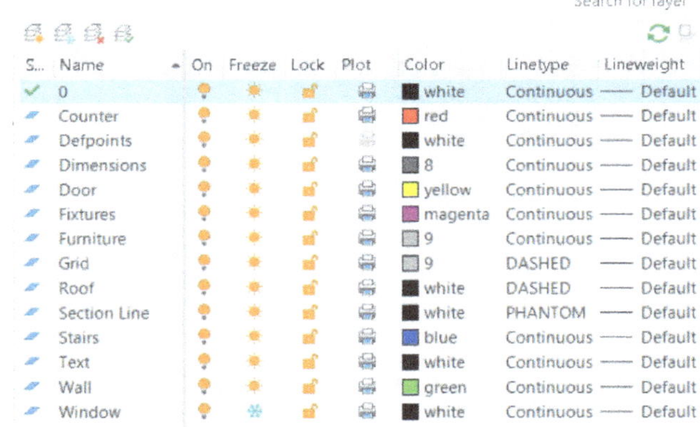

S...	Name	On	Freeze	Lock	Plot	Color	Linetype	Lineweight
✓	0					white	Continuous	Default
	Counter					red	Continuous	Default
	Defpoints					white	Continuous	Default
	Dimensions					8	Continuous	Default
	Door					yellow	Continuous	Default
	Fixtures					magenta	Continuous	Default
	Furniture					9	Continuous	Default
	Grid					9	DASHED	Default
	Roof					white	DASHED	Default
	Section Line					white	PHANTOM	Default
	Stairs					blue	Continuous	Default
	Text					white	Continuous	Default
	Wall					green	Continuous	Default
	Window					white	Continuous	Default

- Close the **Layer Properties Manager**.
- Select the Dining set, cupboard, and bed.
- On the ribbon, click **Home > Layers > Layer drop-down > Furniture**. The selected objects will be transferred to the Furniture layer.

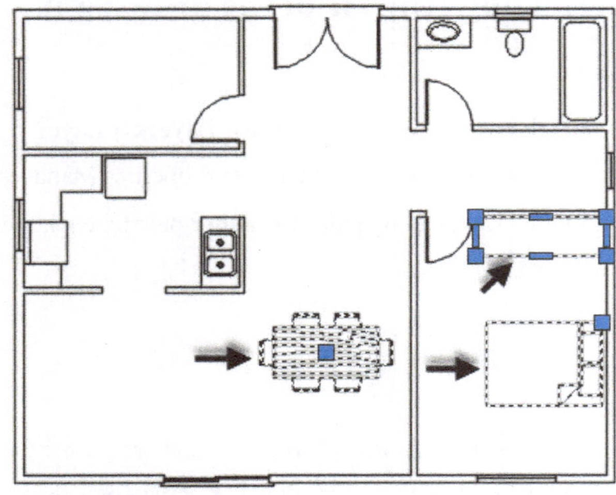

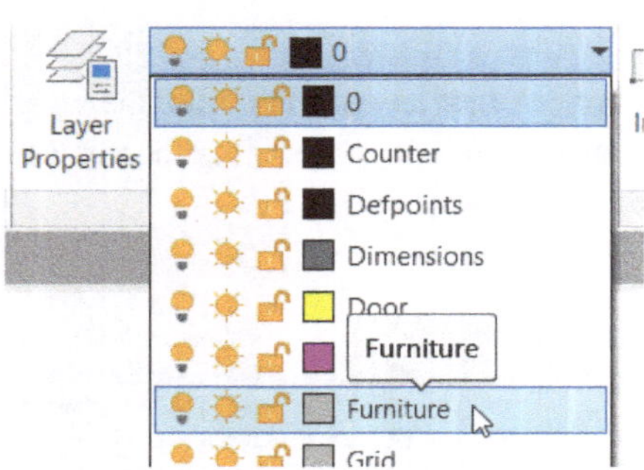

- Press Esc to deselect the selected objects.
- Likewise, transfer the other objects to their respective layers.

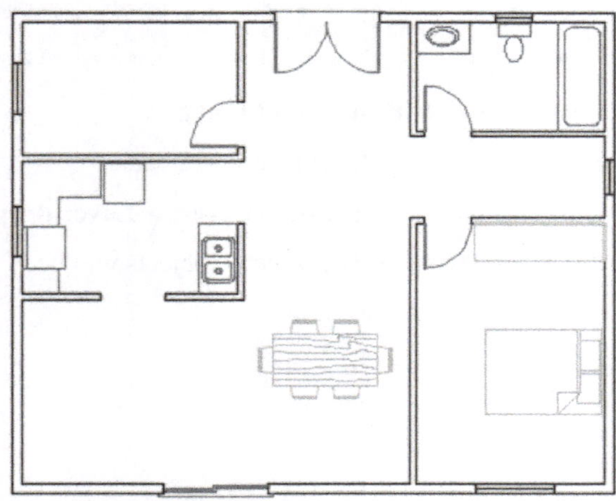

- Open the **Layer Properties Manager** and click the bulb symbols associated with Door, Window, Fixtures, Furniture, and Counter layers. It will hide the corresponding layers.

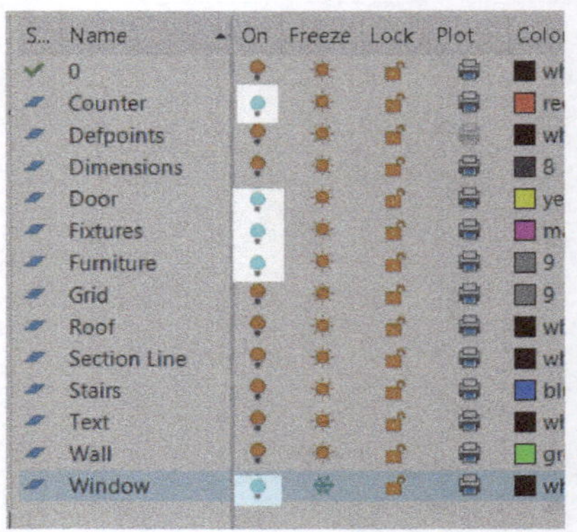

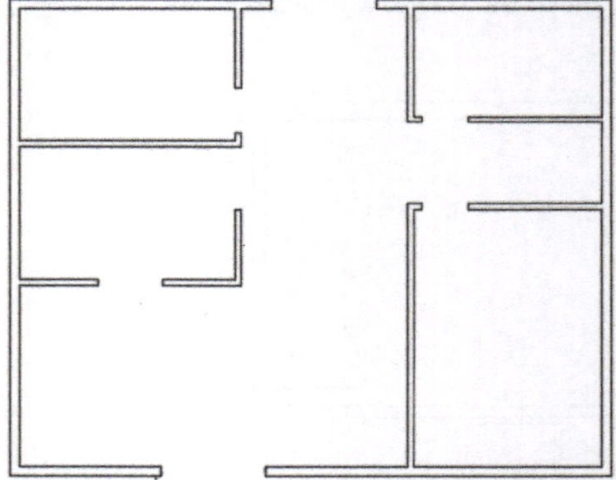

- Create a selection window and select all the walls.
- On the ribbon, click **Home > Layers > Layer drop-down > Wall**. All the walls will be transferred to the **Wall** layer.

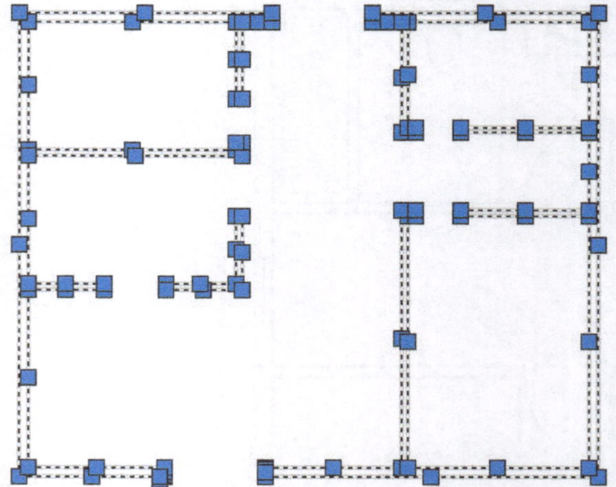

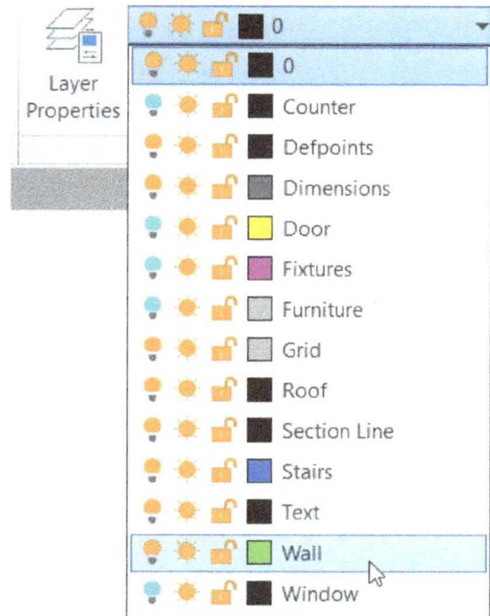

- Now, turn ON the hidden layers by clicking the bulb symbols.

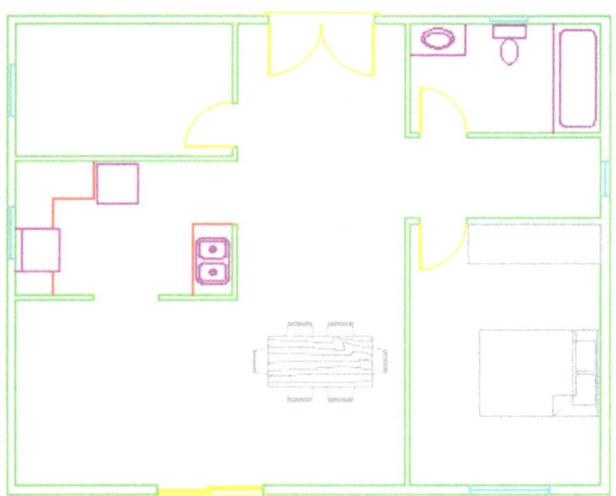

Creating Grid Lines

- On the ribbon, click **Home > Layers > Layer drop-down > Grid**. The Grid layer becomes active.

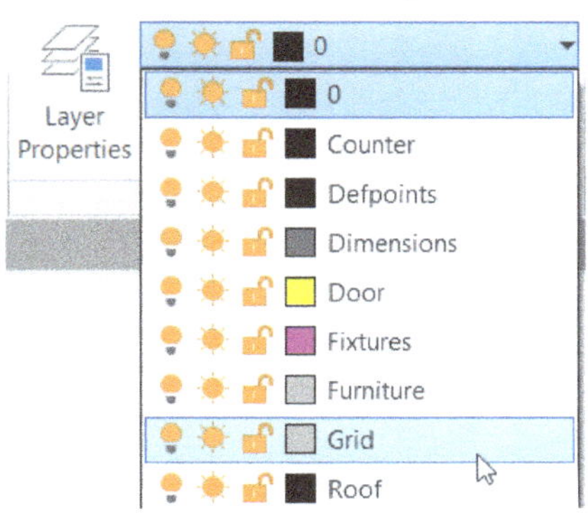

- Activate the **Line** command.
- Press and hold the Shift key and right-click, and then select the **Mid Between 2 Points** option.
- Select the endpoints of the wall, as shown below.

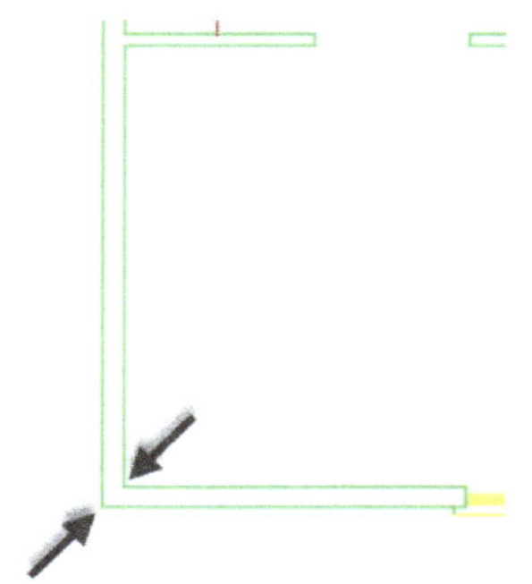

- Move the point upward and click to draw a vertical line of arbitrary length.

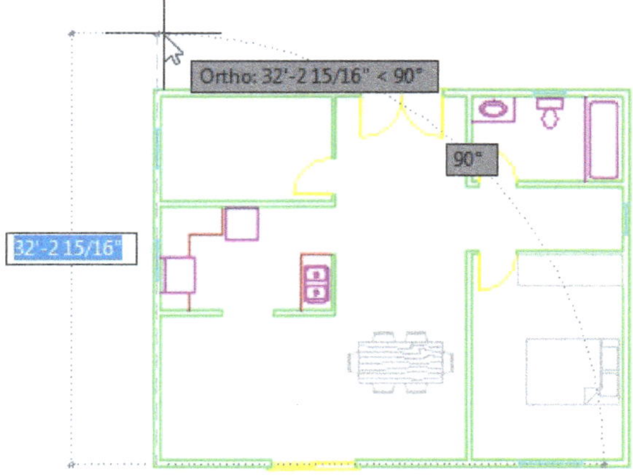

- Press Esc to deactivate the **Line** command.
- Select the line to display grips on it.
- Click the lower end grip and drag the pointer to increase the length of the line.
- Activate the **Offset** command and offset the grid line up to 406.
- Create other grid lines, as shown below.

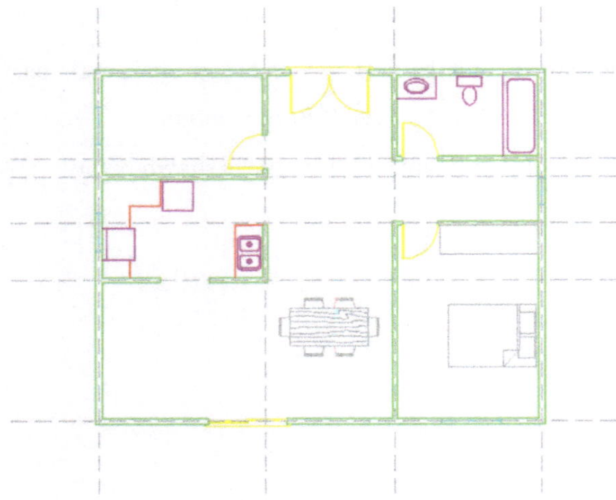

- Create a new layer called **Grid Bubbles** and make it current.

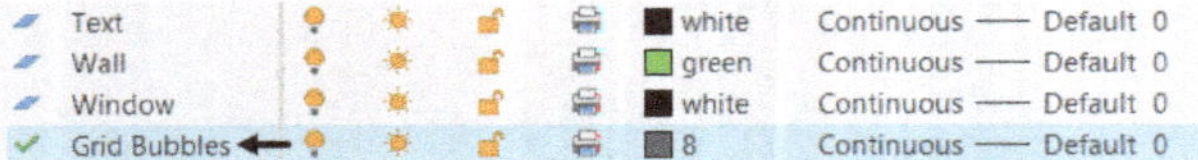

- Create a circle of 12 diameter.
- On the ribbon, click **Insert > Block Definition > Define Attributes**.

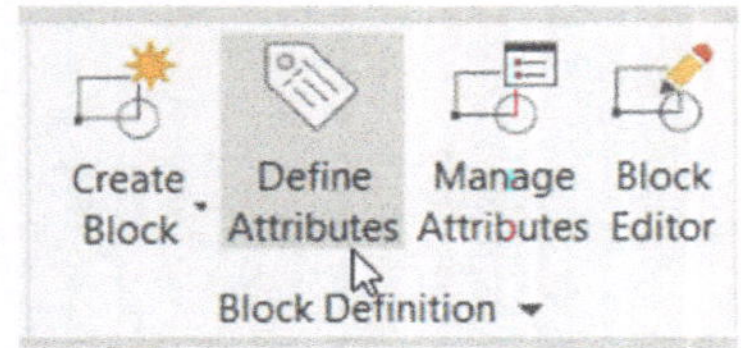

- On the **Attribute Definition** dialog, type-in GRIDBUBBLE in the **Tag** box and select **Justification > Middle center**.
- Type-in 6″ in the **Text height** box and click **OK**.
- Select the center point of the circle. The attribute text will be placed at its center.

GRIDBUBBLE

- On the ribbon, click **Insert > Block Definition > Create Block**.
- Type-in Grid Bubble in the **Name** box and click the **Select objects** button.
- Draw a crossing window to select the circle and attribute. Press Enter to accept the selection.
- Click the **Pick point** icon under the **Base point** section.
- Select the lower quadrant point of the circle to define the base point of the block.

- Uncheck the **Open in block editor** option and click **OK**.
- On the ribbon, click **Insert > Block > Insert > Grid bubble**.

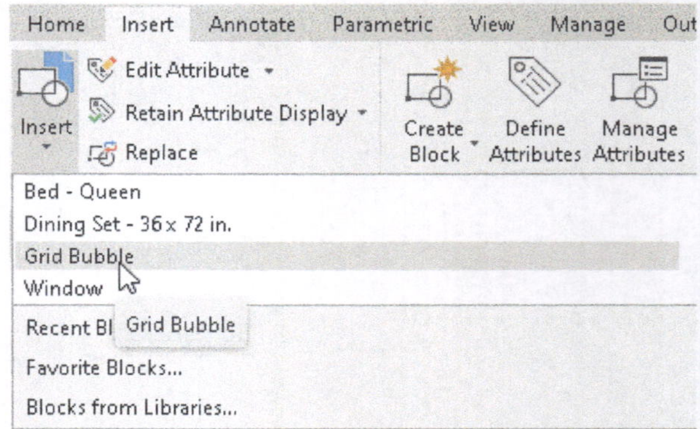

- Select the top endpoint of the first vertical grid line. The **Edit Attributes** dialog pops up.
- Type-in **A** in the GRIDBUBBLE box and click **OK**.

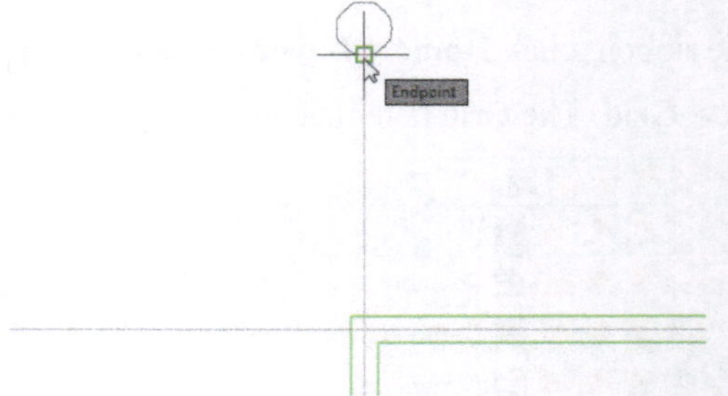

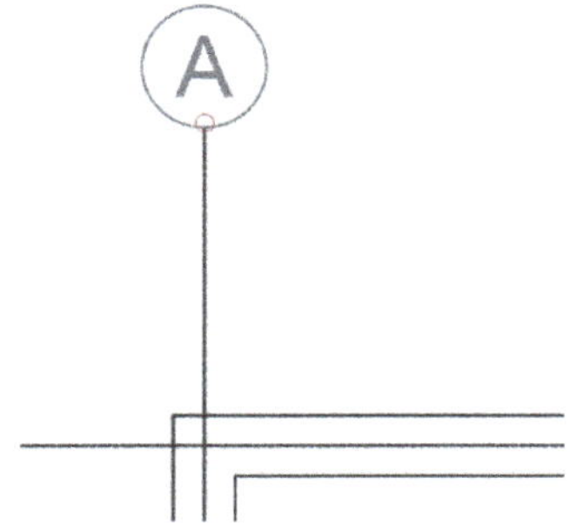

- Likewise, add other grid bubbles to the vertical grid lines.

- Create another block with the name Vertical Grid bubble. Make sure that you select the right quadrant point of the circle as the base point.

- Insert the vertical grid bubbles, as shown below.

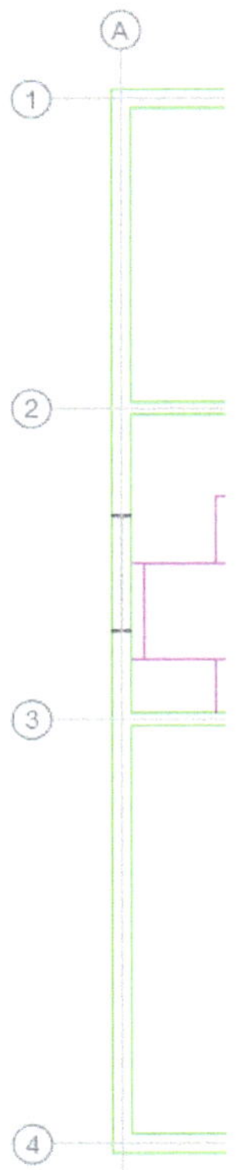

Adding Dimensions

- On the ribbon, click **Home > Layers > Layer drop-down > Dimensions** to make it current.
- Type **D** in the command line and press Enter.
- On the **Dimension Style Manager** dialog, select the **Standard** dimension style and click the **New** button.
- Type-in Floor Plan in the **New Style Name** box and click **Continue**.
- Click the **Primary Units** tab and select **Unit format > Architectural**.
- Set **Precision** to **0′-01/16″.**
- Set **Fraction format** to **Horizontal**.
- Under the **Zero Suppression** section, uncheck the **0 inches** option.
- Click the **Symbols and Arrows** tab.
- Under the **Arrowheads** section, select **First > Architectural tick**. The second arrowhead is automatically changed to **Architectural tick**.
- Select **Leader > Closed Filled** and enter 1/4′ in the **Arrow Size** box.
- Click the **Lines** tab and set **Extend beyond dim lines** and **Offset from origin** to 3″.
- Click the **Text** tab and **Text height** to 6″.
- In the **Text placement** section, set the following settings.

  Vertical-Centered

  Horizontal-Centered

  View Direction-Left-to-Right
- In the **Text alignment** section, select the **Aligned with dimension line** option.
- Click the **Fit** tab and select **Either text or arrows (best fit)** option from the **Fit Options** section.
- In the **Text placement** section, select the **Over dimension line, without Leader** option.
- Click **OK** and click **Set Current** on the **Dimension Style Manager**. Click **Close**.

- On the ribbon, click **Annotate > Dimensions > Dimension**.
- Select the points on the vertical grid lines, as shown below.
- Move the pointer and click to locate the dimension.

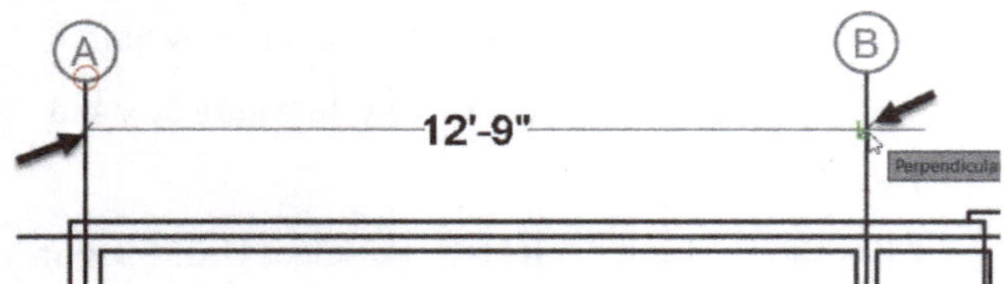

- On the ribbon, click **Annotate > Dimensions > Continue**. You will notice that a dimension is attached to the pointer
- Move the pointer and click on the next grid line.
- Likewise, move the pointer and click on the next grid line.

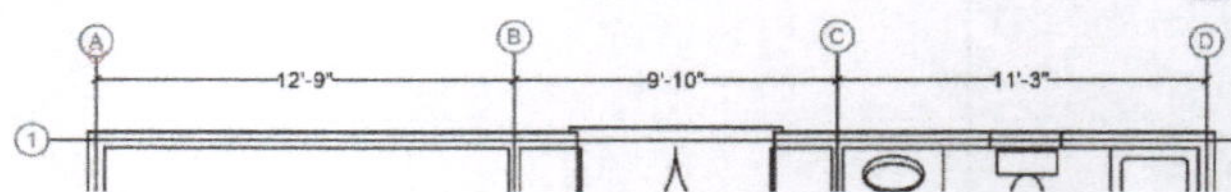

- Activate the **Dimension** command and create the overall horizontal dimension.

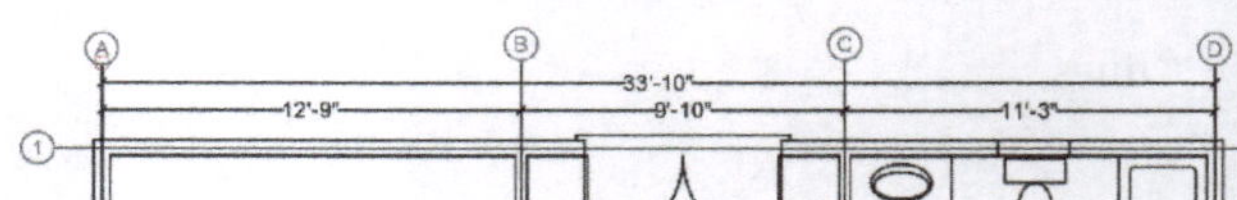

- Likewise, add vertical dimensions to the grid lines.

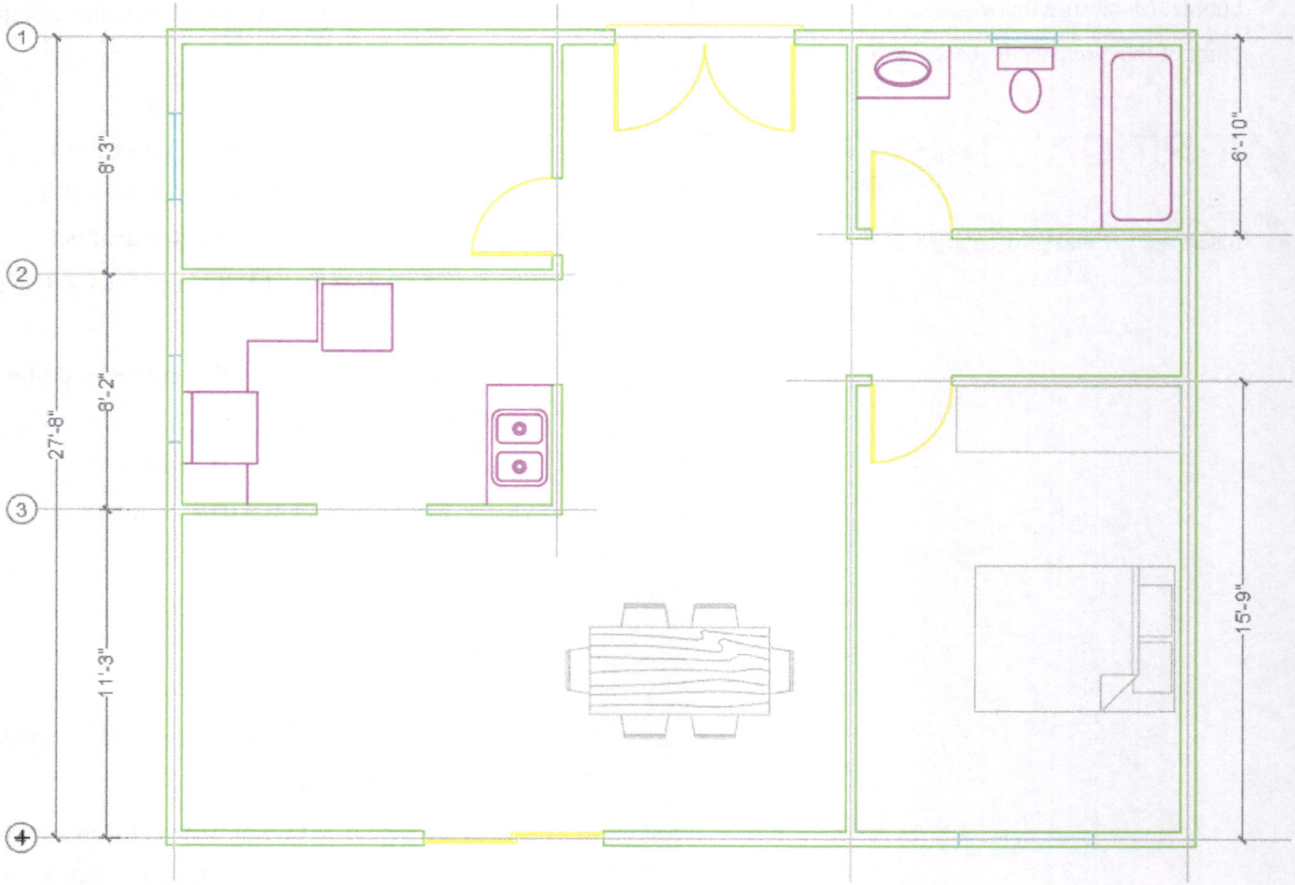

- Complete adding dimensions to the drawing, as shown below.

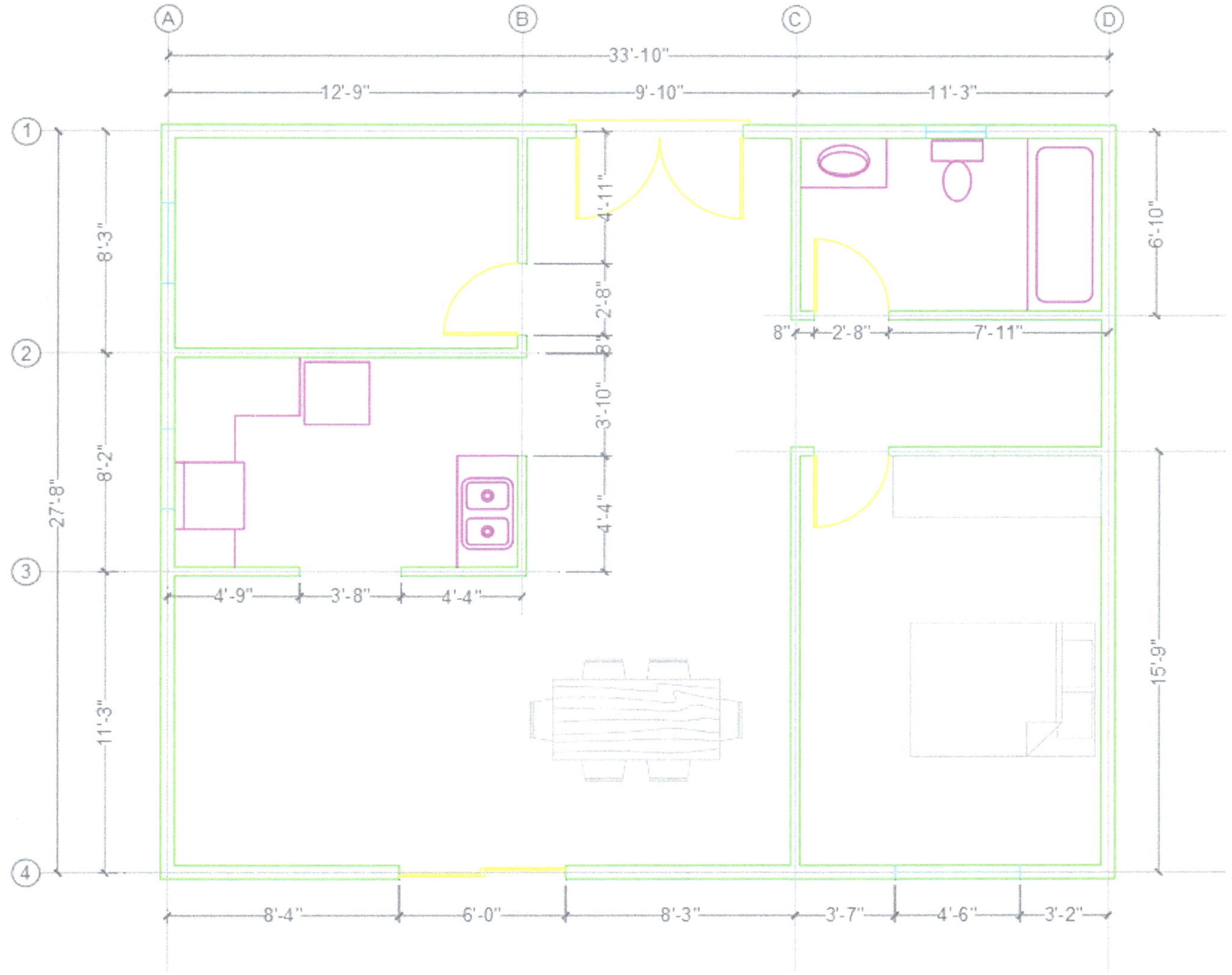

- On the ribbon, click **View > Palettes > Count**.

The **Count** palette appears. It displays the blocks available in the drawing and the count of each block.

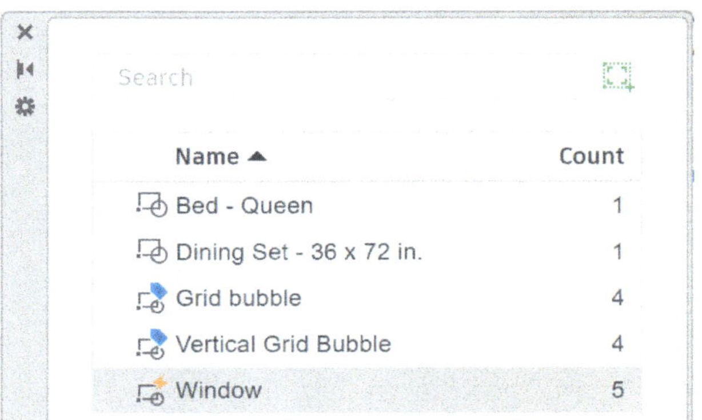

- On the Count palette, click the **Count in a specified area** icon.

- Create a window by specifying the two corner points, as shown.

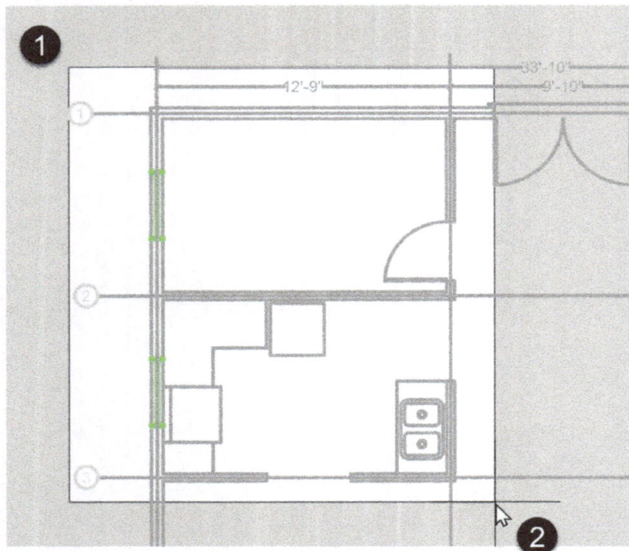

The blocks available in the specified region are displayed on the **Count** palette.

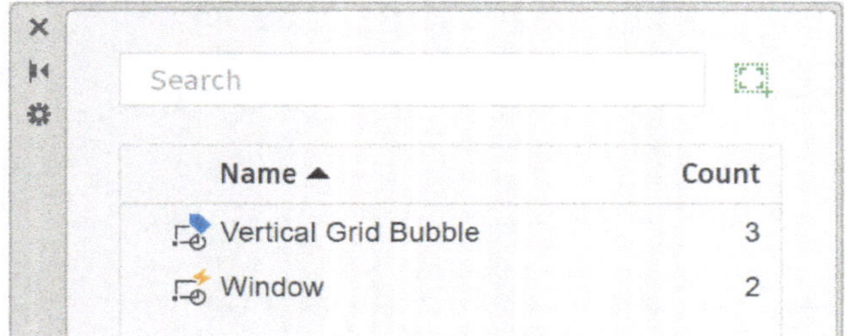

You can click the **Specify Area** icon on the Count toolbar and specify another area in the drawing.

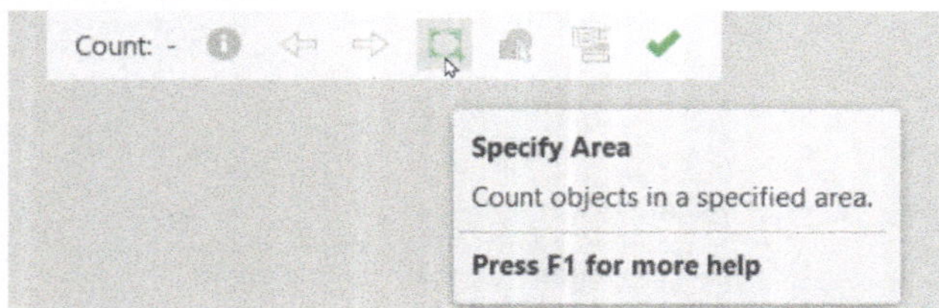

- Select the rectangular selection box and press DELETE.

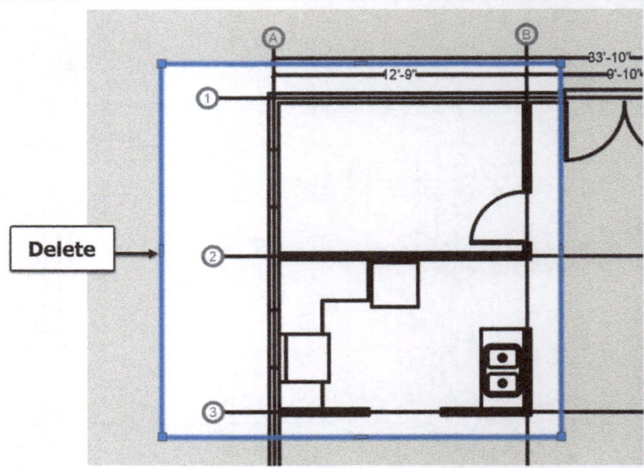

- Click the **Continue and count the entire model space** option on the **Count – Invalid Area** message box.
- On the **Count** palette, click the right mouse button on the **Window** block and select **Review Count Details**.

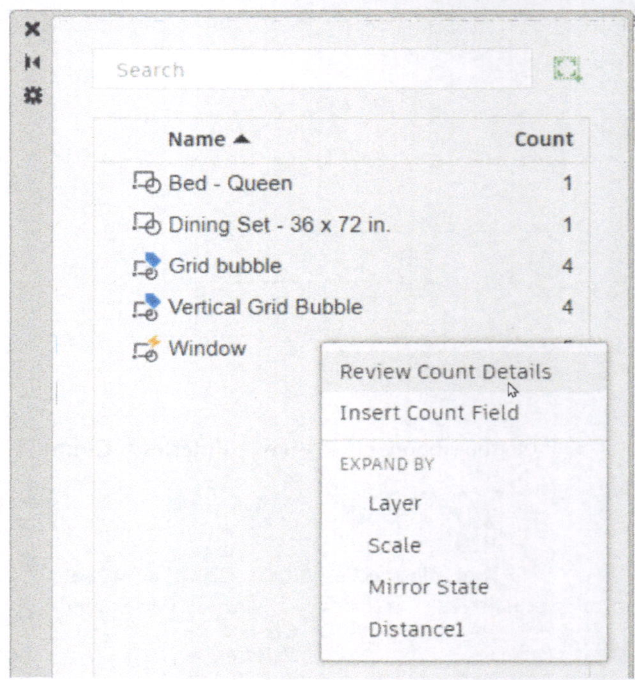

Use the arrows on the **Count** toolbar to zoom to the instances of the Window blocks.

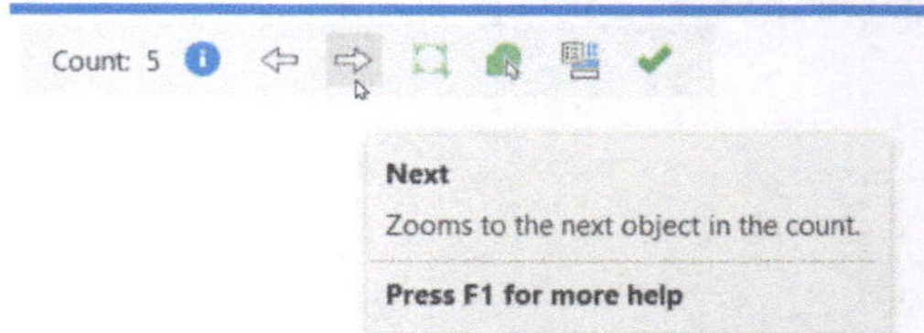

Use the **Insert Count Field** icon on the **Count** toolbar, if you want to insert the count into the drawing.

- Click the **End Count** icon on the **Count** toolbar to close it.
- Click the **Create table** button on the **Count** palette.

- Check the **Select all** box to select all the blocks.

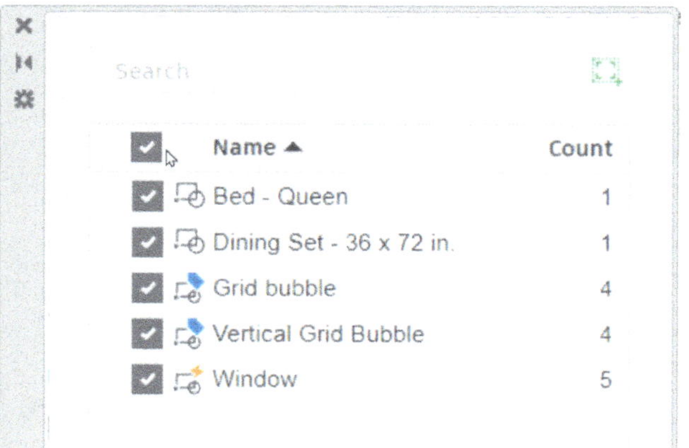

- Click the **Insert** button.
- Click in the graphics window to place the count table.

Item	Count
Window	5
Vertical Grid Bubble	4
Grid bubble	4
Dining Set - 36 x 72 in.	1
Bed - Queen	1

- Save and close the drawing.

Example 2

In this tutorial, you will draw stairs.

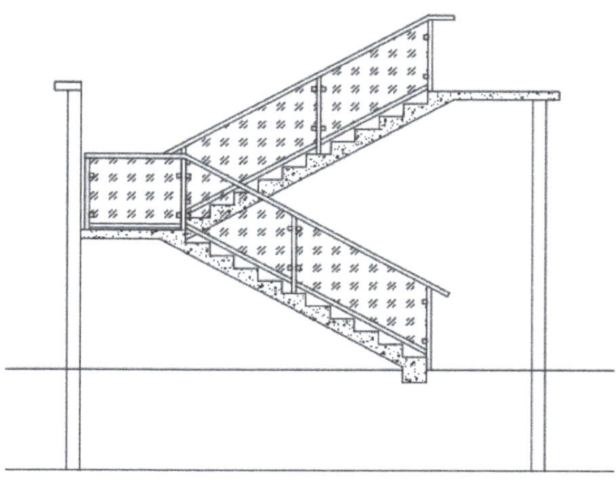

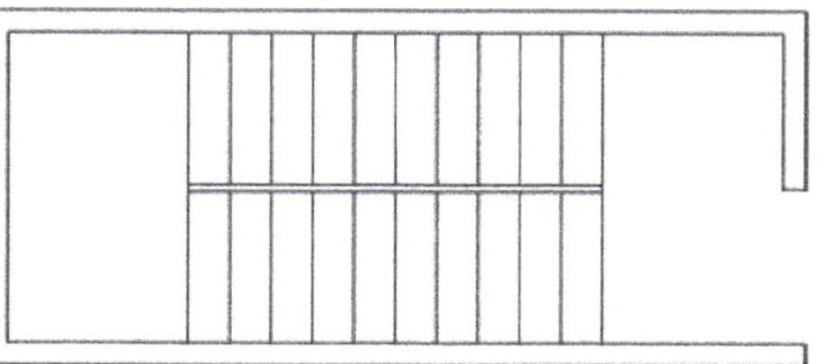

Staircase Nomenclature

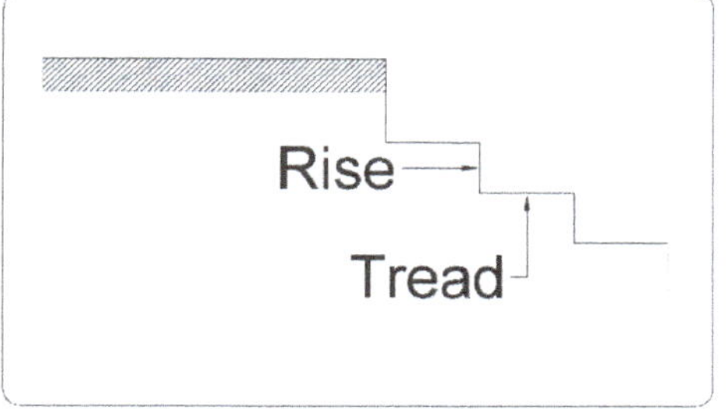

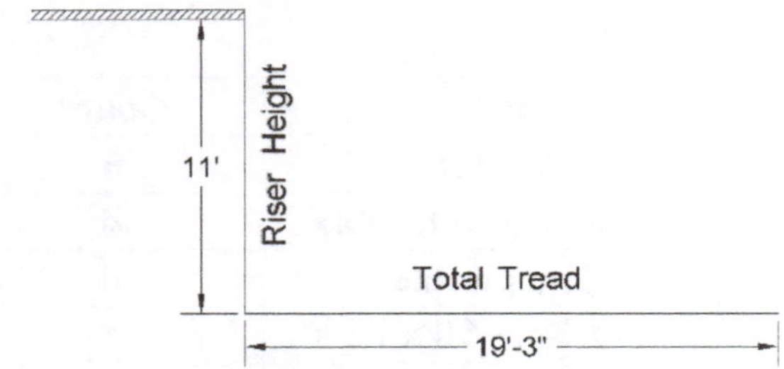

Creating the Stairs

- Start a new AutoCAD file using the **acad.dwt** template.
- Set the **Workspace** to **Drafting & Annotation**.

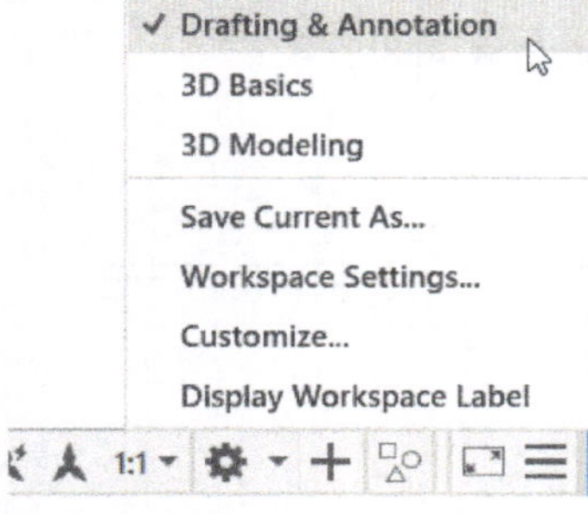

- Type **UN** in the command line and press Enter.
- On the **Drawing Units** dialog, select **Type > Architectural**.
- Select **Precision > 0-01/16**.
- Set the **Insertion Scale** to **Inches**, and click **OK**.
- Type LIMITS in the command line and press Enter.
- Press Enter to accept 0, 0 as the lower limit.
- Type 50', 40' in the command line, and press Enter. The program sets the upper limit of the drawing.
- Click the **Zoom** drop-down > **Zoom All** on the Navigation Bar.
- Turn OFF the grid.
- Click the **ORTHOMODE** icon on the Status bar.
- Click the **Polyline** tool on the **Draw** panel of the **Home** ribbon tab.
- Click to define the start point of the line.
- Move the pointer toward left.
- Type 17'8" and press ENTER.
- Move the pointer upward.
- Type 7' and press ENTER.
- Move the pointer toward the right.
- Type 17'2" and press ENTER.
- Move the pointer downward.
- Type 3'6" and press ENTER.
- Press Esc.

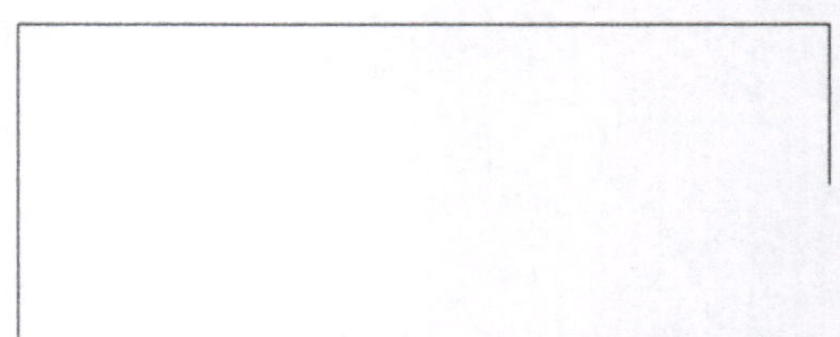

- Click the **Offset** icon on the **Modify** panel of the **Home** ribbon tab.
- Type 6" and press ENTER.
- Select the polyline.
- Move the pointer outward and click.

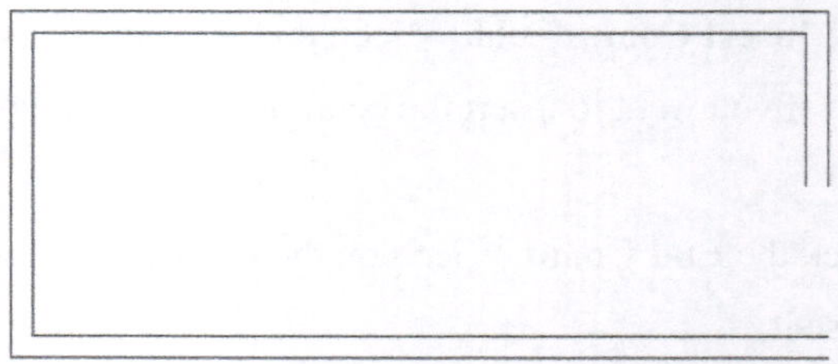

- Click the **Line** tool and close the open ends of the drawing, as shown.

- Create a selection window across all the elements of the drawing.

- Click the **Explode** icon on the **Modify** panel of the **Home** ribbon tab. The polylines are exploded into lines.

- Click the **Offset** icon on the **Modify** panel of the **Home** ribbon tab.
- Type 4' and press ENTER.
- Select the left vertical inner edge.
- Move the pointer toward right and click.

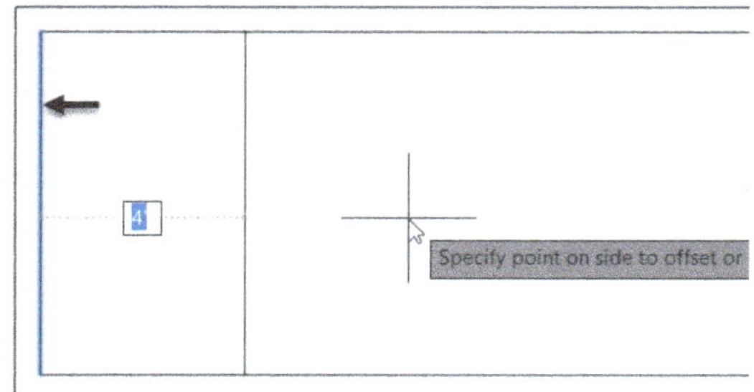

- Press ENTER.
- On the **Home** tab of the ribbon, click **Modify** panel > **Array** drop-down > **Path Array**.
- Select the offset line and press ENTER.
- Click on the inner horizontal line at the location, as shown.

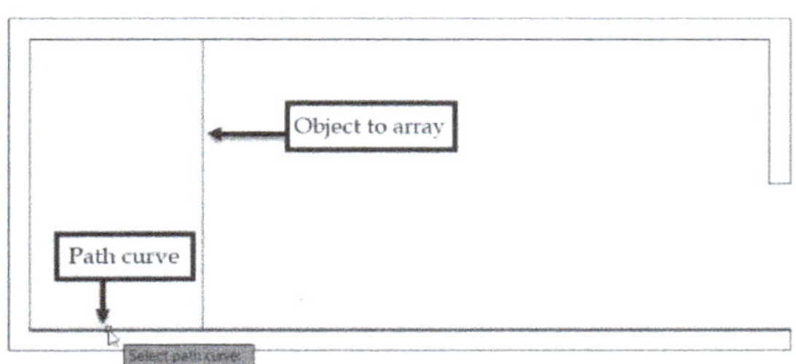

- On the **Array Creation** tab, click the **Items** icon on the **Items** panel.
- Type **11** in the **Items** and **Between** boxes, respectively.

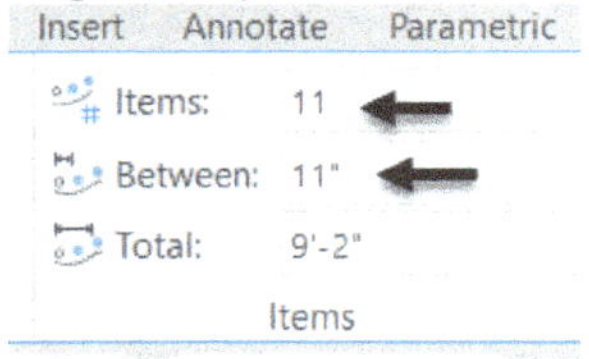

- Deactivate the **Associative** icon on the **Properties** panel.
- Click the **Close Array** icon.

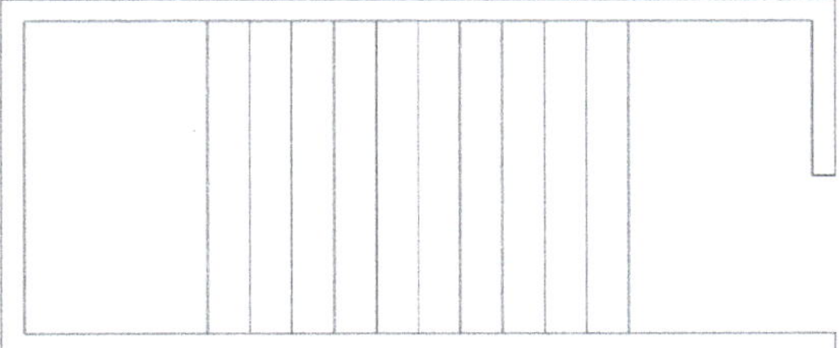

- Offset the inner horizontal lines by 3'5" distance in the inward direction.

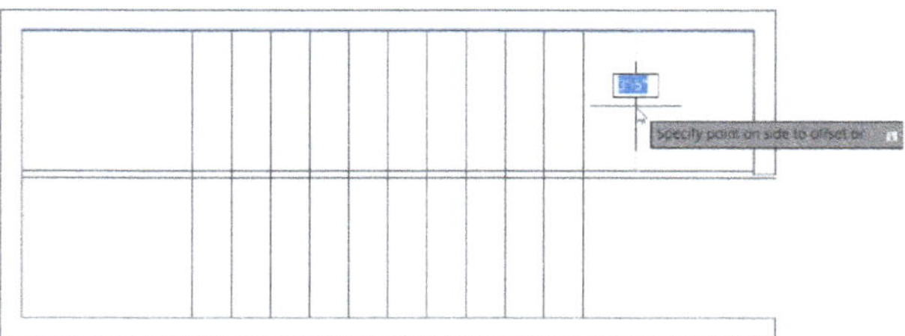

- On the **Home** tab of the ribbon, click **Modify** panel > **Trim/Extend** drop-down > **Trim**.
- Select the portions of the horizontal lines, as shown.

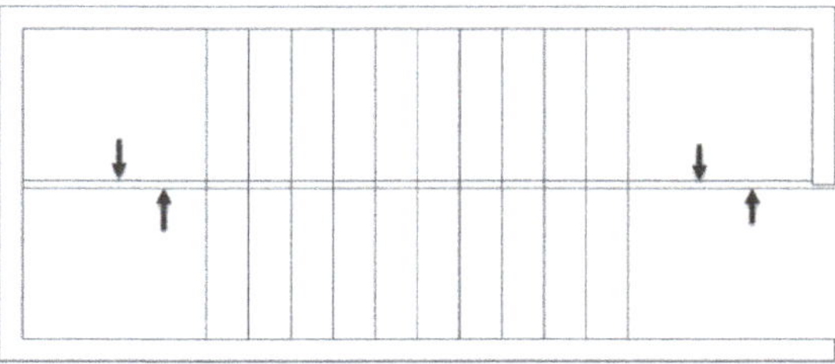

- Zoom-in to the center portion of the stairs.
- Click on the portions of the vertical lines, as shown.

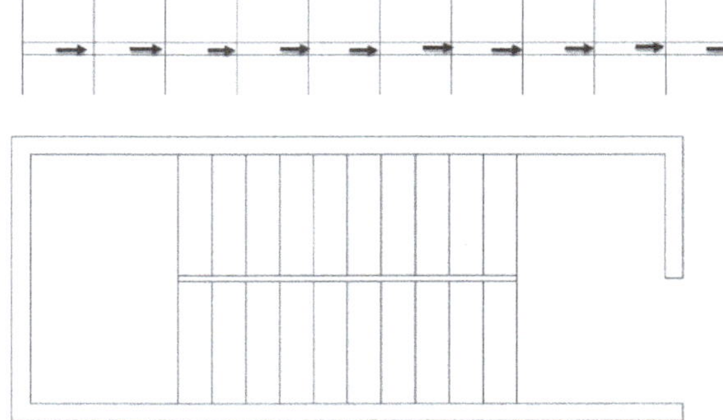

Creating the Section elevation of the Staircase

- Create a horizontal line above the staircase, as shown.

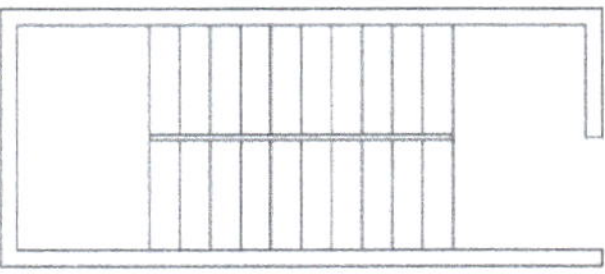

- Click the **Offset** icon on the **Modify** panel of the **Home** ribbon tab.
- Type 4' and press ENTER.
- Select the newly created horizontal line.
- Move the pointer upward and click.

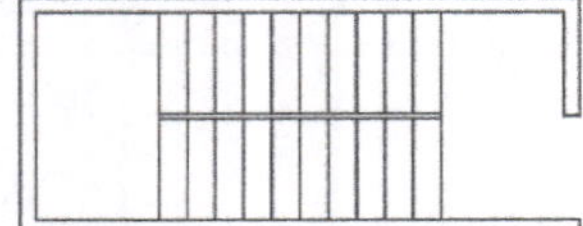

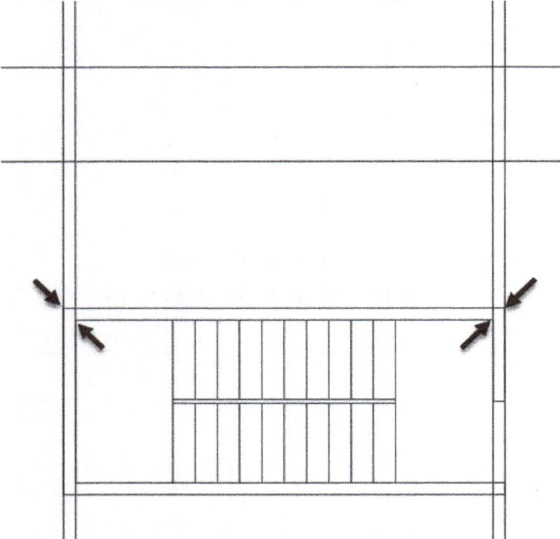

- On the **Home** tab of the ribbon, expand the **Draw** panel and click the **Construction line** icon.
- Select the **Ver** option from the command line.
- Select the corners of the drawing, as shown.

- Click the **Offset** icon on the **Modify** panel of the **Home** ribbon tab.
- Type 11' and press ENTER.
- Select the offset horizontal line.
- Move the pointer upward and click.

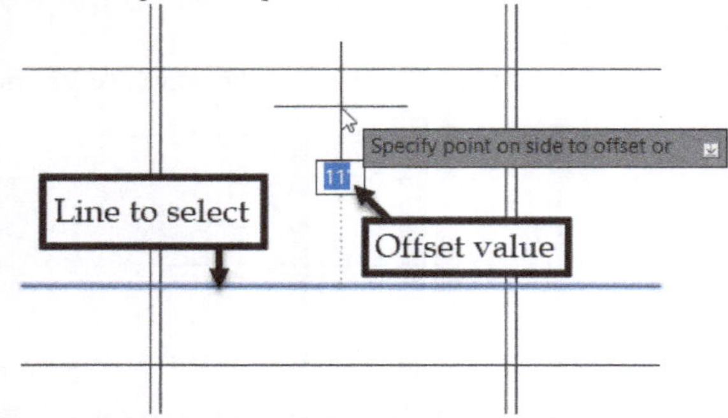

- Press ENTER twice to deactivate the **Offset** tool and then activate it again.
- Type 4" and press ENTER.
- Select the newly created offset line.
- Move the pointer upward and click.

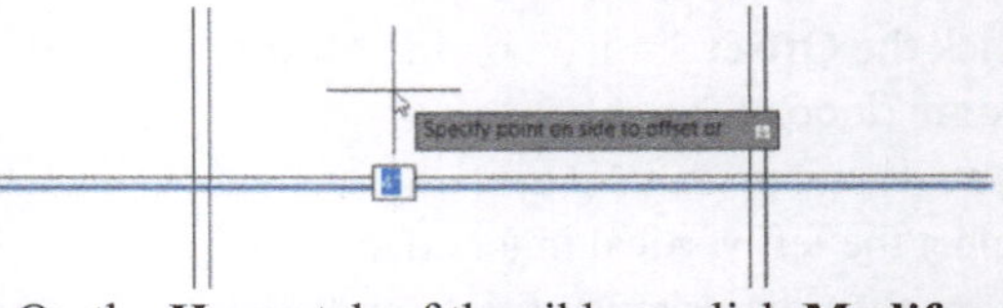

- On the **Home** tab of the ribbon, click **Modify** panel > **Trim/Extend** drop-down > **Trim**.
- Press and hold the left mouse button and drag the pointer across the portions of the construction lines, as shown.

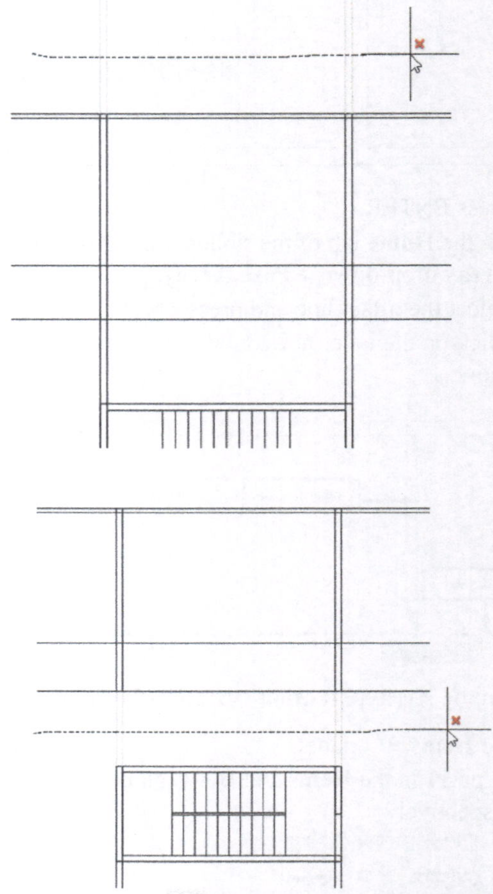

- Click the **Offset** icon on the **Modify** panel of the **Home** ribbon tab.
- Type 6" and press ENTER.
- Offset the outer vertical lines on both sides.

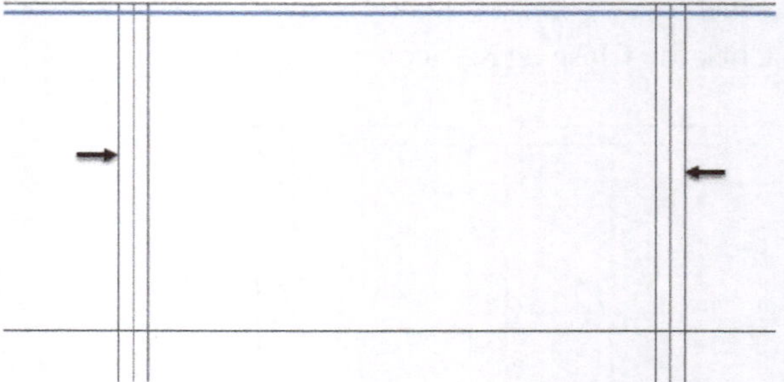

- On the **Home** tab of the ribbon, click **Modify** panel > **Trim/Extend** drop-down > **Trim**.
- Trim the portions of the vertical and horizontal lines, as shown.

- On the **Home** tab of the ribbon, click **Modify** panel > **Trim/Extend** drop-down > **Extend**.
- Select the **Boundary edges** option from the command line.
- Select the lower horizontal edge of the roof, as shown.

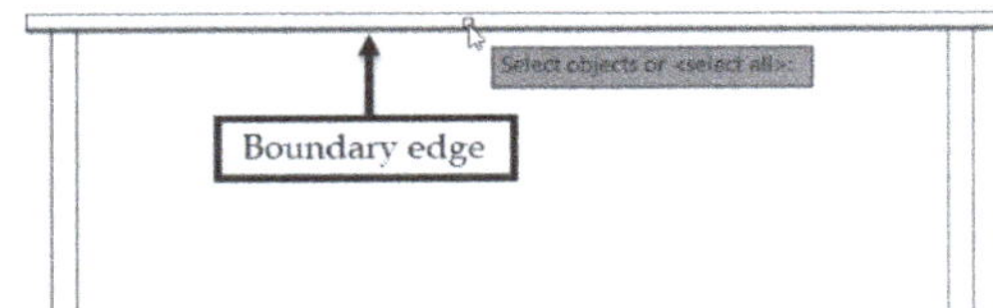

- Press ENTER.
- Press and hold the left mouse button.
- Drag the pointer across the staircase lines; the lines are extended up to the boundary edge.

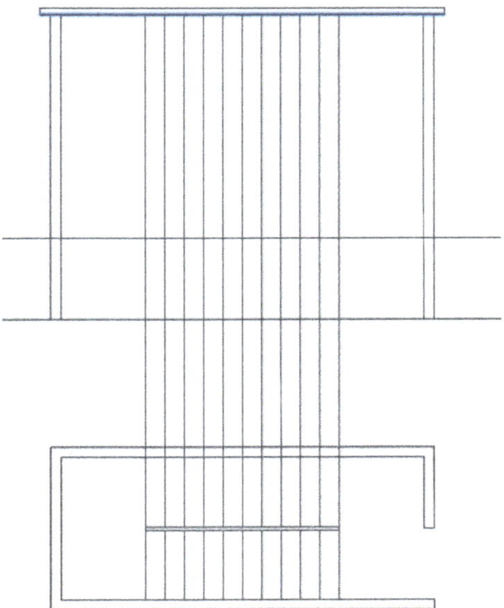

- On the **Home** tab of the ribbon, click **Modify** panel > **Trim/Extend** drop-down > **Trim**.
- Select the **cuTting edges** option from the command line.
- Select the horizontal lines, as shown.
- Press ENTER.

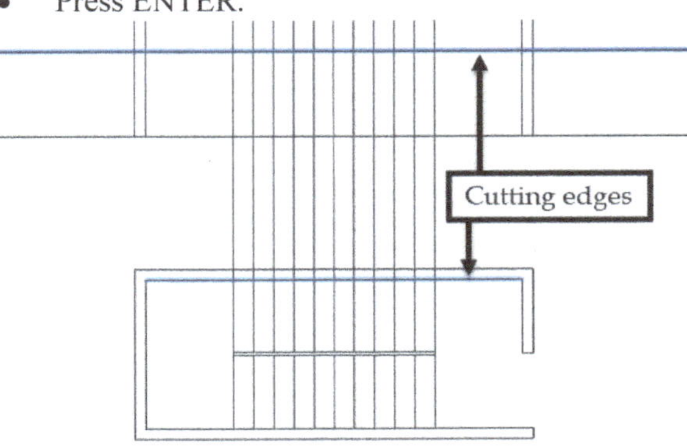

- Click and drag the pointer across the vertical lines, as shown.

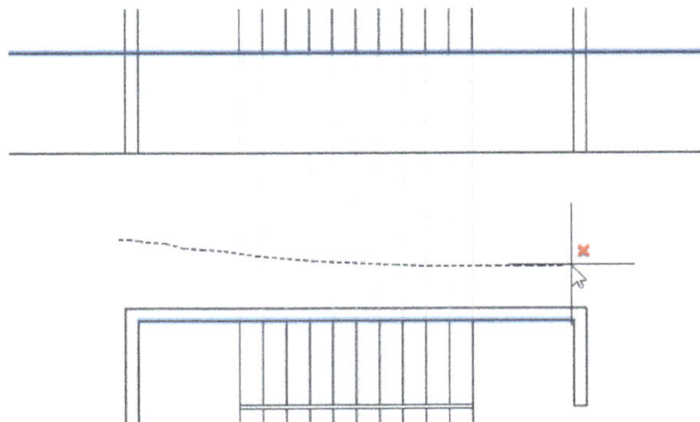

- On the **Home** tab of the ribbon, click **Draw** panel > **Polyline**.
- Select the first point of the polyline.

- Move the pointer upward.
- Type 6" and press ENTER.

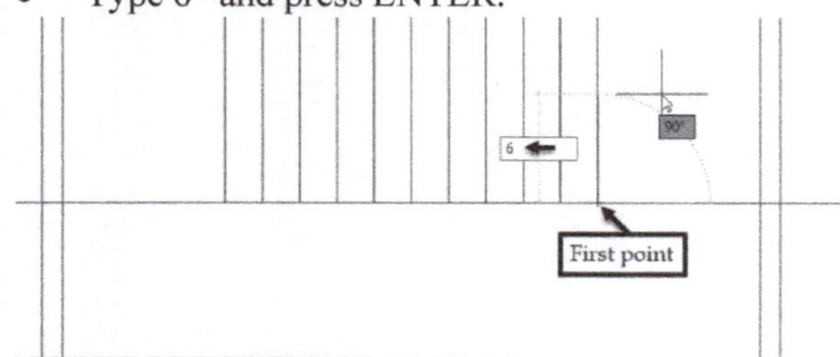

- Move the pointer toward left.
- Type 11" and press ENTER.
- Press Esc.

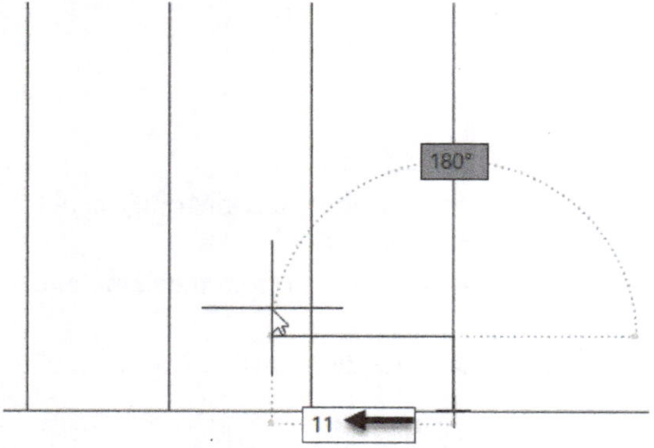

- Select the polyline and click the **Copy** icon on the **Modify** panel of the **Home** ribbon tab.
- Select the first point of the polyline as the base point.

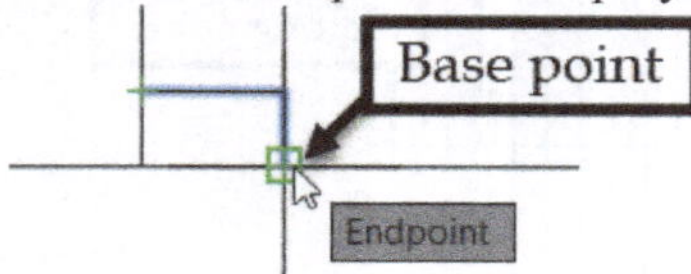

- Move the pointer upward and select the endpoint of the polyline. The polyline is copied.

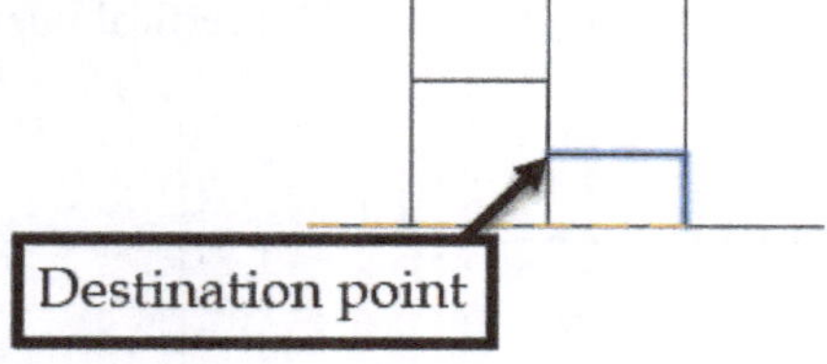

- Likewise, create copies of the polylines, as shown.

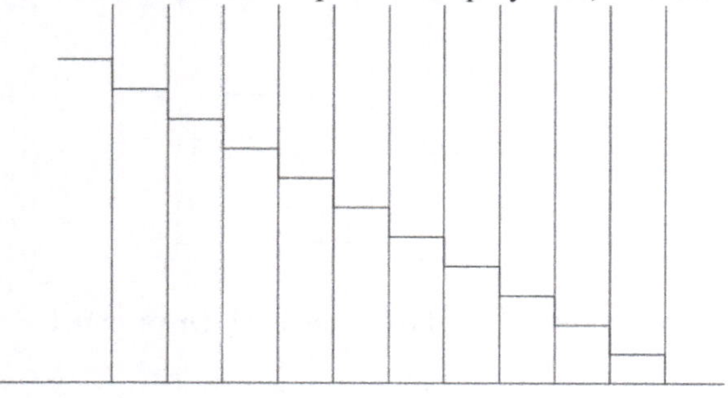

- Expand the **Modify** panel on the **Home** ribbon tab, and then click the **Join** icon.

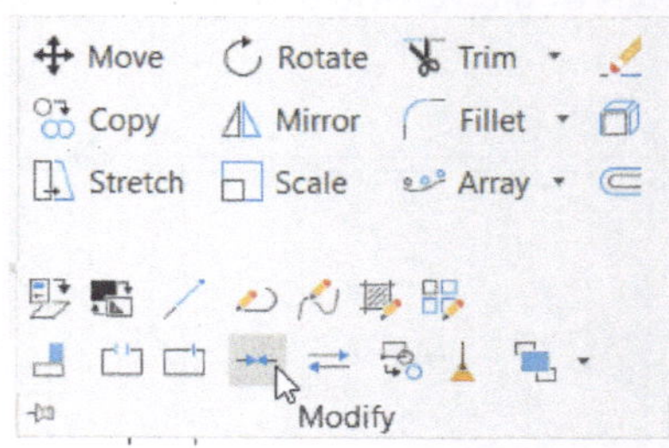

- Select all the polylines and press ENTER. All the polylines are joined.

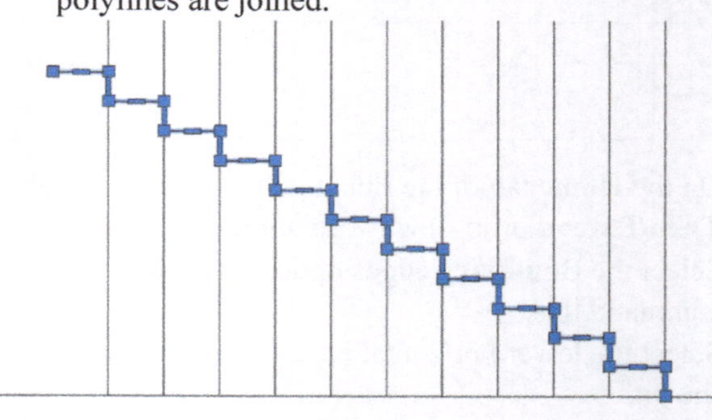

- On the **Home** tab of the ribbon, click **Modify** panel > **Trim/Extend** drop-down > **Extend**.
- Click on the end portion of the polyline to extend it up to the left vertical line.

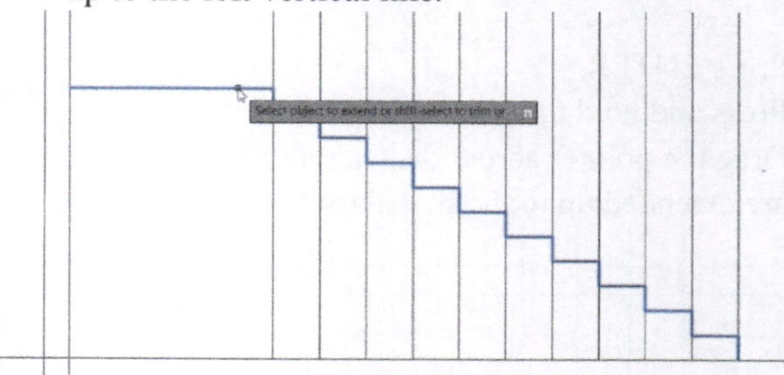

- Press Esc.
- Select the stairs and click the **Mirror** icon on the **Modify** panel of the **Home** ribbon tab.
- Specify the start point of the mirror line, as shown.
- Move the pointer toward the left and specify the endpoint of the mirror line, as shown.

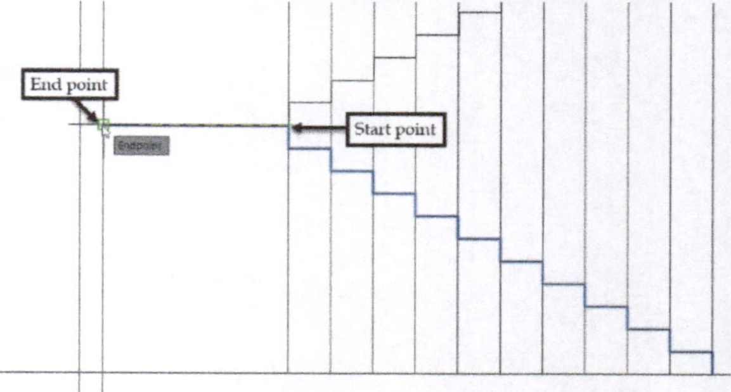

- Select **No** from the command line.
- Select the mirrored polyline.
- Click on the endpoint of the polyline.

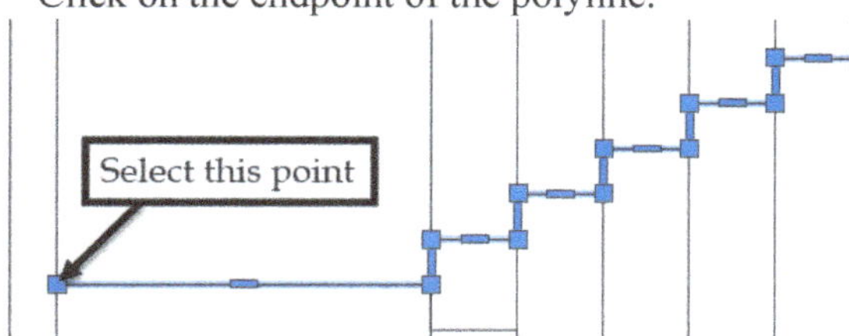

- Move the pointer toward the right and select the vertex point, as shown. The polyline is shortened.

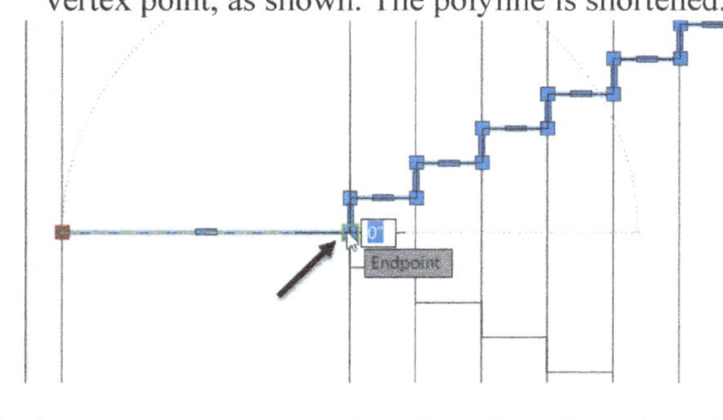

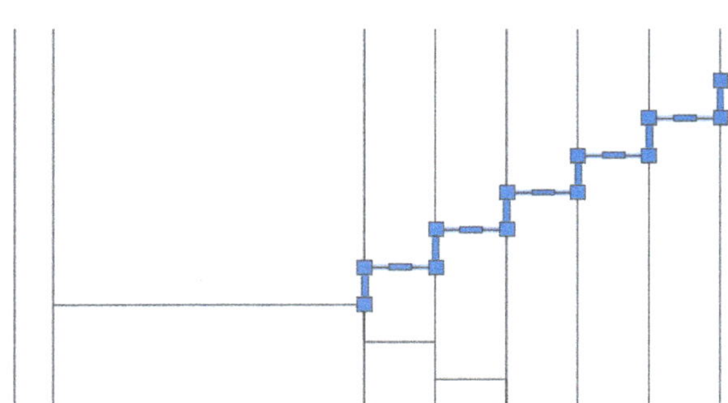

- Press Esc to deselect the mirrored polyline.
- Drag a selection window across the vertical lines from right to left, as shown.

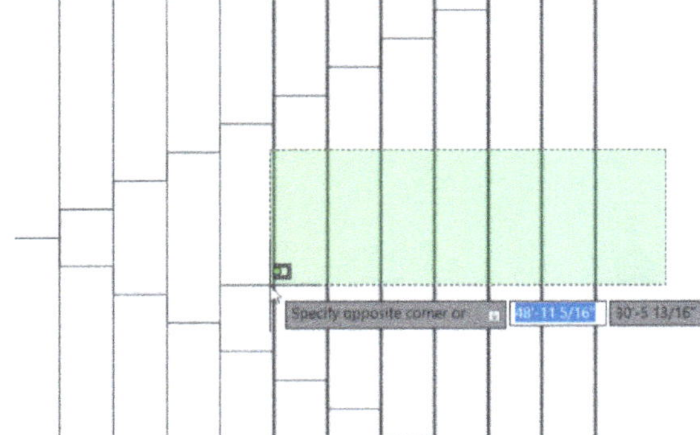

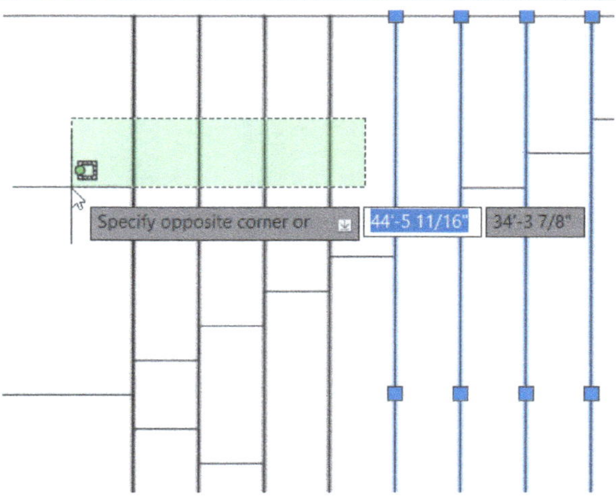

- Press Delete on your keyboard.
- On the **Home** tab of the ribbon, click **Draw** panel >**Line**.
- Select the corner points, as shown.

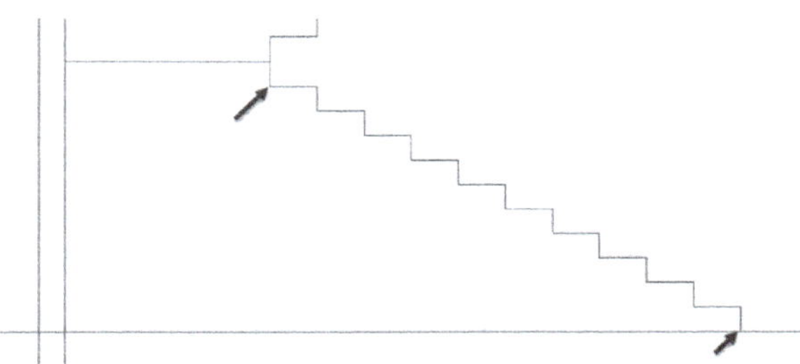

- Offset the newly created line by 4".

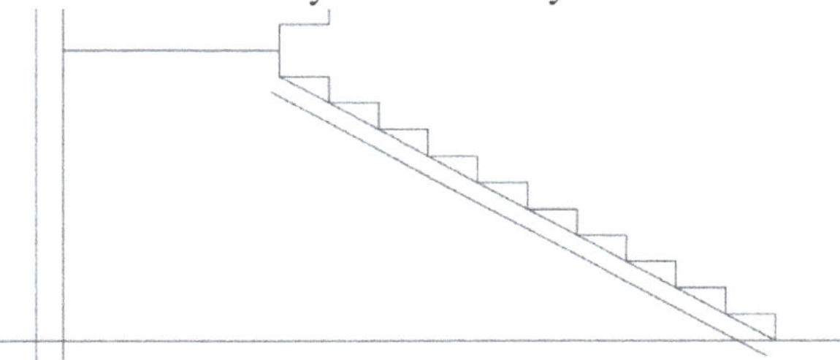

- On the **Home** tab of the ribbon, click **Draw** panel > **Line**.
- Press and hold the Shift key and right click.
- Select the **From** option from the **Object Snap** menu.
- Select the point, as shown.

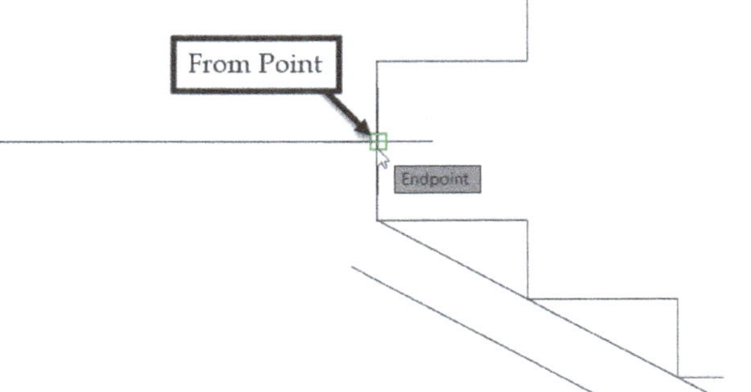

- Move the pointer downward and type 4".
- Press ENTER to specify the start point of the line.
- Move the pointer toward the left and click to create a horizontal line.

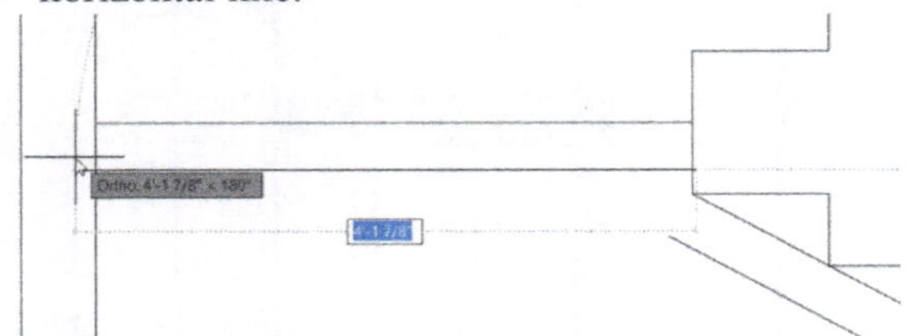

- Click the **Fillet** icon on the **Modify** panel.
- Select the horizontal and inclined lines, as shown.

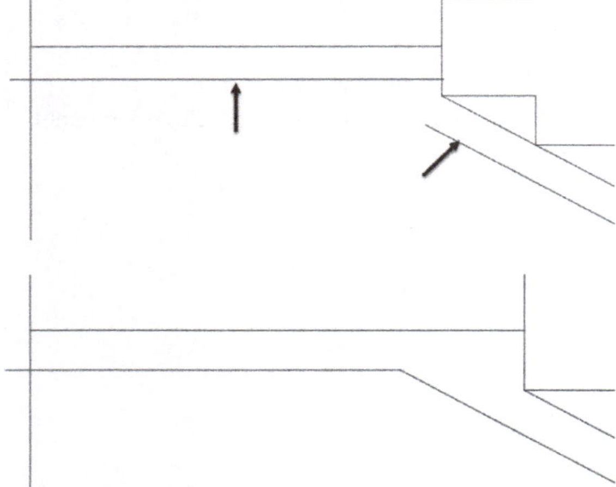

- Delete the inclined line, as shown.

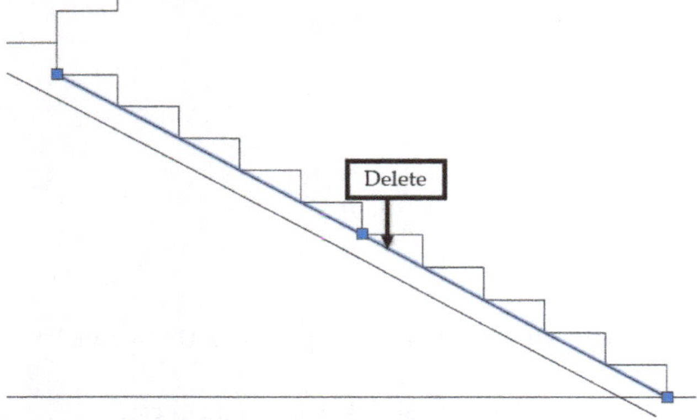

- On the **Home** tab of the ribbon, click **Draw** panel > **Line**.
- Select the corner point of the stair, as shown.

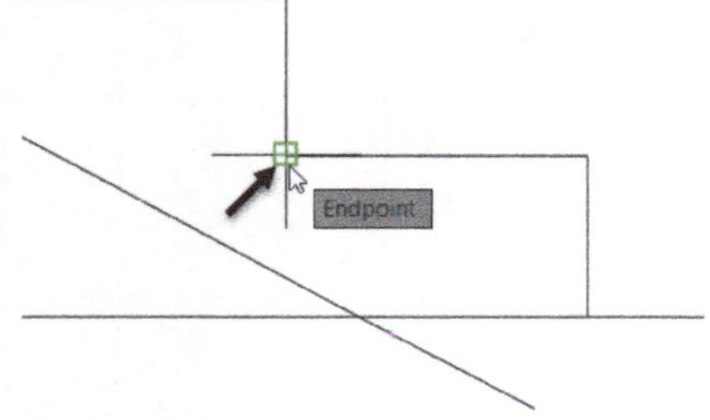

- Move the pointer downward.
- Type 1' and press ENTER.
- Move the pointer toward the right.
- Type 11" and press ENTER.
- Move the pointer upward.
- Type 6" and press ENTER.

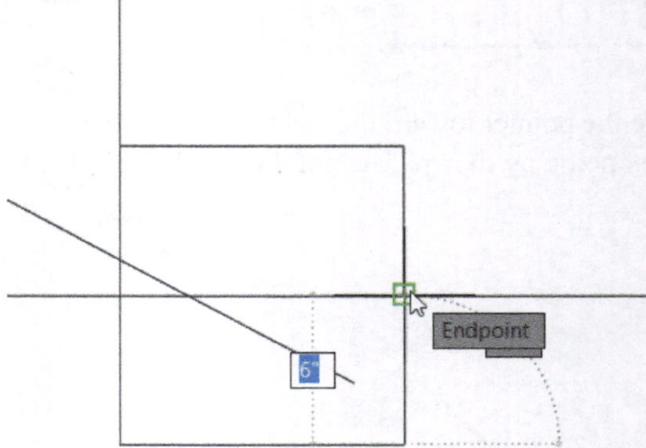

- On the **Home** tab of the ribbon, click **Modify** panel > **Trim/Extend** drop-down > **Trim**.
- Select the portions of the lines, as shown.

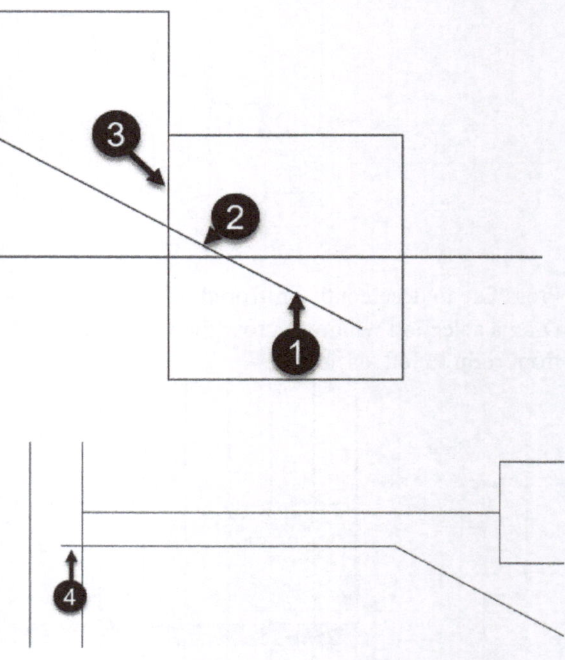

- On the **Home** tab of the ribbon, click **Draw** panel >**Line**.
- Select the corner points, as shown.

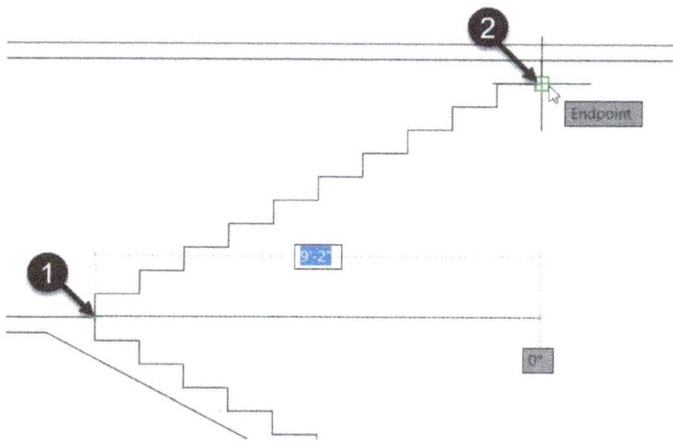

- Offset the newly created line by 4".

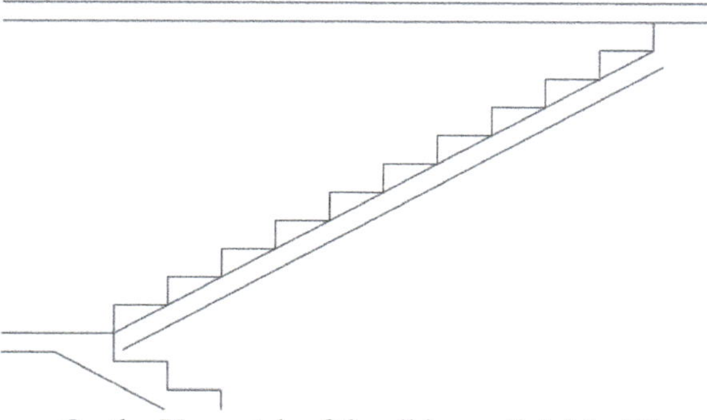

- On the **Home** tab of the ribbon, click **Modify** panel > **Trim/Extend** drop-down > **Extend**.
- Select the inclined line, as shown.

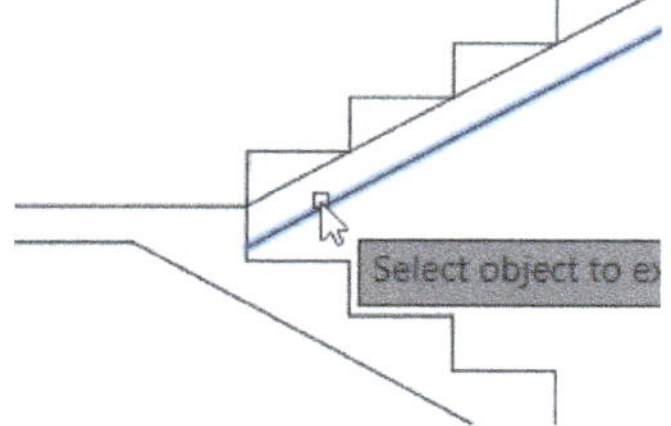

- Extend the inclined line up to the horizontal line, as shown.

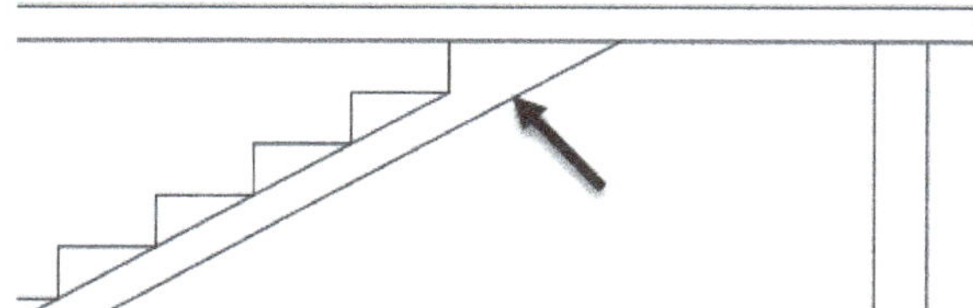

- Extend the vertical lines, as shown. Next, press Esc.

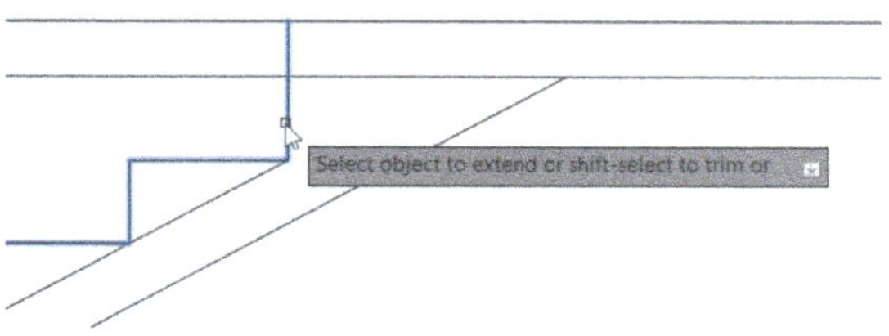

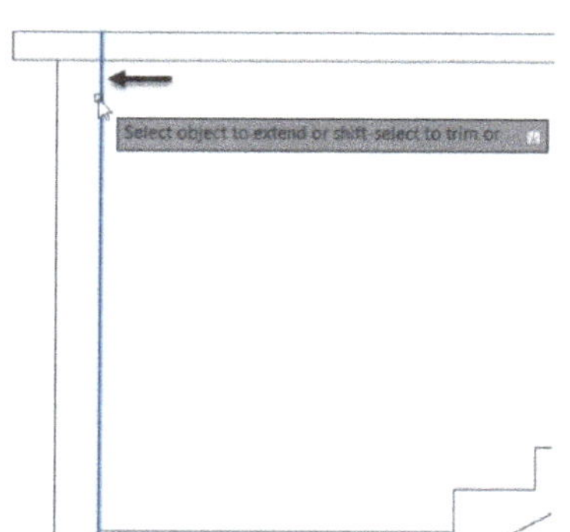

- Select the inclined line, as shown.

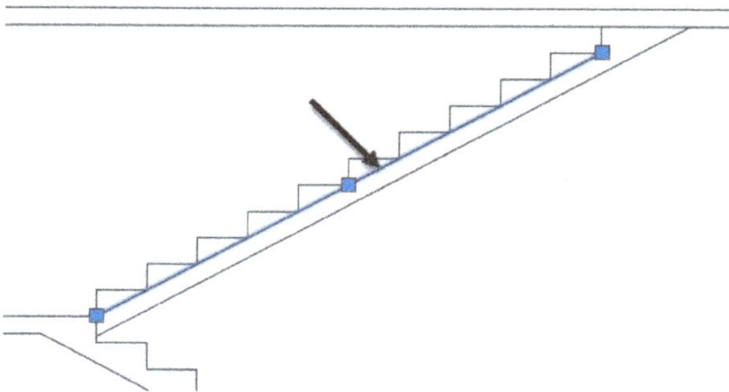

- Press Delete.
- On the **Home** tab of the ribbon, click **Modify** panel > **Trim/Extend** drop-down > **Trim**.
- Select the horizontal lines, as shown.

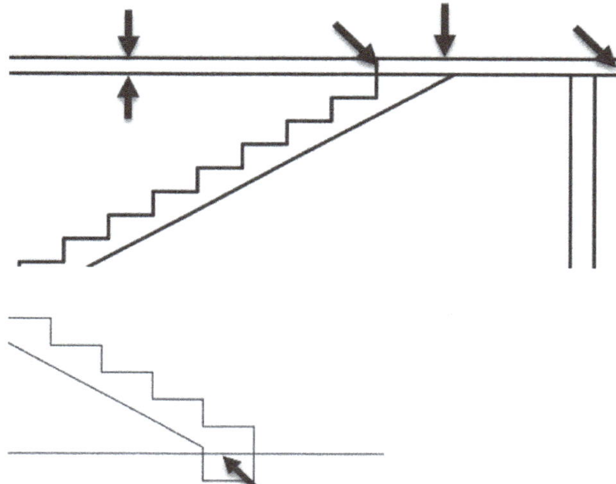

- Offset the horizontal line by 4" in the downward direction.

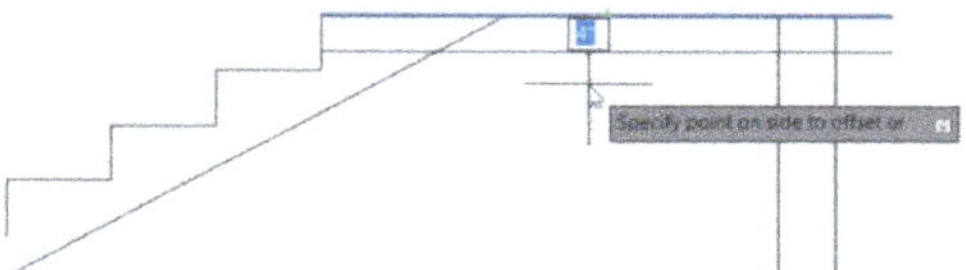

- Trim the elements, as shown.

- Close the end of the offset line.

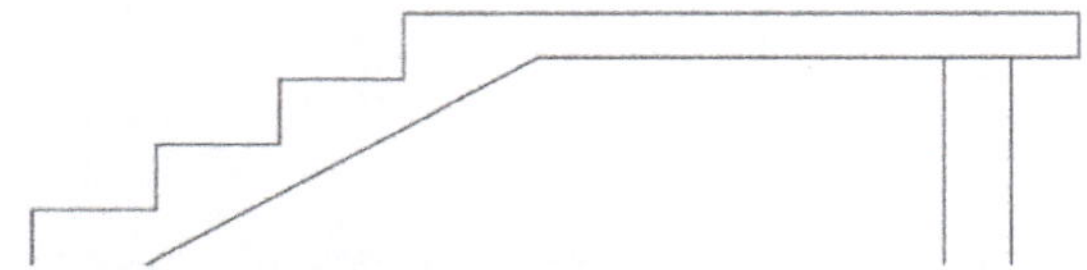

- On the **Home** tab of the ribbon, click **Draw** panel > **Hatch** drop-down > **Hatch**.
- On the **Hatch Creation** tab of the ribbon, expand the **Pattern** gallery, and select the **AR-CONC** pattern.

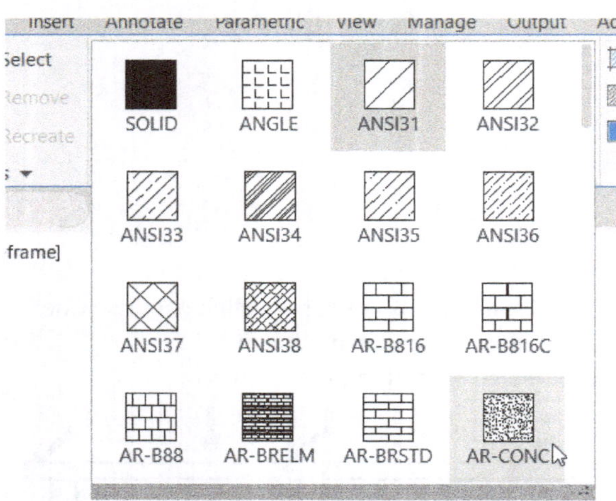

- Pick points in the areas, as shown.

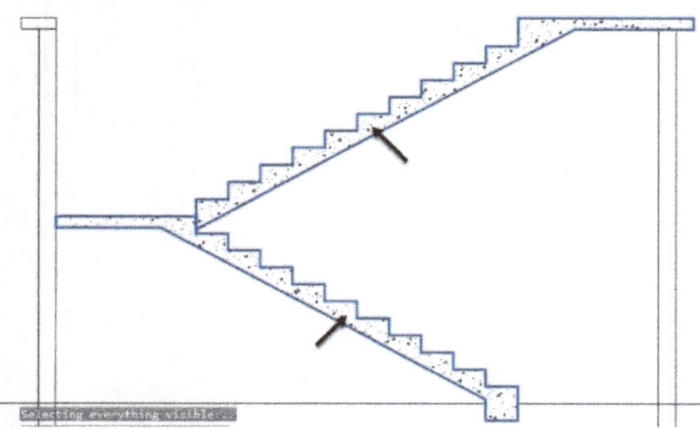

- Click the **Close Hatch Creation** icon on the ribbon.

Creating the Handrail

- On the **Home** tab of the ribbon, click **Draw** panel > **Line**.
- Create two vertical lines of 3' length each at the locations, as shown.

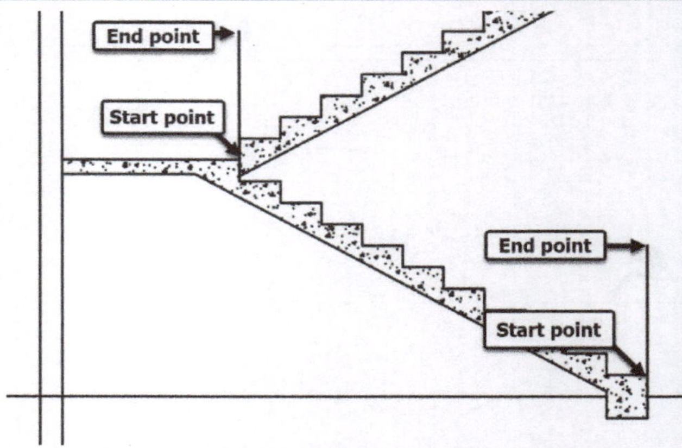

- Activate the **Line** tool and connect the endpoints of the two vertical lines, as shown.

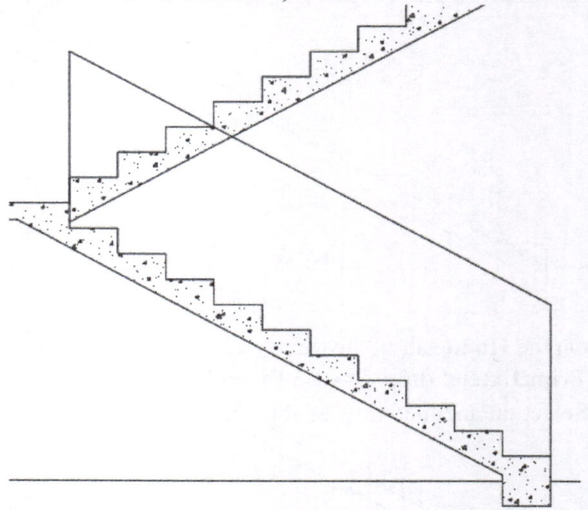

- On the **Home** tab of the ribbon, expand the **Modify** panel and click the **Lengthen** icon.

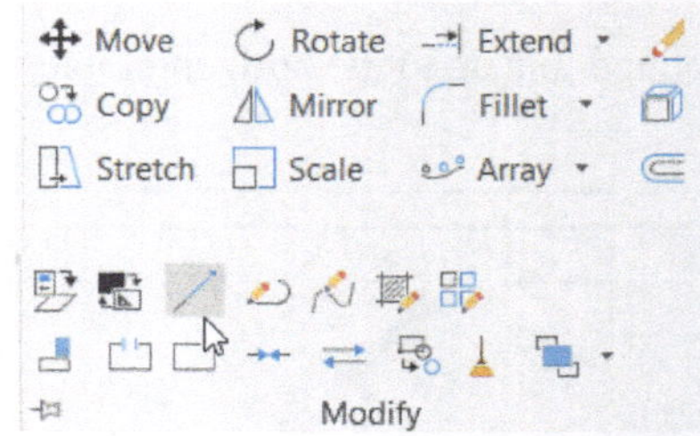

- Select **DElta** from the command line.
- Type 12" and press ENTER.
- Click near the endpoint of the inclined line.

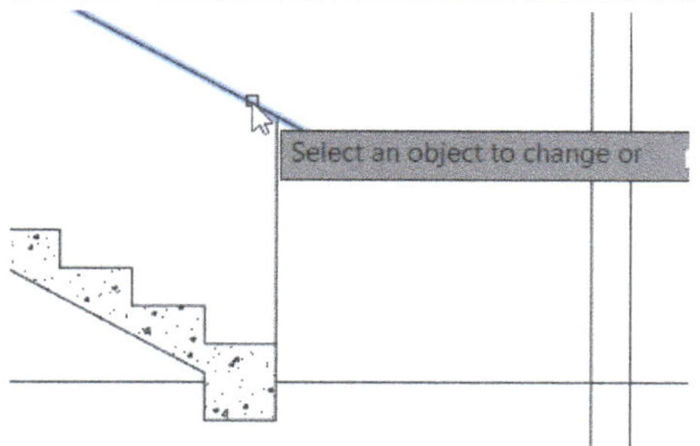

The inclined line is lengthened.

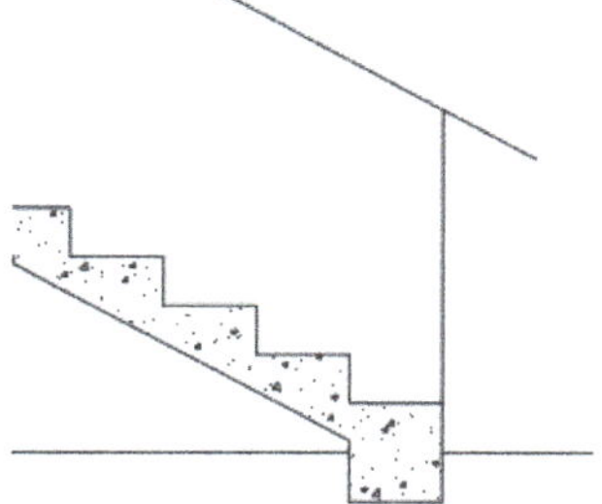

- Create a horizontal line, as shown.

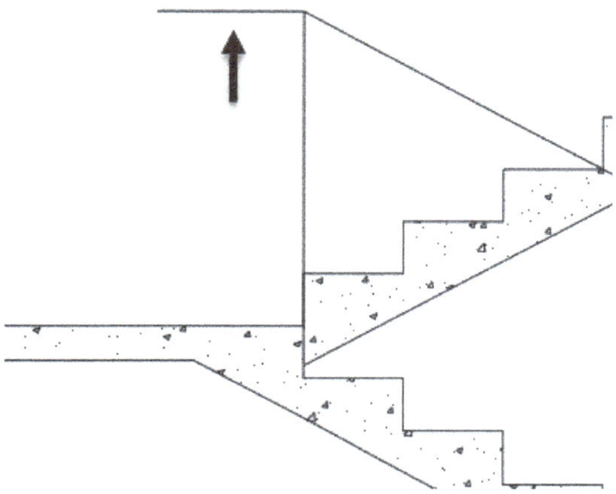

- Click the **Offset** icon on the **Modify** panel.
- Offset the newly created lines by 2" distance, as shown.

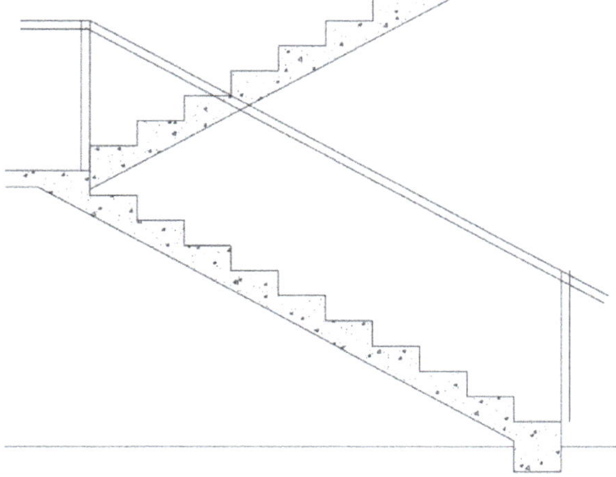

- On the **Home** tab of the ribbon, click **Modify** panel > **Trim/Extend** drop-down > **Extend**.
- Extend the two vertical lines up to the bottom horizontal edge, as shown.

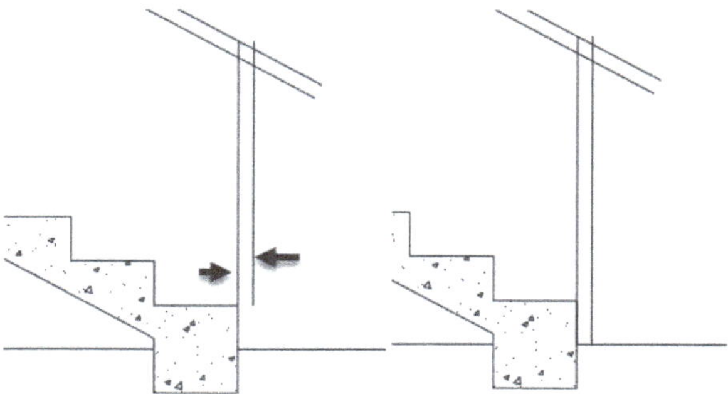

- Activate the **Line** tool.
- Zoom to the fifth step from the top.
- Press and hold the Shift key and right click.
- Select the **Mid between 2 Points** option.
- Select the two endpoints of the step; the start point of the line is specified at the midpoint.

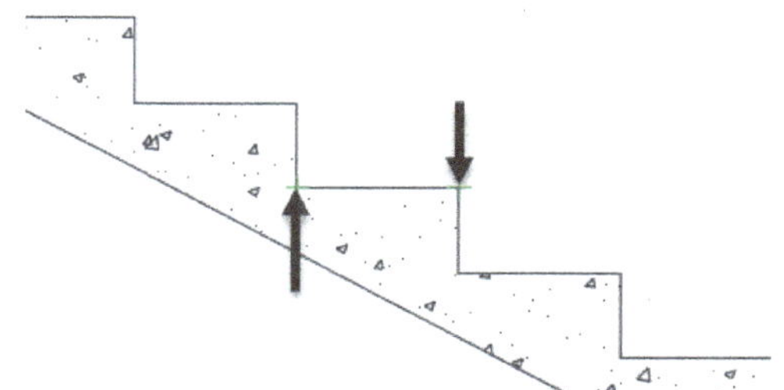

- Move the pointer upward and click.

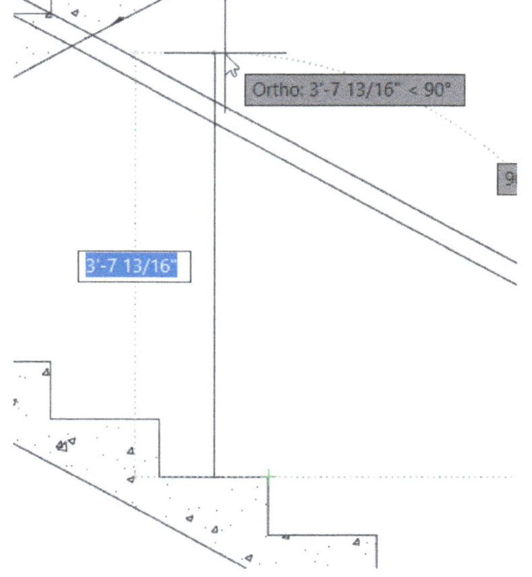

- Click the **Offset** icon on the **Modify** panel.
- Offset the newly created line by 1" on both sides, as shown.
- Delete the middle line.

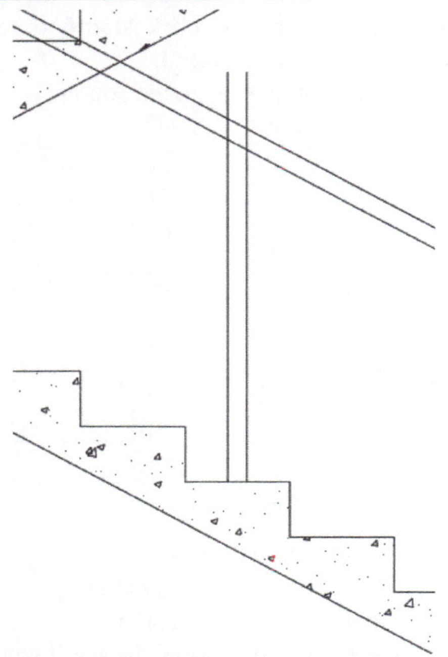

- On the **Home** tab of the ribbon, click **Modify** panel > **Trim/Extend** drop-down > **Trim**.
- Select the **cuTting edges** option from the command line.
- Select the lower inclined and horizontal lines of the handrail.
- Press ENTER.

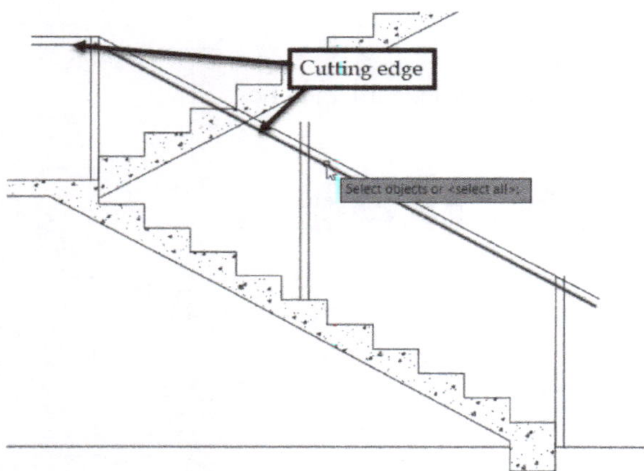

- Trim the portions of the vertical lines, as shown.

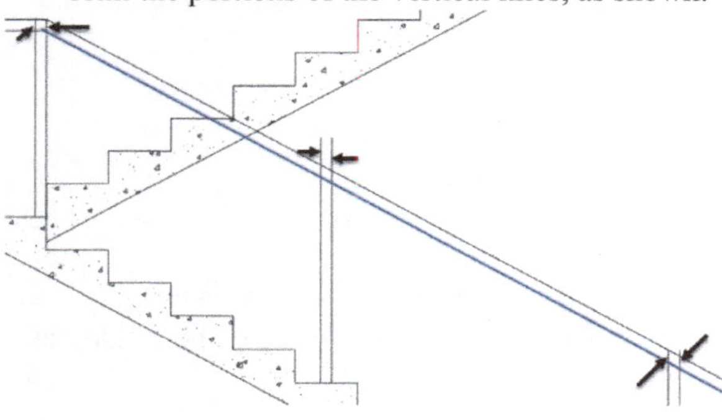

- On the **Home** tab of the ribbon, click **Modify** panel > **Trim/Extend** drop-down > **Trim**.
- Select the unwanted portions of the lines, as shown.

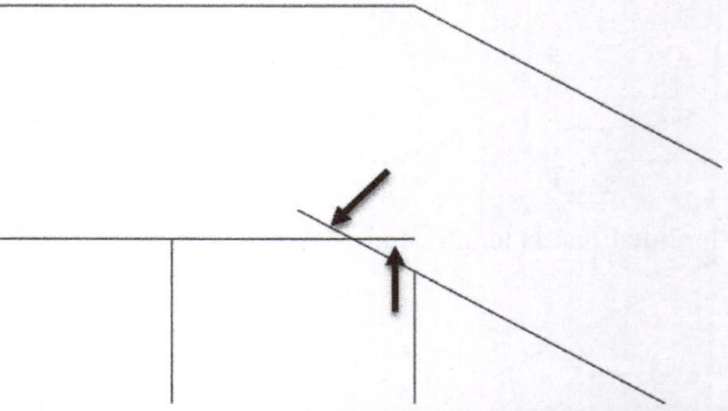

- On the **Home** tab of the ribbon, expand the **Modify** panel and click the **Break at Point** icon.

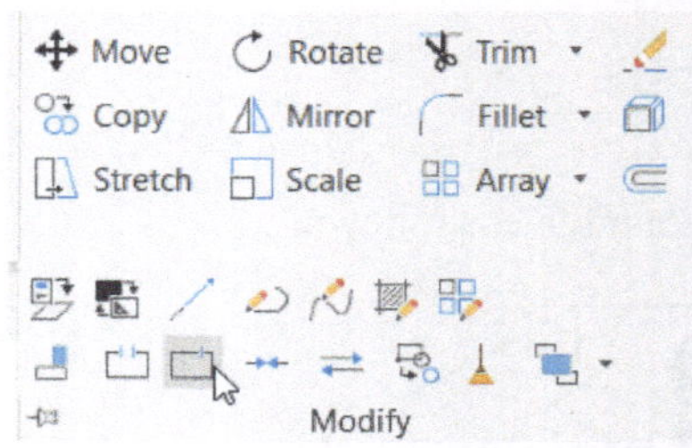

- Select the vertical line, as shown.
- Press and hold the Shift key and right click.
- Select the **Midpoint** option.
- Select the midpoint of the vertical line, as shown. The vertical line is broken at the midpoint.

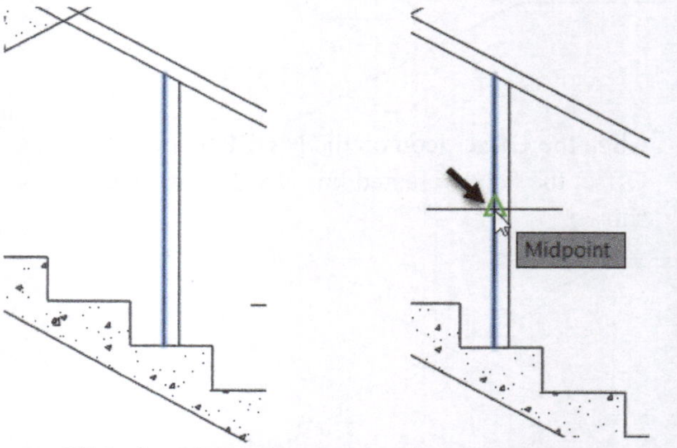

- Click the **Object Snap** drop-down on the status bar and select the **Midpoint** option.
- On the **Home** tab of the ribbon, click **Draw** panel > **Rectangle/Polygon** drop-down > **Rectangle**.
- Select the midpoint of the lower portion of the broken line, as shown.

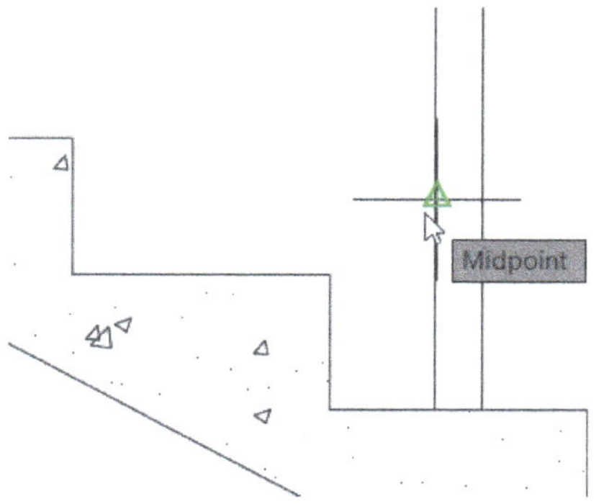

- Select the **Dimensions** option from the command line.
- Type 2 and press ENTER.
- Type 2 and press ENTER.
- Move the pointer upward and click.

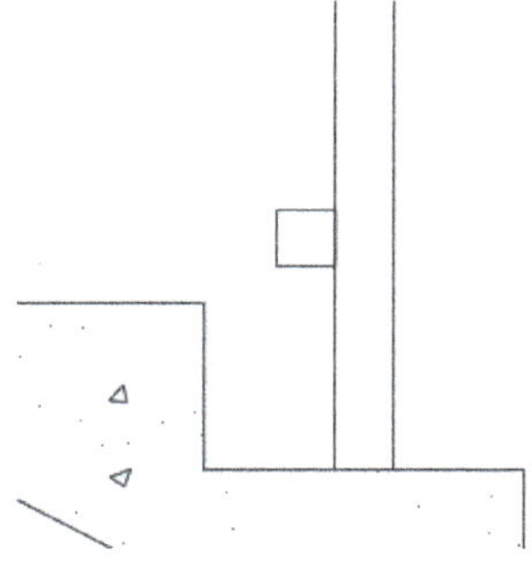

- Select the rectangle and click the **Copy** icon on the **Modify** panel.
- Select the top right corner point of the rectangle.
- Move the pointer upward.
- Select the midpoint of the upper portion of the broken line, as shown.

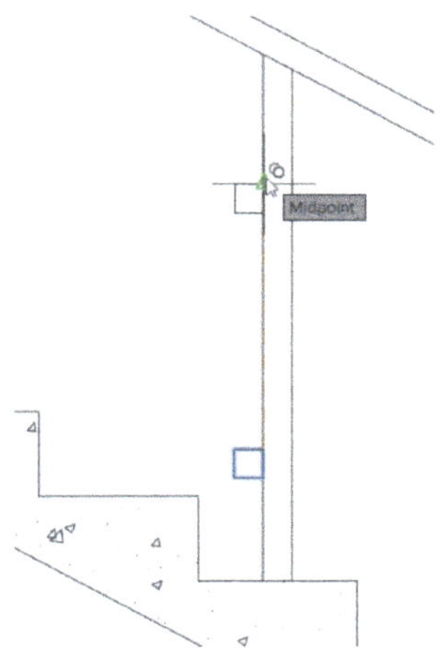

- Select the two rectangles and click the **Copy** icon on the **Modify** panel.
- Select the top right corner point of anyone of the rectangles.

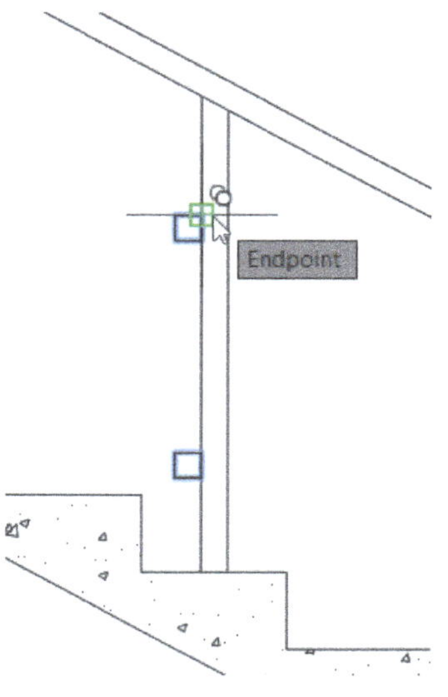

- Move the pointer toward the right.
- Type 4" and press ENTER.

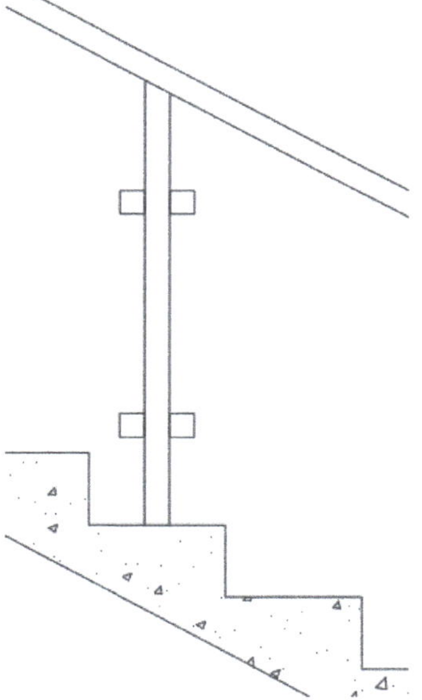

- Press Esc.
- Select the four rectangles and click the **Copy** icon on the **Modify** panel.
- Select the midpoint of the right vertical line.

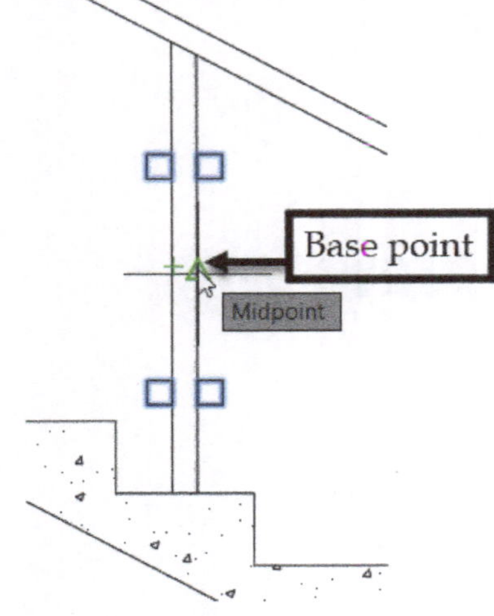

- Move the pointer toward right and select the midpoint of the vertical line, as shown.

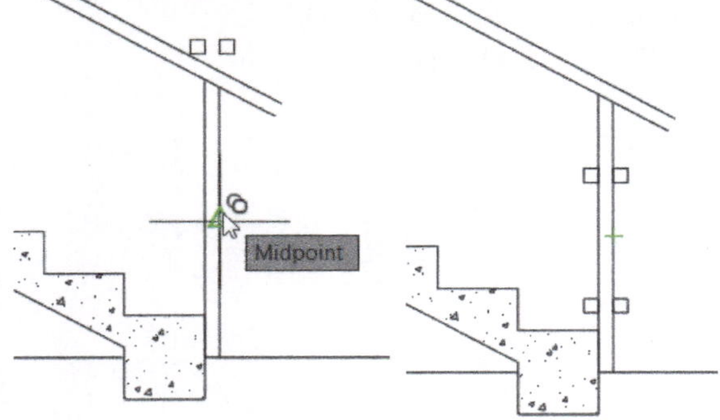

- Move the pointer toward the left and select the midpoint of the vertical line, as shown.

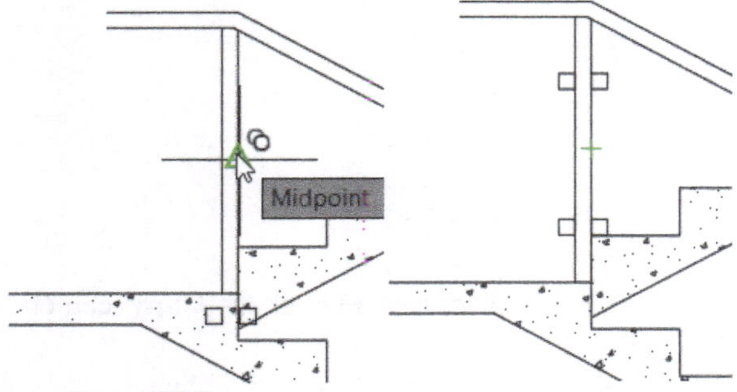

- Press ESC.
- Delete the two rectangles, as shown.

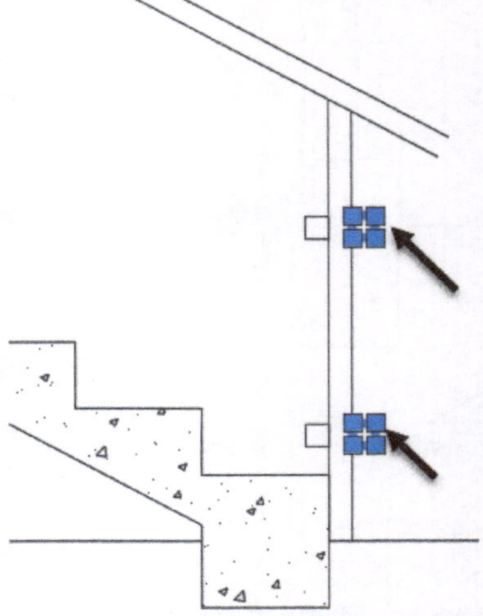

- Select the vertical line and click the **Offset** icon on the **Modify** panel.
- Type 3'8" and press ENTER.
- Move the pointer toward the left and click.

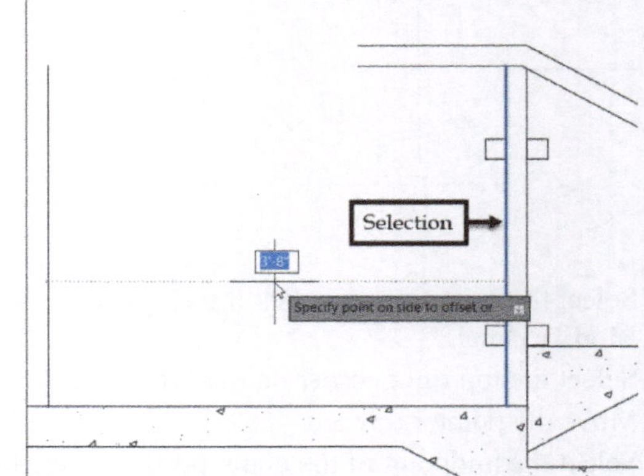

- Press ENTER twice.
- Type 2" and press ENTER.
- Select the newly offset line.
- Move the pointer toward right and click.

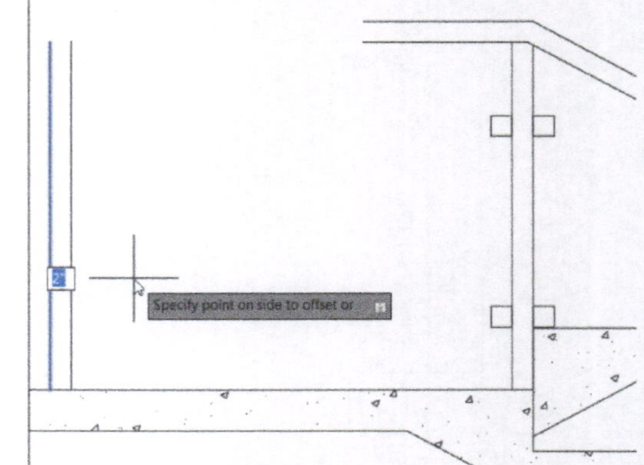

- Click the **Fillet** icon on the **Modify** panel.
- Select the left vertical and top horizontal line, as shown.

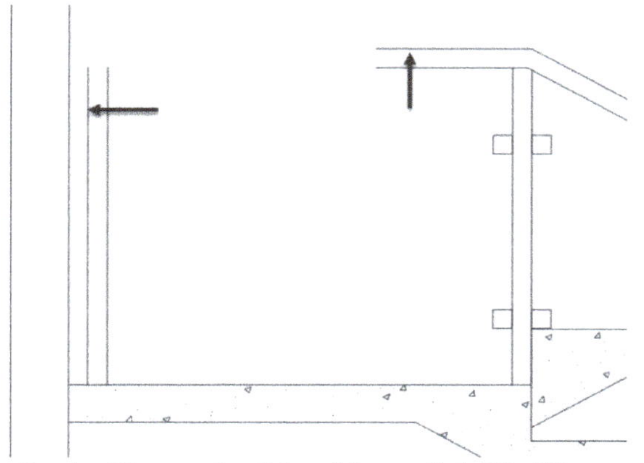

- On the **Home** tab of the ribbon, click **Modify** panel > **Trim/Extend** drop-down > **Extend**.
- Extend the horizontal line up to the vertical line.

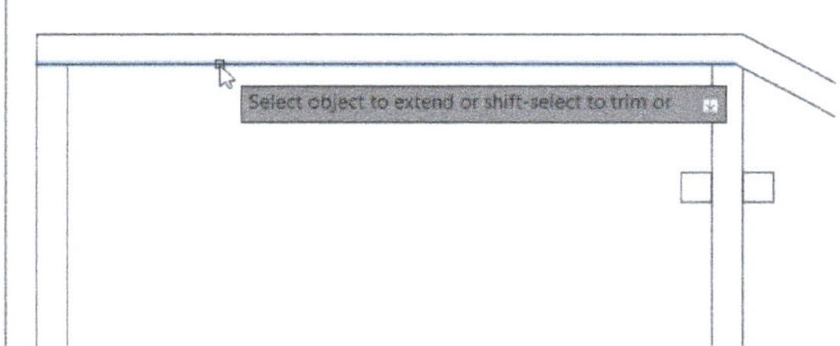

- Press Esc.
- Select the two rectangles and click the **Copy** icon on the **Modify** panel.
- Specify the base and destination points, as shown.

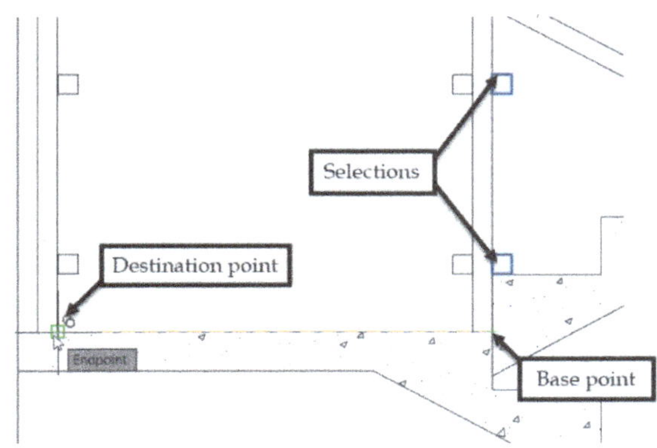

- Click the **Line** icon on the **Draw** panel.
- Select the corner points of the top and bottom steps, as shown.
- Press Esc.

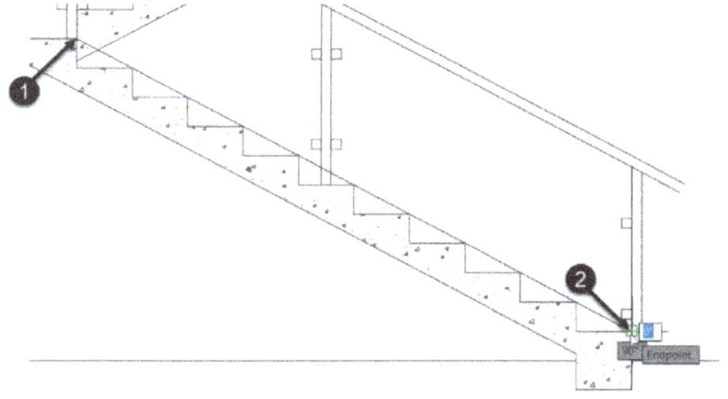

- Select the newly created line and click the **Offset** icon on the **Modify** panel.
- Type 1" and press ENTER.
- Move the pointer upward and click.

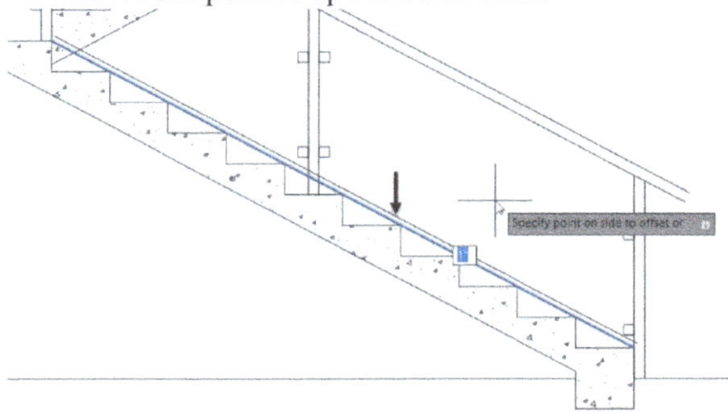

- Press ENTER twice.
- Type 2" and press ENTER.
- Select the newly offset line.
- Move the pointer upward and click.

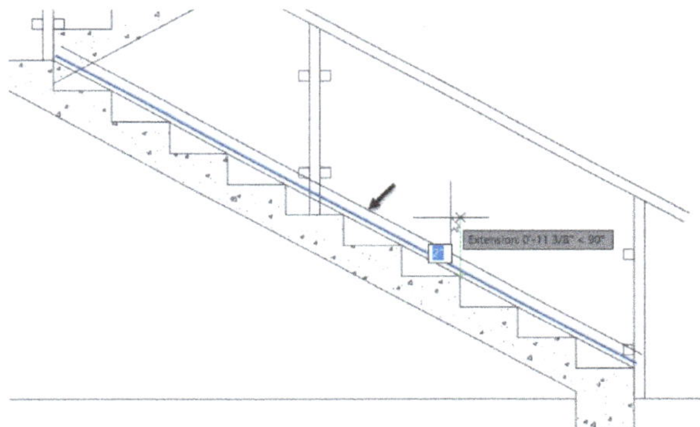

- On the **Home** tab of the ribbon, click **Modify** panel > **Trim/Extend** drop-down > **Trim**.
- Trim the intersecting portions of the lines, as shown.

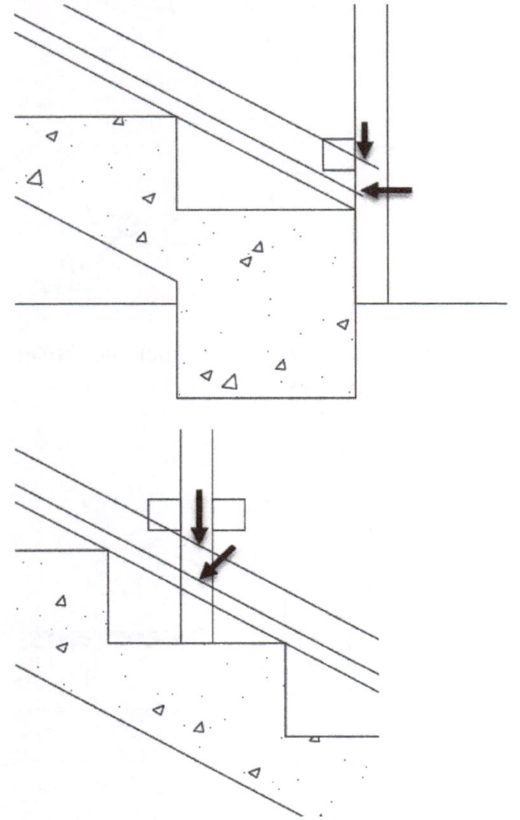

- Press Esc.
- Select the rectangles and click the **Move** icon on the **Modify** panel.
- Specify the base point, as shown.

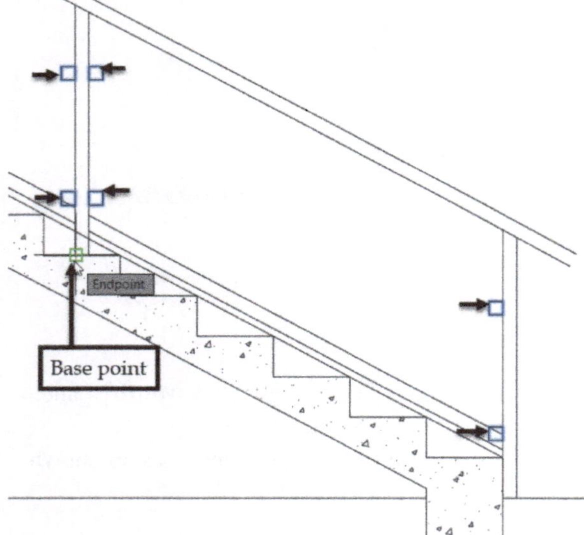

- Move the pointer upward.
- Type 4" and press ENTER.

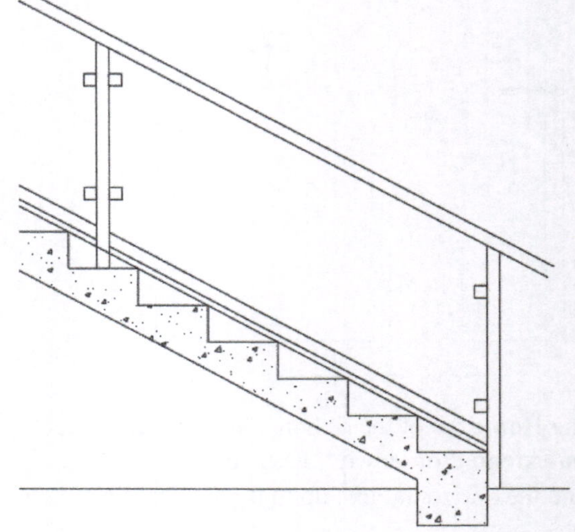

- Select the horizontal line, as shown.
- Click the **Offset** icon on the **Modify** panel.
- Type 33" and press ENTER.
- Move the pointer downward and click.

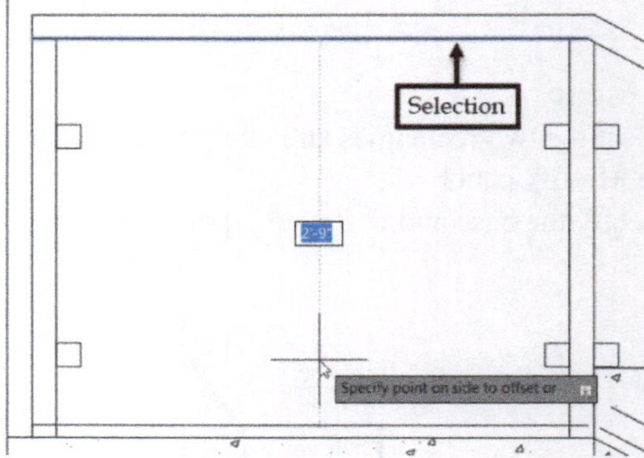

- Press ENTER twice.
- Type 2" and press ENTER.
- Select the newly offset line.
- Move the pointer upward and click.

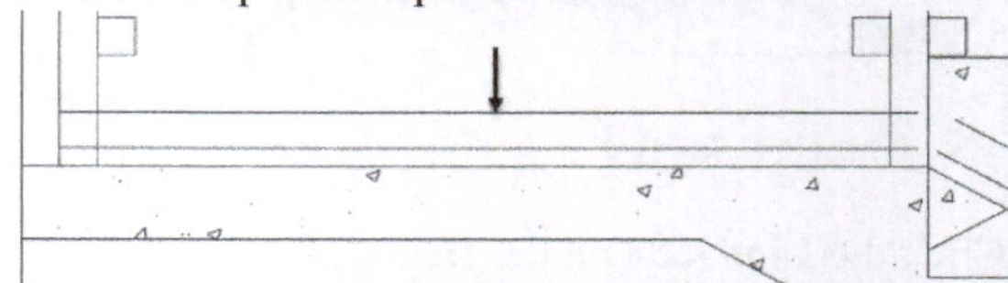

- Click the **Fillet** icon on the **Modify** panel.
- Select the horizontal and inclined lines, as shown.
- Press ENTER and select the horizontal and inclined lines, as shown.

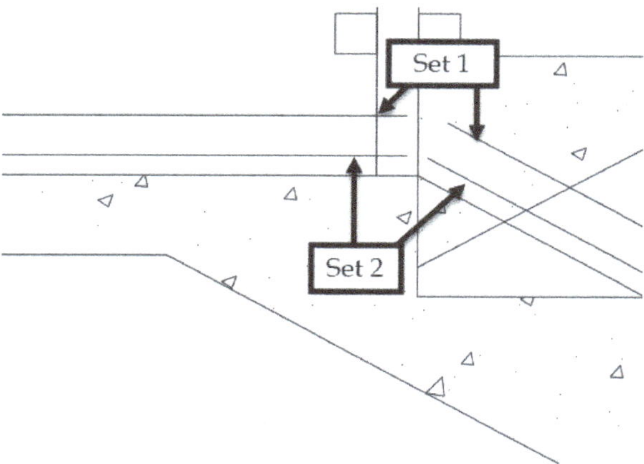

- Click the **Trim** icon on the **Modify** panel.
- Select the portions of the horizontal and inclined line, as shown.

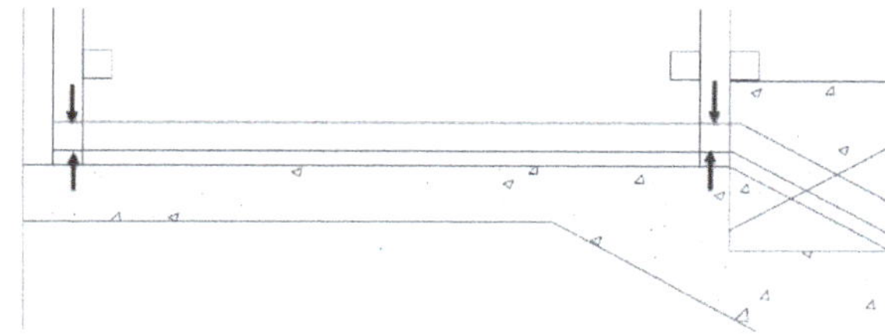

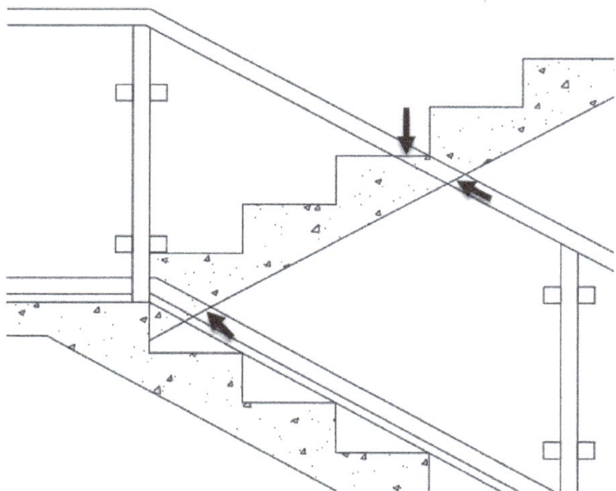

- Select the **eRase** option from the command line.
- Select the inclined line, as shown.
- Press ENTER.

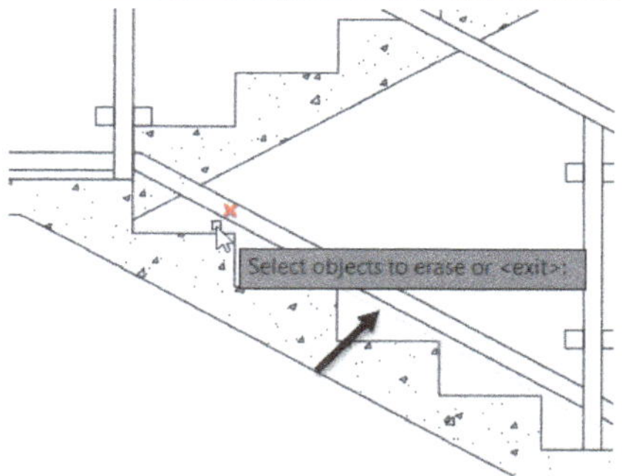

- Close the end of the offset lines using the **Line** tool.

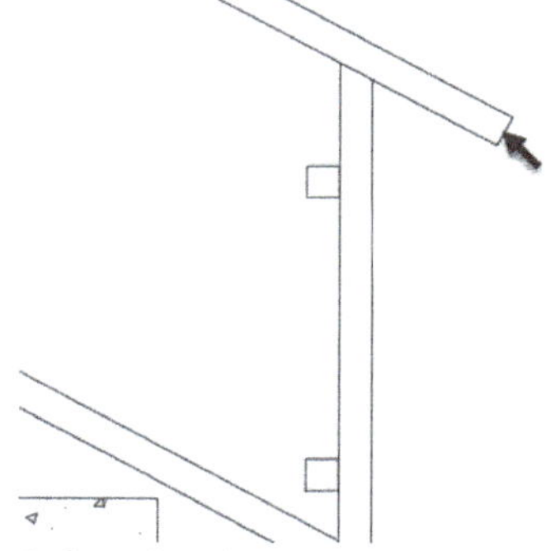

- Select the elements of the handrail.

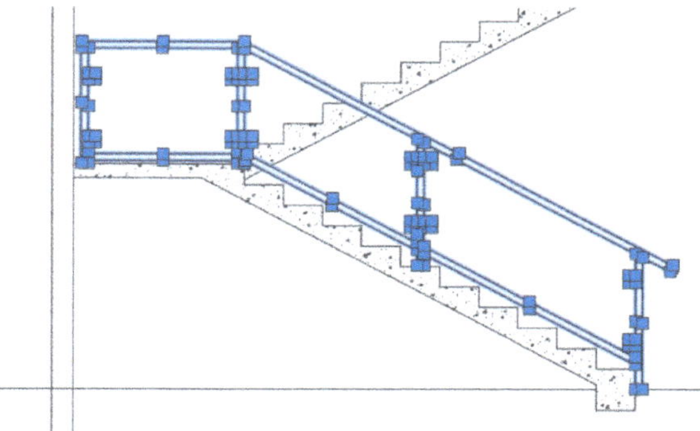

- Click the **Copy** icon on the **Modify** panel.
- Specify the base point, as shown.
- Move the pointer toward the right and click to create the copy.

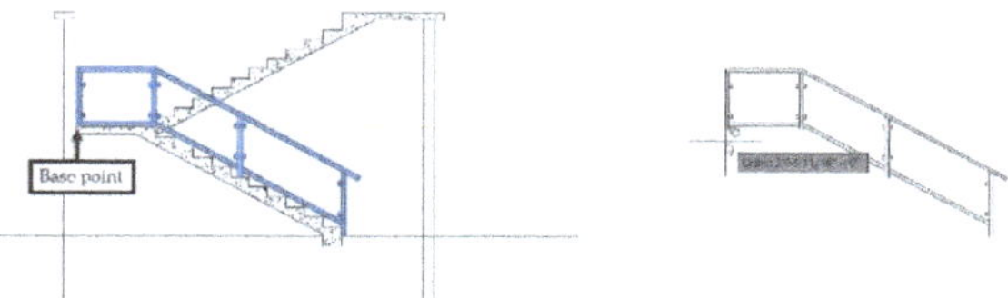

- Click the **Mirror** icon on the **Modify** panel.
- Create a selection window covering all the copied entities of the handrail.

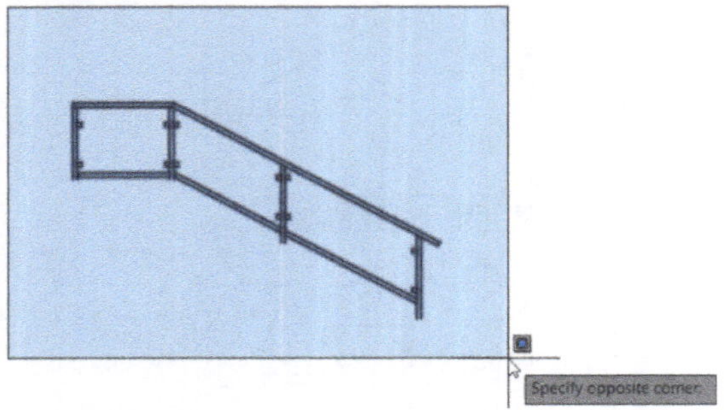

- Press ENTER.
- Select the base point, as shown.
- Move the pointer vertically upward and click.

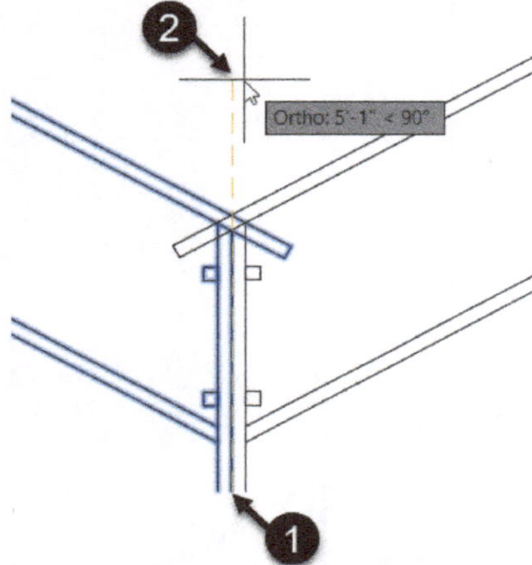

- Select **Yes** from the command line.
- Create a selection window across all the mirrored entities.
- Click the **Move** icon on the **Modify** panel.
- Specify the base point, as shown.

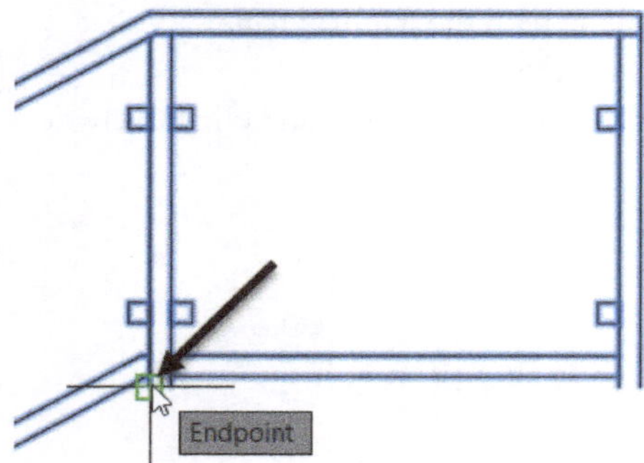

- Move the pointer toward the right and select the destination point.

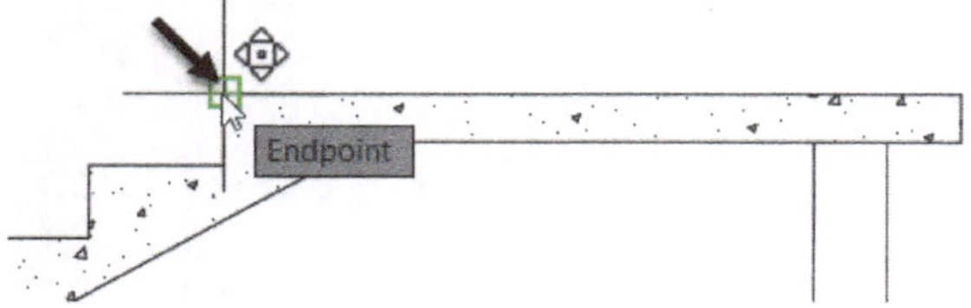

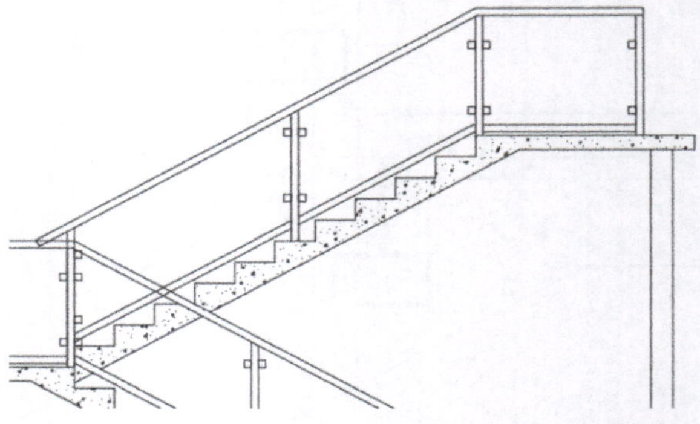

- Select the elements of the handrail, as shown.

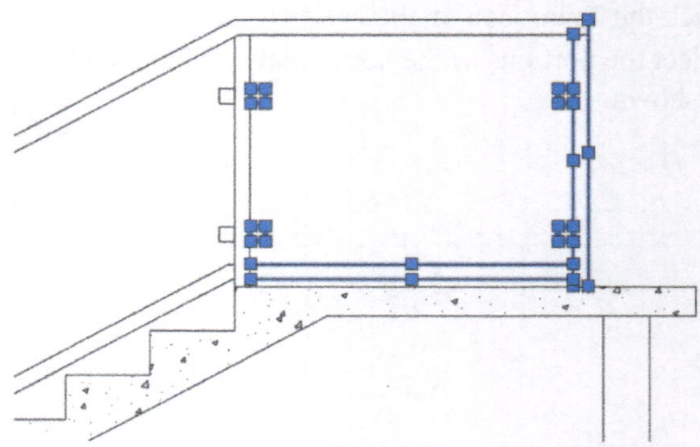

- Press Delete.
- Click the **Line** icon on the **Draw** panel.
- Press and hold the Shift key.
- Right click and select **From** from the **Object Snap** menu.
- Select the corner point of the handrail, as shown.

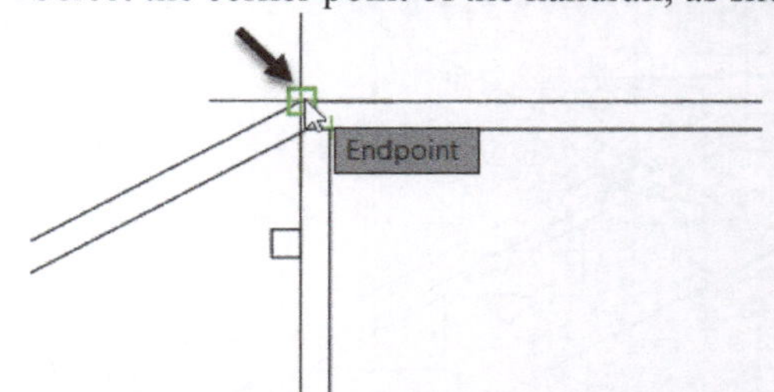

- Move the pointer along the horizontal line, as shown.

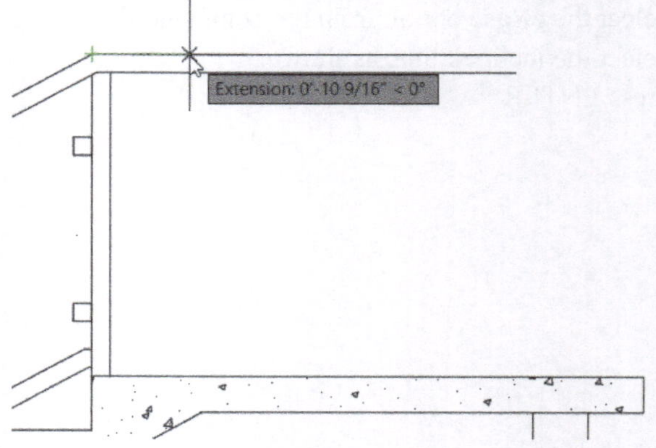

- Type 12 and press ENTER.

- Move the pointer downward and click.
- Click the **Trim** icon on the **Modify** panel.
- Trim the unwanted elements.

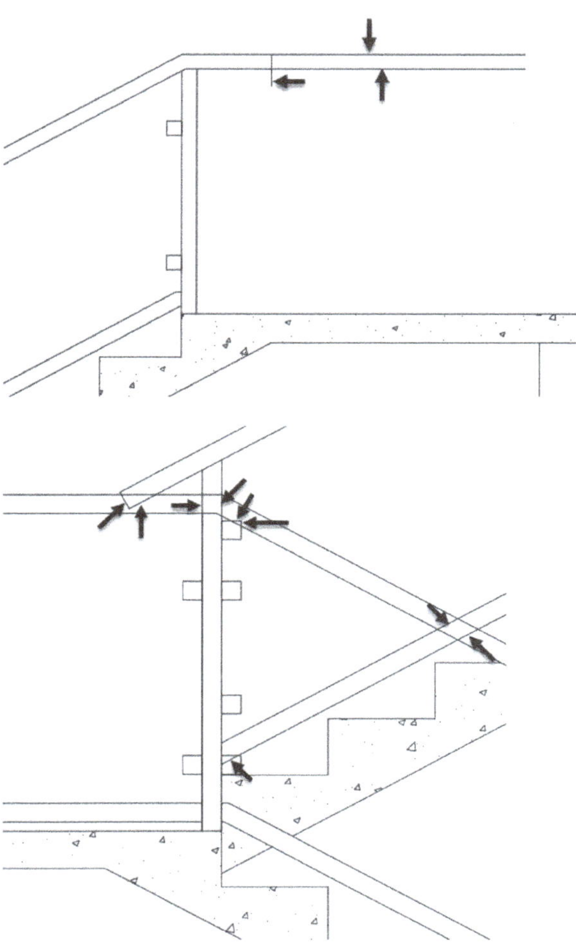

- On the **Home** tab of the ribbon, click **Draw** panel > **Hatch** drop-down > **Hatch**.
- On the **Hatch Creation** tab of the ribbon, expand the **Pattern** gallery, and select the **GOST-GLASS** pattern.

- Pick points in the areas, as shown.

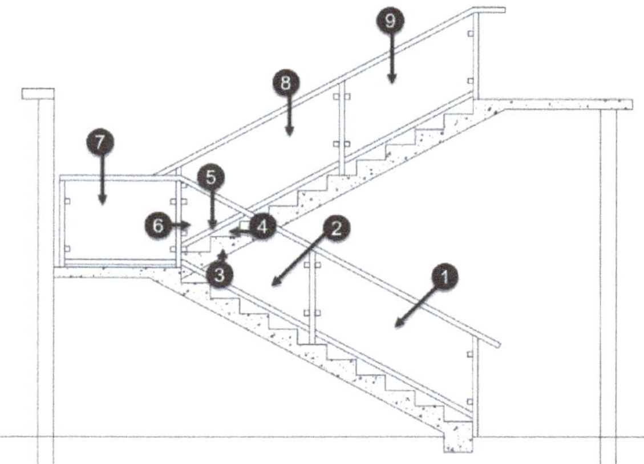

- Right click and select Enter.

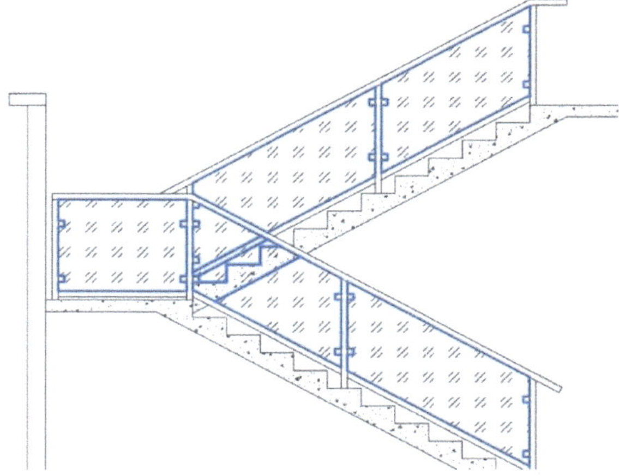

- Delete the construction lines, as shown.

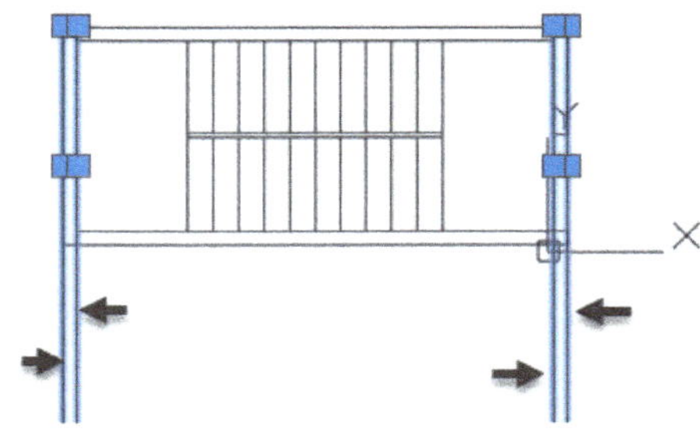

- Save and close the file.

Example 3

In this tutorial, you will create the elevation view using the floor plan.

- Download the Elevation_plan from the companion website and open it.
- Click the **Layer Properties** icon on the **Layers** panel of the **Home** ribbon tab.
- Click the **New Layer** icon on the Layer Properties Manager.
- Type Elevation as the layer name.
- Double-click on the Elevation layer.

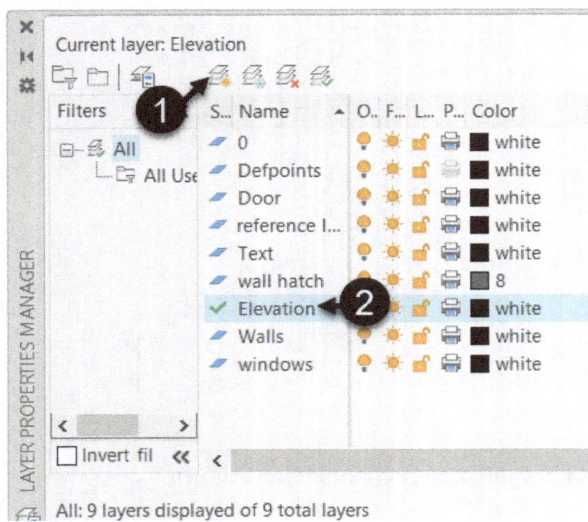

- Close the Layer Properties Manager.
- Draw a horizontal line above the floor plan, as shown.

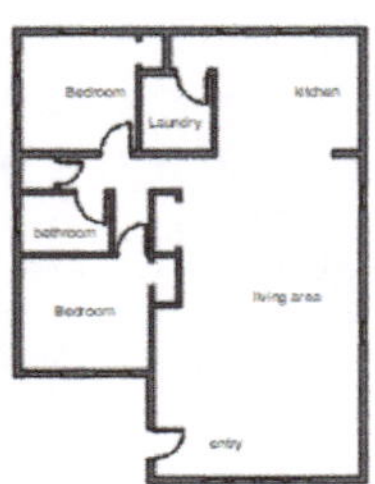

- On the **Home** ribbon tab, expand the **Draw** panel and click the **Ray** tool.

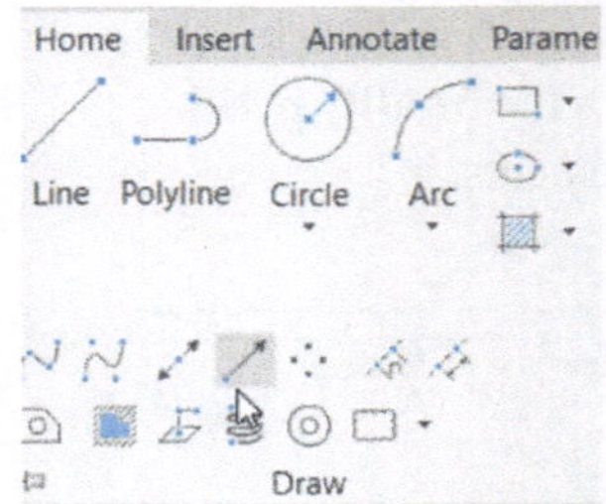

- Select the top-left corner of the floor plan.
- Move the pointer upward and click.

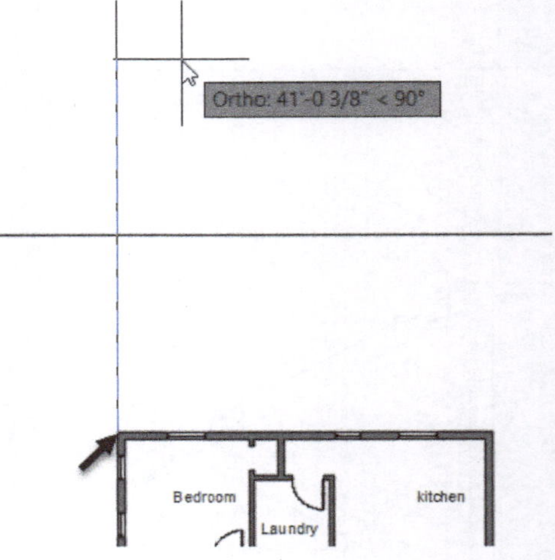

- Press ENTER twice.
- Select the top-right corner of the floor plan.
- Move the pointer upward and click.

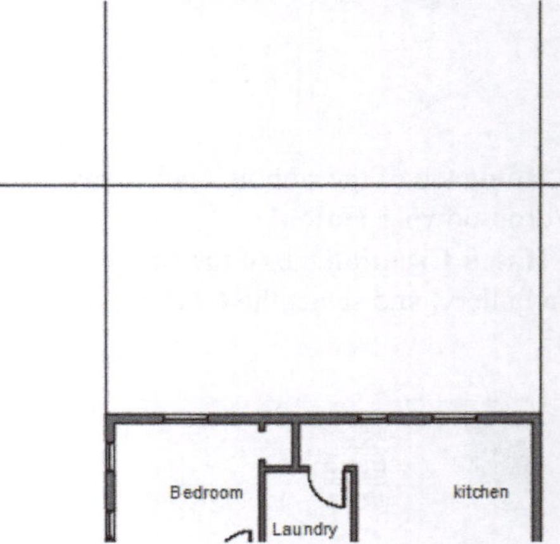

- Click the **Offset** tool on the **Modify** panel of the **Home** ribbon tab.
- Type 6" and press ENTER.
- Select the horizontal line.
- Move the pointer upward and click to create an offset line.

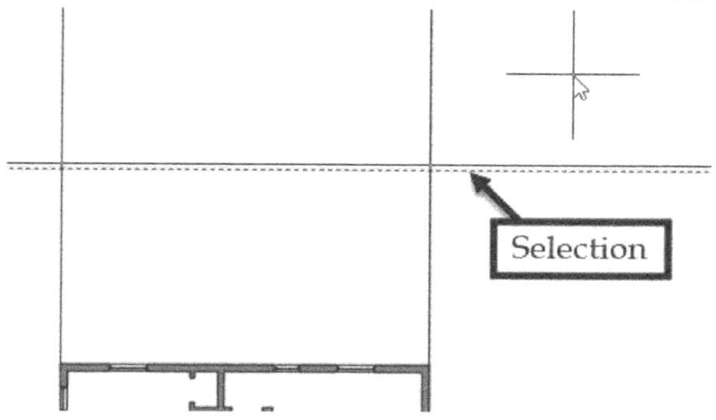

- Press ENTER twice.
- Type 9' as the offset distance. Next, press ENTER.
- Select the offset line created previously.
- Move the pointer upward and click.

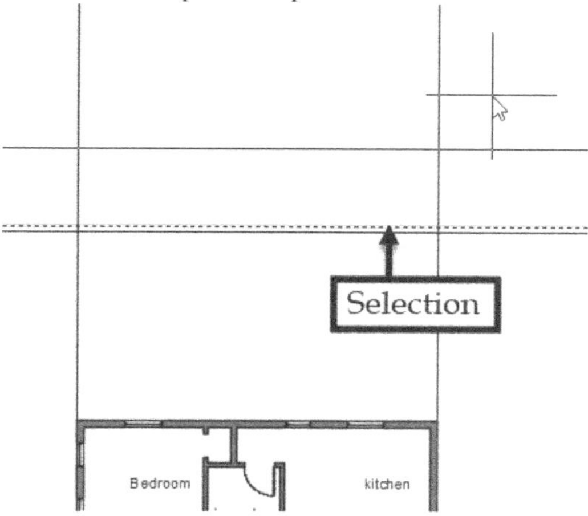

- Press ENTER twice.
- Type 10" as the offset distance. Next, press ENTER.
- Select the offset line created previously.
- Move the pointer upward and click.

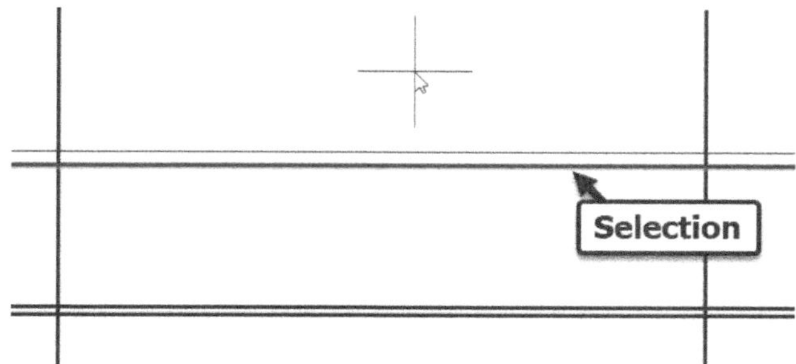

- Press ENTER twice.
- Type 6' as the offset distance. Next, press ENTER.
- Select the offset line created previously.
- Move the pointer upward and click.

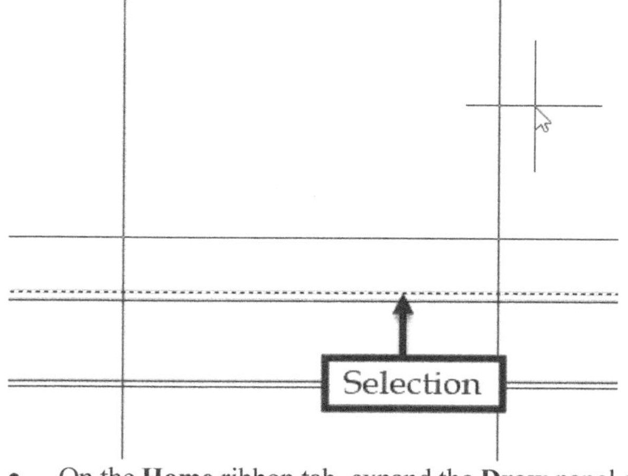

- On the **Home** ribbon tab, expand the **Draw** panel and click the **Construction Line** tool.
- Select the **Offset** option from the command line.
- Type 16" as the offset distance. Next, press ENTER.
- Select the right exterior wall.
- Move the pointer toward right and click.
- Select the left exterior wall.
- Move the pointer toward the left and click.

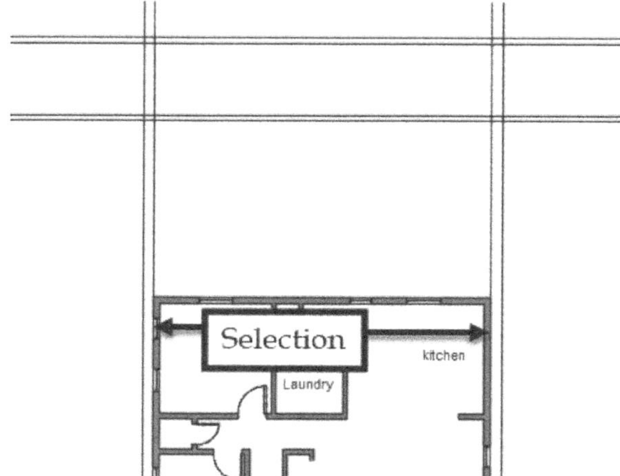

- Press ENTER twice.
- Select the **Ver** option from the command line.
- Select the corner point of the exterior wall, as shown.

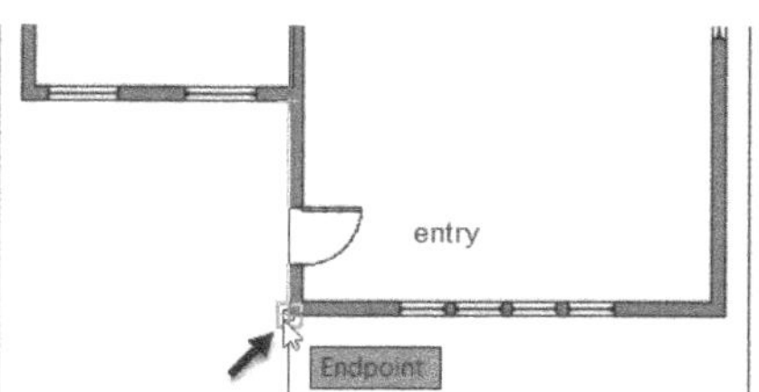

- Press ENTER twice.
- Select the **Offset** option from the command line.
- Type 16" as the offset distance. Next, press ENTER.
- Select the construction line created in the last step.

- Move the pointer toward the left and click.

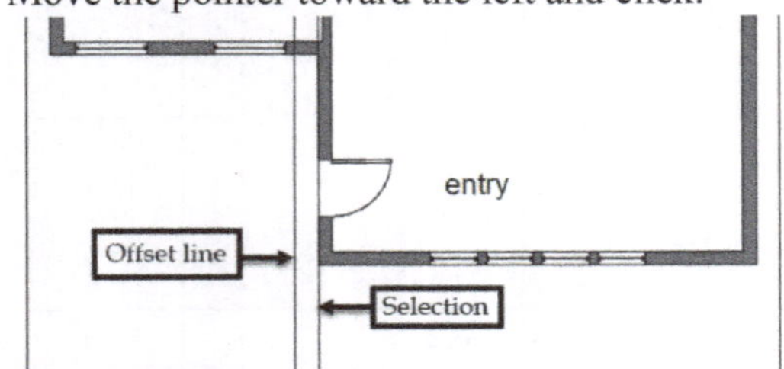

- On the **Home** ribbon tab, expand the **Modify** panel and click the **Break at Point** tool.

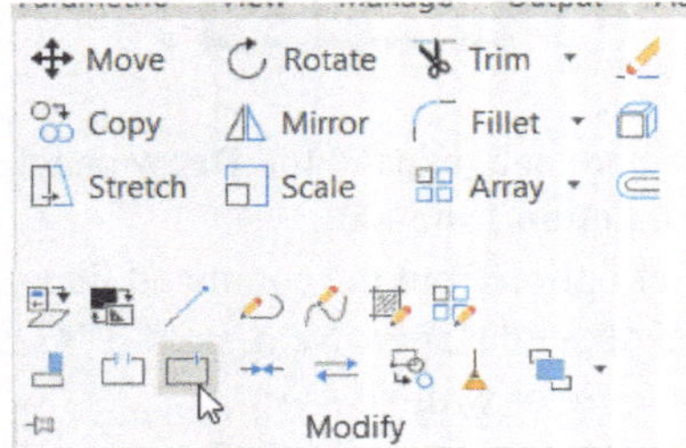

- Select the horizontal line, as shown.

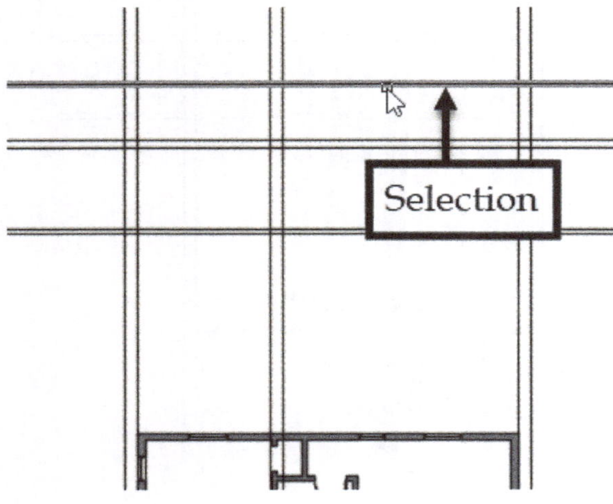

- Select the intersection point between the horizontal and vertical lines, as shown.

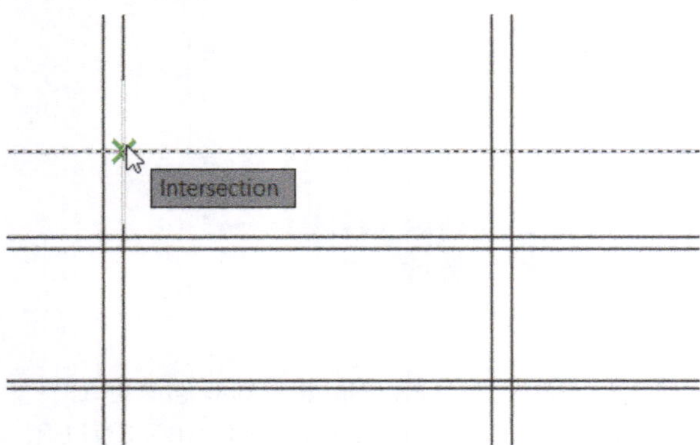

- Press ENTER to activate the **Break at Point** tool.
- Select the right portion of the broken line.
- Select the intersection point between the horizontal and vertical lines, as shown.

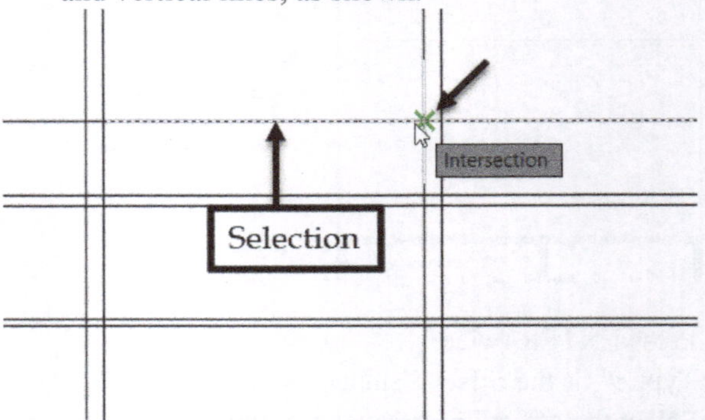

- Click the **Line** tool on the **Draw** panel of the **Home** ribbon tab.
- Select the midpoint of the broken line, as shown.

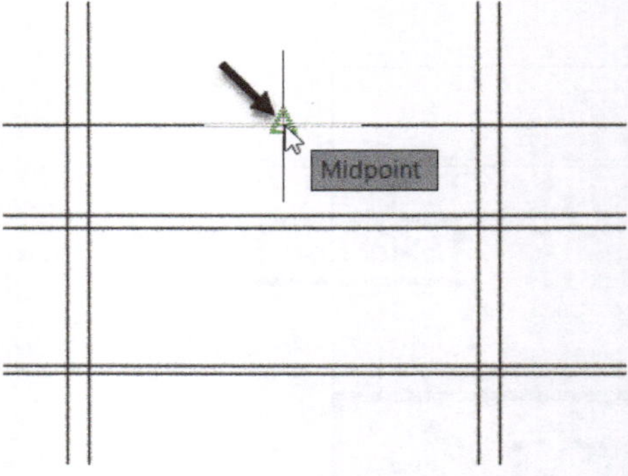

- Select the intersection point between the horizontal and vertical lines, as shown.

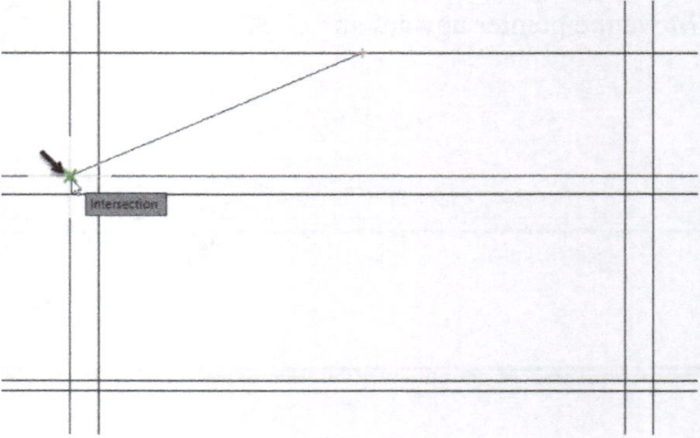

- Press ENTER twice.
- Select the start point of the line created in the last step.
- Select the intersection point between the horizontal and vertical lines, as shown.

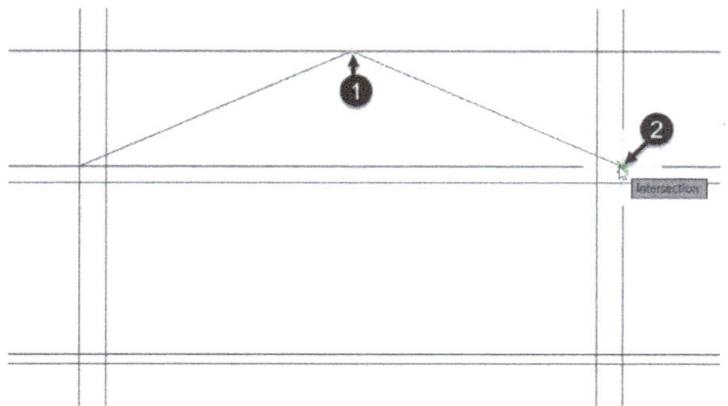

- Click the **Trim** tool on the **Modify** panel of the **Home** ribbon tab.
- Select the portions of the vertical lines, as shown.

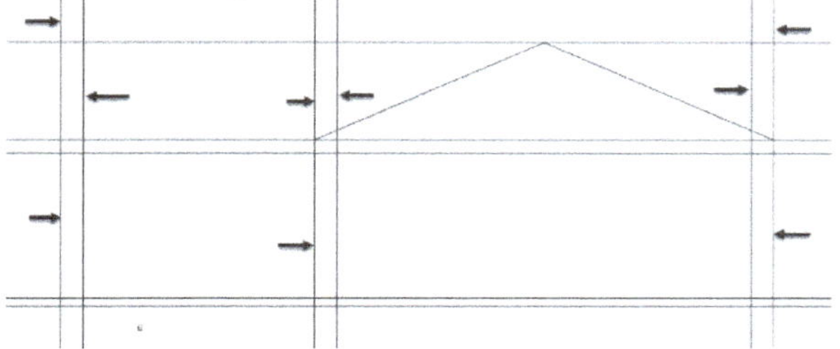

- Click and drag the mouse pointer across the vertical lines, as shown.

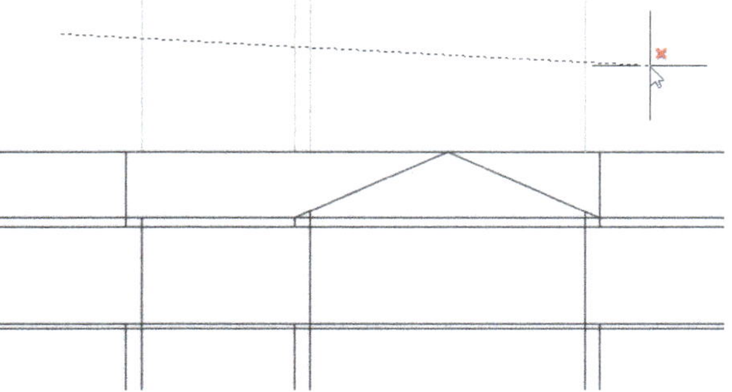

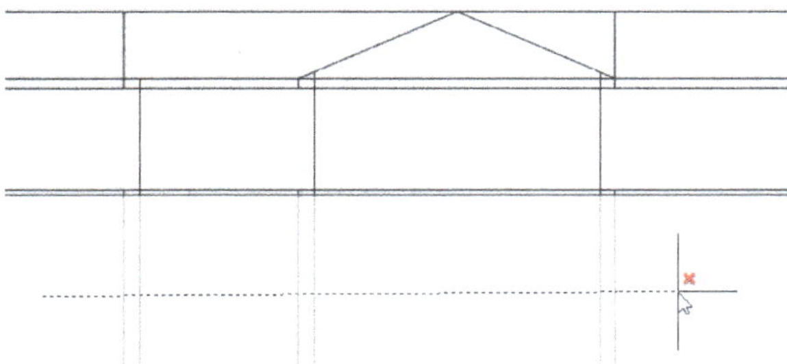

- Trim the horizontal and vertical lines, as shown.

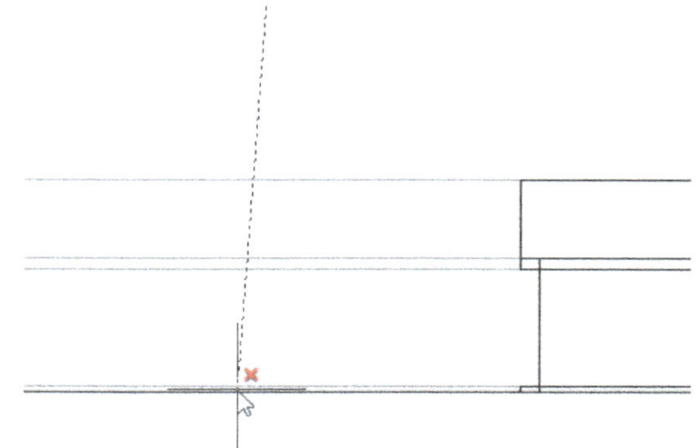

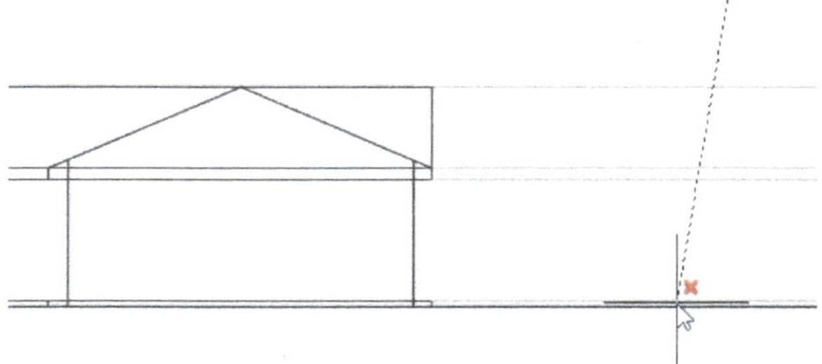

- Trim the small portions, as shown.

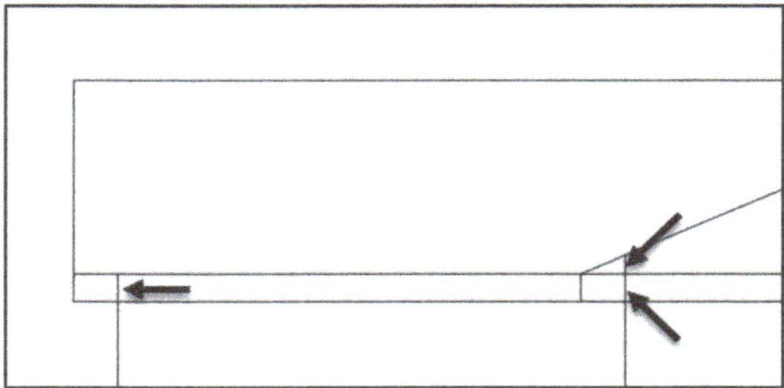

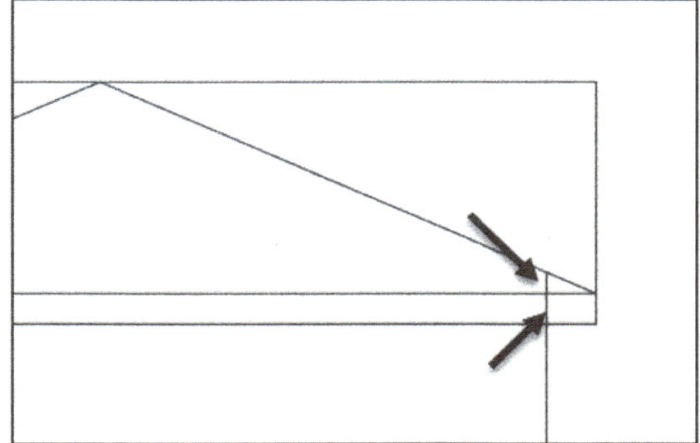

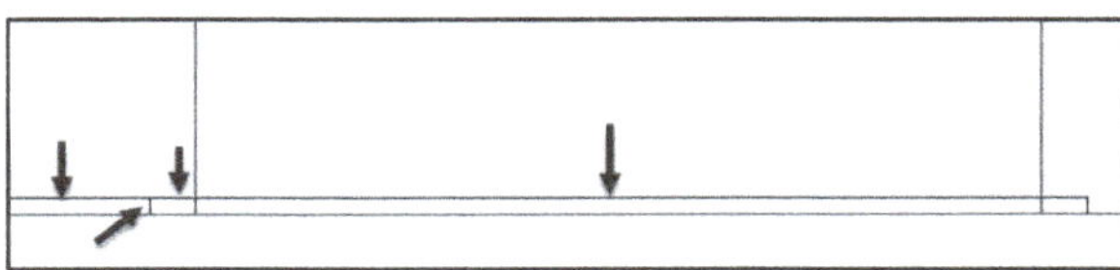

- Press Esc.
- Select the inclined line, as shown.

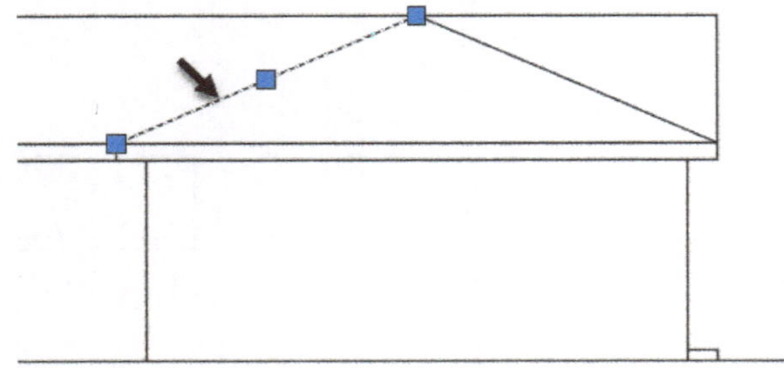

- Click the **Copy** tool on the **Modify** panel of the **Home** ribbon tab.
- Specify the base point, as shown.
- Move the pointer downward and select the intersection point, as shown.

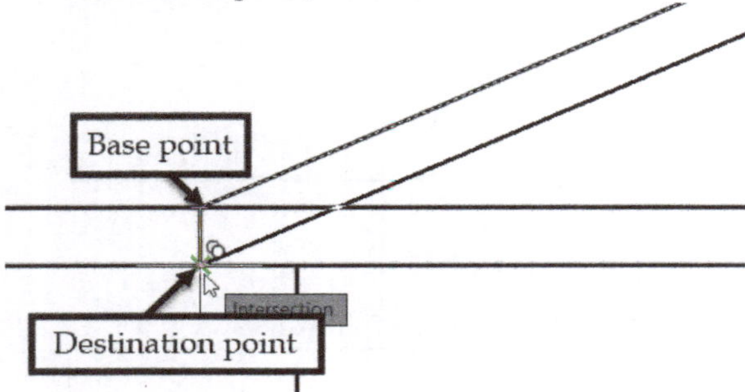

- Press Esc.
- Likewise, copy the other inclined line, as shown.

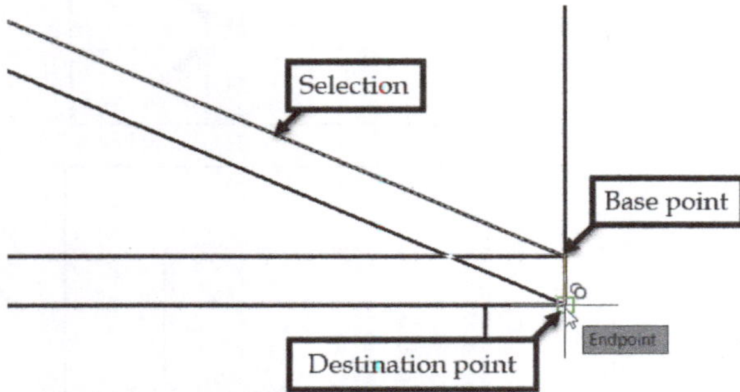

- Click the **Trim** tool on the **Modify** panel of the **Home** ribbon tab.
- Select the portions of the horizontal line, as shown.

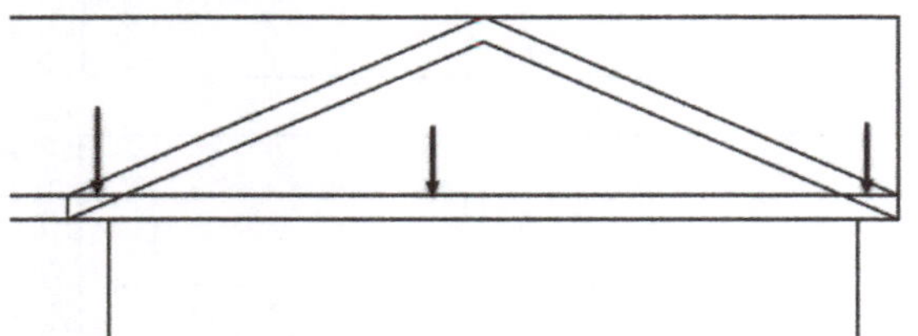

- On the **Home** ribbon tab, expand the **Modify** panel and click the **Break at Point** tool.
- Select the horizontal line, as shown.
- Select the intersection point, as shown.

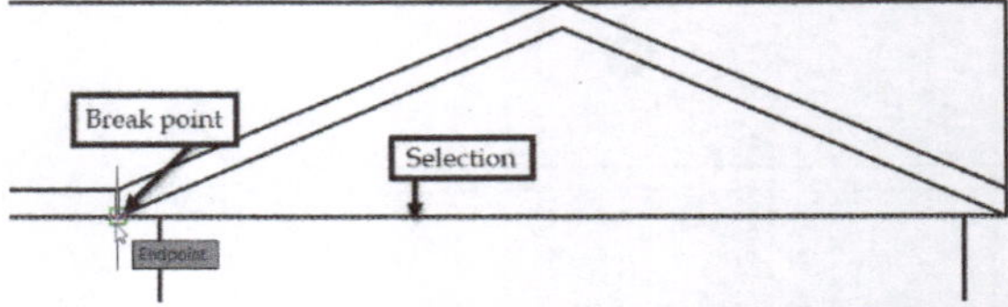

- On the **Home** ribbon tab, expand the **Modify** panel and click the **Break at Point** tool.
- Select the vertical line, as shown.
- Select the intersection point, as shown.

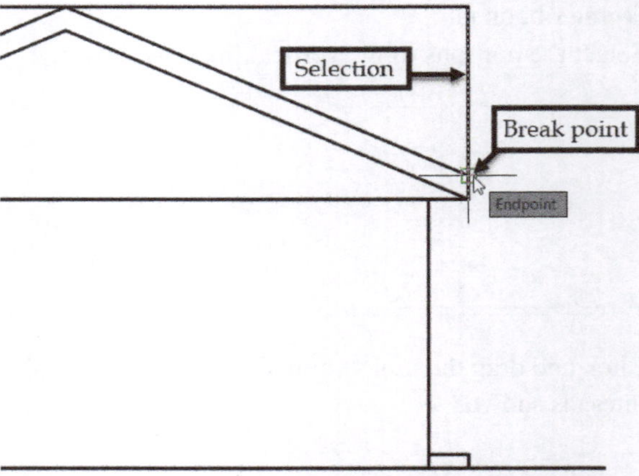

- On the **Home** ribbon tab, expand the **Modify** panel and click the **Break at Point** tool.
- Select the vertical line, as shown.
- Select the intersection point, as shown.

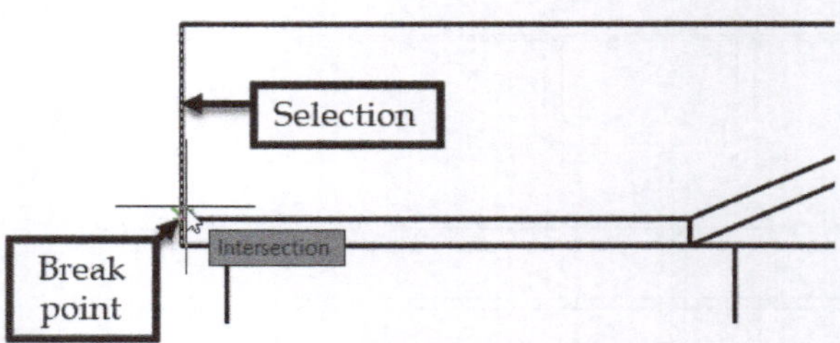

- Select the horizontal line, as shown.
- Press Delete.

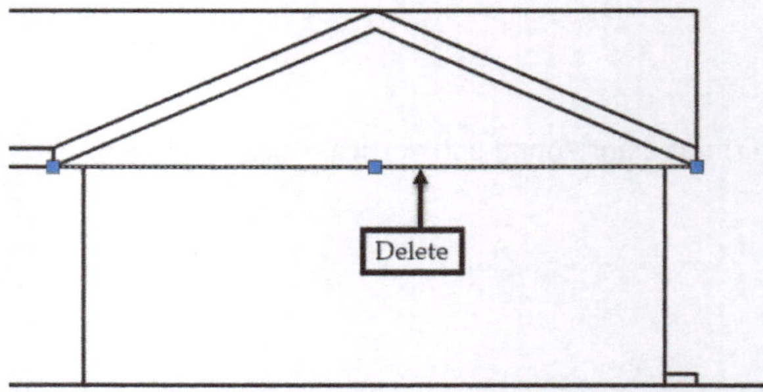

- Click the **Trim** drop-down > **Extend** on the **Modify** panel of the **Home** ribbon tab.
- Select the two vertical lines, as shown.

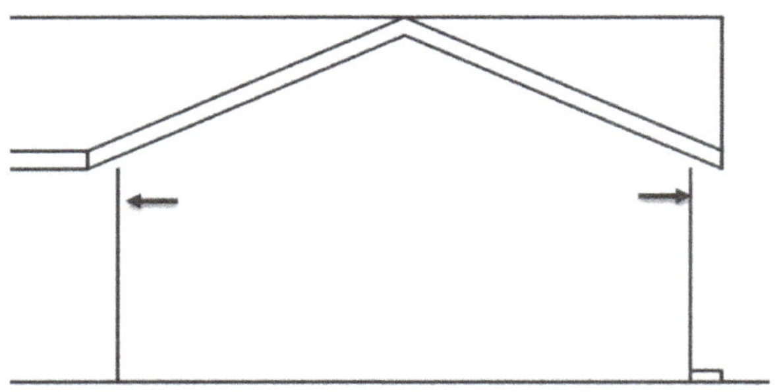

- Create a horizontal line connecting the endpoints of the two vertical lines.

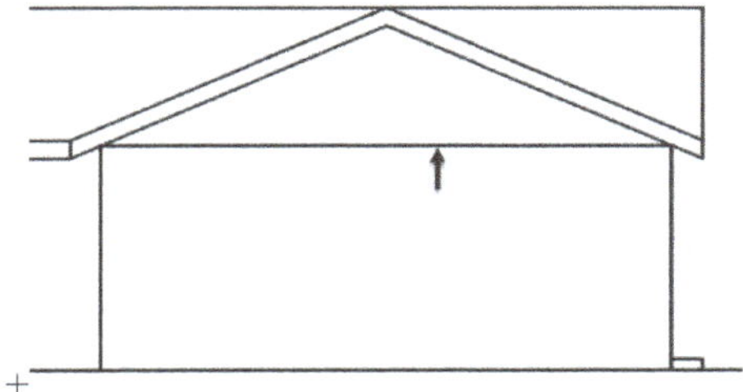

- Select the two construction lines, as shown.
- Press Delete.

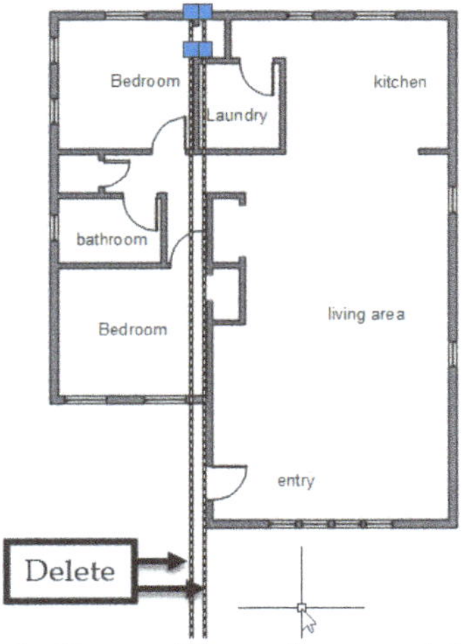

- On the **Home** ribbon tab, expand the **Draw** panel and click the **Construction Line** tool.
- Select the **Ver** option from the command line.
- Zoom to the lower portion of the floor plan and select the vertex points on the window, as shown.

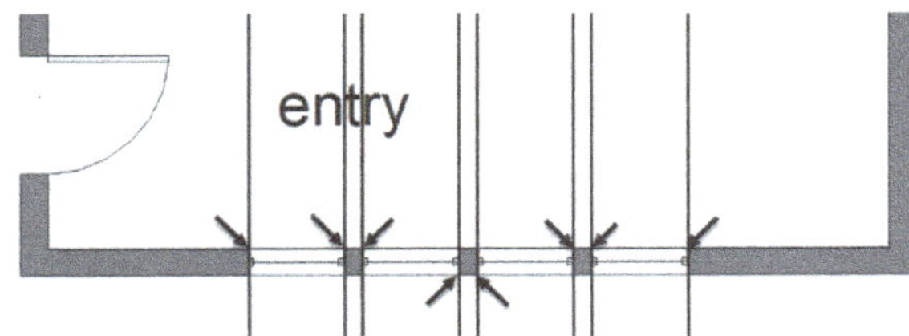

- Click the **Offset** tool on the **Modify** panel of the **Home** ribbon tab.
- Type 3' as the offset distance. Next, press ENTER.
- Select the horizontal line, as shown.
- Move the pointer upward and click.

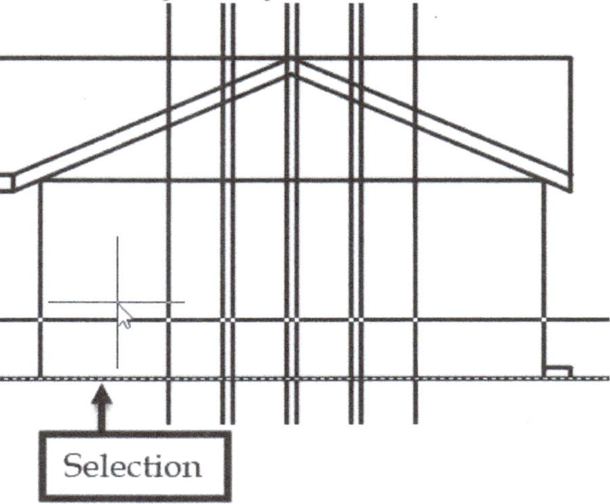

- Click the **Rectangle** tool on the **Draw** panel of the **Home** ribbon tab.
- Select the intersection point, as shown.

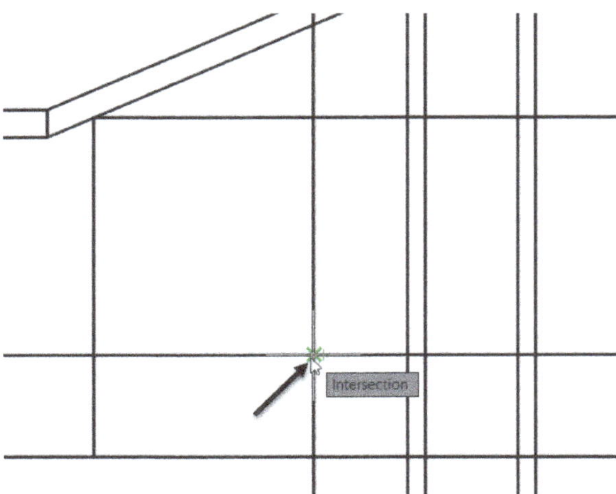

- Select the **Dimensions** option from the command line.
- Select the two intersection points, as shown.

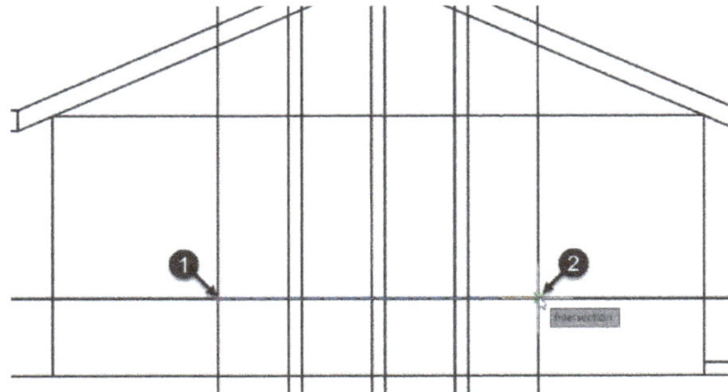

- Type 54 as the rectangle width and press ENTER.
- Move the pointer upward and click.

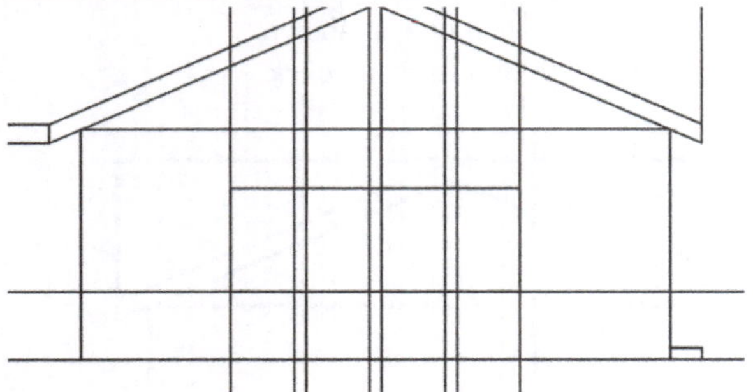

- Click the **Offset** tool on the **Modify** panel of the **Home** ribbon tab.
- Type 4" as the offset distance. Next, press ENTER.
- Select the rectangle.
- Move the pointer outward and click.

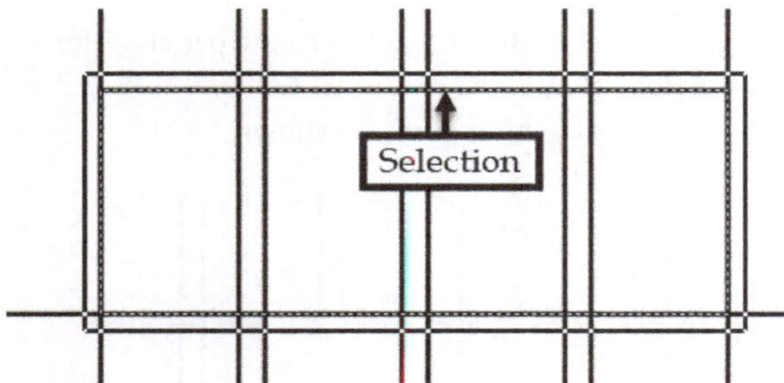

- Click the **Trim** tool on the **Modify** panel of the **Home** ribbon tab.
- Click and drag the pointer across the vertical lines, as shown.

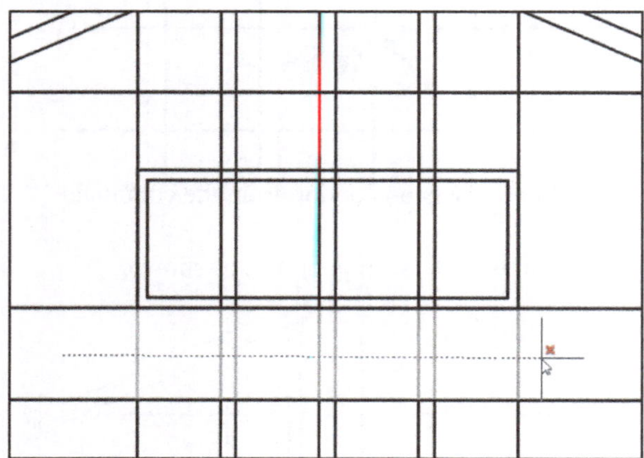

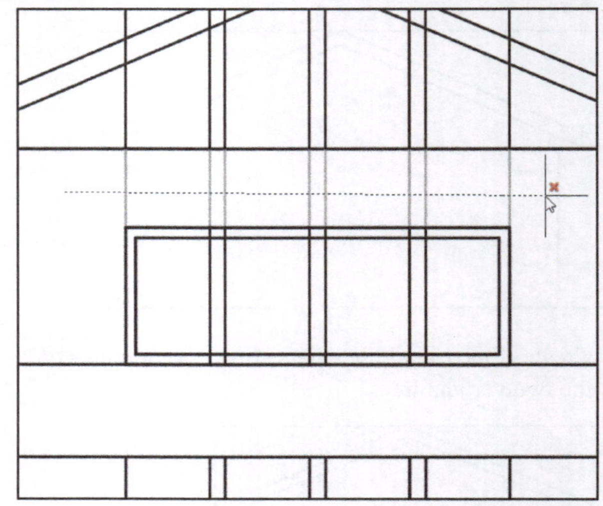

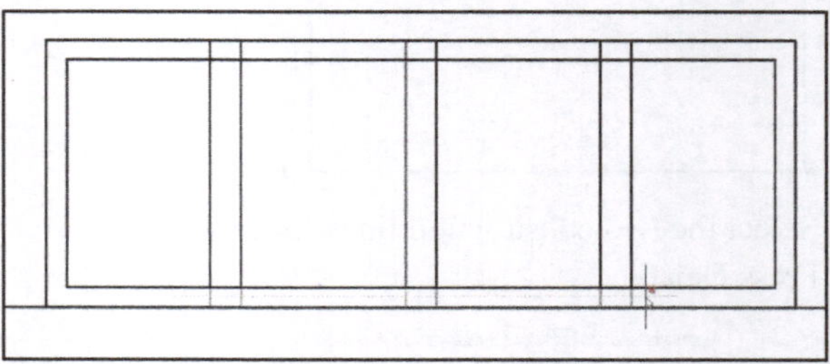

- Press ENTER.
- Select the vertical and horizontal construction lines, and then press Delete.

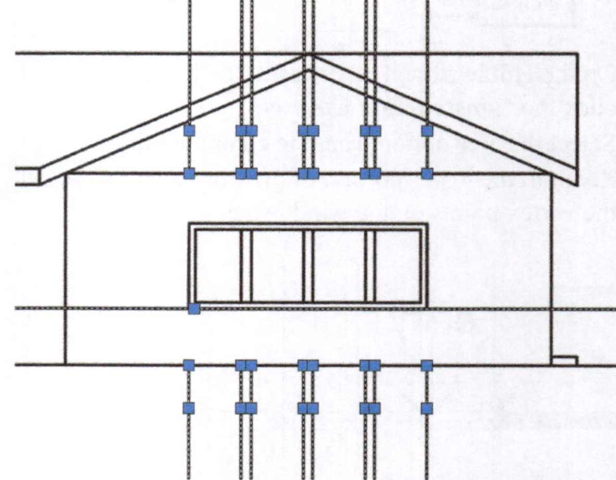

- Select the two rectangles and click the **Explode** tool on the **Modify** panel of the **Home** ribbon tab.

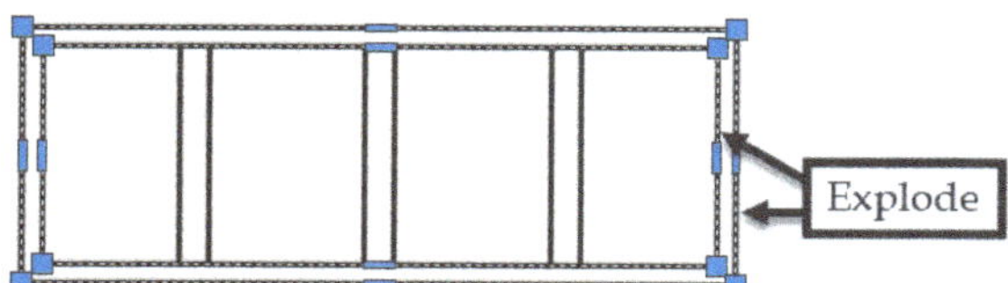

- Type **MLINE** in the command line and press ENTER.
- Select the **Justification** option from the command line.
- Select the **Zero** option from the command line.
- Select the **Scale** option from the command line.
- Type 0.25 and press ENTER.
- Press and hold the Shift key, and then right click.
- Select the **Mid between 2 Points** option.
- Select the two points, as shown.
- Move the pointer downward and click.

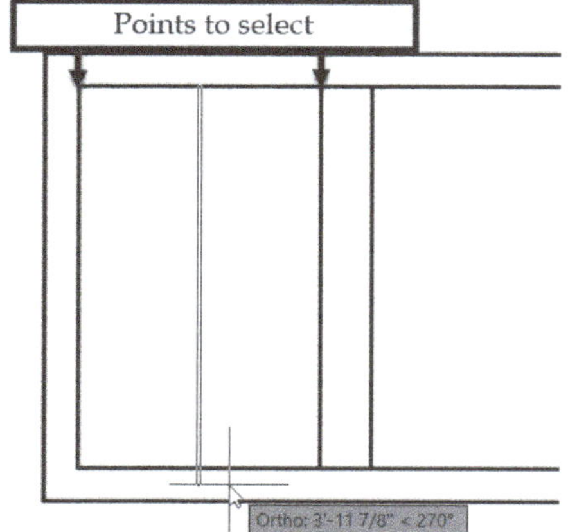

- Press ENTER twice.
- Select the midpoint of the vertical line, as shown.

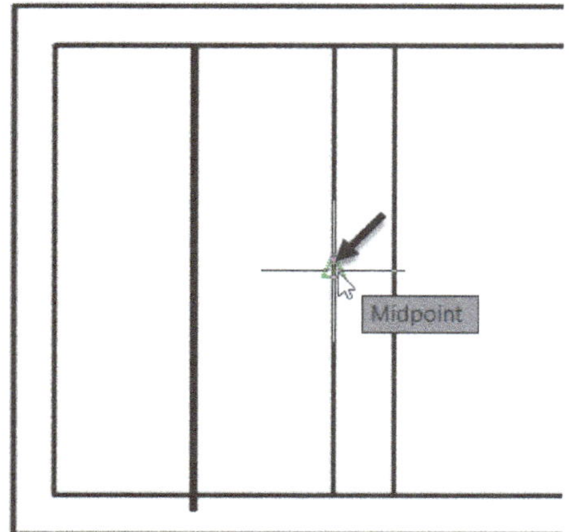

- Move the pointer toward the left and select the midpoint of the vertical line.

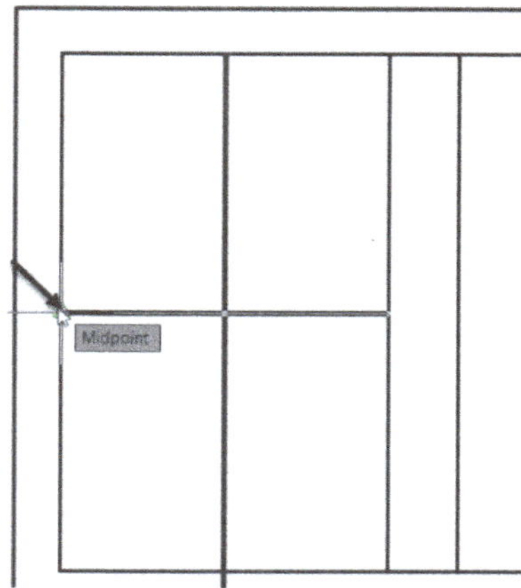

- Press Esc.
- Type **MLEDIT** in the command line and press ENTER.
- Click the **Open Cross** icon on **the Multilines Edit Tools** dialog.

- Select the two multilines; the open cross is created at the intersection of the two multilines.

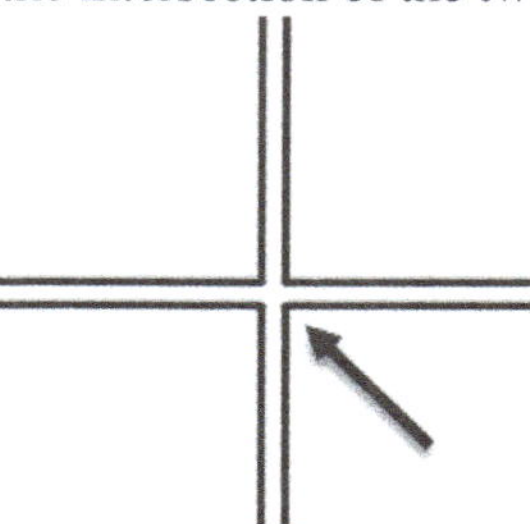

- Click the **Trim** tool on the **Modify** panel of the **Home** ribbon tab.
- Click and drag the pointer across the unwanted portions of the multilines, as shown.

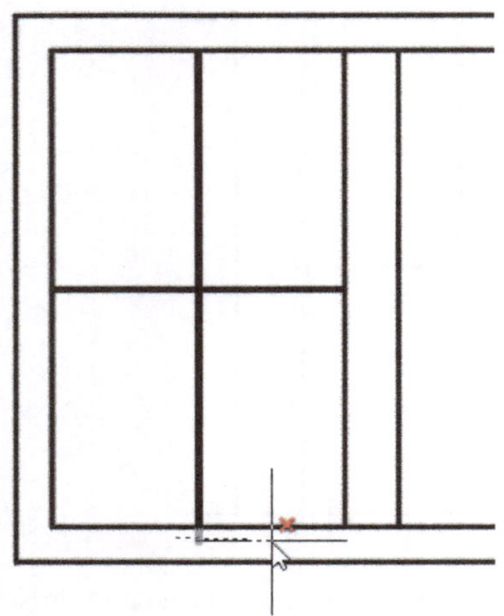

- Click the **Copy** tool on the **Modify** panel of the **Home** ribbon tab.
- Select the two multilines and press ENTER.
- Select the lower-left corner point.
- Move the pointer toward right and select the corner point, as shown.

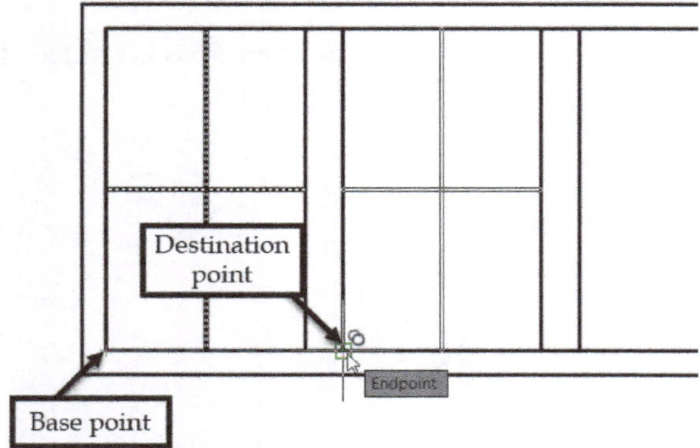

- Likewise, create two more copies of the multi-lines, as shown.

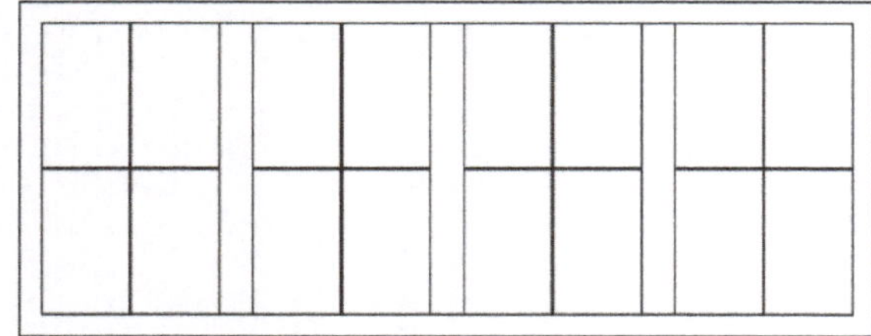

- Likewise, create two windows on the left side.

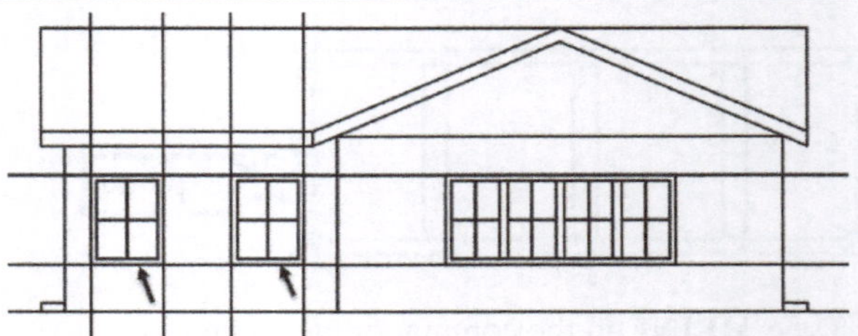

- Delete the construction lines.

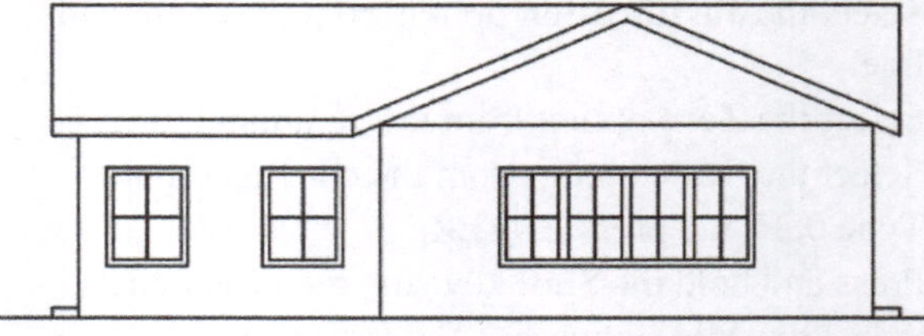

- Click the **Hatch** tool on the **Draw** panel of the **Home** ribbon tab.
- On the **Hatch Creation** tab, click **Pattern** panel > **Pattern** gallery > **AR-RSHKE**.

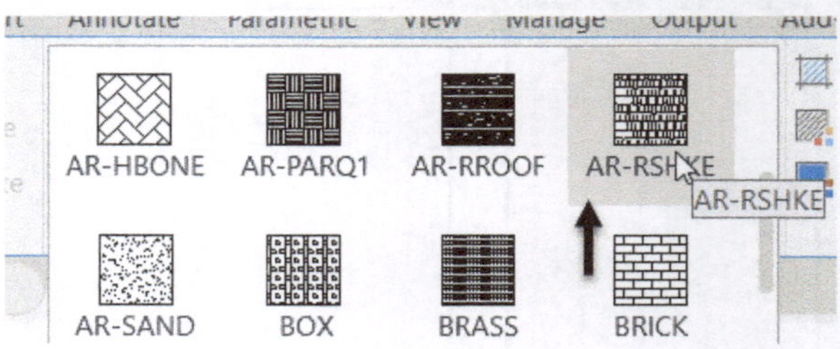

- Click in the regions, as shown.

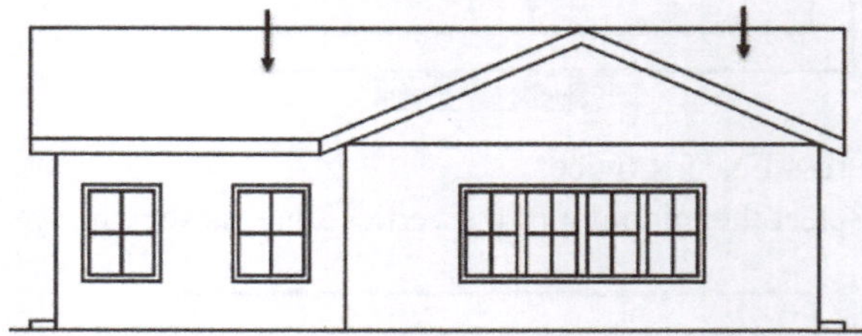

- Press ENTER twice.
- On the **Hatch Creation** tab, click **Pattern** panel > **Pattern** gallery > **AR-B816**.

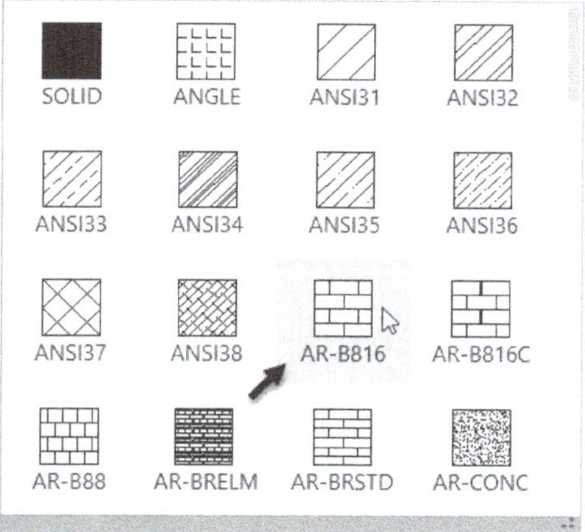

- Click in the regions, as shown.

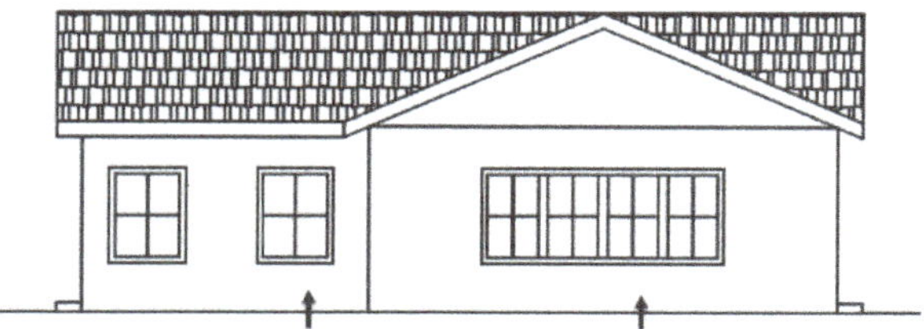

- Press ENTER twice.
- On the **Hatch Creation** tab, click **Pattern** panel > **Pattern** gallery > **ANSI31**.
- On the **Properties** panel, enter **135** and **50** in the **Angle** and **Scale** boxes, respectively.

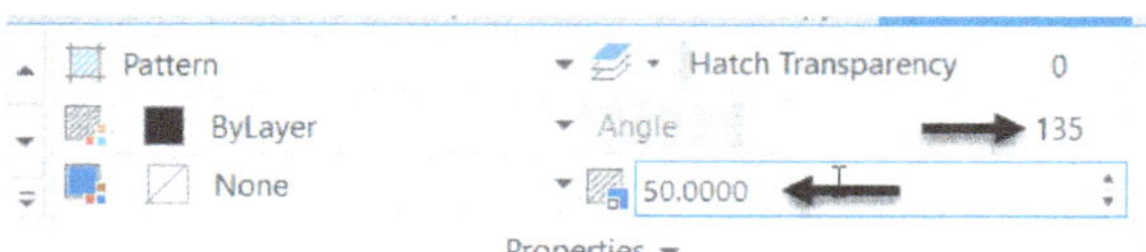

- Click in the region, as shown.

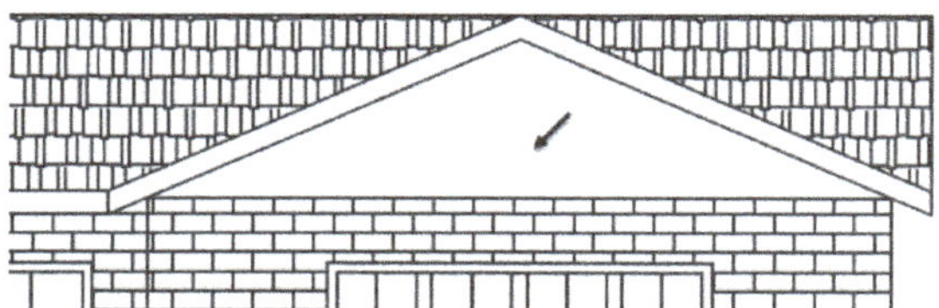

- Press ENTER.
- Save and close the file.

Example 4

In this tutorial, you will create the Roof plan using the floor plan.

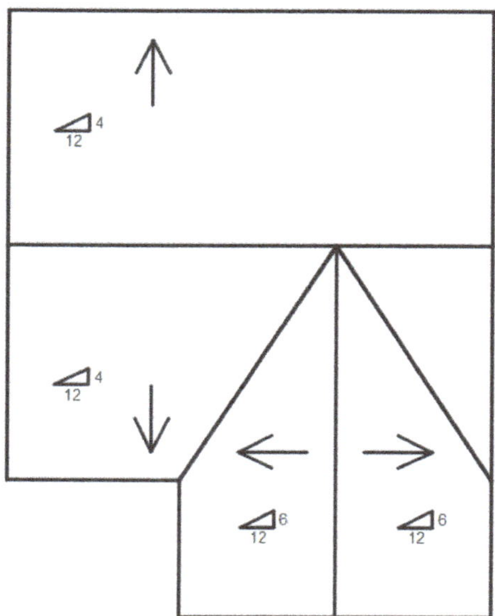

- Download the Floor_plan.dwg from the companion website.
- Start a new drawing using the acad.dwt template.
- Click the **Attach** tool on the **Reference** panel of the **Insert** ribbon tab.
- On the **Select Reference file** dialog, select **Drawing** from the **Files of type** drop-down.
- Go to the location of the Floor_plan.dwg file and double-click on it.
- Click **OK** on the **Attach External Reference** dialog.
- Click in the graphics window.
- Click **Zoom > Zoom All** on the Navigation Bar.
- Click the **Polyline** tool on the **Draw** panel of the **Home** ribbon tab.
- Select the corner points of the floor plan.

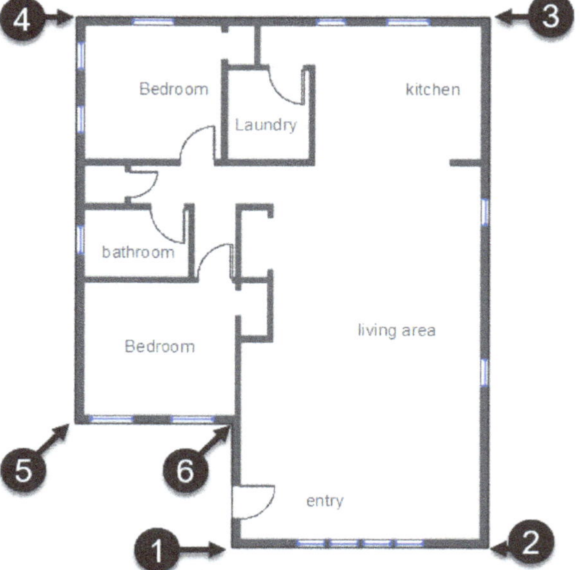

- Select the **Close** option from the command line.
- Click the **Offset** tool on the **Modify** panel of the **Home** ribbon tab.
- Type 16 and press ENTER.
- Select the newly created polyline.
- Move the pointer outward and click.

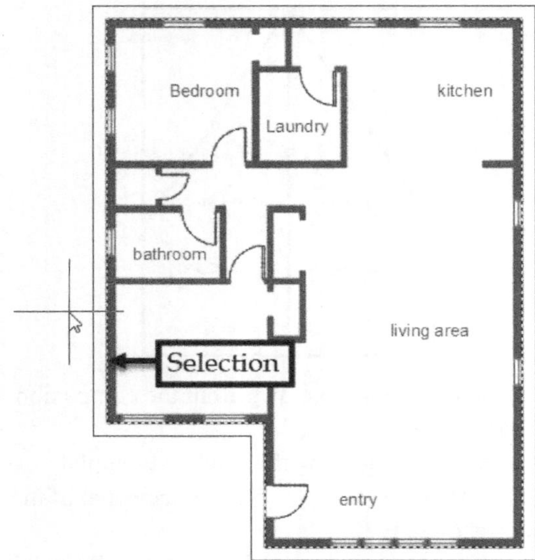

- Click the **Erase** tool on the **Modify** panel of the **Home** ribbon tab.
- Select the polyline used to create the offset.

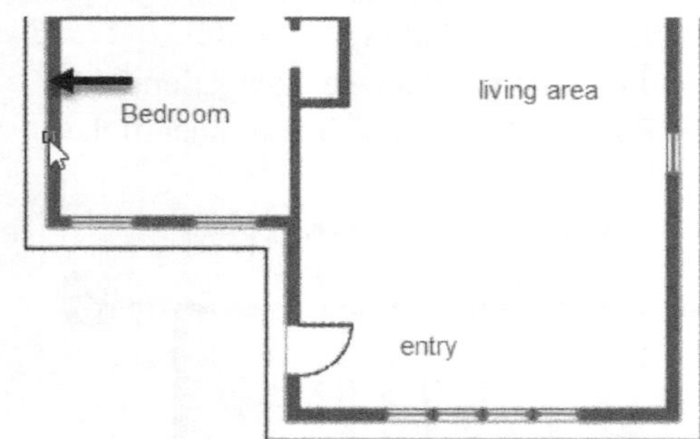

- Press ENTER.
- Click the **Line** tool on the **Draw** panel of the **Home** ribbon tab.
- Select the midpoint of the left vertical line.

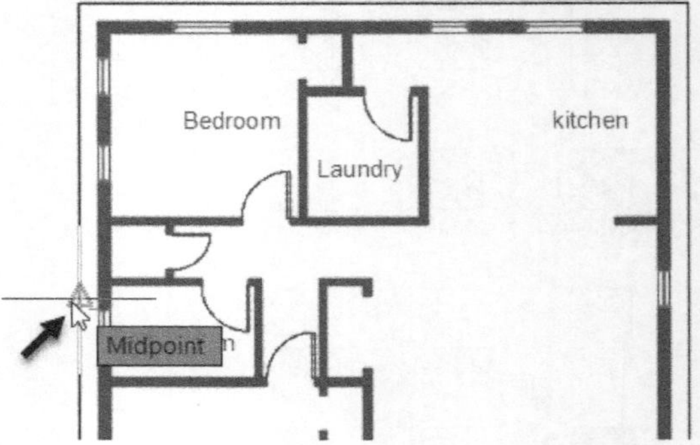

- Make sure the ORTHOMODE icon is active on the status bar.
- Move the pointer toward right and click.

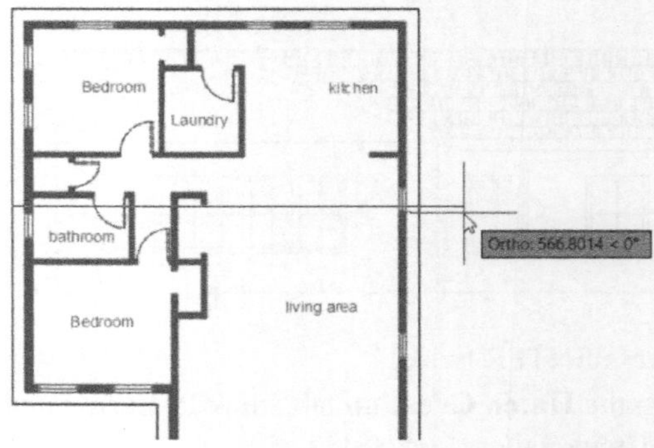

- Press ENTER twice.
- Select the midpoint of the lower horizontal line, as shown.

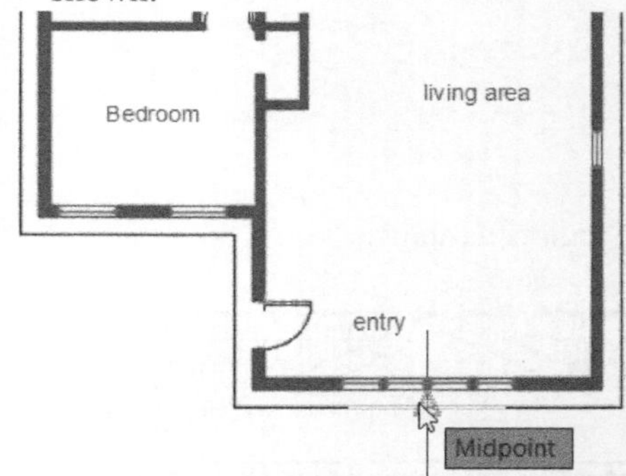

- Move the pointer upward and click.

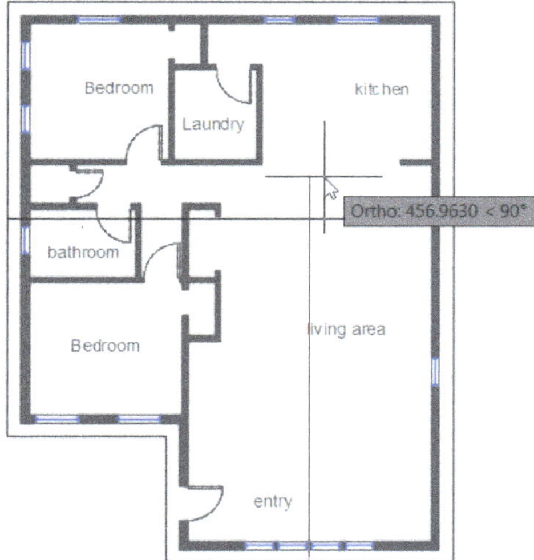

- Click the **Trim** tool on the **Modify** panel of the **Home** ribbon tab.
- Select the unwanted portions of the lines, as shown.

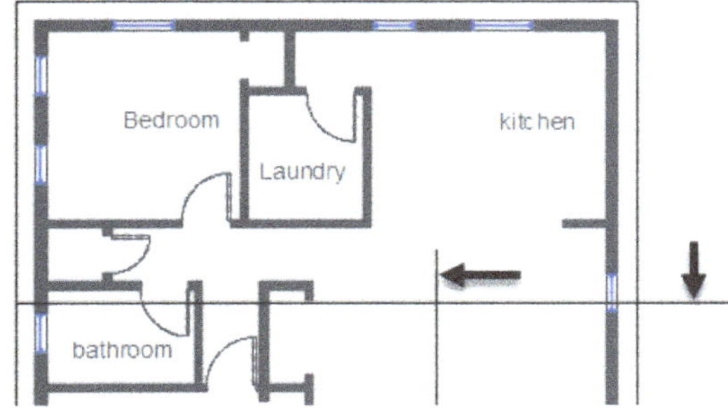

- Click the **Line** tool on the **Draw** panel of the **Home** ribbon tab.
- Select the corner point of the polyline.
- Select the intersection point of the vertical and horizontal lines.

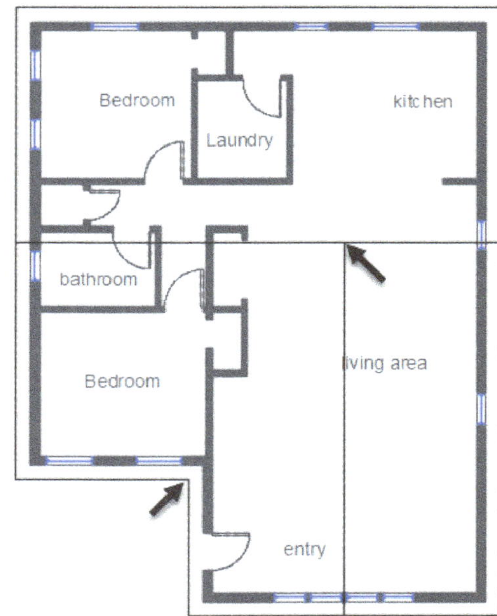

- Press Esc.
- Select the inclined line.
- Click the **Mirror** tool on the **Modify** panel of the **Home** ribbon tab.
- Select the endpoints of the vertical line.
- Select the **No** option from the command line.

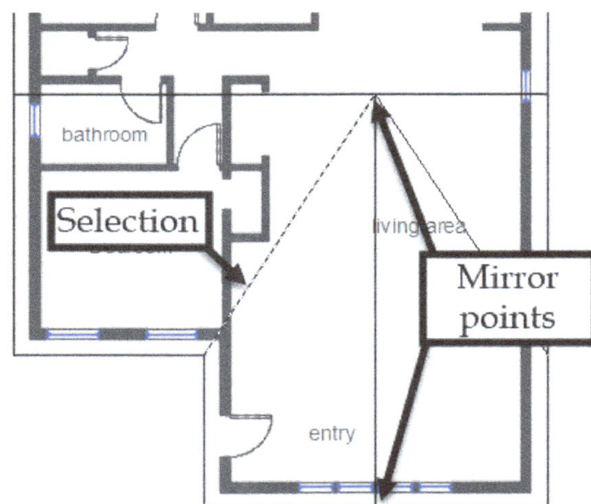

- Select the attachment.
- Click on the origin point of the attachment, and then move the pointer upward.

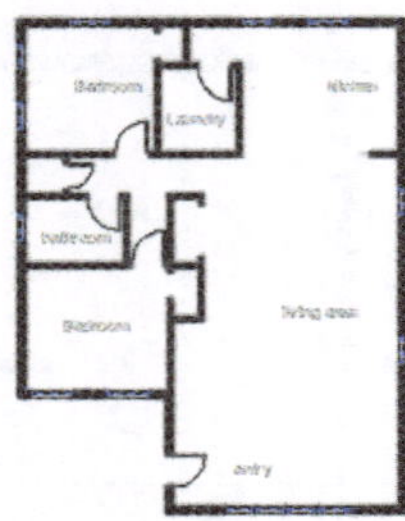

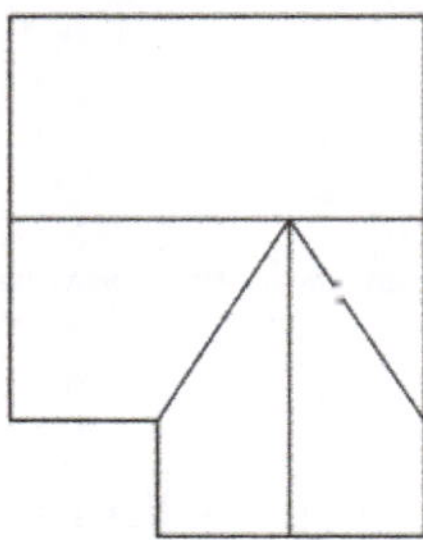

- Click the **Line** tool on the **Draw** panel of the **Home** ribbon tab.
- Click in the graphics window.
- Move the pointer toward the right.
- Type 36 and press ENTER.
- Move the pointer upward.
- Type 18 and press ENTER.

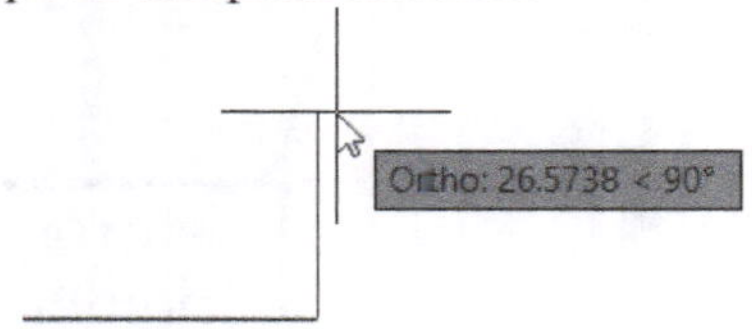

- Select the start point of the horizontal line.

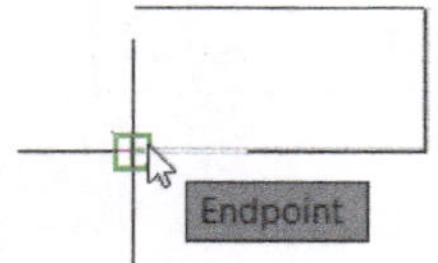

- Press ESC.

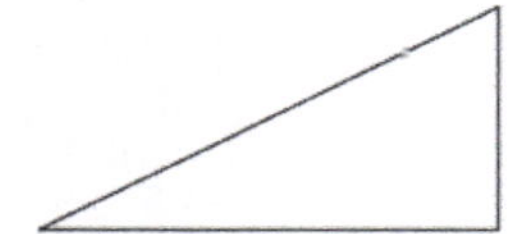

- Click the **Text** drop-down > **Single Line** on the **Annotation** panel of the **Home** ribbon tab.

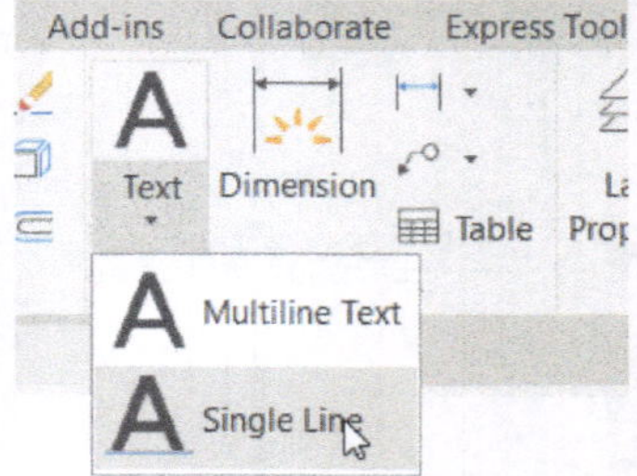

- Click in the graphics window.
- Type **12** as the text height. Next, press ENTER.
- Type 0 as the rotation angle, and then press ENTER.
- Type 4 and click in the graphics window.
- Type 12 and click in the graphics window.
- Press ESC.
- Position the texts at the locations, as shown.

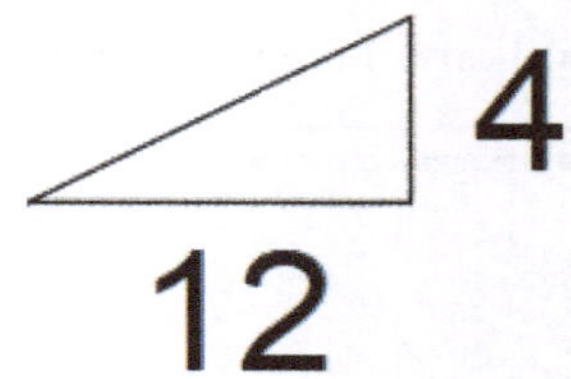

- Create a selection window across the triangle and the texts.
- Click the **Copy** tool on the **Modify** panel of the **Home** ribbon tab.
- Select the base point, as shown.

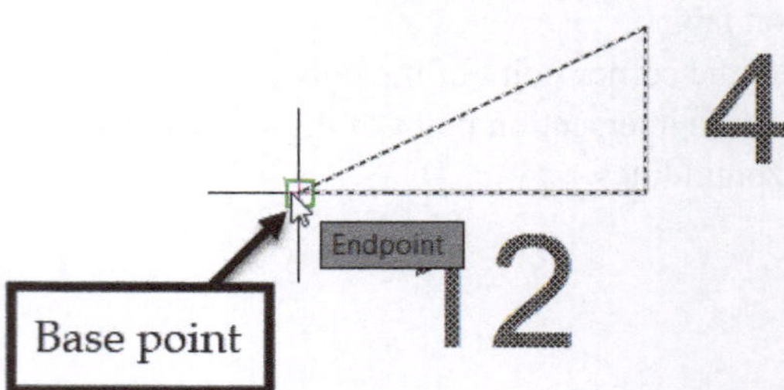

- Place the copies at the locations, as shown.

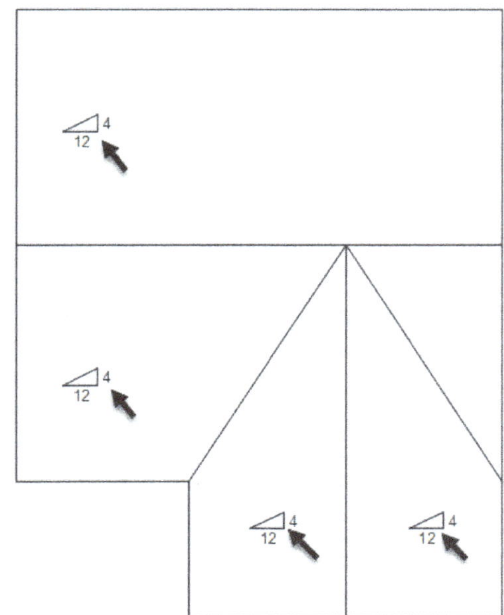

- Zoom the lower portion of the drawing.
- Double-click on 4.
- Type 6 and click in the graphics window.
- Likewise, change the text on the right side to 6.

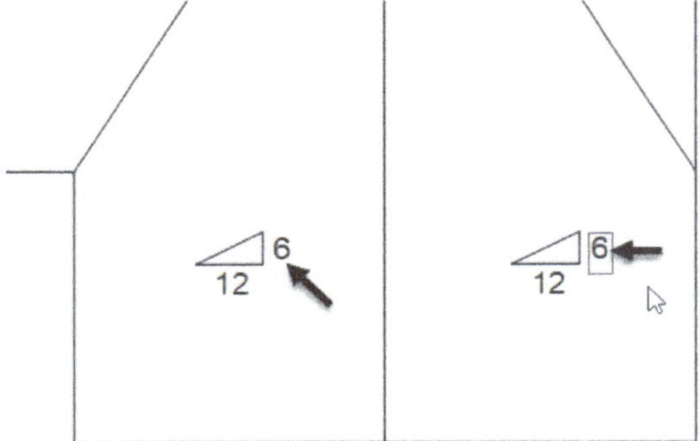

- Press ESC.
- Click the **Rectangle** tool on the **Draw** panel of the **Home** ribbon tab.
- Click at the location, as shown.
- Select the **Dimensions** option from the command line.
- Type 36 and press ENTER.
- Type 72 and press ENTER.
- Move the pointer downward and click.

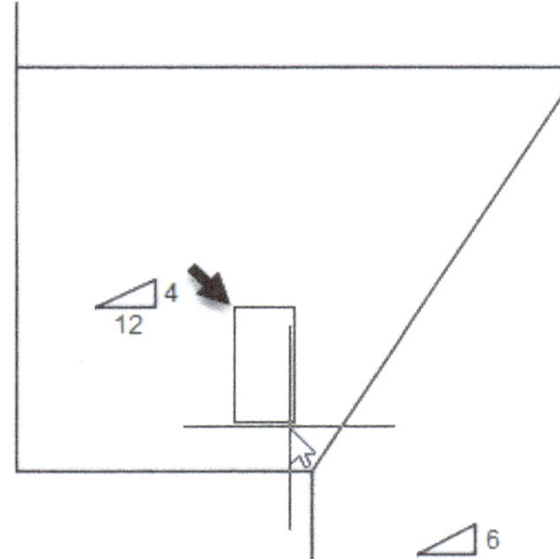

- Click the **Line** tool on the **Draw** panel of the **Home** ribbon tab.
- Select the midpoints of the horizontal edges of the rectangle.

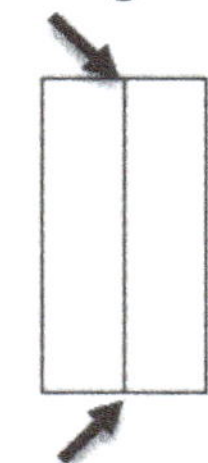

- Press ENTER twice.
- Select the midpoint of the left vertical edge of the rectangle.
- Select the lower endpoint of the vertical line.
- Select the midpoint of the right vertical edge of the rectangle.

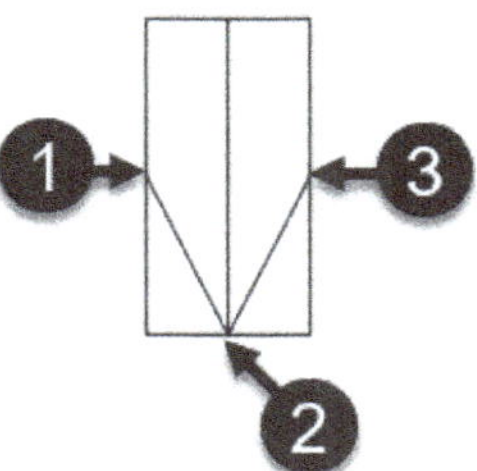

- Press ESC.
- Select the rectangle and press Delete.

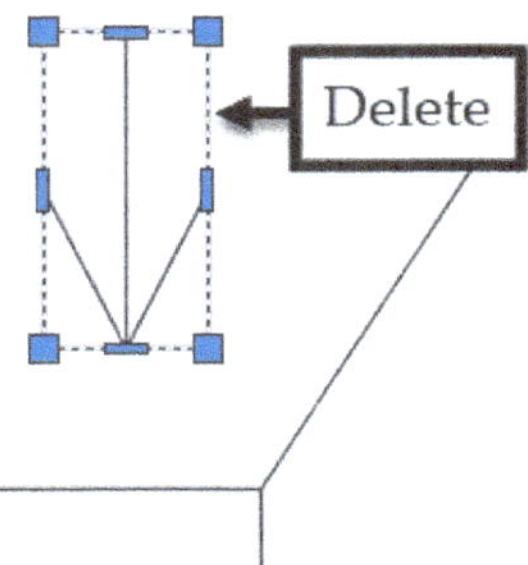

- Create a selection window across all the elements of the arrow.

- Click the **Mirror** tool on the **Modify** panel of the **Home** ribbon tab.
- Select the endpoints of the horizontal line to define the mirror line.

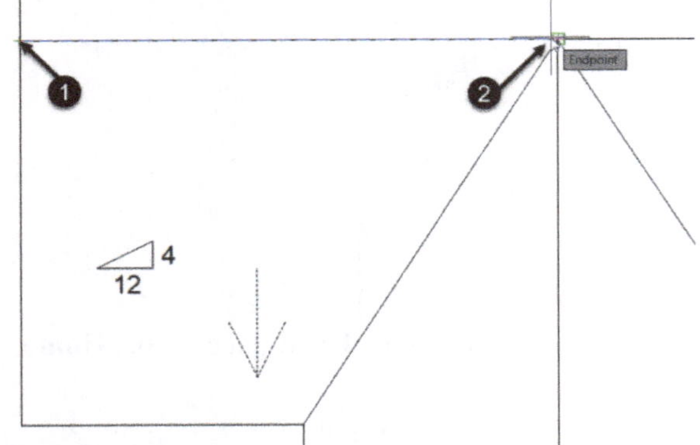

- Select **NO** from the command line.
- Create a selection window across all the elements of the arrow.
- Click the **Rotate** tool on the **Modify** panel of the **Home** ribbon tab.
- Select the base point, as shown.

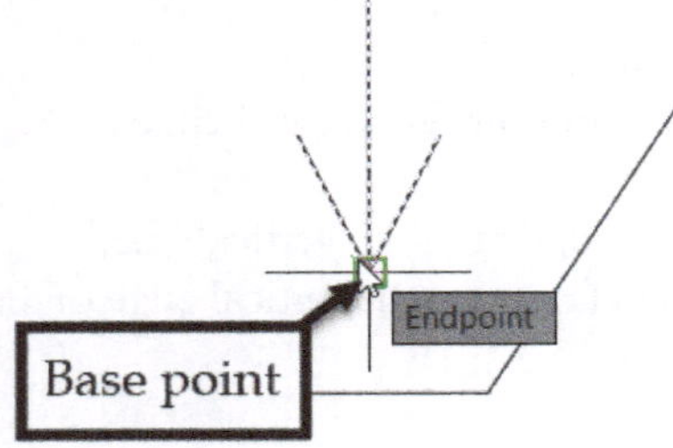

- Select the **Copy** option from the command line.
- Move the pointer vertically downward and click.

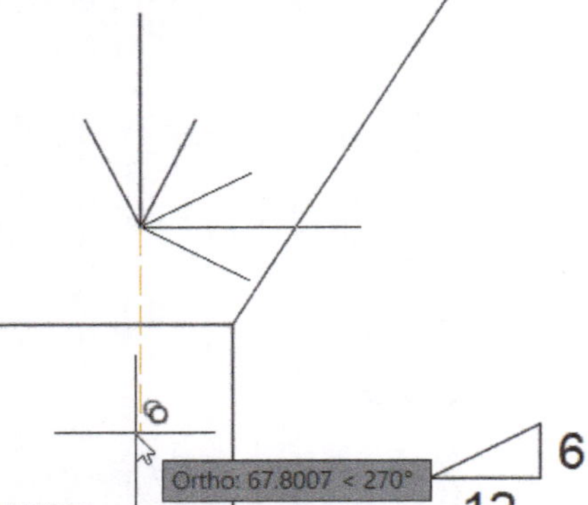

- Select the elements of the rotated arrow.
- Click the **Move** tool on the **Modify** panel of the **Home** ribbon tab.
- Select the base point, as shown.

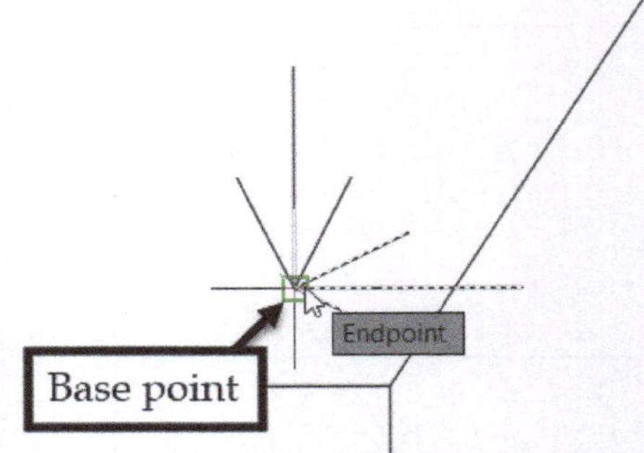

- Move the pointer toward right and click.

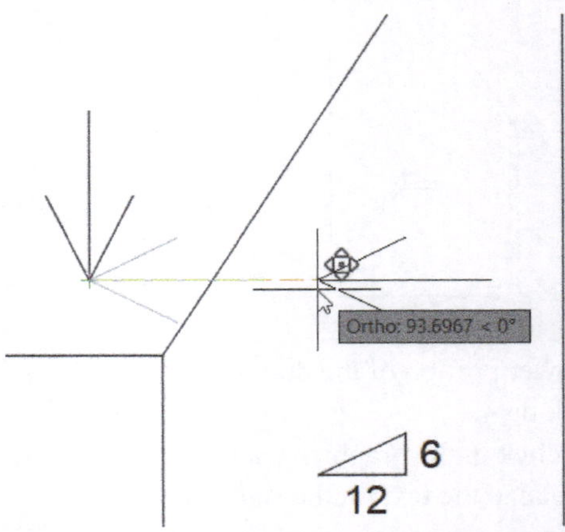

- Select the elements of the arrow moved in the previous step.
- Click the **Mirror** tool on the **Modify** panel of the **Home** ribbon tab.
- Select the endpoints of the vertical line, as shown.

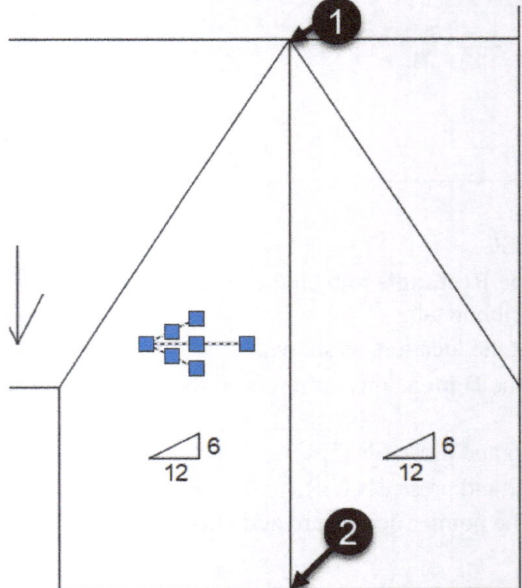

- Select **No** from the command line.
- Save and close the drawing file.

Example 5

In this tutorial, you will create the Roof and wall detail.

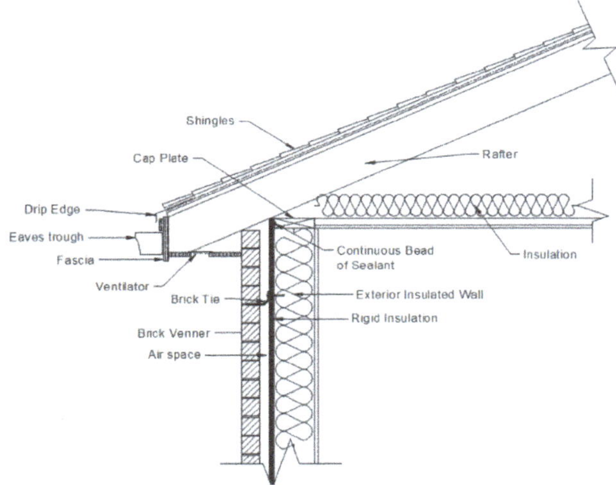

- Download the Roof_&_wall_detail.dwg file and open it.
- Click the **Layer Properties** tool on the **Layers** panel of the **Home** ribbon tab.
- Create a new layer and name it as Roof_wall_detail.
- Make the new layer are current.

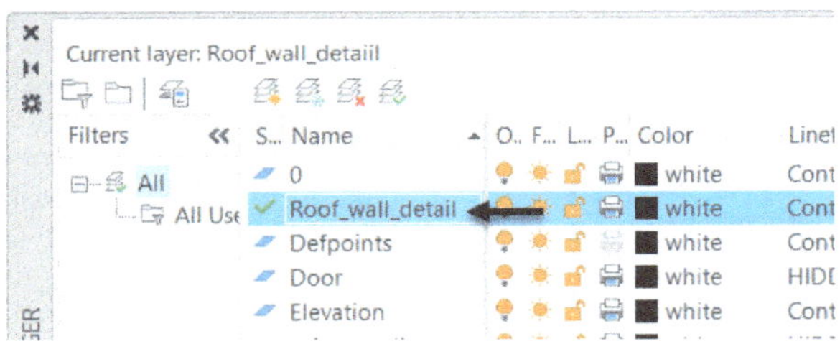

- Close the **Layer Properties Manager** palette.
- On the **Home** ribbon tab, expand the **Draw** panel and click the **Ray** tool.
- Create the projection lines from the elevation view, as shown.

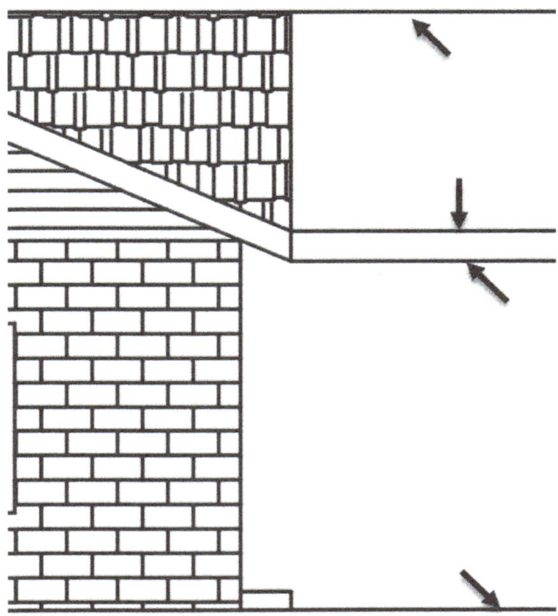

- Create a vertical line, as shown.

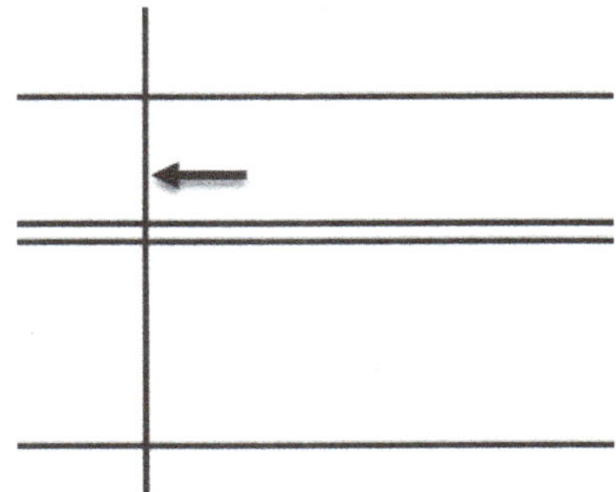

- Select the two inclined lines of the roof, as shown.
- Select the two vertical lines, as shown.

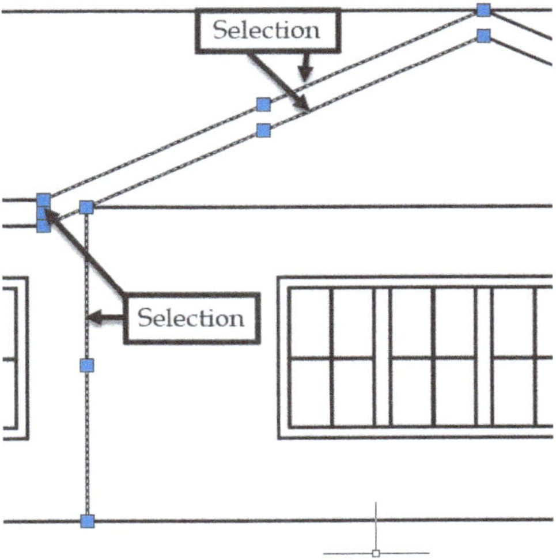

- Click the **Copy** tool on the **Modify** panel of the **Home** ribbon tab.
- Select the base point, as shown.

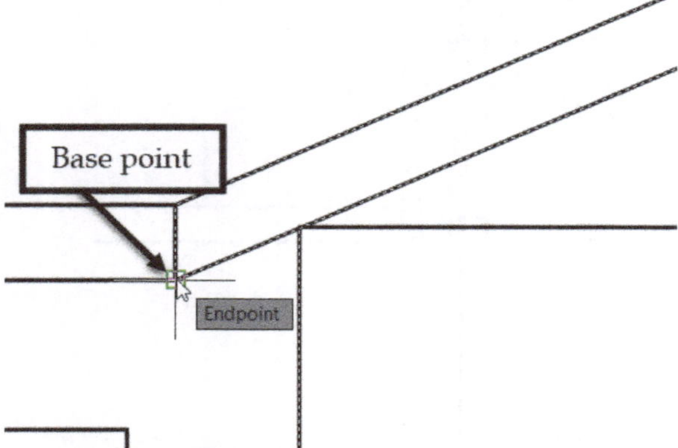

- Move the pointer toward right and select the intersection point, as shown.

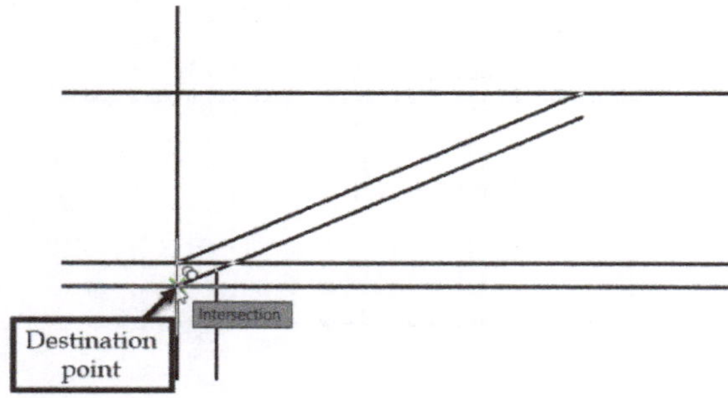

- Delete the reference lines, as shown.

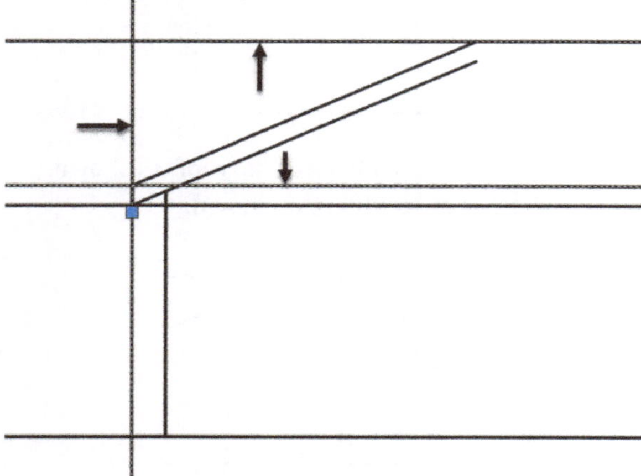

- Click the **Trim** tool on the **Modify** panel of the **Home** ribbon tab.
- Select the portion of the horizontal line, as shown.

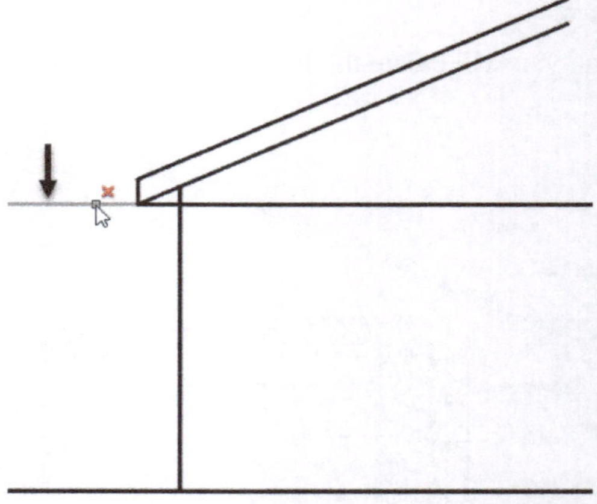

- Click the **Offset** tool on the **Modify** panel of the **Home** ribbon tab.
- Type 4 and press ENTER.
- Select the vertical line.
- Move the pointer toward right and click.

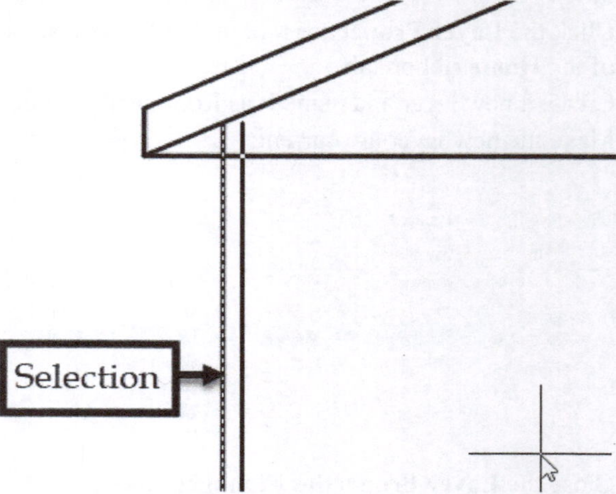

- Press ENTER twice.
- Type 2 and press ENTER.
- Select the offset line.
- Move the pointer toward the right and click.

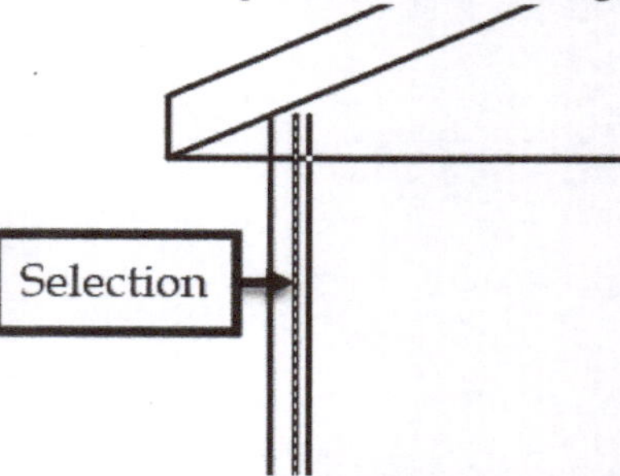

Creating the Brick Venner

- Click the **Rectangle** tool on the **Draw** panel of the **Home** ribbon tab.
- Specify the corner point, as shown.

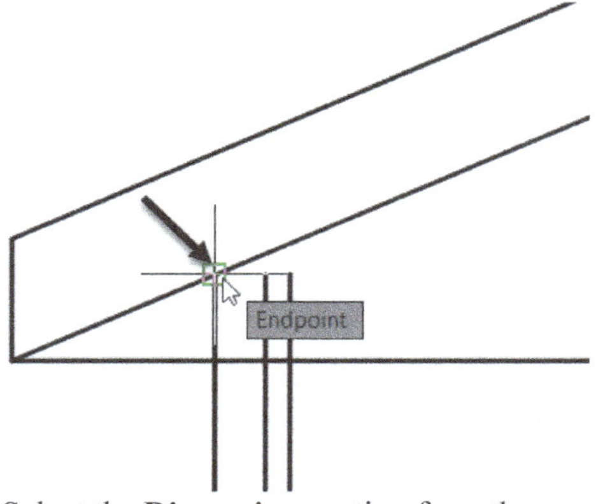

- Select the **Dimensions** option from the command line.
- Type 4 and press ENTER.
- Type 4 and press ENTER.
- Move the pointer downward and click.

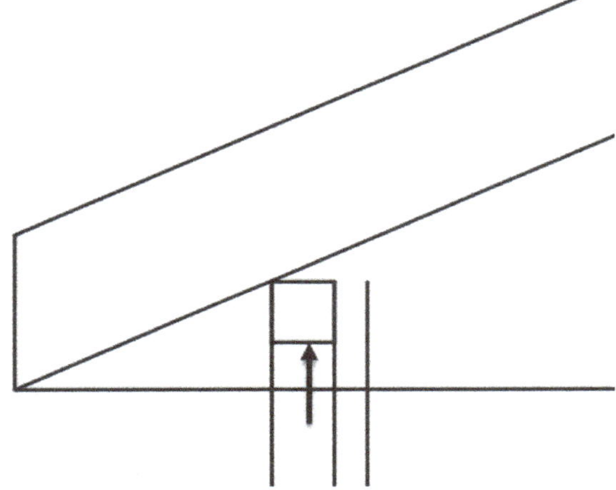

- Select the rectangle and click **Array** drop-down > **Path Array** on the **Modify** panel of the **Home** ribbon tab.

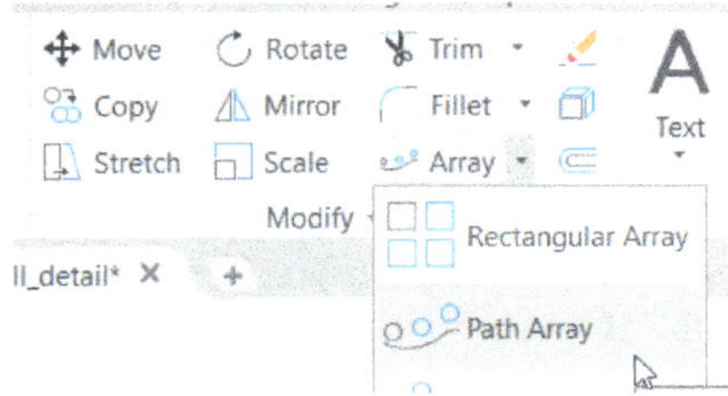

- Select the left vertical line to define the path.

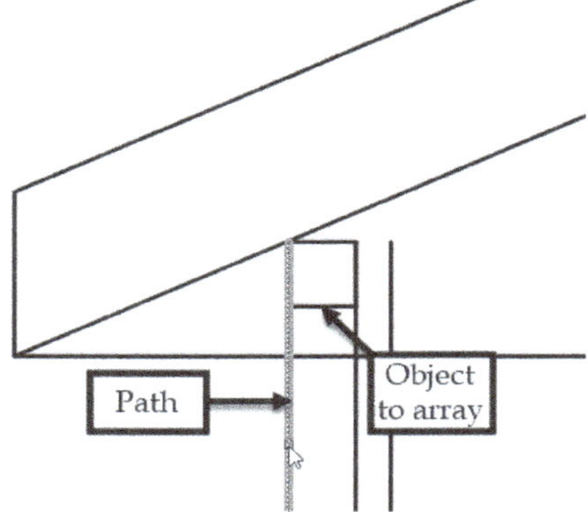

- Click the **Item Count** icon on the **Items** panel of the **Array Creation** ribbon tab.

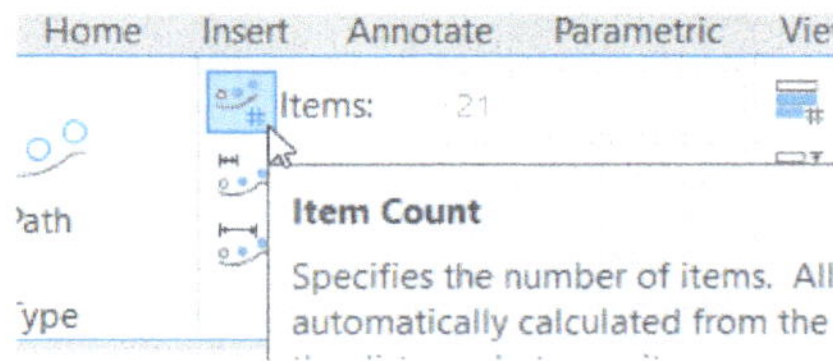

- Type **14** in the **Items** box.
- Type **4.25** in the **Between** box.

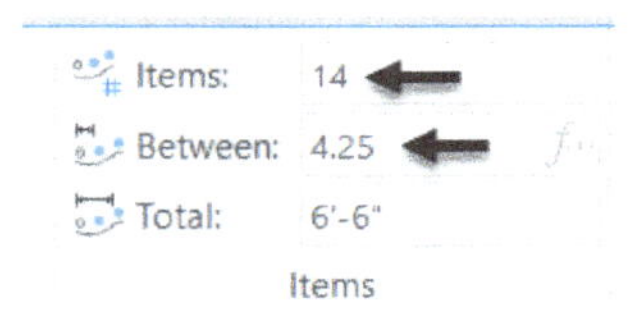

- Deactivate the **Associative** option on the **Properties** panel of the **Array Creation** ribbon tab.

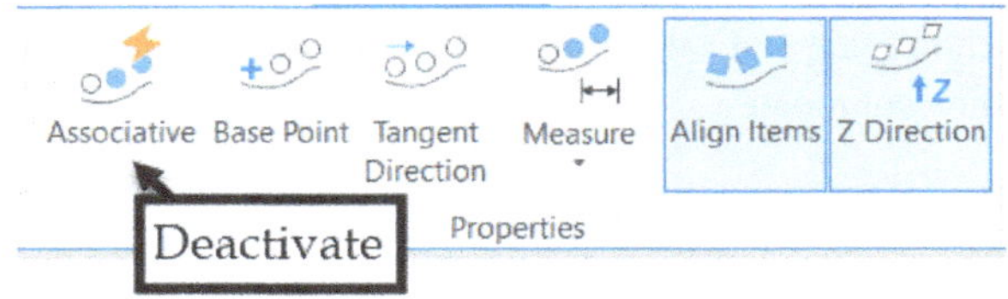

- Click **Close Array** on the ribbon.

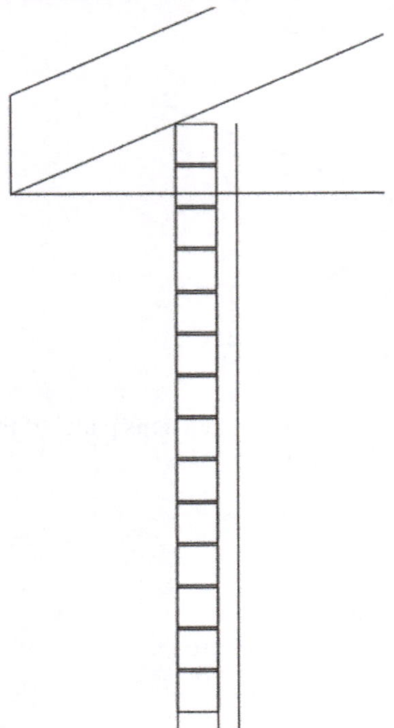

- Select anyone of the rectangles.

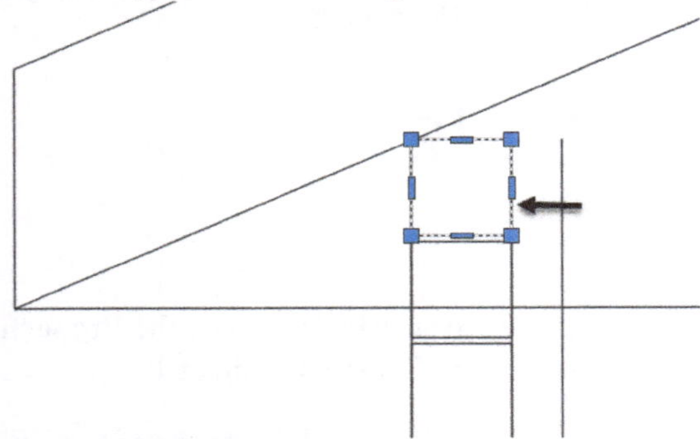

- Right click and select **Select Similar** from the shortcut menu.

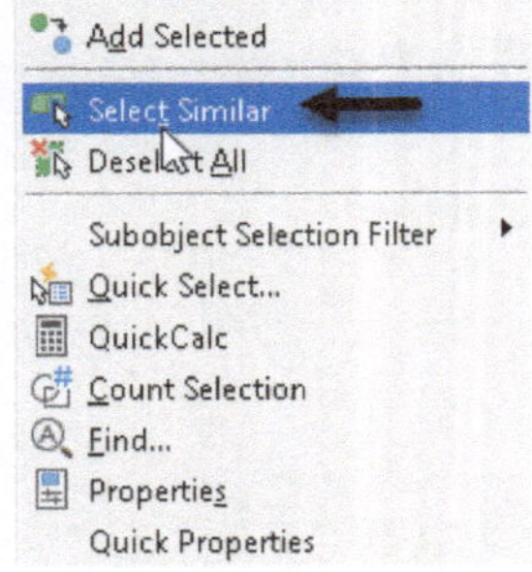

- Click the **Explode** tool on the **Modify** panel of the **Home** ribbon tab. All the rectangles are exploded into individual objects.
- Click the **Line** tool on the **Draw** panel of the **Home** ribbon tab.
- Zoom to the top portion of the drawing.
- Create a line by selecting the two points, as shown.

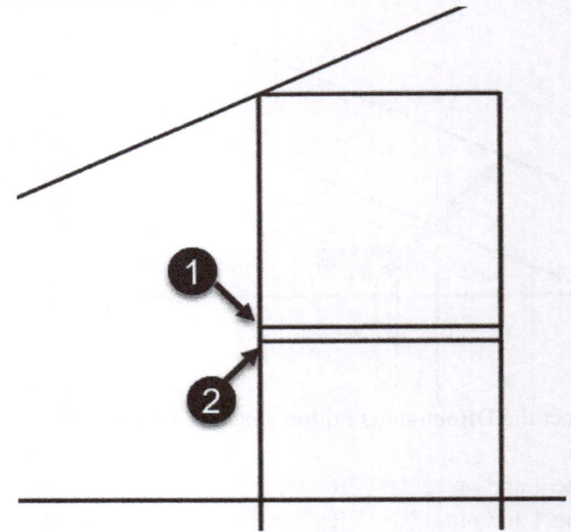

- Press ESC.
- Select the newly created line.
- Click the **Move** tool on the **Modify** panel of the **Home** ribbon tab.
- Select the top endpoint of the selected line.
- Move the pointer toward the right.

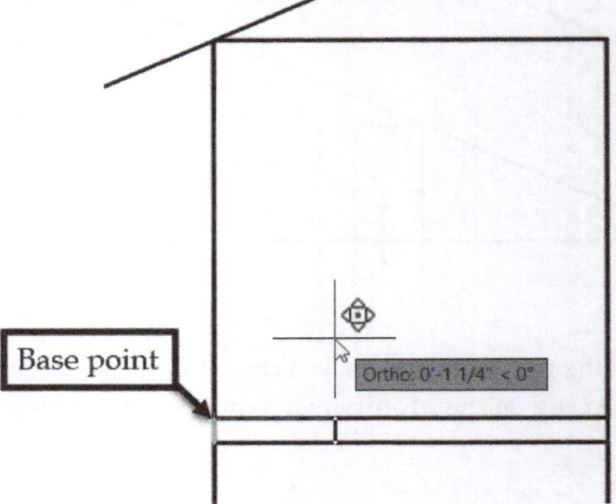

- Type 0.125, and press ENTER.
- Select the line moved in the last step.
- Click **Array** drop-down > **Rectangular Array** on the **Modify** panel of the **Home** ribbon tab.
- On the **Columns** panel, type **2** and **3.75** in the **Columns** and **Between** boxes, respectively.

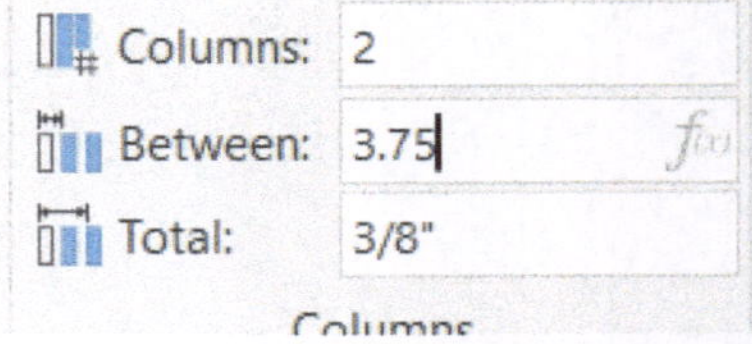

- On the **Rows** panel, type **13** and **-4.25** in the **Rows** and **Between** boxes, respectively.

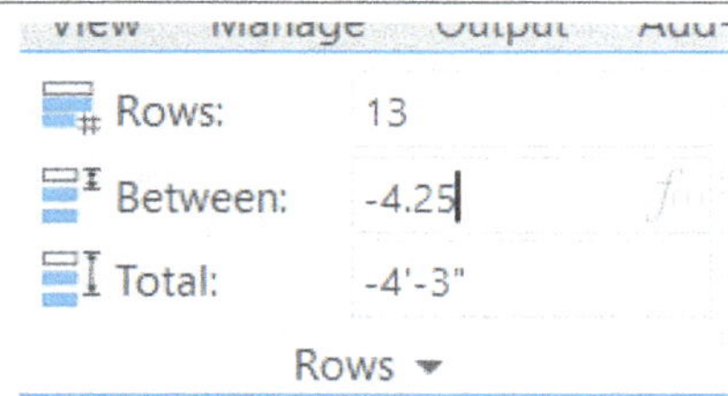

- Deactivate the **Associative** option.
- Click **Close Array** on the ribbon.
- Select the two vertical lines, as shown.

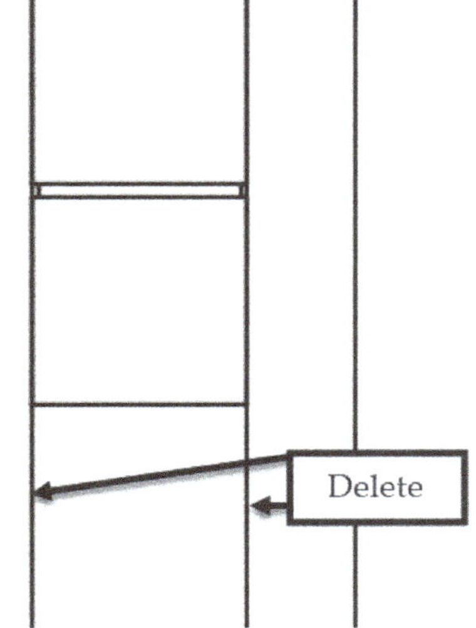

- Press Delete.

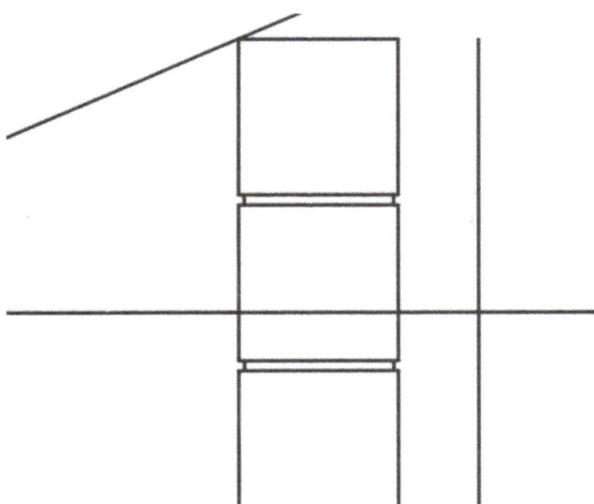

- Click the **Hatch** tool on the **Draw** panel of the **Home** ribbon tab.
- Click the **ANSI31** pattern on the **Pattern** panel.
- Type 8 in the **Hatch pattern Scale** box and press ENTER.

- Click in the regions, as shown.

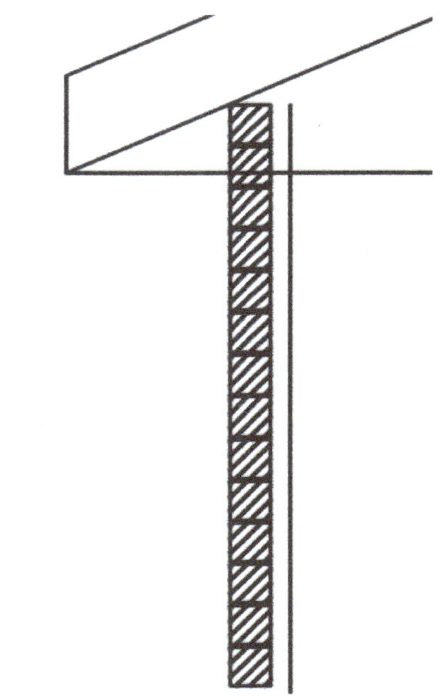

- Click **Close Hatch Creation** on the ribbon.

Creating the Brick Tie

- On the Navigation Bar, click **Zoom** drop-down > **Zoom Window**.
- Create a zoom window at the location, as shown.

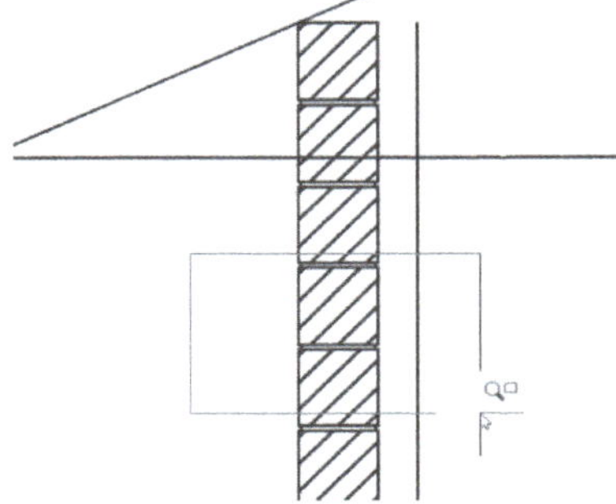

- Click the **Offset** tool on the **Modify** panel of the **Home** ribbon tab.
- Type 0.025, and press ENTER.
- Offset the two horizontal lines in the inward direction.

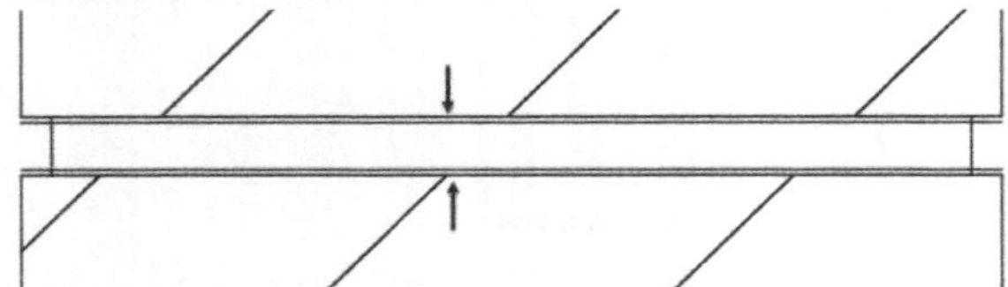

- Press ENTER twice.
- Type 1 and press ENTER.
- Offset the left vertical line towards the right.

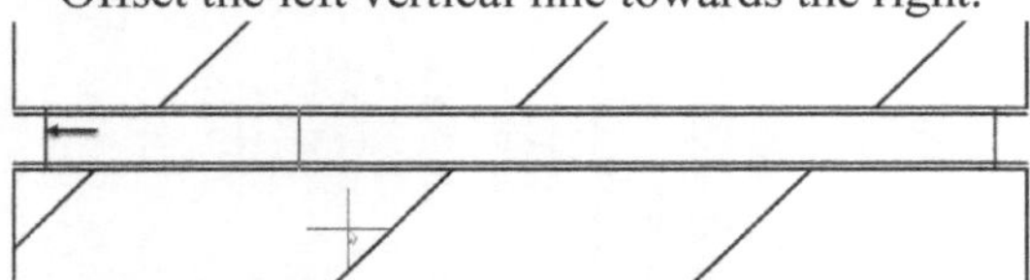

- Click the **Trim** tool on the **Modify** panel of the **Home** ribbon tab.
- Trim the portions of the lines, as shown.

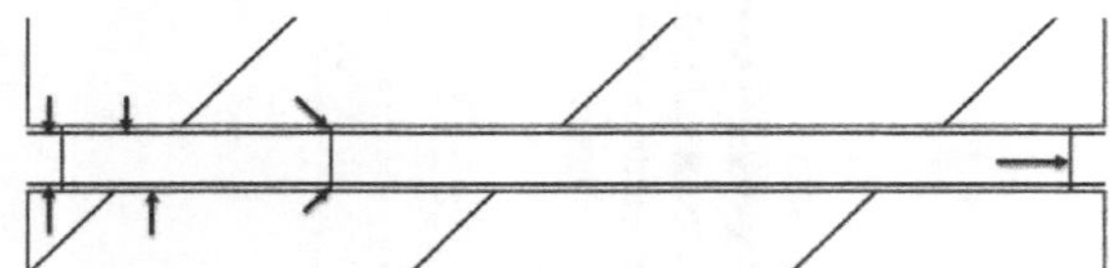

- Click the **Trim** drop-down > **Extend** on the **Modify** panel of the **Home** ribbon tab.
- Select the ends of the horizontal lines, as shown.

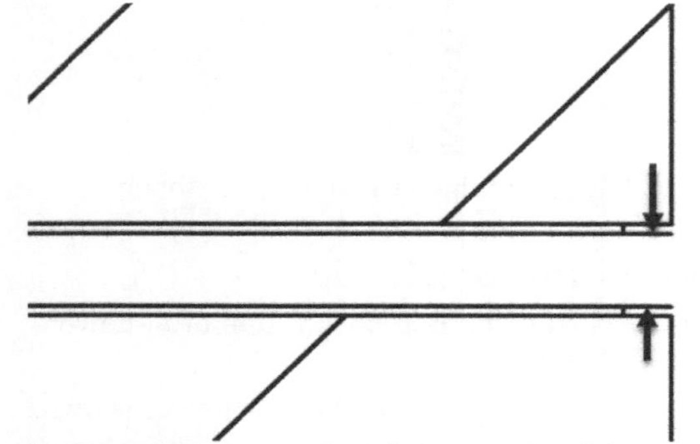

The horizontal lines are extended up to the right vertical line.

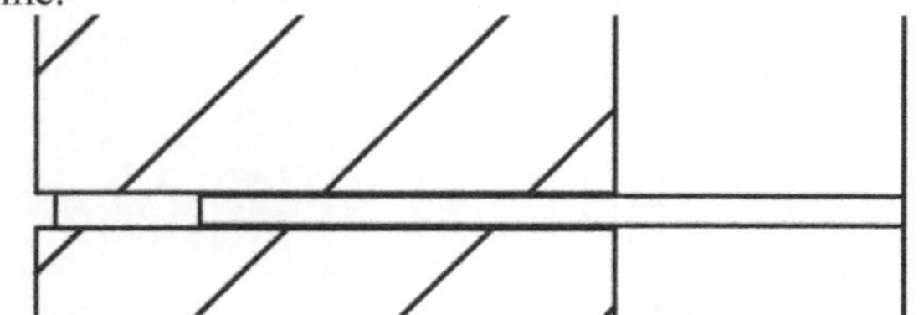

- Click the **Line** tool on the **Draw** panel of the **Home** ribbon tab.
- Select the endpoint of the lower horizontal line.
- Move the pointer vertically upward.
- Type 3 and press ENTER.
- Click the **Offset** tool on the **Modify** panel of the **Home** ribbon tab.
- Type 0.2 and press ENTER.
- Select the vertical line created in the last step.
- Move the pointer toward the left and click.

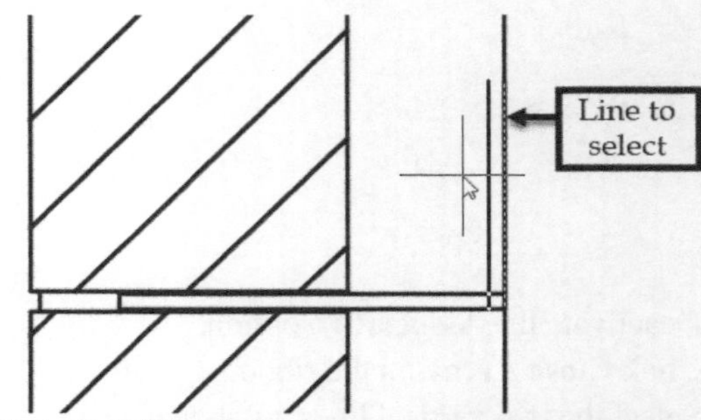

- Click the **Fillet** drop-down > **Chamfer** on the **Modify** panel of the **Home** ribbon tab.
- Select the **Angle** option from the command line.
- Type 1 as the chamfer length.
- Type 45 as the chamfer angle.
- Select horizontal and vertical lines, as shown.

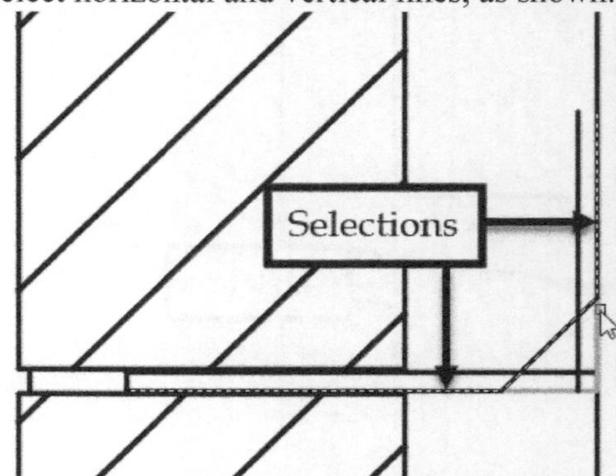

- Likewise, create another chamfer, as shown.

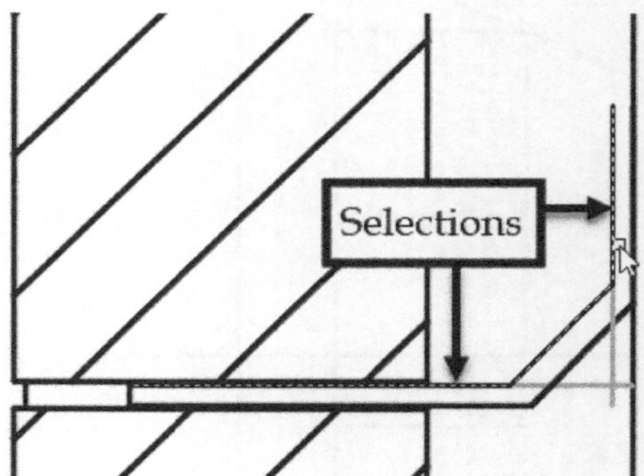

- Click the **Line** tool on the **Draw** panel of the **Home** ribbon tab.
- Cap the ends of the offset lines, as shown.

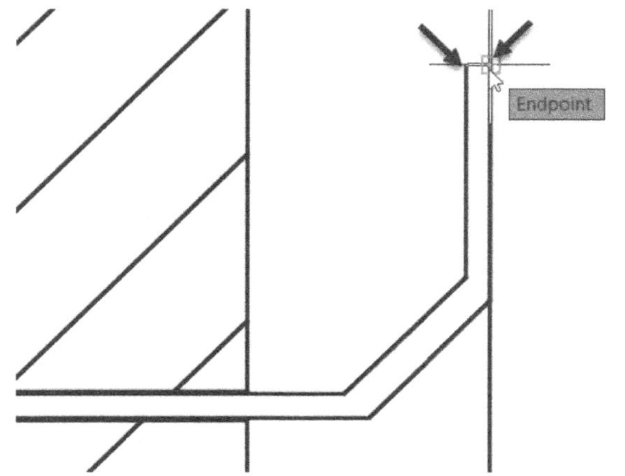

- Click the **Rectangle** tool on the **Draw** panel of the **Home** ribbon tab.
- Click in the graphic window.
- Select the **Dimensions** option from the command line.
- Type 0.2 and press ENTER.
- Type 0.5 and press ENTER.
- Click to create a rectangle.
- Select the rectangle and click the **Move** tool on the **Modify** panel of the **Home** ribbon tab.
- Select the midpoint of the right vertical edge of the rectangle.

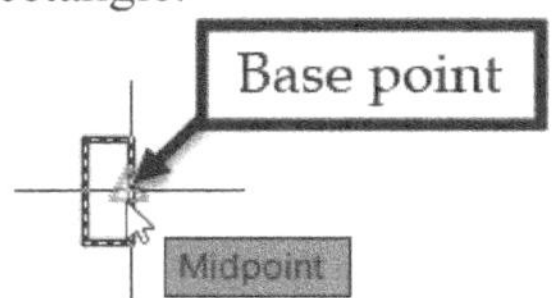

- Move the pointer and select the midpoint of the vertical line, as shown.

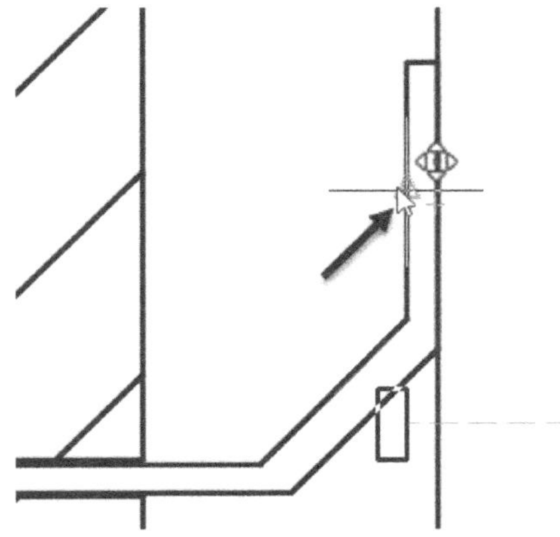

- Click the **Line** tool on the **Draw** panel of the **Home** ribbon tab.
- Select the midpoint of the right vertical edge of the rectangle.
- Move the pointer toward the right.

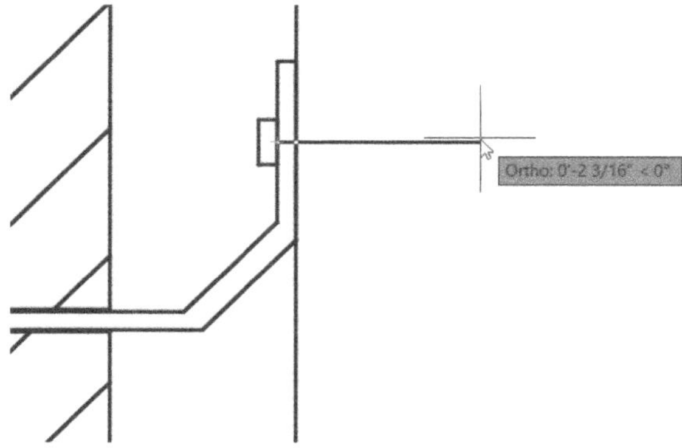

- Type 3.5 and press ENTER.
- Click the **Offset** tool on the **Modify** panel of the **Home** ribbon tab.
- Type 0.1 and press ENTER.
- Offset the newly created horizontal line on both sides.

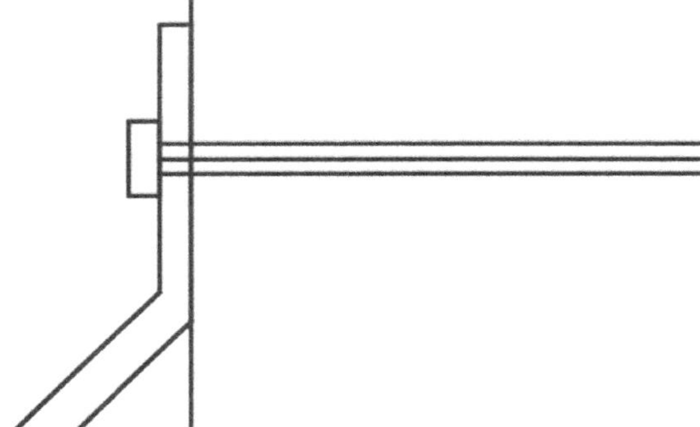

- Press ESC.
- Select the center line and press Delete.
- Click the **Arc** drop-down > **Start, End, Direction** on the **Draw** panel of the **Home** ribbon tab.
- Select the endpoints of the offset lines.
- Move the pointer toward right and click.

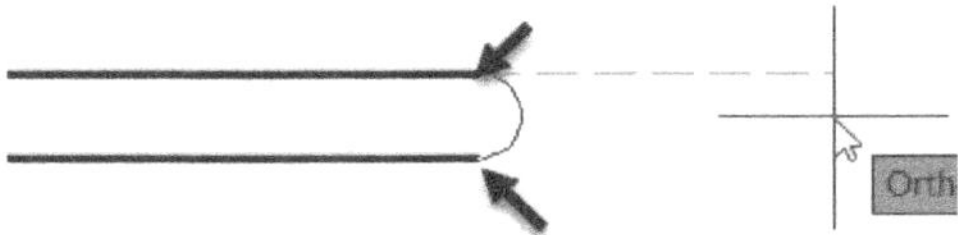

- Click the **Trim** tool on the **Modify** panel of the **Home** ribbon tab.
- Trim the portion of the vertical line, as shown.

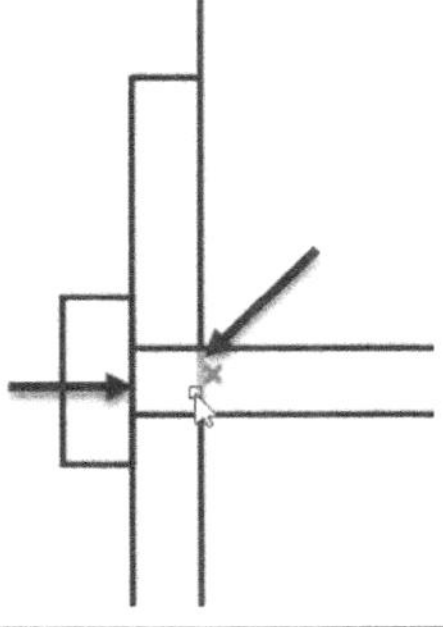

- Click the **Hatch** tool on the **Draw** panel of the **Home** ribbon tab.
- Click the **SOLID** pattern on the **Pattern** panel.

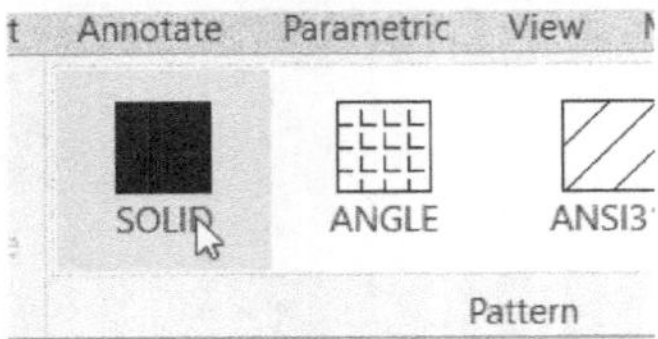

- Click in the regions, as shown.

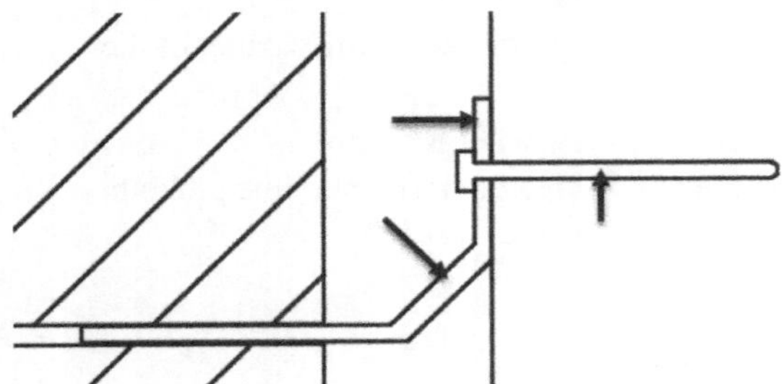

- Click **Close Hatch Creation** on the ribbon.

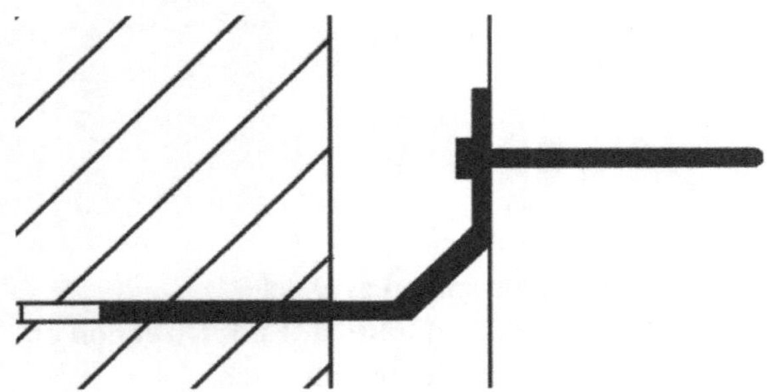

Creating the Insulation

- Click the **Offset** tool on the **Modify** panel of the **Home** ribbon tab.
- Type 1 and press ENTER.
- Select the vertical line, as shown.
- Move the pointer toward right and click.
- Likewise, offset the bottom vertical line.

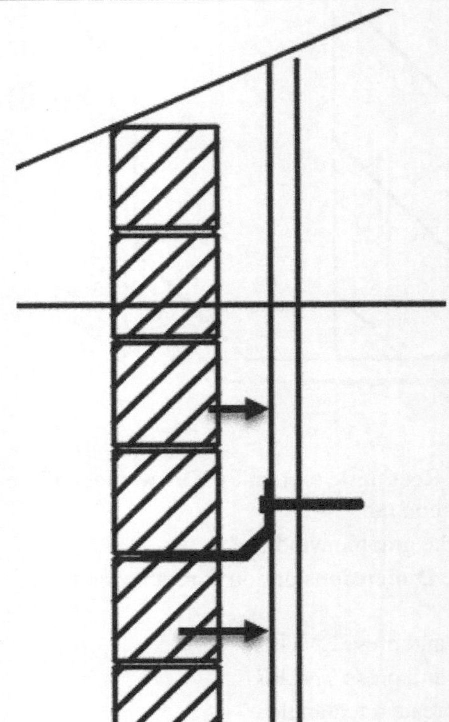

- Press ENTER twice.
- Type 9 and press ENTER.
- Select the offset line.
- Move the pointer toward right and click.

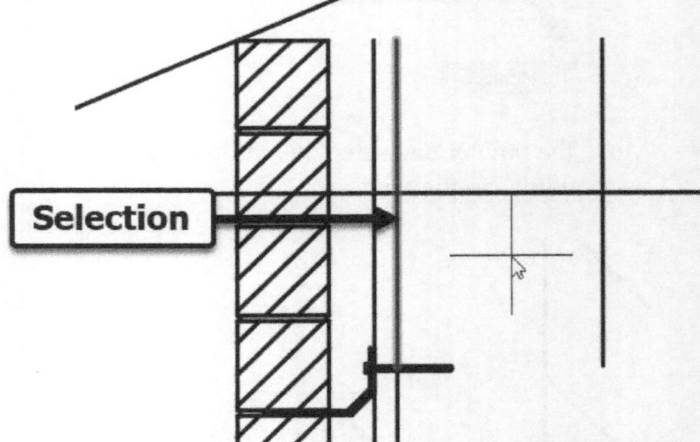

- Click the **Trim** drop-down > **Extend** on the **Modify** panel of the **Home** ribbon tab.
- Click on the lower portion of the newly offset line.

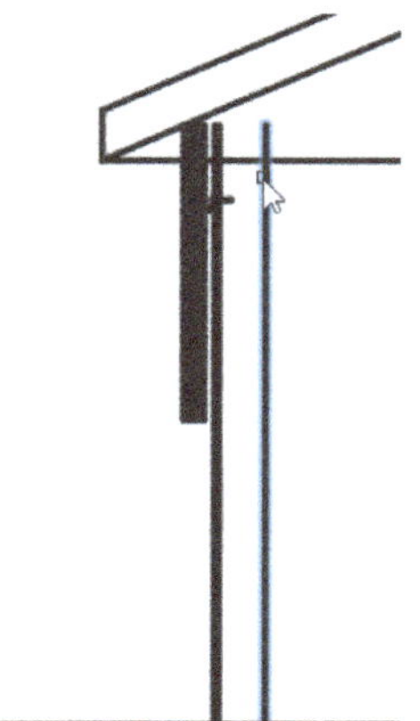

- Click the **Offset** tool on the **Modify** panel of the **Home** ribbon tab.
- Type 0.75 and press ENTER.
- Select the offset line.
- Move the pointer toward right and click.

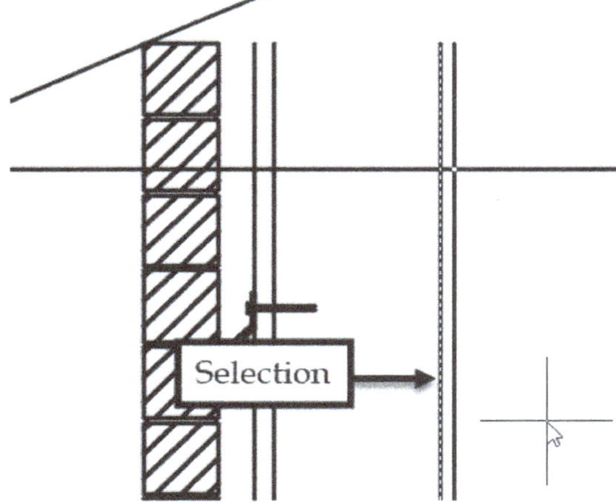

- Click the **Trim** drop-down > **Extend** on the **Modify** panel of the **Home** ribbon tab.
- Click and drag the pointer across the ends of the vertical lines, as shown.

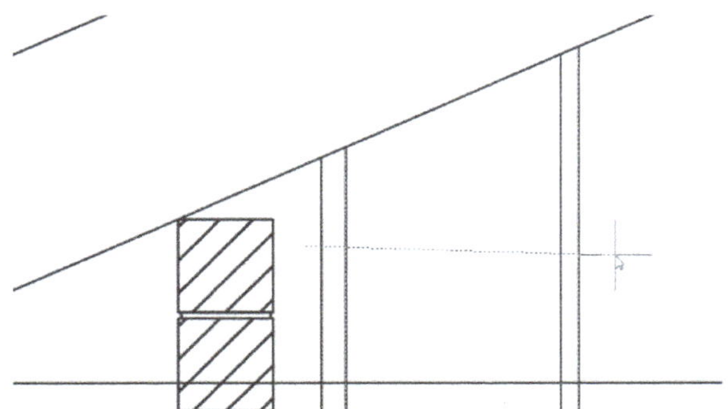

- Create the two horizontal lines, as shown.

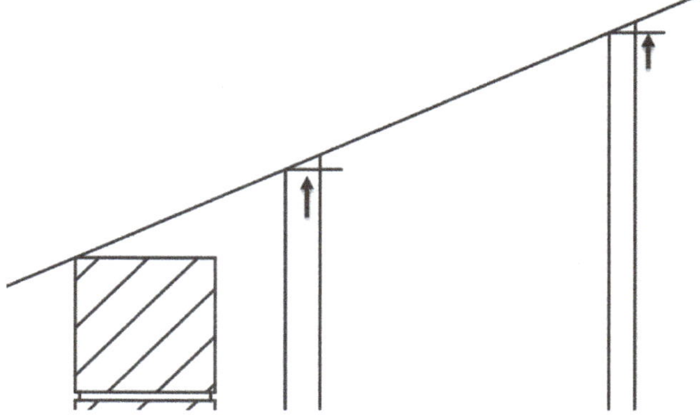

- Trim the extending portions.

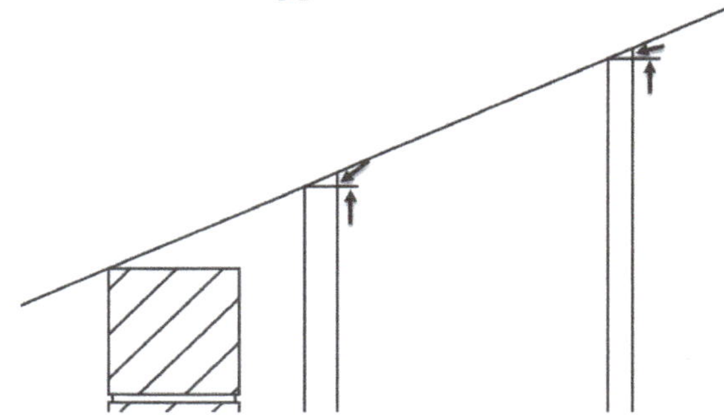

- Click the **Rectangle** tool on the **Draw** panel of the **Home** ribbon tab.
- Select the corner point, as shown.

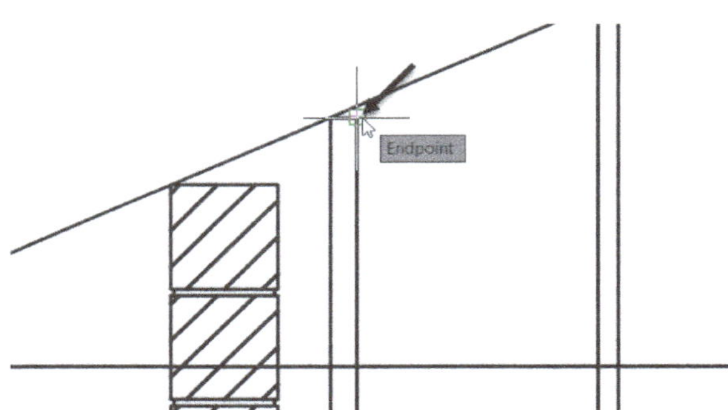

- Select the **Dimensions** option from the command line.
- Type 9 and press ENTER.
- Type 2.25 and press ENTER.
- Move the pointer downward and click.
- Click the **Line** tool on the **Draw** panel of the **Home** ribbon tab.
- Create the diagonal lines by selecting the corner points of the rectangle.

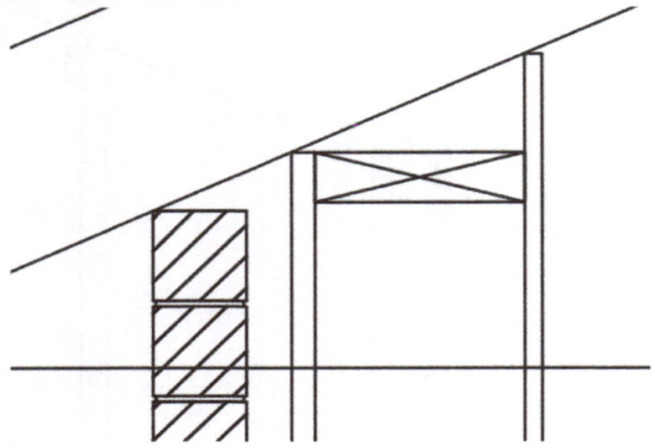

- Expand the **Draw** panel on the **Home** ribbon tab, and then click the **Donut** tool.

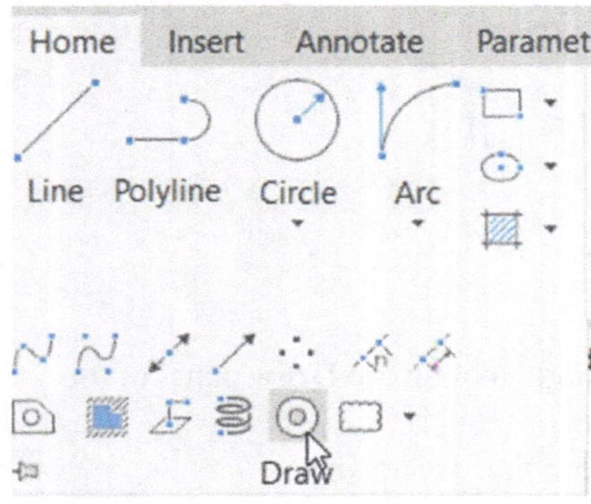

- Type 0 as the internal diameter, and then press ENTER.
- Type 1 as the external diameter, and then press ENTER.
- Select the midpoint of the left vertical edge of the rectangle.

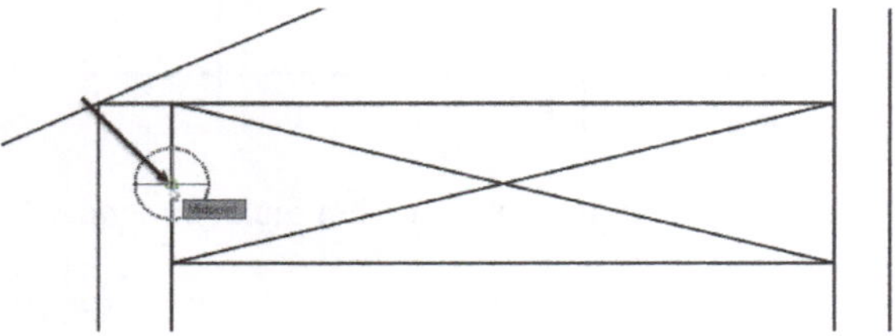

- Press Esc.
- Click the **Line** tool on the **Draw** panel of the **Home** ribbon tab.
- Create a horizontal line, as shown.

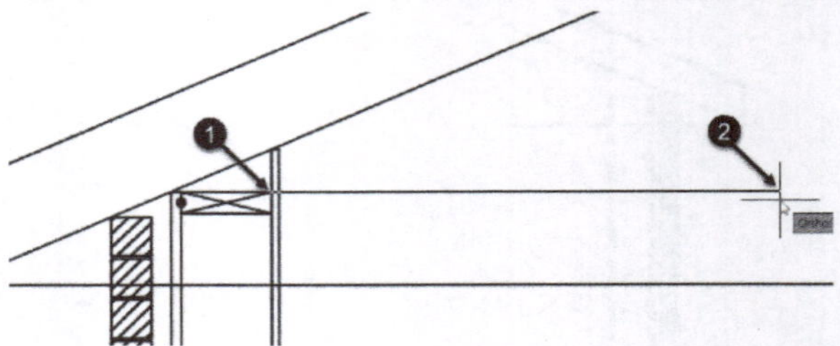

- Click the **Offset** tool on the **Modify** panel of the **Home** ribbon tab.
- Type 1.5 and press ENTER.
- Select the newly created line.
- Move the pointer downward and click.

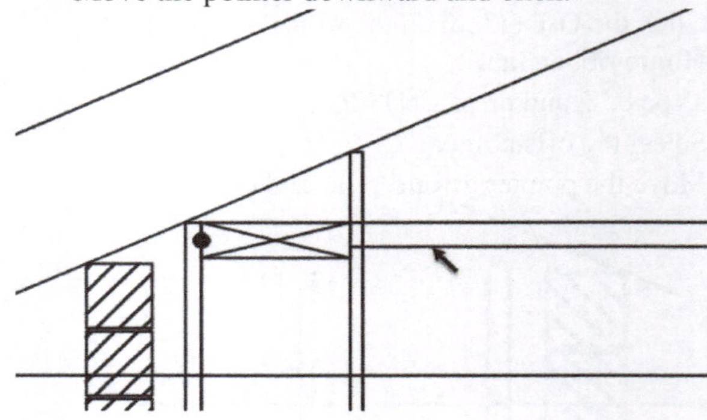

- Press ENTER twice.
- Type 0.75 and press ENTER.
- Select the offset line.
- Move the pointer downward and click.

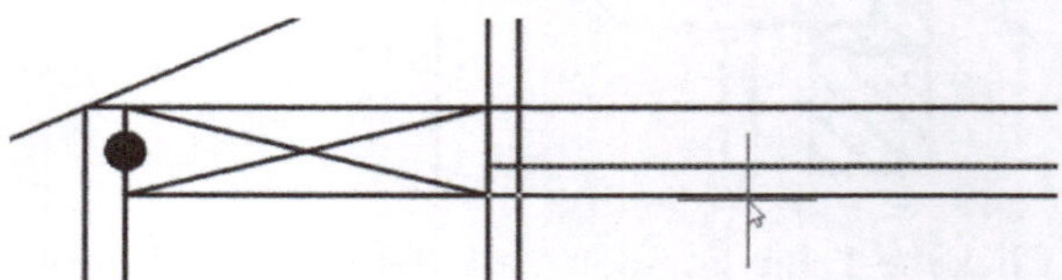

- Press ENTER.
- Click the **Trim** tool on the **Modify** panel of the **Home** ribbon tab.
- Trim the elements, as shown.

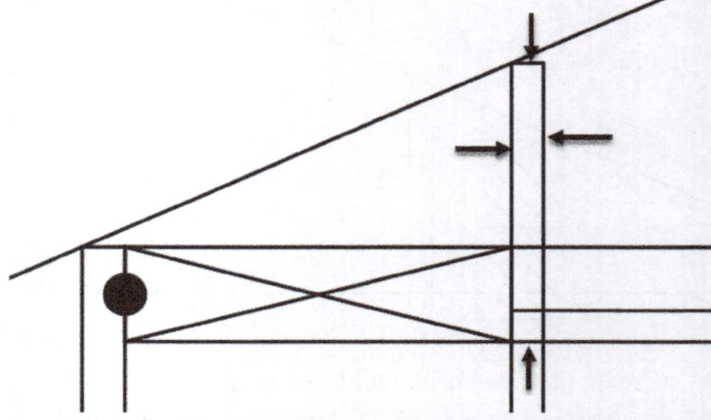

- Click the **Hatch** tool on the **Draw** panel of the **Home** ribbon tab.

- Click the **ANSI37** pattern on the **Pattern** panel.

- Type **2** in the **Hatch Pattern Scale** box on the **Properties** panel.
- Type **45** in the **Angle** box.

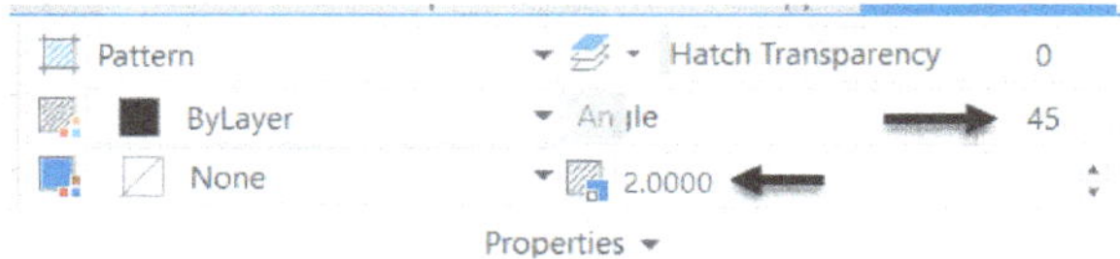

- Click in the regions, as shown.

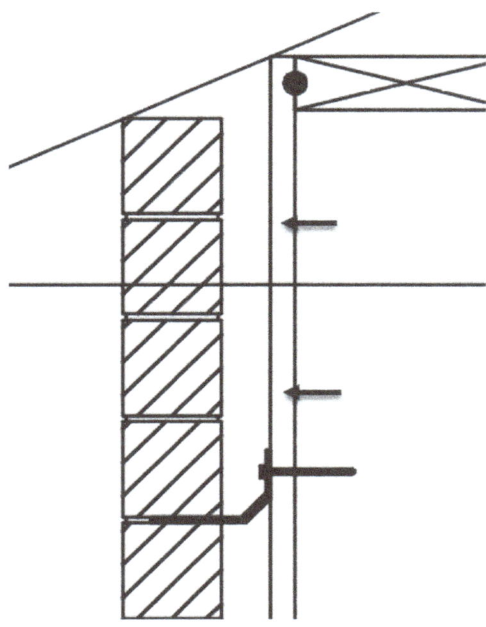

- Click in the area below the brick tie; the **Hatch – Boundary Definition Error** message box appears.

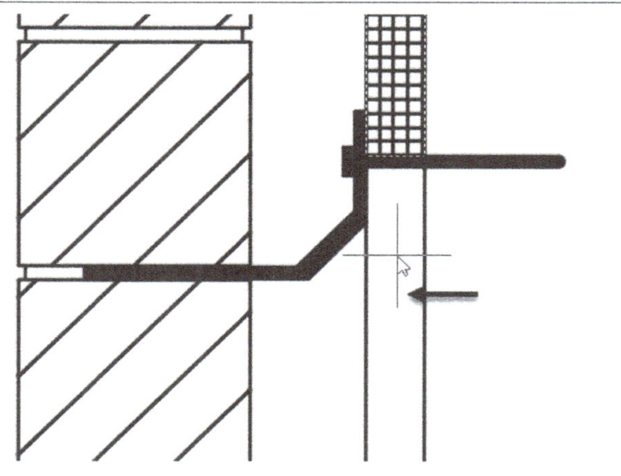

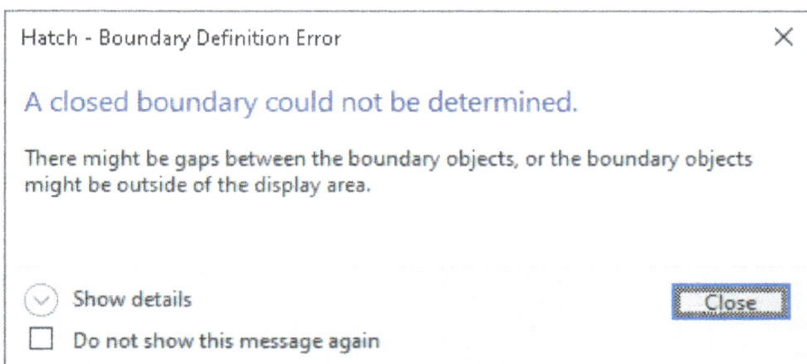

- Click **Close** on the message box.
- Click **Close Hatch Creation** on the ribbon.
- Click the **Line** tool on the **Draw** panel of the **Home** ribbon tab.
- Select the end points of the line, as shown.

- Press ENTER twice.
- Zoom to the lower portion of the and select the lower endpoints of the vertical lines, as shown.

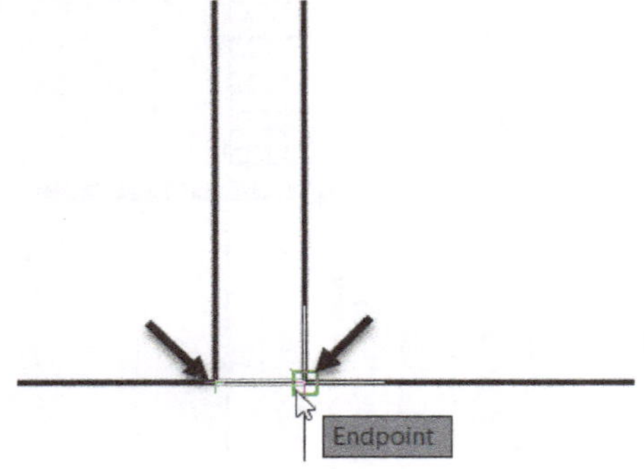

- Expand the **Draw** panel on the **Home** ribbon tab, and then click the **Region** tool.

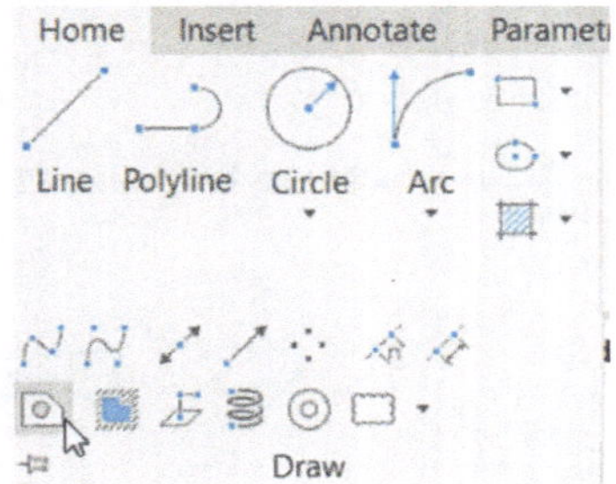

- Select the vertical and horizontal lines, as shown.

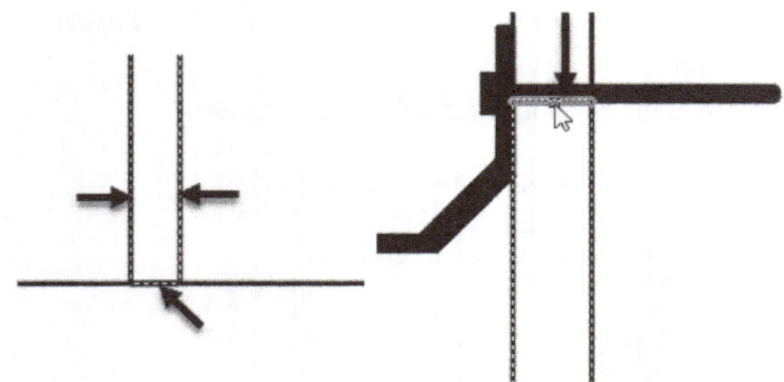

- Press ENTER.
- Click the **Hatch** tool on the **Draw** panel of the **Home** ribbon tab.
- Click in the area below the brick tie.
- Click **Close Hatch Creation** on the ribbon.

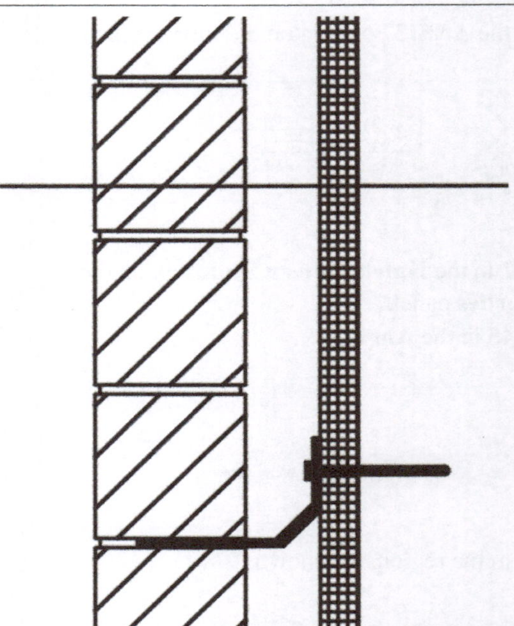

- Click the **Line** tool on the **Draw** panel of the **Home** ribbon tab.
- Select the midpoint of the horizontal line, as shown.
- Move the pointer downward and click.

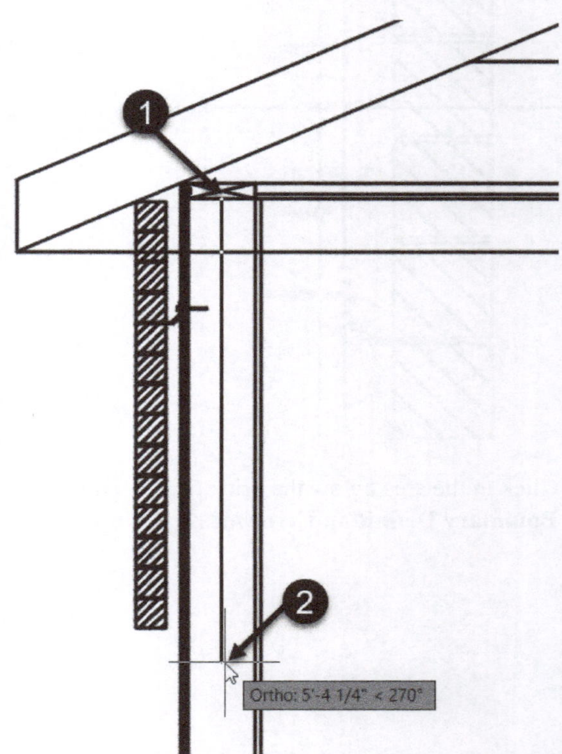

- Click the **Layer Properties** icon on the **Layers** panel of the **Home** ribbon tab.
- Click the **New Layer** icon on the Layer Properties Manager.
- Type **Insulation** as the layer name.

- Click in the **Linetype** column on the Layer Properties Manager.

- Click the **Load** button on the **Select Linetype** dialog.
- Select the **BATTING** linetype from the linetype list.
- Click **OK**.

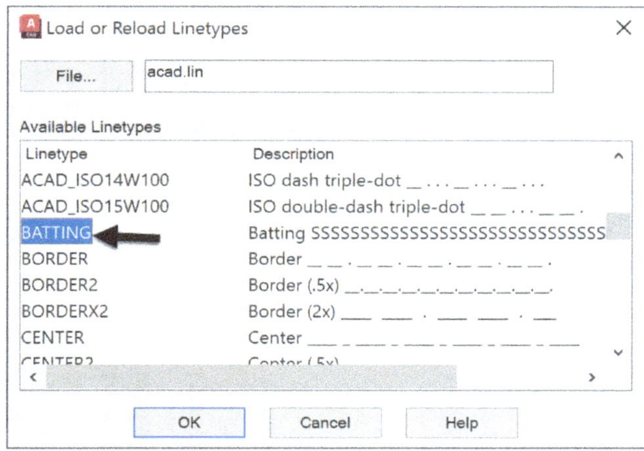

- Select the **BATTING** linetype from the **Select Linetype** dialog.

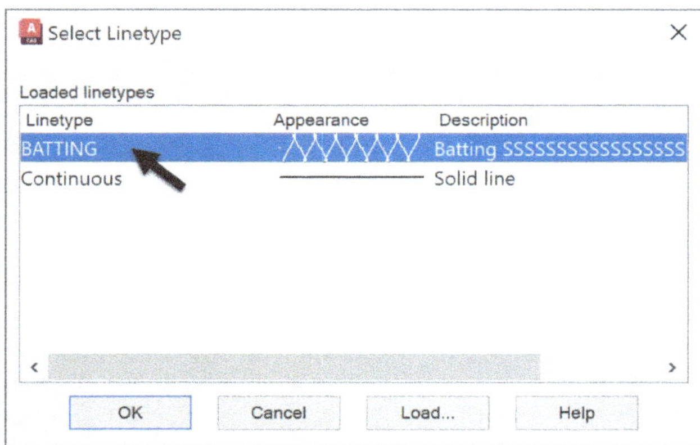

- Click **OK**.
- Select the newly created line.
- Select **Insulation** from the **Layer** drop-down on the **Layers** panel.

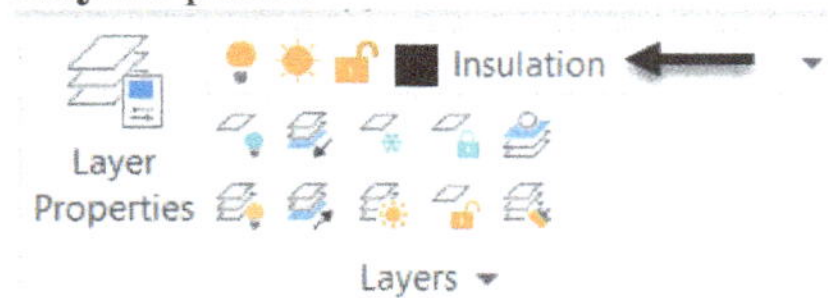

- Select the newly created line.
- Click the **Properties** icon on the **Properties** panel.

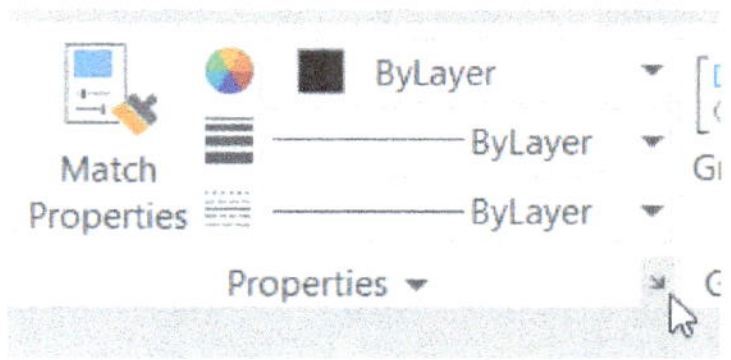

- Type 10 in the **Linetype Scale** box and press ENTER.

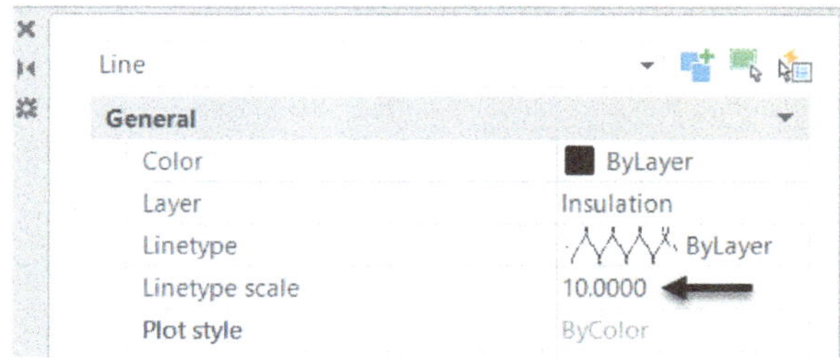

- Press ESC.

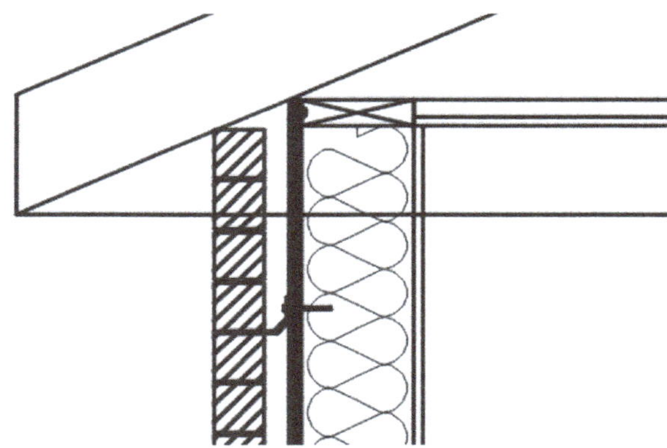

- Click the **Offset** tool on the **Modify** panel of the **Home** ribbon tab.
- Type 3 and press ENTER.
- Select the horizontal line, as shown.
- Move the pointer upward and click.

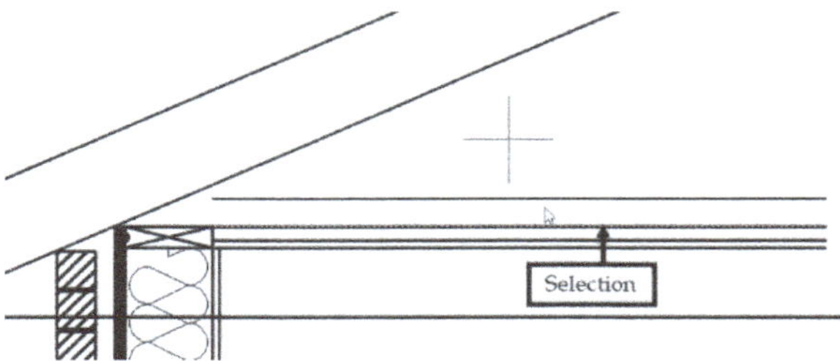

- Press ESC.
- Select the newly created line.
- Select **Insulation** from the **Layer** drop-down on the **Layers** panel.

- Click the **Properties** icon on the **Properties** panel.
- Type 6 in the **Linetype Scale** box and press ENTER.
- Press ESC.

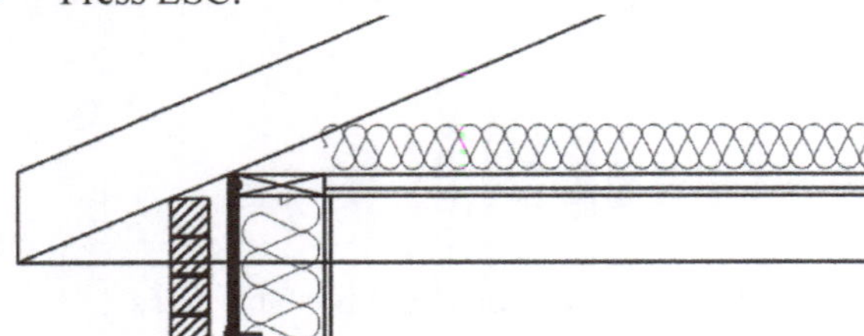

Creating the Roof Detail

- Offset the inclined line by 1".

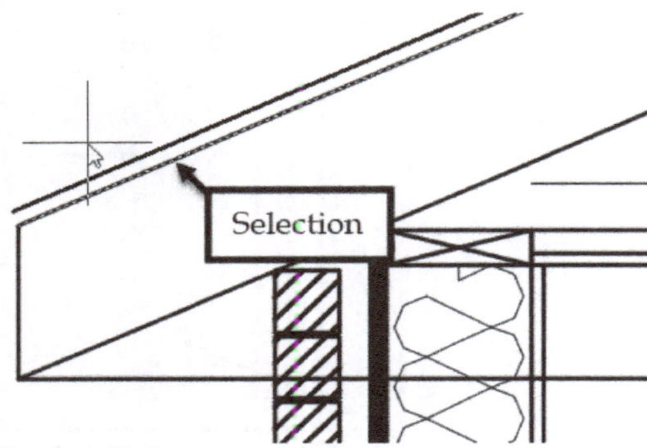

- Close the end of the offset lines.

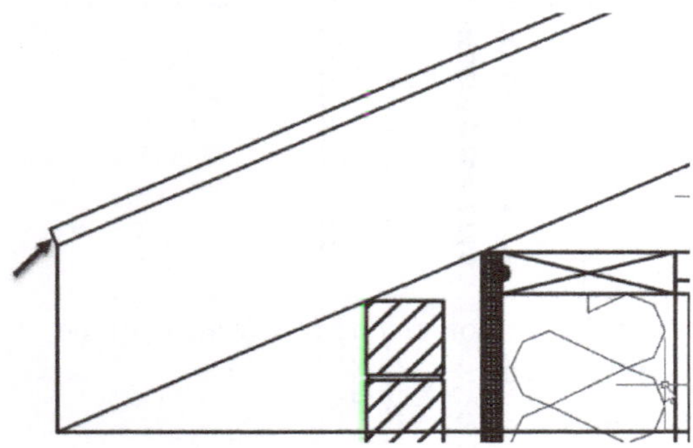

- Offset the new line by 0.5".

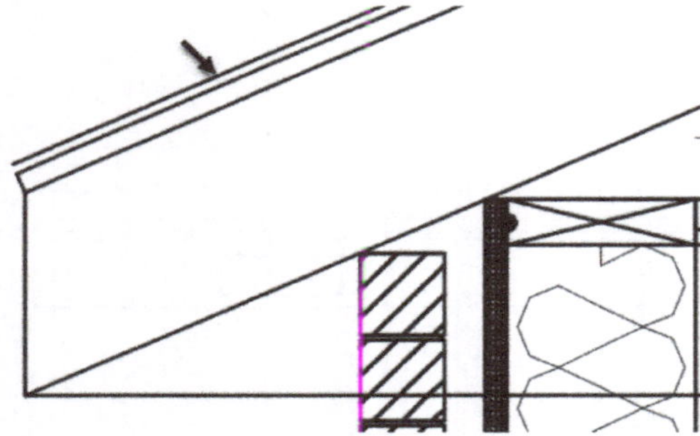

- Select the newly created offset line.
- Select HIDDENX2 from the **Linetype** drop-down on the **Properties** panel.

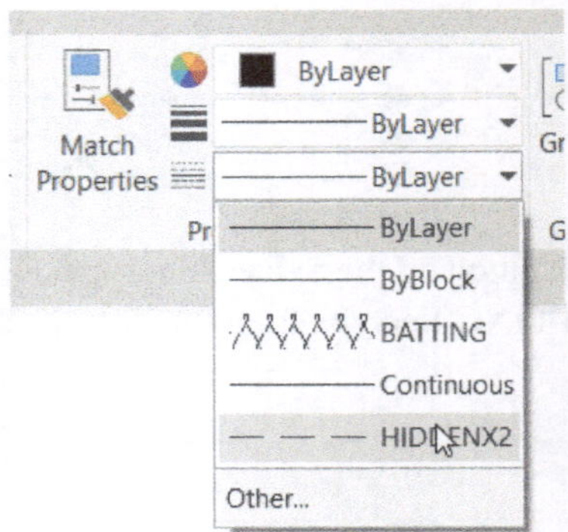

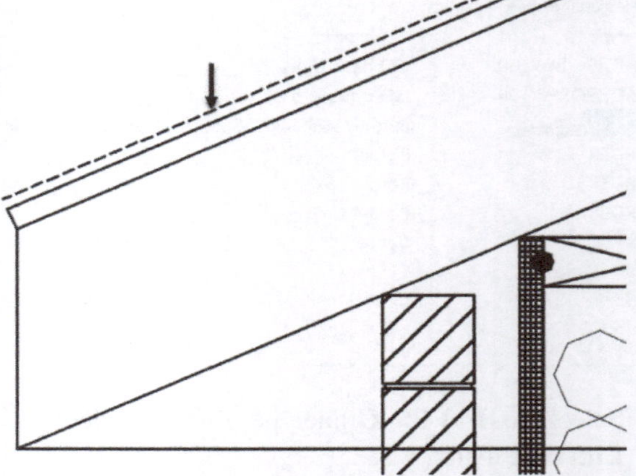

- Create another line with the offset distance of 0.75".

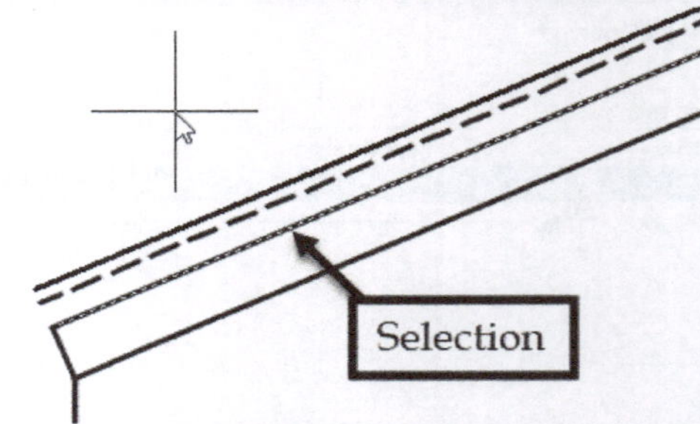

- Expand the **Modify** panel and click the **Lengthen** tool.
- Select the **DElta** option from the command line.
- Type 2.5 and press ENTER.
- Select the newly offset line.

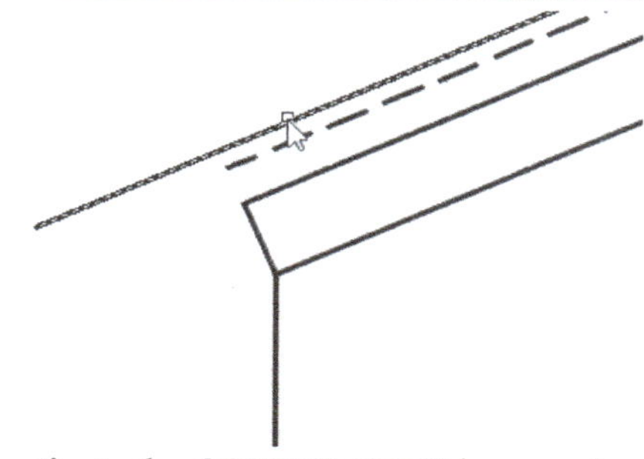

- Deactivate the ORTHOMODE icon on the status bar.
- Click the **Line** tool on the **Draw** panel of the **Home** ribbon tab.
- Select the endpoint of the inclined line, as shown.

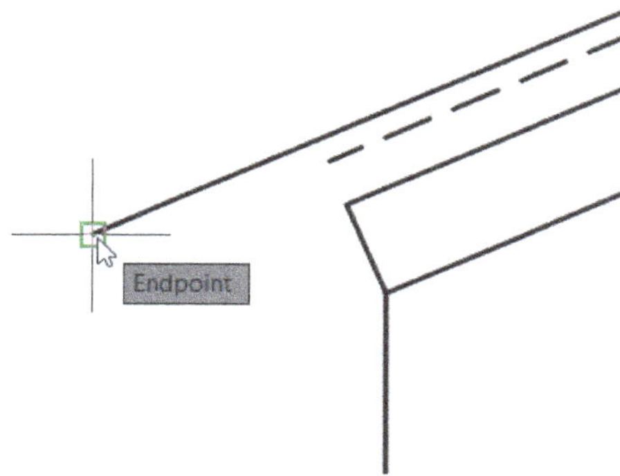

- Press and hold the SHIFT key, and then right click.
- Select the **Parallel** option from the shortcut menu.

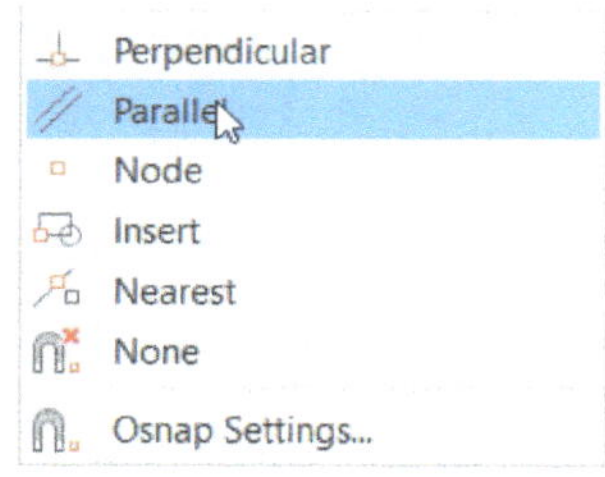

- Move the pointer and hover it on the inclined line, as shown.

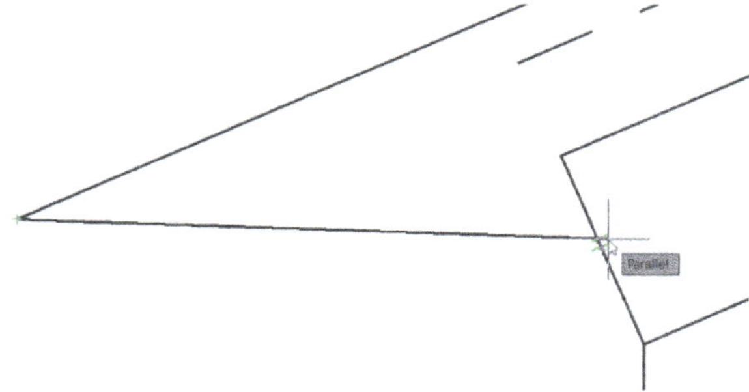

- Move the pointer inline to the start point of the line; a tracking line appears with a parallel symbol, as shown.

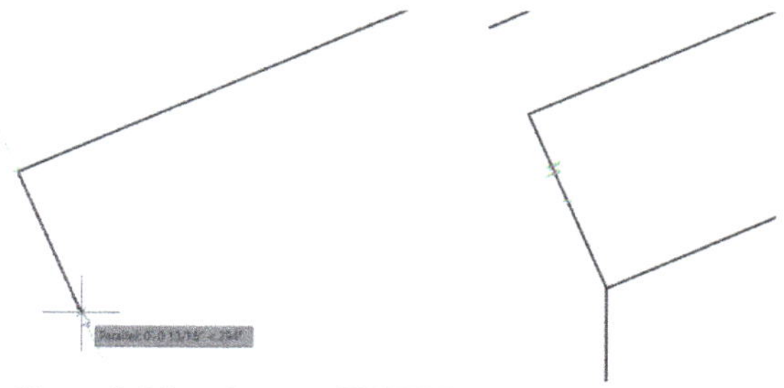

- Type 0.25 and press ENTER.
- Press and hold the SHIFT key, and then right click.
- Select the **Parallel** option from the shortcut menu.
- Move the pointer toward the right and hover the pointer on the endpoint of the hidden line; the tracking line appears from the endpoint.

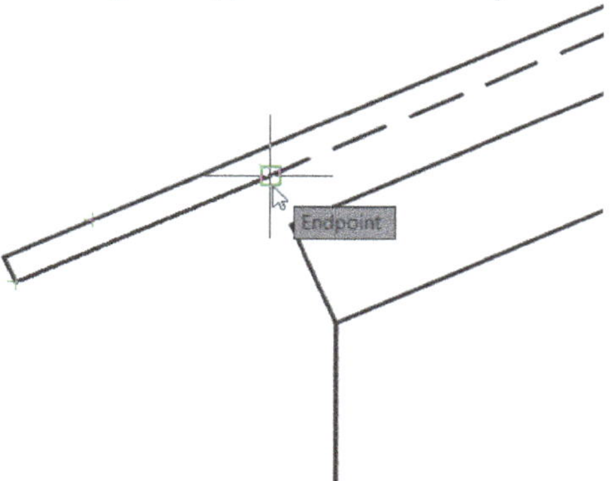

- Type 0.375, and press ENTER.

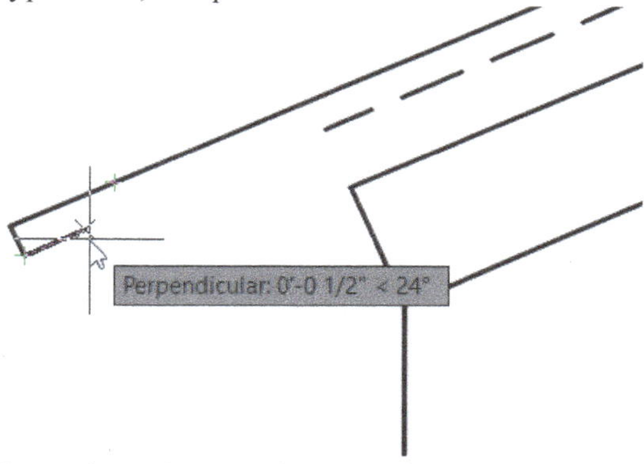

- Hover the pointer on the endpoint of the previous line and move the pointer downward; a tracking line appears from the endpoint.

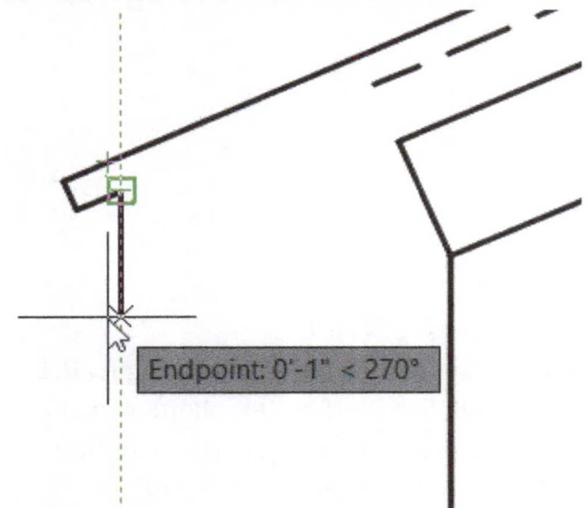

- Type 1.5 and press ENTER.
- Type @.375<225 in the command line and press ENTER.
- Press Esc.

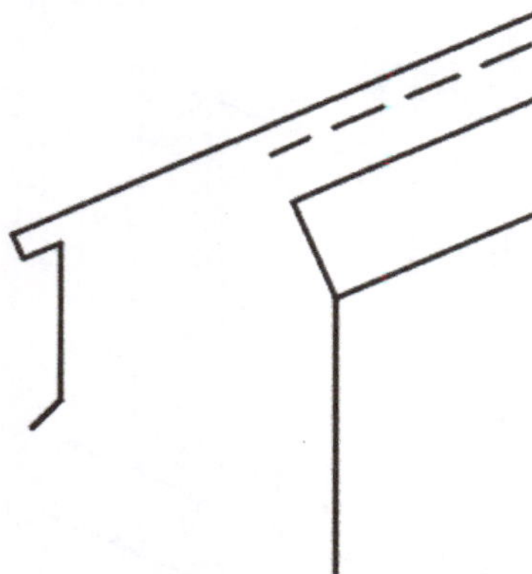

- On the **Draw** panel, click **Arc** drop-down > **Start, End, Direction**.
- Select the end points of the line, as shown.
- Press and hold the Ctrl key and select the endpoint, as shown.

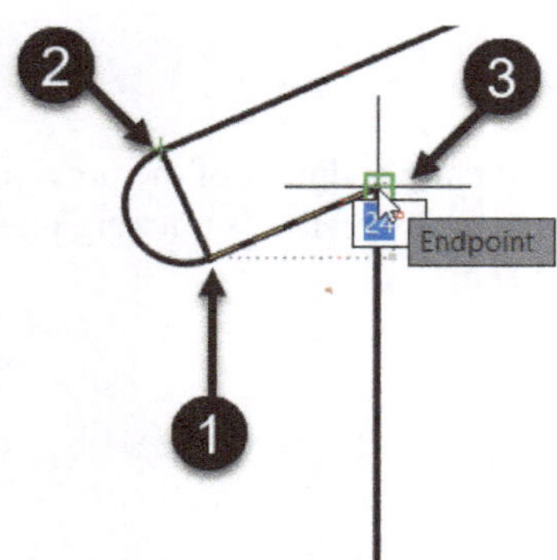

- Delete the line between the endpoints of the arc.

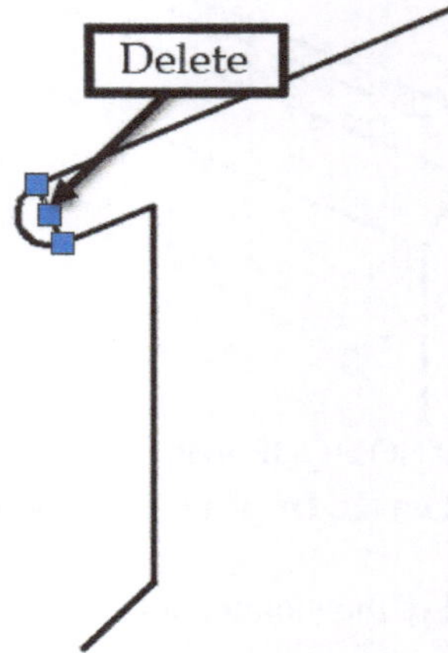

- Expand the **Modify** panel and click the **Break at point** tool.
- Select the inclined line.

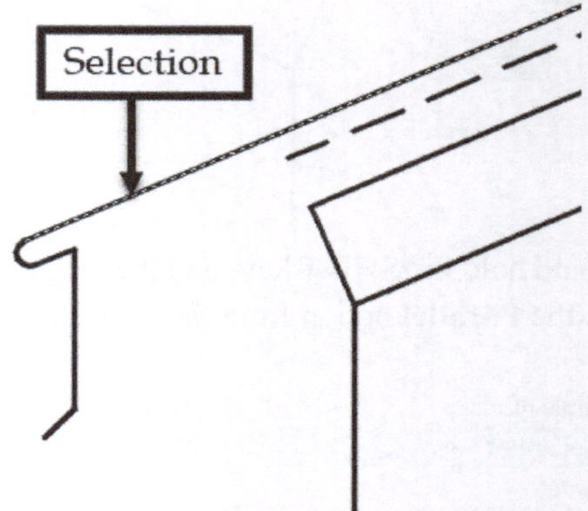

- Press and hold the SHIFT key, and then right click.
- Select the **From** option from the shortcut menu.
- Select the endpoint of the selected line.

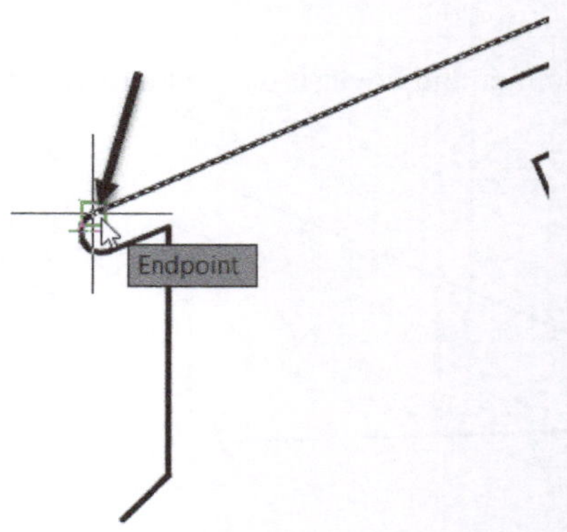

- Move the pointer on the selected line.
- Type 9 and press ENTER; the line is broken at the specified distance.
- Delete the right-side portion of the line.

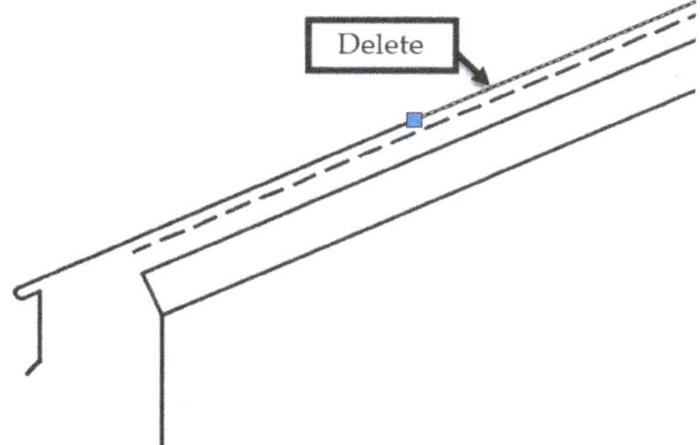

- Click the **Rectangle** tool on the **Draw** panel of the **Home** ribbon tab.
- Select the corner point, as shown.
- Select the **Dimensions** option from the command line.
- Type 1 and press ENTER.
- Type 10 and press ENTER.
- Move the pointer toward the left and click.

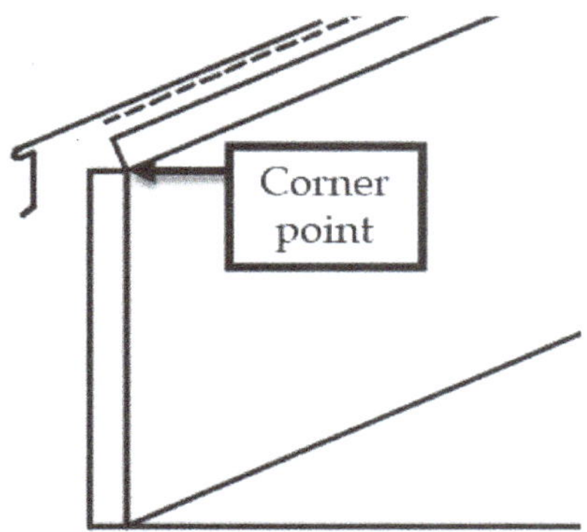

- Click the **Rectangle** tool on the **Draw** panel of the **Home** ribbon tab.
- Select the top-left corner of the rectangle, as shown.
- Select the **Dimensions** option from the command line.
- Type 1 and press ENTER.
- Type 2.5 and press ENTER.
- Move the pointer toward the left and click.

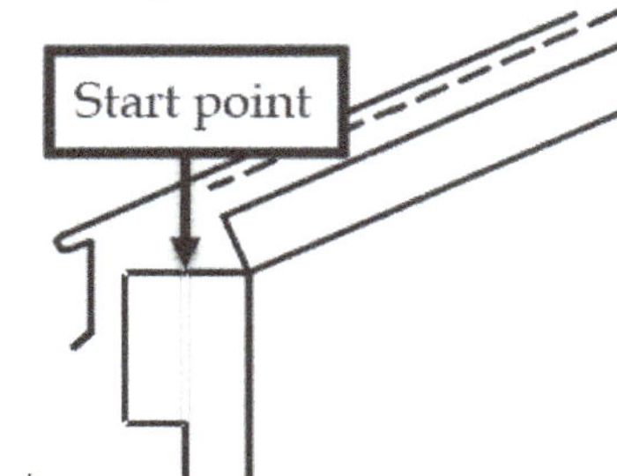

- Move the newly created rectangle downward by 0.5".

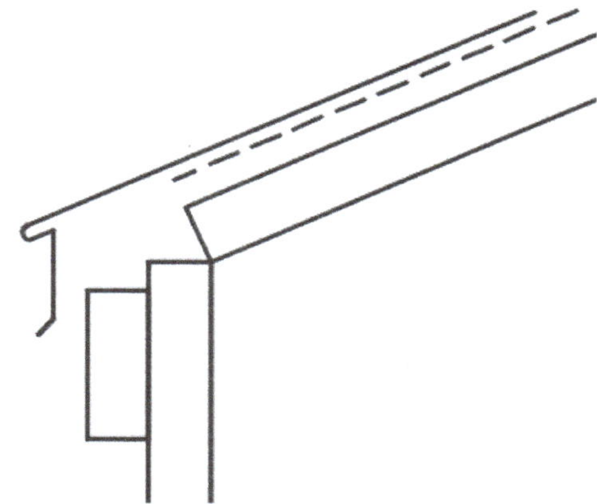

- Click the **Trim** tool on the **Modify** panel of the **Home** ribbon tab.
- Select the **cuTting edges** option from the command line.
- Select the vertical edge of the brick, as shown.
- Press ENTER.
- Click on the horizontal line on the right side.

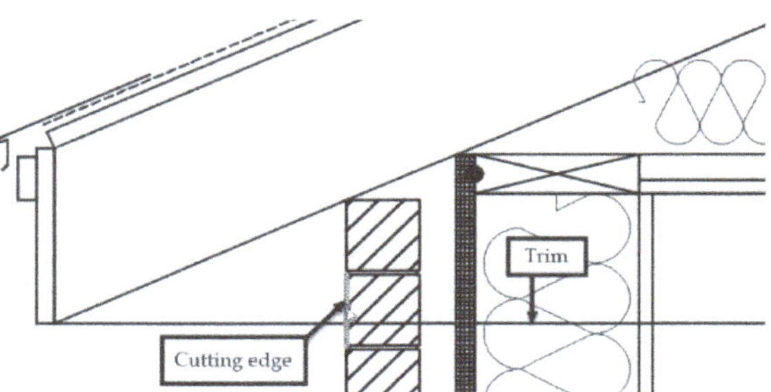

- Click the **Offset** tool on the **Modify** panel of the **Home** ribbon tab.
- Type 1 and press ENTER.
- Select the horizontal line trimmed in the last step.
- Move the pointer upward and click.

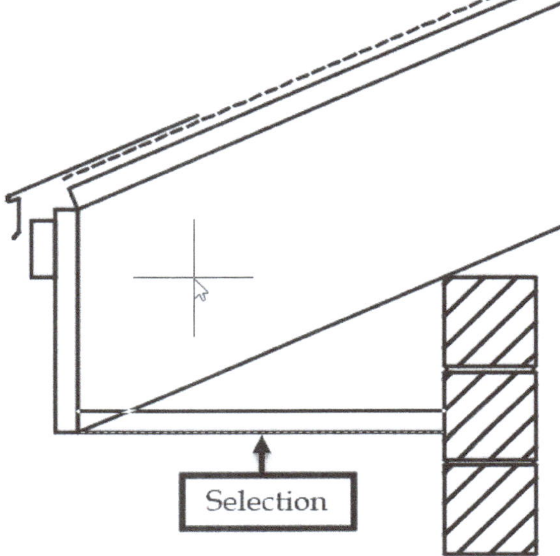

- Select the offset line.
- Move the pointer upward and click.

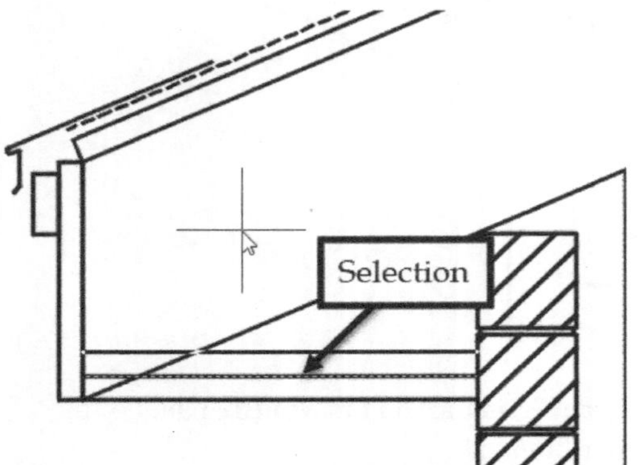

- Click the **Trim** tool on the **Modify** panel of the **Home** ribbon tab.
- Select the portions of the lines, as shown.

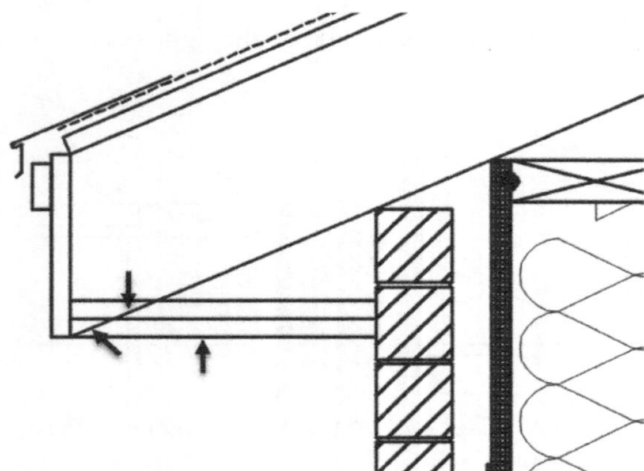

- Click the **Offset** tool on the **Modify** panel of the **Home** ribbon tab.
- Type 7 and press ENTER.
- Select the left vertical line of the veneer brick.
- Move the pointer toward the left and click.

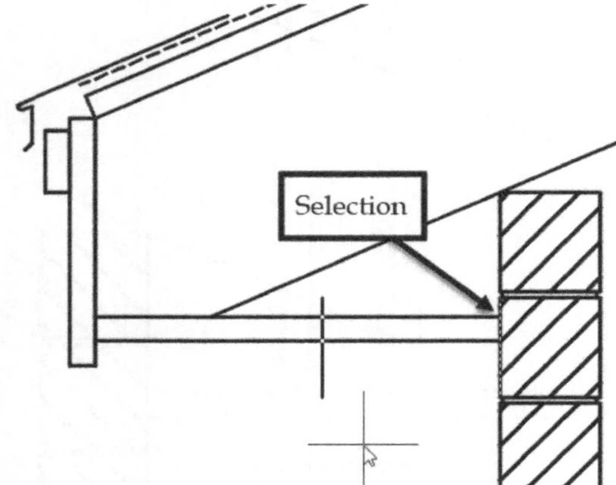

- Press ENTER twice.
- Type 4 and press ENTER.
- Select the offset line.
- Move the pointer toward the left and click.

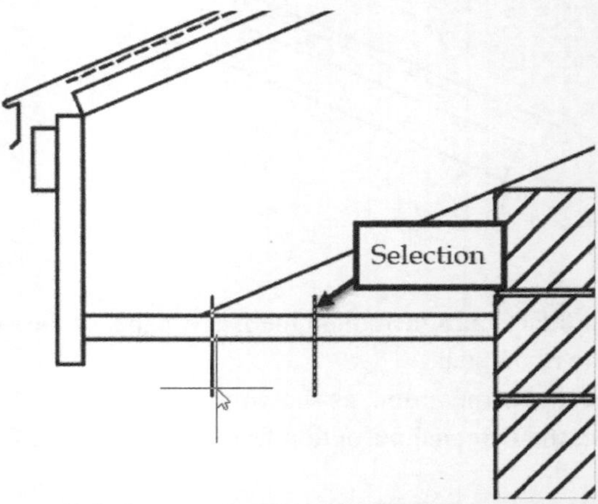

- Click the **Trim** tool on the **Modify** panel of the **Home** ribbon tab.
- Select the portions of the lines, as shown.

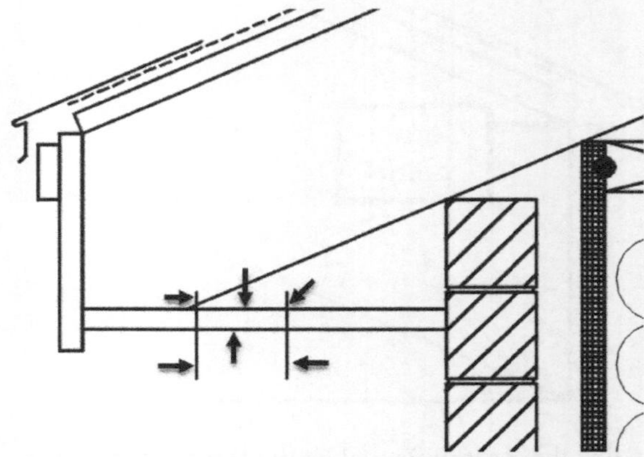

- Create a line by selecting the corner points of the opening, as shown.

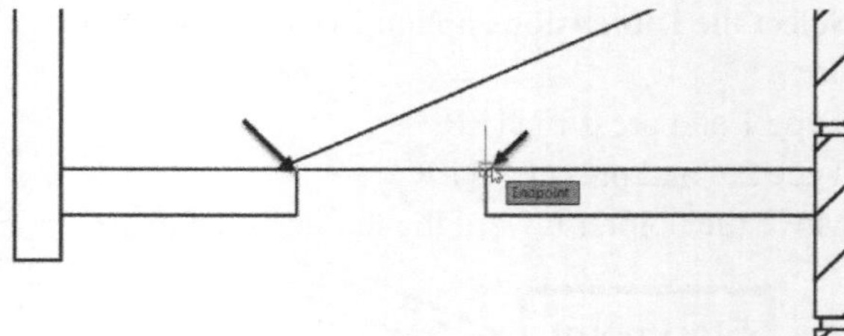

- Press ESC.
- Activate the ORTHOMODE icon on the status bar.
- Select the newly created line and click the **Scale** tool on the **Modify** panel of the **Home** ribbon tab.
- Select the midpoint.

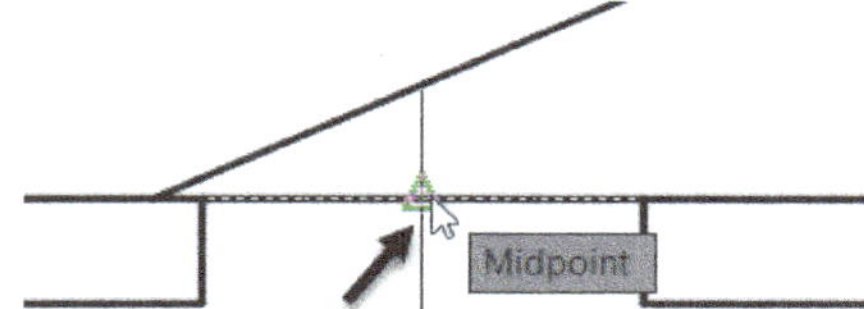

- Move the pointer vertically downward.
- Type 1.25 as the scale factor, and then press ENTER.
- Select the horizontal line and click on its midpoint grip.
- Move the pointer upward.

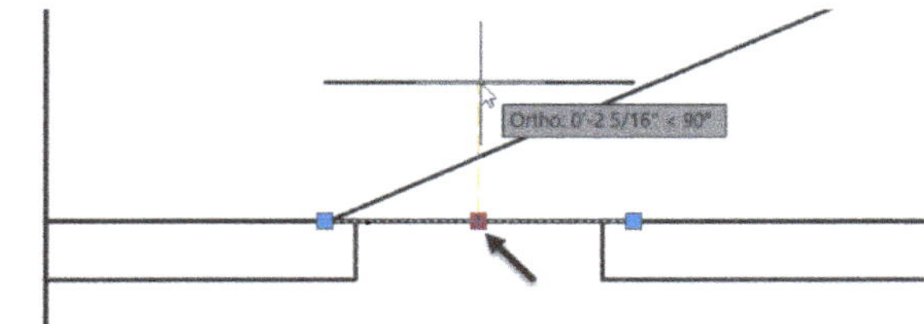

- Type 0.2 and press ENTER.
- Create an inclined line and pattern it, as shown.

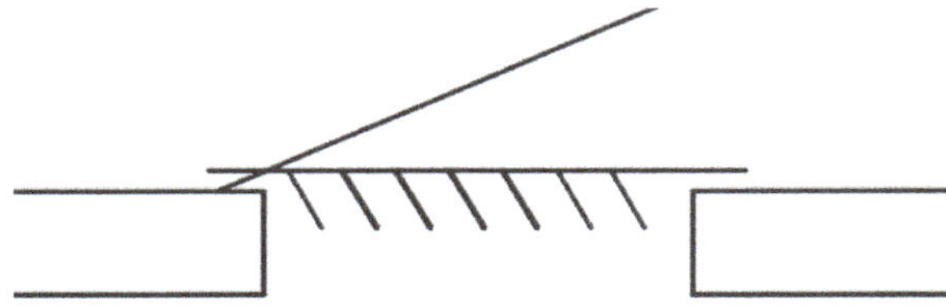

- On the **Home** tab of the ribbon, click **Draw** panel > **Hatch** drop-down > **Hatch**.
- On the **Hatch Creation** tab of the ribbon, expand the **Pattern** gallery, and select the **GOST_WOOD** pattern.

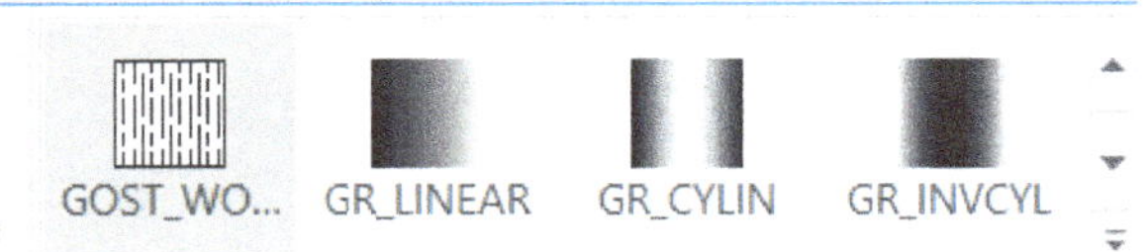

- Pick points in the areas, as shown.

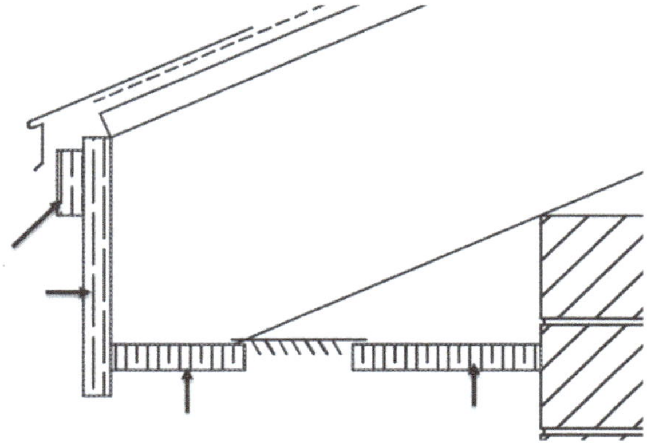

- Type 0.2 in the **Pattern Hatch Scale** box and press ENTER.
- Click the **Close Hatch Creation** icon on the ribbon.
- On the **Home** tab of the ribbon, click **Draw** panel > **Rectangle/Polygon** drop-down > **Rectangle**.
- Select the lower right corner of the rectangle, as shown.
- Select the **Dimensions** option from the command line.
- Type 6 and press ENTER.
- Type 5 and press ENTER.
- Move the pointer downward and click.

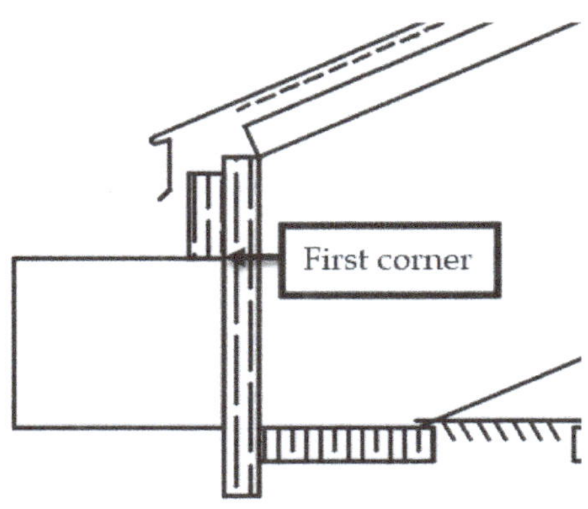

- Select the rectangle and click the **Move** tool on the **Modify** panel.
- Select the top right corner point of the rectangle.
- Move the pointer downward.
- Type 0.5 and press ENTER.
- Select the rectangle and click the **Move** tool on the **Modify** panel.
- Select the top right corner point of the rectangle.
- Move the pointer toward left.
- Type 0.25 and press ENTER.

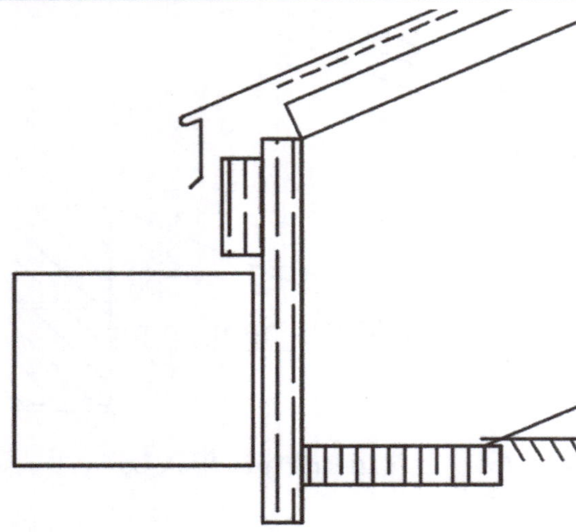

- Click **Circle** drop-down > **Center, Radius** on the **Draw** panel of the **Home** ribbon tab.
- Select the lower-left corner of the rectangle.
- Type 8 and press ENTER.

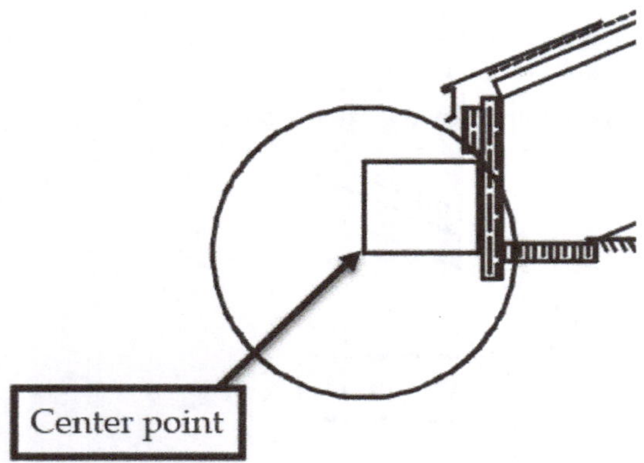

- Select the circle.
- Click on the centerpoint of the circle.
- Move the pointer toward left.
- Type 7 and press ENTER.

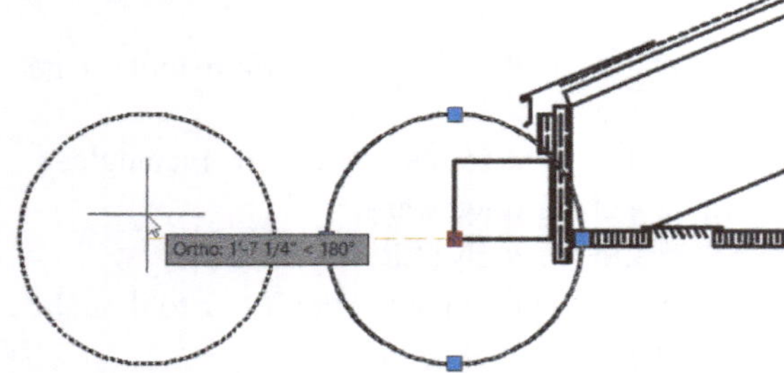

- Click the **Trim** tool on the **Modify** panel.
- Trim the edges of the circle and rectangle, as shown.

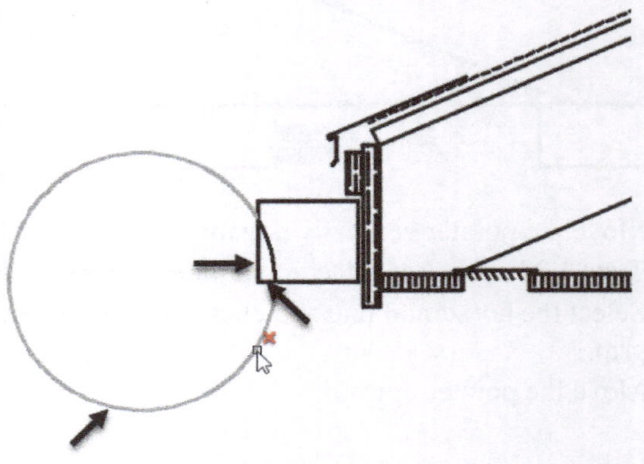

- Press ESC.
- Select the rectangle and click the **Explode** tool on the **Modify** panel.

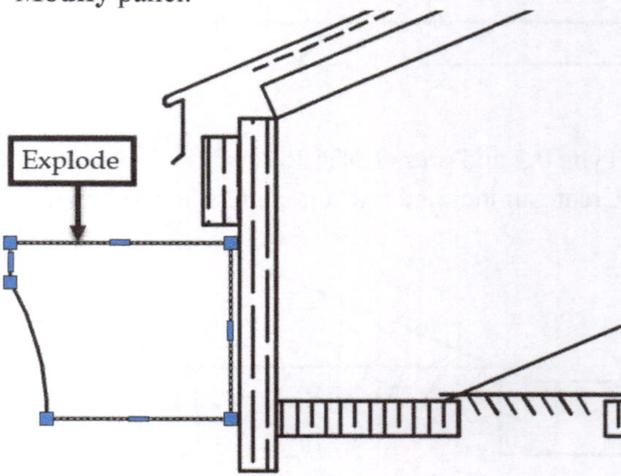

- Select the lines and arcs, as shown in the figure.

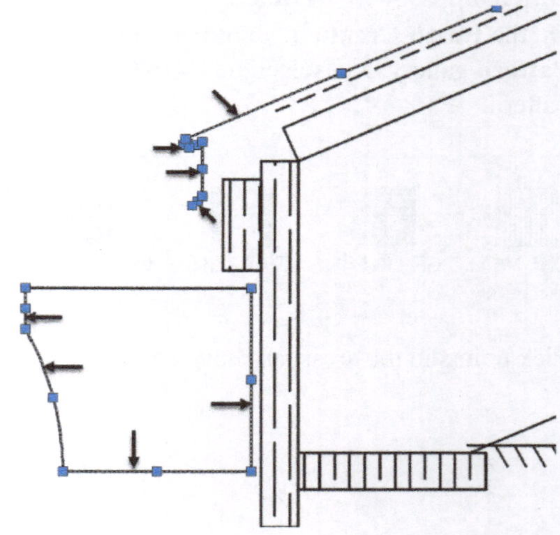

- On the **Home** ribbon tab, click the **Lineweight** drop-down on the **Properties** panel, and then select 0.70.

- Press Esc.
- Click the **Offset** tool on the **Modify** panel.
- Type 1.25 and press ENTER.
- Select the offset line, as shown in the figure.
- Move the pointer upward and click.

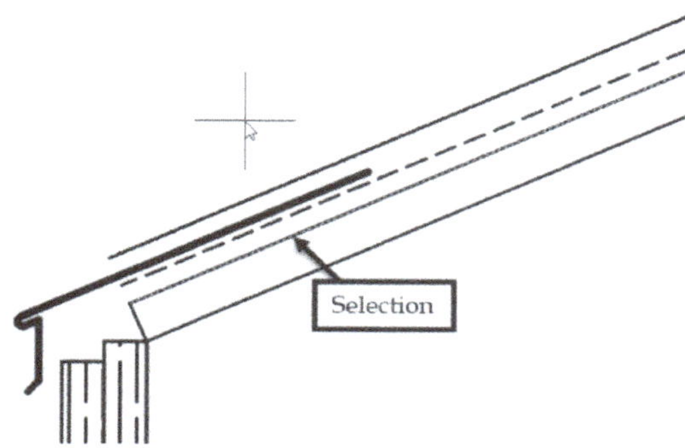

- Deactivate the ORTHOMODE icon on the Status bar.
- Click the **Line** tool on the **Draw** panel.
- Select the endpoint of the offset line.
- Press and hold the SHIFT key and right click.
- Select the **Parallel** option from the shortcut menu.
- Hover the pointer on the line, as shown.

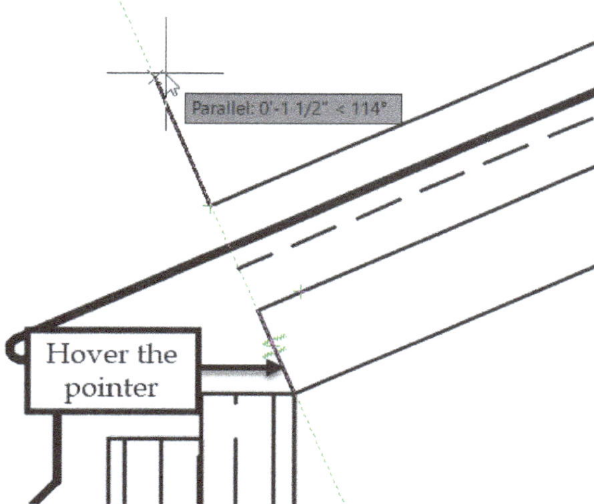

- Move the pointer along the tracking line.
- Type 2 and press ENTER.
- Type @7<20 and press ENTER.
- Press Esc.

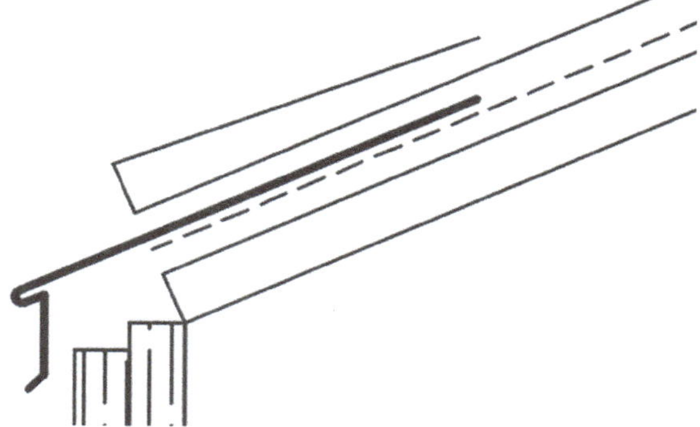

- Select the two newly created lines.
- Click the **Array** drop-down **> Path Array** tool on the **Modify** panel.
- Select the inclined line, as shown.

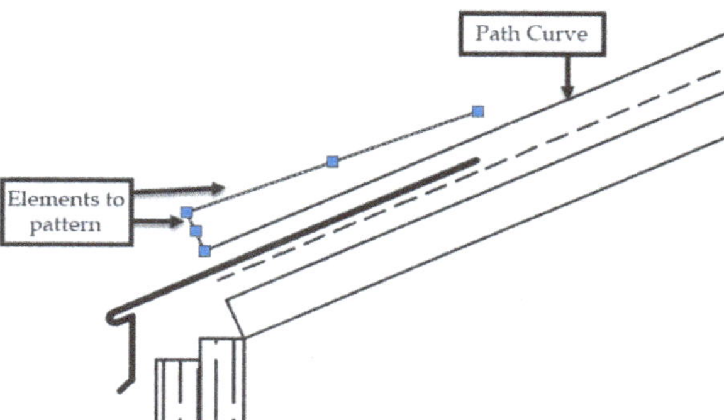

- Type 7 in the **Between** box on the **Items** panel.

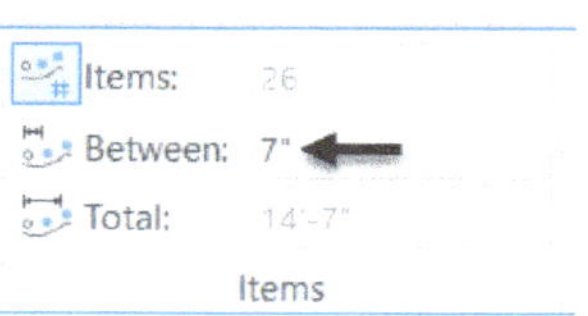

- Click **Close Array** on the ribbon.

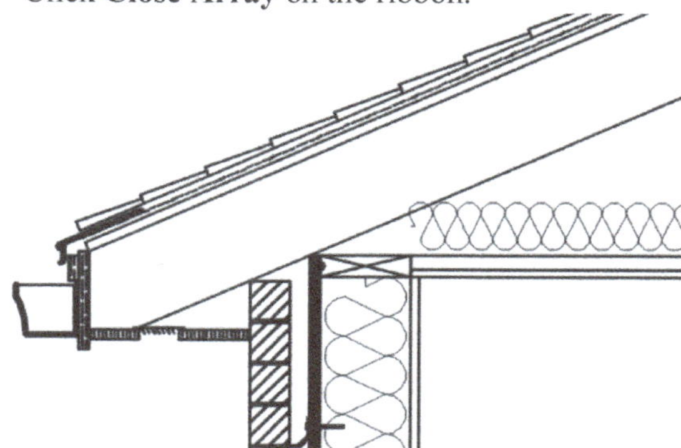

Adding Annotations

- On the **Home** ribbon tab, click **Leader** drop-down > **Leader** on the **Annotation** panel.

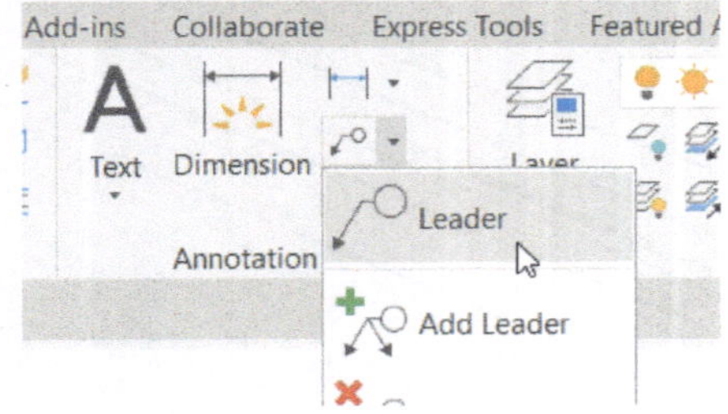

- Deactivate the **ORTHOMODE** icon on the status bar.
- Specify the start point of the leader on the insulation, as shown.
- Move the pointer diagonally toward the bottom right corner, and then click.
- Type **Insulation** and click in the graphics area.

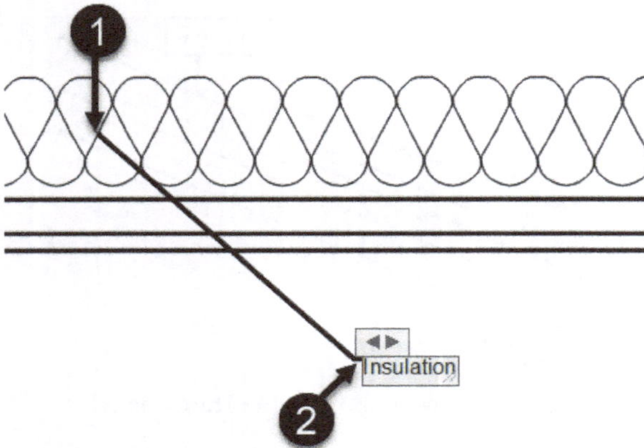

- Select the leader and click the **Properties** icon on the **Properties** panel of the **Home** ribbon tab.

- On the **Properties** palette, scroll to the **Leaders** section and enter 2 in the **Arrowhead size** and **Landing distance** boxes, respectively.

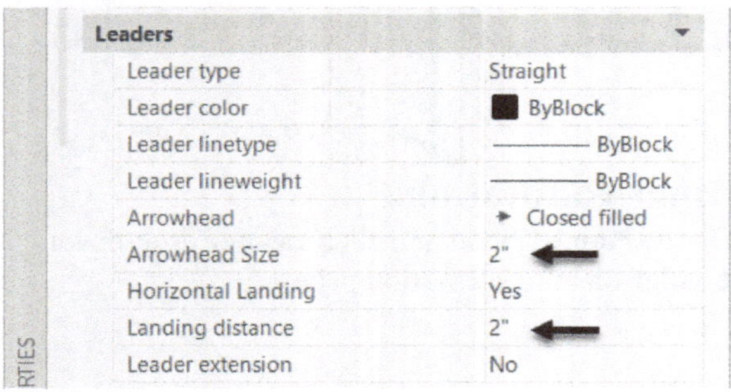

Leaders	
Leader type	Straight
Leader color	ByBlock
Leader linetype	ByBlock
Leader lineweight	ByBlock
Arrowhead	Closed filled
Arrowhead Size	2"
Horizontal Landing	Yes
Landing distance	2"
Leader extension	No

- Scroll to the **Text** section, and then enter **2** and **1** in the **Height** and **Landing gap** boxes, respectively.

Text	
Contents	Insulation
Text style	Standard
Justify	Left
Direction	By style
Width	0"
Height	2"
Rotation	0
Line space factor	1.0000
Line space distance	3 5/16"
Line space style	At least
Background mask	No
Attachment type	Horizontal
Left Attachment	Middle of top line
Right Attachment	Middle of top line
Landing gap	1"
Text frame	No

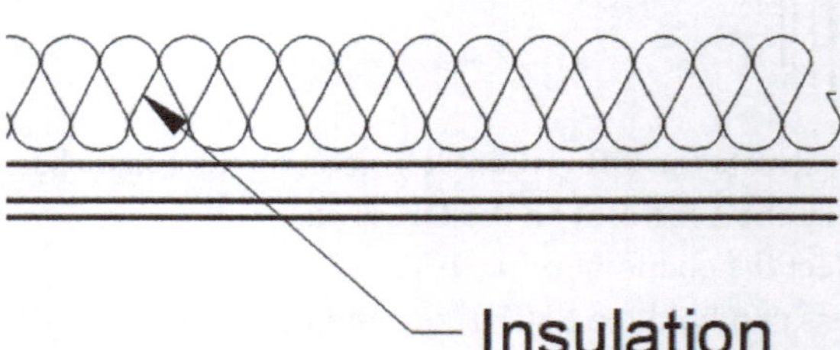

- On the **Home** ribbon tab, click **Leader** drop-down > **Leader** on the **Annotation** panel.
- Specify the start point of the leader on the donut, as shown.
- Move the pointer diagonally toward the bottom right corner, and then click.
- Type Continuous Bead of Sealant and click in the graphics area.

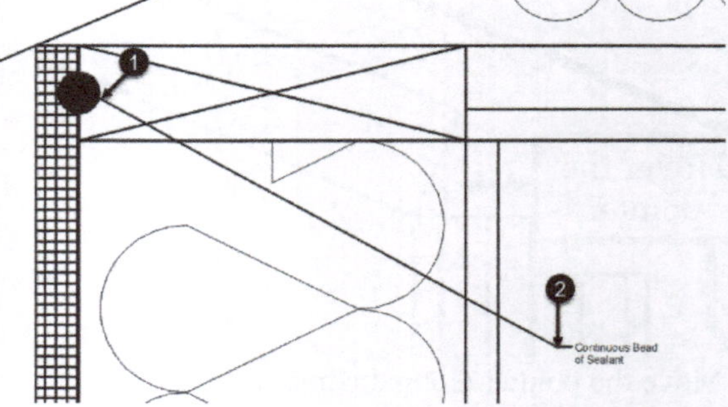

- Click the **Match Properties** tool on the **Properties** panel.

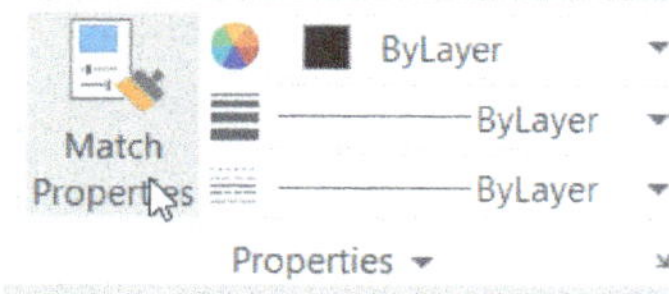

- Select the **Insulation** leader to specify the source object.

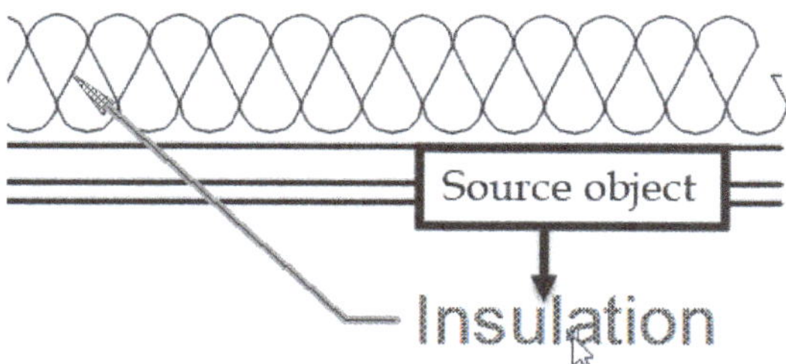

- Select the newly created leader; the properties of the source object are matched with the destination object.

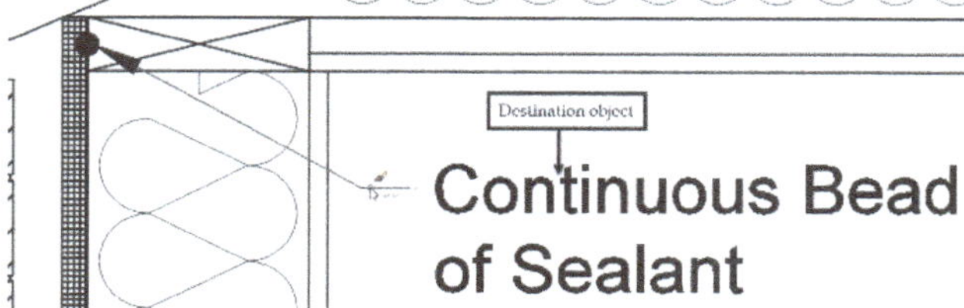

- On the ribbon, click **Annotate > Text > Multiline Text**.
- Click at the location, as shown.

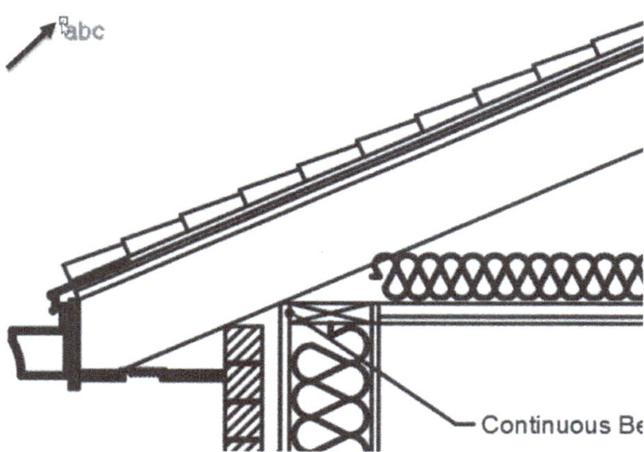

- Select the **Height** option from the command line.
- Type 2" and press ENTER.
- Specify the second corner of the text box, as shown.

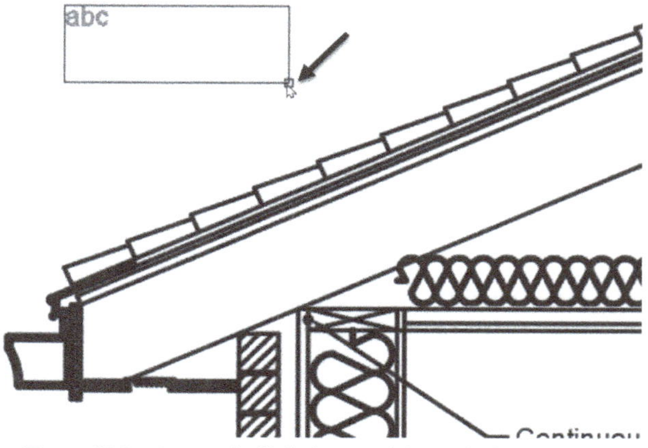

- Type **Shingles** and click in graphics window.
- On the ribbon, click **Annotate > Leaders > Multileader**.
- Select the **select Mtext** option from the command line.
- Select the multiline text from the graphics window.
- Select a point from anyone of the shingles.

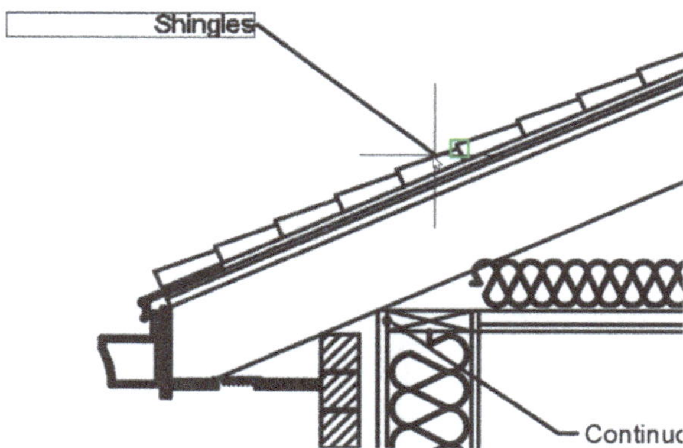

- Click the **Match Properties** tool on the **Properties** panel.
- Select the **Insulation** leader to specify the source object.
- Select the newly created leader.
- Likewise, create the remaining leaders, as shown.

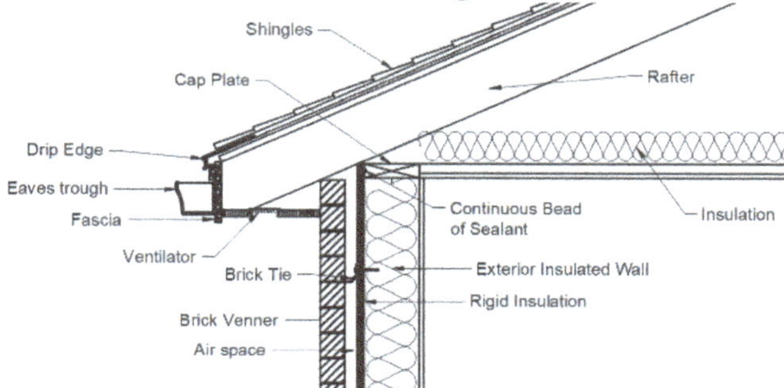

- Activate the **ORTHOMODE** icon on the status bar.
- On the ribbon, click the **Express Tools** tab > **Draw > Break-line Symbol**.

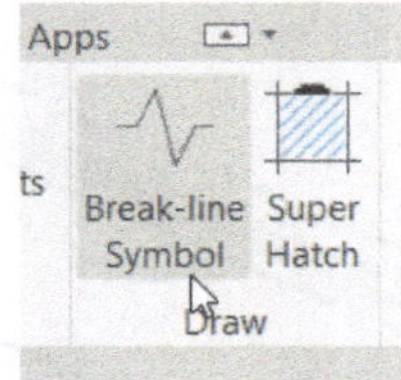

- Select the **Size** option from the command line.
- Type **8** and press ENTER.
- Zoom to the bottom portion and specify the start and endpoints of the break line.

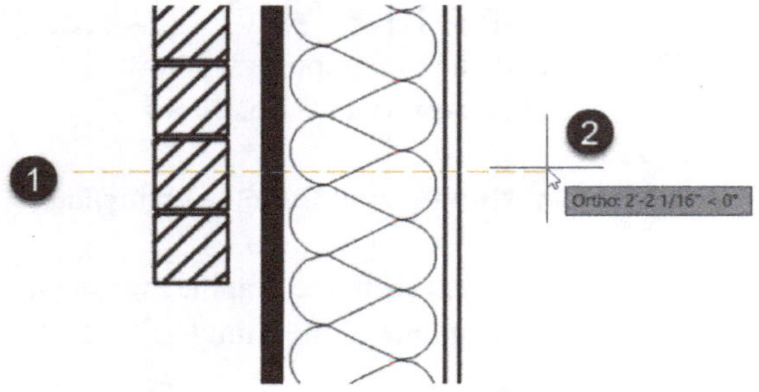

- Select the midpoint of the break line; the break line symbol is placed.

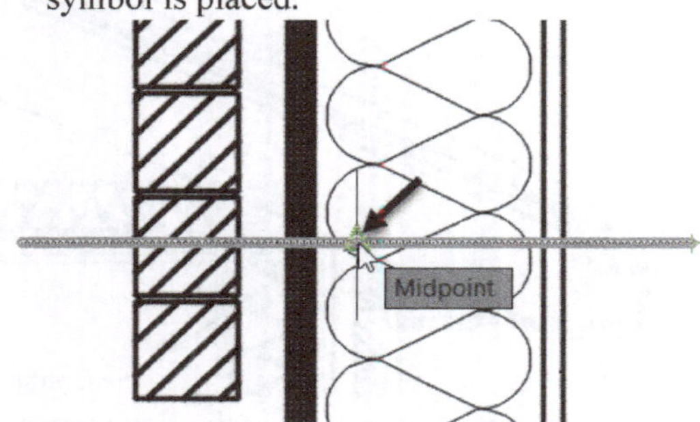

- On the **Home** ribbon tab, click **Trim** drop-down > **Trim** on the **Modify** panel.
- Select the **cuTting edges** option from the command line.
- Select the break line and press ENTER.

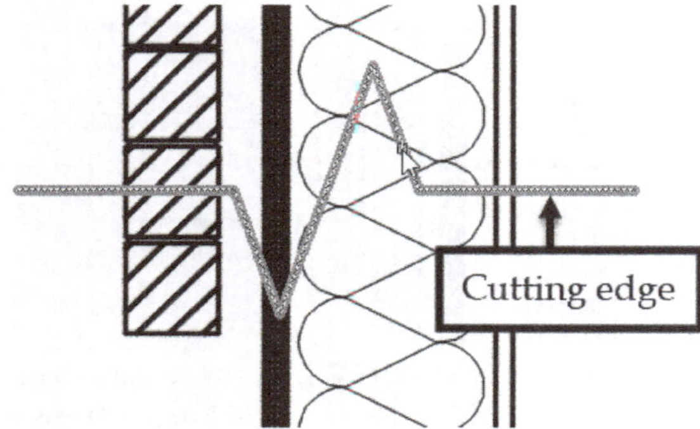

- Trim the elements on the bottom side of the drawing.
- Select the region below the break line and click the **Explode** tool on the **Modify** panel of the **Home** ribbon tab.

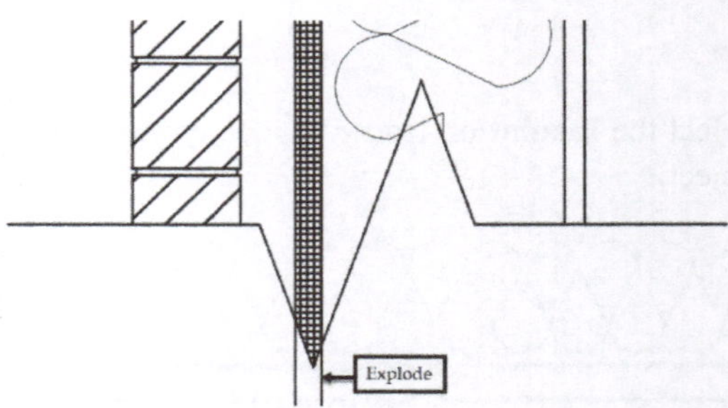

- Trim the vertical lines.

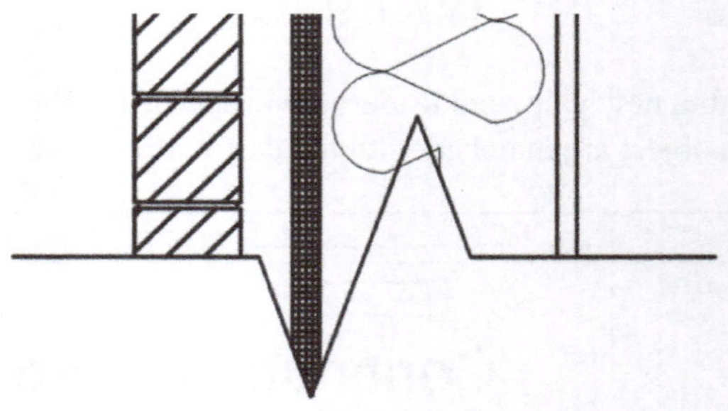

- Likewise, create the break lines and trim the elements, as shown.

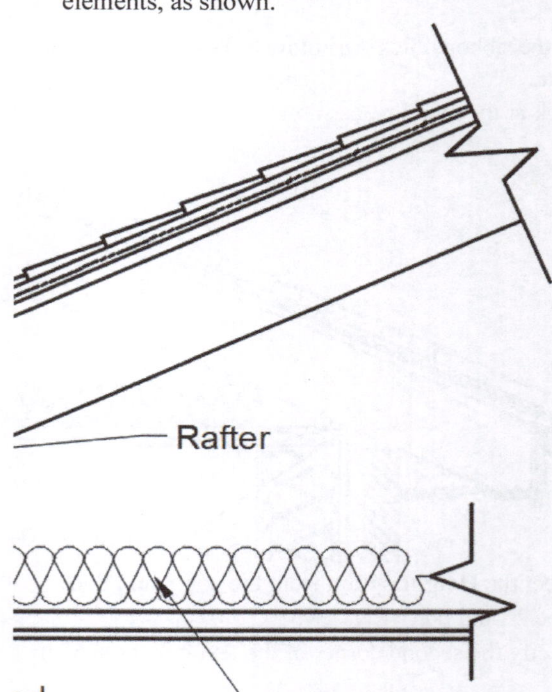

- Save and close the drawing file.

Exercise

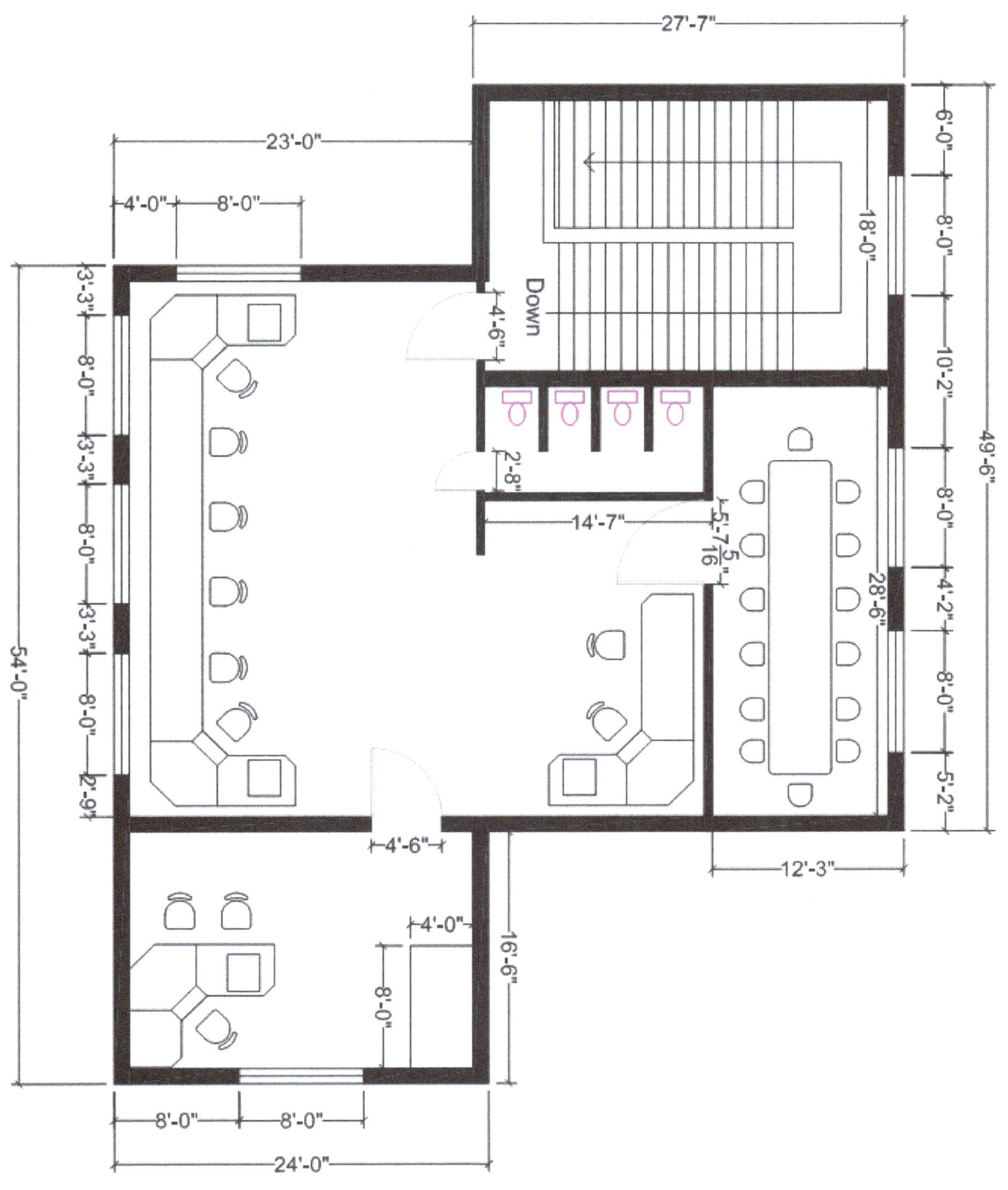

Index

Printed at Repro India Ltd.